Our Living Resources

For sale by the Superintendent of Documents, U.S. Government Printing Office, Washington, DC 20402
Stock #024-010-00708-7

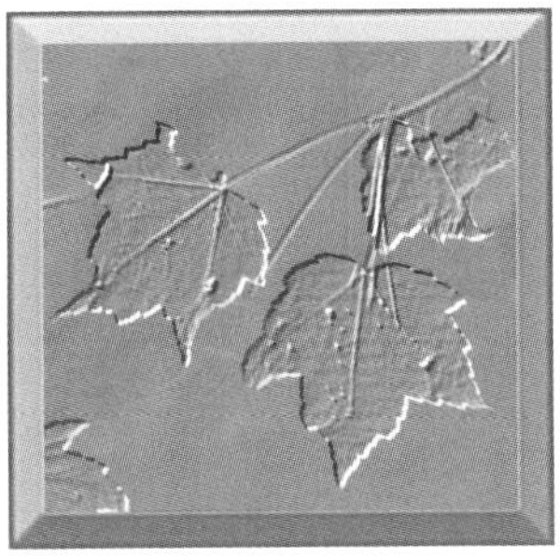

Our Living Resources

A Report to the Nation on the Distribution, Abundance, and Health of U.S. Plants, Animals, and Ecosystems

Senior Science Editor and Project Director
Edward T. LaRoe
National Biological Service

Managing Editor
Gaye S. Farris
National Biological Service

Senior Technical Editor
Catherine E. Puckett
Johnson Controls World Services, Inc., and University of Southwestern Louisiana

Graphics Editor
Peter D. Doran
Bureau of Land Management

Science Editor
Michael J. Mac
National Biological Service

U.S. Department of the Interior — National Biological Service
Washington, DC
1995

National Biological Service Cataloging-in-Publication Data

Our living resources : a report to the nation on the distribution, abundance, and health of U.S. plants, animals, and ecosystems / Edited by Edward T. LaRoe ...[et al.] Washington, D.C. : U.S. Dept. of the Interior, National Biological Service, 1995. 530 p. : ill. ; 28 cm. Includes bibliographical references and index.
1. Biotic communities — United States. 2. Ecology — United States. 3. Animal populations — United States. 4. Plant populations — United States. 5. Natural resources — United States. I. LaRoe, Edward T. II. National Biological Service (U.S.)
QH104 .O98 1995

This report may be cited:

LaRoe, E.T., G.S. Farris, C.E. Puckett, P.D. Doran, and M.J. Mac, eds. 1995. Our living resources: a report to the nation on the distribution, abundance, and health of U.S. plants, animals, and ecosystems. U.S. Department of the Interior, National Biological Service, Washington, DC. 530 pp.

Foreword

The nation's biological resources are the basis of much of our current prosperity and an essential part of the wealth that we will pass on to future generations. Like other forms of wealth, biological diversity constitutes a resource that must be conserved and managed carefully. Proper management of any resource requires (1) inventorying and monitoring the resource, (2) understanding the factors determining its supply and demand, and (3) analyzing options for current and future uses of the resource. Inventory and monitoring is the essential first step in taking stock of the wealth represented in our living resources and planning for their conservation and use.

This report, *Our Living Resources*, is the first product of the Status and Trends Program in the National Biological Service. The report compiles, for scientists, managers, and the lay public, information on many species and the ecosystems on which they depend. As a first step toward a consistent, large-scale understanding of the status and trends of these resources, this report brings together for the first time a host of information about our nation's biological wealth, highlighting causes for both comfort and concern.

The report provides valuable information about causes for the decline of some species and habitats. It also gives insight into successful management strategies that have resulted in recovery of others. The report will also serve as a useful guide for identifying research needs by revealing information gaps that must be filled if we are to achieve a more comprehensive understanding of both current conditions and the anticipated impact of change.

The mission of the National Biological Service is to work with others to provide the information and technologies needed to manage and conserve the nation's biological resources. As the biological science arm of the Department of the Interior—with neither regulatory nor resource-management responsibilities—NBS has as its primary responsibility serving the biological science needs of other Department of the Interior bureaus.

NBS also has a broader role of working with other federal agencies, states, universities, museums, private organizations, and landowners in a "National Partnership" to ensure that a more comprehensive and consistent approach is taken to providing information about the nation's biological resources. All of the players in this new partnership have long and rich histories of collecting and interpreting biological information. The National Biological Service will work with its partners to supplement and integrate this scientific information and make it more accessible.

Our Living Resources is a prime example of NBS's partnership approach. Authors are drawn from more than 15 federal agencies, 15 state agencies, 25 universities, and 13 private organizations. In some cases, individual papers are themselves products of interagency or intergovernmental partnerships.

Statistically reliable information on the status and trends of biological resources is an essential step towards better stewardship of our nation's biological wealth. Equally important is an intensive research program aimed at understanding what factors are responsible for biological changes and the incorporation of that understanding into resource management and policy decisions. NBS works closely with resource managers and other decision makers to analyze how natural forces and human activities affect biological resources and to predict how alternative management and policy decisions might improve or degrade those resources.

NBS is committed to providing better information and making that information easily accessible not only to those who manage and regulate how we use natural resources but also to every American who makes economic use or seeks recreation or simply cherishes the beauty of our living resources. More reliable information and better access to that information will result in better and fairer decisions and a more prosperous future for all Americans.

H. Ronald Pulliam
Director, National Biological Service

Preface

This report is the first of a series of reports on the status and trends of the nation's plants, animals, and ecosystems. It represents an effort to bridge the gap between scientists and resource managers, policy makers, and the general public. Usually, scientists tend to write for scientific journals and communicate with other scientists; this report attempts to collect a great variety of scientific data and interpret it for the nonscientist while maintaining the full credibility of the data.

The articles included represent both invited and contributed papers; that is, where we could identify specific subject experts, we invited them to submit papers, and we also accepted papers contributed by other authors. Following scientific tradition, each article submitted was peer-reviewed, usually by three anonymous scientific reviewers. The articles are often abridged from a complete scientific treatise, but each article contains references and personal contacts if the reader is interested in pursuing the subject in greater depth. Finally, we recognize that this report is incomplete and that more status and trends data exist than we were able to uncover or incorporate into one volume.

In Memoriam

Edward Terhune LaRoe III

Senior science editor Ted LaRoe died of cancer October 19, 1994, having shepherded this report almost to its completion. Had he lived to see *Our Living Resources* published, he would not have lingered to bask in its accomplishment. He would have moved on to new projects, new plateaus, for Ted always had a vision, a sense of where he was going. He also had a vision for the National Biological Service, which he was instrumental in helping to create.

Ted was bright, creative, inquisitive, inspiring, and a man of many accomplishments. His scientific leadership was evident in his active role in issues relating to wetland science, global climate change, coastal resources, ecosystem-based management, and, of course, NBS. Above all, he was a champion of scientific integrity, which, we trust, is evident in this report. We hope he would have been pleased.

Acknowledgments

We extend our sincere appreciation to all who helped produce this report. Especially important were the science editors—Austin K. Andrews, Raymond J. Boyd, Glenn R. Guntenspergen, Russell J. Hall, Michael D. Jennings, Hiram W. Li, Michael J. Mac, William T. Mason, Jr., O. Eugene Maughan, Roy W. McDiarmid, Carole C. McIvor, J. Michael Scott, William K. Seitz, Thomas J. Stohlgren, Benjamin N. Tuggle, Wayne A. Willford, and Gary D. Willson. They served by coordinating reviews, including the peer reviews of articles within their sections. In addition, they each provided an overview to the material in their sections. Assisting with overviews were Gregor T. Auble and B.D. Keeland.

Carl Anderson, Michalann Harthill, Deborah E. Peck, Helen V. Turner, and Sherri L. Hendren each provided tremendous technical support. Contributing expertise in graphics were Nicholas R. Batik, Mary A. Helmerick, Dave Opp, Diane K. Baker, Janine J. Koselak, and M. Jennifer Kapus. Technical editors—Mary Catherine Hager, Beverly Kerr-Mattox, and Kristie A. Weeks—dedicated months to the editing of individual articles. Technical typists Deany M. Cheramie, Dana M. Girod, and Tiffany Alexander Hall assisted by keyboarding, correcting, and proofreading. Technical typist Judy Zabdyr helped in the final stages as did proofreaders Rhonda F. Davis and Lori E. Huckaby, under the direction of editor Beth A. Vairin, who also reviewed the report. Librarian Judy K. Buys performed numerous bibliographic searches to verify citations, and Marilyn Rowland indexed the report. Robert E. Stewart, Director of the National Biological Service's Southern Science Center, graciously allowed the use of his staff, space, and equipment to produce this report, as did Lawrence Bembry, Director of the Bureau of Land Management's Service Center. We are also grateful to all those who gave permission to use their slides and graphics.

Finally, we would like to thank the authors, the peer reviewers, and the state, federal, and private agencies who so willingly gave of their time and data. Without their hard work and cooperation, this report would not have been possible.

Contents

Foreword v
Preface vi
In Memoriam vi
Acknowledgments vii

Part 1 Introduction

Overview 3
Biodiversity: A New Challenge 6
Conservation Landmarks: Bureau of Biological Survey and National Biological Service 7
Activities of the Bureau of Biological Survey 8

Part 2 Distribution, Abundance, and Health

Species

Birds
Overview 15
Breeding Bird Survey: Population Trends 1966-92 17
Winter Population Trends of Selected Songbirds 21
Breeding Productivity and Adult Survival in Nongame Birds 23
Canada Geese in North America 26
Canada Geese in the Atlantic Flyway 28
Arctic Nesting Geese: Alaskan Populations 30
North American Ducks 34
Decline of Northern Pintails 38
Canvasback Ducks 40
Breeding Seabirds in California, Oregon, and Washington 43
Seabirds in Alaska 49
Colonial Waterbirds 53
Shorebirds: East of the 105th Meridian 57
Western North American Shorebirds 60
Raptors 65
Causes of Eagle Deaths 68
Return of Wild Turkeys 70
Mourning Doves 71
Common Ravens in the Southwestern United States, 1968-92 73
Mississippi Sandhill Cranes 75
Piping Plovers 77
California Condors 80
Audubon's Crested Caracara in Florida 82
Puerto Rican Parrots 83
Red-cockaded Woodpeckers 86
Southwestern Willow Flycatchers in the Grand Canyon 89

Mammals
Overview 93
Marine Mammals 94
Indiana Bats 97
Gray Wolves 98
Black Bears in North America 100
Grizzly Bears 103
Black-footed Ferrets 106
American Badgers in Illinois 108
California Sea Otters 110
White-tailed Deer in the Northeast 112
Deer Management at Parks and Refuges 113
North American Elk 115

Reptiles and Amphibians
Overview 117
Turtles 118
Marine Turtles in the Southeast 121

Amphibians 124
A Success Story: The Barton Springs Salamander. 125
American Alligators in Florida 127
Reptiles and Amphibians in the Endangered Longleaf Pine Ecosystem 129
Native Ranid Frogs in California 131
Desert Tortoises in the Mojave and Colorado Deserts 135
Coachella Valley Fringe-toed Lizards 137
Disappearance of the Tarahumara Frog 138

Fishes
Overview 141
Imperiled Freshwater Fishes 142
Southeastern Freshwater Fishes 144
Loss of Genetic Diversity Among Managed Populations 147
Colorado River Basin Fishes. 149
Cutthroat Trout in Glacier National Park, Montana 153
Columbia River Basin White Sturgeon 154

Invertebrates
Overview 159
Diversity and Abundance of Insects 161
Grasshoppers 163
The Changing Insect Fauna of Albany's Pine Barrens 166
Lepidoptera in North America 168
Fourth of July Butterfly Count 171
Species Richness and Trends of Western Butterflies and Moths 172
The Tall-grass Prairie Butterfly Community 174
The Biota of Illinois Caves and Springs 176
Freshwater Mussels: A Neglected and Declining Aquatic Resource 177
Freshwater Mussels in Lake Huron-Lake Erie Corridor. 179
Aquatic Insects As Indicators of Environmental Quality 182
Biodiversity Degradation in Illinois Stoneflies 184

Plants
Overview 189
Microfungi: Molds, Mildews, Rusts, and Smuts. 190
Macrofungi. 192
Truffles, Trees, and Biodiversity 193
Lichens. 194
Bryophytes 197
Floristic Inventories of U.S. Bryophytes 198
Vascular Plants of the United States 200
Environmental Change and the Florida Torreya. 205
Native Vascular Plants 205
Tracking the Mosses and Vascular Plants of New York (1836-1994) 209

Ecosystems

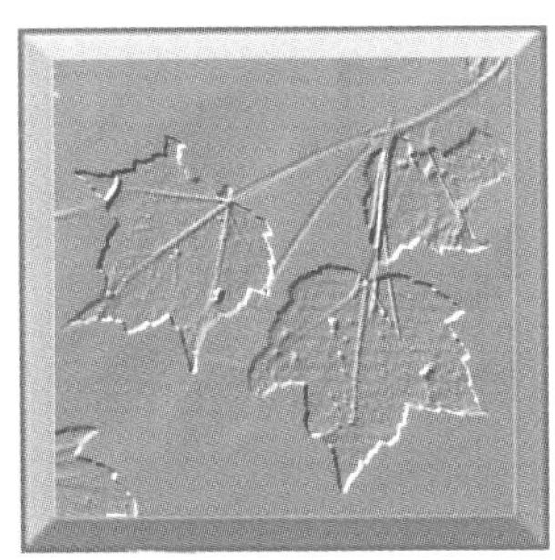

Terrestrial Ecosystems
Overview 213
U.S. Forest Resources. 214
Southern Forested Wetlands 216
Rare Terrestrial Ecological Communities of the United States. 218
Altered Fire Regimes Within Fire-adapted Ecosystems 222
Vegetation Change in National Parks 224
Air Pollution Effects on Forest Ecosystems in North America. 227
Air Quality in the National Park System 228
Whitebark Pine: Ecosystem in Peril. 228
Oak Savannas in Wisconsin 230

Aquatic Ecosystems
Overview 233
Habitat Changes in the Upper Mississippi River Floodplain 234
Biota of the Upper Mississippi River Ecosystem. 236

Fish Populations in the Illinois River . . . 239
Contaminant Trends in Great Lakes Fish . . . 242
Lake Trout in the Great Lakes. . . . 244
Wetlands in Regulated Great Lakes. . . . 247
Decline in the Freshwater Gastropod Fauna in the Mobile Bay Basin . . . 249
Protozoa . . . 252
Marine and Freshwater Algae . . . 255
Freshwater Diatoms: Indicators of Ecosystem Change . . . 256

Coastal and Marine Ecosystems
Overview . . . 259
Nearshore Fish Assemblage of the Tidal Hudson River . . . 260
Natural Resources in the Chesapeake Bay Watershed . . . 263
Florida Manatees . . . 267
Gulf of Mexico Coastal Wetlands: Case Studies of Loss Trends . . . 269
Seagrass Distribution in the Northern Gulf of Mexico . . . 273
Seagrass Meadows of the Laguna Madre of Texas . . . 275
Coastal Barrier Erosion: Loss of Valuable Coastal Ecosystems . . . 277
Reef Fishes of the Florida Keys . . . 279
Coral Reef Ecosystems . . . 280

Riparian Ecosystems
Overview . . . 285
Western Riparian Ecosystems . . . 286
Surface Cover Changes in the Rio Grande Floodplain, 1935-89 . . . 290

Ecoregions

The Great Plains
Overview . . . 295
Declining Grassland Birds . . . 296
Migratory Bird Population Changes in North Dakota . . . 298
Duck Nest Success in the Prairie Potholes . . . 300
Conservation Reserve Program and Migratory Birds in the Northern Great Plains . . . 302
Decline of Native Prairie Fishes . . . 303
The Coyote: An Indicator Species of Environmental Change on the Great Plains . . . 305

Interior West
Overview . . . 309
Ecosystem Trends in the Colorado Rockies . . . 310
The Greater Yellowstone Ecosystem . . . 312
Subalpine Forests of Western North America . . . 314
Southwestern Sky Island Ecosystems . . . 318
Endangered Cui-ui of Pyramid Lake, Nevada . . . 323
Bonytail and Razorback Sucker in the Colorado River Basin . . . 324
Amphibian and Reptile Diversity on the Colorado Plateau . . . 326
Wintering Bald Eagles Along the Colorado River Corridor . . . 328
Mexican Spotted Owls in Canyonlands of the Colorado Plateau . . . 330
Bighorn Sheep in the Rocky Mountain National Parks . . . 332
Desert Bighorn Sheep . . . 333

Alaska
Overview . . . 337
The Arctic Tundra Ecosystem in Northeast Alaska . . . 338
Anadromous Fish of the Central Alaska Beaufort Sea . . . 341
Pacific Salmon in Alaska . . . 343
Wolves and Caribou in Denali National Park, Alaska . . . 347
Kodiak Brown Bears . . . 349
Polar Bears in Alaska . . . 351
Sea Otters in the North Pacific Ocean . . . 353
Pacific Walruses . . . 356
Mentasta Caribou Herd . . . 357
Tundra or Arctic Hares . . . 359

Hawaii

Overview . . . 361
Hawaii Biological Survey . . . 362
Haleakala Silversword . . . 363
Insects of Hawaii . . . 365
Drosophila as Monitors of Change in Hawaiian Ecosystems . . . 368
Birds of Hawaii . . . 372
Hawaii's Endemic Birds . . . 376

Part 3 Special Issues

Global Climate Change

Overview . . . 385
Changes in Winter Ranges of Selected Birds, 1901-89 . . . 386
Changes in Nesting Behavior of Arctic Geese . . . 388
Climate Change in the Northeast . . . 390
Potential Impacts of Climate Change on North American Flora . . . 392

Human Influences

Overview . . . 397
Significance of Federal Lands for Endangered Species . . . 398
Status of U.S. Species: Setting Conservation Priorities . . . 399
Increased Avian Diseases With Habitat Change . . . 401
Captive Propagation, Introduction, and Translocation Programs for Wildlife Vertebrates . . . 405
Raccoon Rabies: Example of Translocation, Disease . . . 406
Contaminants in Coastal Fish and Mollusks . . . 408
Persistent Environmental Contaminants in Fish and Wildlife . . . 413
Wildlife Mortality Attributed to Organophosphorus and Carbamate Pesticides . . . 416
Acidic Deposition ("Acid Rain") . . . 418
Atmospheric Deposition and Solute Transport in a Montane Mixed-Conifer Forest System 421
Agricultural Ecosystems . . . 423

Non-Native Species

Overview . . . 427
Non-native Aquatic Species in the United States and Coastal Waters . . . 428
Nonindigenous Fish . . . 431
Non-native Reptiles and Amphibians . . . 433
Non-native Birds . . . 437
Non-native Animals on Public Lands . . . 440
Exotic Species in the Great Lakes . . . 442
Zebra Mussels in Southwestern Lake Michigan . . . 445
Invasion of the Zebra Mussel in the United States . . . 445
Africanized Bees in North America . . . 448
Bullfrogs: Introduced Predators in Southwestern Wetlands . . . 452
Invasions of the Brown Tree Snake . . . 454
Wild Horses and Burros on Public Lands . . . 456
Purple Loosestrife . . . 458

Habitat Assessments

Overview . . . 461
GAP Analysis: A Geographic Approach to Planning for Biological Diversity . . . 462
Protection Status of Vegetation Cover Types in Utah . . . 463
Biodiversity in the Southwestern California Region . . . 465
Federal Data Bases of Land Characteristics . . . 467
Monitoring Changes in Landscapes from Satellite Imagery . . . 468
Landsat MSS Images . . . 470
The Nation's Wetlands . . . 473

Glossary . . . 479
Index . . . 483

PART 1

Introduction

Terry Donnelly/Tom Stack and Associates

Introduction

by
Edward T. LaRoe
Senior Science Editor and Project Director
National Biological Service

Overview

This report on the distribution, abundance, and health of our nation's biological resources is the first product of the National Biological Service's Status and Trends Program. This information has many potential uses: it can document successful management efforts so resource managers will know what has worked well; it can identify problems so managers can take early action to restore the resource in the most cost-efficient manner; and it can be used to highlight areas where additional research is needed, such as to determine why certain ecological changes are occurring. This report will also be useful to teachers, students, journalists, and citizens in general who are interested in national resource issues.

Another purpose of this report is to help identify gaps in existing resource inventory and monitoring programs. It contains information collected by a variety of existing research and monitoring efforts by scientists in the National Biological Service, other federal and state agencies, academia, and the private sector. The programs that produced the information in this document were not developed in a coordinated fashion to produce an integrated, comprehensive picture of the status and trends of our nation's resources; rather, each was developed for its own particular purpose, usually to help manage a specific resource. Thus, even though articles vary greatly in scope, design, and purpose, this report has identified and attempted to combine many of the existing information sources into a broad picture of the condition of our resources. In the future, these sources will be complemented by additional information from other sources—such as state agencies and other inventory and monitoring studies—to fill in the gaps of knowledge and to provide a more complete understanding of the status of our living resources.

A second report, to be released by the National Biological Service in 1995, will use the information contained in this report and data from other sources to provide a synthesized account of the status and trends of the nation's biodiversity. It will discuss from a historical perspective the factors influencing biodiversity, both natural and human-induced, and provide an integrated description of the status and trends of biological resources on a regional basis.

Status and Trends

The goal of inventory and monitoring programs is to determine the status and trends of selected species or ecosystems. Status studies produce data on the condition of species or ecosystems for a single point in time; trend studies, in contrast, provide a chronological or geographic picture of change in the same resource. Either can measure a number of different biological indicators, such as population size, distribution, health, or physiological factors such as breeding productivity or seed production. Species composition, biodiversity, and age and physical structure are all important indicators of ecosystem status.

Inventory and monitoring programs can provide measures of status and trends to determine levels of ecological success or stress; if such programs are appropriately designed, they can give early warnings of pending problems, allowing resource managers to take remedial action while there are more management options. These earlier options are less severe than if management response is delayed until problems are critical, such as when a species becomes endangered.

One of the challenges resource managers face is to detect long-term trends because such trends are often masked by short-term, random, or undirectional variations (Figure). Plant and animal species often vary greatly in abundance, distribution, or fecundity as a result of forces that include annual or seasonal variations in climate; chance events such as floods and hurricanes; effects from predators or competing species; and even internal physiological processes. Some variations appear totally random; many are cyclic, recurring periodically; and others are long-term in one predominant direction. Scientists have many ways to determine whether apparent changes are biologically and statistically significant, although it is often difficult to detect such trends in their early stages. The design of monitoring programs should address issues such as the number of samples needed, the sampling technique, and the frequency and duration of sampling. All are critical factors in determining the sensitivity of the monitoring program to detect directional change. Data collected in a standard or consistent fashion over many years are especially critical to identify and document trends.

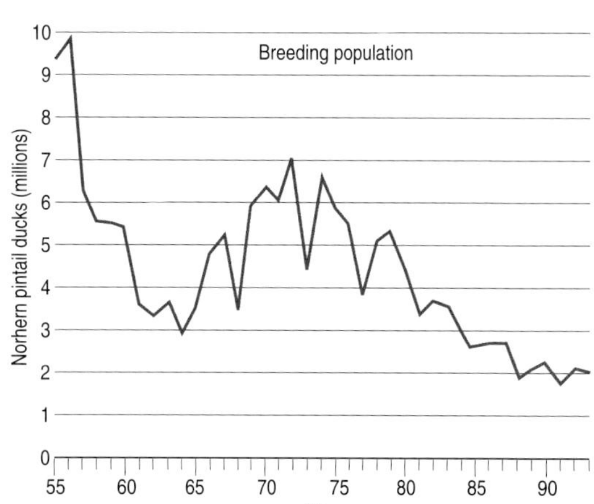

Figure. Northern pintail duck (*Anas acuta*) population data demonstrate the importance of long-term data sets. Annual fluctuations (e.g., 1967-70) reflect yearly fluctuations in breeding success that may have been caused, for example, by differences in rainfall and the abundance of temporary wetlands for nesting habitat. Short-term data sets can give erroneous conclusions; for example, if only data from 1964 to 1972 were available, managers might conclude that populations were increasing. The long-term data, however, describe a statistically and biologically significant population decline. (Source: U.S. Fish and Wildlife Service, 1993. *Status of Waterfowl and Fall Flight Forecast.*)

National Inventory and Monitoring Programs

A number of inventory and monitoring programs have been underway for several to many years in various agencies (Table). Historically, the federal government has been responsible for monitoring the status and trends of migratory species as well as those resident on federal lands. In addition, the federal government monitors habitat conditions on federal lands and, under some circumstances, private lands. Some of the monitoring programs also require international cooperation because many of the migratory species monitored cross international boundaries.

States have monitored resident species and often cooperated in surveys of migratory species. A significant problem with these efforts has been that often the individual agencies or states have used different monitoring procedures and standards, and the results are not comparable from area to area or among different agencies.

The private sector, including particularly The Nature Conservancy, has worked with states to establish Heritage Programs that monitor the distribution and abundance of selected species. This effort has resulted in standardized procedures.

Most inventory and monitoring programs were established for a specific purpose, usually relating to management of natural resources. For example, the efforts to monitor duck populations started 35 years ago to improve the basis for hunting regulations, and the National Wetland Inventory was started in 1979 to determine the condition and rate of wetland loss. Until recently, few, if any, of these programs were intended or have been used to provide broad-based and predictive tools that could help resource managers identify future resource problems.

The National Biological Service has the responsibility for developing information on the status and trends of our nation's plants and animals and the habitats on which they depend. It will achieve this by building on the inventory and monitoring activities existing in the state, federal, and private sectors. The national status and trends effort will continue to depend upon the contributions of these existing programs, and NBS will avoid duplicating programs already under way. Its role will be to coordinate the activities of different agencies into a comprehensive assessment of our living resources, continuing its own contributions, and when necessary, supplementing the current array of activities.

Organization of This Report

In addition to this overview, the report introduction includes articles on the importance of biodiversity and a historical look at biological study in the federal government.

The articles that follow, contributed by a variety of authors and agencies, represent the first effort to pull together information on the status and trends of different groups of biota, ecosystems, and ecoregions as well as related issues. Individual articles in each section are most often arranged from the most general or large scale, to the most specific or small scale. The organization is somewhat arbitrary in that many articles could appear with equally valid justifications in several different locations.

Animals and Plants

Not all groups have received equal treatment, in large part because our current knowledge is not equal among all groups, and inventory and monitoring are focused on comparatively few species. Scientific studies have been greatly assisted in some areas by the work of natural historians and public volunteers. Bird watchers, butterfly collectors, and shell collectors, for example, have provided invaluable scientific information about the geographic ranges of groups in which they are interested. Some of the professional societies today owe their origins to the efforts of amateurs to organize and improve their understanding of biota.

Many of the less visible or charismatic taxa lack the scientific effort or information, much less the volunteer amateur support, to discuss trends in their abundance or distribution. The very title "Animals and Plants" could be viewed as biased by some biologists; although most of the public views mushrooms and other fungi as plants, specialists consider them a separate kingdom, equal both taxonomically as well as in ecological significance to both plants and animals. Despite their significance, plants are simply underrepresented in this report because the data are lacking.

The report begins with birds, the single group for which we have the most data at national and large-scale levels. Because of the significance of birds as important migratory species, there has been a strong role for federal research scientists as well as scientists from state agencies and from Canada and Mexico. Some of the best long-term scientific information on status and trends comes from the Breeding Bird Survey and the Christmas Bird Count.

Table. Selected examples of existing ecological inventory and monitoring programs.

Subject	Institution
Migratory bird surveys	
Breeding birds	National Biological Service (NBS)
Wintering birds	The Audubon Society
Waterfowl surveys	U.S. Fish and Wildlife Service (USFWS), NBS, and states, with international participation
Rare and threatened species	
Listed endangered and threatened species	States, U.S. federal land managers (e.g., National Park Service, Bureau of Land Management (BLM), U.S. Forest Service (USFS)), USFWS, NBS
State Heritage Programs	State agencies and The Nature Conservancy
Endangered marine species	National Marine Fisheries Service (NMFS)
Resident game species	
(e.g., deer, turkey, furbearers)	State fish and wildlife conservation agencies
Habitats and biological communities	
National Wetlands Inventory	USFWS
Gap Analysis Program	NBS in partnership with states and private sector
Environmental Monitoring and Assessment Program	U.S. Environmental Protection Agency
Resources Conservation Act, inventory of wildlife and habitat conditions of farmlands	U.S. Soil Conservation Service (now Natural Resources Conservation Service)
Wildlife and habitat on public lands	
Resources Planning Act assessment of USFS lands	USFS
Federal Land Policy and Management Act assessment of BLM lands	BLM
Contaminants	
Aquatic and terrestrial (Biomonitoring of Environmental Status & Trends National Water Quality Assessment)	NBS with USFWS; U.S. Geological Survey
Marine and coastal	National Oceanic and Atmospheric Administration and NMFS

Ecosystems and Ecoregions

We have also included information on ecosystems and ecoregions. Ecosystems are groups of plants and animals and their nonliving environment such as air and water. For example, one can speak of a coastal wetland ecosystem, whether in North Carolina or Florida, and understand that it includes several specific features or processes shared by all coastal wetland ecosystems.

Ecoregions are geographically defined ecological units, often containing several types of ecosystems, that share common topographic, climatic, and biotic characteristics. Each ecoregion, such as Alaska or Hawaii, can be defined as a single, individual unit on a map, while ecosystems, if mapped, would be scattered about as separate units.

Special Issues

After the status and trends of animals and plants, ecosystems, and ecoregions are presented, a section on related issues follows: global climate change, human influences, non-native species, and methods of habitat assessment. The proliferation of introduced species, both plant and animal, has had a profound influence on the native biota of this country. Many human activities, such as pollution and urbanization, both directly and indirectly affect the health of our living resources. The possibilities of global climate change are examined, followed by a brief overview of national programs such as the Gap Analysis Program, which scientists hope will prove useful in acquiring data to help resource policy makers better protect our resources.

Biodiversity: A New Challenge

by
Edward T. LaRoe
National Biological Service

Resource managers at many state and federal agencies are in the middle of a fundamental change in the practice and objectives of conservation. Traditional management has been directed toward maintaining, usually for harvest purposes, populations of individual species such as ducks, deer, or salmon. Increasingly, however, resource managers are recognizing the critical importance of conserving biological diversity, or biodiversity.

In its simplest terms, biological diversity is the variety of life at all levels: it includes the array of plants and animals; the genetic differences among individuals; the communities, ecosystems, and landscapes in which they occur; and the variety of processes on which they depend. Conserving biological diversity poses dramatic new problems for comprehensive inventory and monitoring: what should be measured or monitored?

Biodiversity is important for many reasons. Its value is often reported in economic terms: for example, about half of all medicinal drugs (Keystone Center 1991; Wilson 1992) come from—or were first found in—natural plants and animals, and therefore these resources are critical for their existing and as yet undiscovered medicinal benefits. Additionally, most foods were domesticated from wild stocks, and interbreeding of different, wild genetic stocks is often used to increase crop yield. Today we use but a small fraction of the food crops used by Native cultures: many of these underused plants may become critical new food sources for the expanding human population or in times of changing environmental conditions.

But biodiversity has an even greater importance: it is the great variety of life that makes existence on earth possible. As a simple example, plants convert carbon dioxide to oxygen during the photosynthetic process; animals breathe this fresh air, releasing energy and providing the second level of the food chain. In turn, animals convert oxygen back to carbon dioxide, providing the building blocks for the formation of sugars during photosynthesis by plants. Microbes (fungi, bacteria, and protozoans) break down the carcasses of dead organisms, recycling the minerals to make them available for new life; along with some algae and lichens, they create soils and improve soil fertility.

Biodiversity provides the reservoir for change in our life-support systems, allowing life to adapt to changing conditions. In a natural population, for example, some individuals will be more resistant to drought or disease or cold; as the environment changes, from season to season, year to year, or over longer periods, and as plagues come and go, these differences among individuals allow at least some members of the population or species to survive and reproduce. This diversity is the basis not only for short-term adaptation to changing conditions, but also for long-term evolution as well.

Like air, water, and soils, biological diversity is part of the capital upon which all life depends. The need for this diversity is greatest in times of environmental stress when plants, animals, and microbes must develop new characteristics or strategies for survival. As we look at the problems of the globe today—global climate change, decreases in the ozone shield and increasing ultraviolet radiation, losses of natural habitats, and pervasive pollution in our streams and oceans—we must recognize that we, as a form of life on earth, need the ability to change in order to cope with new stresses.

Humans cannot survive in the absence of nature. We depend on the diversity of life on earth for about 25% of our fuel (wood and manure in Africa, India, and much of Asia); more than 50% of our fiber (for clothes and construction); almost 50% of our medicines; and, of course, for all our food (Miller et al. 1985). As previously stated, biodiversity produces other benefits: plants produce oxygen for our atmosphere; microbes break down wastes, recycle nutrients, and build the fertility of our soils. One reason our highways are not littered with the carcasses of dead dogs, cats, skunks, armadillos, and deer is biodiversity, in the form of the many scavengers and microbes that we don't often think about, but which play an essential role in the cycle of life. Even species often viewed as "repulsive," such as vultures and maggots, play critical roles in our lives.

Some people believe that because extinction is a natural process, we therefore should not worry about endangered species or the loss of biodiversity. Certainly extinction is natural; it usually occurs as newer forms of life evolve. But under the forces of population growth, technology, and special interests, humans have driven the rate of extinctions today to about 100 times—two orders of magnitude—the natural rate. Even worse, the rate of extinction is still increasing and will be 100 to 1,000 times faster yet in the next 55 years (Miller et al. 1985); scientists today predict that between now and 2030, half the expected lifetime of a child born today, the Earth will lose between a quarter and a third of all existing species. And this is in the absence of new forms of life to replace them.

The last time Earth lost this large a share of its life was 65 million years ago when it may have collided with an asteroid; the impacts of humans on our planet today may have been last equaled by the collision of two heavenly bodies (Wilson 1992).

Scientists cannot honestly say that we need *all* species that exist today for humans to survive; but as a general rule, the more diversity is diminished, the less stable ecosystems become and the greater the fluctuations that occur in plant and animal populations. The more diversity we lose, the more our quality of life and economic potential are diminished, and the greater the risk that we will cause a critical part of the cycle of life to fail.

If humans were allowed to cause the extinction of other species, who would determine which species? If we had been asked 60 years ago what life we could let become extinct, who among us would have insisted that we preserve the lowly mold that was penicillin, the first of the series of antibiotics that have today so changed the quality of our lives? And who, only 5 years ago, would have identified the need to preserve the Pacific yew, which today yields taxol, one of the greatest new hopes in our arsenal against cancer?

References

Keystone Center. 1991. Biological diversity on federal lands, report of a Keystone policy dialogue. The Keystone Center, Keystone, CO. 96 pp.

Miller, K.R., J. Fortado, C. De Klemm, J.A. McNeely, N. Myers, M.E. Soule, and M.C. Trexler. 1985. Issues on the preservation of biological diversity. Pages 337-362 *in* Robert Repetto, ed. The global possible. Yale University Press, New Haven.

Wilson, E.O. 1992. The diversity of life. Belknap Press of Harvard University Press, Cambridge, MA. 424 pp.

Conservation Landmarks: Bureau of Biological Survey and National Biological Service

by

Milton Friend

National Biological Service

A century separates the recent development of the National Biological Service (NBS) and an early predecessor, the Bureau of Biological Survey (BBS). Both organizations were established at critical crossroads for the conservation of the nation's living biological resources and are conservation landmarks of their times. The BBS of the 1920's was described as "a government Bureau of the first rank, handling affairs of great scientific, educational, social, and above all, economic importance throughout the United States and its outlying possessions" (Cameron 1929:144-145). This stature was achieved at a time of great social, economic, and ecological change. BBS had the vision to pioneer new approaches that led to enhanced understanding of the relation between people, other living things, and the environment. The NBS faces similar challenges to address the issues of the 1990's and beyond.

Diminished Natural Resources in a World of Plenty

Early European colonists had an abundance of wildlife to serve subsistence needs. Seemingly endless flocks of ducks, geese, and swans; an abundance of wild turkeys, deer, and bison; green clouds of Carolina parakeets and millions of passenger pigeons; and a bounty of fish and shellfish. This abundance quickly established a viewpoint that the New World's wildlife resources were inexhaustible.

Habitat changes that disrupted the balance of nature soon resulted in economic losses and other hardships because of insect and rodent eruptions. Negative effects of exotic species brought from the Old World further reduced the well-being of many colonists who had come to the New World for a better life. The nation's inexhaustible natural resources and returns from agriculture began to wane significantly. Decimation of previously vast wildlife resources greatly reduced opportunities for cultural and recreational uses of wildlife (Cameron 1929).

Development of the BBS

Roots of the BBS can be traced to the 1883 founding of the American Ornithological Union (AOU) in New York City. Initially, the AOU focused on three subject areas—distribution, biological information and economic impact, and migratory behaviors of birds—all of which became major activities of the BBS. Collaborations and partnerships were developed with numerous ornithologists, field collectors, sportsmen, and observers of nature who were asked to report specific information relative to bird migration. Cooperation also was obtained from the United States Lighthouse Board and the Department of Marine and Fisheries of Canada (Cameron 1929).

Funds for government biological survey programs related to economic ornithology were allocated in 1885 to the Division of Entomology of the U.S. Department of Agriculture. These funds were provided for "the promotion of economic ornithology, or the study of the interrelation of birds and agriculture, an investigation of the food habits, and migration of birds in relation to both insects and plants." The following year additional funds were provided to include the study of mammals and expand the focus

Activities of the Bureau of Biological Survey (Cameron 1929)

Investigation and research

Study of life habits of wild animals
Classification of wild animals
Studies in geographic distribution of wild animals and plants
Life zone investigations of definite areas
Biological surveys of definite areas
Special big game investigations
Investigations for improvement of reindeer in Alaska
Investigations at reindeer experiment station
Investigations of problems of fur farmers
Studies in fur animal disease and parasites
Investigations of problems of rabbit raisers
Studies of rabbit diseases, etc.
Investigations in animal poisons
Studies in bird migration
Bird censuses (general)
Wild fowl censuses
Bird banding
Food habits studies by laboratory examinations of stomach contents of birds, mammals, reptiles, and amphibians
Studies in game bird propagation
Specific studies in covert restocking
Surveys of food resources for waterfowl
Investigations and experiments in predatory animal control
Investigations and experiments in control of injurious rodents
Investigations and experiments in control of other animal pests
Investigations and experiments in control of bird pests

Encouragement of useful forms of wildlife

Advice on game bird and animal propagation methods
Devising of methods for attracting birds about parks, homes, etc.
Encouragement of conservation of wild fur bearers
Advice on small animal production (for pets and laboratory use)
Maintenance and protection of game preserves and birds refuges
Restocking of reservations
Disposal of surplus animals on reservations
Issuance of permits for fur farming on certain Alaskan islands
Administration of Upper Mississippi Wild Life and Fish Refuge Act
Administration of act protecting wildlife on reservations

Repression of undesirable forms of wildlife

Killing of predatory animals
Leadership and demonstration in cooperative effort against predatory animals
Leadership and demonstration in cooperative effort against injurious rodents
Leadership and demonstration in cooperative effort against other animal pests and injurious birds
Processing of poisons and food stuffs for use against predatory and noxious animals

Protection of wildlife

Administration of Migratory Bird Treaty and Lacey acts by warden service and in cooperation with state law enforcement agencies
Issuance of permits for game propagation
Regulation of importation of wild birds and animals
Preparation of regulations under Alaska game law

Dissemination of information

Preparation and editing of publications
Preparation of exhibits and photographs
Answering of inquiries
Addresses by officers (conventions, universities, etc.)

Miscellaneous

Regulation of grazing of domestic stock in certain Alaskan islands

from agriculture and horticulture to the new subject of forestry. At the same time, the work was moved from the Division of Entomology to the new Division of Economic Ornithology and Mammalogy. Dr. C. Hart Merriam became the first division chief in July 1886 (Cameron 1929).

The new division continued to study wildlife food habits, migration, and species distribution. It placed considerable emphasis on educating farmers about birds and animals affecting their interests so that destruction of useful species might be prevented. Dr. Merriam pursued the development of an extensive biological survey, advancing the argument that mapping of faunal and floral areas would benefit farmers by identifying the boundaries of areas fit for the growth of certain crops and those hospitable for certain breeds of livestock. In 1890, the appropriation language for the Department of Agriculture provided for the investigation of "the geographic distribution of animals and plants," causing Dr. Merriam to note that "the division is now in effect a biological survey" (Cameron 1929:27).

The major part of the division's 1891 activities involved an extensive biological survey and biogeographic mapping of the Death Valley region of southern California and southern Nevada. This was followed by additional biological surveys of various areas of the West. Biological surveys also were conducted beyond the continental borders of the United States into Alaska, Canada, and Mexico. In 1896 the Division of Ornithology and Mammalogy became the Division of Biological Survey (Cameron 1929).

Food habit studies, which were continued along with the survey work, emphasized transmitting information to those who could benefit from it. Popular bulletins were prepared on bird migration, the economic impacts of specific wildlife species on agriculture, and the introduction of exotic species. In 1889, the division initiated the more scientific *North American Fauna* series, which included that year a general paper discussing Dr. Merriam's concept of the life zones of North America (Cameron 1929).

The division was elevated to bureau status on July 1, 1905. During the next 34 years, activities expanded to serve the growing U.S. conservation movement. Diverse investigations and research were carried out as well as technical assistance to the public and to game managers; animal damage control; regulatory functions including conservation law enforcement; administration of refuge lands; and public education through publications and exhibits (*see* box). Conservation problems included habitat loss, declining wildlife populations, species extinction, control of exotic species, control of predatory and injurious wildlife, pollution and disease control, and competition between wildlife, agriculture, and forestry.

The BBS was transferred to the Department of Interior on July 1, 1939, and was made part of the U.S. Fish and Wildlife Service (USFWS). In November 1993, the biological research components within the Department of Interior, including those from the USFWS, the National Park Service, the Bureau of Land Management, the Bureau of Reclamation, and the Minerals Management Service were reorganized to form the National Biological Survey. The name was changed to the National Biological Service on January 5, 1995, to more accurately reflect the agency's mission.

Then and Now

Dr. Merriam noted that the chief work of the BBS was to obtain facts, for without a knowledge of facts there can be neither efficient administration nor intelligent regulation of wildlife to meet the needs of the nation (Cameron 1929). That same philosophy is inherent in Secretary of the Interior Bruce Babbitt's remarks about the NBS:

> The National Biological Survey will produce the map we need to avoid the economic and environmental "train wrecks" we see scattered across the country. NBS will provide the scientific knowledge America needs to balance the compatible goals of ecosystem protection and economic progress. . . . [The] National Biological Survey will unlock information about how we protect ecosystems and plan for the future. (National Research Council 1993:181-182).

Land management, regulatory, and law enforcement activities of the BBS remained with the USFWS and other parent bureaus within the Department of Interior when the NBS was formed. Only the biological research components of the department have become part of the NBS. This nonadvocacy biological science program will help the nation to resolve increasingly contentious and challenging issues in managing its biological resources.

References

Cameron, J. 1929. The Bureau of Biological Survey: its history, activities, and organization. John Hopkins Press, New York. 339 pp.

National Research Council (Committee on the Formation of the National Biological Survey). 1993. A biological survey for the nation. National Academy Press, Washington, DC. 205 pp.

For further information:

Milton Friend
National Biological Service
National Wildlife Health Center
6006 Schroeder Rd.
Madison, WI 53711

PART 2

Distribution, Abundance, and Health

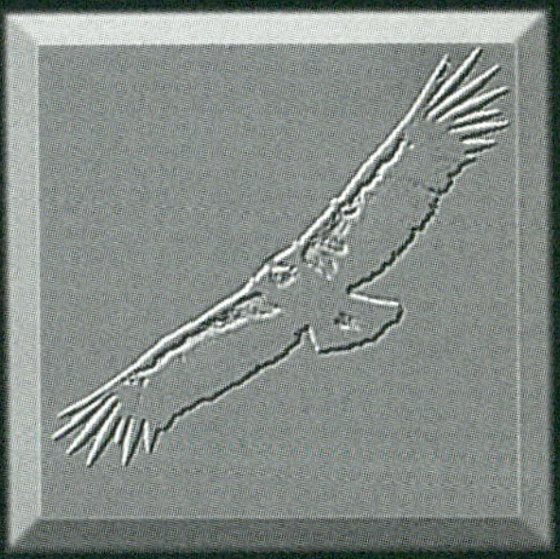

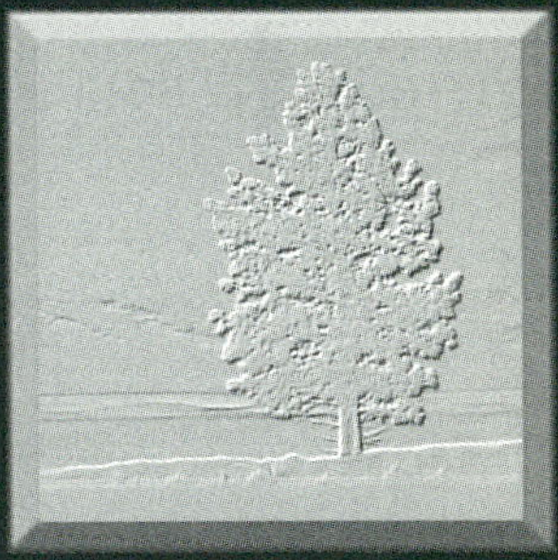

Distribution, Abundance, and Health of *Species*

Birds *15*

Mammals *93*

Reptiles and Amphibians *117*

Fishes *141*

Invertebrates *159*

Plants *189*

Courtesy J.C. Leupold, NBS

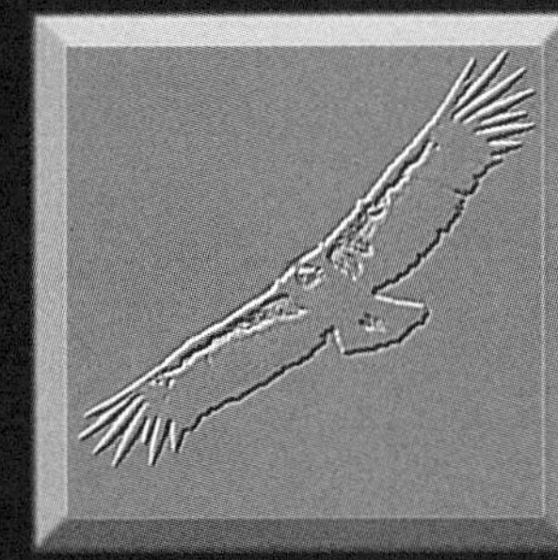

Birds

by
Science Editor
Russell J. Hall
National Biological Service
Division of Research
Washington, DC 20240

Overview

Migratory bird populations are an international resource for which there is special federal responsibility. Moreover, birds are valued and highly visible components of natural ecosystems that may be indicators of environmental quality. Consequently, many efforts have been directed toward measuring and monitoring the condition of North America's migratory bird fauna. The task is not an easy one because the more than 700 U.S. species of migratory birds are highly mobile and may occur in the United States during only part of their annual cycle. Some species annually make round-trip migrations spanning thousands of kilometers or miles, others engage in short or irregular migrations of tens or hundreds of kilometers, and even resident species are capable of moving great distances over short intervals. One often cannot tell whether a bird observed at a given moment is a resident, a migrant, a visitor from another locality, or the same individual seen 10 minutes earlier.

Determining status and trends is further complicated by the fact that each of these species has its own patterns of distribution and abundance, and each species has populations that respond to different combinations of environmental factors. Finally, the sheer abundance of birds estimated at 20 billion individuals in North America at its annual late-summer peak (Robbins et al. 1966) may make it difficult to obtain accurate counts of common species, and the absolute abundance of some may mask important changes in their status.

Biologists have developed many different approaches to determining abundance and trends in abundance, and nearly all of the recognized census methods applicable to birds are represented by the articles in this section. Not surprisingly, trends among the large number of populations treated are mixed.

Results from the nationwide Breeding Bird Survey (Peterjohn et al., this section) and a portion of the large-scale Christmas Bird Count (Root and McDaniel, this section) show that some populations are declining, others increasing, and many show what appears to be normal fluctuations around a more or less stable average. Overall, approximately equal numbers of species appear to be increasing and decreasing over the past two to three decades. Groups of species with the most consistent declines are those characteristic of grassland habitats, apparently reflecting conversion of these habitats to other types of vegetative cover.

Waterfowl populations are monitored closely as a basis for regulating annual harvests at levels consistent with maintenance of populations. Goose populations (Rusch et al.,

Hestbeck's "Canada Geese," Hupp et al., all this section) have shown some impressive gains over the past decades, but most gains have been registered by large-bodied geese, with several smaller species and smaller subspecies of the highly variable Canada goose (*Branta canadensis*) having depressed populations.

Censusing and determining the status of natural Canada goose populations are made more difficult by the widespread introduction and establishment of resident goose populations, which breed outside the traditional Arctic nesting areas and mix with migratory populations on the wintering grounds.

Duck surveys address more than 30 species that might be legally hunted. Even though some species are stable or even increasing, many duck populations have declined in the past decade (Caithamer and Smith, this section). Biologists attribute these declines to losses of breeding and wintering habitats and a long period of drought in breeding areas. Among species receiving special emphasis, canvasbacks (*Aythya valisineria*; Hohman et al., this section) showed a complex pattern with regional changes in distribution and abundance, and pintails (*Anas acuta*; Hestbeck's "Decline of Northern Pintails," this section) showed a widespread and nearly consistent pattern of decline.

Results are preliminary, but two new census programs, the MAPS and BBIRD programs (Martin et al., this section), promise to provide much higher quality information on status and trends by measuring not only the presence of bird populations in breeding areas, but also their success. When fully operational, this approach may offer important clues regarding the causes of observed population changes.

Shorebirds are highly migratory, and status and trends of their populations are largely determined from observations made during periods in their life cycles in which birds congregate in limited breeding, staging, or migratory stopover areas. Populations of eastern (Harrington, this section) and western (Gill et al., this section) species show general patterns of decline, although some species, including those using inland areas, are too poorly studied to detect trends. Apparent dependence on critical breeding and staging areas suggests that populations of many species are vulnerable to habitat loss and disturbance.

Seabirds in the Pacific region (Carter et al., Hatch and Piatt, both this section) include many diverse species that respond differently to factors such as human proximity to nesting areas, oil spills, introduction of predators, depletion of fishery stocks, and availability of human refuse as food. Some species, including certain gulls, brown pelicans (*Pelecanus occidentalis*), and double-crested cormorants (*Phalacrocorax auritus*), have responded positively to recent changes in some areas, whereas others, including murrelets and murres (Family Alcidae) and kittiwakes (Genus *Rissa*), have shown declining trends. Populations of other species appear to fluctuate widely, and information for many species is insufficient to determine long-term trends.

Colonial waterbirds of the continental and east coast regions of the United States (Erwin, this section) show trends related to many of the same factors operating in the Pacific region, with some species recovering from past losses from pesticides while some other species that exploit human refuse are increasing dramatically. Populations of other species, especially certain terns, are declining, probably as a result of habitat loss and degradation or other kinds of human disturbance. Special efforts have been made to determine status and trends of the piping plover (*Charadrius melodus*; Haig and Plissner, this section), a species listed as endangered in certain parts of its range and as threatened in others.

Populations of raptors (Fuller et al., this section) are difficult to census, but ospreys (*Pandion haliaetus*), bald eagles (*Haliaeetus leucocephalus*), and peregrine falcons (*Falco peregrinus*) have increased in numbers as they recover from past effects of pesticides. Populations of most vultures, hawks, and owls are either poorly known or believed to be stable. Notable exceptions are California condors (*Gymnogyps californianus*; Pattee and Mesta, this section), the crested caracara (*Caracara plancus*; Layne, this section), and spotted owls (*Strix occidentalis*), all of which enjoy or have been considered for additional protection. Mortality factors of eagles (Franson et al., this section) have been monitored and, although these data do not directly measure population status, they do indicate trends in the kinds of factors that tend to depress population growth.

The wild turkey (*Meleagris gallopavo*; Dickson, this section) has shown dramatic increases in distribution and abundance in recent decades because of translocations, habitat restoration, and harvest control. Mourning doves (*Zenaida macroura*; Dolton, this section) have shown generally stable populations, although recent population declines in the western states are disturbing. Regional increases of ravens (*Corvus corax*) in the southwest (Boarman and Berry, this section) are primarily of concern because of their potential effects as predators on eggs and young of the desert tortoise (*Gopherus agassizii*).

Populations of severely endangered species, like the California condor (Pattee and Mesta, this section), the Mississippi sandhill crane (*Grus canadensis pulla*; Gee and Hereford, this section), and the Puerto Rican parrot (*Amazona vittata*; Meyers, this section), are reasonably well known. Through censusing these species, biologists have tracked declines, often to a few individuals, and slow recoveries resulting from intensive management activities. Other rare species have populations that are depleted or vulnerable because of recent trends, but which can be censused with far less certainty. For example, willow flycatchers (*Empidonax traillii*; Sogge, this section) breed sparsely in parts of the Grand Canyon where exotic species have displaced natural riparian vegetation; likewise, the status of the red-cockaded woodpecker (*Picoides borealis*) appears closely tied to the decline of the longleaf pine (*Pinus palustris*) ecosystem (Costa and Walker, this section).

Broad-scale programs such as the Breeding Bird Survey, annual waterfowl surveys, and wintering surveys such as the Christmas Bird Count may provide information on status and trends for as many as 75% of U.S. bird species, at least to the extent that they would provide evidence of catastrophic declines. Remaining species may be censused only with difficulty and often with imprecision because they are secretive, rare, highly mobile, or occupy poorly accessible areas. Specialized surveys provide information on some of these groups but, as indicated by the articles in this section, they do so with varying degrees of success. Much work remains to be done on obtaining better information and developing better ways of interpreting available information on difficult-to-census species.

If any overall conclusion is possible on the wide array of information now available on status and trends of bird populations it is this: apparent stability for many species; increases in some species, many of which are generalists adaptable to altered habitats; and decreases in other species, many of which are specialists most vulnerable to habitat loss and degradation.

Reference

Robbins, C.S., B. Bruun, and H.S. Zim. 1966. Birds of North America. Golden Press, New York. 340 pp.

Breeding Bird Survey: Population Trends 1966-92

by
Bruce J. Peterjohn
John R. Sauer
Sandra Orsillo
National Biological Service

The North American Breeding Bird Survey (BBS) was begun in 1966 to collect standardized data on bird populations along more than 3,400 survey routes across the continental United States and southern Canada. The BBS has been used to document distributions and establish continental, regional, and local population trends for more than 250 species.

We summarize here survey-wide patterns in the 1966-92 population trend estimates for 245 species of birds observed on a minimum of 40 routes with a mean relative abundance of 1.0 bird per route. Survey-wide trend estimates are also summarized for six groupings of birds, providing insight into broad geographical patterns of population trends of North American birds.

Methods

The BBS routes are located along secondary roads and surveyed each year during the peak of the breeding season by observers competent in bird identification. Each route is 39.4 km (24.5 mi) long, with 50 stops placed at 0.8-km (0.5-mi) intervals (Robbins et al. 1986). To estimate population change, we used a procedure called route regression, described in greater detail by Geissler and Sauer (1990).

We examined population change in several ways. First, we estimated overall population change for individual species over the entire survey area. Second, we looked for temporal and geographic patterns in individual bird species (e.g., Sauer and Droege 1990).

Additionally, we analyzed overall patterns of population change for several species of particular management interest. Groups of birds were defined by migration status (nonmigratory, short-distance, and Neotropical migrants) or by breeding habitat (grassland, shrubland, or woodland; *see* also Peterjohn and Sauer 1993). For each group, we determined the percentage of species with positive (> 0) trends. If population change is not consistent within the group, about half (50%) of the species should show positive trends. Clearly, some species will show very significant declines (or increases) over the interval, and these species can be identified in the Appendix. However, the percentage of species with positive population trends is a convenient summary of information from all species within the group to demonstrate overall trend patterns.

Finally, to display regional patterns of population change, we calculated the mean trend for the species in each group for each survey route. We used an Arc/Info geographic information system to summarize and display geographic patterns of population change (Isaaks and Srivastava 1989; ESRI 1992).

Trends

Of the 245 species considered, 130 have negative trend estimates, 57 of which exhibit significant declines. Species with negative trend estimates are found in all families, but they are especially prevalent among the mimids (mockingbirds and thrashers) and sparrows. A total of 115 species exhibits positive trends, 44 of which are significant increases. Flycatchers and warblers have the largest proportions of species with increasing populations.

The percentage of increasing species within each group of species having shared life-history traits is summarized in the Table. The most consistent declines are by grassland birds; only 18% have increasing population trends. These declines are most widespread in eastern North America, where few grassland species breed (Fig. 1). Declining populations are also prevalent across the Great Plains, which includes the breeding ranges of most grassland birds. The pattern within western North America is mixed, except for regions of declines along the Pacific coast.

A significant proportion of shrubland and old-field bird species also exhibits population declines (Table). As with grassland birds, regions with declines are most prevalent in eastern North America as well as in the southern Great Plains from Kansas and Missouri south to Texas (Fig. 2). Shrubland species appear to be generally increasing in western North America.

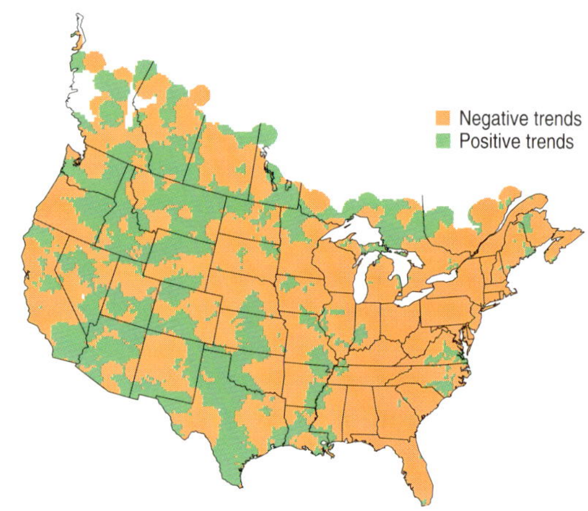

Fig. 1. Geographic patterns in the mean trends for grassland bird species during 1966-92.

Table. Percentage of species with increasing populations for six groups of birds having shared life-history traits. The *P* value indicates the probability that the percentage differs from 50%.

Group	No. of species in each group	Increasing (%)	*P*
Breeding habitats			
Grassland	17	18	0.01
Shrubland	58	34	0.02
Woodland	80	59	0.15
Migration			
Short distance	69	42	0.23
Nonmigratory	41	41	0.35
Neotropical	98	50	0.92
All species	237	47	0.36

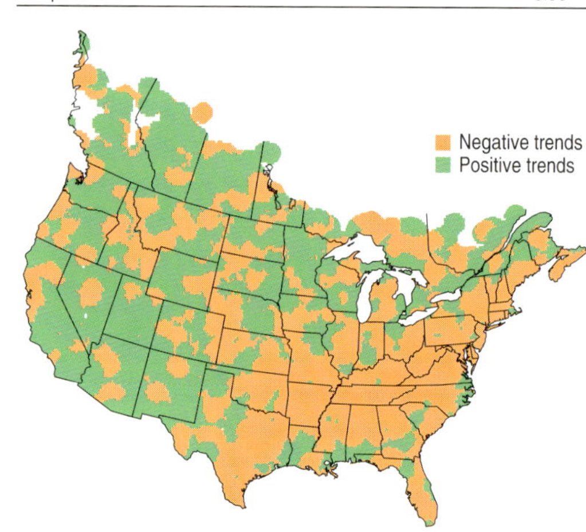

Fig. 2. Geographic patterns in the mean trends for shrubland and old-field bird species during 1966-92.

A majority of woodland bird populations is increasing across most of the continent (Fig. 3). Decreasing populations prevail in a few regions, such as along the Appalachians from West Virginia to northern Alabama, from Arkansas across central Texas, and along the Pacific coast from Oregon to central California. Woodland birds, however, are increasing in more areas than either grassland or early successional species.

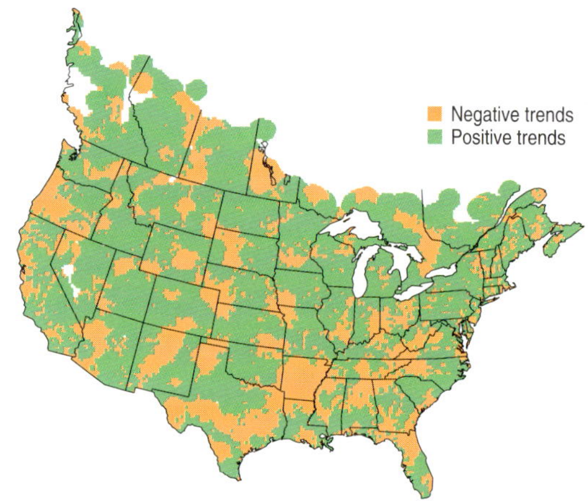

Fig. 3. Geographic patterns in the mean trends for woodland bird species during 1966-92.

Neotropical migrants have received considerable attention in recent years, yet as many species have increased as have decreased during 1966-92 (Table). A region with apparently declining populations extends from the southern Great Plains across the southeastern states and along the Appalachian Mountains to southern New England (Fig. 4). Increasing mean populations prevail across the northern Great Plains and throughout much of western North America. The pattern of population decline in the eastern United States noted by Robbins et al. (1989) occurred after 1978 and is not reflected in these long-term trends.

Short-distance migrants and permanent residents have slightly greater percentages of decreasing species (Table). Both groups have negative mean trends in the southeastern states and from the lower Great Lakes into the Appalachian Mountains, but the patterns elsewhere are mixed (Figs. 5, 6).

These results indicate that grassland and shrubland birds are experiencing the most consistent and widespread declines of any group of species. Whenever possible, appropriate conservation measures should be undertaken to enhance the population trends of these species. While the BBS data indicate the population

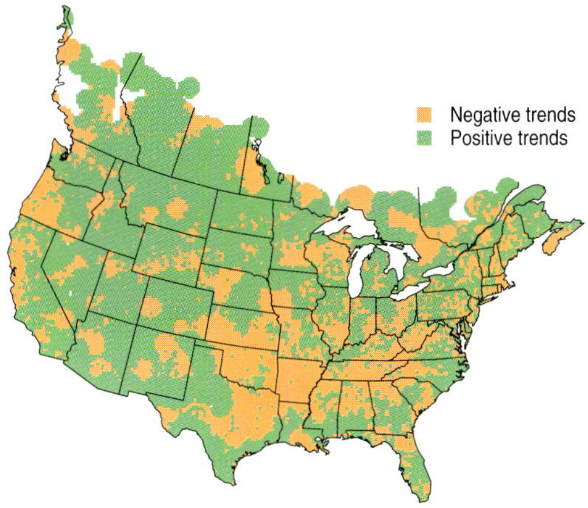

Fig. 4. Geographic patterns in the mean trends for Neotropical migrant bird species during 1966-92.

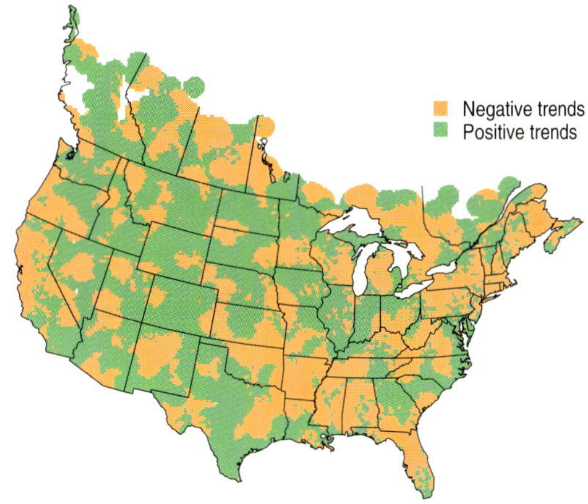

Fig. 5. Geographic patterns in the mean trends for short-distance migrant bird species during 1966-92.

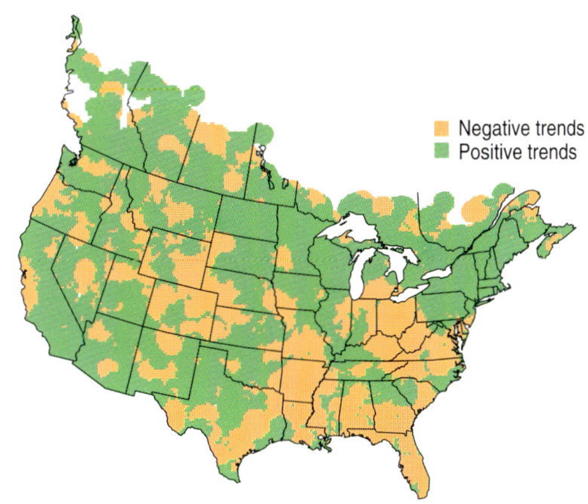

Fig. 6. Geographic patterns in the mean trends for permanent resident bird species during 1966-92.

trends for breeding birds, these data are not designed to identify the factors responsible for these trends. To understand how bird populations are responding to the changing habitat conditions in North America, additional studies are needed that would combine the BBS results with regional data on land-use changes, weather conditions, and other variables.

References

ESRI. 1992. Understanding GIS: the Arc/Info method. Environmental Systems Research Institute, Inc., Redlands, CA. 416 pp.

Geissler, P.H., and J.R. Sauer. 1990. Topics in route-regression analysis. Pages 53-56 *in* J.R. Sauer and S. Droege, eds. Survey designs and statistical methods for the estimation of avian population trends. U.S. Fish and Wildlife Service Biological Rep. 90(1).

Isaaks, E.H., and R.M. Srivastava. 1989. An introduction to applied geostatistics. Oxford University Press, New York. 561 pp.

Peterjohn, B.G., and J.R. Sauer. 1993. North American Breeding Bird Survey annual summary 1990-1991. Bird Populations 1:52-67.

Robbins, C.S., D. Bystrak, and P.H. Geissler. 1986. The Breeding Bird Survey: its first fifteen years, 1965-1979. U.S. Fish and Wildlife Service Resour. Publ. 157. 196 pp.

Robbins, C.S., J.R. Sauer, R.S. Greenberg, and S. Droege. 1989. Population declines in North American birds that migrate to the Neotropics. Proceedings of the National Academy of Science USA 86:7658-7662.

Sauer, J.R., and S. Droege. 1990. Recent population trends of the eastern bluebird. Wilson Bull. 102:239-252.

Appendix. Population trends of birds from the North American Breeding Bird Survey. To appear in this list, the species must have been seen on > 40 routes at an average count of > 1 bird/route. We present trends (%/year), probability (*P*), and the number of routes on which the species was seen. *See* Peterjohn and Sauer 1993 for group classification.

Species	Scientific name	Trend	*P**	No. of routes
American white pelican	*Pelecanus erythrorhynchos*	3.60	0.00	152
Great egret	*Casmerodius albus*	1.5	ns	513
Little blue heron	*Egretta caerulea*	-1.45	ns	429
Cattle egret	*Bubulcus ibis*	2.09	ns	475
White ibis	*Eudocimus albus*	3.17	ns	173
White-faced ibis	*Plegadis chihi*	32.27	0.01	61
Canada goose	*Branta canadensis*	7.05	0.00	1,090
Mottled duck	*Anas fulvigula*	-5.27	0.03	64
Mallard	*A. platyrhynchos*	0.98	ns	1,890
Northern pintail	*A. acuta*	-5.65	0.00	502
Blue-winged teal	*A. discors*	-0.92	ns	814
Northern shoveler	*A. clypeata*	0.18	ns	379
Gadwall	*A. strepera*	3.76	0.00	389
Lesser scaup	*Aythya affinis*	2.08	ns	263
Red-breasted merganser	*Mergus serrator*	-9.57	0.02	53
Black vulture	*Coragyps atratus*	1.72	ns	540
Turkey vulture	*Cathartes aura*	0.37	ns	1,691
Ring-necked pheasant	*Phasianus colchicus*	-1.24	0.10	1,263
Northern bobwhite	*Colinus virginianus*	-2.43	0.00	1,338
Scaled quail	*Callipepla squamata*	-3.31	0.00	104
Gambel's quail	*C. gambelii*	0.90	ns	82
California quail	*C. californica*	-0.04	ns	264
Mountain quail	*Oreortyx pictus*	1.37	ns	112
American coot	*Fulica americana*	-0.51	ns	620

Species	Scientific name	Trend	*P**	No. of routes
Sandhill crane	*Grus canadensis*	4.30	0.00	259
Killdeer	*Charadrius vociferus*	-0.38	ns	2,692
Black-necked stilt	*Himantopus mexicanus*	0.63	ns	119
Willet	*Catoptrophorus semipalmatus*	-0.72	ns	295
Upland sandpiper	*Bartramia longicauda*	3.28	0.00	687
Long-billed curlew	*Numenius americanus*	-1.61	ns	234
Marbled godwit	*Limosa fedoa*	0.71	ns	188
Common snipe	*Gallinago gallinago*	0.14	ns	1,011
Laughing gull	*Larus atricilla*	6.01	0.00	125
Franklin's gull	*L. pipixcan*	-5.95	ns	231
Ring-billed gull	*L. delawarensis*	7.43	0.02	684
California gull	*L. californicus*	-1.27	ns	230
Herring gull	*L. argentatus*	-2.06	0.09	474
Glaucous-winged gull	*L. glaucescens*	3.85	0.09	40
Great black-backed gull	*L. marinus*	-1.47	ns	125
Black tern	*Chlidonias niger*	-4.51	0.00	368
Rock dove	*Columba livia*	1.04	0.06	2,255
Band-tailed pigeon	*C. fasciata*	-3.69	0.00	189
White-winged dove	*Zenaida asiatica*	0.03	ns	78
Mourning dove	*Z. macroura*	0.02	ns	2,726
Common ground dove	*Columbina passerina*	-3.13	0.01	194
Yellow-billed cuckoo	*Coccyzus americanus*	-1.30	0.00	1,637
Lesser nighthawk	*Chordeiles acutipennis*	5.08	0.03	118
Common nighthawk	*C. minor*	-0.34	ns	1,609

Species	Scientific name	Trend	P*	No. of routes
Chuck-will's-widow	*Caprimulgus carolinensis*	-0.78	ns	522
Black swift	*Cypseloides niger*	1.61	ns	79
Chimney swift	*Chaetura pelagica*	-0.84	0.08	1,789
White-throated swift	*Aeronautes saxatalis*	-3.38	ns	189
Broad-tailed hummingbird	*Selasphorus platycercus*	0.42	ns	115
Rufous hummingbird	*S. rufus*	-3.38	0.00	188
Red-headed woodpecker	*Melanerpes erythrocephalus*	-1.84	0.00	1,236
Acorn woodpecker	*M. formicivorus*	0.98	ns	138
Golden-fronted woodpecker	*M. aurifrons*	-1.86	ns	56
Red-bellied woodpecker	*M. carolinus*	0.59	ns	1,246
Yellow-bellied sapsucker	*Sphyrapicus varius*	-0.85	ns	605
Nuttall's woodpecker	*Picoides nuttallii*	1.44	ns	95
Downy woodpecker	*P. pubescens*	0.14	ns	2,214
Yellow-shafted flicker	*Colaptes auratus*	-2.75	0.00	2,062
Red-shafted flicker	*C. cafer*	-0.87	ns	689
Olive-sided flycatcher	*Contopus borealis*	-2.52	0.00	736
Western wood-pewee	*C. sordidulus*	-0.39	ns	637
Eastern wood-pewee	*C. virens*	-1.64	0.00	1,719
Yellow-bellied flycatcher	*Empidonax flaviventris*	3.58	0.01	263
Acadian flycatcher	*E. virescens*	0.50	ns	854
Alder flycatcher	*E. alnorum*	1.30	0.04	788
Willow flycatcher	*E. traillii*	-0.62	ns	1,152
Least flycatcher	*E. minimus*	-0.55	ns	1,150
Hammond's flycatcher	*E. hammondii*	1.50	ns	221
Dusky flycatcher	*E. oberholseri*	0.72	ns	265
Pacific-slope flycatcher	*E. difficilis*	1.47	ns	218
Eastern phoebe	*Sayornis phoebe*	0.64	ns	1,650
Ash-throated flycatcher	*Myiarchus cinerascens*	2.38	0.01	370
Great crested flycatcher	*M. crinitus*	0.03	ns	1,804
Brown-crested flycatcher	*M. tyrannulus*	6.15	0.00	47
Cassin's kingbird	*Tyrannus vociferans*	-1.74	ns	138
Western kingbird	*T. verticalis*	1.51	0.01	942
Eastern kingbird	*T. tyrannus*	-0.10	ns	2,267
Scissor-tailed flycatcher	*T. forficatus*	-0.08	ns	244
Horned lark	*Eremophila alpestris*	-0.65	ns	1,750
Purple martin	*Progne subis*	0.71	ns	1,623
Tree swallow	*Tachycineta bicolor*	1.27	0.04	1,707
Violet-green swallow	*T. thalassina*	0.76	ns	511
Northern rough-winged swallow	*Stelgidopteryx serripennis*	0.95	ns	2,119
Bank swallow	*Riparia riparia*	-0.48	ns	1,318
Cliff swallow	*Hirundo pyrrhonota*	0.98	ns	1,737
Barn swallow	*H. rustica*	0.37	ns	2,701
Gray jay	*Perisoreus canadensis*	-1.28	ns	350
Steller's jay	*Cyanocitta stelleri*	0.39	ns	328
Blue jay	*C. cristata*	-1.81	0.00	1,986
Scrub jay	*Aphelocoma coerulescens*	1.27	0.04	272
Pinyon jay	*Gymnorhinus cyanocephalus*	-1.65	ns	132
Black-billed magpie	*Pica pica*	-1.34	0.05	577
American crow	*Corvus brachyrhynchos*	0.85	0.06	2,578
Fish crow	*C. ossifragus*	2.93	0.01	466
Chihuahuan raven	*C. cryptoleucus*	-2.48	ns	87
Common raven	*C. corax*	3.66	0.00	1,202
Black-capped chickadee	*Parus atricapillus*	1.89	0.00	1,433
Carolina chickadee	*P. carolinensis*	-0.67	ns	862
Mountain chickadee	*P. gambeli*	0.07	ns	291
Chestnut-backed chickadee	*P. rufescens*	-1.54	ns	133
Plain titmouse	*P. inornatus*	-2.30	0.01	180
Tufted titmouse	*P. bicolor*	0.64	ns	1,289
Black-crested titmouse	*P.b. atricristatus*	2.06	0.03	64
Verdin	*Auriparus flavicep*	-1.38	ns	97
Bushtit	*Psaltriparus minimus*	-1.13	ns	250
Red-breasted nuthatch	*Sitta canadensis*	2.48	0.00	872
Brown-headed nuthatch	*S. pusilla*	-1.30	ns	290
Cactus wren	*Campylorhynchus brunneicapillus*	-0.89	ns	136
Rock wren	*Salpinctes obsoletus*	-1.68	0.04	509
Carolina wren	*Thryothorus ludovicianus*	1.01	0.03	1,118
Bewick's wren	*Thryomanes bewickii*	-0.35	ns	594
House wren	*Troglodytes aedon*	1.55	0.00	1,924
Winter wren	*T. troglodytes*	2.25	ns	659
Golden-crowned kinglet	*Regulus satrapa*	-0.01	ns	541
Ruby-crowned kinglet	*R. calendula*	-1.31	ns	656

Species	Scientific name	Trend	P*	No. of routes
Blue-gray gnatcatcher	*Polioptila caerulea*	1.03	ns	1,233
Black-tailed gnatcatcher	*P. melanura*	-0.22	ns	57
Eastern bluebird	*Sialia sialis*	2.52	0.00	1,633
Mountain bluebird	*S. currucoides*	0.56	ns	422
Veery	*Catharus fuscescens*	-1.06	0.06	964
Gray-cheeked thrush	*C. minimus*	-4.46	ns	43
Swainson's thrush	*C. ustulatus*	0.00	ns	707
Hermit thrush	*C. guttatus*	2.10	0.01	912
Wood thrush	*Hylocichla mustelina*	-1.88	0.00	1,510
American robin	*Turdus migratorius*	1.03	0.01	2,588
Varied thrush	*Ixoreus naevius*	2.16	0.06	148
Wrentit	*Chamaea fasciata*	-1.39	ns	113
Gray catbird	*Dumetella carolinensis*	-0.42	ns	1,941
Northern mockingbird	*Mimus polyglottos*	-0.98	0.03	1,694
Sage thrasher	*Oreoscoptes montanus*	1.16	ns	244
Brown thrasher	*Toxostoma rufum*	-1.19	0.01	1,917
Curve-billed thrasher	*T. curvirostre*	-3.59	0.00	100
California thrasher	*T. redivivum*	-4.06	0.05	83
Sprague's pipit	*Anthus spragueii*	-3.52	0.02	140
Cedar waxwing	*Bombycilla cedrorum*	2.36	0.00	1627
Phainopepla	*Phainopepla nitens*	2.53	0.05	104
Loggerhead shrike	*Lanius ludovicianus*	-3.20	0.00	1364
European starling	*Sturnus vulgaris*	-0.99	0.02	2727
White-eyed vireo	*Vireo griseus*	-0.15	ns	945
Solitary vireo	*V. solitarius*	3.28	0.00	954
Warbling vireo	*V. gilvus*	1.31	0.01	1740
Philadelphia vireo	*V. philadelphicus*	1.50	ns	191
Red-eyed vireo	*V. olivaceus*	1.39	0.01	2020
Tennessee warbler	*Vermivora peregrina*	4.21	ns	341
Orange-crowned warbler	*V. celata*	-0.71	ns	346
Nashville warbler	*V. ruficapilla*	1.35	ns	673
Northern parula	*Parula americana*	0.82	ns	970
Yellow warbler	*Dendroica petechia*	0.94	0.05	2161
Chestnut-sided warbler	*D. pensylvanica*	-0.60	ns	788
Magnolia warbler	*D. magnolia*	2.80	0.00	527
Cape May warbler	*D. tigrina*	2.95	ns	239
Myrtle warbler	*D. coronata*	1.41	0.09	575
Audubon's warbler	*D.c. auduboni*	0.08	ns	386
Black-throated gray warbler	*D. nigrescens*	2.32	0.07	190
Townsend's warbler	*D. townsendi*	1.63	ns	145
Hermit warbler	*D. occidentalis*	0.79	ns	82
Black-throated green warbler	*D. virens*	-0.45	ns	637
Blackburnian warbler	*D. fusca*	0.87	ns	511
Pine warbler	*D. pinus*	2.12	0.00	797
Prairie warbler	*D. discolor*	-2.15	0.00	773
Bay-breasted warbler	*D. castanea*	-0.04	ns	216
Blackpoll warbler	*D. striata*	-0.33	ns	178
Black-and-white warbler	*Mniotilta varia*	0.91	ns	1126
American redstart	*Setophaga ruticilla*	-0.58	ns	1299
Ovenbird	*Seiurus aurocapillus*	0.55	ns	1278
Northern waterthrush	*S. noveboracensis*	0.49	ns	615
Kentucky warbler	*Oporornis formosus*	-0.77	ns	685
Mourning warbler	*O. philadelphia*	0.15	ns	538
MacGillivray's warbler	*O. tolmiei*	-0.58	ns	309
Common yellowthroat	*Geothlypis trichas*	-0.48	ns	2361
Hooded warbler	*Wilsonia citrina*	1.49	ns	608
Wilson's warbler	*W. pusilla*	0.53	ns	525
Canada warbler	*W. canadensis*	-0.73	ns	504
Yellow-breasted chat	*Icteria virens*	-0.43	ns	1273
Summer tanager	*Piranga rubra*	-0.19	ns	761
Scarlet tanager	*P. olivacea*	0.22	ns	1257
Western tanager	*P. ludoviciana*	-0.31	ns	472
Northern cardinal	*Cardinalis cardinalis*	-0.21	ns	1591
Pyrrhuloxia	*C. sinuatus*	-0.73	ns	61
Rose-breasted grosbeak	*Pheucticus ludovicianus*	-0.19	ns	1146
Black-headed grosbeak	*P. melanocephalus*	-0.32	ns	509
Blue grosbeak	*Guiraca caerulea*	1.86	0.00	1014
Lazuli bunting	*Passerina amoena*	0.14	ns	417
Indigo bunting	*P. cyanea*	-0.57	ns	1725
Painted bunting	*P. ciris*	-3.21	0.00	269
Dickcissel	*Spiza americana*	-1.58	0.02	791
Green-tailed towhee	*Pipilo chlorurus*	0.41	ns	212
Rufous-sided towhee	*P. erythrophthalmus*	-1.99	0.00	1951

Species	Scientific name	Trend	P*	No. of routes
California towhee	*P. californicus*	-0.22	ns	113
Brown towhee	*P. fuscus*	-2.67	0.00	83
Cassin's sparrow	*Aimophila cassinii*	-2.85	0.00	171
Chipping sparrow	*Spizella passerina*	-0.04	ns	2,300
Clay-colored sparrow	*S. pallida*	-1.31	0.02	444
Brewer's sparrow	*S. breweri*	-3.68	0.00	376
Field sparrow	*S. pusilla*	-3.25	0.00	1,581
Vesper sparrow	*Pooecetes gramineus*	-0.25	ns	1,488
Lark sparrow	*Chondestes grammacus*	-3.42	0.00	935
Black-throated sparrow	*Amphispiza bilineata*	-3.78	0.02	225
Sage sparrow	*A. belli*	-2.43	ns	210
Lark bunting	*Calamospiza melanocorys*	-2.86	0.03	359
Savannah sparrow	*Passerculus sandwichensis*	-0.57	ns	1,461
Baird's sparrow	*Ammodramus bairdii*	-1.52	ns	134
Grasshopper sparrow	*A. savannarum*	-4.48	0.00	1,479
Fox sparrow	*Passerella iliaca*	0.44	ns	224
Song sparrow	*Melospiza melodia*	-0.80	0.09	2,079
Lincoln's sparrow	*M. lincolnii*	3.99	0.02	420
Swamp sparrow	*M. georgiana*	0.50	ns	783
White-throated sparrow	*Zonotrichia albicollis*	-1.44	0.01	635
White-crowned sparrow	*Z. leucophrys*	-1.91	0.01	274
Slate-colored junco	*Junco hyemalis*	-0.47	ns	545
Oregon junco	*J.h. oregonus*	-1.23	0.08	341
Gray-headed junco	*J.h. caniceps*	2.04	ns	50
McCown's longspur	*Calcarius mccownii*	8.32	0.00	68
Chestnut-collared longspur	*C. ornatus*	0.62	ns	153
Bobolink	*Dolichonyx oryzivorus*	-1.33	0.01	1,147

Species	Scientific name	Trend	P*	No. of routes
Red-winged blackbird	*Agelaius phoeniceus*	-1.06	0.01	2,760
Tricolored blackbird	*A. tricolor*	4.83	ns	69
Eastern meadowlark	*Sturnella magna*	-2.18	0.00	1,742
Western meadowlark	*S. neglecta*	-0.56	ns	1,334
Yellow-headed blackbird	*Xanthocephalus xanthocephalus*	1.53	ns	649
Brewer's blackbird	*Euphagus cyanocephalus*	-1.15	0.06	1,006
Great-tailed grackle	*Quiscalus mexicanus*	7.40	0.00	198
Boat-tailed grackle	*Q. major*	2.52	0.05	118
Common grackle	*Q. quiscula*	-1.44	0.00	2,196
Bronzed cowbird	*Molothrus aeneus*	-1.12	ns	55
Brown-headed cowbird	*M. ater*	-0.88	0.06	2,780
Orchard oriole	*Icterus spurius*	-1.38	0.03	1,313
Baltimore oriole	*I. galbula*	0.26	ns	1,594
Bullock's oriole	*I.g. bullockii*	-0.81	ns	614
Scott's oriole	*I. parisorum*	2.26	ns	113
Pine grosbeak	*Pinicola enucleator*	6.36	0.01	152
Purple finch	*Carpodacus purpureus*	-1.19	0.05	921
Cassin's finch	*C. cassinii*	1.27	ns	235
House finch	*C. mexicanus*	-0.14	ns	1,420
Red crossbill	*Loxia curvirostra*	2.13	ns	438
White-winged crossbill	*L. leucoptera*	-5.52	0.09	155
Pine siskin	*Carduelis pinus*	0.38	ns	778
Lesser goldfinch	*C. psaltria*	-1.22	ns	281
American goldfinch	*C. tristis*	-1.06	0.03	2,165
Evening grosbeak	*Coccothraustes vespertinus*	-0.37	ns	596
House sparrow	*Passer domesticus*	-1.65	0.00	2,557

*ns--not significant.

For further information:

Bruce G. Peterjohn
National Biological Service
Patuxent Environmental Science Center
Laurel, MD 20708

Winter Population Trends of Selected Songbirds

by

Terry L. Root
University of Michigan

Larry McDaniel
National Center for Atmospheric Research

Many studies have found significant changes, primarily declines, in populations of breeding birds throughout the United States. Most studies have focused on birds that migrate to the Neotropics for winter. Speculations about causes of observed declines have primarily implicated habitat fragmentation and loss (e.g., deforestation) in Central and South America. The National Audubon Society's Christmas Bird Counts (CBC), begun in the winter of 1900-01, provide the data needed to discern consistent population trends in birds wintering throughout the United States.

For this study we used the CBC data to examine population trends of songbirds with ranges that apparently are limited by lower temperatures in the North. We chose these species to track populations of birds that could be in peril in the future. These birds potentially will be more quickly affected by changing climate than other birds, and we need baseline information on them to document possible consequences of global climatic change. The species that are indeed declining need to be monitored because the possible synergistic effects of declining populations and changing climate could result in local and even regional extinctions.

Methods

We examined 30 years of CBC data (winters of 1959-60 to 1988-89) for 50 songbirds whose northern range edges are associated with January minimum temperatures (Root 1988b). For each songbird species or subspecies at each count site, we calculated the number of individuals seen per counting effort (e.g., hours of observation). Yearly averages for each of the conterminous states were determined from these values for each species. Data were used from all count sites that were censused at least 25 of the 30 years. For details on the method we used to calculate population trends, *see* Geissler and Noon (1981) and contact us. All of our conclusions rest on very conservative analyses.

Trends

Of the 50 songbirds examined, 27 (54%) exhibited a statistically and biologically significant trend in at least one state (Fig. 1). Of these 27 species, 16 (59%) had populations declining in more states than states in which they were increasing; 12 exhibited only declines and 4 had a population increase in at least one state. Ten (37%) of the 27 species had populations increasing in more states than states exhibiting declines, with 7 exhibiting only population increases. One (4%) species had populations increasing and decreasing in the same number of states.

In general, the populations of birds that eat seeds from grasses and forbs (e.g., sparrows and meadowlarks) seem to be declining more frequently than those of birds that eat seeds from shrubs and trees, or berries (e.g., tufted titmouse [*Parus bicolor*] and cedar waxwing [*Bombycilla*

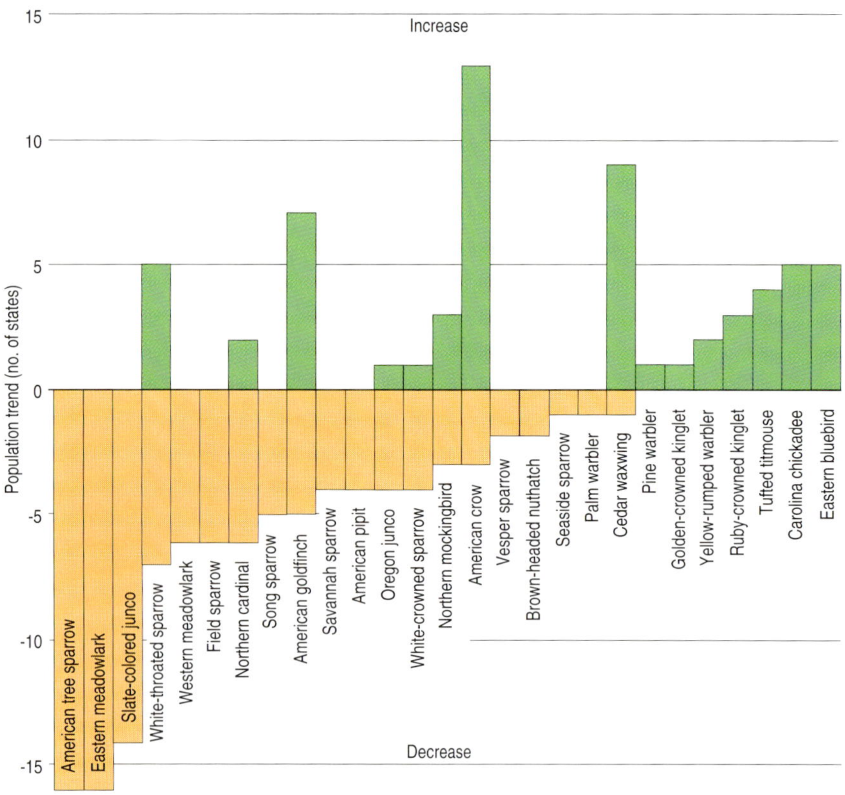

Fig. 1. Number of states with population trends either declining or increasing for 27 songbirds.

cedrorum]) (Fig. 1). This situation may be due to the fact that the grassland and early successional habitats are being modified, while ornamental fruiting bushes, shrubs, and trees planted in urbanized areas may be benefiting the increasing species (Beddall 1963). The explanation, however, is certainly more complex than this, given that some birds do not fit the pattern. For example, the American pipit (*Anthus rubescens*), which eats berries, crustaceans, and mollusks (Ehrlich et al. 1988), is decreasing in four states and increasing in none (Fig. 1).

To evaluate the areas of the conterminous states showing increases or decreases in their bird populations, we counted the number of species showing a population change in each state and then calculated the percentage with respect to the number of the 27 species occurring in each state (Fig. 2). A total of 24 (50%) of the states has greater than 5% of these wintering bird species showing positive population trends, while 32 (67%) show declines of similar magnitudes.

Mapping the percentages (Fig. 3) indicates that the largest increase is in South Carolina, with the far western states, those in the north-central region, and a scattering of states in the eastern portion of the conterminous states showing positive population trends.

The largest decreases (Fig. 3) are in South Carolina, Georgia, Florida, Alabama, Louisiana, and Delaware. The Pacific states, those in the

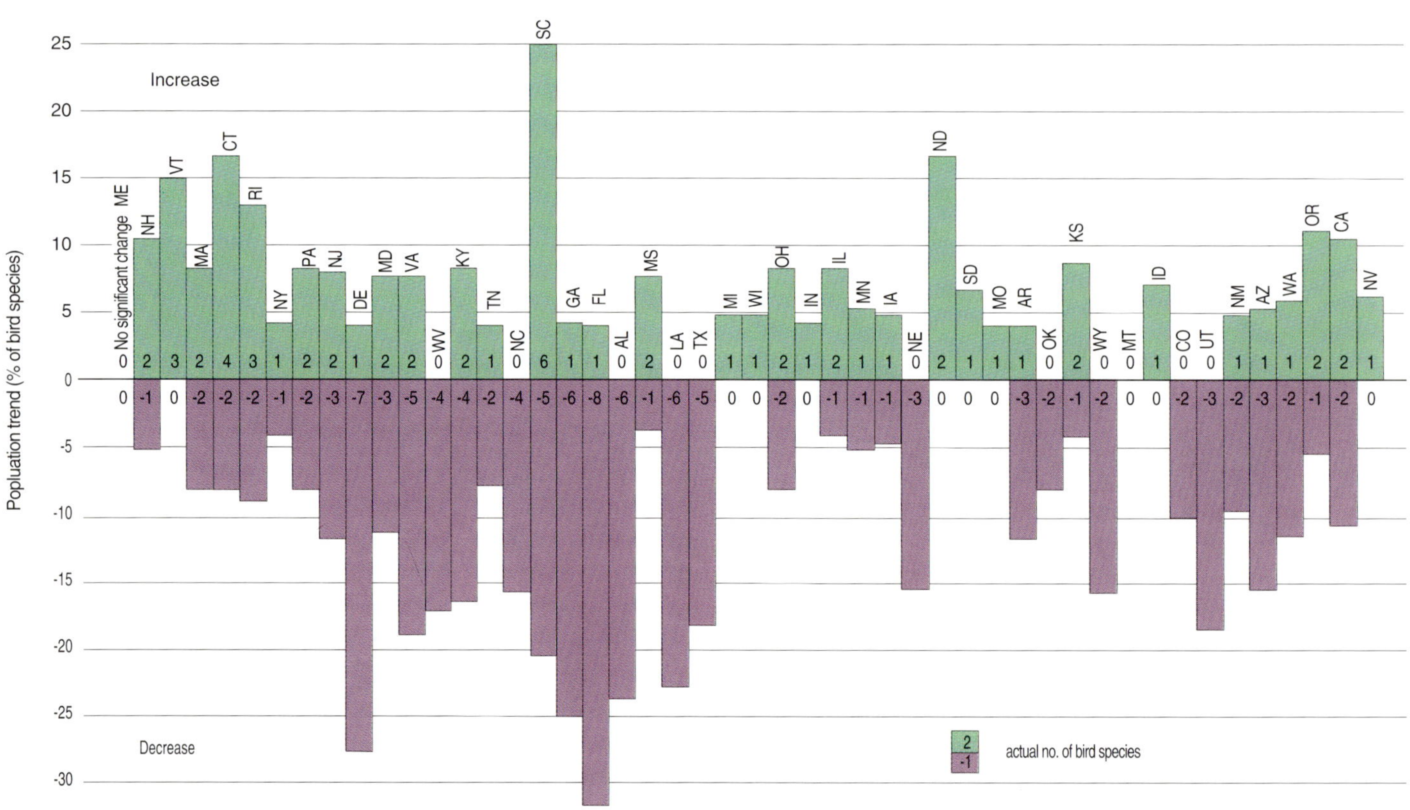

Fig. 2. Number and percentage of 27 birds showing declining and increasing population trends.

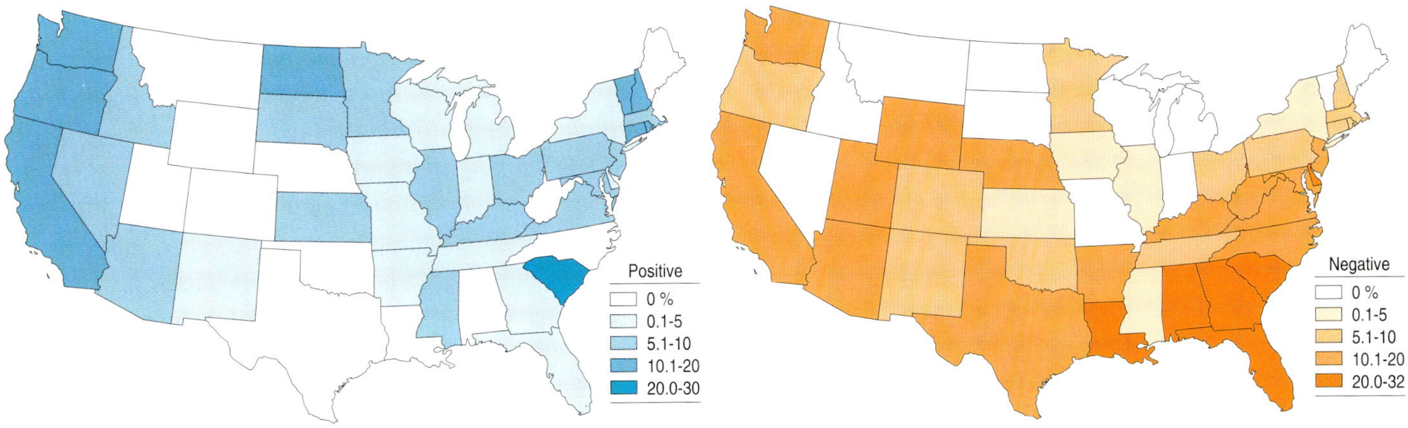

Fig. 3. Percentage of 27 birds showing positive and negative trends.

Great Plains, and the southeastern portion of the conterminous states generally show the greatest declines, though the actual reasons for these population changes will need to be examined in more detail. Certainly, the pattern of extensive declines in most of the southern coastal states is quite alarming.

Additionally, regions of the country that could be particularly influenced by global climatic change are the southern coasts (because of increased storms and degradation of coastal wetlands; IPCC 1990), and the Great Plains (owing to a significant decline in soil moisture; Leatherman 1992). Hence, the populations of birds in these areas need to be closely monitored to ensure preservation actions are taken before the combined effects of population declines and climate change result in extinctions. More studies and monitoring are warranted to understand the possible consequences of these patterns.

The analyses presented here can also be used to investigate population trends of target species across the country. Compare, for instance, the trends by state for the American tree sparrow (*Spizella arborea*; one of the most declining birds examined) and the cedar waxwing (one of the most increasing birds) with maps of their winter range and abundance patterns (Root 1988a). This comparison reveals that significant population trends, whether positive or negative, seem to occur primarily along these species' northern range boundaries and in many coastal states. Such analyses could help target specific regions of the country where population trends of key (e.g., threatened) species need watching.

References

Beddall, B.G. 1963. Range expansions of the cardinal and other birds in the northeastern states. Wilson Bull. 75:140-158.

Ehrlich, P.R., D.S. Dobkin, and D. Wheye. 1988. The birder's handbook. Simon and Schuster, New York. 785 pp.

Geissler, P.H., and B.R. Noon. 1981. Estimates of avian population trends from the North America Breeding Bird Survey. Pages 42-51 *in* C.J. Ralph and J.M. Scott, eds. Estimating the numbers of terrestrial birds. Studies in Avian Biology 6.

IPCC. 1990. Climate change: the IPCC scientific assessment. (also *see* Climate change 1992: the supplementary report to the IPCC scientific assessment.) Intergovernmental Panel on Climate Change. Cambridge University Press, New York, NY. 364 pp.

Leatherman, S.P. 1992. Sea level rise: implications and responses. Pages 256-263 *in* S.K. Majumdar, L.S. Kalkstein, B. Yarnal, E.W. Miller, and L.M. Rosenfeld, eds. Global climate change: implications, challenges and mitigation measures. Pennsylvania Academy of Science, Phillipsburg, NJ.

Root, T.L. 1988a. Atlas of wintering North American birds. University of Chicago Press, IL. 312 pp.

Root, T.L. 1988b. Environmental factors associated with avian distributional boundaries. Journal of Biogeography 15:489-505.

For further information:

Terry L. Root
University of Michigan
School of Natural Resources and Environment
430 E. University
Ann Arbor, MI 48109

Breeding Productivity and Adult Survival in Nongame Birds

Populations of many North American landbirds, including forest-inhabiting species that winter in the Neotropics, seem to be declining (Robbins et al. 1989; Terborgh 1989). These declines have been identified through broad-scale, long-term survey programs that identify changes in abundance of species, but provide little information about causes of changes in abundance or the health of specific populations in different geographic locations.

Population health is a measure of a population's ability to sustain itself over time as determined by the balance between birth and death rates. Indices of population size do not always provide an accurate measure of population health because population size can be maintained in unhealthy populations by immigration of recruits from healthy populations (Pulliam 1988). Poor population health across many populations in a species eventually results in the decline of that species. Early detection of population declines allows managers to correct problems before they are critical and widespread.

Demographic data (breeding productivity and adult survival) provide the kind of early warning signal that allows detection of

by
Thomas E. Martin
National Biological Service
David F. DeSante
The Institute for Bird Populations
Charles R. Paine
Montana Cooperative Wildlife Research Unit
Therese Donovan
University of Montana
Randall Dettmers
Ohio Cooperative Research Unit
James Manolis
Minnesota Cooperative Fish and Wildlife Research Unit
Kenneth Burton
The Institute for Bird Populations

unhealthy populations in terms of productivity or survival problems (Martin and Guepel 1993). In addition, demographic data can help determine whether population declines are the result of low breeding productivity or low survival in migration or winter. Breeding productivity data also can help identify habitat conditions associated with successful and failed breeding attempts. Such information is critical for developing habitat- and land-management practices that will maintain healthy bird populations (Martin 1992). Here, we provide examples of the kinds of information that can be obtained by broad-scale demographic studies.

Demographic Programs

The Monitoring Avian Productivity and Survivorship (MAPS) and Breeding Biology Research and Monitoring Database (BBIRD) programs were developed to gather the demographic data needed to provide early and locality-specific warning signals of population problems. MAPS uses large, stationary mistnets to capture and examine young and adult birds for between-year changes and to determine long-term trends in adult population size, productivity, and adult survival. BBIRD locates and monitors bird nests to study changes in nesting success, determine causes of nesting failure (e.g., weather, habitat, nest predation, or nest parasitism), and identify habitat conditions associated with successful reproduction. Though both programs are new, they are growing rapidly. We present example data to demonstrate initial results and burgeoning potential of these programs for the future.

MAPS

Initiated in 1989 and coordinated by The Institute for Bird Populations, MAPS is a cooperative effort among federal and state agencies, private organizations, and bird banders to operate a standardized continent-wide network of mist-netting and banding stations during the breeding season (DeSante 1992; DeSante et al. 1993a, 1993b). A typical MAPS station involves about ten 12-m (39-ft) mistnets over a 20-ha (49-acre) area. All birds captured throughout the breeding season are identified to species, age, and sex, and are banded with U.S. Fish and Wildlife Service bands.

As of 1992, 170 stations were in operation and more than 94,000 captures of more than 200 bird species were recorded. The number of adult birds captured is used as an index of adult population size while the proportion of young provides an index of postfledgling productivity (Baillie et al. 1993).

BBIRD

The BBIRD program, initiated in 1992, provides detailed information on nesting productivity and habitat needs of nongame birds at a national scale. BBIRD is a cooperative effort among biologists studying nesting productivity at local sites across the country. Participants follow a standard field protocol to obtain raw data on nesting productivity, causes of reproductive failure, vegetation measures at several spatial scales, and point counts (bird counts). Data from each local site are overseen by individual independent investigators who can obtain comparative information from other sites. In addition, overview analyses to identify national and regional trends are conducted at the Montana Cooperative Wildlife Research Unit.

BBIRD study sites are in large forested blocks to minimize fragmentation effects and provide baseline information on productivity in undisturbed habitats as well as in auxiliary sites that have no habitat restrictions (e.g., grazed, fragmented, or logged sites). The BBIRD program now includes 23 sites in 17 states. Over 8,000 nests of more than 150 bird species were monitored during the first 2 years of the program.

Variation in Productivity

The data provided by MAPS and BBIRD suggest that weather may be an important influence on population dynamics at large and even continental scales. Prior data from constant-effort mist-netting in scrub habitat on the west coast have suggested that avian productivity may peak during average weather conditions and may be depressed when weather conditions deviate from average (DeSante and Geupel 1987). These facts are especially important because one of the most important ecological results of global climate change may be a greater annual variability in both local and large-scale weather conditions.

Changes in indices of adult population size and postfledging productivity from the first 4 years of MAPS are presented for all species pooled and for each target species caught at 10 or more stations in 1992 in the Northeast and Northwest regions. These data indicate that productivity varied greatly from year to year, presumably a result of large-scale weather conditions (e.g., precipitation and temperature) just before and during the breeding seasons. Productivity was poor across most of North America, but especially in the eastern third of the continent in 1990. Adult population sizes declined significantly in the East in 1991, presumably a result of the poor productivity in 1990. In 1992 productivity was poor again in the East but good in the West. These results suggest that productivity in a given year may influence population sizes and population dynamics in subsequent years for many species over a large area.

BBIRD data likewise suggest that weather may substantially affect nesting productivity. Unusually wet weather conditions were reported at 6 of 14 BBIRD sites in 1992 when nest success of several species, including wood thrush (*Hylocichla mustelina*) and red-eyed vireo (*Vireo olivaceus*), was lower in 1992 than in 1993 (Table 1). These same two species also had reduced breeding productivity based on MAPS data. They produced fewer young per successful nest in 1992 than in 1993, a fact which also may be related to weather; some research suggests that clutch size as well as fledging success can be affected by weather conditions and may even provide a particularly sensitive measure of a species' tolerance to changing climatic conditions (e.g., Rotenberry and Wiens 1989). Further research may show that climatic variability is an important influence on the population trends of species.

Table 1. Wood thrush and red-eyed vireo nest success based on Mayfield (1961, 1975) estimates at midwestern BBIRD sites during 1992 and 1993 (numbers of nests are in parentheses).

State	Wood thrush		Red-eyed vireo	
	1992	1993	1992	1993
Ohio	23.0 (52)	33.1 (194)	6.6 (19)	33.7 (83)
Arkansas	45.6 (11)	58.0 (15)	35.3 (35)	42.1 (36)
Minnesota			19.0 (51)	23.0 (25)

Habitat-specific Differences

Forest fragmentation, where large forest blocks are cut and interspersed with open habitat, is believed to be particularly detrimental for breeding nongame birds. For example, BBIRD data show that fragmentation was associated with lower nest success in several species at midwestern BBIRD sites. Ovenbirds (*Seiurus aurocapillus*) were particularly sensitive to fragmentation effects; their reduced nest success resulted primarily from increased predation, although the parasitism rates of brown-headed cowbird (*Molothrus ater*) were also higher in fragments. No clear effect of fragmentation was noted for red-eyed vireos, although nest success differed substantially among unfragmented sites, potentially reflecting more subtle differences in habitat suitability or landscape-level effects (Table 2).

Adult Survival in Two Eastern Thrushes

Analysis of 3 years (1990-92) of MAPS data for veery (*Catharus fuscescens*) and wood thrush indicated low and substantially different

State	Ovenbird		Red-eyed vireo	
	Fragmented	Unfragmented	Fragmented	Unfragmented
Ohio	13.7 (35)	33.1 (45)	30.0 (52)	24.6 (50)
Wisconsin	19.8 (30)	42.6 (51)	26.4 (13)	50.8 (13)
Arkansas		51.9 (41)		38.7 (71)
Minnesota		44.5 (159)		21.0 (76)

Table 2. Ovenbird and red-eyed vireo nest success based on Mayfield (1961, 1975) estimates at fragmented and unfragmented midwestern BBIRD sites during 1992 and 1993 (numbers of nests are in parentheses).

($P < 0.06$) adult survival probabilities from 1990 to 1991. According to Breeding Bird Survey data, veery populations declined by 1.0% per year between 1966 and 1991, while wood thrush populations showed a statistically greater decline of 2.0% per year (Peterjohn and Sauer 1993). This difference in population declines is mirrored by survival indices; MAPS estimates of wood thrush survival are half that of the veery, possibly because of differences in adult survival over winter. This possibility is especially interesting because wood thrushes winter in Mexico and Central America where a greater proportion of the tropical forests have been cleared than in South America where veeries winter. Differences in estimated survival of the two species, however, could simply reflect different life-history traits (e.g., wood thrushes having lower adult survival associated with higher fertility; Martin in press). Estimated survival differences could also result from differences in breeding-site fidelity, which is related to nest success; a variety of evidence shows that birds disperse more in breeding seasons that follow nesting failure, potentially biasing survival estimates. Further nest-monitoring data from North America and survivorship data from both North America and the Neotropics are needed to identify causes of population declines in these and other Neotropical migratory landbirds.

Trends

Preliminary results from the MAPS and BBIRD programs suggest that population trends of nongame landbirds are influenced by

Courtesy T. Martin, NBS

Monitoring of nests, such as this one belonging to a red-faced warbler (*Cardellina rubrifrons*), provides information on breeding productivity.

weather-induced productivity problems, survival problems during migration or winter, and degradation of breeding habitat. These results emphasize the importance of national programs such as MAPS and BBIRD in providing baseline information on both continental and local habitat-specific processes that influence avian population dynamics. Ultimately, these data on breeding productivity and adult survival and their underlying environmental determinants will provide information critical for managing North American landbirds.

References

Baillie, S.R., R.E. Green, M. Body, and S.T. Buckland. 1993. An evaluation of the constant effort sites scheme. British Trust for Ornithology, Thetford. 103 pp.

DeSante, D.F. 1992. Monitoring Avian Productivity and Survivorship (MAPS): a sharp, rather than blunt, tool for monitoring and assessing landbird populations. Pages 511-521 *in* D.C. McCullough and R.H. Barrett, eds. Wildlife 2001: populations. Elsevier Applied Science, London.

DeSante, D.F., and G.R. Geupel. 1987. Landbird productivity in central coastal California: the relationship to annual rainfall, and a reproductive failure in 1986. Condor 89:636-653.

DeSante, D.F., K.M. Burton, and O.E. Williams. 1993a. The Monitoring Avian Productivity and Survivorship (MAPS) program second annual report (1990-1991). Bird Populations 1:68-97.

DeSante, D.F., O.E. Williams, and K.M. Burton. 1993b. The Monitoring Avian Productivity and Survivorship (MAPS) program: overview and progress. Pages 208-222 *in* D.M. Finch and P.W. Stangel, eds. Status and management of Neotropical migratory birds. Gen. Tech. Rep. RM-229. U.S. Forest Service, Rocky Mountain Forest and Range Experiment Station, Fort Collins, CO.

Martin, T.E. 1992. Breeding productivity considerations: what are the appropriate habitat features for management? Pages 455-473 *in* J.M. Hagan and D.W. Johnston, eds. Ecology and conservation of Neotropical migrants. Smithsonian Institution Press, Washington, DC.

Martin, T.E. Variation and covariation of life history traits of birds in relation to nest sites, nest predation, and food. Ecological Monographs. In press.

Martin, T.E., and G.R. Guepel. 1993. Nest-monitoring plots: methods for locating nests and monitoring success. Journal of Field Ornithology 64:507-519.

Mayfield, H. 1961. Nesting success calculated from exposure. Wilson Bull. 73:255-261.

Mayfield, H. 1975. Suggestions for calculating nest success. Wilson Bull. 87:456-466.

Peterjohn, B.G., and J.R. Sauer. 1993. North American Breeding Bird Survey annual summary 1990-1991. Bird Populations 1:52-67.

Pulliam, H.R. 1988. Sources, sinks, and population regulation. American Naturalist 132:652-661.

Robbins, C.S., J.R. Sauer, R.S. Greenberg, and S. Droege. 1989. Population declines in North American birds that migrate to the Neotropics. Proceedings of the National Academy of Science 86:7658-7662.

Rotenberry, J.T., and J.A. Wiens. 1989. Reproductive biology of shrubsteppe passerine birds: geographical and temporal variation in clutch size, brood size, and fledging success. Condor 91:1-14.

Terborgh, J. 1989. Where have all the birds gone? Essays on the biology and conservation of birds that migrate to the American tropics. Princeton University Press, NJ. 207 pp.

For further information:

Thomas E. Martin
National Biological Service
Cooperative Wildlife Research Unit
University of Montana
Missoula, MT 59812

Canada Geese in North America

by
Donald H. Rusch
Richard E. Malecki
National Biological Service
Robert Trost
U.S. Fish and Wildlife Service

Canada geese (*Branta canadensis*) are probably more abundant now than at any time in history. They rank first among wildlife watchers and second among harvests of waterfowl species in North America. Canada geese are also the most widely distributed and phenotypically (visible characteristics of the birds) variable species of bird in North America. Breeding populations now exist in every province and territory of Canada and in 49 of the 50 United States. The size of the 12 recognized subspecies ranges from the 1.4-kg (3-lb) cackling Canada goose *(B.c. minima)* to the 5.0-kg (11-lb) giant Canada goose (*B.c. maxima*; Delacour 1954; Bellrose 1976).

Market hunting and poor stewardship led to record low numbers of geese in the early 1900's, but regulated seasons including closures, refuges, and law enforcement led to restoration of most populations. Winter surveys were begun to study population trends and set responsible harvest regulations for these long-lived and diverse birds. Winter surveys begun in 1936-37 probably represent the oldest continuing index of migratory birds in North America.

Surveys

Sporadic counts of migrating and wintering Canada geese from the ground were supplemented by regular tallies from the air in the early 1950's. Winter surveys began because the subarctic and arctic nesting areas of many subspecies were still unknown and aerial surveys of these remote areas were impractical.

The well-designed spring surveys of Canada geese that began in the 1970's with the Eastern Prairie population have now expanded to include several others (Office of Migratory Bird Management 1993). Spring surveys estimate numbers of each population at the time of year when subspecies are reproductively isolated and geographically separated. The smaller subspecies of Canada geese nest farther north (arctic and subarctic regions of Alaska and Canada), and most winter farther south (gulf states and Mexico) than do the larger subspecies.

Status and Trends

Most aggregations of wintering geese were overharvested in the early 1900's. Those subspecies that nested in temperate regions closer to humans were most heavily hunted. By 1930 the giant Canada geese, which nested in the northern parts of the deciduous forest and tall-grass prairie, were believed extirpated. Numbers of the large geese that nested in the Great Plains and Great Basin (*B.c. moffitti*) were also severely reduced. Small Canada geese from the remote arctic and subarctic breeding ranges fared somewhat better, possibly because of less exposure to unregulated exploitation, but were also reduced in number.

Although hunting depleted numbers of Canada geese, human activity also created new habitats for these birds. Agriculture led to the clearing of forests and the plowing of prairies, creating the open landscapes preferred by geese. Cereal grains and pastures provided new food sources for geese, and the development of mechanical combines and pickers created an increased supply of waste grain (Hine and Schoenfeld 1968). In addition, uniform hunting regulations and improved wildlife law enforcement curtailed goose harvests after the signing of the Migratory Bird Treaty in 1916, and most goose populations increased over the next several decades (Figure). National wildlife refuges provided key sanctuaries and further assisted recovery of Canada goose numbers.

The giant Canada goose was "rediscovered" by Harold C. Hanson, a biologist of the Illinois Natural History Survey; the publication of his book *The Giant Canada Goose* in 1965 initiated a restoration effort that became one of the great success stories of wildlife management. These large geese were restored to their former range in the Mississippi and Central flyways and now breed in all states east of the Mississippi River.

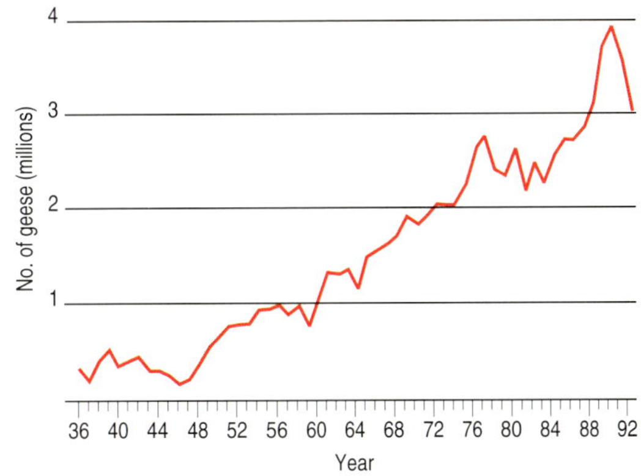

Figure. Total numbers of Canada geese counted on winter surveys, 1936-93.

Research and improved scientific management led to better understanding of diversity, distribution, and population dynamics of Canada geese in the 1970's. Awareness of differences in distribution and migration among the subspecies allowed managers to effectively control goose harvests. Improved management led to stable or increasing numbers of Canada geese in most populations (Table). The Mississippi Flyway Giant, Hi-line, Rocky Mountain, and Western Prairie/Great Plains populations, all composed mainly of large subspecies (*B.c. maxima* and *moffitti*), grew at about twice the rate of other populations that contained mainly smaller subspecies. The population numbers of the large geese that breed in the states of the Atlantic Flyway have also increased dramatically, but this trend was masked by declining numbers of geese in

Table. Canada goose population indices (in 1,000's) based on surveys conducted during fall and winter, 1969-93.

Year	Population*											
	AP	SJBP	MVP	Max(MF)	EPP	WP/GP	TGPP	SGPP	H-LP	RMP	DSKY	CCG
1969-70	775.2	106.9	324.7	50.8	106.6			151.2	44.2	25.8	22.5	
1970-71	675.0	127.3	292.3	64.4	126.3		133.2	148.5	40.5	25.3	19.8	
1971-72	700.2	117.6	293.9	55.8	157.4		160.9	160.9	31.4	36.6	17.9	
1972-73	712.0	101.3	295.9	54.2	181.4		148.4	259.4	35.6	37.1	15.8	
1973-74	760.2	136.0	277.9	57.6	205.8		160.5	153.6	24.5	42.8	18.6	
1974-75	819.3	101.0	304.4	57.0	197.1		133.5	123.7	41.2	46.7	26.5	
1975-76	784.5	115.5	304.9	62.1	204.4		203.7	242.5	55.6	51.6	23.0	
1976-77	923.6	129.8	478.5	58.5	254.2		171.3	210.0	67.6	54.3	24.1	
1977-78	833.2	180.4	575.5	60.1	270.2		215.5	134.0	65.1	59.0	24.0	
1978-79	823.6	142.7	434.5	77.1	207.2		187.6	163.7	33.8	62.7	25.5	
1979-80	780.1	127.0	394.9	86.4	171.8		165.9	213.0	67.3	77.3	22.0	64.1
1980-81	955.0	120.3	367.4	102.9	150.9		257.7	168.2	94.4	93.8	23.0	127.4
1981-82	702.6	118.5	250.9	107.6	145.3	175.0	284.7	156.0	81.9	64.3	17.7	87.1
1982-83	888.7	129.9	303.7	149.9	213.4	242.0	171.8	173.2	75.9	68.2	17.0	54.1
1983-84	822.4	129.9	352.8	103.9	163.1	150.0	279.9	143.5	39.5	55.5	10.1	26.2
1984-85	814.2	129.3	477.2	151.7	168.4	230.0	207.0	179.1	76.4	90.3	7.5	25.8
1985-86	905.4	158.0	618.9	180.1	169.0	115.0	198.2	181.0	69.8	68.3	12.2	32.1
1986-87	754.8	129.8	514.6	231.9	183.4	324.0	163.2	190.9	98.1	71.5		51.4
1987-88	737.9	158.8	564.6	225.9	228.5	272.1	315.8	139.1	66.8	71.4	12.2	54.8
1988-89	660.7	170.2	734.6	252.2	184.5	330.3	224.2	284.8	100.1	73.9	11.8	69.9
1989-90	733.8	159.4	1098.2	284.3	324.9	271.0	159.0	378.1	105.9	102.4	11.7	76.8
1990-91	706.9	142.2	939.7	345.1	218.4	390.0	315.5	508.5	116.6	86.7		110.2
1991-92	654.5	107.2	766.8	234.8	189.4	341.9	280.4	620.2	140.5	115.7	18.0	104.6
1992-93	569.2	104.4	673.4	282.6	146.4	318.0	238.7	328.2	118.5	99.5	16.6	149.3

*Populations are Atlantic (AP), Southern James Bay (SJBP), Mississippi Valley (MVP), Mississippi Flyway Giant (Max[MF]), Eastern Prairie (EPP), Western Prairie/Great Plains (WP/GP), Tall-grass Prairie (TGPP), Short-grass Prairie (SGPP), Hi-line (H-LP), Rocky Mountain (RMP), Dusky (DSKY), and Cackling Canada Goose (CCG).

Canada's eastern subarctic regions.

Although small geese with long migrations have generally not fared as well as large geese with short migrations, some small geese have responded well to intensive management. Introduced Arctic foxes (*Alopex lagopus*) depleted populations of the Aleutian Canada goose (*B.c. leucopareia*), and the subspecies was nearly extinct by 1940. About 300 were rediscovered in the Aleutians on Buldir Island in 1962 (Jones 1963). Subsequent removal of foxes and translocation of wild geese have led to increases to about 750 geese in 1975 and more than 11,000 in 1993.

Heavy hunting caused numbers of cackling Canada geese to plummet to record lows in the early 1980's, but intensive research (Raveling and Zezulak 1992) and harvest control have brought about a sustained recovery (Table).

Recent genetic studies of Canada geese support the existence of two major groups that last shared a common ancestor about 1 million years ago. The large-bodied group (*B.c. canadensis, interior, maxima, moffitti, fulva, occidentalis*) is mainly continental in distribution, while the small-bodied group (*hutchinsii, taverneri, minima, leucopareia*) breeds in coastal Alaska and Arctic Canada (Rusch et al. in press).

The future of these diverse stocks of Canada geese depends upon information adequate to permit simultaneous protection of rare forms, responsible subsistence and recreational hunting of abundant populations, and control of nuisance Canada geese in urban and suburban environments. Delineation of breeding ranges and spring surveys that monitor numbers of pairs and their productivity offer the most realistic approach to population management and the conservation of this remarkable diversity of geese.

Ranges of most populations have been described, and spring surveys are in place for some. Development and continuation of spring surveys for each subspecies of Canada geese are prerequisites for their conservation and management. The species can no doubt be perpetuated without spring surveys, but without continued monitoring, management, and conservation, it is likely that rare forms will disappear, opportunities for subsistence and recreational hunting will diminish, and nuisance problems caused by large geese living near humans will increase.

References

Bellrose, F.C. 1976. Ducks, geese and swans of North America. Stackpole, Harrisburg, PA. 544 pp.

Delacour, J.T. 1954. The waterfowl of the world. Vol. 1. Country Life, Ltd., London. 251 pp.

Hanson, H.C. 1965. The giant Canada goose. Southern Illinois University Press, Carbondale. 226 pp.

Hine, R.L., and C. Schoenfeld, eds. 1968. Canada goose management. Denbar Educational Research Services, Madison, WI. 194 pp.

Jones, R.D., Jr. 1963. Buldir Island, site of a remnant population of Aleutian Canada geese. Wildfowl 14:80-84.

Office of Migratory Bird Management. 1993. Status of waterfowl and fall flight forecast. U.S. Fish and Wildlife Service, Washington, DC. 37 pp.

Raveling, D.G., and D.S. Zezulak. 1992. Changes in distribution of cackling Canada geese in autumn. California Fish and Game 78:65-77.

Rusch, D.H., D.D. Humburg, M.D. Samuel, and B.D. Sullivan, eds. 1994. Biology and management of Canada geese. Proceedings of the 1991 International Canada Goose Symposium. In press.

For further information:

Donald H. Rusch
National Biological Service
Wisconsin Cooperative Wildlife Research Unit
University of Wisconsin
Madison, WI 53706

Canada Geese in the Atlantic Flyway

by
Jay B. Hestbeck
National Biological Service

Large changes have occurred in the geographic wintering distribution and subspecies composition of the Atlantic Flyway population of Canada geese (*Branta canadensis*) over the last 40 years. The Atlantic Flyway can be thought of as being partitioned into four regions: South, Chesapeake, mid-Atlantic, and New England. Wintering numbers have declined in the southern states (North Carolina, South Carolina, Georgia, Florida), increased then decreased in the Chesapeake region (Delaware, Maryland, Virginia), and increased markedly in the mid-Atlantic region (New York, New Jersey, Pennsylvania, West Virginia) (Serie 1993; Fig. 1). In the New England region (Maine, New Hampshire, Vermont, Massachusetts, Rhode Island, Connecticut), wintering numbers increased from around 6,000 during 1948-50 to between 20,000 and 30,000 today (Serie 1993).

Overall, the total number of wintering geese reached a peak of 955,000 in 1981 and has since declined 40% to 569,000 in 1993. Compounding these distributional changes in wintering numbers, the subspecies composition has also changed. The Canada goose population is composed of migrant geese (primarily *B.c.*

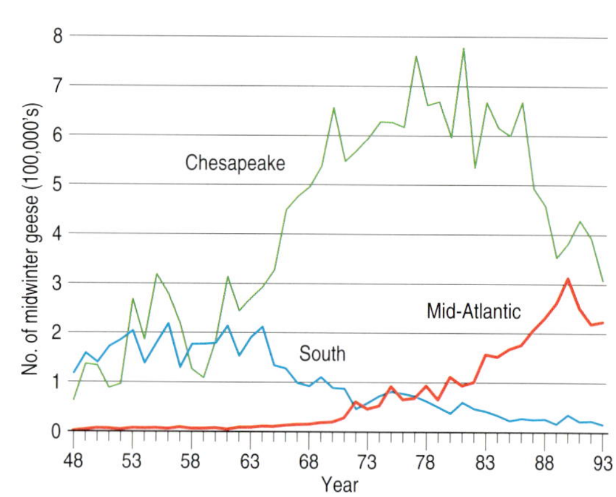

Fig. 1. Midwinter number of Canada geese in mid-Atlantic, Chesapeake, and South regions of the Atlantic Flyway, 1948-93 (Midwinter Survey, U.S. Fish and Wildlife Service, Office of Migratory Bird Management).

canadensis and *B.c. interior*) that breed in the subarctic regions of Canada and resident geese (primarily *B.c. maxima* and *B.c. moffitti*) that breed in southern Canada and the United States (Stotts 1983). The number of resident geese in Maine to Virginia has increased considerably from maybe 50,000 to 100,000 in 1981 (Conover and Chasko 1985) to an average of 560,000 in 1992-93 (H. Heusman, Massachusetts Division of Fisheries and Wildlife, personal communication). This rapid increase in resident geese suggests that the migrant population has declined more than the 40% decline observed in total wintering geese from 1981 to 1993.

Flying neck-banded goose (*Branta canadensis*).

Population Changes

Changes in population numbers result from changes in production, survival, and movement, acting singly or in combination. Consequently, understanding the reason for population changes involves detecting variation in survival, production, and movement over time and relating that variation to changes in wintering numbers. During the 1970's, the decrease of wintering geese in the South and increase in the Chesapeake region appeared to result from increased survival of geese in the Chesapeake and possibly from movement or short-stopping of geese from the South to the Chesapeake (Trost et al. 1986). Short-stopping occurs when migrant geese winter in a more northern location than their traditional, more southern, migration terminus.

During the 1980's, the decrease of wintering geese in the Chesapeake appeared to result from an 11% decrease in average survival from 1963-74 to 1984-88 (Hestbeck 1994a). This decrease in survival corresponded to a 36% increase in average harvest rate for the Atlantic Flyway from 1963-74 to 1984-88 (Fig. 2). Overall, the flyway harvest rate, as a 3-year average, increased from 19% in 1962-64 to 34% in 1982-84, and then slowly declined to 31% by 1990-92. The eastern Canada harvest rate has slowly increased from 4.2% in 1968-70 to 8.1% in 1990-92. The slight decline in the harvest rate in the flyway since 1982-84 has been partially offset by harvest rate increases in eastern Canada.

The decrease in number of geese wintering in the Chesapeake region in the 1980's was not related to changes in production. Production for migrants, measured from the Canadian data, remained constant over the period of population decline in the Chesapeake (Fig. 3). Average production recently declined during 1991-92 for geese harvested in Quebec. I also used harvest age ratios for the mid-Atlantic and Chesapeake regions to test for differences in production between these regions (Hestbeck 1994b). If the changes in wintering number resulted from changes in production, the average annual change in the age ratios would be higher for the mid-Atlantic region than for the Chesapeake region. The average annual changes were not different between these regions, however, indicating that regional production differences were not present.

The decrease in number of geese wintering in the Chesapeake region in the 1980's was not caused by migrant geese short-stopping in the mid-Atlantic instead of returning to the Chesapeake. From neck-band data, the probability of returning or moving to the different regions was estimated and indicated that, although geese traditionally returned to the same wintering area, they also changed wintering areas from year to year (Hestbeck 1994b). In years with harsher winters, geese wintered farther south than during milder winters (Hestbeck et al. 1991). Overall, the probability of returning or moving to the Chesapeake region was higher than the probability of returning or moving to any other region. When population size, survival, and movement were combined to estimate net movement among regions, the estimated net movements among regions were small and did not correspond to the changes in numbers of wintering geese. Taken

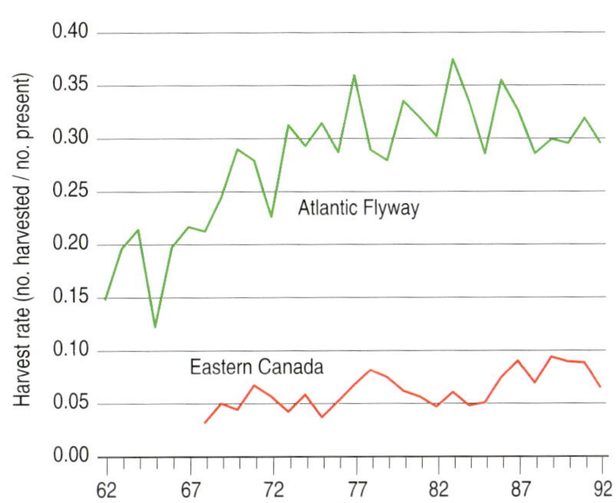

Fig. 2. Harvest rate of Canada geese in the Atlantic Flyway, 1962-92 (Harvest and Midwinter Surveys, U.S. Fish and Wildlife Service, Office of Migratory Bird Management) and eastern Canada, 1968-92 (Harvest Survey, Canadian Wildlife Service, National Wildlife Research Centre).

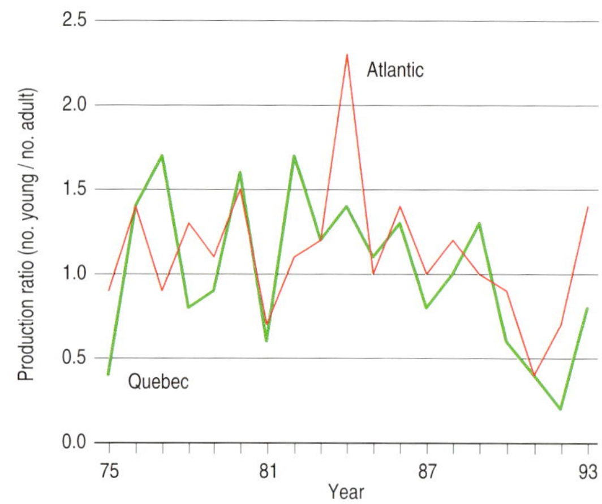

Fig. 3. Production ratio of Canada geese in Quebec and Atlantic regions of eastern Canada, 1975-93 (Waterfowl Parts Collection Survey, Canadian Wildlife Service, Atlantic Region, Sackville, N.B.).

together, these results suggested that the increases in the number of wintering geese in the mid-Atlantic region did not result from short-stopping of geese.

The increase of wintering geese in the mid-Atlantic most likely resulted from expanding resident populations. Resident geese generally have larger body sizes, allowing them to winter farther north than smaller-bodied migrant geese (Lefebvre and Raveling 1967). Resident and migratory-resident geese may selectively remain in the mid-Atlantic region. In addition, the resident population may be increasing faster than the migrant population because survival and production appear higher for residents than for migrants. Residents survive better partly because they are familiar with areas of food and refuge and may avoid hunting areas (Johnson and Castelli 1994). Production may be higher for resident than migrant geese because the climate is less variable and milder with a longer growing season in southern Canada and the United States than in the subarctic. Resident geese may also reach reproductive age earlier than migrant geese because the southerly growing season is longer, providing greater food resources.

References

Conover, M.R., and G.G. Chasko. 1985. Nuisance Canada goose problems in the eastern United States. Wildlife Society Bull. 13:228-233.

Hestbeck, J.B. 1994a. Survival of Canada geese banded in winter in the Atlantic Flyway. Journal of Wildlife Management 58(4): 748-756.

Hestbeck, J.B. 1994b. Changing number of Canada geese wintering in different regions of the Atlantic Flyway. *In* D.H. Rusch, D.D. Humburg, M.D. Samuel, and B.D. Sullivan, eds. Proceedings of the 1991 International Canada Goose Symposium. Milwaukee, WI. In press.

Hestbeck, J.B., J.D. Nichols, and R.A. Malecki. 1991. Estimates of movement and site fidelity using mark-resight data of wintering Canada geese. Ecology 72:523-533.

Johnson, F.A., and P.M. Castelli. 1994. Demographics of Canada geese breeding in southeastern Canada and the northeastern United States. *In* D.H. Rusch, D.D. Humburg, M.D. Samuel, and B.D. Sullivan, eds. Proceedings of the 1991 International Canada Goose Symposium. Milwaukee, WI. In press.

Lefebvre, E.A., and D.G. Raveling. 1967. Distribution of Canada geese in winter as related to heat loss at varying environmental temperatures. Journal of Wildlife Management 31:538-546.

Serie, J. 1993. Waterfowl harvest and population survey data. U.S. Fish and Wildlife Service, Office of Migratory Bird Management, Laurel, MD. 68 pp.

Stotts, V.D. 1983. Canada goose management plan for the Atlantic Flyway, 1983-95. Part 2. History and current status. The Atlantic Flyway Waterfowl Council. mimeo.

Trost, R.E., R.A. Malecki, L.J. Hindman, and D.C. Luszcz. 1986. Survival and recovery rates of Canada geese from Maryland and North Carolina 1963-1974. Proceedings of the Annual Conference of the Southeastern Association of Fish and Wildlife Agencies 40:454-464.

For further information:

Jay B. Hestbeck
National Biological Service
Massachusetts Cooperative Fish and Wildlife Research Unit
University of Massachusetts
Box 34220
Amherst, MA 01003

Arctic Nesting Geese: Alaskan Populations

by
Jerry Hupp
Robert Stehn
Craig Ely
Dirk Derksen
National Biological Service

North American populations of most goose species have remained stable or have increased in recent decades (USFWS and Canadian Wildlife Service 1986). Some populations, however, have declined or historically have had small numbers of individuals, and thus are of special concern. Individual populations of geese should be maintained to ensure that they provide aesthetic, recreational, and ecological benefits to the nation. Monitoring and management efforts for geese should focus on individual populations to ensure that genetic diversity is maintained (Anderson et al. 1992).

Alaska is the only state with viable breeding populations of arctic geese. Five species (11 subspecies) nest in Alaska, and although these species also breed in arctic regions of Canada or Russia, most geese of the Pacific Flyway originate in Alaska or use Alaskan habitats during migration. Alaskan geese are often hunted for subsistence by Alaskan Natives.

While data for some areas are lacking, populations of greater white-fronted geese (*Anser albifrons frontalis*) and medium-sized Canada geese (*Branta canadensis*) in interior and northern Alaska appear stable or have increased (King and Derksen 1986). Although only a small number of lesser snow geese (*Chen caerulescens caerulescens*) nest in Alaska, substantial populations occur in Canada and Russia. Populations of Pacific black brant (*B. bernicla nigricans*), emperor geese (*C. canagica*), greater white-fronted geese, and cackling Canada geese (*B.c. minima*) on the Yukon-Kuskokwim Delta (YKD) of western Alaska have declined from their historical numbers and are the focus of special management efforts (USFWS 1989). In addition, populations of tule white-fronted geese (*A.a. gambeli*), Aleutian Canada geese (*B.c. leucopareia*), Vancouver

Canada geese (*B.c. fulva*), and dusky Canada geese (*B.c. occidentalis*) are of special concern because of their limited geographic distributions and small numbers.

Inventory of Arctic Geese

An annual index of the Pacific black brant population has been obtained since 1964 by the U.S. Fish and Wildlife Service (USFWS) during aerial surveys of wintering areas along the Pacific coast (Bartonek 1994a). Population trends of cackling Canada geese and greater white-fronted geese from 1965 to 1979 were based on surveys conducted by USFWS and state agency biologists on migration areas in the Klamath Basin of Oregon and California. Population trends of those two species from 1980 to 1993 were based on coordinated surveys on wintering areas (Bartonek 1994b).

Emperor geese have been inventoried by USFWS biologists during aerial surveys of spring and fall migration areas on the Alaska Peninsula and the YKD since 1980 (Bartonek 1992). We used the highest count within a year to determine the population trend for emperor geese. Population indices for tule white-fronted geese were obtained from surveys on wintering and migration areas in the Pacific Flyway in intermittent years since 1978. Aleutian Canada geese have been counted on a spring staging area in northern California since 1975. Dusky Canada geese have been inventoried on their wintering areas in the Pacific Flyway since 1953. There are no data on population trends of Vancouver Canada geese; however, the winter population in the northern portion of southeastern Alaska was estimated by USFWS biologists in 1986.

Status of Alaskan Geese

Yukon-Kuskokwim Delta Geese

Most geese on the YKD nest within 30 km (15-20 mi) of the Bering Sea but winter in diverse areas. Pacific black brant primarily winter along the Pacific coast of Mexico while greater white-fronted geese and cackling Canada geese primarily winter in the Central Valley of California. In recent years, increasing numbers of cackling Canada geese have wintered in Oregon. Most emperor geese winter in the Aleutian Islands.

These four species experienced sharp population declines (30%-50%) between the early 1960's and mid-1980's (Fig. 1). The declines were likely due to the combined effects of subsistence harvest of breeding birds and eggs on the YKD, excessive sport harvest on the wintering areas, poor weather during nesting, and fox predation of nests (USFWS 1989). In 1984, the USFWS, Yupik Natives, state wildlife agencies, and sport hunters cooperated to reduce sport and subsistence harvest. Since then populations of cackling Canada geese and greater white-fronted geese have begun to recover while emperor geese and black brant remain near historical lows (Fig. 1). Poor winter survival of juvenile emperor geese may be slowing recovery of that species (Schmutz et al. 1994). Winter survival of cackling Canada geese has improved since the reduction in sport hunting; however, there is no evidence that their survival in summer has improved (Raveling et al. 1992).

Tule White-fronted Geese

The only known nesting area for tule white-fronted geese is in Upper Cook Inlet (Timm et al. 1982) and the adjacent Susitna in south-central Alaska. Tule geese may also occur on the Innoko National Wildlife Refuge in western Alaska. The numbers of tule geese counted on wintering areas in the Central Valley of California in recent years are higher than during the late 1970's (Fig. 2). It is unclear if the increase is due to population growth or because of improved understanding of the winter distribution.

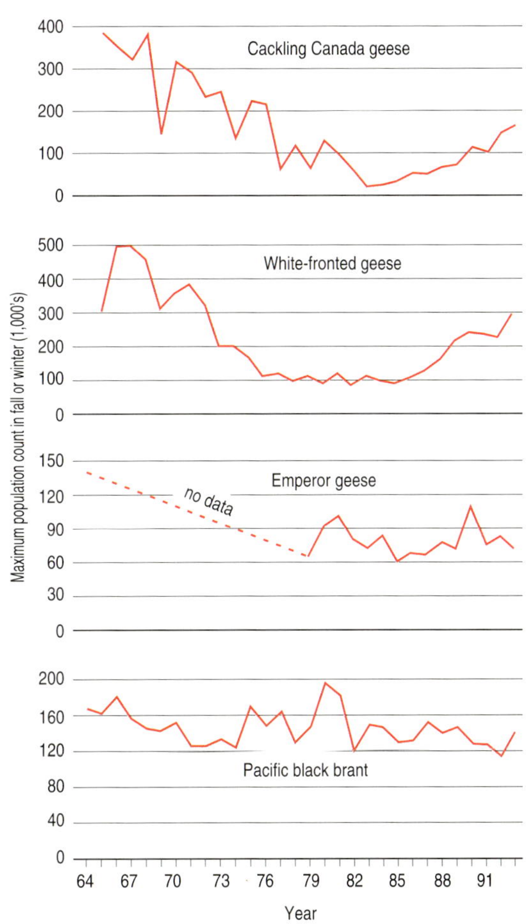

Fig. 1. Population trends of arctic geese that nest on the Yukon-Kuskokwim Delta, Alaska (1964-93).

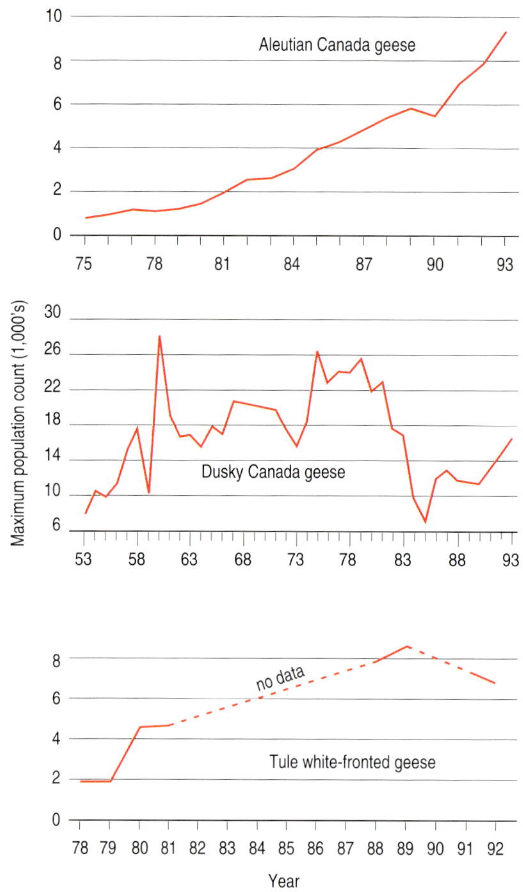

Fig. 2. Population trends of Aleutian Canada geese (1975-93), dusky Canada geese (1953-93), and tule white-fronted geese (1978-89).

Dusky Canada Geese

Dusky Canada geese primarily nest on the Copper River Delta of south-central Alaska, the islands of Prince William Sound, and Middleton Island in the Gulf of Alaska. They winter in the Willamette Valley of Oregon and the lower Columbia River. The population was stable or increased between the 1950's and 1970's. During the early 1980's, however, the population declined, then stabilized at a lower level in the mid-1980's (Fig. 2). The decline was largely due to reduced nesting success as a result of habitat changes on the nesting area following the 1964 Alaska earthquake. Invasion of shrubs and loss of wet meadow habitats resulted in more mammalian predators and greater nest predation (Subcommittee on Dusky Canada Geese 1992).

Aleutian Canada Geese

Although once abundant on the Aleutian, Commander, and Kuril islands, the numbers of Aleutian Canada geese were greatly reduced by foxes and dogs introduced to nesting islands by commercial fur farmers before World War II (Byrd and Woolington 1983). The subspecies was classified as endangered in 1967, and by the mid-1970's fewer than 800 individuals remained (USFWS 1991). Sport harvest on migration and wintering areas in Oregon and California was stopped in 1975, and fox control was initiated on nesting islands. Geese were also transplanted to fox-free islands. The population of Aleutian Canada geese responded to recovery efforts and has grown to more than 9,000 individuals (Fig. 2). The status of the subspecies was changed from endangered to threatened in 1991.

Vancouver Canada Geese

Vancouver Canada geese nest and use brood-rearing areas in southeastern Alaska (Lebeda and Ratti 1983) and winter on coastal wetlands near the breeding areas. Few data on breeding numbers exist because Vancouver Canada geese nest in coastal forests and are difficult to survey. About 10,000 Vancouver Canada geese wintered in the northern portion of southeastern Alaska in 1986 (Hodges and Conant 1986). Wintering sites are scattered among coastal wetlands and have not been consistently surveyed. Consequently, population trends of this subspecies are not known. Population trends are likely influenced by environmental variables because sport and subsistence harvest are minimal (King and Derksen 1986).

Status of Habitats of Special Concern

Yukon-Kuskokwim Delta

The YKD (Fig. 3) is the primary waterfowl nesting area in Alaska (King and Dau 1981); it provides critical nesting and brood-rearing habitat for more than 400,000 geese. In addition, the entire population of Wrangel Island lesser snow geese uses the YKD during fall staging (Ely et al. 1993). While much of the YKD is within the Yukon Delta National Wildlife Refuge, it is also a region where more than 17,000 Yupik Natives live in 40 Native villages. Large private inholdings, primarily Native corporation lands, exist within the refuge and contain important waterfowl nesting habitat. Meeting the subsistence needs of Native people while maintaining or enhancing waterfowl populations on the YKD requires close coordination among the Yupik Natives and federal and state agencies. Management of subsistence waterfowl harvest on the YKD has been difficult because of cultural differences and constraints imposed by the Migratory Bird Treaty Act. Coordinated management efforts will be especially important in the future as Native populations increase.

Izembek Lagoon

Nearly the entire world population of more than 120,000 Pacific black brant uses Izembek

Lagoon (Fig. 3) as a fall staging area for about 2 months. Although Izembek Lagoon is protected as a national wildlife refuge and state game refuge, it is near offshore oil leases in Bristol Bay. Should oil development proceed, increased aircraft activity over Izembek Lagoon could result in a significant increase in disturbance that could prevent brant from accumulating sufficient body fat for their nonstop flight to wintering areas in Mexico. This lack of sufficient body fat could result in increased mortality (Ward et al. 1994).

Bristol Bay Lowlands

Estuaries on the north side of the Alaska Peninsula (Fig. 3) provide critical migration habitat for cackling Canada geese, Taverner's Canada geese (*B. c. taverneri*), and emperor geese, and nesting habitat for a unique group of greater white-fronted geese. Part of this area is protected in State Critical Habitat Areas managed by the Alaska Department of Fish and Game. At least 5,265 ha (13,000 acres) of important habitat, however, is state land that may be subject to resource development.

Teshekpuk Lake Special Areas

Up to 32,000 Pacific black brant (25% of the world population) and 30,000 individuals of other goose species molt annually on Teshekpuk Lake Special Area (TLSA) (Fig. 3) on the National Petroleum Reserve in Alaska (Derksen 1978; King 1984). The area is managed by the Bureau of Land Management, and special regulations govern resource development on the TLSA to minimize adverse impacts to wildlife. Energy development in adjacent areas, though, may result in increased aircraft activity that could disturb molting geese and reduce their ability to secure forage needed for feather replacement (Jensen 1990).

Interior Wetlands

Greater white-fronted and Canada goose-nesting and brood-rearing habitats occur in interior wetlands near the Yukon, Tanana, Kuskokwim, Koyukuk, Susitna, and Innoko rivers (King and Lensink 1971). National wildlife refuges encompass much of the important habitat, although some areas are managed by the state of Alaska, private landowners, and the Bureau of Land Management. At present, there is relatively little human-related disturbance in these areas, although placer mining, oil exploration and development, timber harvest, and military training could affect some areas.

Upper Cook Inlet

About 100,000 geese and swans use Upper Cook Inlet (Fig. 3) as spring migration habitat. In addition, this inlet is one of two nesting areas of tule white-fronted geese. Development of oil and gas, coal, timber, and mineral deposits has either been proposed or is ongoing in Upper Cook Inlet and may affect coastal wetlands used by migratory waterfowl. Most of the important waterfowl habitats in this area are state game refuges or Critical Habitat Areas managed by the Alaska Department of Fish and Game.

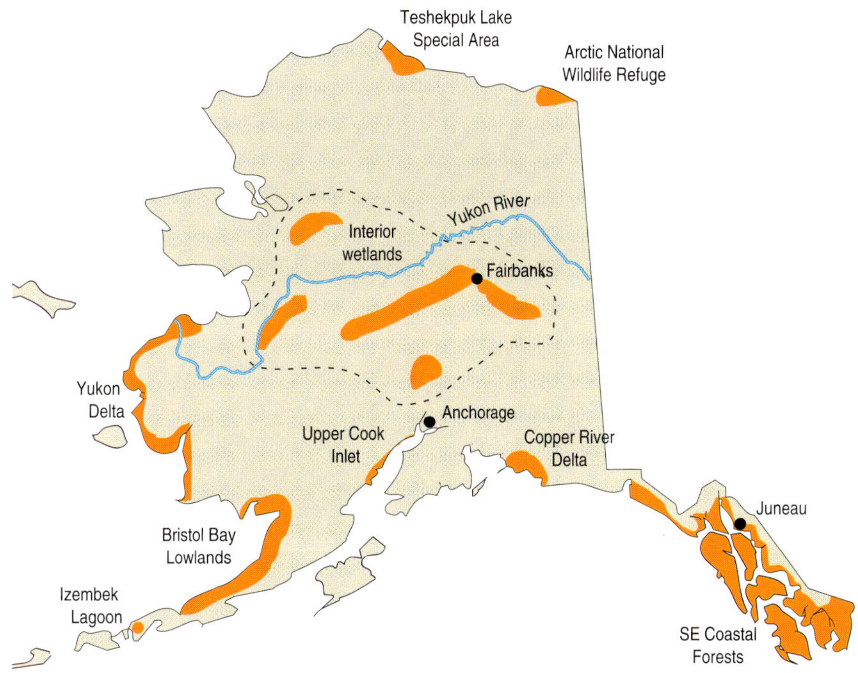

Fig. 3. Alaskan habitats of special importance to geese.

Alaska Coastal Forests

Some nesting and brood-rearing areas of Vancouver Canada geese (Fig. 3) occur in areas of commercially harvestable timber (Lebeda and Ratti 1983). Logging activities on U.S. Forest Service land on the Tongass National Forest could affect these habitats. In addition, timber harvest on Native corporation lands may restrict opportunities to transplant Vancouver Canada geese into areas of suitable habitat or may limit natural expansion of the subspecies range (King and Derksen 1986). Use of tidal areas to store harvested timber before shipping can affect wintering habitat of Vancouver Canada geese and migration habitats of other waterbirds.

Arctic National Wildlife Refuge

As many as 300,000 lesser snow geese and an unknown number of greater white-fronted geese stage on the Arctic National Wildlife Refuge (Fig. 3) before fall migration. During staging, geese feed intensively and build fat reserves for migration. Proposed petroleum leases on the refuge would result in increased aircraft activity that could disrupt feeding behavior of geese, displace birds from feeding habitats, and reduce their ability to accumulate body fat before migration (Brackney et al. 1987). Diminished fat reserves could reduce survival during migration.

References

Anderson, M.G., J.M. Rhymer, and F.C. Rohwer. 1992. Philopatry, dispersal, and the genetic structure of waterfowl populations. Pages 365-395 *in* B.D.J. Batt, A.D. Afton, M.G. Andersen, C.D. Ankney, D.H. Johnson, J.A. Kadlec, and G.L. Krapu, eds. Ecology and management of breeding waterfowl. University of Minnesota Press, Minneapolis.

Bartonek, J.C. 1992. 1992 Pacific Flyway briefing material. U.S. Fish and Wildlife Service, Portland, OR. 90 pp.

Bartonek, J.C. 1994a. Pacific brant Midwinter Survey January 1993. U.S. Fish and Wildlife Service, Portland, OR. 4 pp.

Bartonek, J.C. 1994b. 1993 Pacific Flyway fall and winter goose surveys. U.S. Fish and Wildlife Service, Portland, OR. 8 pp.

Brackney, A.W., R.M. Platte, J.M. Morton, and D. Whiting. 1987. Ecology of lesser snow geese staging on the Coastal Plain of the Arctic National Wildlife Refuge, fall 1985. Pages 372-392 *in* G.W. Garner and P.E. Reynolds, eds. 1985 Update report, baseline study of the fish, wildlife, and their habitats. Vol. 1. U.S. Fish and Wildlife Service, Anchorage, AK.

Byrd, G.V., and D.W. Woolington. 1983. Ecology of Aleutian Canada Geese at Buldir Island, Alaska. U.S. Fish and Wildlife Service Special Scientific Report-Wildlife, 253. 18 pp.

Derksen, D.V. 1978. Summary of Teshekpuk Lake aerial goose surveys (1976-1978). U.S. Fish and Wildlife Service, Anchorage, AK. 20 pp.

Ely, C.E., J.Y. Takekawa, and M.L. Wege. 1993. Distribution, abundance, and productivity of Wrangel Island lesser snow geese *Anser caerulescens* during autumn migration on the Yukon-Kuskokwim Delta, Alaska. Wildfowl 44:24-32.

Hodges, J.I., and B. Conant. 1986. Experimental Canada goose survey northern portion of southeast Alaska. U.S. Fish and Wildlife Service, Juneau, AK. 9 pp.

Jensen, K.C. 1990. Responses of molting Pacific black brant to experimental aircraft disturbance in the Teshekpuk Lake Special Area, Alaska. Ph.D. dissertation, Texas A & M University, College Station. 72 pp.

King, J.G., and D.V. Derksen. 1986. Alaska goose populations: past, present, and future. Transactions of the North American Wildlife and Natural Resources Conference 51:464-479.

King, J.G., and C.P. Dau. 1981. Waterfowl and their habitats in the eastern Bering Sea. Pages 739-753 *in* D.W. Hood and J.A. Calder, eds. The eastern Bering Sea shelf: oceanography and resources. Vol. 2. University of Washington Press, Seattle.

King, J.G., and C.J. Lensink. 1971. An evaluation of Alaskan habitat for migratory birds. U.S. Fish and Wildlife Service, Washington, DC. 45 pp.

King, R.J. 1984. Results of the 1982 and 1983 aerial goose surveys at Teshekpuk Lake, Alaska. U.S. Fish and Wildlife Service, Fairbanks, AK. 10 pp.

Lebeda, C.S., and J.T. Ratti. 1983. Reproductive biology of Vancouver Canada geese on Admiralty Island, Alaska. Journal of Wildlife Management 47:297-306.

Raveling, D.G., J.D. Nichols, J.E. Hines, D.S. Zezulak, J.G. Silveira, J.C. Johnson, T.W. Aldrich, and J.A. Weldon. 1992. Survival of cackling Canada geese, 1982-1988. Journal of Wildlife Management 56:63-73.

Schmutz, J.A., S.E. Cantor, and M.R. Petersen. 1994. Seasonal and annual survival of emperor geese. Journal of Wildlife Management 58:525-535.

Subcommittee on Dusky Canada Geese. 1992. Pacific Flyway management plan for dusky Canada geese. Pacific Flyway Study Committee, U.S. Fish and Wildlife Service, Portland, OR. 28 pp.

Timm, D.E., M.L. Wege, and D.S. Gilmer. 1982. Current status and management challenges for tule white-fronted geese. Transactions of the North American Wildlife and Natural Resources Conference 47:453-463.

USFWS and Canadian Wildlife Service. 1986. North American Waterfowl Management Plan. U.S. Fish and Wildlife Service, Washington, DC. 19 pp.

USFWS. 1989. Comprehensive management plans for arctic nesting geese in Alaska. U.S. Fish and Wildlife Service, Anchorage, AK. 107 pp.

USFWS. 1991. Aleutian Canada goose recovery plan. U.S. Fish and Wildlife Service, Anchorage, AK. 55 pp.

Ward, D.H., R.A. Stehn, and D.V. Derksen. 1994. Response of staging brant to disturbance at the Izembek Lagoon, Alaska. Wildlife Society Bull. 22:220-228.

For further information:

Jerry Hupp
National Biological Service
Alaska Science Center
1011 E. Tudor Road
Anchorage, AK 99503

North American Ducks

by
David F. Caithamer
Graham W. Smith
U.S. Fish and Wildlife Service

Increased predation and habitat degradation and destruction coupled with drought, especially on breeding grounds, have caused the declines of some duck populations. More than 30 species of ducks breed in North America, in areas as diverse as the arctic tundra and the subtropics of Florida and Mexico. For many of these species, however, the Prairie Pothole region of the north-central United States and south-central Canada is the most important breeding area (Fig. 1), although migratory behavior and the life histories of different species lead them to use many wetland habitats.

Numerous sources of information are available on the status of duck populations in North America. The two most comprehensive and reliable sources are the Breeding Population and Habitat Survey, conducted since 1955 and encompassing the Prairie Pothole region, boreal forests, and tundra habitats from South Dakota to Alaska (Caithamer et al. 1993; Fig. 1), and the Midwinter Survey, encompassing the United States and portions of Canada and

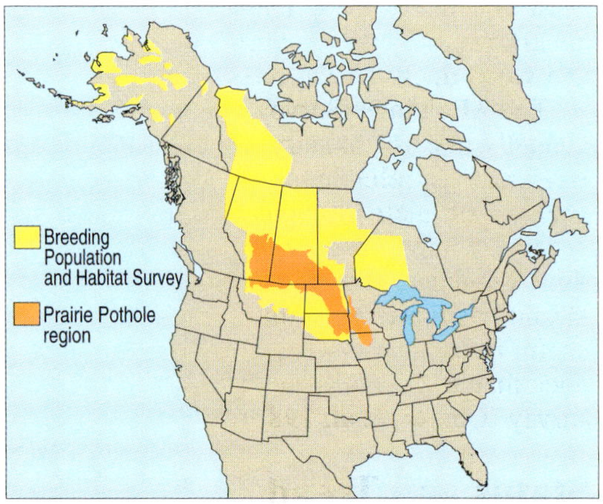

Fig. 1. The Prairie Pothole region and areas sampled in the Breeding Population and Habitat Survey.

Mexico at regular intervals. Results from these surveys are the basis for this article.

The Breeding Population and Habitat Survey is conducted during May and June when most species occupy their breeding ranges. Pilot-biologists and observers in airplanes identify and count ducks on a sample of transects. Not all ducks are visible from the air, so some transects are resurveyed more thoroughly with a helicopter or from the ground to obtain complete counts. These data are used to correct the air counts and obtain unbiased estimates of duck densities in these areas. Estimates of number of pairs of ducks are expanded to provide population estimates for the entire surveyed area. This survey, conducted by the Canadian Wildlife Service and the U.S. Fish and Wildlife Service (USFWS), is among the most extensive and comprehensive surveys conducted annually for any group of animals anywhere in the world. Survey estimates are the major determinant governing the regulation-setting process for the sport harvesting of ducks by both Canadian and United States provincial, state, and federal governments.

The Breeding Population and Habitat Survey is most reliable for mallards (*Anas platyrhynchos*), gadwall (*A. strepera*), American wigeon (*A. americana*), green-winged teal (*A. crecca*), blue-winged teal (*A. discors*), northern shoveler (*A. clypeata*), redhead (*Aythya americana*), canvasback (*A. valisineria*), and scaup (*A. affinis* and *A. marila*). Researchers and managers are trying to expand the geographic range of this survey in the Pacific Flyway, eastern Canada, and the northwestern United States.

The breeding survey, however, poorly monitors species such as whistling ducks (*Dendrocygna* spp.), mottled ducks (*Anas fulvigula*), American black ducks (*A. rubripes*), most sea ducks and mergansers (*Lophodytes cucullatus, Mergus merganser, M. serrator*), and wood ducks (*Aix sponsa*).

The Midwinter Survey has been conducted annually in early January since the mid-1940's. It is not as reliable as the breeding survey because of methodological shortcomings and because winter is a poor time to survey population abundance (Eggeman and Johnson 1989). Despite its limitations, this survey does provide useful information on such species as the black duck that are not well surveyed by the breeding survey (Conroy et al. 1988).

Status and Trends

Population estimates of all ducks from the breeding survey have varied from 26.5 to 42.8 million since 1955 (Fig. 2). Generally, breeding populations were high in the 1950's and 70's and low in the 60's, 80's, and 90's. The 1993 estimate of 28.0 million was 20% below the 1955-92 average.

Estimates of ducks from the Midwinter Survey also have varied since 1955 (Fig. 2). The 1993 estimate of 10.3 million ducks was the lowest recorded, and 44% below the 1955-92 average.

The Breeding Population and Habitat Survey provides reliable estimates for seven species of dabbling ducks, while the Midwinter Survey provides estimates for eight. The breeding population of total dabbling ducks in 1993 was 20% below the 1955-92 average. Compared with the 1955-92 average, 1993 breeding population estimates suggest population declines for mallards, American wigeon, blue-winged teal, and northern pintail. Population estimates were unchanged for green-winged teal and increased for gadwall and northern shoveler (Figs. 3-5). During the most recent 10-year period, the breeding population of northern pintail decreased, gadwall populations increased, and populations of six other species were stable (Table). Midwinter estimates of all species of dabbling ducks were stable or increased during 1984-93 (Table).

Midwinter estimates are the only long-term data available for black ducks. Apparent differences in population trends between the breeding and midwinter surveys (Table) are a function of differences in the quality of the surveys and in the populations monitored by the surveys. For example, breeding mallards have increased in recent years in the Atlantic Flyway, which is outside the breeding survey area. The breeding survey indicates a stable trend for mallards while the winter survey indicates an increasing trend; the two surveys monitor different portions of the total continental population.

Five species of diving ducks are monitored by breeding and winter surveys. Because lesser scaup are not distinguished from greater scaup in the surveys, these species have been combined. Breeding populations of diving ducks in 1993 were 18% below the 1955-92 average. Redhead and scaup breeding populations were lower than average, whereas the canvasback population was near average, and the ring-necked duck (*Aythya collaris*) population was above average (Figs. 4, 6). From 1984 to 1993, the breeding population of scaup declined while the breeding population of ring-necked ducks increased (Table). The Midwinter Survey also indicated an increasing population of ring-necks during this period (Table).

Fourteen species of sea ducks, mergansers, and their allies were monitored by the breeding survey. These 14 species plus the harlequin duck (*Histrionicus histrionicus*) were monitored during the Midwinter Survey. Because

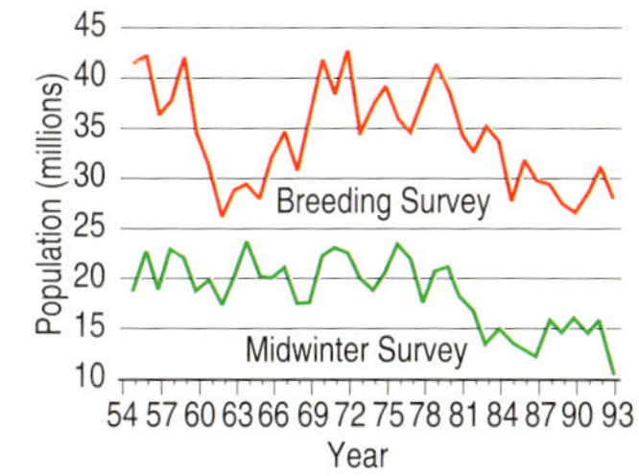

Fig. 2. Duck populations in North America, 1955-93, from the Breeding Population and Habitat Survey and the Midwinter Survey.

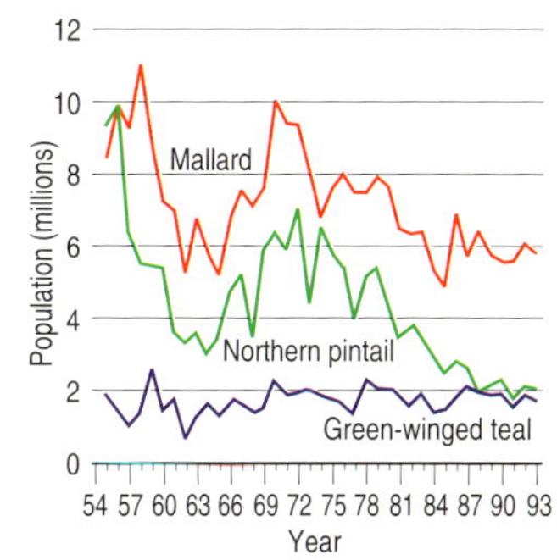

Fig. 3. Mallard, northern pintail, and green-winged teal breeding population estimates, 1955-93.

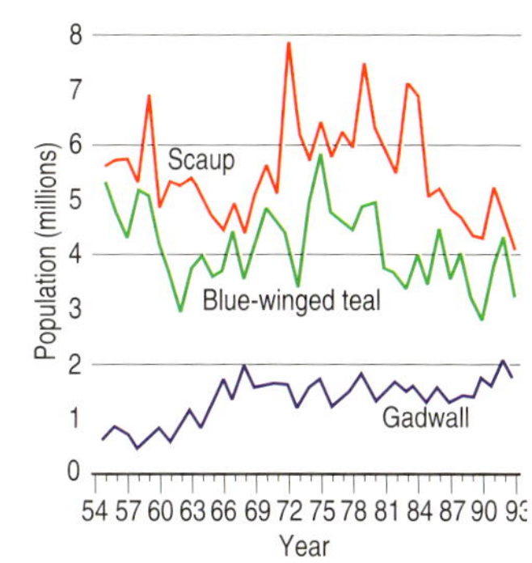

Fig. 4. Scaup, blue-winged teal, and gadwall breeding population estimates, 1955-93.

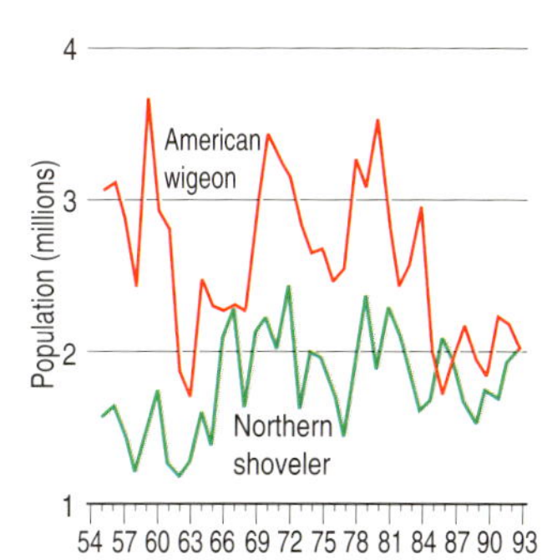

Fig. 5. American wigeon and northern shoveler breeding population estimates, 1955-93.

Table. Estimated annual numbers (in thousands) and recent trends (1984-93) of ducks based on the survey areas monitored by breeding and midwinter surveys.

Tribe and species	Breeding		Wintering	
	No.	Trend	No.	Trend
Perching ducks				
Wood duck			33	Stable
Dabbling ducks				
American black duck			278	Stable
American wigeon	2,053	Stable	1,088	Stable
Blue-winged teal	3,192	Stable		
Blue-winged and cinnamon teal			166	Increasing
Gadwall	1,755	Increasing	1,168	Increasing
Green-winged teal	1,694	Stable	2,086	Increasing
Mallard	5,708	Stable	4,994	Increasing
Mottled duck			129	Increasing
Northern pintail	2,053	Decreasing	2,241	Stable
Northern shoveler	2,046	Stable	638	Stable
Diving ducks				
Canvasback	472	Stable	298	Stable
Greater and lesser scaup	4,080	Decreasing	1,070	Stable
Redhead	485	Stable	336	Stable
Ring-necked duck	868	Increasing	421	Increasing
Sea ducks and mergansers				
Bufflehead	869	Increasing	126	Increasing
Eiders[a]	8	Decreasing	132	Stable
Goldeneye[b]	592	Stable	122	Stable
Harlequin duck			<1	Stable
Mergansers[c]	528	Stable	264	Increasing
Oldsquaw	174	Decreasing	10	Decreasing
Scoters[d]	1,006	Decreasing	160	Stable
Stifftails				
Ruddy duck	387	Stable	110	Decreasing

[a]Eiders include common eider (*Somateria mollissima*), king eider (*S. spectabilis*), spectacled eider (*S. fischeri*), and Steller's eider (*Polysticta stelleri*).
[b]Goldeneye include Barrow's goldeneye (*Bucephala islandica*) and common goldeneye (*B. clangula*).
[c]Mergansers include hooded merganser (*Laphodytes cucallatus*), red-breasted merganser (*Mergus serrata*), and common merganser (*M. merganser*).
[d]Scoters include black scoter (*Melanitta nigra*), surf scoter (*M. perspcillata*), and white-winged scoter (*M. fusca*).

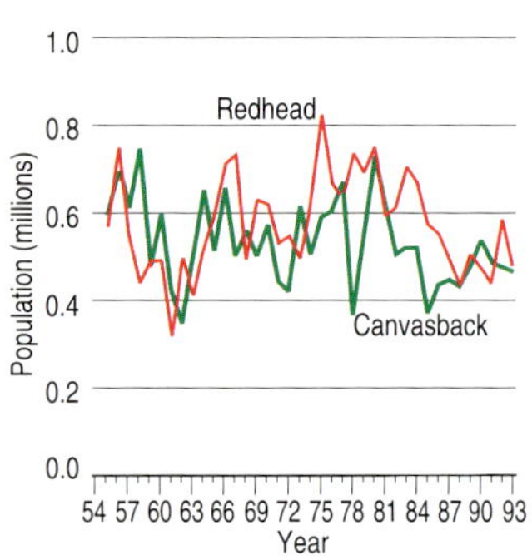

Fig. 6. Redhead and canvasback breeding population estimates, 1955-93

some of these species are difficult to identify during aerial surveys, or are encountered rarely, they are combined with related species (*see* Table).

Collectively, breeding populations of mergansers and their allies were 9% lower in 1993 compared to the 1955-92 average. Merganser, oldsquaw (*Clangula hyemalis*), eider, and scoter breeding populations in 1993 were all lower than their 1955-92 averages (*see* Table for species). The breeding population of goldeneye in 1993 was similar to the 1955-92 average, whereas the bufflehead (*Bucephala albeola*) breeding population was higher than the long-term average. During the last 10 years, breeding populations of eiders, oldsquaw, and scoters decreased, bufflehead increased, and goldeneye and mergansers were stable (Table). Winter population estimates during 1983-92 decreased for oldsquaw, increased for bufflehead and mergansers, and were stable for other species in the sea duck tribe (Table).

In the United States and Canada, wood ducks are the only representative of the tribe Cairinina and ruddy ducks (*Oxyura jamaicensis*) are the only representative of the Oxyurini tribe. Wood ducks are hard to survey because they inhabit forested wetlands where it is difficult to obtain reliable counts. Their current population, however, is greater than in the early 1900's (Bellrose 1980). Midwinter counts of wood ducks during 1983-92 indicated a stable population (Table). Ruddy duck breeding populations in 1993 were similar to the 1955-92 average.

Factors Affecting Population Status

Duck population changes occur on breeding, staging, and wintering habitats, with the changes on breeding habitats having the greatest effect on populations. Degradation and destruction of wetlands over the last 200 years have diminished duck populations; wetland alteration and degradation continue. The rate of wetland loss has been greatest in prime agricultural areas such as the Prairie Pothole region (Fig. 1), and lowest in northern boreal forests and tundra. Thus, species such as dabbling ducks that mostly nest in the severely altered Prairie Potholes have been harmed more than species like sea ducks and mergansers that nest farther north (Bellrose 1980; Johnson and Grier 1988).

Because most dabbling ducks need grassy cover for nesting (Kaminski and Weller 1992), conversion of native grasslands to agricultural production, including pastures, has reduced available nesting cover and contributed to a reduced nesting success for dabblers. This condition is especially true in the Prairie Pothole region of the United States and Canada (Fig. 1). In addition, highly variable precipitation in the Prairie Potholes has changed the number of wetlands available for nesting. For example, in 1979 there were 6.3 million wetlands in the surveyed portion of the Prairie Pothole region, but by the next spring, wetlands in the same area had decreased 55% to 2.9 million. Two years later they increased more than 100% to 4.2 million. These annual changes can temporarily mask the long-term declining trend in wetland abundance across the Prairie Pothole region.

The changing availability of wetland habitats in the Prairie Potholes region causes substantial fluctuations in some duck populations. During periods of high precipitation, larger wetland basins are full or overflowing, and shallow wetlands are abundant. Species such as the northern pintail, which tend to use shallow or ephemeral wetlands for feeding, produce more young when wetland numbers increase (Smith 1970; Hochbaum and Bossenmaier 1972). Consequently, population numbers increase as they did during the 1970's.

During the driest periods, however, such as those in the 1980's, only the deepest and most permanent wetlands retain water, causing population declines in species such as pintails that rely primarily on shallow wetlands. Population numbers are more stable for species like the canvasback, which rely on deeper marshes, and are therefore less affected by annual changes in wetland numbers because deeper marshes consistently retain water, providing ample habitat in most years (Stewart and Kantrud 1973).

Nest success in the Prairie Pothole region has declined in recent years largely because of increased nest predation caused by the range expansion of some predators and by reduced nesting habitat (Sargeant and Raveling 1992). Fewer and smaller areas of nesting habitat concentrate duck nests, enhancing the ability of predators to find nests. Predators such as raccoons (*Procyon lotor*) have expanded their range northward, probably because they can den in buildings, rock piles, and other human-made sites during winter.

Although wetland drainage, urbanization, and other human-caused changes have resulted in wintering habitat losses, these losses have been offset, at least for dabbling ducks, by increased fall and winter food from waste grain left in stubble fields. In addition, the national wildlife refuge system has protected and managed many staging and wintering areas for the benefit of waterfowl.

Modern duck-hunting regulations are believed to keep recreational harvest at levels compatible with the long-term welfare of duck populations. The proportion of ducks harvested varies regionally and by species, age, and sex. In 1992, 2%-12% of the adult mallards from the Prairie Pothole region were killed by hunters. Harvest rates of other species were generally lower. These conservative harvest rates are unlikely to cause population declines (Blohm 1989).

Conclusions

Changes in duck populations reflect changes in quality and quantity of waterfowl habitats. Long-term declines in populations have been caused by extensive habitat alterations. By contrast, short-term changes primarily reflect weather and resultant availability of wetland habitats. Maintenance of the current monitoring system and initiatives to improve our monitoring capability are essential for effective duck management.

Maintaining or increasing the quality and quantity of waterfowl habitat is needed to stabilize or increase duck populations. Agricultural policies and practices can profoundly affect habitat availability in Canada and the United States. For example, the Conservation Reserve Program, in which certain agricultural areas were set aside and planted in grasses, has added much-needed dabbling duck nesting habitat and therefore has improved their productivity in the U.S. portion of the Prairie Pothole region (R.E. Reynolds, USFWS, personal communication). The North American Waterfowl Management Plan, through its regional joint ventures, is striving to increase the habitat available for waterfowl and to improve monitoring of some populations.

References

Bellrose, F.C. 1980. Ducks, geese and swans of North America. 3rd ed. Stackpole Books, Harrisburg, PA. 540 pp.

Blohm, R.J. 1989. Introduction to harvest: understanding surveys and season setting. Pages 118-129 *in* K.H. Beattie, ed. Sixth International Waterfowl Symposium, Washington, DC.

Caithamer, D.F, J.A. Dubovsky, F.A. Johnson, J.R. Kelley, Jr., and G.W. Smith. 1993. Waterfowl: status and fall flight forecast. Administrative Rep., U.S. Fish and Wildlife Service, Washington, DC. 37 pp.

Conroy M.J., J.R. Goldsberry, J.E. Hines, and D.B. Stotts. 1988. Evaluation of aerial transect surveys for wintering American black ducks. Journal of Wildlife Management 52:694-703.

Eggeman, D.R., and F.A. Johnson. 1989. Variation in effort and methodology for the midwinter waterfowl inventory in the Atlantic Flyway. Wildlife Society Bull. 17:227-233.

Hochbaum, G.S., and E.F. Bossenmaier. 1972. Response of pintail to improved breeding habitat in southern Manitoba. Canadian Field-Naturalist 86:79-81.

Johnson, D.H., and J.W. Grier. 1988. Determinants of breeding distributions of ducks. Wildlife Monograph 100:1-37.

Kaminski, R.M., and M.W. Weller. 1992. Breeding habitats of nearctic waterfowl. Pages 568-589 *in* B.J. Batt, A.D. Afton, M.G. Anderson, C.D. Ankney, D.H. Johnson, J.A. Kadlec, and G.L. Krapu, eds. Ecology and management of breeding waterfowl. University of Minnesota Press, Minneapolis.

Sargeant, A.B., and D.G. Raveling. 1992. Mortality during the breeding season. Pages 396-422 *in* B.J. Batt, A.D. Afton, M.G. Anderson, C.D. Ankney, D.H. Johnson, J.A. Kadlec, and G.L. Krapu, eds. Ecology and management of breeding waterfowl. University of Minnesota Press, Minneapolis.

Smith, R.I. 1970. Response of pintail breeding populations to drought. Journal of Wildlife Management 34:943-946.

Stewart, R.E., and H.A. Kantrud. 1973. Ecological distribution of breeding waterfowl populations in North Dakota. Journal of Wildlife Management 37:39-50.

For further information:

David F. Caithamer
U.S. Fish and Wildlife Service
Office of Migratory Bird Management
Henshaw Laboratory
11500 American Holly Dr.
Laurel, MD 20708

Decline of Northern Pintails

by
Jay B. Hestback
National Biological Service

The size of the continental breeding population of northern pintail (*Anas acuta*) has greatly varied since 1955, with numbers in surveyed areas ranging from a high of 9.9 million in 1956 to a low of 1.8 million in 1991. This variation results primarily from differences in the numbers of breeding pintails in the prairie region of Canada and the United States (Fig. 1); these numbers ranged from 8.6 million in 1956 to 0.5 million in 1991; numbers in the northern regions from Alaska to northern Alberta and northern Manitoba varied primarily between 1 and 2 million.

Breeding pintails prefer seasonal shallow-water habitats without tall emergent aquatic vegetation (Smith 1968). The proportions and distribution of breeding pintails on the prairies vary annually depending on the amount of annual precipitation and the resulting increase or decrease in the availability of suitable breeding habitat (Smith 1970; Johnson and Grier 1988).

Changes in the size of the continental pintail population result from changes in production, survival, or both. Consequently, understanding population changes involves detecting variation in survival and production over time and relating that variation to changes in population size. Once the cause of the decline is determined, appropriate management strategies can be developed to reverse it.

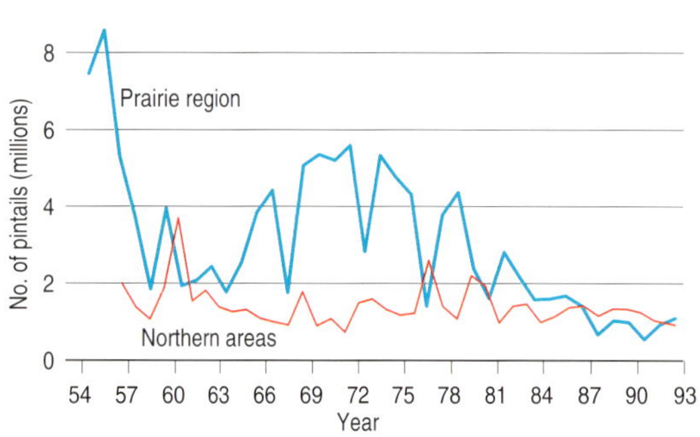

Fig. 1. Number of pintails in northern areas from Alaska to northern Alberta and northern Manitoba and in the prairie region from southern Alberta and central Montana to southern Manitoba and the Dakotas from 1955 to 1993 (Breeding Population and Habitat Survey, U.S. Fish and Wildlife Service, Office of Migratory Bird Management).

Status and Trends

I arbitrarily partitioned the population data into periods of relative growth, stability, and decline to help explain changes in the continental breeding population, which declined from 1955 to 1962, increased from 1963 to 1970, remained at a high stable level from 1971 to 1979, and declined from 1980 to 1992. I also partitioned the continental population into flyways based on data from recoveries of winter-banded pintails. This data indicated that pintails exhibit a high fidelity to the winter-banding region and flyway (Hestbeck 1993). Data from recoveries of summer-banded pintails were used to associate birds between breeding and wintering areas.

Pintails (*Anas acuta*).

Data on the pintail population were obtained through various surveys conducted by the United States and Canada. The Breeding Population and Habitat Survey provided estimates for the number of breeding pintails and for the total number of ponds. The total number of ponds was used as an index of breeding-habitat availability where the availability increased as the number of ponds increased. Annual survival rates were estimated from legband recoveries of summer-banded pintails.

I estimated average survival rates for the previously listed time periods for all areas with banding data. As an index of production, I used the number of young females divided by the number of adult females (i.e., age-ratio) harvested annually in each flyway reported in the Waterfowl Parts Collection Survey (U.S. Fish and Wildlife Service, Office of Migratory Bird Management). Because of possible harvest differences among flyways and large variation in annual ratios, I estimated the average age-ratio for each flyway for the above time periods.

Changes in the continental population can be addressed by studying changes in flyway populations because pintails from different summer breeding areas were associated with certain wintering areas. Generally, pintails wintering in the Pacific Flyway were associated with breeding areas in the western states and provinces from Alaska to Saskatchewan and central Montana. Pintails in the Central Flyway were primarily associated with breeding areas in Saskatchewan, eastern Montana, Manitoba, and the Dakotas. Pintails in the Mississippi Flyway were primarily associated with breeding areas from Saskatchewan and Minnesota to James Bay. Pintails in the Atlantic Flyway were primarily associated with breeding areas from James Bay to the Canadian Maritimes.

If 1980-92 population declines were caused by poor reproduction, production would be lower. Production, however, remained relatively constant over periods of population growth (1963-70), stability (1971-79), and decline

(1980-92) for the Atlantic, Mississippi, and Central flyways (Fig. 2). Production in the Pacific Flyway exhibited a substantial decline from 2.40 in 1963-70, to 1.78 in 1971-79, and to 1.60 in 1980-92.

Likewise, survival would be lower during 1980-92 if population declines were caused by declines in survival. Comparisons of average survival rates between 1980-92 and earlier periods were possible for only a limited number of areas because few pintails were banded in many regions. In the area encompassing northern Alberta, northeastern British Columbia, and southwestern Northwest Territories, average survival during 1980-92 was higher than the average for earlier periods for adult males (80% versus 68%), young males (68% versus 53%), and adult females (69% versus 64%). In southern Alberta, average survival during 1980-92 was higher than the average for earlier periods for adult males (74% versus 70%) and young females (86% versus 55%). Survival remained constant between 1980-92 and earlier periods for all age-classes of pintails banded in southern Saskatchewan and southern Manitoba. In the Dakotas, average survival during 1980-92 was higher for only adult males (77% versus 66%).

These data reveal that possible declines in pintail survival did not cause the population declines observed during the 1980's. Overall, survival was higher during 1980-92 than during earlier periods for adult males that winter in the Pacific, Central, and Mississippi flyways and for young females that winter in the Pacific Flyway. Survival remained constant between time periods for adult females and young males in the Pacific, Central, and Mississippi flyways.

Given the small changes in production and survival, pintail numbers should stabilize in the Central and Mississippi flyways and possibly the Atlantic Flyway. In the Pacific Flyway, however, the survival increases of young females has not compensated for the overall decrease in production.

During the 1980's the Canadian prairies on the average received less precipitation, resulting in reduced availability of pintail breeding habitat. Hopes for increased pintail population size have been based, in part, on the expectation that increased precipitation in the western Canadian prairies would result in increased breeding habitat and production. Female-based age-ratio data suggest, though, that increased production is unlikely to occur even with increased precipitation because pintail production remained low even when water was plentiful. Average age-ratios for the Pacific Flyway when water in the western Canadian prairies was above average (total May ponds for southern Alberta and southern Saskatchewan exceeding 2.68 million) steadily declined from 3.11 in the 1960's, to 2.03 in the 1970's, and 1.86 in the 1980's.

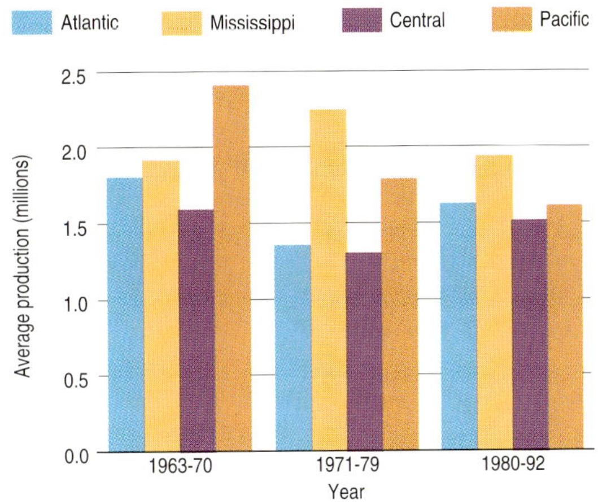

Fig. 2. Average production of pintails in Atlantic, Mississippi, Central, and Pacific flyways for 1963-70, 1971-79, and 1980-92 (Waterfowl Parts Collection Survey, U.S. Fish and Wildlife Service, Office of Migratory Bird Management).

Consequently, a fundamental change appears to have occurred in pintail productivity on western Canadian prairies, meaning that we cannot base pintail management on the hope that increased precipitation will result in a return to the higher levels of production experienced in the 1960's.

Researchers suspect that the production decline may be related to the fact that the shallow-water breeding habitat favored by pintails is most susceptible to agricultural drainage. By 1989, 78% of the pothole margins (the transition zone where potholes meet farmland) and 22% of wet basins were degraded by agricultural activity in prairie Canada (F.D. Caswell and A. Didiuk, Canadian Wildlife Service, personal communication). Increased intensification of agriculture may also contribute to lower production on the prairies through increased grazing and cropping, increased nest destruction, and increased use of agricultural chemicals (Ducks Unlimited 1990). Further research on the western Canadian prairies is necessary to determine specific causes of production declines in pintails and to determine methods to increase production.

References

Ducks Unlimited. 1990. Sprig: population recovery strategy for the northern pintail. Ducks Unlimited, Inc., Long Grove, IL. 30 pp.

Hestbeck, J.B. 1993. Overwinter distribution of northern pintail populations in North America. Journal of Wildlife Management 57:582-589.

Johnson, D.H., and J.W. Grier. 1988. Determinants of breeding distributions of ducks. Wildlife Monograph 100. 37 pp.

Smith, R.I. 1968. The social aspects of reproductive behavior in the pintail. Auk 85:381-396.

Smith, R.I. 1970. Response of pintail breeding populations to drought. Journal of Wildlife Management 34:943-946.

For futher information:

Jay B. Hestbeck
National Biological Service
Cooperative Fish and Wildlife
Research Unit
University of Massachusetts
Amherst, MA 01003

Canvasback Ducks

by
William L. Hohman
G. Michael Haramis
Dennis G. Jorde
Carl E. Korschgen
John Takekawa
National Biological Service

Canvasbacks (*Aythya valisineria*) are unique to North America and are one of our most widely recognized waterfowl species. Unlike other ducks that nest and feed in uplands, diving ducks such as canvasbacks are totally dependent on aquatic habitats throughout their life cycle. Canvasbacks nest in prairie, parkland, subarctic, and Great Basin wetlands; stage during spring and fall on prairie marshes, northern lakes, and rivers; and winter in Atlantic, Pacific, and Gulf of Mexico bays, estuaries, and some inland lakes. They feed on plant and animal foods in wetland sediments. Availability of preferred foods, especially energy-rich subterranean plant parts, is probably the most important factor influencing geographic distribution and habitat use by canvasbacks.

In spite of management efforts that have included restrictive harvest regulations and frequent hunting closures in all or some of the flyways (Anderson 1989), canvasback numbers declined from 1955 to 1993 and remain below the population goal (540,000) of the North American Waterfowl Management Plan (USFWS and Canadian Wildlife Service 1994). Causes for this apparent decline are not well understood, but habitat loss and degradation, low rates of recruitment, a highly skewed sex ratio favoring males, and reduced survival of canvasbacks during their first year are considered important constraints on population growth.

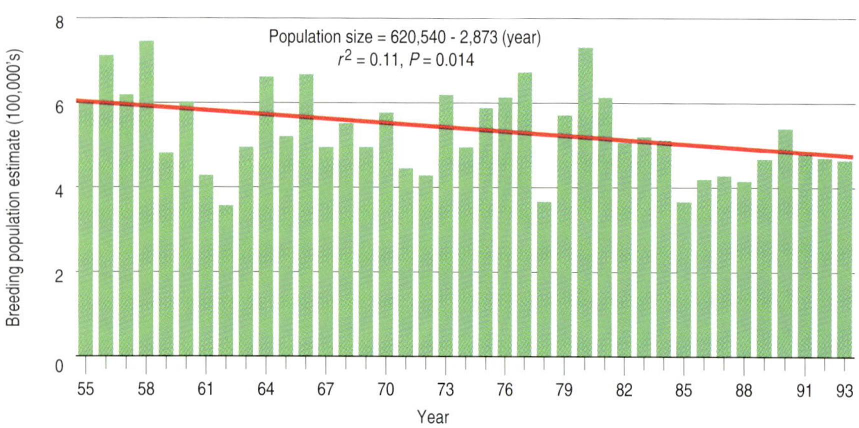

Figure. Estimated breeding population of canvasbacks, 1955-93 (data from the U.S. Fish and Wildlife Service, Office of Migratory Bird Management).

Status and Trends

Canvasback population trends are monitored by means of annual Breeding Waterfowl and Habitat Surveys and Midwinter Waterfowl Inventories (MWI). Readers should refer to cited literature for additional information regarding methods.

Canvasback Numbers and Distribution

Between 1955 and 1993 population indices for canvasbacks fluctuated between 353,700 and 742,400 and averaged 534,000 ducks (Figure). The population showed a general rate of decline of 0.6% per year during the period; however, because population estimates are imprecise, annual differences are difficult to detect. For example, a population change of more than 30% would be needed to detect a significant difference between years with 90% confidence.

The winter distribution of canvasbacks has changed since the 1950's, when most canvasbacks (79%) were found wintering in the Atlantic or Pacific flyways. The proportion of the continental population wintering in the Central and Mississippi flyways increased from 21% in 1955-69 to 44% in 1987-92 as a result of declines in canvasback numbers at Chesapeake Bay and San Francisco Bay and increases in the Gulf of Mexico region. Only about 23,000 canvasbacks winter in Mexico, but numbers may be increasing (Office of Migratory Bird Management, unpublished data). Shifts in winter distribution probably reflect regional differences in habitat availability, but may also indicate differences in survival and recruitment.

Sex Ratios

Canvasbacks have a highly skewed sex ratio favoring males. Sex ratios of wintering canvasbacks in Louisiana (1.6-1.8 males:female; Woolington 1993) and San Francisco Bay (2.2 males:female; J. Takekawa, unpublished data) are lower than those observed in the Atlantic Flyway (2.9-3.2 males:female), but sex ratios apparently decreased in two mid-Atlantic states between 1981 and 1987 (Haramis et al. 1985, 1994). Based on recent (1987-92) MWI and sex ratio data, we calculated that the continental sex ratio for canvasbacks likely lies between 2.0 and 2.5 males:female.

Survival

Annual survival rates of female canvasbacks (56%-69%) are lower than those of males (70%-82%; Nichols and Haramis 1980). Survival rates also vary geographically (survival is greater in the Pacific Flyway than in the Atlantic; Nichols and Haramis 1980) and are positively related to body mass in early winter (Haramis et al. 1986). Survival of females in their first year probably is reduced relative to that of adults. Assuming that all surviving females return to their natal areas to breed, return rates for female canvasbacks breeding in southwestern Manitoba suggest that only 21% of hens survive their first year compared to 69% annual survival of older hens (Serie et al. 1992).

Nichols and Haramis (1980) found no association between canvasback harvest regulations and survival. However, an analysis of return rates for female canvasbacks in southwestern Manitoba indicated that survival of immatures was significantly related to harvest (M.G. Anderson, Ducks Unlimited-Canada, unpublished data). The canvasback season was closed in the Atlantic, Central, and Mississippi flyways during 1986-93, but about 8,000 birds were harvested annually in Canada and 10,000 in the Pacific Flyway. There is also a substantial illegal harvest of canvasbacks at some sites (Haramis et al. 1993; Korschgen et al. 1993; W.L. Hohman, unpublished data). However, the current level of hunting-related mortality is probably not limiting population growth. Rather, annual variation in recruitment and degradation and loss of breeding, migrational, and wintering habitats are more likely influencing population size.

Courtesy W.L. Hohman, NBS

Canvasback (*Aythya valisineria*).

Time-specific Survival Rates and Sources of Mortality

Survival rates for adults in spring and summer are unknown. In spite of a nationwide ban on the use of lead shot by waterfowl hunters, ingestion of spent lead shotgun pellets by waterfowl is common and likely will remain so for many years. More than 50% of spring-migrating canvasbacks captured at a major staging area on the Mississippi River had elevated blood lead levels (Havera et al. 1992). Lead-exposed birds have reduced body mass, fat, and protein (Hohman et al. 1990), so their subsequent survival and ability to reproduce and perform activities such as courtship, migration, or molt, may be compromised.

Nest success (i.e., embryonic survival) of canvasbacks is highly variable, especially for birds nesting on the prairies. For example, nest success in southwestern Manitoba in wet years was 54%-60%, but in dry years averaged only 17% (Serie et al. 1992). In spite of habitat loss and degradation, ranges in nest success observed in southwestern Manitoba were similar in 1961-72 (21%-62%; Stoudt 1982) and 1974-80 (17%-60%; Serie et al. 1992). Mammalian predation, especially by mink (*Mustela vison*) and raccoon (*Procyon lotor*), is an important factor affecting the nest success of prairie-nesting canvasbacks.

Mortality of prefledged ducklings is high, especially during the first 10 days (C.E. Korschgen, unpublished data). In northwestern Minnesota, estimated survival rates for ducklings up to 10 days old ranged from near zero to 70%, but differed between sexes during the first 25 days of life (male > female; C.E. Korschgen, unpublished data). Predation and weather were the primary sources of duckling mortality. Survival of young between fledging and fall migration is unknown; however, production estimates calculated from harvest information (0.16-1.07 young:adult) suggest that recruitment rates for canvasbacks generally are low compared to other ducks.

Survival rates for fall-migrating canvasbacks have not been studied, but survival rates have been estimated at several major wintering sites. Adult and immature females had high winter survival at Chesapeake Bay (83%-100%; Haramis et al. 1993) and coastal Louisiana (≥ 95%; Hohman et al. 1993). Winter survival was lower at Catahoula Lake, Louisiana (57%-92%), where canvasbacks were not only shot illegally but where substantial numbers of birds were also exposed to lead (W.L. Hohman, unpublished data).

Habitat Trends

Historically, climate, grazing, and fire were major factors affecting habitats of prairie-nesting waterfowl. Since settlement, however, human activities, especially those related to agriculture, have had a major impact on the quantity and quality of breeding habitats. Nationwide, over 53% of original wetlands have been lost. Wetland losses in states where canvasbacks historically nested range from less than 1% (Alaska) to 89% (Iowa); however, deeper wetlands preferred by nesting canvasbacks probably have been drained to a lesser extent than shallower wetlands.

Northern lakes used by canvasbacks for molting and staging before fall migration probably have been least affected by human and natural perturbations. Nonetheless, disturbances related to commercial and recreational activities, nutrient enrichment of lakes resulting from sewage discharges and agricultural runoff, introductions of herbivorous fish, and alteration of lake levels for generation of hydroelectric power have reduced the suitability and use of some traditional staging areas in the southern boreal forest region.

Most of the traditional stopover habitats used by migrating canvasbacks no longer provide suitable feeding and resting opportunities (Kahl 1991). For example, of the more than 40 former migration stopover areas in the upper portion of the Mississippi Flyway, only Lake Christina in west-central Minnesota, two pools on the Upper Mississippi River, and two areas on the Great Lakes have peak populations of more than 5,000 canvasbacks (Korschgen 1989). Restoration efforts begun in 1987 at Lake Christina were successful in reestablishing submersed aquatic vegetation and canvasback use. Habitat on the Upper Mississippi River increased in extent from the mid-1960's to the late 1980's. However, record drought in 1988-89 and extensive flooding in 1993 in the Upper Mississippi River basin have caused major declines in habitat quality and abundance.

In the Great Lakes region, increased bird use of Lake St. Clair and Long Point on Lake Erie coincided with improved water quality and increased production of submersed aquatic plants, especially wildcelery (*Vallisneria americana*). These improvements are attributed to regulation of water discharges into the Great Lakes and perhaps the proliferation of zebra mussels (*Dreissena polymorpha*).

In the Pacific Flyway, coastal habitats used by migrating canvasbacks have not changed greatly since the 1950's, although development has increased in some areas (e.g., Puget Sound). Whereas use of some inland sites (e.g., Great Salt Lake, Utah; Malheur National Wildlife Refuge (NWR), Oregon; and Stillwater NWR, Nevada) declined during the 1970's or 1980's, canvasback use of Klamath Basin NWR, Oregon-California, and Pyramid Lake, Nevada, has increased.

Degradation of water quality in the Chesapeake Bay caused by nutrient enrichment, turbidity, and sedimentation reduced the abundance of aquatic plant and animal foods most important to canvasbacks in winter (Haramis 1991). Declining availability of plant foods caused canvasbacks to shift to mostly animal foods. Canvasback numbers declined in response to loss of aquatic plants in the Chesapeake Bay, but increased in North Carolina and Virginia where preferred plant foods were still abundant (Lovvorn 1989). Aquatic plants are now declining in the coastal areas of North Carolina and other wintering areas throughout the Atlantic Flyway. Unless the widespread decline of aquatic plant foods is reversed, the number of canvasbacks wintering in the Atlantic Flyway is not likely to increase.

San Francisco Bay is the most important wintering area for canvasbacks in the Pacific Flyway. Urban development there has greatly reduced available habitat. In remaining habitats, canvasbacks are exposed to high levels of environmental contaminants (Miles and Ohlendorf 1993). Canvasbacks make extensive use of salt evaporation ponds in northern San Francisco Bay (Accurso 1992). These ponds recently came under public ownership, but their management as tidal salt marshes will probably reduce their use by canvasbacks. Increasing numbers of canvasbacks have been observed recently on wetland easements and sewage lagoons in the northern San Joaquin Valley.

Increased numbers of canvasbacks are wintering in the Gulf of Mexico region, especially at Catahoula Lake, where, since 1985, peak numbers (up to 78,000 birds) have equaled or exceeded counts on traditional wintering areas such as Chesapeake Bay and San Francisco Bay. Birds appear to be attracted to Catahoula Lake because of its abundant plant foods and stable flooding regime (Woolington and Emfinger 1989). These birds are at risk of lead poisoning, however, because of the high density of spent lead shot contained in lake sediments.

Information Gaps

Information needs for improved management of canvasbacks include banding or radiotelemetry data sufficient to provide habitat information and estimates of region-specific rates of survival, band recovery, and recruitment; survival rates of immature birds between hatch and arrival on wintering areas; and cross-seasonal effects of winter nutrition and contaminant exposure on reproduction.

References

Accurso, L.M. 1992. Distribution and abundance of wintering waterfowl on San Francisco Bay, 1988-1990. M.S. thesis, Humboldt State University, Acadia, CA. 252 pp.

Anderson, M.G. 1989. Species closures — a case study of wintering waterfowl on San Francisco Bay, 1988-1990. M.S. thesis, Humbolt State University, Acadia, CA. 252 pp.

Haramis, G.M. 1991. Canvasback. Pages 17.1-17.10 *in* S.L. Funderburk, J.A. Mihursky, S.J. Jordan, and D. Riley, eds. Habitat requirements for Chesapeake Bay living resources. 2nd ed. Living Resources Subcommittee, Chesapeake Bay Program, Annapolis, MD.

Haramis, G.M., E.L. Derleth, and W.A. Link. 1994. Flock sizes and sex ratios of canvasbacks in Chesapeake and North Carolina. Journal of Wildlife Management 58:123-130.

Haramis, G.M., J.R. Goldsberry, D.G. McAuley, and E.L. Derleth. 1985. An aerial photographic census of Chesapeake and North Carolina canvasbacks. Journal of Wildlife Management 49:449-454.

Haramis, G.M., D.G. Jorde, and C.M. Bunck. 1993. Survival of hatching-year female canvasbacks wintering on Chesapeake Bay. Journal of Wildlife Management 57:763-770.

Haramis, G.M., J.D. Nichols, K.H. Pollock, and J.E. Hines. 1986. The relationship between body mass and survival of wintering canvasbacks. Auk 103:506-514.

Havera, S.P., R.M. Whitton, and R.T. Shealy. 1992. Blood lead and ingested shot in diving ducks during spring. Journal of Wildlife Management 56:539-545.

Hohman, W.L., R.D. Pritchert, J.L. Moore, and D.O. Schaeffer. 1993. Survival of female canvasbacks wintering in coastal Louisiana. Journal of Wildlife Management 57:758-762.

Hohman, W.L., R.D. Pritchert, R.M. Pace, D.W. Woolington, and R. Helm. 1990. Influence of ingested lead on body mass of wintering canvasbacks. Journal of Wildlife Management 54:211-215.

Kahl, R. 1991. Restoration of canvasback migrational staging habitat in Wisconsin. Tech. Bull. 172. Department of Natural Resources, Madison, WI. 47 pp.

Korschgen, C.E. 1989. Riverine and deepwater habitats for diving ducks. Pages 157-189 *in* L.M. Smith, R.L. Pederson, and R.M. Kaminski, eds. Habitat management for migrating and wintering waterfowl in North America. Texas Tech University Press, Lubbock.

Korschgen, C.E., K. Kenow, J. Nissen, and J. Wetzel. 1993. Final report: canvasback hunting mortality and hunter education efforts on the Upper Mississippi River National Wildlife and Fish Refuge. U.S. Fish and Wildlife Service, LaCrosse, WI. 50 pp.

Lovvorn, J.R. 1989. Distributional responses of canvasbacks to weather and habitat change. Journal of Applied Ecology 26:113-130.

Miles, A.K., and H.M. Ohlendorf. 1993. Environmental contaminants in canvasbacks wintering on San Francisco Bay, California. California Fish and Game 79:28-38.

Nichols, J.D., and G.M. Haramis. 1980. Inferences regarding survival and recovery rates of winter-banded canvasbacks. Journal of Wildlife Management 4:164-173.

Serie, J.R., D.L. Trauger, and J.E. Austin. 1992. Influence of age and selected environmental factors on reproductive performance of canvasbacks. Journal of Wildlife Management 56:546-555.

Stoudt, J.H. 1982. Habitat use and productivity of canvasbacks in southwestern Manitoba, 1961-72. U.S. Fish and Wildlife Service Special Sci. Rep. Wildlife 248. 31 pp.

Woolington, D.W. 1993. Sex ratios of wintering canvasbacks in Louisiana. Journal of Wildlife Management 57:751-757.

Woolington, D.W., and J.W. Emfinger. 1989. Trends in wintering canvasback populations at Catahoula Lake, Louisiana. Proceedings of the Annual Conference of the Southeastern Association of Fish and Wildlife Agencies 43:396-403.

USFWS and Canadian Wildlife Service. 1994. North American waterfowl management plan 1994 update, expanding the commitment. U.S. Fish and Wildlife Service, Washington, DC. 40 pp.

For further information:

William L. Hohman
National Biological Service
Southern Science Center
700 Cajundome Blvd.
Lafayette, LA 70506

Breeding Seabirds in California, Oregon, and Washington

by

Harry R. Carter
David S. Gilmer
National Biological Service
Jean E. Takekawa
Roy W. Lowe
Ulrich W. Wilson
U.S. Fish and Wildlife Service

More than two million seabirds of 29 species nest along the west coasts of California, Oregon, and Washington, including three species listed on the federal list of threatened and endangered species: the brown pelican (*Pelecanus occidentalis*), least tern (*Sterna antillarum*), and marbled murrelet (*Brachyramphus marmoratus*). The size and diversity of the breeding seabird community in this region reflect excellent nearshore prey conditions; subtropical waters within the southern California Bight area; complex tidal waters of Strait of Juan de Fuca and Puget Sound in Washington; large estuaries at San Francisco Bay, Columbia River, and Grays Harbor-Willapa bays; and the variety of nesting habitats used by seabirds throughout the region, including islands, mainland cliffs, old-growth forests, and artificial structures.

Breeding seabird populations along the west coast have declined since European settlement began in the late 1700's because of human occupation of, commercial use of, and introduction of mammalian predators to seabird nesting islands. In the 1900's, further declines occurred in association with rapid human population growth and intensive commercial use of natural resources in the Pacific region. In particular, severe adverse impacts have occurred from partial or complete nesting habitat destruction on islands or the mainland, human disturbance of nesting islands or areas, marine pollution, fisheries, and logging of old-growth forests (Ainley and Lewis 1974; Bartonek and Nettleship 1979; Hunt et al. 1979; Sowls et al. 1980; Nettleship et al. 1984; Speich and Wahl 1989; Ainley and Boekelheide 1990; Sealy 1990; Ainley and Hunt 1991; Carter and Morrison 1992; Carter et al. 1992; Vermeer et al. 1993).

Methods

Population status of breeding seabirds on the west coast has been measured primarily through the determination of and trends in population size, based on counts of birds and nests at nesting colonies (e.g., Sowls et al. 1980). At-sea surveys also have been used to approximate population sizes for breeding and nonbreeding populations and species as well as their foraging distribution alongshore and offshore (e.g., Briggs et al. 1987). Rather than just monitoring small plots of nests on a few accessible islands to determine status and trends, relatively accurate and standardized censuses of entire coastal seabird breeding populations (except for certain nesting areas of difficult-to-census species) have been conducted annually or periodically to determine the overall status of many species breeding on the west coast (Figs. 1-4). However, we have considered census accuracy, natural variability, trends at well-studied colonies (e.g., Farallon National Wildlife Refuge) and many other factors in assessing population status and trends.

Status and Trends

Storm-petrels (Hydrobatidae)

Increasing numbers of Leach's storm-petrels (*Oceanodroma leucorhoa*) have been documented recently in Oregon (R.W. Lowe, USFWS, unpublished data), although this

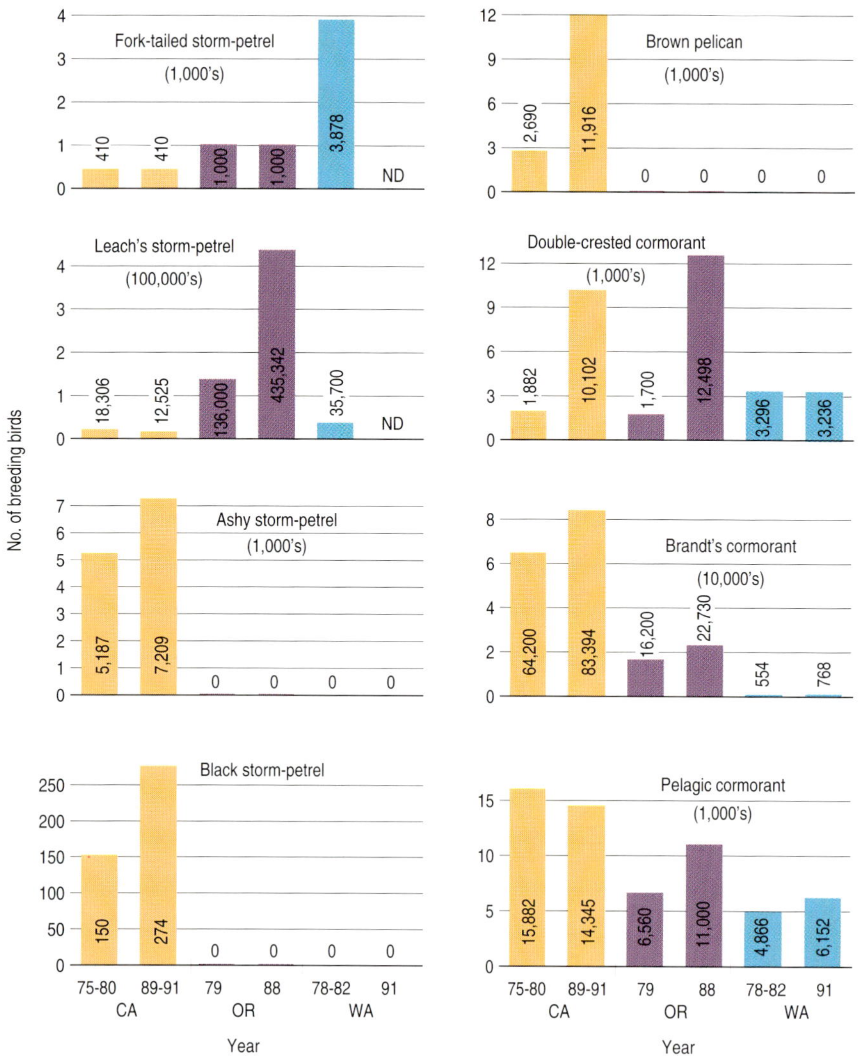

Fig. 1. Status and trends of breeding populations of storm-petrels, pelicans, and cormorants on the west coasts of California, Oregon, and Washington. Data for small inland populations of white pelicans and double-crested cormorants are not included. ND — no data available; 0 — no coastal nesting. Sources: CA (Hunt et al. 1979; Sowls et al. 1980; Carter et al. 1992); OR (Varoujean and Pitman 1979; R.W. Lowe, unpublished data); and WA (Speich and Wahl 1989; U.W. Wilson, unpublished data). Also *see* Carter et al. (in press) for double-crested cormorant.

increase probably represents greater survey effort (Fig. 1). They have declined in northern California because of the loss of burrow-nesting habitats due to soil erosion and defoilation by nesting cormorants (Carter et al. 1992). Ashy storm-petrels (*O. homochroa*) have declined recently at the world's largest known colony at the South Farallon Islands, possibly because of high gull predation (W.J. Sydeman, Point Reyes Bird Observatory, unpublished data). This decline is of concern because the small world population of this species (fewer than 10,000 breeding birds) nests entirely in California. Greater numbers of ashy and black storm-petrels (*O. melania*) have been documented recently in southern California, although this probably reflects greater survey efforts (Carter et al. 1992). In Fig. 1, similar numbers of fork-tailed storm-petrels (*O. furcata*) are indicated over the past decade in Oregon and California because survey efforts confirmed very small numbers. Declines in California are suspected (Carter et al. 1992), but further work is required to establish trends.

Pelicans (Pelecanidae)

Brown pelicans have increased recently at the only two remaining colonies (West Anacapa and Santa Barbara islands) in the Channel Islands in southern California (Fig. 1), following severe pre-1975 declines primarily due to eggshell thinning from marine pollutants (Anderson et al. 1975; Anderson and Gress 1983; Carter et al. 1992; F. Gress and D.W. Anderson, University of California-Davis, personal communication). Breeding success is still low and limited recovery may involve immigration of birds out of Mexico. Concern exists for adverse effects of continuing low levels of marine pollutants, commercial fisheries, and the 1990 *American Trader* oil spill. Although the brown pelican has shown recent population increases, white pelicans have been extirpated from parts of interior California and have declined at inland colonies in northern California because of low reproduction related to water developments and drought (Carter et al. 1992; P. Moreno and D.W. Anderson, University of California-Davis, personal communication). Small colonies still exist at Sheepy Lake and Clear Lake in the Klamath Basin area. These conditions also exist at other inland areas in Oregon, Washington, and Nevada, but problems seem fewer farther east.

Cormorants (Phalacrocoracidae)

Double-crested cormorants (*Phalacrocorax auritus*) have increased dramatically in coastal regions of California and Oregon (Fig. 1) because of reduced human disturbance, reduced levels of marine pollutants in southern California, and recent use of artificial nesting areas in San Francisco Bay and Columbia River estuaries (Gress et al. 1973; Carter et al. 1992). They have not increased in Puget Sound because of high human disturbance and predation by bald eagles (*Haliaeetus leucocephalus*), which has caused colony abandonments (Henny et al. 1989; Speich and Wahl 1989; Carter et al. in press; U.W. Wilson, unpublished data). Declines have been reported at interior colonies in California, Oregon, and Washington due to water developments, human disturbance at colonies, and large-scale shooting of birds at hatcheries (during smolt releases) and at aquacultural facilities (Carter et al. in press; R.W. Lowe, unpublished data; R. Bayer, personal communication; P. Moreno, unpublished data). Brandt's and pelagic cormorant (*P. penicillatus* and *P. pelagicus*) populations have fluctuated in response to El Niño conditions (Ainley and Boekelheide 1990; Ainley et al. 1994). At the South Farallon Islands, these cormorants appear very sensitive to El Niño conditions, which result in quite poor reproduction and mortality of subadult and adult birds (Boekelheide and Ainley 1989; Ainley and Boekelheide 1990). Overall, numbers have remained stable or increased in most areas in the region (e.g., Carter et al. 1992), whereas these birds now occur at lower abundance than previously at the South Farallon Islands (Ainley et al. 1994). Numbers have increased in southern California, but the birds have suffered from gill-net and oil-spill mortality as well as human disturbance at colonies (H.R. Carter, unpublished data).

Gulls, Terns, and Skimmers (Laridae and Rynchopidae)

The predominant nesting gull on the west coast is the western gull (*Larus occidentalis*). Numbers have increased, especially in California (Fig. 2), probably because of the bird's use of human and fishing refuse and reduced human disturbance. Numbers have reached saturation at the world's largest colony at the South Farallon Islands (Ainley et al. 1994); however, expansion is occurring at other major colonies in central and southern California (Carter et al. 1992). Glaucous-winged gulls (*L. glaucescens*) have remained stable or increased in Puget Sound (U.W. Wilson, unpublished data).

California gulls (*L. californicus*) have recently expanded from interior colonies to nest in San Francisco Bay (Fig. 2; Carter et al. 1992; P. Woodin, San Francisco Bay Bird Observatory, unpublished data). They face serious threats at inland colonies in interior California because of water developments. At the world's largest colony at Mono Lake, low water levels have resulted in the formation of land bridges to nesting islands, allowing access by coyotes (*Canis latrans*) in certain years (Jones and Stokes Associates 1993). Similar problems exist at other northern California colonies for many seabird and colonial waterbird species (W.D. Shuford, Point Reyes Bird Observatory, unpublished data).

The status of California gulls at inland colonies in Oregon and Washington is not well known. Status and trends of inland colonies of ring-billed gulls (*L. delawarensis*) in California, Oregon, and Washington are not known, although problems related to low water levels may occur at many colonies. Many thousands have nested recently in northern California (W.D. Shuford, unpublished data). Small numbers (< 500 breeding birds) also nest along the Washington coast (Speich and Wahl 1989). Small numbers (< 10 breeding birds) of Heermann's gulls (*L. heermanni*) nested in the early 1980's along the central California coast but none are known to do so now. Franklin's gulls (*L. pipixcan*) recently nested in small numbers (< 100 breeding birds) at Lower Klamath Lake, California, but their status in the region is not known.

Low thousands of Caspian, Forster's, least, and elegant terns (*Sterna caspia, S. forsteri, S. antillarum, S. elegans*) and black skimmers (*Rynchops niger*) now occur in the region through increases (especially along the southern California coast) due to colony protection and use of artificial nesting sites (Speich and Wahl 1989; Carter et al. 1992). Certain tern colonies have been eliminated or shifted (especially in San Francisco Bay) because of human disturbance and red fox (*Vulpes vulpes*) or other mammalian predation (P. Woodin, unpublished data). Overall, least tern colonies in California appear somewhat stable because of extensive management. They undoubtedly occur at lower

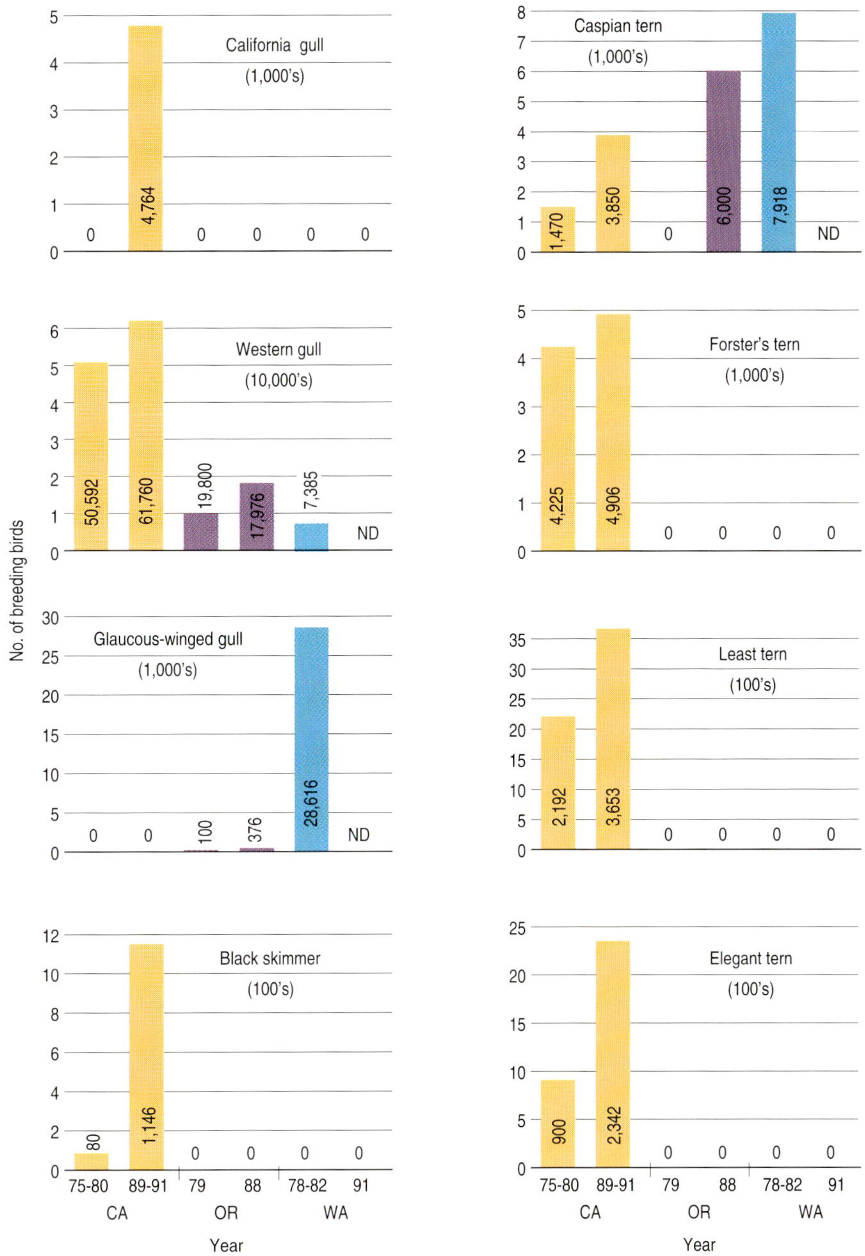

Fig. 2. Status and trends of breeding populations of gulls, terns, and skimmers. Small coastal populations of gulls (Heermann's and ring-billed) and royal terns, as well as large or small inland populations of gulls (ring-billed and California), terns (black, gull-billed, Caspian, and Forster's), and black skimmers are not included. ND — no data available; 0 — no coastal nesting. Sources: CA (Hunt et al. 1979; Sowls et al. 1980; Carter et al. 1992); OR (Varoujean and Pitman 1979; R.W. Lowe, unpublished data); and WA (Speich and Wahl 1989; U.W. Wilson, unpublished data). Also *see* Carter et al. (in press) for double-crested cormorant.

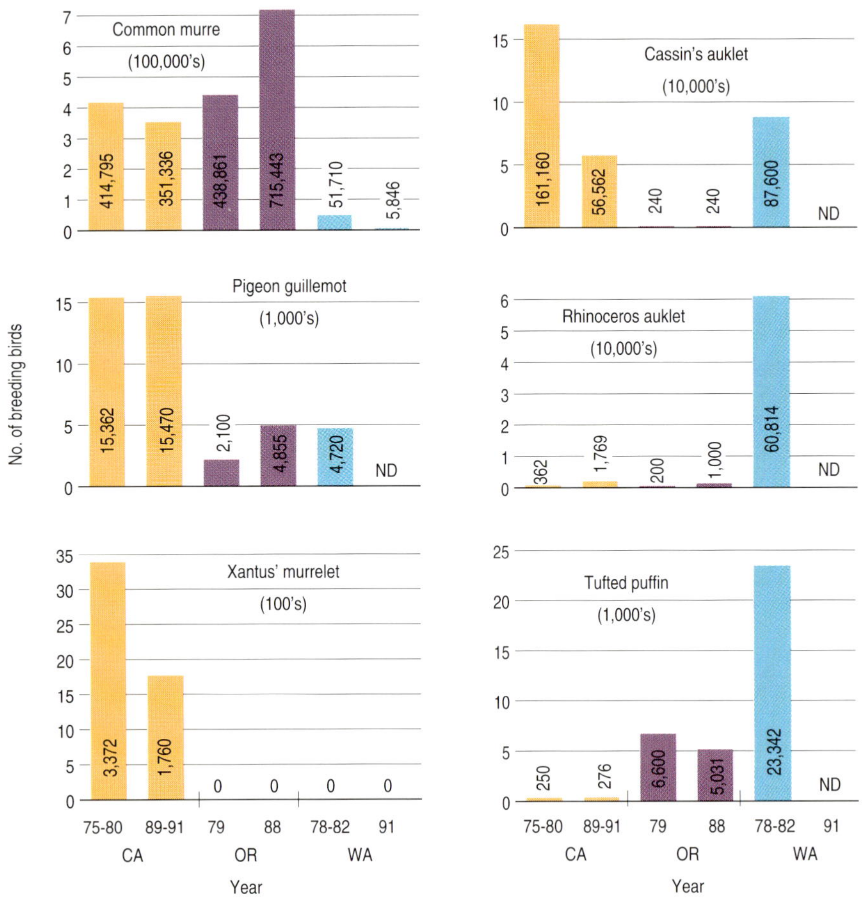

Fig. 3. Status and trends of breeding populations of several alcids in California, Oregon, and Washington. Data for marbled murrelets and historical nesting by ancient murrelets are not included. ND — no data available; 0 — no coastal nesting. Sources: CA (Hunt et al. 1979; Sowls et al. 1980; Carter et al. 1992); OR (Varoujean and Pitman 1979; R.W. Lowe, unpublished data); and WA (Speich and Wahl 1989; U.W. Wilson, unpublished data). Also *see* Carter et al. (in press) for double-crested cormorant.

than historical levels because of loss of nesting habitat, which continues to be threatened (Carter et al. 1992; R. Jurek, California Department of Fish and Game, personal communication). Low numbers (< 100 breeding birds) of arctic terns (*S. paradisaea*) have nested in coastal Washington in the past but not now (Speich and Wahl 1989). Small numbers (< 100 breeding birds) of gull-billed and royal terns (*S. nilotica* and *S. maxima*) recently colonized the southern California coast, although gull-billed terns have nested inland at the Salton Sea for a few decades. The status of black terns (*Chlidonias niger*) is not known.

Alcids (Alcidae)

Pigeon guillemot (*Cepphus columba*) populations have remained stable overall (Fig. 3), but major fluctuations have occurred in response to El Niño events at the South Farallon Islands and on the Oregon coast (Hodder and Graybill 1985; Ainley and Boekelheide 1990). A significant population and new nesting areas have been found recently in southern California, although higher numbers reflect both better survey techniques and population increases (Carter et al. 1992). Ancient murrelets (*Synthliboramphus antiquus*) nested on the Washington coast in the early 1900's but no longer do (Speich and Wahl 1989). Cassin's auklets (*Ptychoramphus aleuticus*) have declined at the largest known colony in the region at the South Farallon Islands, probably because of high gull predation and loss of burrow-nesting habitat from soil erosion (Carter et al. 1992; W.J. Sydeman, unpublished data). However, lower numbers also were found at Prince Island in southern California where numbers of nesting gulls are lower. Differences in survey techniques probably account for part of the lower numbers found recently, but other data on soil conditions, densities of nesting gulls, and gull predation support a decline at the South Farallon Islands (W.J. Sydeman, unpublished data). Hundreds also were killed in the 1984 *Puerto Rican* and 1986 *Apex Houston* oil spills (Ford et al. 1987; Page et al. 1990).

Rhinoceros auklets (*Cerorhinca monocerata*) have increased throughout the region. Largest numbers occur at Protection and Destruction islands, but burrow occupancy has fluctuated widely between years (Wilson and Manuwal 1986; U.W. Wilson, unpublished data). The South Farallon Islands were recolonized after a 100-year absence in the early 1970's (Ainley and Lewis 1974) and reached saturation levels by the late 1980's (Carter et al. 1992; Ainley et al. 1994). Nesting has recently extended to the Channel Islands (Carter et al. 1992). Thousands of rhinoceros auklets were killed in the 1986 *Apex Houston* oil spill (Page et al. 1990).

The largest tufted puffin (*Fratercula cirrhata*) populations occur along the west coast of the Olympic Peninsula (Speich and Wahl 1989), but their status there is not well known. In Puget Sound, this species has declined substantially (U.W. Wilson, unpublished data). At small colonies in Oregon and California, their numbers appear stable (Carter et al. 1992; Fig. 3), despite impacts due to El Niño at the South Farallon Islands (Ainley and Boekelheide 1990; Ainley et al. 1994). They have recently recolonized southern California where they have not nested since the early 1900's (Carter et al. 1992).

Common murres (*Uria aalge*) are the dominant member of the breeding seabird community on the west coast but they have declined substantially in central California and Washington (Figs. 3, 4) because of the combined effects of high mortality from gill-net fishing and oil spills plus poor reproduction during intense El Niño events. In central California, large historical declines in the late 1800's and early 1900's almost led to the extinction of this population (Ainley and Lewis 1974). Population growth occurred, however, between the 1950's and the 1970's, producing about 230,000 breeding birds by 1980-82 (Takekawa et al. 1990). Over 70,000 murres were estimated to have been killed in gill nets in central California between 1979 and 1987, before heavy fishing restrictions were imposed in 1987 to stop mortality (Takekawa et al. 1990). Additional mortality (10,000+ murres) occurred during the 1984 *Puerto Rican* and 1986 *Apex Houston* oil spills (Ford et al. 1987; Page et al. 1990). At the South Farallon Islands, reproductive success was almost nil during intense El Niño events in 1983 and 1992 (Ainley and Boekelheide 1990; W.J. Sydeman, unpublished data). Because of these and other factors, the central California population declined by over 60% from 1982 to 1989 and has not recovered (Fig. 4; Takekawa et al. 1990; Carter et al. 1992; Ainley et al. 1994; H.R. Carter, unpublished data).

In Washington, murre numbers crashed during the 1982-83 El Niño (Wilson 1991), although there was heavy mortality from gill nets at this time; mortality from gill nets still continues in Puget Sound. In addition, certain colonies have been disturbed by low-flying aircraft, especially near military bases. Numbers of breeding murres in Washington are lower than indicated in Figs. 3 and 4 because many birds counted in colonies in recent years (and used to derive estimates) do not appear to be breeding (U.W. Wilson, unpublished data). Significant mortality occurred during the 1984 *Arco Anchorage*, 1988 *Nestucca*, and 1991 *Tenyo Maru* oil spills. In the *Nestucca* spill alone, about 30,000 murres were estimated to have died (Ford et al. 1991). The Washington population of murres has been almost extirpated over the last decade and has not recovered.

In contrast, murre populations in Oregon and northern California have been stable or increasing to date, despite human disturbance at several colonies (Takekawa et al. 1990; R.W. Lowe, unpublished data) and some losses of Oregon birds from oil spills and the use of gill nets in Washington. In addition, these areas were known to experience lower productivity through colony abandonment during intense El Niño conditions in 1993 (Fig. 4; H.R. Carter, unpublished data; J.E. Takekawa and R.W. Lowe, unpublished data). Thus, it appears clear that decline and lack of recovery of populations in central California and Washington have resulted primarily from human causes, especially gill nets and oil spills.

Marbled murrelets probably have declined substantially throughout the region largely because of the direct loss of most (90%-95%) of their old-growth forest nesting habitat to large-scale logging since the mid-1800's (Carter and Morrison 1992; FEMAT 1994). About 10,000-20,000 birds remain. In addition, hundreds of murrelets have been killed in gill nets and oil spills in central California, Puget Sound, and off the Olympic Peninsula (Carter and Morrison 1992; H.R. Carter, unpublished data). Murrelets appear to have very low reproductive rates (based on nests examined and at-sea counts of juveniles), probably because of high avian nest predation in fragmented forests and possibly lower breeding success during intense El Niño events. This species was listed as threatened in California, Oregon, and Washington in 1992, and is being considered carefully with regard to the future of old-growth forests and the timber industry in this region. Small populations in California, Oregon, and southwestern Washington are isolated and susceptible to extinction from various potential disturbances in the future.

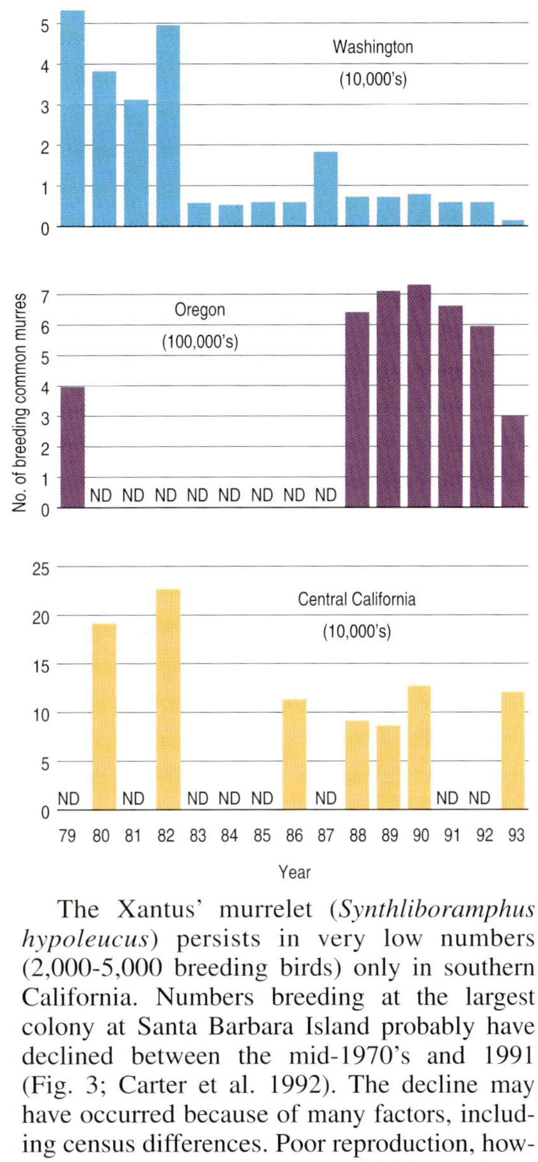

Fig. 4. Status and trends of breeding populations of common murres in Washington, Oregon, and central California. ND — no data available. Sources: WA (Wilson 1991; U.W. Wilson, unpublished data); OR (Varoujean and Pitman 1979; R.W. Lowe, unpublished data); and Central CA (Takekawa et al. 1990; Carter et al. 1992; H.R. Carter and J.E. Takekawa, unpublished data).

The Xantus' murrelet (*Synthliboramphus hypoleucus*) persists in very low numbers (2,000-5,000 breeding birds) only in southern California. Numbers breeding at the largest colony at Santa Barbara Island probably have declined between the mid-1970's and 1991 (Fig. 3; Carter et al. 1992). The decline may have occurred because of many factors, including census differences. Poor reproduction, however, has occurred because of high levels of avian and mammalian predation and has probably led to this decline. Other smaller colonies may disappear because of mortality from oil spills from offshore platforms in Santa Barbara Channel and oil tanker traffic into Los Angeles harbor and other factors. Larger numbers of nesting birds are now suspected in southern California (H.R. Carter, unpublished data). A significant portion of the small world population of this species nests in southern California while the remainder nests on the northwest coast of Baja California, Mexico. This candidate species may be considered for federal and state listing in the near future.

Future Efforts

Because of the continuing decline of and threats to seabirds on broad regional and local levels along the west coast, efforts to determine status and trends of seabirds must be extended beyond current levels. Long-term efforts must be shared among many federal and state agencies, universities, and private groups, including (1) the development of a coordinated long-term monitoring and research program, including data-base development and maintenance; (2) extending monitoring to all coastal and inland areas and species; (3) developing new methodologies for surveying nocturnal species of murrelets, auklets, and storm-petrels; (4) conducting studies of specific conservation problems such as loss of nesting habitats (e.g., old-growth forests), gill-net mortality (e.g., Puget Sound), oil-spill mortality, human disturbance, water developments, and agricultural practices; (5) restoring lost or depleted seabird colonies and habitats; and (6) examining the possible long-term effects of human fisheries and global climate change on seabird prey resources and nesting habitats.

References

Ainley, D.G., and R.J. Boekelheide, eds. 1990. Seabirds of the Farallon Islands: ecology, dynamics, and structure of an upwelling-system community. Stanford University Press, Stanford, CA. 450 pp.

Ainley, D.G., and G.W. Hunt, Jr. 1991. The status and conservation of seabirds in California. Pages 103-114 *in* J.P. Croxall, ed. Seabird status and conservation: a supplement. International Council of Bird Preservation Tech. Bull. 11.

Ainley, D.G., and T.J. Lewis. 1974. The history of Farallon Island marine bird population, 1854-1972. Condor 76:432-446.

Ainley, D.G., W.J. Sydeman, S.A. Hatch, and U.W. Wilson. 1994. Seabird population trends along the west coast of North America: causes and extent of regional concordance. Studies in Avian Biology 15:119-133.

Anderson, D.W., and F. Gress. 1983. Status of a northern population of California brown pelicans. Condor 85:79-88.

Anderson, D.W., J.R. Jehl, R.W. Risebrough, L.A. Woods, L.R. DeWeese, and W.G. Edgecomb. 1975. Brown pelicans: improved reproduction off the southern California coast. Science 190:806-808.

Bartonek, J.C., and D.N. Nettleship, eds. 1979. Conservation of marine birds of northern North America. U.S. Fish and Wildlife Service, Wildlife Res. Rep. 11. 319 pp.

Boekelheide, R.J., and D.G. Ainley. 1989. Age, resource availability, and breeding effort in Brandt's cormorant. Auk 106:389-401.

Briggs, K.T., W.B. Tyler, D.B. Lewis, and D.R. Carlson. 1987. Bird communities at sea off California: 1975 to 1983. Studies in Avian Biology 11. 74 pp.

Carter, H.R., and M.L. Morrison, eds. 1992. Status and conservation of the marbled murrelet in North America. Proceedings of a 1987 Pacific Seabird Group Symposium. Proceedings of the Western Foundation of Vertebrate Zoology 5. Camarillo, CA. 134 pp.

Carter, H.R., G.J. McChesney, D.L. Jaques, C.S. Strong, M.W. Parker, J.E. Takekawa, D.L. Jory, and D.L. Whitworth. 1992. Breeding populations of seabirds in California, 1989-1991. Vol 1. U.S. Fish and Wildlife Service, Dixon, CA. [Unpublished final report.]

Carter, H.R., A.L. Sowls, M.S. Rodway, U.W. Wilson, R.W. Lowe, F. Gress, and D.W. Anderson. Status of the double-crested cormorant on the west coast of North America. Colonial Waterbirds. In press.

Ford, R.G., G.W. Page, and H.R. Carter. 1987. Estimating mortality of seabirds from oil spills. Pages 747-751 *in* Proceedings of the 1987 Oil Spill Conference. American Petroleum Institute, Washington, DC.

Ford, R.G., D.H. Varoujean, D.R. Warrick, W.A. Williams, D.B. Lewis, C.L. Hewitt, and J.L. Casey. 1991. Seabird mortality resulting from the *Nestucca* oil spill incident winter 1988-89. Ecological Consulting Incorporated, Portland, OR. [Unpublished report.]

Forest Ecosystem Management Assessment Team (FEMAT). 1994. Forest ecosystem management: an ecological, economic, and social assessment. U.S. Departments of Agriculture and Interior, Washington, DC.

Gress, F., R.W. Risebrough, D.W. Anderson, L.F. Kiff, and J.R. Jehl, Jr. 1973. Reproductive failures of double-crested cormorants in southern California and Baja California. Wilson Bull. 85:197-208.

Henny, C.J., L.J. Blus, S.P. Thompson, and U.W. Wilson. 1989. Environmental contaminants, human disturbance and nesting of double-crested cormorants in northwestern Washington. Colonial Waterbirds 12:198-206.

Hodder, J., and M.R. Graybill. 1985. Reproduction and survival of seabirds in Oregon during the 1982-83 El Niño. Condor 87:535-541.

Hunt, G.L., R.L. Pitman, M. Naughton, K. Winnett, A. Newman, P.R. Kelly, and K.T. Briggs. 1979. Reproductive ecology and foraging habits of breeding seabirds. Pages 1-399 *in* Summary of marine mammal and seabird surveys of the southern California Bight area 1975-1978. Vol. 3—Investigators' reports. Part 3. Seabirds—Book 2. University of California-Santa Cruz. For U.S. Bureau of Land Management, Los Angeles, CA. Contract AA550-CT7-36. [Unpublished report.]

Jones and Stokes Associates. 1993. Environmental impact report for the review of Mono Basin water rights of the City of Los Angeles. Draft. May. (JSA 90-171.) Prepared for California State Water Resources Control Board, Division of Water Rights, Sacramento, CA.

Nettleship, D.N., G.A. Sanger, and P.F. Springer, eds. 1984. Marine birds: their feeding ecology and commercial fisheries relationships. Proceedings of the Pacific Seabird Group Symposium. Canadian Wildlife Service Spec. Publ., Ottawa, Ontario. 220 pp.

Page, G.W., H.R. Carter, and R.G. Ford. 1990. Numbers of seabirds killed or debilitated in the 1986 *Apex Houston* oil spill in central California. Pages 164-174 *in* S.G. Sealy, ed. Auks at sea. Proceedings of an International Symposium of the Pacific Seabird Group. Studies in Avian Biology 14.

Sealy, S.G., ed. 1990. Auks at sea. Proceedings of an International Symposium of the Pacific Seabird Group. Studies in Avian Biology 14. 180 pp.

Sowls, A.L., A.R. DeGange, J.W. Nelson, and G.S. Lester. 1980. Catalog of California seabird colonies. U.S. Fish and Wildlife Service, FWS/OBS 37/80. 371 pp.

Speich, S.M., and T.R. Wahl. 1989. Catalog of Washington seabird colonies. U.S. Fish and Wildlife Service, Biological Rep. 88(6). 510 pp.

Takekawa, J.E., H.R. Carter, and T.E. Harvey. 1990. Decline of the common murre in central California, 1980-1986. Pages 149-163 *in* S.G. Sealy, ed. Auks at sea. Proceedings of an International Symposium of the Pacific Seabird Group. Studies in Avian Biology 14.

Varoujean, D.H., and R.L. Pitman. 1979. Oregon seabird survey 1979. U.S. Fish and Wildlife Service, Portland, OR. Unpublished report.

Vermeer, K., K.T. Briggs, K.H. Morgan, and D. Siegel-Causey, eds. 1993. The status, ecology, and conservation of marine birds of the North Pacific. Proceedings of a Pacific Seabird Group Symposium. Canadian Wildlife Service Spec. Publ. Ottawa, Ontario. 263 pp.

Wilson, U.W. 1991. Responses of three seabird species to El Niño events and other warm episodes on the Washington coast, 1979-1990. Condor 93:853-858.

Wilson, U.W., and D.A. Manuwal. 1986. Breeding biology of the rhinoceros auklet in Washington. Condor 88:143-155.

For further information:

Harry R. Carter
National Biological Service
California Pacific Science Center
6924 Tremont Rd.
Dixon, CA 95620

Seabirds in Alaska

by
Scott A Hatch
John F. Piatt
National Biological Service

About 100 million seabirds reside in marine waters of Alaska during some part of the year. Perhaps half this population is composed of 50 species of nonbreeding residents, visitors, and breeding species that use marine habitats only seasonally (Gould et al. 1982). Another 30 species include 40-60 million individuals that breed in Alaska and spend most of their lives in U.S. territorial waters (Sowls et al. 1978). Alaskan populations account for more than 95% of the breeding seabirds in the continental United States, and eight species nest nowhere else in North America (USFWS 1992).

Seabird nest sites include rock ledges, open ground, underground burrows, and crevices in cliffs or talus. Seabirds take a variety of prey from the ocean, including krill, small fish, and squid. Suitable nest sites and oceanic prey are the most important factors controlling the natural distribution and abundance of seabirds.

The impetus for seabird monitoring is based partly on public concern for the welfare of these birds, which are affected by a variety of human activities like oil pollution and commercial fishing. Equally important is the role seabirds serve as indicators of ecological change in the marine environment. Seabirds are long-lived and slow to mature, so parameters such as breeding success, diet, or survival rates often give earlier signals of changing environmental conditions than population size itself. Seabird survival data are of interest because they reflect conditions affecting seabirds in the nonbreeding season, when most annual mortality occurs (Hatch et al. 1993b).

Techniques for monitoring seabird populations vary according to habitat types and the breeding behavior of individual species (Hatch and Hatch 1978, 1989; Byrd et al. 1983). An affordable monitoring program can include but a few of the 1,300 seabird colonies identified in Alaska, and since the mid-1970's, monitoring efforts have emphasized a small selection of surface-feeding and diving species, primarily kittiwakes (*Rissa* spp.) and murres (*Uria* spp.). Little or no information on trends is available for other seabirds (Hatch 1993a). The existing monitoring program occurs largely on sites within the Alaska Maritime National Wildlife Refuge, which was established primarily for the conservation of marine birds. Data are collected by refuge staff, other state and federal agencies, private organizations, university faculty, and students.

Status of Monitored Birds

Kittiwakes

Kittiwakes are small, pelagic (open sea) gulls that range widely at sea and feed on a variety of small fish and plankton, which they capture at the sea surface. Black-legged kittiwakes (*Rissa tridactyla*) have been studied intensively because they are widely distributed and easy to observe. Among 10 locations for which population trend data are available, 3 show significant declines since the mid-1970's, 3 show increases, and 4 show no consistent trends (Fig. 1). The overall trend is unknown, although widespread declines are anticipated because of a downward

Courtesy S. Hatch, NBS

Dense colonies of common murres (*Uria aalge*) breed on bare cliff ledges—here on the Semidi Islands, western Gulf of Alaska.

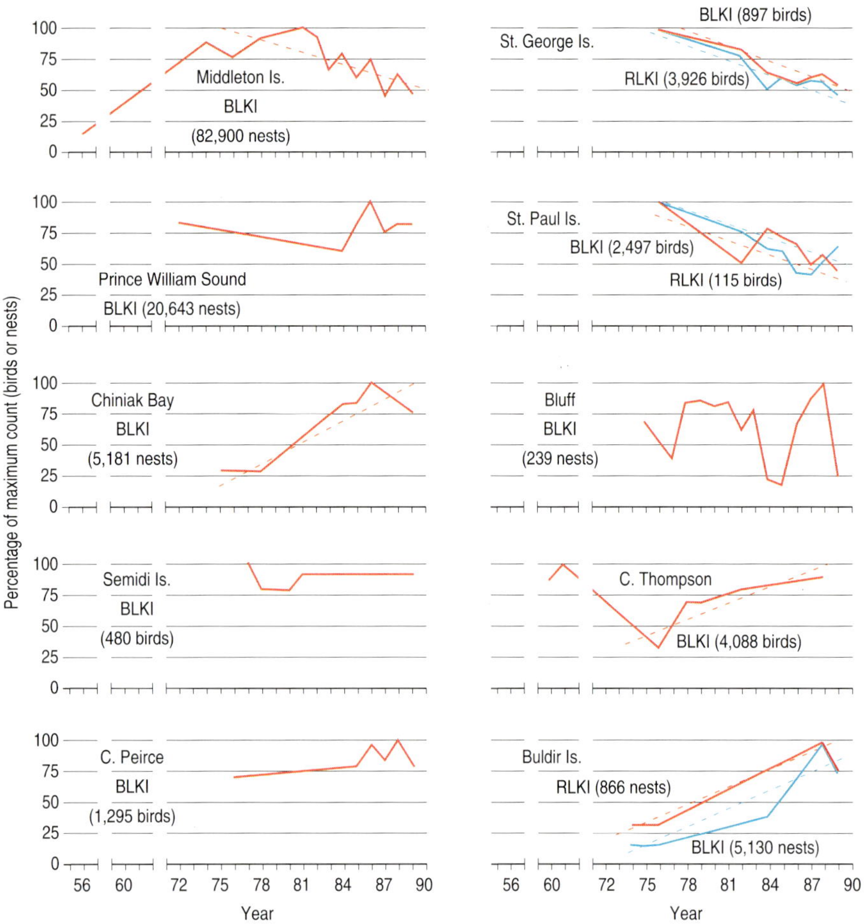

Fig. 1. Population trends of black-legged kittiwakes (BLKI) and red-legged kittiwakes (RLKI) at selected colonies in Alaska. The maximum count of birds or nests is indicated for each location. Dashed lines indicate significant regressions ($P < 0.05$) of data collected since 1970 (P is a measure of the confidence that the decline or increase is statistically reliable. $P < 0.05$ indicates a high probability that the population trend depicted actually occurred). *See* Hatch et al. 1993a and references cited therein for data sources.

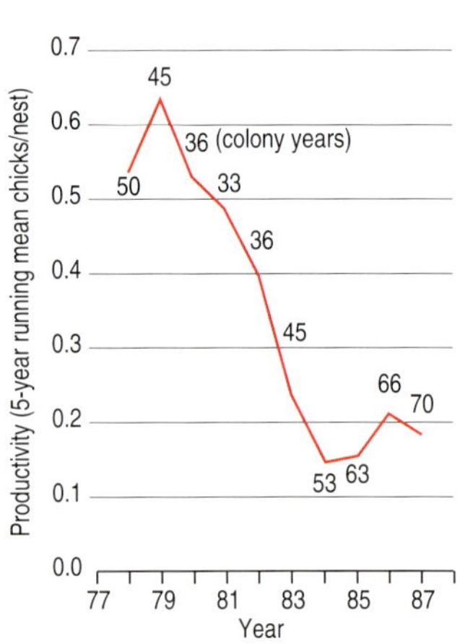

Fig. 2. Productivity (chicks fledged per nest built) of black-legged kittiwakes in Alaskan colonies, 1976-89. The number of colony-years included in each mean is indicated. *See* Hatch et al. 1993a for raw data.

trend in the production of offspring (Fig. 2); some large colonies fail chronically. On Middleton Island, for example, breeding has been a total or near-total failure in 10 of the last 12 years (1983-94; Hatch et al. 1993a; Hatch, unpublished data). The colony is declining at an average rate of 7% per year (equal to adult mortality), suggesting there is no recruitment (Hatch et al. 1993b). If survival estimates obtained on Middleton apply generally, the near-term future of kittiwakes is unfavorable because average productivity of 0.2 chicks per pair (Fig. 2) is inadequate to maintain populations.

Where red-legged kittiwakes (*R. brevirostris*) have been monitored, they show population trends similar to black-legged kittiwakes (Fig. 1). In 1989 their population was down by 50% in the Pribilof Islands, but they were more numerous at Buldir Island than in the mid-1970's (Byrd and Williams 1993). Because most of the world population of red-legged kittiwakes breeds in the Pribilofs (75% on St. George Island), their decline at that location is cause for concern.

Murres

Murres are large-bodied, abundant, and wide-ranging seabirds that feed mostly on schools of fish they pursue by diving underwater, sometimes to depths of 100-200 m (330-650 ft). Repeated counts of one or both murre species (common murre, *Uria aalge*, and thick-billed murre, *U. lomvia*) are available for 12 locations in Alaska (Fig. 3). Since 1970 common murres have declined significantly at two colonies, and thick-billed murres have declined at one. Murres (species not distinguished) increased at two colonies over the same period. Between the 1950's and the 1970's, murres increased at one location (Middleton Island) and declined at another (Cape Thompson), but they have since been relatively stable at both colonies. In 1989 the *Exxon Valdez* oil spill killed substantial numbers of common murres at several colonies in the Gulf of Alaska (Piatt et al. 1990a).

Available data are insufficient to identify overall trends. Murres are relatively consistent producers of young, averaging 0.5-0.6 chicks per pair annually in both species (Byrd et al. 1993).

Threatened and Endangered Species

No breeding seabirds are currently listed as threatened or endangered in Alaska. The short-tailed albatross (*Diomedea albatrus*), with fewer than 1,000 individuals surviving, breeds in Japan but visits Alaskan waters during most months of the year. The species is vulnerable to incidental take by commercial fishing gear, especially gill nets and longlines (Sherburne 1993).

Three species that breed in Alaska were recently listed as category 2 (possibly qualifying for threatened or endangered status, but more information is needed for determination): the red-legged kittiwake, marbled murrelet (*Brachyramphus marmoratus*), and Kittlitz's murrelet (*B. brevirostris*). As noted previously, red-legged kittiwakes have declined substantially on the Pribilof Islands (Fig. 1). Marine bird surveys conducted in Prince William Sound in 1972-73 and 1989-93 suggest a significant decline of marbled murrelets in that area (Klosiewski and Laing 1994). This finding is corroborated by Audubon Christmas Bird Counts from coastal sites in Alaska, which reveal a downward trend since 1972 (Piatt, unpublished data). Kittlitz's murrelet also showed a decline in the Prince William Sound surveys (Klosiewski and Laing 1994). With an estimated population of fewer than 20,000 birds range-wide (van Vliet 1993), this species is one of the rarest of auks (Family Alcidae). Both murrelets were adversely affected by the *Exxon Valdez* oil spill (Piatt et al. 1990a).

Other Species

Scant information is available to assess numerical changes for most seabird species in Alaska. We know that some species were seriously reduced or locally extirpated by foxes introduced to islands in the 1800's and early 1900's. About 450 islands from southeastern Alaska to the western Aleutians were used as release sites for arctic (*Alopex lagopus*) and red foxes (*Vulpes vulpes*) (Bailey 1993). The species most affected included open-ground nesters such as gulls (*Larus* spp.), terns (*Sterna* spp.), and fulmars (*Fulmarus glacialis*), and burrowing birds like ancient murrelets (*Synthliboramphus antiquus*), Cassin's auklets (*Ptychoramphus aleuticus*), tufted puffins (*Fratercula cirrhata*), and storm-petrels (*Oceanodroma* spp.). In spite of natural die-offs and eradication efforts, foxes remain on about 50 islands to which they were introduced (Bailey 1993).

Recent counts suggest that fulmars are increasing at two of their major colonies (Semidi Islands and Pribilof Islands), and several small colonies have been established since the mid-1970's (Hatch 1993b). Counts of least and crested auklets (*Aethia pusilla* and *A. cristatella*) also indicate possible increases at two colonies in the Bering Sea (Piatt et al. 1990b; Springer et al. 1993).

Red-faced cormorants (*Phalacrocorax urile*) declined about 50% on the Semidi Islands between 1978 and 1993, while pelagic cormorants (*P. pelagicus*) increased on Middleton Island between 1956 and the mid-1970's (Hatch, unpublished data). Glaucous-winged gulls (*Larus glaucescens*) increased on Middleton from none breeding in 1956 to more than 20,000 birds in 1993 (Hatch, unpublished data); this species has also shown marked increases following removal of introduced foxes at several sites in the Aleutian Islands (Byrd et al. 1994). Marine bird surveys in Prince William Sound (Klosiewski and Laing 1994) suggest that arctic terns (*Sterna paradisaea*), glaucous-winged gulls, pelagic cormorants, horned puffins (*Fratercula corniculata*), and pigeon guillemots (*Cepphus columba*) have all declined in that area. Terns and guillemots have recently increased on several Aleutian Islands following fox removal (Byrd et al. 1994).

Factors Affecting Seabirds

Alaskan seabirds are killed incidentally in drift gill nets used in high seas (DeGange et al. 1993), and oil pollution poses a significant threat, as demonstrated by the *Exxon Valdez* spill. There is little doubt, however, that the introduction of exotic animals, especially foxes, but also rats, voles, ground squirrels, and rabbits has been the most damaging source of direct mortality associated with human activity (Bailey 1993). Unlike one-time catastrophes, introduced predators exert a continuous negative effect on seabird populations.

Changes in food supply, whether natural or related to human activity, are another important influence on seabird populations. The postwar period from 1950 to the 1990's has seen explosive growth and constant change in commercial fisheries of the northeastern Pacific (Alverson 1992). Driving these changes, or in some cases possibly driven by them, are major shifts in the composition of marine fish stocks. In the Gulf of Alaska, for example, a shift occurred in the late 1970's and early 1980's toward greater abundance of groundfish (cod, *Gadus macrocephalus*; various flatfishes; and especially walleye pollock, *Theragra chalcogramma*), possibly at the expense of small forage species such as herring (*Clupea harengus*), sandlance (*Ammodytes hexapterus*), and capelin (*Mallotus villosus*; Alverson 1992) (Fig. 4). Coincident with these changes, diets of a variety of seabirds such as murres, murrelets, and kittiwakes have shifted from being predominantly capelin-based

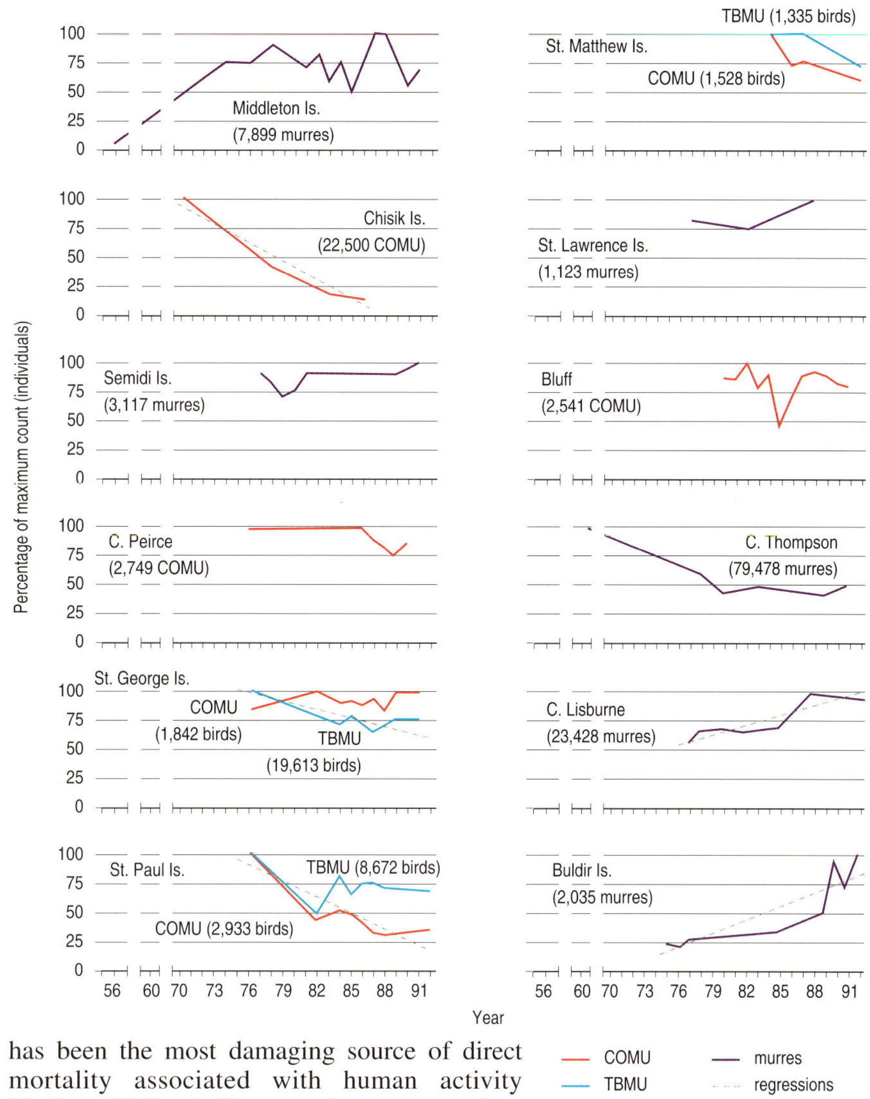

Fig. 3. Population trends of common murres (COMU) and thick-billed murres (TBMU) at selected colonies in Alaska. Counts of "murres" included unspecified numbers of common and thick-billed murres. The maximum count of individuals is indicated for each location. Dashed lines indicate significant regressions ($P < 0.05$) of data collected since 1970. *See* Hatch 1993a for data sources.

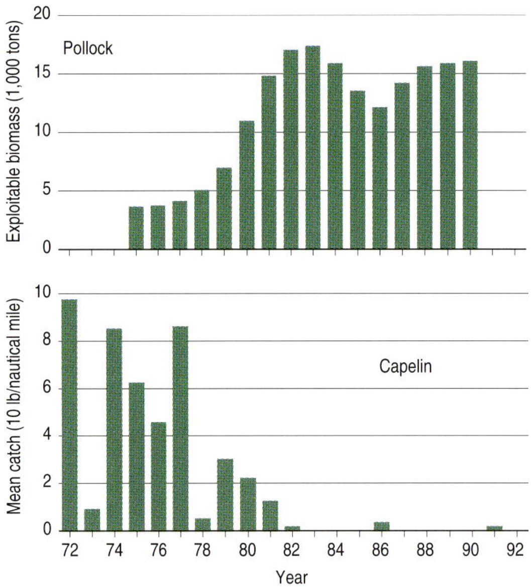

Fig. 4. Temporal changes in marine fish stocks of the Gulf of Alaska: total pollock biomass (age 2+) from stock assessment surveys by the National Marine Fisheries Service, 1975-90 (above; Marasco and Aron 1991), and catch per unit effort of capelin in midwater trawls in Pavlov Bay, western Gulf of Alaska, 1972-92 (below; P. Anderson, NMFS, unpublished data).

to pollock-based (Piatt, unpublished data). Seabird declines and breeding failures correspond to the shift, as do drastic declines in harbor seals (*Phoca vitulina*) and northern sea lions (*Eumetopias jubatus*) in the Gulf of Alaska (Merrick et al. 1987; Pitcher 1990).

The wholesale removal of large quantities of fish biomass from the ocean is likely to have major, if poorly known, effects on the marine ecosystem. An emerging issue is whether fish harvests are altering marine ecosystems to the detriment of seabirds and other consumers like pinnipeds and whales.

The relative role of fishing and natural environmental variation in regulating these systems is another matter for long-term research. In any case, seabird monitoring will continue to provide valuable insights into marine food webs, especially changes that affect the ocean's top-level consumers, including humans.

For further information:

Scott A. Hatch
National Biological Service
Alaska Science Center
Anchorage, AK 99503

References

Alverson, D.L. 1992. A review of commercial fisheries and the Steller sea lion (*Eumetopias jubatus*): the conflict arena. Reviews in Aquatic Sciences 6:203-256.

Bailey, E.P. 1993. Fox introductions on Alaskan islands: history, impacts on avifauna, and eradication. U.S. Fish and Wildlife Service, Resour. Publ. 193. 54 pp.

Byrd, G.V., R.H. Day, and E.P. Knudson. 1983. Patterns of colony attendance and censusing of auklets at Buldir Island, Alaska. Condor 85:274-280.

Byrd, G.V., and J.C. Williams. 1993. Red-legged kittiwake (*Rissa brevirostris*). Pages 1-12 *in* The birds of North America, 60. A. Poole and F. Gill, eds. The Academy of Natural Sciences, Philadelphia; The American Ornithologists' Union, Washington, DC.

Byrd, G.V., E.C. Murphy, G.W. Kaiser, A.Y. Kondratyev, and Y.V. Shibaev. 1993. Status and ecology of offshore fish-feeding alcids (murres and puffins) in the North Pacific. Pages 176-186 *in* K. Vermeer, K.T. Briggs, K.H. Morgan, and D. Siegel-Causey, eds. The status, ecology, and conservation of marine birds of the North Pacific. Canadian Wildlife Service, Ottawa.

Byrd, G.V., J.L. Trapp, and C.F. Zeillemaker. 1994. Removal of introduced foxes: a case study in restoration of native birds. Transactions of the 59th North American Wildlife and Natural Resources Conference. Wildlife Management Institute, Washington, DC. In press.

DeGange, A.R., R.H. Day, J.E. Takekawa, and V.M. Mendenhall. 1993. Losses of seabirds in gill nets in the North Pacific. Pages 204-211 *in* K. Vermeer, K.T. Briggs, K.H. Morgan, and D. Siegel-Causey, eds. The status, ecology, and conservation of marine birds of the North Pacific. Canadian Wildlife Service, Ottawa.

Gould, P.J., D.J. Forsell, and C.J. Lensink. 1982. Pelagic distribution and abundance of seabirds in the Gulf of Alaska and eastern Bering Sea. U.S. Fish and Wildlife Service FWS/OBS-82/48. 294 pp.

Hatch, S.A. 1993a. Population trends of Alaskan seabirds. Pacific Seabird Group Bull. 20:3-12.

Hatch, S.A. 1993b. Ecology and population status of northern fulmars *Fulmarus glacialis* of the North Pacific. Pages 82-92 *in* K. Vermeer, K.T. Briggs, K.H. Morgan, and D. Siegel-Causey, eds. The status, ecology, and conservation of marine birds of the North Pacific. Canadian Wildlife Service, Ottawa.

Hatch, S.A., G.V. Byrd, D.B. Irons, and G.L. Hunt, Jr. 1993a. Status and ecology of kittiwakes (*Rissa tridactyla* and *R. brevirostris*) in the North Pacific. Pages 140-153 *in* K. Vermeer, K.T. Briggs, K.H. Morgan, and D. Siegel-Causey, eds. The status, ecology, and conservation of marine birds of the North Pacific. Canadian Wildlife Service, Ottawa.

Hatch, S.A., and M.A. Hatch. 1978. Colony attendance and population monitoring of black-legged kittiwakes on the Semidi Islands, Alaska. Condor 90:613-620.

Hatch, S.A., and M.A. Hatch. 1989. Attendance patterns of murres at breeding sites: implications for monitoring. Journal of Wildlife Management 53:483-493.

Hatch, S.A., B.D. Roberts, and B.S. Fadely. 1993b. Adult survival of black-legged kittiwakes *Rissa tridactyla* in a Pacific colony. Ibis 135:247-254.

Klosiewski, S.P., and K.K. Laing. 1994. Marine bird populations of Prince William Sound, Alaska, before and after the *Exxon Valdez* oil spill. U.S. Fish and Wildlife Service, Anchorage, AK. 89 pp.

Marasco, R., and W. Aron. 1991. Explosive evolution—the changing Alaska groundfish fishery. Reviews in Aquatic Sciences 4:299-315.

Merrick, R.L., T.R. Loughlin, and D.G. Calkins. 1987. Decline in abundance of the northern sea lion, *Eumetopias jubatus*, in Alaska, 1956-86. Fishery Bull. 85:351-365.

Piatt, J.F., C.J. Lensink, W. Butler, M. Kendziorek, and D.R. Nysewander. 1990a. Immediate impact of the "Exxon Valdez" oil spill on marine birds. Auk 107:387-397.

Piatt, J.F., B.D. Roberts, and S.A. Hatch. 1990b. Colony attendance and population monitoring of least and crested auklets on St. Lawrence Island, Alaska. Condor 92:97-106.

Pitcher, K.W. 1990. Major decline in number of harbor seals, *Phoca vitulina richardsi*, on Tugidak Island, Gulf of Alaska. Marine Mammal Science 6:121-134.

Sherburne, J. 1993. Status report on the short-tailed albatross *Diomedea albatrus*. U.S. Fish and Wildlife Service, Ecological Services, Anchorage, AK. 58 pp.

Sowls, A.L., S.A. Hatch, and C.J. Lensink. 1978. Catalog of Alaskan seabird colonies. U.S. Fish and Wildlife Service FWS/OBS-78/78. 254 pp.

Springer, A.M., A.Y. Kondratyev, H. Ogi, Y.V. Shibaev, and G.B. van Vliet. 1993. Status, ecology, and conservation of *Synthliboramphus* murrelets and auklets. Pages 187-201 *in* K. Vermeer, K.T. Briggs, K.H. Morgan, and D. Siegel-Causey, eds. The status, ecology, and conservation of marine birds of the North Pacific. Canadian Wildlife Service, Ottawa.

USFWS. 1992. Alaska seabird management plan. U.S. Fish and Wildlife Service, Division of Migratory Birds, Anchorage, AK. 102 pp.

van Vliet, G. 1993. Status concerns for the "global" population of Kittlitz's murrelet: is the "glacier murrelet" receding? Pacific Seabird Group Bull. 20:15-16.

Colonial Waterbirds

by
R. Michael Erwin
National Biological Service

Colonial waterbirds, that is, seabirds (gulls, terns, cormorants, pelicans) and wading birds (herons, egrets, ibises), have attracted the attention of scientists, conservationists, and the public since the turn of the century when plume hunters nearly drove many species to extinction. The first national wildlife refuge at Pelican Island, Florida, was founded to conserve a large nesting colony of the brown pelican (*Pelecanus occidentalis*). The National Audubon Society also established a game warden system to monitor and protect important waterbird colonies. These efforts helped establish federal laws to protect migratory birds and their nesting habitats in North America.

Although the populations of many species rebounded in the early part of the 20th century, major losses and alteration of coastal wetlands still threaten the long-term sustainability of many colonial waterbirds. A national, coordinated monitoring program is needed to monitor population status and trends in colonial waterbirds (Erwin et al. 1993). The Canadian Wildlife Service has recently established a national seabird monitoring program (D. Nettleship, CWS, personal communication). In addition, better coordination and cooperation for monitoring waterbirds are needed on both their breeding grounds in North America and their wintering grounds in Latin America where wetland loss is also a critical problem (Erwin et al. 1993). This article summarizes the status and trends of selected waterbird species in North America, but excludes Alaska, Hawaii, and the Pacific coast, which are described elsewhere.

Population Surveys

Data on the population status of colonial waterbirds come from many sources. The Breeding Bird Survey (Peterjohn and Sauer 1993) is useful as a visual index for the more widely distributed species that occur along coasts and across the interior of the United States and Canada (e.g., great blue herons [*Ardea herodias*] and herring gulls [*Larus argentatus*]), but it is not effective for many waterbird species that nest in wetlands.

Recently, Christmas Bird Count (CBC) data have been analyzed, providing an index to numbers of wintering birds (J.R. Sauer, National Biological Service, personal communication). For waterbirds, these counts must be used with caution since water conditions can have a major effect on the feeding distribution of waterbirds during the count period in December. Thus, trends in CBC counts may indicate more about trends in wetland conditions than trends in populations of any particular waterbird species.

More precise estimates of species' populations at colony sites have been conducted over the years by state, federal, and private organizations. Although a few states (e.g., Florida, Illinois, Massachusetts, Texas, and Virginia) have conducted annual surveys over a long period for at least some species, there is little consistency among their methods and the frequency of surveys (Erwin et al. 1985). Many data on breeding populations are kept at the state level, but these data seldom predate 1980, precluding assessment of long-term trends in many of these long-lived species.

Even though more than 50 species of colonial waterbirds breed in the United States, Canada, and Mexico, this article focuses on the 22 species for which sufficient data are available to indicate population changes, at least at a regional level.

Courtesy D. Smith

Common tern (*Sterna hirundo*).

Pelecaniformes

Pelicans and their allies (cormorants, anhingas) suffered from DDT use, and their numbers plummeted to the point where the eastern and California brown pelicans became endangered. The eastern subspecies, however, was recently removed from the threatened list because of its rapid numerical and range increases (Table).

The American white pelican (*Pelecanus erythrorhynchos*) has shown similar sharp increases in the western regions of Canada and the United States (Evans and Knopf 1993). Double-

Table. Regional, national, and continental population status and trends of selected colonial waterbirds in the United States[a] as reported by the Breeding Bird Survey, Christmas Bird Counts, and other sources.

		Population status		BBS/CBC trend[b]			
Species	Region	Early period	Recent period	% change	% +/-routes	Years	References[c]
Pelecaniformes							
American white pelican	Continent			+5.3**		1966-91	BBS
	U.S.	17,872 nests (1964)	22,299 nests (1980-81)				1
	Canada	14,103 (1967-69)	53,345 (1985-86)				1
	Mexico	Sporadic (100 nests)					1
	U.S.			+3.8***		1966-89	CBC (winter)
Eastern brown pelican	U.S.	7,800-8,300 (1970-76)	26,461 (1989)	NA			2
Double-crested cormorant	Continent			+6.5***	+0.61***	1966-91	BBS
	U.S.			+2.3**	+0.61***		
	Canada			+11.5***	+0.64*		
	U.S.			+8.2***		1966-89	CBC (winter)
Ciconiiformes							
Great blue heron	Continent			+1.5**	0.60***	1966-91	BBS
	U.S.			+1.9	0.61***		
	Canada			+0.7 ns	0.54*		
Great blue heron	U.S.			+2.2***		1966-89	CBC
Snowy egret	U.S.			+2.0**		1966-89	CBC
Reddish egret	U.S. Gulf coast	1,700-2,200 pr. (1976-78)	1,370-1,900 pr. (1989-90)				3,4,5,6
Black-crowned night-heron	U.S.			+2.8		1966-89	CBC
White ibis	Southeast U.S.	40,000-80,000 pr.(1967-71)	22,000-50,000 pr. (1987-93)				7
	U.S.			+5.0**		1966-89	CBC
White-faced ibis	Western U.S.	4,500-5,500 pr. (1967-75)	13,000-13,500 pr. (1985)				8
	U.S.			+7.6**		1966-89	CBC
Wood stork	Southeast U.S.	2,500-5,200 pr. (1976-82)	6,729 pr. (1993)				9,10
	U.S.			+1.3 ns		1966-89	CBC
Charadriiformes							
Razorbill	N. Gulf St. Lawrence, Can.	16,200 birds (1960)	2,380 birds (1982)				11
Atlantic puffin	Canada (Witless Bay)	300,000-340,000 pr.(1973)	225,000 pr.(1978-80)				11
	U.S.	125 pr. (1977)	135 pr. (1993)				12
Great black-backed gull	U.S.			+3.6**		1966-89	CBC
Herring gull	Atlantic coast U.S.	110,000 pr.(1978-82)	< 100,000 pr. (1988-92)				13
	U.S.			+0.5 ns		1966-89	CBC
Ring-billed gull	Continent			+7.9**	+0.60***	1966-91	BBS
	U.S.			+16.5**	+0.58***		
	Canada			+5.7*	+0.62***		
	U.S.			+4.2***		1966-89	CBC
Franklin's gull	Continent			-6.0 ns		1966-91	BBS
	U.S.			-19.2***			
	Canada			-1.2 ns			
Gull-billed tern	Mid-Atlantic U.S. (VA-SC)	1,100-1,600 pr. (1977)	1,125-1625 pr. (1993)				14,15,16
	Gulf coast U.S. (TX-AL)	1,200-2,100 pr. (1977)	3,000 pr. (1990)				3,4,5
	U.S.			-1.5*		1966-89	CBC
Forster's tern	Continent			-2.4 ns	-0.58*		BBS
	U.S.			-3.2 ns	-0.60**		BBS
	Canada			Insuff. data			
	U.S.			+4.3***			CBC
Common tern	Great Lakes U.S.	1,691 nests (1977)	1,916 nests (1989)				17, 18
Roseate tern	N. Atlantic U.S.	2,855-3,285 pr. (1976-80)	3,200 pr. (1993)				19, 20
	U.S. Caribbean	Uncertain pre-1975	1,900-2,500 pr. (1975-80)				13,18
Least tern (interior ssp.)	Mississippi River	4,100-4,700 birds (1986-87)	6,833 birds (1991)			1986-91	21
Black tern	Continent			-3.9**	-0.59***	1966-92	BBS
	U.S.			-5.6***	-0.64***		
	Canada			-3.4	-0.52 ns		

[a]Excluding Alaska, Hawaii, and the Pacific coast states.

[b]Breeding Bird Survey trends statistically test for an annual (% change) trend (H_0: trend = 0) and % of increasing (+) or decreasing (-) routes (H_0: no. routes + = no. routes -). Probability levels: *$P < 0.10$;** $P < 0.05$; *** $P < 0.01$. A lower P value means there is more confidence that the trend is real. A population trend change at the $P < 0.10$ level is considered statistically significant; ns = not significant. Christmas Bird Count trends are conducted similar to annual BBS trend (J.R. Sauer, NBS, unpublished data).

[c]Sources: numbers refer to literature reference number; BBS = Breeding Bird Survey results (J.R. Sauer and B. Peterjohn, NBS, personal communication); CBC = Christmas Bird Count trend results (J.R. Sauer, personal communication).

1—Evans and Knopf 1993; 2—P. Wilkerson, South Carolina Wildlife and Marine Resources Department; 3—Lange, in press; 4—Portnoy 1978; 5—Martin and Lester 1990; 6—Runde 1991; 7—P. Frederick, University of Florida, unpublished data; 8—D. Manry, unpublished data; 9—Ogden et al. 1987; 10—J. Ogden and M. Coulter, National Park Service, unpublished data; 11—Nettleship and Birkhead 1985; 12—B. Allen, Maine Department of Inland Fisheries and Wildlife, unpublished data; 13—Nisbet, in press; 14—Spendelow and Patton 1988; 15—Erwin 1979; 16—J. Parnell and P. Wilkerson, University of North Carolina and South Carolina Wildlife and Marine Resources Department, unpublished data; 17—Scharf et al. 1992; 18—Blokpoel and Tessier 1993; 19—Gochfeld 1983; 20—J.A. Spendelow, NBS, unpublished data; 21—E. Kirsch and J. Sidle, NBS, unpublished data.

crested cormorant (*Phalacrocorax auritus*) populations also declined during the 1940-70 period, probably because of DDT and other pesticides; however, this species has increased dramatically across Canada and the northern United States (Table). In the Great Lakes and elsewhere, this species' increases have attracted considerable attention because of the negative effects on fisheries and on the aquaculture industry (Blokpoel and Scharf 1991; Blokpoel and Tessier 1991; Nettleship and Duffy, in press).

Ciconiiformes

Heron, egret, and ibis nesting colonies were reduced along much of the U.S. coastline in the early 1900's as a result of the millinery trade; however, the species have all recovered their former ranges. Great blue herons are the most abundant and ubiquitous of the wading birds in North America; all indications suggest that their populations have increased, especially in the United States (Butler 1992; Table). One reason for this trend may be that winter survival has increased as herons feed heavily at aquaculture facilities in the southern United States.

The reddish egret (*Egretta rufescens*) is listed as a species of management concern to the USFWS (OMBM 1994). It nests in small numbers along the gulf coast and in southern Florida (Table). Reddish egrets seem to have declined some in Texas (Lange, in press) and Louisiana (Portnoy 1978; Martin and Lester 1990; Figure), but the data are not adequate in Florida to assess trends.

Snowy egrets (*E. thula*) were prized by plume traders at the turn of the century, and the species suffered dramatic population declines; however, by the 1970's these egrets had recovered their former range. More recently, their populations declined in some Atlantic regions such as Virginia (Williams et al. 1990) and southern Florida (Robertson and Kushlan 1974; Ogden 1978; Table).

The black-crowned night-heron (*Nycticorax nycticorax*), which occurs across all of North America, may be declining in parts of Canada, south to Texas (Davis 1993) and perhaps Virginia (Williams et al. 1990; Table).

Ibises are more nomadic in their breeding distribution than are other wading birds. White ibis (*Eudocimus albus*) have declined markedly in southern Florida as a result of hydrologic changes in the Everglades (Robertson and Kushlan 1974; Ogden 1978). Their breeding distribution has shifted northward, and large colonies exist in Georgia and the Carolinas (Ogden 1978; Bildstein 1993). Over the entire southeastern United States the species may not have undergone major changes, although state estimates have been erratic (twofold changes in 2-3 years; Table).

The white-faced ibis (*Plegadis chihi*) was formerly (1987) on the USFWS management concern list, but is not on the 1994 national list (OMBM 1994). Population data for the central and western populations (noncoastal) indicate a marked increase in the numbers of these ibis from the early 1970's to 1985 (D. Manry, personal communication; Table).

Wood storks (*Mycteria americana*), which have been federally listed as endangered since 1984, nest from Florida north to South Carolina in the United States, in Cuba, and in enormous numbers in the river deltas of eastern Mexico, especially the Usumacinta-Grijalva Delta. Stork colonies have shifted north from the Everglades to central and northern Florida, Georgia, and South Carolina since the 1970's (Robertson and Kushlan 1974; Ogden 1978; Ogden et al. 1987). Recent inventories of nesting populations in the United States indicate a modest increase in numbers over the past 10-15 years (Table; Figure).

Because of the mobility of wood storks and ibis, monitoring them requires a regional approach to ensure standardization in survey timing and methods. Individual state inventories are inadequate to address many highly mobile species.

Charadriiformes

This order of colonial-nesting waterbirds includes the alcids (murres, puffins, auks), shorebirds, gulls, terns, and black skimmers (*Rynchops niger*). Although some species of alcids and terns were nearly extirpated by hunters or millinery traders during the early 1900's, they rebounded well in many areas.

Alcid populations are rare in the eastern United States. In maritime Canada, however, alcid numbers are substantial (Nettleship and Birkhead 1985; Erskine 1992), though there is concern over Canada's razorbill (*Alca torda*) populations, which declined by more than 75% from 1960 to 1982 (Nettleship and Birkhead 1985). These declines may be the result of conflicts with commercial fisheries.

Canadian populations of Atlantic puffins (*Fratercula arctica*) have declined a great deal in some areas. The largest Atlantic puffin colony in North America is at Witless Bay, Newfoundland (61% of continental breeding total); this colony has declined by 25%-35% from 1973 to 1980 (Nettleship and Birkhead 1985). Again, competition between birds and commercial fisheries (capelin) may be causing much of the decline. In Maine, a successful transplant program has been in effect for more than a decade to reintroduce nesting Atlantic puffins onto several coastal islands (Kress and Nettleship 1988); numbers remain small, however (Table).

Gull populations have increased substantially from the middle part of the century to the present (Buckley and Buckley 1984; Nisbet, in press). Great black-backed gulls (*Larus marinus*) have increased in some mid-Atlantic states but have probably declined in Maine (Nisbet, in press; Table). Herring gull populations probably peaked around 1980 at about 110,000 pairs along the northeastern U.S. coastline, but populations may have declined during the 1980's

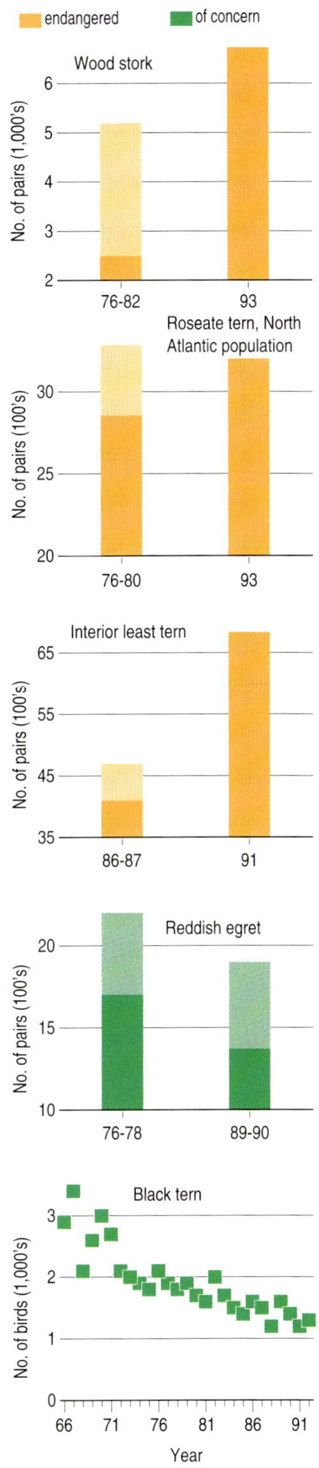

Figure. Trends of selected colonial waterbirds either endangered or on the U.S. Fish and Wildlife Service's list of species of management concern in the lower 48 states (excluding the Pacific coast). Black tern trends are count indices from the Breeding Bird Survey (mean or average number of birds per route). Lighter color shows range of variation in estimates.

(Nisbet, in press); BBS and CBC data do not show any change (Table). Changes in landfill practices that have reduced food supplies along the northeastern coast may have reduced winter survival and slowed the population growth of this species. In the Great Lakes, however, herring gulls have shown a dramatic increase since the late 1970's.

Ring-billed gulls (*L. delawarenis*) continue to increase across the northern tier of states, Canada, and the Great Lakes (Blokpoel and Scharf 1991; Blokpoel and Tessier 1991; Table). The BBS and CBC data suggest significant increases in the United States and Canada (Table). Refuge and resource managers are concerned over the reported decline in the Franklin's gull (*L. pipixcan*), an interior, marsh-nesting species that may be vulnerable to agricultural pesticides (White and Kolbe 1985). The BBS trends indicate that the numbers of this species significantly declined in the United States from 1966 to 1991. However, adding 1992 and 1993 data indicates a nonsignificant decline in the United States, which raises the question of the value of BBS data for this flock-feeding species (J.R. Sauer, personal communication).

Gull-billed terns (*Sterna nilotica*) are a species of special concern to many coastal states and were on the former (1987) USFWS management list. Recent population figures from Texas (Lange, in press), Louisiana (Martin and Lester 1990), and the mid-Atlantic region (Virginia to South Carolina) suggest that the population is reasonably stable over the long term but erratic from year to year (Table).

The Forster's tern (*S. forsteri*) nests both along coasts and across the interior of the northern tier of states and Canadian provinces. State surveys do not suggest declines in most states from New Jersey (C.D. Jenkins, New Jersey Division of Fish, Game and Wildlife, personal communication) to Virginia (Erwin 1979). Data are insufficient in the Great Lakes to assess trends. The trends from the BBS and CBC are contradictory, with breeding trends indicating declines and wintering trends a significant increase. This species is erratic in its nesting and probably not sampled well by either of these surveys.

Common terns (*S. hirundo*), while abundant and increasing along the U.S. northeastern coast (Buckley and Buckley 1984), are considered endangered, threatened, or a species of special concern in six Great Lakes states and Ontario (Blokpoel and Scharf 1991; Scharf et al. 1992). Even though tern numbers increased from 1977 to 1989 in the U.S. Great Lakes (Table), the number of their colony sites has declined from 31 to 23. Competition with the ring-billed gull is a major factor in this decline (Scharf et al. 1992).

The roseate tern (*S. dougallii*) is an endangered species (since 1987) and breeds in two populations in the western Atlantic. The western North Atlantic population includes the maritime provinces south to Long Island, New York (with a few possibly from New Jersey to Georgia); the U.S. Neotropical population is confined to Puerto Rico, the Virgin Islands, and southern Florida. In the northern population, the number of breeding pairs ranged from 2,855 to 3,285 pairs during the 1976-80 period (Gochfeld 1983) to 3,200 estimated pairs in 1993 (J. Spendelow, National Biological Service, personal communication; Table; Figure). In the southern U.S. population, pair estimates from the 1976-79 period range from about 1,900 (Gochfeld 1983) to about 2,600 pairs in the Florida Keys, Puerto Rico, and the Virgin Islands (Blokpoel and Tessier 1993; Table). Earlier records are sparse in this region, making trends difficult to determine.

The least tern (*S. antillarum*) is divided into three subspecies in the United States and Canada; the interior (*S.a. athalassos*) and California (*S.a. browni*) subspecies are listed as endangered. In the Mississippi River drainages, the interior least tern seems to have increased from the 1986-87 period to 1991 (E. Kirsch and J. Sidle, NBS, unpublished data; Table; Figure). The 1993 floods probably prevented recent nesting in many river stretches.

The black tern (*Chlidonias niger*) is listed as either endangered or a species of concern in many northern states, including New York, Iowa, Illinois, Wisconsin, Ohio, and Indiana. Its population has decreased at the BBS continental and U.S. levels from 1966 to 1992 (Table; Figure). From 1982 to 1991, BBS data indicate a significant increase in Canada with continued decrease in the United States. This suggests a species' displacement to the north, possibly a result of changes in wetland conditions in the northern tier of the United States. A confounding factor may also be that the Canadian surveys have been more intensive for this species in recent years.

References

Bildstein, K.L. 1993. White ibis: wetland wanderer. Smithsonian Institution Press, Washington, DC. 242 pp.

Blokpoel, H., and W.C. Scharf. 1991. Status and conservation of seabirds nesting in the Great Lakes of North America. Pages 17-41 *in* J. Croxall, ed. Status and conservation of the world's seabirds. International Council for Bird Preservation, Cambridge, England.

Blokpoel, H., and G.D. Tessier. 1991. Distribution and abundance of colonial waterbirds nesting in the Canadian portions of the lower Great Lakes system in 1990. Canadian Wildlife Service Tech. Rep. Series 117. 16 pp.

Blokpoel, H., and G.D. Tessier. 1993. Atlas of colonial waterbirds nesting on the Canadian Great Lakes, 1989-1991. Part 1. Cormorants, gulls, and island-nesting terns

on Lake Superior in 1989. Canadian Wildlife Service Tech. Rep. Series 181. 96 pp.
Buckley, P.A., and F.G. Buckley. 1984. Seabirds of the north and middle Atlantic coasts of the United States: their status and conservation. Pages 101-134 *in* J.P. Croxall, P.G.H. Evans, and R.W. Schreiber, eds. International Council for Bird Preservation Tech. Publ. 2. Cambridge, England.
Butler, R.W. 1992. Great blue heron (*Ardea herodias*). Pages 1-20 *in* A. Poole and F. Gill, eds. The birds of North America, No. 25. Academy of Natural Sciences, Philadelphia, PA, and The American Ornithologists' Union, Washington, DC.
Davis, W.E., Jr. 1993. Black-crowned night-heron (*Nycticorax nycticorax*). Pages 1-20 *in* A. Poole and F. Gill, eds. The birds of North America, No. 74. Academy of Natural Sciences, Philadelphia, PA, and The American Ornithologists' Union, Washington, DC.
Erskine, A.J. 1992. Atlas of breeding birds of the maritime provinces. McCurdy Printing, Halifax, Nova Scotia. 270 pp.
Erwin, R.M. 1979. Coastal waterbird colonies: Cape Elizabeth, Maine to Virginia. U.S. Fish and Wildlife Service FWS/OBS-79/10. 212 pp.
Erwin, R.M., P.C. Frederick, and J.L. Trapp. 1993. Monitoring of colonial waterbirds in the United States: needs and priorities. Pages 18-22 *in* M. Moser, R.C. Prentice, and J. van Vessem, eds. Waterfowl and wetland conservation in the 1990's — a global perspective. Proceedings of the International Waterfowl and Wetlands Research Bureau Symposium, St. Petersburg Beach, FL. International Waterfowl and Wetlands Research Bureau Special Publ. 26, Slimbridge, UK.
Erwin, R.M., P.H. Geissler, M.L. Shaffer, and D.A. McCrimmon, Jr. 1985. Colonial waterbird monitoring: a strategy for regional and national evaluation. Pages 342-357 *in* W. McComb, ed. Proceedings of a Workshop on Management of Nongame Species and Ecological Communities. University of Kentucky Agricultural Experiment Station, Lexington.
Evans, R.M., and F.L. Knopf. 1993. American white pelican (*Pelacanus erythrorhynchos*). Pages 1-24 *in* A. Poole and F. Gill, eds. The birds of North America, No. 57. Academy of Natural Sciences, Philadelphia PA, and The American Ornithologists' Union, Washington, DC.
Gochfeld, M. 1983. The roseate tern: world distribution and status of a threatened species. Biological Conservation 25:103-125.
Kress, S.W., and D.N. Nettleship. 1988. Re-establishment of Atlantic puffin (*Fratercula arctica*) at a former breeding site in the Gulf of Maine. Journal of Field Ornithology 59:161-170.
Lange, M. Texas coastal waterbird colonies: 1973-1990 census summary, atlas, and trends. U.S. Fish and Wildlife Service and Texas Parks and Wildlife Department, Angleton, TX. 209 pp. In press.
Martin, R.P., and G.D. Lester. 1990. Atlas and census of wading bird and seabird colonies in Louisiana, 1990. Louisiana Department of Wildlife and Fisheries and Louisiana Natural Heritage Program, Special Publ. 3, Baton Rouge. 182 pp.
Nettleship, D.N., and T.R. Birkhead, eds. 1985. The Atlantic Alcidae. Academic Press, New York. 574 pp.
Nettleship, D.N., and D.C. Duffy, eds. The double-crested cormorant: biology, conservation, and management. Colonial Waterbirds 17 (Special Publ.). In press.
Nisbet, I.C.T. Seabirds of the eastern United States: status and conservation. *In* D.N. Nettleship, ed. Seabirds of North America: status and conservation requirements. A special science publication in association with Stichting Greenpeace Council, Amsterdam, The Netherlands. In press.
OMBM. 1994. Migratory nongame birds of management concern in the United States: the 1994 list. U.S. Fish and Wildlife Service, Office of Migratory Bird Management, Washington, DC.
Ogden, J.C. 1978. Recent population trends of colonial wading birds on the Atlantic and Gulf Coastal Plains. Pages 137-154 *in* A. Sprunt, IV, J.C. Ogden, and S. Winckler, eds. Wading birds. National Audubon Society Res. Rep. 7, New York.
Ogden, J.C., D.A. McCrimmon, Jr., G.T. Bancroft, and B.W. Patty. 1987. Breeding populations of the wood stork in the southeastern United States. Condor 89:752-759.
Peterjohn, B.G., and J.R. Sauer. 1993. North American Breeding Bird Survey annual summary, 1990-91. Bird Populations 1:1-15.
Portnoy, J.W. 1978. Nesting colonies of seabirds and wading birds-coastal Louisiana, Mississippi, and Alabama. U.S. Fish and Wildlife Service FWS/OBS-77/07. 102 pp.
Robertson, W.B., Jr., and J.A. Kushlan. 1974. The south Florida avifauna. Pages 414-452 *in* Memoir 2 environments of south Florida: present and past. Miami Geological Society report, FL.
Runde, D. 1991. Trends in wading bird nesting populations in Florida, 1976-78 and 1986-89. Final performance report, Division of Wildlife, Florida Game and Freshwater Fish Commission, Tallahassee.
Scharf, W.C., G.W. Shugart, and J.L. Trapp. 1992. Distribution and abundance of gull, tern and cormorant colonies of the U.S. Great Lakes, 1989 and 1990. U.S. Fish and Wildlife Service, Office of Migratory Bird Management, Washington, DC. 89 pp.
Spendelow, J.A., and S.R. Patton. 1988. National atlas of coastal waterbird colonies in the contiguous United States: 1976-82. U.S. Fish and Wildlife Service Biological Rep. 88(5). 326 pp.
White, D.H., and E.J. Kolbe. 1985. Secondary poisoning of Franklin's gulls in Texas by monocrotophos. Journal of Wildlife Diseases 21:76-78.
Williams, B., J.W. Akers, J.W. Via, and R.A. Beck. 1990. Longitudinal surveys of the beach nesting and colonial waterbirds of the Virginia barrier islands, 1975-1987. Virginia Journal of Science 41:381-388.

For further information:

R. Michael Erwin
National Biological Service
Patuxent Environmental Science Center
11410 American Holly Dr.
Laurel, MD 20708

Shorebirds: East of the 105th Meridian

The North American group of shorebirds includes 48 kinds of sandpipers, plovers, and their allies, many of which live for most of the year in coastal marine habitats; others live principally in nonmarine habitats including grasslands, freshwater wetlands, and even second-growth woodlands. Most North American shorebirds are highly migratory, while others are weakly migratory, or even nonmigratory in some parts of their range. Here we discuss shorebirds east of the 105th meridian (roughly east of the Rocky Mountains). Historically, populations of many North American species were dramatically reduced by excessive gunning (Forbush 1912). Most populations recovered after the passage of the Migratory Bird Treaty Act of 1918, although some species never recovered and others have declined again.

High proportions of entire populations of shorebirds migrate by visiting one or a small number of "staging sites," areas where the birds accumulate fat to provide fuel before continuing

by
Brian A. Harrington
Manomet Observatory for Conservation Sciences

with their long-distance, nonstop flights to the next site (Morrison and Harrington 1979; Senner and Howe 1984; Harrington et al. 1991). Growing evidence (Schneider and Harrington 1981) indicates that staging areas are unusually productive sites with highly predictable but seasonally ephemeral "blooms" of invertebrates, which shorebirds use for fattening. In some cases, especially for "obligate" coastal species, specific sites are traditionally used; even other species sites may shift between years. Because of this, conservationists believe some species are at risk through loss of strategic migration sites (Myers et al. 1987). Other species are threatened by the loss of breeding and wintering habitats (Page et al. 1991; Haig and Plissner 1993; B. Leachman and B. Osmundson, U.S. Fish and Wildlife Service, unpublished data).

The predicted consequences of global warming, such as sea-level change, will also strongly affect the intertidal marine habitats, which many species of shorebirds depend upon. Some of the strongest warming effects will be at high latitudes, including those where many shorebirds migrate to breed, as well as south temperate latitudes, where many of them winter.

Population Trend Data

Information on population trends in North American shorebirds comes largely from studies designed for other purposes, except in the case of a few species that breed within latitudes covered by the Breeding Bird Survey (BBS) and one game species, the American woodcock (*Scolopax minor*). We divide these studies into two types, those based on surveys during breeding and nonbreeding seasons.

Population trend data from breeding seasons come mostly from studies of declining or threatened species such as piping plovers (*Charadrius melodus*; Haig and Plissner 1993), mountain plovers (*C. montanus*; Graul and Webster 1976; F.L. Knopf, U.S. Fish and Wildlife Service, unpublished data), and snowy plovers (*C. alexandrinus*; Page et al. 1991). Additional data come from the BBS and from special survey efforts on game species such as American woodcock (Sauer and Bortner 1991).

Nonbreeding season data come mostly from aerial surveys of migrants on Delaware Bay during spring (Clark et al. 1993), of migrants by the International Shorebird Surveys (ISS) during spring and fall (Harrington et al. 1989), and by the Maritimes Shorebird Surveys (MSS) in eastern Canada during fall (Morrison et al. 1994). Although none of these projects was designed principally to gather data for population trend monitoring, they are the only data bases on migrant species that have been systematically compiled through a period of years. The Christmas Bird Counts are an exception; they are conducted when most shorebirds are south of the United States.

Largely voluntary efforts of the ISS of Manomet Observatory, the MSS of the Canadian Wildlife Service, the BBS of the National Biological Service, and surveys on Delaware Bay (DELBAY) coordinated by New Jersey and Delaware state wildlife agencies have produced rough data useful for trend analysis. Because the BBS is conducted during the breeding season and is based on roadside surveys, its value is greatest in analyzing population change of broadly distributed shorebirds common in temperate latitudes where survey effort is greatest. The ISS, MSS, and DELBAY projects have focused on migration season counts and, therefore, are the best (though not ideal) available resources for monitoring northern-breeding shorebirds, which include most species in North America.

Plovers

Three of the eight species of plover that regularly occur east of the 105th meridian (snowy plover, piping plover, and mountain plover) are species of concern (endangered, threatened, or candidate species); killdeer (*C. vociferus*) and perhaps black-bellied plover (*Pluvialis squatarola*) are in decline (Table). In North America, all of these except the black-bellied plover are distributed principally in temperate latitudes; snowy, piping, and mountain plovers breed in special, localized habitats (principally sandy beaches, salt lakes, and salt flats for snowy and piping plovers, short-grass prairie for mountain plovers). There has been no evaluation of trends for Wilson's plover (*Charadrius wilsonia*), typically a beach-nesting species in southern North America. There are no statistically significant population changes in American golden- (*P. dominica*) and semipalmated plovers (*C. semipalmatus*).

Oystercatchers, Avocets, and Stilts

No significant population changes have been detected in the three species of these groups east of the 105th meridian (Table).

Sandpipers

This is the largest family of shorebirds. Five species of this family listed in the Table—willet (*Catoptrophorus semipalmatus*), upland sandpiper (*Bartramia longicauda*), long-billed curlew (*Numenius americanus*), marbled godwit (*Limosa fedoa*), and American woodcock—commonly breed in the contiguous 48 United States. Two others, the long-billed curlew, which nest principally in short-grass prairie, and the American woodcock found in second-growth woodland, show significant population declines. Upland sandpipers (tall-grass habitats, including croplands) show a significant increase. The remaining sandpiper species breed principally north of the contiguous 48 states. Six of these—ruddy turnstone (*Arenaria interpres*), red knot (*Calidris canutus*), sanderling (*C. alba*), white-rumped sandpiper (*C. fuscicollis*), Baird's sandpiper (*C. bairdii*), and buff-breasted sandpiper (*Tryngites subruficollis*)—are principally high-latitude breeders; two (red knot and sanderling) of the three species for which trend analysis data are available are in decline (Table). The remaining species can be grouped as taiga or middle Arctic breeders; seven of these have not been evaluated for population trend change; five species—whimbrel (*Numenius phaeopus*), semipalmated sandpiper (*Calidris pusilla*), least sandpiper (*C. minutilla*), short-billed dowitcher (*Limnodromus griseus*), and common snipe (*Gallinago gallinago*)—were in significant decline (Table), and four species—greater and lesser yellowlegs (*Tringa melanoleuca* and *T. flavipes*), spotted sandpiper (*Actitis macularia*), and dunlin (*C. alpina*)—showed no significant change (Table). No species showed significantly increased population trends.

Phalaropes

Only one (Wilson's phalarope; *Phalaropus tricolor*) of the three species of North American phalaropes has been evaluated for population change, and it showed significant declines (Table).

Summary and Recommendations

Population trend evaluation has been conducted for 27 of 41 shorebird species common in the United States east of the 105th meridian. Of the 27 species for which trend data are available, 12 show no change, 1 increased, and 14 decreased (Table). There were no clear correlations with habitat.

It is important that shorebird populations are monitored nationally, yet most species are hard to monitor because they inhabit regions that are difficult to access for much of the year. Migration seasons appear to be the most practical time for monitoring most species. Unfortunately, sampling for population monitoring during nonbreeding seasons presents a group of unresolved analytical challenges. Additional work on existing data can help identify how or whether broad, voluntary, or professional networks can collect data that will better meet requirements for monitoring population change.

References

Clark, K.E., L.J. Niles, and J. Burger. 1993. Abundance and distribution of migrant shorebirds in Delaware Bay. Condor 95:69-705.

Forbush, E.H. 1912. Game birds, wild-fowl, and shore birds. Massachusetts State Board of Agriculture, Boston. 622 pp.

Graul, W.D., and L.E. Webster. 1976. Breeding status of the mountain plover. Condor 78:265-267.

Haig, S., and J.H. Plissner. 1993. Distribution and abundance of piping plovers: results and implications of the 1991 international census. Condor 95:145-156.

Harrington, B.A., F.J. Leeuwenberg, S. Lara Resende, R. McNeil, B.T. Thomas, J.S. Grear, and E.F. Martinez. 1991. Migration and mass change of white-rumped sandpipers in North and South America. Wilson Bull. 103:621-636.

Table. Species, major habitats, and population change in North American breeding shorebirds in the United States east of the 105th meridian.*

Scientific name	Common name	Habitat	Reference and status	Significance
Pluvialis squatarola	Black-bellied plover	Coastal	a- d+	*P* < 0.10(a) ns(d)
P. dominica	American golden-plover	Upland	d-	ns
Charadrius alexandrinus	Snowy plover	Coastal	g threatened	
C. wilsonia	Wilson's plover	Coastal	unknown	
C. semipalmatus	Semipalmated plover	Mixed	a- d+	ns(a) ns(d)
C. melodus	Piping plover	Coastal	c threatened	
C. vociferus	Killdeer	Upland	b-	*P* < 0.05
C. montanus	Mountain plover	Upland	b+	ns
Haematopus palliatus	American oystercatcher	Coastal	unknown	
Himantopus mexicanus	Black-necked stilt	Fresh water	b-	ns
Recurvirostra americana	American avocet	Fresh water	b-	ns
Tringa melanoleuca	Greater yellowlegs	Mixed	a-	ns
T. flavipes	Lesser yellowlegs	Mixed	a+	ns
T. solitaria	Solitary sandpiper	Fresh water	unknown	
Catoptrophorus semipalmatus	Willet	Coastal	a± b+ d-	ns(a) ns(b) ns(d)
Actitis macularia	Spotted sandpiper	Fresh water	b+	ns
Bartramia longicauda	Upland sandpiper	Upland	b+	*P* < 0.05
Numenius phaeopus	Whimbrel	Coastal	a- d+	*P* < 0.01(a) ns(d)
N. americanus	Long-billed curlew	Upland	b-	*P* < 0.05
Limosa haemastica	Hudsonian godwit	Coastal	unknown	
L. fedoa	Marbled godwit	Mixed	b+	ns
Arenaria interpres	Ruddy turnstone	Coastal	a- d+ e-	ns(a) ns(d) ns(e)
Calidris canutus	Red knot	Coastal	a- d- e-	ns(a) *P* < 0.10(d) ns(e)
C. alba	Sanderling	Coastal	a- d- e-	*P* < 0.01(a) ns(d) *P* < 0.01(e)
C. pusilla	Semipalmated sandpiper	Mixed	a- d- e-	ns(a) *P* < 0.02(d) *P* < 0.05(e)
C. mauri	Western sandpiper	Mixed	unknown	
C. minutilla	Least sandpiper	Mixed	a+ d-	ns(a) *P* < 0.05(d)
C. fuscicollis	White-rumped sandpiper	Mixed	unknown	
C. bairdii	Baird's sandpiper	Fresh water	unknown	
C. melanotos	Pectoral sandpiper	Fresh water	unknown	
C. maritima	Purple sandpiper	Coastal	unknown	
C. alpina	Dunlin	Mixed	d- e±	ns(d) ns(e)
C. himantopus	Stilt sandpiper	Fresh water	unknown	
Tryngites subruficollis	Buff-breasted sandpiper	Upland	unknown	
Limnodromus griseus	Short-billed dowitcher	Coastal	a- d- e+	*P* < 0.05(a) *P* < 0.08(d) *P* = 0.12(e)
L. scolopaceus	Long-billed dowitcher	Fresh water	unknown	
Gallinago gallinago	Common snipe	Fresh water	b-	*P* < 0.05
Scolopax minor	American woodcock	Special	b- f-	*P* < 0.05(b) *P* < 0.05(f)
Phalaropus tricolor	Wilson's phalarope	Fresh water	b-	*P* < 0.05
P. lobatus	Red-necked phalarope	Special	unknown	
P. fulicaria	Red phalarope	Special	unknown	

* In the "reference and status" column and the "significance" column, "a" through "g" refer to a reference in footnote **. The reference footnotes also give the years the survey was conducted. If "+" follows the letter in the "reference and status" column, the population is increasing. If "-" follows the letter, the population is declining. In the "significance" column, "ns" means population increase or decrease is not significant. "*P*" is a measure of the confidence that the decline or increase is actually significant. A lower *P* value means there is more confidence that the trend is real. A population trend change at the *P* < 0.10 level is considered statistically significant.

** a — Howe et al. (1989) for 1972-83.
b — B.G. Peterjohn, NBS, unpublished analysis, National Biological Service, Breeding Bird Survey, 1982-91.
c — Haig and Plissner 1993.
d — Morrison et al., in press 1974-91.
e — Clark et al. 1993 for 1986-92.
f — Sauer and Bortner 1991.
g — U. S. Fish and Wildlife Service, Office of Endangered Species, unpublished data.

Harrington, B.A., J.P. Myers, and J.S. Grear. 1989. Coastal refueling sites for global bird migrants. Pages 4293-4307 *in* O.T. Magoon, H. Converse, D.Miner, L.T. Tobin, and D. Clark, eds. Proceedings of the Sixth Symposium on Coastal and Ocean Management. American Society of Civil Engineers, New York.

Howe, M.A., P.H. Geissler, and B.A. Harrington. 1989. Population trends of North American shorebirds based on the International Shorebird Survey. Biological Conservation 49:185-200.

Morrison, R.I.G., C. Downes, and B. Collins. 1994. Population trends of shorebirds on fall migration in eastern Canada, 1974-1991. Wilson Bull. 106. In press.

Morrison, R.I.G., and B.A. Harrington. 1979. Critical shorebird resources in James Bay and eastern North America. Pages 498-507 *in* Transactions of the 44th North American Wildlife and Natural Resources Conference. Wildlife Management Institute, Washington, DC.

Myers, J.P., R.I.G. Morrison, P.Z. Antas, B.A. Harrington, T.E. Lovejoy, M. Sallaberry, S.E. Senner, and A. Tarak. 1987. Conservation strategy for migratory species. American Scientist 75:19-26.

Page, G., L.E. Stenzel, W.D. Shuford, and C.R. Bruce. 1991. Distribution and abundance of the snowy plover on its western North American breeding grounds. Journal of Field Ornithology 62:245-255.

Sauer, J.R., and J.B. Bortner. 1991. Population trends from the American woodcock singing-ground survey, 1970-88. Journal of Wildlife Management 55:300-312.

Schneider, D.C., and B.A. Harrington. 1981. Timing of shorebird migration in relation to prey depletion. Auk 98:197-220.

Senner, S.E., and M.A. Howe. 1984. Conservation of Nearctic shorebirds. Pages 379-421 *in* J. Burger and B. Olla, eds. Shorebirds: breeding behavior and populations. Plenum Press, New York.

For further information:

Brian A. Harrington
Manomet Observatory for
Conservation Sciences
Manomet, MA 02345

Western North American Shorebirds

by
Robert E. Gill, Jr.
Colleen M. Handel
National Biological Service
Gary W. Page
Point Reyes Bird Observatory

Shorebirds are a diverse group that includes oystercatchers, stilts, avocets, plovers, and sandpipers. They are familiar birds of seashores, mudflats, tundra, and other wetlands, but they also occur in deserts, high mountains, forests, and agricultural fields. Widespread loss and alteration of these habitats, especially wetlands and grasslands during the past 150 years, coupled with unregulated shooting at the turn of the century, resulted in population declines and range contractions of several species throughout North America. In the western portion of the continent, efforts to monitor the status and trends of shorebirds have been in effect for only the past 15-25 years and for only a few species. Methods exist to monitor population trends for most shorebirds, but only broadscale, international efforts, relying largely on volunteer help, will accomplish this.

In this article we address shorebirds primarily in western North America, the region west of the Continental Divide from northern Alaska to southern Mexico. The 12 states, a Canadian province and territory, and the western portion of Mexico within this region represent about 25% of the North American landmass (Fig. 1). Western North America includes portions of three broad ecological domains: the Polar Domain, encompassing the tundra and boreal forests that cover most of Greenland, Canada, and Alaska; the Humid Temperate Domain, including the humid midlatitude forests and shrublands within the United States, southern portions of the Canadian prairie provinces, and along the west coast of North America; and the Dry Domain, encompassing the short-grass prairies, sagebrush provinces, and deserts (Fig. 1; Bailey 1978, 1989).

Sources of Data

We derived seasonal distribution of shorebirds within these ecological domains from numerous sources, mostly range maps in field guides, books, and our familiarity with the birds within the region (AOU 1983; Robbins et al. 1983; Hayman et al. 1986; Godfrey 1987; National Geographic Society 1987; Paulson 1993).

No continent-wide protocol exists for monitoring the status and trends of North American shorebirds. Current information has largely been acquired through independent programs sponsored by a combination of federal, state, and private conservation agencies. Efforts have mostly been regional, including broadscale monitoring directed primarily at birds during the nonbreeding season (Howe et al. 1989; Gill and Handel 1990; Page et al. 1992; Skagen and Knopf 1993; Morrison et al. 1994) or have focused on individual species (Handel and Dau 1988; Gill et al. 1991; Page et al. 1991; Haig 1992; Handel and Gill 1992a; Knopf 1994; F.L. Knopf, USFWS, unpublished report). We have relied primarily on this information and that of our ongoing studies to summarize the status and trends of shorebirds in western North America.

Shorebirds of the Region

Breeding

Among the 51 species that regularly breed in North America, 47 (92%) do so within western North America (Table). Within this region, the Polar Domain supports the greatest number of breeding species (37), including 5 that breed nowhere else on the continent. The Humid

Temperate Domain provides breeding areas for 20 species while only 12 breed in the Dry Domain (Fig. 1). The number of species breeding within the domains in the West generally exceeds those breeding east of the Continental Divide, even though the eastern area is much larger.

Western North American shorebirds nest in a variety of habitats, although most species (53%) are restricted to either coastal or interior wetlands (Page and Gill 1994). About a third of the species nest primarily on uplands, especially Arctic and subarctic tundra and dry temperate grasslands.

Wintering

Thirty-six (70%) of the continent's breeding species winter in western North America, including seven that are restricted to the region (Table). The continental distribution of species shifts southward in winter, but numbers are still higher in the West than in the East (Fig. 1). Only 4 of the 37 species breeding in the Polar Domain of western North America remain there during winter. About 30 species spend the winter in the Humid Temperate and Dry domains. Populations of 12 (25%) of western North America's breeding species spend the winter entirely on other continents or throughout Oceania (*see* glossary; Table).

Most shorebirds use a much broader range of habitats during winter than during the breeding period. All species use one or more coastal habitats in winter and two-thirds of the species also use interior habitats (Page and Gill 1994). Wetlands, the single-most important habitat both along the coast and in the interior of western North America, are used by about 80% of all species. Sandy and rocky shorelines along the Pacific coast are also important habitats and are used by about a quarter of the species (Page and Gill 1994).

Migrating

All species of North American shorebirds are migratory to some degree, with the possible exception of both species of oystercatchers and Wilson's plover; they are not migratory in the true sense but do make short, local movements. Shorebirds migrate in spring and fall over three broadly defined corridors encompassing the western, central, and eastern portions of the continent to wintering areas in North, Central, and South America (Morrison and Myers 1989). Other migratory corridors funnel Arctic breeders from western North America across the Pacific Ocean to wintering areas in Asia, Australasia, and Oceania (*see* glossary; Gill and Handel 1981; Handel and Gill 1992b; Page and Gill 1994). The distances traveled between breeding and wintering grounds vary greatly within and among species, often exceeding 8,000 km (5,000 mi) for such species as Hudsonian and bar-tailed godwits.

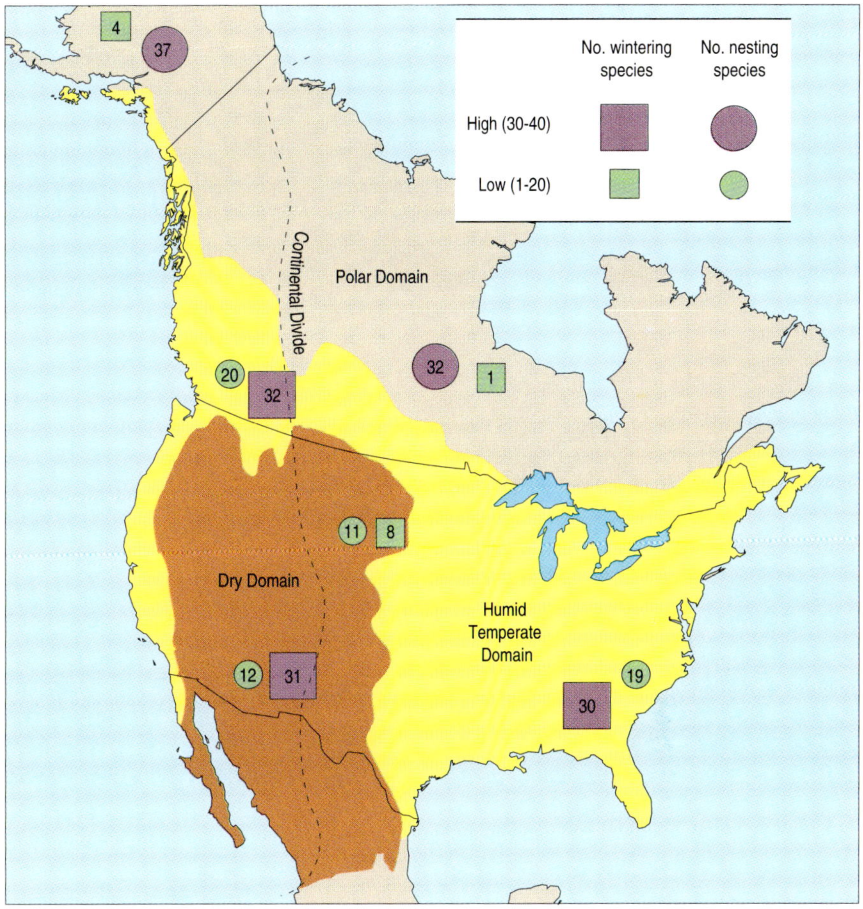

Fig. 1. Number of shorebird species regularly breeding and wintering within three broad ecological domains of North America west and east of the Continental Divide (dashed line).

Wetlands are the most important habitat used by shorebirds during spring and fall migrations. Throughout western North America about 140 discrete wetlands and several additional wetland complexes (e.g., Central Valley of California) have been identified as being important to shorebirds during these periods (Fig. 2). Most staging areas (85%) host populations of 1,000-10,000 birds, but 18 sites support 100,000-1 million shorebirds during the peak of migration (Fig. 2). Because shorebirds use different migration pathways and strategies during spring and fall, the locations of critical staging areas shift between the two seasons (Fig. 2).

Status and Trends

Size of Populations

Population estimates exist for only about a quarter of the species that breed or winter in western North America (Table), and even these few vary widely in terms of statistical rigor and precision. These estimates range between

Table. Seasonal occurrence and status and trends of populations of shorebirds in North America west and east of the Continental Divide.

Species	Occurrence		Population*	
	Breeding	Wintering	Size	Trend
Black-bellied plover (*Pluvialis squatarola*)	Both regions	Both regions	Unknown	Unknown
Pacific golden-plover (*P. fulva*)	Western	Western	Unknown	Unknown
American golden-plover (*P. dominica*)	Both regions	Absent	Unknown	Unknown
Snowy plover (*Charadrius alexandrinus*)	Both regions	Both regions	18,500 b	Population decline and range contraction past century; significant decline in western region past 25 yr (Page et al. 1991; Page and Gill 1994)
Wilson's plover (*C. wilsonia*)	Both regions	Both regions	Unknown	Unknown
Common ringed plover (*C. hiaticula*)	Eastern	Absent	Unknown	Unknown
Semipalmated plover (*C. semipalmatus*)	Both regions	Both regions	Unknown	Unknown
Piping plover (*C. melodus*)	Eastern	Mostly eastern	4,700 b	Declining (Haig 1992)
Killdeer (*C. vociferus*)	Both regions	Both regions	Unknown	Decline of 2.0% per yr in western region past 25 yr; 5.3% decline per yr last 10 yr (J. Sauer and S. Droege, NBS, unpublished data)
Mountain plover (*C. montanus*)	Both regions	Mostly western	5,000-15,000 b	Population decline and range contraction past century; continental decline 3.6% per yr past 25 yr (Knopf 1994; F.L. Knopf, USFWS, unpublished report)
American oystercatcher (*Haematopus palliatus*)	Both regions	Both regions	Unknown	Unknown
Black oystercatcher (*H. bachmani*)	Western	Western	7,600 b	Unknown (Page and Gill 1994)
Black-necked stilt (*Himantopus mexicanus*)	Both regions	Both regions	25,000 w	Population decline and range contraction past century; no significant change in population size past 25 yr (Page and Gill 1994; J. Sauer and S. Droege, NBS, unpublished data)
American avocet (*Recurvirostra americana*)	Both regions	Both regions	100,000+ w	Population decline past century; decline of 3.6% per yr in western region past 10 yr (Page and Gill 1994; G. Page, unpublished data; J. Sauer and S. Droege, NBS, unpublished data)
Greater yellowlegs (*Tringa melanoleuca*)	Both regions	Both regions	Unknown	Unknown
Lesser yellowlegs (*T. flavipes*)	Both regions	Both regions	Unknown	Unknown
Solitary sandpiper (*T. solitaria*)	Both regions	Both regions	Unknown	Unknown
Willet (*Catoptrophorus semipalmatus*)	Both regions	Both regions	Unknown	Population decline and range contraction past century; population increase of 2.8% per yr in United States past 25 yr; 0.5% increase in West in past 10 yr (Page and Gill 1994; J. Sauer and S. Droege, NBS, unpublished data)
Wandering tattler (*Heteroscelus incanus*)	Western	Western	Unknown	Unknown
Spotted sandpiper (*Actitis macularia*)	Both regions	Both regions	Unknown	Population stable over continent and western region past 25 yr (J. Sauer and S. Droege, NBS, unpublished data)
Upland sandpiper (*Bartramia longicauda*)	Both regions	Absent	Unknown	Population decline and range contraction past century; 3.6% annual increase on continental basis past 25 yr; no significant trend in western region (Page and Gill 1994; J. Sauer and S. Droege, NBS, unpublished data)
Eskimo curlew (*Numenius borealis*)	Both regions	Absent	25-50 b	Almost extirpated over past century; may be extinct (Gollop et al. 1986; Alexander et al. 1991)
Whimbrel (*N. phaeopus*)	Both regions	Both regions	Unknown	Unknown
Bristle-thighed curlew (*N. tahitiensis*)	Western	Absent	7,000 b	Unknown (Gill and Redmond 1992; C. Handel, unpublished data)
Long-billed curlew (*N. americanus*)	Both regions	Both regions	Unknown	Population decline and range contraction over past century; annual decrease of 3.0% on continental basis past 10 yr (J. Sauer and S. Droege, NBS, unpublished data)
Hudsonian godwit (*Limosa haemastica*)	Both regions	Absent	Unknown	Unknown
Bar-tailed godwit (*L. lapponica*)	Western	Absent	25,000-40,000 b	Unknown (Page and Gill 1994; R. Gill, unpublished data)
Marbled godwit (*L. fedoa*)	Both regions	Both regions	100,000+ w	Population decline and range contraction over past century; no significant trend throughout continent or western region past 25 yr (Page and Gill 1994; G. Page, unpublished data; J. Sauer and S. Droege, NBS, unpublished data)
Ruddy turnstone (*Arenaria interpres*)	Both regions	Both regions	Unknown	Unknown
Black turnstone (*A. melanocephala*)	Western	Western	61,000-99,000 b	Unknown (Handel and Gill 1992a)
Surfbird (*Aphriza virgata*)	Western	Western	50,000-70,000 s	Unknown (Page and Gill 1994)
Red knot (*Calidris canutus*)	Both regions	Both regions	Unknown	Unknown
Sanderling (*C. alba*)	Both regions	Both regions	Unknown	Unknown
Semipalmated sandpiper (*C. pusilla*)	Both regions	Absent	Unknown	Unknown
Western sandpiper (*C. mauri*)	Western	Both regions	Unknown	Unknown
Least sandpiper (*C. minutilla*)	Both regions	Both regions	Unknown	Unknown
White-rumped sandpiper (*C. fuscicollis*)	Both regions	Absent	Unknown	Unknown
Baird's sandpiper (*C. bairdii*)	Both regions	Absent	Unknown	Unknown
Pectoral sandpiper (*C. melanotos*)	Both regions	Absent	Unknown	Unknown
Purple sandpiper (*C. maritima*)	Eastern	Eastern	Unknown	Unknown
Rock sandpiper (*C. ptilocnemis*)	Western	Western	Unknown	Unknown
Dunlin (*C. alpina*)	Both regions	Both regions	450,000-600,000 w	Unknown (Page and Gill 1994)
Stilt sandpiper (*C. himantopus*)	Both regions	Both regions	Unknown	Unknown
Buff-breasted sandpiper (*Tryngites subruficollis*)	Both regions	Absent	Unknown	Unknown
Short-billed dowitcher (*Limnodromus griseus*)	Both regions	Both regions	Unknown	Unknown
Long-billed dowitcher (*L. scolopaceus*)	Both regions	Both regions	Unknown	Unknown
Common snipe (*Gallinago gallinago*)	Both regions	Both regions	Unknown	Population stable past 25 yr but 3.9% decline per yr on continental basis during past 10 yr; decline in western region not significant past 10 yr (J. Sauer and S. Droege, NBS,unpublished data)
American woodcock (*Scolopax minor*)	Eastern	Eastern	Unknown	Unknown
Wilson's phalarope (*Phalaropus tricolor*)	Both regions	Both regions	1,500,000 f	Population decline past century; expansion of range past 50 yr; annual population declines of 7.5% throughout United States and 8.1% in central region during past 10 yr (Jehl 1988; Page and Gill 1994; J. Sauer and S. Droege, NBS, unpublished data)
Red-necked phalarope (*P. lobatus*)	Both regions	Absent	Unknown	Unknown
Red phalarope (*P. fulicaria*)	Both regions	Western	Unknown	Unknown

*Population estimates for Western North America unless otherwise stated (*see* Fig. 1). Sources for estimates of population size given with trend information. Population size — estimated number of individual birds for b — breeding season, f — fall, w — winter, and s — spring. Geographic regions under population trend are defined in Robbins et al. (1986).

10,000 and 100,000 individuals for most populations, but number from as few as 25 birds for the endangered Eskimo curlew to about 500,000 for the Pacific race of the dunlin (*Calidris alpina pacifica*). A few other species for which some data are available, such as western sandpiper and Wilson's phalarope, have populations that exceed a million (Page and Gill 1994).

Population Trends

For most species, reliable quantitative data on population trends are either not available or too recent to assess trends. Assessment of long-term population trends is based largely on historical accounts of relative abundance and distribution and knowledge of habitat alteration within breeding and wintering ranges. Nonetheless, populations of several species of western North American shorebirds have declined significantly over the past 150 years (Page and Gill 1994). One Arctic breeder, the Eskimo curlew, is on the verge of extinction (Gollop et al. 1986; Alexander et al. 1991). Conversion of native grasslands for agriculture, loss of wetlands, and market hunting before the turn of the century have been attributed as factors primarily responsible for these declines. No species is known to have increased in overall population size over this period.

Information on more recent population trends comes primarily from the North American Breeding Bird Survey (BBS), a system of roadside surveys designed primarily to monitor populations of breeding landbirds. The BBS does not sample most western shorebird breeding populations very well because of its sporadic coverage, poor sampling of wetland habitats, and lack of coverage of the most important shorebird breeding grounds in the Arctic, which are roadless. Despite these limitations, BBS does provide valuable trend information, particularly for grassland species in the temperate zone. Additional information on population trends can also be obtained from surveys that target species of concern, such as the snowy plover (Page et al. 1991), or particular habitats of concern, such as the Arctic Coastal Plain of Alaska (D. Troy Ecological Assoc. and British Petroleum Exploration, unpublished report; Andres 1994).

Recent survey data show a mixture of declining, increasing, and apparently stable population trends (Table). Over the past 25 years, western populations of willet and upland sandpiper appear to have been rebounding (J. M. Sauer and S. Droege, unpublished data). Numbers of several other species, such as the black-necked stilt, marbled godwit, and spotted sandpiper, appear to have stabilized (J.M. Sauer and S. Droege, unpublished data). Western populations of several other species, however, have significantly declined over the past 25 years, including the snowy plover, killdeer, mountain plover, American avocet, long-billed curlew, common snipe, and Wilson's phalarope (Table). Such relatively short-term trends among wetland species are difficult to interpret, however, as they may reflect changes in distribution in response to drought conditions rather than absolute declines in population size (Page and Gill 1994).

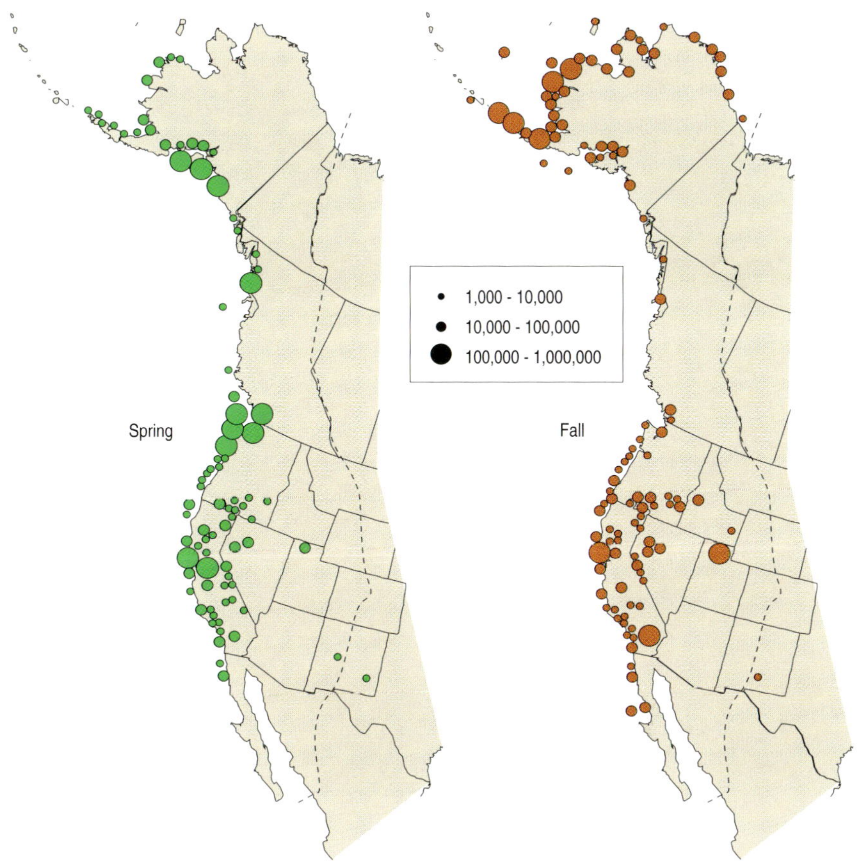

Fig. 2. Location of important staging areas in western North America used by shorebirds during spring and fall migration. Size of dot indicates the estimated peak number of shorebirds at each site.

Most changes in populations appear linked to habitat alteration. For example, since 1970 the snowy plover, heavily dependent on coastal habitats, has disappeared as a breeding species from over 60% of its historic California nesting sites (Page and Stenzel 1981). Introducing plants to stabilize sand dunes, increasing recreational use of beaches, and heavy nest predation by feral foxes threaten to reduce coastal populations further (Page and Gill 1994). Fluctuating water levels in interior wetlands result in unpredictable changes in availability of nesting habitat away from the coast (Page et al. 1991). The breeding range of the mountain plover has contracted markedly in several western states and the continental population has declined significantly during the past 25 years, probably because of habitat degradation on wintering grounds in central and southern California (Knopf 1994; F.L. Knopf, NBS, unpublished report). Given the substantial loss of wetlands throughout all western states except Alaska (median loss of 37%; Page and Gill 1994) and a similar loss of native grasslands (Knopf 1994), it is likely that other species of temperate-breeding shorebirds for which we have no trend data have also suffered population declines.

Shorebirds breeding throughout the remote and sparsely populated Polar Domain have been least affected by loss of breeding habitats. Most of these species, however, are dependent on wetlands and other greatly altered habitats outside this region during winter and migration. Information from long-term studies in Europe suggests that populations of Arctic-breeding shorebirds can be affected by conditions on the wintering grounds as well as by those on the breeding grounds (Goss-Custard and Moser 1988; Moser 1988). Arctic breeders such as the buff-breasted sandpiper, upland sandpiper, and American golden-plover winter primarily in grassland habitats of the pampas in South America. These habitats have been virtually eliminated by agricultural development (Bucher and Nores 1988; Blanco et al. 1993). The bristle-thighed curlew, unique among shorebirds because of its flightlessness during molt (Marks 1993), is threatened by problems associated with increasing human populations on wintering grounds in Oceania, including the introduction of mammalian predators (Marks et al. 1990; Gill and Redmond 1992).

Courtesy R.E. Gill, Jr., NBS

Surfbirds (*Aphiza virgata*) and black turnstones (*Arenaria melanocephala*).

In long-term studies of shorebirds nesting at Prudhoe Bay on Alaska's North Slope between 1981 and 1992, considerable annual variation in nesting density and nest success has been found in several species of shorebirds (D. Troy, Troy Ecological Associates and British Petroleum Exploration, unpublished report). Much of this variation has been attributed to predation and environmental factors such as snow cover and temperature at the start of the breeding season. However, among eight species of intensively monitored shorebirds, only dunlin (*Calidris alpina articola*) have exhibited a general, but not significant, downward trend in nesting density over this 10-year period.

Detecting Future Trends

To conserve the tremendous biodiversity of our shorebird resources in western North America, we suggest a two-tiered monitoring program that addresses trends in both habitat availability and shorebird population size. In this program we should:

- Identify and map the current geographic extent and quality of breeding, staging, and wintering habitats important to shorebirds, particularly those species with relatively small populations or restricted habitat requirements;
- Monitor the extent and quality of these habitats, evaluating them at periodic intervals;
- Develop cooperative, international programs to monitor trends in shorebird populations;
- Monitor a representative sample of shorebird populations and evaluate trends in comparison with changes in critical habitats; and
- Establish cooperative, international agreements to protect critical breeding, staging, and wintering habitats, with priority given to those species with low numbers, specific habitat requirements, and immediate threats.

Recently developed technology and continental habitat mapping now provide the tools to identify and map the current extent of wetlands and other habitats important to shorebirds of western North America. By coupling this with current information on shorebird distribution and habitat requirements, we will be able to identify areas critical for shorebirds. The same technology can be used to monitor changes in these habitats over time.

Several existing programs can be adapted or modified to provide reliable information on trends in size of several shorebird populations. Each species needs to be evaluated individually to determine where it could be monitored most cost-effectively—breeding grounds, staging areas, or wintering grounds. Programs such as the International Shorebird Survey, Breeding Bird Survey, and Christmas Bird Count can be used to coordinate efforts of large numbers of volunteers to simultaneously collect information on several species of shorebirds. For many other species like the snowy plover, buff-breasted sandpiper, and bristle-thighed curlew—of particular concern or difficult to monitor with these programs—specific surveys need to be designed and repeated periodically to effectively monitor population trends.

References

Alexander, S.A., R.S. Ferguson, and K.J. McCormick. 1991. Key migratory bird terrestrial habitat sites in the Northwest Territories. 2nd ed. Canadian Wildlife Service Occasional Papers 71. Ottawa, Canada. 184 pp.

Andres, B.A. 1994. Coastal zone use by postbreeding shorebirds in northern Alaska. Journal of Wildlife Management 58:206-213.

AOU. 1983. Check-list of North American birds. 6th ed. American Ornithologists' Union, Washington, DC. 877 pp.

Bailey, R.G. 1978. Description of the ecoregions of the United States. U.S. Forest Service, Intermountain Region, Ogden, UT. 77 pp.

Bailey, R.G. 1989. Explanatory supplement to ecoregions of the continent. Environmental Conservation 16:307-309.

Blanco, D., R. Banchs, P. Canevari, and M. Oesterheld. 1993. Critical sites for the Eskimo curlew (*Numenius borealis*), and other Nearctic grassland shorebirds in Argentina and Uruguay. U.S. Fish and Wildlife Service and Wetlands for the Americas, Manomet, MA. 89 pp. Unpublished report.

Bucher, E.H., and M. Nores. 1988. Present status of birds in steppes and savannas of northern and central Argentina. International Council for Bird Preservation Tech. Publ. 7:71-79.

Gill, R.E., Jr., and C.M. Handel. 1981. Shorebirds of the eastern Bering Sea. Pages 719-738 *in* D.W. Hood and J.A. Calder, eds. The eastern Bering Sea shelf: oceanography and resources. Vol. 2. University of Washington Press, Seattle.

Gill, R.E., Jr., and C.M. Handel. 1990. The importance of subarctic intertidal habitats to shorebirds: a study of the central Yukon-Kuskokwim Delta, Alaska. Condor 92:702-725.

Gill, R.E., Jr., R.L. Lanctot, J. Mason, and C.M. Handel. 1991. Observations on habitat use, breeding chronology and parental care in bristle-thighed curlews on the Seward Peninsula, Alaska. Wader Study Group Bull. 61:28-36.

Gill, R.E., Jr., and R.L. Redmond. 1992. Distribution, numbers, and habitat of bristle-thighed curlews *Numenius tahitiensis* on Rangiroa Atoll. Notornis 39:17-26.

Godfrey, W.E. 1987. Birds of Canada. National Museum of Canada Bull. 203. Ottawa. 595 pp.

Gollop, J.B., T.W. Berry, and E.H. Iverson. 1986. Eskimo curlew: a vanishing species? Saskatchewan Natural History Society Special Publ. 17. 160 pp.

Goss-Custard, J.D., and M.E. Moser. 1988. Rates of change in the numbers of dunlin, *Calidris alpina*, wintering in British estuaries in relation to the spread of *Spartina angelica*. Journal of Applied Ecology 25:95-109.

Haig, S.M. 1992. Piping plover. *In* A. Poole, P. Stettenheim, and F. Gill, eds. The birds of North America, No. 2. Academy of Natural Sciences, Philadelphia, and American Ornithologists' Union, Washington, DC.

Handel, C.M., and C.P. Dau. 1988. Seasonal occurrence of migrant whimbrels and bristle-thighed curlews on the Yukon-Kuskokwim Delta, Alaska. Condor 90:782-790.

Handel, C.M., and R.E. Gill, Jr. 1992a. Breeding distribution of the black turnstone. Wilson Bull. 104:122-135.

Handel, C.M., and R.E. Gill, Jr. 1992b. Roosting behavior of premigratory dunlins *Calidris alpina*. Auk 109:57-72.

Hayman, P., J. Marchant, and T. Prater. 1986. Shorebirds: an identification guide to the waders of the world. Houghton Mifflin, Boston, MA. 412 pp.

Howe, M.A., P.H. Geissler, and B.A. Harrington. 1989. Population trends of North American shorebirds based on the International Shorebird Survey. Biological Conservation 49:185-199.

Jehl, J.R., Jr. 1988. Biology of the eared grebe and Wilson's phalarope in the non-breeding season: a study of adaptations to saline lakes. Studies in Avian Biology 12.

Knopf, F.L. 1994. Avian assemblages on altered grasslands. Studies in Avian Biology 15:247-257.

Marks, J.S. 1993. Molt of bristle-thighed curlews in the northwestern Hawaiian Islands. Auk 110:573-587.

Marks, J.S., R.L. Redmond, P. Hendricks, R.B. Clapp, and R.E. Gill, Jr. 1990. Notes on longevity and flightlessness in bristle-thighed curlews. Auk 107:779-781.

Morrison, R.I.G., C. Downes, and B. Collins. 1994. Population trends of shorebirds on fall migration in eastern Canada, 1974-1991. Wilson Bull. 106:431-447.

Morrison, R.I.G., and J.P. Myers. 1989. Shorebird flyways in the New World. Pages 85-96 *in* H. Boyd and J.-Y. Pirot, eds. Flyways and reserve networks for waterbirds. Special Publ. 9. International Waterfowl and Wetlands Research Bureau, Slimbridge, U.K.

Moser, M.E. 1988. Limits to the numbers of grey plovers *Pluvialis squatarola* wintering in British estuaries: an analysis of long-term population trends. Journal of Applied Ecology 25:473-485.

National Geographic Society. 1987. A field guide to the birds of North America. National Geographic Society, Washington, DC. 464 pp.

Page, G.W., and R.E. Gill, Jr. 1994. Shorebirds in western North America: late 1800s to late 1900s. Studies in Avian Biology 15:147-160.

Page, G.W., W.D. Shuford, J.E. Kjelmyr, and L.E. Stenzel. 1992. Shorebird numbers in wetlands of the Pacific Flyway: a summary of counts from April 1988 to January 1992. Point Reyes Bird Observatory, Stinson Beach, CA. 42 pp. Unpublished report.

Page, G.W., and L.E. Stenzel, eds. 1981. The breeding status of the snowy plover in California. Western Birds 12:1-40.

Page, G.W., L.E. Stenzel, W.D. Shuford, and C.R. Bruce. 1991. Distribution and abundance of the snowy plover on its western North American breeding grounds. Journal of Field Ornithology 62:245-255.

Paulson, D. 1993. Shorebirds of the Pacific Northwest. University of Washington Press, Seattle. 406 pp.

Robbins, C.S., B. Bruun, and H.S. Zim. 1983. Birds of North America: a guide to field identification. Golden Press, New York. 360 pp.

Robbins, C.S., D. Bystrak, and P.H. Geissler. 1986. The Breeding Bird Survey: its first fifteen years, 1965-1979. U.S. Fish and Wildlife Service Resour. Publ. 157. 196 pp.

Skagen, S.K., and F.L. Knopf. 1993. Toward conservation of midcontinental shorebird migrations. Conservation Biology 7:533-541.

For further information:

Robert E. Gill, Jr.
National Biological Service
Alaska Science Center
1011 E. Tudor Rd.
Anchorage, AK 99503

Raptors

by
Mark R. Fuller
Charles J. Henny
Petra Bohall Wood
National Biological Service

Raptors, or birds of prey, which include the hawks, falcons, eagles, vultures, and owls, occur throughout North American ecosystems. As predators, most of them kill other vertebrates for their food. Compared to most other animal groups, birds of prey naturally exist at relatively low population levels and are widely dispersed within their habitats. The natural scarcity of raptors, combined with their ability to move quickly, the secretive behavior of many species, and the difficulties of detecting them in rugged terrain or vegetation, all make determining their population status difficult.

As top predators, raptors are key species for our understanding and conservation of ecosystems. Changes in raptor status can reflect changes in the availability of their prey species, including population declines of mammals, birds, reptiles, amphibians, and insects. Changes in raptor status also can be indicators of more subtle detrimental environmental changes such as chemical contamination and the occurrence of toxic levels of heavy metals (e.g., mercury, lead). Consequently, determining and monitoring the population status of raptors are necessary steps in the wise management of our natural resources.

Methods

We did not compile summary statistics or analyze data for any species; rather, we only have summarized the interpretations and analyses of others. Our summary of raptor status draws largely on the biological literature and on state and federal government reports. Much of this information is summarized in Johnsgard (1988), Palmer (1988), and White (1994) and in proceedings sponsored by the National Wildlife Federation (NWF 1988, 1989a, 1989b, 1990, 1991). Other information is from unpublished data (S.W. Hoffman, HawkWatch International; J.C. Bednarz, Arkansas State University; and W.R. DeRagon, U.S. Army Corps of Engineers).

Interpretations and analyses to determine raptor status and trends can be characterized in four general types: impressions of biologists and of other serious observers of wildlife; impressions or nonstatistical analyses of organized searches or of tallies of birds seen (e.g., Christmas Bird Counts); statistical analyses of intensive quantitative status surveys; and statistical analyses of standardized counts, incorporating estimates of the survey effort (e.g., number of persons, time expended, area covered).

Our conclusion about the status of each species (Table) is usually applied on a nationwide scale, but often must be qualified because of local or regional concerns. These reflect habitat modification or contamination for which we did not have information on a broader scale. We used statistical results when available, but usually our conclusions are based on impressions or qualitative analyses because only that is available on a scale across the species' range, or the United States.

Northern goshawk (*Accipiter gentilis*) in adult plumage, is an example of a raptor species for which there is concern about status.

Courtesy M.R. Fuller

Selected Species

Ospreys

Nesting ospreys (*Pandion haliaetus*) are concentrated along the Atlantic coast, Great Lakes, the northern Rocky Mountains, and in the Pacific Northwest. Most regional populations declined through the early 1970's, but the magnitude of decline varied, with the North Atlantic coast and Great Lakes being most severe. After the 1972 nationwide ban of the insecticide DDT, raptor productivity improved and population numbers increased in most areas. Ospreys also benefited from reservoir construction, especially in the West. Osprey numbers generally are stable, but in some areas they are still increasing. The large stick nests of ospreys, like those of bald eagles (*Haliaeetus leucocephalus*), are relatively conspicuous, thus aiding counts of occupied nests, which are used as a measure of population size. Counts from most states in the early 1980's provided an estimate of about 8,000 nesting pairs. Also, because several osprey populations were studied for many years, a general knowledge of their population dynamics permits a greater understanding of this species' status.

Snail Kite

The endangered snail kite (*Rostrhamus sociabilis*) breeds in central and southern Florida, the northern extent of the species' range, where it is associated with wetlands that are affected by management of water levels. From 1900 to 1960 the population declined; however, it then increased, and now remains stable with fluctuations from 300 to 800 birds (R.E. Bennetts, University of Florida, personal communication).

Bald Eagles

Many local bald eagle populations showed sharp declines (25% to 100%) from 1950 to the 1970's. Populations were adversely affected by shooting, habitat destruction, and organochlorine pesticides (primarily DDT). The bird was protected by the Bald Eagle Protection Act of 1940. In 1978 it was reclassified as endangered in 43 states and threatened in 5. With the documented effects of DDT on reproduction, early studies emphasized locating breeding pairs and monitoring reproductive success.

After the nationwide ban of DDT in 1972, bald eagle reproduction improved and populations began increasing. In 1981 about 1,300 pairs nested in the United States outside Alaska. The active protection of nesting habitat and release of hand-reared eagles aided the population increase. In 1993 at least 4,016 pairs of bald eagles nested in the contiguous United States, with an estimated additional 20,000-25,000 pairs in Alaska. Bald eagles nesting along the shorelines of Lakes Superior, Michigan, Huron, and Erie have lower reproductive rates and relatively high concentrations of the toxic DDE and PCB compounds (Bowerman 1993). Bald eagles nesting in Maine also have low reproductive success, probably because of environmental contaminants.

Habitat loss remains a threat in many areas. Historically there was a continuous (though scattered) distribution of bald eagles in the Southwest, south into Sonora and Baja California, Mexico, where now only a remnant population exists. Because population increases were not uniform throughout the range, the U.S. Fish and Wildlife Service has proposed downlisting this species from endangered to threatened in certain geographic areas.

Hawks

Populations of sharp-shinned hawks (*Accipiter striatus*) in the Midwest might be increasing, but analyses of eastern hawk migration count stations reveal a drop in numbers of juveniles, and blood samples collected from sharp-shinned hawks in the Northeast contained high DDE pesticide concentrations. Many other factors could be involved in a population decline, however. The sharp-shinned hawk provides an example of how monitoring can warn researchers of a potential, long-term decline in a regional population.

Similarly, the northern goshawk (*A. gentilis*) counts of eastern migrants suggest a stable population, but analyses of counts from the West reveal a decline. There is no widespread standardized design for surveying goshawks during the breeding season.

Habitat loss has reduced the number of Harris' hawks (*Parabuteo unicinctus*), whose northern range extent is the southwestern United States. Searches reveal that Harris' hawks have been extirpated from some areas such as the Colorado River Valley, California and Arizona, and that clearing of brush for agriculture likely has led to more than 50% reduction in Texas in the winter.

The biological status of the ferruginous hawk (*Buteo regalis*) remains uncertain because it is stable in some areas (e.g., Great Plains), but declining in other areas (e.g., half the western

Table. Status and trends of raptors in the United States.

Species	Status/trend	Comment*
Black vulture (*Coragyps atratus*)	Stable	Population estimation difficult because of flocking and wide-ranging behavior, secretive nesting
Turkey vulture (*Cathartes aura*)	Stable	
California condor (*Gymnogyps californianus*)	Endangered; extirpated from wild, 1987	Captive propagation and release underway
Osprey (*Pandion haliaetus*)	Increasing	Good information
Hook-billed kite (*Chondrohierax uncinatus*)	Unknown	Extreme northern range limit
American swallow-tailed kite (*Elanoides forficatus*)	Stable	Greatly reduced from historical range
White-tailed kite (*Elanus caeruleus*)	Increasing	Recent range expansion
Snail kite (*Rostrhamus sociabilis*)	Endangered, stable	Northern range limit
Mississippi kite (*Ictinia mississippiensis*)	Increasing	Range expansion
Bald eagle (*Haliaeetus leucocephalus*)	Threatened or endangered in contiguous U.S.; increasing	Status reassessment underway
Northern harrier (*Circus cyaneus*)	Stable	Nomadic, no standard survey; local concern
Sharp-shinned hawk (*Accipiter striatus*)	Stable	Regional differences
Cooper's hawk (*A. cooperii*)	Stable	
Northern goshawk (*A. gentilis*)	Unknown	C2; petition to list *A.g. laingi*; threatened
Common black hawk (*Buteogallus anthracinus*)	Stable	Limited distribution
Harris' hawk (*Parabuteo unicinctus*)	Stable	Fragmented distribution, northern range limit
Gray hawk (*Buteo nitidus or Austurina plagiata*)	Stable	C2; limited distribution, northern range limit
Hawaiian hawk (*B. solitarius*)	Endangered	Difficult to survey, limited distribution
Red-shouldered hawk (*B. lineatus*)	Stable	Local concern
Broad-winged hawk (*B. platypterus*)	Stable	Migration count decline in 1980's
Puerto Rican broad-winged hawk (*B.p. brunnescens*)	Unknown	C2; limited distribution
Short-tailed hawk (*B. brachyurus*)	Stable	Northern range limit; about <500 birds in U.S.
Swainson's hawk (*B. swainsoni*)	Unknown	C3; local concern
White-tailed hawk (*B. albicaudatus*)	Stable	Northern range limit, about 200-400 birds in U.S.
Zone-tailed hawk (*B. albonotatus*)	Stable	Northern range limit, about 100 pairs in U.S.
Red-tailed hawk (*B. jamaicensis*)	Stable	Local increases; Breeding Bird Survey data
Ferruginous hawk (*B. regalis*)	Unknown	C2
Rough-legged hawk (*B. lagopus*)	Stable	
Golden eagle (*Aquila chrysaetos*)	Stable	
Crested caracara (*Caracara plancus*)	Unknown	Northern range limit
American kestrel (*Falco sparverius*)	Stable	Breeding Bird Survey data
American kestrel, Florida (*F.s. paulus*)	Declining	C2
Merlins (*F. columbarius*)	Stable	
Aplomado falcon (*F. femoralis septentrionalis*)	Endangered	Northern range limit; captive propagation and release underway
American peregrine falcon (*F. peregrinus anatum*)	Endangered; increasing	
Arctic peregrine falcon (*F.p. tundrius*)	Threatened	Proposed to delist
Gyrfalcon (*F. rusticolus*)	Stable	
Prairie falcon (*F. mexicanus*)	Stable	
Barn owls (*Tyto alba*)	Stable	Local concern
Flammulated owl (*Otus flammeolus*)	Unknown	Recent surveys reveal more birds, larger range
Virgin Islands screech-owl (*O. nudipes newtoni*)	Unknown	C2; limited distribution
Eastern screech-owl (*O. asio*)	Stable	
Western screech-owl (*O. kennicottii*)	Stable	
Whiskered screech-owl (*O. trichopsis*)	Unknown	Northern range limit
Great horned owl (*Bubo virginianus*)	Stable	
Snowy owl (*Nyctea scandiaca*)	Stable	U.S. breeding, AK only
Northern hawk owl (*Surnia ulula*)	Unknown	U.S. breeding, AK, northern Minnesota
Northern pygmy owl (*Glaucidium gnoma*)	Unknown	Current survey efforts
Ferruginous pygmy owl (*G. brasilianum cactorum*)	Unknown	C2; northern range limit
Elf owl (*Micrathene whitneyi*)	Unknown	Current survey efforts
Burrowing owl (*Athene cunicularia*)	Declining	Local concern
Northern spotted owl (*Strix o. caurina*)	Threatened	Current survey efforts
Mexican spotted owl (*S.o. lucida*)	Threatened	Current survey efforts
California spotted owl (*S.o. occidentalis*)	Unknown	C2; current survey efforts
Barred owl (*S. varia*)	Stable	Western range expansion
Great gray owl (*S. nebulosa*)	Stable	
Long-eared owl (*Asio otus*)	Stable	Local concern
Short-eared owl (*A. flammeus*)	Stable	Local concern
Boreal owl (*Aegolius funereus*)	Stable	Population estimation difficult
Northern saw-whet owl (*A. acadicus*)	Stable	Concern in southeast AK

*Category 2 (C2)—Proposal to list is possibly appropriate but available data are not conclusive for threatened or endangered status.
Category 3 (C3)—Proven more abundant or widespread than previously believed or not subject to identifiable threat.

Causes of Eagle Deaths

by
J. Christian Franson
Lou Sileo
Nancy J. Thomas
National Biological Service

The U.S. Department of the Interior has investigated the deaths of more than 4,300 bald and golden eagles (*Haliaeetus leucocephalus* and *Aquila chrysaetos*) since the early 1960's as part of an ongoing effort to monitor causes of wildlife mortality. The availability of dead eagles for study depends on finding carcasses in fair to good condition and transporting them to the laboratory. Such opportunistic collection and the fact that recent technological advances have enhanced our diagnostic capabilities, particularly for certain toxins, mean that results reported here do not necessarily reflect actual proportional causes of death for all eagles in the United States throughout the 30-year period. This type of sampling does, however, identify major or frequent causes of death.

Most diagnosed deaths of eagles in our study resulted from accidental trauma, gunshot, electrocution, and poisoning (Fig. 1). Accidental trauma, such as impacts with vehicles, power lines, or other structures, was the most frequent cause of death in both eagle species (23% of bald and 27% of golden). Gunshot killed about 15% of each species. Electrocution was twice as frequent in golden (25%) than in bald eagles (12%), probably because of the preference of golden eagles for prairie habitats and their use of utility poles as perches.

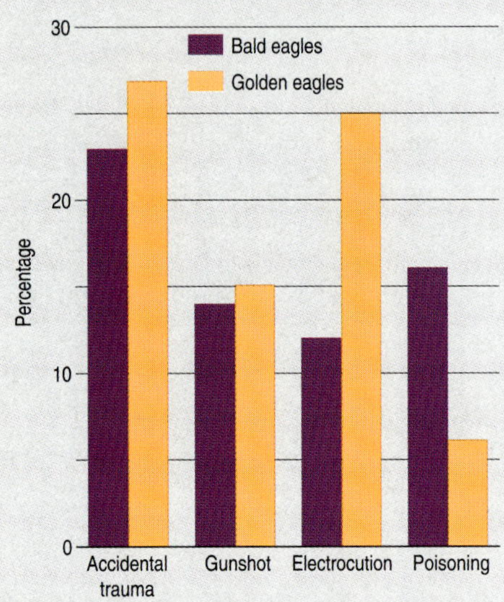

Fig. 1. Causes of mortality of bald and golden eagles over the past 30 years.

Lead poisoning was diagnosed in 338 eagles from 34 states (Fig. 2). Eagles become poisoned by lead after consuming lead shot and, occasionally, bullet fragments present in food items. Agricultural pesticides accounted for most remaining poisonings; organophosphorus and carbamate compounds killed 139 eagles in 25 states (Fig. 3). Eagles are exposed to these chemicals in a variety of ways, often by consuming other animals that died of direct poisoning or from baits placed to deliberately kill wildlife.

Overall, poisonings were more frequent in bald eagles (16%) than golden eagles (6%). The reasons for this are unclear, but may be related to factors that influence submission of carcasses for examination or differences in species' preferences for agricultural, rangeland, and wetland habitats.

Necropsy examination of a bald eagle at the National Wildlife Health Center, Madison, Wisconsin.

For further information:

J. Christian Franson
National Biological Service
National Wildlife Health Center
6006 Schroeder Rd.
Madison, WI 53711

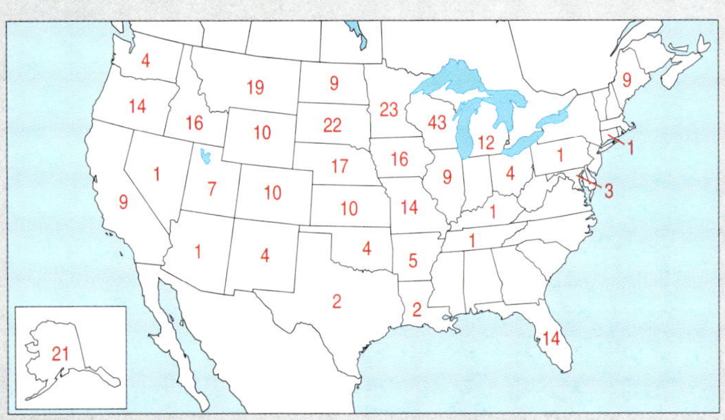

Fig. 2. Nationwide distribution of lead-poisoned eagles.

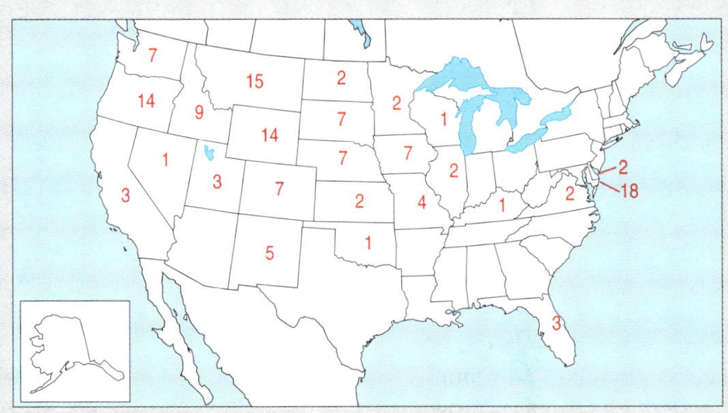

Fig. 3. Nationwide distribution of eagle poisonings caused by organophosphorus and carbamate pesticides.

states). Status determination is complicated by the low density of nesting birds and fluctuations in breeding associated with cycles of prey abundance. It remains in Category 2, i.e., possibly appropriate to propose to list but available data are not conclusive for threatened or endangered status.

Falcons

American peregrine falcon (*Falco peregrinus anatum*) populations declined as a result of contamination by DDT and other organochlorine pesticides. The species was extirpated as a breeding bird in the eastern United States and declared endangered elsewhere. Peregrine recovery has been accomplished in the eastern United States and supplemented in the West (except Alaska) by release of hundreds of peregrines bred in captivity. Now several generations originating from released peregrines have survived and produced young in the wild. In some locales (e.g., parts of California), however, young are still not produced at normal rates. In Alaska nesting numbers of the Arctic subspecies increased naturally, and it was downlisted to threatened status in 1984. Now the Artic peregrine falcon is proposed for removal from the Endangered Species List.

Owls

The distribution of the ferruginous pygmy owl (*Glaucidium brasilianum cactorum*) extends north only into southern Arizona and southern Texas, and concern exists about its status because of the fragmentation and loss of deciduous riparian woodlands and remnant mesquite habitat. The subspecies occurring there, the cactus ferruginous pygmy owl, was elevated from Category 2 as of March 1993 and is being considered for listing as threatened.

The spotted owl (*Strix occidentalis*) is being surveyed extensively and studied because the northern and Mexican subspecies are threatened. In the Pacific Northwest the threat to these owls is loss of old-growth forest, and in the Southwest, general loss of forest habitat. The attention focused on spotted owls has resulted in the only standardized, broad-scale survey of an owl species. Since 1968 the number of known owl nesting areas in Oregon has increased from 27 records (9 sightings, 18 specimens) to about 2,700 separate sites known to be occupied by pairs or single birds sometime within the last 5 years (E. Forsman, U.S. Forest Service, personal communication). This does not reflect an increase in owls; rather, it reflects our ignorance of owl numbers and distribution, largely resulting from lack of survey effort.

Conclusions

Raptors, as top predators, naturally occur at low densities relative to many other organisms. As a group, raptors are poorly surveyed and there are few quantitative data with which to determine their population status and trends. A summary of our assessment of the status and population trends of the 60 species and subspecies of raptors we considered (Table) includes the following: 2 are declining in numbers and 5 are increasing; 16 (27%) are thought to be stable; 19 (32%) are classified as stable, but this assessment is qualified because of local or regional concerns or poor information; the information for 12 (20%) is so poor that we could not determine their status; 7 (12%) of these species or subspecies are endangered or threatened; and 9 (15%) are in Category 2 or 3, reflecting recent concern that they might be endangered or threatened.

We must learn more about the distribution and population dynamics of all our raptor species. With knowledge of their status and trends and information about their distribution and habitat requirements, we can avoid expensive, disruptive, last-resort management of these birds. With knowledge of their ecology, we can conserve biodiversity.

References

Bowerman, W.W. 1993. Regulation of bald eagles (*Haliaeetus leucocephalus*) productivity in the Great Lakes Basin: an ecological and toxological approach. Ph.D. dissertation, Michigan State University, East Lansing. 291 pp.

Johnsgard, P.A. 1988. North American owls: biology and natural history. Smithsonian Institution Press, Washington, DC. 295 pp.

NWF. 1988. Proceedings of the Southwest Raptor Management Symposium and Workshop. Scientific and Tech. Series 11, National Wildlife Federation, Washington, DC. 395 pp.

NWF. 1989a. Proceedings of the Western Raptor Management Symposium and Workshop. Scientific and Tech. Series 12, National Wildlife Federation, Washington, DC. 320 pp.

NWF. 1989b. Proceedings of the Northeast Raptor Management Symposium and Workshop. Scientific and Tech. Series 13, National Wildlife Federation, Washington, DC. 356 pp.

NWF. 1990. Proceedings of the Southeast Raptor Management Symposium and Workshop. Scientific and Tech. Series 14, National Wildlife Federation, Washington, DC. 248 pp.

NWF. 1991. Proceedings of the Midwest Raptor Management Symposium and Workshop. Scientific and Tech. Series 15, National Wildlife Federation, Washington, DC. 290 pp.

Palmer, R.S., ed. 1988. Handbook of North American birds. Vols. 4 and 5. Yale University Press, New Haven, CT. 433 pp. and 462 pp.

White, C.M. 1994. Population trends and current status of selected western raptors. Studies in Avian Biology 15:161-172.

For further information:

Mark R. Fuller
National Biological Service
Raptor Research and Technical Assistance Center
3948 Development Ave.
Boise, ID 83705

Return of Wild Turkeys

by
James G. Dickson
U.S. Forest Service

The wild turkey (*Meleagris gallopavo*) is a large gallinaceous bird characterized by strong feet and legs adapted for walking and scratching, short wings adapted for short rapid flight, a well-developed tail, and a stout beak useful for pecking. These birds probably originated some 2 to 3 million years ago in the Pliocene epoch. Molecular data suggest this genetic line diverged from pheasant-like birds about 11 million years ago. There are two species in the genus, the wild turkey of the United States, portions of southern Canada, and northern Mexico; and the ocellated turkey (*M. ocellata*) in the Yucatan region of southern Mexico, Belize, and northern Guatemala. This article focuses on the return of the wild turkey.

Sources of Information

Historical information on turkeys comes from documented accounts of early explorers, which have been summarized by Mosby and Handley (1943) and Schorger (1966). Recent national population estimates are composite figures obtained from individual state wildlife management agencies. Researchers use many survey techniques including harvest estimates, brood counts, winter flock surveys, and hunter and landowner observations. Kennamer et al. (1992) recently summarized state estimates. At present, there is no consistent, widespread monitoring technique.

Wild turkey (*Meleagris gallopavo*).

Life History

According to most accounts, wild turkeys were quite abundant at the time of European colonization of North America. Wild turkeys became a major food of these settlers as they moved westward across the forested eastern United States. Turkeys were also used for clothing, ornamentation, and food by many Native American tribes. As the nation grew in the 1800's, wild turkey numbers dwindled. The birds were harvested without restraint and marketed for human consumption. In addition, their forest habitat was cleared for agriculture and wood products. In the early 1900's, population numbers continued to decline. By 1920, wild turkeys were extirpated from 18 of the 39 states of their ancestral range (Mosby and Handley 1943).

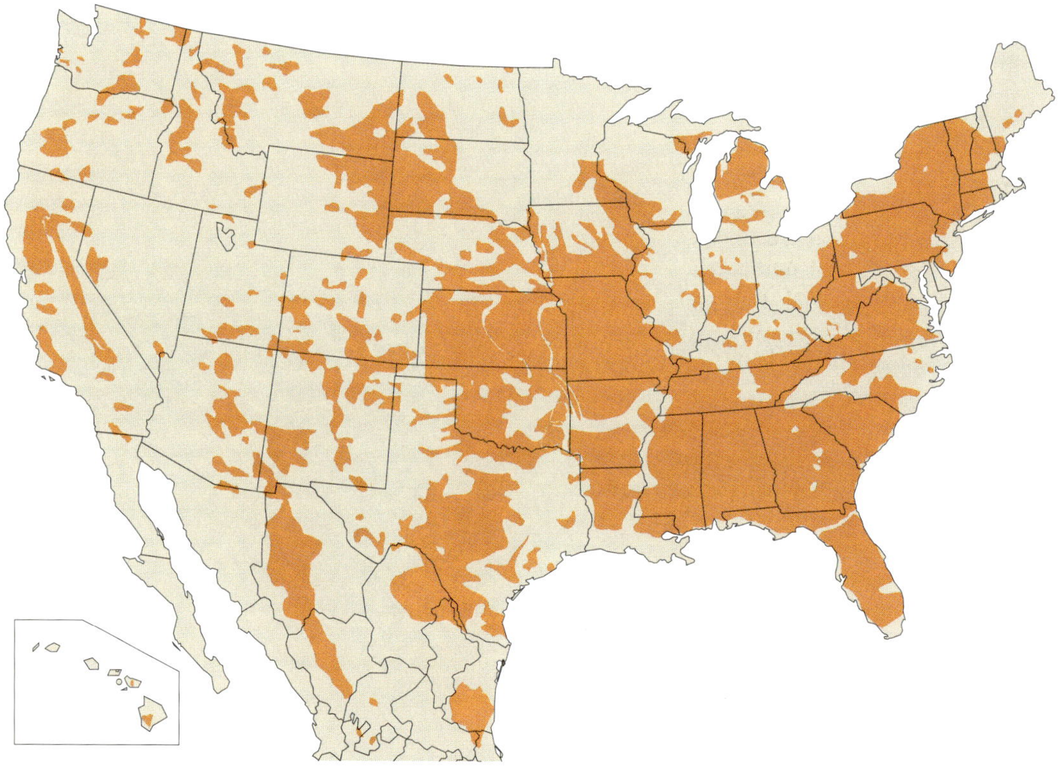

Fig. 1. Distribution of the wild turkey in the United States and Mexico in 1989 (Stangel et al. 1992).

After the early 1900's little change occurred in wild turkey distribution and populations until after World War II when resources were directed to restoring and managing the nation's wildlife populations, including the wild turkey. A technique that many state agencies believed to be promising, but did not work, was artificial propagation of game-farm or pen-raised turkeys. Turkeys raised in captivity were not properly imprinted on (recognition and attachment) wild hens and did not have the experience and survival skills necessary to live and reproduce in the wild.

Restoration through trapping wild turkeys in the wild and relocating them was the proper solution, but this technique was not easily accomplished with the wary bird. Development of the rapidly propelled cannon net, originally designed for capturing waterfowl, was a major factor in relocating large numbers of wild turkeys for restoration. Thousands of wild turkeys were captured or moved with this technique or variations of it; in addition, drop nets and immobilizing drugs were used.

Several other factors contributed to the return of the wild turkey: the maturing of the eastern forests, which had been almost eliminated; increased knowledge from research; spread of sound management practices; and better protection of new flocks vulnerable to poaching.

The restoration of the wild turkey is a great wildlife management success story. In the early part of this century only tens of thousands of wild turkeys were found in a few remote areas. By 1959 the total population approached one-half million (Kennamer et al. 1992), and by 1994 almost all of the forested eastern United States and much of the forested West had been restocked (Fig. 1), with the total population now probably approaching 4 million (Fig. 2). At present, there are viable wild turkey populations with hunting seasons in every state but Alaska, and the annual harvest exceeds one-half million turkeys. The state wildlife management agencies, aided by the National Wild Turkey Federation and supported by sportmen's dollars, undertook a tremendous task and achieved dramatically successful results (Dickson 1992). Turkey hunting continues to be pursued by millions of dedicated hunters.

Future population expansion is expected to be somewhat limited. Most suitable turkey habitat has been stocked, and, generally, populations in these areas have already gone through their high-productivity phase. Population expansion is also limited because appropriate habitat will be lost as the human population expands.

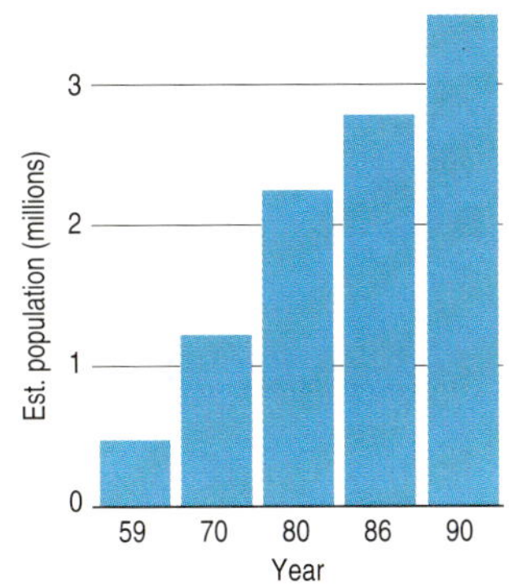

Fig. 2. Estimated U.S. wild turkey population, 1959-90 (from Kennamer et al. 1992).

References

Dickson, J.G., ed. 1992. The wild turkey: biology and management. Stackpole Books, Harrisburg, PA. 463 pp.

Kennamer, J.E., M. Kennamer, and R. Brenneman. 1992. History. Pages 6-17 *in* J.G. Dickson, ed. The wild turkey: biology and management. Stackpole Books, Harrisburg, PA.

Mosby, H.S., and C.O. Handley. 1943. The wild turkey in Virginia: its status, life history and management. Virginia Division of Game and Inland Fisheries, Richmond. P-R Project. 281 pp.

Schorger, A.W. 1966. The wild turkey: its history and domestication. University of Oklahoma Press, Norman. 625 pp.

Stangel, P.W., J.I. Smith, and P.L. Leberg. 1992. Systematics and population genetics. Pages 18-28 *in* J.G. Dickson, ed. The wild turkey: biology and management. Stackpole Books, Harrisburg, PA.

For further information:

James G. Dickson
U.S. Forest Service
Wildlife Habitat Laboratory
PO Box 7600, SFA Station
Nacogdoches, TX 75962

Mourning Doves

by
David D. Dolton
U.S. Fish and Wildlife Service

The mourning dove (*Zenaida macroura*) is one of the most widely distributed and abundant birds in North America (Droege and Sauer 1990). It is also the most important U.S. game bird in terms of numbers harvested. The U.S. fall population of mourning doves has been estimated to be about 475 million (Tomlinson et al. 1988; Tomlinson and Dunks 1993).

The breeding range of the mourning dove extends from the southern portions of the Canadian Provinces throughout the continental United States into Mexico, the islands near Florida and Cuba, and scattered areas in Central America (Aldrich 1993; Fig. 1). Although some mourning doves are nonmigratory, most migrate south to winter in the United States from northern California to Connecticut, south throughout most of Mexico and Central America to western Panama.

Within the United States, three areas contain breeding, migrating, and wintering mourning dove populations that are largely independent of each other (Kiel 1959). In 1960 three areas were established as separate management units: the Eastern (EMU), Central (CMU), and Western (WMU; Fig. 1).

The two main tools used to manage mourning doves are an annual breeding population survey (known as the Mourning Dove Call-count Survey; Dolton 1993a, b) and harvest surveys. The Call-count Survey provides an annual index to population size as well as data for determining long-term trends in dove populations. State harvest surveys and the National Migratory Bird Harvest Information Program, begun in 1992, estimate dove harvest. In addition, recoveries from banded doves have provided vital information for managing the species (Hayne 1975; Dunks et al. 1982; Tomlinson et al. 1988).

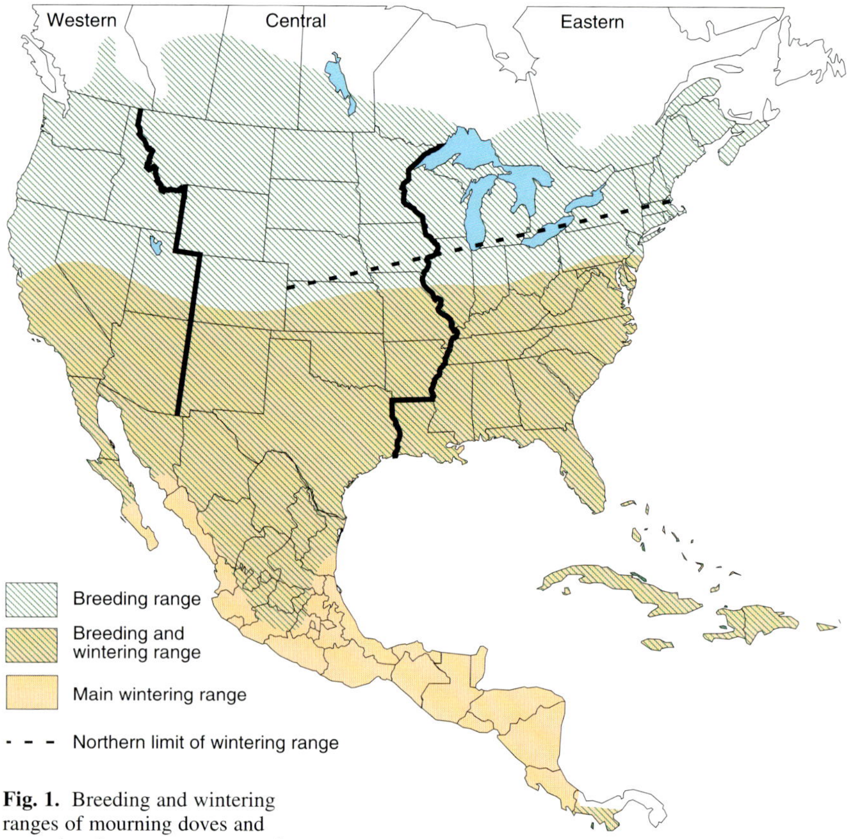

Fig. 1. Breeding and wintering ranges of mourning doves and mourning dove management units in the United States.

Status and Trends

The Eastern Management Unit includes 27 states—30% of the U.S. land area. The 1993 population indices were 18.3 doves heard and 14.9 doves seen per route (Dolton 1993b; Fig. 2). Both estimates are above the long-term trend estimates. Between 1966 and 1993, the population has been relatively stable. Dove harvest in the EMU was relatively constant from 1966 to 1987, with between 27.5 million and 28.5 million birds taken. The latest estimate, a 1989 survey, indicated that the harvest had dropped to about 26.4 million birds shot by an estimated 1.3 million hunters (Sadler 1993).

The Central Management Unit consists of 14 states containing 46% of the U.S. land area. Of the three units, the CMU has the highest mourning dove population index. The 1993 index for the unit of 23.9 doves heard per route is slightly below the long-term trend estimate (Dolton 1993b; Fig. 2). For doves seen, the estimate of 26.8 is also below what was expected. Even though there appears to be an increase in doves seen and a slight decrease in doves heard between 1966 and 1993, in statistical terms there is no significant trend indicated for either count. Although hunting pressure and harvest varied widely among states, dove harvest in the CMU generally increased between 1966 and 1987 to an annual average of about 13.5 million birds. In 1989 almost 11 million doves were taken by about 747,000 hunters (Sadler 1993).

The Western Management Unit comprises seven states and represents 24% of the land area in the United States. The 1993 population indices of 9.3 doves heard and 8.5 doves seen per route are slightly above their long-term trend estimates (Dolton 1993b; Fig. 2). Significant downward trends in numbers of doves heard and seen for the unit occurred between 1966 and 1993. From 1987 to 1993, however, a significant positive trend occurred in the unit although the indices were still below those of the 1960's. After a decline in the dove breeding population, dove harvest in the WMU declined significantly. In the early 1970's, about 7.3 million doves were taken by an estimated 450,000 hunters. By 1989, the harvest had dropped to about 4 million birds shot by about 285,000 hunters (Sadler 1993).

In summary, mourning dove populations in the EMU and CMU are relatively stable. Although the population of doves in the WMU declined from a high in the mid-1960's, it appears that it stabilized during the past 7-10 years. U.S. dove harvest appears to be decreasing. The mourning dove remains an extremely important game bird, however, especially since more doves are harvested than all other migratory game birds combined. A 1991 survey indicated that the mourning dove provided about 9.5 million days of hunting recreation for 1.9 million people (USFWS and U.S. Bureau of Census 1993).

Year-to-year population changes are normal and expected. Although populations are relatively stable in the Eastern and Central Management units, declining long-term trends in the past two decades are cause for concern in the Western Unit and in local areas elsewhere. A combination of factors may have been detrimental to dove populations in some areas: habitat and agricultural changes including loss of nesting habitat through reclamation and industrial and urban development, changes in agricultural practices that may have reduced food sources, and possibly overharvest of doves in local areas. In California, for example, many live oak trees have been cut for wood products resulting in a loss of nesting habitat. Reclamation projects or lowered water tables eliminated thousands of acres of mesquite nesting habitat in Arizona. Since many doves from the WMU winter in Mexico during a 5- to 6-month period each year, agricultural changes there may negatively affect doves.

In the CMU, agricultural changes were evaluated and compared with dove population trends in the eastern group of states (R.R. George, Texas Parks and Wildlife Department, unpublished data); mourning dove population indices appeared to be most closely correlated with changes in number of farms (positive) or farm size (negative). In addition, an analysis identified number of farms and acres of soybeans, oats, and sorghum over time as good indicators of the number of doves heard.

Early records indicate that mourning doves were present, although not abundant, when the United States was settled by colonists (Reeves and McCabe 1993). The resulting clearing of forests, introduction of new food plants, grazing and trampling by livestock that promoted seed-producing plants used by doves, and the creation of stock ponds providing more widely distributed drinking water in the arid West all benefited the mourning dove so that they are probably more numerous now than in colonial times.

These birds are quite adaptable and readily nest and feed in urban and rural areas. The mourning dove has recently even expanded its range northward.

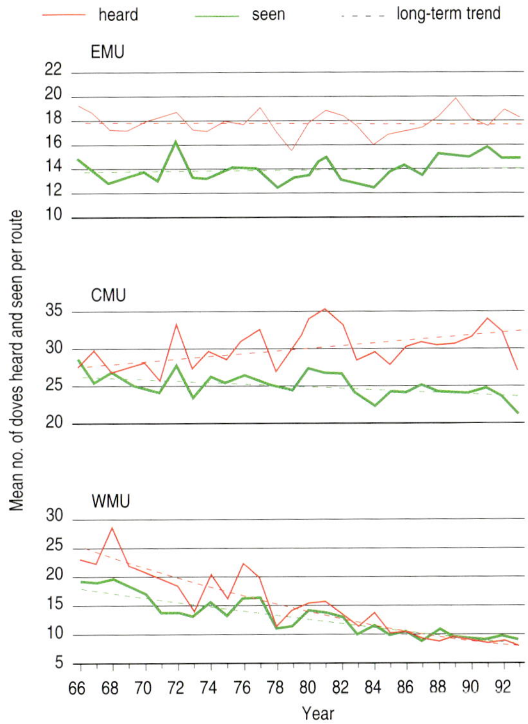

Fig. 2. Population indices of mourning doves in the Eastern (EMU), Central (CMU), and Western (WMU) Management units, 1966-93.

References

Aldrich, J.W. 1993. Classification and distribution. Pages 47-54 *in* T.S. Baskett, M.W. Sayre, R.E. Tomlinson, and R.E. Mirarchi, eds. Ecology and management of the mourning dove. Stackpole Books, Harrisburg, PA.

Dolton, D.D. 1993a. The Call-count Survey: historic development and current procedures. Pages 233-252 *in* T.S. Baskett, M.W. Sayre, R.E. Tomlinson, and R.E. Mirarchi, eds. Ecology and management of the mourning dove. Stackpole Books, Harrisburg, PA.

Dolton, D.D. 1993b. Mourning dove breeding population status, 1993. U.S. Fish and Wildlife Service, Laurel, MD. 16 pp.

Droege, S., and J.R. Sauer. 1990. North American Breeding Bird Survey annual summary 1989. U.S. Fish and Wildlife Service Biological Rep. 90(8). 16 pp.

Dunks, J.H., R.E. Tomlinson, H.M. Reeves, D.D. Dolton, C.E. Braun, and T.P. Zapatka. 1982. Migration, harvest, and population dynamics of mourning doves banded in the Central Management Unit, 1967-77. U.S. Fish and Wildlife Service Special Sci. Rep.—Wildlife 249. 128 pp.

Hayne, D.W. 1975. Experimental increase of mourning dove bag limit in Eastern Management Unit, 1965-72. Southeastern Association of the Game and Fish Commissioners Tech. Bull. 2. 56 pp.

Kiel, W. H., Jr. 1959. Mourning dove management units—a progress report. U.S. Fish and Wildlife Service Special Sci. Rep.—Wildlife 42. 24 pp.

Reeves, H.M., and R.E. McCabe. 1993. Historical perspective. Pages 7-46 *in* T.S. Baskett, M.W. Sayre, R.E. Tomlinson, and R.E. Mirarchi, eds. Ecology and management of the mourning dove. Stackpole Books, Harrisburg, PA.

Sadler, K.C. 1993. Mourning dove harvest. Pages 449-458 *in* T.S. Baskett, M.W. Sayre, R.E. Tomlinson, and R.E. Mirarchi, eds. Ecology and management of the mourning dove. Stackpole Books, Harrisburg, PA.

Tomlinson, R.E., D.D. Dolton, H.M. Reeves, J.D. Nichols, and L.A. McKibben. 1988. Migration, harvest, and population characteristics of mourning doves banded in the Western Management Unit, 1964-77. U.S. Fish and Wildlife Service Tech. Rep. 13. 101 pp.

Tomlinson, R.E., and J.H. Dunks. 1993. Population characteristics and trends in the Central Management Unit. Pages 305-340 *in* T.S. Baskett, M.W. Sayre, R.E. Tomlinson, and R.E. Mirarchi, eds. Ecology and management of the mourning dove. Stackpole Books, Harrisburg, PA.

USFWS and U.S. Bureau of the Census. 1993. 1991 National survey of fishing, hunting, and wildlife-associated recreation. U.S. Government Printing Office, Washington, DC. 124 pp.

For further information:

David D. Dolton
U.S. Fish and Wildlife Service
Office of Migratory Bird Management
11500 American Holly Dr.
Laurel, MD 20708

Common Ravens in the Southwestern United States, 1968-92

by
William I. Boarman
Kristin H. Berry
National Biological Service

The common raven (*Corvus corax*) is a large black passerine bird found throughout the northern hemisphere including western and northern North America. Ravens are scavengers that frequently feed on road-killed animals, large dead mammals, and human refuse. They kill and eat prey including rodents, lambs (Larsen and Dietrich 1970), birds, frogs, scorpions, beetles, lizards, and snakes. They also feed on nuts, grains, fruits, and other plant matter (Knight and Call 1980; Heinrich 1989). Their recent population increase is of concern because ravens eat agricultural crops and animals whose populations may be depleted.

Ravens are closely associated with human activities, frequently visiting solid-waste landfills and garbage containers at parks and food establishments, being pests of agricultural crops, and nesting on many human-made structures. In two recent surveys in the deserts of California (FaunaWest Wildlife Consultants 1989; Knight and Kawashima 1993), ravens were more numerous in areas with more human influences, and were often indicators of the degree to which humans affect an area.

Annual Breeding Bird Surveys (BBS) conducted nationwide by the U.S. Fish and Wildlife Service (USFWS) indicated that raven populations in several parts of the country significantly increased during 1965-79 (Robbins et al. 1986). This increase concerns resource managers because ravens feed on agricultural crops and animal species of interest to humans. For instance, in the deserts of the southwestern United States, ravens prey on young desert tortoises (*Gopherus agassizii*; Berry 1985; Fig. 1), which in the Mojave and Colorado deserts are listed as a threatened species by the USFWS (Federal Register 1990). Because of high levels of raven predation on tortoises, the Bureau of Land Management has taken action to reduce this predation (BLM 1990, 1994). We report here on a 24-year trend in raven abundance along roadsides in the deserts of the southwestern United States and surrounding regions, where increasing raven populations interest resource management agencies (BLM 1990; USFWS 1994).

Fig. 1. Juvenile desert tortoise shell found beneath an active raven nest. The hole in the shell was probably pecked open by a raven to eat the organs.

Our analysis of BBS 1968-92 data focuses on arid lands and neighboring habitats in California, Nevada, Utah, and Arizona. We used data from 137 39.2-km (24.5-mi) routes within the following BBS strata: Great Basin Desert; mountain highlands of Arizona; Sonoran-Colorado Desert; Mojave Desert; basins and ranges, including portions of the northern Mojave and Great Basin deserts; Central Valley; and southern California grasslands, California foothills (southern California routes only), and Los Angeles ranges combined into one (coastal southern California).

Status and Trends

Between 1968 and 1992, the latest year for which data were available, raven populations increased significantly ($P \leq 0.01$) throughout the study area (Fig. 2), in spite of relatively high variances among routes. Raven sightings increased 76-fold in the Central Valley of California, 14-fold in the Sonoran-Colorado Desert, and 10-fold in the Mojave Desert over the 24-year period. Statistically significant but lower increases in raven populations were experienced in the heavily urbanized coastal southern California strata. The results for the mountain highlands stratum are questionable because of a low number of routes ($n = 7$; B. Peterjohn, NBS, personal communication).

In three studies, raven numbers were highest along powerlines, intermediate along highways, and lowest in open desert areas (Austin 1971; FaunaWest Wildlife Consultants 1989; Knight and Kawashima 1993). These reports and observations of raven use of human-based resources for food, water, and nesting substrate (Knight and Call 1980; FaunaWest Wildlife Consultants 1989; Heinrich 1989) suggest that high raven populations are a result of human subsidies (Boarman 1993).

Increased raven populations may be a concern for threatened and endangered species if increased numbers of ravens result in greater predation. In California alone, there are 96 threatened or endangered species, some of which are or may be at risk of increased raven predation if raven populations continue to grow. On San Clemente Island, ravens are a predator of the endangered San Clemente Island loggerhead shrike (*Lanius ludovicianus mearnsi*), and along coastal California they prey on endangered populations of the California least tern (*Sterna antillarum browni*; Belluomini 1991). The carcasses of 11 chuckwallas (*Sauromalus obesus*), a candidate species for listing as threatened or endangered by the USFWS, were recently found beneath one raven nest (personal observation). This finding may be a rare occurrence, but if raven populations continue to increase, more ravens may begin to prey on chuckwallas. We are conducting more research to understand the foraging ecology and population biology of ravens and their effects on their prey populations. This research will help us determine how much of a threat ravens pose to the region's biodiversity and learn how to reduce these effects.

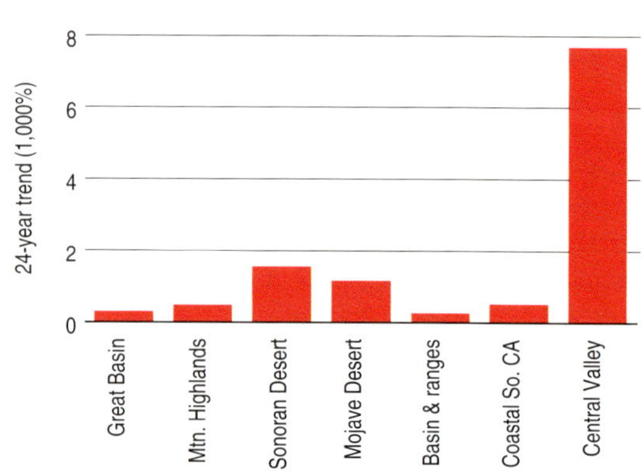

Fig. 2. A 24-year trend in the average (mean) number of raven sightings within each stratum studied.

References

Austin, G.T. 1971. Roadside distribution of the common raven in the Mohave Desert. California Birds 2:98.

Belluomini, L.A. 1991. The status of the California least tern at Camp Pendleton, California during the breeding season of 1990. Natural Resources Management Branch, Southwest Division Naval Facilities Engineering Command, San Diego, CA. 77 pp.

Berry, K.H. 1985. Avian predation on the desert tortoise in California. Report to Southern California Edison, Co. Bureau of Land Management, Riverside, CA. 20 pp.

BLM. 1990. Draft raven management plan for the California Desert Conservation Area. Bureau of Land Management, Riverside, CA. 59 pp.

BLM. 1994. Environmental assessment for 1994 experimental program to shoot ravens. Bureau of Land Management, Riverside, CA. 8 pp.

Boarman, W.I. 1993. When a native predator becomes a pest: a case study. Pages 191-206 *in* S.K. Majumdar, E.W. Miller, D.E. Baker, E.K. Brown, J.R. Pratt, and R.F. Schmalz, eds. Conservation and resource management. Pennsylvania Academy of Sciences, Easton.

FaunaWest Wildlife Consultants. 1989. Relative abundance and distribution of the common raven in the deserts of southern California and Nevada during spring and summer of 1989. Bureau of Land Management, Riverside, CA. 60 pp.

Federal Register. 1990. Endangered and threatened wildlife and plants; determination of threatened status for the Mojave population of the desert tortoise. Federal Register 55:12178-12191.

Heinrich, B. 1989. Ravens in winter. Summit Books, New York. 379 pp.

Knight, R., and M. Call. 1980. The common raven. Bureau of Land Management Tech. Note 344. 61 pp.

Knight, R., and J. Kawashima. 1993. Responses of raven and red-tailed hawk populations to linear right-of-ways. Journal of Wildlife Management 57:266-271.

Larsen, K.H., and J.H. Dietrich. 1970. Reduction of a raven population on lambing grounds with DRC-1339. Journal of Wildlife Management 34:200-204.

Robbins, C.S., D. Bystrak, and P.H. Geissler. 1986. The Breeding Bird Survey: its first fifteen years, 1965-1979. U.S. Fish and Wildlife Service Resour. Publ. 157. 196 pp.

USFWS. 1994. Desert tortoise (Mojave population) recovery plan. U.S. Fish and Wildlife Service, Portland, OR. 73 pp.

Common raven (*Corvus corax*).

For further information:

William I. Boarman
National Biological Service
Desert Tortoise Research Project
6221 Box Springs Blvd.
Riverside, CA 92507

Mississippi Sandhill Cranes

by

George F. Gee
National Biological Service

Scott G. Hereford
U.S. Fish and Wildlife Service

Resident sandhill cranes formed a continuous population in Georgia and Florida and widely separated populations along the Gulf Coastal Plain of Texas, Louisiana, Mississippi, and Alabama (Figure). The Mississippi sandhill crane (*Grus canadensis pulla*) was one of the widely separated populations on the Coastal Plain that bred in pine savannas in southeastern Mississippi, just east of the Pascagoula River to areas just west of the Jackson County line, south to Simmons Bayou, and north to an east-west line 8-16 km (5-10 mi) north of VanCleave.

Agricultural and industrial development including World War II ship building, fire suppression, and forestry practices destroyed much of the sandhill crane's habitat in Jackson County, Mississippi. The U.S. Fish and Wildlife Service (USFWS) added the Mississippi sandhill crane to the endangered species list in 1973 and established the Mississippi Sandhill Crane National Wildlife Refuge in 1974. The USFWS began captive breeding at the Patuxent Wildlife Research Center (PWRC) in 1965 to protect the subspecies during habitat restoration and to provide stock for reintroduction.

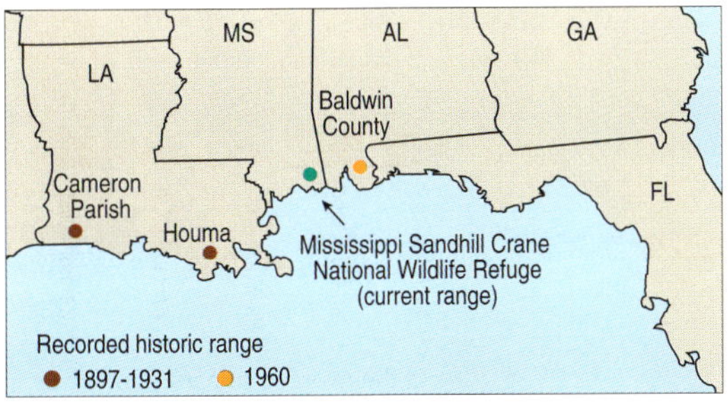

Figure. Range of Mississippi sandhill cranes.

Morphological, physiological, and genetic differences exist among crane subspecies (Aldrich 1972). Mississippi birds mature earlier and begin egg production about 6 weeks later than Florida sandhill cranes. Genetic studies (Dessauer et al. 1992; Jarvi et al. 1994) show a level of heterozygosity (*see* glossary) in the wild Mississippi population about half that in other sandhill cranes. As in other small populations, cranes seem to have genetic weaknesses. In the captive population, for example, 17% of all birds die from detectable heart murmurs and when released to the wild, 36% with heart murmur and 83% without heart murmurs survive for 1 year after release.

Status and Trends

Population Decline

In the 1800's the species was abundant enough for farmers to consider it a pest. Although population studies only started recently, it appears the population has been small for most of this century (Table 1).

Table 1. Estimated numbers of Mississippi sandhill cranes on the Mississippi Sandhill Crane National Wildlife Refuge, 1929-93.

Year	Wild	Captive released	Total
1929			50 -100
1949			50+
1969			50-60
1975			30-50
1978			40-50
1979			40-50
1980			50
1981	41	9	50
1982	41	9	50
1983	34	9	43
1984	27	13	40
1985	13	19	32
1986	23	18	41
1987	17	16	33
1988	21	23	44
1989	21	33	54
1990	24	49	73
1991	19	73	92
1992	20	88	108
1993	20	115	135

Until the 1940's, the human population in Jackson County was small, and the remnant population of Mississippi sandhill cranes remained stable. The suitable pine savanna habitat shrunk from over 40,500 ha (100,000 acres) in 1940 to 10,530 ha (26,000 acres) in the 1960's, which were designated as critical habitat by the USFWS. The USFWS requested a population study in 1960 when Mississippi proposed building Interstate Highway 10 through the last of the crane habitat. The Nature Conservancy, the U.S. Department of Transportation, and State of Mississippi donated land to the refuge.

Recent Reintroductions

The first releases of hand-reared birds failed. Thus, releases of Mississippi sandhills on the refuge during the 1980's were birds raised by their parents or surrogate parents. These parent-reared birds proved wilder than the hand-reared birds and adapted well to the pine savanna. Unfortunately, the parent-rearing technique reduced production and increased expenses.

The PWRC developed a new hand-rearing technique that visually isolated chicks from humans and imprinted them on adult sandhill cranes in the chick-rearing area. Caretakers dressed in sheets to hide their human form when handling birds, and encounters with cranes were limited. Juveniles were placed in socialization pens in the fall to form three cohorts (parent-reared, hand-reared, and a mixed group). A gentle release on the refuge allowed the birds to leave the release pen when ready and to return for food for a period after release. Surprisingly, a greater percentage of hand-reared birds has survived than the parent-reared birds, although both groups have paired and produced fertile eggs. The releases increased the refuge population from 44 in 1988 to 135 in 1993 (Table 1).

Status in Jackson County, Mississippi

The population decline of the Mississippi sandhill crane reflects the loss of the mesic and hydric pine savanna once abundant in the area. Savannas occur on coastal terraces, elevated ridges, and uplands. Fire frequency and intensity, combined with soil type and hydrology, provide successional regulation of the savanna. Woody, forested communities replace the savanna without fire. Before ditching, the flat topography of the terraces allowed sheet flow of water across the terraces and supported extensive areas of open savanna. When the refuge was established, about 75% of the crane savannas had been destroyed (by residential or commercial development) or changed to one of several different forest types. Only 5% of the original savanna type that supported the cranes remains on the Gulf Coastal Plain. For this reason, Mississippi sandhill cranes now occur only on the refuge and adjacent private lands in southeastern Mississippi.

The Mississippi sandhill crane population nests only on the 7,813-ha (19,300-acre) refuge. The only other large tract of remnant savanna that might be suitable nesting habitat exists southeast of the refuge on the proposed Grand Bay National Wildlife Refuge. Savanna used by the Mississippi sandhill crane exists as highly fragmented remnants that the refuge must manage to provide nesting, foraging, and roosting sites (Table 2).

Mortality and natural recruitment may also restrict population viability. Predation (primarily mammalian) causes high mortality during the first year of life. Other factors that may limit populations include tumors, contaminants, microbial pathogens, and parasites. The prevalence of tumors in the wild Mississippi sandhill crane population far exceeds that expected in other birds and mammals.

Table 2. Mississippi sandhill crane nesting sites on refuge, by habitat.

Type of habitat	Number	Percentage
Open savanna	82	49
Swamp edges	62	38
Pine plantations	12	7
Forest edges	8	5
Cleared lands	2	1

Research Needs

Research needs include assessing the effects of prescribed burns and other mechanical techniques on habitat restoration and crane use; assessing the effects of water levels, water-level fluctuations, and hydrology on crane nesting and fledging success; determining the level of propagation and captive release conditioning needed to maintain population size during restoration; developing genetic management to protect the gene pool; and determining disease and contaminant sources for tumors and poor reproductive success in captive and wild flocks.

References

Aldrich, J. 1972. A new subspecies of sandhill cranes from Mississippi. Proceedings of the Biological Society of Washington 85(5):63-70.

Dessauer, H.C., G.F. Gee, and J.S. Rogers. 1992. Allozyme evidence for crane systematics and polymorphisms within populations of sandhill, sarus, Siberian, and whooping cranes. Molecular Phylogenetics and Evolution 1(4):279-288.

Jarvi, S.I., G.F. Gee, M.M. Miller, and W.E. Briles. 1994. Detection of haplotypes of the major histocompatibility complex in Florida sandhill cranes. Journal of Heredity. In press.

For further information:

George F. Gee
National Biological Service
Patuxent Environmental Science Center
Laurel, MD 20708

Piping Plovers

by
Susan Haig
National Biological Service
Jonathan H. Plissner
University of Georgia

Editor's note: This paper is largely a synopsis of a paper by Haig and Plissner (1993) in *Condor*.

The piping plover (*Charadrius melodus*) is a wide-ranging, beach-nesting shorebird whose population viability continues to decline as a result of habitat loss from development and other human disturbance (Haig 1992). In 1985 the species was listed as endangered in the Great Lakes Basin and Canada and threatened in the northern Great Plains and along the U.S. Atlantic coast. The U.S. Fish and Wildlife Service (USFWS) is proposing that birds in the northern Great Plains also be listed as endangered.

Each year, many breeding areas are censused and some winter surveys are conducted. In 1991 biologists from Canada, the United States, Mexico, and various Caribbean nations carried out a simultaneous census of piping plovers at all known breeding and wintering sites. Census goals were to establish baseline population levels for all known piping plover sites and to census additional potential breeding and wintering sites (Figure).

Status

This census covered 2,099 sites, resulting in the highest number of breeding and wintering piping plovers ever recorded. It will be repeated three or four more times over the next 15-20 years for more accurate assessment of population trends.

Winter Census

The total number of wintering birds (3,451) reported constituted 63% of the breeding birds (5,486) counted (Tables 1, 2). Most birds (55%; $N = 1{,}898$) were found along the Texas coast where the census concentrated on birds in previously uncensused stretches of Laguna Madre's back bays. The highest concentration of birds in local sites was also reported in Texas (Haig and Plissner 1993). Although the 1991 census discovered more wintering birds than had been previously reported, a large proportion of piping plovers were not seen in the winter census.

Better census efforts in Louisiana, northern Cuba, and on many of the smaller Caribbean islands may reveal additional winter sites. Previous reviews of their distribution did not indicate that birds moved farther south than the Caribbean (Haig and Oring 1985). Relatively few birds are seen on the Atlantic coast in winter, a contrast to the 36% of plovers that breed along the Atlantic coast. Thus, the largest gap in our understanding of piping plover distribution during winter appears to be in locating winter sites for Atlantic coast breeders.

Breeding Census

All known piping plover breeding sites were censused in 1991 (Table 2). Piping plovers were widely distributed in small populations across their breeding range (Figure); most adults (63.2%) bred in the northern Great Plains and prairies of the United States and Canada. Thirty-six percent were found on the Atlantic coast and

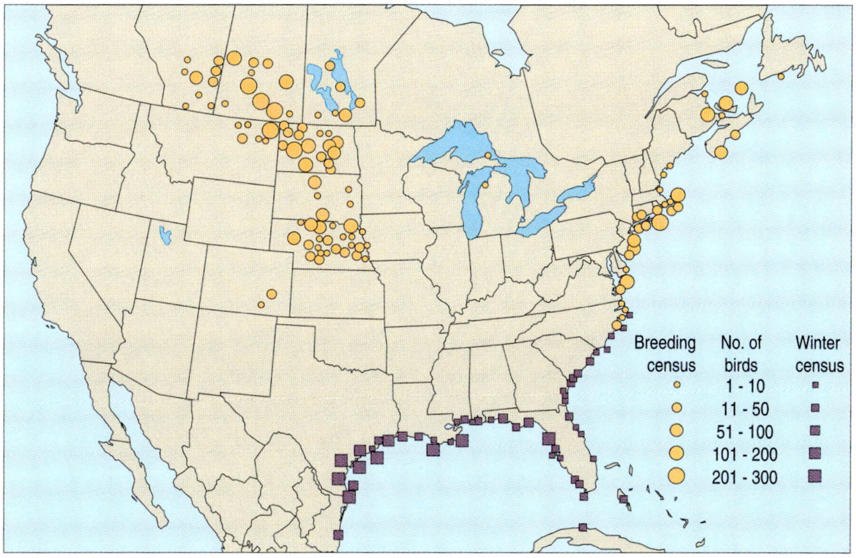

Figure. Distribution of piping plovers throughout the annual cycle in 1991.

Table 1. Numbers of wintering piping plovers and sites where birds occurred in 1991.

Location	Birds	Sites
U.S. Atlantic		
North Carolina	20	7
South Carolina	51	8
Georgia	37	6
Florida	70	9
Total	178	30
U.S. Gulf		
Florida	481	31
Alabama	12	1
Mississippi	59	7
Louisiana	750	23
Texas	1,904	64
Total	3,206	126
Mexico Gulf	27	4
Caribbean		
Bahamas	29	1
Turks and Caicos	0	0
Cuba	11	1
Jamaica	0	0
Puerto Rico	0	0
Cayman Islands	0	0
Total	40	2
Combined total	3,451	162

less than 1% occurred on the Great Lakes. Sites with the highest concentrations of breeding birds also were found in the northern Great Plains (also known in Canada as the Great Prairie); however, each local population consisted of only a small (less than 8%) proportion of the total breeding population. Local populations were even smaller on the Atlantic coast.

Migration Areas

Atlantic coast piping plovers are commonly seen on east coast beaches during spring and fall migration. Migration routes of inland birds are poorly understood, however. Only a few occurrences of piping plovers have been reported at seemingly appropriate inland migration sites such as Kirwin National Wildlife Refuge in Kansas, Cheyenne Bottoms Wildlife Management Area in Kansas, and Great Salt Plains National Wildlife Refuge in Oklahoma. It appears that inland birds may fly nonstop to gulf coast sites.

Trends

Because simultaneous, species-wide censuses were not conducted in the past, assessing population trends is difficult. Examination of long-term census data at specific sites is useful in some cases. Most midcontinent sites that have been monitored for 10 years or more have experienced a decline (Table 3). The cumulative effects of problems in the prairies have been modeled, and results indicate that piping plovers in the Great Plains are now declining by 7% annually (Ryan et al. 1993), a devastating trend for the species. Atlantic coast numbers remain stable; however, there has been unprecedented effort to protect piping plovers along the U.S. Atlantic coast. Results from previous censuses (Table 3) should be considered rough population estimates; as is true with many bird species, we have little information regarding the intensity of census efforts in those population estimates.

Threats

In the northern Great Plains, water-level regulation policies on the major rivers (e.g., Platte, Missouri) serve as a direct source of chick mortality and an indirect source of habitat loss through vegetation encroachment and flooding (Schwalbach 1988; Sidle et al. 1992). We know that because 20% of northern Great Plains (Great Prairie) birds use river sites, loss of productivity on rivers such as the Missouri can

Table 2. Piping plover breeding census, 1991.

Location	Adults	Sites where piping plovers occurred
Atlantic Coast		
Canada		
New Brunswick	203	24
Newfoundland	7	1
Nova Scotia	113	34
Prince Edward Island	110	20
Quebec	76	11
St. Pierre/Miquelon	4	2
Canada Atlantic total	513	92
U.S.		
Maine	38	8
Massachusetts	293	50
Rhode Island	47	7
Connecticut	67	7
New York	338	69
New Jersey	280	22
Delaware	10	3
Maryland	35	1
Virginia	270	14
North Carolina	86	14
South Carolina	2	1
U.S. Atlantic total	1,466	196
Atlantic total	1,979	288
Great Lakes		
Duluth, MN	0	0
Wisconsin	1	1
Michigan	39	14
Long Point, Ontario	0	0
Great Lakes total	40	15
Northern Great Plains/Prairie		
Canada Prairie		
Alberta	180	27
Saskatchewan	1,172	71
Manitoba	80	12
Lake of Woods, Ontario	5	1
Canada Prairie total	1,437	111
U.S. Great Plains		
Montana	308	39
North Dakota	992	115
South Dakota	293	47
Lake of Woods, MN	13	1
Colorado	13	4
Nebraska	398	106
Iowa	13	2
Kansas	0	0
Oklahoma	0	0
U.S. Great Plains total	2,030	314
Combined totals		
Canada	1,950	203
United States	3,536	525
Total	5,486	728

significantly affect annual productivity for the species. A similar threat to piping plovers occurs on Lake Diefenbaker in Saskatchewan, the largest piping plover breeding site in the world, where each year water levels are raised soon after parents have laid their clutches, resulting in a loss of all nests.

Avian and mammalian predation is a problem throughout the species' breeding range, although population numbers appear to be stabilizing on the Atlantic coast and the Great Lakes as a result of using predator exclosures over nests (Rimmer and Deblinger 1990; Mayer and Ryan 1991; Melvin et al. 1992). Human disturbance continues to be a problem on the Atlantic coast (Strauss 1990), and in the Great Lakes, piping plovers may also be suffering from a lack of viable habitat (Nordstrom 1990). Comparison of food availability at northern Great Plains sites with Great Lakes sites indicated lower diversity and abundance of invertebrates on the Great Lakes. Finally, recent evidence suggests that Great Lakes birds may be suffering from high levels of toxins (i.e., PCB's), which may be a prime factor in low productivity and population growth (USFWS, East Lansing, Michigan, personal communication).

The discovery of the high proportion of wintering piping plovers on algal and sand flats has significant implications for future habitat protection. Current development of these areas on Laguna Madre in Texas and Mexico, increased dredging operations, and the continuous threat of oil spills in the Gulf of Mexico will result in serious loss of piping plover wintering habitat.

Courtesy S. Haig

Piping plover (*Charadrius melodus*).

In summary, piping plovers suffer from many factors that may cause their extinction in the next 50 years. Most devastated are the Great Lakes and northern Great Plains birds whose viability is severely threatened. Unfortunately, recovery is hindered by a lack of knowledge about the winter distribution, status of winter sites, adequate water-management policy in western breeding sites, and direct human disturbance on the Atlantic coast.

Table 3. Changes in numbers of piping plovers at specific breeding areas.*

Location	1st est.		2nd est.		1991 census	% Change 1st est. 1991	% Change 2nd est. 1991
	Year	No.	Year	No.			
Atlantic Coast							
Newfoundland	1968	30	1984	4	7	-72	+75
Cadden Beach, Nova Scotia	1976	56	1983	28	20	-64	-29
Maine	1976	48	1982	12	38	-21	+217
Rhode Island	1945	80	1983	20	47	-41	+135
Connecticut	1980	40	1983	34	67	+68	+97
Long Island, NY	1939	1,000	1983	200	338	-66	+69
New Jersey	1980	118	1983	64	280	+137	+338
Delaware	1978	80	1984	18	10	-88	-44
Maryland	1972	85	1984	25	35	-59	+40
Great Lakes							
Michigan	1979	77	1982	14	39	-49	+179
Wisconsin	1900	140	1983	6	1	-99	-83
Northern Great Plains/Prairie							
Big Quill Lake, Saskatchewan	1978	210	1984	186	151	-28	-19
Chain Lakes, Alberta	1976	50	n.a.	n.a.	9	-72	n.a.
Lake Manitoba, Manitoba	1980	27	1984	9	3	-89	-67
Lake of the Woods, MN	1982	44	1986	32	13	-70	-59
Niobrara River, NE 1978	1981	92	1985	100	110	+20	+10

*Sources are listed in Haig and Oring (1985) and Haig and Plissner (1993).

References

Haig, S.M. 1992. The piping plover. Pages 1-18 *in* A. Poole, P. Stettenheim, and F. Gill, ed. Birds of North America. American Ornithologists' Union, Washington, DC.

Haig, S.M., and L.W. Oring. 1985. Distribution and status of the piping plover throughout the annual cycle. Journal of Field Ornithology 56:334-345.

Haig, S.M., and J.H. Plissner. 1993. Distribution and abundance of piping plovers: results and implications of the 1991 International Census. Condor 95:145-156.

Mayer, P.M., and M.R. Ryan. 1991. Electric fences reduce mammalian predation on piping plover nests and chicks. Wildlife Society Bull. 19:59-62.

Melvin, S.M., L.H. MacIvor, and C.R. Griffin. 1992. Predator exclosures: a technique to reduce predation at piping plover nests. Wildlife Society Bull. 20:143-148.

Nordstrom, L.H. 1990. Assessment of habitat suitability for reestablishment of piping plovers on the Great Lakes National Seashores. M.S. thesis, University of Missouri, Columbia. 36 pp.

Rimmer, D.W., and R.D. Deblinger. 1990. Use of predator exclosures to protect piping plover nests. Journal of Field Ornithology 61:217-223.

Ryan, M.R., B.G. Root, and P.M. Mayer. 1993. Status of the piping plover in the Great Plains of North America: a demographic simulation model. Conservation Biology 7:581-591.

Schwalbach, M.J. 1988. Conservation of least terns and piping plovers along the Missouri River and its major tributaries in South Dakota. M.S. thesis, South Dakota State University, Brookings. 43 pp.

Sidle, J.G., D.E. Carlson, E.M. Kirsch, and J.J. Dinan. 1992. Flooding: mortality and habitat renewal for least terns and piping plovers. Colonial Waterbirds 15:132-136.

Strauss, E. 1990. Reproductive success, life history patterns, and behavioral variation in a population of piping plovers subjected to human disturbance. Ph.D. dissertation, Tufts University, Boston, MA. 123 pp.

For further information:

Susan M. Haig
National Biological Service
Forest and Range Ecosystem Science Center
Oregon State University
3200 SW Jefferson Way
Corvallis, OR 97331

California Condors

by
Oliver H. Pattee
National Biological Service
Robert Mesta
U.S. Fish and Wildlife Service

The California condor (*Gymnogyps californianus*) is a member of the vulture family. With a wingspan of about 3 m (9 ft) and weighing about 9 kg (20 lb), it spends much of its time in soaring flight visually seeking dead animals as food. The California condor has always been rare (Wilbur 1978; Pattee and Wilbur 1989). Although probably numbering in the thousands during the Pleistocene epoch in North America, its numbers likely declined dramatically with the extinction of most of North America's large mammals 10,000 years ago. Condors probably numbered in the hundreds and were nesting residents in British Columbia, Washington, Oregon, California, and Baja California around 1800. In 1939 the condor population was estimated at 60-100 birds, and its home range was reduced to the mountains and foothills of California, south of San Francisco and north of Los Angeles.

Conservation to halt the condor's decline included establishing the Sisquoc (1937) and Sespe (1947) condor sanctuaries within the Los Padres National Forest, obtaining fully protected status under California Fish and Game Code (1953), placement on California's first state endangered species list (1971), and, finally, being listed by the federal government under the Endangered Species Act of 1973 (Wilbur 1978). The success of these efforts could not be judged, however, because verifiable status and trends data did not become available until 1982. By using these data, we confirmed the decline in condor numbers over the past 50 years was even greater than thought.

Population estimates before 1939 were based entirely on guesswork and interpretation of the fossil record, historical accounts, museum collections, or anecdotal observations by early naturalists and scholars. We believed there were fewer condors because they were no longer seen in many areas where they were once commonly observed. The condor's plight generated widespread interest among conservationists to know the actual population size and its rate of decline.

Koford (1953) conducted the first major life-history study of the California condor and provided the first documented enumeration of the species. His count was based on numbers seen in the largest single flocks with an unspecified adjustment for condors not seen. Another estimate in 1965 (Miller et al. 1965) compared flock sizes seen in the late 1950's and early 1960's with those reported by Koford.

A yearly survey was begun by volunteers in 1965 and continued through 1981 (except for 1979). This survey used multiple observers at strategic sites who counted all condors seen for a 2-day period in October (Mallette and Borneman 1966; Wilbur 1980). The yearly population estimates of this October survey were quite different from year to year and failed to provide any statistical measures of variability, although results did show a gradual downward trend in condor numbers.

California condor (*Gymnogyps californianus*).

The annual October survey was replaced in 1982 by a counting method (Snyder and Johnson 1985) using photographs of soaring condors to recognize differences in feather patterns. This method allowed individuals to be identified and counted. Although an improvement over previous techniques, this method is time consuming and only works when there are few animals. The photographic census was discontinued after 1985 because all condors had been marked with uniquely colored and numbered tags and radio transmitters.

Trends

Data used to determine the population size of California condors before 1982 (Figure) were biased for many reasons. Foremost was the fact that no surveyors could explain how they used the number of condors they saw to estimate how many condors actually existed. Nor could they say how sure they were of being right. Consequently, the severity of the decline and number of condors dying were grossly underestimated. Because management was unaware of the severity of the decline and urgency of the crisis, critical decisions to save the condors were delayed. For example, the ability to recognize individuals based on methods that started in 1982 (Table) allowed us to realize we had lost five adult condors (about 30% of the wild population) during winter 1984-85. Understanding the critical nature of this loss ultimately led to the decision to capture the remaining wild birds.

As of January 1994 there were 66 birds, and the future of the captive population appears bright. The World Center for Birds of Prey in Boise, Idaho, became the third captive site in September 1993, joining the San Diego Wild Animal Park and the Los Angeles Zoo. The George Miksch Avian Research Center in Bartlesville, Oklahoma, is scheduled to become the fourth captive breeding facility in 1994. We expect all captive flocks to do well and continue to increase, providing young birds for release in California as well as yet-to-be selected sites in Arizona and New Mexico.

Timely and accurate status and trends data will continue to be important to the condor recovery program as more birds are released. Not only will these data be needed to monitor the success of the release, but also they are essential for identifying problems, which is especially critical because no known or suspected mortality factors in California have been significantly reduced, much less eliminated. The relocation of all released California condors to a site near the Sisquoc Sanctuary after the death of the fourth bird (three lost to powerline collisions) reflects the close monitoring necessary to ensure that appropriate actions can be taken as quickly as possible.

With the wild population consisting of only nine young birds with a restricted range and still dependent on artificial feeding stations, conventional radiotelemetry and tagging have been adequate. As the number of birds increase and their territories expand, however, conventional methods for monitoring and locating birds will be unable to fulfill the recovery program's needs. For the release program to succeed, we will need to identify and remove or avoid key mortality factors such as the powerline collision hazard at the first site. To accomplish this, we will need to monitor and locate dozens of individual condors scattered over a million or more hectares. Equipment to do this exists but has not been modified or adequately tested for use on condors. Eventually a simple, inexpensive survey procedure will be needed to track the wild condor population as it increases and starts reproducing. Developing these procedures now is essential.

Table. Status of the wild and captive California condor populations, 1982-93.

Year	No. of captive birds	No. of wild birds	Total
1982	3	21	24
1983	9	16	25
1984	16	11	27
1985	21	6	27
1986	25	2	27
1987	27	0	27
1988	28	0	28
1989	32	0	32
1990	40	0	40
1991	52	0	52
1992	56	7	63
1993	66	9	75

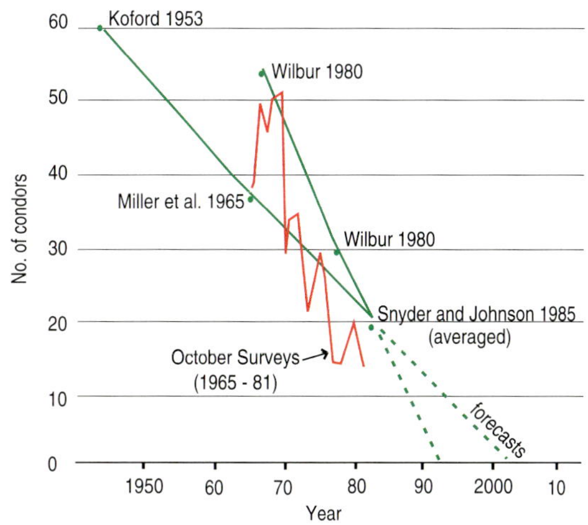

Figure. Estimates of the California condor population, 1945-82 (Snyder and Johnson 1985). Used with permission from the *Condor*©.

References

Koford, C.B. 1953. The California condor. National Audubon Society Res. Rep. 4. 154 pp.

Mallette, R.D., and J.C. Borneman. 1966. First cooperative survey of the California condor. California Fish and Game 52:185-203.

Miller, A.H., I. McMillan, and E. McMillan. 1965. The current status and welfare of the California condor. National Audubon Society Res. Rep. 6. 61 pp.

Pattee, O.H., and S.R. Wilbur. 1989. Turkey vulture and California condor. Pages 61-65 *in* Proceedings of the Western Raptor Management Symposium and Workshop. National Wildlife Federation, Washington, DC. 320 pp.

Snyder, N.F.R., and E.V. Johnson. 1985. Photographic censusing of the 1982-1983 California condor population. Condor 87:1-13.

Wilbur, S.R. 1978. The California condor, 1966-76: a look at its past and future. U.S. Fish and Wildlife Service North American Fauna 72. 136 pp.

Wilbur, S.R. 1980. Estimating the size and trend of the California condor population, 1965-1978. California Fish and Game 66:40-48.

Courtesy D. Clendenen

California condors have a wingspan of about 3 m or 9 ft.

For further information:

Oliver H. Pattee
National Biological Service
Patuxent Environmental Science Center
1510 American Holly Dr.
Laurel, MD 20708

Audubon's Crested Caracara in Florida

by
James N. Layne
Archbold Biological Station

Audubon's crested caracara (*Caracara plancus audubonii*) is a species characteristic of the grassland ecosystems of central Florida and is one of the state's most distinctive birds. The Florida population is threatened and widely separated from the main species' range, which extends from extreme southwestern Louisiana, southern Texas, and southern Arizona to the tip of South America, including Tierra del Fuego and the Falkland Islands. Another isolated population occurs on Cuba and the Isle of Pines.

Audubon's crested caracara (*Caracara plancus audubonii*) in Florida.

The number of Florida caracaras is believed to have undergone a substantial decline from the early historic level in the 1950's and 1960's (Layne in press), with the total state population estimated at 250 in the early 1950's (Sprunt 1954) and fewer than 100 birds in the late 1960's (Heinzman 1970). Based on the apparent continuing decrease in its numbers, Florida's population of Audubon's crested caracara was federally listed as threatened in 1987 (Federal Register 1987). As part of a general study of the life history, ecology, and behavior of the caracara in Florida, I monitored its distribution and population status from 1972 to 1991.

Information was obtained from road and off-road searches in all parts of the known range; systematic roadside and aerial surveys in a 5,116-km^2 (1,975-mi^2) area within the core portion of the range; published records; museum specimens; and sighting reports from over 500 cooperators. Logistical limitations prevented surveying the entire potential Florida range thoroughly enough in any given year to obtain a reasonably accurate picture of the distribution and total population. Thus, estimates of the statewide distribution and numbers were based on records combined over 5-year periods: 1972-76, 1977-81, 1982-86, and 1987-91. Searches were most intensive from 1972 to 1981 and in the final period 1987-91. Because areas along public roads were surveyed more intensively than those remote from highways, there was a lower probability of detecting caracaras whose territories did not overlap roads than those whose territories included roads. This bias appeared to be at least partially compensated for by a tendency of caracaras to concentrate along highways because of the attraction of roadkills as a food source.

Status and Trends

The breeding range of Audubon's crested caracara in Florida (Fig. 1), based on records from the most recent 5-year period of the study (1987-91), did not differ significantly from that during 1973-76 (Layne 1978). Caracaras were documented in 20 counties in central peninsular Florida, with most locations in the same 5-county area as in the earlier years. Counties with 10% or more of the 183 estimated locations during 1987-91 included (number of locations in parentheses) Glades (41), Highlands (34), Okeechobee (23), and Osceola (18). The data indicate no obvious change has occurred in the overall range or core area of the distribution of the caracara in Florida from that shown by Howell (1932). As there had been relatively little alteration of the natural habitats of the state up to that time, Howell's range map is assumed to reflect the early historical distribution.

The estimated number of adult caracaras during 5-year intervals from 1972 to 1991 ranged from 196 to 312 (Fig. 2). The variation between periods reflects differences in sampling effort rather than changes in actual numbers. Thus, the adult population over the 20-year period appears to have been stable with a minimum of about 300 individuals in 150 territories. Further evidence that the population remained generally stable between 1972 and

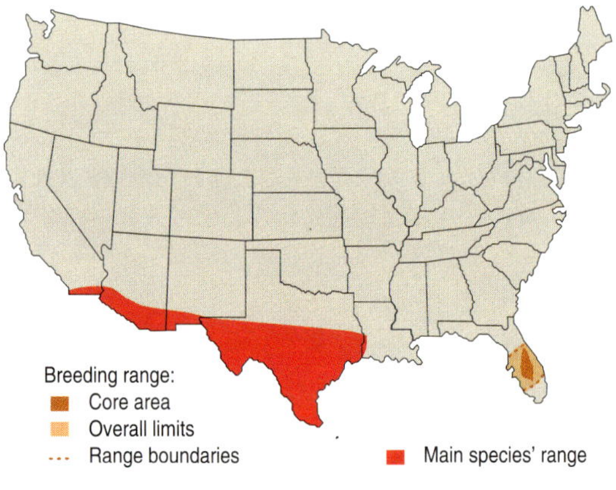

Fig. 1. Breeding range of Audubon's crested caracara in Florida based on records from 1987 to 1991; range boundaries shown by Howell (1932), and main species' range in western United States.

1991 is the similarity in adult-immature age ratios during this interval (Fig. 2). Although immatures could not be censused accurately because they tend to wander individually or in aggregrations after the break up of family groups, they are believed to have numbered between 100 and 200 in any one year, giving a total statewide population of 400-500.

The estimate of the minimum adult population includes single adults observed in an area only once during a 5-year interval as representing a pair on an established territory. Assuming that such individuals were actually unmated transients reduces the estimated adult population to about 230 individuals. Regardless of which estimate of the adult population during 1972-91 is accepted, it is highly unlikely that the Florida population was reduced to fewer than 100 birds between 1967 and 1970 (Heinzman 1970).

Although the range of Audubon's crested caracara in Florida appears to have remained unchanged for the past 60 years and numbers have been stable over at least the past 20 years, the future status of the population is still of concern. Most birds occur on private ranchlands subject to habitat degradation or loss from intensification of agricultural practices or other development. The most immediate threat is large-scale conversion of native range and improved pasture habitats to citrus groves.

A decline in the Florida caracara population within the next 10 years appears likely if citrus conversion and other habitat losses continue at the present rate. Because caracaras are relatively long-lived and strongly attached to their territories, residents may persist in a territory despite unfavorable changes, but may not be replaced by new individuals when they finally leave or die. The result may be a significant time lag before the effects of deleterious habitat changes are reflected in an actual population decline. The magnitude of the time lag in detection of any trend in the Florida distribution and population of Audubon's crested caracara also will depend upon the effectiveness of future monitoring efforts.

References

Federal Register. 1987. Endangered and threatened wildlife and plants: threatened status for the Florida population of Audubon's crested caracara. Federal Register 52(128):25229-25232.

Heinzman, G. 1970. The caracara survey, a four year report. Florida Naturalist 43:149.

Howell, A.H. 1932. Florida bird life. Florida Department of Game and Freshwater Fish, Tallahassee. 479 pp.

Layne, J.N. 1978. Audubon's caracara (*Caracara cheriway auduboni*). Pages 34-36 *in* H.W. Kale II, ed. Birds, rare and endangered biota of Florida. Vol. 2. University Press of Florida, Gainesville.

Layne, J.N. Audubon's crested caracara (*Polyborus plancus audubonii*). *In* J.A. Rodgers, H.W. Kale II, and H. Smith, eds. Birds, rare and endangered biota of Florida. 2nd ed. University Press of Florida, Gainesville. In press.

Sprunt, A., Jr. 1954. Florida bird life. Coward-McCann, New York. 527 pp.

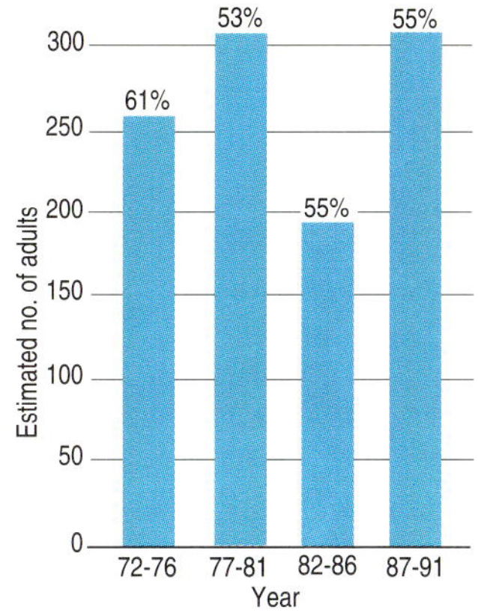

Fig. 2. Estimated numbers of adult Audubon's crested caracaras in Florida over 5-year intervals from 1972 to 1991, based on the assumption that localities where adults were recorded represent territories occupied by an adult pair. Percentage of locations that had immature birds versus those that had adults are given above bars.

For further information:

James N. Layne
Archbold Biological Station
PO Box 2057
Lake Placid, FL 33862

Puerto Rican Parrots

by

J. Michael Meyers
National Biological Service

Since the arrival of Columbus in Puerto Rico, the Taino Indian has disappeared and the parrot has just barely survived (Wadsworth 1949; Snyder et al. 1987). The Puerto Rican parrot (*Amazona vittata*) had shared its habitat with the peaceful Taino Indians for centuries before the arrival of European settlers in the Caribbean.

Status and Trends

Upon arrival of the Spanish in 1493, the Puerto Rican parrot lived in all major habitats of Puerto Rico and the adjacent smaller islands of Culebra, Mona, Vieques, and possibly the Virgin Islands (Snyder et al. 1987). Parrots occupied eight major climax or old-growth forest types (Little and Wadsworth 1964) that covered Puerto Rico and were interspersed only by small, scattered, sandy, or marshy areas near the coast (Snyder et al. 1987). Parrots nested in cavities of large trees that were plentiful throughout the forests. Fertile, moist lowland forests in the coastal plain as well as forested mountain valleys contained much of the fruits and seeds necessary to feed a thriving parrot population. The forests of Puerto Rico probably supported a parrot population of 100,000-1,000,000 at the end of the 15th century (Snyder et al. 1987; Wiley 1991).

Little habitat change occurred in Puerto Rico during the first 150 years of European settlement. By 1650 the Spanish population had increased to 880 (Snyder et al. 1987); parrots still occupied all major habitats and were plentiful (Fig. 1). During the next two centuries the human population soared to almost 500,000 (Fig. 1), and clearing for agriculture, especially in the lowlands, eradicated forests in Puerto Rico (Wadsworth 1949). By 1836 reports by

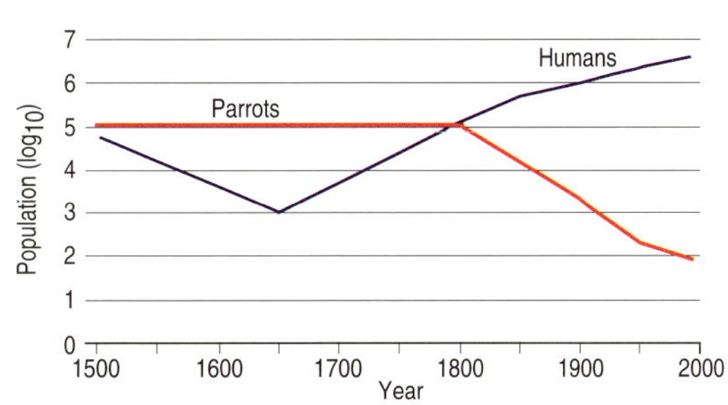

Fig. 1. Population trends of humans and Puerto Rican parrots since 1500 (Snyder et al. 1987 and U.S. Census data; all data for the year 2000 are projected). Populations are converted to $\log_{10}$ for showing trends.

Moritz, a German naturalist, indicated that the Puerto Rican parrot population had begun to decline (Snyder et al. 1987).

By 1900 the human population had doubled to a million (Fig. 1). About 76% of the land area of Puerto Rico had been converted from forest to agriculture (Snyder et al. 1987); less than 1% of the old-growth forest remained after more than 400 years of European civilization. At this time, the parrot population must have been low, but no data exist. By 1937 U.S. Forest Service (USFS) rangers estimated the Puerto Rican parrot population at about 2,000 birds (Wadsworth 1949). A few years later, parrots were found only in the Luquillo Mountains, formerly a forest reserve of the Spanish Crown and now managed by the USFS. This area contained the last forest habitat suitable for Puerto Rican parrots.

Population surveys of the Puerto Rican parrot were not conducted until the 1950's. Early estimates of the parrot population in Puerto Rico are based on few written records and general observations (Snyder et al. 1987), knowledge of the parrot's biology, and extrapolation of population surveys conducted by Rodríguez-Vidal (1959). During the 1950's, Rodríguez-Vidal of the Puerto Rico Department of Agriculture and Commerce conducted the first extensive study of the Puerto Rican parrot. He reported a population of 200 Puerto Rican parrots by the mid-1950's (Fig. 2). About 20 years later the population had dwindled to 14 individuals that inhabited an isolated rain forest of the Luquillo Mountains.

Courtesy J.M. Meyers, NBS

Puerto Rican parrot (*Amazona vittata*).

In 1968 Kepler, U.S. Fish and Wildlife Service (USFWS), organized parrot surveys by placing observers at strategic sites, including overlooks from prominent rocks, road-cuts, and building roofs. Snyder et al. (1987) improved the survey method in 1972 by constructing 10 treetop lookouts in areas of major parrot use. Parrot surveys are conducted from these platforms during the breeding season and pre- and postbreeding season (Snyder et al. 1987). Observers collect information on parrot numbers, directions, and their distance from the platform by the time of day. By 1993 this treetop lookout system was expanded to 38 platforms (Vilella and García 1994).

In 1968 implementation of the Puerto Rican Parrot Recovery Plan began; it is a cooperative effort of scientists and managers of the Puerto Rico Department of Environmental and Natural Resources, USFS (Caribbean National Forest and International Institute of Tropical Forestry), USFWS Puerto Rican Parrot Field Office, and the National Biological Service. After the recovery program began, the parrot population increased to 47 birds by 1989 (Wiley 1980; Lindsey et al. 1989; Meyers et al. 1993); however, about 50% of the population was destroyed by Hurricane Hugo that same year. A small population of 22-24 individuals remained in late 1989 (Fig. 2). Since then, the population recovered to 38-39 by early 1994 (F.J. Vilella, USFWS, personal communication). After the hurricane, the number of successful nesting pairs increased from a maximum of 5 to 6 pairs from 1991 to 1993 (Meyers et al. 1993; Vilella and García 1994).

Research and Management

Puerto Rican parrots declined in relation to the increasing human population (Fig. 1). Conversion of forests to agriculture and loss of forest habitat, on which the species depended for food and nest cavities, was the primary cause for decline. Shooting parrots for food or protection of crops and capture for pets were secondary causes for decline. The remnant parrot population in the Luquillo Mountains was further stressed when trails and roads were created and when human uses of the forest timber were encouraged in the early 1900's (Snyder et al. 1987). Storms before the arrival of Europeans probably had little effect on the parrot population because the population was more widespread, and hurricanes tend to affect only a small geographic area. Severe hurricanes in 1898, 1928, 1932, and 1989 reduced small, now-isolated populations even further. The apparent ability of the population to rebound after these storms is suggested by increases in the parrot population and in nesting pairs after

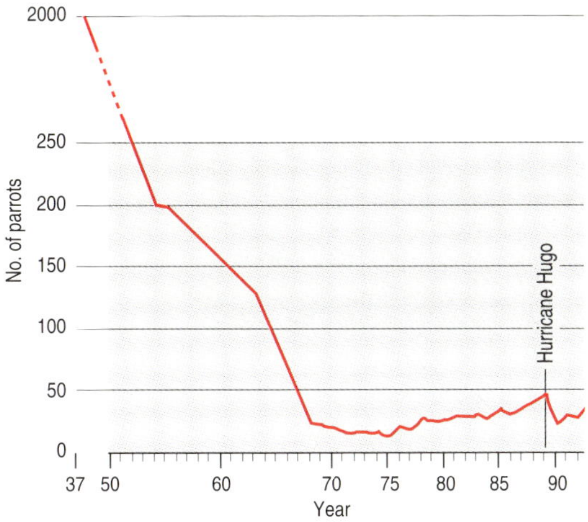

Fig. 2. Population trends of the Puerto Rican parrot in the 20th century.

Hurricane Hugo hit the island in 1989 (Meyers et al. 1993).

Intense research and management strategies during the last 27 years have prevented the extinction of the Puerto Rican parrot. Much of the effort to rebuild the population has involved research and management of nesting sites (Wiley 1980; Snyder et al. 1987; Lindsey et al. 1989; Wiley 1991). Predators, such as black rats (*Rattus rattus*) and pearly-eyed thrashers (*Margarops fuscatus*), have been controlled (Snyder et al. 1987). Bot fly (*Philornis* spp.) infestations of nestlings are still a minor problem (Lindsey et al. 1989). Management of nests by fostering captive-reared young into wild nests, guarding nests, controlling honey bees (*Apis mellifera*), improving and maintaining existing nest cavities, and creating enhanced nesting cavities should increase the population of the Puerto Rican parrot (Wiley 1980; Lindsey et al. 1989; Wiley 1991; Lindsey 1992; Vilella and García 1994).

Hurricanes will continue to threaten the wild population of the Puerto Rican parrot. Researchers estimate that storms equal to the intensity of Hugo (sustained winds of 166 km/h or 104 mi/h) occur at least every 50 years in northeastern Puerto Rico (Scatena and Larsen 1991). The risk of extinction caused by hurricanes will be reduced by establishing a geographically separated wild population (USFWS 1987).

Introduced parrots and parakeets are common in Puerto Rico, including some of the genus *Amazona*. Monitored populations of these non-native birds have increased from 50% to 250% during 1990-93 (J.M. Meyers, National Biological Service, unpublished data). If they expand their ranges to include older forests, these populations may pose a threat to the Puerto Rican parrot by introducing diseases and by competing for resources. At present, none of the introduced *Amazona* populations are found near the Luquillo Mountains; however, orange-fronted parakeets (*Aratinga canicularis*) have foraged and nested in these mountains at lower elevations (J.M. Meyers, NBS, unpublished data).

As the Puerto Rican parrot population increases, it is possible that suitable nesting sites may limit population growth. Before this occurs, research and management should concentrate on increasing the wild population. The ability of the Puerto Rican parrot to expand its population in a manner similar to the exotic parrots in Puerto Rico, in a variety of natural and human-altered environments, should not be underestimated and may be the key to its recovery.

References

Lindsey, G.D. 1992. Nest guarding from observational blinds: strategy for improving Puerto Rican parrot nest success. Journal of Field Ornithology 63:466-472.

Lindsey, G.D., M.K. Brock, and M.H. Wilson. 1989. Current status of the Puerto Rican parrot conservation program. Pages 89-99 *in* Wildlife management in the Caribbean islands. Proceedings of the Fourth Meeting of Caribbean Foresters. U.S. Department of Agriculture, Institute of Tropical Forestry, Río Piedras, Puerto Rico.

Little, E.L., Jr., and F.H. Wadsworth. 1964. Common trees of Puerto Rico and the Virgin Islands (reprint). Agriculture Handbook 249. U.S. Department of Agriculture, Washington, DC. 556 pp.

Meyers, J.M., F.J. Vilella, and W.C. Barrow, Jr. 1993. Positive effects of Hurricane Hugo: record years for Puerto Rican parrots nesting in the wild. Endangered Species Tech. Bull. 27:1,10.

Rodríguez-Vidal, J.A. 1959. Puerto Rican parrot study. Monographs of the Department of Agriculture and Commerce 1. San Juan, Puerto Rico. 15 pp.

Scatena, F.N., and M.C. Larsen. 1991. Physical aspects of Hurricane Hugo in Puerto Rico. Biotropica 23:317-323.

Snyder, N.R.F., J.W. Wiley, and C.B. Kepler. 1987. The parrots of Luquillo: natural history and conservation of the Puerto Rican parrot. Western Foundation of Vertebrate Zoology, Los Angeles, CA. 384 pp.

USFWS. 1987. Recovery plan for the Puerto Rican parrot, *Amazona vittata*. U.S. Fish and Wildlife Service, Atlanta, GA. 69 pp.

Vilella, F.J., and E.R. García. 1994. Post-hurricane management of the Puerto Rican parrot. *In* J.A. Bissonette and P.R. Krausman, eds. International Wildlife Management Congress Proceedings. The Wildlife Society, Washington, DC. In press.

Wadsworth, F.H. 1949. The development of the forested land resources of the Luquillo Mountains, Puerto Rico. Ph.D. dissertation, University of Michigan, Ann Arbor. 481 pp.

Wiley, J.W. 1980. The Puerto Rican parrot (*Amazona vittata*): its decline and the program for its conservation. Pages 133-159 *in* R. E. Pasquier, ed. Conservation of new world parrots. International Council for Bird Preservation Tech. Publ. 1. Smithsonian Institute Press, Washington, DC.

Wiley, J.W. 1991. Status and conservation of parrots and parakeets in the Greater Antilles, Bahama Islands, and Cayman Islands. Bird Conservation International 1:187-214.

For further information:

J. Michael Meyers
National Biological Service
Patuxent Environmental
Science Center
PO Box N
Palmer, Puerto Rico 00721-0501
USA

Red-cockaded Woodpeckers

by
Ralph Costa
U.S. Fish and Wildlife Service
Joan L. Walker
U.S. Forest Service

The red-cockaded woodpecker (RCW; *Picoides borealis*) is a territorial, nonmigratory, cooperative breeding species (Lennartz et al. 1987). Ecological requirements include habitat for relatively large home ranges (34 to about 200 ha or 84 to about 500 acres; Connor and Rudolph 1991); old pine trees with red-heart disease for nesting and roosting (Jackson and Schardien 1986); and open, parklike forested landscapes for population expansion, dispersal (Connor and Rudolph 1991), and necessary social interactions.

Historically, the southern pine ecosystems, contiguous across large areas and kept open with recurring fire (Christensen 1981), provided ideal conditions for a nearly continuous distribution of RCWs throughout the South. Within this extensive ecosystem red-cockaded woodpeckers were the only species to excavate cavities in living pine trees, thereby providing essential cavities for other cavity-nesting birds and mammals, as well as some reptiles, amphibians, and invertebrates (Kappes 1993). The loss of open pine habitat since European settlement precipitated dramatic declines in the bird's population and led to its being listed as endangered in 1970 (Federal Register 35:16047).

We obtained historic RCW distribution data, arranged by state and county, from published sources (Jackson 1971; Hooper et al. 1980), and interviews with various red-cockaded woodpecker experts. Current distribution and abundance data were obtained from natural resource agencies and knowledgeable biologists. Most records were reported between January 1993 and March 1994, and most represent direct census data. Specific references are available from R. Costa (Table).

Several terms are used to describe red-cockaded woodpecker abundance. "Group" refers to birds that cooperate to rear the young from a single nest. It usually consists of a breeding male and female, and zero to four helpers, usually the group's male offspring from previous breeding seasons. For reporting purposes, single bird groups (usually male) are tallied. The collection of cavity trees used by a group for nesting and roosting is the "cluster." Although single tree clusters do occur, typically each cluster consists of 2 to more than 15 cavity trees and may occupy 2 to more than 4 ha (5 to more than 10 acres). Each group normally occupies and defends only one cluster. "Population" refers to the aggregation of groups that are more distant than 29 km (18 mi) from the nearest group. A single isolated group may constitute a population.

Table. Number of red-cockaded woodpecker active clusters, by state and land ownership category, for various years between 1990-94.*

	Ownership			
State	Federal	State	Private	Total
Alabama	150	8	25	183
Arkansas	35	0	121	156
Florida	1,063	128	94	1,285
Georgia	431	2	218	651
Kentucky	5	0	0	5
Louisiana	422	10	73	505
Mississippi	152	0	22	174
North Carolina	408	162	163	733
Oklahoma	0	9	1	10
South Carolina	456	39	186	681
Tennessee	1	0	0	1
Texas	218	26	61	305
Virginia	0	0	5	5
Total	3,341	384	969	4,694

*For information on references, contact R. Costa.

Red-cockaded woodpecker (*Picoides borealis*).

Historical Distribution and Abundance

The historical range of this species covered southeast Virginia to east Texas and north to portions of Tennessee, Kentucky, southeast Missouri, and eastern Oklahoma (Figure). The range included the entire longleaf pine ecosystem, but the birds also inhabited open shortleaf, loblolly, and Virginia pine forests, especially in the Ozark-Ouachita Highlands and the southern tip of the Appalachian Highlands.

Red-cockaded woodpecker abundance was described variously as fairly common (Woodruff 1907), locally common (Murphey 1939), common (Chapman 1895), or abundant (Audubon 1839). Occasional occurrences were noted for New Jersey (Hausman 1928), Pennsylvania (Gentry 1877), Maryland (Meanly 1943), and Ohio (Dawson and Jones 1903).

The distribution map (Figure) displays only counties for which specimens or reliable sources can be cited. The gaps in the distribution undoubtedly contained red-cockaded woodpeckers in the past. Most counties without documented occurrences are found in the longleaf pine-shortleaf pine-loblolly pine-hardwoods transition areas in the east gulf region (Figure), where richer soils and rolling topographies were associated with intense agriculture and interrupted fire regimes. Such areas possibly supported smaller populations that were quickly lost with the forest clearing and therefore were never recorded.

Status and Causes of Decline

Red-cockaded woodpeckers survive as very small (1-5 groups) to large (groups of 200 or more) populations. There are at least small populations in most states with historical occurrences (Table). Except for a population of about 90 groups in southern Arkansas and northern Louisiana, the largest populations are found within the historical longleaf pine ecosystem. Other populations outside the longleaf pine range consist of fewer than 20 groups in single or several adjacent counties. Within the longleaf range, there are 4 populations with more than 200 groups and 11 populations with more than 100 groups; all but one are found on federal lands. The remaining longleaf pine-associated populations are small and isolated. Such small populations are threatened by adverse effects of demographic isolation, increased predation and cavity competition, and stochastic (random) natural events such as hurricanes.

The decline of the red-cockaded woodpecker coincided with the loss of the longleaf ecosystem. As forests were cleared, birds were isolated in forest tracts where unmerchantable trees were left. Aerial and ground photographs from the 1930's show that scattered medium to large trees (0.4-2 per ha or 1-5 per acre) were left in many stands. The culled trees (undoubtedly including red-cockaded woodpecker cavity trees) provided residual nesting and foraging habitat for the birds. In some places these trees remain and are used by red-cockaded woodpeckers today.

Since the 1950's, on lands managed for forest products, the forest structure and composition changed in conjunction with clearcutting, short timber rotations, conversion of longleaf stands to other pine species, and "clean" forestry practices (removal of cavity, diseased, or defective trees). These practices eliminated much of the remaining red-cockaded woodpecker habitat. Additionally, aggressive fire suppression promoted the development of a hardwood midstory in pine forests. The adverse impacts of a dense midstory on RCW populations are well-documented (Connor and Rudolph 1989; Costa and Escano 1989).

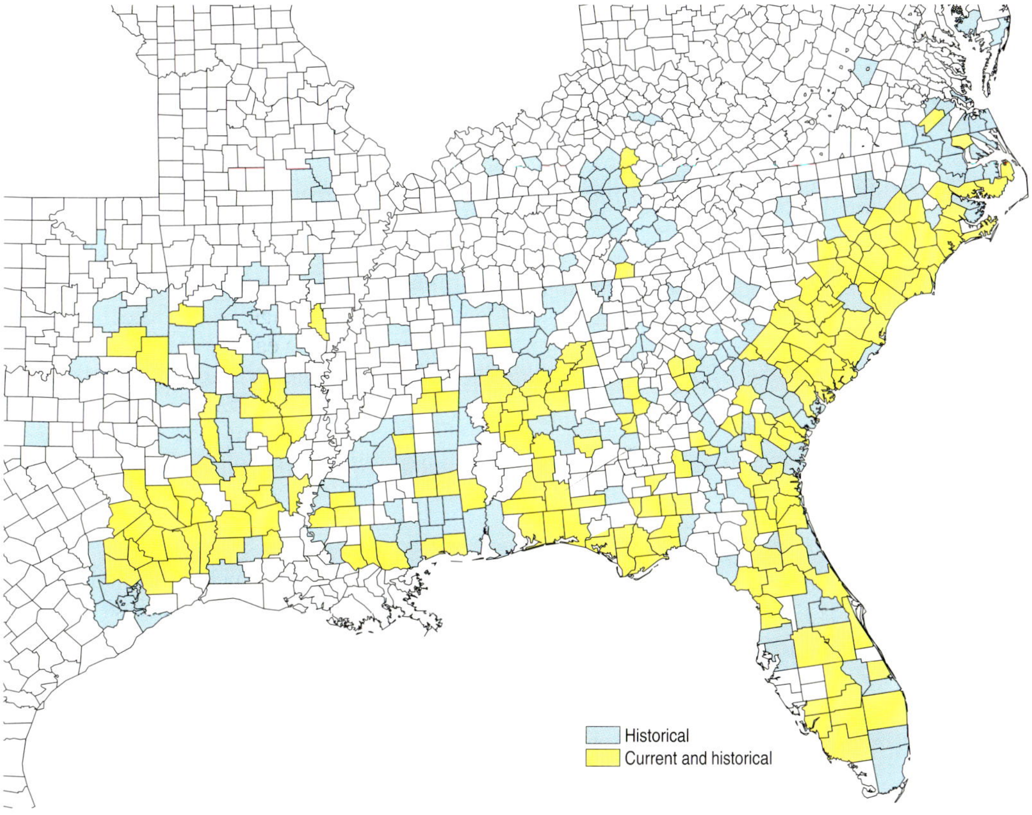

Figure. Distribution of red-cockaded woodpeckers by county and state. Most historical RCW records are cited from Jackson 1971 and Hooper et al. 1980. For information on references, contact R. Costa.

Recent Developments and the Future

The Red-cockaded Woodpecker Recovery Plan (USFWS 1985) specifies that rangewide recovery will be achieved when 15 viable populations are established and protected by adequate habitat management programs. The recovery populations are to be distributed across the major physiographic provinces and within the major forest types that can be managed to sustain viable populations. Each recovery population will likely require 400 breeding pairs (or 500 active clusters, as some clusters are occupied by single birds or contain nonbreeding groups) to ensure long-term population viability (Reed et al. 1993; Stevens, in press). At a density of 1 group/80-120 ha (200-300 acres; USFWS 1985; USFS 1993), landscapes of at least 40,000 ha (100,000 acres) will be needed to support viable populations. Most forested pine areas large enough to supply this habitat are on public, mostly federal, lands.

With two exceptions (Hooper et al. 1991; USFS, Apalachicola National Forest, FL, unpublished data), there is no evidence that red-cockaded woodpecker populations can expand to viable levels without considerable human intervention. Conversely, numerous population extirpations have been documented (Baker 1983; Costa and Escano 1989; Cox and Baker, in press). Ensuring the survival of the species, even in the short term (50 years), will require landscape-scale habitat and population management to provide the forest structure and composition needed for nesting and foraging habitat and population expansion; and to manage limiting factors (primarily a lack of suitable cavity trees, cavity competition, and demographic isolation) that can extirpate small populations. Both strategies are part of management guidelines drafted by several federal land stewards (USFS 1993; U.S. Army 1994; USFWS 1994).

These ecosystem management plans promote practices that minimize landscape fragmentation, retain suitable numbers of potential cavity trees well distributed throughout the landscape, and restore the original forest cover by planting the appropriate pine species. They recommend the use of growing-season fires to control hardwoods, create open forest conditions, and begin to restore the understory plant communities of the pine ecosystems. Stabilization and growth of small high-risk populations will be aided by creating artificial red-cockaded woodpecker cavities (Copeyon 1990) and translocating juvenile birds from stable larger populations into small ones (Rudolph et al. 1992). Technologies that minimize or eliminate predation and competition problems are available (Carter et al. 1989).

During the past 4-7 years, several populations have stabilized or increased (Gaines et al., in press; Richardson and Stockie, in press) as a result of implementing conservation biology principles—that is, integrating available technology with the species' life history and ecological requirements. The limited number of juvenile birds, however, may hinder recovery progress in all populations simultaneously.

References

Audubon, J.J. 1839. Ornithological biography. Vol. 5. A. and C. Black, Edinburgh.

Baker, W.W. 1983. Decline and extirpation of a population of red-cockaded woodpeckers in northwest Florida. Pages 44-45 *in* D.A. Wood., ed. Red-cockaded Woodpecker Symposium II Proceedings. Florida Game and Freshwater Fish Commission, Tallahassee.

Carter, J.H., III, J.R. Walters, S.H. Everhart, and P.D. Doerr. 1989. Restrictors for red-cockaded woodpecker cavities. Wildlife Society Bull. 17:68-72.

Chapman, F.M. 1895. Handbook of birds of eastern North America. D. Appleton and Co., New York. 431 pp.

Christensen, N.L. 1981. Fire regimes in southeastern ecosystems. U.S. Forest Service Gen. Tech. Rep. WO-26: 112-136.

Connor, R.N., and D.C. Rudolph. 1989. Red-cockaded woodpecker colony status and trends on the Angelina, Davy Crockett and Sabine National forests. U.S. Forest Service, Southern Forest Experiment Station, Res. Paper SO-250. 15 pp.

Connor, R.N., and D.C. Rudolph. 1991. Forest habitat loss, fragmentation, and red-cockaded woodpecker population. Wilson Bull. 103 (3): 446-457.

Copeyon, C.K. 1990. A technique for constructing cavities for the red-cockaded woodpecker. Wildlife Society Bull. 18:303-311.

Costa, R., and R.E. Escano. 1989. Red-cockaded woodpecker status and management in the Southern Region in 1986. U.S. Forest Service Southern Region Tech. Publ. R8-TP 12. 71 pp.

Cox, J., and W.W. Baker. In press. Distribution and status of the red-cockaded woodpecker in Florida: 1992 update. Red-cockaded Woodpecker Symposium III: species recovery, ecology and management. Stephen F. Austin State University, Nacogdoches, TX.

Dawson, W.L., and L. Jones. 1903. The birds of Ohio. Vol. 1. The Wheaton Publishing Co., Columbus, OH. 671 pp.

Gaines, G.D., W.L. Jarvis, and K. Laves. In press. Red-cockaded woodpecker management on the Savannah River Site: a management/research success story. Red-cockaded Woodpecker Symposium III: species recovery, ecology and management. Stephen F. Austin State University, Nacogdoches, TX.

Gentry, T.G. 1877. Life-histories of birds of eastern Pennsylvania. Vol. 2. J.H. Choate, Salem, MA.

Hausman, L.A. 1928. Woodpeckers, nuthatches, and creepers of New Jersey. New Jersey Agricultural Experiment Station Bull. 470:1-48.

Hooper, R.G., D.L. Krusac, and D.L. Carlson. 1991. An increase in a population of red-cockaded woodpeckers. Wildlife Society Bull. 19:277-286.

Hooper, R.G., L.J. Niles, R.F. Harlow, and G.W. Wood. 1982. Home ranges of red-cockaded woodpeckers in coastal South Carolina. Auk 99(4):675-682.

Hooper, R.G., A.F. Robinson, and J.A. Jackson. 1980. The red-cockaded woodpecker: notes on life history and management. U.S. Forest Service, Southeastern Area, State and Private Forestry, Gen. Rep. SA-GR 9. 8 pp.

Jackson, J.A. 1971. The evolution, taxonomy, distribution, past populations and current status of the red-cockaded woodpecker. Pages 4-29 *in* R.L. Thompson, ed. The

Ecology and Management of the Red-cockaded Woodpecker, Proceedings of a Symposium. Bureau of Sport Fisheries and Wildlife and Tall Timbers Research Station, Tallahassee, FL.

Jackson, J.A., and B.J. Schardien. 1986. Why do red-cockaded woodpeckers need old trees? Wildlife Society Bull. 14:318-322.

Kappes, J.J. 1993. Interspecific interactions associated with red-cockaded woodpecker cavities at a north Florida site. M.S. thesis, University of Florida, Gainesville. 75 pp.

Lennartz, M.R., R.G. Hooper, and R.F. Harlow. 1987. Sociality and cooperative breeding of red-cockaded woodpeckers (*Picoides borealis*). Behavioral Ecology and Sociobiology 20:77-88.

Meanly, R.M. 1943. Red-cockaded woodpecker breeding in Maryland. Auk 60:105.

Murphey, E.E. 1939. *Dryobates borealis* (Vieillot), in A.C. Bent, Life histories of North American woodpeckers. Smithsonian Institution U.S. National Museum Bull. 174:72-79.

Reed, J.M., J.R. Walters, T.E. Emigh, and D.E. Seaman. 1993. Effective population size in red-cockaded woodpeckers: population and model differences. Conservation Biology 7(2):302-308.

Richardson, D., and J. Stockie. In press. Intensive management of a small red-cockaded woodpecker population at Noxubee National Wildlife Refuge. Red-cockaded Woodpecker Symposium III: Species Recovery, Ecology and Management. Stephen F. Austin State University, Nacogdoches, TX.

Rudolph, D.C., R.N. Connor, D.K. Carrie, and R.R. Schaefer. 1992. Experimental reintroduction of red-cockaded woodpeckers. Auk 109(4):914-916.

Stevens, E.E. In press. Population viability for red-cockaded woodpeckers. Red-cockaded Woodpecker Symposium III: Species Recovery, Ecology and Management. Stephen F. Austin State University, Nacogdoches, TX.

U.S. Army. 1994. Management guidelines for the red-cockaded woodpecker on army installations. U.S. Army Legal Services Agency, Arlington, VA. 19 pp.

USFS. 1993. Draft environmental impact statement for the management of the red-cockaded woodpecker and its habitat on national forests in the Southern Region. U.S. Forest Service, Southern Region, Atlanta, GA. 460 pp.

USFWS. 1985. Red-cockaded woodpecker recovery plan. U.S. Fish and Wildlife Service, Atlanta, GA. 88 pp.

USFWS. 1994. Draft strategy and guidelines for the recovery and protection of the red-cockaded woodpecker on national wildlife refuges. U.S. Fish and Wildlife Service, Atlanta, GA. 50 pp.

Woodruff, E.S. 1907. Some interesting records from southern Missouri. Auk 24:348-349.

For further information:

Ralph Costa
U.S. Fish and Wildlife Service
Red-cockaded Woodpecker Field Office
Department of Forest Resources
Clemson University
261 Lehotsky Hall
Box 341003
Clemson, SC 29634

Southwestern Willow Flycatchers in the Grand Canyon

by
Mark K. Sogge
National Biological Service

The southwestern willow flycatcher (*Empidonax traillii extimus*) occurs, as its name implies, throughout most of the southwestern United States (Fig. 1). It is a Neotropical migrant songbird, i.e., one of many birds that return to the United States and Canada to breed each spring after migrating south to the Neotropics (Mexico and Central America) to winter in milder climates. In recent years, there has been strong evidence of declines in many Neotropical migrant songbirds (e.g., Finch and Stangel 1993), including the southwestern willow flycatcher (Federal Register 1993). The flycatcher appears to have suffered significant declines throughout its range, including total loss from some areas where it historically occurred. These declines, as well as the potential for continued and additional threats, prompted the U.S. Fish and Wildlife Service (USFWS) to propose listing the southwestern willow flycatcher as an endangered species (Federal Register 1993).

The southwestern willow flycatcher is one of four distinct races of willow flycatchers that breed in North America. All races breed in shrubby or woodland habitats, usually adjacent to, or near, surface water or saturated soil. Riparian areas—woodland and shrub areas along streams and rivers—are particularly favored. In fact, the southwestern willow flycatcher is a riparian obligate, breeding only in riparian vegetation. It prefers tall, dense willows and cottonwood habitat where dense vegetation continues from ground level to the tree canopy. Southwestern willow flycatchers appear to breed in stands of the exotic and invasive tamarisk (*Tamarix* spp.) only at locations above 625 m (2,051 ft) elevation (Federal Register 1993), and where the tamarisk stands have suitable structural characteristics (Fig. 2). Thus, many areas dominated by tamarisk are not suitable flycatcher habitat. Being a riparian obligate, the southwestern willow flycatcher is particularly sensitive to the alteration and loss of riparian habitat (including tamarisk invasion), which is a widespread and pervasive problem throughout the Southwest.

Because of the decline and precarious status of southwestern willow flycatchers, it is important to document the status of the species, where it occurs, how many individuals are present, and where they are successfully breeding. Information on trends is also important in managing and protecting the species. Grand Canyon

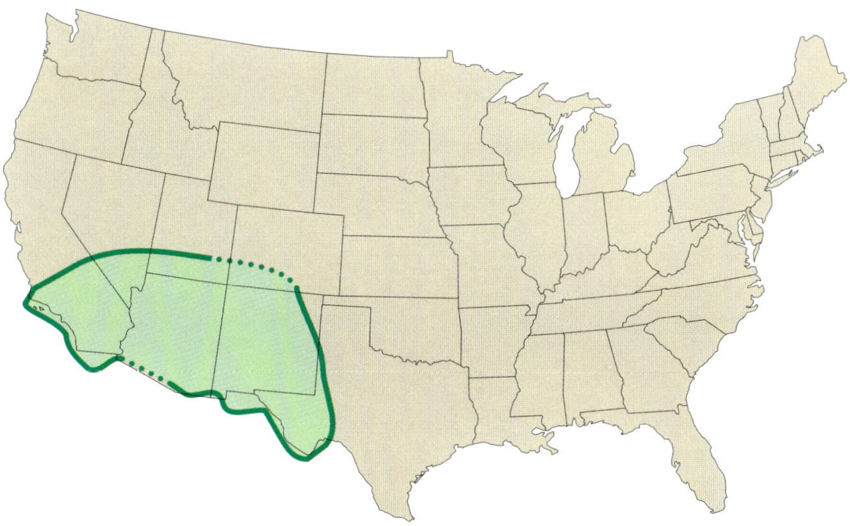

Fig. 1. Breeding distribution of the southwestern willow flycatcher. Dotted line represents areas where distribution is uncertain.

Courtesy M.K. Sogge, NBS

Fig. 2. Southwestern willow flycatcher breeding territory in tamarisk habitat along the Colorado River in the Grand Canyon.

National Park, the USFWS, and the U.S. Bureau of Reclamation have been regularly monitoring the status of the southwestern willow flycatcher in the Grand Canyon since 1982. The National Biological Service's Colorado Plateau Research Station at Northern Arizona University has conducted this monitoring since 1992. The Grand Canyon is one of the few areas with such a long record of willow flycatcher population data; the only others are the Santa Margarita and Kern rivers in southern California.

Methods

Our monitoring program involved intensive surveys of about 450 km (280 mi) of the Colorado River in Arizona between Glen Canyon Dam (Lake Powell) and upper Lake Mead. This portion of the river flows from elevation 945 m (3,100 ft) at the dam to 365 m (1,200 ft) at Lake Mead. We walked through or floated along all potential southwestern willow flycatcher habitat patches along the river corridor and looked and listened for willow flycatchers. Although willow flycatchers look very similar to several other flycatchers, they can be readily identified by their distinctive "fitz-bew" song. To increase the chance of detecting resident flycatchers, we played a tape recording of willow flycatcher songs and calls (Fig. 3) as we moved through our survey areas. This technique usually elicits a response from any resident southwestern willow flycatchers that may be present (Tibbitts et al. 1994). We conducted surveys from May through July at about 160 habitat patches each year (1992 and 1993), and made repeated trips to each site (Sogge et al. 1993).

Courtesy M.K. Sogge, NBS

Fig. 3. Surveyor broadcasting taped vocalizations and looking for response from willow flycatchers.

Status and Trends

Surveys conducted between 1982 and 1991 looked only at the upper 114 km (71 mi) of the river and counted primarily singing males. Within this same stretch, we detected only two singing male willow flycatchers in 1992, and three in 1993. These willow flycatchers were found only in the dense riparian habitat dominated by tamarisk, but including some willows along the river corridor above 860 m (2,800 ft) elevation. The breeding population of southwestern willow flycatchers in the Grand Canyon was very low: we found only one nest in 1992, and only three in 1993. Worse yet, each of the three 1993 willow flycatcher nests was brood-parasitized by brown-headed cowbirds (*Molothrus ater*), and none produced young willow flycatchers. With such a small breeding population, and the potential for severe loss of breeding effort due to cowbirds, there is concern over the continued survival of the species within Grand Canyon.

Based on comparison with past willow flycatcher surveys in the Grand Canyon (river mi 0-71; Brown 1988, 1991), willow flycatchers have declined since the mid-1980's (Fig. 4). Because we could conduct more surveys and our methods were more likely to detect flycatchers than the pre-1992 surveys (conducted without using tape playback), the population decline of the southwestern willow flycatcher in Grand Canyon may be even more dramatic than our data indicate.

We did find willow flycatchers in areas of the river corridor where surveys had not been previously conducted: three in 1992 and five in 1993. Two other willow flycatchers were also found during separate bird studies on the river corridor. These birds were found in tamarisk (above 530 m; 1,900 ft) or willow (below 530 m; 1,900 ft) habitats. None of these willow flycatchers established territories or bred, however, and most were probably migrants simply passing through the area (Sogge et al. 1993).

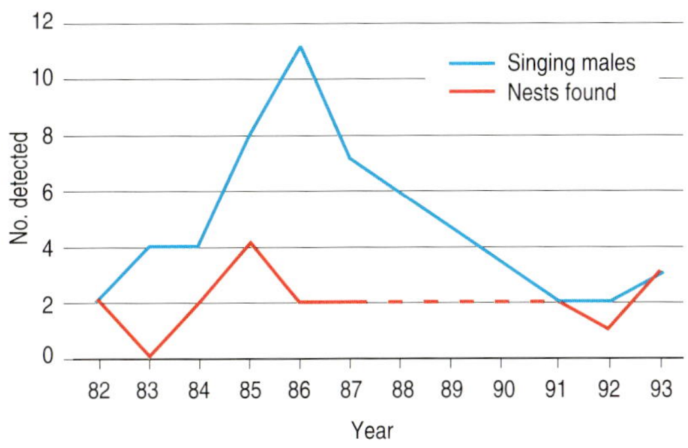

Fig. 4. The numbers of singing male southwestern willow flycatchers and flycatcher nests detected in the Grand Canyon (river mi 0 to 71), 1982-93. Dotted lines represent years when surveys were not conducted.

The low breeding population, historical declines, and potentially limited productivity in the Grand Canyon reflect the plight of the southwestern willow flycatcher throughout its range. Declines have been noted virtually everywhere the flycatcher occurs, and threats to its survival are widespread and immediate. As human activities such as urbanization, water diversion, agriculture, and grazing in riparian areas continue in the Southwest, so do the loss and alteration of riparian habitat. Vital wintering habitat in Mexico and Central America is also being lost to similar human activities.

Brood parasitism by brown-headed cowbirds is another significant threat to southwestern willow flycatchers within the Grand Canyon and in many other areas. In fact, cowbirds may be one of the greatest threats in areas where breeding habitat is protected, such as the Grand Canyon and other national parks and protected areas. Cowbirds lay their eggs in the nests of other birds (the host), who subsequently abandon the nests or raise the cowbird chicks. Female cowbirds will sometimes remove or destroy host eggs, and cowbird chicks often monopolize the parental care of the hosts. Thus, cowbird parasitism can reduce the number of host young produced, and in some cases, cowbirds may be the only young successfully raised by the host. Such effects have been recorded for southwestern willow flycatchers in the Grand Canyon and in other areas as well (Federal Register 1993). Conversely, control and removal of cowbirds have resulted in local increases in southwestern willow flycatchers and other songbirds. Cowbird brood parasitism is related to riparian loss and fragmentation because cowbird parasitism is highest in fragmented habitats.

The southwestern willow flycatcher is a unique and valuable part of the riparian community in the Southwest. Although recent and planned future surveys provide important status and distributional information on the flycatcher in the Grand Canyon and a few other areas within Arizona, there is a critical need for basic surveys and ecological research (including the effect of brown-headed cowbirds) on this species throughout most of its range, particularly in New Mexico, southern Utah, and Colorado. As a riparian obligate species whose continued existence is directly tied to the future of our remaining riparian habitats, its precarious status and historic decline help illustrate the need for riparian preservation and management. Such management is important not only for the southwestern willow flycatcher, but also for all plant and animal species that make up and depend on these valuable riparian areas.

Southwestern flycatcher (*Empidonax traillii extimus*).

References

Brown, B.T. 1988. Breeding ecology of a willow flycatcher population in Grand Canyon, Arizona. Western Birds 19(1):25-33.

Brown, B.T. 1991. Status of nesting willow flycatchers along the Colorado River from Glen Canyon Dam to Cardenas Creek, Arizona. U.S. Fish and Wildlife Service Endangered Species Rep. 20. 34 pp.

Federal Register. 1993. Proposal to list the southwestern willow flycatcher as an endangered species, and to designate critical habitat. U.S. Fish and Wildlife Service 23 July 1993. Federal Register 58:39495-39522.

Finch, D.M., and P.W. Stangel. 1993. Status and management of Neotropical migratory birds; 1992 September 21-25; Estes Park, CO. Gen. Tech. Rep. RM-229. U.S. Forest Service, Rocky Mountain Forest and Range Experiment Station, Fort Collins, CO. 422 pp.

Sogge, M.K., T.J. Tibbitts, and S.J. Sferra. 1993. Status of the southwestern willow flycatcher along the Colorado River between Glen Canyon Dam and Lake Mead—1993. Summary report. National Park Service Cooperative Park Studies Unit, Northern Arizona University, U.S. Fish and Wildlife Service, and Arizona Game and Fish Department. 69 pp.

Tibbitts, T.J., M.K. Sogge, and S.J. Sferra. 1994. A survey protocol for the southwestern willow flycatcher (*Empidonax traillii extimus*). National Park Service Tech. Rep. NPS/NAUCPRS/NRTR-94/04. 24 pp.

For further information:

Mark K. Sogge
National Biological Service
Colorado Plateau Research Station
Northern Arizona University
Box 5614
Flagstaff, AZ 86011

Thomas Kitchin/Tom Stack & Associates©

Mammals

by
Science Editor
Benjamin N. Tuggle
U.S. Fish and Wildlife Service
Chicago Illinois Field Office
Barrington, Illinois 60010

Overview

Many mammalian population studies have been initiated to determine a species' biological or ecological status because of its perceived economic importance, its abundance, its threatened or endangered state, or because it is viewed as our competitor. As a result, data on mammalian populations in North America have been amassed by researchers, naturalists, trappers, farmers, and land managers for years.

Inventory and monitoring programs that produce data about the status and trends of mammalian populations are significant for many reasons. One of the most important reasons, however, is that as fellow members of the most advanced class of organisms in the animal kingdom, the condition of mammal populations most closely reflects our condition. In essence, mammalian species are significant biological indicators for assessing the overall health of advanced organisms in an ecosystem.

Habitat changes, particularly those initiated by humans, have profoundly affected wildlife populations in North America. Though Native Americans used many wildlife species for food, clothing, and trade, their agricultural and land-use practices usually had minimal adverse effects on mammal populations during the pre-European settlement era. In general, during the post-Columbian era, most North American mammalian populations significantly declined, primarily because of their inability to adapt and compete with early European land-use practices and pressures.

Habitat modification and destruction during the settlement of North America occurred very slowly initially. Advances in agriculture and engineering accelerated the loss or modification of habitats that were critical to many species in climax communities. These landscape transformations often occurred before we had any knowledge of how these environmental changes would affect native flora and fauna. Habitat alterations were almost always economically driven and in the absence of land-use regulations and conservation measures many species were extirpated.

In addition to rapid and sustained habitat and landscape changes from agricultural and silvicultural practices, other factors such as unregulated hunting and trapping, indiscriminate predator and pest control, and urbanization also contributed significantly to the decline of once-bountiful mammalian populations. These practices, individually and collectively, have been directly correlated with the decline or extinction of many sensitive species.

The turn of the century brought a new focus on conservation efforts in this country. Populations of some species, such as the white-tailed deer (*Odocoileus virginianus*), showed marked recovery after regulatory and conservation strategies began. Ardent wildlife management and conservation programs, started primarily for game species, have increased our knowledge and understanding of species and habitat interactions. Conservation programs have also positively affected many species that share habitat with the target species the programs are designed to aid. To complement these efforts, however, integrated regulatory legislation and conservation policies that specifically help sustain nontarget species and their habitats are still imperative.

The increased emphasis on the importance of managing for biological diversity and adopting an ecosystem approach to management has enhanced our efforts to move from resource-management practices that are oriented to single species to strategies that focus on the long-term conservation of native populations and their natural habitats. Thus, an integrated and comprehensive inventory and monitoring program that coordinates data on the status and trends of our natural resources is critical to successfully manage habitats that support a diverse array of plant and animal species.

This section provides knowledge on the status and trends of some higher vertebrate species that occupy some of this country's most diverse ecosystems. Many articles discuss historical and present species distribution, while others discuss the need for further research to fill our gaps of knowledge regarding the species. The articles cover a range of mammal species, some that have benefited greatly from past conservation efforts, and others that are now threatened or endangered, with the effort to recover them just beginning. Some species have been directly affected by habitat loss or modification, others by past hunting and trapping pressures.

We should not forget that our survival depends on wildlife, particularly higher vertebrates, nor should we forget that the status of wildlife populations serves as an advance indicator of overall environmental quality.

Marine Mammals

by
Anne Kinsinger
National Biological Service

Summarized from National Oceanic and Atmospheric Administration (1994)

At least 35 species of marine mammals are found along the U.S. Atlantic coast and in the Gulf of Mexico: 2 seal species, 1 manatee, and 32 species of whales, dolphins, and porpoises (*see* Table 1 for status of selected species). Seven of these species are listed as endangered under the Endangered Species Act (ESA). At least 50 species of marine mammals are found in U.S. Pacific waters: 11 species of seals and sea lions; walrus; polar bear; sea otter; and 36 species of whales, dolphins, and porpoises; 11 species are listed as endangered or threatened under the ESA (*see* Table 2 for the status of selected species).

Table 1. Status of selected Atlantic and Gulf of Mexico coast species of marine mammals.

Species and geographic area	Abundance	Status	Trends	Official status in designated U.S. waters
Fin whale, NE U.S.	5,200	Unknown	Unknown	Endangered*
Humpback whale, NW Atlantic	5,100 (2,888-8,112)	Possibly 65% of 1850 population	Unknown	Endangered*
Northern right whale, NW Atlantic	350	Probably <5% of original number	Unknown	Endangered*
Pilot whales, NE U.S.	Unknown	Unknown	Unknown	
Bottlenose dolphin				
NE U.S. coastal type	Unknown	Possibly down by 50% 1987-88	Unknown	Depleted**
NE U.S. offshore type	10,000-13,000	Unknown		
Gulf of Mexico (offshore and coastal types)	35,000-45,000	Possibly down by 50% 1987-88	Unknown	
Whitesided dolphin, NE U.S.	27,600	Unknown	Unknown	
Spotted dolphin, NE U.S.	200	Unknown	Unknown	
Harbor porpoise, Gulf of Maine	47,200	Unknown	Unknown	Proposed as threatened*
Harbor seal, NE U.S.	26,000	Unknown	Increasing	
Beaked whales (six species in U.S. waters)	Unknown	Unknown	Unknown	

*Endangered Species Act.
**Marine Mammal Protection Act.

NMFS Assessments

The National Marine Fisheries Service (NMFS), an agency within the National Oceanic and Atmospheric Administration (NOAA), conducts research and status studies on many of these marine mammals under the authorities of the Magnuson Fisheries Conservation and Management Act, the Marine Mammal Protection Act (MMPA), and the ESA. The results of the status surveys include information required by the MMPA and the ESA on abundance (population size); status (as compared with historical levels or current viability); trends (changes in abundance); and status in U.S. waters. These results, published annually by NOAA, are the basis for this summary (NOAA 1994).

Estimates of abundance in U.S. waters are available for many, though not all, marine mammal species. Information on status and trends, however, is extremely limited because so little is known of the basic life history of many marine mammal species that scientists can determine neither status nor whether a population estimate represents a healthy, sustainable population. Moreover, long-term trends in many populations cannot be determined because historical population data are not available.

The NMFS provides assessments for 139 stocks (i.e., populations of species or groups of species that are treated together for management) of marine mammals; the status of 120 stocks is unknown, and trend data are only available for 19 stocks. The recently reauthorized MMPA requires the NMFS to conduct periodic assessments of marine mammal stocks that occur in U.S. waters. For this reason, better status and trends data are likely to become available over the next few years.

Abundance and status data for selected marine mammals are summarized in Table 1 (Atlantic species) and Table 2 (Pacific species). Trend data are mixed, but a number of conservation success stories have come from marine mammals. The bowhead and grey whales have shown significant population increases, as have California sea lions, the northern elephant seal, harbor seals in California, Oregon, Washington, and the Northeast, and the southern sea otter. These increases are largely the result of prohibition of commercial whaling by the International Whaling Commission (IWC) and by protection enacted under the MMPA and ESA. Other marine mammal populations, such as the Steller sea lion and the common dolphin in the eastern tropical Pacific, are still declining. Causes of decline in marine mammal populations include bycatch associated with commercial fishing, illegal killings, strandings, entanglement, disease, ship strikes, altered food sources, and possibly exposure to contaminants.

Population Trends

Whales

The eastern North Pacific stock of grey whale (*Eschrichtius robustus*) is rising (Fig. 1) and is one success story in species restoration. The NMFS estimates that the historical populations of grey whales in 1896 were around 15,000-20,000. While current population levels are below the estimated carrying capacity of 24,000, they appear higher than historical levels and represent a substantial gain. The population growth rate between 1968 and 1988 was 3.3% per year. After 3 years of review, on 15 June 1994, this species was removed from protection (delisted) under the ESA, an indication of successful management.

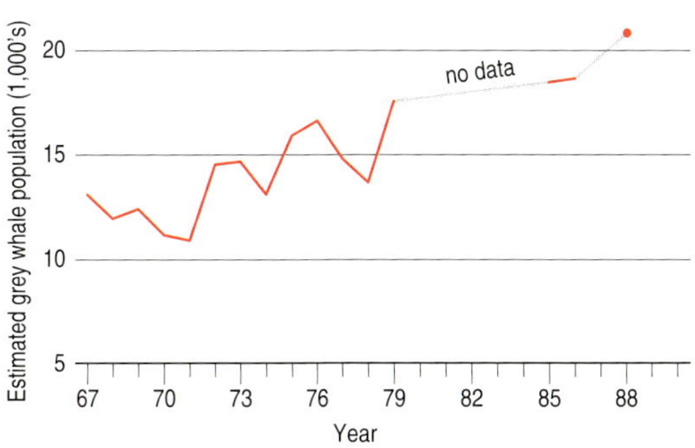

Fig. 1. Estimated population of grey whales, 1967-90 (NOAA 1994).

Table 2. Status of selected Pacific coast species of marine mammals.

Species and area	Abundance	Status	Trends	Official status in designated U.S. waters
Fin whale	935	Unknown	Unknown	Endangered*
Humpback whale, E Pacific	~1,400	Probably less than 15% of 1850 population	Unknown	Endangered*
Northern right whale	Unknown	Unknown	Unknown	Endangered*
Bowhead whale, W. Arctic	7,500	About 40% of 1848 population size	Increasing at 3.1%/yr, 1978-88	Endangered*
Grey whale	20,869 (19,200-22,700)	Recovered to historical 1845 abundance levels	Increasing at 3.3%/yr, 1968-88	Removed from ESA listing June 1994
E. tropical Pacific dolphins				
NE spotted	731,000	Depleted	Declining	
W/S spotted	1,298,000	Unknown	Stable	
Coastal spotted	30,000	Unknown	Stable	
E spinner	631,800	Depleted, 44% of late 1950's population	Stable	Depleted**
Whitebelly spinner	1,019,000	Unknown	Stable	
N common	476,300	Unknown	Declining	
Central common	406,100	Unknown	Stable	
S common	2,210,900	Unknown	Stable	
Common (pooled)	3,093,300	Unknown	Stable	
Striped	1,918,000	Unknown	Stable	
Harbor porpoise				
SE Alaska	2,052			
W Gulf of Alaska	1,273			
N California	10,000			
Central California	3,806			
Inland Washington	3,298			
Oregon/Washington	23,701			
Hawaiian monk seal	1,550	Declined 50% since 1950's	Unknown, pup counts declining to variable	Endangered*
California sea lion (CA, OR, WA)	111,016	Unknown	Increasing 10.2%/yr since 1983	
Harbor seal		Unknown	Increasing?	
Alaska	63,000		Declining	
California	23,113		Increasing	
Oregon/Washington	45,713		Increasing	
Northern fur seal				
Pribilof Islands	982,000	< 40% of 1950's population	No significant trend since 1983 on St. Paul Is.	Depleted**
San Miguel	6,000		Increasing	
Steller sea lion Northern Pacific	116,000	< 22% of late 1950's population	Declined 73% since 1960	Threatened*

*Endangered Species Act.
**Marine Mammal Protection Act.

The bowhead whale (*Balaena mysticetus*) is an endangered species that has shown a significant increase since the IWC adopted new rules in 1980 regulating its harvest for subsistence purposes by Native Americans (Fig. 2). The total prewhaling (before the mid-1800's) population of the bowhead whale is believed to have been 12,000-18,000; NMFS estimates that by 1900 it was probably in the low thousands. The current population of 7,500 is about 40% of its estimated 1848 population level (Table 2), more than 3 times the population low reached in 1980. The bowhead whale population has been growing by about 3% per year since 1978.

The endangered western North Atlantic population of right whales (*Eubalaena glacialis*) is considered by NMFS to be the only northern hemisphere right whale population with a significant number of individuals, about 300-350 animals (Table 1). Other stocks are considered virtually extinct: only five to seven sightings have been made in the last 25 years. Estimates of the pre-18th century population are as high as

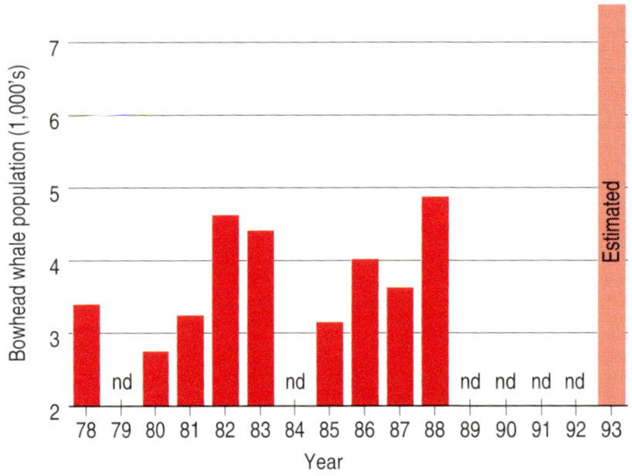

Fig. 2. Actual counts of bowhead whales, 1978-90 (NOAA 1994).

10,000. NMFS believes that human influences such as ship strikes and net entanglements are affecting about 60% of the population. The agency notes that the annual loss of even a single right whale has measurable effect on the population, by greatly inhibiting recovery of the species.

Dolphins and Porpoises

The coastal migratory stock of Atlantic bottlenose dolphin (*Tursiops truncatus*) is listed as depleted under the MMPA (Table 1). This coastal stock incurred a loss of up to 50% during a 1987-88 die-off. Long-term trends are unknown, but the stock may require as many as 50 years to recover.

Harbor porpoises (*Phocoena phocoena*) occur on both U.S. coasts and are faring relatively well. The northwestern Atlantic harbor porpoise is found from Newfoundland, Canada, to Florida. The NMFS 1991-92 population estimate of the Gulf of Maine population is 47,200 (Table 1), but estimates of abundance for other populations do not exist. NMFS has found that harbor porpoise mortality from sink gill-net fisheries along the east coast of North America from Canada to North Carolina appears large compared with the species' natural reproduction rates. Management actions are being taken to address this issue, but long-term trends are unknown. On the west coast, NMFS's combined population estimate for northern California, Oregon, and Washington coastal stocks is 45,713.

The NMFS assesses 10 stocks of eastern tropical Pacific dolphins. Although population trends for most populations cannot be detected, the northeastern stocks of offshore spotted dolphin and the common dolphin may be declining (Table 2). These two stocks, as well as the eastern spinner and the striped dolphin, are incidentally taken in the international fishery for yellowfin tuna in the tropical Pacific waters off Mexico and Central America. Although mortality has been reduced in recent years, populations are still declining, or at best not increasing.

Seals and Sea Lions

According to the NMFS, harbor seal (*Phoca vitulina*) populations have increased recently throughout much of their range because of protection by the MMPA. Recent NMFS surveys estimate that at least 26,000 harbor seals inhabit the Gulf of Maine (Table 1). Populations of California harbor seals are also increasing; a recent survey resulted in a count of about 23,000 harbor seals residing in the Channel Islands and along the California mainland (Table 2), an increase from about 12,000 in 1983. The population of harbor seals in Oregon and Washington has been estimated at 45,700, and is also increasing. Harbor seal counts in the Central Gulf of Alaska, however, have declined significantly in the past two decades; numbers are currently estimated by NOAA at 63,000 seals.

The northern fur seal (*Callorhinus ursinus*) is considered depleted under the MMPA. Production on one of its major breeding areas, Alaska's Pribilof Islands, dropped more than 60% between 1955 and 1980, but has since stabilized. The current population is less than 40% of the mid-1950's level; no significant trend in the Pribilof Islands population has been noted since 1983 (Table 2).

The northern sea lion or Steller sea lion (*Eumetopias jubatus*) is listed as threatened under the ESA. Species numbers have declined sharply throughout its range in the last 34 years (Table 2). The number of adults and juveniles in U.S. waters dropped from 154,000 in 1960 to 40,000 in 1992, a reduction of 73%. Most of this decline occurred in Alaska waters, and is believed due to a combination of factors, including incidental kills, illegal shooting, changes in prey availability and biomass, and perhaps other unidentified factors.

The U.S. population of California sea lions (*Zalophus californianus*) is increasing at a rate of about 10% annually. In 1990, NMFS estimated that the U.S. population was 111,000 individuals (Table 2). A number of human-related interactions, such as incidental take during fishing, entanglement, illegal killing, and pollutants, result in sea lion deaths.

Reference

NOAA. 1994. Our living ocean: report on the status of U.S. living marine resources, 1993. NOAA Tech. Memorandum NMFS-F/SPO-15. National Oceanic and Atmospheric Administration, National Marine Fisheries Service, Silver Spring, MD. 136 pp.

For further information:

Michael Payne
National Marine Fisheries Service
Office of Protected Resources
F/PR2
1335 East-West Highway
Silver Spring, MD 20910

Indiana Bats

by
Ronald D. Drobney
National Biological Service
Richard L. Clawson
Missouri Department of Conservation

The Indiana bat (*Myotis sodalis*) is an endangered species that occurs throughout much of the eastern United States (Fig. 1). Although bats are sometimes viewed with disdain, they are of considerable ecological and economic importance. Bats consume a diet consisting largely of nocturnal insects and thereby are a natural control for both agricultural pests and insects that are annoying to humans. Furthermore, many forms of cave life depend upon nutrients brought into caves by bats in the form of guano or feces (Missouri Department of Conservation 1991).

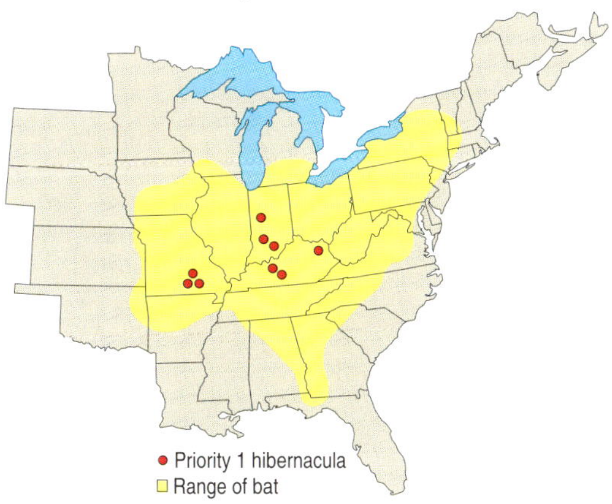

Fig. 1. Range of the Indiana bat and locations of Priority 1 hibernacula (*see* text for definitions).

Indiana bats use distinctly different habitats during summer and winter. In winter, bats congregate in a few large caves and mines for hibernation and have a more restricted distribution than at other times of the year. Nearly 85% of the known population winters in only seven caves and mines in Missouri, Indiana, and Kentucky, and approximately one-half of the population uses only two of these hibernacula.

In spring, females migrate north from their hibernacula and form maternity colonies in predominantly agricultural areas of Missouri, Iowa, Illinois, Indiana, and Michigan. These colonies, consisting of 50 to 150 adults and their young, normally roost under the loose bark of dead, large-diameter trees throughout summer; however, living shagbark hickories (*Carya ovata*) and tree cavities are also used occasionally (Humphrey et al. 1977; Gardner et al. 1991; Callahan 1993; Kurta et al. 1993).

As a consequence of their limited distribution, specific summer and winter habitat requirements, and tendency to congregate in large numbers during winter, Indiana bats are particularly vulnerable to rapid population reductions resulting from habitat change, environmental contaminants, and other human disturbances (Brady et al. 1983). Additionally, because females produce only one young per year, recovery following a population reduction occurs slowly. Concerns arising from the high potential vulnerability and slow recovery rate have led to a long-term population monitoring effort for this species.

Bat Census Design

The first rangewide census of wintering Indiana bats was made in 1975. All subsequent population data were gathered according to standardized cave census techniques established by the Indiana Bat Recovery Team in 1983 (Brady et al. 1983). Data presented in this article are based upon counts made at 2-year intervals at Priority 1 hibernacula, which are caves where winter populations exceeding 30,000 bats have been recorded. We chose to use data only from Priority 1 caves because they contain the majority of bats in the population. During midwinter cave censuses, bats hanging singly and in small clusters of up to 25 were counted individually. The number of bats in larger clusters was determined by multiplying the surface area of the cluster by bat density (Fig. 2).

Bat Populations: Trends and Recovery Prospects

Before the 1970's, the population status of Indiana bats was poorly understood because the locations of many of their winter hibernacula were unknown, and the counts that were conducted were made irregularly and inconsistently. The 1975 census established a benchmark of nearly 450,000 bats using Priority 1 hibernacula. Since 1983 the number of bats tallied has declined significantly, reaching a low of 347,890 during the most recent census in 1993 (Fig. 3).

Courtesy R. L. Clawson

Fig. 2. Hibernating cluster of Indiana bats.

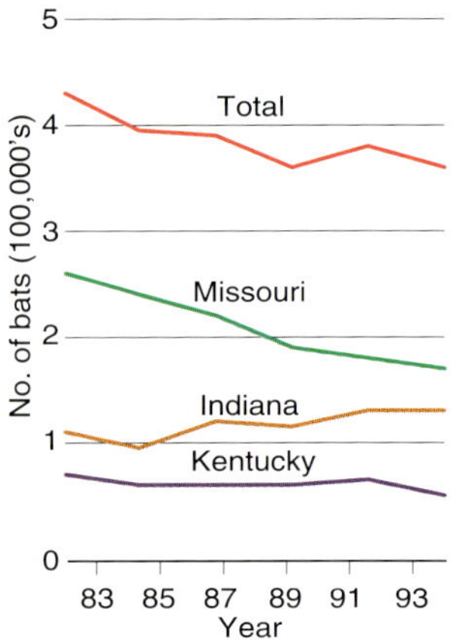

Fig. 3. State and national trends for Indiana bats, 1983-93.

Although the national trend indicates a 22% decline during the past 10 years, this decrease has not been consistent across the species' winter range (Fig. 3). Most of the decrease in the 10-year national census results can be accounted for by a precipitous 34% decline in the number of bats counted in Missouri. A more favorable pattern has been noted in Indiana, where numbers have increased, and in Kentucky, where the population has remained relatively stable.

Recovery efforts have included placing gates or fences across cave entrances to eliminate disturbances to hibernating bats. These exclusion devices have not halted population declines, suggesting that other factors are negatively influencing bat populations.

Another potential threat is the loss of habitat used by maternity colonies. Maternity roost sites in dead trees exposed to sunlight and located in upland forests and near streams are particularly important. Losses of these sites through streamside deforestation and stream channelization pose significant threats to population recovery.

Pesticides and other environmental contaminants represent additional hazards. Indiana bats are exposed to lingering residues of chlorinated hydrocarbon pesticides such as aldrin and heptachlor. These products have been banned since the 1970's, but persist in the soil and in insects upon which bats feed. Potential detrimental effects of the new generation of pesticides, including organophosphates, are unknown.

The long-term prognosis for Indiana bat populations is uncertain. The fact that wintering populations appear to be increasing in Indiana and are remaining relatively stable in Kentucky provides the basis for some optimism. A better understanding of their summer habitat requirements and factors affecting survival and reproduction is needed so that more effective recovery efforts can be formulated. It is important to recognize, however, that even if the factors that are negatively influencing Indiana bat populations are removed, recovery will occur slowly because this species has a low reproductive rate.

For further information:

Ronald D. Drobney
National Biological Service
Missouri Cooperative Fish and Wildlife Research Unit
112 Stephens Hall
University of Missouri
Columbia, MO 65211

References

Brady, J.T., R.K. LaVal, T.H. Kunz, M.D. Tuttle, D.E. Wilson, and R.L. Clawson. 1983. Recovery plan for the Indiana bat. U.S. Fish and Wildlife Service, Washington, DC. 94 pp.

Callahan, E.V. 1993. Indiana bat summer habitat requirements. M.S. thesis, University of Missouri, Columbia. 74 pp.

Gardner, J.E., J.D. Garner, and J.E. Hofmann. 1991. Summer roost selection and roosting behavior of *Myotis sodalis* (Indiana bat) in Illinois. Final report. Illinois Natural History Survey, Illinois Department of Conservation, Champaign. 56 pp.

Humphrey, S.R., A. R. Richter, and J.B. Cope. 1977. Summer habitat and ecology of the endangered Indiana bat, *Myotis sodalis*. Journal of Mammalogy 58:334-346.

Kurta, A., D. King, J.A. Teramino, J.M. Stribley, and K.J. Williams. 1993. Summer roosts of the endangered Indiana bat (*Myotis sodalis*) on the northern edge of its range. American Midland Naturalist 129:132-138.

Missouri Department of Conservation. 1991. Endangered bats and their management in Missouri. Missouri Department of Conservation, Jefferson City. 8 pp.

Gray Wolves

by
L. David Mech
National Biological Service
Daniel H. Pletscher
University of Montana
Clifford J. Martinka
National Biological Service

The gray wolf (*Canis lupus*) originally occupied all habitats in North America north of about 20° north latitude (in Mexico), except for the southeastern United States, where the red wolf (*C. rufus*) lived. By 1960 the wolf was exterminated by federal and state governments from all of the United States except Alaska and northern Minnesota. Until recently, 24 subspecies of the gray wolf were recognized for North America, including 8 in the contiguous 48 states. After the gray wolf was listed as an endangered species in 1967, recovery plans were developed for the eastern timber wolf (*C.l. lycaon*), the northern Rocky Mountain wolf (*C.l. irremotus*), and the Mexican wolf (*C.l. baileyi*). The other subspecies in the contiguous United States were considered extinct.

The Eastern Timber Wolf Recovery Plan (U.S. Fish and Wildlife Service 1992) set as criteria for recovery the following conditions: a viable wolf population in Minnesota consisting of at least 200 animals, and either a population of at least 100 wolves in the United States within 160 km (100 mi) of the Minnesota population, or a population of at least 200 wolves if farther than 160 km (100 mi) from the Minnesota population. The Northern Rocky Mountain Wolf Recovery Plan (U.S. Fish and Wildlife Service 1987) defined recovery as when at least 10 breeding pairs of wolves inhabit each of three specified areas in the northern Rockies for 3 successive years. The Mexican Wolf Recovery Plan (U.S. Fish and Wildlife Service 1982) called for a self sustaining population of at least 100 Mexican wolves in a 12,800-km^2 (4,941-mi^2) range.

A recent revision of wolf subspecies in North America (Nowak 1994), however, reduced the number of subspecies originally occupying the contiguous 48 states from eight to four. It classified the wolf currently inhabiting northern Montana as being *C.l. occidentalis*, primarily a Canadian and Alaskan wolf. It considered *C.l. nubilus* to be the wolf remaining in most of the range of the former northern Rocky Mountain wolf and the present range of the eastern timber wolf; this leaves the eastern timber wolf extinct in its former U.S. range, surviving now only in southeastern Canada. The new classification may have implications for the recovery criteria propounded by the Eastern Timber Wolf and Northern Rocky Mountain Wolf recovery plans. The reclassification did not change the status of the Mexican wolf.

This article is based on a review of the literature and recent personal communications. Most of the studies cited depended primarily on the use of aerial radio-tracking and observation (Mech 1974; Mech et al. 1988).

Population Status by Region

Lake Superior Region

After wolves were protected in 1974 by the Endangered Species Act of 1973, their numbers and distribution in Minnesota increased, and individuals began recolonizing Wisconsin (Mech and Nowak 1981). The population increased in Wisconsin and began recolonizing Michigan (Hammill 1993). The Minnesota population increased at about 3% per year (Fuller et al. 1992); its distribution continues to increase (Paul 1994). The best estimate of its current size is 1,740-2,030 wolves. Wisconsin and mainland Michigan each supported an estimated 50+ wolves in early 1994 (A.P. Wydeven, Wisconsin Department of Natural Resources, personal communication; J. Hammill, Michigan Department of Natural Resources, personal communication), and Isle Royale National Park about 14 wolves (Peterson 1994).

As wolves increased in Minnesota, they also began dispersing westward into North and South Dakota (Licht and Fritts 1994). The only records from these states involve 10 wolves killed from 1981 through 1992, but the possibility remains that small populations may occur in some of the more remote areas. Sufficient prey certainly exist there, so if dispersing wolves from Minnesota and Manitoba are not killed by humans, they should be able to breed and start populations.

Western United States

Wolves were virtually absent in the western United States (other than an occasional animal that disperses from Canada) from the mid-1930's through 1980 (Ream and Mattson 1982). The nearest breeding population through this period was probably in Banff National Park, Alberta. Wolves were completely protected in extreme southeastern British Columbia in the 1960's (Pletscher et al. 1991). This led to recolonization of the area and adjacent northwestern Montana, and in 1986 a den was documented in Glacier National Park, Montana (Ream et al. 1989). This population, which straddles the Canadian border, has since grown to four packs and about 45 wolves.

Three breeding packs have been reported elsewhere in western Montana (Fritts et al. 1994), all probably founded by animals that dispersed from Glacier National Park. Additionally, an animal that dispersed from Glacier is in northeastern Idaho, and a wolf shot in 1992 just south of Yellowstone National Park was genetically related to Glacier wolves (Fritts et al. 1994). Animals that have dispersed, primarily from the Glacier area, have begun backfilling the area between Glacier National Park and Jasper National Park, Alberta (Boyd et al. 1994). This connection to larger wolf populations in Canada will enhance the viability of the U.S. population.

Courtesy L.D. Mech, NBS

Gray wolf (*Canis lupus*).

Although occasional wolves have been sighted in Wyoming and Washington and numerous sightings have been reported from central Idaho, no reproduction has been documented in these states, with the possible exception of litters in Washington in 1990 (S.H. Fritts, U.S. Fish and Wildlife Service, personal communication). An environmental impact statement on the reintroduction of wolves to Yellowstone and central Idaho was completed in early 1994.

Factors Impeding Wolf Recovery

In small populations, the death of any individual can seriously impede recovery, meaning that factors that may not affect larger populations may hinder recovery of smaller ones. Such factors hindering the recovery of wolves include illegal and accidental killing of wolves by humans, canine parvovirus (Mech and Goyal 1993; Johnson et al. 1994; Wydeven et al. 1994), sarcoptic mange (A.P. Wydeven et al., Wisconsin Department of Natural Resources, personal communication), possibly Lyme disease (Thieking et al. 1992), and heartworm (*Dirofilaria immitis*; Mech and Fritts 1987). Of these, only killing by humans is subject to human control.

Future Outlook

All wolf populations in the contiguous 48 states are increasing. Minnesota wolves occupy all suitable areas there and even have been colonizing agricultural regions where the Eastern Timber Wolf Recovery Team felt they should not be (U.S. Fish and Wildlife Service 1992). Thus, in 1993, the Department of Agriculture's Animal Damage Control Program destroyed a record 139 wolves for livestock depredation control (Paul 1994). As wolf populations continue to grow in other newly colonized areas, there may be an increasing need for control of those wolves preying on livestock (Fritts 1993). Because the public has so strongly supported wolf recovery and reintroduction, it may be difficult for many to understand the need for control. Thus, strong efforts at public education will be required.

References

Boyd, D.K., P.C. Paquet, S. Donelon, R.R. Ream, D.H. Pletscher, and C.C. White. 1994. Dispersal characteristics of a recolonizing wolf population in the Rocky Mountains. *In* L.D. Carbyn, S.H. Fritts, and D.R. Seip, eds. Ecology and conservation of wolves in a changing world. Canadian Circumpolar Institute, Edmonton, Alberta. In press.

Fritts, S.H. 1993. The downside of wolf recovery. International Wolf 3(1): 24-26.

Fritts, S.H., E.E. Bangs, J.A. Fontaine, W.G. Brewster, and J.F. Gore. 1994. Restoring wolves to the northern Rocky Mountains of the United States. *In* L.D. Carbyn, S.H. Fritts, and D.R. Seip, eds. Ecology and conservation of wolves in a changing world. Canadian Circumpolar Institute, Edmonton, Alberta. In press.

Fuller, T.K., W.E. Berg, G.L. Radde, M.S. Lenarz, and G.B. Joselyn. 1992. A history and current estimate of wolf distribution and numbers in Minnesota. Wildlife Society Bull. 20:42-55.

Hammill, J. 1993. Wolves in Michigan: a historical perspective. International Wolf 3:22-23.

Johnson, M.R., D.K. Boyd, and D.H. Pletscher. 1994. Serology of canine parvovirus and canine distemper in relation to wolf (*Canis lupus*) pup mortalities. Journal of Wildlife Diseases 30:270-273.

Licht, D.S., and S.H. Fritts. 1994. Gray wolf (*Canis lupus*) occurrences in the Dakotas. American Midland Naturalist 132: 74-81.

Mech, L.D. 1974. Current techniques in the study of elusive wilderness carnivores. Pages 315-322 *in* Proceedings of the 11th International Congress of Game Biologists. National Swedish Environment Protection Board, Stockholm.

Mech, L.D., and S.H. Fritts. 1987. Parvovirus and heartworm found in Minnesota wolves. Endangered Species Tech. Bull. 12(5-6): 5-6.

Mech, L.D., S.H. Fritts, G. Radde, and W.J. Paul. 1988. Wolf distribution in Minnesota relative to road density. Wildlife Society Bull. 16:85-88.

Mech, L.D., and S.M. Goyal. 1993. Canine parvovirus effect on wolf population change and pup survival. Journal of Wildlife Diseases 29:330-333.

Mech, L.D., and R.M. Nowak. 1981. Return of the gray wolf to Wisconsin. American Midland Naturalist 105:408-409.

Nowak, R.M. 1994. Another look at wolf taxonomy. *In* L.D. Carbyn, S.H. Fritts, and D.R. Seip, eds. Ecology and conservation of wolves in a changing world. Canadian Circumpolar Institute, Edmonton, Alberta. In press.

Paul, W.J. 1994. Wolf depredation on livestock in Minnesota: annual update of statistics 1993. U.S. Department of Agriculture, Animal Damage Control, Grand Rapids, MN. 10 pp.

Peterson, R.O. 1994. Out of the doldrums for Isle Royale wolves? International Wolf 4(2):19.

Pletscher, D.H., R.R. Ream, R. Demarchi, W.G. Brewster, and E.E. Bangs. 1991. Managing wolf and ungulate populations in an international ecosystem. Transactions of the North American Wildlife and Natural Resources Conference 56:539-549.

Ream, R.R., M.W. Fairchild, D.K. Boyd, and A.J. Blakesley. 1989. First wolf den in western U.S. in recent history. Northwestern Naturalist 70:39-40.

Ream, R.R., and U.I. Mattson. 1982. Wolf status in the northern Rockies. Pages 362-381 *in* F.H. Harrington and P.C. Paquet, eds. Wolves of the world. Noyes Publishing, Park Ridge, NJ.

Thieking, A., S.M. Goyal, R.F. Berg, K.L. Loken, L.D. Mech, and R.P. Thiel. 1992. Seroprevalence of Lyme disease in Minnesota and Wisconsin wolves. Journal of Wildlife Diseases 28:177-182.

U.S. Fish and Wildlife Service. 1982. Mexican wolf recovery plan. USFWS, Albuquerque, NM. 103 pp.

U.S. Fish and Wildlife Service. 1987. Northern Rocky Mountain wolf recovery plan. USFWS, Denver, CO. 119 pp.

U.S. Fish and Wildlife Service. 1992. Recovery plan for the eastern timber wolf. USFWS, Twin Cities, MN. 73 pp.

Wydeven, A.P., R.N. Schultz, and R.P. Thiel. 1994. Gray wolf monitoring in Wisconsin — 1979-1991. *In* L.D. Carbyn, S.H. Fritts, and D.R. Seip, eds. Ecology and conservation of wolves in a changing world. Canadian Circumpolar Institute, Edmonton, Alberta. In press.

For further information:

L. David Mech
National Biological Service
North Central Forest Experiment Station
1992 Folwell Ave.
St. Paul, MN 55108

Black Bears in North America

by
Michael R. Vaughan
National Biological Service
Michael R. Pelton
University of Tennessee

Habitat loss, habitat fragmentation, and unrestricted harvest have significantly changed the distribution and abundance of black bears (*Ursus americanus*) in North America since colonial settlement. Although bears have been more carefully managed in the last 50 years and harvest levels are limited, threats from habitat alteration and fragmentation still exist and are particularly acute in the southeastern United States. In addition, the increased efficiency in hunting techniques and the illegal trade in bear parts, especially gall bladders, have raised concerns about the effect of poaching on some bear populations. Because bears have low reproductive rates, their populations recover more slowly from losses than do those of most other North American mammals.

Black bear populations are difficult to inventory and monitor because the animals occur in relatively low densities and are secretive by nature. Black bears are an important game species in many states and Canada and are an important component of their ecosystems. It is important that they be continuously and carefully monitored to ensure their continued existence.

Black Bear Survey Data

Information on the distribution and status of black bears in North America came from several unpublished reports and scientific publications. *Traffic USA* (McCracken et al. 1995) reports periodically on the status of black bears in North America. Two reports on the status and conservation of the bears of the world were presented at meetings of the International Conference on Bear Research and Management in 1970 and 1989 (Cowan 1972; Servheen 1990). Finally, much of the information for this report is from data collected by survey for a report by the International Union for the Conservation of Nature and Natural Resources/Species Survival Commission (IUCN/SSC) Bear Specialist Group (Pelton et al. 1994).

Range and Status

Black bears historically ranged over most of the forested regions of North America, including all Canadian provinces, Alaska, all states in the conterminous United States, and significant portions of northern Mexico (Hall 1981; Fig. 1). Their current distribution is restricted to relatively undisturbed forested regions (Pelton 1982; Pelton et al. 1994; Fig. 2). Black bears can still be found throughout Canada with the exception of Prince Edward Island (extirpated in 1937), and in at least 40 of the 50 states; their status in Mexico is uncertain (Leopold 1959; Fig. 2).

In the eastern United States black bear range is continuous throughout New England but becomes increasingly fragmented from the mid-Atlantic down through the Southeast (Maehr 1984). In the Southeast, most populations are now restricted to the Appalachian mountain chain or to coastal areas intermittently in all states from Virginia to Louisiana (J. Wooding, Florida Freshwater Fish and Game Commission, unpublished data).

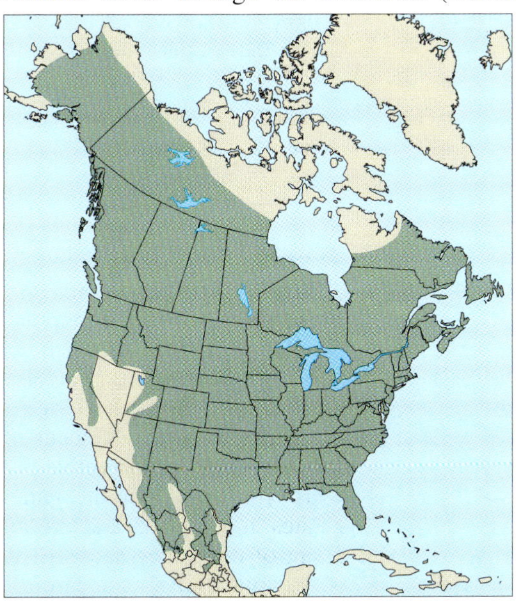

Fig. 1. Historical distribution of the American black bear (modified from Hall 1981).

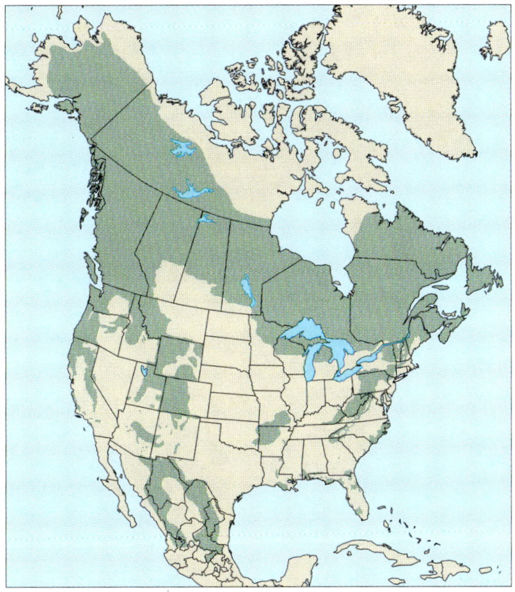

Fig. 2. Present distribution of the American black bear, based on survey responses from provinces and states (Pelton 1994) and research projects in Mexico (D. Doan, Texas A & I University, personal communication).

Recently, 11 Canadian provinces and territories reported stable black bear populations, and 10 provinces and territories estimated population sizes totaling about 359,000-373,000 (Pelton et al. 1994; McCracken et al. 1995; Table 1). Bears are legally harvested in all Canadian provinces and territories; total annual mortality from all sources (e.g., hunting, road kills, nuisance kills) is estimated at more than 23,000 (Pelton et al. 1994).

Table 1. Population estimates and trends of American black bears in Canada (adapted from Pelton et al. 1994).

Province	Population estimate	Trend
Alberta	39,600	Stable
British Columbia	121,600	Stable
Manitoba	25,000	Stable
New Brunswick	Unknown	Stable/declining*
Newfoundland	6,000-10,000	Stable
Northwest Territories	5,000+**	Stable
Nova Scotia	3,000	Stable
Ontario	65,000-75,000	Stable/increasing
Quebec	60,000	Stable
Saskatchewan	24,000**	Stable
Yukon	10,000	Stable
Total	359,200-373,200	

* Stable — East and Northeast; declining — West and Central.
**1991 or 1992 estimates from McCracken et al. (1995).

Thirty-eight of 40 states responding to a 1993 survey (Pelton et al. 1994) reported stable or increasing populations; only Idaho and New Mexico reported decreasing populations (Table 2). Based on data from 38 states, the total population estimate for black bears in the United States ranges from about 307,000 to 332,000 (excluding South Dakota and Wyoming). Black bears are listed as threatened or endangered in Florida, Louisiana, Mississippi, South Dakota, and Texas; rare in Missouri; and protected in Kentucky. They are unclassified in Connecticut. The remainder of the 40 states responding to the survey classify black bears as a game species (Table 2). In 1970 Arizona and Nevada listed black bears as a protected species and Texas listed them as game (Cowan 1972); thus the current classifications (Table 2) represent an upgrade in status for Arizona and Nevada and a downgrade for Texas. The status of bears in all remaining states covered in both surveys remained essentially unchanged.

The Southern Appalachian Region (Tennessee, North Carolina, South Carolina, and Georgia) is an area of special concern, and bear populations there have been routinely monitored since the late 1960's by the Southern Appalachian Bear Study Group. Initial estimates placed the population at 2,000-2,500 bears. The establishment of a network of black bear sanctuaries in the 1970's, scattered throughout the national forests in North Carolina, Tennessee, and Great Smoky Mountains National Park, provided protection for bears in the region, and estimates remain at 2,000-2,500 bears.

Courtesy J. Kasbohm

Black bear (*Ursus americanus*).

Two of 16 recognized subspecies of black bears (Hall 1981) require special mention: the Louisiana bear (*U.a. luteolus*), with a range of east Texas, all of Louisiana, and southern Mississippi; and the Florida bear (*U.a. floridanus*), with a range of Florida and southern Alabama. The U.S. Fish and Wildlife Service was petitioned in 1987 and 1990 to list the Louisiana bear and the Florida bear, respectively, as endangered species under the Endangered Species Act of 1973. In 1992 the Louisiana bear was officially placed on the federal endangered species list as a threatened species, and the Florida bear was placed in a "warranted but precluded" category. This latter category indicates that although biological evidence supports listing, several other species of higher priority are awaiting listing and will be listed before the Florida bear. At present, the U.S. Fish and Wildlife Service is considering listing bears in southern but not northern Florida.

Given the data available, the total minimum population of black bears reported in North America approaches 650,000-700,000. Total annual mortality (mostly from hunting) for the United States (more than 19,000) and Canada (more than 23,000) exceeds 42,000, which is less than 10% of the known population. Many state wildlife agencies accept that bear populations can sustain 20%-25% annual harvest mortality, with the understanding that some areas are more sensitive to overharvest than others. Thus, except for those in the southeastern United States and in Idaho and New Mexico, North American black bear populations appear stable or on the increase. Only concentrated research on isolated populations of bears remaining in the southeastern United States will answer questions concerning the long-term viability of those populations.

Table 2. Population estimates and trends of American black bears in the United States (adapted from Pelton et al. 1994).

State	Estimated population size	Population trend	Status
Alabama	<50	Stable	Game
Alaska	100,000*	Stable	Game
Arizona	2,500	Stable	Game
Arkansas	2,200	Slightly increasing	Game
California	20,000	Slightly increasing	Game
Colorado	8,000-12,000	Unknown	Game
Connecticut	15-30	Increasing	Unclassified
Florida	1,000-2,000	Stable	Threatened
Georgia	1,700	Slightly increasing	Game
Idaho	20,000-25,000*	Slightly decreasing	Game
Kentucky	<200	Increasing	Protected
Louisiana	200-400	Slightly increasing	Threatened
Maine	19,500-20,500	Stable	Game
Maryland	175-200	Slightly increasing	Game
Massachusetts	700-750	Slightly increasing	Game
Michigan	7,000-10,000	Slightly increasing	Game
Minnesota	15,000	Increasing	Game
Mississippi	<50	Slightly increasing	Endangered
Missouri	50-130	Increasing	Rare
Montana	15,000-20,000	Stable	Game
Nevada	300	Increasing	Game
New Hampshire	3,500	Increasing	Game
New Jersey	275-325	Increasing	Game
New Mexico	3,000	Decreasing	Game
New York	4,000-5,000	Slightly increasing	Game
North Carolina	6,100	Increasing	Game
Oklahoma	120	Increasing	Game
Oregon	25,000	Increasing	Game
Pennsylvania	7,500	Stable	Game
South Carolina	200	Slightly increasing	Game
South Dakota	Unknown	Unknown	Threatened
Tennessee	750-1,500	Increasing	Game
Texas	50 *	Increasing	Threatened
Utah	800-1,000	Slightly increasing	Game
Vermont	2,300	Stable	Game
Virginia	3,000-3,500	Slightly increasing	Game
Washington	27,000-30,000	Increasing	Game
West Virginia	3,500	Increasing	Game
Wisconsin	6,200	Slightly increasing	Game
Wyoming	Unknown	Stable	Game
Total	306,935-331,805		

*1991 estimates from McCracken et al. (1995).

References

Cowan, I.M. 1972. The status and conservation of bears (Ursidae) of the world—1970. International Conference on Bear Research and Management 2:343-367.

Hall, E.R. 1981. The mammals of North America. 2nd ed. John Wiley and Sons, New York. 1,181 pp.

Leopold, A.S. 1959. Wildlife of Mexico. University of California Press, Berkeley. 608 pp.

Maehr, D.S. 1984. Distribution of black bears in eastern North America. Eastern Workshop on Black Bear Research and Management 7:74.

McCracken, C., K.A. Johnson, and D. Rose. 1995. Status, management, and commercialization of the American black bear (*Ursus americanus*). Traffic USA. Washington, DC. In press.

Pelton, M.R. 1982. Black bear. Pages 504-514 *in* J.A. Chapman and G.A. Feldhamer, eds. Wild mammals of North America; biology, management, and economics. John Hopkins University Press, Baltimore and London.

Pelton, M.R., F. vanManen, A. Coley, K. Weaver, J. Pedersen, and T. Eason. 1994. Black bear conservation action plan—North America. IUCN/SSC Bear Specialist Group Tech. Rep. In press.

Servheen, C. 1990. The status and conservation of the bears of the world. International Conference on Bear Research and Management, Monograph Series 2. 32 pp.

For further information:

Michael R. Vaughan
National Biological Service
Virginia Cooperative Fish and Wildlife Research Unit
Virginia Polytechnic Institute and State University
Blacksburg, VA 24061

Grizzly Bears

by
David J. Mattson
R. Gerald Wright
Katherine C. Kendall
Clifford J. Martinka
National Biological Service

Grizzly bears (*Ursus arctos*) once roamed over most of the western United States from the high plains to the Pacific coast (Fig. 1). In the Great Plains, they seem to have favored areas near rivers and streams, where conflict with humans was also likely. These grassland grizzlies also probably spent considerable time searching out and consuming bison that died from drowning, birthing, or winter starvation, and so were undoubtedly affected by the elimination of bison from most of the Great Plains in the late 1800's. They are potential competitors for most foods valued by humans, including domesticated livestock and agricultural crops, and under certain limited conditions are also a potential threat to human safety. For these and other reasons, grizzly bears in the United States were vigorously sought out and killed by European settlers in the 1800's and early 1900's.

Between 1850 and 1920 grizzlies were eliminated from 95% of their original range, with extirpation occurring earliest on the Great Plains and later in remote mountainous areas (Fig. 1a). Unregulated killing of grizzlies continued in most places through the 1950's and resulted in a further 52% decline in their range between 1920 and 1970 (Fig. 1b). Grizzlies survived this last period of slaughter only in remote wilderness areas larger than 26,000 km^2 (10,000 mi^2). Altogether, grizzly bears were eliminated from 98% of their original range in the contiguous United States during a 100-year period.

Because of this dramatic decline and the uncertain status of grizzlies in areas where they had survived, their populations in the contiguous United States were listed as threatened under the Endangered Species Act in 1975. High levels of grizzly bear mortality in the Yellowstone area during the early 1970's were also a major impetus for this listing. Grizzly bears persist as identifiable populations in five areas (Fig. 1b): the Northern Continental Divide, Greater Yellowstone, Cabinet-Yaak, Selkirk, and North Cascade ecosystems. All these populations except Yellowstone's have some connection with grizzlies in southern Canada, although the current status and future prospects of Canadian bears are subject to debate. The U.S. portions of these five populations exist in designated recovery areas, where they receive full protection of the Endangered Species Act.

Grizzlies potentially occur in two other areas: the San Juan Mountains of southern Colorado and the Bitterroot ecosystem of Idaho and Montana. There are no plans for augmenting or recovering grizzlies in the San Juan Mountains, and serious consideration is being given to reintroducing grizzlies into the Bitterroots as an "experimental nonessential" population.

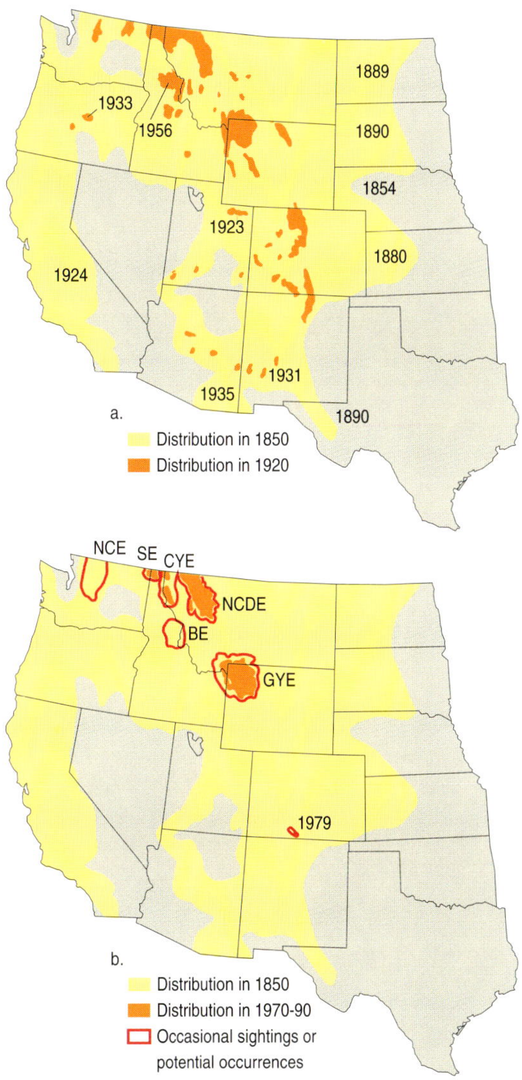

Fig. 1. Approximate distribution of grizzly bears in 1850 compared to 1920 (a; Merriam 1922) and 1970-90 (b). Local extinction dates, by state, appear in (a). Populations identified in (b) are NCE — North Cascades ecosystem, SE — Selkirk ecosystem, CYE — Cabinet-Yaak ecosystem, BE — Bitterroot ecosystem, NCDE — Northern Continental Divide ecosystem, GYE — Greater Yellowstone ecosystem. As indicated in (b), a grizzly was killed in the San Juan Mountains of southern Colorado in 1979.

Status and Trends

Recent research in the Northern Continental Divide, Yellowstone, and Selkirk ecosystems has produced growth and size estimates for these grizzly bear populations. Study results, however, have been compromised by either small sample sizes, incomplete coverage, or possibly unrepresentative samples. These types of studies are also relatively expensive and require the capture and radio tagging of bears, although without the aid of radio tagging, it is even more difficult to directly count or otherwise monitor grizzly bear populations in their extensive, typically forested, ranges.

Because of these difficulties, we have only rough estimates of size for U.S. grizzly bear populations. Many grizzlies exist only in the Northern Continental Divide and Yellowstone ecosystems. We can be confident that there are at least 175 bears in the Northern Continental Divide ecosystem and 142 in the Yellowstone ecosystem, and a minimum of about 360 in the entire contiguous United States (Table). On the other hand, it is unlikely that more than 75 animals inhabit each of the Cabinet-Yaak, Selkirk, and North Cascade populations.

We have few reliable estimates of population trends for the same reasons that we have few reliable estimates of population size. In most cases we do not have any information on trends or the populations are so small (as in the Selkirks) that the death of only a few individuals can turn a growing population into a declining one (Table). Current best estimates for the Northern Continental Divide and Yellowstone areas suggest that these largest populations have been stable or slightly increasing in recent years. Even for these relatively well-studied populations, however, obtaining a reliable estimate of trends is difficult because of large and diverse study areas, small samples, and potentially biased observations.

Long-term viability of a population or species is achieved when there are enough animals and sufficient secure and productive habitat to ensure that the population or species will survive for the indefinite future. Certainly, direct mortality that accompanied the arrival of European settlers had catastropic consequences for grizzly bears. Other catastrophes related to disease, climate change, and changes in human values could yet be visited upon grizzlies.

Viability analysis is not an exact science, yet there are some rules of thumb that can be used to identify populations at substantially greater risk of extinction than others. For example, among existing isolated populations of brown bears (also *U. arctos)* and grizzly bears worldwide, only populations that were reduced to no fewer than about 450 bears responded with rapid growth when given protection. Conversely, even with protection, populations smaller than 200 continued to decline (Mattson and Reid 1991). All of these smaller populations also occupied areas less than 10,000 km^2 (3,900 mi^2) at the time they were given legal protection. This relationship between range size and vulnerability is consistent with the fact that only North American grizzly populations occupying areas larger than 26,000 km^2 (10,500 mi^2) in 1920 survived to the present. The Selkirk and Cabinet-Yaak ecosystems are about 5,200 km^2 (about 2,000 mi^2) and the remaining ecosystems are about 24,600-29,500 km^2 (about 9,500-11,400 mi^2). We expect populations with current ranges less than 29,500 km^2 (11,400 mi^2) to be at substantially greater risk of extinction.

Exchange of genes among individuals and populations is also important to survival of populations. Allendorf et al. (1991) estimated that populations of about 500 interbreeding grizzlies may be required to maintain normal levels of genetic diversity. This genetically effective population size equates to total population sizes of around 2,000 because not all bears breed. Given that the maximum documented movement of grizzly bears away from their mother's range is 45-105 km (28-65 mi; Blanchard and Knight 1991), it is unlikely that populations separated by a greater distance exchange breeding animals. Furthermore, bear movement across these gaps is entirely dependent upon their surviving often hostile conditions.

No grizzly bear population in the contiguous United States could be considered robust by our

Table. Recent population and trend estimates for areas in the contiguous United States occupied or potentially occupied by grizzly bears (NCDE — Northern Continental Divide ecosystem, GYE — Greater Yellowstone ecosystem, CYE — Cabinet-Yaak ecosystem, C — Cabinet portion only (95% confidence interval), SE — Selkirk ecosystem, NCE — North Cascades ecosystem, BE — Bitterroot ecosystem, SJE — San Juan ecosystem).

	Minimum population estimate		Population estimate assuming 60% sightability[a]			
Area	Average (mean)	Range (95% CI)	Average (mean)	Range (95% CI)	Trend estimate[b]	Long-term viability
NCDE[c]	242[d]	175-308	404	293-514	Stable to slightly +[e]	?
NCDE[f]	302[d]	219-384	502	365-640		
GYE[g]	197[d]	142-252	329	237-420	ca. +0.01[h]	?
CYE[i]	<15		29	9-55 [j]	?	Not viable
SE[k]	26-36 [l]		?		0 to +0.02[k], recently-[m]	Not viable
NCE[n]	10-20 [l]		<50[l]		?	Not viable
BE	0		Possible presence		?	Not viable
SJE	0		Possible presence		?	Not viable

[a]Based on results from Aune and Kasworm (1989) suggesting that 60% of adult females were observed in their study area. Accordingly, minimum population estimates are divided by 0.6.
[b]Expressed as an increasing (+) or decreasing (-) population, where available in terms of per capita rate of increase or decrease per year. A "?" indicates populations for which there are no substantive or reliable estimates of trend.
[c]Data from USFWS (1993) and MFWP (1993).
[d]Mean and 95% confidence intervals for 3-year sums of "unduplicated" adult females observed in an area (n = 4 years, except for CYE n = 3 years) minus known adult female mortality for the corresponding 3-year period, divided by 0.284 (the assumed proportion of adult females in the population) for NCDE and GYE.
[e]From Keating (1986), Aune and Kasworm (1989), McLellan (1989), and MFWP (1993).
[f]Using 22.8% adult females in the population and assuming a 1:1 adult sex ratio, based on the upper 95% confidence interval for estimates of percentage of adults in grizzly bear populations from the NCDE (MFWP 1993).
[g]Data from Knight et al. (1993).
[h]From Knight et al. (1988).
[i]Data for 1986-90 from MFWP (1993). Minimum population estimate is for the Cabinet portion only. Data from USFWS (1993).
[j]The lower confidence interval = 0, but 9 bears were radio-marked and known to be alive.
[k]From Wielgus (1993).
[l]Including bears in adjacent Canada.
[m]From R.B. Wielgus, University of British Columbia, Vancouver, personal communication (1994).
[n]From Almack et al. (1991).

rules of thumb for population viability. Clearly, the small populations of the North Cascade, Selkirk, and Bitterroot ecosystems, the San Juan Mountains, and the U.S. portion of the Cabinet-Yaak ecosystem are not viable. Although the North Cascade ecosystem is close to 26,000 km^2 (10,000 mi^2), its prospects are compromised by its isolation, even from populations in Canada. Similarly, although the Cabinet-Yaak and Selkirk populations can potentially receive bears that have dispersed from other populations, their 5,200-km^2 (2,000-mi^2) ranges are within the size boundaries of many U.S. populations that went extinct between 1920 and 1970 (Fig. 2) and are similar to those of European populations that appear to be declining toward extinction.

Prospects for the larger Northern Continental Divide and Greater Yellowstone populations are better but still uncertain. The Yellowstone population is probably no larger than 420 animals (Table) and is very isolated, making its long-term status tenuous. The Northern Continental Divide population probably has the best prospects because it is the largest population, in the largest area, and within the range of movement of other grizzly bear populations. Nonetheless, even this population is near the thresholds of 450 animals and the 26,000-km^2 (10,000-mi^2) range size historically associated with persistence of grizzlies in the United States and Europe.

The prognosis for the Selkirk, Cabinet-Yaak, and Northern Continental Divide populations might be improved if their connections with Canadian grizzly populations were considered. These southern Canadian grizzlies, however, do not have protection comparable to the U.S. Endangered Species Act and, outside of national parks, they are all hunted. There is also serious debate over the status of Canadian grizzly populations, especially in southwest Alberta and the northern Selkirks. Thus, there is no evidence that Canadian grizzlies will guarantee the long-term survival of neighboring U.S. populations.

Implications

Since listing of the species under the Endangered Species Act in 1975, populations have probably stabilized in the Yellowstone and Northern Continental Divide ecosystems. Little if any of the former range has been reoccupied, however, and five of seven potential or existing populations do not have optimistic prospects, and even the two largest populations remain at risk.

About 88% of all grizzly bears that have been studied and died within the United States during the last 20 years were killed by humans, both legally and illegally. Humans remain the almost exclusive source of grizzly mortality, despite protection under the Endangered Species Act. Improved protection of these populations is accordingly dependent upon reducing the frequency of contact between grizzly bears and humans, primarily by managing levels of human activity in areas where we want grizzly bears to survive.

The Selkirk and Cabinet-Yaak grizzly bear populations may also need to be augmented by management if they are to survive beyond the next 100 years, whereas the North Cascade, Bitterroot, and San Juan populations will require the import of bears from elsewhere if they are to grow or persist even in the short term. The Yellowstone and Northern Continental Divide populations will need at least existing levels of protection, along with reliable monitoring and timely management.

References

Allendorf, F.W., R.B. Harris, and L.H. Metzgar. 1991. Estimation of effective population size of grizzly bears by computer simulation. Pages 650-654 *in* E.C. Dudley, ed. The unity of evolutionary biology: Proceedings of the Fourth International Congress of Systematic and Evolutionary Biology. Vol. 2. Dioscorides Press, Portland, OR.

Almack, J.A., W.L. Gaines, P.H. Morrison, J.R. Eby, R.H. Naney, G.F. Wooten, S.H. Fitkin, and E.R. Garcia. 1991. North Cascades grizzly bear ecosystem evaluation: final report. Interagency Grizzly Bear Committee, Denver, CO. 146 pp.

Aune, K., and W. Kasworm. 1989. Final report: East Front grizzly studies. Montana Department of Fish, Wildlife, and Parks, Helena. 332 pp.

Blanchard, B.M., and R.R. Knight. 1991. Movements of Yellowstone grizzly bears. Biological Conservation 58:41-67.

Keating, K.A. 1986. Historical grizzly bear trends in Glacier National Park, Montana. Wildlife Society Bull. 14:83-87.

Knight, R., J. Beecham, B. Blanchard, L. Eberhardt, L. Metzgar, C. Servheen, and J. Talbott. 1988. Report of the Yellowstone grizzly bear population task force: equivalent population size for 45 adult females. Interagency Grizzly Bear Committee, Denver, CO. 8 pp.

Knight, R.R., B.M. Blanchard, and D.J. Mattson. 1993. Yellowstone grizzly bear investigations: annual report of the Interagency Study Team 1992. National Park Service, Bozeman, MT. 26 pp.

Mattson, D.J., and M.W. Reid. 1991. Conservation of the Yellowstone grizzly bear. Conservation Biology 5:364-372.

McLellan, B.N. 1989. Dynamics of a grizzly bear population during a period of industrial resource extraction. 3. Natality and rate of increase. Canadian Journal of Zoology 67:1865-1868.

Merriam, C.H. 1922. Distribution of grizzly bears in U.S. Outdoor Life (December):405-406.

MFWP. 1993. Five year update of the programmatic environmental impact statement: the grizzly bear in northwestern Montana, 1986-1990. Montana Department of Fish, Wildlife, and Parks, Helena. 228 pp.

USFWS. 1993. Grizzly bear recovery plan. U.S. Fish and Wildlife Service, Missoula, MT. 181 pp.

Wielgus, R.B. 1993. Causes and consequences of sexual habitat segregation in grizzly bears. Ph.D. thesis, University of British Columbia, Vancouver. 88 pp.

For further information:

David J. Mattson
National Biological Service
Cooperative Park Studies Unit
Department of Fish and Wildlife Resources
University of Idaho
Moscow, ID 83843

Black-footed Ferrets

by
Dean Biggins
Jerry Godbey
National Biological Service

The black-footed ferret (*Mustela nigripes*) was a charter member of endangered species lists for North America, recognized as rare long before the passage of the Endangered Species Act of 1973. This member of the weasel family is closely associated with prairie dogs (*Cynomys* spp.) of three species, a specialization that contributed to its downfall. Prairie dogs make up 90% of the ferret diet; in addition, ferrets dwell in prairie dog burrows during daylight, venturing out mostly during darkness. Trappers captured black-footed ferrets during their quests for other species of furbearers. Although the species received increased attention as it became increasingly rare, the number of documented ferrets fell steadily after 1940 (Fig. 1), and little was learned about the animals before large habitat declines made studies of them difficult. These declines were brought about mainly by prairie dog control campaigns begun before 1900 and reaching high intensity by the 1920's and 1930's.

Black-footed ferrets, almost extinct by 1985, are being reintroduced from captive breeding but still lack genetic diversity.

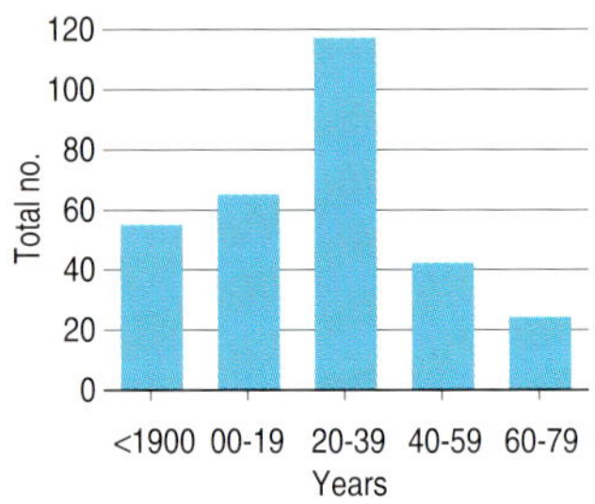

Fig. 1. Black-footed ferrets collected before 1980.

Much of what is known about black-footed ferret biology was learned from research during 1964-74 on a remnant population in South Dakota (Linder et al. 1972; Hillman and Linder 1973), and from 1981 to the present on a population found at Meeteetse, Wyoming, and later transferred to captivity (Biggins et al. 1985; Forrest et al. 1988; Williams et al. 1988). Nine ferrets from the sparse South Dakota population (only 11 ferret litters were located during 1964-72) were taken into captivity from 1971 to 1973, and captive breeding was undertaken at the U.S. Fish and Wildlife Service's Patuxent Wildlife Research Center in Maryland (Carpenter and Hillman 1978). Although litters were born there, no young were successfully raised. The last of the Patuxent captive ferrets died in 1978, and no animals were located in South Dakota after 1979.

Black-footed ferrets were "rediscovered" in prairie dog complexes at Meeteetse in 1981, giving conservationists what seemed a last chance to learn about the species and possibly save it from extinction. That population remained healthy (70 ferret litters were counted from 1982 to 1986) through 1984 (Fig. 2), a period when much was learned about ferret life history and behavior. In 1985, sylvatic plague, a disease deadly to prairie dogs, was confirmed in the prairie dogs at Meeteetse, creating fear that the prairie dog habitat vital for ferrets would be lost. In addition, field biologists were reporting a substantial decrease in the number of ferrets detected. The fear of plague was quickly overshadowed by the discovery of canine distemper in the ferrets themselves. It is a disease lethal to ferrets.

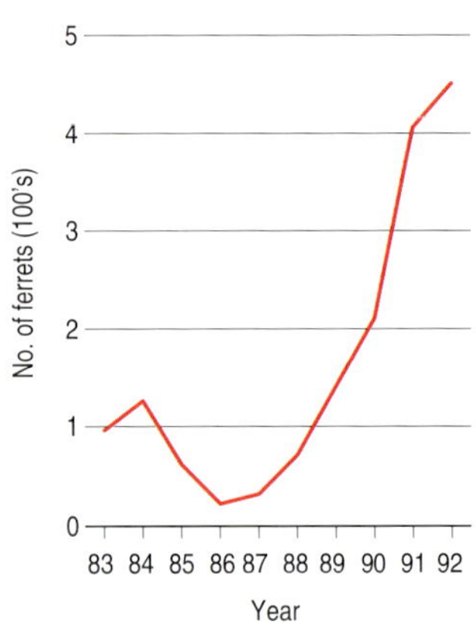

Fig. 2. Black-footed ferret population from Meeteetse, Wyoming, 1983-92 (all captive from 1986 to present).

In 1985 six ferrets were captured to begin captive breeding, but two of them brought the distemper virus into captivity, and all six died (Williams et al. 1988). A plan was formulated to place more animals from Meeteetse into captivity to protect them from distemper and to start the breeding program. By December 1985, only 10 ferrets were known to exist, 6 in captivity and 4 at Meeteetse. The following year, the surviving free-ranging ferrets at Meeteetse produced only two litters, a number thought too small to sustain the wild population. Because both the Meeteetse and captive populations were too small to sustain themselves, all remaining ferrets were removed from the wild, resulting in a captive population of 18 individuals by early 1987.

Captive breeding of ferrets eventually became successful (Fig. 2). Although the captive population is growing, researchers fear the consequences of low genetic diversity (already documented by O'Brien et al. 1989) and of inbreeding depression (*see* glossary). A goal of the breeding program is to retain as much genetic diversity as possible, but the only practical way to increase diversity is to find more wild ferrets. In spite of intensive searches of the remaining good ferret habitat and investigations of sighting reports, no wild ferrets have been found.

The captive breeding program now is producing sufficient surplus ferrets for reintroduction into the wild; 187 ferrets were released into prairie dog colonies in Shirley Basin, Wyoming, during 1991-93. Challenges facing the black-footed ferret reintroduction include low survivorship of released ferrets due to high dispersal and losses to other predators; unknown influence of low genetic diversity; canine distemper hazard; indirect effect of plague on prairie dogs and possible direct effect on ferrets; and low availability of suitable habitat for reintroduction. The scarcity of habitat reflects a much larger problem with the prairie dog ecosystem and needs increased attention.

At the turn of this century, prairie dogs reportedly occupied more than 40 million ha

(100 million acres) of grasslands, but by 1960 that area had been reduced to about 607,500 ha (1.5 million acres; Marsh 1984). Much reduction was attributed to prairie dog control programs, which continue. For example, in South Dakota in the late 1980's, $6.2 million was spent to apply toxicants to prairie dog colonies on Pine Ridge Indian Reservation (Sharps 1988). At least two states (Nebraska and South Dakota) have laws prohibiting landowners from allowing prairie dogs to flourish on their properties; if the land manager does not "control" the "infestation," the state can do so and bill expenses to the owner (Clarke 1988).

Sylvatic plague also has been devastating to prairie dogs and was the likely cause of the dramatic decline in prairie dogs at Meeteetse. Although the Meeteetse complex recently supported the densest and most vigorous population of black-footed ferrets ever known, it cannot be considered as ferret habitat now because of plagued-induced losses of prairie dogs. Plague is present in most of the monitored white-tailed prairie dog (*Cynomys leucurus*) complexes, including the Shirley Basin ferret reintroduction site (Table). The plague's persistence could be responsible for the generally lower densities of white-tailed prairie dogs (averaging fewer than seven prairie dogs per hectare or fewer than three per acre).

Several prairie dog complexes have been evaluated as sites for reintroduction of black-footed ferrets (Table). The evaluation involves grouping clusters of colonies separated by fewer than 7 km (4.3 mi) into complexes, based on movement capabilities of ferrets (Biggins et al. 1993); these areas include some of the best prairie dog complexes remaining in the states. Nevertheless, other extensive prairie dog complexes were not considered for ferret reintroduction.

Most of the original range of the black-footed ferret was associated with black-tailed prairie dog (*Cynomys ludovicianus*) complexes, which now exhibit the highest population densities of all prairie dogs (Table). Black-footed ferret reintroductions recently began at black-tailed prairie dog complexes near Malta, Montana, and Badlands National Park, South Dakota (Table). At present, the best example of a large complex of black-tailed prairie dogs is near Nuevos Casas Grandes, Chihuahua, Mexico (Table). It supports an impressive associated fauna and is a potential reintroduction site for black-footed ferrets.

Ramifications of a healthy prairie dog ecosystem extend well beyond black-footed ferrets. The prairie dog is a keystone species of the North American prairies. It is an important primary consumer, converting plants to animal biomass at a higher rate than other vertebrate herbivores of the short-grass prairies, and its burrowing provides homes for many other species of animals and increases nutrients in surface soil. This animal also provides food for many predators. We estimated it takes 700-800 prairie dogs to annually support a reproducing pair of black-footed ferrets and a similar biomass of associated predators (Biggins et al. 1993), suggesting that large complexes of prairie dog colonies are necessary to support self-sustaining populations of these second-order consumers.

The 98% loss of the productive prairie dog ecosystem has not yet motivated legal protection or plans for management. There is no federal legislation directly promoting the welfare of the prairie dog ecosystem (even on public lands), and the only existing state legislation promotes poisoning.

To develop a plan for remedial action, several immediate research needs are apparent in the prairie dog ecosystem: determine the relative diversity and abundance of invertebrates and

Three species of prairie dogs make up 90% of the black-footed ferret's diet; prairie dog burrows are also used by the ferrets during the day.

Table. Prairie dog complexes evaluated for black-footed ferret reintroductions. (Some data from Black-footed Ferret Interstate Coordinating Committee.)

State	Site	Prairie dog species*	Complex size (ha)	Hectares of prairie dogs	Prairie dogs estimate	Prairie dogs/ha colony
United States						
Arizona	Aubrey Valley	Gunnison's	44,167	7,390	34,067	4.61
Colorado	Little Snake	White-tailed	252,075	31,624	36,875	1.17
	Wolf Creek	White-tailed	65,607	3,174	20,009	6.30
	Sterling	Black-tailed	57,824	2,366	16,786	7.10
Montana	Custer Creek	Black-tailed	38,879	425	16,750	39.39
	Malta Bureau of Land Management	Black-tailed	583,430	7,600	167,299	22.01
	Charles M. Russell Refuge	Black-tailed	28,508	896	22,371	25.00
North Dakota	Roosevelt National Park	Black-tailed	14,126	594	39,270	66.11
	Marmarth National Park	Black-tailed	7,257	548	21,208	38.70
	Fort Yates	Black-tailed	6,739	579	20,823	35.96
South Dakota	Badlands National Park	Black-tailed	17,016	1,669	74,081	44.39
Wyoming	Meeteetse	White-tailed	53,846	5,111	1,299	0.25
	Shirley Basin	White-tailed	48,987	20,612	75,155	3.65
	Medicine Bow	White-tailed	74,958	27,235	24,492	0.90
	Recluse	Black-tailed	98,802	7,181	59,895	8.34
	Bolton Ranch	White-tailed	28,068	4,420	7,858	1.78
	Kinney Rim	White-tailed	43,509	7,220	608	0.08
Mexico						
Chihuahua	Nuevos Casas Grandes	Black-tailed	87,866	54,541	994,986	18.24

*Gunnison's prairie dog (*Cynomys gunnisoni*), white-tailed prairie dog (*C. leucurus*), and black-tailed prairie dog (*C. ludovicianus*).

Courtesy B. Miller

Prairie dog control campaigns, like this one in Arizona, circa 1913, contributed to the decline of the black-footed ferret.

small- to medium-sized vertebrates on prairie dog complexes, as well as the degree of dependence on prairie dogs of selected associated species; examine the effect of complex size, as well as constituent colony sizes, numbers, and juxtaposition on diversity and abundance of associated species; investigate the recent history of plague on selected complexes to determine the relation between complex size (and morphology) and resistance to decimation by plague; and develop methods for reestablishing prairie dog colonies and reconstructing complexes in suitable areas where prairie dogs have been extirpated.

The black-footed ferret cannot be reestablished on the grasslands of North America in viable self-sustaining populations without large complexes of prairie dog colonies. The importance of this system to other species is not completely understood, but large declines in some of its species should serve as a warning. The case of the black-footed ferret provides ample evidence that timely preventive action would be preferable to the inefficient "salvage" operations. Furthermore, there is considerable risk of irreversible damage (e.g., genetic impoverishment) with such rescue efforts.

References

Biggins, D.E., B.J. Miller, L. Hanebury, R. Oakleaf, A. Farmer, R. Crete, and A. Dood. 1993. A technique for evaluating black-footed ferret habitat. Pages 73-88 *in* J.L. Oldemeyer, D.E. Biggins, B.J. Miller, and R. Crete, eds. Management of prairie dog complexes for the reintroduction of the black-footed ferret. U.S. Fish and Wildlife Service Biological Rep. 13.

Biggins, D.E., M.H. Schroeder, S.C. Forrest, and L. Richardson. 1985. Movements and habitat relationships of radio-tagged black-footed ferrets. Pages 11.1-11.17 *in* S.H. Anderson and D.B. Inkley, eds. Proceedings of the Black-footed Ferret Workshop. Wyoming Game and Fish Department, Cheyenne.

Carpenter, J.W., and C.N. Hillman. 1978. Husbandry, reproduction, and veterinary care of captive ferrets. Proceedings of the American Association of Zoo Veterinarians Workshop, Knoxville, TN. 1979:36-47.

Clarke, D.C. 1988. Prairie dog control—a regulatory viewpoint. Pages 119-120 *in* D.W. Uresk and G. Schenbeck, eds. Eighth Great Plains Wildlife Damage Control Workshop Proceedings. U.S. Forest Service Gen. Tech. Rep. RM-154.

Forrest, S.C., D.E. Biggins, L. Richardson, T.W. Clark, T.M. Campbell III, K.A. Fagerstone, and E.T. Thorne. 1988. Population attributes for the black-footed ferret (*Mustela nigripes*) at Meeteetse, Wyoming, 1981-1985. Journal of Mammalogy 69(2):261-273.

Hillman, C.N., and R.L. Linder. 1973. The black-footed ferret. Pages 10-20 *in* R.L. Linder and C.N. Hillman, eds. Proceedings of the Black-footed Ferret and Prairie Dog Workshop. South Dakota State University Publications, Brookings.

Linder, R.L., R.B. Dahlgren, and C.N. Hillman. 1972. Black-footed ferret-prairie dog interrelationships. Pages 22-37 *in* Proceedings of the Symposium on Rare and Endangered Wildlife of the Southwestern U.S. New Mexico Department of Game and Fish, Santa Fe.

Marsh, R.E. 1984. Ground squirrels, prairie dogs and marmots as pests on rangeland. Pages 195-208 *in* Proceedings of the Conference for Organization and Practice of Vertebrate Pest Control. ICI Plant Protection Division, Fernherst, England.

O'Brien, S.J., J.S. Martenson, M.A. Eichelberger, E.T. Thorne, and F. Wright. 1989. Biochemical genetic variation and molecular systematics of the black-footed ferret, *Mustela nigripes*. Pages 21-33 *in* Conservation biology and the black-footed ferret. Yale University Press, New Haven, CT.

Sharps, J. 1988. Politics, prairie dogs, and the sportsman. Pages 117-118 *in* D.W. Uresk and G. Schenbeck, eds. Eighth Great Plains Wildlife Damage Control Workshop Proceedings. U.S. Forest Service Gen. Tech. Rep. RM-154.

Williams, E.S., E.T. Thorne, M.J.G. Appel, and D.W. Belitsky. 1988. Canine distemper in black-footed ferrets (*Mustela nigripes*) from Wyoming. Journal of Wildlife Diseases 24:385-398.

For further information:

Dean E. Biggins
National Biological Service
Midcontinent Ecological Science Center
4512 McMurry Ave.
Fort Collins, CO 80525

American Badgers in Illinois

by
Barbara Ver Steeg
Illinois Natural History Survey
Richard E. Warner
University of Illinois

The American badger (*Taxidea taxus*) is a medium-sized carnivore found in treeless areas across North America, such as the tallgrass prairie (Lindzey 1982). Badgers rely primarily on small burrowing mammals as a prey source; availability of badger prey may be affected by changes in land-use practices that alter prey habitat. In the midwestern United States most native prairie was plowed for agricultural use beginning in the mid-1800's (Burger 1978). In the past 100 years, Midwest agriculture has shifted from a diverse system of small farms with row crops, small grains, hay, and livestock pasture to larger agricultural operations employing a mechanized and chemical approach to cropping. The result is a more uniform agricultural landscape dominated by two primary row crops, corn and soybeans. The effects of such land-use alterations on badgers are unknown. In addition, other human activities such as hunting and trapping have no doubt had an impact on native vertebrates such as the badger. Our ongoing study was initiated to determine the distribution and status of badgers in Illinois.

Trends in carnivore abundance are difficult to evaluate because most species are secretive or visually cryptic. Trapping records, one of the earliest historical data sources for furbearers, are virtually nonexistent for badgers in the 1800's (Obbard et al. 1987). In Illinois, badgers have been protected from harvest since 1957. Furthermore, population estimates derived from furbearer harvest data are complicated by market price bias (Erickson 1982). Thus, data for estimating long-term population trends in Illinois badgers are few and flawed. Our approach is to document and evaluate current population parameters, behavior, and habitat use in the context of present and historical habitat quality and availability.

Most research on badgers has been limited to the western United States. Although results have varied somewhat among these studies, average densities (estimated subjectively from mark-recapture and home range data) have ranged from 0.38 to 5 badgers/km^2 (0.98-12.95 badgers/mi^2). We use radio telemetry to collect intensive data at a field site in west-central Illinois. Preliminary results suggest that individual badger home range size in Illinois is an order of magnitude larger than that of western badgers, implying that badger density in Illinois is much lower. The home range size estimates of two badgers in Minnesota were also larger than those reported for western states (Sargeant and Warner 1972; Lampe and Sovada 1981).

More than 65% of the Illinois landscape is under intensive row-crop agriculture (Neely and Heister 1987). Although badger prey exist throughout Illinois, available prey in row crops is limited to small species such as the deer mouse (*Peromyscus maniculatus*), which occur in low uniform densities. Important prey species reported in the West, such as ground squirrels (*Spermophilus* spp.), have average densities similar to Illinois deer mice, but they are much larger animals and may be concentrated into easily hunted loose colonies (Messick and Hornocker 1981; Minta 1990).

In Illinois, badgers appear to use most frequently cover types that are relatively undisturbed by plowing, including hayfields, pastures, and linear habitats such as roadsides and fencelines. These habitats offer the greatest concentration of small mammalian prey and the lowest frequency of agricultural disturbance. If badgers are limited by available prey, it is possible that the current badger population density is lower than when native prairie and its accompanying prey species' populations dominated the landscape.

Although badgers are legally protected in Illinois, human-induced mortality such as vehicle collisions and agricultural accidents take a toll on populations. Large predators that might prey on adult badgers, such as the black bear (*Ursus americanus*), gray wolf (*Canis lupus*), and mountain lion (*Felis concolor*), have been extirpated since the 19th century (Hoffmeister 1989). However, our study shows that predation by coyotes (*Canis latrans*) and domestic dogs significantly affects juvenile badgers; fewer than 70% of juveniles survive to dispersal, reducing overall recruitment.

The badger's range may be expanding eastward from its former boundaries within the Midwest; observations of range expansion in Missouri, southern Illinois, Indiana, and Ohio suggest that agricultural practices have converted previously forested acres to more suitable badger habitat (Moseley 1934; Leedy 1947; Mumford 1969; Hubert 1980; Mumford and Whitaker 1982; Long and Killingley 1983; Gremillion-Smith 1985; Whitaker and Gammon 1988).

Courtesy M. Georgi, Illinois Natural History Survey

American badger (*Taxidea taxus*).

Our study revealed that badgers are distributed and breeding throughout Illinois. The dynamics of badger range expansion are difficult to pinpoint, in part because of the cryptic nature of the species. In Illinois and probably the agricultural Midwest in general, individual badgers move over such large areas that live sightings or indications of badger presence are few and far between. Opportunistic observations to evaluate local badger distribution underestimate geographic range; thus, a focused regionwide attempt to evaluate badger range in the Midwest might demonstrate a wider distribution than expected.

Badgers in Illinois appear to be a species with intermediate status: though they are neither abundant nor of high economic value, they are widely distributed and have adapted to a greatly altered environment. Understanding what factors cause a species such as the badger to become more or less abundant is vitally important in conservation biology and wildlife management.

References

Burger, G.V. 1978. Agriculture and wildlife. Pages 89-107 *in* H.P. Brokaw, ed. Wildlife and America. Council on Environmental Quality, Washington, DC.

Erickson, D.W. 1982. Estimating and using furbearer harvest information. Pages 53-66 *in* G.C. Sanderson, ed. Midwest furbearer management. Central Mountains and Plains Section of The Wildlife Society, Wichita, KS.

Gremillion-Smith, C. 1985. Range extension of the badger (*Taxidea taxus*) in southern Illinois. Transactions of the Illinois Academy of Science 78:111-114.

Hoffmeister, D.F. 1989. Mammals of Illinois. University of Illinois Press, Urbana, IL. 348 pp.

Hubert, G.F., Jr. 1980. Badger status evaluation. Illinois Department of Conservation, Job Completion Report, Federal Aid Project W-49-R-34, Study XII. 12 pp.

Lampe, R.P., and M.A. Sovada. 1981. Seasonal variation in home range of a female badger (*Taxidea taxus*). Prairie Naturalist 15:55-58.

Leedy, D.L. 1947. Spermophiles and badgers move eastward in Ohio. Journal of Mammalogy 28:290-292.

For further information:

Barbara Ver Steeg
Illinois Natural History Survey
607 E. Peabody Dr.
Champaign, IL 61820

Lindzey, F.G. 1982. The North American badger. Pages 653-663 *in* J.A. Chapman and G.A. Feldhammer, eds. Wild mammals of North America. Johns Hopkins University Press, Baltimore, MD.

Long, C.A., and C.A. Killingley. 1983. The badgers of the world. Charles C. Thomas Publishers, Springfield, IL. 404 pp.

Messick, J.P., and M.G. Hornocker. 1981. Ecology of the badger in southwestern Idaho. Wildlife Monograph 76. 53 pp.

Minta, S.C. 1990. The badger, *Taxidea taxus*, (Carnivora: Mustelidae): spatial-temporal analysis, dimorphic territorial polygyny, population characteristics, and human influences on ecology. Ph.D. thesis, University of California, Davis. 317 pp.

Moseley, E.L. 1934. Increase of badgers in northwestern Ohio. Journal of Mammalogy 15:156-158.

Mumford, R.E. 1969. Distribution of the mammals of Indiana. Indiana Academy of Science Monograph 1. 114 pp.

Mumford, R.E., and J.O. Whitaker, Jr. 1982. Mammals of Indiana. Indiana University Press, Bloomington, IN. 537 pp.

Neely, R.D., and C.G. Heister, compilers. 1987. The natural resources of Illinois: introduction and guide. Illinois Natural History Survey Special Publ. 6. 224 pp.

Obbard, M.E., J.G. Jones, R. Newman, A. Booth, A.J. Satterthwaite, and G. Linscombe. 1987. Furbearer harvests in North America. Pages 1007-1038 *in* M. Novak, J.A. Baker, M.E. Obbard, and B. Malloch, eds. Wild furbearer management and conservation in North America. The Ontario Trappers Association and the Ministry of Natural Resources, Toronto, Ontario.

Sargeant, A.B., and D.W. Warner. 1972. Movements and denning habits of a badger. Journal of Mammalogy 53:207-210.

Whitaker, J.O., Jr., and J.R. Gammon. 1988. Endangered and threatened vertebrate animals of Indiana; their distribution and abundance. Indiana Academy of Science Monograph 5. 122 pp.

California Sea Otters

by
James A. Estes
Ronald J. Jameson
James L. Bodkin
David R. Carlson
National Biological Service

Information on the size, distribution, and productivity of the California sea otter population is broadly relevant to two federally mandated goals: removing the population's listing as threatened under the Endangered Species Act (ESA) and obtaining an "optimal sustainable population" under the Marine Mammal Protection Act. Except for the population in central California, sea otters (*Enhydra lutris*) were hunted to extinction between Prince William Sound, Alaska, and Baja California (Kenyon 1969). Wilson et al. (1991), based on variations in cranial morphology, recently assigned subspecific status (*E. l. nereis*) to the California sea otter. Furthermore, mitochondrial DNA analysis has revealed genetic differences among populations in California, Alaska, and Asia (NBS, unpublished data).

In 1977, the California sea otter was listed as threatened under the ESA, largely because of its small population size and perceived risks from such factors as human disturbance, competition with fisheries, and pollution. Because of unique threats and growth characteristics, the California population is treated separately from sea otter populations elsewhere in the North Pacific.

Courtesy D. Buchich

Sea otter (*Enhydra lutris*).

Survey Design

Data on the size and distribution of the California sea otter population have been gathered for more than 50 years. In 1982 we developed a survey technique in which individuals in most of the California sea otter's range are counted from shore by groups of two observers using binoculars and spotting scopes. Supplemental data for each sighting include group size, activity, number and size of pups, and habitat. Areas that cannot be counted from shore are surveyed from a low-flying aircraft. Rangewide surveys are done in late spring and mid-autumn.

Population Trends, 1914-93

The California sea otter population has increased steadily through most of the 1900's (Fig. 1). Rate of increase was about 5% per year until the mid-1970's. Although only one survey was completed between 1976 and 1982, the collective data suggest that population growth had ceased by the mid-1970's, and that the population may have declined by as much as 30% between the mid-1970's and early 1980's. Counts made since 1983 have increased at about 5%-6% per year. In spring 1993, 2,239 California sea otters were counted.

The California sea otter's lineal range (distance along the 9-m [5-fathom] isobath between the northernmost and southernmost sightings) has also increased, although more slowly and erratically than the population size (data summarized by Riedman and Estes 1990). The

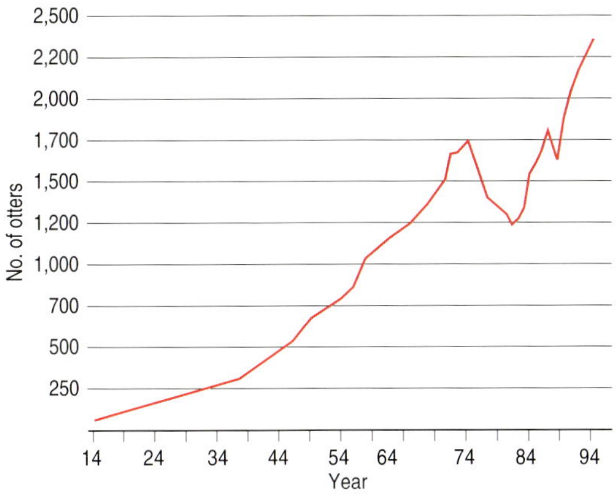

Fig. 1. Trends in abundance of the California sea otter population, 1914-93.

direction of range expansion was predominately southward before 1981, but northward thereafter. Comparison between spring surveys conducted in 1983 and 1993 (Fig. 2) is sufficient to draw several conclusions. First, the population's range limits changed little during this 10-year period, even though large numbers of individuals accumulated near the range peripheries. Second, population density increased throughout this time, although rates of increase were lowest near the center of the range. Finally, the relative abundance of individuals has remained largely unchanged (compare Fig. 2a [1983] with Fig. 2b [1993], noting the similarity in forms of distributions for kilometer segments 10-21).

Although the number of dependent pups counted in spring surveys almost doubled between 1983 and 1993, the geographic range within which these pups were born has changed very little (Fig. 2). Rate of annual pup production ranged from 0.14 to 0.28, but in most years it varied between 0.18 and 0.21. There are no obvious trends in rate of annual pup production between 1983 and 1993. Although the incremental change in the population from one year to the next appeared positively related to the annual number of births, this relationship cannot be shown to be statistically significant.

Implications

From the mid-1970's to the early 1980's, the California sea otter population ceased growing and probably declined. Entanglement mortality in a coastal set-net fishery was the likely cause of this decline (Wendell et al. 1985). Restrictions were imposed on the fishery in 1982, and the population apparently responded by resuming its prior rate of increase.

The maximum rate of increase for sea otter populations is about 20% per year. Except for the California otters, all increasing populations for which data are available have grown at about this rate (Estes 1990). These patterns, coupled with the absence of any size- or density-related reduction in growth rates, make the relatively slow rate of increase in the California population perplexing.

Although the ultimate reason for disparate growth rates among sea otter populations is unknown, we believe that causes relate more to increased mortality than diminished reproduction. While it is difficult to compare population-level reproductive rates between sea otters in Alaska and California, longitudinal studies of

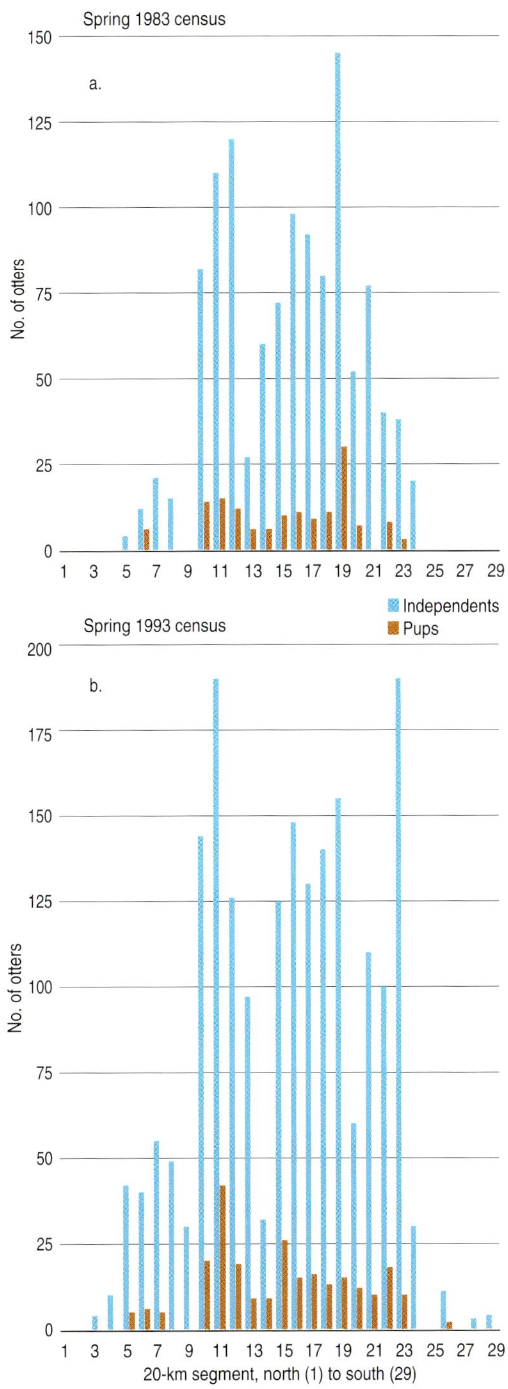

Fig. 2. Distribution and abundance of California sea otters in 1983 (a) and 1993 (b). Data are from the spring surveys.

marked individuals in the two regions indicate that both age of first reproduction and annual birth rate of adult females are similar. Furthermore, the close similarity between the theoretical maximum rate of increase and observed rates of population increase for sea otters in Washington, Canada, and portions of Alaska suggests that mortality from birth to senescence in these populations is quite low. In contrast, rates of mortality in the California sea otter are comparatively high, with an estimated 40%-50% of newborns lost before weaning (Siniff and Ralls 1991; Jameson and Johnson 1993; Riedman et al. 1994). This alone would significantly depress a population's potential rate of increase. Furthermore, the age composition of beach-cast carcasses in California indicates that most postweaning deaths occur well in advance of physiological senescence (Pietz et al. 1988; Bodkin and Jameson 1991). These patterns likely explain the depressed rate of increase in the California sea otter population.

Although the demographic patterns of mortality in California sea otters are becoming clear, the causes of deaths remain uncertain. There is growing evidence for the importance of predation by great white sharks (*Carcharodon carcharias*). Contaminants may also be having a detrimental effect on California sea otters, although as yet there is no direct evidence for this. However, polychlorinated biphenyl (PCB) and DDT levels, known to be high in the California Current, are also high in the liver and muscle tissues of California sea otters (Bacon 1994). Of particular concern are that average PCB levels in California sea otters approach those that cause reproductive failure in mink, which are in the same family as otters; and preweaning pup losses are especially high in primiparous (*see* glossary) females. This latter point may be significant because environmental contaminants that accumulate in fat can be transferred via milk in extraordinarily high concentrations, especially to the first-born young in species such as the sea otter which has prolonged sexual immaturity.

References

Bacon, C.E. 1994. An ecotoxicological comparison of organic contaminants in sea otters (*Enhydra lutris*) among populations in California and Alaska. M.S. thesis, University of California, Santa Cruz. 55 pp.

Bodkin, J.L., and R.J. Jameson. 1991. Patterns of seabird and marine mammal carcass deposition along the central California coast, 1980-1986. Canadian Journal of Zoology 69(5):1149-1155.

Estes, J.A. 1990. Growth and equilibrium in sea otter populations. Journal of Animal Ecology 59:385-401.

Jameson, R.J., and A.M. Johnson. 1993. Reproductive characteristics of female sea otters. Marine Mammal Science 9(2):156-167.

Kenyon, K.W. 1969. The sea otter in the eastern Pacific Ocean. North American Fauna 68:1-352.

Pietz, P., K. Ralls, and L. Ferm. 1988. Age determination of California sea otters from teeth. Pages 106-115 *in* D.B. Siniff and K. Ralls, eds. Population status of California sea otters. Final report to the Minerals Management Service, U.S. Department of the Interior 14-12-001-3003.

Riedman, M.L., and J.A. Estes. 1990. The sea otter (*Enhydra lutris*): behavior, ecology, and natural history. U.S. Fish and Wildlife Service Biological Rep. 90(14). 126 pp.

Riedman M.L., J.A. Estes, M.M. Staedler, A.A. Giles, and D.R. Carlson. 1994. Breeding patterns and reproductive success of California sea otters. Journal of Wildlife Management 58:391-399.

Siniff D.B., and K. Ralls. 1991. Reproduction, survival, and tag loss in California sea otters. Marine Mammal Science 7(3):211-229.

Wendell, F.E., R.A. Hardy, and J.A. Ames. 1985. Assessment of the accidental take of sea otters, *Enhydra lutris*, in gill and trammel nets. Marine Research Branch, California Department of Fish and Game. 30 pp.

Wilson, D.E., M.A. Bogan, R.L. Brownell, Jr., A.M. Burdin, and M.K. Maminov. 1991. Geographic variation in sea otters, *Enhydra lutris*. Journal of Mammalogy 72(1):22-36.

For further information:

J.A. Estes
National Biological Service
University of California
Santa Cruz, CA 95064

White-tailed Deer in the Northeast

by
Gerald L. Storm
National Biological Service
William L. Palmer
Pennsylvania Game Commission

Populations of white-tailed deer (*Odocoileus virginianus*) have changed significantly during the past 100 years in the eastern United States (Halls 1984). After near extirpation in the eastern states by 1900, deer numbers increased during the first quarter of this century. The effects of growing deer populations on forest regeneration and farm crops have been a concern to foresters and farmers for the past 50 years.

In recent years, deer management plans have been designed to maintain deer populations at levels compatible with all land uses. Conflicts, however, between deer and forest management or agriculture still exist in the Northeast. Areas that were once exclusively forests are now a mixture of forest, farm, and urban environments that create increased interactions and conflicts between humans and deer, including deer-vehicle collisions. Management of deer near urban environments presents a unique challenge for local resource managers (Porter 1991).

This report describes trends in abundance of white-tailed deer in the northeastern United States, relationships between harvest and population estimates, and conflicts between deer and other resources.

Data Surveys

We contacted biologists in each of 13 northeastern states to acquire estimates of deer population size, harvest, and deer-vehicle collisions. We featured harvest data for antlered deer from all 13 states to describe deer abundance during 1983-92, as well as data from selected states to describe relations between deer harvests and population size.

Biologists in the northeastern states also provided information on trends in reported conflicts between deer and land use and other natural resources. We determined the proportion of states that expressed conflicts for particular categories such as deer and agriculture, deer and forestry, or deer and other resources.

Population Estimates and Management Implications

White-tailed deer populations have increased in all 13 northeastern states during 1983-92, based on either population estimates or number of antlered deer harvested. Population estimates for nine states indicated an increase from less than 1.5 million in the early 1980's to 1.8 million in the early 1990's (Fig. 1). Deer density in the deer range of these states

Deer Management at Parks and Refuges

Even though states are responsible for managing deer within their boundaries, they do not control all land areas. The level of management for a state may be an ecological or political unit. However, states usually lack data on deer and their habitats for small units such as municipalities, parks, refuges, or military facilities, and they are not directly responsible for management of these special areas. Presented here are examples of two state parks, two national parks and a national historic site, and three national wildlife refuges.

Parks

Ridley Creek and Tyler state parks in Pennsylvania provide two examples of where attempts have been made to manage high deer densities in and around urban areas. Such high densities pose significant problems because of deer feeding on ornamental plants and deer-vehicle collisions. At Ridley Creek State Park, a 1,052-ha (2,600-acre) area near Philadelphia, hunters harvested 97-344 deer per year during eight controlled hunts held between 1983 and 1992. From 160 to 491 deer were observed during annual counts made from helicopters (no count was made in 1990). A count of 491 in 1983 indicated that the deer density was in excess of 46.7 deer/km^2 (121 deer/mi^2) in the park. Hunter harvests resulted in a significant herd reduction, as 160 deer were counted in 1992 compared to 491 in 1983.

Controlled hunts were conducted during 4 years—1987, 1988, 1989, and 1991—at Tyler State Park in eastern Pennsylvania. The hunts in December 1987 and January 1988 yielded a kill of 487 deer; this number equates to 70.3 deer harvested per km^2 (182 deer/mi^2) on the 692-ha (1,710-acre) park. During 1987, 455 deer were counted during aerial surveys compared to 49 during 1992, indicating that controlled hunts resulted in a significant reduction in deer abundance at Tyler State Park.

National Parks

The 2,335-ha (5,770-acre) Catoctin Mountain National Park, administered by the National Park Service in Maryland, has been noticeably affected by deer since at least 1981. Estimates of deer density indicated an increase from 9.6 to 23.5 deer/km^2 (25 to 61 deer/mi^2) between 1986 and 1989. The presence of deer at this density has led to concern over the effect of deer on native plants, including rare species. The National Park Service is preparing an environmental assessment to review various management alternatives and to select a strategy to manage deer at Catoctin Mountain Park. Unlike in state parks, harvest of deer from National Park Service lands is difficult, if not illegal, to implement; hence, management options are more limited.

Estimates of deer abundance at Gettysburg National Military Park and Eisenhower National Historic Site from 1987 through 1992 indicated an increase from 721 to 1,018 deer on a 2,862-ha (7,072-acre) area near Gettysburg in Adams County, Pennsylvania (Storm et al. 1992; Tzilkowski and Storm 1993). The 1992 population equates to a density of 35.5 deer/km^2 (92 deer/mi^2), which is 10 times higher than that prescribed by the Pennsylvania Game Commission for Adams County. The deer herd at Gettysburg has been associated with high levels of damage to farm crops and forest plant communities, as well as deer-vehicle collisions. An environmental impact statement is being prepared to develop a strategy for managing the Gettysburg deer population.

Refuges

The number of deer harvested by hunters increased twofold between 1983 and 1992 at each of the three national wildlife refuges examined. During 1992, the number of deer taken by hunters was 165 (17.8/km^2 [46/mi^2]) for Eastern Neck, 210 (7.7/km^2 [20/mi^2]) for Great Swamp, and 109 (4.2/km^2 [11/mi^2]) at Montezuma. Although we did not obtain estimates of prehunt populations at these three refuges, if we assume that 35% of the population was killed, the prehunt herd size at the Great Swamp Refuge was 600 deer, which equates to 22 deer/km^2 (57 deer/mi^2).

Harvests by hunters appear to control deer at national wildlife refuges, despite the fact that each refuge manager has a unique set of cultural and biological attributes to consider in deer management. Although hunting is a viable deer management alternative for most refuges, there is still a need to monitor the size of deer herds, determine the most suitable technique to survey deer at each refuge and the most useful demographic data, and monitor plant communities to assess the effect of feeding by deer on plant resources.

Courtesy The Pennsylvania Game Commission

White-tailed deer fawn.

References

Storm, G.L., D.F. Cottam, R.H. Yahner, and J.D. Nichols. 1992. A comparison of two techniques for estimating deer density. Wildlife Society Bull. 20:197-203.

Tzilkowski, W.M., and G.L. Storm. 1993. Detecting change using repeated measures analysis: white-tailed deer abundance at Gettysburg National Military Park. Wildlife Society Bull. 21:411-414.

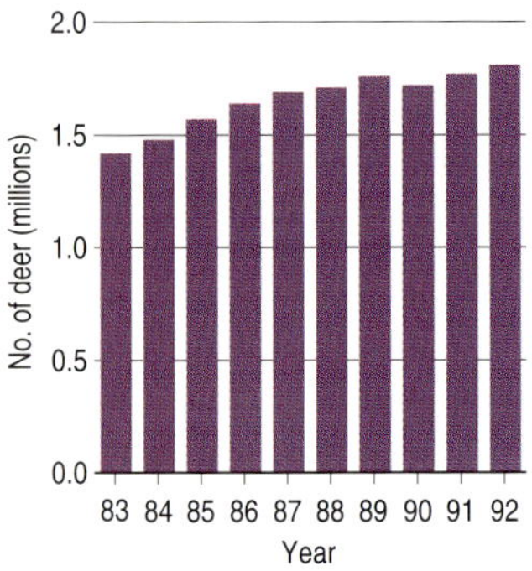

Fig. 1. The trend in the size of the white-tailed deer population in nine northeastern states (Connecticut, Delaware, Maine, Massachusetts, New Hampshire, New York, Pennsylvania, Rhode Island, Vermont), 1983-92.

has increased from 4.3 deer/km^2 (11.1 deer/mi^2) in 1983 to 5.5 deer/km^2 (14.2 deer/mi^2) in 1992. Density estimates ranged from 2.7 deer/km^2 (7.1 deer/mi^2) in Rhode Island to 9.7 deer/km^2 (25.1 deer/mi^2) in Pennsylvania. The total 1992 population of white-tailed deer in the Northeast (including estimates provided by personal communication with biologists from Maryland, New Jersey, Virginia, and West Virginia) was estimated at about 3.0 million.

The total antlered (Fig. 2) and antlerless harvest for all 13 states was estimated at 600,000 in 1983 and 900,000 in 1992. Managers manipulate the harvest of antlered to antlerless deer to obtain a desired population (i.e., appropriate age and sex ratios). During the past decade, deer populations in the Northeast have continued to increase except in states that harvested markedly more antlerless than antlered deer. In Pennsylvania, for example, the deer population increased until the harvest of antlerless deer reached levels necessary to curb the upward trend in the population. In contrast, Massachusetts has consistently harvested more antlered than antlerless deer and the population

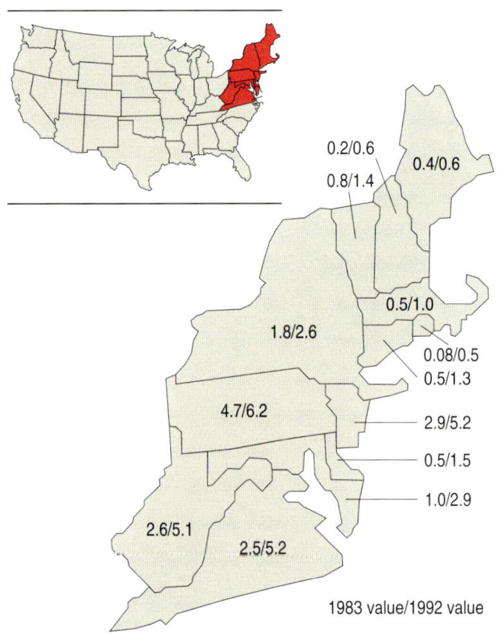

Fig. 2. The harvest of antlered white-tailed deer (number per square mi or 259 ha of deer range) in 13 northeastern states in 1983 (first value) and in 1992 (second value); estimates for Virginia and West Virginia include young-of-the-year males (button bucks).

continues to increase. These two examples illustrate how a prescribed harvest of antlerless deer can be used to achieve a population response that is consistent with each state's management objective. The magnitude of the antlerless and antlered deer harvest is a key factor for adjusting populations. The actual female-male ratio in the population, reproductive rates, and the sex-specific mortality caused by nonhunting factors also affect the population trends of each state.

Ten of 13 states responded to the request for

White-tailed deer (*Odocoileus virginianus*).

information on deer conflicts during the past decade; only two of these indicated no conflict between current deer populations and land use or other natural resources. Four of the eight states with conflicts indicated increasing trends in agriculture-deer conflicts. Conflicts increased between deer and urban habitats in eight states, and vehicle-deer collisions increased in seven of the states. Seven states indicated they had problems between deer and forest regeneration, and two of these states indicated the problem was becoming commoner. Seven states reported deer conflicts with parks and refuges; such problems included lack of forest regeneration as well as deer feeding on ornamental shrubs on private property. Four of these states indicated increasing trends in these kinds of problems.

Conclusions and Present Outlook

The trends in abundance of deer in northeastern states are largely a function of regulated harvests by hunters. A significant amount of information on annual harvest by hunters and deer demographics is available in each northeastern state. Thus, the process of managing white-tailed deer may serve as a model to evaluate monitoring techniques, population dynamics, and effects of wildlife on cultural and other natural resources.

Managers of parks and refuges need better information to predict trends in regeneration and development of forests and the role of deer in forest regeneration. This will require the use of new and appropriate survey techniques (Wiggers and Beckerman 1993) and the ability to evaluate, interpret, and manage data acquired during long-term monitoring of deer and habitats used by deer (Tzilkowski and Storm 1993). Management goals can only be achieved through knowledge of trends in deer abundance and a better understanding of public attitudes toward natural resources.

References

Halls, L.K., ed. 1984. White-tailed deer: ecology and management. Stackpole Books, Harrisburg, PA. 870 pp.

Porter, W.F. 1991. White-tailed deer in eastern ecosystems: implications for management and research in national parks. Natural Resources Report NPS/NRSUNY/NRR-91/05, National Park Service, Denver, CO. 57 pp.

Tzilkowski, W.M., and G.L. Storm. 1993. Detecting change using repeated measures analysis: white-tailed deer abundance at Gettysburg National Military Park. Wildlife Society Bull. 21:411-414.

Wiggers, E.P., and S.F. Beckerman. 1993. Use of thermal infrared sensing to survey white-tailed deer populations. Wildlife Society Bull. 21:263-268.

For further information:

Gerald L. Storm
National Biological Service
Pennsylvania Cooperative Fish and Wildlife Research Unit
University Park, PA 16802

North American Elk

by
James M. Peek
University of Idaho

North American elk or wapiti (*Cervus elaphus*) represent how a wildlife species can recover even after heavy exploitation of populations and habitats around the turn of the century. This species is highly prized by wildlife enthusiasts and by the hunting public, which has provided the various state wildlife agencies with ample support to restore populations to previously occupied habitats and to manage populations effectively. Additionally, the Rocky Mountain Elk Foundation, founded in 1984, has promoted habitat management, acquisition, and proper hunting ethics among many segments of the hunting public.

Current population size is estimated at 782,500 animals for the entire elk range (Rocky Mountain Elk Foundation 1989). Projections of population trends for the national forests and for the entire U.S. elk range are for continued increases through the year 2040 (Flather and Hoekstra 1989).

This species occupies more suitable habitat than at any time in the century, and populations are at all-time highs (Figure). Elk populations in the United States primarily occupy federally managed lands, including national forests, public lands, national parks, and several wildlife refuges. Substantial populations occur on private holdings, including large ranches and reservations owned by Native Americans. Populations have been introduced into Michigan and Pennsylvania and recently have expanded in Nevada and California. In Canada, elk have increased their range into northern British Columbia since 1950 and occupy crown lands in Alberta, British Columbia, and Manitoba. Elk populations in the mountain parks of Jasper, Yoho, Kootenay, and Banff are an important part of the fauna, and the populations in Elk Island National Park and Riding Mountain National Park have been extensively investigated. In Alberta and the western United States, an industry centered around ranching elk has proliferated in recent years.

Perhaps the most spectacular improvement in elk populations is in California, where one population that originally consisted of about 600 individuals in the Owens Valley has now grown to over 2,500 Tule elk in 22 different populations (Phillips 1993). Aquiring habitat and reintroducing elk are the major reasons for the increase.

Problems associated with elk management include the reduced life expectancies of males, which in some areas are attributable to hunting. This problem has been aggravated by increased access to formerly inaccessible habitat, allowing more bulls to be hunted. Additionally, elk have moved into more accessible habitats that provide less cover during hunting seasons. In some cases, hunting has increased enough to lower bull elk life expectancies even in areas where access has not increased. Means to address these issues include reductions in season lengths, quotas on bulls either through hunter registration or limited-entry permit hunts, closures of extensive areas to vehicle access during the hunting season, and more integrated management of timber harvest to accommodate the needs of elk for escape cover.

Such restrictions vary in their effectiveness, depending upon numbers and distribution of hunters, other human disturbances, and the amount and kind of forest involved. In open pine forests, for example, restricting access may be less effective than in denser fir forests, making other hunting regulations, such as limited-entry hunts, necessary. Elk occupying open rangelands where conifer cover is poorly distributed are largely subject to limited-entry hunting. Elk are sensitive to human activity

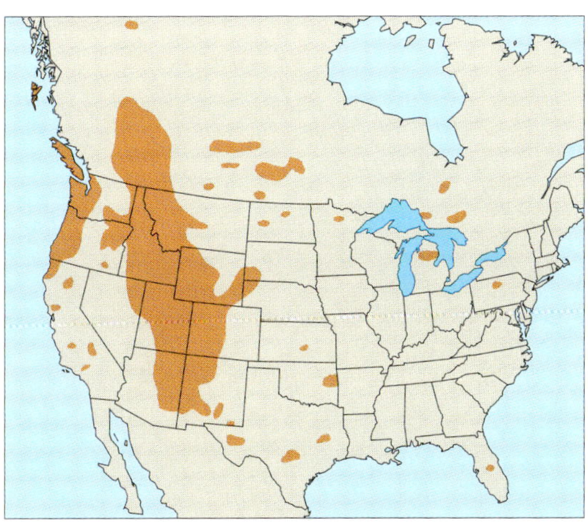

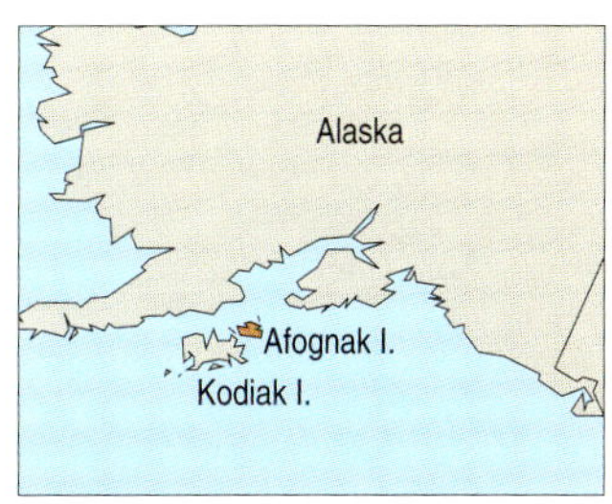

Figure. Distribution of elk in North America as of 1978, based on information provided by provincial and state wildlife agencies (modified from Thomas and Toweill 1982, used with permission, Wildlife Management Institute).

even in national parks where they are not hunted and may become partially conditioned to human presence. Recreational, logging, grazing, seismic, and mining activities must be restricted to times and places where animals are least affected.

As elk numbers have increased in farming areas, depredation on cash crops has also increased. Efforts to address this issue include special "depredation" hunts designed to move animals away from problem areas or to reduce populations, planting less palatable crops, fencing hay and valuable crops to prevent access by elk, feeding elk, and hazing to discourage use. An integrated and specially tailored approach is often necessary to address this important problem.

Whether the high densities of elk that occur within Yellowstone National Park are perceived to be a problem depends upon one's viewpoint. Current research on the condition of park plant communities heavily used by wintering elk suggests that factors interact to influence these communities. Grasslands that have been protected for more than 30 years did not exhibit changes in productivity when compared with grazed grasslands (Coughenour 1991). On the other hand, when protected stands are compared with stands open to browsing, it appears that woody plants may have been adversely altered through prolonged heavy grazing (Chadde and Kay 1991). Past actions that affected plants include fire protection, concentrated grazing pressure by bison (*Bison bison*) in some areas, and altered grizzly bear (*Ursus arctos*) feeding behavior. Within Yellowstone Park, the prospective restoration of wolf (*Canis lupus*) populations and changes in grizzly bear populations since the elimination of artifical food sources will undoubtedly affect elk populations that exist primarily within the park.

Natural changes in habitat across the western elk range have largely benefited elk. Efforts to improve range conditions by modifying livestock grazing practices will provide more forage for elk, even if losses in woody plants may reduce the habitat quality for deer. Better livestock management should also mean accommodating elk habitat use by providing ungrazed pastures within grazing allotments and by manipulating livestock grazing so plants retain their palatability to elk. As livestock is managed more effectively across western public lands, forage plants that wildlife use will benefit, thus also benefiting elk.

On the other hand, some traditional high-quality elk winter habitats, which contain seral (*see* glossary) shrub ranges that developed after large fires earlier this century, are now growing into conifer stands. Some conifers like Douglas fir (*Pseudotsuga menziesii*) are palatable and highly digestible for elk, and even pole-size stands can provide needed cover during severe winters or hunting seasons. As conifers dominate a larger proportion of the winter ranges and associated spring habitats, however, they shade out other species and habitat quality may deteriorate, eventually hurting elk populations. These long-term changes are not easily dealt with in short-term management efforts.

Nevertheless, the future of elk populations in North America seems secure. Demand for hunting as well as the nonconsumptive values of elk will ensure the success of substantial populations. Elk populations will benefit from improved habitat conditions on arid portions of the range, improved livestock management, more effective integrated management of forested habitats, and continued implementation of fire management policies in the major wilderness areas and national parks.

References

Chadde, S.W., and C.E. Kay. 1991. Tall willow communities on Yellowstone's northern range: a test of the "natural regulation" paradigm. Pages 231-262 *in* R.B. Keiter and M.S. Boyce, eds. The greater Yellowstone ecosystem. Yale University Press, New Haven, CT.

Coughenour, M.B. 1991. Biomass and nitrogen responses to grazing of upland steppe on Yellowstone's northern winter range. Journal of Applied Ecology 28:71-82.

Flather, C.H., and T.W. Hoekstra. 1989. An analysis of the wildlife and fish situation in the United States: 1989-2040. U.S. Department of Agriculture Forest Service Gen. Tech. Rep. RM-178. 147 pp.

Phillips, B. 1993. Good news for tules: Destanella Flat. Bugle 10:21-31.

Rocky Mountain Elk Foundation. 1989. Wapiti across the West. Bugle 6:138-140.

Thomas, J.W., and D.E. Toweill, eds. 1982. Elk of North America. Stackpole Books, Harrisburg, PA. 698 pp.

For further information:

James M. Peek
University of Idaho
Department of Fish and Wildlife Resources
Moscow, ID 83843

Courtesy John Moran, The Gainesville Sun ©

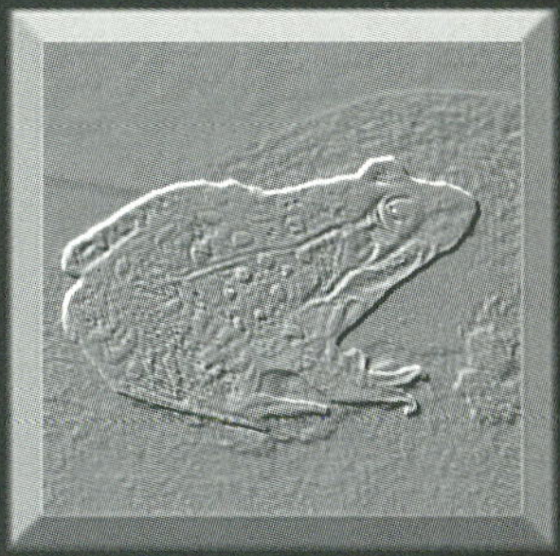

Reptiles and Amphibians

by
Science Editor
Roy W. McDiarmid
National Biological Service
National Museum of Natural History
Washington, DC 20560

Overview

Amphibians and reptiles are important elements of our national biological heritage and deserve special attention. They are crucial to the natural functioning of many ecological processes and key components of important ecosystems. In some areas certain species are economically consequential; others are aesthetically pleasing to many people, and as a group they represent significant segments of the evolutionary history of North America. Knowledge gained from past study of amphibian development and metamorphosis has contributed immensely to our understanding of basic biological processes and has directly benefited humans.

The native herpetofauna of the continental United States includes about 230 species of amphibians (about 62% of which are salamanders and 38% frogs) and some 277 species of reptiles (about 19% turtles, 35% lizards, 45% snakes, and less than 1% crocodilians). If the list were expanded to include native species from Puerto Rico and the U.S. Virgin Islands in the Caribbean, Hawaii, the Trust Territory of the Pacific Islands, and the U.S. Territories in the Pacific, the amphibian list would increase by about 20 native species (all frogs) and another 5 non-native frog species. If the reptile inventory were expanded similarly, the list would increase by 2 turtles, 83 lizards, 18 snakes, and 1 crocodilian. Another 2 species of turtles, 17 lizards, 2 snakes, and 1 crocodilian have been introduced. An updated summary of this information is scheduled for publication later this year (McDiarmid, unpublished data).

Many U.S. reptile and amphibian checklists and field guides have been written over the past 50 years. The data for such summaries come from researchers working with various aspects of the biology of amphibians and reptiles and are found in many scientific publications. These summary field guides give the impression that the herpetofauna of the United States is well known and well studied. When we realize how little is known of the herpetofauna of comparable areas in South America, such an assumption is valid. A cursory review of U.S. data, however, provides a somewhat different view. Since 1978 the total herpetofaunal diversity of the United States has increased by almost 12%, from 454 to 507 species. Much of that increase, though, has resulted from a new knowledge of complex groups of species (e.g., eastern plethodontid salamanders) through the application of molecular techniques to gain a better understanding of the patterns of species formation and of the phylogenetic (evolutionary) history of certain groups. New species are still being discovered in relatively populated parts of the country (e.g., salamanders from California; D. Wake, Museum of Vertebrate Zoology, University of California, Berkeley, personal communication).

Baseline information of the status and health of U.S. populations of amphibians and reptiles is remarkably sparse. No national program of monitoring populations of amphibians and reptiles, comparable to the North American Breeding Bird Survey (now coordinated by the National Biological Service), is operational. Programs in some states (e.g., Kansas, Illinois, Maryland, Wisconsin) have been moderately successful in monitoring amphibians, but clearly a national program is needed. Long-term data (more than 10 years) from specific sites in many habitats in different parts of the country were and are essential to detect continental or global patterns of change in the distribution and abundance of species' populations. A recent publication (Heyer et al. 1994) recommended standard guidelines and techniques for monitoring amphibian populations and habitats; a similar volume on reptiles is planned. What remains is to establish a national program for such monitoring studies; the Declining Amphibian Populations Task Force, a part of the Species Survival Commission of the World Conservation Union, together with the National Biological Service, should play major roles in establishing such programs for amphibians. Similarly, organizations that deal with the conservation of turtles and crocodilians need to be expanded to develop an effective national monitoring program for reptiles.

Habitat degradation and loss seem to be the most important factors adversely affecting amphibian and reptile populations in North America. The drainage and loss of small aquatic habitats and their associated wetlands have had a major adverse effect on many amphibian species and some reptiles.

Many other factors in the decline of reptiles and amphibians have been implicated; most, perhaps all, are human-caused. For example, non-native species of gamefish introduced for sport have been implicated in the decline of frog populations in mountainous areas of some western states. Similarly, the introduction, accidental or intentional, of other non-native species (e.g., bullfrogs in western states, anoline lizards in south Florida, and snakes in Guam) has harmed native species in other parts of the country. Although populations of a few species have been severely impacted for diverse reasons (*see* the articles on California native frogs and the Tarahumara frog [*Rana tarahumarae*]), it is not too late to prevent the extirpation of others. Certain management and conservation decisions based on adequate scientific data and careful planning have proven successful (*see* articles on Coachella Valley fringe-toed lizard [*Uma inornata*] and the American alligator [*Alligator mississippiensis*]), but too often these initiatives are reactive and occur only after a species is in trouble.

Clearly, a better coordinated national program that looks at all species of amphibians and reptiles is desirable. Local and state programs to monitor amphibian and reptile populations are beginning; these efforts need to be expanded nationally. It is obvious that early detection of problems is crucial to successful remedial action. In many ways, a national program of monitoring amphibian and reptile populations is like preventive medicine; the earlier a problem is detected, the greater the likelihood of successful treatment and the lower the cost. A proactive national program based on standardized scientific methodology and applied across all species and habitats will go a long way toward ensuring that amphibians and reptiles remain a healthy component of our national biological heritage. They are too important overall to receive anything less.

Reference

Heyer, W.R., M.A. Donnelly, R.W. McDiarmid, L.-A.C. Hayek, and M.S. Foster, eds. 1994. Measuring and monitoring biological diversity: standard methods for amphibians. Smithsonian Institution Press, Washington, DC. 364 pp.

Turtles

by
Jeffrey E. Lovich
National Biological Service

Turtles have existed virtually unchanged for the last 200 million years. Unfortunately, some of the same traits that allowed them to survive the ages often predispose them to endangerment. Delayed maturity and low and variable annual reproductive success make turtles unusually susceptible to increased mortality through exploitation and habitat modifications (Brooks et al. 1991; Congdon et al. 1993).

In general, turtles are overlooked by wildlife managers in spite of their ecological significance and importance to humans. Turtles are, however, important as scavengers, herbivores, and carnivores, and often contribute significant biomass to ecosystems. In addition, they are an important link in ecosystems, providing dispersal mechanisms for plants, contributing to environmental diversity, and fostering symbiotic associations with a diverse array of organisms. Adults and eggs of many turtles have been used as a food resource by humans for centuries (Brooks et al. 1988; Lovich 1994). As use pressures and habitat destruction increase, management that considers the life-history traits of turtles will be needed.

Documenting Turtle Population Status

I reviewed the population trends of turtles in the United States by examining most references (Ernst et al. 1994) that document the trends of turtle species and populations. Because few long-term studies (lasting more than one generation of the species examined) have focused on turtles, data on population fluctuations over time are generally unavailable (but *see* Gibbons 1990; Congdon et al. 1993). Techniques for conducting population studies of turtles and analyzing the data are summarized in Gibbons (1990).

Although we know less than desired about the actual extent of population fluctuations in most turtle populations, we do know that many turtles in the United States are at great risk of decline and extinction. Of the 55 native turtle species in the United States and its offshore waters, 25 (45%) require conservation, and 21 (38%) are protected or are candidates for protection under the Endangered Species Act. Of the 11 species and subspecies listed as candidates for protection under the ESA, 4 are considered declining, and 7 have unknown population statuses (Table). All tortoises and marine turtles require conservation action. Of the remaining 46 turtle species (aquatic and semi-aquatic forms), 16 (35%) require conservation action. The percentage of U.S. turtles requiring conservation action (45%) is similar to that of the world (41%; IUCN/SSC Tortoise and Freshwater Turtle Specialist Group 1991).

Although no turtles in the United States are known to have become extinct since European colonization (Honegger 1980), many species have experienced significant declines in numbers and distribution during the last 100 years. For example, several bog turtle (*Clemmys muhlenbergii*) populations in western New York, and all populations in western Pennsylvania, are apparently extirpated (Collins 1990; Ernst et al. 1994). Some populations of the spotted turtle (*C. guttata*) have also shown dramatic declines (Lovich 1989). Even wide-ranging, formerly common species such as the common box turtle (*Terrapene carolina*; Ernst et al. 1994), desert tortoise (*Gopherus agassizii;* USFWS 1993), gopher tortoise (*G. polyphemus*; McCoy and Mushinsky 1992), common slider (*Trachemys scripta*; Warwick 1986), and the alligator snapping turtle (*Macroclemys temminckii*; Pritchard 1989) have declined significantly, underscoring the importance of monitoring "common"

Table. U.S. turtle species in need of conservation.

Family and species	Common name	Status*
Cheloniidae	Sea turtles	
Caretta caretta	Loggerhead	Threatened
Chelonia mydas	Green sea turtle	Endangered or threatened according to population or geographic area
Eretmochelys imbricata	Hawksbill	Endangered
Lepidochelys kempii	Kemp's ridley	Endangered
L. olivacea	Olive ridley	Endangered or threatened according to population or geographic area
Chelydridae	Snapping turtles	
Macroclemys temminckii	Alligator snapping turtle	Unknown but vulnerable; C2 candidate
Dermochelyidae	Leatherback sea turtles	
Dermochelys coriacea	Leatherback	Endangered
Emydidae	Semi-aquatic pond turtles	
Clemmys insculpta	Wood turtle	May become threatened if trade not brought under control
C. marmorata	Western pond turtle	Declining; C2 candidate
C. muhlenbergii	Bog turtle	Unknown; are or may be threatened by international trade; C2 candidate
Emydoidea blandingii	Blanding's turtle	Declining; C2 candidate
Graptemys barbouri	Barbour's map turtle	Unknown; C2 candidate
G. caglei	Cagle's map turtle	Unknown
G. flavimaculata	Yellow-blotched map turtle	Threatened, but insufficiently known; may be threatened by international trade
G. oculifera	Ringed map turtle	Threatened; restricted distribution
Malaclemys terrapin	Diamondback terrapin	Some populations unknown, others declining; C2 candidate; listing applies to population or geographic area
Pseudemys alabamensis	Alabama red-bellied turtle	Endangered; restricted distribution
P. rubriventris	Red-bellied turtle	Endangered, according to population or geographic area
Kinosternidae	Mud and musk turtles	
Kinosternon flavescens	Yellow mud turtle	Unknown; C2 candidate; listing applies to population or geographic area
K. hirtipes	Mexican rough-footed mud turtle	Unknown; C2 candidate; listing applies to population or geographic area
Sternotherus depressus	Flattened musk turtle	Threatened
Testudinidae	Tortoises	
Gopherus agassizii	Desert tortoise	Some populations threatened, others are C2 candidates; may become threatened if trade not brought under control. Status of Sonoran Desert population unknown
G. berlandieri	Texas tortoise	May become threatened if trade not brought under control Receiving some conservation action
G. polyphemus	Gopher tortoise	Declining. Some populations threatened, others are C2 candidates; may become threatened if trade not controlled. Receiving some conservation action
Trionychidae	Softshell turtles	
Apalone spinifera	Spiny softshell turtle	Are or may be affected by international trade

*C2 — Possibly qualifying for threatened or endangered status, but more information is needed for determination.

Courtesy J.E. Lovich from the Al Redmond collection

Barbour's map turtle (*Graptemys barbouri*) is restricted to the Apalachicola River system of Alabama, Florida, and Georgia. The species is a candidate for listing under the Endangered Species Act.

species (Dodd and Franz 1993). The alarming decline of marine turtle populations is discussed later in this section.

Perhaps the best data on long-term population changes in turtles are for the diamondback terrapin (*Malaclemys terrapin*), a species exploited heavily during the 19th century as a gourmet food (McCauley 1945; Carr 1952). Terrapin populations declined rapidly, causing some states to set seasons and limits for their protection as early as 1878. The market for terrapin meat eventually waned, and terrapin populations recovered somewhat because their habitat remained largely intact. Unfortunately, some terrapin populations may be declining again because of renewed regional harvesting (Garber 1988), increased habitat destruction, mortality from vehicles, and drowning in crab traps (Ernst et al. 1994).

Some turtle species, such as members of the map turtle genus *Graptemys,* have restricted ranges (Lovich and McCoy 1992) that place them at extreme risk of extinction. In addition, the popularity of many species, particularly tortoises, as pets, contributes to the decline of wild populations (IUCN/SSC 1989; Ernst et al. 1994). Disease also appears to contribute to population declines in some turtles (Balazs 1986; Dodd 1988; Jacobson et al. 1991) and even seems a major challenge to the recovery of the federally threatened desert tortoise (USFWS 1993).

Because of individual longevity, delayed maturity, and long generation times of turtles, long-term studies are required to monitor the dynamics of turtle populations (Gibbons 1990); recovery of most threatened species will be slow. Programs in which hatchlings are propagated in captivity and later released into the wild will do little to assist the recovery of turtles until the ultimate causes of decline are corrected (Frazer 1992).

Efforts to conserve turtles in the United States should be concentrated in areas of high species diversity, where many species have limited distributions, and where populations are at great risk. Notable high-risk areas include shallow wetlands inhabited by freshwater turtles and coastal zones occupied by sea turtles. The most significant area of turtle endemism in the United States is along the Coastal Plain of the Gulf of Mexico (Lovich and McCoy 1992). Eleven species of turtles in the southeastern United States, where diversity is high (Iverson and Etchberger 1989; Iverson 1992), require conservation action, adding to the importance of implementing immediate conservation programs in that region.

References

Balazs, G.H. 1986. Fibropapillomas in Hawaiian green turtles. Marine Turtle Newsletter 39:1-3.

Brooks, R.J., G.P. Brown, and D.A. Galbraith. 1991. Effects of a sudden increase in natural mortality of adults on a population of the common snapping turtle (*Chelydra serpentina*). Canadian Journal of Zoology 69:1314-1320.

Brooks, R.J., D.A. Galbraith, E.G. Nancekivell, and C.A. Bishop. 1988. Developing guidelines for managing snapping turtles. Pages 174-179 *in* R.C. Szaro, K.E. Severson, and D.R. Patton, tech. coords. Management of amphibians, reptiles, and small mammals in North America. U.S. Forest Service Gen. Tech. Rep. RM-166.

Carr, A.F. 1952. Handbook of turtles. The turtles of the United States, Canada, and Baja California. Comstock Publishing Associates, Cornell University Press, Ithaca, NY. 542 pp.

Collins, D.E. 1990. Western New York bog turtles: relicts of ephemeral islands or simply elusive? Pages 151-153 *in* R.S. Mitchell, C.J. Sheviak, and D.J. Leopold, eds. Ecosystem management: rare species and significant habitats. Proceedings of the Fifteenth Annual Natural Areas Conference, New York State Museum Bull. 471.

Congdon, J.D., A.E. Dunham, and R.C. Van Loben Sels. 1993. Delayed sexual maturity and demographics of Blanding's turtles (*Emydoidea blandingii*): implications for conservation and management of long-lived organisms. Conservation Biology 7:826-833.

Dodd, C.K., Jr. 1988. Disease and population declines in the flattened musk turtle *Sternotherus depressus*. American Midland Naturalist 119:394-401.

Dodd, C.K., Jr., and R. Franz. 1993. The need for status information on common herpetofaunal species. Herpetological Review 24:47-49.

Ernst, C.H., J.E. Lovich, and R.W. Barbour. 1994. Turtles of the United States and Canada. Smithsonian Institution Press, Washington, DC. In press.

Frazer, N.B. 1992. Sea turtle conservation and halfway technology. Conservation Biology 6:179-184.

Garber, S.B. 1988. Diamondback terrapin exploitation. Plastron Papers 17(6):18-22.

Gibbons, J.W., ed. 1990. Life history and ecology of the slider turtle. Smithsonian Institution Press, Washington, DC. 368 pp.

Honegger, R.E. 1980. List of amphibians and reptiles either known or thought to have become extinct since 1600. Biological Conservation 19:141-158.

IUCN/SSC. 1989. The conservation biology of tortoises. Occasional Papers of the International Union for the Conservation of Nature and Natural Resources, Species Survival Commission (SSC) 5. 204 pp.

IUCN/SSC Tortoise and Freshwater Turtle Specialist Group. 1991. Tortoises and freshwater turtles: an action plan for their conservation. 2nd ed. International Union for the Conservation of Nature and Natural Resources, Gland, Switzerland. 48 pp.

Iverson, J.B. 1992. Global correlates of species richness in turtles. Herpetological Journal 2:77-81.

Iverson, J.B., and C.R. Etchberger. 1989. The distributions of the turtles of Florida. Florida Scientist 52:119-144.

Jacobson, E.R., J.M. Gaskin, M.B. Brown, R.K. Harris, C.H. Gardiner, J.L. LaPoite, H.P. Adams, and C. Reggiardo. 1991. Chronic upper respiratory tract disease of free-ranging desert tortoises (*Xerobates agassizii*). Journal of Wildlife Diseases 27:296-316.

Lovich, J.E. 1989. The spotted turtles of Cedar Bog, Ohio: historical analysis of a declining population. Pages 23-28 *in* R.C. Glotzhober, A. Kochman, and W.T. Schultz, eds. Proceedings of Cedar Bog Symposium II. Ohio Historical Society.

Lovich, J.E. 1994. Biodiversity and zoogeography of non-marine turtles in Southeast Asia. Pages 381-391 *in* S.K. Majumdar, F.J. Brenner, J.E. Lovich, J.F. Schalles, and E.W. Miller, eds. Biological diversity: problems and challenges. Pennsylvania Academy of Science, Easton, PA. In press.

Lovich, J.E., and C.J. McCoy. 1992. Review of the *Graptemys pulchra* group (Reptilia, Testudines, Emydidae), with descriptions of two new species. Annals of Carnegie Museum 61:293-315.

McCauley, R.H. 1945. The reptiles of Maryland and the District of Columbia. Privately printed, Hagerstown, MD. 194 pp.

McCoy, E.D., and H.R. Mushinsky. 1992. Studying a species in decline: changes in populations of the gopher tortoise on federal lands in Florida. Florida Scientist 55:116-124.

Pritchard, P.C.H. 1989. The alligator snapping turtle: biology and conservation. Milwaukee Public Museum, WI. 104 pp.

USFWS. 1993. Draft recovery plan for the desert tortoise (Mojave population). U.S. Fish and Wildlife Service, Portland, OR. 170 pp.

Warwick, C. 1986. Red-eared terrapin farms and conservation. Oryx 20:237-240.

For further information:

Jeffrey E. Lovich
National Biological Service
Midcontinent Ecological Science Center
Palm Springs Field Station
63-500 Garnet Ave.
PO Box 2000
North Palm Springs, CA
92258

Marine Turtles in the Southeast

by

C. Kenneth Dodd, Jr.
National Biological Service

Five species of marine turtles frequent the beaches and offshore waters of the southeastern United States: loggerhead (*Caretta caretta*), green (*Chelonia mydas*), Kemp's ridley (*Lepidochelys kempii*), leatherback (*Dermochelys coriacea*), and hawksbill (*Eretmochelys imbricata*). All five are reported to nest, but only the loggerhead and green turtle do so in substantial numbers. Most nesting occurs from southern North Carolina to the middle west coast of Florida, but scattered nesting occurs from Virginia through southern Texas. The beaches of Florida, particularly in Brevard and Indian River counties, host what may be the world's largest population of loggerheads.

Marine turtles, especially juveniles and subadults, use lagoons, estuaries, and bays as feeding grounds. Areas of particular importance include Chesapeake Bay, Virginia (for loggerheads and Kemp's ridleys); Pamlico Sound, North Carolina (for loggerheads); and Mosquito Lagoon, Florida, and Laguna Madre, Texas (for greens). Offshore waters also support important feeding grounds such as Florida Bay and the Cedar Keys, Florida (for green turtles), and the mouth of the Mississippi River and the northeast Gulf of Mexico (for Kemp's ridleys). Offshore reefs provide feeding and resting habitat (for loggerheads, greens, and hawksbills), and offshore currents, especially the Gulf Stream, are important migratory corridors (for all species, but especially leatherbacks).

Most marine turtles spend only part of their lives in U.S. waters. For example, hatchling loggerheads ride oceanic currents and gyres (giant circular oceanic surface currents) for many years before returning to feed as subadults in southeastern lagoons. They travel as far as Europe and the Azores, and even enter the Mediterranean Sea, where they are susceptible to longline fishing mortality. Adult loggerheads may leave U.S. waters after nesting and spend years in feeding grounds in the Bahamas and Cuba before returning. Nearly the entire world population of Kemp's ridleys uses a single Mexican beach for nesting, although juveniles and subadults, in particular, spend much time in U.S. offshore waters.

The biological characteristics that make sea turtles difficult to conserve and manage include a long life span, delayed sexual maturity, differential use of habitats both among species and life stages, adult migratory travel, high egg and juvenile mortality, concentrated nesting, and vast areal dispersal of young and subadults. Genetic analyses have confirmed that females of most species return to their natal beaches to nest (Bowen et al. 1992; Bowen et al. 1993). Nesting assemblages contain unique genetic markers showing a tendency toward isolation from other assemblages (Bowen et al. 1993); thus, Florida green turtles are genetically different from green turtles nesting in Costa Rica and Brazil (Bowen et al. 1992). Nesting on warm sandy beaches puts the turtles in direct conflict with human beach use, and their use of rich offshore waters subjects them to mortality from commercial fisheries (National Research Council 1990).

Marine turtles have suffered catastrophic declines since European discovery of the New World (National Research Council 1990). In a relatively short time, the huge nesting assemblages in the Cayman Islands, Jamaica, and Bermuda were decimated. In the United States, commercial turtle fisheries once operated in south Texas (Doughty 1984), Cedar Keys, Florida Keys, and Mosquito Lagoon; these fisheries collapsed from overexploitation of the mostly juvenile green turtle populations. Today, marine turtle populations are threatened worldwide and are under intense pressure in the Caribbean basin and Gulf of Mexico, including Cuba, Mexico, Hispaniola, the Bahamas, and Nicaragua. Subadult loggerheads are captured extensively in the eastern Atlantic Ocean and Mediterranean Sea. Thus, marine turtles that hatch or nest on U.S. beaches or migrate to U.S. waters are under threats far from U.S. jurisdiction. Marine turtles can be conserved only through international efforts and cooperation.

Information on the status and trends of southeastern marine turtle populations comes from a variety of sources, including old fishery records, anecdotal accounts of abundance, beach surveys for nests and females, and trawl and aerial surveys for turtles offshore. Surveys for marine turtles are particularly difficult because most of their lives are spent in habitats that are not easily surveyed. Hence, most status and trends information comes from counting females and nests. Few systematic long-term (more than 10-20 years) surveys have been conducted; the most notable are the nesting surveys at Cumberland Island and adjacent barrier islands in Georgia (T.H. Richardson, University of Georgia, unpublished data), and beaches south of Melbourne in Brevard County, Florida (Ehrhart et al. 1993). Beach monitoring is fairly widespread in many areas of the Southeast, but coverage varies considerably among beaches and field crews. The only long-term sampling of lagoonal or bay populations occurs at Mosquito Lagoon and Chesapeake Bay, although short-duration surveys have sampled Florida Bay, Pamlico Sound, and Laguna Madre. Trawl surveys of inlets and ship channels and aerial surveys of offshore waters have been undertaken periodically.

Loggerhead and Green Turtles

The number of turtles nesting fluctuates substantially from one year to the next, making interpretation of beach counts difficult. The Florida nesting populations of loggerheads and green turtles appear stable based on 12 years of data from east-central Florida (Ehrhart et al. 1993; Fig. 1). The green turtle nesting population may be increasing because of protective measures over the last 20 years or so, although the number of nesting females is still low (assuming 3-5 nests per female). North of Florida, nesting loggerhead numbers are declining 3%-9% a year in Georgia and South Carolina (National Research Council 1990). The main cause of mortality is drowning in shrimp and fish nets (National Research Council 1990), although turtle excluder devices (TEDs; Fig. 2a) have helped reduce mortality (Fig 2b; Henwood et al. 1992). Large juveniles are most susceptible to drowning, and this is a critical life stage in the population dynamics of sea turtles (Crouse et al. 1987).

Few data are available for lagoonal turtles, although similar numbers have been captured in Mosquito Lagoon and Chesapeake Bay from one year to the next. Loggerhead and green turtle populations, both adult and subadult, have undoubtedly declined from historical levels because of beach development and disturbance, the collection of eggs, and destructive fishing

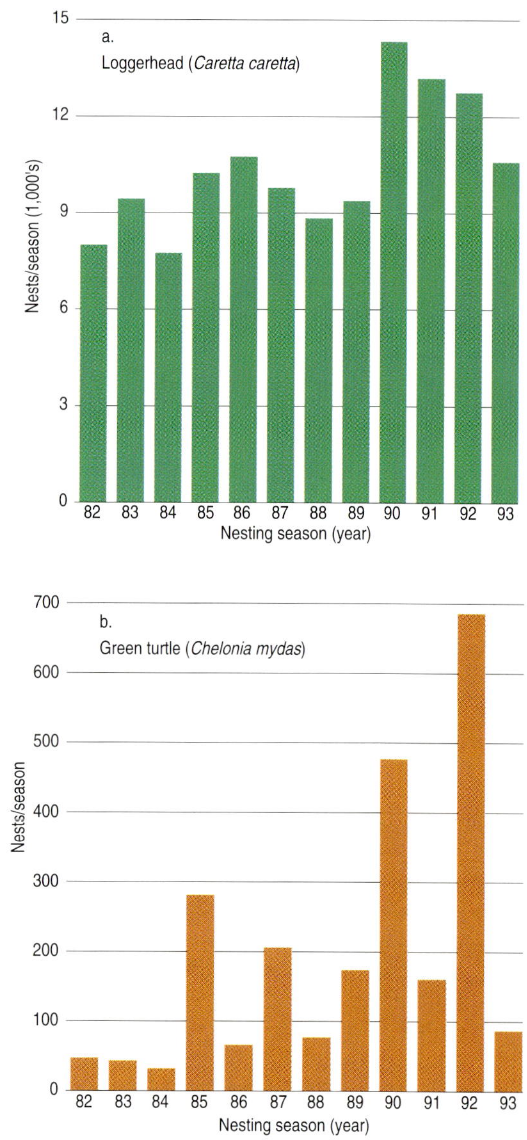

Fig. 1 a. Loggerhead nest totals in south Brevard County, Florida, 1982-93. **b.** Green turtle nest totals in south Brevard County, Florida, 1982-93. From Ehrhart et al. (1993).

practices. Most high-level nesting occurs on the remaining undeveloped or lightly developed beaches. Even there, plans for development and disorientation from lights pose serious and continuing problems.

Kemp's Ridley

At one time, more than 40,000 females nested in a single mass nesting (termed "arribada") in Tamaulipas, Mexico. Several arribadas probably occurred each year. Since 1947 a drastic reduction in the number of nesting females caused the near extinction of this species (Ross et al. 1989). Today only about 400-500 turtles nest each year despite stringent protection of the nesting beach. The principal threat to this species is incidental take during shrimp fishing.

Leatherback and Hawksbill

The leatherback and hawksbill are rare nesters in the southeastern United States, but offshore waters are important for feeding, resting, and as migratory corridors. The status and trends of these species in U.S. offshore waters are unknown, although they are severely threatened throughout the Caribbean. Leatherbacks are taken by trawlers or are otherwise entangled in nets. Hawksbills are sought, especially in Cuba, for their shell, which is used for jewelry and similar items. The solitary nesting habits of hawksbills make them particularly difficult to monitor.

Summary

Sea turtles are threatened by beach development, light pollution, ocean dumping, incidental take in trawl and longline fisheries, disease (especially fibropapillomas), and many other variables. Because sea turtles are long-lived species, trends are difficult to monitor. Present methods of beach monitoring are extremely labor-intensive, expensive, and biased toward one segment of the population. Very little is known about marine turtle life-history and habitat requirements away from nesting beaches, and virtually nothing is known about male turtles. Because the effectiveness of measures aimed at protecting turtles may not be seen for decades, known conservation strategies should be favored over unproven mitigation schemes. Acquiring nesting habitat should be encouraged. One of the most important management measures to protect sea turtles, especially of the juvenile and subadult size class, in the southeastern United States, Caribbean, and western Atlantic Ocean is the use of TEDs to minimize drowning in commercial fisheries. Mature females should also be protected because of their importance to future reproduction. Researchers need to identify migratory routes, feeding and developmental habitat, and ways to minimize adverse impacts during all life-history stages.

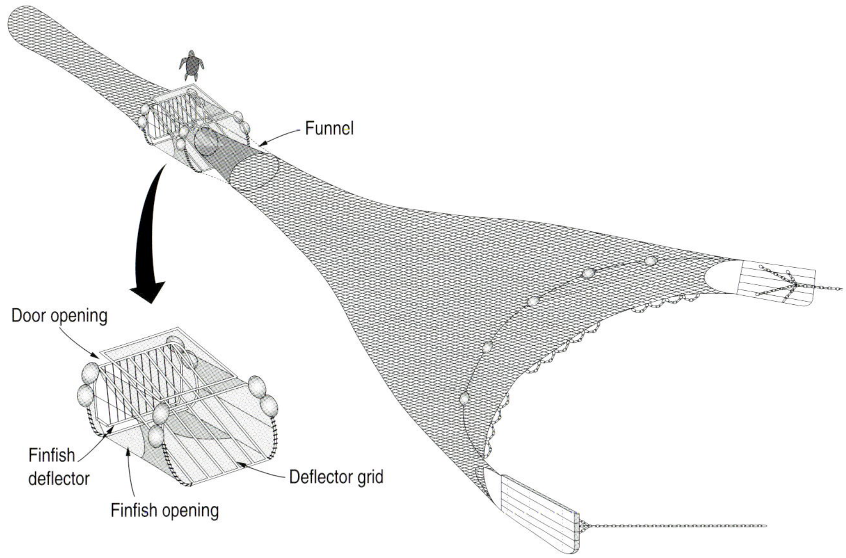

Fig. 2a. Schematic of a turtle excluder device (TED). From Watson et al. (1986).

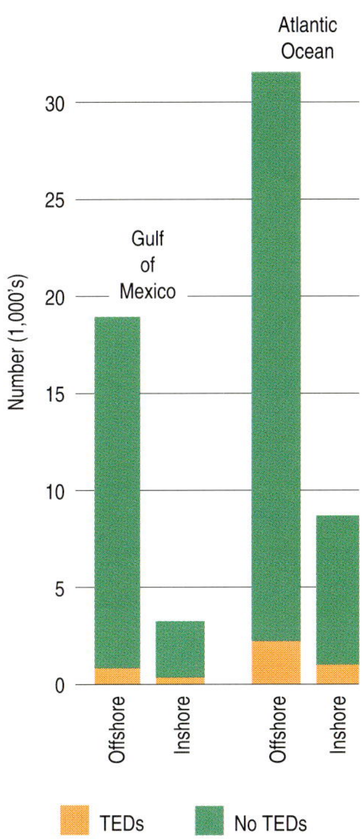

Fig. 2b. Incidental capture of sea turtles in inshore and offshore waters of the United States before and after regulations requiring the use of TEDs on the U.S. shrimp fleet. From Henwood et al. (1992).

References

Bowen, B., J.C. Avise, J.I. Richardson, A.B. Meylan, D. Margaritoulis, and S.R. Hopkins-Murphy. 1993. Population structure of loggerhead turtles (*Caretta caretta*) in the northwestern Atlantic Ocean and Mediterranean Sea. Conservation Biology 7:834-844.

Bowen, B.W., A.B. Meylan, J.P. Ross, C.J. Limpus, G.H. Balazs, and J.C. Avise. 1992. Global population structure and natural history of the green turtle (*Chelonia mydas*) in terms of matriarchal phylogeny. Evolution 46:865-881.

Crouse, D.T., L.B. Crowder, and H. Caswell. 1987. A stage-based population model for loggerhead sea turtles and implications for conservation. Ecology 68:1412-1423.

Doughty, R.W. 1984. Sea turtles in Texas: a forgotten commerce. Southwestern Historical Quarterly 88:43-70.

Ehrhart, L.H., W.E. Redfoot, R.D. Owen, and S.A. Johnson. 1993. Studies of marine turtle nesting beach productivity in central and south Brevard County, Florida, in 1993. Report to Florida Department of Environmental Protection, Institute of Marine Research, St. Petersburg. 20 pp.

Henwood, T., W. Stuntz, and N. Thompson. 1992. Evaluation of U.S. turtle protective measures under existing TED regulations, including estimates of shrimp trawler related mortality in the wider Caribbean. National Oceanic and Atmospheric Administration Tech. Memorandum NMFS-SEFSC-303. 15 pp.

National Research Council. 1990. Decline of the sea turtles. Causes and prevention. National Academy Press, Washington, DC. 259 pp.

Ross, J.P., S. Beavers, D. Mundell, and M. Airth-Kindree. 1989. The status of Kemp's ridley. Center for Marine Conservation, Washington, DC. 51 pp.

Watson, J.W., J.F. Mitchell, and A.K. Shah. 1986. Trawling efficiency device: a new concept for selective shrimp trawling gear. Marine Fisheries Review 48:1-9.

For further information:

C. Kenneth Dodd, Jr.
National Biological Service
Southeastern Biological Science Center
7920 N.W. 71st St.
Gainesville, FL 32653

Amphibians

by
R. Bruce Bury
P. Stephen Corn
C. Kenneth Dodd, Jr.
Roy W. McDiarmid
Norman J. Scott, Jr.
National Biological Service

Amphibians are ecologically important in most freshwater and terrestrial habitats in the United States: they can be numerous, function as both predators and prey, and constitute great biomass. Amphibians have certain physiological (e.g., permeable skin) and ecological (e.g., complex life cycle) traits that could justify their use as bioindicators of environmental health. For example, local declines in adult amphibians may indicate losses of nearby wetlands. The aquatic breeding habits of many terrestrial species result in direct exposure of egg, larval, and adult stages to toxic pesticides, herbicides, acidification, and other human-induced stresses in both aquatic and terrestrial habitats. Reported declines of amphibian populations globally have drawn considerable attention (Bury et al. 1980; Bishop and Petit 1992; Richards et al. 1993; Blaustein 1994; Pechmann and Wilbur 1994).

Approximately 230 species of amphibians, including about 140 salamanders and 90 anurans (frogs and toads) occur in the continental United States. Because of their functional importance in most ecosystems, declines of amphibians are of considerable conservation interest. If these declines are real, the number of listed or candidate species at federal, state, and local levels could increase significantly. Unfortunately, because much of the existing information on status and trends of amphibians is anecdotal, coordinated monitoring programs are greatly needed.

Faunal Comparisons

North American amphibian species exhibit two major distributional patterns, endemic and widespread. Endemic species (Figure) tend to have small ranges or are restricted to specific habitats (e.g., species that occur only in one cave or in rock talus on a single mountainside). Declines are documented best for endemic species, partly because their smaller ranges make monitoring easier. Populations of endemics are most susceptible to loss or depletions because of localized activities (Bury et al. 1980; Dodd 1991). Examples of endemic species affected by different local impacts include the Santa Cruz long-toed salamander (*Ambystoma macrodactylum croceum*) in California, the Texas blind salamander (*Typhlomolge rathbuni*) in Texas, and the Red Hills salamander (*Phaeognathus hubrichti*) in Alabama; these three species are listed as federally threatened or endangered.

The number of endemic species that have suffered losses or are suspected of having severe threats to their continued existence has increased in the last 15 years (Table). In part, the increase reflects descriptions of new species with restricted ranges, but the accelerating pace of habitat alteration is the primary threat.

Table. The number of amphibian species showing documented or perceived declines in 1980 (Bury et al. 1980) and 1994.

Distribution pattern	Number of species	
	1980	1994
Endemic or relict	33	52
Widespread	5	33

The ranges of most endemics in the western states (26 species) are widely dispersed across the landscape. In contrast, endemics in the eastern and southeastern states (25 species) tend to be clustered in centers of endemism, such as in the Edwards Plateau (Texas), Interior (Ozark) Highlands (Arkansas, Oklahoma), Atlantic Coastal Plain (Texas to Virginia), and uplands or mountaintops in the Appalachian Mountains (West Virginia to Georgia).

Widespread species often are habitat generalists. Many were previously common, but have shown regional or rangewide declines (Hine et al. 1981; Corn and Fogelman 1984; Hayes and Jennings 1986; Table). Reported declines of widespread species often lack explanation, perhaps because these observations have only recently received general attention or because temporal and spatial variations in population sizes of many amphibians are not well understood. Some reports are for amphibians in relatively pristine habitats where human impacts are not apparent.

A few examples of declines in widespread species illustrate the threats they face across the country:

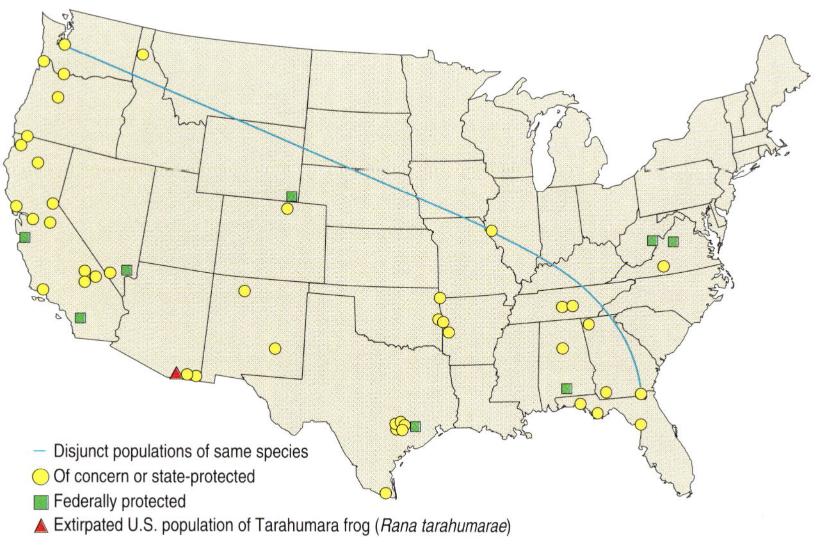

Figure. Distribution of U.S. endemic amphibian species; those west of the 100th meridian tend to be more broadly dispersed.

- Amphibians predominate in small forest streams of the Pacific Northwest. Because timber is harvested without adequate streamside protection, many populations of the tailed frog (*Ascaphus truei*) and torrent salamanders (*Rhyacotriton* spp.) have been severely affected; some populations soon will warrant consideration for listing.
- The western toad (*Bufo boreas*) once was common in the Rocky Mountains, but now occurs at fewer than 20% of known localities from southern Wyoming to northern New Mexico.
- Many salamander and frog populations in the southeastern United States have been negatively affected, some severely, because of degradation of stream habitats (e.g., the hellbender, *Cryptobranchus alleganiensis*) and conversion of natural pinewood and hardwood forests and associated wetlands (e.g., gopher frog, *Rana capito*) to plantation forestry, agriculture, and urban uses.
- Leopard frogs (*Rana* spp.), which are used in teaching and research institutions, were once abundant in most of the United States. Populations in this diverse group have declined, sometimes significantly, in midwestern, Rocky Mountain, and southwestern states.

Western toad (*Bufo boreas*).

Causes of Declines

No single factor has been identified as the cause of amphibian declines, and many unexplained declines likely result from multiple causes. Human-caused factors may intensify natural factors (Blaustein et al. 1994b) and produce declines from which local populations cannot recover and thus they go extinct. Known or suspected factors in those declines include destruction and loss of wetlands (Bury et al. 1980); habitat alteration, such as impacts from timber harvest and forest management (Corn and Bury 1989; Dodd 1991; Petranka et al. 1993); introduction of non-native predators, such as sportfish and bullfrogs, especially in western states (Hayes and Jennings 1986; Bradford 1989); increased variety and use of pesticides and herbicides (Hine et al. 1981); effects of acid precipitation, especially in eastern North America and Europe (Freda 1986; Beebee et al. 1990; Dunson et al. 1992); increased ultraviolet radiation reaching the ground (Blaustein et al. 1994a); and diseases resulting from decreased immune system function (Bradford 1991; Carey 1993; Pounds and Crump 1994).

A Success Story: *The Barton Springs Salamander*

A success story from the Edwards Plateau in Texas illustrates the importance of baseline ecological data, current science, and the types of partnerships essential for conservation of amphibians. The recently described Barton Springs salamander (*Eurycea sosorum*) occurs only in three springs within about 300 m (984 ft) of each other within the city limits of Austin. This salamander has one of the smallest known distributions of any North American vertebrate.

Pools associated with the two primary springs had been developed as municipal swimming and wading pools, and standard cleaning procedures had eliminated most salamanders. With cooperation of city authorities and local volunteers, pool maintenance practices detrimental to the salamander were modified, and populations of the salamander seem to be increasing and expanding their ranges within the spring system.

Barton Springs salamander (*Eurycea sosorum*).

Amphibian populations also may vary in size because of natural factors, particularly extremes in the weather (Bradford 1983; Corn and Fogelman 1984). The size of amphibian populations may vary, sometimes dramatically, from year to year, so what is perceived as a decline may be part of long-term fluctuations (Pechmann et al. 1991). The effect of global climate change on amphibians is speculative, but it has the potential for causing the loss of many species.

Monitoring Needs

A profound need exists for national coordination of regional inventories and population studies, including a national effort to monitor amphibians on parks, forests, wildlife refuges, and other public lands. Only through long-term studies will better data on population changes through time and between sites become available. Such data are essential to evaluating the status and trends of amphibian species in the United States. Some regional surveys and inventories exist but only for a few species; these studies should be expanded into a coordinated effort with long-term monitoring of populations at many sites across the country as the goal.

In addition, more research is needed to determine the impact of natural and human-caused factors on the different life-history stages and environments of amphibians. Also, the assumption that amphibians are good indicators needs to be tested rigorously (Pechmann and Wilbur 1994). Likewise, understanding the dynamics of populations between habitats and regions, and the roles amphibians play in aquatic and terrestrial ecosystems is essential. Detailed work on the ecology of species and the factors implicated in declines needs to continue.

References

Beebee, T.J.C., R.J. Flower, A.C. Stevenson, S.T. Patrick, P.G. Appleby, C. Fletcher, C. Marsh, J. Natkanski, B. Rippey, and R.W. Battarbee. 1990. Decline of the natterjack toad *Bufo calamita* in Britain: palaeoecological, documentary, and experimental evidence for breeding site acidification. Biological Conservation 53:1-20.

Bishop, C.A., and K.E. Petit, eds. 1992. Declines in Canadian amphibian populations: designing a national monitoring strategy. Canadian Wildlife Service Occasional Paper 76:1-120.

Blaustein, A.R. 1994. Chicken Little or Nero's fiddle? A perspective on declining amphibian populations. Herpetologica 50:85-97.

Blaustein, A.R., P.D. Hoffman, D.G. Hokit, J.M. Kiesecker, S.C. Walls, and J.B. Hays. 1994a. UV repair and resistance to solar UV-B in amphibian eggs: a link to population declines? Proceedings of the National Academy of Sciences 91:1791-1795.

Blaustein, A.R., D.B. Wake, and W.P. Sousa. 1994b. Amphibian declines: judging stability, persistence, and susceptibility of populations to local and global extinction. Conservation Biology 8:60-71.

Bradford, D.F. 1983. Winterkill, oxygen relations, and energy metabolism of a submerged dormant amphibian, *Rana muscosa*. Ecology 64:1171-1183.

Bradford, D.F. 1989. Allotopic distribution of native frogs and introduced fishes in high Sierra Nevada lakes of California: implication of the negative effect of fish introductions. Copeia 1989:775-778.

Bradford, D.F. 1991. Mass mortality and extinction in a high-elevation population of *Rana muscosa*. Journal of Herpetology 5:174-177.

Bury, R.B., C.K. Dodd, Jr., and G.M. Fellers. 1980. Conservation of the Amphibia of the United States: a review. U.S. Fish and Wildlife Service Res. Publ. 134:1-34.

Carey, C. 1993. Hypothesis concerning the causes of the disappearance of boreal toads from the mountains of Colorado. Conservation Biology 7:355-362.

Corn, P.S., and R.B. Bury. 1989. Logging in western Oregon: responses of headwater habitats and stream amphibians. Forest Ecology and Management 29:39-57.

Corn, P.S., and J.C. Fogelman. 1984. Extinction of montane populations of the northern leopard frog (*Rana pipiens*) in Colorado. Journal of Herpetology 18:147-152.

Dodd, C.K., Jr. 1991. The status of the Red Hills salamander *Phaeognathus hubrichti*, Alabama, USA, 1976-1988. Biological Conservation 55:57-75.

Dunson, W.A., R.L. Wyman, and E.S. Corbett. 1992. A symposium on amphibian declines and habitat acidification. Journal of Herpetology 16:349-352.

Freda, J. 1986. The influence of acidic pond water on amphibians: a review. Water, Air, and Soil Pollution 30:439-450.

Hayes, M.P., and M.R. Jennings. 1986. Decline of ranid frog species in western North America: are bullfrogs (*Rana catesbeiana*) responsible? Journal of Herpetology 20:490-509.

Hine, R.L., B.L. Les, and B.F. Hellmich. 1981. Leopard frog populations and mortality in Wisconsin, 1974-76. Wisconsin Department of Natural Resources Tech. Bull. 122:1-39.

Pechmann, J.H.K., D.E. Scott, R.D. Semlitsch, J.P. Caldwell, L.J. Vitt, and J.W. Gibbons. 1991. Declining amphibian populations: the problem of separating human impacts from natural fluctuations. Science 253:892-895.

Pechmann, J.H.K., and H.M. Wilbur. 1994. Putting declining amphibian populations in perspective: natural fluctuations and human impacts. Herpetologica 50:65-84.

Petranka, J.W., M.E. Eldridge, and K.E. Haley. 1993. Effects of timber harvesting on southern Appalachian salamanders. Conservation Biology 7:363-370.

Pounds, J.A., and M.L. Crump. 1994. Amphibian declines and climate disturbance: the case of the golden toad and harlequin frog. Conservation Biology 8:72-85.

Richards, S.J., K.R. McDonald, and R.A. Alford. 1993. Declines in populations of Australia's endemic tropical rainforest frogs. Pacific Conservation Biology 1:66-77.

For further information:

R. Bruce Bury
National Biological Service
200 S.W. 35th St.
Corvallis, OR 97333

American Alligators in Florida

by
Allan R. Woodward
Florida Game and Fresh Water Fish Commission
Clinton T. Moore
U.S. Fish and Wildlife Service

The American alligator (*Alligator mississippiensis*) is an integral component of wetland ecosystems in Florida. Alligators also provide aesthetic, educational, recreational, and economic benefits to humans. Because of the commercial value of alligator hides for making high-quality leather products, alligator hunting was a major economic and recreational pursuit of many Floridians from the mid-1800's to 1970. The Florida alligator population varied considerably during the 1900's in response to fluctuating hunting pressure caused by unstable markets for luxury leather products.

The declining abundance of alligators during the late 1950's and early 1960's led to the 1967 classification of the Florida alligator population as endangered throughout its range. Federal and international regulations imposed during the 1970's and 1980's helped control trade of alligator hides, and illegal hunting of alligators was checked. The Florida alligator population responded immediately to protection and was reclassified as threatened in 1977 and as threatened because of its similarity in appearance to the American crocodile (*Crocodylus acutus*) in 1985 (Neal 1985).

Assessments of Florida's alligator population were based on sporadic surveys before 1974 (Wood et al. 1985). The Florida Game and Fresh Water Fish Commission implemented annual night-light surveys that used spotlights to detect alligator eyeshine in 1974 to provide a more objective basis for assessing population trends (Wood et al. 1985). Although all areas were not sampled every year, these data are the best available for alligator populations in Florida and are useful for estimating population trends (Woodward and Moore 1990). Because survey areas were not a random sample of all alligator habitat in Florida, trend results are applicable only to deepwater habitats and navigable wetlands.

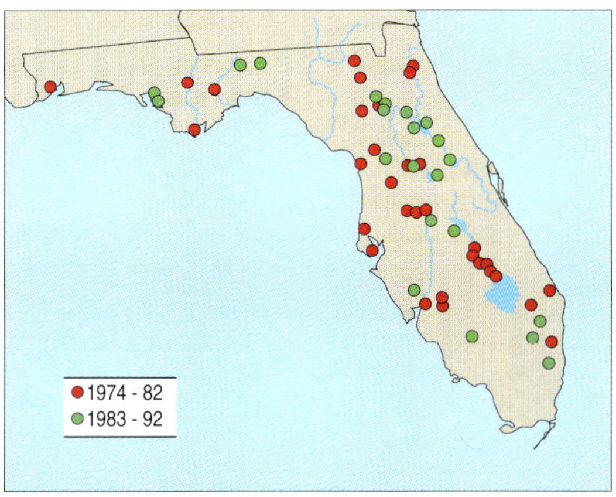

Fig. 1. Locations of survey areas for night-light counts of alligators in Florida, 1974-92.

Design of Alligator Surveys, 1974-92

We conducted night-light counts (Woodward and Marion 1978) with high-intensity spotlights from boats on 54 areas throughout Florida (Fig. 1) during 1974-92 (Woodward and Moore 1990). The number of areas surveyed in any year ranged from 7 in 1974 to 43 in 1980. In 1983 the number of areas surveyed was reduced to 22 to allow observers to conduct replicate counts on areas each year (Fig. 1). Eighteen of the 22 areas were subjected to alligator harvests of some type.

Courtesy John Moran, The Gainesville Sun ©

Alligators at dusk, Payne's Prairie State Preserve, Florida.

We analyzed observed densities of alligators per kilometer (0.62 mile) of shoreline to estimate trends for each area during the periods 1974-92 and 1983-92. Size classes corresponded to the overall population, juveniles (0.3-1.2 m [1-4 ft]), harvestable sizes (1.2 m or longer [4 ft or longer]), and adults (1.8 m or longer [6 ft or longer]; hatchlings less than 0.3 m long [1 ft] were excluded from trend analysis).

Count densities represent only alligators observed during the survey. Most (more than 65%) alligators were submerged during surveys and not detected (Murphy 1977; Brandt 1989; Woodward and Linda 1993). Alligators in wetlands adjacent to surveyed areas may have been undetected (Woodward and Linda 1993). Counts, however, do provide a relative measure of alligator abundance that is useful for estimating population trends, provided that rates of detection do not vary annually.

Status and Trends

From 1974 to 1992, the density of alligators on surveyed wetlands increased an average 41%

or 1.9% annually. Average annual densities of harvestable alligators increased 2.7%, while average annual densities of adults increased 2.5%. The 0.5% average annual increase in counts of juvenile alligators during 1974-92 was not significant. These trends confirm that the Florida alligator population increased during the apparent recovery of the 1970's and 1980's (Neal 1985). We observed cyclic patterns in abundance over time for all size classes (Fig. 2). Cyclic population levels may represent varying availability of counted alligators due to fluctuations in water level not fully accounted for in our analyses. They may also reflect population changes brought about by periodic droughts or, to a lesser extent, severe winters.

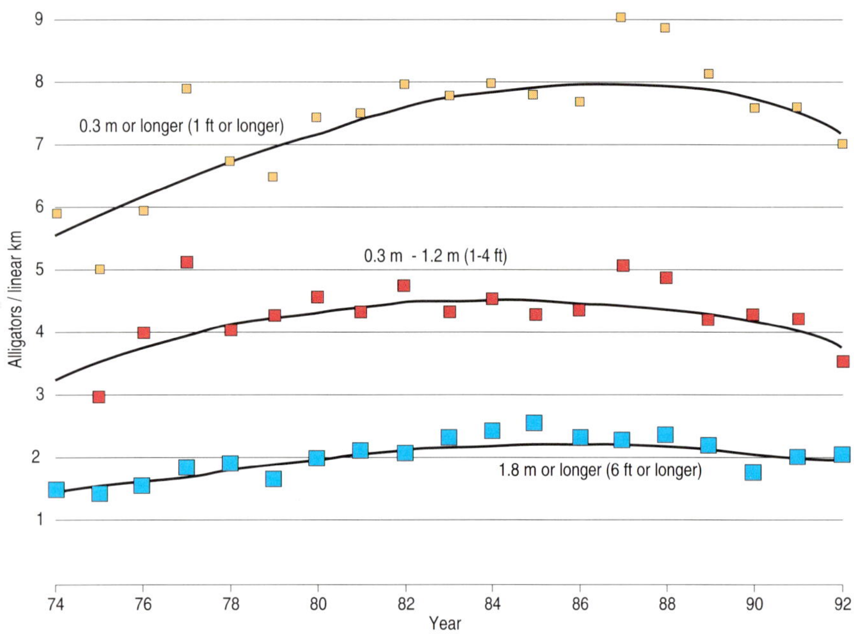

Fig. 2. Annual indices (mean number of alligators detected per linear kilometer [0.62 mi] of survey route) and smoothed trend estimates (Cleveland 1979) for three size classes of the statewide alligator population in Florida, 1974-92.

From 1983 to 1992, observed densities of adult alligators declined 3.2% per year, but we did not detect such trends in other size classes (Fig. 2). It is too early to draw conclusions concerning the influence of harvests on alligator populations since legal harvesting began in 1987 because of the variable nature of night-light alligator counts and the uncertain effects of wariness. Relatively stable populations of juveniles and harvestable alligators indicate that hatchling recruitment (replenishment) is sufficient to replace alligators lost through harvest. Consequently, alligator harvests do not seem to have negatively affected the Florida alligator population as a whole.

Historically, the Florida alligator population was threatened by habitat loss and excessive illegal hunting (Hines 1979), but recently environmental contamination has been associated with population declines. Wetland drainage and alteration during the 1900's destroyed alligator habitat and permanently reduced alligator populations in some wetlands, particularly in freshwater marshes (Neal 1985). State legislation, most recently the Wetlands Protection Act of 1984 (Florida Statutes 403.91), has significantly protected remaining wetlands, but alteration and loss of wetlands persist. Between the mid-1970's and mid-1980's, 10,542 ha (26,030 acres) of wetlands per year were lost to agriculture and other development (Frayer and Hefner 1991). Thus, habitat loss remains a threat to alligator populations.

Illegal hunting is now negligible and has been replaced by regulated, managed harvests. Florida implemented a nuisance alligator control program in 1978 in response to increasing problem alligators during the 1970's (Hines and Woodward 1980). Because the nuisance alligator program targets individual alligators, the removal of these animals is unlikely to measurably affect alligator populations (Hines and Woodward 1980; Jennings et al. 1989). The state game commission introduced managed harvests of alligators and their eggs in 1987 to create conservation incentives by enhancing economic value of wild alligators (Wiley and Jennings 1990). Studies of the effects of harvest on alligator populations demonstrated that harvests are sustainable at certain rates (Jennings et al. 1988; Woodward et al. 1992). Annual monitoring and effective control of harvest rates ensure that populations will not suffer long-term depletion.

More recently, environmental toxins have been implicated in the sharp decline of the alligator population on Lake Apopka, Florida's third-largest lake (Woodward et al. 1993; Guillette et al. 1994). Widespread pollution of wetlands by potentially toxic petrochemicals and metals may threaten the long-term viability of other alligator populations within Florida. For the present, the status of the Florida alligator population is secure; however, continued habitat loss and toxic contamination will negatively affect alligator populations and may eventually compromise their conservation.

References

Brandt, L.A. 1989. The status and ecology of the American alligator (*Alligator mississippiensis*) in Par Pond, Savannah River Site. M.S. thesis, Florida International University, Fort Lauderdale. 89 pp.

Cleveland, W.S. 1979. Robust locally weighted regression and smoothing scatterplots. Journal of the American Statistical Association 74:829-836.

Frayer, W.E., and J.M. Hefner. 1991. Florida wetlands: status and trends, 1970's to 1980's. U. S. Fish and Wildlife Service, Atlanta. 33 pp.

Guillette, L.J., Jr., T.S. Gross, G.R. Masson, J.M. Matter, H.F. Percival, and A.R. Woodward. 1994. Developmental abnormalities of the gonad and abnormal sex hormone concentrations in juvenile alligators from contaminated and control lakes in Florida. Environmental Health Perspectives 102:680-688.

Hines, T.C. 1979. The past and present status of the American alligator in Florida. Proceedings of the Annual Conference of the Southeastern Association of Fish and Wildlife Agencies 33:224-232.
Hines, T.C., and A.R. Woodward. 1980. Nuisance alligator control in Florida. Wildlife Society Bull. 8:234-241.
Jennings, M.L., H.F. Percival, and A.R. Woodward. 1988. Evaluation of alligator hatchling and egg removal from three Florida lakes. Proceedings of the Annual Conference of the Southeastern Association of Fish and Wildlife Agencies 42:283-294.
Jennings, M.L., A.R. Woodward, and D.N. David. 1989. Florida's nuisance alligator control program. Pages 29-36 *in* S.R. Craven, ed. Proceedings of the Fourth Eastern Wildlife Damage Control Conference, Madison, WI.
Murphy, T.M. 1977. Distribution, movement, and population dynamics of the American alligator in a thermally altered reservoir. M.S. thesis, University of Georgia, Athens. 64 pp.
Neal, W. 1985. Endangered and threatened wildlife and plants; reclassification of the American alligator in Florida to threatened due to similarity of appearance. Federal Register 50(119):25,672-25,678.
Wiley, E.N., and M.L. Jennings. 1990. An overview of alligator management in Florida. Pages 274-285 *in* Proceedings of the Tenth Working Meeting Crocodile Specialist Group. IUCN The World Conservation Union, Gland, Switzerland.
Wood, J.M., A.R. Woodward, S.R. Humphrey, and T.C. Hines. 1985. Night counts as an index of American alligator population trends. Wildlife Society Bull. 13:262-273.
Woodward, A.R., and S.B. Linda. 1993. Alligator population estimation. Final Report, Florida Game and Fresh Water Fish Commission, Tallahassee. 36 pp.
Woodward, A.R., and W.R. Marion. 1978. An evaluation of night-light counts of alligators. Proceedings of the Annual Conference of the Southeastern Association of Fish and Wildlife Agencies 32:291-302.
Woodward, A.R., and C.T. Moore. 1990. Statewide alligator surveys. Final Report, Florida Game and Fresh Water Fish Commission, Tallahassee. 24 pp.
Woodward, A.R., C.T. Moore, and M.F. Delany. 1992. Experimental alligator harvest. Final Report, Florida Game and Fresh Water Fish Commission, Tallahassee. 118 pp.
Woodward, A.R., H.F. Percival, M.L. Jennings, and C.T. Moore. 1993. Low clutch viability of American alligators on Lake Apopka. Florida Scientist 56:52-63.

For further information:

Allan R. Woodward
Florida Game and Fresh Water Fish Commission
4005 S. Main St.
Gainesville, FL 32601

Reptiles and Amphibians in the Endangered Longleaf Pine Ecosystem

by
C. Kenneth Dodd, Jr.
National Biological Service

The Coastal Plain of the southeastern United States contains a rich diversity of reptiles and amphibians (herpetofauna). Of the 290 species native to the Southeast, 170 (74 amphibians, 96 reptiles) are found within the range of the remnant longleaf pine (*Pinus palustris*) ecosystem (Fig. 1). Many of these species are not found elsewhere, particularly those amphibians that require temporary ponds for reproduction. Many Coastal Plain species are listed federally or by states as endangered or threatened or are candidates for listing (Fig. 1). Examples include the flatwoods salamander (*Ambystoma cingulatum*), striped newt (*Notophthalmus perstriatus*), Carolina and dusky gopher frogs (*Rana capito capito* and *R.c. sevosa*), eastern indigo snake (*Drymarchon corais couperi*), gopher tortoise (*Gopherus polyphemus*), eastern diamondback rattlesnake (*Crotalus adamanteus*), and Florida pine snake (*Pituophis melanoleucus mugitus*).

Studies in the Southeast

Information on the status and trends of the Coastal Plain herpetofauna comes from limited studies of selected species or populations, mostly within the last decade. The only intensive long-term quantitative and community-based studies have been at the Savannah River Site on the upper Coastal Plain of South Carolina. Most other studies have been distributional surveys for species such as Red Hills salamanders (*Phaeognathus hubrichti*), gopher frogs, striped newts, flatwoods salamanders, gopher tortoises, and Florida scrub lizards (*Sceloporus woodi*). Few studies have reported detailed habitat requirements for suspected declining species throughout their range. Surveys generally range 1-2 years in duration. Other trend information is derived from studies conducted by university scientists, private organizations, or state resource agencies. Concern for the future of the entire herpetofaunal community in the Southeast rests mostly on the well-documented loss of the old-growth longleaf pine ecosystem, although few community-based herpetofaunal surveys have been undertaken in this habitat.

Status

The fire-adapted longleaf pine community once stretched from southeastern Virginia to eastern Texas (Fig. 2). At present, less than 14% of the historical 282,283 km^2 (70 million acres) longleaf pine forest remains (Means and Grow 1985; Noss 1989), and most of it is on private land. Less than 1% is old-growth forest. Conversion of longleaf pine forests for agriculture, timber plantations, and urban needs (Ware et al. 1993) is accelerating (Fig. 3) and probably threatens the continued existence of many amphibian and reptile species, particularly in southern Georgia and Florida. For example, longleaf pine forests in Florida declined from 30,756 km^2 (7.6 million acres) in 1936 to only 3,845 km^2 (0.95 million acres) in 1989, an 88% decrease (Cerulean 1991). In southeastern Georgia the longleaf pine forest declined 36% (to 931 km^2 [230,000 acres]) between 1981 and 1988 (Johnson 1988). Most of this conversion has been from second- or third-growth longleaf pine stands to slash or loblolly pine plantation forestry.

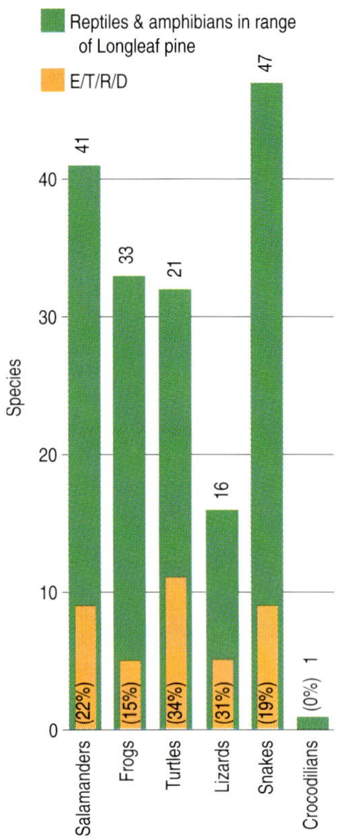

Fig. 1. Reptiles and amphibians within the southeastern Coastal Plain. Green bars = total number; Gold bars = number of species in need of conservation and management. E = endangered, T = threatened, R = rare, D = declining.

The effects of the loss of the longleaf pine ecosystem on the herpetofaunal community have never been assessed directly, but several species are known to have been affected. For example, the number of gopher tortoises, a key species within the longleaf pine ecosystem, has declined by an estimated 80% during the last 100 years (Auffenberg and Franz 1982). More than 300 invertebrates and 65 vertebrates use gopher tortoise burrows (Jackson and Milstrey 1989; Fig. 4), so an 80% reduction in gopher tortoises could represent a substantial reduction in the biodiversity of the longleaf pine ecosystem.

Amphibians that breed in temporary ponds have been particularly affected both because of direct habitat destruction and the slower loss of wetland breeding sites by ditching. Breeding, foraging, and overwintering sites are also affected by certain types of forest plantation site preparation. Only five populations of striped newts remain in Georgia (Dodd 1993; L. LaClaire, USFWS, personal communication); the flatwoods salamander has disappeared from the eastern section of its range; gopher frogs are nearly extirpated in North Carolina, Alabama, and Mississippi; and dusky salamanders (*Desmognathus* spp.) appear to have declined or disappeared in coastal South Carolina and peninsular Florida.

On the other hand, the long-term community studies at the Savanna River Site, where the destructive effects of plantation forestry are not prevalent, do not reveal declining trends, although some amphibian populations there fluctuate widely from one year to the next in both numbers and reproductive output (Pechmann et al. 1991). A 5-year study on a north Florida biological preserve disclosed declining amphibian numbers, but the study coincided with a severe regional drought (Dodd 1992). In west-central Florida, amphibian communities have changed composition because of

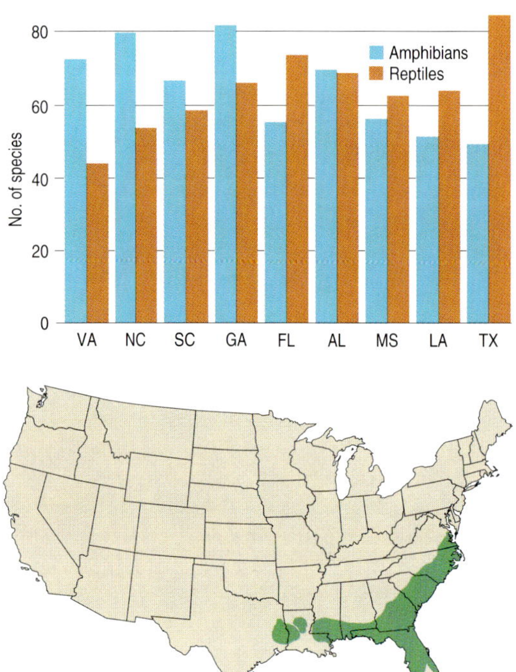

Fig. 2. Historical distribution of the longleaf pine ecosystem in the southeastern Coastal Plain. Chart shows the present total number of species of amphibians and reptiles in various southeastern states.

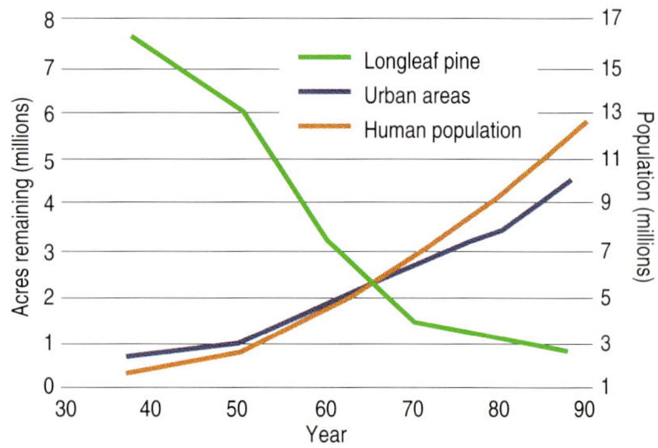

Fig. 3. Trend in loss of longleaf pine forest in relation to urban development and increases in human population in Florida, 1930-90 (Cerulean 1991; used with permission from The Nature Conservancy).

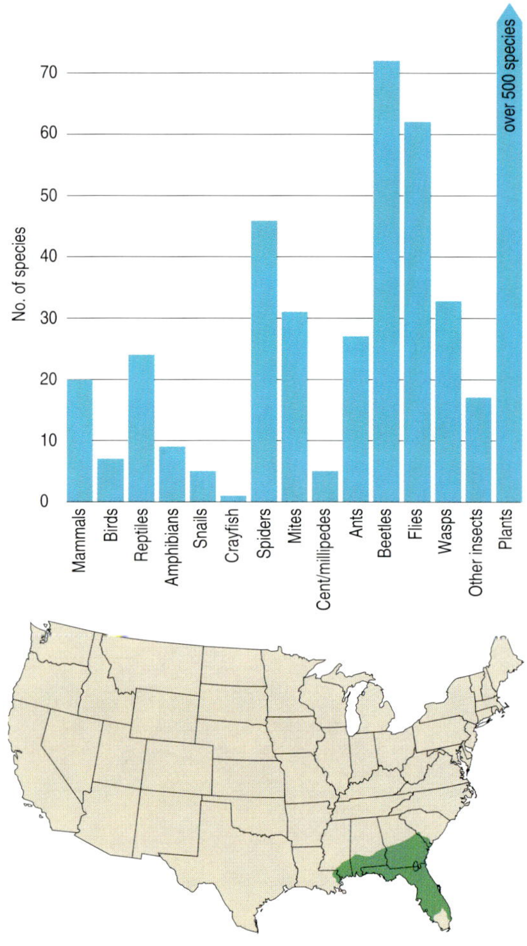

Fig. 4. The distribution of the gopher tortoise (*Gopherus polyphemus*) in the southeastern United States. The chart shows the number of species of various taxa known to use its burrow and the number of plant taxa described from the longleaf pine-wiregrass ecosystem.

urbanization (Delis 1993), but the long-term effects of the change are unknown. The overall status of the Red Hills salamander (federal threatened list) remained the same from 1976 to 1988 (Dodd 1991), although habitat loss continued from plantation forestry. Virtually no data exist for terrestrial reptile populations or communities except for the gopher tortoise. Anecdotal information for all terrestrial reptiles suggests population declines, particularly in areas affected by imported red fire ants (*Solenopsis invicta*).

Local centers of amphibian and reptile diversity need to be identified within the remaining longleaf pine community. Surveys, basic life-history studies of sensitive species, and long-term monitoring of amphibian and reptile populations need to be initiated. Many species that are restricted to wetland and upland habitats appear to be declining, but precise baseline data are lacking. Factors impeding the identification of population trends include the longevity of many species, the effects of periodic natural events such as drought, and what appear to be random population fluctuations. At the same time, when the known extent of habitat loss is coupled with declining trends elsewhere (Blaustein and Wake 1990; Wyman 1990) that result from unknown or hypothesized causes (UVB light, acidity, heavy metals, estrogen-mimicking compounds, roads, habitat fragmentation), the study and monitoring of amphibian and reptile populations in remnant southeastern longleaf pine forests will become especially imperative.

References

Auffenberg, W., and R. Franz. 1982. The status and distribution of the gopher tortoise (*Gopherus polyphemus*). Pages 95-126 *in* R.B. Bury, ed. North American tortoises: conservation and ecology. U.S. Fish and Wildlife Service Res. Rep. 12.

Blaustein, A.R., and D.B. Wake. 1990. Declining amphibian populations: a global phenomenon? Trends in Ecology and Evolution 5:203-204.

Cerulean, S.I. 1991. The preservation 2000 report. Florida's natural areas—what have we got to lose? The Nature Conservancy, Winter Park, FL. 74 pp.

Delis, P.R. 1993. Effects of urbanization on the community of anurans of a pine flatwood habitat in west central Florida. M.S. thesis, University of South Florida, Tampa. 47 pp.

Dodd, C.K., Jr. 1991. The status of the Red Hills salamander *Phaeognathus hubrichti*, Alabama, USA, 1976-1988. Biological Conservation 55:57-75.

Dodd, C.K., Jr. 1992. Biological diversity of a temporary pond herpetofauna in north Florida sandhills. Biodiversity and Conservation 1:125-142.

Dodd, C.K., Jr. 1993. Distribution of striped newts (*Notophthalmus perstriatus*) in Georgia. Report to U.S. Fish and Wildlife Service, Jacksonville, FL. 52 pp.

Jackson, D.R., and E.G. Milstrey. 1989. The fauna of gopher tortoise burrows. Florida Nongame Wildlife Program Tech. Rep. 5:86-98.

Johnson, T.G. 1988. Forest statistics for southeast Georgia, 1988. USDA Forest Service Resour. Bull. SE-104. 53 pp.

Means, D.B., and G. Grow. 1985. The endangered longleaf pine community. ENFO (Florida Conservation Foundation) Sept:1-12.

Noss, R.F. 1989. Longleaf pine and wiregrass: keystone components of an endangered ecosystem. Natural Areas Journal 9:211-213.

Pechmann, J.H.K., D.E. Scott, R.D. Semlitsch, J.P. Caldwell, L.J. Vitt, and J.W. Gibbons. 1991. Declining amphibian populations: the problem of separating human impacts from natural fluctuations. Science 253:892-895.

Ware, S., C. Frost, and P.D. Doerr. 1993. Southern mixed hardwood forest: the former longleaf pine forest. Pages 447-493 *in* W.H. Martin, S.G. Boyce, and A.C. Echternacht, eds. Biodiversity of the southeastern United States. Lowland terrestrial communities. John Wiley and Sons, New York.

Wyman, R.L. 1990. What's happening to the amphibians? Conservation Biology 4:350-352.

For further information:

C. Kenneth Dodd, Jr.
National Biological Service
Southeastern Biological Science Center
7920 N.W. 71st St.
Gainesville, FL 32653

Native Ranid Frogs in California

by
Mark R. Jennings
National Biological Service

Many recent declines and extinctions of native amphibians have occurred in certain parts of the world (Wake 1991; Wake and Morowitz 1991). All species of native true frogs have declined in the western United States over the past decade (Hayes and Jennings 1986). Most of these native amphibian declines can be directly attributed to habitat loss or modification, which is often exacerbated by natural events such as droughts or floods (Wake 1991). A growing body of research, however, indicates that certain native frogs are particularly susceptible to population declines and extinctions in habitats that are relatively unmodified by humans (e.g., wilderness areas and national parks in California; Bradford 1991; Fellers and Drost 1993; Kagarise Sherman and Morton 1993). To understand these declines, we must document the current distribution of these species over their entire historical range to learn where they have disappeared.

In 1988 the California Department of Fish and Game commissioned the California Academy of Sciences to conduct a 6-year study on the status of the state's amphibians and reptiles not currently protected by the Endangered Species Act. The study's purpose was to determine amphibians and reptiles most vulnerable to extinction and provide suggestions for future research, management, and protection by state, federal, and local agencies (Jennings and Hayes 1993). This article describes the distribution and status of all native true frogs in California as determined by the California Fish and Game study.

Status

All species studied have suffered declines in distribution and abundance, largely because of habitat loss or modification from farming, grazing, logging, urban development, suppression of brush fires, and flood-control or water-development projects. The species have also been affected by the widespread introduction of vertebrate and invertebrate aquatic predators.

Northern Red-legged Frog (*Rana aurora aurora*)

This frog, restricted to lower elevations (300 m [984 ft]) of the north coast region of California (Fig. 1), has disappeared from about 15% of its historical range in California. It is not in danger of extinction in the state.

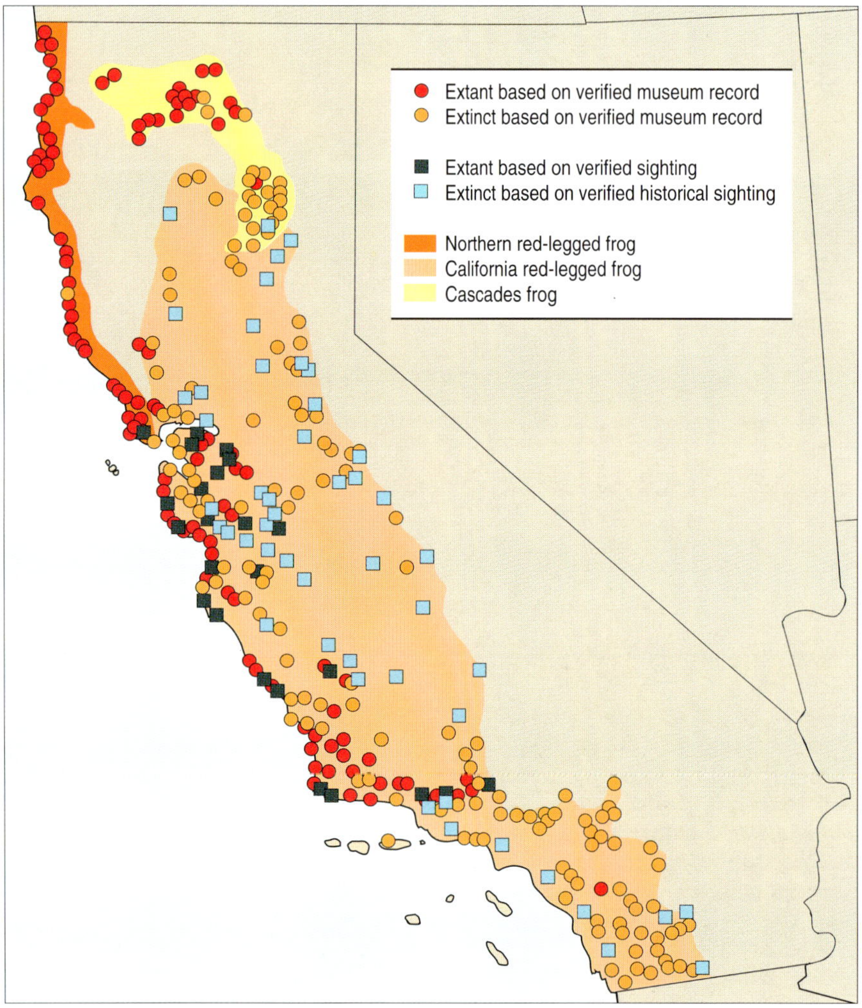

Fig. 1. Historical and current distribution of the northern red-legged frog, California red-legged frog, and Cascades frog in California based on 2,068 museum records and 302 records from other sources. Dots indicate locality records based on verified museum specimens. Squares indicate locality records based on verified sightings (e.g., field notes, photographs, published papers). Red dots and green squares denote localities where native frogs are extant. Gold dots and blue squares indicate where native frogs are presumed extinct. Figure modified from Jennings and Hayes (1993).

California Red-legged Frog (*R.a. draytonii*)

This frog was originally found over most of California below 1,524 m (500 ft) and west of the deserts and the Sierra Nevada crest (Fig. 1). Although the California red-legged frog has now disappeared from about 75% of its historical range in the state, around the turn of the century it was abundant enough to support an important commercial fishery in the San Francisco fish markets (Jennings and Hayes 1984). California red-legged frogs have almost completely disappeared from the Central Valley and southern California since 1970 and are currently proposed for listing as endangered by the U.S. Fish and Wildlife Service (Federal Register 1994).

Cascades Frog (*R. cascadae*)

The Cascades frog was originally found in northern California above 230 m (755 ft; Fig. 1), where it was historically very abundant. Since the mid-1970's, the species extensively declined, disappearing from about 50% of its range in the state. No habitat loss hypothesis adequately explains why this frog survived with current land-use practices for over 50 years before its decline. It is still abundant in California only in the northern third of its range on lands under federal ownership.

Foothill Yellow-legged Frog (*R. boylii*)

This frog was originally found over most of California below 1,829 m (6,000 ft), west of the deserts and the Sierra-Cascade crest (Fig. 2). In many locations before 1970, populations contained hundreds of individuals (Zweifel 1955), but the frog has now completely disappeared from southern California and from about 45% of its historical range over the entire state. Most populations were apparently healthy until the mid-1970's, when a population crash occurred in southern California and the Sierra Nevada foothills after several years of severe floods and drought, which may have been responsible for the declines, although it is not certain. Because this species was an important component of the food web in many streamside ecosystems, its loss has probably negatively affected several organisms, such as garter snakes (*Thamnophis* spp.), which historically relied upon it as a major food source.

Spotted Frog (*R. pretiosa*)

The spotted frog was historically recorded only from scattered localities in the extreme northeastern part of California below 1,372 m (4,500 ft), where it was apparently restricted to large marshy areas filled by warmwater (more than 20°C [68°F]) springs (Fig. 2). It has now disappeared from about 99% of its range, and is only known from one location in the state. It appears to be on the verge of extinction in California.

Yavapai Leopard Frog (*R. yavapaiensis*)

This frog was originally found along the Colorado River and in the Coachella Valley of southeastern California (Fig. 2). It has not been seen in the state since the mid-1960's and now seems to be extinct at all sites examined. This leopard frog has been replaced in California by the introduced bullfrog (*R. catesbeiana*) and the Rio Grande leopard frog (*R. berlandieri*), which are able to thrive in human-modified reservoirs and canals in the Yavapai leopard frog's original range (Jennings and Hayes 1994).

Mountain Yellow-legged Frog (*R. muscosa*)

This species was historically abundant in the Sierra Nevada at elevations largely above 1,829 m (6,000 ft), and also in the San Gabriel, San Bernardino, and San Jacinto mountains of southern California above 369 m (1,210 ft; Fig. 3). The mountain yellow-legged frog has disappeared from about 50% of its historical range in the Sierra Nevada and about 99% of its historical range in southern California. Some researchers believe that the widespread introduction of non-native trout into high-elevation lakes is the major reason for the decline of this species in the Sierra Nevada (Bradford 1989; Bradford et al. 1993). The species, however, experienced massive die-offs in many parts of its range during the 1970's (Bradford 1991) after several years of severe floods and drought, and continues to decline in relatively pristine areas such as wilderness areas and national parks.

Such observations indicate that present land-management practices of setting aside large tracts of land for the "protection of biodiversity" may not be adequate for ensuring the continued survival of this species. Already, the loss of this frog over large areas has negatively affected organisms such as the western terrestrial garter snake (*Thamnophis elegans*), which relied upon it as a major food source (Jennings et al. 1992). To keep these populations from extinction, resource managers may need to initiate active management efforts for mountain yellow-legged frogs (such as fish eradication programs in selected high-elevation lakes, fencing of riparian zones to exclude livestock grazing, and relocating hiking trails and campgrounds away from sensitive riparian habitats).

Northern Leopard Frog (*R. pipiens*)

This frog was historically recorded from scattered localities below 1,981 m (6,500 ft) in

Northern red-legged frog (*Rana aurora aurora*).

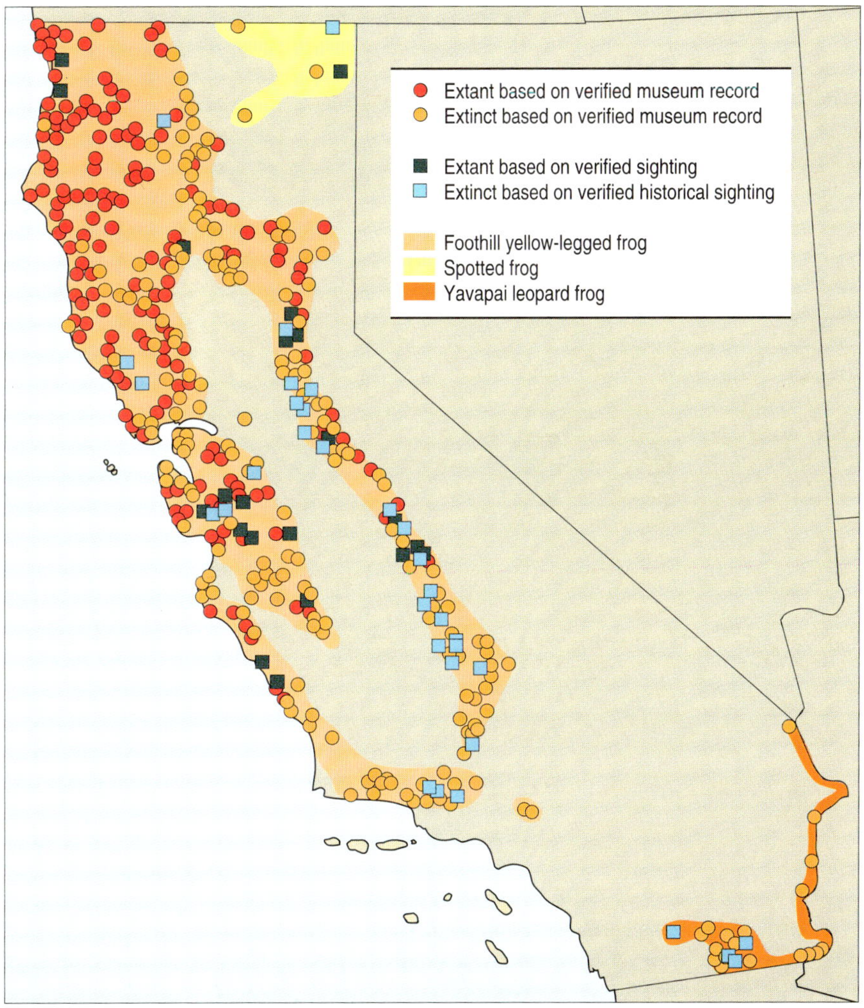

Fig. 2. Historical and current distribution of the foothill yellow-legged frog, spotted frog, and Yavapai leopard frog in California based on 3,316 museum records and 171 records from other sources. Dots indicate locality records based on verified museum specimens. Squares indicate locality records based on verified sightings (e.g., field notes, photographs, published papers). Red dots and green squares denote localities where native frogs are extant. Gold dots and blue squares indicate where native frogs are presumed extinct. Figure modified from Jennings and Hayes (1993).

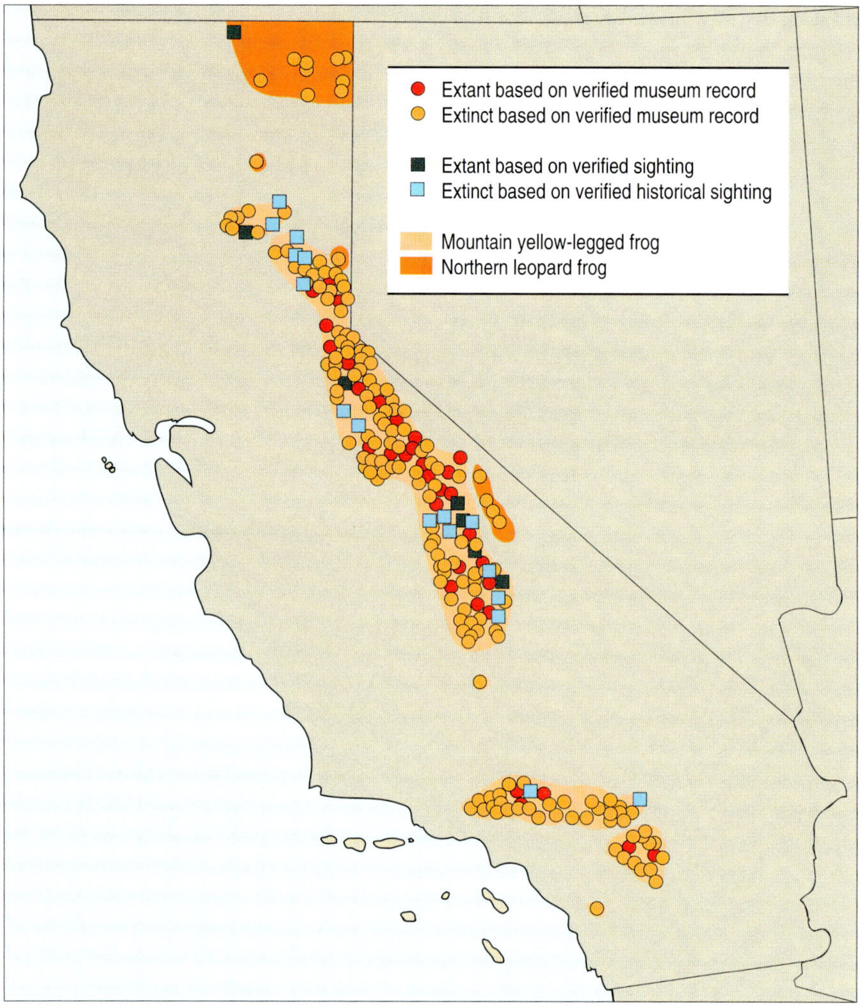

Fig. 3. Historical and current distribution of the mountain yellow-legged frog, and presumed native populations of the northern leopard frog in California based on 2,565 museum records and 673 records from other sources. Dots indicate locality records based on verified museum specimens. Squares indicate locality records based on verified sightings (e.g., field notes, photographs, published papers). Red dots and green squares denote localities where native frogs are extant. Gold dots and blue squares indicate where native frogs are presumed extinct. Figure modified from Jennings and Hayes (1993).

the eastern part of California (Fig. 3). Some populations were introduced into the state within the past 100 years (Jennings and Hayes 1993), most around the turn of the century (Storer 1925). This species has disappeared from about 95% of its range in California and is now found only in one national wildlife refuge near the Oregon border. Most localities where this frog was historically found have not changed appreciatively during the past 50 years, so the reasons for the species' decline and disappearance remain a mystery.

For further information:

Mark R. Jennings
National Biological Service
Alaska Science Center
Piedras Blancas Field Station
PO Box 70
San Simeon, CA 93452

References

Bradford, D.F. 1989. Allotopic distribution of native frogs and introduced fishes in the high Sierra Nevada lakes of California: implication of the negative effects of fish introductions. Copeia 1989(3):775-778.

Bradford, D.F. 1991. Mass mortality and extinction in a high elevation population of *Rana muscosa*. Journal of Herpetology 25(2):174-177.

Bradford, D.F., D.M. Graber, and F. Tabatabai. 1993. Isolation of remaining populations of the native frog, *Rana muscosa*, by introduced fishes in Sequoia and Kings Canyon National Parks, California. Conservation Biology 7(4):882-888.

Federal Register. 1994. Endangered and threatened wildlife and plants; proposed endangered status for the California red-legged frog. Federal Register 59(22):4888-4895.

Fellers, G.M., and C.A. Drost. 1993. Disappearance of the Cascades frog *Rana cascadae* at the southern end of its range, California, USA. Biological Conservation 65(2):177-181.

Hayes, M.P., and M.R. Jennings. 1986. Decline of ranid frog species in western North America: are bullfrogs (*Rana catesbeiana*) responsible? Journal of Herpetology 20(4):490-509.

Jennings, M.R., and M.P. Hayes. 1984. Pre-1900 overharvest of the California red-legged frog (*Rana aurora draytonii*): the inducement for bullfrog (*Rana catesbeiana*) introduction. Herpetologica 41(1):94-103.

Jennings, M.R., and M.P. Hayes. 1993. Amphibian and reptile species of special concern in California. Final report submitted to the California Department of Fish and Game, Inland Fisheries Division, Rancho Cordova, under Contract (8023). 336 pp.

Jennings, M.R., and M.P. Hayes. 1994. Decline of native ranid frogs in the desert southwest. *In* P.R. Brown and J.W. Wright, eds. Proceedings of the Conference on the Herpetology of the North American Deserts. Southwestern Herpetologists Society, Spec. Publ. 5. In press.

Jennings, W.B., D.F. Bradford, and D.F. Johnson. 1992. Dependence of the garter snake *Thamnophis elegans* on amphibians in the Sierra Nevada of California. Journal of Herpetology 26(4):503-505.

Kagarise Sherman, C., and M.L. Morton. 1993. Population declines of Yosemite toads in the eastern Sierra Nevada of California. Journal of Herpetology 27(2):186-198.

Storer, T.I. 1925. A synopsis of the amphibia of California. University of California Publications in Zoology 27:1-343.

Wake, D.B. 1991. Declining amphibian populations. Science 253(5022):860.

Wake, D.B., and H.J. Morowitz. 1991. Declining amphibian populations—a global phenomenon? Findings and recommendations. Alytes 9(1):33-42.

Zweifel, R.G. 1955. Ecology, distribution, and systematics of frogs of the *Rana boylei* group. University of California Publications in Zoology 54(4):207-292.

Desert Tortoises in the Mojave and Colorado Deserts

by
Kristin H. Berry
Philip Medica
National Biological Service

The desert tortoise (*Gopherus agassizii*) is a widespread species of the southwestern United States and Mexico. Within the United States, desert tortoises live in the Mojave, Colorado, and Sonoran deserts of southeastern California, southern Nevada, southwestern Utah, and western Arizona (Fig. 1). A substantial portion of the habitat is on lands administered by the U.S. Department of the Interior.

The U.S. government treats the desert tortoise as an indicator or umbrella species to measure the health and well-being of the ecosystems it inhabits. The tortoise functions well as an indicator because it is long-lived, takes 12-20 years to reach reproductive maturity, and is sensitive to changes in the environment. In 1990 the U.S. Fish and Wildlife Service listed the species as threatened in the northern and western parts of its geographic range (Fig. 1) because of widespread population declines and overall habitat loss, deterioration, and fragmentation.

Because some populations exhibit significant genetic, morphologic (*see* glossary), and behavioral differences, the Desert Tortoise Recovery Team identified six distinctive population segments (Fig. 1) for critical habitat protection and long-term conservation within the Mojave and Colorado deserts (e.g., Lamb et al. 1989; USFWS 1994). The population segments are representative of distinctive climatic, floristic, and geographic regions.

Surveys

The primary sources of information on status and trends of desert tortoise populations are from study plots established by the U.S. Bureau of Land Management and state fish and game agencies. More than 30 permanent study plots, each of which is 2.6 km^2 or larger (1 mi^2 or more), are surveyed at intervals ranging from 2 to 10 years. Study plots provide data on population characteristics, including density, size-age class structure, sex ratios and numbers of breeding females, recruitment of juveniles into the adult population, causes of death, and mortality rates (Berry 1990). Researchers use mark-recapture techniques to conduct 60-day surveys in spring for live and dead tortoises.

Trends for habitat condition on study plots are measured by using quantitative data on native and exotic annual and perennial vegetation (Berry 1990). Associated data on past and recent human activities or influences include numbers of visitors per season; density of dirt roads, trails, and vehicle tracks; levels and types of livestock grazing; and acreage disturbed by mining and mineral development and utility corridors.

The data base for the six population segments varies considerably; some segments contain several plots that have been sampled for 11-17 years, whereas others have few plots that have been sampled only 1 or 2 years (Berry 1990; USFWS 1994).

Trends

Condition and trends in tortoise populations vary within and between population segments. One measure of population condition is change in density. Examples of changes in density for nine study plots in California and Nevada are shown in Fig. 2 (Berry 1990; D.B. Hardenbrook, Nevada Division of Wildlife, and S. Slone, Bureau of Land Management, personal communication). The greatest declines in densities, for all size classes and for breeding females (up to 90%), occurred in the western Mojave segment between the 1970's and 1990's. Similar declines (30%-60%) also occurred in the eastern Colorado Desert segment between 1979 and 1992, with the greatest declines registered at the Chuckwalla Bench plot (Fig. 2). Moderate declines of 20%-25% were reported from some sites in the eastern Mojave Desert segment (Piute Valley and Goffs). The northeastern Mojave also exhibited declines on some plots (e.g., Ivanpah Valley and Gold Butte). In contrast, the northern Colorado Desert population segment showed indications of growth in the breeding adults at one plot (Ward Valley), and the upper Virgin River segment appears stable (USFWS 1994).

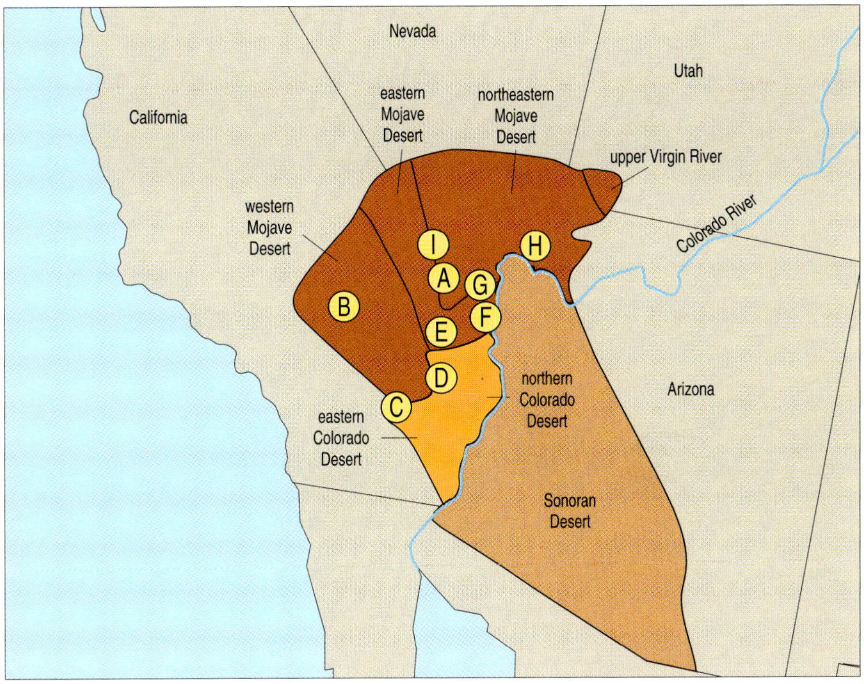

Fig. 1. U.S. range of the desert tortoise (*Gopherus agassizii*). The six population segments for desert tortoises federally listed as threatened occur in parts of the Mojave and Colorado deserts that lie north and west of the Colorado River.

Desert tortoise (*Gopherus agassizii*).

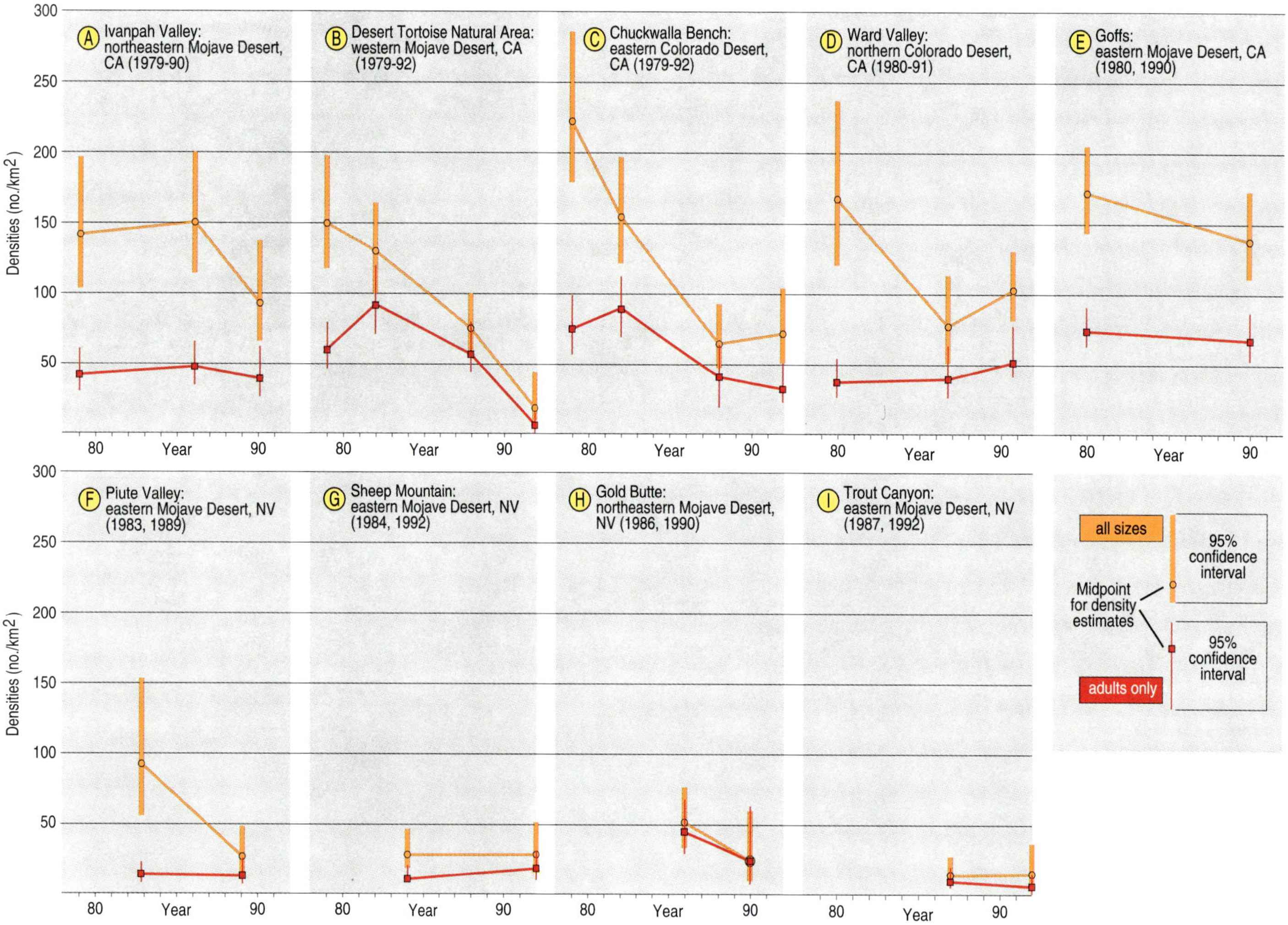

Fig. 2. Examples of changes in desert tortoise population densities at nine study sites in California and Nevada. The midpoint for density estimates of all sizes of tortoises (orange line) is shown by a dot on a bar representing the 95% confidence interval (CI); the midpoint for density estimates for adult tortoises only (red lines) is depicted by a square on a bar representing the 95% CI. Causes of declines vary by site.

Causes of population declines differed somewhat within and between population segments, but were primarily related to human activities. Higher than normal losses or mortality rates were attributed to many causes, such as illegal collecting, vandalism, upper respiratory tract disease or shell disease, predation by common ravens, crushing by vehicles both on and off roads, and trampling by livestock (BLM 1988; USFWS 1994). For example, 14.6%-28.9% of desert tortoise carcasses collected from western Mojave plots in the 1970's and early 1980's showed signs of gunshots (tortoises were shot while still alive), but only 0%-3.1% of carcasses from the less-visited eastern Mojave and northern Colorado deserts showed such signs (Berry 1986). Deaths from vehicles on paved roads were also highest in the western Mojave, where densities of dirt roads and vehicle trails are higher than elsewhere.

Of particular concern is the recent appearance of a highly infectious and usually fatal upper respiratory tract disease caused by the bacterium *Mycoplasma agassizii*. The disease, apparently introduced through the release of captive tortoises (Jacobson 1993), has caused the deaths of thousands of wild tortoises in the Mojave Desert during the last few years (K.H. Berry, unpublished data).

Fragmented and deteriorated habitats also affect population vitality. Populations in areas with high levels of exotic annual plants are declining at substantially higher rates than those in less disturbed areas.

In summary, tortoise populations occurring in relatively undisturbed and remote areas with little vehicular access and low human visitation generally were stable, or exhibited lower rates of decline than tortoise populations in areas with high levels of disturbance, high vehicular access, and high human visitation.

References

Berry, K.H. 1986. Incidence of gunshot deaths in desert tortoises in California. Wildlife Society Bull. 14:127-132.

Berry, K.H. 1990. The status of the desert tortoise in California in 1989 (with amendments to include 1990-1992 data sets). Draft report from U.S. Bureau of Land Management to U.S. Fish and Wildlife Service, Region 1, Portland, OR.

BLM. 1988. Desert tortoise habitat management on the public lands: a rangewide plan. U.S. Bureau of Land Management, Washington, DC. 23 pp.

Jacobson, E.R. 1993. Implications of infectious diseases for captive propagation and introduction programs of threatened/endangered reptiles. Journal of Zoo and Wildlife Medicine 24(3):245-255.

Lamb, T., J. Avise, and J.W. Gibbons. 1989. Phylogeographic patterns in mitochondrial DNA of the desert tortoise (*Xerobates agassizi*) and evolutionary relationships among the North American gopher tortoises. Evolution 43(1):76-87.

USFWS. 1994. Desert Tortoise (Mojave Population) Recovery Plan. U.S. Fish and Wildlife Service, Portland, OR. 77 pp. + appendices.

For further information:

Kristin H. Berry
National Biological Service
Riverside Field Station
6221 Box Springs Blvd.
Riverside, CA 92507

Coachella Valley Fringe-toed Lizards

by

Cameron Barrows
The Nature Conservancy

Allan Muth

Mark Fisher
University of California, Boyd Deep Canyon Desert Research Center

Jeffrey Lovich
National Biological Service

Fringe-toed lizards (*Uma* spp.) inhabit many of the scattered windblown sand deposits of southeastern California, southwestern Arizona, and northwestern Mexico. These lizards have several specialized adaptations: elongated scales on their hind feet ("fringes") for added traction in loose sand, a shovel-shaped head and a lower jaw adapted to aid diving into and moving short distances beneath the sand, elongated scales covering their ears to keep sand out, and unique morphology (form or structure) of internal nostrils that allows them to breathe below the sand without inhaling sand particles.

While these adaptations enable fringe-toed lizards to successfully occupy sand dune habitats, the same characteristics have restricted them to isolated sand "islands." Three fringe-toed lizard species live in the United States: the Mojave (*U. scoparia*), the Colorado Desert (*U. notata*), and the Coachella Valley (*U. inornata*). Of the three, the Coachella Valley fringe-toed lizard has the most restricted range and has been most affected by human activities. In 1980 this lizard was listed as a threatened species by the federal government.

In 1986 the Coachella Valley Preserve system was established to protect habitat for the Coachella Valley fringe-toed lizard. This action set several precedents: it was the first Habitat Conservation Plan established under the revised (1982) Endangered Species Act and the newly adopted Section 10 of the act, it established perhaps the only protected area in the world set aside for a lizard, and its design was based on a model of sand dune ecosystem processes, the sole habitat for this lizard. Three disjunct sites in California, each with a discrete source of windblown sand, were set aside to protect fringe-toed lizard populations: Thousand Palms, Willow Hole, and Whitewater River. Collectively, the preserves protect about 2% of the lizards' original range.

Eight years after the establishment of the preserve system, few Coachella Valley fringe-toed lizards exist outside the boundaries of the three protected sites. Barrows (author, unpublished data) recently identified scattered pockets of windblown sand occupied by fringe-toed lizards in the hills along the northern fringe of the valley, but only at low densities. Fringe-toed lizard populations within the protected sites have been monitored yearly since 1986. During this period, California experienced one of its most severe droughts, which ended in spring 1991. Numbers of fringe-toed lizards within the Thousand Palms and Willow Hole sites declined during the drought, but rebounded after 1991 (Fig. 1). By 1993, after three wet springs, lizard numbers had increased substantially.

Lizards at the Whitewater River site were intensively monitored since 1985 by using mark-recapture methods to count the population on a 2.25-ha (5.56-acre) plot. In 1986 this site had the highest population density of the three protected sites. As with the other two sites, the Whitewater River population declined throughout the drought, but only increased slightly after the drought broke in 1991 (Fig. 2). Compounding the drought effect, much of the fine sand preferred by fringe-toed lizards was blown off the site during the dry years. This condition was unique to the Whitewater River

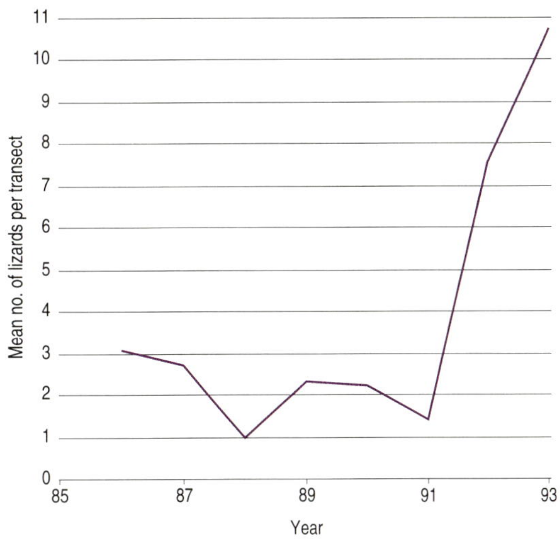

Fig. 1. The mean number of lizards per transect at the Thousand Palms and Willow Hole sites, 1986-93. Data were pooled from five 10 x 1,000 m (32.8 x 3,281 ft) transects. All transects were sampled six times each year, and all sampling was conducted within a 6-week span in the late spring of each year.

Courtesy C. Barrows

Coachella Valley fringe-toed lizard (*Uma inornata*).

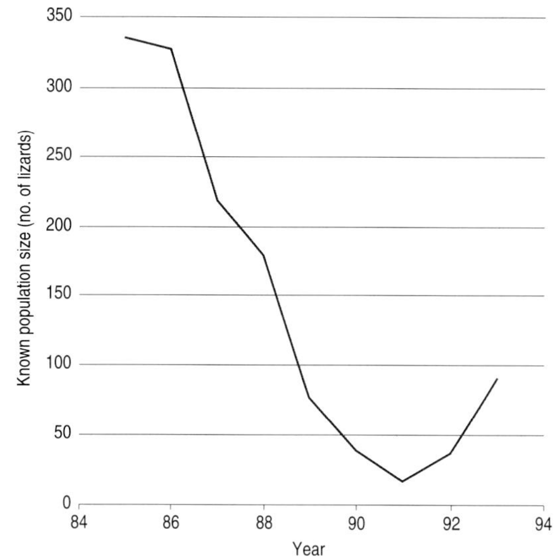

Fig. 2. The known population size of a Coachella Valley fringe-toed lizard population on a 2.25-ha (5.56-acre) study plot on the Whitewater River preserve.

site; the other two protected sites have much deeper sand deposits and are less susceptible to wind erosion. New windblown sand was deposited on the Whitewater River site in 1993 after a period of high rainfall. The population appears to be increasing in response to these favorable conditions.

The decline in fringe-toed lizards during the monitoring period appears to be the result of responses to natural fluctuations in habitat. The dynamic nature of sand dune systems, coupled with the lizards' apparent sensitivity to drought, underlines the importance of preserve design. Appropriate designs anticipate the effect of natural habitat fluctuation.

The ecological model that governed the design of the Coachella Valley Preserve system was reevaluated in 1993 with one disturbing result. A primary sand source was identified that supplies the sand dunes at the Thousand Palms site, but was not emphasized sufficiently in the original model and design. Fortunately, the sand source and its path to the existing preserve have not been affected severely by human development at this time, so options for correcting the design's shortcomings are still available. The fringe-toed lizard population sustained by this sand source has been the largest of the three sites for the past few years. Monitoring the lizards without investigating ecosystem processes would not have identified the design error until it was too late to correct.

For further information:

Cameron Barrows
The Nature Conservancy
53277 Avenida Diaz
La Quinta, CA 92253

Disappearance of the Tarahumara Frog

by
S.F. Hale
Herpetologist
C.R. Schwalbe
National Biological Service
J.L. Jarchow
Sonora Pet Hospital, Tucson, AZ
C.J. May
Pima Community College
C.H. Lowe
University of Arizona
T.B. Johnson
Arizona Game and Fish Department

In the spring of 1983 the last known Tarahumara frog in the United States was found dead. Overall, the species seems to be doing well in Mexico, although the decline of more northern populations are of concern. The Tarahumara frog (*Rana tarahumarae*) inhabits seasonal and permanent bedrock and bouldery streams in the foothills and main mountain mass of the Sierra Madre Occidental of northwestern Mexico. It ranges from northern Sinaloa, through western Chihuahua and eastern and northern Sonora, and until recently into extreme south-central Arizona (Fig. 1). Arizona localities, all in Santa Cruz County, include three drainages in the Atascosa-Pajarito Mountains (Campbell 1931; Little 1940; Williams 1960) and three in the Santa Rita Mountains (Hale et al. 1977).

Population Estimates, 1975-93

We have drawn our review from museum records, the published literature, and reports, journal entries, and personal observations by the authors, other biologists, and knowledgeable persons. From May 1975 through June 1977, we conducted an ecological, demographic, and life-history study of the population at Big Casa Blanca Canyon (Santa Rita Mountains).

Between 1980 and 1993, we visited 22 of 30 historical Tarahumara frog localities. We surveyed 43 additional streams with potential habitat and found Tarahumara frogs at 25 new localities in Mexico. Localities were extensively searched, often both day and night, sometimes repeatedly. Frogs and tadpoles were counted, size-classed, and sexed when possible. Time, streamwater pH, air, substrate and water temperatures, habitat description and condition, and relative abundances of other aquatic vertebrates were noted.

During the summers of 1982-83, rain samples were collected at The Nature Conservancy's Sonoita Creek and Canelo Hills preserves for pH determination and heavy metal analysis. Both sites are within 22-56 km (14-35 mi) of declining frog populations and 64-129 km (40-80 mi) north and northwest of copper smelters. Streamwater samples from sites of declining populations in Sycamore and Big Casa Blanca canyons in Arizona and Carabinas Canyon in northeastern Sonora were also collected for pH and heavy metal analyses.

Decline of Populations

In April 1974, 27 dead and dying Tarahumara and leopard frogs were observed at Sycamore Canyon, Atascosa-Pajarito Mountains, the best-known and most frequently visited Tarahumara frog population. The last sightings of Tarahumara frogs in that range were in the summer of 1974.

The decline of the Santa Rita Mountains population began in 1977 (Fig. 2). Total numbers of frogs (adults and juveniles) captured plummeted from 252 in 1976 to 46 in 1977; estimated total population size fell from a maximum 1,020 frogs to 625 (Hale and May 1983). In June 1977 some captured frogs became unresponsive and often died, apparently from the stress of capture, a response not previously observed. In 1978 no frogs marked in prior years, nor tiny larvae attributable to that year's breeding, were found. Larger tadpoles from 1976-77 persisted. Twenty newly metamorphosed frogs were observed in 1978 and 40 in 1979; from 1980 to 1982 we saw one to three frogs attributable to those frogs. In spring 1983 the last known Tarahumara frog in the United States was found dead. Repeated visits (some times yearly) to all former Arizona localities have yielded no additional sightings.

Three of seven populations studied from 1981 to 1986 in northern Sonora appeared healthy, with adult and juvenile frogs as well as both small and large larvae, suggesting a stable, reproductive population. Frogs were not seen at three other sites where they had been found in the 1970's and early 1980's. The last population, in Carabinas Canyon, Sierra El Tigre, which contained numerous frogs and tadpoles, was in the initial stages of a major decline when first observed in fall 1981. Within a year all frogs had disappeared from the downstream end of this population, but frogs in the upper portion of the drainage appeared to have suffered no decline in numbers through our most recent visit to the site in 1986.

Carabinas Canyon frogs displayed clinical signs suggestive of heavy metal poisoning, including irregular muscular activity and failure of muscular coordination (ataxia), partial paralysis of the hind legs, dilated pupils unresponsive to light, and a loss of the righting response. The skin was often dry on the head and back. Symptoms were amplified by the stress of capture and handling. Frogs displaying obvious signs of heavy metal poisoning were already dying.

Field examinations of dead frogs showed no evidence of gross pathological disorders. Skin cultures showed no common pathogens; species representing probable normal skin flora and opportunistic secondary pathogens attacking a debilitated host were present. Histopathological examinations of five dying frogs (E. Jacobson, J. Hillis Miller Health Center, College of Veterinary Medicine, Gainesville, Florida) revealed no gross pathologies (Hale and May 1983; Hale and Jarchow 1988).

Populations of Chiricahua and Yavapai leopard frogs (*Rana chiricahuensis* and *R. yavapaiensis*) declined with the Tarahumara frog where they occurred together, although leopard frogs were not eliminated from most Tarahumara frog sites. In Sycamore Canyon, Chiricahua leopard frogs have managed to maintain a small but viable population near Yank Spring, but numbers decrease downstream in previously favorable leopard frog habitat. The Chiricahua leopard frog has experienced catastrophic declines elsewhere, and is in danger of disappearing from most of its range (Clarkson and Rorabaugh 1989).

Rain collected at the Sonoita Creek and Canelo Hills preserves in the summers of 1982

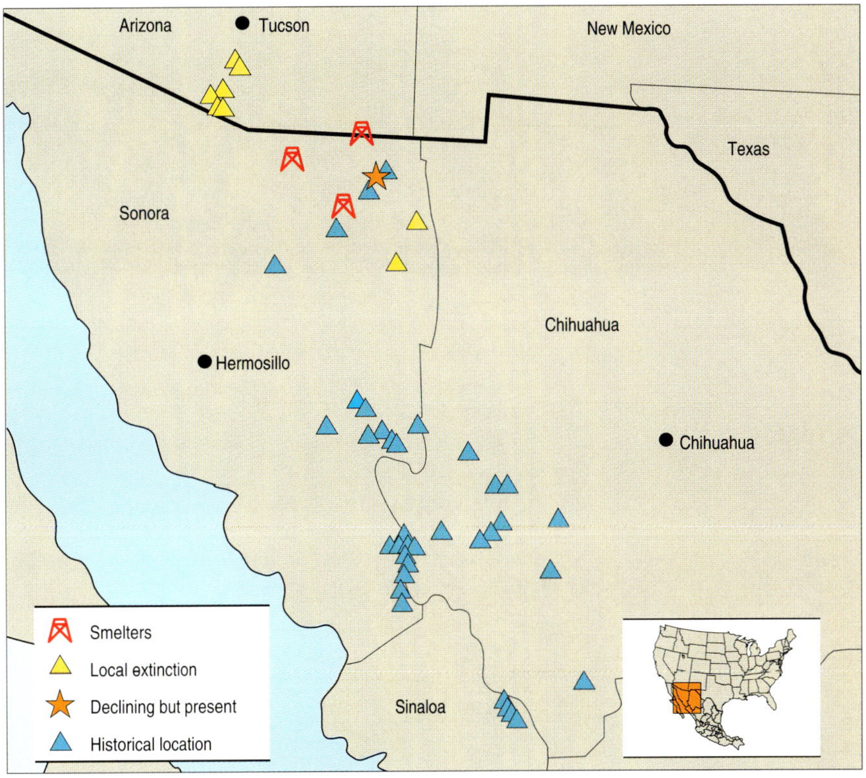

Fig. 1. Range of the Tarahumara frog, *Rana tarahumarae*. Copper smelters are at Douglas, AZ (now closed), and Cananea and Nacozari, Sonora. Historical locations include both surveyed populations that appeared stable, and unvisited historical localities (Campbell 1931; Little 1940; Williams 1960; Hale et al. 1977; Hale and May 1983; Hale and Jarchow 1988).

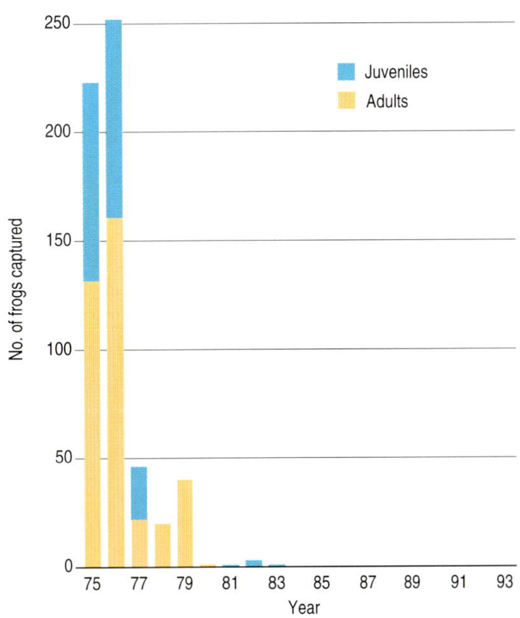

Fig. 2. Number of Tarahumara frogs captured 1975-93, Big Casa Blanca Canyon, Santa Rita Mountains, Santa Cruz County, AZ (Hale and May 1983).

Tarahumara frog (*Rana tarahumarae*) in Mexico.

and 1983 was consistently very acidic, attributed primarily to particulates produced by a copper smelter in Douglas, Arizona (Blanchard and Stromberg 1987), which has since been shut down. The alkaline soils in the area may buffer the streams from the acid rain; stream pH values were always slightly basic.

Analyses of water from affected streams showed consistently elevated levels of cadmium, a toxic metal, especially in relation to levels of the essential metal, zinc. In several species of vertebrates, sensitivity to cadmium toxicity is reduced with zinc supplementation (Supplee 1963; Webb 1972). At Sycamore Canyon and Big Casa Blanca Canyon localities, frogs survived longest near springs where zinc concentrations were highest. Levels of arsenic in streamwater were occasionally elevated (Hale and Jarchow 1988).

Although the proximity of operating copper smelters is correlated with population declines in Tarahumara and leopard frogs, exact causes of declines are not clear. No declines in frogs were noted until the 1970's, yet copper smelter emissions were much higher in the areas of declines in the early 1900's than recently. One of our hypotheses that accounts for the timing of the declines relates them to a long-term leaching of acid-soluble zinc from canyon walls, accumulation of insoluble cadmium in stream sediment, and sediment accumulation in stream pools from infrequent heavy rains before declines.

In southern and central Sonora, ranid frog populations appeared stable and reproductive at least through 1986; no population declines or extirpations were noted, either of Tarahumara or leopard frogs. Populations visited since 1986 do not appear to be declining.

Conclusions

We are confident that the Tarahumara frog no longer occurs in the United States, based upon repeated surveys of historical and potential habitat in southern Arizona. Although repeated surveys since 1983 in Mexico have not been as extensive as in the United States, sites visited in central and southern Sonora apparently continue to support healthy frog populations.

We conclude that the Tarahumara frog is not threatened with extinction throughout its range at this time, although the sudden declines and local extirpations in northern populations, coincident with declines of leopard frogs, are a serious concern.

State and federal resource management agencies in both Arizona and Sonora, Mexico, with independent biologists and the Arizona-Sonora Desert Museum (ASDM) and Centro Ecologico de Sonora have formed the Tarahumara Frog Reestablishment Oversight Group. This group proposes to reestablish the Tarahumara frog in selected historical sites and maintain captive frog populations at ASDM and elsewhere to provide stock for additional reintroduction. By intensively monitoring reintroduced populations and measuring important environmental variables we hope to determine the cause of declines in native ranid frogs in this area. Rain, streamwater, and air quality will be assessed continuously at each site, including pH, heavy metals, solar radiation (especially ultraviolet), and air particulates. Stream bottom substrate and tissue samples from frogs and frog prey and predator species will be sampled for heavy metals. Only after the causes of the declines have been identified and corrected can we expect long-term reestablishment of Tarahumara frogs and recovery of leopard frogs.

For further information:

S.F. Hale
Herpetologist
139 W. Suffolk Dr.
Tucson, AZ 85704

References

Blanchard, C.L., and M. Stromberg. 1987. Acidic precipitation in southeastern Arizona: sulfate, nitrate and trace-metal deposition. Atmospheric Environment 21:2375-2381.

Campbell, B. 1931. *Rana tarahumarae*, a frog new to the United States. Copeia 1931:164.

Clarkson, R.W., and J.C. Rorabaugh. 1989. Status of leopard frogs (*Rana pipiens* complex: Ranidae) in Arizona and southeastern California. Southwestern Naturalist 34:531-538.

Hale, S.F., and J.L. Jarchow. 1988. The status of the Tarahumara frog (*Rana tarahumarae*) in the United States and Mexico: Part 2. C.R. Schwalbe and T.B. Johnson, eds. Report to Arizona Game and Fish Department, Phoenix, and Office of Endangered Species, U.S. Fish and Wildlife Service, Albuquerque, NM. 101 pp.

Hale, S.F., and C.J. May. 1983. Status report for *Rana tarahumarae* Boulenger. Arizona Natural Heritage Program, Tucson. Report to Office of Endangered Species, U.S. Fish and Wildlife Service, Albuquerque, NM. 99 pp.

Hale, S.F., F. Retes, and T.R. Van Devender. 1977. New populations of *Rana tarahumarae* (Tarahumara frog) in Arizona. Journal of the Arizona Academy of Science 11:134-135.

Little, E.L., Jr. 1940. Amphibians and reptiles of the Roosevelt Reservoir area, Arizona. Copeia 1940:260-265.

Supplee, W.C. 1963. Antagonistic relationship between dietary cadmium and zinc. Science 139:119-120.

Webb, M. 1972. Protection by zinc against cadmium toxicity. Biochemical Pharmacology 21:2767-2771.

Williams, K.L. 1960. Taxonomic notes on Arizona herpetozoa. Southwestern Naturalist 5:25-36.

Courtesy N.M. Burkhead, NBS

Fishes

Science Editor
O. Eugene Maughan
National Biological Service
Arizona Cooperative Fish and Wildlife Research Unit
University of Arizona
Tucson, AZ 85721

Overview

The inescapable conclusion from the data presented in this section is that within historical time, native fish communities have undergone significant and adverse changes. These changes generally tend toward reduced distributions, lowered diversity, and increased numbers of species considered rare. These changes have been more inclusive and more dramatic in the arid western regions where there are primarily endemic (native) species, but similar, though more subtle changes, have occurred throughout the country. These trends are the same whether one focuses on faunas (Johnson; Starnes; and Walsh et al., this section) or on populations or genetic variation within a single species (Marnell; Miller et al.; and Philipp and Claussen, this section). Changes in fish communities may be indicative of the overall health of an aquatic system; some species have narrow habitat requirements.

The fact that fish populations have changed over historical time should not come as any great surprise. We have massively modified fish habitat through the very water demands that define our society (domestic, agricultural, and industrial water supplies; waste disposal; power generation; transportation; and flood protection). All of these activities have resulted in controlling or modifying the flow or degrading the quality of natural waters. In addition, almost all contaminants ultimately find their way into the aquatic system. Species of fishes that have evolved under the selection pressures imposed by natural cycles have often been unable to adapt to the changes imposed on them as a result of human activities.

Physical and chemical changes in their habitats are not the only stresses that fishes have encountered over time. Through fish management programs, the aquarium trade, and accidental releases, many aquatic species have been introduced to new areas far beyond their native ranges. Although these introductions were often done with the best of intentions, they have sometimes subjected native fish species to new competitors, predators, and disease agents that they were ill-equipped to withstand.

The data presented by Philipp and Claussen (this section) further suggest that managed fish populations (hatchery-stocked populations) have a lower genetic diversity than unmanaged populations. In other words, theoretically, the smaller the gene pool, the less likely a species may be able to adapt to changing environmental conditions.

It appears unlikely that the forces that have led to these changes in our fish fauna will lessen significantly in the immediate future. Therefore, if we are to preserve the diversity and adaptive potential of our fishes, we must understand much more of their ecology. Vague generalizations about habitat requirements or the results of biotic interactions are no longer enough. We must know quantitatively and exactly how fishes use habitat and how that use changes in the face of biotic pressures. Only when armed with such information are we likely to reduce the current trends among our native fishes.

Imperiled Freshwater Fishes

by
James E. Johnson
National Biological Service

The United States is blessed with perhaps 800 species of native freshwater fishes (Lee et al. 1980; Moyle and Cech 1988; Warren and Burr 1994). These fishes range from old, primitive forms such as paddlefish, bowfin, gar, and sturgeon, to younger, more advanced fishes, such as minnows, darters, and sunfishes. They are not equally distributed across the nation, but tend to concentrate in larger, more diverse environments such as the Mississippi River drainage (375 species; Robison 1986; Warren and Burr 1994). Drainages that have not undergone recent geological change, such as the Tennessee and Cumberland rivers, are also rich in native freshwater fishes (250 species; Starnes and Etnier 1986). Fewer native fishes are found in isolated drainages such as the Colorado River (36 species; Carlson and Muth 1989). More arid states west of the 100th meridian average about 44 native fish species per state, while states east of that boundary average more than three times that amount (138 native species; Figure).

Extinction, dispersal, and evolution are naturally occurring processes that influence the kinds and numbers of fishes inhabiting our streams and lakes. More recent human-related impacts to aquatic ecosystems, such as damming of rivers, pumping of aquifers, addition of pollutants, and introductions of non-native species, also affect native fishes, but at a more rapid rate than natural processes. Some fishes are better able to withstand these rapid changes to their environments or are able to find temporary refuge in adjacent habitats; fishes that lack tolerance or are unable to retreat face extinction.

In 1979 the Endangered Species Committee of the American Fisheries Society (AFS) developed a list of 251 freshwater fishes of North America judged in danger of disappearing (Deacon et al. 1979), 198 of which are found in the United States. A decade later, AFS updated the list (Williams et al. 1989), noting 364 taxa of fishes in some degree of danger, 254 of which are native to the United States. Both AFS lists used the same endangered and threatened categories defined in the Endangered Species Act of 1973, and added a special concern category to include fishes that could become threatened or endangered with relatively minor disturbances to their habitat. These imperiled native fishes are the first to indicate changes in our surface waters; thus their status provides us with a method of judging the health of our streams and lakes. This article compares the two AFS data sets to assess the trends in the status of freshwater fishes in the United States over the past decade.

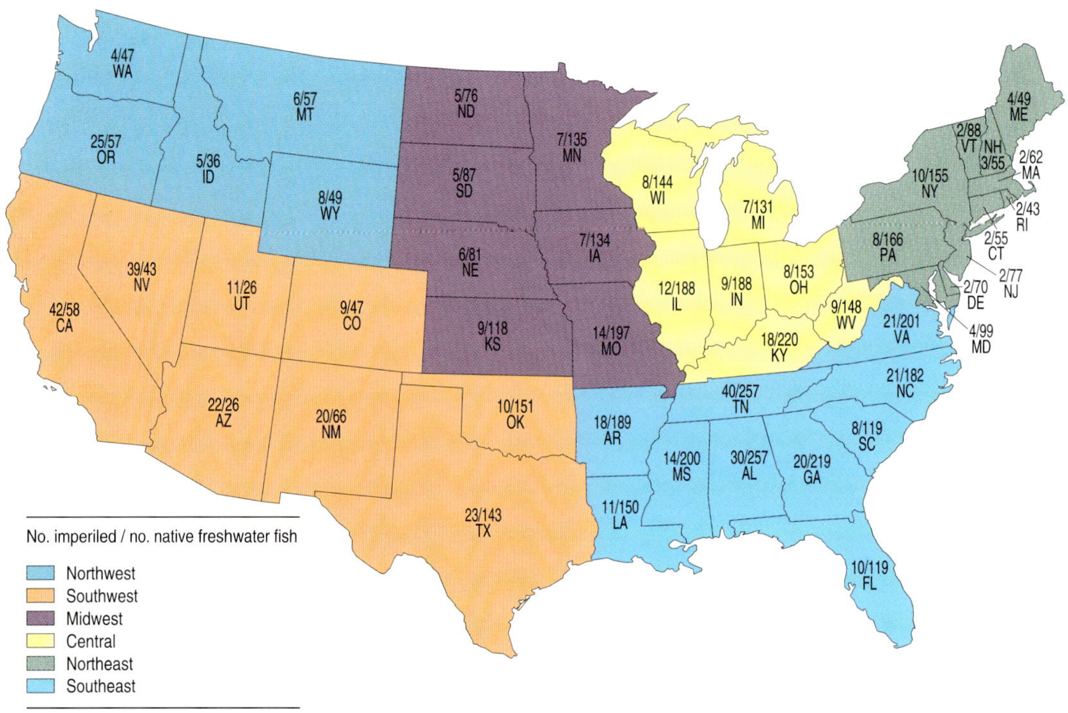

Figure. Number of fishes considered imperiled and number of native freshwater fishes of the contiguous United States by state (redrawn from Warren and Burr 1994).

Basis of the American Fisheries Society Listings

The 1979 and 1989 AFS listings were based entirely on biological considerations throughout the geographic range of the taxon and ignored jurisdictional or political considerations. For example, the johnny darter (*Etheostoma nigrum*) is a small darter found in clear streams from the East coast to the Continental Divide; the species reaches the western periphery of its range in Colorado. Johnny darters are rare in Colorado, which recognizes the species' rarity (Johnson 1987). Throughout most of its range, however, the johnny darter is common and thus was not included in the AFS listing. Only those taxa that appear imperiled are included in the lists; populations were not considered unless they were distinct enough to be recognized as subspecies.

The preliminary 1979 AFS listing was obtained by asking knowledgeable fishery scientists which fishes should be included. Those taxa were added to a 1972 listing of protected fishes (Miller 1972) that was then sent out to every state and to selected federal agencies for review.

The native fish faunas of some areas of the country are better studied than others and may therefore be better represented in the listing. The 1989 listing used knowledgeable biologists but not extensive agency review to build upon the 1979 listing. These two data bases provide the best information presently available on rare native fishes of the United States.

Changes in the Status of Native Freshwater Fishes, 1979-89

Analysis of the 1989 list provides some basic information on the status and trends of the native fishes of the United States. About one-fourth of our native freshwater fishes are perceived to be imperiled. Ninety-three percent of imperiled species are in trouble because of the deteriorating quality of the aquatic habitats on which they depend; this deterioration results from physical, chemical, and biological effects to our surface waters and underground aquifers. Overuse, introduction of non-native species, disease, and other problems that also affect our native fishes cause much less endangerment than habitat destruction.

The increase of taxa of fishes between the 1979 (189 taxa) and 1989 (254 taxa) AFS listings does not include 19 taxa that were removed from the 1989 listing because of extinction, taxonomic revisions, or better information on status. Seventy-five imperiled taxa that did not appear in the 1979 AFS listing were added to the 1989 AFS listing, an increase of 38% in a single decade. In addition, the status of 39 fishes was changed: 7 taxa improved (e.g., changed from threatened to special concern), 22 taxa declined, and 10 taxa were recognized as extinct (Table). No fish was removed from the 1989 AFS listing because of successful recovery efforts, indicating that our freshwater fishes continue to decline overall, and factors causing those changes appear difficult to reverse.

Species	Population trend
Pallid sturgeon (*Scaphirhynchus albus*)	Declined
Longjaw cisco (*Coregonus alpenae*)	Extinct
Deepwater cisco (*C. johannae*)	Extinct
Blackfin cisco (*C. nigripinnis*)	Extinct
Alvord cutthroat trout (*Oncorhynchus clarki* ssp.)	Extinct
Fish Creek Springs tui chub (*Gila bicolor euchila*)	Improved
Independence Valley tui chub (*G.b. isolata*)	Extinct
Thicktail chub (*G. crassicauda*)	Extinct
Chihuahua chub (*G. nigrescens*)	Improved
Least chub (*Iotichthys phlegethontis*)	Declined
White River spinedace (*Lepidomeda albivallis*)	Declined
Cape Fear shiner (*Notropis mekistocholas*)	Declined
Blackmouth shiner (*N. melanostomus*)	Declined
Oregon chub (*Oregonichthys crameri*)	Declined
Blackside dace (*Phoxinus cumberlandensis*)	Declined
Loach minnow (*Rhinichthys cobitis*)	Declined
White River sucker (*Catostomus clarki intermedius*)	Declined
Zuni bluehead sucker (*C. discobolus yarrowi*)	Improved
Shortnose sucker (*Chasmistes brevirostris*)	Declined
June sucker (*C. liorus mictus*)	Declined
Lost River sucker (*Deltistes luxatus*)	Declined
Razorback sucker (*Xyrauchen texanus*)	Declined
Pygmy madtom (*Noturus stanauli*)	Declined
Alabama cavefish (*Speoplatyrhinus poulsoni*)	Declined
Preston springfish (*Crenichthys baileyi albivallis*)	Declined
White River springfish (*C.b. baileyi*)	Declined
Moorman springfish (*C.b. thermophilus*)	Declined
Railroad Valley springfish (*C. nevadae*)	Declined
Devils Hole pupfish (*Cyprinodon diabolis*)	Improved
Desert pupfish (*C. macularius*)	Declined
Amistad gambusia (*Gambusia amistadensis*)	Extinct
San Marcos gambusia (*G. georgei*)	Extinct
Gila topminnow (*Poeciliopsis occidentalis*)	Improved
Spring pygmy sunfish (*Elassoma* sp.)	Improved
Sharphead darter (*Etheostoma acuticeps*)	Improved
Amber darter (*Percina antesella*)	Declined
Blue pike (*Stizostedion vitreum glaucum*)	Extinct
Utah Lake sculpin (*Cottus echinatus*)	Extinct
Shoshone sculpin (*C. greenei*)	Declined

Table. Population trends for endangered, threatened, and special concern freshwater fishes of the United States whose status changed between 1979 and 1989 (Williams et al. 1989).

The relation between declining aquatic habitats and fishes facing extinction is not as simple as might be expected. Species with limited distributions are more likely to be jeopardized by changes in their local aquatic habitats than are species with extensive ranges. Many fishes on the lists have local distributions, and a few, such as the Clear Creek gambusia (*Gambusia heterochir*) and Devils Hole pupfish (*Cyprinodon diabolis*), are limited to a single spring. These unique fishes could be lost by a single, isolated event. Some of the widespread species included in the listings—such as paddlefish (*Polyodon spathula*) and six taxa of sturgeons—depend on large rivers, and their inclusion indicates widespread threats to these extensive habitats.

States with the most listed (imperiled) species include California (42), Tennessee (40), and Nevada (39). Somewhat fewer listed fishes are found in Alabama (30), Oregon (25), Texas (23), Arizona (22), Virginia (21), North

Carolina (21), New Mexico (20), and Georgia (20; Figure). Regionally, the Southwest has the highest mean number of fish species listed per state (22.5), closely followed by the Southeast (19.3); the northeastern states have the lowest mean number of native fish species in trouble (3.7). Nearly half (48%) of the southwestern native fishes are jeopardized, followed by fishes of the Northwest (19%), the Southeast (10%), the Midwest (6.4%), the central states (5.9%), and the Northeast (4.3%; Warren and Burr 1994).

The AFS will likely update its listing of native fishes in peril toward the end of this decade, thus providing us with more than 20 years of information on the status of these fishes, a short time in the overall life of a species but a good data base upon which to evaluate the environmental health of our streams and lakes. If the trend over the last decade continues, we can expect a further decline in the richness of our native fishes. In addition, as aquatic habitat deterioration becomes more extensive, we can expect to see an increase in the listing of widespread fishes.

For further information:

James E. Johnson
National Biological Service
Arkansas Cooperative Research Unit
Department of Biological Sciences
University of Arkansas
Fayetteville, AR 72701

References

Carlson, C.A., and R.T. Muth. 1989. The Colorado River: lifeline of the American Southwest. Pages 220-239 *in* D. P. Dodge, ed. Proceedings of the International Large Rivers Symposium. Canadian Journal of Fisheries and Aquatic Sciences, Special Publ. 106.

Deacon, J.E., G. Kobetich, J.D. Williams, S. Contreras, et al. 1979. Fishes of North America endangered, threatened, or of special concern: 1979. Fisheries 4(2):30-44.

Johnson, J.E. 1987. Protected fishes of the United States and Canada. American Fisheries Society, Bethesda, MD. 42 pp.

Lee, D.S., C.R. Gilbert, C.H. Hocutt, R.E. Jenkins, D.E. McAllister, and J.R. Stauffer, Jr. 1980. Atlas of North American freshwater fishes. North Carolina State Museum of Natural History. 254 pp. (Reissued 1981 with appendix; 867 pp.)

Miller, R.R. 1972. Threatened freshwater fishes of the United States. Transactions of the American Fisheries Society 101(2):239- 252.

Moyle, P.B., and J.J. Cech, Jr. 1988. Fishes: an introduction to ichthyology. Prentice Hall, Englewood Cliffs, NJ. 559 pp.

Robison, H.W. 1986. Zoogeographic implications of the Mississippi River basin. Pages 267-285 *in* C.H. Hocutt and E.O. Wiley, eds. The zoogeography of North American freshwater fishes. John Wiley and Sons, Inc., New York.

Starnes, W.C., and D.A. Etnier. 1986. Drainage evolution and fish biogeography of the Tennessee and Cumberland rivers drainage realm. Pages 325-361 *in* C.H. Hocutt and E.O. Wiley, eds. The zoogeography of North American freshwater fishes. John Wiley and Sons, New York.

Warren, M.L., and B.M. Burr. 1994. Status of freshwater fishes of the United States: overview of an imperiled fauna. Fisheries 19(1):6-18.

Williams, J.E., J.E. Johnson, D.A. Hendrickson, S. Contreras-Balderas, J.D. Williams, M. Navarro-Mendoza, D.E. McAllister, and J.E. Deacon. 1989. Fishes of North America endangered, threatened, or of special concern: 1989. Fisheries 14(6):2-20.

Southeastern Freshwater Fishes

by
Stephen J. Walsh
Noel M. Burkhead
James D. Williams
National Biological Service

North America has the richest fauna of temperate freshwater fishes in the world, with about 800 native species in the waters of Canada and the United States. The center of this diversity is in the southeastern United States, where as many as 500 species may exist (62% of the continental fauna north of Mexico). Many coastal marine species also enter fresh waters of the Southeast, and at least 34 foreign fish species are established in the region.

Although freshwater fishes of the United States are better studied than any fish fauna of comparable scope in the world (Lee et al. 1980; Hocutt and Wiley 1986; Matthews and Heins 1987; Page and Burr 1991; Mayden 1992), large gaps exist in scientific knowledge about the biology and ecology of most species. New species are still being discovered, and the taxonomy of other species is being refined.

Seriously declining populations of freshwater fishes in the United States concern the scientific community (Deacon et al. 1979; Williams et al. 1989; Moyle and Leidy 1992; Warren and Burr 1994). This article briefly summarizes the current conservation status of southeastern freshwater fishes; the Southeast is emphasized because of its important fish biodiversity and to focus attention on the growing problem of adverse human impacts on the region's aquatic habitats (Mount 1986; Burkhead and Jenkins 1991; Etnier and Starnes 1991; Warren and Burr 1994).

Courtesy N.M. Burkhead, NBS

Principal causes of declining fish resources in the Southeast are due to habitat perturbations, such as loss of forested stream cover, mining activities, and impoundments, as at this site in northern Georgia.

Hydrologic Regions

The southeastern United States as defined here is delimited on the north and west by the Ohio and Mississippi rivers. The following hydrologic regions (Fig. 1) are defined on the basis of common geophysical characteristics and similar fish faunas of the drainages within each region (Hocutt and Wiley 1986): (a) Atlantic Slope—coastal waters from the Roanoke River (Virginia) southward to the Altamaha River (Georgia); (b) Peninsular—waters from the Satilla River (Georgia) to the Ochlockonee River (Florida); (c) Lower Apalachicola Basin—waters from the Apalachicola River (Florida) westward to the Perdido River (Alabama); (d) Lower Mobile Basin—lowland portions of the Tombigbee and Alabama rivers and tributaries (Alabama and Mississippi); (e) Lower Mississippi—the Mississippi River and its eastern tributaries below the Ohio River (Mississippi, Tennessee, and Kentucky); (f) Interior Plateau—upland waters of the middle and lower Ohio River and southern tributaries, including the lower Cumberland and Tennessee rivers (Kentucky and Tennessee); and (g) Southern Appalachians—upland waters of the mountains in the geological provinces known as the Cumberland Plateau, Valley and Ridge, Blue Ridge, and Piedmont, south of the Kanawha (West Virginia) and Roanoke rivers. Many fishes are widely distributed in the Southeast and occur in two or more hydrologic regions.

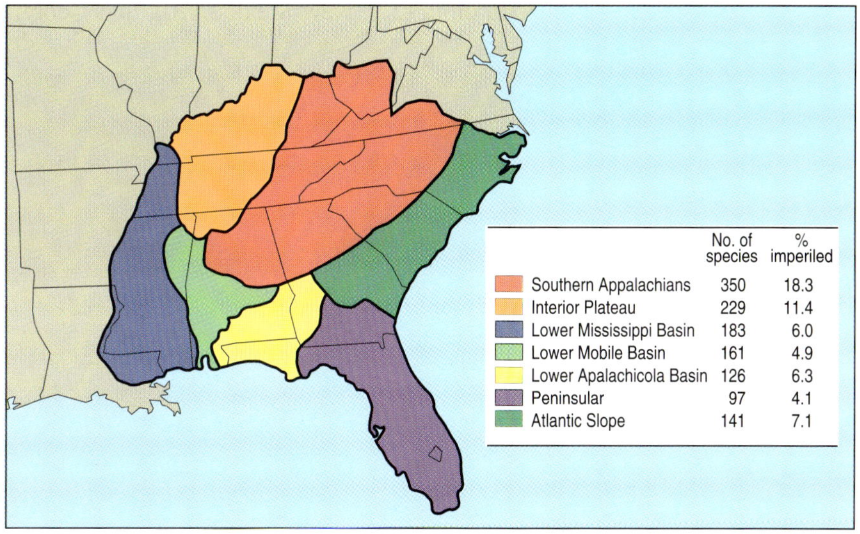

Fig. 1. Total numbers of freshwater fishes and percentage imperiled by hydrographic region of the southeastern United States.

Imperiled Freshwater Fishes

The Southeast has about 485 known species of native freshwater fishes, representing 27 families. Most of the diversity of the southeastern fish fauna is in five families: the darters and perches (family Percidae; 31.3%); the minnows (family Cyprinidae; 29.7%); the madtoms and bullhead catfishes (family Ictaluridae; 6.8%); the suckers (family Catostomidae; 6.6%); and the sunfishes and basses (family Centrarchidae; 5.8%). The greatest diversity is in the Appalachian Mountains and Interior Plateau (Fig. 1), but other regions of the Southeast also harbor many more species than do similar-sized geographic areas elsewhere in the United States.

As of January 1994 the U.S. Fish and Wildlife Service (USFWS) had designated 15 southeastern fish species as endangered and 12 as threatened, representing 6% of the entire regional fish fauna. Ninety-three fish taxa (19%) are imperiled (endangered, threatened, or of special concern) in the Southeast, including proposed listings and those recognized by other authors (Williams et al. 1989). During the past 25 years, only seven species were upgraded by the USFWS, mainly because of discovery of new populations, inadequate knowledge at the time of listing, or invalid taxonomy. No endangered or threatened species have been delisted. A steady upward trend in designation of imperiled southeastern fishes has occurred in the last 20 years (Fig. 2); the number of species considered imperiled by the USFWS increased from 3 (less than 1%) in 1974 to 84 (17%) in 1994 (USFWS listings only). During the 10-year period from 1979 to 1989, the number of species considered imperiled by the American Fisheries Society increased from 63 (13%) to 81 (17%; Fig. 2).

An alarming 21% of the nearly 300 species of minnows and darters are imperiled in the Southeast. Considered alone, more than 30% of the 150 species of darters are in trouble, representing the highest total number of species in any one family. Madtom catfishes (genus *Noturus*) are also disproportionally imperiled among large families of more than 30 species (Etnier and Starnes 1991; Warren and Burr 1994). Among smaller groups of fishes, the most severe status is among the sturgeons and paddlefish, where seven of the eight (86%) southeastern species are in jeopardy. In terms of ecological requirements, most imperiled species are those that live in small to large creeks and small rivers, are closely associated with clean stream-bottom substrates, or are isolated in spring and cave environments.

On a regional scale, the greatest number of imperiled species occurs in the highland areas of the Appalachians and Interior Plateau,

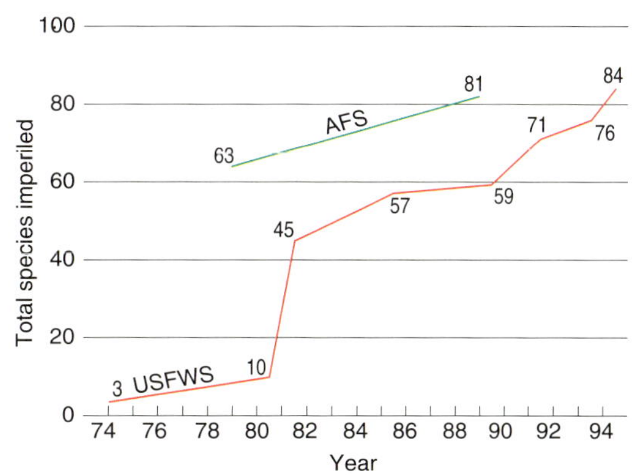

Fig. 2. Total numbers of imperiled fishes in the Southeast during the last 20 years, as recognized by the American Fisheries Society (AFS) and the U.S. Fish and Wildlife Service (USFWS). Numbers represent imperiled species during years of listing activity.

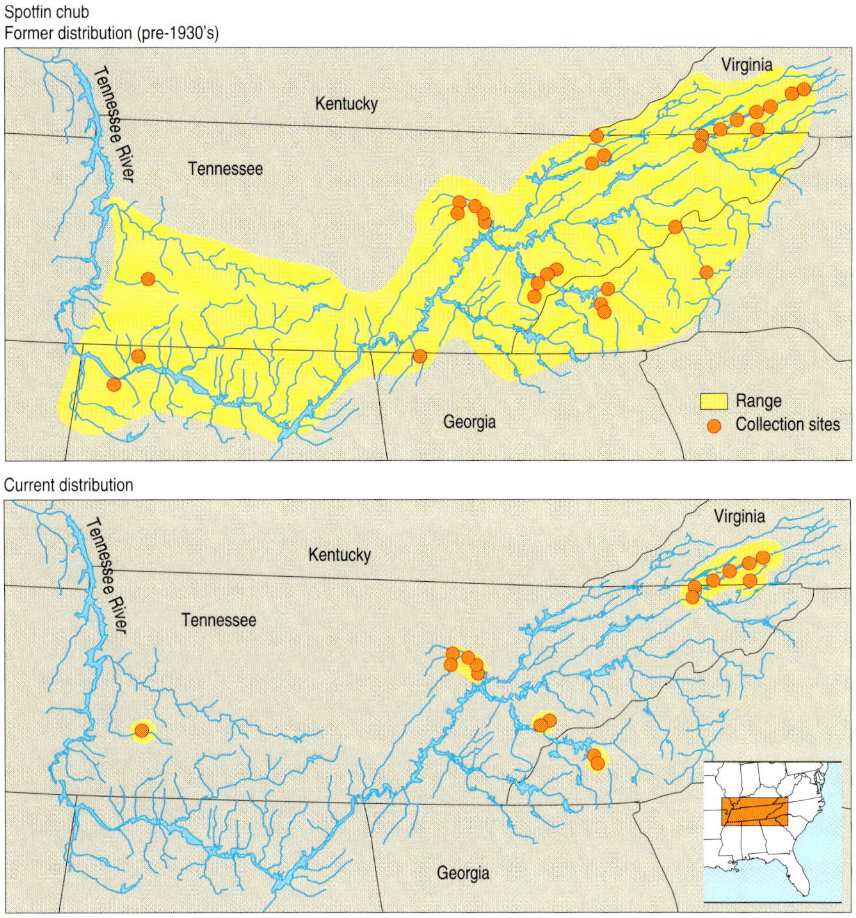

Fig. 3. An example of habitat fragmentation, decline, and isolation of populations of a southeastern freshwater fish, the endangered spotfin chub (*Cyprinella monacha*). Former (pre-1930's) and present range in yellow.

Courtesy N.M. Burkhead, NBS

Tangerine darter (*Percina aurantiaca*).

Courtesy N.M. Burkhead, NBS

Mountain redbelly dace (*Phoxinus oreas*).

followed by the Coastal Plain subregions (Fig. 1). This geographic trend is correlated with both a high level of diversity in the respective hydrologic regions and the quite localized or endemic distributions of many species. Especially important are a number of watersheds that harbor many species confined within those drainages; these watersheds include the Tennessee River, the Mobile Basin, the Cumberland River, and the Roanoke and James rivers (Warren and Burr 1994). Most jeopardized species have restricted distributions, but the number of more geographically widespread species that are disappearing from large portions of their ranges is increasing.

Two species of southeastern fishes have become extinct in the last century: the harelip sucker (*Moxostoma lacerum*) and the whiteline topminnow (*Fundulus albolineatus*). At least one other species, the least darter (*Etheostoma microperca*), has disappeared from the southern portion of its range that falls within the region covered here. The slender chub (*Erimystax cahni*) has not been seen since 1987 and may be near extinction. Two other species peripheral to the Southeast are feared extinct: the Scioto madtom (*Noturus trautmani*) and the Maryland darter (*Etheostoma sellare*; Etnier 1994).

The declining status of freshwater fishes among divergent taxonomic groups and across broad habitat types and geographic areas is interpreted as evidence for widespread and pervasive threats to the entire North American fish fauna (Moyle and Leidy 1992; Warren and Burr 1994). In the Southeast, fish declines are the result of the same factors that cause global deterioration of aquatic resources, primarily habitat loss and degraded environmental conditions. The principal causes of freshwater fish imperilment in the Southeast and other areas of the United States are dams and channelization of large rivers, urbanization, agriculture, deforestation, erosion, pollution, introduced species, and the cumulative effects of all these factors (Moyle and Leidy 1992; Warren and Burr 1994). The most insidious threat to southeastern fishes is sedimentation and siltation resulting from poor land-use patterns that eliminate suitable habitat required by many bottom-dwelling species. Cumulative effects of physical habitat modifications have caused widespread fragmentation of many fish populations in the Southeast (Fig. 3), presenting difficult challenges for those trying to reverse and restore diminished fish stocks.

Aquatic resources are often resilient and capable of recovery, given favorable conditions. Conservation of southeastern fishes will require significant changes in land management and socioeconomic factors (Moyle and Leidy 1992; Warren and Burr 1994), but such changes are necessary to stem future losses of biodiversity. The first step required is to improve public education on the value and status of native aquatic organisms. For resource managers and policy makers, increased efforts must be made to assume proactive management of entire watersheds and ecosystems; establish networks of aquatic preserves; restore degraded habitats; establish long-term research, inventory, and monitoring programs on fishes; and adopt improved environmental ethics concerning aquatic ecosystems (Warren and Burr 1994). The southeastern fish fauna is a national treasure of biodiversity that is imminently threatened. If this precious heritage is to be passed on, its stewardship must be improved through cooperative actions of all public and private sectors within the region.

References

Burkhead, N.M., and R.E. Jenkins. 1991. Fishes. Pages 321-409 *in* Virginia's Endangered Species: Proceedings of a Symposium. McDonald and Woodward Publishing Co., Blacksburg,VA. 672 pp.

Deacon, J.E., G. Kobetich, J.D. Williams, S. Contreras, et al. 1979. Fishes of North America endangered, threatened, or of special concern: 1979. Fisheries 4(2):30-44.

Etnier, D.A. 1994. Our southeastern fishes —what have we lost and what are we likely to lose. Southeastern Fishes Council Proceedings 29:5-9.

Etnier, D.A., and W.C. Starnes. 1991. An analysis of Tennessee's jeopardized fish taxa. Journal of the Tennessee Academy of Science 66(4):129-133.

Hocutt, C.H., and E.O. Wiley, eds. 1986. The zoogeography of North American freshwater fishes. John Wiley and Sons, Inc., New York. 866 pp.

Lee, D.S., C.R. Gilbert, C.H. Hocutt, R.E. Jenkins, D.E. McAllister, and J.R. Stauffer, Jr. 1980. Atlas of North American freshwater fishes. North Carolina State Museum of Natural History, Raleigh. 854 pp. (Reissued in 1981 with appendix; 867 pp.)

Matthews, W.J., and D.C. Heins, eds. 1987. Community and evolutionary ecology of North American stream fishes. University of Oklahoma Press, Norman. 310 pp.

Mayden, R.L., ed. 1992. Systematics, historical ecology, and North American freshwater fishes. Stanford University Press, CA. 969 pp.

Mount, R.H., ed. 1986. Vertebrate animals of Alabama in need of special attention. Alabama Agricultural Experiment Station, Auburn University. 124 pp.

Moyle, P.B., and R.A. Leidy. 1992. Loss of biodiversity in aquatic ecosystems: evidence from fish faunas. Pages 127-169 *in* P.L. Fiedler and S.K. Jain, eds. Conservation biology: the theory and practice of nature conservation, preservation and management. Chapman and Hall, New York. 507 pp.

Page, L.M., and B.M. Burr. 1991. A field guide to freshwater fishes of North America north of Mexico. Peterson field guide series. Houghton Mifflin Co., Boston, MA. 432 pp.

Warren, M.L., Jr., and B.M. Burr. 1994. Status of freshwater fishes of the United States: overview of an imperiled fauna. Fisheries 19(1):6-18.

Williams, J.E., J.E. Johnson, D.A. Hendrickson, S. Contreras-Balderas, J.D. Williams, M. Navarro-Mendoza, D.E. McAllister, and J.E. Deacon. 1989. Fishes of North America endangered, threatened, or of special concern: 1989. Fisheries 14(6):2-20.

For further information:

Stephen J. Walsh
National Biological Service
Southeastern Biological Science Center
7920 NW 71st St.
Gainesville, FL 32653

Loss of Genetic Diversity Among Managed Populations

by
David P. Philipp
Julie E. Claussen
Illinois Natural History Survey

Species are composed of genetically divergent units usually interconnected by some (albeit low) level of gene flow (Soulé 1987). Because of this restriction in gene flow, natural selection can genetically tailor populations to their environments through the process of local adaptation (Wright 1931).

Because freshwater and anadromous (i.e., adults travel upriver from the sea to spawn) fishes are restricted by the boundaries of their aquatic habitats, genetic subdivisions may be more pronounced for these vertebrates than for others. Consequently, managers of programs for these species must realize that the stock (i.e., local discrete populations), and not the species as a whole, must be the units of primary management concern (Kutkuhn 1981).

Genetic variability in a species occurs both among individuals *within* populations as well as *among* populations (Wright 1978). Variation *within* populations is lost through genetic drift (*see* glossary; Allendorf et al. 1987), a process increased when population size becomes small. Variation *among* populations is lost when previously restricted gene flow between populations is increased for some reason (e.g., stocking, removal of natural barriers such as waterfalls); differentiation between populations is lost as a result of the homogenization of two previously distinct entities (Altukhov and Salmenkova 1987; Campton 1987).

Beyond this loss of genetic variation, mixing two groups can result in outbreeding depression, which is the loss of fitness in offspring that results from the mating of two individuals that are too distantly related (Templeton 1987). This loss in fitness is caused by the disruption of the process that produced advantageous local adaptations through natural selection. Inbreeding depression, on the other hand, is the loss of fitness produced by the repeated crossing of related organisms. The area of optimal relatedness occurs between inbreeding depression and outbreeding depression.

Loss of Genetic Integrity Through Stocking

Many sportfish populations are managed by using a combination of harvest regulation, habitat manipulation, and stocking. Jurisdiction for these activities falls to federal, state, tribal, and local governments, as well as private citizens. Many resource managers in the past were unaware of the long-term consequences that stocking efforts would have on the genetic integrity of local populations (Philipp et al. 1993).

Fish introductions can be classified into three types: non-native introductions, in which a given species of fish is introduced into a body of water outside its native range (regardless of any political boundaries); stock transfers, in which fish from one stock are introduced into a water body in a different geographic region inhabited by a different stock of that same species, yet are still within their native range; and genetically compatible introductions, in which fish are removed from a given water body and they, or more often their offspring, are introduced back into that water body or another water body that is still within the boundaries of the genetic stock serving as the hatchery brood source (Philipp et al. 1993).

Although non-native introductions may often cause ecological problems for the environments in which they are introduced, they can also cause genetic problems if they hybridize with closely related native species. Examples of this are the hybridization of introduced smallmouth bass (*Micropterus dolomieu*) and spotted bass (*M. punctulatus*) with native Guadalupe bass (*M. treculi*) in Texas (Morizot et al. 1991), and the hybridization of introduced rainbow trout (*Oncorhynchus mykiss*) with native Apache trout (*O. apache*; Carmichael et al. 1993). The greatest degree of genetic damage, that is, the loss of genetic variation among populations, is caused by stock transfers, a common

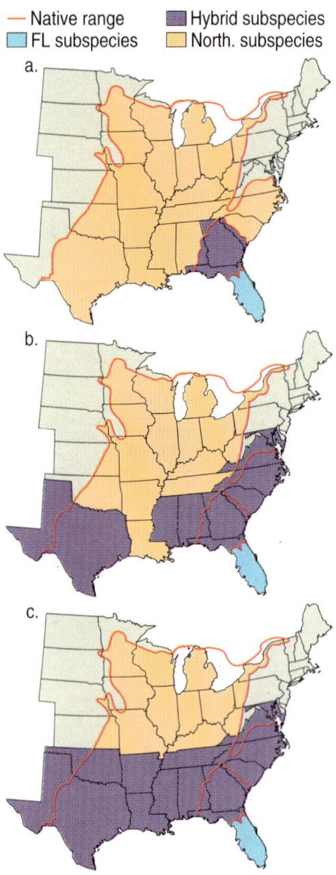

Figure. Loss of genetic variation among largemouth bass populations. **a.** The native range of the largemouth bass (*Micropterus salmoides*) is delineated by the red lines (MacCrimmon and Robbins 1975). As first described by Bailey and Hubbs (1949), the Florida subspecies, *M.s. floridanus*, was restricted to peninsular Florida (blue); the northern subspecies, *M.s. salmoides,* covered most of the rest of the range of the species; and there was a relatively small intergrade zone between the two resulting from some indeterminable combination of natural hybridization and human-caused mixing of stocks. **b.** The expansion of the intergrade by 1980 was described by Philipp et al. (1983). Because detailed ranges were not explored in all states, and because this intergrade zone expansion was likely caused by state stocking programs, entire states are classified according to whether the intergrade zone was expanded. **c.** The current intergrade zone is now even larger because of the addition of more states in which largemouth bass containing at least some *M.s. floridanus* genes are being introduced either by the state fish and game agencies themselves or by private groups. Notice that the entire southern and eastern portion of the original range of the northern subspecies, *M.s. salmoides*, is at risk of being inundated with *M.s. floridanus* genes.

practice among fisheries management agencies and the private sector.

Largemouth Bass

Largemouth bass (*Micropterus salmoides*) exemplify how introduction programs cause the loss of genetic diversity. The original range of the largemouth bass was restricted to parts of the central and southeastern United States (Figure), extending northward into some of southern Ontario (MacCrimmon and Robbins 1975). Bailey and Hubbs (1949), however, described two subspecies. The Florida subspecies, *M.s. floridanus*, was formerly restricted to much of peninsular Florida (Figure, a), whereas the range of the northern subspecies, *M.s. salmoides*, extended north and west of an intergrade zone that included parts of South Carolina, Georgia, Alabama, and northern Florida. It is likely, though, that the intergrade zone had already been expanded from the original natural hybrid zone as a result of early fish stocking programs.

Since 1949, however, much more serious stocking efforts have extended this intergrade zone. A survey of largemouth bass populations conducted in the late 1970's (Philipp et al. 1983) revealed that the intergrade zone had grown considerably larger through the deliberate stocking efforts of the involved state agencies (Figure, b). Additional introductions of *M.s. floridanus* since that genetic survey have now spread the genes of that subspecies across the entire southern range of *M.s. salmoides* (Figure, c).

This introduction of the Florida largemouth has compromised the genetic integrity of all the populations of the northern largemouth bass into which the species has been introduced (populations in Texas, Oklahoma, Arkansas, Louisiana, Mississippi, Tennessee, Alabama, Georgia, South Carolina, North Carolina, Virginia, and Maryland, at a minimum). Those now-genetically mixed populations have lost much of their distinctness because of the loss of among-population genetic variation that accompanies this type of homogenization. Populations other than those in the water bodies actually stocked will be affected as well because of inevitable gene flow into and between other connected populations. As a result, genetic integrity is now at risk for all populations of this important sportfish species throughout the southern and eastern portions of its native range.

In addition, because the two subspecies have quite different characteristics (Cichra et al. 1982; Fields et al. 1987; Kleinsasser et al. 1990), these massive stock transfers will likely result in outbreeding depression. More specifically, the Florida subspecies exhibits significantly poorer survival, growth, and reproductive success in Illinois than does the northern subspecies (Philipp 1991; Philipp and Whitt 1991). Also, the offspring resulting from crossing the two subspecies (in either direction) are less fit in Illinois than are the offspring of the pure northern subspecies (Philipp 1991). These results extend to populations of the northern subspecies across its range from Texas to Minnesota (unpublished data).

Conclusions

The genetic integrity of largemouth bass stocks, and likely of many other managed fish species as well, is eroding as a result of management programs that inadvertently permit or deliberately promote stock transfers. This causes not only the loss of genetic variation among populations, but through outbreeding depression it is also probably negatively affecting the fitness of many native stocks involved. We need to address genetic integrity when restoring native populations.

References

Allendorf, F.W., N. Ryman, and F.M. Utter. 1987. Genetics and fishery management past, present, and future. Pages 1-19 *in* N. Ryman and F.M. Utter, eds. Population genetics and fishery management. Washington Sea Grant Program, Seattle.

Altukhov, Y.P., and E.A. Salmenkova. 1987. Stock transfer relative to natural organization, management and conservation of fish populations. Pages 333-344 *in* N. Ryman and F.M. Utter, eds. Population genetics and fishery management. Washington Sea Grant Program, Seattle.

Bailey, R.M., and C.L. Hubbs. 1949. The black basses (*Micropterus*) of Florida, with description of a new species. University of Michigan Museum of Zoology Occasional Papers 516:1-40.

Campton, D.E. 1987. Natural hybridization and introgression in fishes: methods of detection and interpretation. Pages 161-192 *in* N. Ryman and F.M. Utter, eds. Population genetics and fishery management. Washington Sea Grant Program, Seattle.

Carmichael, G.J., J.N. Hanson, M.E. Schmidt, and D.C. Morizot. 1993. Introgression among Apache, cutthroat, and rainbow trout in Arizona. Transactions of the American Fisheries Society 122:121-130.

Cichra, C.E., W.H. Neill, and R.L. Noble. 1982. Differential resistance of northern and Florida largemouth bass to cold shock. Proceedings of the Southeastern Association of Fish and Wildlife Agencies 34(1980):19-24.

Fields, R., S.S. Lowe, C. Kaminski, G.S. Whitt, and D.P. Philipp. 1987. Critical and chronic thermal maxima of northern and Florida largemouth bass and their reciprocal F_1 and F_2 hybrids. Transactions of the American Fisheries Society 116:856-863.

Kleinsasser, L.J., J.H. Williamson, and B.G. Whiteside. 1990. Growth and catchability of northern Florida and F_1 hybrid largemouth bass in Texas ponds. North American Journal of Fisheries Management 10:462-468.

Kutkuhn, J.H. 1981. Stock definition as a necessary basis for cooperative management of Great Lakes fish resources. Canadian Journal of Fisheries and Aquatic Sciences 38:1476-1478.

MacCrimmon, H.R., and W.H. Robbins. 1975. Distribution of the black basses in North America. Pages 56-66 *in*

R.H. Stroud and H. Clepper, eds. Black bass biology and management. Sport Fishing Institute, Washington, DC.
Morizot, D.C., S.W. Calhoun, L.L. Clepper, M.E. Schmidt, J.H. Williamson, and G.J. Carmichael. 1991. Multispecies hybridization among native and introduced centrarchid basses in central Texas. Transactions of the American Fisheries Society 120:283-289.
Philipp, D.P. 1991. Genetic implications of introducing Florida largemouth bass, *Micropterus salmoides floridanus*. Canadian Journal of Fisheries and Aquatic Sciences 48(1):58-65.
Philipp, D.P., W.F. Childers, and G.S. Whitt. 1983. A biochemical genetic evaluation of the northern and Florida subspecies of largemouth bass. Transactions of the American Fisheries Society 112:1-20.
Philipp, D.P., J.M. Epifanio, and M.J. Jennings. 1993. Conservation genetics and current stocking practices: are they compatible? Fisheries 18:14-16.
Philipp, D.P., and G.S. Whitt. 1991. Survival and growth of northern, Florida, and reciprocal F_1 hybrid largemouth bass in central Illinois. Transactions of the American Fisheries Society 120:156-178.
Soulé, M.E., ed. 1987. Conservation biology: the science of scarcity and diversity. Sinauer Associates, Inc., Sunderland, MA. 584 pp.
Templeton, A.R. 1987. Coadaptation and outbreeding depression. Pages 105-116 *in* M.E. Soulé, ed. Conservation biology: the science of scarcity and diversity. Sinauer Associates, Inc., Sunderland, MA.
Wright, S. 1931. Evolution in Mendelian populations. Genetics 16:97-159.
Wright, S. 1978. Evolution and the genetics of populations. Vol. 4. Variability within and among natural populations. University of Chicago Press, IL.

For further information:

David P. Philipp
Illinois Natural History Survey
Center for Aquatic Ecology
607 E. Peabody Dr.
Champaign, IL 61820

Colorado River Basin Fishes

by
Wayne C. Starnes
Smithsonian Institution

The Colorado River and its tributaries have undergone drastic alterations from their natural states over the past 125 years. These alterations include both physical change or elimination of aquatic habitats and the introductions of numerous non-native species, particularly fish. Ironically, several more species occur at most localities today than were historically present before these alterations. This situation complicates the use of biodiversity as a litmus test for monitoring trends of either the deterioration or the health of an aquatic ecosystem.

An Altered Ecosystem

Over its entire basin (Figure), the Colorado River has been changed from its natural state perhaps as much as any river system in the world. The demands for water and power in the arid West have drastically altered the system by impoundments, irrigation diversions, diking, channelization, pollutants, and destruction of bank habitats by cattle grazing and other practices. Some reaches, ranging from desert spring runs to main rivers, have been completely dewatered or, seasonally, their flows consist almost entirely of irrigation return laden with silt and chemical pollutants. The Gila River of Arizona, one of the Colorado's largest tributaries, has not flowed over its lower 400 km (248 mi) since the early 1900's. These alterations and their effects on the fish fauna have been discussed by several authors (Miller 1961; Minckley and Deacon 1968; Stalnaker and Holden 1973; Carlson and Muth 1989; Minckley and Deacon 1991). Only a few small tributaries, mostly at higher elevations, retain most of their natural characteristics.

Courtesy J.N. Rinne

Captive bonytail (*Gila elegans*), rarest of the larger river species in the Colorado River Basin.

Native Fish Fauna

Despite the expansive drainage basin (631,960 km^2 [243,937 mi^2]) of the Colorado River, the system supported only a relatively small number of native fish species compared with basins of much smaller size east of the Continental Divide. The Colorado Basin's native fauna, however, was nearly unique. If two former marine invaders are removed from the 51 native taxa known from the system (Table 1), 42 of the 49 that remain (86%) are considered endemic to the system. The greatest diversity of taxa (44) was distributed in the Lower Basin downstream of the Arizona-Utah border, in a variety of habitats that include mainstem rivers, smaller tributaries, and isolated springs. The Upper Basin was much less diverse, containing 14 species, including a subset of the Lower Basin fauna plus 4 headwater species that occur in cooler water and a warm spring endemic. Basinwide, about 5 species occurred mostly in mainstem river or larger tributary habitats, 37 were restricted to smaller, in some cases isolated, habitats, and 7 were more generally distributed among different habitat types.

Trends

As a consequence of habitat alterations, the prevailing trend among native fish populations in the Colorado River Basin has been drastic

Figure. Colorado River Basin.

reductions that include decreased abundance in all or part of their ranges, overall range reductions, or virtual or actual extinctions (Tables 1 and 2). Presently, 40 of the 49 strictly freshwater, native species are considered either possibly or actually jeopardized or are extinct (Table 1). Of the 40, 12 are of special concern, 25 are considered endangered or threatened, and 3 are believed extinct.

In the Lower Basin, only 3 of the 10 native species that inhabited the mainstem of the lower Colorado River remained by the 1940's but by the 1960's, none remained. In the lower Salt River portion of the Gila River system, the original complement of 14 taxa was also reduced to 3 by the 1940's and to 2 by the 1960's; today, they are probably extirpated. In the early 1900's, the isolated springs of the Pluvial White River system in southern Nevada harbored 17 endemic taxa; today, 1 of those taxa is extinct, 9 endangered, 3 threatened, and the remainder of

Table 1. Native fish taxa of the Colorado River Basin including currently recognized subspecies. Taxa denoted by * may eventually prove genetically distinct from populations outside the Colorado River Basin. Those denoted "(m)" are marine invaders. Status of jeopardized and extinct species appears in parentheses: E = endangered; T = threatened; SC = special concern; X = extinct (based, in part, on Carlson and Muth 1989; Williams et al. 1989; and the National Biological Service's Category 2 list). Common names bracketed with quotation marks indicate that those species are undescribed and not officially named.

Scientific name	Common name
Family Elopidae	
Elops affinis (m)	Machete
Family Cyprinidae	
Agosia chrysogaster	Longfin dace
Gila cypha (E)	Humpback chub
G. elegans (E)	Bonytail
G. intermedia (SC)	Gila chub
G. robusta jordani (E)	Pahranagat chub
G. robusta robusta (SC)	Roundtail chub
G. seminuda (E)	Virgin chub
Lepidomeda albivallis (E)	White River spinedace
L. altivelis (X)	Pahranagat spinedace
L. mollispinis mollispinis (T)	Virgin spinedace
L.m. pratensis (E)	Big Spring spinedace
L. vittata (T)	Little Colorado spinedace
Meda fulgida (T)	Spikedace
Moapa coriacea (E)	Moapa dace
Plagopterus argentissimus (E)	Woundfin
Ptychocheilus lucius (E)	Colorado squawfish
Rhinichthys cobitis (T)	Loach minnow
R. deaconi (X)	Las Vegas dace
R. osculus osculus	Speckled dace
R. osculus ssp. (SC)	"Preston speckled dace"
R. osculus ssp. (SC)	"Meadow Valleys speckled dace"
R. osculus ssp. (SC)	"White River speckled dace"
R.o. thermalis (SC)	Kendall Warm Springs dace
R.o. velifer (SC)	Pahranagat speckled dace
Family Catostomidae	
Catostomus clarki clarki	Desert sucker
C.c. intermedius (E)	White River sucker
C. clarki ssp. (E)	"Meadow Valley sucker"
C. discobolus discobolus	Bluehead sucker
C.d. yarrowi (SC)	Zuni sucker
C. insignis	Sonora sucker
C. latipinnis (SC)	Flannelmouth sucker
C. platyrhynchus	Mountain sucker
C. sp. (SC)	"Little Colorado sucker"
Xyrauchen texanus (E)	Razorback sucker
Family Salmonidae	
Oncorhynchus apache (T)	Apache trout
O. clarki pleuriticus (SC)	Colorado cutthroat trout
O. gilae (T)	Gila trout
*Prosopium williamsoni**	Mountain whitefish
Family Goodeidae	
Crenichthys baileyi albivallis (E)	Preston springfish
C.b. baileyi (E)	White River springfish
C.b. grandis (E)	Hiko springfish
C.b. moapae (T)	Moapa springfish
C.b. thermophilus (T)	Moorman springfish
C. nevadae (T)	Railroad Valley springfish
Family Cyprinodontidae	
Cyprinodon macularius macularius (E)	Desert pupfish
C. sp. (X)	"Monkey Springs pupfish"
Family Poeciliidae	
Poeciliopsis occidentalis (SC)	Gila topminnow
Family Cottidae	
*Cottus bairdi**	Mottled sculpin
*C. beldingi**	Paiute sculpin
Family Mugilidae	
Mugil cephalus (m)	Striped mullet

special concern. On the other hand, a few small tributaries, by virtue of their isolation, rare intermittent flows in lower reaches, and physical barriers, have been spared significant alterations or invasions by non-native species and retain an intact native fauna (e.g., Redfield Canyon, Arizona, Table 2).

In the larger rivers of the Upper Basin, such as the Green, lower Yampa, and most of the upper Colorado, most native taxa are extant but one or two (razorback sucker [*Xyrauchen texanus*], possibly bonytail [*Gila elegans*]), are represented by very rare individuals that may not be reproducing; all native fishes are greatly exceeded in numbers and kind by non-native taxa. In smaller tributaries of that region, varied numbers of native taxa persist; in the worst affected streams (e.g., most Green River tributaries in Utah), most taxa have been replaced by non-native taxa (author's observation).

Case studies of two endangered Colorado River species, which are hallmarks to conservationists, further elucidate patterns of decline among these fishes. They are large, long-lived (20-50 years) species that inhabit larger streams. The Colorado squawfish (*Ptychocheilus lucius*) is a highly migratory (Tyus 1990) predatory minnow. Perhaps because of fragmentation or impediment of migratory routes, its original extensive range has been reduced by roughly two-thirds, and it is uncommon where it remains. The last confirmed report in the Gila River was in 1950 and the last in the Lower Basin in 1975 (Miller 1961; Minckley 1973; Maddux et al. 1993).

The fourth species, the humpback chub (*Gila cypha*), is strictly a denizen of turbulent canyon reaches so difficult to sample that it was not discovered until 1946; it ranged from Boulder Canyon on the lower Colorado throughout canyon reaches of the Upper Basin well into Wyoming. Today, it occurs only in Grand Canyon, Arizona (Maddux et al. 1993), near the confluence of the Colorado and Little Colorado rivers, and in five Upper Basin canyon areas (rare in three), although the genetic "purity" of the Upper Basin populations is questioned. Recovery plans are in place for these fish as well as the bonytail and the razorback sucker. These fish are all easily propagated in captivity. It is otherwise difficult to find anything positive in the history of these or other Colorado Basin native fishes over the past several decades.

Non-native Species

Concomitant with the pervasive physical alteration of the Colorado River ecosystem has been both purposeful and accidental introductions of at least 72 non-native fish taxa (Maddux et al. 1993), including those indigenous to other North American basins and more exotic species. Alterations of the ecosystem's natural characteristics have apparently tipped the ecologic balance in favor of many of the non-native species that now vastly outnumber natives in numbers of species (Table 2), population density, and often biomass at most localities. There is evidence that some, such as the extremely pervasive red shiner (*Cyprinella lutrensis*), displace native taxa (Douglas et al. 1994) while others, such as channel and flathead catfish (*Ictalurus punctatus* and *Pylodictis olivaris*), are known predators on larval and juvenile native species (several references in Maddux et al. 1993). The introduced white sucker (*Catostomus commersoni*) is hybridizing extensively with native suckers throughout much of the Upper Basin (author's observation), possibly threatening the genetic integrity of those taxa. These and other interactions between non-native and native taxa may have significant negative effects on native fishes. The dominance held by non-native fishes may be symptomatic of the overall degree of alteration of the Colorado River ecosystem and could potentially confound future studies of biodiversity.

Table 2. Overall and relative abundance of native and non-native fishes from various localities in the Colorado River Basin. Numbers for 1800's represent original complements of native taxa. For subsequent years, total abundance is followed by ratio of non-native to native taxa in parentheses. Sources: Miller 1961; Taba et al. 1965; Vanicek et al. 1970; Stalnaker and Holden 1973; Cross 1975; Holden and Stalnaker 1975a,b; Suttkus et al. 1976; Carlson et al. 1979; Miller et al. 1982; Valdez et al. 1982; Valdez 1984,1990; Wick et al. 1985; Platania and Bestgen 1988; Griffith and Tiersch 1989.

Locality	Survey date				
	1800's	1940's	ca. 1965	ca. 1975	ca.1985
Yampa-Green River area, CO-UT	10	-	21(12/9)	22(13/9)	24(15/9)
White River, CO-UT	9	-	-	13(7/6)	12(5/7)
Dolores River, CO	9	-	-	11(7/4)	16(12/4)
Colorado River, Lake Powell,UT to Gunnison River, CO	10	-	15(9/6)	29(19/10)	31(23/8)
San Juan River, NM	9	-	-	-	18(12/6)
Colorado River, Grand Canyon, AZ	10	-	-	19(15/4)	-
Virgin River, AZ-UT	6	-	-	19(13/6)	-
Lower Colorado River, AZ-CA	10	12(9/3)	11(11/0)	-	-
Salt River near Phoenix, AZ	14	9(6/3)	22(20/2)	-	-
Redfield Canyon, San Pedro River system, AZ	5	-	-	-	5(0/5)

Altered Species Diversity and Biodiversity Studies

While native taxa have declined, there have actually been two- to threefold increases in the number of species at most localities in the Colorado Basin because of the success of introduced taxa (Table 2). If future biodiversity monitoring is to truly gauge positive and negative shifts in the health of the Colorado River ecosystems, then an accurate baseline is necessary. A baseline describing unaltered native fauna might be an ideal but unattainable goal. That line could be approached, however, by divesting faunal lists of all non-native taxa and determining, as much as possible, the true extent of diversity of that which remains. In fish, it is practical to do so to the level of distinctive populations through studies of genetic variability. With luck, it is even possible to include extirpated populations through DNA studies of museum specimens if historic material is available.

Once a baseline is determined, researchers and managers can know where to try to "hold the line" in maintaining diversity through management and protection. Of course, on a systemwide basis, the baseline diversity of a pristine system can never be reattained because genetically unique populations have already been lost. On a more local basis, however, positive increments and recovery of the habitat are indicated if monitoring reveals increased diversity resulting from the successful reestablishment of taxa which were conserved in other, less altered, portions of the system.

For monitoring purposes, when non-native species are added to biodiversity determinations, we must carefully tease out the cause of shifts toward or from the "desired baseline" which, in the case of the Colorado River, is probably a value far less than the present overall number of species. Thus, "desirable" outcomes may be indicated by overall decreases in diversity caused by the disappearance of non-native taxa as an indicator of habitat "healing," but not so by the loss of native taxa. Conversely, actual increases may yet be positive if caused by reestablishment of native taxa, but may be an indicator of further degradation if caused by success of additional non-natives. Realistically, monitoring will have to include, in addition to determinations of diversity, attention to shifts in dominance among native and non-native species, which can be indicative of both positive and negative trends.

References

Carlson, C.A., and R. Muth. 1989. The Colorado River: lifeline of the West. Pages 220-239 *in* D.P. Dodge, ed. Proceedings of the International Large River Symposium. Canadian Special Publ. in Fishery and Aquatic Sciences 106.

Carlson, C.A., C.G. Prewitt, D.E. Snyder, and E.J. Wick. 1979. Fishes and macroinvertebrates of the White and Yampa rivers, Colorado. Bureau of Land Management, Biological Sciences Series 1. Denver, CO. 276 pp.

Cross, J.N. 1975. Ecological distribution of the fishes of the Virgin River. M.S. thesis, University of Nevada, Las Vegas. 187 pp.

Douglas, M.E., P.C. Marsh, and W.L. Minckley. 1994. Indigenous fishes of western North America and the hypothesis of competitive displacement: *Meda fulgida* (Cyprinidae) as a case study. Copeia 1994:9-19.

Griffith, J.S., and T.R. Tiersch. 1989. Ecology of fishes in Redfield Canyon, Arizona, with emphasis on *Gila robusta intermedia*. Southwestern Naturalist 34:131-164.

Holden, P.B., and C.B. Stalnaker. 1975a. Distribution of fishes in the Dolores and Yampa river systems of the Upper Colorado Basin. Southwestern Naturalist 19:403-412.

Holden, P.B., and C.B. Stalnaker. 1975b. Distribution and abundance of mainstream fishes of the Middle and Upper Colorado River basins, 1967-1973. Transactions of the American Fisheries Society 104:217-231.

Maddux, H.R., L.A. Fitzpatrick, and W.R. Noonan. 1993. Colorado River endangered fishes critical habitat draft biological support document. U.S. Fish and Wildlife Service, Salt Lake City, UT. 225 pp.

Miller, R.R. 1961. Man and the changing fish fauna of the American Southwest. Papers of the Michigan Academy of Science, Arts, and Letters 46:365-404.

Miller, W.H., D. Archer, H.M. Tyus, and R.M. McNatt. 1982. Yampa River fishes study, final report. Colorado River Fishery Project. U.S. Fish and Wildlife Service, Salt Lake City, UT. 79 pp.

Minckley, W.L. 1973. Fishes of Arizona. Arizona Game and Fish Department, Phoenix. 293 pp.

Minckley, W.L., and J.E. Deacon. 1968. Southwestern fishes and the enigma of endangered species. Science 159:1424-1432.

Minckley, W.L., and J.E. Deacon, eds. 1991. Battle against extinction: native fish management in the American West. University of Arizona Press, Tucson. 517 pp.

Platania, S.P., and K.R. Bestgen. 1988. An interim report on the fishes of the lower San Juan River, New Mexico, 1987. New Mexico Department of Game and Fish, Santa Fe. 61 pp.

Stalnaker, C.B., and P.B. Holden. 1973. Changes in native fish distribution in the Green River system, Utah-Colorado. Utah Academy of Science, Arts, Letters Proceedings 50:25-32.

Suttkus, R.D., G.H. Clemmer, C. Jones, and C.R. Shoop. 1976. Survey of fishes, mammals and herpetofauna of the Colorado River and adjacent riparian areas of the Grand Canyon National Park. Grand Canyon National Park Colorado River Research Serial Contribution 34. 48 pp.

Taba, S.S., J.R. Murphy, and H.R. Frost. 1965. Notes on the fishes of the Colorado River near Moab, Utah. Utah Academy of Science, Arts, Letters Proceedings 42:280-283.

Tyus, H.M. 1990. Potamodromy and reproduction of Colorado River squawfish (*Ptychocheilus lucius*). Transactions of the American Fisheries Society 119:1035-1047.

Valdez, R. 1984. A survey of fish collections in the White River, Utah. Report for White River Shale Corporation, Ecosystem Research Institute, Logan, UT. 39 pp.

Valdez, R. 1990. The endangered fish of Cataract Canyon. Bio/West Report 134-3 for Bureau of Reclamation, Salt Lake City, UT. 94 pp.

Valdez, R., P. Mangan, M. McInerny, and R.P. Smith. 1982. Tributary report: fishery investigations of the Gunnison and Dolores rivers. Pages 322-365 *in* Colorado River Fisheries Project Final Report of Field Investigations, Part 2. U.S. Fish and Wildlife Service, Salt Lake City, UT.

Vanicek, C.D., R.H. Kramer, and D.R. Franklin. 1970. Distribution of Green River fishes in Utah and Colorado following closure of Flaming Gorge Dam. Southwestern Naturalist 14:297-315.

Wick, E.J., J.A. Hawkins, and C.A. Carlson. 1985. Colorado squawfish and humpback chub habitat monitoring, 1981-1982. Colorado State University Larval Fish Lab and Colorado Division of Wildlife Rep. SE 3-6. 41 pp.

Williams, J.E., J.E. Johnson, D.A. Hendrickson, S. Contreras-Balderas, J.D. Williams, M. Navarro-Mendoza, D.E. McAllister, and J.E. Deacon. 1989. Fishes of North America endangered, threatened, or of special concern: 1989. Fisheries 14:2-80.

For further information:

Wayne C. Starnes
Smithsonian Institution
Division of Fishes MRC 159
Washington, DC 20560

Cutthroat Trout in Glacier National Park, Montana

by
Leo F. Marnell
National Biological Service

The indigenous fishery of Glacier National Park has been radically altered from its pristine condition during the past half-century through introductions of non-native fishes and the entry of non-native species from waters outside the park. These introductions have adversely affected the native westslope cutthroat trout (*Oncorhynchus clarki lewisi*; Fig. 1) throughout much of its park range.

The effects of non-native fishes on indigenous fisheries have been reviewed by Taylor et al. (1984), Marnell (1986), and Moyle et al. (1986). Effects of fish introductions in Glacier National Park include establishment of non-native trout populations in historically fishless waters, genetic contamination (i.e., hybridization) of some native westslope cutthroat trout stocks, and ecological interferences with various life-history stages of native trout.

Research conducted in the park during the 1980's addressed the genetic effects of fish introductions on native trout. Of 47 lakes known or suspected to contain cutthroat trout or trout hybrids, 32 lakes contained viable populations of cutthroat trout, rainbow trout (*O. mykiss*), or hybrids. Trout introduced in the other waters were evidently unable to sustain themselves through natural reproduction.

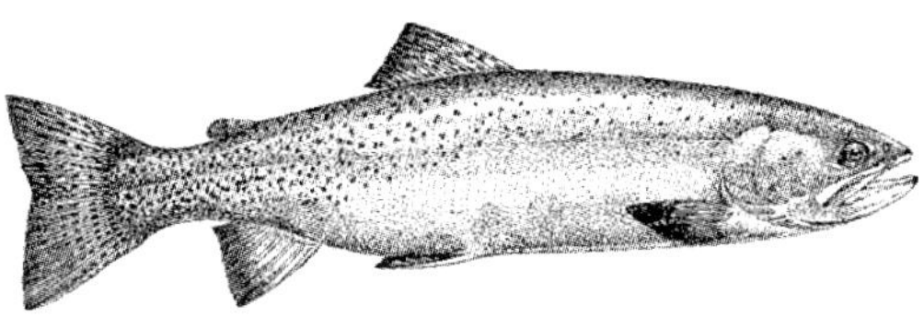

Fig. 1. Westslope cutthroat trout (*Oncorhynchus clarki lewisi*).

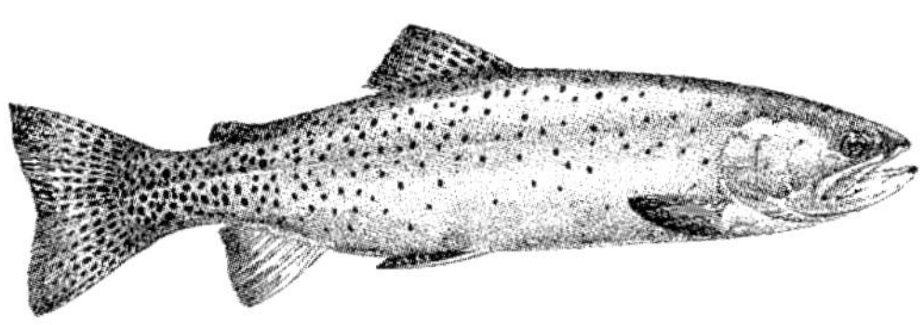

Fig. 2. Yellowstone cutthroat trout (*Oncorhynchus clarki bouvieri*).

About 30 trout sampled from each lake underwent laboratory genetic analyses. Close agreement of the results from two analytical procedures yielded a high degree of confidence in the conclusions (Marnell et al. 1987). Genetic classifications in Tables 1 and 2 reflect the combined results of the analyses.

Fourteen pure strain populations of westslope cutthroat trout persist in 15 lakes (i.e., some interconnected lakes contain a single trout population) in the North and Middle Fork drainages of the Flathead River; the species was historically present in these waters (labeled as "stable" populations in Table 1).

Pure strain native trout also inhabit four other Middle Fork lakes (i.e., Avalanche, Snyder, and Upper and Lower Howe lakes), but it is unclear whether they are indigenous or were transplanted from other park waters. Recent findings from sediment paleolimnology studies suggest that trout have been present in at least one of these lakes for more than 300 years (D. Verschuren, University of Minnesota, and author, unpublished data). Hence, trout populations in these four lakes are tentatively classified as indigenous (Table 1).

Introduced populations of Yellowstone cutthroat trout (*O. clarki bouvieri*; Fig. 2) and trout hybrids including cutthroat-rainbow trout (*O. clarki* spp. x *O. mykiss*) occur in 13 lakes distributed among the three continental drainages

Table 1. Status and trends of cutthroat trout and their hybrids in the North and Middle Fork, Flathead River drainages of Glacier National Park, Montana.

Lake	Area (ha)	Trout classification*	Population status**
North Fork, Flathead R.			
Akokala	9	WCT	Stable
Arrow	23	WCT x YCT	Hybrid
Bowman	691	WCT	Unstable
Camas	8	YCT	Non-native
Cerulean	20	WCT	Stable
Evangeline	28	YCT	Non-native
Grace	32	WCT x YCT	Hybrid
Kintla	688	WCT	Unstable
Logging	444	WCT	Unstable
Quartz	349	WCT	Stable
Lower Quartz	67	WCT	Stable
Middle Quartz	19	WCT	Stable
Trout	86	WCT	Stable
Middle Fork, Flathead R.			
Avalanche	23	WCT	Stable
Fish	3	WCT x YCT	Hybrid
Harrison	101	WCT	Unknown
Hidden	110	YCT	Non-native
Lincoln	14	WCT	Stable
Lower Howe	12	WCT	Stable
Lower Isabel	17	WCT	Stable
McDonald	2,760	WCT	Unstable
Ole	2	WCT	Stable
Snyder	2	WCT	Stable
Upper Howe	3	WCT	Stable
Upper Isabel	6	WCT	Stable

* WCT — pure strain westslope cutthroat trout.
YCT — the introduced Yellowstone cutthroat trout.
x — two or more species have hybridized.
**Stable — native population exists in a pristine environment.
Unstable — declining condition resulting from presence of competing non-native species.
Hybrid and non-native populations — classified without regard to population condition.

Table 2. Status and trends of non-native and hybrid trout populations in the South Saskatchewan and Missouri river drainages of Glacier National Park, Montana.

Lake	Area (ha)	Trout classification*	Population status**
South Saskatchewan River			
Lower Slide	15	YCT x RBT	Hybrid
Otokomi	9	YCT x RBT	Hybrid
Red Eagle	55	YCT x WCT x RBT	Hybrid
Upper Slide	5	YCT x RBT	Hybrid
Upper Missouri River Drainage			
Katoya	4	YCT	Non-native
Morning Star	4	YCT	Non-native
Old Man	17	YCT	Non-native

*YCT— introduced Yellowstone cutthroat trout.
x — two or more species have hybridized.
RBT— rainbow trout.
WCT — westslope cutthroat trout.
**Hybrid and non-native populations are classified without regard to population condition.

that form their headwaters in Glacier National Park (Tables 1 and 2). Native cutthroat trout were not found east of the Continental Divide in the Missouri River or South Saskatchewan River drainages within the park.

In addition to genetic concerns, ecological disturbances associated with the presence of introduced fishes have compromised the native westslope cutthroat fishery. Fish are no longer stocked in park waters; however, several waters, including some that contain undisturbed native fisheries, remain vulnerable to invasion by non-native migratory species. Introduced kokanee salmon (*O. nerka*), a specialized planktivore, are believed to be competing with juvenile stages of native trout in some waters, especially during periods of winter ice cover when plankton may be limited. Predation by introduced lake trout (*Salvelinus namaycush*) has also been implicated in the decline of native cutthroat trout in several large glacial lakes in the North and Middle Fork drainages (Marnell 1988). Native cutthroat trout have been compromised by fish introductions and invasions throughout about 84% of their historic range in Glacier National Park (Marnell 1988).

Although native cutthroat trout have been adversely affected throughout a large portion of their park range, the species has not been lost from any water where it was historically present. Glacier National Park remains one of the last strongholds of genetically pure strains of lacustrine (i.e., lake-adapted) westslope cutthroat trout. This fact could have important implications for reestablishment of this unique subspecies throughout the central Rocky Mountains, where this trout has disappeared from most of its original range.

References

Marnell, L.F. 1986. Impacts of hatchery stocks on wild fish populations. Pages 339-347 *in* R.H. Stroud, ed. Fish culture in fisheries management. American Fisheries Society, Bethesda, MD.

Marnell, L.F. 1988. Status of the westslope cutthroat trout in Glacier National Park, Montana. American Fisheries Society Symposium 4:61-70. Bethesda, MD.

Marnell, L.F., R.J. Behnke, and F.W. Allendorf. 1987. Genetic identification of cutthroat trout (*Salmo clarki*) in Glacier National Park, Montana. Canadian Journal of Fisheries and Aquatic Sciences 44:1830-1839.

Moyle, P.B., H. Li, and B.A. Barton. 1986. The Frankenstein effect: impact of introduced fishes in North America. Pages 415-426 *in* R.H. Stroud, ed. Fish culture in fisheries management. American Fisheries Society, Bethesda, MD.

Taylor, J.N., W.R. Courtenay, Jr., and J.A. McCann. 1984. Known impacts of exotic fishes in the continental United States. Pages 322-373 *in* W.R. Courtenay, Jr., and J.R. Stauffer, eds. Distribution, biology, and management of exotic fishes. Johns Hopkins University Press, Baltimore, MD.

For further information:

Leo F. Marnell
National Biological Service
Midcontinent Ecological Science Center
Glacier Field Station
Glacier National Park
West Glacier, MT 59936

Columbia River Basin White Sturgeon

by
Allen I. Miller
Timothy D. Counihan
Michael J. Parsley
Lance G. Beckman
National Biological Service

White sturgeon (*Acipenser transmontanus*), the largest freshwater fish in North America, live along the west coast from the Aleutian Islands to central California (Scott and Crossman 1973). Genetically similar reproducing populations inhabit three major river basins: Sacramento-San Joaquin, Columbia, and Fraser. The greatest number of white sturgeon are in the Columbia River Basin.

Historically, white sturgeon inhabited the Columbia River from the mouth upstream into Canada, the Snake River upstream to Shoshone Falls, and the Kootenai River upstream to Kootenai Falls (Scott and Crossman 1973; Figure). White sturgeon also used the extreme lower reaches of other tributaries, but not extensively. Current populations in the Columbia River Basin can be divided into three groups: fish below the lowest dam, with access to the ocean (the lower Columbia River); fish isolated (functionally but not genetically) between dams; and fish in several large tributaries.

The Columbia River has supported important commercial, treaty, and recreational white sturgeon fisheries. A commercial fishery that began in the 1880's peaked in 1892 when 2.5 million kg (5.5 million lb) were harvested (Craig and Hacker 1940). By 1899 the population had been severely depleted, and annual harvest was very low until the early 1940's, but the population recovered enough by the late 1940's that the commercial fishery expanded. A 1.8-m (6-ft) maximum size restriction was enacted to prevent another population collapse. Total harvest doubled in the 1970's and again in the 1980's because of increased treaty and recreational fisheries. From 1983 to 1994, 15 substantial regulatory changes were implemented on the mainstem Columbia River downstream from McNary Dam as a result of increased fishing. Columbia River white sturgeon are still economically important. Recreational, commercial, and treaty fisheries in the Columbia River downstream from McNary Dam were valued at $10.1 million in 1992 (Tracy 1993).

Several factors make white sturgeon relatively vulnerable to overexploitation and changes in their environment. The fish may live more than 100 years (Rieman and Beamesderfer 1990), and overexpolitation is well documented for long-lived, slow-growing fish (Ricker 1963). Female white sturgeon are slow to reach sexual maturity; in the Snake River they mature at age 15-32 (Cochnauer 1981). Mature females in the Columbia Basin only spawn every 2-11 years (Stockley 1981; Cochnauer 1983; Welch and Beamesderfer 1993). Sustainable harvest levels vary for impoundments in the Columbia River. Several impoundments are managed as groups, making overexploitation more likely in impoundments with low sustainable harvest levels.

White sturgeon populations in free-flowing and inundated reaches of the Columbia River Basin have been negatively affected by the abundant hydropower dams in most of the mainstem Columbia and Snake rivers (Rieman and Beamesderfer 1990). These dams have altered the magnitude and timing of discharge, water depths, velocities, temperatures, turbidities, and substrates, and have restricted sturgeon movement within the basin. Sturgeons in other river basins have declined in response to dam-induced habitat alterations (Artyukhin et al. 1978).

Mainstem Columbia River

Abundance and growth of white sturgeon are greatest in the lower Columbia River (Figure). These fish use estuarine and marine habitats as well as riverine habitats, allowing them to feed on anadromous prey fishes (those fishes traveling upriver from the sea to spawn; Tracy 1993). Although the lower Columbia River population may be the only one in this basin that is abundant and stable, even it is at some risk of collapse (Rieman and Beamesderfer 1990). Of the 11 populations isolated between dams upstream, white sturgeon are known to be relatively abundant in only 3 (Figure). White sturgeon densities in three of the remaining eight populations are much lower than in the abundant populations. Data are sparse for the remaining five populations, although Zinicola and Hoines (1988) reported that in 1988 fewer than 10 white sturgeon were harvested in each of four of these impoundments and only 34 in another.

Although the lower Columbia River population probably declined during the 1980's, adoption of more restrictive harvest regulations appears to have stabilized the population (Tracy 1993). Successful spawning occurs each year in this reach (McCabe and Tracy 1993). Catch-per-unit-effort of most size groups in the three populations for which data are available declined considerably from 1987 to 1991; fisheries there have collapsed and the populations are at risk of collapse (Beamesderfer and Rien 1993). Recruitment in some populations appears limited to years with high river discharges in spring (Miller and Beckman 1993). Although most of the mainstem populations appear unstable, their genetic similarity to the stable lower Columbia River population has excluded them from consideration for listing under the federal Endangered Species Act.

Overexploitation and poaching have reduced population size (Beamesderfer and Rien 1993), and impoundments and altered hydrographs caused by development of the hydropower system have altered critical spawning habitat (Parsley et al. 1993). Because the factors identified as causing declines in other white sturgeon populations are present to varying degrees in each of the other eight upstream impoundments, these populations are likely declining as well.

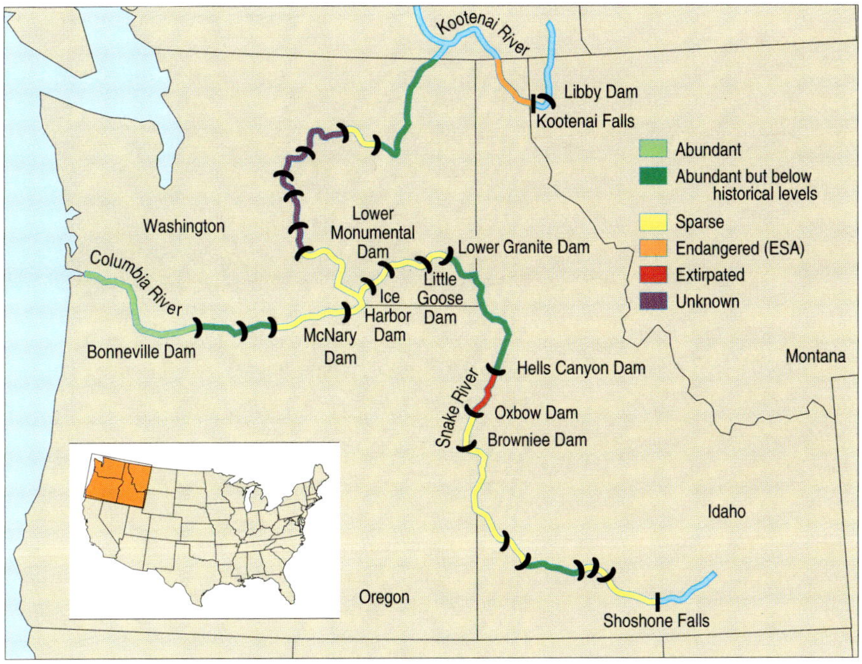

Figure. Distribution and status of white sturgeon in the U.S. portion of the Columbia River Basin.

Kootenai River

Current research on white sturgeon in the Kootenai River indicates that this population is unstable and declining. The U.S. Fish and Wildlife Service listed the Kootenai River population as endangered in 1994.

This population has declined to fewer than 1,000 fish, about 80% of which are more than 20 years old. Apperson and Anders (1990) concluded that virtually no recruitment has occurred since 1974, soon after Libby Dam began regulating flows, thereby altering historical discharge patterns of the river. This altering of discharge patterns is thought to be a major causal factor limiting recruitment into this unique sturgeon population. Research on the Kootenai River is examining the effects of increased discharge on the spawning behavior of white sturgeon. During 1993 increased discharges resulted in the collection of only three white sturgeon eggs despite intensive efforts to collect early lifestages of white sturgeon (Marcuson 1994).

Fishing for white sturgeon in the Kootenai River has been regulated in Idaho since 1944, in Montana since 1957, and in British Columbia since 1952, indicating that overharvesting may have been affecting population size. Fishing for white sturgeon has been closed in Montana since 1979, and catch and release angling

Courtesy Oregon Historical Society

Fifteen-hundred-pound white sturgeon caught near Payette, Idaho, circa 1911.

restrictions have been in place since 1984 in Idaho and 1990 in British Columbia.

Snake River

The Snake River has 12 dams from its mouth upstream to Shoshone Falls in Idaho. White sturgeon are believed to exist in small numbers in the lower three pools on the Snake River formed by Ice Harbor, Lower Monumental, and Little Goose dams (Zinicola and Hoines 1988). Of the nine impoundments upstream from Little Goose Dam, white sturgeon are relatively abundant in two, present at low numbers in six, and are absent in another (PSMFC 1992).

Although little is known about the early life history and spawning habitat requirements of white sturgeon in the Snake River, the construction and operation of the river's dams are likely to have the same effects as the impoundments on the Columbia and Kootenai rivers. White sturgeon appear more abundant in regions of the Snake River where free-flowing river habitat exists (PSMFC 1992), such as between Lower Granite and Hells Canyon dams where 76% of the river is free-flowing. Conversely, white sturgeon are not present in the impoundments created by Hells Canyon Dam and not abundant in the impoundment created by Oxbow Dam, which constitute two continuous slackwater regions (Welsh and Reid 1971).

While free-flowing sections of the Snake River exist in varying proportions between the dams, impoundments upstream of these sections influence both water temperature and the annual discharge pattern. At least 28 sturgeon died during July 1990 because of low dissolved-oxygen levels in Brownlee Pool (PSMFC 1992). Sturgeon production in the Snake River also appears limited by dewatering from irrigation diversions (Lukens 1981) and small spawning populations (Cochnauer et al. 1985).

Harvest of white sturgeon from the Snake River has had a definite negative impact on these populations, but the magnitude of the effect is unknown. Commercial fishing was permitted on the Snake River until 1943; then increasingly restrictive regulations were implemented from 1944 to 1969. In 1970 catch and release regulations were imposed on the entire river. A recommendation has been made that 3 of the 12 reaches of the Snake River discussed in this article be completely closed to fishing (Cochnauer et al. 1985).

Summary

Habitat changes (e.g., decreased discharges resulting in decreased spawning habitat) caused by development of the hydropower system have contributed to white sturgeon population declines in the Columbia River Basin; spawning habitat has been particularly affected by dams.

Overharvest of white sturgeon has caused population declines in several Columbia River Basin populations, both historically and in the past two decades. Recent management changes have helped alleviate overharvest in much of the Columbia River Basin, but refinement of management strategies is still needed in some areas.

The status of the 25 Columbia River Basin white sturgeon populations varies considerably: 1 is stable and abundant; 5 are relatively abundant, but probably at lower levels than in the past; 12 are sparse and many are declining; 5 have unknown status but creel data suggest they are sparse; 1 is sparse, declining, and listed under the Endangered Species Act; and white sturgeon have probably been extirpated from another. Conditions that have contributed to stock declines in other white sturgeon populations are present in populations whose status is unknown, suggesting that populations with unknown status may also be declining.

References

Apperson, K.A., and P.J. Anders. 1990. Kootenai River white sturgeon investigations and experimental culture, annual progress report 1989. Bonneville Power Administration, Division of Fish and Wildlife Project 88-65. 50 pp.

Artyukhin, Y.K., A.D. Sukhoparova, and L.G. Fimukhira. 1978. The gonads of sturgeon, *Acipenser güldenstädti*, in the littoral zone below the dam of the Volograd water engineering system. Journal of Ichthyology 18:912-923.

Beamesderfer, R.C., and T.A. Rien. 1993. Dynamics and potential production of white sturgeon populations in three lower Columbia River reservoirs. Pages 175-206 *in* R.C. Beamesderfer and A.A. Nigro, eds. Status and habitat requirements of the white sturgeon populations in the Columbia River downstream from McNary Dam. Vol 1. Final Report to Bonneville Power Administration, Portland, OR.

Cochnauer, T.G. 1981. Survey status of white sturgeon populations in the Snake River, Bliss Dam to C.J. Strike Dam. Idaho Department of Fish and Game, River and Stream Investigations, Job Performance Rep., Project F-73-R-2, Job I-b, Boise. 25 pp.

Cochnauer, T.G. 1983. Abundance, distribution, growth, and management of white sturgeon *Acipenser transmontanus* in the middle Snake River, Idaho. Ph.D. dissertation, University of Idaho, Moscow. 52 pp.

Cochnauer, T.G., J.R. Lukens, and F.E. Partridge. 1985. Status of white sturgeon, *Acipenser transmontanus*, in Idaho. Pages 127-133 *in* F.P. Binkowski and S.I. Doroshov, eds. North American sturgeons: biology and aquaculture potential. Dr. W. Junk, Dordrecht, The Netherlands.

Craig, J.A., and R.L. Hacker. 1940. The history and development of the fisheries of the Columbia River. U.S. Bureau of Fisheries Bull. 49(32):132-216.

Lukens, J.R. 1981. Snake River sturgeon investigations: Bliss Dam upstream to Shoshone Falls. Idaho Department of Fish and Game, Boise. 24 pp.

Marcuson, P. 1994. Kootenai River white sturgeon investigations, annual progress report 1993. Bonneville Power Administration, Division of Fish and Wildlife Project. 88-65. 67 pp.

McCabe, G.T., Jr., and C.A. Tracy. 1993. Spawning characteristics and early life history of white sturgeon *Acipenser transmontanus* in the Lower Columbia River. Pages 19-49 *in* R.C. Beamesderfer and A.A. Nigro, eds. Status and habitat requirements of the white sturgeon populations in the Columbia River downstream from McNary Dam. Vol. 1. Final Report to Bonneville Power Administration, Portland, OR.

Miller, A.I., and L.G. Beckman. 1993. A recruitment index for white sturgeon in the Columbia River downstream from McNary Dam. Pages 141-166 *in* R.C. Beamesderfer and A.A. Nigro, eds. Status and habitat requirements of the white sturgeon populations in the Columbia River downstream from McNary Dam. Vol. 2. Final Report to Bonneville Power Administration, Portland, OR.

PSMFC. 1992. White sturgeon management framework plan. Pacific States Marine Fisheries Commission, Portland, OR. 201 pp.

Parsley, M.J., L.G. Beckman, and G.T. McCabe, Jr. 1993. Spawning and rearing habitat use by white sturgeons in the Columbia River downstream from McNary Dam. Transactions of the American Fisheries Society 122(2):217-227.

Ricker, W.E. 1963. Big effects from small causes: two examples from fish population dynamics. Journal of the Fisheries Research Board of Canada 20:257-264.

Rieman, B.E., and R.C. Beamesderfer. 1990. White sturgeon in the lower Columbia River: is the stock overexploited? North American Journal of Fisheries Management 10:388-396.

Scott, W.B., and E.J. Crossman. 1973. Freshwater fishes of Canada. Fisheries Research Board of Canada Bull. 184, Ottawa. 966 pp.

Stockley, C. 1981. Columbia River sturgeon. Washington Department of Fisheries Prog. Rep. 150, Olympia. 28 pp.

Tracy, C.A. 1993. Status of white sturgeon resources in the mainstem Columbia River. Final Report. Dingell/Johnson-Wallop/Breaux Project F-77-R, Washington Department of Fisheries, Battleground. 16 pp.

Welch, D.W., and R.C. Beamesderfer. 1993. Maturation of female white sturgeon in lower Columbia River impoundments. Pages 89-108 *in* R.C. Beamesderfer and A.A. Nigro, eds. Status and habitat requirements of the white sturgeon populations in the Columbia River downstream from McNary Dam. Vol. 2. Final Report to Bonneville Power Administration, Portland, OR.

Welsh, T.L., and W.W. Reid. 1971. Hells Canyon fisheries investigations: 1970 annual report. Idaho Department of Fish and Game, Boise. 56 pp.

Zinicola, T., Jr., and L.J. Hoines. 1988. Washington state sport catch report. Washington Department of Fisheries, Olympia. 83 pp.

For further information:

Allen I. Miller
National Biological Service
Northwest Biological Science Center
Columbia River Research Laboratory
M.P. 5.48L Cook-Underwood Rd.
Cook, WA 98605

Courtesy J.R. Bruschwein

Invertebrates

by
Science Editor
William T. Mason, Jr.
National Biological Service
Southeastern Biological Science Center
7920 NW 71st St.
Gainesville, FL 32653

Overview

Invertebrates are impressive in abundance and diversity, living on land and in water and air. Many species are borne to distant places on air and water currents, and via modern transportation.

Of the millions of species of animals worldwide, about 90% are invertebrates, that is, animals without backbones (Opler, Powell, this section). The arthropods, or jointed-leg invertebrates such as beetles, account for 75% of this total. More than 90,000 described insect species inhabit North America (Hodges, Powell, this section); the Lepidoptera (butterflies and moths) alone account for about 11,500 of these (Powell, this section).

Within an acre of land and water, hundreds of different invertebrates form an ecological web of builders, gatherers, collectors, predators, and grazers, all interacting with each other and each a necessary component of a healthy ecosystem. The large macroscopic invertebrates—like bees, beetles, butterflies, grasshoppers, snails, and earthworms—are well known, but other invertebrates are almost invisible because they are extremely tiny or camouflaged for protection. We have just begun to understand the ecology of some commercially important species, but we understand very little about the behavior, communication, and function of many other invertebrates within various ecosystems.

Each individual invertebrate is a highly complex, specialized animal. Some molt (change or metamorphose) into several distinct life stages. For example, some insects transform from egg to larva, then to pupa, and finally emerge as a terrestrial winged adult. Some aquatic invertebrates do not have pupal stages, and the larvae (nymphs or naiads) grow progressively larger by molts. Earthworms bear cocoons that each contain about six miniature juveniles; they also reproduce by fragmentation (architomy).

Changes to the environment can disrupt basic interactions of invertebrate species, thereby affecting other organisms in the food chain. Disruptions of natural food cycles may cause drastic changes in the community structure and ecological web of life. This is especially true of the fauna that dwell in fragile ecosystems like caves and springs (Webb, *see* box). Eventually even humans are affected by changes to food webs and destruction of beneficial habitat for wildlife.

Most invertebrates can survive extreme natural events like severe storms, blizzards, and flooding. When confronted by unnatural disturbances, however, such as excessive siltation from urban and highway developments, eutrophication (excessive nutrients) by runoff from agricultural lands, and contamination of aquatic habitats by toxic substances and acids, invertebrate populations can be severely damaged. Airborne toxicants like acid rain are harmful to the long-term well-being of insects. If disturbances are sufficient, natural fauna may be extirpated (removed or lost) and replaced by more tolerant kinds. This "unbalanced" situation usually results in a population explosion of a few species (e.g., Tubificidae: Oligochaeta and red-blooded Chironomidae: Diptera). Such a biological reaction makes these aquatic invertebrates excellent bioindicators of overall environmental conditions (Bartsch and Ingram 1959). The use of aquatic invertebrates for bioassay (testing the toxicity of substances to "standard" test organisms) has greatly helped to minimize adverse effects of contaminants on aquatic life.

Butterflies and moths are particularly susceptible to environmental disturbances (Opler, this section), although their responses to mild disturbances and changes may be slow, lasting decades (Otte, Swengel, and Swengel and Swengel, this section). McCabe (this section) concludes that some of the flux in biodiversity is likely due to the "edge effect" at the interface from one habitat to another, and not necessarily to anthropogenic (human-caused) disturbances.

In the aquatic realm, organic chemicals and other toxic substances, acids and alkalis, and mine drainage can quickly decimate populations of mussels, mayflies, and stoneflies, whereas reduced water flow and introduction of pollutants like silt and excessive nutrients (Mason et al., Webb, this section) cause a slow, relentless destruction of the indigenous fauna.

Courtesy V. Krantz

Leaf miner moth (*Acrocercops arbutella*).

In the past 50 years, nearly 72% of the United States' 297 native mussel species have become endangered, threatened, or of special concern (Williams and Neves, this section). Their populations have been damaged because of siltation, point and nonpoint source pollution, and outright habitat destruction.

The zebra mussel (*Dreissena polymorpha*) and some other nonindigenous species represent "biological pollution" (Schloesser and Nalepa, this section), and should be considered much like toxic pollution for control and treatment. Non-native zebra mussels lack predators and have invaded nearly the full length of the Mississippi River and its major tributaries, threatening the native mussel fauna of the eastern United States (Williams and Neves, this section). The impact of the zebra mussel and other nonindigenous species is covered in greater detail in the "Non-native Species" section of this report.

Courtesy R.J. Gill

Citophilus mealybug (*Pseudococcus calceolariae*).

Historical data bases (e.g., Otte, this section) have traditionally focused on commercially important invertebrate species such as clams and oysters. In contrast, little information exists on the status and trends of nonconsumptive, indigenous invertebrate life, and existing data are often not in formats for use in modern decision-making tools (Messer et al. 1991).

An important, often-overlooked problem with providing scientifically credible data involves the taxonomy and systematics (identification and classification) of organisms. Today, our museum collections of invertebrates are often old and worn out, and there are few trained taxonomists to renew archival materials. In fact, many "type" specimens used for original species' descriptions in the early 1900's are unusable, making comparisons of recently collected specimens impossible.

Canada has been doing continuous biomonitoring for several decades, which has now resulted in status and trends analyses of subtle perturbations like acidification (Chmielewski and Hall 1993). It is clear that the success of future assessments in the United States will greatly depend on availability of and access to high-quality data; stop-gap measures are unlikely to prove successful because of inconsistencies caused by differing collection methods, taxonomy, and reporting units.

This section is organized by general articles on invertebrates and followed by terrestrial and aquatic case studies. The authors drew on original data, often unpublished, and therefore, although some of the studies may appear outdated, this does not detract from the usefulness of the examples.

Basic research on the taxonomy and ecology of species and communities is urgently needed as groundwork for future status and trend assessments. Complex ecological relationships are poorly understood. Only a few working ecosystem models (e.g., Chesapeake Bay) are sufficiently developed to allow semi-quantitative predictions about cause-effect relationships between some biologic components (e.g., plankton) and abiotic conditions. Other biological components need to be added to the modeling framework, especially as related to food web interactions. Future status and trends information gathering should be supportive of ecosystem model development wherever possible.

References

Bartsch, A.F., and W.M. Ingram. 1959. Stream life and the pollution environment. Public Works 90:104-110.

Chmielewski, C.M., and R.J. Hall. 1993. Changes in the emergence of blackflies (Diptera: Simuliidae) over 50 years from Algonquin Park streams: is acidification the cause? Canadian Journal of Fisheries and Aquatic Sciences 50:1517-1529.

Messer, J.J., R.A. Linthurst, and W.S. Overton. 1991. An EPA program for monitoring ecological status and trends. Environmental Monitoring and Assessment 17:67-78.

Diversity and Abundance of Insects

by
Ronald W. Hodges
U.S. Department of Agriculture Systematic Entomology Laboratory

Insects are the most diverse group of organisms (Wheeler 1990); potentially they are highly indicative of environmental change through close adaptation to their environment; they represent the majority of links in the community foodchain; and they likely have the largest biomass of the terrestrial animals (Holden 1989). Thus, knowledge about them is fundamental to studying the environment.

The 34 orders of insects have 90,968 described species and an estimated 72,500+ undescribed species in 653 families and 12,578 genera (Arnett 1985; Kosztarab and Schaefer 1990) in America north of Mexico. Of the described species 71,931 are in the orders Coleoptera (beetles, 23,640), Diptera (flies, 19,562), Hymenoptera (ants, bees, wasps, and sawflies, 17,429), and Lepidoptera (moths and butterflies, 11,300). Undescribed species are distributed mainly among Homoptera (aphids, leafhoppers, scale insects, and allies, 4,334), Coleoptera (2,627), Diptera (41,622), Hymenoptera (18,571), and Lepidoptera (2,700; Kosztarab and Schaefer 1990).

Some aspects of the immature stages of 8,668 species are known (Kosztarab and Schaefer 1990); however, very few are fully known (i.e., documented with voucher specimens and publications with illustrations of eggs, each larval instar, and pupae). Detailed knowledge of the immature stages is important because insects often are present as adults for a short period during the year, but are present as eggs, larvae, or pupae during most of the year.

Taxonomic literature useful for identifying described species is available for less than 30% of them in the adult stage. No major order has been subjected to revisionary study at the specific level, and only two such projects are under way, Lepidoptera (Dominick 1971+) and Diptera (Griffiths 1980+). Some smaller orders, some families, and many genera have been revised for North America (e.g., bethylid wasps [Evans 1978], cerambycid beetles [Linsley and Chemsak 1961-84], chrysidid wasps [Bohart and Kimsey 1982], dragonflies [Odonata; Needham and Westfall 1955], grasshoppers [Orthoptera; Otte 1981, 1983], lady beetles [Gordon 1985], springtails [Collembola; Christiansen and Bellinger 1980-81], thrips of Illinois [Thysanoptera; Stannard 1968], and beetles of the Pacific Northwest [Hatch 1953-71]). Several family or ordinal groups have been revised for Canada and the northern United States. Diptera (Stone et al. 1965; Systematic Entomology Laboratory, U.S. Department of Agriculture, unpublished), Heteroptera (Henry and Froeschner 1988), and Hymenoptera (Krombein et al. 1979) have been cataloged. The Lepidoptera have a checklist (Hodges et al. 1983). A nomenclatorial data base (BIOTA, Biosystematic Information on Terrestrial Arthropods, available via Internet or on CD-ROM) for terrestrial arthropods (less Crustacea) is being developed and coordinated by the Systematic Entomology Laboratory, USDA (Hodges 1994).

For all major orders much revisionary work is needed to define and discriminate among species, genera, and higher taxa in a broad sense and with recognition of variation in nearly all characters. From these works field guides and identification manuals must be developed. Literature for lay workers and students should provide identification to the species level by state or region as this information is necessary for conducting surveys (*Keys to British Insects*—a continuing publication series of the Royal Entomological Society, London—is an excellent example).

Several states have programs to document their fauna with publications and voucher material: California Insect Survey, Florida State Collection of Arthropods, Illinois Natural History Survey, New York State Natural History Survey, and Insects of Virginia, Blacksburg. Few state faunal lists exist; the few that do are outdated or limited: all insects of New York (Leonard 1926) and North Carolina (Brimley 1938, 1942; Wray 1950, 1967); Lepidoptera of Florida (Kimble 1965), Maine (Brower 1974, 1983, 1984), New York (Forbes 1923, 1948, 1954, 1960), and Pennsylvania (Tietz 1952). Checklists or faunal lists of Odonata exist for 39 states and provinces (Westfall 1984). Surveys by county are under way for Kentucky (Lepidoptera, University of Louisville, unpublished data), Maryland (scattered orders and families, Maryland Entomological Society), Missouri (moths, J.R. Heitzman, unpublished data), Ohio (Lepidoptera, Ohio Lepidopterists; Metzler 1980; Iftner et al. 1992; Rings et al. 1992; unpublished data), and the western United States (butterflies; Stanford and Opler 1993). Extensive data have been collected on the distribution of Alaskan butterflies by the Alaska Lepidoptera Survey (Philip, University of Alaska, unpublished data).

Courtesy G. Zolnerowich

Leaf-footed bug *(Thasus* sp.).

No site in North America has been fully surveyed for all insects; however, the Mount Desert Island, Maine, survey (Procter 1946) was an early attempt to do so. Of an estimated 6,000 species, 3,400 have been reported from the H.J. Andrews Experimental Forest, Oregon (Parsons et al. 1991). Craters of the Moon National Monument, Idaho (Horning and Barr 1970); Deep Creek in San Bernardino County, California (S.I. Frommer, University of California, Riverside, unpublished data); and Pawnee Grasslands, Colorado (Kumar et al.

Courtesy J.R. Bruschwein

Mantidfly *(Mantispa interrupta).*

Courtesy C.E. Williams

Robber fly (*Diogmites symmachus*).

Courtesy J. Heraty

Spurge hawkmoth (*Hyles euphorbiae*).

1976) have been intensively surveyed for insects but none of the surveys approaches completion. A constant problem has been the inability to identify all the numerous taxa.

Sampling for taxa, except for aquatics, is based mainly on adults; results are highly variable, depending on the competency of the samplers, knowledge of habits of organisms, weather during sampling periods, and phases of the moon and wavelength of light (for those species attracted to light). With exceptions (aquatic insects; Merritt et al. 1984), sampling techniques to estimate species diversity within an area have not been developed or are preliminary for limited taxa.

Identification of adults within large orders depends on highly trained, experienced taxonomists who have access to good collections and libraries; very few taxonomists exist relative to the number of taxa. Identifications in collections must be held suspect unless the taxa have been revised in contemporary terms and the specimens studied and vouchered by the revisor (Hodges 1976) or other specialist.

Individuals capable and willing to provide authoritative identifications are becoming fewer each year. Many have retired or could retire. There has been a significant redirection of systematists from basic revisionary work to other research areas. Nearly 30% fewer persons are entering the field because the likelihood of obtaining a position upon completion of training is extremely poor (Lutz 1994). Technical and monetary support for systematists and curators always has been limited and is becoming more restricted.

Collections vary in size from small private collections to the 30+ million specimens in the National Insect Collection in the National Museum of Natural History. Many state universities, particularly in the Midwest and on the West Coast, have collections of 1+ million specimens, as do several private and public institutions. Despite this large number, many species are represented by few specimens and almost none with comprehensive representation by county and by state.

Surveys of many taxa are possible but require individuals to initiate them; sufficient taxonomic literature and research to enable recognition of taxa; curatorial support for preparing, sorting, and identifying specimens and potential supervision of the surveys; and adequate collection and library facilities for species recognition and permanent storage of voucher specimens.

These comments are meant to provide perspective on the status of systematic entomology and thus the role insects may have in the work of the National Biological Service.

References

Arnett, R.H. 1985. American insects: a handbook of the insects of America north of Mexico. Van Nostrand Reinhold Co., New York. 850 pp.

Bohart, R.M., and S.S. Kimsey. 1982. A synopsis of the Chrysididae in America north of Mexico. Memoirs of the American Entomological Institute 33:1-266.

Brimley, C.S. 1938. The insects of North Carolina. North Carolina Department of Agriculture, Raleigh. 560 pp.

Brimley, C.S. 1942. Supplement to insects of North Carolina. North Carolina Department of Agriculture, Raleigh. 39 pp.

Brower, A.E. 1974. A list of the Lepidoptera of Maine—Part 1, the macrolepidoptera. Life Sciences and Agriculture Experiment Station Tech. Bull. 66:1-136.

Brower, A.E. 1983. A list of the Lepidoptera of Maine—Part 2. Maine Department of Conservation and the Department of Entomology Tech. Bull. 109:1-60.

Brower, A.E. 1984. A list of the Lepidoptera of Maine—Part 2 (concluded). Maine Department of Conservation and the Department of Entomology Tech. Bull. 114:1-70.

Christiansen, K.A., and P. Bellinger. 1980-81. The Collembola of North America north of the Rio Grande, a taxonomic analysis. Grinnell College, Grinnell, IA. 4 vols.

Dominick, R.B., ed. 1971+. The moths of America north of Mexico, including Greenland. The Wedge Entomological Research Foundation, Washington, DC. [19 parts]

Evans, H.E. 1978. The Bethylidae of America north of Mexico. Memoirs of the American Entomological Institute 27:1-332.

Forbes, W.T.M. 1923. The Lepidoptera of New York and neighboring states. Part 1. Cornell University Agricultural Experiment Station Memoir 68:1-729.

Forbes, W.T.M. 1948. The Lepidoptera of New York and neighboring states. Part 2. Cornell University Agricultural Experiment Station Memoir 274:1-263.

Forbes, W.T.M. 1954. The Lepidoptera of New York and neighboring states. Part 3. Cornell University Agricultural Experiment Station Memoir 329:1-433.

Forbes, W.T.M. 1960. The Lepidoptera of New York and neighboring states. Part 4. Cornell University Agricultural Experiment Station Memoir 371:1-188.

Gordon, R.D. 1985. The Coccinellidae (Coleoptera) of America north of Mexico. Journal of the New York Entomological Society 93(1):1-912.

Griffiths, C.D.G., ed. 1980+. Flies of the Nearctic Region. E. Schweizerbart'sche Verlagsbuchhandlung. Stuttgart. [19 parts]

Hatch, M.H. 1953-71. The beetles of the Pacific Northwest. Part 1: Introduction and Adephaga, 340 pp.; Part 2: Staphyliniformia, 384 pp.; Part 3: Pselaphidae and Diversicornia I, 503 pp.; Part 4: Macrodactyles, Palpicornes, and Heteromera, 268 pp.; Part 5: Rhipiceroidea, Sternoxi, Phylophaga, Rhynchophora, and Lamellicornia, 662 pp. University of Washington Press, Seattle.

Henry, T.J., and R.C. Froeschner. 1988. Catalog of the Heteroptera, or true bugs, of Canada and the continental United States. E.J. Brill, Leiden, The Netherlands. 958 pp.

Hodges, R.W. 1976. Presidential address 1976—what insects can we identify? Journal of the Lepidopterists' Society 30:245-251.

Hodges, R.W. 1994. BIOTA. Association of Systematics Collection Newsletter 21(6):72.

Hodges, R.W., et al., eds. 1983. Check list of the Lepidoptera of America north of Mexico. E.W. Classey Ltd. and the Wedge Entomological Research Foundation, London, England. 284 pp.

Holden, C. 1989. Entomologists wane as insects wax. Science 246:754-756.

Horning, D.S., Jr., and W.F. Barr. 1970. Insects of Craters of the Moon National Monument, Idaho. University of Idaho College of Agriculture Miscellaneous Series 8:1-118.

Iftner, D.C., J.A. Shuey, and J.V. Calhoun. 1992. Butterflies and skippers of Ohio. Ohio Biological Survey Bull. 9(1):1-212.

Kimble, C.P. 1965. Lepidoptera of Florida. Arthropods of Florida and Neighboring Land Areas 1:1-363.

Kosztarab, M., and C.W. Schaefer. 1990. Conclusions. Pages 241-247 *in* M. Kosztarab and C.W. Schaefer, eds. Systematics of the North American insects and arachnids: status and needs. Virginia Agricultural Experiment Station Information Series 90-1.

Krombein, K.V., P.D. Hurd, Jr., D.R. Smith, and B.D. Burks. 1979. Catalog of the Hymenoptera in America north of Mexico. Smithsonian Institution Press, Washington, DC. 2,735 pp.

Kumar, R., R.J. Lavigne, J.E. Lloyd, and R.E. Pfadt. 1976. Insects of the Central Plains Experimental Range Pawnee National Grassland. University of Wyoming Agricultural Experiment Station Science Monograph 32:1-74.

Leonard, M.D., ed. 1926. A list of the insects of New York with a list of the spiders and certain other allied groups. Cornell University Agricultural Experiment Station Memoir 101:1-1,121.

Linsley, E.G., and J.A. Chemsak. 1961-84. The Cerambycidae of North America. Parts 1-7. University of California Press, Berkeley.

Lutz, D. 1994. More than stamp collecting. American Scientist 82(2):120-121.

Merritt, R.W., K.W. Cummins, and V.H. Resch. 1984. Collecting, sampling, and rearing methods for aquatic insects. Pages 11-26 *in* R.W. Merritt and K.W. Cummins, eds. An introduction to the aquatic insects of North America. Kendall/Hunt, Dubuque, IA.

Metzler, E.H. 1980. Annotated checklist and distribution maps of the royal moths and giant silkworm moths (Lepidoptera: Saturniidae) in Ohio. Ohio Biological Survey Biological Notes 14:1-10.

Needham, J.G., and M.J. Westfall, Jr. 1955. A manual of the dragonflies of North America (Anisoptera). University of California Press, Berkeley. 615 pp.

Otte, D. 1981. North American grasshoppers. Vol. 1. Gomphocerinae and Acridinae. Harvard University Press, Cambridge, MA. 275 pp.

Otte, D. 1983. North American grasshoppers. Vol. 2. Oediodinae. Harvard University Press, Cambridge, MA. 366 pp.

Parsons, G.R., G. Cassis, A.R. Moldenke, J.D. Lattin, et al. 1991. Invertebrates of the H.J. Andrews Experimental Forest, Western Cascade Range, Oregon. V: An annotated list of insects and other arthropods. U.S. Forest Service General Tech. Rep. 290:1-168.

Procter, W. 1946. Biological survey of the Mount Desert Region. Part 7. Wistar Institute, Philadelphia, PA. 566 pp.

Rings, R.W., E. H. Metzler, F. J. Arnold, and D.H. Harris. 1992. The owlet moths of Ohio, order Lepidoptera family Noctuidae. Ohio Biological Survey Bull. 9(2):1-222.

Stanford, R.E., and P.A. Opler. 1993. Atlas of western U.S.A. butterflies, including adjacent parts of Canada and Mexico. Published by authors, Denver, CO. 275 pp.

Stannard, L.J. 1968. The thrips, or Thysanoptera, of Illinois. Illinois Biological Monographs 25:1-200.

Stone, A., C.W. Sabrosky, W.W. Wirth, et al. 1965. A catalog of the Diptera of America north of Mexico. Agriculture Handbook 276:1-1696.

Tietz, H.M. 1952. The Lepidoptera of Pennsylvania. Pennsylvania State College, School of Agriculture, Agriculture Experiment Station. State College, PA. 194 pp.

Westfall, M.J., Jr. 1984. Odonata. Pages 126-176 *in* R.W. Merritt and K.W. Cummins, eds. An introduction to the aquatic insects of North America. Kendall/Hunt, Dubuque, IA.

Wheeler, Q.D. 1990. Insect diversity and cladistic constraints. Annals of the Entomological Society of America 83(6):1031-1047.

Wray, D.L. 1950. Insects of North Carolina: second supplement. North Carolina Department of Agriculture, Raleigh. 59 pp.

Wray, D.L. 1967. Insects of North Carolina: third supplement. North Carolina Department of Agriculture, Raleigh. 181 pp.

Tiger beetle (*Cicindela hirticollis*).

For further information:

Ronald W. Hodges
Systematic Entomology Laboratory, USDA
c/o National Museum of Natural History, MRC-168
Washington, DC 20560

Grasshoppers

by

Daniel Otte
Academy of Natural Sciences, Philadelphia

Grasshoppers (Orthoptera:Acrididae) are perhaps the most important grazing herbivores in the nation's grasslands, which from a human standpoint, are the most important food-producing areas. The damage that grasshoppers do to plants varies with the species. A few dozen species at most are highly injurious to crops, while those that feed on economically unimportant plants may have no measurable impact, and those that feed on detrimental plants are highly beneficial. Given such differences, it becomes important to distinguish properly between harmful and beneficial species. Grasshopper abundance in all kinds of grasslands means they are an important factor in the ecological equation. Their economic importance—positive and negative—means that they must be included in all studies of grassland and desert-grassland communities.

Taxonomic Status

More than 1,000 species of grasshoppers have been described from the United States (Otte 1976, 1994, unpublished data base). Taxonomic revisions at the Academy of Natural Sciences in Philadelphia (ANSP) reveal that approximately 20% of the U.S. species represented in the existing ANSP collection are undescribed (Otte 1981; unpublished data). Most new species belong to the very large genus *Melanoplus*, which contains some of the most injurious grasshopper species known. A considerable number of undescribed species are from the eastern states, from approximately central Texas to New England. New species are turning up even in extremely well-studied areas such as Michigan and Florida. It is expected that at least tens of species remain to be discovered in the coastal ranges of California, and many other mountain peaks in the western states should have species unique to them. Much of the academy's collecting efforts have been directed to investigating the grasshopper faunas of such mountain peaks ("sky islands").

Natural Range Increases

Great Lakes Region

Documenting natural range changes requires that comparable collections be made at several points in time. The only such case involving grasshoppers that I am aware of involves the ranges of two grasshopper species along the Great Lakes shores. *Trimerotropis huroniana* and *T. maritima* displace one another on the dunes surrounding the Great Lakes, with *T. huroniana* occupying the northern shores and *T. maritima* the southern shores (Otte 1970). The boundary between these two species has shifted in the last seven decades. The two species may well be competitive on four different lakefronts, on the north-south shores of Lakes Michigan and Huron.

Prairie Peninsula

In southern Michigan the bandwing grasshopper (*Pardalophora haldemani*) was abundant in 1943 and the related species, *P. apiculata*, was rare (Cantrall 1943). By 1968 *P. haldemani* had been completely replaced by *P. apiculata*, probably because subtle habitat changes gave *P. apiculata* an advantage over the strictly prairie species *P. haldemani*.

Unnatural Range Increases

Precise documentation of range changes in grasshoppers could be achieved if historical collecting sites could be resurveyed today. We are reasonably certain, though, that the cutting of eastern forests (mainly during the last century) opened up habitats for numerous species adapted to grasslands and forest edges. Numerous prairie margin species now occur widely in the eastern United States in areas that were almost completely covered by forests. By colonizing roadsides, other species have become extremely widely distributed. The Carolina locust (*Dissosteira carolina*), for example, is a ubiquitous roadside species that is now found in previously heavily forested regions. Whether the overall range (outer limits of the range) has changed is debatable because the species inhabits river margins and small natural eroded areas within the eastern forest region.

In the western United States, certain species do well in eroded habitats that often result from overgrazing. Thus, the ranges of species specializing on eroded ground probably increased along with increases in grazing. The clear-winged grasshopper (*Camnula pellucida*), a pest species from the northern Great Plains that greatly damages crops in the northern United States and western Canada, is now extremely abundant in overgrazed mountain meadows in the western states and is a good indicator of meadow degradation there.

Many pest species specialize on agricultural fields; their ranges have increased because of irrigation and the planting of crops in normally desert habitats (e.g., migratory grasshopper [*Melanoplus sanguinipes*], two-striped grasshopper [*M. bivittatus*], and differential grasshopper [*M. differentialis*]). Ball et al. (1942) documented numerous cases of grasshoppers moving into areas altered by agricultural practices.

Range Reductions and Extinctions: Case Studies

The Rocky Mountain Locust

Although it was the most abundant species during much of the last century in western North America, the Rocky Mountain locust (*M. spretus*) is now extinct; no specimens have been collected in this century. This species spread its destruction over many western states and was the source of great difficulty for early farmers east of the Rocky Mountains. The complete disappearance of the species has puzzled biologists for decades. The most reasonable hypothesis is that this species reproduced mainly along river valleys in Montana and Idaho and that with the heavy grazing of these habitats, beginning in the last part of the 1800's, these breeding areas were so heavily disturbed that breeding was disrupted (Lockwood and DeBrey 1990). Today, frozen remains of this species can still be found in glaciers in Montana.

California Coastal Ranges

The ANSP collections revealed that two undescribed species of *Melanoplus* were collected only in what is now downtown San Francisco. Recent revisions of the Marginatus group of the genus *Melanoplus* (Otte 1981, 1994, unpublished data base) reveal that the coastal ranges of California contain numerous members of this group, but that the ranges of many of the species are extremely limited. A subgroup of the Marginatus group speciated around the San Francisco Bay area, two species are known from the Berkeley area, two from San Francisco proper, and several from the north side of San Francisco Bay. The San Francisco species were collected in the first decade of this century when some natural vegetation still existed in San Francisco. South of San Francisco these species are replaced by related species. Several other species in the group are known only from the Monterey Bay region, and one species only from a single locality.

Two Rare Species

Two individuals of an extremely rare grasshopper species (*Eximacris superbum* Hebard) were collected in south Texas. Repeated efforts to collect the species have not met with success, although the species possibly breeds only during winter or in the early spring and might still turn up when an effort is made to collect it in early spring. The only relative in this genus (*E. phenax* Otte) is known from a single male collected at Big Cedar in the Kiamichi Mountains of Oklahoma. Searches for this species have also been unsuccessful; again, it is possible that the species overwinters in the adult stage and therefore is not present during normal grasshopper breeding times.

Mountain Islands

Some species of grasshopper are known only from mountain tops (sky islands). In the East, some *Melanoplus* species are known from single balds (grassy mountain summits) in the Appalachian region (three new species are presently being described; Otte, unpublished data). In the western United States members of the Montanus group are also known only from single localities (A.B. Gurney, unpublished data). Surveys of mountains in Colorado, New Mexico, Arizona, Utah, and Nevada showed that some of these species have not yet been described and are believed to occur only on single mountains.

Within their limited ranges on mountains, the grasshopper species are further limited by environmental disturbance. I have encountered many overgrazed mountain meadows, sometimes even highly isolated ones surrounded by forest. These differ from ungrazed meadows chiefly in the height of the vegetation and the number of plant species there, and consequently in the incidence of short-winged grasshoppers. Collections from high mountain passes, where meadows are partly protected from cattle by fences along the road, show a clear effect of vegetation length on diversity: in the protected areas, nonflying grasshopper species are present, sometimes in large numbers, but are absent in grazed areas, while flying species, which have wide distributions (weedy species), are common. The principal reason for the difference appears to be that short-winged, nonflying species are highly vulnerable to bird predation, and without protective vegetation are unable to survive.

Pleistocene Islands in Northern Florida

The northern half of Florida contains a number of habitats that remained exposed as islands during interglacial periods. Several grasshopper groups have species associated with these former islands and species' ranges are highly restricted (Hubbell 1932). These areas are also ideal for farming and therefore have been greatly altered during the last 50-80 years. It is extremely likely that some species never collected were lost. It remains to be seen which species collected earlier this century still exist.

Management Implications

Large differences exist in range sizes between species that can fly and those that cannot (Otte 1979). In the latter group are numerous species known from a single or a few localities. Most of these inhabit island habitats (isolated bogs, prairie openings within the eastern forests, balds on the Appalachian range, mountain meadows on western mountaintops, hammocks in Florida, and perhaps coastal islands along the East coast). Many species in these regions have probably already been lost. Others can be saved by creating new sanctuaries and properly modifying existing ones. Within such regions it should be possible to set aside small sanctuaries or strings of sanctuaries from which cattle and other grazing mammals are excluded. Such sanctuaries already exist along highways where cattle are kept away from roadsides and railways.

References

Ball, E.D., E.R. Tinkham, R. Flock, and C.T. Vorhies. 1942. The grasshoppers and other Orthoptera of Arizona. Tech. Bull. Arizona College of Agriculture 93:275-373.

Cantrall, I.J. 1943. The ecology of the Orthoptera and Dermaptera of the George Reserve, Michigan. Miscellaneous Publ. of the University of Michigan Museum of Zoology 54:1-184.

Hubbell, T.H. 1932. A revision of the *puer* group of the North American genus *Melanoplus*, with remarks on the taxonomic value of the concealed male genitalia in the Cyrtacanthacridinae (Acrididae). University of Michigan Museum of Zoology Miscellaneous Publ. 23:1-64.

Lockwood, J.A., and L.D. DeBrey. 1990. A solution for the sudden and unexplained extinction of the Rocky Mountain locust, *Melanoplus spretus* (Walsh). Environmental Entomology 19:1194-1205.

Otte, D. 1970. A comparative study of communicative behavior in grasshoppers. University of Michigan Museum of Zoology Miscellaneous Publ. 141:1-168.

Otte, D. 1976. Species richness patterns of New World desert grasshoppers in relation to plant diversity. Journal of Biogeography 3:197-207.

Otte, D. 1979. Biogeographic patterns in flight capacity of Nearctic grasshoppers (Orthoptera, Acrididae). Entomological News 90(4):153-158.

Otte, D. 1981. The North American grasshoppers. Vol 1. Acrididae: Gomphocerinae and Acridinae. Harvard University Press, Cambridge, MA. 275 pp.

Otte, D. 1994. The Orthoptera species file: computer catalog to the genera and species of world grasshoppers. Academy of Natural Sciences of Philadelphia. Unpublished data base.

For further information:

Daniel Otte
Academy of Natural Sciences
Department of Entomology
1900 Benjamin Franklin Parkway
Philadelphia, PA 19103

The Changing Insect Fauna of Albany's Pine Barrens

by
Tim L. McCabe
New York State Museum

Sand plains and similar inland sand deposits are desertlike islands in a sea of moist land. Because of rapid drainage of rainwater, sand plains are modern-day refugia that represent drier conditions that have existed off and on during the past 10,000 years. Drainage makes for drier soils that mimic prairie conditions and consequently harbor prairie relicts; thus these communities support a specialized flora and fauna. Sand barrens abound with rare or endemic forms, many of which are endangered.

The Albany pine barrens is a sand plains community and one of a relatively few scrub-oak (*Quercus ilicifolia*), pitch-pine (*Pinus rigida*) communities. Around the turn of the century, the Albany pine barrens was the site of intensive collecting by museum entomologists. Consequently, it has a historically well-documented and diverse insect fauna, making it possible to compare the fauna after a century of transition. Today, the region is heavily urbanized, and only 15.5 km^2 (6 mi^2) of the original 104 km^2 (40 mi^2) of natural barrens remain. As this habitat has been lost, 31 species of butterflies, moths, and skippers (Lepidoptera) have become locally extinct during the last century (McCabe et al. 1993). The past two decades have witnessed the most rapid change to the Albany pine barrens as well as the most dramatic decline of its resident insects.

Table. Insect species historically recorded from the Albany pine barrens but now extirpated (modified after McCabe et al. 1993).

Order	Family	Species
Coleoptera	Cicindelidae	*Cicindela patruela*
Diptera	Asilidae	*Cyrtopogon laphriformis*
		Promachus bastardii
Odonata	Libellulidae	*Williamsonia lintneri*
Lepidoptera	Hesperiidae	*Poanes viator*
		Erynnis brizo
		E. persius
	Nymphalidae	*Speyeria idalia*
		Phyciodes batesii
	Noctuidae	*Acronicta lanceolaria*
		A. radcliffei
		Agroperina lutosa
		Anomogyna badicollis
		Argyrostrotis quadrifiliaris
		Catocala pretiosa
		Eugraphe subrosea
		Homohadena badistriga
		Lithophane georgii
		L. lepida
		L. semiusta
		L. thaxteri
		Platypolia anceps
		Psectraglaea carnosa
		Pyreferra ceromatica
		Xylena cineritia
		X. thoracica
		Xylotype capax
	Geometridae	*Metarrhanthis apiciaria*
		Brephos infans
		Semiothisa eremiata
	Megalopygidae	*Megalopyge crispata*
	Saturniidae	*Citheronia sepulchralis*
		C. imperialis
	Sphingidae	*Hemaris gracilis*
		Darapsa versicolor

Insect Surveys and Data Collection

A general survey of all insect species included collections made using malaise traps, light traps, and netting. Because pine barren communities require regular disturbance regimes (e.g., fire) to maintain the unique open habitats that characterize them, we evaluated insects in areas that had been recently burned. Postburn sites of 1, 5, 12, and 30 years of age were sampled. I gathered published records from numerous sources and, through comparison with recent catalogs and museum holdings, I attempted to identify those species that have a restricted distribution or at least are unusual for New York State (McCabe et al. 1993). The population of the Karner blue butterfly (*Lycaeides melissa samuelis*) was the focus of an intensive monitoring program using a visual transect method (Higgins et al. 1991; McCabe et al. 1993; Meyer and McCabe 1993). Better known and easily identified groups have also been evaluated (McCabe and Huether 1985 [1986]; McCabe 1985; McCabe et al. 1993; McCabe and C. Weber, unpublished data).

Changes in Species Composition

The group for which I am able to make the most reliable comparisons is the Lepidoptera, particularly the owlet moths (Noctuidae). Unfortunately, I began investigations too late (1980) to witness the extirpation of many of the species historically recorded from Albany's pine barrens (Table). Of the 31 species of Lepidoptera extirpated from the pine barrens, 5 are partial to wetland habitats, which have suffered severely in the pine barrens: a skipper (*Poanes viator*), a sphinx (*Darapsa versicolor*), and three owlets (*Agroperina lutosa*, *Eugraphe subrosea*, and *Argyrostrotis quadrifiliaris*). Two owlet species (*Xylena cineritia* and *Acronicta lanceolaria*) are known to cycle in and out of an area in unpredictable patterns; thus their recent absence is thought temporary. Most of the remaining species now have distributions to the south (owlets: *Catocala pretiosa, Pyreferra ceromatica,* and *Xylotype capax*; flannel moth: *Megalopyge crispata*; giant silkworm moths: *Citheronia sepulchralis* and *C. imperialis*) or to the north (owlets: *Xestia (Anomogyna) badicollis, Lithophane georgii, L. lepida, L. semiusta, L. thaxteri, Platypolia anceps,* and *Xylena thoracica*; geometer: *Brephos infans*; sphinx: *Hemaris gracilis*). Many of these species are not restricted to pitch-pine barrens, but the Albany pine barrens represents important habitat at the extreme edges of their ranges.

The species at the margins of their distribution in Albany have witnessed losses almost equally divided between north and south, suggesting that regular "pulses" of an insect species' distribution account for more species losses than can be attributed to the nearly sevenfold loss of habitat. It therefore seems appropriate to look closer at those species whose decline is most relevant to habitat loss.

The owlets *Psectraglaea carnosa* and *Chaetaglaea cerata* are usually found in coastal heath habitats but have been recorded from Albany. *Chaetaglaea cerata* is at precariously low levels in Albany, and *P. carnosa* is now locally extinct. Recent records of *C. cerata* and the last reports of *P. carnosa* were from an area of the pine barrens adjacent to the current landfill.

Another once-common species in the pine barrens is the owlet *Homohadena badistriga*, but I have observed this moth only once during the last 4 years (1989-93). *Homohadena badistriga* caterpillars show a marked preference for the native shrub *Lonicera dioica* over all other *Lonicera* in the area. This shrub species, which appears to be a favorite browse of deer (personal observation), has become far less abundant in the past 12 years (J. Mattox, Bard College, personal communication). None of 27 bushes of *L. dioica* I had visited in 1982 exist today.

The owlet *Agrotis stigmosa,* which favors the periphery of open dunes, has a simpler story. The two most substantial open dunes in the Albany barrens have recently been developed, and *A. stigmosa* has subsequently been rarely encountered and may soon be lost.

The Karner blue butterfly (*Lycaeides melissa samuelis*), now listed as an endangered species, has markedly declined in the Albany barrens (Figure). This species appears to be a barrens relict that has been losing ground over all of the Northeast. Its larvae feed on *Lupinus perennis* (lupine). Another lycaenid butterfly, *Incisalia irus*, also dependent on lupine, has suffered a similar decline. The continued decline of *L. melissa samuelis* on the Albany pine plains (Figure) is illustrated by using both recent data (Higgins et al. 1991; Meyer and McCabe 1993) and earlier population estimates of Cryan (1980) and Schweitzer (1988, 1990). This downward trend continues even though some sites now support more lupine than a decade ago and appear to be well protected.

Pine Barrens Management

Native pine barrens plants such as pitch pines, New Jersey tea, and lupine are very difficult to establish successfully. Seedlings are shaded out by scrub oak. Young pitch pines are heavily browsed by deer and severely attacked by an introduced pine sawfly; younger plants are completely defoliated. Lupines are devoured by cottontail rabbits. Most characteristic pine barrens plants require open, disturbed sites.

Fire has been scientifically employed as a management practice on the Albany barrens only quite recently. Scrub oak successfully regenerates after burns, as does the locust, *Robinia pseudoacacia*, a tree introduced from the Southeast for fence posts. One of the pine barrens rarities is *Chytonix sensilis*, a fungus-feeding moth. The year after a burn, fire-blackened trunks support luxurious growths of this fungus. Despite this, *C. sensilis* was most abundant in 12-year-old burn sites where hardly any fungus had been present. In areas unburned for more than 30 years, only *C. sensilis* females were collected. One 12-year-old site is the same one that supports *Chaetaglaea cerata* and had supported *P. carnosa*, suggesting that a burn frequency of at least 12 years is best to promote some of the choicest pine barrens associates. I trapped moths extensively in postburn sites of 1, 5, 12, and 30 years of age. No site was available with a postburn age between 12 and 30 years; an optimal burn frequency will likely fall somewhere within this range. A frequent burn schedule would be highly detrimental to insect species very susceptible to fires, such as one of the elfin butterflies, *Incisalia henrici*.

Species on the periphery of their range may not be reliable indicators of habitat quality. Natural fluctuation in range limits appears more significant than formerly considered. This can be attested to by the extirpation of 31, and the addition of 32, moth species. The decline of characteristic pine barrens species has to be examined on a case-by-case basis.

The Albany pine barrens has also been adversely affected by vehicular traffic, windbreaks created by roads and buildings, development of open dunes, introductions of exotic species, and even the frequency of fires, which promote some and compromise other pine barren rarities. Cutting to create oak openings should be considered as a management practice. In addition, open dunes may have to be artificially maintained where artificial windbreaks interfere.

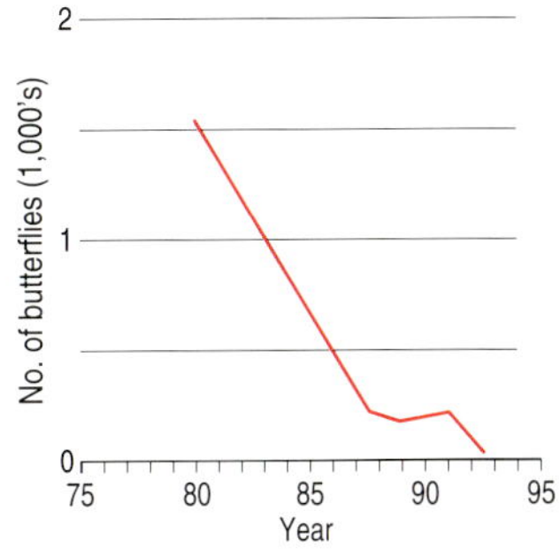

Figure. Decline of the Karner blue butterfly at Willow Avenue site in Albany (McCabe et al.1993).

References

Cryan, J.F. 1980. The Karner blue butterfly (*Lycaeides melissa samuelis* Nabokov) in the Hudson Valley sand belt of New York. Part 2. An annotated list of Hudson Valley sand belt populations and their status. A report for the New York State Department of Environmental Conservation.

Higgins, L.E., T.L. McCabe, A. Meyer, and M. Rusch. 1991. Albany pine bush preserve—1991 entomological report. Prepared for The Nature Conservancy and the City of Albany, NY.

For further information:

Tim L. McCabe
New York State Museum
The State Education Department
University of the State of New York
Albany, NY 12230

McCabe, T.L. 1985. An annotated list of pine bush caddis (Insecta: Trichoptera). Skenectada 3:17-18.

McCabe, T.L., and J.P. Huether. 1985[1986]. An annotated list of pine bush Cerambycidae (Insecta: Coleoptera). Skenectada 3:19-23.

McCabe, T., A. Meyer, C. Weber, and L. Higgins. 1993. Albany pine bush project 1991-1992 entomological report. Submitted to the Eastern New York Chapter of The Nature Conservancy. 111 pp.

Meyer, A.M., and T.L. McCabe. 1993. Albany pine bush project. Karner blue butterfly report. The Nature Conservancy. Unpublished report.

Schweitzer, D.F. 1988. Supplement to 1988 Karner blue population studies: the Crossgates Mall population. A consultant's report to the City of Albany, NY. Unpublished report. 5 pp.

Schweitzer, D.F. 1990. The 1990 status of selected Karner blue remnants in Saratoga and Albany counties, New York, with a discussion of monitoring methods. New York Department of Environmental Conservation, Endangered Species Unit.

Lepidoptera Inventories in the Continental United States

by
Jerry A. Powell
University of California, Berkeley
Essig Museum of Entomology

Lepidoptera (butterflies and moths) make up about 13% of the described and named 90,000 insect species of North America (11,500 named) and are among the better known large orders, although no complete inventory of Lepidoptera species exists for any state, county, or locality in North America.

The rationale for local or regional inventory of insects is related to their importance in biodiversity. Insects make up 75% of all described animals, and in natural communities their species outnumber those of all other higher organisms combined. Thus interrelationships between insects and other organisms form the most prevalent and comprehensive elements of the fabric of biological communities.

Lepidoptera are the major group of plant-feeding insects, and local inventories of Lepidoptera can help indicate the stability and diversity of plant communities. When we have several reasonably complete local inventories of Lepidoptera in different regions of the country, we will be able to make predictions about overall insect—and therefore biological—diversity, and about relationships between plant and insect species richness on a local or regional basis. Such knowledge will lead to more efficient methods of assessing the health and loss of biological diversity.

Once a baseline inventory is done, monitoring of changes in species richness and abundance to assess the ecological health of the community can be carried out. Inventory of a diverse group of insects such as the Lepidoptera must involve various approaches and collecting procedures. This article summarizes the status of local and state inventories of Lepidoptera and suggests a model for planning comparable faunal inventories of major insect groups.

Lepidoptera Surveys

To gather information on the status of current inventories, I mailed a one-page questionnaire to 25 lepidopterists thought to be developing local or state lists. Nearly all responded, and several are conducting more than one census. Early in 1993 I published a request for information on inventories in the *News of The Lepidopterists' Society*, which is distributed to about 1,000 members in North America. The responses were fewer than I had expected; there may be many more inventories in progress than those reported to me. For completed local and state lists, I searched the literature, but the results are likely to be incomplete because such lists are lengthy and often are published in obscure literature not well referenced by abstracting services.

A thorough local inventory must depend upon diverse methods: daytime searches for butterflies and diurnal moths, nighttime collections of moths attracted to ultraviolet (UV) or mercury vapor lights, and rearing caterpillars (larvae) to the adult stage. In some regions a fourth approach, "sugaring," the attraction of moths to sweet, fermenting bait, is effective for many species not readily attracted to lights. Generating an inventory for a large group of insects such as Lepidoptera is difficult because the season that each species can be found may be short; species abundance varies widely from year to year; several techniques and specialists' experience are needed to complete a thorough census; and, beginning early in the survey, individuals of vagrant species are encountered.

A major problem in compiling an inventory is the identification of species. This is easily accomplished for butterflies (6% of the Lepidoptera), and there are hundreds of local and state lists (Field et al. 1974). Identifications are accessible for the larger moths (macrolepidoptera), including inchworm moths (Geometridae), giant silkworm moths (Saturniidae), hawk moths (Sphingidae), owlet moths (Noctuidae), and related families. However, for many so-called "microlepidoptera" (primitive suborders, leaf miner and leaf roller moths, etc.), 10%-90% of the local species in some families are undescribed. As a result, most state and local lists have dealt only with macrolepidoptera or have treated the microlepidoptera species only cursorily.

Inventories and Trends

There are published Lepidoptera lists or surveys in progress from at least 30 states, and local inventories of at least macrolepidoptera for 35 or more reserves, counties, or islands in the continental United States. The tendency of lepidopterists to compile state and local lists, which had been expressed primarily by faunal studies of butterflies (Field et al. 1974), increasingly has encompassed moths. Half of the state lists and 85% of the local inventories have been published since 1964, and there are an even larger number in progress. More of these include microlepidoptera than before probably because of considerable progress in the descriptive taxonomy of most families during the past 35 years (e.g., Covell 1984).

State Lists

The older and more comprehensive state lists are in the eastern United States (Fig.1). The most complete state lists of Lepidoptera are those for New York (Forbes 1923-60), New Jersey (Smith 1910; Muller 1965-76), and Maine (Brower 1974-86), although these lists have many identification problems. The most active are in Ohio, Kentucky, Mississippi, Florida (Kimball 1965), and Texas. There are lists primarily or only of macrolepidoptera for some states, including Arizona (Bailowitz et al. 1990), Michigan (Moore 1955), Pennsylvania (Tietz 1952), and Maryland (D.C. Ferguson et al., National Museum of Natural History, unpublished data). Lists of described species for the western states are now being done (Fig. 1).

Local Inventories

Thirty-five local inventories have been published or are in progress (Fig. 1). These vary greatly in moth families included, geographic size, and number of years in progress. Several inventories, including those of Martha's Vineyard and Nantucket, Massachusetts (Jones and Kimball 1943); Mount Desert Island, Maine (Proctor 1946); Welder Wildlife Refuge, Texas (Blanchard et al. 1985); Ash Canyon, Arizona (N. McFarland, Sierra Vista, AZ, unpublished data); and three in California (McFarland 1965; Powell, unpublished data) span 10-50 years and are estimated to be 85%-95% complete (Table).

Unfortunately, no two inventories can be meaningfully compared because they vary in important parameters. Many have recorded only macrolepidoptera, often only one sampling approach was emphasized, inventories are made of sites that vary greatly in size, inventory duration ranges considerably (Table), and the methods of recording data are often inconsistent.

A Model Inventory

We have been conducting inventories in California to document species discovery rates and other comparative data. The most comprehensive inventory is at the University of California Big Creek Reserve in coastal Monterey County, an area of diverse habitats and elevations. The census has been carried out primarily by specialists' visits. We have sampled in all months, on 175 dates, recording every species in each sample; we spent 180 personnel-days for diurnal species, recorded more than 260 UV light samples, and processed 1,350 larval collections and their rearing. The census (more than 900 species) is believed more than 90% complete, with 3% or fewer of the species in each three-date sample new to the list during dates 155-175 (Fig. 2). Butterflies and diurnal moths make up 16% of the total, and microlepidoptera recorded only as larvae make up another 9%.

The species discovery rate was slow because we could not sample the whole reserve during each visit, and most of the effort followed a consummate fire in the fourth year of our 12-year inventory; many species were first collected in year 9 or 10. Nevertheless, the results provide a realistic idea of the effort required in a complex community to achieve a reliable species accumulation curve (Fig. 2).

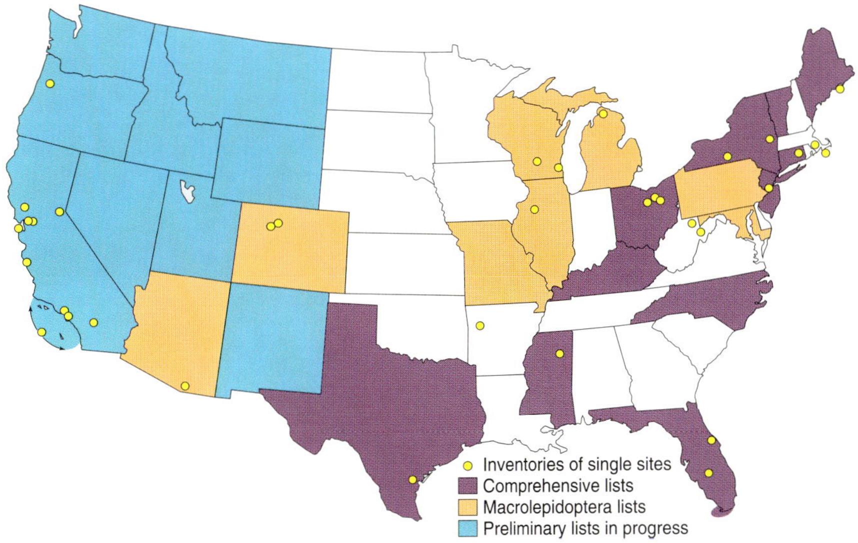

Fig. 1. Distribution of state and local inventories of Lepidoptera in the contiguous 48 United States. States having comprehensive lists (all families) published or in progress, those with macrolepidoptera lists, and those with preliminary lists in progress are indicated. Dots indicate locations of 35 local inventories of single sites, reserves, and islands, either published or in progress.

Table. Comparison of size, duration, estimated percentage completed, and numbers of species recorded in local inventories of Lepidoptera, listed by state.

State*	Area (km^2)	Duration (years)	% est. censused	No. of species
Arizona macro	< 10	13	> 95	900 +
California macro	< 10	10	> 95	278
California micro	< 10	25	80-85	160
California micro	16	12	85-90	376
Florida macro	< 10	2	80-90	318
Illinois micro	200	50	90-95	945
Maine macro	< 10	4	< 70	349
New Jersey macro	< 10	5	90	410
New York macro	100	30 +	> 95	872
Oregon macro	< 10	1.5	70-80	360
Texas macro	30	24 +	50-70 ?	481
West Virginia macro	< 10	6	90	400

* Macro — macrolepidoptera
Micro — microlepidoptera

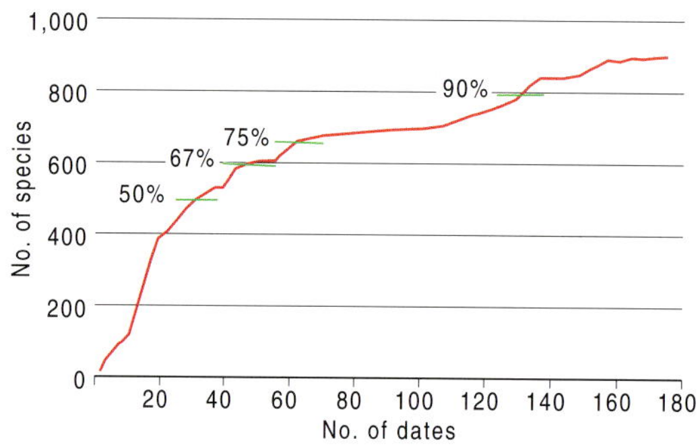

Fig. 2. Species discovery curve for all Lepidoptera at the Big Creek Reserve, Monterey County, California, based on collections during 1980-93. The total (910 species) is believed to be more than 90% of the resident fauna. Points along the curve are indicated when 50%, 67%, 75%, and 90% of the recorded total were reached.

The data from Big Creek and other inventories (e.g., Butler and Kondo 1991) demonstrate that short-term effort is inadequate to inventory insects. We cannot determine faunal composition from a few visits to a site or even comprehensive sampling over one season. If a group under study is relatively uniform in biology, one sampling or trapping technique may be adequate and a steeper species accumulation curve can be attained. At Big Creek, all Lepidoptera accumulation did not reach 50% until 25 dates, or 75% until 65 dates (Fig. 2).

Planning Inventories

A comprehensive inventory should employ diverse sampling approaches, as outlined previously. Light trapping alone may be expected to recover about 75% of the species after extended effort. If monitoring changes in populations is a goal, a subset of the fauna (e.g., one or a few well-known families) should be the focus, with sampling standardized by method (e.g., light trap), site, frequency, and so forth, so as to be repeatable. To make local inventories comparable, data should be identified in several ways: (1) results should be recorded by standardized subsets of the area; (2) sampling effort should be quantified and reported (e.g., number of person-hours or days, dates, UV samples); (3) first records for each species should be recorded to document species discovery rates; (4) voucher specimens should be preserved, especially for small moths, because detailed study by a specialist may be necessary to distinguish species. Ideally, every specimen can be bar-coded to the data base, a rapid process if carried out in tandem with data entry initially as is being done in Costa Rica (Janzen 1992).

We do not know how many species of moths and butterflies live in any state, county, or locality in North America. We need baseline inventories that are standardized by area or sampling effort by which different parts of the continent or tropical faunas can be compared to extrapolate patterns in regional, national, or world biodiversity of Lepidoptera.

References

Bailowitz, R.A. et al. 1990. A checklist of the Lepidoptera of Arizona. Utahensis 10:13-32.

Blanchard, A. et al. 1985. Checklist of Lepidoptera of the Rob and Bessie Welder Wildlife Refuge near Sinton, Texas. The Southwestern Entomologist 10:195-214.

Brower, A.E. 1974-86. A list of the Lepidoptera of Maine. Parts 1, 2. Maine Agricultural Experiment Station, Tech. Bull. 66:1-136; 109:1-60; 114:1-70.

Butler, L., and V. Kondo 1991. Macrolepidopterous moths collected by blacklight trap at Cooper's Rock State Forest, West Virginia: a baseline study. Agriculture and Forestry Experiment Station, West Virginia University, Bull. 705:1-25.

Covell, C.V. 1984. A guide to the moths of eastern North America. Houghton-Mifflin Co., Boston, MA. 496 pp.

Field, W.D., C.F. dos Passos, and J.H. Masters. 1974. A bibliography of the catalogs, lists, faunal and other papers on the butterflies of North America north of Mexico arranged by state and province (Lepidoptera: Rhopalocera). Smithsonian Contributions to Zoology 157:1-104.

Forbes, W.T.M. 1923-60. Lepidoptera of New York and neighboring states, I-IV. Cornell University Agricultural Experiment Station Memoirs 68:1-729; 274:1-263; 329:1-433; 371:1-188.

Janzen, D.H. 1992. Information on the bar code system that INBio uses in Costa Rica. Insect Coll. News 7:24.

Jones, F.M., and C.P. Kimball. 1943. The Lepidoptera of Nantucket and Martha's Vineyard. Publ. Nantucket Maria Mitchell Assoc. 4:1-217.

Kimball, C.P. 1965. The Lepidoptera of Florida. Division of Plant Industry, Florida Department of Agriculture, Gainesville. 363 pp.

McFarland, N. 1965. The moths (Macroheterocera) of a chaparral plant association in the Santa Monica Mountains of southern California. Journal of Res. on Lepidoptera 4:43-74.

Moore, S. 1955. An annotated list of the moths of Michigan, exclusive of Tineoidea (Lepidoptera). Museum of Zoology, University of Michigan, Ann Arbor, Miscellaneous Publ. 88:1-87.

Muller, J. 1965-76. Supplemental list of Macrolepidoptera of New Jersey. Journal of the New York Entomological Society 73:63-77; 76:303-306; 81:66-71; 84:197-201.

Proctor, W. 1946. Biological survey of the Mount Desert region. Part VII. The insect fauna. Wistar Institute of Anatomy and Biology, Philadelphia, PA. 566 pp.

Smith, J.B. 1910. Report of the New Jersey State Museum for 1909. Trenton, NJ. 887 pp.

Tietz, H.M. 1952. Lepidoptera of Pennsylvania. A manual. Pennsylvania State College, Agricultural Experiment Station, State College, PA. 194 pp.

For further information:

Jerry A. Powell
Essig Museum of Entomology
University of California, Berkeley
201 Wellman Hall
Berkeley, CA 94720

Fourth of July Butterfly Count

by
Ann B. Swengel
International Count Co-editor

The Xerces Society started the Fourth of July Butterfly Count (FJC) in 1975, sponsoring it annually until 1993, when the North American Butterfly Association (NABA) assumed administration. The general methods of the butterfly count are patterned after the highly successful Christmas Bird Count (CBC), founded in 1900 and sponsored by the National Audubon Society (Swengel 1990).

Courtesy A.B. Swengel

Painted lady (*Vanessa cardui*) nectaring on showy milkweed (*Asclepias speciosa*).

The results of the FJC, including butterfly data, count-site descriptions, and weather information on count day, are published annually. The count was designed as an informal program for butterfly enthusiasts and the general public. These counts can never substitute for more formal scientific censusing because data sets from the counts have flaws that impair scientific analysis. Nevertheless, the FJC program does provide data that, with considerable caution, can be useful for science and conservation (Swengel 1990). FJC data have been used to study the biology, status, and trends of both rare and widely distributed species (Swengel 1990; Nagel et al. 1991; Nagel 1992; Swengel, unpublished data).

Analysis and Application

I reviewed FJC count reports and other publications for applications of FJC data to monitor the status and trends of North American butterfly species. These studies varied considerably in sample size, amount of data manipulation and statistical analysis, and degree of variable control. Different methods of using FJC data include, in order of ascending statistical refinement: presence or absence of a species in a subset of counts; highest observed number of a species on a single count; individuals of a species per count for a subset of counts in a given year; and individuals of a species per count hours or per count miles. The subset of counts used to supply data for analysis also varied from a single count to all counts in a certain region or all counts ever reporting a given species during the study period. The sample subset and statistical approach are best determined by the nature and extent of available data.

The rationales, methodologies, shortcomings, and validity of analyzing FJC data have been detailed elsewhere (Swengel 1990), but are based on the substantial ornithological literature regarding the scientific use of CBC and other types of survey data. As ornithologists have clearly indicated, these kinds of data sets must be used with great care because (1) the sample sites and dates depend on when and where volunteer observers choose to conduct a count; (2) the quality of sampling and accuracy of data vary among counts; (3) only certain species are sampled adequately enough to allow data interpretation; and (4) the species complex can vary somewhat from year to year. Even with such constraints, these data sets are valuable because of the numerous sites surveyed, their wide geographic scope, and the relatively low cost of data acquisition.

Interpreting Count Data

For the first 11 years of the count program (1975-85), only a few dozen counts were held annually, but since then the number of annual counts has increased steadily to 209 in North America in 1993. Each FJC annual report since 1982 has provided a table that details how many counts reported each species and which single count found the most individuals of each species. Although informal, this table indicates the frequency and abundance of butterfly species as observed in the counts.

Several rare species with federal status under the Endangered Species Act have been sampled in the counts, as reviewed in the introduction to the 1993 FJC annual report (Opler and Swengel 1994). A researcher using FJC to study rare butterflies must be careful in interpreting the data, however. Unless a number of FJC counts are specifically designed to sample rare species well, it is unlikely that rare species will be sampled adequately enough to allow scientific analysis of status and trends. Even in these cases, however, site data for rare species reported in FJC remain useful as leads to follow in status surveys of extant populations for these species (Opler and Swengel 1994). Most likely, the data should be considered as augmenting additional, more formal scientific study and should be confirmed, either by alternative survey means or by contacting the counters for documentation.

Because of the larger sample size, FJC data may better demonstrate the population trends of more abundant and widespread species. For example, the painted lady (*Vanessa cardui*) is a

Courtesy A.B. Swengel

Monarch (*Danaus plexippus*) nectaring on dwarf blazingstar (*Liatris cylindracea*).

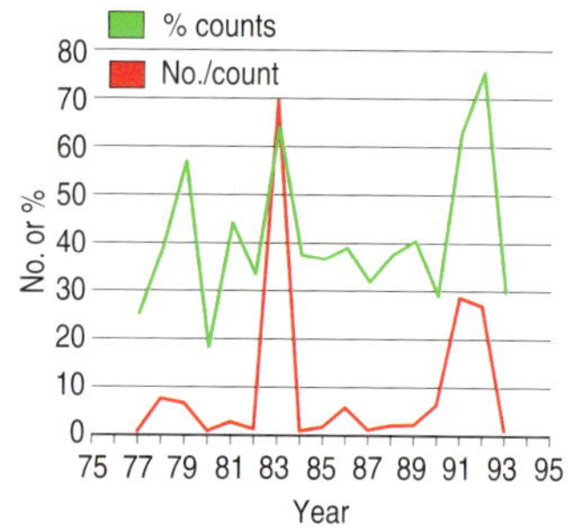

Fig. 1. Number of painted ladies (*Vanessa cardui*) per count and percentage of counts reporting this species, for all counts in North America north of Mexico, 1977-93.

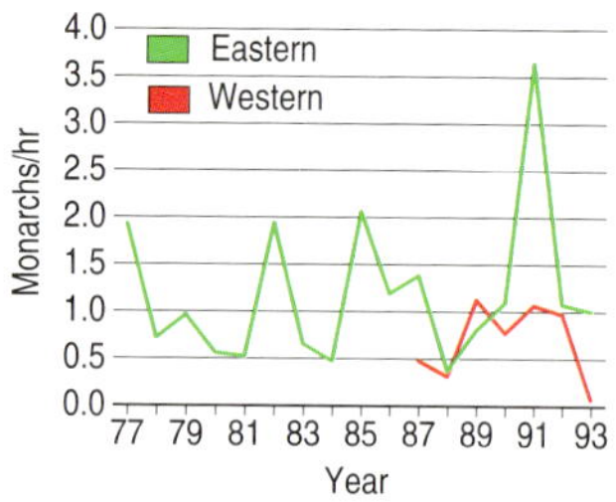

Fig. 2. Mean number of monarchs (*Danaus plexippus*) per party-hour for counts reporting the species east of the Rocky Mountains, 1977-93, and west of the Rocky Mountains, 1987-93.

For further information:

Ann B. Swengel
North American Butterfly Association
909 Birch St.
Baraboo, WI 53913

subtropical species with a tendency to wander (immigrate) outside its residential range into temperate regions, with periodic years of massive invasions. FJC data clearly reflect this aspect of the species' natural history by showing dramatic fluctuations in painted lady frequency and abundance in the counts in 1979, 1983, and 1992 (Swengel 1993; Fig. 1). These outbreaks may correlate with weather perturbations in the species' residential range (Myres 1985; Swengel 1993).

FJC data have also been used to document fluctuations in other immigrant species (Swengel 1990), but especially to monitor population trends of the migrant monarch butterfly (*Danaus plexippus*; Swengel 1990 and unpublished data) that breeds in temperate North America and overwinters in Mexico and coastal California. The number of these butterflies fluctuates considerably (Fig. 2); fluctuations tend to correlate with major climatic perturbations such as the El Niño Southern Oscillation and major volcanic eruptions (Swengel, unpublished data). Monarchs and painted ladies often show dramatic fluctuations in the same years (e.g., 1978-79, 1982-83, 1991-92), but usually they vary in opposite directions (Figs. 1 and 2), suggesting that the same widespread climatic phenomena tend to affect both species in different ways. Because conservationists are concerned about threats to the overwintering habitat of monarchs, long-term data sets such as FJC are valuable to check for persistent downward trends.

While FJC cannot replace more formal and intensive scientific surveying, it does offer a readily available and ever-enlarging data set that, with caution, is useful for science and conservation because of its relative continuity, inexpensiveness, large size, and widespread sampling.

Courtesy A.B. Swengel

Karner blue (*Lycaeides melissa samuelis*) male basking on grass.

References

Myres, M.T. 1985. A southward return migration of painted lady butterflies, *Vanessa cardui*, over southern Alberta in the fall of 1983, and biometeorological aspects of their outbreaks into North America and Europe. The Canadian Field-Naturalist 99:147-155.

Nagel, H. 1992. The link between Platte River flows and the regal fritillary butterfly. The Braided River 4:10-11.

Nagel, H.G., T. Nightengale, and N. Dankert. 1991. Regal fritillary butterfly population estimation and natural history on Rowe Sanctuary, Nebraska. Prairie Naturalist 23:145-152.

Opler, P.A., and A.B. Swengel. 1994. NABA-Xerces Fourth of July butterfly counts 1993 report. North American Butterfly Association, Morristown, NJ. 72 pp.

Swengel, A.B. 1990. Monitoring butterfly populations using the 4th of July butterfly count. American Midland Naturalist 124:395-406.

Swengel, A. 1993. Permutations of painted ladies. American Butterflies1(2):34.

Species Richness and Trends of Western Butterflies and Moths

by
Paul A. Opler
National Biological Service

Butterflies and large moths are among the best-sampled insects and as such are excellent indicators of ecological conditions or environmental change. Because the caterpillars of most Lepidoptera are herbivorous, their species richness is most often a reflection of plant diversity (Brown and Opler 1990).

Management or restoration of invertebrate diversity requires comprehensive data about the status and occurrence of species. I present the species richness of butterflies and three moth families in the 17 western conterminous states and five smaller subareas in the West.

Data Collection

The species richness of western butterflies and moths (Lepidoptera) was determined by using four county-level atlases and counting the number of species recorded in each state or region (Peigler and Opler 1993; Smith 1993; Stanford and Opler 1993; Opler, unpublished data). The county atlases were developed by using specimen data from field surveys, private collections, museums, and scientific monographs. The records analyzed include all historical data; thus the map for a particular species may not represent its current status.

Butterflies (superfamilies Papilionoidea and Hesperioidea), hawkmoths (Sphingidae), silkmoths (Saturniidae), and tiger moths (Arctiidae) are relatively well-sampled groups and therefore give a good preliminary indication of the geographic patterns of species richness. Populations of the selected butterflies and moths in the 17 conterminous western states and five subregions were selected as sampling units.

The five subregions are the lower Rio Grande Valley of South Texas, the Big Bend region of Texas, the Colorado Front Range, the isolated mountains of southeastern Arizona and adjacent New Mexico (the so-called "sky islands"), and southern California south of the Transverse Ranges (Fig. 1). They were selected based on *a priori* knowledge of species richness and patterns of endemic species occurrence.

The number of resident butterflies was determined by counting the number of species recorded for each state or region. Species known to be nonresidents (vagrants or sporadic residents) in a particular state or region were excluded. For the three moth families, all species recorded in a particular state or region, including vagrants, were included in the counts.

The reader should be aware that the quantity and quality of the data are not sufficient to analyze temporal trends for individual species. In addition, all geographic units have not been sampled with equal intensity.

Status and Trends

In the 17 western states, 915 species of butterflies and moths in the studied groups are recorded. The number of species ranges from 181 (20% of total species count) for North Dakota to 520 (57% of total species count) for Texas (Table 1). In general, there are fewer species of butterflies and moths in more northern states and in states with less topographic diversity, which creates less variety in terrain. Of course, larger states tend to have more species than smaller states, since large states, on average, have more diverse habitats and topography. These trends are similar to those of other organisms as well.

The patterns for butterflies and the three moth families are similar, except that species richness of hawkmoths is unexpectedly high in Nebraska and Oklahoma (Table 1), most likely because of the immigration of nonbreeding tropical species (Smith 1993).

Each of the five subregions is smaller than Washington, the smallest western state, yet species richness is greater in all subregions (except the Lower Rio Grande) than in nearly two-thirds (11 out of 17) of the states studied (Table 2). The richest subregion, with 273 species, is the sky islands of southeastern Arizona and southwestern New Mexico. Species richness is second highest in the Front Range of Colorado, which straddles the Continental Divide and includes a large elevational range and diverse habitats ranging from prairie to alpine tundra. The relatively small Lower Rio Grande Valley has the fewest species of the five subregions, but still has more species than some states that are almost 15 times as large. Moreover, the best remaining native habitats in this subregion amount to only a few thousand hectares. Sampling intensity is relatively high for the Front Range, sky islands, and southern California, but increased sampling efforts in the Lower Rio Grande Valley and Big Bend might add significant numbers of species.

Each subregion has a distinct butterfly and moth fauna that includes many endemics—20 or more are potential candidates for listing as endangered species. Each of the four subregions that adjoin the Mexican border also hosts from a few to many Mexican species that occur nowhere else in the United States.

Table 1. Number of species of selected Lepidoptera by state.

State	Area km² (mi²)	Hawk-moths	Silk-moths	Tiger moths	Butter-flies	Total
Arizona	294.0 (113.5)	49	31	111	246	437
California	404.8 (156.2)	30	17	52	225	324
Colorado	268.3 (103.6)	32	18	71	230	351
Idaho	210.9 (81.4)	16	7	24	154	201
Kansas	211.9 (81.8)	23	9	34	133	199
Montana	376.6 (145.4)	10	6	27	184	227
Nebraska	198.4 (76.6)	36	10	38	170	254
Nevada	284.6 (109.9)	18	9	28	181	236
New Mexico	314.2 (121.3)	31	24	83	272	410
North Dakota	179.5 (69.3)	30	3	16	132	181
Oklahoma	177.9 (68.7)	39	13	30	146	228
Oregon	249.2 (96.2)	23	9	28	154	214
South Dakota	196.6 (75.9)	12	7	32	149	200
Texas	678.6 (261.9)	69	34	127	290	520
Utah	212.6 (82.1)	24	14	46	197	281
Washington	172.2 (66.5)	17	8	27	140	192
Wyoming	251.2 (97.0)	18	7	49	197	271
Totals for western U.S.	4,681.5 (1,807.1)	99	68	219	529	915

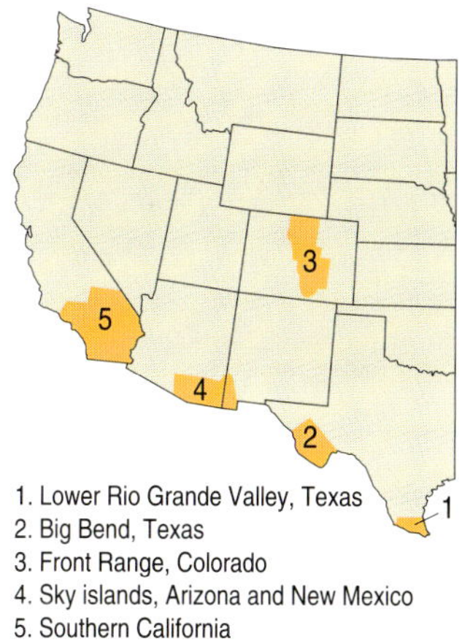

Fig. 1. Western United States showing five subregions of high species richness.

Table 2. Number of species of selected Lepidoptera by subregion.

Regions	Area km²(mi²)	Hawk-moths	Silk-moths	Tiger moths	Butter-flies	Total
Lower Rio Grande[a]	14.0 (5.4)	33	11	11	115	170
Big Bend[b]	37.3 (14.4)	28	10	10	151	199
Front Range[c]	71.5 (27.6)	26	11	11	176	224
Sky islands[d]	41.4 (16.0)	41	25	25	182	273
Southern California[e]	116.6 (45.0)	23	14	33	159	229

[a]Lower Rio Grande Valley includes all of Cameron, Hidalgo, Willacy, and Starr counties, TX.
[b]Big Bend includes all of Brewster, Jeff Davis, and Presidio counties, TX .
[c]The Colorado Front Range includes all of Boulder, Clear Creek, Custer, Douglas, El Paso, Fremont, Gilpin, Grand, Huerfano, Jackson, Jefferson, Larimer, Park, Pueblo, Summit, and Teller counties.
[d]Sky islands include all of Cochise and Santa Cruz counties, the eastern half of Pima County, AZ, and all of Hidalgo County, NM.
[e]Southern California includes all of Imperial, Los Angeles, Orange, Riverside, San Bernardino, and San Diego counties.

Two tailed swallow-tail (*Papilio multicaudata*).

The highest species richness of western Lepidoptera is in the Southwest, usually in areas that adjoin the Mexican border. Invertebrates are seldom considered in management plans for parks, preserves, or refuges, and their management needs are often not the same as those for vertebrate wildlife or plants. Processes unfavorable to Lepidoptera diversity include overgrazing, overuse of controlled burns, urbanization, and excessive modification or recreational use of selected specialized ecosystems such as wetlands and dunes. Because invertebrates account for more than 90% of animal species, it makes good sense for managers to address the health and populations of these species in planning and in making management decisions. Management which favors high Lepidoptera species richness is usually similar to that which favors natural ecosystem processes and the maintenance of extensive native plant populations.

Tiger moth (*Gnophaela vermiculata*).

For further information:

Paul A. Opler
National Biological Service
Information Transfer Center
1201 Oak Ridge Dr.
Suite 200
Fort Collins, CO 80525

References

Brown, J.W., and P.A. Opler. 1990. Patterns of butterfly species density in peninsular Florida. Journal of Biogeography 17:615-622.

Peigler, R.S., and P.A. Opler. 1993. Moths of western North America. 1. Distribution of Saturniidae of western North America. Contributions of the C.P. Gillette Insect Biodiversity Museum, Colorado State University, Fort Collins. 24 pp.

Smith, M.J. 1993. Moths of western North America. 2. Distribution of Sphingidae of western North America. Contributions of the C.P. Gillette Insect Biodiversity Museum, Colorado State University, Fort Collins. 34 pp.

Stanford, R.E., and P.A. Opler. 1993. Atlas of western USA butterflies, including adjacent parts of Canada and Mexico. Published by authors, Denver, CO. 275 pp.

Courtesy P.A. Opler, NBS

Sphinx moth (*Proserpinus juanita*).

The Tall-grass Prairie Butterfly Community

by
Ann B. Swengel
Scott R. Swengel
Baraboo, Wisconsin

The prairie biome is a plant community dominated by grasses and nongrassy herbs (wildflowers or "forbs"), with some woody shrubs and occasional trees. Prairie is classified into three major types by rainfall and consequent grass composition. The easternmost and moistest division is the tall-grass prairie (Risser et al. 1981). Although tall-grass prairie once broadly covered the middle of the United States (Fig. 1), this biome is now estimated to be at least 99% destroyed from presettlement by pioneers, who converted it for agricultural uses. Prairie loss continues through plowing, extreme overgrazing, and development, but at varying degrees. Prairie is also lost passively because the near-total disruption of previous ecological processes causes shifts in floristic composition and structure.

As a result of this habitat destruction, butterflies and other plants and animals that are obligate to the prairie ecosystem are rare and primarily restricted to prairie preserves. The Dakota skipper (*Hesperia dacotae*) and the regal fritillary (*Speyeria idalia*) are federal candidates for listing under the Endangered Species Act, and additional prairie butterfly species are on state lists as officially threatened or endangered. Patches of original prairie vegetation remain in preserves, parks, unintensively used farmlands such as hayfields and pastures, and in unused land. These remnants of prairie, however, are isolated and often in some state of ecological degradation.

The existence of prairie depends on the occurrence of certain climatic conditions and disturbance processes such as animal herbivory and fire. These natural processes, however, are severely disrupted today because of the destruction and fragmentation of the prairie biome. Without management intervention, the vegetational composition and structure of prairie sites are altered through invasion of woody species and smothering under dead plant matter. Prairie usually requires active management to maintain the ecosystem and its biodiversity, but it is difficult to know exactly which processes once naturally maintained the prairie ecosystem. Frequent fire, whether caused by lightning or set by native peoples, is usually considered the dominant prehistoric process that maintained prairie; thus management for tall-grass prairie in most states relies primarily or solely on frequent fire (e.g., Sauer 1950; Hulbert 1973; Vogl 1974). Other researchers (e.g., England and DeVos 1969), however, assert that prairie was the result of grazing by large herds of ungulates as in the Serengeti in Africa.

Despite this scientific conflict, it appears certain that successful management for maintaining the prairie landscape and its native species should be based on these natural processes, whatever they were. The vast diversity and specificity of insects to certain plants and habitat features make them fine-tuned ecological indicators. Thus, butterfly conservation is useful not only for maintaining these unique species, but also for helping us monitor and learn about the soundness of our general ecosystem management.

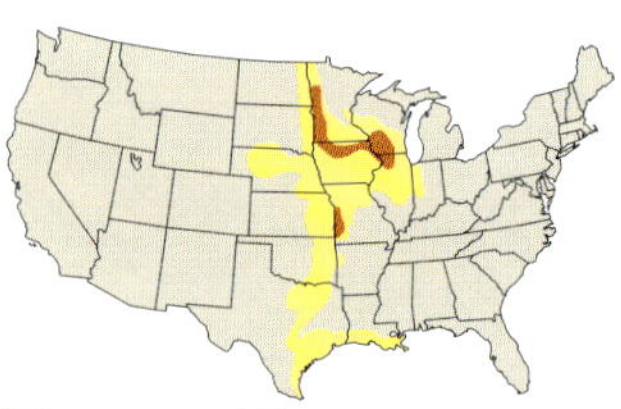

Fig. 1. Original boundaries of the tall-grass prairie biome in the United States (Risser et al. 1981) and locations of study sites (A.B. Swengel, unpublished data).

Survey and Classification

We counted 90 butterfly species and 80,906 individuals in surveys from 1988 to 1993 at 93 prairies varying from 1 to 445 ha (3 to 1,100 acres) in the Upper Midwest (Illinois, Iowa, Minnesota, Wisconsin) and southwestern Missouri (Fig. 1). Most sites are managed principally with fire, with burns averaging about 25% (range 0-99% or more) of the prairie patch per year. Many Missouri sites are managed primarily with summer haying along with a little burning and cattle grazing. The vegetation in each survey unit was relatively uniform.

Any species observed 100 or more times was designated a study species. Before analyzing the results, we classified the study species by habitat niche breadth: prairie specialist, grassland, generalist, and invader. We used population indices (individuals observed per hr in each unit) to identify which units had relatively greater densities of particular species and which factors might account for this variation. Details regarding the survey and statistical methodologies are provided elsewhere (A.B. Swengel, unpublished data).

Management and Distribution

The overwhelming destruction of prairie habitat has had disastrous consequences for prairie-specialist butterflies, not just because of the outright loss of appropriate living space but also because of habitat fragmentation. Because prairie-specialist butterflies are rarely encountered outside of these fragmented prairie patches, populations at different sites may have minimal gene flow and are rarely able to recolonize sites of local extinctions. For example, the regal fritillary is the most widespread prairie butterfly species, but it requires larger habitat patches or connected networks of habitat patches to maintain populations. The arogos skipper *(Atrytone arogos iowa*) and ottoe skipper (*Hesperia ottoe*) also occur widely in the prairie biome but are more restricted in their habitat requirements, resulting in more localized and spotty distributions. The Dakota skipper and poweshiek skipper *(Oarisma poweshiek*) are most restricted in range, occurring only in northern prairie, and have further habitat restrictions within that range. As a result, the northern Midwest (northwestern Iowa, western Minnesota, and the eastern Dakotas) is the region where tall-grass prairie conservation has the most potential for maintaining the greatest diversity of prairie-specialist butterflies.

Regal fritillaries (*Speyeria idalia*) mating on pale purple coneflower (*Echinacea pallida*).

Our surveys show that the management occurring at a prairie critically affects whether prairie-specialist butterflies exist at the site and at what abundance. Although each butterfly species has its own response to fire, the prairie specialists show a pronounced and statistically significant decline after fire; this decline persists 4 or more years (A.B. Swengel, unpublished data). Species with the broadest habitat adaptation (invaders) are most abundant in recently burned units and least abundant in units left unburned the longest. Species of intermediate adaptations (grasslands, generalists) showed milder, intermediate trends.

Unintensive haying management (a single annual or biennial cutting with removal of the clipped vegetation) is more favorable for butterfly diversity. Such haying is more beneficial for butterflies sooner after treatment and causes a less pronounced variation in butterfly abundance between different treatment years. In general, butterflies are more abundant in the first years after haying than after burning; specialists account for much of this difference (Fig. 2). Our limited opportunities to test light grazing show that it may also serve specialist butterflies better than fire.

Prairie-specialist butterflies apparently respond to different management types because of varying degrees of mortality (e.g., fire causes more direct mortality than haying or grazing) and because of differences in continuity of required habitat resources (e.g., fire removes all cover but is followed by regrowth of thick cover, while unintensive haying and grazing can more consistently maintain moderate cover). Management also indirectly affects butterfly populations by altering the abundance and occurrence of plants they depend on as well as the vegetational structure and physical features they require.

These results are consistent with butterfly conservation experience around the world, particularly in Europe and Australia (Butterflies Under Threat Team 1986; Kirby 1992; New 1993). Thus, simply preserving habitat is not sufficient to conserve insect biodiversity; suitable management approaches and land uses compatible with the habitat's native biodiversity must be preserved. It is possible to maintain plants successfully without protecting the associated animals, but it is impossible to maintain the associated animals successfully without protecting the plants.

It appears desirable for managers to aim for diversity and patchiness in prairie-management approaches within and among sites rather than broadly applying a single management formula for prairie everywhere. Whether or not a site is managed specifically to conserve insects, declines and extirpations of insects specialized

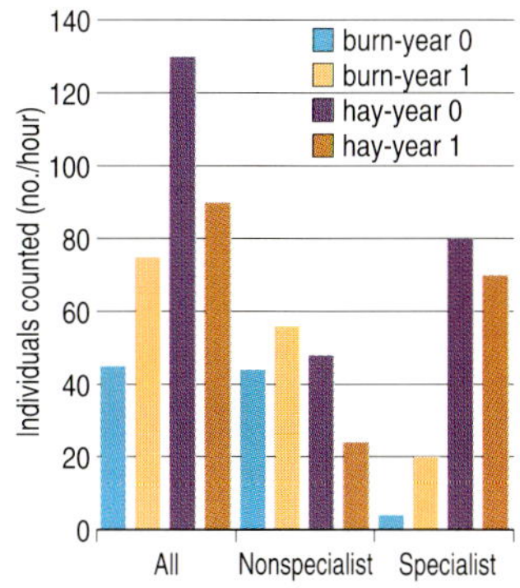

Fig. 2. Abundance of all, nonspecialist (grassland, generalists, and invaders), and prairie-specialist study species in the first years of fire and hay management, Missouri study sites.

to the habitat indicate that ecological degradation has already occurred there, while maintenance of these species indicates success in ecosystem conservation. Because we found that management with mechanical cutting or light grazing appears most effective for maintaining both the prairie habitat and its associated specialist insects (seeming to indicate an ecosystem adaptation to herbivory), we recommend that these methods should have a primary role in modern prairie management for the conservation of biodiversity. There is cause for optimism, however, because no known prairie butterfly species have gone extinct, despite their rarity. Instead, these species have persisted on habitat remnants, showing that appropriate habitat preservation and management should translate into readily measurable conservation successes.

For further information:

Ann B. Swengel
909 Birch St.
Baraboo, WI 53913

References

Butterflies Under Threat Team. 1986. The management of chalk grassland for butterflies. Joint Nature Conservation Committee, Peterborough, U.K. 80 pp.

England, R.E., and A. DeVos. 1969. Influence of animals on pristine conditions on the Canadian grasslands. Journal of Range Management 22:87-94.

Hulbert, L.C. 1973. Management of Konza Prairie to approximate pre-whiteman influences. Pages 14-19 *in* L.C. Hulbert, ed. Third Midwest prairie conference proceedings. Kansas State University, Manhattan.

Kirby, P. 1992. Habitat management for invertebrates: a practical handbook. Royal Society for the Protection of Birds, Bedfordshire, U.K. 149 pp.

New, T.R. 1993. Conservation biology of Lycaenidae (butterflies). IUCN, Gland, Switzerland. 173 pp.

Risser, P.G., E.C. Birney, H.D. Blocker, S.W. May, W.J. Parton, and J.A. Wiens. 1981. The true prairie ecosystem. Hutchinson Ross Publishing Co., Stroudsburg, PA. 557 pp.

Sauer, C. 1950. Grassland climax, fire and management. Journal of Range Management 3:16-20.

Vogl, R.J. 1974. Effect of fire on grasslands. Pages 139-194 *in* T.T. Kozlowski and C.E. Ahlgren, eds. Fire and ecosystems. Academic Press, New York.

The Biota of Illinois Caves and Springs

by
Donald W. Webb
Illinois Natural History Survey

Caves and springs tend to be inhabited by a highly specialized and intolerant diversity of vertebrate and invertebrate species. Ongoing research on the aquatic and terrestrial macroinvertebrates and terrestrial vertebrates inhabiting 105 springs and caves in Illinois (Figure) surveyed from 1990 to 1993 has verified the uniqueness of this biota and highlighted the very fragile ecosystem in which these organisms survive. Data on more than 8,000 invertebrate specimens, representing 4 phyla, 11 classes, and 32 orders, have been collected and the data entered into a data base. More than 2,500 specimens and 27 species of vertebrates (3 fishes, 7 salamanders, 4 frogs, 1 turtle, 4 birds, 1 raccoon, and 7 bats) were observed in caves, dominated by the salamanders and bats.

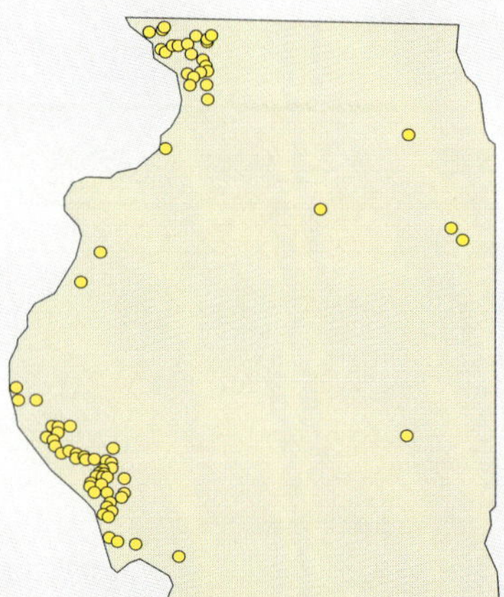

Figure. Distribution of springs and caves in Illinois.

The water chemistry of the Illinois springs and cave streams was typical of most hardwater springs, although nitrate levels in one spring and one cave stream in the karst region of Monroe County exceeded the Illinois Pollution Control Board's Maximum Contamination Level of 10 mg/L (10 ppm), raising concern over the effects of agricultural runoff on the biota of Illinois cave streams. The detection of mercury in the tissue of amphipods and isopods was noted, although no detectable level of mercury was determined in any of the water samples tested.

Karst limestone regions have sinks, underground streams, and caves. Qualitative collections of invertebrates and observations of vertebrates were made to determine species richness and the spatial distribution of each species. In caves, habitat selection and cave preference (entrance, twilight, and dark zones) were examined for aquatic invertebrates and terrestrial vertebrates and invertebrates.

The aquatic macroinvertebrates were dominated in abundance and diversity by noninsect arthropods, several of which are currently on federal and state endangered species lists (e.g., the amphipod *Gammarus acherondytes*). In terms of abundance, the amphipods *Gammarus minus* and *G. pseudolimneaus* and the turbellarian *Phagocata gracilis* dominated surface springs, while the amphipod *G. troglophilus* dominated cave streams. The diversity of oligochaete worms, with 24 taxa, proved to be the most surprising feature of the study, especially because several unidentified taxa of worms were collected that may be new species. *Varichaetadrilus angustipenis,* although previously collected only rarely in Illinois, was recorded from numerous springs. The collection of *Allonais paraguayensis* in Old Driver Spring was the most interesting find; this species has been reported only from a locality in Louisiana and an aquarium in New York. The presence of *A. paraguayensis* in Illinois represents a significant range extension for this species. The occurrence of unidentifiable taxa of Lumbricidae and Lumbriculidae also poses interesting systematic questions.

Aquatic macrophytes were scarce in most springs examined, although the moss *Leptodictyum riparium* was abundant in the spring head of Old Driver Spring and the forb *Mentha piperita* plugged the upper reaches of the outflow channel of Old Driver and Rose springs.

The terrestrial fauna of the cave was dominated by insects (heleomyzid and mycetophilid flies, collembolans, carabid and staphylinid beetles, and camel crickets), amphibians (seven species), and bats (seven species). The federally listed endangered gray bat (*Myotis grisescens*) was observed in one cave, and the Indiana bat (*M. sodalis*) in six caves. The state-listed endangered southeastern bat (*M. austroriparius*) was observed in two caves. The federally listed endangered Pleistocene disc snail (*Discus macclintocki*) is known from one cave in northwestern Illinois.

Implications of Surveys

In Illinois, the biota of springs and cave streams typifies the hypothesis that hardwater springs in eastern North America are dominated by noninsect macroinvertebrates (Glazier 1991). Although amphipods and turbellarians were the most abundant organisms in surface springs, it was the diversity evident within the oligochaete worms that proved the most exciting feature of surface springs. Twenty-four taxa, four of which may prove new to science, and several new state records were found. Several new localities for the spring cavefish *Forbesichthys agassiz* were also discovered. In the cave streams, the amphipods were the most diverse and abundant macroinvertebrates, in particular the troglobitic amphipod *Gammarus troglophilus*. Six state-endangered macroinvertebrates are known from Illinois caves.

References

Glazier, D.S. 1991. The fauna of North American temperate cold springs: patterns and hypotheses. Freshwater Biology 26:527-542.

For Further Reading on Cave and Spring Biota

Colbo, M.H. 1991. A comparison of the spring-inhabiting genera of Chironomidae from the Holarctic with those from natural and manmade springs in Labrador, Canada. Memoirs of the Entomological Society of Canada 155:169-179.

Erman, N.A., and D.C. Erman. 1990. Biogeography of caddisfly (Trichoptera) assemblages in cold springs of the Sierra Nevada (California, USA). Contribution 200, California Water Resources Center, University of California, Riverside. 28 pp.

Forbes, S.A. 1882. The blind cave fishes and their allies. American Naturalist 16(1):1-5.

Forester, R.M. 1991. Ostracode assemblages from springs in the western United States: implications for paleohydrology. Memoirs of the Entomological Society of Canada 155:181-201.

Gardner, J.E. 1991. Illinois caves: a unique resource. Pages 447-452 *in* L.M. Page and M.R. Jeffords, eds. Our living heritage: the biological resources of Illinois. Bull. of the Illinois Natural History Survey 34(4):357-475.

Glazier, D.S., and J.L. Gooch. 1987. Macroinvertebrate assemblages in Pennsylvania (U.S.A.) springs. Hydrobiologia 150:33-43.

Gooch, J.L., and D.S. Glazier. 1991. Temporal and spatial patterns in mid-Appalachian springs. Memoirs of the Entomological Society of Canada 155:29-49.

Peck, S.B., and K. Christiansen. 1990. Evolution and zoogeography of the invertebrate cave faunas of the driftless area of the upper Mississippi River valley of Iowa, Minnesota, Wisconsin, and Illinois, USA. Canadian Journal of Zoology 68(1):73-88.

Peck, S.B., and J.J. Lewis. 1977. Zoogeography and evolution of the subterranean invertebrate faunas of Illinois and southeastern Missouri. Bull. of the National Speleological Society 40(2):39-63.

Pritchard, G. 1991. Insects in thermal springs. Memoirs of the Entomological Society of Canada 155:89-106.

Roughley, R.E., and D.J. Larson. 1991. Aquatic Coleoptera of springs in Canada. Memoirs of the Entomological Society of Canada 155:125-140.

Smith, I.M. 1991. Water mites (Acari: Parasitengona: Hydrachnida) of spring habitats in Canada. Memoirs of the Entomological Society of Canada 155:141-167.

Webb, D.W. 1993. Status survey for a cave amphipod, *Gammarus acherondytes*, Hubricht and Mackin (Crustacea: Amphipoda) in southern Illinois. Illinois Natural History Survey, Center for Biodiversity Tech. Rep. 1993(9):1-8.

Webb, D.W., P.C. Reed, and M.J. Wetzel. 1992. The springs of Illinois: a report on the fauna, flora, and hydrogeology of six basic-water springs in southern Illinois. Report to the Illinois Nature Preserves Commission. Unpublished. 41 pp.

Webb, D.W., S.J. Taylor, and J.K. Krejca. 1993. The biological resources of Illinois caves and other subterranean environments. Determination of the diversity, distribution, and status of the subterranean faunas of Illinois caves and how these faunas are related to groundwater quality. Illinois Natural History Survey, Center for Biodiversity Tech. Rep. 1993 (8):1-157.

Weise, J.G. 1957. The spring cave-fish, *Chologaster papilliferus*, in Illinois. Ecology 38:195-204.

Williams, D.D. 1991. Life history traits of aquatic arthropods in springs. Memoirs of the Entomological Society of Canada 155:63-87.

Williams, N.E. 1991. Geographical and environmental patterns in caddisfly (Trichoptera) assemblages from coldwater springs in Canada. Memoirs of the Entomological Society of Canada 155:107-124.

For further information:

Donald W. Webb
Illinois Natural History Survey
Center for Biodiversity
607 East Peabody Dr.
Champaign, IL 61820

Freshwater Mussels: A Neglected and Declining Aquatic Resource

by
James D. Williams
Richard J. Neves
National Biological Service

The United States has the greatest diversity of freshwater mussels in the world. Of the five families and roughly 1,000 species occurring globally, nearly 300 species and subspecies in the families Unionidae and Margaritiferidae reside here (Turgeon et al. 1988). The number of mussels historically known for each state varies tremendously (Fig. 1), but the diversity of freshwater mussels in just the Southeast is unmatched by any other area in the world.

Mussels were an important natural resource for Native Americans, who used them for food, tools, and jewelry. During the late 1800's and early 1900's, mussel shells supported an important commercial fishery; shells were used to manufacture pearl buttons until the advent of plastic buttons in the 1940's. Today the commercial harvest of freshwater mussel shells is exported to Asia for the production of spherical beads that are inserted into oysters, freshwater mussels, and other shellfish to produce pearls.

There are no federal regulations on the harvest of mussels, except those species on the federal list of endangered or threatened species. Several states, however, regulate size, species, gear used, and season that mussels can be taken. Japanese demand for the high-quality U.S. mussel shells in recent years pushed the price to $13/kg ($6/lb) in 1991. Shell exports peaked in 1991 at more than 8 million kg (9,000 tons), but demand declined in 1992 and 1993 and has leveled off to about 4 million kg (4,500 tons; Baker 1993).

Determining Status

In reviewing the conservation status of freshwater mussels, we included all species and subspecies recognized in the American Fisheries Society list of common and scientific names of mollusks from the United States and Canada

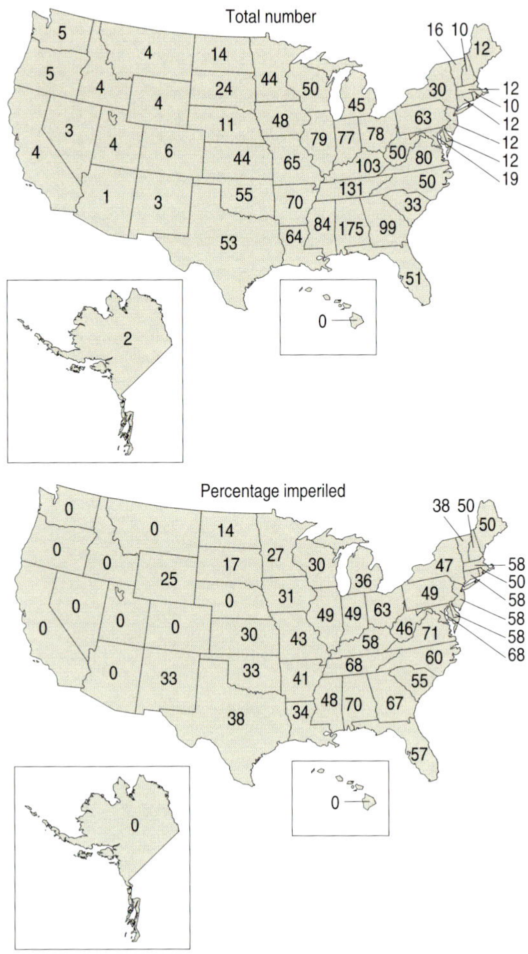

Fig. 1. Number of species and subspecies of freshwater mussels historically known to occur within each state and the percentage now classified as imperiled.

(Turgeon et al. 1988). Distribution data and conservation status were obtained from research publications, books, original data from biologists, and a recent synopsis by Williams et al. (1993).

The status categories were based on information for each species throughout its geographic range. The conservation status categories for a mussel species were defined as follows: endangered—in danger of extinction throughout all or a significant portion of its range; threatened—is likely to become endangered throughout all or a significant portion of its range; special concern—may become threatened or endangered by relatively minor disturbances to its habitat; undetermined—historical and current distribution and abundance have not been evaluated recently; and currently stable—distribution and abundance are seemingly stable, or may have declined in portions of range but not in need of immediate conservation.

Freshwater mussels from the Tombigbee River at Memphis Landing, Pickens County, Alabama. Southern combshell (*Epioblasma penita*); female, top, male, bottom.

Decline of Mussels

The decline of freshwater mussels, which began in the late 1800's, has resulted from various habitat disturbances, most significantly, modification and destruction of aquatic habitats by dams and pollution. Freshwater habitats suffer not only from direct alterations by humans but indirectly from abuse of terrestrial habitats, such as from siltation, especially evident if one compares the levels of imperilment of aquatic versus terrestrial species. Master (1990) recognized 55% of North America's mussels as extinct or imperiled, compared to only 7% of the continent's bird and mammal species.

Aquatic habitat loss comes in a variety of forms such as from effects of dams, dredging, and channelization, or from more subtle effects of siltation and contaminants associated with construction and agriculture. Dams, with their altered flow regimes and attendant reservoirs, have caused the extirpation of 30%-60% of the native mussel species in selected U.S. rivers (Williams et al. 1992; Layzer et al. 1993). Siltation resulting from deforestation, poor agricultural and land-use practices, and removal of riparian vegetation can destabilize the stream bottom and eliminate benthic organisms such as mollusks (Ellis 1931). Many streams that look healthy can be polluted by contaminants like heavy metals, pesticides, and acid mine drainage. The effects of pollution and habitat alteration on mussels were reviewed by Fuller (1974).

Competition from non-native mollusks also has contributed to the loss of native mussel populations. The Asian clam (*Corbicula fluminea*), introduced to the U.S. west coast in the 1930's, has invaded nearly every watershed nationwide (McMahon 1983). Local population explosions of the Asian clam have adversely affected some, but not all, native mussels (Belanger et al. 1990; Leff et al. 1990). The recently introduced zebra mussel (*Dreissena polymorpha*) appears poised to decimate many of the remaining mussel populations. Zebra mussels were discovered in the United States at Lake St. Clair in 1988 and spread rapidly throughout the Great Lakes. In 1991 they were found in the Illinois River, and by late 1991 had spread to the Tennessee River (Nalepa and Schloesser 1992). They are now found throughout the Mississippi River and portions of its major tributaries, even to southern Louisiana. During the next 10-20 years, zebra mussels will most likely spread throughout most of the United States and southern Canada.

The adverse modification and destruction of aquatic habitats, along with the introduction of nonindigenous species, have resulted in the decline of freshwater mussels. The percentage of imperiled mussel species for eastern states is high (Fig. 1). Of the 297 native mussel species in the United States, 71.7% are considered endangered, threatened, or of special concern (Fig. 2), including 21 mussels that are endangered and presumed extinct. Seventy species (23.6%) are considered to have stable populations (Fig. 2), although many of these also have declined in abundance and distribution. Many species in the latter group occur in larger rivers and reservoirs and are projected to suffer severe declines as the zebra mussel invades these ecosystems.

The rapid decline of mussels during this century went almost unnoticed until the past 30 years. Although most of the described threats to survival of mussels have existed for more than a century, the increased geographic area covered by these threats and the cumulative effects of human expansion and development have now overwhelmed aquatic systems.

The demise in both populations and species diversity of our mussel fauna is likely occurring in other freshwater mollusks (especially snails) and aquatic organisms, but too few surveys have been conducted to record such trends. Conservation and restoration should focus on the ecosystem and watershed level instead of directing concerns to the individual species. To effectively carry out such a broad recovery effort will require an unparalleled level of cooperation and coordination of private, state, and federal agencies. Perhaps even more critical to the success of ecosystem and watershed conservation is the involvement of the general public, conservation organizations, and private corporations. If the decline of aquatic mollusks continues, we will witness the greatest extinction of these organisms experienced in modern times.

References

Baker, P.M. 1993. Resource management: a shell exporter's perspective. Pages 69-71 *in* K.S. Cummings, A.C. Buchanan, and L.M. Koch, eds. Conservation and Management of Freshwater Mussels. Proceedings of a symposium. Illinois Natural History Survey, Champaign.

Belanger, T.V., C.G. Annis, and D.D. VanEpps. 1990. Growth rates of the Asiatic clam, *Corbicula fluminea*, in the upper and middle St. Johns River, Florida. Nautilus 104:4-9.

Ellis, M.M. 1931. Some factors affecting the replacement of the commercial fresh-water mussels. U.S. Bureau of Fisheries Fishery Circular 7:1-10.

Fuller, S.L.H. 1974. Clams and mussels (Mollusca: Bivalvia). Pages 215-273 *in* C.W. Hart and S.L.H. Fuller, eds. Pollution ecology of freshwater invertebrates. Academic Press, New York.

Layzer, J.B., M.E. Gordon, and R.M. Anderson. 1993. Mussels: the forgotten fauna of regulated rivers. A case study of the Caney Fork River. Regulated Rivers: Research and Management 8:63-71.

Leff, L.G., J.L. Burch, and J.V. McArthur. 1990. Spatial distribution, seston removal, and potential competitive interactions of the bivalves *Corbicula fluminea* and *Elliptio complanata*, in a coastal plain stream. Freshwater Biology 24:409-416.

Master, L. 1990. The imperiled status of North American aquatic animals. Biodiversity Network News 3:1-2, 7-8.

McMahon, R.F. 1983. Ecology of an invasive pest bivalve, *Corbicula*. Pages 505-561 *in* W.D. Russell-Hunter, ed. The Mollusca. Vol. 6. Ecology. Academic Press, New York.

Nalepa, T.F., and D.W. Schloesser, eds. 1992. Zebra mussels: biology, impacts, and control. Lewis Publishers, Boca Raton, FL.

Turgeon, D.D., A.E. Bogan, E.V. Coan, W.K. Emerson, W.G. Lyons, W.L. Pratt, C.F.E. Roper, A. Scheltema, F.G. Thompson, and J.D. Williams. 1988. Common and scientific names of aquatic invertebrates from the United States and Canada: Mollusks. American Fisheries Society Special Publ. 16. 277 pp.

Williams, J.D., S.L.H. Fuller, and R. Grace. 1992. Effects of impoundment on freshwater mussels (Mollusca: Bivalvia: Unionidae) in the main channel of the Black Warrior and Tombigbee rivers in western Alabama. Bull. Alabama Museum of Natural History 13:1-10.

Williams J.D., M.L. Warren, Jr., K.S. Cummings, J.L. Harris, and R.J. Neves. 1993. Conservation status of freshwater mussels of the United States and Canada. Fisheries 18(9):6-22.

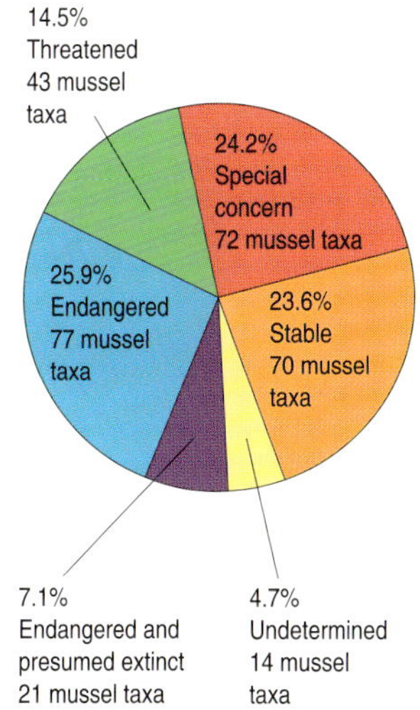

Fig. 2. The percentage of the U.S. mussel fauna classified by conservation status category: undetermined, endangered and presumed extinct, endangered, threatened, special concern, and stable.

For further information:

James D. Williams
National Biological Service
Southeastern Biological Science Center
7920 N.W. 71st St.
Gainesville, FL 32653

Freshwater Mussels in the Lake Huron-Lake Erie Corridor

by
Don W. Schloesser
National Biological Service
Thomas F. Nalepa
National Oceanic and Atmospheric Administration

An early indicator of adverse human effects on large open-water systems in North America was western Lake Erie, part of the Lake Huron-Lake Erie corridor of the Laurentian Great Lakes (Fig. 1). Local pollution of tributaries of western Lake Erie was recognized as early as 1890, when populations of whitefish (Salmonidae) and lake herring (*Coregonus artedi*) in the Detroit River declined (Beeton 1961). Waters of western Lake Erie stopped yielding whitefish and herring in the 1920's-30's, but not until the 1950's, after extensive biological investigations, were the open waters of western Lake Erie believed to have been polluted by human "local" activities (National Academy of Sciences 1970). Eutrophication (the addition of nutrients) of western Lake Erie created unsuitable conditions (primarily low dissolved oxygen concentrations) for fish and other animals in a major portion of Lake Erie—the world's 12th largest lake. By the early 1960's, Lake Erie was declared "biologically dead" (Burns 1985).

Among the many ecosystem components affected by human-induced changes to western Lake Erie (Burns 1985) is the native mussel fauna (Bivalvia: Unionidae). Reduced mussel populations that survived degraded conditions of the 1950's have been used in status and trends studies to evaluate traditional forms of pollution in western Lake Erie. Studies in the 1990's have focused on evaluating the effects of exotic species on mussel populations in the Lake Huron-Lake Erie corridor. Exotic species have recently been characterized as "biological pollution," a new concept in evaluating status and trends data. Our study shows both historical, long-term effects from human activities and recent, dramatic effects from exotic species on mussel populations in waters of the Great Lakes.

Sampling Populations

The Lake Huron-Lake Erie corridor receives water from three of the five Laurentian Great Lakes, the largest freshwater system in the world

(Fig. 1). Relatively pristine water enters the St. Clair River, passes through Lake St. Clair and the Detroit River, and enters western Lake Erie.

Freshwater native mussels were collected by scuba divers in the Lake Huron-Lake Erie corridor (Fig. 1) at 46 stations during six sampling periods from 1961 to 1992. In Lake St. Clair, mussels were collected at 29 stations in 1986, 1990, and 1992. Ten replicate quadrate samples (0.5 m^2 each [5.4 ft^2]) were obtained at each station and sampling date. In western Lake Erie, mussels were collected four times at one index station in 1989-91 and once at 17 historically sampled stations in 1961, 1982, and 1991. Sampling at the index station was performed with an epibenthic sled (46 x 25 cm [18 x 63 in]). Sampling at the 17 historically sampled stations was performed with a Ponar grab sampler. Three replicate Ponar (0.05 m^2 [0.5 ft^2]) samples of the substrate were collected at each station. Mussels were identified following Clarke (1981) and comparisons with bivalve taxonomic reference collections. Taxonomic nomenclature follows Turgeon et al. (1988) and Williams et al. (1993).

Fig. 1. The Lake Huron-Lake Erie corridor, including Lake St. Clair and western Lake Erie (in red).

Table 1. Species of native mussels historically found in the Lake Huron-Lake Erie corridor of the Great Lakes (modified from Clarke and Stansbery 1988).

Species
Mucket (*Actinonaias ligamentina [carinata]*)
Elktoe (*Alasmidonta marginata*)
Slippershell mussel (*A. viridis*)
Threeridge (*Amblema plicata plicata*)
Cylindrical papershell (*Anodontoides ferussacianus*)
Purple wartyback (*Cyclonaias tuberculata*)
Spike (*Elliptio dilatata*)
Northern riffleshell (*Epioblasma torulosa rangiana*)
Snuffbox (*E. triquetra*)
Wabash pigtoe (*Fusconaia flava*)
Wavy-rayed lampmussel (*Lampsilis fasciola*)
Pocketbook (*L. ovata*)
Eastern lampmussel (*L. siliquoidea*)
White heelsplitter (*Lasmigona complanata complanata*)
Creek heelsplitter (*L. compressa*)
Fluted-shell (*L. costata*)
Fragile papershell (*Leptodea fragilis*)
Eastern pondmussel (*Ligumia nasuta*)
Black sandshell (*L. recta*)
Threehorn wartyback (*Obliquaria reflexa*)
Hickorynut (*Obovaria olivaria*)
Round hickorynut (*O. subrotunda*)
Round pigtoe (*Pleurobema coccineum*)
Ohio pigtoe (*P. cordatum*)
Pink heelsplitter (*Potamilus alatus*)
Pink papershell (*P. ohiensis*)
Kidneyshell (*Ptychobranchus fasciolaris*)
Giant floater (*Pyganodon grandis*)
Mapleleaf (*Quadrula quadrula*)
Pimpleback (*Q. pustulosa pustulosa*)
Salamander mussel (*Simpsonaias ambigua*)
Squawfoot (*Strophitus undulatus*)
Lilliput (*Toxolasma parvus*)
Fawnsfoot (*Truncilla donaciformis*)
Deertoe (*T. truncata*)
Pondhorn (*Uniomerus tetralasmus*)
Paper pondshell (*Utterbackia imbecillis*)
Rayed bean (*Villosa fabalis*)
Rainbow (*V. iris*)

Table 2. Number of species of native mussels and average (mean) density (number/m^2) in Lake St. Clair and western Lake Erie of the Lake Huron-Lake Erie corridor, 1961-92.

Lake/year	Total no. of species	Average (mean) no/m^2
Lake St. Clair		
1986	18	2
1990	16	2
1992	12	<1
Western Lake Erie		
1961	8	10
1982	5	4
1991	0	0

Historical Status

Around 1900 the Lake Huron-Lake Erie corridor was characterized as having one of the most abundant freshwater mussel faunas in North American lakes (Goodrich and van der Schalie 1932; Mackie et al. 1980): 39 species (Table 1).

Before 1990 mussel populations existed in most areas of the Lake Huron-Lake Erie corridor (Fig. 2). In Lake St. Clair, mussel populations were similar in 1986 and 1990 (Table 2). Numbers of mussels per unit area were relatively low (2/m^2 [0.2/ft^2]), but consistent, and there were 16-18 species found throughout the lake in 1990. The relatively healthy populations of mussels are attributed to the pristine water flowing into the lake from Lake Huron (Herdendorf et al. 1986).

In western Lake Erie, mussel populations that had survived low water quality in the 1950's declined between 1961 and 1982 (Table 2). Numbers declined from 10/m^2 (0.9/ft^2) to 4/m^2 (0.4/ft^2), and the number of species

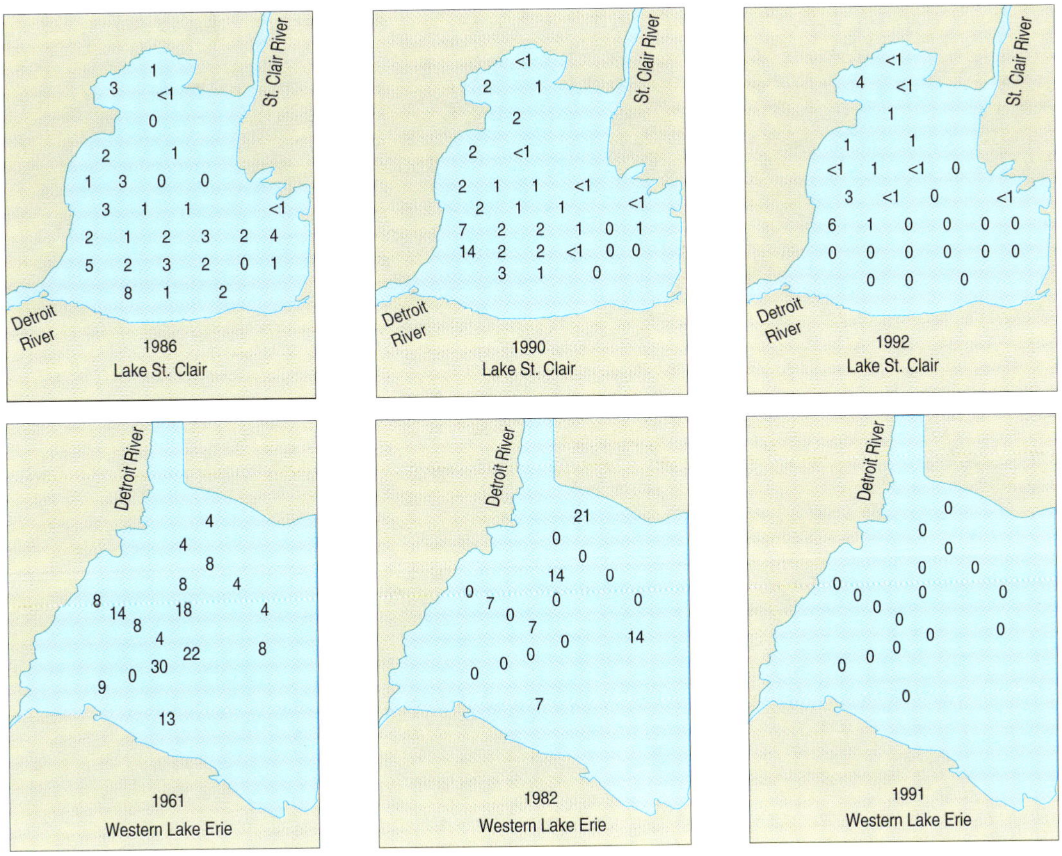

Fig. 2. Average (mean) densities (number/m^2) of native mussels in Lake St. Clair and western Lake Erie of the Lake Huron-Lake Erie corridor of the Great Lakes, 1961-92.

declined from eight to five between 1961 and 1982. The declining populations of native mussels are attributed to pollution that originated from tributary rivers of the lake prior to the 1970's. In the mid-1970's, pollution-abatement programs were begun, and water and substrate quality began to improve in western Lake Erie by the mid-1980's. By the late 1980's, environmental quality improved dramatically and pollution-sensitive indicators such as burrowing mayflies (*Hexagenia* spp.) began to return to western Lake Erie (Farara and Burt 1993; Schloesser, unpublished data).

Current Status

In the early 1990's, however, native mussel populations declined dramatically in the Lake Huron-Lake Erie corridor, despite improvements in water and substrate quality (Fig. 2; Table 2). In Lake St. Clair, substantial declines of mussels were documented between 1990 and 1992. Numbers and species of mussels were about half those found only 2 years earlier. Most changes in mussel populations in Lake St. Clair occurred in the southern portion of the lake, where mussels are no longer found (Fig. 2). In Lake Erie, mussel populations virtually disappeared in offshore waters between 1982 and 1991 (Fig. 2; Table 2).

Recent changes in native mussel populations in the Lake Huron-Lake Erie corridor are attributed to mortality caused by the exotic zebra mussel (*Dreissena polymorpha*); these exotics attach to the surface of mussels in such high numbers that native mussels are unlikely to be able to breathe and eat (Fig. 3). Intensive sampling indicated that native mussel populations declined rapidly between September 1989 and May-June 1990 (Fig. 4). Zebra mussels became abundant the summer of 1989, when infestation on clams increased from 24 mussels to 7,000 mussels per clam (Schloesser and Kovalak 1991; Nalepa and Schloesser 1992).

Erosion caused by deforestation, poor agricultural practices, and destruction of riparian zones, and organic and inorganic pollution have long been recognized as other causes for mussel mortality (Williams et al. 1993). Our knowledge of the zebra mussel, however, and its colonization on native mussels indicates that native mussel mortalities in the 1990's are attributable to

Fig. 3. Typical native mussel (*Potamilus alatus*) uncolonized (left) and colonized (right) by the exotic zebra mussel (*Dreissena polymorpha*) in the Lake Huron-Lake Erie corridor of the Great Lakes.

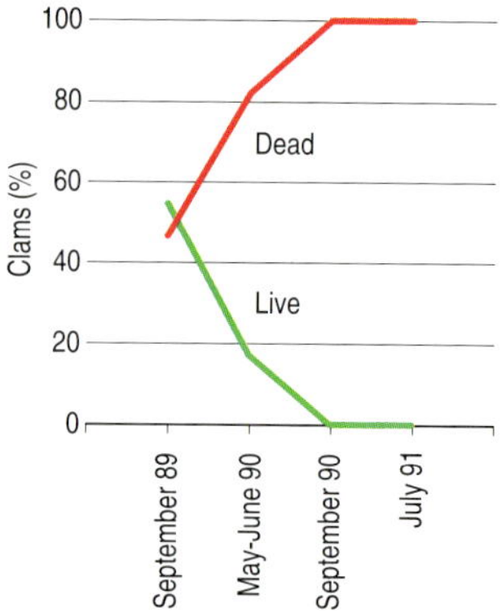

Fig. 4. Percentage of live and dead native mussels collected at an index station in western Lake Erie of the Lake Huron-Lake Erie corridor of the Great Lakes, 1989-91.

For further information:

Don W. Schloesser
National Biological Service
Great Lakes Science Center
Ann Arbor, MI 48105

biological pollution. Exotic species such as zebra mussels are being recognized as new and widespread threats to ecosystem stability throughout North America (Office of Technology Assessment 1993).

References

Beeton, A.M. 1961. Environmental changes in Lake Erie. Transactions of the American Fisheries Society 90:153-159.

Burns, N.M. 1985. Erie: the lake that survived. Rowman and Allanheld, Totowa, NJ. 320 pp.

Clarke, A.H. 1981. The freshwater molluscs of Canada. National Museum of Natural Sciences, Ottawa. 446 pp.

Clarke, A.H., and D.H. Stansbery. 1988. Are some Lake Erie mollusks products of post-Pleistocene evolution? Pages 85-92 *in* J.F. Downhower, ed. The biogeography of the island region of western Lake Erie. Ohio State University Press, Columbus. 208 pp.

Farara, D.G., and A.J. Burt. 1993. Environmental assessment of western Lake Erie sediments and benthic communities—1991. Ontario Ministry of Environment and Energy, Water Resources Branch, Great Lakes Section by Beak Consultants Limited, Brampton, Ontario. 193 pp.

Goodrich, C., and H. van der Schalie. 1932. The naiad species of the Great Lakes. Occasional Papers of the Museum of Zoology University of Michigan 238:8-14.

Herdendorf, C.E., C.N. Raphael, and E. Jaworski. 1986. The ecology of Lake St. Clair wetlands: a community profile. U.S. Fish and Wildlife Service Biological Rep. 85(7.7). 187 pp.

Mackie, G.L., D.S. White, and T.W. Zdeba. 1980. A guide to freshwater mollusks of the Laurentian Great Lakes with special emphasis of the genus *Pisidium*. U.S. Environmental Protection Agency EPA-600/3-80-068. Duluth, MN. 144 pp.

Nalepa, T.F., and D.W. Schloesser, eds. 1992. Zebra mussels: biology, impacts, and control. CRC Press, Inc., Boca Raton, FL. 810 pp.

National Academy of Sciences. 1970. Eutrophication: causes, consequences, correctives. Proceedings of a symposium. Washington, DC. 661 pp.

Office of Technology Assessment (U.S. Congress). 1993. Harmful non-indigenous species in the United States. OTA-F-565. U.S. Government Printing Office, Washington, DC. 390 pp.

Schloesser, D.W., and W.P. Kovalak. 1991. Infestation of unionids by *Dreissena polymorpha* in a power plant canal in Lake Erie. Journal of Shellfish Res. 10(2): 355-359.

Turgeon, D.D., A.E. Bogan, E.V. Coan, W.K. Emerson, W.G. Lyons, W.L. Pratt, C.F.E Roper, A. Scheltema, F.G. Thompson, and J.D. Williams. 1988. Common and scientific names of aquatic invertebrates from the United States and Canada: mollusks. American Fisheries Society Special Publ. 16. 277 pp.

Williams, J.D., M.L. Warren, Jr., K.S. Cummings, J.L. Harris, and R.J. Neves. 1993. Conservation status of freshwater mussels of the United States and Canada. Fisheries 18(9):6-22.

Aquatic Insects As Indicators of Environmental Quality

by
William T. Mason, Jr.
National Biological Service
Calvin R. Fremling
Winona State University
Alan V. Nebeker
U.S. Environmental Protection Agency

Aquatic insects are among the most prolific animals on earth, but are highly specialized and represent less than 1% of the total animal diversity (Pennak 1978). Most people know the 12 orders and about 11,000 species of North American aquatic insects (Merritt and Cummins 1984) only by the large adults that fly around or near wetlands.

Aquatic insects are excellent overall indicators of both recent and long-term environmental conditions (Patrick and Palavage 1994). The immature stages of aquatic insects have short life cycles, often several generations a year, and remain in the general area of propagation. Thus, when environmental changes occur, the species must endure the disturbance, adapt quickly, or die and be replaced by more tolerant species. These changes often result in an overabundance of a few tolerant species, and the communities become destabilized or "unbalanced."

Members of the order Diptera, or true flies, are especially good "bioindicators" of aquatic environmental conditions because, in addition to the attributes of other aquatic insects, they occupy the full spectrum of habitats and conditions (Paine and Gaufin 1956; Roback 1957; Mason 1975; Hudson et al. 1990).

Although considerable information on aquatic insects and other macroinvertebrates has been collected since the 1950's, most studies have been abbreviated surveys. There are few good examples of long-term biomonitoring of aquatic insects in the United States because of the discontinuance of most routine biomonitoring in the 1980's. We present ongoing and past examples of surveillance monitoring of aquatic insects of the Ohio and Mississippi rivers. Our interest here centers on the immature stages of aquatic insects that, although usually unnoticed, are part of the framework of natural ecosystems (Fig. 1).

Ohio River Aquatic Insects

During 1963-67, aquatic insects (primarily midges [Diptera], caddisflies [Trichoptera], mayflies [Ephemeroptera], and stoneflies [Plecoptera]) and other benthic invertebrates were monitored at 80-161 km (50-100 mi) increments along the 1,582 km (963 mi) of the mainstem Ohio River from Pittsburgh, Pennsylvania, to Cairo, Illinois (Mason et al. 1971). Rock-filled basket samplers were a preliminary collection device in addition to Ponar substrate grab collections.

In the upper Ohio River from river mile 0 to 260 (418 km) at Addison, Ohio, during 1965-67, the aquatic insect diversity (Fig. 1a) and individuals (Fig. 1b) in rock-filled basket samplers were low compared with collections from downriver sites. The macroinvertebrate fauna consisted mostly of pollution-tolerant midge larvae and worms, indicating poor to fair water quality. In the lower reach from Louisville to Evansville (distance of about 200 river miles or 322 km) the fauna was double to triple that of the upriver stations and contained facultative and some clean-water taxa, indicating improved water quality.

Although the total aquatic insect diversity in the baskets at river mile 601 (968 km) in 1965 exceeded those at river mile 788 (1,268 km) by about one-third, during the next 2 years the diversity at Evansville increased over that at Louisville by 30%-40%. This significant increase was probably caused by environmental changes (e.g., increased eutrophication that provided more foods for these insects) that favored Chironomidae nonbiting midges and Hydropsychidae net-spinning caddisflies. During the 3-year period, pollution-tolerant species replaced some of the clean-water "green" species.

Aquatic insects are also useful indicators of contamination of the sediments and waters that may have gone unnoticed by routine physicochemical measurements. Uptake of toxic substances, such as heavy metals and organochlorine compounds, causes various kinds of deformities of the larval and pupal Chironomidae (Hamilton and Saether 1971; Lenat 1993). Depending on the severity of the pollution, these deformed individuals do not reach maturity and the populations are eventually reduced (van Urk et al. 1992). During the 1963-67 Ohio River monitoring program, Mason and Lewis observed larval deformities in samples taken from the sediments from the upper reaches of the Ohio River near Pittsburgh, Pennsylvania, the lower Monongahela River, and Kanawha River (Mason, unpublished data).

Management Implications

There is a need to establish long-term monitoring and reporting on macroinvertebrate populations such as that carried out during 1963-67. A monitoring program could evaluate the success of pollution clean-up and identify biological indicators to help balance water uses among urban centers, transportation, industry, and fishing and other recreation. Water chemistry and physical measurements alone are not sufficient to determine subtle shifts in aquatic populations.

Locating point sources of contaminants or thermal wastes so that they discharge directly to trout streams and lakes usually results in loss of stonefly populations, which, in turn, adversely affects fisheries. The effects of aerial spraying and other types of insecticide applications on stonefly and other sensitive aquatic organisms should be considered during site-preparation planning. Natural resource managers often recommend set backs, or buffer strips of untilled land adjacent to streams, as an effective way to minimize harm from pollution runoff.

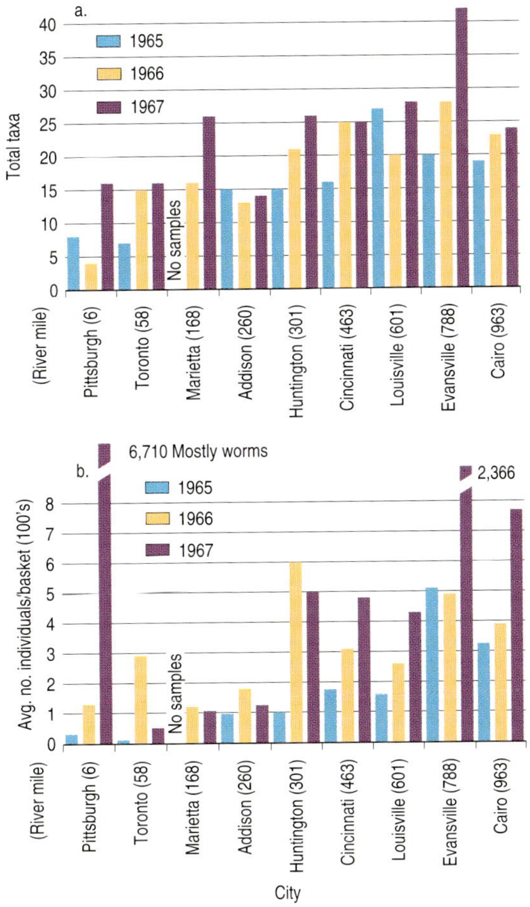

Fig. 1. (a) Total number of aquatic insect and other macroinvertebrate taxa and (b) average number of individuals collected in basket samplers in the Ohio River, 1965-67 (Mason et al. 1971).

Mississippi River Ephemeroptera

The nymphs of burrowing mayflies (Ephemeroptera) live in U-shaped tubes in the silt bottoms of shallow, slow-moving waters (Berner and Pescador 1988). Although mass emergences of adult burrowing mayflies in the Upper Mississippi River have been considered a nuisance (Fremling 1968), their abundance represents a wealth of fish food biomass; their abundance also reflects environmental health.

During 1957-69, three species of burrowing mayflies (*Hexagenia bilineata, H. limbata*, and *Pentagenia vittigera*) were monitored in the 3,218-km (2,000-mi) reach of the Mississippi River from Minneapolis, Minnesota, to New Orleans, Louisiana (Fremling 1964, 1970). In the 1930's, 29 navigation dams were built in the upper reaches of the Mississippi River, and burrowing mayflies became abundant in the slow-moving silted shallows. The insects were much less abundant downstream from St. Louis, Missouri, where no dams existed. The surveys of Mississippi River mayflies continue today.

During the years 1957-69 and 1976, about 1,300 collections of *Hexagenia* showed that most of the navigation pools and impoundments upstream from Minneapolis and St. Paul, Minnesota, supported large populations of burrowing mayflies. Both *Hexagenia* species were conspicuously rare in the 48-km (30-mi) reach downstream from the Twin Cities. There, a heavy pollutant load caused low dissolved oxygen levels on the river bottom for much of the year. The mayflies were also rare in the upper reach of Lake Pepin, a large (32-km [20-mi]) natural impoundment farther downstream, where they had been abundant in years past. Apparently Lake Pepin was a settling basin for pollutants and decaying algae caused by overfertilization from the Twin Cities area.

A 1986 mayfly survey revealed that recent pollution abatement measures in the Twin Cities created favorable conditions for mayflies to return to densities of the 1950's-60's. The distribution of *Hexagenia* species reflects the status of aquatic life inhabiting a large river that was otherwise difficult to monitor effectively or economically by standard chemical testing (Fremling 1989, 1990).

Management Implications

As with the Ohio River insects, there is a need to maintain a network of routine monitoring stations along the 3,218 km (2,000 mi) of the Mississippi River to learn when atypical emergences of mayflies and other aquatic insects occur. This information will allow public officials and administrators to pinpoint more intensive and detailed analytical surveys that could determine causes of the emergences.

Today, the greatest future threat to the burrowing mayflies in the Mississippi River lies in accelerated siltation and subsequent filling of the navigation pools. These filled areas are rapidly becoming floodplain forests, a conversion that eliminates them as burrowing mayfly habitat, thereby reducing food stocks for fisheries.

References

Berner, L., and M.L. Pescador. 1988. The mayflies of Florida. University of Florida Press, Gainesville. 415 pp.

Fremling, C.R. 1964. Mayfly distribution indicates water quality on the Upper Mississippi River. Science 146(3648):1164-1166.

Fremling, C.R. 1968. Documentation of mass emergence of *Hexagenia* mayflies from the Upper Mississippi River. Transactions of the American Fisheries Society 97(3):278-281.

Fremling, C.R. 1970. Mayfly distribution as a water quality index. Water Pollution Control Research Series 16030 DQH 11/70. U.S. Environmental Protection Agency, Water Quality Office, Washington, DC. 39 pp.

Fremling, C.R. 1989. *Hexagenia* mayflies: biological monitors of water quality in the Upper Mississippi River. Journal of the Minnesota Academy of Science 55(1):139-143.

Fremling, C.R. 1990. Recurrence of *Hexagenia* mayflies demonstrates improved water quality in Pool 2 and Lake Pepin, Upper Mississippi River. Pages 243-248 *in* I.C. Campbell, ed. Mayflies and stoneflies. Kluwer Academic Publishers, The Netherlands.

Hamilton, A.L., and O.A. Saether. 1971. The occurrence of characteristic deformities in the chironomid larvae of several Canadian lakes. Canadian Entomologist 103:363-368.

Hudson, P.L., D.R. Lenat, B.A. Caldwell, and D. Smith. 1990. Chironomidae of the southeastern United States: a checklist of species and notes on biology, distribution, and habitat. U.S. Fish and Wildlife Service Fish and Wildlife Res. 7. 46 pp.

Lenat, D.R. 1993. Using mentum deformities of *Chironomus* larvae to evaluate the effects of toxicity and organic loading in streams. Journal of the North American Benthological Society 12:265-269.

Mason, W.T., Jr. 1975. Chironomidae (Diptera) as biological indicators of water quality. Pages 40-51 *in* C.C. King and L.E. Elfner, eds. Organisms and biological communities as indicators of environmental quality. Circular 8. Ohio Biological Survey, Columbus.

Mason, W.T., Jr., P.A. Lewis, and J.B. Anderson. 1971. Macroinvertebrate collections and water quality monitoring in the Ohio River Basin 1963-1967. Office of Technical Programs, Ohio Basin Region, and Analytical Quality Control Laboratory, U.S. Environmental Protection Agency, Cincinnati, OH. 52 pp.

Merritt, R.W., and K.W. Cummins. 1984. Introduction. Pages 1-3 *in* R.W. Merritt and K.W. Cummins, eds. An introduction to the aquatic insects of North America. 2nd ed. Kendall/Hunt Publishing Company, Dubuque, IA.

Paine, G.W., and A.R. Gaufin. 1956. Aquatic Diptera as indicators of pollution in a midwestern stream. The Ohio Journal of Science 56:291-304.

Patrick, R., and D.M. Palavage. 1994. The value of species as indicators of water quality. Proceedings of the Academy of Natural Sciences Philadelphia 145:55-92.

Pennak, R.W. 1978. Fresh-water invertebrates of the United States. 2nd ed. John Wiley & Sons, Inc. New York. 803 pp.

Roback, S.S. 1957. The immature tendipedids of the Philadelphia area. Academy of Natural Sciences of Philadelphia Monograph 9. 152 pp.

van Urk, G., F.C.M. Kerkum, and H. Smit. 1992. Life cycle patterns, density, and frequency of deformities in *Chironomus* larvae (Diptera: Chironomidae) over a contaminated sediment gradient. Canadian Journal of Fisheries and Aquatic Sciences 49:2291-2299.

For further information:

William T. Mason, Jr.
National Biological Service
Southeastern Biological Science Center
7920 N.W. 71st St.
Gainesville, FL 32653

Biodiversity Degradation in Illinois Stoneflies

by

Donald W. Webb
Illinois Natural History Survey

Preliminary analysis of the recent collections of Illinois stoneflies indicates a reduction in the species richness in Illinois, a reduction in the spatial distribution of many species, the dominance of more generalist species more tolerant to environmental perturbations, and the extirpation of several species.

These general trends can be expanded for all of the central United States. The reduction in stream flow through the construction of locks and dams and the resulting effect of increased sedimentation have severely affected the habitat and niche selection available to species such as stoneflies that require rapidly flowing streams. This situation has been compounded by the erosional effects of deforestation and agricultural practices, which are maximizing the amount of land put into cultivation, as well as the increased problems related to nonpoint pollution from agricultural pesticides and fertilizers. To properly delineate these trends, the status of stoneflies and most other groups of aquatic organisms in the central United States needs to be evaluated.

In Illinois, stoneflies (Insecta: Plecoptera) were collected extensively from 1926 through 1940 by T.H. Frison (Frison 1929, 1935, 1937, 1942), with additional winter-emerging stoneflies collected from 1960 to 1970 by H.H. Ross's "Winter Stonefly Club" (Ricker and Ross 1968, 1969; Ross and Ricker 1971). From the thousands of specimens collected, the Illinois Natural History Survey has an exceptional record of species diversity and spatial distribution of Illinois stoneflies.

In 1990 we began a reevaluation of the species richness and spatial distribution of Illinois stoneflies (Webb and Harris 1993). The focus of this study was to compare current species richness and distribution patterns with those determined by Frison, Ross, and Ricker. To do this, we developed a data base for the Illinois specimens in the collections of the Illinois Natural History Survey, and we extensively resurveyed stoneflies in each of the 25 major drainages within the state (Figure).

Status

We evaluated the status of each stonefly species on the basis of the locality information and classified each species as rare, uncommon, or common (Table). This evaluation revealed that 39% of the 61 species reported were known from three localities or fewer. In addition, we developed a checklist of the Illinois species and updated their varied nomenclature.

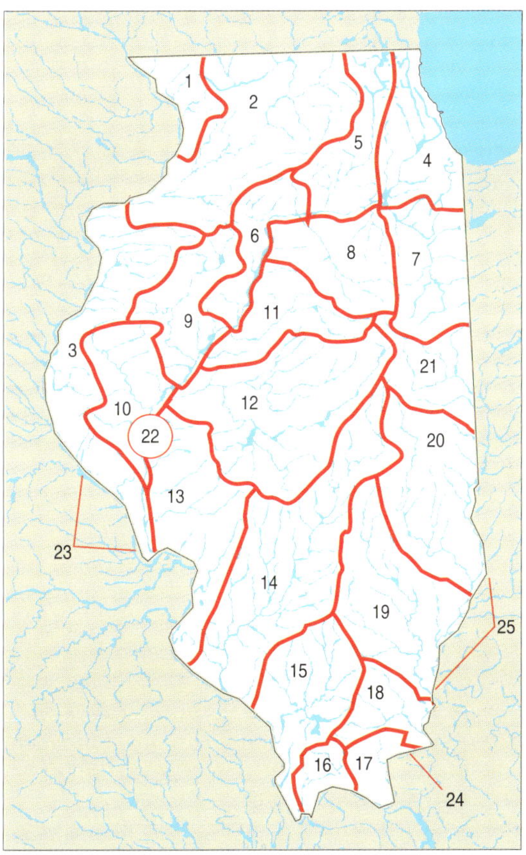

Figure. Twenty-five major river drainages in Illinois.

After 4 years of collecting we consider 13 Illinois stonefly species rare (*Acroneuria filicis* Frison; *A. perplexa* Frison; *Allocapnia nivicola* [Fitch]; *A. smithi* Ross and Ricker; *Haploperla brevis* [Banks]; *Isoperla burksi* Frison; *Nemoura trispinosa* Claassen; *Paragnetina media* [Walker]; *Prostoia completa* [Walker]; *Shipsa rotunda* [Claassen]; *Soyedina vallicularia* [Wu]; *Zealeuctra fraxina* Ricker and Ross; and *Z. narfi* Ricker and Ross). We found that 6 have been extirpated from Illinois (*Allocapnia illinoensis* Frison; *Alloperla roberti* Surdick; *Amphinemura nigritta* [Provancher]; *Isoperla conspicua* Frison; *I. marlynia* [Needham and Claassen]; *Leuctra tenuis* [Pictet]); 4 species have possibly been extirpated (*Isogenoides varians* [Walsh]; *Leuctra sibleyi* Claassen; *Nemocapnia carolina* Banks; *Paracapnia angulata* Hanson), and 1 rare species (*Alloperla caudata* Frison) is common. One species, *Soyedina vallicularia* [Wu], has been added to the state list.

Data from over 50,000 Illinois stonefly specimens in the collections of the Illinois Natural History Survey are being analyzed to determine the species richness and spatial distribution of Illinois stoneflies by drainage basin. This assessment will be based separately on earlier data (Frison 1929, 1935, 1937, 1942; Ricker and Ross 1968, 1969; Ross and Ricker 1971) and will evaluate these data relative to collections since 1990.

Apparent Trends

In reevaluating the current status of Illinois stoneflies, our first concern was the status of so many "rare" species in Illinois. We wanted to determine if the limited locality records for these species reflect actual rare distribution in Illinois, inadequate sampling, or an accidental occurrence (i.e., the species is not normally a part of the indigenous Illinois fauna). It is now apparent that 13 of these species are truly rare in Illinois; many of these are at the edge of their distributions. The eastern deciduous forest with its gravel- and cobble-bottomed streams extends only slightly into Illinois and several of these rare species are found only in these habitats. Similarly, the limestone and sandstone outcroppings of the Shawnee Hills in southern Illinois offer another area of high-quality streams and are home for several rare species of Illinois stoneflies. To a very limited extent, springs in Illinois are a refugia for a few rare species. For only one species, *Alloperla caudata*, does it appear that inadequate sampling during April and May produced a biased picture of

Courtesy M.R. Jeffords, Illinois Natural History Survey

Pteronarcys pictetii, one of the largest stoneflies in Illinois, is common to big rivers. In the nymphal stage, this species serves as an important food for fish.

Table . Relative abundance of Illinois stoneflies. R — rare (1-3 localities); U — uncommon (4-14 localities); and C — common (more than 14 localities). Surnames within or outside parentheses refer to the authors of the species name.

Species	Relative abundance
Group Euholognatha	
Capniidae	
Capniinae	
Allocapnia	
forbesi Frison	U
granulata (Claassen)	C
illinoensis Frison	R
mystica Frison	C
nivicola (Fitch)	R
recta (Claassen)	U
rickeri Frison	C
smithi Ross & Ricker	R
vivipara (Claassen)	C
Nemocapnia	
carolina Banks	R
Paracapnia	
angulata Hanson	R
Leuctridae	
Leuctrinae	
Leuctra	
rickeri James	U
sibleyi Claassen	R
tenuis (Pictet)	R
Zealeuctra	
claasseni (Frison)	U
fraxina Ricker & Ross	R
narfi Ricker & Ross	R
Nemouridae	
Amphinemurinae	
Amphinemura	
delosa (Ricker)	U
nigritta (Provancher)	R
varshava (Ricker)	U
Nemourinae	
Nemoura	
trispinosa Claassen	R
Prostoia	
completa (Walker)	R
Shipsa	
rotunda (Claassen)	R
Soyedina	
vallicularia (Wu)	R
Taeniopterygidae	
Brachypteryinae	
Strophopteryx	
fasciata (Burmeister)	C
Taeniopteryginae	
Taeniopteryx	
burksi Ricker & Ross	C
lita Frison	U
metequi Ricker & Ross	U
nivalis (Fitch)	C
parvula Banks	U
Group Systellognatha	
Chloroperlidae	
Chloroperlinae	
Alloperla	
caudata Frison	R
roberti Surdick	R
Haploperla	
brevis (Banks)	R
Perlidae	
Perlinae	
Acroneuriinae	
Acroneuriini	
Acroneuria	
abnormis (Newman)	C
evoluta Klapalek	U
filicis Frison	R
frisoni Stark and Brown	C
internata (Walker)	U
perplexa Frison	R
Attaneuria	
ruralis (Hagen)	C
Perlesta	
decipiens (Walsh)	C
Perlinella	
drymo (Newman)	U
ephrye (Newman)	U
Perlinae	
Neoperlini	
Neoperla	
clymene (Newman) complex	C
Perlini	
Agnetina	
flavescens (Walsh)	C
Paragnetina	
kansensis (Banks)	U
media (Walker)	R
Perlodidae	
Isoperlinae	
Clioperla	
clio (Newman)	C
Isoperla	
bilineata (Say)	C
burksi Frison	R
conspicua Frison	R
decepta Frison	C
longiseta Banks	U
marlynia (Needham & Claassen)	R
mohri Frison	R
nana (Walsh)	C
richardsoni Frison	U
Perlodinae	
Perlodini	
Hydroperla	
crosbyi (Needham & Claassen)	U
fugitans (Needham & Claassen)	U
Isogenoides	
varians (Walsh)	R
Pteronarycidae	
Pteronarcyinae	
Pteronacrcyini	
Pteronarcys	
pictetii Hagen	C

its distribution, as this species is found commonly across the Shawnee Hills. The apparent or possible extirpation of 10 species of stoneflies from Illinois creates serious concern. None of the species were collected in Illinois during 1990-94, although we are still unsure about the extirpation of four of these species, since they are large-river species and we are not satisfied with our collecting techniques for these habitats.

In evaluating the changes in spatial distribution of Illinois stoneflies, the "winter stoneflies," in particular, the genus *Allocapnia,* offer the best examples (Frison 1929, 1935, 1937, 1942; Ricker and Ross 1968, 1969; Ross and Ricker 1971; Webb and Harris 1993). One striking example is *A. granulata.* In the 1920's, 30's, and 60's, this species was distributed and abundant from southeastern to northern Illinois. A drastic reduction in the distribution of this species has occurred since 1970, and in the past 4 years, this species has only been collected within the Rock River drainage of northern

Illinois. As yet, no cause for this reduced distribution has been proposed. This is the most spectacular example we have discovered, but similar distribution patterns have been noted in other species, particularly within the genus *Acroneuria*. Our recent collections reveal that generalist species—those tolerant of a variety of environmental perturbations—apparently are becoming the dominant species in Illinois. *Allocapnia vivipara, Taeniopteryx burksi,* and *Isoperla bilineata* are examples of this trend; all are widespread throughout Illinois in many ecological habitats.

References

Frison, T.H. 1929. Fall and winter stoneflies, or Plecoptera, of Illinois. Bull. of the Illinois Natural History Survey 18:345-409.

Frison, T.H. 1935. The stoneflies, or Plecoptera, of Illinois. Bull. of the Illinois Natural History Survey 20:281-471.

Frison, T.H. 1937. Studies of Nearctic insects. II. Descriptions of Plecoptera, with special reference to the Illinois species. Bull. of the Illinois Natural History Survey 21:78-99.

Frison, T.H. 1942. Studies of North American Plecoptera, with special reference to the fauna of Illinois. Bull. of the Illinois Natural History Survey 22:235-355.

Ricker, W.E., and H.H. Ross. 1968. North American species of *Taeniopteryx* (Plecoptera: Insecta). Journal of the Fisheries Research Board Canada 25:1423-1439.

Ricker, W.E., and H.H. Ross. 1969. The genus *Zealeuctra* and its position in the family Leuctridae (Plecoptera, Insecta). Canadian Journal of Zoology 47:1113-1127.

Ross, H.H., and W.E. Ricker. 1971. The classification, evolution, and dispersal of the winter stonefly genus *Allocapnia*. University of Illinois Biology Monographs 45:1-162.

Webb, D.W., and M.A. Harris. 1993. Survey of 21 rare species of stoneflies (Plecoptera) in Illinois. Illinois Natural History Survey, Center for Biodiversity Tech. Rep. 1993(3):1-14.

For further information:

Donald W. Webb
Illinois Natural History Survey
Center for Biodiversity
607 East Peabody Dr.
Champaign, IL 61820

Courtesy R.J. Helmerick

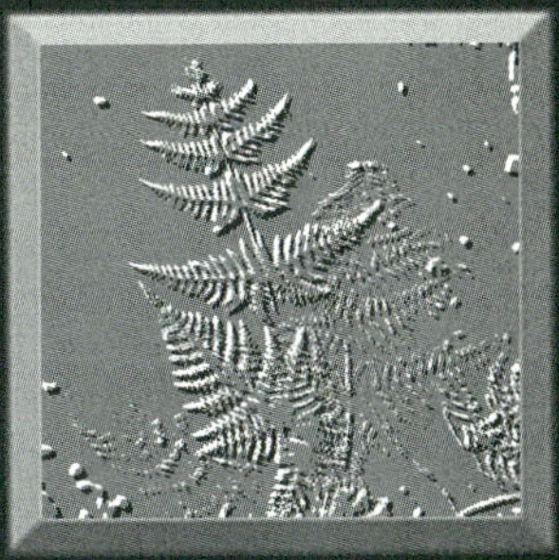

Plants

by
Science Editor
Glenn R. Guntenspergen
National Biological Service
Southern Science Center
700 Cajundome Boulevard
Lafayette, LA 70506

Overview

This section describes trends in two of the major kingdoms of life on earth: the green plants of the Kingdom Plantae and the molds, lichens, and mushrooms of the Kingdom Fungi. Members of the plant and fungal kingdoms have both economic and ecological importance. Plants transform solar energy into usable economic products essential in our modern society and provide the basis for most life on earth by generating oxygen as a product of photosynthesis. Fungi not only mediate critical biological and ecological processes including the breakdown of organic matter and recycling of nutrients, but they also play important roles in mutualistic associations with plants and animals. Members of the Kingdom Fungi also produce commercially valuable substances including antibiotics and ethanol, while other fungi are pathogenic and cause damage to crops and forest trees. Because fungi and plants play such fundamental roles in our lives, it is important to have a comprehensive knowledge of the taxa comprising these groups. However, at a time when we are increasingly recognizing the importance of these groups, we are impoverishing our biological heritage. Rates of species loss are reaching alarming levels as ecosystems are degraded and habitat is lost. This erosion of biological diversity threatens the maintenance of long-term sustainable development and protection of the earth's biosphere.

Questions involving biological diversity are now of major concern to scientists, the general public, and government agencies with mandates for natural resource protection. Much of this concern has been directed toward tropical forest systems because of their high levels of biodiversity, although other regions, including the United States, deserve our immediate attention. Certainly, a first step toward conserving biological diversity must be based on a firm knowledge of the numbers and distribution of existing species. Developing good estimates of species diversity is also important in describing historical and current trends of species dynamics. Unfortunately, despite the existence of various state and regional surveys, the efforts of taxonomists and natural historians, and the publication of various floras, we still do not have precise estimates of the status of plant and fungal taxa in the United States. Estimates for vascular plant taxa in the United States range upward from 17,000 species (Morin, Morse et al., this section). In contrast to this well-studied group, only 5%-10% of an estimated 1.5 million fungal species have been described worldwide (Rossman, this section).

Even though the bulk of information about our native vascular flora was collected in the 19th and early 20th centuries, significant data about the status of plants in the United States continue to be collected as species expand their ranges, as other species thought locally extirpated are rediscovered, as poorly surveyed areas are explored, and as species become extinct. Even in states like New York, which has a long and currently active program of botanical exploration, additional species of vascular plants continue to be documented as poorly surveyed areas are given more comprehensive coverage (Miller and Mitchell, this section).

Herbaria and museums continue to be important repositories for this information because collecting by their personnel represents a significant effort at inventorying plant and fungal species in this country (Morin, this section). Unfortunately, their role is increasingly at risk as support for collecting declines. In other cases, a shortage of trained specialists will prevent an adequate inventory of biotic diversity. Although many regional checklists exist as well as excellent manuals that cover bryophyte systematics, floristic inventories of bryophytes have been hampered primarily by a lack of trained professionals (Merrill, this section).

The flora of the American countryside has been much changed since European settlement. Over the past 20 years alone, more than 200 species of non-native vascular plants have been recorded in New York state; these species represent an important risk to native plant communities (Miller and Mitchell, this section). Human activities are responsible for the introduction of these invasive exotics as well as the extinction of some species with small geographic ranges or those restricted to unique habitats.

If current trends in land-use continue, however, even species with more widespread distributions will be at risk. For example, lichens as a group are declining in many areas from the effects of air pollution. It is estimated that as much as 80%-90% of the original lichen flora has disappeared from urbanized areas (Bennett, this section). Likewise, marked declines in macrofungi have been documented in Europe although similar trends in this country have not been published because, in part, of the incomplete inventory and lack of monitoring of these groups in the United States (Mueller, this section). Among the more completely documented vascular plants, The Nature Conservancy reports that 9.8% of native species have been lost from at least one state, more than 200 native species have become extinct in the United States, and an additional 403 native plant taxa need protection under the United States Endangered Species Act (Morse et al., this section).

The articles in this section represent an important step in describing the status of the plant and fungal taxa in this country. They provide a snapshot illustrating our knowledge of past and current distributions of plants; the importance of developing a more comprehensive data base for various groups, especially the fungi; and the need to develop a comprehensive inventory of the continually changing and evolving flora of the United States. If we are to understand the causes underlying the changes in patterns of diversity and make predictions about the threats of anthropogenic (human-caused) activities, we must have a quantitative understanding about the nature and distribution of the taxa composing our flora.

Microfungi: Molds, Mildews, Rusts, and Smuts

Amy Y. Rossman
U.S. Department of Agriculture

Fungi are a group of organisms that exist as a vast network of tiny threads growing in and out of all kinds of organic matter. As they grow, the threads secrete enzymes that break down the substances around them, releasing nutrients into the environment. Without fungi, the world would be completely covered with organic debris that would not rot, and nutrients would not be available for plant growth. All plants would die.

Microfungi include the organisms that are called molds and mildews as well as rusts and smuts, which cause plant diseases. They grow in all substrates, including plants, soil, water, insects, cows' rumen (*see* glossary), hair, and skin. Microfungi are said to be small because only part of the fungus is visible at one time, if at all. The visible parts produce thousands of tiny spores that are carried by the air, spreading the fungus. Most of the fungal body consists of microscopic threads extending through the substrate in which it grows. The invisible fungal structure may be extremely large, often extending for miles as, for example, the "humongous fungus" occurring in the north-central United States (Rensberger 1992).

Among the multitudinous molds are humble servants such as *Penicillium notatum*, the source of penicillin, and *Tolyposporium niveum*, a producer of cyclosporin, the immune-system suppressant used for organ transplant operations. In sustainable agriculture the fungal performers are agents of biological control and crop nutrition, helping the environment through the reduced use of chemical pesticides and fertilizers. Fungi can stop a hoard of locusts by attacking the chitinous insect exoskeleton or control nematodes that destroy the roots of crop plants (CAB 1993). Although strains of fungi can degrade plastics and break down hazardous wastes such as dioxin (Jong and Edwards 1991), only a fraction of these fungi have been screened as beneficial organisms.

Microfungi can also be harmful, causing the irritating human affliction known as athlete's foot as well as disastrous diseases of crops and trees. The potato famine in Ireland during the mid- to late 1800's was caused by a fungus called *Phytophthora infestans* that rotted the potato crops for several years (Large 1962). Because of this disease, many Irish immigrated to the United States. Once the nature of the disease was determined, a solution based on fungus control was found. Knowing what fungi exist, where they occur, and what they do is essential.

Diversity of Microfungi

The microfungi are the most diverse group of all the fungi but the least understood or documented. Only about 5%-10% of all fungal species have been described, much less characterized and put to use or controlled. Investigations to explore the diversity of microfungi have shown that they are much more diverse than previously thought. Very small samples of tropical rainforest leaf litter yielded up to 145 different species of microfungi (Bills and Polishook 1994). About 200,000 fungal species have been described worldwide (Reed and Farr 1993), yet an estimated 1-1.5 million species may exist (Hawksworth 1991; Rossman 1994).

Within the United States, information has been published about 13,000 species of microfungi on plants or plant products (Farr et al. 1989), probably only a fraction of the species thought to exist. Specimens of microfungi are housed in the U.S. National Fungus Collections and other institutions that serve as reservoirs of information and documentation about our nation's natural heritage. By comparing the species reported in the literature with those represented in the collections, one can estimate the number of microfungi known in the United States at 29,000 species (Farr et al. 1989). In areas of the world where fungi have been well studied, the ratio of vascular plants to fungi is about 6 to 1, suggesting that there may actually be 120,000 species of fungi within the United States.

Internet Information

Although the numbers and kinds of fungi in the United States are not known, information about the microfungi associated with plants and plant products in the United States is available over Internet at this telnet address: FUNGI.ARS-GRIN.GOV. After the word *OK* appears on the screen, type *login user;* when prompted for a password, type *user*. By doing this, anyone can find out what fungi might occur on the flowers in his or her own backyard. Data can also be consulted on accurate scientific names of microfungi, recent literature on plant-associated fungi, specimens in the U.S. National Fungus Collections, and records of microfungi on plants throughout the world. In an instant, reports of fungi can be consulted by those making land-management decisions or helping a farmer control a disease.

Survey and Inventory Needs

Knowing which microfungi occur within the United States provides information upon which plant quarantine decisions are made. A wrong decision allowing entry of a harmful pathogen can profoundly affect this nation's biological resources. In the eastern United States, a devastating disease called chestnut blight, caused by *Cryphonectria parasitica* and unknowingly imported from Europe on logs, killed virtually all the towering chestnut trees that once dominated our forests in the last century (Anagnostakis 1987). Now on the forest floor only skeletons of the trees can be seen with decay fungi rotting the bleached "bones" of these fallen giants.

Another disease, dogwood anthracnose, occurs on flowering dogwood trees in both the eastern and western United States. The causal fungus, *Discula destructiva*, was unknown until 1991 (Redlin 1991). Still unknown is whether this fungus was imported or was already present in the United States before its appearance as dogwood anthracnose. Because microfungi are small, their existence may not be noticed until they cause serious diseases.

A program to inventory and monitor microfungi in the United States does not exist at present; thus it is impossible to determine if species of microfungi are increasing or declining. Efforts to document the biodiversity of microfungi in the United States are limited to reports by plant pathologists who encounter disease-causing organisms or search for useful biological-control organisms. Information about the occurrence and biology of microfungi will increase the ability to make accurate decisions about the importation of agricultural products, to control microfungi already present, and to determine if beneficial microfungi are being lost because of habitat destruction. With increased knowledge the unexplored world of microfungi can be put to work to solve our most pressing environmental and agricultural problems.

References

Anagnostakis, S. 1987. Chestnut blight: the classical problem of an introduced pathogen. Mycologia 79(1):23-37.

CAB. 1993. Locust project enters phase two. Commonwealth Agricultural Bureau (CAB) International News. June. p. 4.

Bills, G.F., and J.D. Polishook. 1994. Abundance and diversity of microfungi in leaf litter of a lowland rain forest in Costa Rica. Mycologia 86:187-198.

Farr, D., G. Bills, G. Chamuris, and A. Rossman 1989. Fungi on plants and plant products in the United States. American Phytopathological Society Press, St. Paul, MN. 1,152 pp.
Hawksworth, D.L. 1991. The fungal dimension of biodiversity: magnitude, significance, and conservation. Mycological Res. 95:641-655.
Jong, S.C., and M.J. Edwards. 1991. American type culture collection catalogue of filamentous fungi, 18th ed. Rockville, MD. 667 pp.
Large, E.C. 1962. Advance of the fungi. Dover, New York. 488 pp.
Redlin, S.C. 1991. *Discula destructiva* sp. nov., cause of dogwood anthracnose. Mycologia 83:633-642.
Reed, C.F., and D.F. Farr. 1993. Index to Saccardo's Sylloge Fungorum. Volumes I-XXVI IN XXIX 1882-1972. Reed Herbarium, Darlington, MD. 884 pp.
Rensberger, B. 1992. Underground goliath. Michigan mushroom over 1,500 years old. Washington Post, April 2.
Rossman, A.Y. 1994. A strategy for an all-taxa inventory of fungal biodiversity. *In* C.I. Peng, ed. Biodiversity and terrestrial ecosystems. Bull. of the Academy Sinica Institute Botany. In press.

For further information:

Amy Y. Rossman
U.S. Department of Agriculture
U.S. National Fungus Collections
Beltsville, MD 20705

Macrofungi

by
Gregory M. Mueller
The Field Museum, Chicago

Courtesy G.M. Mueller, The Field Museum

Fig. 1. *Entoloma salmoneum.* The salmon-colored entoloma is a common recycler of forest litter in North American forests.

Macrofungi are a diverse, commonly encountered, and ecologically important group of organisms. Like most fungi, the major part of these organisms consists of a mass of thin, microscopic threads (termed mycelium) growing in soil, decomposing leaves, and other substrate. They differ from other fungi by forming large, macroscopic fruitbodies at some time in their life; the mushrooms sold in grocery stores are an example of these fruitbodies. This group of fungi includes all mushrooms (Fig. 1), morels, puffballs, bracket fungi, and cup fungi.

Macrofungi are vitally significant in forests; many species help break down dead organic material, such as dead tree trunks and leaves, into simple compounds usable by growing plants. Thus, they act as nature's recyclers, without which forests could not function. Some species are major plant pathogens (causes of disease) that cause millions of dollars of damage to U.S. forests each year. Still other species enter into a necessary, mutually beneficial association with trees such as oaks, pines, firs, and spruces. In this association (Fig. 2), termed mycorrhizae, the mycelium of the fungus brings water and nutrients to the tree in return for taking excess food from the tree. Neither the tree nor the fungus can survive without the other. Finally, some of these fungi form an important part of the diet of many small mammals and insects. For example, small truffle-like fungi are a major food source of the northern flying squirrel (*Glaucomys sabrinus; see* box). Because macrofungi are an indispensable component of the forest ecosystem, information on which fungi occur in the forests and on the specific role that they play is necessary for management and maintenance of our forests.

Courtesy G.M. Mueller, The Field Museum

Fig. 2. Mycorrhizae formed between ponderosa pine and *Laccaria laccata* in the laboratory. Note the branched pine roots and threadlike fungal hyphae.

Macrofungi also directly affect people. Though some fungi are deadly poisonous, others are prized as edibles. Commercial mushroom harvesting is a multimillion-dollar-a-year business in the United States; for example, the industry added an estimated $40 million to the Oregon economy in 1993 alone. Additionally, several thousand amateur mushroom hunters in the United States collect solely for their own enjoyment.

Number of Species

Considering the human, ecological, and economic importance of these organisms, it is somewhat surprising that there is not a good estimate of the number of species of macrofungi that occur in North America. Because there are neither checklists of North American mushrooms and their relatives nor comprehensive regional treatments, the best estimates of North American diversity are based on comparisons with numbers of these organisms reported from Europe. More than 3,000 species of mushrooms and their relatives are reported from western Europe (Moser 1983), but most scientists who study fungi (mycologists) would estimate that far more species occur in North America. For example, more than twice as many species of *Lactarius, Amanita*, and *Clitocybe* are reported from the continental United States (Hesler and Smith 1979; Bigelow 1982, 1985; Jenkins 1986) than from western Europe (Moser 1983).

Better estimates exist for species diversity of the other groups of North American macrofungi. Gilbertson and Ryvarden (1986, 1987) treated more than 400 species of polypore fungi, Smith et al. (1981) listed nearly 300 species of puffballs and relatives, and Seaver (1942, 1951) covered more than 350 species of cup fungi and other macro ascomycetes. Based on these data, it is reasonable to predict that there are 5,000-10,000 species of macrofungi in the United States. A compilation of herbarium records in U.S. and Canadian museums and universities would provide a good first step in predicting the diversity of these organisms.

Fig. 3. *Cantharellus cibarius*. The chanterelle is one of the important fungi forming mycorrhizae with pines and oaks in North American forests.

Declining Fungi

Change in the frequency of occurrence of macrofungi in Europe is well documented; many species that form ectomycorrhizae (a kind of mycorrhizae; *see* glossary) are showing a marked decline, and some species involved with wood decay show a marked increase in fruiting. More than 50% of the reported species of mushrooms in Europe occur on at least one country's "Red List" (*see* glossary: "red data book") (Arnolds and de Vries 1993), and once-common species such as *Hydnum repandum* and some of the chanterelles (Fig. 3) appear lost from some countries. Air pollution, particularly acid rain, has been implicated in this observed decline in ectomycorrhizal fungi fruiting frequency and diversity in Europe (Fellner 1993; Pegler et al. 1993). Intensive collecting of edible fungi such as chanterelles, *Hydnum*, and boletes might also be negatively affecting fruiting patterns of these fungi, but additional data are needed to document this. In any case, the observed change in fungal fruiting is correlated with a decline in forest health, but cause and effect are hard to document. Rigorous studies to determine if similar trends in macrofungi fruiting patterns have occurred in the United States do not exist.

Current Studies of Diversity

The baseline data necessary for estimating fungal diversity and for investigating trends in fruiting patterns and frequencies of macrofungi in the United States and Canada are not yet available although various methods are beginning to be used to obtain these necessary data. For example, studies of species diversity and frequency of particular fungi in Pacific Northwest old-growth forests have documented that while a single season of collecting will uncover most of the decomposer macrofungi, mycorrhizal fungi fruit much more erratically (Vogt et al. 1992). Thus, to develop a reasonable

---

Truffles, Trees, and Biodiversity

by
Robert Fogel
University of Michigan

Most Americans identify truffles as expensive, Epicurean delights from Europe, found with the aid of pigs. Because truffles are produced belowground, we remain ignorant of the rich diversity and importance of truffles in North America. Truffles (ascomycetes) and the similar-appearing false truffles (basidiomycetes) play a major role in determining the structure and function of forest ecosystems by providing nutrients to many economically valuable trees in exchange for carbohydrates from the trees. This mycorrhizal (fungus root) symbiosis is obligate; that is, truffles and trees, especially conifers, cannot survive without each other. One of the problems in reforesting large areas of the Southwest is identifying ectomycorrhizal fungi suitable for inoculation of tree seedlings destined for sites with calcareous soils.

Truffles and false truffles are food items for many animals, including many endangered or threatened species. In old-growth Douglas-fir (*Pseudotsuga menziesii*) forests, truffles not only provide soil nutrients to the trees controlling forest structure, but they also are an important link in the food web supporting the endangered spotted owl. Northern flying squirrels (*Glaucomys sabrinus*) glide down to the forest floor at night to feed on truffles. While feeding on truffles, flying squirrels become vulnerable to predation from the northern spotted owl (*Strix occidentalis caurina*), coyotes (*Canis latrans*), bobcats (*Lynx rufus*), and other predators.

Given the undeniably important role of truffles in determining the structure and function of forest ecosystems, how much is known about the distribution of truffles and false truffles? The paucity of information and potential impact of surveys on our knowledge base can be illustrated by an ongoing National Science Foundation-funded survey of the Great Basin, an area of 712,250 km^2 (275,000 mi^2) between the Sierra Nevada and Wasatch mountains and including most of Nevada and parts of California, Idaho, Utah, Wyoming, and Oregon. No truffles or false truffles had been reported from the area before the survey. Over three summers, the survey produced 1,119 collections of truffles and false truffles from 40 mountain ranges.

In addition, the survey produced evidence for extinction of many truffles in the Great Basin. A few truffles obligately associated with a single tree species outside the Great Basin have switched within the Great Basin to new tree species, providing supporting evidence for extinction of local tree species. New endemic species have been found and the geographic ranges of some species greatly expanded. Populations of some endemic species are restricted to a single mountain range.

Knowledge of truffles is important to the biodiversity in the United States. Without such knowledge, there is a danger of losing or degrading ecosystems through ignorance about the status of keystone fungal species. If ecosystems are lost, then species dependent on specific ecosystems will also be lost.

For more information:

Robert Fogel
University of Michigan
Ann Arbor, MI 48109

estimate of species richness and dominance, researchers must sample over several years. These studies also have documented that certain collecting techniques work better for some fungi than others, which emphasizes the need to develop standardized sampling protocols for collecting data on fungal species' richness and fruiting patterns.

Satellite imagery has been combined with a long-term mapping program of fungal fruitbodies to assess the relative health and growth of particular tree-mycorrhiza fungus pairs in southern Mississippi (Cibula and Ovrebo 1988). This approach shows great promise for directly investigating the effect of certain fungi on tree health. These data, however, are based only on aboveground information, and there is still some question about how well the appearance of fruitbodies growing under a particular tree predicts what fungi are forming mycorrhizae with that tree at that time. To address this question, researchers have developed molecular techniques using DNA amplification procedures to compare the mycorrhizae on the roots of certain trees with fungal fruitbodies occurring near the tree (Bruns and Gardes 1993). The preliminary data documented that there is not always a one-to-one correlation between fruitbodies and mycorrhizae, and that caution must be used when using fruitbodies alone.

Further Studies

The studies mentioned in this article illustrate the range of work in the United States on assessing diversity and determining possible changes in fruiting patterns of macrofungi. More work is needed to document the status and trends of macrofungi in North America. These data are vital because of the integral role that macrofungi play in forest systems as decomposers and recyclers, plant pathogens, mutualists, and food for small mammals, and because of the growing commercial importance of these fungi.

For further information:

Gregory M. Mueller
The Field Museum
Department of Botany
Chicago, IL 60605

References

Arnolds, E., and B. de Vries. 1993. Conservation of fungi in Europe. Pages 211-234 *in* D.N. Pegler, L. Boddy, B. Ing, and P.M. Kirk, eds. Fungi of Europe: investigation, recording and conservation. The Royal Botanic Gardens, Kew, U.K.

Bigelow, H.E. 1982. North American species of *Clitocybe*. Part 1. Beihefte zur Nova Hedwigia 72:5-280.

Bigelow, H.E. 1985. North American species of *Clitocybe*. Part 2. Beihefte zur Nova Hedwigia 81:281-471.

Bruns, T., and M. Gardes. 1993. Molecular tools for the identification of ectomycorrhizal fungi: taxon-specific oligonucleotide probes for suilloid fungi. Molecular Ecology 2:233-242.

Cibula, W.G., and C.L. Ovrebo. 1988. Mycosociological studies of mycorrhizal fungi in two loblolly pine plots in Mississippi and some relationships with remote sensing. Pages 268-307 *in* J.D. Greer, ed. Remote sensing for resource inventory, planning and monitoring. Proceedings of the Second Forest Service Remote Sensing Application Conference. American Society for Photogrammetry and Remote Sensing, Falls Church, VA.

Fellner, R. 1993. Air pollution and mycorrhizal fungi in central Europe. Pages 239-250 *in* D.N. Pegler, L. Boddy, B. Ing, and P.M. Kirk, eds. Fungi of Europe: investigation, recording and conservation. The Royal Botanic Gardens, Kew, U.K.

Gilbertson, R.L., and L. Ryvarden. 1986. North American Polypores. Vol. 1. Fungiflora A/S, Oslo. 433 pp.

Gilbertson, R.L., and L. Ryvarden. 1987. Pages 437-885 *in* North American Polypores. Vol. 2. Fungiflora A/S, Oslo.

Hesler, L.R., and A.H. Smith. 1979. North American species of *Lactarius*. The University of Michigan Press, Ann Arbor. 841 pp.

Jenkins, D.T. 1986. *Amanita* of North America. Mad River Press, Eureka, CA. 198 pp.

Moser, M. 1983. Keys to the Agarics and Boleti (Polyporales, Boletales, Agaricales, Russulales). Roger Phillips, London. 535 pp.

Pegler, D.N., L. Boddy, B. Ing, and P.M. Kirk, eds. 1993. Fungi of Europe: investigation, recording and conservation. The Royal Botanic Gardens, Kew, U.K. 322 pp.

Seaver, F.J. 1942. The North American cup fungi (Operculates). Rev. ed. Seaver, New York. Reprinted by Lubrecht and Cramer. 377 pp. + 74 plates.

Seaver, F.J. 1951. The North American cup fungi (Inoperculates). Seaver, New York. 428 pp.

Smith, A.H., H.V. Smith, and N.S. Weber. 1981. How to know the non-gilled fleshy fungi. 2nd ed. William C. Brown, Dubuque, IA. 324 pp.

Vogt, K.A., J. Bloomfield, J.F. Ammirati, and S.R. Ammirati. 1992. Sporocarp production by Basidiomycetes, with emphasis on forest ecosystems. Pages 563-581 *in* G.C. Carroll and D.T. Wicklow, eds. The fungal community. Marcel Dekker, Inc., New York.

Lichens

by
James P. Bennett
National Biological Service

Lichens are a unique life form because they are actually two separate organisms, a fungus and an alga, living together in a symbiosis. Lichens seem to reproduce sexually, but what appears to be a fruiting structure is actually that of the fungal component. Consequently, lichens are classified by botanists as fungi, but are given their own lichen names.

Lichens are small plant-like organisms that grow just about everywhere: soils, tree trunks and branches, rocks and artificial stones, roofs, fences, walls, and even underwater. They are famous for surviving climatic extremes and are even the dominant vegetation in those habitats. Some lichens, however, are only found in very specialized habitats. The diversity of lichens in an area, therefore, is highly dependent on habitat diversity. Many special habitats across the United States are declining or disappearing because of human activities, and some lichen species are consequently in decline.

Lichens are very diverse in form: some grow flat and appressed to a substrate, others are more leaf-like and grow free of the substrate, and yet others have complex filamentous and blade-like forms.

Lichens are unique botanically because they lack any outside covering, or cuticle, and consequently are directly exposed to the atmosphere, which they depend upon for their nutrients and water, neither of which is derived from their hosts. Moistened lichen tissues act as blotters, soaking up chemicals or materials deposited on their surfaces. Unfortunately, this feature has also made them highly susceptible to air pollutants; lichens are perhaps the plant species most susceptible to sulfur dioxide, heavy metals, and acid rain.

Lichens play important roles in ecosystems. They break down rocks and form soil by excreting weak acids, or in arid ecosystems like deserts, they help bind the soil surface by forming crusts. They are important food sources for invertebrates and vertebrates, including reindeer that eat reindeer "moss," which is actually a lichen (Fig. 1). In addition, some birds depend on certain lichens for nest-building materials. Finally, some lichens can fix nitrogen from the atmosphere and contribute a significant portion of this to certain forest ecosystems (e.g., the Pacific Northwest).

A rich lichen flora in a region indicates a lack of disturbance in the area for two reasons. First, lichens can only appear in an area if both the fungus and alga are propagated there and coincide. Isolation of an area so that propagules (*see* glossary) cannot reach the area will slow down recolonization significantly. Second, lichens grow slowly, usually only a few millimeters a year. Thus, colonization of an area by lichen species typically does not occur even over the span of one human generation.

Status

The best estimates of the number of U.S. lichen species are between 3,500 and 4,000, grouped in about 400 genera. The current checklist for the United States and Canada is probably in excess of 3,600 (Egan 1987).

Some species are cosmopolitan and are found from coast to coast. Most species, however, are more limited in their geographic distributions. The percentage of species that are rare nationally is high: about one-third of more than 400 lichens described by Hale (1979) are rare, and this ratio could probably be applied to the total number for the United States. Thirty-eight percent of the lichen flora of Hawaii is considered endemic. Five lichen species have been nominated for federal threatened and endangered listing (Pittam 1991), and several states (e.g., California, Minnesota, and Missouri) have listed some species as threatened or endangered.

No state has a complete lichen flora published. Incomplete floras or checklists are known for Alaska, California, Colorado, Connecticut, Florida, Hawaii, Louisiana, Michigan, Minnesota, North Carolina, New Mexico, New York, South Dakota, Tennessee, Texas, and Washington. Most of the rest of the country's lichen flora remains unexplored. Species for these partial checklists number in the several hundreds, with the exception of California with 999 taxa. Nationally, centers of diversity for lichens include the Pacific Northwest, California, the southern Appalachians, Florida, and Maine. On a more local scale, wetlands and floodplains tend to have higher numbers of lichen species than more arid areas. The presence of a bog or a rocky outcropping in an area will typically double the number of species present.

Courtesy J.P. Bennett

Fig. 1. *Cladina mitis* and *C. rangiferina* (reindeer moss), Voyageurs National Park, MN.

There are about 10 lichen herbarium collections with active lichen taxonomists in the United States, and about 5 in Canada. Many of these collections are poorly funded, not computerized, and stored in inadequate or outdated facilities. Fewer than two dozen practicing lichenologists work in the United States and Canada, and very few graduate students are being trained in lichenology. Most academic botany or biology departments do not have lichenologists.

Trends

About 100 years ago, lichens had disappeared from many cities in Europe and Great Britain and the term "lichen desert" was coined to describe the phenomenon; these lichen deserts were caused by air pollution. Here in the United States, lichen deserts are well known in our cities and nearby rural areas, but are unfortunately poorly documented. Most information is anecdotal, but some studies have shown lichen deserts in eastern Pennsylvania (Nash 1975), the Cuyahoga Valley in Ohio (Wetmore 1989), northern Indiana on Lake Michigan (Wetmore 1988), Cedar Rapids, Iowa (Saunders 1976), Los Angeles (Sigal and Nash 1983), Seattle, Washington (Johnson 1979),

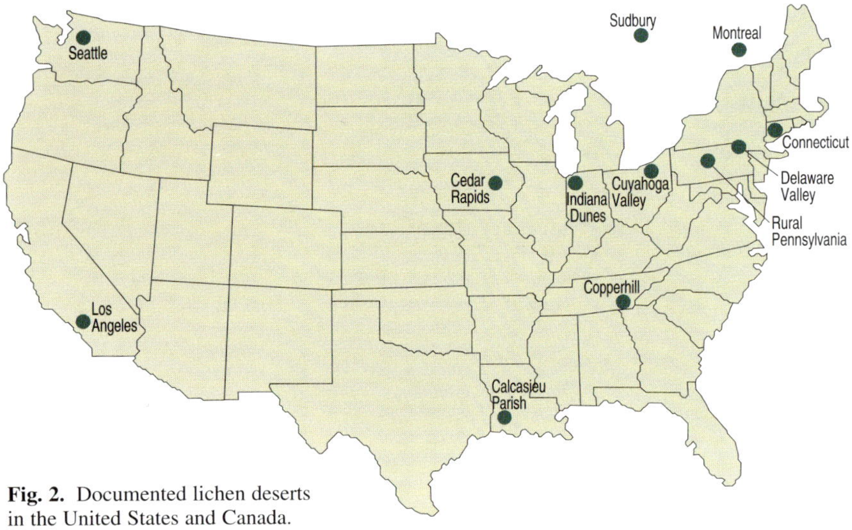

Fig. 2. Documented lichen deserts in the United States and Canada. Strong anecdotal evidence exists that lichen deserts also occur in most major cities.

Copperhill, Tennessee (Mather 1978), and in Canada in Montreal (LeBlanc and De Sloover 1970) and Sudbury (LeBlanc et al. 1972) (Fig. 2). In some of these areas, researchers estimate that as much as 80%-90% of the original lichen flora is gone (Nash 1975; Wetmore 1989). Acid rain has diminished lichen diversity in remote rural areas such as north-central Pennsylvania (Showman and Long 1992), central and southwestern Connecticut (Metzler 1980), and southwestern Louisiana (Thompson et al. 1987). Sensitive species must be studied and monitored to determine the effects of air pollutants.

Some lichens are unique to old-growth forests. *Usnea longissima*, which only grows in old-growth spruce forests, has vanished from many sites in western Europe (Esseen et al. 1992) and may be repeating this pattern in parts of the United States. Other old-growth forest lichens, including *Alectoria sarmentosa*, *Lobaria scrobiculata*, and *Ramalina thrausta*, are now quite rare in the eastern United States because of habitat destruction and loss.

In addition, scientific overcollecting may become a problem for lichens. One species, *Gymnoderma lineare*, was overcollected in Great Smoky Mountains National Park, Tennessee, in the late 1970's, and is now proposed for federal listing as endangered. Collecting is no longer allowed in certain areas (e.g., some national parks and nature preserves), and both the American Bryological and Lichenological Society and the British Lichenological Society do not always permit collecting at popular sites during their annual forays. Some hobby overcollecting of lichens for dye materials or architectural tree models is thought to be a problem in a few areas, but is not well documented.

Trends in lichenology in this country are not encouraging and are at odds with trends in the rest of world (Galloway 1992). With fewer universities offering training in the discipline, fewer surveys and lists of floras being done, less literature being published, and at the same time lichens disappearing from our ecosystems, it is clear that the science is heading the opposite direction of what is needed. Other countries, including England, Canada, the Netherlands, and Japan, are increasing funding for lichenology, training more students, publishing more literature, and conserving their lichen flora. Given the problems confronting lichen habitats, the size of the United States, and the potential flora it may have, lichen science needs more attention. A reasonable start would be a preliminary checklist for every state and an identification of priority areas for future surveys.

References

Egan, R.S. 1987. A fifth checklist of the lichen-forming, lichenocolous and allied fungi of the continental United States and Canada. Bryologist 90:77-173.

Esseen, P.-A., B. Ehnstrom, L. Ericson, and K. Sjoberg. 1992. Boreal forests—the focal habitats of fennoscandia. Chapter 7 *in* L. Hansson, ed. Ecological principles of nature conservation. Elsevier, London.

Galloway, D.J. 1992. Biodiversity: a lichenological perspective. Biodiversity and Conservation 1:312-323.

Hale, M.E. 1979. How to know the lichens. 2nd ed. Wm. C. Brown Publishers, Dubuque, IA. 246 pp.

Johnson, D.W. 1979. Air pollution and the distribution of corticolous lichens in Seattle, Washington. Northwest Science 53(4):257-263.

LeBlanc, F., and J. De Sloover. 1970. Relation between industrialization and the distribution and growth of epiphytic lichens and mosses in Montreal. Canadian Journal of Botany 48:1485-1496.

LeBlanc, F., D.N. Rao, and G. Comeau. 1972. The epiphytic vegetation of *Populus balsamifera* and its significance as an air pollution indicator in Sudbury, Ontario. Canadian Journal of Botany 50:519-528.

Mather, T.C. 1978. Lichens as indicators of air pollution in the vicinity of Copperhill, Tennessee. Georgia Journal of Science 36:127-139

Metzler, K.J. 1980. Lichens and air pollution: a study in Connecticut. Report of Investigations 9. State Geological and Natural History Survey of Connecticut. 30 pp.

Nash, T.H., III. 1975. Influence of effluents from a zinc factory on lichens. Ecological Monographs 45:183-198.

Pittam, S.K. 1991. The rare lichens project, a progress report. Evansia 8:45-47.

Saunders, J.R. 1976. The influence of SO_2 on corticolous lichens in Cedar Rapids, Iowa. Submitted in partial fulfillment for College Honors (Biology) at Coe College, Cedar Rapids, IA, May 1976. 111 pp.

Showman, R.E., and R.P. Long. 1992. Lichen studies along a wet sulfate deposition gradient in Pennsylvania. Bryologist 95:166-170.

Sigal, L.L., and T.H. Nash III. 1983. Lichen communities on conifers in southern California mountains: an ecological survey relative to oxidant air pollution. Ecology 64:1343-1354.

Thompson, R.L., G.J. Ramelow, J.N. Beck, M.P. Langley, and J.C. Young. 1987. A study of airborne metals in Calcasieu Parish, Louisiana using the lichens, *Parmelia praesorediosa* and *Ramalina stenospora*. Water, Air and Soil Pollution 36:295-309.

Wetmore, C.M. 1988. Lichens and air quality in Indiana Dunes National Lakeshore. Mycotaxon 33:25-39.

Wetmore, C.M. 1989. Lichens and air quality in Cuyahoga Valley National Recreation Area, Ohio. Bryologist 92:273-281.

For further information:

James P. Bennett
National Biological Service
Wisconsin Cooperative
Research Unit
University of Wisconsin-Madison
Madison, WI 53705

Bryophytes

by
Gary L. Smith Merrill
The Field Museum, Chicago

Bryophytes (mosses, liverworts, and hornworts) are small green plants that reproduce by means of spores (or vegetatively) instead of seeds. Most are only a few centimeters high, although some mosses attain a half meter (20 in) or more. Although often small and inconspicuous, bryophytes are remarkably resilient and successful. They are sensitive indicators of air and water pollution, and play important roles in the cycling of water and nutrients and in relationships with many other plants and animals. Information about bryophytes and their ecology is essential to develop comprehensive conservation and management policies and to restore degraded ecosystems.

There are three main groups of bryophytes: mosses (Musci); liverworts, also known as hepatics (Hepaticae); and hornworts (Anthocerotae). Bryophytes rank second (after the flowering plants) among major groups of green land plants, with an estimated 15,000-18,000 species worldwide. In North America north of Mexico, there are 1,320 species of mosses in 312 genera (Anderson et al. 1990), and 525 species of hepatics and hornworts in 119 genera (Stotler and Crandall-Stotler 1977), or somewhat more than 10% of the world's bryophyte species.

Mosses are most abundant and conspicuous in moist habitats, but are also found in grasslands and deserts, where they endure prolonged dry periods. Hepatics also include some arid-adapted species, but most are plants of humid environments. In mosses and leafy hepatics, the conspicuous plant body is leafy; in some liverworts and all hornworts, the plant is a flattened, ribbon-like "thallus" that lies flat on the ground. Bryophytes have no roots but are anchored by slender threads called rhizoids, which also play a role in the absorption of water and mineral nutrients.

Bryophytes have successfully exploited many environments, perhaps partly because they are rarely in direct competition with higher plants (Anderson 1980). For such small organisms, the climate near the ground (microclimate) is often very different from conditions recorded by standard meteorological methods, and shifts in temperature and humidity are often extreme. A remarkable adaptation of bryophytes is their ability to remain alive for long periods without water, even under high temperatures, then resume photosynthesis within seconds after being moistened by rain or dew.

Ecological Roles

Most bryophytes appear to absorb water and mineral nutrients directly into leaves and stems, a fact that makes them extremely vulnerable to airborne pollutants in solution (*see* references in Longton 1980). Where abundant, bryophytes may constitute an important sink for moisture and nutrients. Mosses are reliable indicators of soil conditions because they tend to accumulate chemical elements somewhat indiscriminately. The analysis of concentrations of pollutants in older bryophyte specimens could be used to document increases in pollution levels over time.

Bryophytes are also closely associated with organisms as diverse as protozoa, rotifers (microscopic aquatic animals), nematodes, earthworms, mollusks, insects, and spiders (Gerson 1982), as well as plants and fungi. Direct interactions of bryophytes include providing food, shelter, and nesting materials for small mammals and invertebrates; indirectly, they serve as a matrix for a variety of interactions between organisms.

Bryophytes occur in all types of environments, except salt water. They occur on both shaded and exposed soil and rocks, the bark of living trees, and on decaying logs and litter in humid forests (evergreen and deciduous). Many are subaquatic in swamps, bogs, and fens, and some grow submerged or emergent in streams. There are no marine bryophytes, but a few grow on coastline rocks and can tolerate exposure to salt spray.

Courtesy G.L. Merrill, The Field Museum

The newly discovered moss genus and species, *Ozobryum ogalalense*, is known only from four localities in northwest Kansas and adjacent Nebraska (Merrill 1992). The species forms soft, compact cushions on exposed lime-rich outcrops in native prairie pastures. The outcrops are porous and charged with moisture, making them a magnet for several species of bryophytes in an otherwise hostile environment. *Ozobryum* underscores the fact that discoveries can still be made in areas of the country where bryophytes are poorly known.

In the moss-carpeted rainforests of the Pacific Northwest, bryophytes make up a significant proportion of the biomass. Peat moss (*Sphagnum*) is a dominant organism in northern peatland communities and is of some economic importance in horticulture and as an energy source. Bryophytes of arid grasslands and deserts are few, but there are mosses that appear adapted to prairies and to the periodic intense disturbance of grazing and fire (Merrill 1991).

Floristics and Distribution

Basic information on the distribution of bryophytes is available for at least the northeastern United States, eastern Canada, and the Pacific

Northwest. Some parts of the continent are less well known, chiefly remote areas of the Rockies, the arid Southwest, and the Great Plains. Much information about the bryophytes of the interior plains may be "irretrievably lost since most of the natural grassland, with whatever mosses it may have sheltered, is under cultivation" (Schofield 1980, p. 131), but fieldwork can still yield important discoveries (Merrill 1992) as well as basic distributional information.

A much-improved picture of bryophyte distribution in North America will emerge as the result of the preparation of treatments for Volume 13 of *Flora North America* (scheduled for publication in 1996), but much of the necessary distributional information is simply not available now.

Status

Some bryophyte species appear to thrive in disturbed habitats (both "naturally" disturbed and those due to human activity). Many bryophytes, however, are quite rare, have extremely local distributions, and are at risk. Changes in land use and loss of habitat represent the greatest threat to bryophyte diversity. Cutting forests, draining bogs and wetlands, and destroying rock faces by quarrying and road building are especially destructive.

Most bryophytes are unlikely to be picked for their own sake, but where mosses are particularly abundant, as in the Pacific Northwest, commercial harvesting for horticultural purposes can have a significant effect. The loss of bryophyte habitat is likely to have a ripple effect, since other organisms closely associated with them are also likely to be lost. Efforts at habitat restoration must take into account the difficulty of re-creating the specialized conditions that many bryophytes require.

Future Needs and Priorities

Basic floristic inventories are an essential part of any assessment of the role of bryophytes in natural ecosystems. While checklists are available that cover the whole of North America (as well as many states), and floristic works are available that make the task of identifying species easier, these do not provide information on the status of individual species. Inventories are needed to identify areas where many rare bryophytes occur; these areas should be given priority in establishing conservation reserves. In addition, trained specialists are scarce, and their numbers are decreasing. The advent of modern electronic data-base technology makes it possible to capture important distributional information contained in existing collections, but this also is time-intensive and expensive. Priorities are to support basic floristic research on bryophytes (and the training of new bryologists and information specialists needed to deal with the formidable task of documenting bryophyte diversity) and to provide support to institutions that maintain the major national resource collections of bryophytes.

References

Anderson, L.E. 1980. Cytology and reproductive biology of mosses. Pages 37-76 *in* R.J. Taylor and A.E. Leviton, eds. The mosses of North America. Pacific Division of the American Association for the Advancement of Science, San Francisco.

Anderson, L.E., H.A. Crum, and W.R. Buck. 1990. List of the mosses of North America north of Mexico. Bryologist 93:448-499.

Gerson, U. 1982. Bryophytes and invertebrates. Pages 291-332 *in* A.J.E. Smith, ed. Bryophyte ecology. Chapman and Hall, New York.

Longton, R.E. 1980. Physiological ecology of mosses. Pages 77-113 *in* R.J. Taylor and A.E. Leviton, eds. The mosses of North America Symposium. Pacific Division, American Association for the Advancement of Science, San Francisco.

Merrill, G.L. Smith. 1991. Bryophytes of Konza Prairie Research Natural Area, Kansas. Bryologist 94:383-391.

Merrill, G.L. Smith. 1992. *Ozobryum ogalalense* (Pottiaceae), a new moss genus and species from the American Great Plains. Novon 2:256-258.

Schofield, W.B. 1980. Phytogeography of the Mosses of North America (north of Mexico). Pages 131-170 *in* R.J. Taylor and A.E. Leviton, eds. The mosses of North America. Pacific Division of the American Association for the Advancement of Science, San Francisco.

Stotler, R., and B. Crandall-Stotler. 1977. A checklist of the liverworts and hornworts of North America. Bryologist 80:405-428.

For further information:

Gary L. Smith Merrill
The Field Museum
Department of Botany
Roosevelt Rd. at Lakeshore Dr.
Chicago, IL 60605

Floristic Inventories of U.S. Bryophytes

by
Alan Whittemore
Bruce Allen
Missouri Botanical Garden, St. Louis

Few floristic inventories of bryophytes have been made in the United States, primarily because of lack of trained personnel. The publication of modern manuals for the eastern United States for mosses (Crum and Anderson 1981) and liverworts and hornworts (Schuster 1966-92) has improved the situation. The paucity of manuals in the western United States is especially critical because of the uniqueness of the western North American flora. Eighty percent of the genera of bryophytes known to be endemic to temperate North America are confined to the area west of the 110th meridian (approximately the Rocky Mountains and west), but very few bryologists work there (Schofield 1980; Schuster 1984).

Mosses

Mosses are the best known of the three bryophyte groups and have the most species: 1,320 species in 312 genera (Anderson et al. 1990). The only manual of mosses that treats all of North America north of Mexico is by A.J. Grout (1928-40), but is now outdated. Although this flora is unreliable for the mosses in the

Courtesy D. Kearns, Missouri Botanical Garden

The moss *Leucolepis acanthoneuron*.

midcontinent, it covers the mosses from the eastern United States and the west coast regions well.

The eastern forest region is the strongest area for moss floristics in the United States. The United States east of the Mississippi is covered well by Crum and Anderson's (1981) flora. Most states there have recent checklists of mosses. In addition, several regional floras cover parts of more than one state (e.g., Crum [1983] for upper Michigan and nearby areas and Redfearn [1983] for the Ozark region).

The distribution of mosses in other parts of the country is not as well known. There are checklists of mosses for nearly every U.S. state (Pursell 1982), although many were published 30-40 years ago and are outdated. The Southeast has the fewest checklists; the northern parts of Mississippi, Alabama, and Georgia and the southern parts of Arkansas are poorly known.

The Southwest is also one of the least known U.S. areas. It has great diversity of habitats including mountains, grasslands, and desert habitats. Although checklists have been published for all of the states and a flora has been published for Utah (Flowers 1973), the mosses of Nevada, Arizona, New Mexico, and parts of Texas are probably still the least known in the country. The recent publication of the moss flora of Mexico (Sharp et al. 1994) will considerably aid workers in this region, but much basic floristic work needs to be done.

Good state checklists exist for the Great Plains and the Pacific Northwest, which has checklists for the entire region as well as a regional flora (Lawton 1971). The Great Plains is reasonably well covered with checklists and two regional floras for all of the midcontinent. Moss diversity in this region is low, and many of the mosses are members of the eastern moss flora. But the mosses in this region have not been extensively surveyed, and the area continues to yield surprises such as *Ozobryum ogalalense*, a new genus (Merrill 1993).

Alaska has a checklist and work has begun on a synoptic flora that will cover the Arctic area (Mogensen 1985). Floristically, however, the Arctic areas of Alaska are fundamentally different from the rest of the United States. A portion of flora can be named by using Arctic European floras; otherwise, the flora can be named only by specialists with access to the scattered literature and a good herbarium.

Liverworts and Hornworts

No part of the United States can be considered well-inventoried for liverworts or hornworts. The eastern half of the country is much better known than the West. The preparation of Schuster's manual of the liverworts and hornworts of eastern North America (1966-92), which resulted in the publication of several dozen new species (mostly from the southern Appalachians and Florida), has improved our knowledge of these plants in the East. Many taxonomic problems still need serious study, however, and known ranges of distribution are still incomplete.

Our knowledge of the liverwort and hornwort floras in the western half of the country has improved recently because of a series of local checklists (mostly of national parks and similar small floristic units) for the Pacific Northwest. For large parts of the northwestern United States, however, we still rely on a few pioneering studies from 1890 to 1940.

Courtesy D. Kearns, Missouri Botanical Garden

The liverwort *Asterella echinella*.

The most poorly known part of the country is undoubtedly the interior Southwest (New Mexico, Arizona, and surrounding regions). Data from this area are so scanty and inadequate that it is difficult to assess the regional liverwort and hornwort floras in any meaningful way. Recent studies, though, describe several new taxa and some range extensions. For instance, *Mannia fragrans*, which seems widespread in the mountains of the western United States, was not reported from any state west of Colorado before 1987. Likewise, Bischler's (1979) revision of the xerophytic liverwort genus *Plagiochasma* increased the number of species known from the United States from three to five (adding two widespread Mexican species from

Texas and Arizona). Numerous additions to the flora can be expected from this part of the country if intensive fieldwork is conducted.

Study of these plants has been handicapped by the lack of identification manuals over much of the continent. The completion of Schuster's manual (1966-92) has improved the situation in eastern North America, but there is still almost no usable literature from the western half of the country. Since the first half of the century, there have been no floristic treatments with identification aids of any kind published for any area west of the 110th meridian, with the single exception of the brief checklist of the liverworts and hornworts of Olympic National Park by Hong et al. (1989). In the whole of this large area, which makes up more than half of the country, specimens can only be identified reliably by specialists with access to rare and often outdated literature. Even in the well-studied extreme Northeast (i.e., New England and New York), new taxa continue to be found (for example, *Pellia megalospora* Schust. was not described until 1981). Further collection and study will surely provide many more range extensions. Likewise, the very distinctive endemic genus *Schofieldia* Godfrey was not described from western Washington until 1976, even though it is without close relatives and is rather common in subalpine sites from northwestern Washington north through the central part of the Alaska panhandle.

For further information:

Alan Whittemore
Missouri Botanical Garden
PO Box 299
St. Louis, MO 63166

References

Anderson, L.E., H.A. Crum, and W.R. Buck. 1990. List of mosses of North America north of Mexico. Bryologist 93:448-499.

Bischler, H. 1979. *Plagiochasma* Lehm. et Lindenb. IV. Les taxa americains. Revue Bryologique et Lichenologique 45:255-334.

Crum, H. 1983. Mosses of the Great Lakes forest. 3rd ed. University of Michigan, Ann Arbor. 417 pp.

Crum, H., and L.E. Anderson. 1981. Mosses of eastern North America. 2 vols. Columbia University Press.

Flowers, S. 1973. Mosses: Utah and the West. Brigham Young University Press, Provo, UT. 567 pp.

Grout, A.J. 1928-40. Moss flora of North America. 3 vols. Self-published, New Fane, VT.

Hong, W.S., K. Flander, D. Stockton, and D. Trexler. 1989. An annotated checklist of the liverworts and hornworts of Olympic National Park, Washington. Evansia 6:33-52.

Lawton, E. 1971. Moss flora of the Pacific Northwest. Hattori Botanical Laboratory, Nichinan, Japan. 362 pp.

Merrill, G.S. 1993. *Ozobryum ogalalense* (Pottiaceae), a new genus and species from the American Great Plains. Novon 2:255-258.

Mogensen, G.S. 1985. Illustrated moss flora of Arctic North America and Greenland. Bioscience 17:1-8.

Pursell, R.A. 1982. A synopsis of moss floristics in the eastern United States. Beihefte zur Nova Hedwigia 71:451-454.

Redfearn, P.L., Jr. 1983. Mosses of the Interior Highlands of North America. Revision Missouri Botanical Garden, St. Louis. 104 pp.

Schofield, W.B. 1980. Phytogeography of the mosses of North America (north of Mexico). Pages 131-170 *in* R.J. Taylor and A.E. Leviton, eds. The Mosses of North America Symposium. Pacific Division, American Association for the Advancement of Science, San Francisco.

Schuster, R.M. 1966-92. The Hepaticae and Anthocerotae of North America, east of the hundredth meridian. Vols. 1-4, Columbia University Press, New York. Vols. 5-6, The Field Museum, Chicago.

Schuster, R.M. 1984. Phytogeography of the Bryophyta. Pages 463-626 *in* R.M. Schuster, ed. New manual of bryology. Vol. 1. Hattori Botanical Laboratory, Nichinan, Japan.

Sharp, A.J., H. Crum, and P.M. Eckel, eds. 1994. The moss flora of Mexico. Memoirs of the New York Botanical Garden 69. 1113 pp.

Vascular Plants of the United States

by

Nancy Morin
Missouri Botanical Garden

Information on the plants of the United States can be found in floras, monographs, and various lists and reports. Herbarium collections provide an invaluable record of past and current distributions of U.S. plants and form the basis for published accounts of the plants such as floras and checklists. Properly understanding and managing U.S. plant resources depend on having physical samples that document the characteristics and distributions of plants and on the scientific studies of the relationships, characteristics, distributions, and physical requirements of the plants. Although such documentation exists for some areas of the country, many areas are still poorly known, and authoritative references are still lacking for some.

About 17,000 species of vascular plants (i.e., flowering plants, gymnosperms, and ferns) occur in the contiguous United States and Alaska (Flora of North America Editorial Committee 1993); Hawaii is home to more than 1,800 species of flowering plants (Wagner et al. 1990), few of which are found on the North American mainland. Trees have been most completely documented, followed by shrubs and showy herbaceous plants. Known distributions of rare plants are generally available in computerized data bases, often maintained by state Natural Heritage Programs. Nationwide database files for rare plants are maintained by The Nature Conservancy.

Non-natives and inconspicuous natives are often overlooked by plant collectors and thus are less well documented. In much of the continent, and especially in highly populated areas, however, the native flora has been altered so completely by humans that "native" or "natural" vegetation is almost beyond conception. Because of this, the historical portrait of plant distribution that can be drawn based on herbarium specimens is extremely valuable to understand the pre-Columbian composition of our flora and the relation of plants to their environment. Modern collecting still brings many new species to light. Between 1975 and 1989, for example, 725 new taxa of vascular plants were reported from the conterminous United States alone (Hartman 1990).

The following discussions indicate what published plant information and data bases exist and describe the level of current and historical plant collecting in the United States.

Major Plant Groups

Few families or genera in the United States have been studied comprehensively throughout their range during the past 50 years, and until now there has been no source that brings together the best existing knowledge of U.S. plant taxa. To provide such a resource, plant taxonomists from the United States and Canada have established the Flora of North America project. Scientific information on the names, relationships, characteristics, and distributions of all plants that grow outside of cultivation in North America north of Mexico will be published in 14 volumes and in an online data base over the next 8 years. To date, two volumes have been published (Flora of North America Editorial Committee 1993). As information is synthesized and published, research needs can be evaluated. Checklists of North American plants are currently available (Soil Conservation Service 1982; Kartesz 1994), and the Soil Conservation Service maintains a data base of Plant List of Attributes, Nomenclature, Taxonomy, and Symbols (PLANTS) for North America.

Pteridophytes

About 500 species of ferns and fern allies are found in the United States, excluding Hawaii where about 200 occur. The most recent treatment of ferns for North America is in Volume 2 of *Flora of North America* (Flora of North America Editorial Committee 1993). Recent studies involving DNA analysis, isozyme work, and modern statistical analyses have significantly improved our understanding of genetic relationships among groups of ferns (Wagner and Smith 1993). Fern groups in the dry areas of the Southwest especially need study.

Gymnosperms

Gymnosperms, with 118 species (none native to Hawaii), include the economically important conifers. Tremendous research has been done on conifers, including detailed population studies of individual species. The most recent treatment of gymnosperms for North America is Volume 2 of *Flora of North America* (Flora of North America Editorial Committee 1993). The *Atlas of United States Trees* (Little 1971), although somewhat outdated, is still the best source for precise distributional information for conifers.

Angiosperms

Most vascular plant species in the United States are angiosperms, those plants bearing what are commonly recognized as flowers. The large sunflower family has been intensively studied over the past several decades, although work on this family is hampered by its complexity and the difficulty of identifying individual plants. In addition, more extensive surveying of the Southwest is needed to understand the family. An account of Asteraceae for the southeastern United States was published in *The Vascular Flora of the Southeastern United States* (Radford et al. 1980); Great Basin species are treated in Volume 5 of the *Intermountain Flora* (Cronquist et al. 1972-94), and Asteraceae will appear as the final published volume of *Flora of North America*.

The grass family is the most agriculturally important family in the United States, both for its forage value and as a source for crop and rangeland weeds. Researchers coordinated by Utah State University are revising the *Manual of the Grasses of the United States* (Hitchcock and Chase 1950).

Much work on the complex legume family has been done by researchers in the U.S. Department of Agriculture. Genera such as *Astragalus*, with more than 325 species, still require tremendous work to understand; it is extremely difficult to identify individual species. An international program to develop a checklist of species in this family, with distribution, growth habit, and economic information, is being carried out by the International Legume Data Information System (ILDIS); the Missouri Botanical Garden is the center for North American information for this project.

The sedge family includes ecologically important species, especially in wetlands where sedges dominate. Although sedges are being intensively studied, individual species can be difficult to identify; *Carex* alone contains more than 400 species. Cyperaceae specialists have been collaborating on common solutions to taxonomic problems in this group; volume 11 of *Flora of North America* will synthesize the best information available on the family.

Regional Floras

Hawaii

The *Manual of the Flowering Plants of Hawaii* (Wagner et al. 1990) gives excellent coverage for flowering plants. Two fern floras

Courtesy G. Yatskievych, MO Dept. Conserv.

Scleria cilata, a member of the sedge family, Cyperaceae.

All drawings by Laurie Lange; Flora of North America Editorial Committee©

The fern *Cyrtomium falcatum*.

Courtesy G. Yatskievych, MO Dept. Conserv.

Bouteloua gracilis, a member of the grass family, Poaceae (Gramineae).

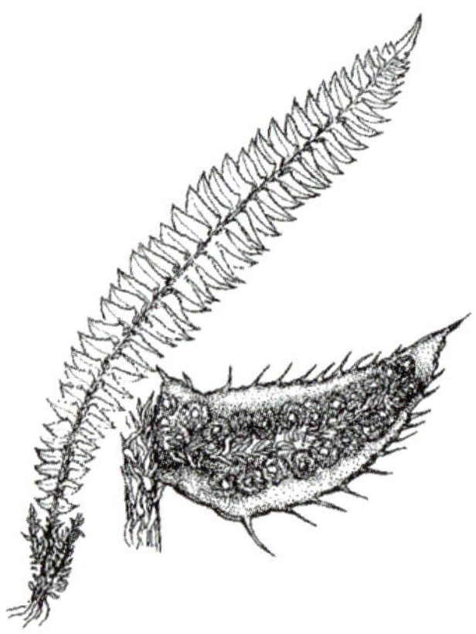

The fern *Polystichum lonchitis*.

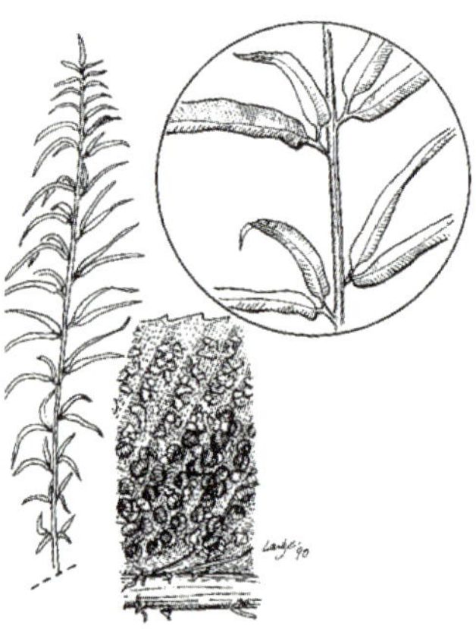

The fern *Pityrogramma trifoliata*.

are in progress. In addition, the Bishop Museum, the National Tropical Botanical Garden, and the National Museum of Natural History, Smithsonian Institution, are collaboratively creating a data base for their flowering plant specimens from Hawaii, a project to be completed by 1996. The Bishop Museum has a checklist data base of native and cultivated plants in Hawaii, but additional collecting is needed to document native plants, particularly on Molokai and Kauai. Collecting is needed throughout the islands to document the introduction and spread of alien plants. Scientists at the National Tropical Botanical Garden have discovered 20 new taxa from Kauai alone since 1990, and some 200 species of naturalized plants have been discovered in Hawaii in the past 5 years.

Alaska

Alaska has such a huge area of wilderness that basic botanical exploration is essential; *Flora of Alaska and Neighboring Territories* (Hulten 1968) is a useful work. In addition, a data base for Alaskan plants is maintained at the University of Alaska Museum in Fairbanks. Rare plants are tracked by the University of Alaska, the Alaska Natural Heritage Program, and the Alaska Rare Plant Working Group (an ad hoc group of botanists from state and federal agencies, the university, and nongovernmental organizations).

The West

The western region of the continental United States is probably the least known. Some areas (mostly near cities with universities, along highways, and popular camping sites) are relatively well known, but in less populated areas not near paved roads, much remains to be explored.

Three excellent floras treat the plants of the west coast: *Vascular Plants of the Pacific Northwest* (Hitchcock et al. 1955-69); *Intermountain Flora* (Cronquist et al. 1972-94); and *The Jepson Manual: Higher Plants of California* (Hickman 1993). State floras for Oregon (Peck 1961), Washington (Piper 1906), and Idaho (Davis 1952) are out of date and need to be revised. A revised checklist for Oregon is in preparation (A. Liston, Oregon State University, personal communication). Specimen data bases are being developed for California, Oregon, and Idaho. California herbaria have developed a model project (Specimen Management System for California Herbaria, SMASCH) to computerize data from all their California specimens. Specimens (including lichens and fungi) in Oregon herbaria are being put into a data base.

The Klamath-Siskiyou area of California and Oregon, mid-elevation Sierra, and the intermountain portion of California are the most poorly known regions. For instance, a showy new species of the genus *Neviusia*, the snow-wreath, previously known from only one species in the southeastern United States, was recently discovered in 1992 in an accessible area of northern California (Shevock et al. 1992). In addition, the rich flora of southwestern Oregon is poorly represented in herbaria, as are several counties in north-central Oregon (A. Liston, personal communication).

Intermountain Area

The number of collections from the Intermountain region has doubled in the past 20 years. *The Intermountain Flora* (Cronquist et al. 1972-94), which treats Utah, most of Nevada, southeastern Oregon, southern Idaho, and eastern California, is comprehensive; five of seven volumes have been published. An unpublished flora of Nevada exists (Kartesz 1987).

Nevada is one of the most poorly explored and documented states. Recent collectors have concentrated activity in the Great Basin mountains of Nevada and the Colorado Plateau of Utah. Even in areas seemingly well-collected, such as Zion National Park in southwestern Utah, a number of new species have been discovered and described since 1975 (Hartman 1990). *A Utah Flora* (Welsh 1993) and *Atlas of the Vascular Plants of Utah* (Albee et al. 1988) are modern and thorough treatments.

The Southwest

Although many local floras have been prepared for the Rocky Mountain areas, few have been published. Data bases on distribution of species are also being developed for individual states at the University of New Mexico, Utah State University, Colorado State University, the University of Colorado, and the University of Wyoming. A computerized checklist is being prepared for New Mexico at New Mexico State University in Las Cruces. Most of Arizona and New Mexico have been poorly collected, but these two states are thought to be the floristically richest areas in the United States, and new and surprising species are being discovered yearly. References for New Mexico (Martin and Hutchins 1980-81) are outdated or poor. In New Mexico, for instance, even frequently visited sites like the Chiricahuas still reveal treasures, such as *Apacheria*, a new genus discovered in 1973 (Mason 1975).

Northern Arizona University maintains a data base on conifers and grasses of the state; the remainder of its Arizona holdings are also being entered. In addition, the University of Arizona has a major data-base project. Areas needing more collection in Arizona include north of the Colorado River and parts of the

Colorado Plateau (L.R. Landrum, Arizona State University, and T.J. Ayers, Northern Arizona University, personal communication).

Although much of Colorado is also poorly known, all of Wyoming will have been surveyed by 1998, with recent collection data fully computerized (R. Hartman, University of Wyoming, personal communication).

The Great Plains

The Flora of the Great Plains (Great Plains Flora Association 1986) and its associated *Atlas of the Flora of the Great Plains* (Great Plains Flora Association 1977) are the result of careful study of the region in the 1960's and 1970's. The University of Kansas herbarium contains specimens representative of the entire flora; these specimens have been recently annotated by experts. This herbarium, in combination with those at the University of Nebraska, Kansas State University in Manhattan, North Dakota State University in Fargo, and the University of Minnesota (which has specimen data online), probably has fully covered this region and has current, active collecting programs. These herbaria are collaborating to develop a Central United States Plant Inventory Database (CUSPID). South Dakota and the eastern half of Montana have been undercollected.

Great Lakes

Many poorly known and interesting species are restricted to the Great Lakes region, and other typically more northern species occur here (The Nature Conservancy 1994). Recent floras are available or are being prepared for Illinois, Michigan, Minnesota, and Ohio. The floras of Indiana and Wisconsin need to be updated. Information from specimens treated in recent volumes of the *Michigan Flora* (Voss 1972) is being entered into a data base, and the Kent State University herbarium is computerizing its collection.

The Eastern Forest

The region covered by the eastern forest has been settled longer than any other area in the United States. Habitats here have undergone tremendous alteration and many introduced species now dominate the landscape. These plants should be regularly inventoried to document the occurrence and spread of alien species and to monitor the effects of environmental change. For instance, in 1950, 20% of the species in the northeastern United States were non-native (Fernald 1950); in 1986, 36% of the flora of New York was non-native (Mitchell 1986).

Regional, statewide, and local floristic studies and publications are traditional in the Northeast, but the older work is sometimes taxonomically and nomenclaturally outdated, and many areas remain inadequately inventoried. Two standard references for the vascular plants of the Northeast are *Gray's Manual of Botany* (Fernald 1950) and the recently revised *Manual of Vascular Plants of Northeastern United States and Adjacent Canada* (Gleason and Cronquist 1991). Seymour's (1982) *The Flora of New England* is also useful.

Botanists in Maine, Vermont, New Hampshire, Connecticut, and Massachusetts are updating checklists or older floras or preparing new ones. In New York, an active collaborative flora project has produced 10 illustrated installments, plus a checklist (Mitchell 1986) and an atlas of county records (New York Flora Association 1990). For Pennsylvania, Rhoads and Klein's (1993) recent atlas is available.

A book on the aquatic plants of northeastern North America is soon to be published (G.E. Crow, University of New Hampshire, and C.B. Hellquist, North Adams State College, Massachusetts, personal communication). In addition, the Association of Northeastern Herbaria, organized in 1991, is coordinating the preparation of specimen-based electronic data bases and the sharing of data. Specimen data from herbaria at the University of Massachusetts (Amherst), the Buffalo Museum of Science, the New York State Museum, and the University of Maine are partly or completely stored electronically. A large computer-stored data base also exists for Pennsylvania plants.

The South

The *Manual of the Vascular Plants of Texas* (Correll and Johnston 1970) is being updated. A number of regional floras and checklists have been published within the last two decades, but there are no regional floras for the Rolling Plains or the Trans-Pecos areas. Specimen records at the University of Texas at El Paso have been computerized, and type specimens at the University of Texas at Austin are computerized and online.

In general, local floras, checklists, and atlases are more commonly available for southeastern states than are complete state floras. In the southeast, Alabama, Arkansas, and Mississippi are the most poorly known, and northern Florida, Georgia, northwestern Louisiana, and eastern Oklahoma need considerably more study. In Alabama, in particular, the poorly collected areas are the Coastal Plain north of Mobile and Baldwin counties, north to the Cumberland Plateau. For overviews, see *The Vascular Flora of the Southeastern United States* (Radford et al. 1980), of which two of the five projected volumes have been published. *A Generic Flora of the Southeastern United States* (Wood and Miller 1958-90), which includes

Courtesy D. Boufford, Harvard

Fabaceae (Leguminosae), *Baptisa australis,* a member of the legume family.

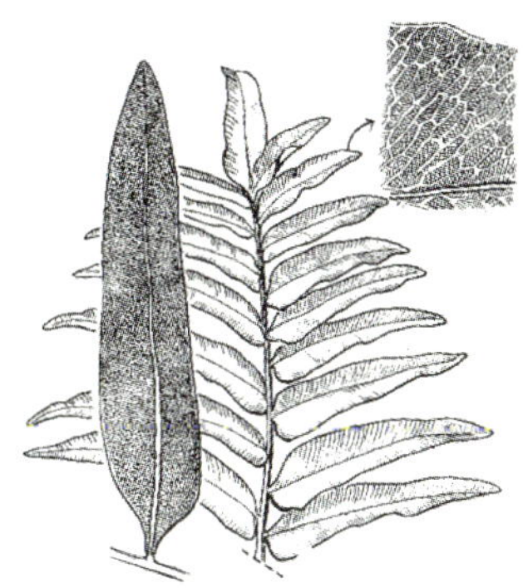

The fern *Acrostichum danaeifolium.*

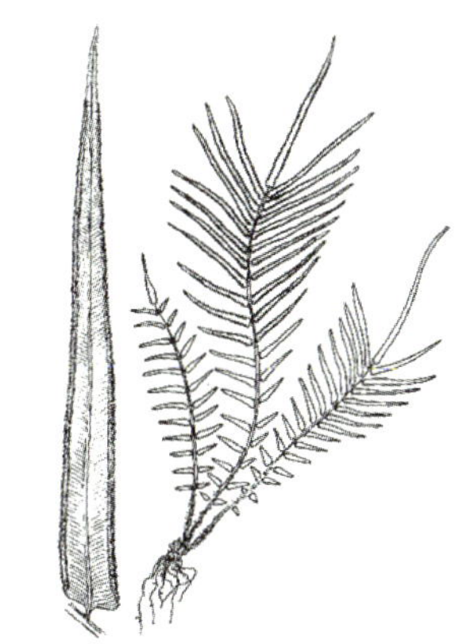

The fern *Pityrogramma vittata.*

Courtesy G. Straley, University of British Columbia

Picea sitchensis, a member of the pine family, Pinaceae.

keys to genera and discussions about species and their distributions in the Southeast, is about 80% finished. The latest complete flora is Small's (1933) manual. The *Manual of the Vascular Flora of the Carolinas* (Radford et al. 1968) is a standard and reliable reference. A flora of Florida and atlas of the vascular plants of Florida are under way (R.P. Wunderlin, University of South Florida, personal communication). In addition, extensive computerized data bases on distribution, literature, and nomenclature of Florida plants exist at the University of South Florida.

In Florida, the specimen coverage is incomplete in sparsely populated areas (e.g., several eastern Panhandle counties and northeastern counties). At least one new species per year is described from Florida and these mostly have limited distributions and are in imperiled habitats.

The fern *Phanerophlebia auriculata.*

Herbaria in the southeastern United States have formed a consortium (Southeastern Regional Floral Information System) to computerize specimen records in all southeastern herbaria. The information from this project is available online at the University of Alabama.

Invasion of weedy species is one of the most serious threats to native vegetation in the southeastern United States. Much better documentation of the occurrence and spread of these species is needed to control these invaders.

Collecting and Monitoring

Active collecting programs document and monitor changes in distribution of native and introduced species. Introduced plants and plant migrations often affect the distribution and health of native plants. At present, it can take as long as 20 years after an introduction to collect and record the species in the literature.

Long-term care of these national collections is vital; many regional herbaria no longer have curatorial support, and some have been or are in danger of being abandoned by their institutions, which will limit resources and information for studies.

The fern *Ctenitis sloanei.*

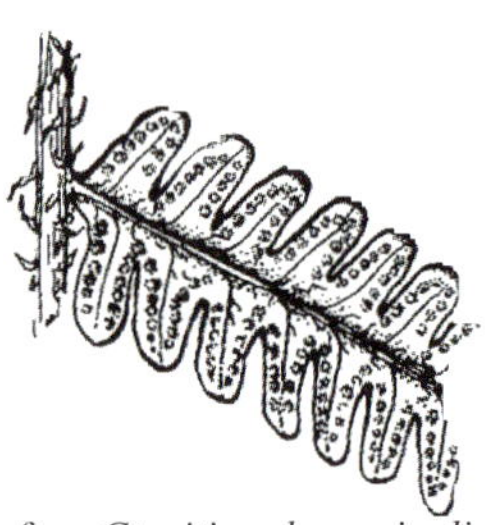
The fern *Ctenitis submarginalis.*

References

Albee, B., L.M. Shultz, and S. Goodrich. 1988. Atlas of the vascular plants of Utah. Utah Museum of Natural History, Salt Lake City. 670 pp.

Correll, D.S., and M.C. Johnston. 1970. Manual of the vascular plants of Texas. Texas Research Foundation, Renner. 188 pp.

Cronquist, A., A.H. Holmgren, N.H. Holmgren, J.L. Reveal, P.K. Holmgren, and R.C. Barneby. 1972-94+. Intermountain flora: vascular plants of the Intermountain West, U.S.A. 5+ vols. Hafner Publishing Co., New York. (Vol. 5-Composite. Columbia University Press.)

Davis, R. 1952. Flora of Idaho. W.C. Brown, Dubuque, IA. 828 pp.

Fernald, M.L. 1950. Gray's manual of botany. 8th ed. American Book Co., New York.

Flora of North America Editorial Committee. 1993. Flora of North America: north of Mexico. Vols. 1 and 2. Oxford University Press, New York.

Gleason, H.A., and A. Cronquist. 1991. Manual of vascular plants of northeastern United States and adjacent Canada. 2nd ed. New York Botanical Garden, New York. 910 pp.

Great Plains Flora Association. 1977. Atlas of the flora of the Great Plains. University Press of Kansas, Lawrence. 600 pp.

Great Plains Flora Association. 1986. Flora of the Great Plains. University Press of Kansas, Lawrence. 1,392 pp.

Hartman, R.L. 1990. New taxa described from the conterminous United States, 1975-1989. Unpublished report.

Hickman, J., ed. 1993. The Jepson manual: higher plants of California. University of California, Berkeley. 1,400 pp.

Hitchcock, A.S., and A. Chase. 1950. Manual of the grasses of the United States. 2 vols. Dover Publications, New York.

Hitchcock, C.L., A. Cronquist, M. Ownbey, and J.W. Thompson. 1955-69. Vascular plants of the Pacific Northwest. 5 vols. University of Washington Press, Seattle.

Hulten, E. 1968. Flora of Alaska and neighboring territories: a manual of the vascular plants. Stanford University Press, Stanford, CA. 1,008 pp.

Kartesz, J.T. 1987. A flora of Nevada. Ph.D. dissertation, University of Nevada, Reno.

Kartesz, J.T. 1994. A synonymized checklist of the vascular flora of the United States, Canada, and Greenland. 2 vols. Timber Press, Portland, OR.

Little, E.L., Jr. 1971. Atlas of United States trees. Vol. 1. Conifers and important hardwoods. U.S.D.A. Miscellaneous Publ. 1146. Washington, DC.

Martin, W.C., and C.R. Hutchins. 1980-81. A flora of New Mexico. 2 vols. J. Cramer, Vaduz, NM.

Mason, C.T., Jr. 1975. *Apacheria chiricahuensis*: a new genus and species from Arizona. Madrono 23:105-108.

Mitchell, R.S. 1986. A checklist of New York State plants. New York State Museum Bull. 458. 272 pp.

New York Flora Association. 1990. Preliminary vouchered atlas of New York State flora. New York State Museum Institute, Albany. 496 pp.

Peck, M.E. 1961. A manual of the higher plants of Oregon. Binfords & Mort, Portland, OR. 936 pp.

Piper, C.V. 1906. Flora of the state of Washington. Contributions from the United States National Herbarium 11:1-637.

Radford, A.E., H.E. Ahles, and C.R. Bell. 1968. Manual of the vascular flora of the Carolinas. University of North Carolina Press, Chapel Hill. 1,183 pp.

Radford, A.E., J.W. Hardin, J.R. Massey, E.L. Core, and L.S. Radford, eds. 1980. Vascular flora of the southeastern United States. 3 vols. University of North Carolina Press, Chapel Hill.

Rhoads, A.F., and W.M. Klein, Jr. 1993. The vascular flora of Pennsylvania: annotated checklist and atlas. American Philosophical Society, Philadelphia. 636 pp.

Seymour, F.C. 1982. The flora of New England: a manual for the identification of all vascular plants including ferns and fern allies growing without cultivation in New England. Phytologia Memoirs 5. Moldenke, Plainfield, NJ. 611 pp.

Shevock, J.R., B. Ertter, and D.W. Taylor. 1992. *Neviusia cliftonii* (Rosaceae: Kerrieae), an intriguing new relict species from California. Novon 2:285-289.

Small, J.K. 1933. Manual of the southeastern flora. University of North Carolina Press, Chapel Hill. 1,554 pp.

Soil Conservation Service. 1982. National list of scientific plant names. 2 vols. U.S. Department of Agriculture, Washington, DC.

The Nature Conservancy. 1994. The conservation of biological diversity in the Great Lakes ecosystem: issues and opportunities. (Unpublished report.)

Voss, E.G. 1972. Michigan flora: a guide to the identification and occurrence of the native and naturalized seed plants of the state. 2 vols. Cranbrook Institute of Science, Bloomfield Hills, MI.

Wagner, W.L., D.R. Herbst, and S.H. Sohmer. 1990. Manual of the flowering plants of Hawaii. 2 vols. University of Hawaii Press and Bishop Museum Press, Honolulu.

Wagner, W.H., Jr., and A.R. Smith. 1993. Pteridophytes. Flora of North America. Vol. 1:247-266. Oxford University Press, New York.

Welsh, S. L. 1993. A Utah flora. Brigham Young University, Provo, UT.

Wood, C.E., Jr., and N. G. Miller, eds. 1958-90. A generic flora of the southeastern United States. Journal of the Arnold Arboretum Supplementary Series, 199+.

For further information:

Nancy Morin
Missouri Botanical Garden
PO Box 299
St. Louis, MO 63166

Environmental Change and the Florida Torreya

by
Mark W. Schwartz
University of California-Davis
Sharon M. Hermann
Tall Timbers Research Station

Many of the best-known cases of catastrophic decline in trees are linked to introduced pathogens that circumvent the natural defenses of their adopted host, leaving it vulnerable to attack. Notable examples of such declines include Dutch elm disease and the chestnut blight. Similarly, numerous studies have linked environmental degradation (e.g., acid rain, ozone depletion, and global warming) to altered interactions among species. In the case of plants and their pathogens, environmental degradation may result in increased disease susceptibility and mortality as is true for the general forest declines in Europe and the widespread decline of red spruce (*Picea rubens*) in the northeastern United States. Identifying the specific mechanisms for increased mortality in nonspecific tree declines is often very difficult, and debate ensues as to which sources of mortality are primary disease agents and which are merely opportunistic.

Both introduced pathogens and altered environmental conditions have been hypothesized as contributing to the decline of *Torreya taxifolia*, a narrowly restricted endemic conifer. The range of the Florida torreya spans an area of less than 400 km^2 (154 mi^2) along the Apalachicola River in northern Florida and adjacent Georgia. In the 1950's this mid-sized tree species was struck by a catastrophic decline that has left it on the verge of extinction in the wild. High mortality is reducing the population by an estimated 5% per year. Formerly a common tree within its range, there are fewer than 1,500 trees left in the wild.

The average height of a Florida torreya is currently less than 1 m (3.3 ft). The average age of the oldest stem on trees is less than 15 years. While a handful of trees produces pollen, there have been no sexually mature females observed in the wild for at least 15 years. Symptoms of disease include needle spots, needle necrosis, and stem cankers. Primary stem mortality has reduced the average height of trees by 10 cm (4 in) during the past 3 years. Thus, the Florida torreya has shown no sign of recovery or stabilization during the 35 years subsequent to the onset of the species' decline. If current patterns persist, the Florida torreya is destined for extinction in the wild.

The search for a cause for the decline of the Florida torreya began in the 1960's when a team of pathologists studying the case could find no introduced fungal pathogens. Pathologists studying the problem during the 1990's have shown that (1) there does not appear to be any viral or bacterial pathogens associated with *T. taxifolia*; (2) a very common native fungal endophyte (*Pestalotia natans*), often pathogenic in other plants, does not appear virulent on *T. taxifolia*; and (3) the less common *Scytalidium* sp., not typically noted for its pathogenicity, produces pathogenic symptoms on *T. taxifolia* and was likely introduced to the region during the late 1950's, when slash pine plantations were planted from nursery stock. Finally, growth experiments have suggested that environmental stress triggers episodes of mortality in the trees. Greenhouse experiments on Florida torreya trees derived from cuttings also suggest the likelihood that structural changes in the slope forests along the Apalachicola that have resulted in lower light levels have also stressed wild populations of Florida torreya.

The current hypothesis is that the decline of Florida torreya is a result of facultative (*see* glossary) pathogens attacking trees under conditions of increased environmental stress. Several potential stress factors, including fire suppression, climatic changes such as temperature extremes and drought, and altered hydrologic regimes in ravine forests and resultant changes in nutrient flow have also been hypothesized as contributing to the species' decline.

Despite extensive research to find a link between disease agents and environmental stress, the mechanisms for forest decline remain rather speculative. *Torreya taxifolia* has such a narrow distribution that a decline in the populations in the Apalachicola basin has brought the species to near extinction. With the increasing magnitude of abiotic environmental changes, we may expect more cases that are similar to the decline of *T. taxifolia*. Unfortunately, the lack of identification of specific disease agents and specific mechanisms has hindered action to correct potential problems that cause forest declines. Given the difficulty in delineating mechanisms for declines, we typically cannot ascertain exact mechanisms until it may be too late. Waiting to be absolutely certain of the triggers for particular forest declines before corrective action is taken is likely to be a costly strategy.

For further information:

Mark W. Schwartz
Center for Population Biology
University of California-Davis
Davis, CA 95616

Native Vascular Plants

Most of the familiar flora of the American landscape, such as trees, shrubs, herbs, vines, grasses, and ferns, are known as vascular plants. These plants have systems for transporting water and photosynthetic products and are differentiated into stems, leaves, and roots. Nonvascular plants—the algae, fungi, and mosses and lichens—are considered in other articles in this volume. Except in Arctic and alpine areas, vascular plants dominate nearly all of North America's natural plant communities. About 17,000 species of vascular plants are native to one or more of the 50 U.S. states, along with several thousand additional native subspecies, varieties, and named natural hybrids (Kartesz 1994).

Human activities have expanded the geographical distributions of many plant species, particularly farm crops, timber trees, garden plants, and weeds. When a non-native plant

by
Larry E. Morse
The Nature Conservancy
John T. Kartesz
North Carolina Botanical Garden
Lynn S. Kutner
The Nature Conservancy

species is found growing outside cultivation, it is considered an exotic species in that area. About 5,000 exotic species are known outside cultivation in the United States. While many exotic plant species are desirable in some contexts (such as horticulture), hundreds of invasive non-natives have become major management problems when established in places valued as natural areas (McKnight 1991; U.S. Congress 1993). A few particularly troublesome non-natives are regulated under specific federal or state laws as noxious weeds.

Geographic Distribution

Western and southern states have the largest numbers of native vascular plant species in the country. (Fig. 1, revised from Kartesz 1992). California, with more than 5,000 native vascular plant species, has almost one-third of the total number for the entire United States. Texas, with about 4,500 native species, ranks second. Arizona, Florida, Georgia, New Mexico, and Oregon all have over 3,000 native species.

Hawaii, as a remote oceanic island archipelago, has relatively few native species (Carlquist 1970), but nearly all (89%) of the native Hawaiian flowering (angiosperm) species are endemic to that region (Wagner et al. 1990). A small number of vascular plants, including a species of lycopod (*Huperzia haleakalae*), are native to both Hawaii and the North American mainland.

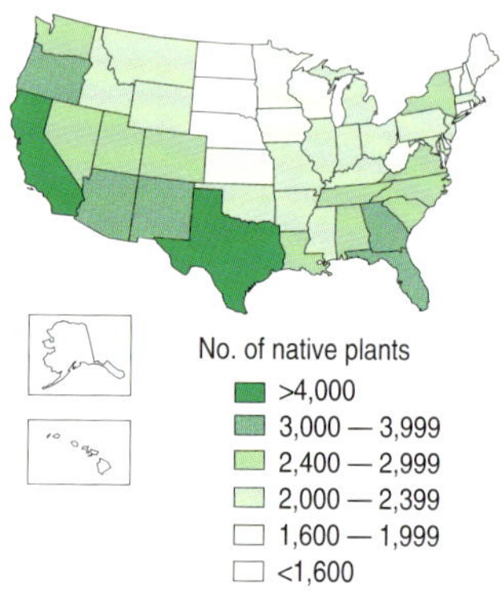

Fig. 1. The number of native vascular plant species in each state.

In every state, hundreds of plant species are established as exotics. States with coastal areas, major agricultural regions, and large cities generally have the highest numbers of non-native plants. A modest number of native U.S. species, such as the northern catalpa (*Catalpa speciosa*), have also spread from cultivation beyond their native ranges. Some familiar mainland species, like a wild blackberry (*Rubus argutus*) and a grass known as broomsedge (*Andropogon virginicus*), have become problem weeds in Hawaii (Smith 1989).

Rare Species

As of February 1994, 403 native U.S. species, subspecies, or varieties of vascular plants and one nonvascular plant have been formally protected under the provisions of the U.S. Endangered Species Act of 1973 (USFWS 1994). Nearly half the 822 native U.S. federally listed species are plants. The U.S. Fish and Wildlife Service considers an additional 1,953 kinds of plants as candidates for such listing (Federal Register 1993).

The first U.S. national lists of rare plants depended largely on nominations from specialists already familiar with various rare species and omitted many potential candidates. Many state-level rare plant lists were also developed in the 1970's; these generally addressed species considered rare in a particular area regardless of abundance elsewhere.

The Nature Conservancy and the network of Natural Heritage Programs use a consistent methodology to inventory natural diversity and to assess rarity and endangerment for all currently recognized species of vascular plants in North America, Hawaii, and portions of Latin America (Jenkins 1985). By using a five-level scale from 1 (rarest and most vulnerable—typically five or fewer existing occurrences) to 5 (demonstrably widespread, abundant, and secure), a global or rangewide rank (G1 to G5) is determined for each species. With the use of the same five-level scale, conservation priority ranks are assigned for national (N1 to N5) and subnational or state (S1 to S5) status. Ranks are used conservatively throughout the Natural Heritage Network and are assigned after careful review of a species' status. Additional ranks are used to indicate species that occurred historically within a jurisdiction, but which are not presently known. A species is presumed extinct if efforts to relocate it are unsuccessful, if no suitable habitat remains, or if the loss has been well documented. Species are considered "historic" (possibly extinct) if there is reliable evidence from biological surveys that the species occurred within the past few centuries in a given area (Snyder 1993).

The Natural Heritage Network has documented the status of thousands of rare species. At the same time, plant surveys have shown that a comparable number of plants are substantially more common than previously believed. Species status information from all 50 U.S. State Natural Heritage Programs is combined with national and rangewide data in the Natural Heritage Network's Central Scientific Databases maintained by The Nature Conservancy. The inventories and data bases of the Natural Heritage Network continuously gather, organize, and revise information on species rarity and distribution as it becomes available.

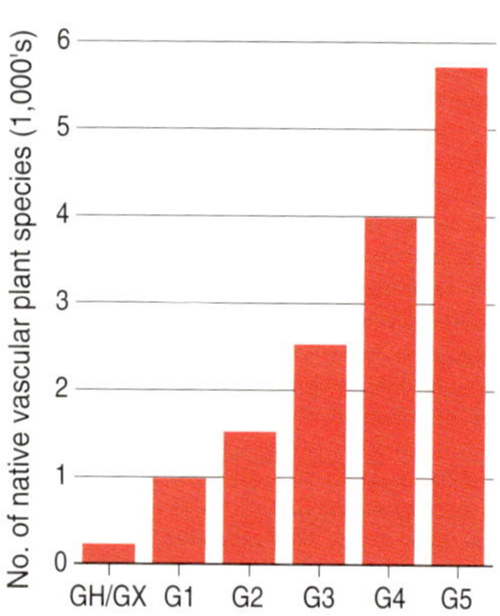

Fig. 2. The number of native vascular plant species in the United States in each global rank. GH/GX means species is potentially extinct; G1 to G5 rank the species from rarest (G1) to most common (G5).

The number of species in the United States in each global rank is presented in Fig. 2. For example, more than 4,850 species (about 28%) of the native U.S. vascular plants are considered globally rare (ranked G1, G2, or G3) by The Nature Conservancy and the Natural Heritage Network. Of these, about 960 species are ranked G1 and occur at fewer than five sites globally or are comparably imperiled.

Globally rare native species of vascular plants are concentrated in the western and southern states (Fig. 3), with greatest proportions in Arizona, California, Florida, Georgia, Hawaii, Nevada, New Mexico, Texas, and Utah.

In addition to these globally rare species, about 4,500 other species of widespread or more common vascular plants (ranked G4 or G5) are being actively inventoried in at least one state where they are rare.

Loss of Species

The patterns and causes of plant species' losses are often important components of state-level conservation studies. The loss, or suspected loss, of a species from a portion of the landscape is referred to as "extirpation."

A recent study (Kutner and Morse, unpublished report) of the losses of U.S. native vascular plants revealed that about 1,772 (9.8%) of these species have been lost from at least one state. Of these species, 438 (25%) may be lost from the floras of two or more states. The proportion of species potentially extirpated from each state varies dramatically across the nation (Fig. 4), with the largest losses reported from northeastern states and from Hawaii. Delaware has experienced the proportionately highest loss from its flora, with more than 15% of its species potentially extirpated. Many of the northeastern and mid-Atlantic states have lost more than 5% of their native vascular plants. This region of the United States has experienced hundreds of years of human development and includes many of the most densely populated and intensely developed states. Many plants that have been lost from these states may now be similarly threatened in portions of their remaining ranges.

About 28% of the native flora is considered globally rare (ranked G1, G2, or G3) by the Natural Heritage Network, but only 12% of the potentially extirpated species are globally rare. Most potentially extirpated species have been lost from one or two states and are currently globally common (ranked G4 or G5). In the United States, 110 of these globally common species have been lost from three or more states, and more than 35 species have been lost from four or more states. Of the most common species (global rank G5), about 285 have been lost from two or more states. Common species that have been lost from many states may not be as secure from imperilment as previously believed. Additionally, the effect of species' losses on other plants and animals in a community is often unknown. Rangewide analyses could indicate species that would benefit from further research and a better understanding of potential threats, thus helping prevent subsequent losses.

Many species that are endangered, threatened, or formal candidates for federal protection have also lost parts of their ranges. Nearly 6% of listed and proposed endangered species and 20% of listed and proposed threatened species are reported extirpated from at least one state. About 16% of the category 1 candidate species (top candidates for listing as endangered or threatened) and almost 11% of the category 2 candidate species (possibly qualifying for threatened or endangered status, but more information is needed) have been similarly affected.

Some currently rare species had widespread historical distributions. For example, American chaffseed (*Schwalbea americana*) is a federally listed endangered species with a Natural Heritage rank of G2. The historical range of this species extended from Mississippi to Massachusetts; the plant is currently known from about 20 populations in five states, mostly in South Carolina. The most significant threat to this species is fire suppression, which allows plant succession to proceed to the point where there is not enough light for the plant to compete successfully. Habitat loss has also caused the extirpation of several *Schwalbea* populations. For rare species such as *S. americana*, further state-level extirpations could seriously affect the species' survival.

Wetland Species

Although there are fewer than 7,000 native wetland vascular plant species in the United States, plants that occur mostly in wetlands are more likely to be extirpated from at least one state. Based on the USFWS National Wetlands Inventory (Reed 1988), about half of the potentially extirpated species are either obligate (*see* glossary) or facultative (*see* glossary) wetland species.

Wetlands and aquatic ecosystems have been severely affected in the United States; approximately 53% of these ecosystems have been destroyed in the 48 contiguous states (Dahl 1990). Aquatic species frequently have specific habitat requirements and can be threatened by both habitat loss and changes in local hydrology. In the mid-Atlantic region, several intertidal vascular plants have been extirpated from the Delaware River system because of habitat alteration (Ferren and Schuyler 1980).

Possibly Extinct Species

About 90 mainland U.S. and 110 Hawaiian vascular plant species may be extinct, according to records of the USFWS and The Nature Conservancy (Russell and Morse 1992). For

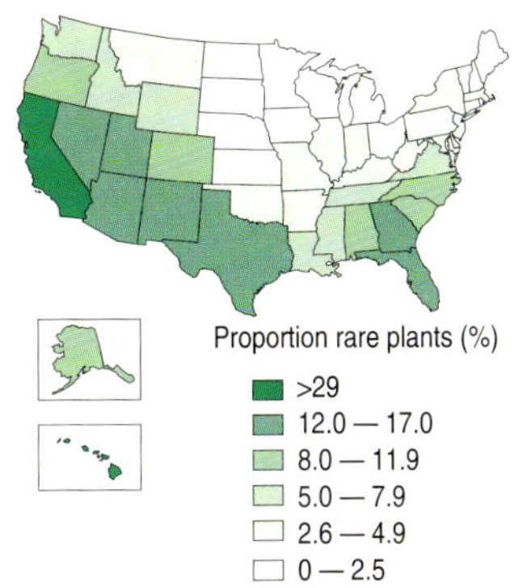

Fig. 3. The proportion of globally rare vascular plant species (ranked G1, rarest, to G3, more common) in the native flora of each U.S. state.

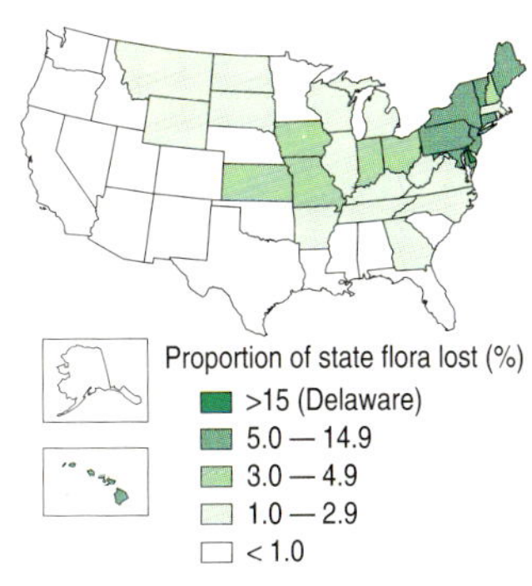

Fig. 4. The proportion of species reported potentially extirpated from the native flora of each U.S. state.

example, Nuttall's mudwort (*Micranthemum micranthemoides*) has been recorded from Delaware, the District of Columbia, Maryland, New Jersey, New York, Pennsylvania, and Virginia, but despite searches, it has not definitely been seen since September 1941.

Several species of U.S. plants are extirpated from the wild, but still exist in cultivation. Most familiar of these is the Franklinia (*Franklinia alatamaha*), a small tree known historically only from the Altamaha River in southeastern Georgia, but which is now widely cultivated as an ornamental in eastern states.

Ongoing fieldwork has resulted in the rediscovery of many species. The running buffalo clover (*Trifolium stoloniferum*) was rediscovered in West Virginia in 1983 (Bartgis 1985) and has been found subsequently in Indiana, Kentucky, Missouri, and Ohio. In Oregon, a population of *Lomatium peckianum* was located in 1983 for the first time in more than 50 years. The discovery of additional populations has changed the species' federal status from a category 1 candidate to a former candidate (Kagan and Vrilakas 1993). In Montana, several recent rediscoveries have occurred, including a 1985 rediscovery of *Trifolium microcephalum*, a species of clover not seen since it was first collected by Meriwether Lewis in 1805 or 1806 (Hoy 1993). Likewise, during the 1991 field season the yellow passionflower (*Passiflora lutea*) was located at two sites in Delaware for the first time since the early 1800's (Clancy 1993). These examples illustrate the importance of ongoing inventories as well as the dynamic nature of local and regional floras.

Threats to Diversity

Habitat alteration and incompatible land use are the major threats to most rare U.S. plant species. Apart from certain species of cacti, ginseng, and various showy wildflowers, relatively few rare U.S. plants are primarily threatened by overcollecting. Global climate change (Peters and Lovejoy 1992; Morse et al. 1993) and sea-level rise (Reid and Trexler 1991) may pose additional threats to some native U.S. plant species.

Species at higher risk of extinction usually include those having small geographic ranges, narrow habitat requirements, unusual life histories, or vulnerability to exotic pests or diseases. In addition, reduced biodiversity of local floras is of high concern, even if plants lost from a particular geographical region are common and secure elsewhere. Finally, depletion of even widespread species can occur if exploitation or habitat destruction occurs beyond a sustained-yield rate.

Assessment of the causes and patterns of species losses in the United States, combined with ongoing documentation of natural diversity and studies of rarity, endangerment, and threats, will refine conservation priorities by identifying species or areas that will most benefit from further protection and research. Analyses of ongoing inventory and monitoring work could provide early warnings of widespread threats to biological diversity, thereby perhaps improving the protection of both rare and more common plants and allowing the development and implementation of conservation strategies before crises occur.

References

Bartgis, R.L. 1985. Rediscovery of *Trifolium stoloniferum* Muhl. ex A. Eaton. Rhodora 87:425-429.

Carlquist, S. 1970. Hawaii: a natural history. Natural History Press, Garden City, NY. 463 pp.

Clancy, K. 1993. The yellow passionflower, *Passiflora lutea* L., rediscovered in Delaware. Castanea 58:153-155.

Dahl, T.E. 1990. Wetlands losses in the United States 1780s to 1980s. U.S. Fish and Wildlife Service, Washington, DC. 13 pp.

Federal Register. 1993. Plant taxa for listing as endangered or threatened species: notice of review. Federal Register 58:51144-51190.

Ferren, W.R., and A.E. Schuyler. 1980. Intertidal vascular plants of river systems near Philadelphia. Proceedings of the Academy of Natural Sciences of Philadelphia 132:86-120.

Hoy, J. 1993. Rediscovering lost species. Kelseya [Montana Native Plant Society] 6:5.

Jenkins, R.E. 1985. Information methods: why the Heritage programs work. The Nature Conservancy News 35:21-23.

Kagan, J., and S. Vrilakas. 1993. Extinct and extirpated plants from Oregon. Kalmiopsis [Native Plant Society of Oregon] 3:12-16.

Kartesz, J.T. 1992. Preliminary counts for native vascular plant species of U.S. states and Canadian provinces. Biodiversity Network News [The Nature Conservancy] 5:6.

Kartesz, J.T. 1994. A synonymized checklist of the vascular flora of the United States, Canada, and Greenland. 2nd ed. Timber Press, Portland, OR. 622 pp.

McKnight, B.N., ed. 1991. Biological pollution: the control and impact of invasive exotic species. Indiana Academy of Science, Indianapolis. 261 pp.

Morse, L.E., L.S. Kutner, G.D. Maddox, J.T. Kartesz, L.L. Honey, C.M. Thurman, and S.J. Chaplin. 1993. The potential effects of climate change on the native vascular flora of North America: a preliminary climate-envelopes analysis. Report TR-103330, Electric Power Research Institute, Palo Alto, CA. 120 pp.

Peters, R.L., and T.L. Lovejoy, eds. 1992. Global warming and biological diversity. Yale University Press, New Haven, CT. 386 pp.

Reed, P.B., Jr. 1988. National list of plant species that occur in wetlands: 1988 national summary. U.S. Fish and Wildlife Service, Washington, DC. 244 pp.

Reid, W.V., and M.C. Trexler. 1991. Drowning the national heritage: climate change and U.S. coastal biodiversity. World Resources Institute, Washington, DC. 48 pp.

Russell, C.A., and L.E. Morse. 1992. Plants. Biodiversity Network News [The Nature Conservancy] 5:4.

Smith, Clifford W. 1989. Non-native plants. Pages 60-69 *in* C.P. Stone and D.B. Stone, eds. Conservation biology in Hawai'i. University of Hawaii Cooperative National Park Resources Study Unit, Honolulu.

Snyder, D.B. 1993. Extinct, extant, extirpated, or historical? Or in defense of historical species. Bartonia 57, Supplement:50-57.
U.S. Congress, Office of Technology Assessment. 1993. Harmful non-indigenous species in the United States. U.S. Government Printing Office OTA-F-565 (September). 391 pp.
USFWS 1994. Box score: listings and recovery plans. U.S. Fish and Wildlife Service Endangered Species Tech. Bull. 19:24.
Wagner, W.L., D.R. Herbst, and S.H. Sohmer. 1990. Manual of the flowering plants of Hawai'i. University of Hawaii Press, Honolulu. 1,853 pp.

For further information:

Larry E. Morse
The Nature Conservancy
1815 N. Lynn St.
Arlington, VA 22209

Tracking the Mosses and Vascular Plants of New York (1836-1994)

by
Norton G. Miller
Richard S. Mitchell
New York State Museum

New York, the third most populous state, has highly varied topography, geology, soils, and climate, and a complex history of land use, all of which are reflected in a rich flora of native, introduced, and opportunistic species. Large parts of the state support beech-maple, oak-chestnut (now modified as a result of the elimination of chestnut), or hemlock-northern hardwood forest, and there are extensive tracts of red spruce-balsam fir forest in the Adirondack and Catskill mountains. Alpine tundra is present on the highest Adirondack peaks at elevations above about 1,372 m (4,500 ft), while salt marshes, freshwater ponds, and dwarf pine barrens occur at or near sea level on Long Island. Almost all land in the state has been glaciated and therefore available for plant occupation no longer than 18,000 years. In 1880 nearly 78% of the state's land was in farms or farm woodlots, but by 1980, 61% of New York was classified as forested.

The flora of New York is an economically important resource and the foundation of healthy sustainable environmental systems. The state's flora and its composition have been studied since the early 1800's, allowing researchers to present trends in the numbers of vascular plant and moss species. In our work, we have emphasized the study of voucher (*see* glossary) specimens, which allow us and our successors to verify identifications and evaluate the application of species concepts of other researchers.

Status

Organized study of the New York flora began in 1836 with a botanical survey that was a part of the New York State Geological and Natural History Survey. This survey led to the publication of John Torrey's *A Flora of the State of New-York* (Torrey 1843). The state's plant resources continued to be investigated at the New York State Museum under government sponsorship that began in 1867 and continues to the present. The regionally significant herbarium and extensive data collections that have resulted from this research and exploration provide the documentation for this article as well as our ongoing work and information from other important botanical collections.

Totals for the major groups of mosses and vascular plants (as of February 1994) are given in Table 1, and increases in the numbers of known species are listed in Table 2. Torrey's 1843 flora enumerated 1,452 species, while a 1994 compendium (R.S. Mitchell, unpublished data) lists 3,451, an increase of 58%. The differences, in part, are due to a dramatic increase in the number of reported non-native species, of which 77% (1,122 of 1,449) are naturalized (naturally reproducing and spreading). The differences are also due to a significant rise in the number of species recognized as indigenous to the state (an increase of 711). Native species from other parts of the United States are listed in the tables as such, even if they are also known to have been introduced into New York. Although the number of known native plant species has steadily increased, the apparent decrease in the number of native species from House's to Mitchell's list (Table 2) was the result of taxonomic reinterpretation that reduced many taxa (especially species of *Rubus* and *Crataegus*) into synonymy over the latter half of the 20th century.

Table 1. Current tally of New York flora.

Plants	Native	Non-native	Total
Vascular plants			
Pteridophytes	113	7	120
Gymnosperms	18	13	31
Angiosperms	1,871	1,429	3,300
Total	2,002	1,449	3,451
Persisting	1,933	1,122	3,055
Mosses			
Sphagnidae	52	-	52
Andreaeidae	3	-	3
Bryidae	417	4	421
Total	472	4	476
Persisting	469	4	473

The data reflect both an intensification of botanical exploration during the 19th and 20th centuries and the arrival of numerous plant waifs (nonpersistent alien species), mainly from Eurasia, many of which became naturalized as population centers, commerce, and transportation networks enlarged. In addition, a few native species apparently continue to expand their ranges northward, as exemplified by discoveries in 1993 of the large floating bladderwort (*Utricularia inflata* Walter) and beakgrain (*Diarrhena americana obovata* [Gleason] Brandenb.) in southeastern New York state. The list of mosses (Table 2) grew most dramatically between 1866 and 1957 as a result of field and herbarium study. Miller's 1994 synopsis of the state's bryophyte flora (unpublished data) shows that many new discoveries continue to be made. Several non-native moss species have been recognized near nurseries and botanical gardens, the

Courtesy R.S. Mitchell

Spreading globeflower (*Trollius laxus* Salisbury), a threatened species in New York.

Table 2. Historical documentation of New York flora.

Plants	No. of species by reference			
Vascular plants	Torrey (1843)	House (1924)	Mitchell (1986)	Mitchell (1994 tally)*
Pteridophytes	59	94	110	120
Gymnosperms	14	25	29	31
Angiosperms	1,379	2,825	3,252	3,300
Native	1,291	2,133**	1,995***	2,002
Non-native	161	811	1,296	1,449
Total	1,452	2,944	3,291	3,451
Mosses	Peck (1866-1912)	Ketchledge (1957)	Ketchledge (1980)	Miller (1994 tally)*
Sphagnidae	15	29	46	52
Andreaeidae	3	2	3	3
Bryidae	295	413	414	421
Total	313	444	463	476

* R.S. Mitchell, unpublished data; N.G. Miller, unpublished data.
** Of this number, some 300 species are now placed in synonymy in the light of modern taxonomic research.
*** This number has been reduced by four, to reflect species eliminated from the list as a result of changes in taxonomic status, discovery of incorrectly identified plants, and faulty literature reports.

likely points of introduction, and more adventive (*see* glossary) mosses will almost certainly be discovered as field exploration continues.

The numbers of vascular plants and mosses considered rare in New York are substantial. In conformity with New York State Heritage Program designations, we tallied the number of species in the following categories: S1 (5 or fewer sites), S2 (6-20 sites), and SH (no site verified within the past 15 years). By these criteria, roughly a fourth of New York's native vascular plants (435 species: 22%) and mosses (119 species: 26%) are rare. Of the native species, we consider 69 of the vascular plants and 3 of the mosses extirpated because most have not been observed within New York this century.

Trends

For the past 70 years, an average of 11 species of vascular plants per year were newly documented for New York. Since 1980 the number of native vascular plants added to the flora has been 1 per year, while the number of exotic species has been over 200. For mosses, a less well-known group of plants, one additional native species per year on average was discovered between 1957 and 1994. Although the steepest increase in knowledge of both groups occurred in the 1800's and early 1900's, significant information on plant diversity continues to accumulate at a steady rate, as the ranges of species in the state become better known.

Although there is a long history of botanical exploration in New York state, many areas still have not been surveyed adequately. Poorly known regions include parts of the Allegheny Upland of central and western New York, the Champlain Valley, and portions of the Adirondack Mountain region and adjacent districts. The Hudson Highlands area, previously poorly explored, is being intensively studied by botanists from the New York State Biological Survey.

In the last decade, 11 New York plant species considered extirpated have been discovered at new sites, including *Sphagnum angermanicum* Melin, a rare peat moss, and prairie smoke (*Geum triflorum* Pursh), an herb thought extirpated by Torrey in 1843 and rediscovered in the 1980's. Nine additional species thought extirpated and over 50 species designated "critically imperiled" by New York State Heritage Program criteria were reclassified into less sensitive categories as new information became available, thereby lessening the urgency of conservation measures. Of about 70 extirpated species, some have been lost because of expanding population centers, but many have been lost because of wetland drainage and increased forest cover that altered their specialized, often calcareous habitats.

We predict that one or two native species per year will continue to be added to the vascular plant and moss floras of New York state through new discoveries. By contrast, previously undocumented non-native vascular plants will be added at an annual rate some 20-fold greater than that of the native flora. Inventories and assessments of liverworts, fungi, lichens, and terrestrial, aquatic, and marine algae, are much less advanced in New York than those for the vascular plants and mosses, which deserve attention as well.

References

House, H.D. 1924. Annotated list of the ferns and flowering plants of New York state. Bull. of the New York State Museum 254. 759 pp.

Ketchledge, E.H. 1957. Checklist of the mosses of New York State. New York State Museum Bull. 363. 55 pp.

Ketchledge, E.H. 1980. Revised checklist of the mosses of New York State. New York State Museum Bull. 440. 19 pp.

Mitchell, R.S. 1986. A checklist of New York State plants. New York State Museum Bull. 458. 272 pp.

Peck, C.H. 1866. List of the mosses of the state of New-York. Pages 42-70 *in* 19th Annual Report of the Regents of the University of the State of New York on the Cabinet of Natural History. [Additions published in succeeding reports through 1912.]

Torrey, J. 1843. A flora of the state of New-York. 2 vols. Carroll and Cook Printers, Albany. 484 and 572 pp.

For further information:

Norton G. Miller
New York State Museum
Biological Survey
Albany, NY 12230

Distribution, Abundance, and Health of

Ecosystems

Terrestrial *213*

Aquatic *233*

Coastal and Marine *259*

Riparian *285*

Courtesy S.L. Grace, NBS

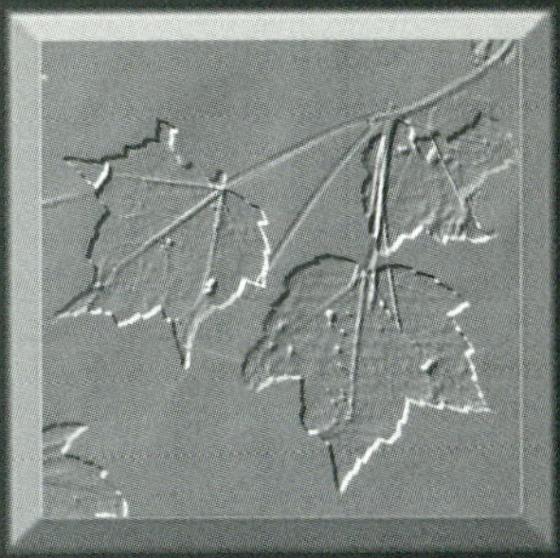

Terrestrial Ecosystems

Science Editor
Raymond J. Boyd
Bureau of Land Management
Service Center
Denver Federal Center, Bldg. 50
Denver, CO 80225

Text by
B.D. Keeland
National Biological Service
Southern Science Center
700 Cajundome Blvd.
Lafayette, LA 70506

Overview

Whereas several other sections in this volume cover individual species or locations, the articles in this section address the integration of individual species into communities and ecosystems (*see* glossary). Terrestrial ecosystems include a rich variety of community types and cover a range extending from nearly aquatic wetlands along our coasts and myriad rivers, lakes, and streams, to mountain tops and arid, desert locations. The diversity of these ecosystems offers both challenge and opportunity. The challenge stems from the sheer number of potential ecosystems to be analyzed. Grossman and Goodin (this section) discuss 371 imperiled and critically imperiled communities, and state that this number represents only 10%-15% of all terrestrial communities. This implies a minimum of 2,500-3,500 individual terrestrial community types. Obviously, a single report cannot hope to address more than a few of these many terrestrial communities and ecosystems.

Discussions of biological diversity have traditionally revolved around the protection of individual species. More recently, we have begun to realize that protection of community or ecosystem diversity is equally important. Patchwork conversions of natural landscapes for agriculture, silviculture, and development result in a fragmentation that leaves small remnant areas of natural ecosystems (Burgess and Sharpe 1981). As these natural patches become smaller and more isolated, their ability to maintain healthy populations of many plant and animal species is reduced (Harris 1984). As individual species are lost from each fragment, the community changes and both species and ecosystem diversity are reduced. Thus, large numbers of natural ecosystems are now in danger.

Kendall (this section) discusses one such imperiled ecosystem. The whitebark pine (*Pinus albicaulis*) ecosystem of the western mountains is endangered because of the combined effects of an introduced disease and fire suppression. The effects of introduced diseases on natural species and ecosystems have been well documented. Several species, such as the American chestnut (*Castanea dentata*), have been virtually eliminated and other species have been greatly reduced by introduced diseases. The effects on ecosystems where these species were previously found have been dramatic (Shugart and West 1977).

Alteration of natural fire regimes has played a major role in the reshaping of natural ecosystems. In many systems a reduction in fire frequency can lead to invasion by fire-intolerant species and eventual loss of the original ecosystem.

This is shown by Henderson and Epstein (this section) in their discussion of how fire supression and other factors caused tremendous losses of oak savannas throughout the Midwest. In other systems, an increase in fire frequency can also lead to changes in ecosystem structure and function. Although we now realize that fire is a natural and necessary part of many ecosystems, it was not until after the devastating fires of Yellowstone National Park that the general public was alerted to the benefits of such fires (Elfring 1989). An effective fire-suppression program can allow accumulation of vast amounts of detritus (dead organic material such as leaves, branches, and stems). If this material is not consumed periodically by small fires burning along the forest floor, it will accumulate to the point of providing raw materials for an exceptionally intense fire that can burn tree crowns and destroy the existing forest. Ferry et al. (this section) discuss four fire-adapted ecosystems that have been affected by modified fire regimes and conclude, "Managers must balance the suppression program with a program of prescribed fire applied on a landscape scale if we are to meet our stewardship responsibilities."

Numerous variables in addition to disease and fire affect our natural resources. These variables include pollution (Peterson, this section; Nash et al., this section), conversions to other uses, harvesting activities such as logging, and global climate change. Cole (this section) demonstrates that over the past 5,000 years change has been a natural part of our terrestrial ecosystems. Within a given ecosystem some species decline in importance while others increase over time, resulting in a change in the overall character of the ecosystem. A key feature to stand out in the 5,000-year chronology developed by Cole is that current rates of change are about 10 times higher than presettlement rates. Human intervention in one form or another is now the principal agent of change. Darr (this section) provides a review of U.S. Forest Service data and discusses changes being brought about by forestry-management practices. At a reduced spatial scale, Keeland et al. (this section) discuss changes within the forested wetlands of the southeastern United States. Forested wetlands have been especially reduced and fragmented as a result of land-use conversions, predominantly to agricultural activities.

A common thread here, as in all sections in this report, is that if unchecked, human activities will continue to result in an upset balance of species interactions, alteration of ecosystems, and extensive habitat loss.

References

Burgess, R.L., and D.M. Sharpe, eds. 1981. Forest island dynamics in man-dominated landscapes. Springer-Verlag, New York. 310 pp.

Elfring, C. 1989. Yellowstone: fire storm over fire management. BioScience 39(10):667-672.

Harris, L.D. 1984. The fragmented forest. The University of Chicago Press, Chicago. 211 pp.

Shugart, H.H., and D.C. West. 1977. Development of an Appalachian deciduous forest succession model and its application to assessment of the impact of the chestnut blight. Journal of Environmental Management 5:161-179.

U.S. Forest Resources

by
David R. Darr
U.S. Forest Service

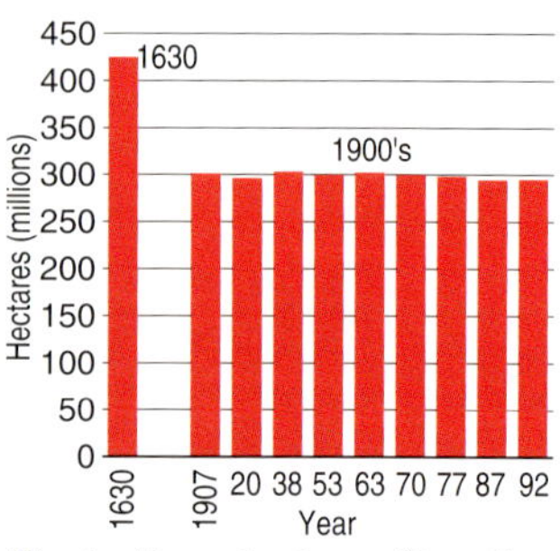

Fig. 1. Forest land area (Powell et al. 1993).

The Secretary of Agriculture is directed by law to make and keep current a comprehensive inventory and analysis of the present and prospective conditions of and requirements for the renewable resources of U.S. forests and rangelands. This inventory includes all forests and rangelands, regardless of ownership. The work is carried out by people in the Forest Inventory and Analysis program of the U.S. Department of Agriculture Forest Service (USFS).

Inventories provide key forest resource information for planners and policy makers. Increasingly, people turn to these inventories for information on biological diversity, forest health, and developmental decisions.

Information is collected from over 130,000 permanent sample plots selected to assure statistical reliability. Vegetation on the plots is measured on average about every 10 years. Characteristics of the vegetation and land are measured, including ownership, productivity for timber production, the kinds and sizes of trees, how fast trees are growing, whether any trees have died from natural causes, and whether any trees have been cut (USFS 1992).

Characteristics of Forest Land

Over the years, the U.S. forest cover has changed because of the way people use and manage forest land. Today, about 33% of the U.S. land area, or 298 million ha (737 million acres), is forest land, about two-thirds of the forested area in 1600 (Fig. 1). Since 1600, some 124 million ha (307 million acres) of forest land have been converted to other uses, mainly agricultural. More than 75% of this conversion occurred in the 19th century, but by 1920, clearing forests for agriculture had largely halted.

Some 34% of all forest land is federally owned and managed by the U.S. Forest Service, the Bureau of Land Management, and other federal agencies. The rest is owned by nonfederal public agencies, forest industry, farmers, and other private individuals. About 19 million ha (47 million acres; 6% of all U.S. forest land) are reserved from commercial timber harvest in wilderness, parks, and other land classifications.

Forest land is widely but unevenly distributed. North Dakota has the smallest percentage of forest cover (1%) and Maine has the greatest (89%). Forest areas vary greatly from sparse scrub forests of the arid interior West to the highly productive forests of the Pacific coast and the South, and from pure hardwood forests to multispecies mixtures and coniferous forests. In total, 52% of the forest land is east of the Great Plains states. In the East, the oak-hickory forest type group is most common, while in the West, the category referred to as "other softwoods" is most common.

U.S. forests provide wildlife habitat and thereby support biodiversity; take carbon out of the air and thus serve as carbon sinks; and provide the outdoor environments desired by many people for recreation.

Timberland forests are logged for lumber, plywood, and paper products. This timberland is generally the most productive and capable of producing at least 1.4 m^3 of industrial wood per hectare a year (20 ft^3/acre) and is not reserved from timber harvest (Powell et al. 1993). Two-thirds of the nation's forested ecosystems (198 million ha or 490 million acres) are classed as timberland. Because of historical interest in timber production, more information is available for the characteristics of timber inventories on timberland than for other forest land.

Timberland ownership patterns vary throughout the United States. For the country as a whole, 73% of all timberland is owned by private individuals and firms. The remaining 27% is in federal, state, and other public ownerships. Much of the privately owned land is in the East and much of the national forest land is in the West (Fig. 2). Most of the publicly owned land is managed according to plans that account for the various uses and values provided by forest cover. Forest industry lands are generally managed with timber production being the main interest. Other private forest lands are managed for a variety of interests, reflecting the divergent views of the some 6 million owners in this category.

The nation's timberland contains an estimated 24.3 billion m^3 (858 billion ft^3) of timber, of which 92% is in growing stock—live, sound trees suited for roundwood (timber) products. Softwoods such as pine are generally used to make lumber and plywood for use in construction. Hardwoods, such as oak, are used in making furniture and pallets. Both softwoods and hardwoods are used in manufacturing paper products. The nation's softwood growing stock volume amounts to 57% of the total, with about 66% of this volume in the West. Total softwood growing stock volume has been slightly declining recently (Fig. 3). By contrast, hardwood growing stock volume increased 7% between 1987 and 1992. More than 90% of all hardwood timber volume is in the eastern United States.

Mortality, Growth, Harvest

Mortality is the result of natural causes such as insects, disease, fire, and windthrow. Between 1962 and 1986, mortality averaged 122 million m^3 (4.3 billion ft^3) per year. Mortality increased to 155 million m^3 (5.5 billion ft^3) in 1991, but was still less than 1% of the U.S. growing stock volume.

Net annual growth, which already has mortality subtracted out, totaled 612 million m^3 (21.6 billion ft^3) in 1991—about 2.7% of the growing stock inventory. Total growing stock growth declined about 2% between 1986 and 1991 (Fig. 4), the first decline in net annual growth since 1952. The decline between 1986 and 1991 occurred with softwoods, which declined 4.4% to 339 million m^3 (12 billion ft^3). Net annual growth for hardwoods increased 0.9%.

Removals from timber inventories are losses by other than natural causes (mortality) and include harvest of roundwood products. Timber removals from growing stock inventory in 1991 totaled 461.5 million m^3 (16.3 billion ft^3) or 2.1% of the inventory. Average timber removals have risen each decade since the 1950's. Almost 55% of all timber removals came from the forests of the South, up from 45% in 1970. Twenty-three percent of all removals came from Pacific coast forests, 17% from the North, and 5% from forests in the Rocky Mountains. Softwoods accounted for two-thirds of all growing stock removals in 1991. Timber removals continued to be concentrated on private land in 1991.

The growth-removals ratio for the United States is greater than one for all species (1.3), for softwoods (1.1), and for hardwoods (1.8), which indicates that the timber inventory is increasing. In the 1920's, timber growth was about one-half the rate of harvest. By the 1940's, improved forest growth rates (partly because of forest protection from fire), as well as declines in harvest rates, resulted in timber growth and harvest coming into approximate balance.

References

Powell, D.S., J.L. Faulkner, D.R. Darr, Z. Zhu, and D.W. MacCleery. 1993. Forest resources of the United States, 1992. Gen. Tech. Rep. RM-234. U.S. Forest Service, Rocky Mountain Forest and Range Experiment Station, Fort Collins, CO. 132 pp.

USFS. 1992. Forest Service resource inventories: an overview. Forest inventory, economics, and recreation research. U.S. Forest Service, Washington, DC. 39 pp.

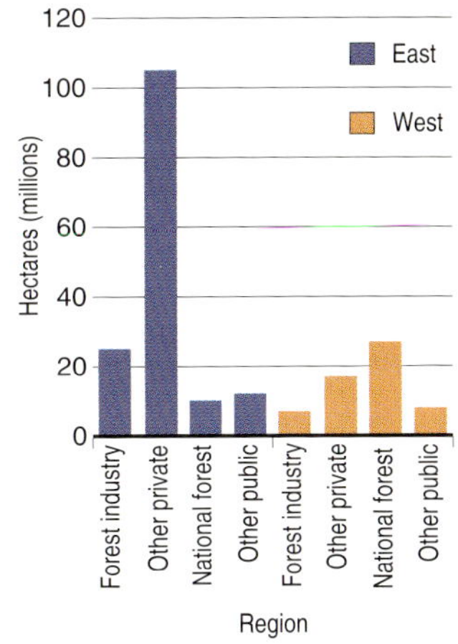

Fig. 2. Timberland ownership patterns by regions, 1992 (Powell et al. 1993).

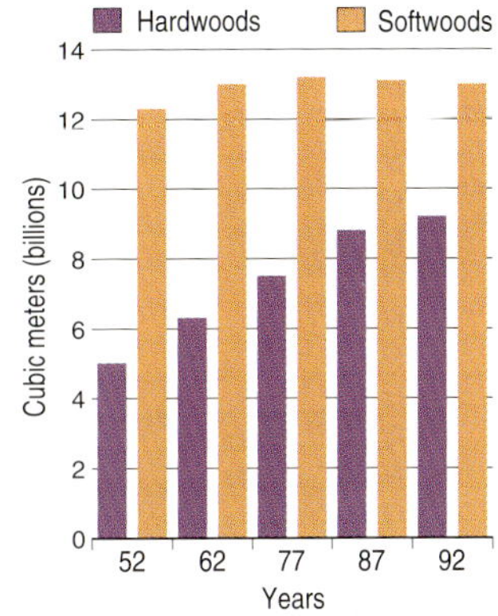

Fig. 3. Softwood and hardwood growing stock volume, selected years (Powell et al. 1993).

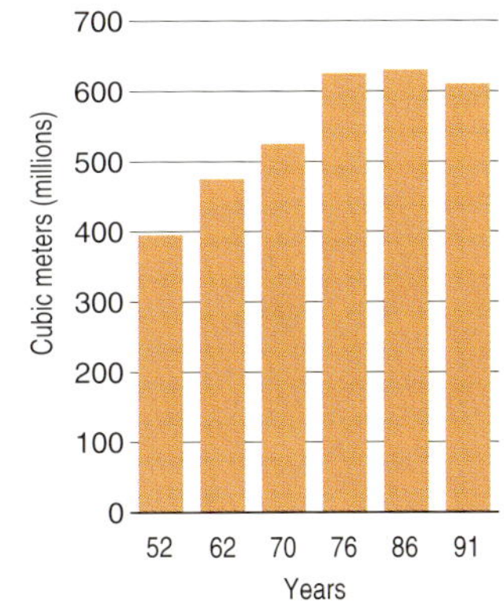

Fig. 4. Net annual growth, selected years (Powell et al. 1993).

For further information:

David R. Darr
U.S. Forest Service
PO Box 96090
Washington, DC 20090

Southern Forested Wetlands

by
B.D. Keeland
James A. Allen
Virginia V. Burkett
National Biological Service

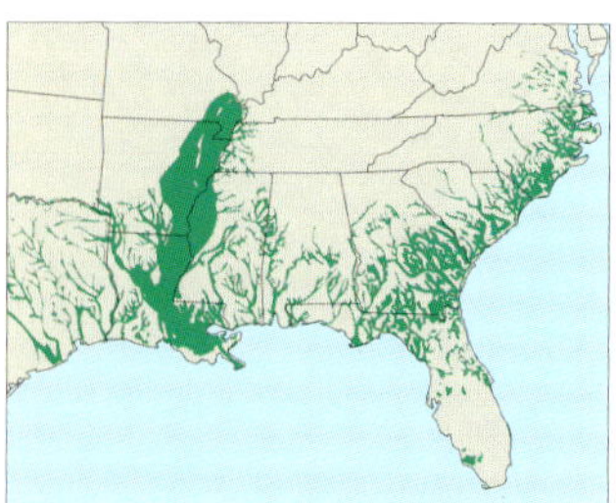

Fig. 1. Approximate distribution of forested wetlands along rivers and streams in the southeastern United States prior to European colonization (Putnam et al. 1960).

European settlers in many parts of the southern United States encountered a landscape largely comprising forested wetlands. These wetlands were a major feature of river floodplains and isolated depressions or basins from Virginia to Florida, west to eastern Texas and Oklahoma, and along the Mississippi River to southern Illinois (Fig. 1). Based on the accounts of pre-20th-century naturalists such as Audubon, Banister, John and William Bartram, Brickell, and Darby, the flora and fauna of many wetlands were unusually rich even by precolonial standards (Wright and Wright 1932). These early travelers described vast unbroken forests of oaks, ashes, maples (*Quercus, Fraxinus, Acer*), and other tree species, many with an almost impassable understory of saplings, shrubs, vines, switch cane, and palmetto. Low swampy areas with deep, long-term flooding were dominated by baldcypress (*Taxodium distichum)* and tupelo (*Nyssa aquatica* or *N. sylvatica* var *biflora*) and typically had sparse understories.

Most southern forested wetlands fall in the broad category of bottomland hardwoods, characterized and maintained by a natural hydrologic regime of alternating annual wet and dry periods and soils that are saturated or inundated during a portion of the growing season. Variations in elevation, hydroperiod, and soils result in a mosaic of plant communities across a floodplain. Wharton et al. (1982) classified bottomland hardwoods into 75 community types, including forested wetland types such as Atlantic white cedar bogs (*Chamaecyparis thyoides*), red maple (*Acer rubrum* var *drummondii*) and cypress-tupelo swamps, pocosins, hydric hammocks, and Carolina bays.

Realistic estimates of the original extent of forested wetlands are not available because accurate records of wetlands were not maintained until the early 20th century, and many accounts of wetland size were little more than speculation (Dahl 1990). Klopatek et al. (1979) estimated the precolonial forested wetland area of the United States to be about 27.2 million ha (67.2 million acres), but Abernathy and Turner (1987) suggested that this figure was low because it ignored small isolated wetlands.

Status

Estimates of the current forested wetland area vary. Shaw and Fredine (1956) estimated that as of the mid-1950's, the United States had about 19.1 million ha (47.2 million acres) of forested wetlands. Frayer et al. (1983) reported a similar total, 20.1 million ha (49.7 million acres), as of the mid-1970's. Between 1940 and 1960, the area of southern bottomland hardwoods increased from about 14.8 to 15.1 million ha (36.6 to 37.3 million acres) but declined to 12.5 to 13.1 million ha (30.9 to 32.4 million acres) by the mid-1970's (Hefner and Brown 1985; Turner et al. 1988). By the mid-1980's, an additional 1.4 million ha (3.5 million acres) of forested wetlands were lost, mostly from the southeastern United States.

The Southeast (including Alabama, Arkansas, Florida, Georgia, Kentucky, Louisiana, Mississippi, North Carolina, South Carolina, and Tennessee) makes up only 16% of the surface area of the conterminous United States yet accounts for about 47% of the total wetland area and 65% of the forested wetland area (Hefner and Brown 1985). Fifteen percent of the land surface of the Southeast can be categorized as wetlands, whereas only 5% of the land surface on a national basis is wetlands.

Before the mid-1970's, about 54% of palustrine wetland losses on a national basis were in forested areas. Palustrine wetlands include all nontidal wetlands dominated by trees, shrubs, persistent emergents, emergent mosses or lichens, and all such wetlands that occur in tidal areas where salinity due to ocean-derived salts is below 0.5 ppt (Cowardin et al. 1979). Between the mid-1950's and the mid-1970's, more than 2.2 million ha (5.4 million acres) of palustrine forested wetlands were lost within the Southeast, accounting for 92% of the national loss for this wetland type (Hefner and Brown 1985). Since the mid-1970's, loss of forested wetlands has accounted for 95% of all palustrine wetland losses (Dahl et al. 1991).

Despite dramatic losses since the beginning of the colonial period, southern forested wetlands currently account for about 36% of all wetlands and 60% to 65% of all forested wetlands in the conterminous United States (Hefner and Brown 1985; Dahl et al. 1991). Although loss rates have declined recently, most wetland acreage lost every year in the United States is from southern forested wetlands (Alig et al. 1986).

The most dramatic wetland loss in the entire nation has occurred in the forested wetlands of the Lower Mississippi River Alluvial Floodplain (LMRAF). This vast wetland extends nearly 1,000 km (621 mi) from the confluence of the Mississippi and Ohio rivers to the Gulf of Mexico and originally covered more than 10.1 million ha (25.0 million acres; Hefner and Brown 1985). About 8 million ha (19.8 million acres) of this area were forested wetlands in Arkansas, Louisiana, and Mississippi. Recent estimates reveal that fewer than 2 million ha (4.9 million acres) of forested wetlands remain in the LMRAF (The Nature Conservancy 1992), and the remaining portions of the original area are extremely fragmented (Fig. 2) and have lost many of their original functions (Mitsch and Gosselink 1993). Also, alterations in hydrology and poor timber management practices have resulted in a degraded condition of many of the remaining forests (Alig et al. 1986).

Turner et al. (1988) reported annual loss rates of 3.1% for forested wetlands in Arkansas, 0.9% for Louisiana, and 0.5% for Mississippi from 1960 to 1975. Recent U.S. Forest Service inventories indicate continued annual loss rates of 0.7% and 1.0% for the oak-gum-cypress forest type in the Louisiana and Mississippi portions of the LMRAF (May and Bertelson 1986; Kelly and Sims 1989; Vissage et al. 1992).

Causes of Loss

Since colonial times, wetlands have been regarded as a menace and a hindrance to land development: wastelands that were valuable only if drained. During the mid-19th century, Congress passed the Swamp Lands Acts of 1849, 1850, and 1860, granting swamp and periodically flooded bottomlands to the states. Five southern states received 16.7 million ha (41.3 million acres) for draining. By 1960, over 40 million ha (98 million acres) of former wetland area in the United States were under drainage (Turner et al. 1988). Most wetlands were drained for conversion to agriculture; such conversions account for 87% of our national wetland losses.

Large-scale federal navigation, flood-control, and drainage projects have played a large role in these conversions by making previously flood-prone lands dry enough for planting crops (USDI 1988). Other losses have resulted from construction of flood-control structures and reservoirs, mining and petroleum extraction, and urban development. A 40% increase in the population of the South between 1960 and 1980 (Alig et al. 1986) has accelerated wetland losses.

Future Prospects

A significant future threat is global climate change; in particular, sea-level rise represents a direct threat to thousands of hectares of coastal wetlands (Titus et al. 1984). Although the main effects of sea-level rise would be seen in coastal marshes, extensive areas of bottomland hardwood and swamp forest in Florida and Louisiana could be affected by increased flooding and saltwater intrusion (Titus et al. 1984; Pezeshki et al. 1987; Conner and Brody 1989).

Legislation such as the Clean Water Act and the "Swampbuster" provision of the 1985 Public Law 100-233 "Farm Bill" has slowed, but not completely prevented, the loss of forested wetlands. In the future, however, the amount of new losses of forested wetlands may be of less concern than the fragmentation and degradation of the few remaining large wetland areas.

While the amount of forested wetlands in the South is expected to continue declining, there are good prospects for restoration in some areas. Recognition of the scale and effects of bottomland hardwood losses has resulted in interest in restoration techniques. Serious restoration began in the mid-1980's, when state and federal agencies began reforesting former agricultural lands (Haynes and Moore 1988; Savage et al. 1989; Newling 1990). The pace of reforestation picked up rapidly following the establishment of the Conservation Reserve Program (CRP) and later the Wetland Reserve Program, two federal agricultural programs that provide payment to private landowners who plant trees on a portion of their land. The combined efforts of the agencies and these two agricultural programs have resulted in the planting of about 65,000 ha (160,615 acres) of bottomland hardwood forests in the southern United States since 1985. Most restoration has occurred in the LMRAF.

Prospects for a similar rate of reforestation over the coming decade appear excellent. Federal and state natural resource agencies continue to reforest their lands. In addition, they have become heavily involved in promoting reforestation on private lands through initiatives such as the Wetland Reserve Program, the U.S. Fish and Wildlife Service's Partners for Wildlife Program, and the North American Waterfowl Management Plan.

Partnerships are being sought between the forest industry, individual landowners, universities, and several state and federal agencies. Examples of such partnerships include Scott Paper Company's enrollment of 27,500 ha (67,952 acres) near Mobile, Alabama, in the Gulf Coast Joint Venture of the North American Waterfowl Management Plan, and a reforestation research project being initiated in west-central Mississippi that involves International Paper Company, the National Council of the Pulp and Paper Industry for Air and Stream Improvement, six federal agencies, and two universities.

Although there is growing concern that many reforestation projects have not been fully successful, it is clear that when properly done, reforestation can yield impressive results in the LMRAF region (Allen 1990). The technical feasibility of reforestation, along with the current environment of federal, state, and private cooperation in much of the region, suggests that the LMRAF may be one of the best areas of the country to seriously attempt a net gain of wetlands.

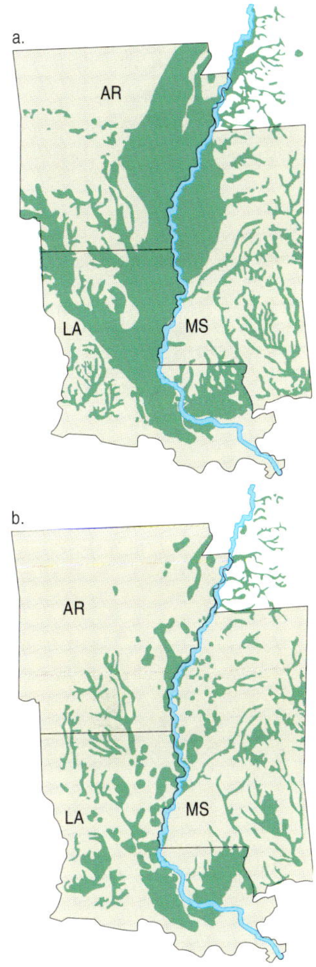

Fig. 2. Distribution of forested wetlands along the Lower Mississippi River: (a) Precolonial extent based on Putnam et al. (1960); (b) recent extent based on 1982 data (data source: U.S. Fish and Wildlife Service, Vicksburg, Mississippi).

References

Abernathy, Y., and R.E. Turner. 1987. US forested wetlands: 1940-1980. BioScience 37:721-727.

Alig, R.J., H.A. Knight, and R.A. Birdsey. 1986. Recent area changes in southern forest ownerships and cover types. U.S. Forest Service, Southern Forest Experiment Station Res. Paper. SE-260. 10 pp.

Allen, J.A. 1990. Establishment of bottomland oak plantations on the Yazoo National Wildlife Refuge complex. Southern Journal of Applied Forestry 14:206-210.

Conner, W.H., and M. Brody. 1989. Rising water levels and the future of southeastern Louisiana swamp forests. Estuaries 12:318-323.

Cowardin, L.M., V. Carter, F.C. Golet, and E.T. LaRoe. 1979. Classification of wetlands and deepwater habitats of the United States. U.S. Fish and Wildlife Service FWS/OBS-79/31. 103 pp.

Dahl, T.E. 1990. Wetlands losses in the United States 1780's to 1980's. U.S. Fish and Wildlife Service, Washington, DC. 21 pp.

Dahl, T.E., C.E. Johnson, and W.E. Frayer. 1991. Status and trends of wetlands in the conterminous United States, mid-1970's to mid-1980's. U.S. Fish and Wildlife Service, Washington, DC. 28 pp.

Frayer, W.E., M.J. Monahan, D.C. Bowden, and F.A. Graybill. 1983. Status and trends of wetlands and deepwater habitats in the conterminous United States, 1950's to 1970's. Colorado State University, Fort Collins. 31 pp.

Haynes, R.J., and L. Moore. 1988. Reestablishment of bottomland hardwoods within national wildlife refuges in the Southeast. Pages 95-103 *in* Proceedings of a conference: increasing our wetland resources. National Wildlife Federation—Corporate Conservation Council, Washington, DC.

Hefner, J.M., and J.D. Brown. 1985. Wetland trends in the southeastern United States. Wetlands 4:1-11.

Kelly, J.F., and M. Sims. 1989. Forest resources of Mississippi. USDA Forest Service, Southern Forest Experiment Station, Res. Bull. SO-147. 63 pp.

Klopatek, J.M., R.J. Olson, C.J. Emerson, and J.L. Jones. 1979. Land use conflicts with natural vegetation in the United States. Environmental Conservation 6:192-200.

May, D.M., and D.F. Bertelson. 1986. Forest statistics for Louisiana parishes. USDA Forest Service, Southern Forest Experiment Station, Res. Bull. SO-115. 59 pp.

Mitsch, W.J., and J.G. Gosselink. 1993. Wetlands, 2nd ed. Van Nostrand Reinhold, New York. 722 pp.

Newling, C.J. 1990. Restoration of bottomland hardwood forests in the Lower Mississippi Valley. Restoration and Management Notes 8:23-28.

Pezeshki, S.R., R.D. Delaune, and W.H. Patrick, Jr. 1987. Response of baldcypress (*Taxodium distichum* L. var. *distichum*) to increases in flooding salinity in Louisiana's Mississippi River Deltaic Plain. Wetlands 7:1-10.

Putnam, J.A., G.M. Furnival, and J.S. McKnight. 1960. Management and inventory of southern hardwoods. U.S. Forest Service, Agriculture Handbook 181. 102 pp.

Savage, L., D.W. Pritchett, and C.E. Depoe. 1989. Reforestation of a cleared bottomland hardwood area in northeast Louisiana. Restoration and Management Notes 7:88.

Shaw, S.P., and C.G. Fredine. 1956. Wetlands of the United States. Their extent and their value to waterfowl and other wildlife. U.S. Fish and Wildlife Service Circ. 39. 67 pp.

The Nature Conservancy. 1992. The forested wetlands of the Mississippi River, an ecosystem in crisis. The Nature Conservancy, Baton Rouge. 24 pp.

Titus, J.G., T.R. Henderson, and J.M. Teal. 1984. Sea level rise and wetlands loss in the United States. National Wetlands Newsletter 6(5):3-6.

Turner, R.E., S.W. Forsythe, and N.J. Craig. 1988. Bottomland hardwood forest land resources of the southeastern United States. Pages 13-28 *in* J.R. Clark and J. Benforado, eds. Wetlands of bottomland hardwood forests. Elsevier, Amsterdam, The Netherlands.

USDI. 1988. The impact of federal programs on wetlands, Vol. 1. The Lower Mississippi Alluvial Plain and the Prairie Pothole region. U.S. Department of the Interior, Washington, DC. 114 pp.

Vissage, J.S., P.E. Miller, and A.J. Hartsell. 1992. Forest statistics for Louisiana parishes–1991. U.S. Forest Service, Southern Forest Experiment Station Resour. Bull. SO-168. 65 pp.

Wharton, C.H., W.M. Kitchens, and T.W. Sipe. 1982. The ecology of bottomland hardwood swamps of the Southeast: a community profile. U.S. Fish and Wildlife Service. FWS/OBS-81/37. 133 pp.

Wright, A.H., and A.A. Wright. 1932. The habitats and composition of the vegetation of Okefinokee Swamp, Georgia. Ecological Monographs 2:109-232.

For further information:

B.D. Keeland
National Biological Service
Southern Science Center
700 Cajundome Blvd.
Lafayette, LA 70506

Rare Terrestrial Ecological Communities of the United States

by
Dennis H. Grossman
Kathleen Lemon Goodin
The Nature Conservancy

Federal agencies and conservation organizations have shifted their focus from managing individual species to managing entire ecosystems to protect biological diversity and conserve natural resources. Although ecological communities provide a more appropriate level of biological organization for characterizing ecosystems than individual species, the lack of a standard ecological community classification has impeded progress for ecosystem protection and management.

The Nature Conservancy and the Association of Natural Heritage Programs and Conservation Data Centers (Natural Heritage Network) have developed a framework for the classification of ecological communities. The first product from this effort is a preliminary list of rare terrestrial communities across the conterminous United States. This list was completed for the U.S. Fish and Wildlife Service (Grossman et al. 1994). This article provides a summary of the information from the Grossman et al. report, including a review of the status of information concerning rare communities of the United States, an analysis of regional patterns of rarity, and a discussion of the application of this information toward protection efforts. The use of ecological communities as a coarse conservation unit promotes conservation of the underlying ecological processes and biotic interactions that sustain the ecosystems across the landscape and ensures protection of biological diversity and rare species.

The Nature Conservancy/Natural Heritage Network Classification System

The basic goal of the community classification effort is to provide a complete listing of *all* communities that represent the variation in ecological systems. The classification hierarchy for terrestrial communities is based on the biological characteristics of existing vegetation types. These types range from early successional through climax associations and include seral stages that are maintained by natural and human-induced management and disturbance regimes.

The classification hierarchy is partitioned into terrestrial, aquatic, and subterranean "systems." The upper levels of the terrestrial system have been derived through the modification of United Nations Educational, Scientific and Cultural Organization (1973) and Driscoll et al. (1984) and refer to the physiognomic attributes (structural) of the vegetation. The two finest levels of the classification hierarchy are based on floristic analysis and are determined through the identification of diagnostic species (Westhoff and van der Maarel 1973).

Ranking System

The Nature Conservancy and Natural Heritage Network rank all elements of natural biological diversity according to their relative rarity and vulnerability to aid in ranking critical areas for conservation. The community ranks are consistent with the overall conservation ranking approach applied to all elements of natural diversity within The Nature Conservancy/Natural Heritage Network methodology (Master 1991). The communities described in this report have been ranked G1 and G2 according to The Nature Conservancy/Natural Heritage Network ranking system (*see* Table 1).

Table 1. The Nature Conservancy/Natural Heritage Network conservation ranks for rare communities.

Rank	Definition
G1	Critically imperiled globally because of extreme rarity. Generally five or fewer occurrences or less than about 800 ha (or 2,000 acres) or because of some factor making the community particularly vulnerable to extinction.
G2	Imperiled globally because of extreme rarity. Generally 6-20 occurrences or 800-4,000 ha (2,000-10,000 acres) or because of some factor making the community very vulnerable to extinction throughout range.

Listing Globally Rare Community Types

The development of the list of rare communities proceeded from the identification of rare communities at the state level, to the production of regional classifications of the rare state types, and finally to the generation of a consistent list of rare communities at the national level. Most state heritage programs have developed a classification system at the state level; these systems are based on available data and literature, input from experts, and field verification. State conservation ranks have been assigned to most of these communities based on the analysis of existing information.

A preliminary list of G1 and G2 communities was compiled by each of the Nature Conservancy's science regions (Table 2) through a detailed evaluation of all rare state types reported by the state heritage programs in each region. Each rare state type was reclassified to conform to the classification and nomenclature standards of the national framework. Rare communities that cross regional boundaries were identified and re-classified as necessary to produce the national list of G1 and G2 communities.

Table 2. The Nature Conservancy science regions.

East	Southeast	Midwest	West
Connecticut	Alabama	Illinois	Alaska*
Delaware	Arkansas	Indiana	Arizona
Maine	Florida	Iowa	California
Maryland	Georgia	Kansas	Colorado
Massachusetts	Kentucky	Michigan	Hawaii*
New Hampshire	Louisiana	Minnesota	Idaho
New Jersey	Mississippi	Missouri	Montana
New York	North Carolina	Nebraska	Nevada
Pennsylvania	Oklahoma	North Dakota	New Mexico
Rhode Island	South Carolina	Ohio	Oregon
Vermont	Tennessee	South Dakota	Utah
Virginia	Texas	Wisconsin	Washington
West Virginia			Wyoming

*Not included in this report.

Patterns of Community Rarity

Within the lower 48 United States, 371 globally rare terrestrial vegetated communities have been documented (Grossman et al. 1994). Preliminary evaluation of the proportion of G1 and G2 types indicates that these will account for about 10%-15% of all terrestrial communities. It is premature to attempt detailed national analysis and synthesis of existing data because of the preliminary nature of the overall classification and the unevenness in available community information among regions. We can, however, provide a preliminary examination of the relative proportion of rare communities in each physiognomic class within each region.

Eastern Region

Fourteen percent of the nationally rare communities occur in the eastern region (Anderson et al. 1994). Many new community types are still being identified. Most of the rare communities reported from the eastern region were forest, followed by sparse woodland and herbaceous types (Figure). The rarity of these communities is either related to the suitability of

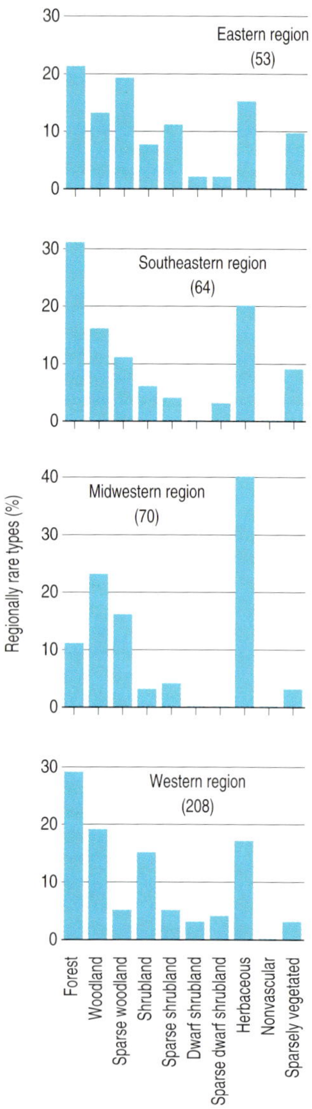

Figure. Rare terrestrial communities by region and physiognomic class. Numbers in parentheses refer to number of community types.

these habitats for land conversion or due to association with naturally rare habitats.

Southeastern Region

The rare communities occurring in the southeastern region account for about 18% of the nationally listed types (Patterson et al. 1994). This region is dominated by forest, and numerous diverse and intact rare forest associations remain. Most of the rare types fall within the forest class, followed by the herbaceous and woodland classes (Figure). The communities within the herbaceous class remain poorly defined throughout this region, but this class still represents a large portion of the rare types. We believe that the total number of types and the number of rare types within the herbaceous class will increase as additional information becomes available. Fire suppression has threatened many of the woodland types, and the actual number of rare woodland types is also presumed higher than now reported.

Midwestern Region

About 19% of the communities in the list of nationally rare communities occur in the midwestern region (Ambrose et al. 1994). Although the proportion of rare community types in this region is relatively small because of the historically coarse level of classification for this region, the magnitude of land conversion to agricultural production is staggering. The herbaceous class accounts for 40% of the rare types in the midwestern region, and the woodland and sparse woodland types make up another 38% (Figure). The rare herbaceous types reflect the remnant patches of the once-extensive prairie province. The woodland and sparse woodland communities have been heavily affected by the disruption of historical fire regimes and agricultural development.

Western Region

Most rare and threatened types identified in the national list of rare communities (about 56%) occur in the western region (Reid et al. 1994). This reflects the region's rich base of ecological and biological data and the consistent application of a detailed level of community classification, as well as a high level of natural diversity in this large region. Most rare types in the western region occur within the forest class, followed by the woodland, herbaceous, and shrubland classes (Figure). Fire suppression as a widespread forest-management practice over many decades has pushed many forest types to this status of rarity. Flood-control and water-diversion projects have similarly affected many of the forest and woodland riparian types. The rarity of the herbaceous communities across the western region is reported to be primarily the result of overgrazing and, to a lesser degree, direct agricultural conversion.

No regions reported rare communities in the nonvascular class and few were documented within the dwarf shrubland, sparse dwarf shrubland, and sparsely vegetated class. This result may not reflect the actual status of rare communities in these classes throughout the United States but rather the shortage of available information.

Knowledge Gaps

The rare communities for several states are not documented at this time. This does not mean there are no rare communities in those states but instead indicates the lack of available information. These knowledge gaps were documented during the listing of rare communities. Information gaps at the state level included incomplete or overly coarse classifications, lack of conservation ranks, and the lack of time and support for field verification. Those states where significant work remains are listed in Table 3.

Many communities recognized as rare still require additional work to complete their classification, ranking, and description process. The number of communities in this group presently totals 482.

Limitations

The number of rare communities varies among regions, reflecting unevenness in the quantity and quality of community information among the regions, along with varying levels of classification development and subtle differences in procedures for conservation ranking. To some extent, the regional variation also reflects the actual differences in ecological and biological diversity, the results of landscape fragmentation, and land-cover conversion.

While rarity of ecological communities is critical information for biodiversity conservation and management, appropriate protection and management activities should be determined for each individual rare community. Communities assigned a rank of G1 or G2 are very rare and occur generally within a restricted range of environmental conditions. These ranks do not reflect why a particular community is rare; such analysis, however, is fundamental to setting guidelines for protection and long-term management.

Some communities are naturally rare because of their association with an uncommon habitat. For example, the rarity of the inland salt marsh association (*Scirpus maritimus-Atriplex patula-Eleocharis parvula* herbaceous vegetation) has been documented, but the community is not noticeably rarer than it was 100 years ago.

This kind of community occurs on saturated saline mud flats associated with rare inland salt springs in Illinois, Michigan, and New York (Ambrose et al. 1994; Anderson et al. 1994). The environmental characteristics that support this biological association have similarly restricted the use of this habitat for agricultural production and most other types of land conversion, although some communities have been degraded by salt-extraction operations. Though this community is unlikely to disappear because of human-induced disturbance, individual communities should be protected from degradation due to incompatible land use.

In contrast, the mesic tall-grass prairie association (*Andropogon gerardii-Sorghastrum nutans-Sporobolus heterolepis* [*Liatrus* spp.-*Silphium laciniatum*] herbaceous vegetation) in the Midwest was common a century ago but is very rare today. The existing occurrences of this association type represent remnants of a community whose acreage has rapidly declined because of the value of its habitat for agricultural production (Ambrose et al. 1994). It has also suffered from the large-scale alteration of historical fire regimes. Rare communities such as this are quite threatened and require immediate protection and management.

Future

The list of nationally rare communities will help ensure their recognition and set priorities for their protection, an important step for conservation. Even if the list of rare communities were complete, however, it would still be insufficient to conserve and manage biological diversity. A comprehensive national conservation strategy for all communities, including common ones, is necessary to protect and manage the full spectrum of biological diversity and ecological systems.

The development of a standard community classification system has dramatically increased our capability to make better informed conservation and ecosystem management decisions at multiple geographic scales. The synthesis of existing community data on nationally rare types has identified the strengths and weaknesses of the existing information base, information that will help us decide how to accumulate and analyze data to fill critical gaps in our knowledge.

The acquisition and management of the ecological and biological data needed to complete the national classification represent a major challenge. The success of many ecosystem management initiatives will depend upon this information. A concerted cooperative effort is necessary to conserve and manage our biological and ecological resources.

Table 3. Knowledge gaps related to state community classification, ranking, and inventory.

State classification not completed	No state ranks	Coarse state classification	Limited inventory
Maryland	California	California	Alabama
Rhode Island	Georgia	Florida	Georgia
South Dakota	Maryland	North Dakota	North Dakota
West Virginia	Rhode Island	Texas	South Dakota
	South Dakota	Virginia	
	Tennessee		
	West Virginia		

References

Ambrose, D., J. Drake, and D. Faber-Langendoen. 1994. Rare plant communities of the United States: midwestern region. *In* D.H. Grossman, K.L. Goodin, and C.L. Reuss, eds. Rare plant communities of the conterminous United States: an initial survey. The Nature Conservancy, Arlington, VA. In press.

Anderson, M.G., L.A. Sneddon, and K.J. Metzler. 1994. Rare plant communities of the United States: eastern region. *In* D.H. Grossman, K.L. Goodin, and C.L. Reuss, eds. Rare plant communities of the conterminous United States: an initial survey. The Nature Conservancy, Arlington, VA. In press.

Driscoll, R.S., D.L. Merkel, R.L. Radloff, D.E. Snyder, and J.S. Hagihara. 1984. An ecological land classification framework for the United States. U.S. Forest Service Miscellaneous Publ. 1439.

Grossman, D.H., K.L. Goodin, and C.L. Reuss, eds. 1994. Rare plant communities of the conterminous United States: an initial survey. The Nature Conservancy, Arlington, VA. In press.

Master, L. 1991. Assessing threats and setting priorities for conservation. Conservation Biology 5(4):559-563.

Patterson, K.D, D.J. Allard, and S. Landaal. 1994. Rare plant communities of the United States: southeastern region. *In* D.H. Grossman, K.L. Goodin, and C.L. Reuss, eds. Rare plant communities of the conterminous United States: an initial survey. The Nature Conservancy, Arlington, VA. In press.

Reid, M., L. Engelking, and P. Bourgeron. 1994. Rare plant communities of the United States: western region. *In* D.H. Grossman, K.L. Goodin, and C.L. Reuss, eds. Rare plant communities of the conterminous United States: an initial survey. The Nature Conservancy, Arlington, VA. In press.

United Nations Educational, Scientific and Cultural Organization (UNESCO). 1973. International classification and mapping of vegetation. Series 6, Ecology and conservation. United Nations Educational, Scientific and Cultural Organization, Paris. 93 pp.

Westhoff, V., and E. van der Maarel. 1973. The Braun-Blanquet approach. Pages 626-828 *in* R.H. Whittaker, ed. Handbook of vegetation science. Part V. Ordination and classification of communities. Junk, The Hague.

For further information:

Dennis H. Grossman
The Nature Conservancy
1815 N. Lynn St.
Arlington, VA 22209

Altered Fire Regimes Within Fire-adapted Ecosystems

by
Gardner W. Ferry
Robert G. Clark
Roy E. Montgomery
Bureau of Land Management
Robert W. Mutch
U.S. Forest Service
Willard P. Leenhouts
U.S. Fish and Wildlife Service
G. Thomas Zimmerman
National Park Service

Fires ignited by people or through natural causes have interacted over evolutionary time with ecosystems, exerting a significant influence on numerous ecosystem functions (Pyne 1982). Fire recycles nutrients, reduces biomass, influences insect and disease populations, and is the principal change agent affecting vegetative structure, composition, and biological diversity. As humans alter fire frequency and intensity, many plant and animal communities are experiencing a loss of species diversity, site degradation, and increases in the size and severity of wildfires. This article examines the role fire plays in the ecological process around which most North American ecosystems evolved.

The five plant communities selected for study were the sagebrush steppe, juniper woodlands, ponderosa pine forest, lodgepole pine forest, and the southern pineland (Fig. 1). Status and trends of altered fire regimes in fire-adapted ecosystems highlight the role that fire plays in wildland stewardship. Fire regimes are considered as the total pattern of fires over time that is characteristic of a region or ecosystem (Kilgore and Heinselman 1990).

Sagebrush-grass Plant Communities

Greater frequency of fire has seriously affected the sagebrush steppe during the last 50 years (Table). One such community, the semi-arid intermountain sagebrush (*Artemisia* species) steppe, encompasses about 45 million ha (112 million acres). After repeated fires, non-native European annual grasses such as cheatgrass (*Bromus tectorum*) and medusahead (*Taeniatherum caput-medusae*) now dominate the sagebrush steppe (West and Hassan 1985). It is unclear whether cheatgrass invasion, heavy grazing pressure, or shorter fire return intervals initiated the replacement of perennial grasses and shrubs by the non-native annual grasses. It is clear, however, that wildfires aid in replacing native grasses with cheatgrass, as well as causing the loss of the native shrub component (Whisenant 1990). Inventories show that cheatgrass is dominant on about 6.8 million ha (17 million acres) of the sagebrush steppe and that it could expand into an additional 25 million ha (62 million acres) in the sagebrush steppe and the Great Basin sagebrush type (Pellant and Hall 1994).

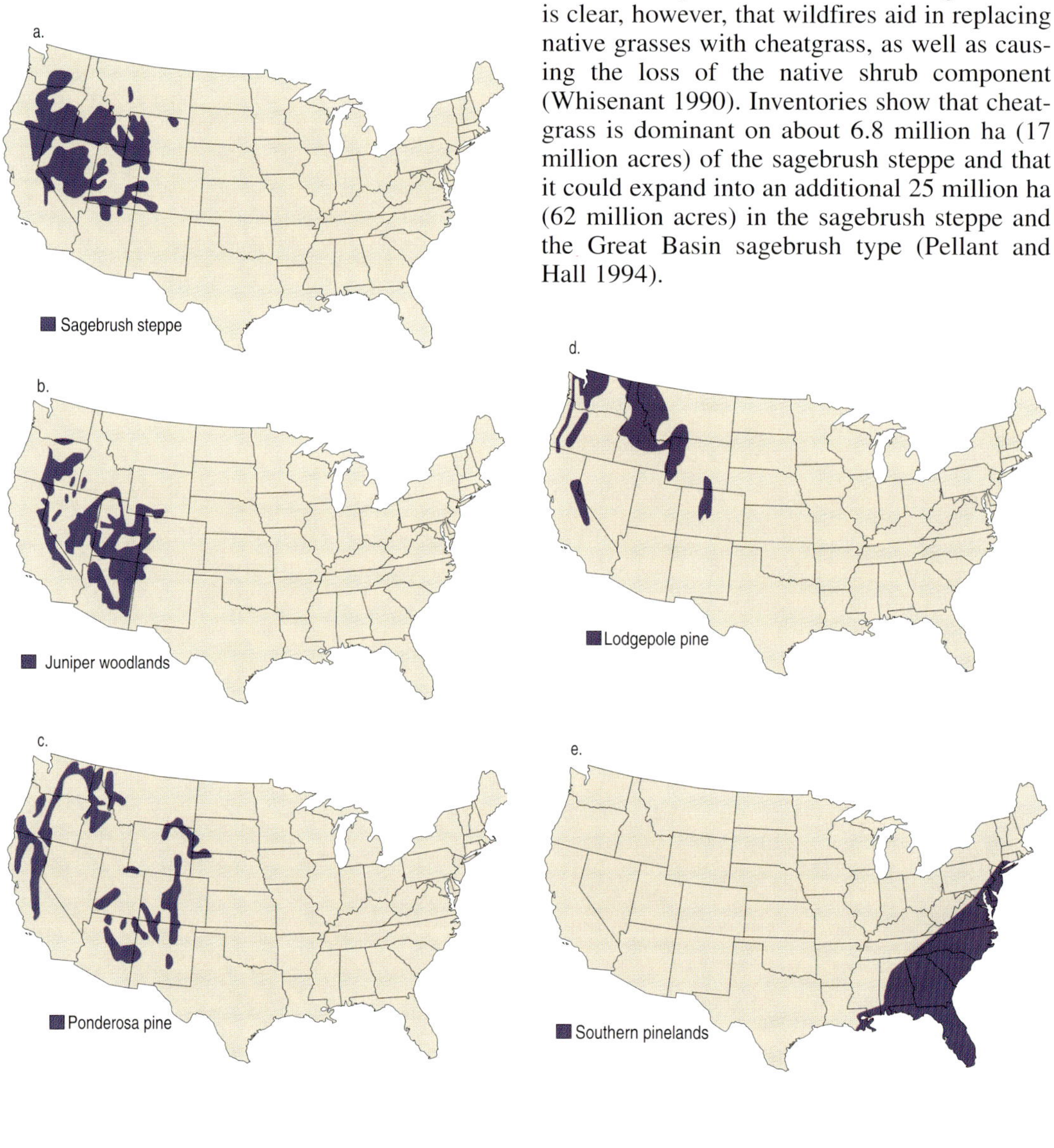

Fig. 1. Range of: a —sagebrush steppe; b — juniper woodlands; c — ponderosa pine; d — lodgepole pine; and e — southern pineland communities in the United States.

Table. Increase in the number of wildfires and area burned on sagebrush steppe in Idaho (data from the Bureau of Land Management, Idaho State Office, Boise).

	1950-59	1960-69	1970-79	1980-89
Number of wildfires	Data incomplete	1,344	1,406	2,334
Area burned (ha)	751,000	663,000	900,000	1,316,000

Western Juniper Woodlands

Juniper woodlands occupy 17 million ha (42 million acres) in the Intermountain region (West 1988). Juniper species common to this region are western juniper (*Juniperus occidentalis*), Utah juniper (*J. osteosperma*), single-seeded juniper (*J. monosperma*), and Rocky Mountain juniper (*J. scopulorum*). Presettlement juniper woodlands were usually savanna-like or confined to rocky outcrops not typically susceptible to fire (Nichol 1937).

Juniper woodlands began increasing in both density and distribution in the late 1800's (R.F. Miller, Eastern Oregon Agricultural Research Center, unpublished data; Fig. 2) because of climate, grazing, and lack of fire (Miller and Waigand 1994). Warm and wet climate conditions then were ideal for juniper and grass seed production. Fire frequency had decreased because the grazing of domestic livestock had greatly reduced the grasses and shrubs that provided fuel, and relocation of Native Americans eliminated an important source of ignition. Continued grazing and 50 years of attempted fire exclusion have allowed juniper expansion to go unchecked.

Ponderosa Pine Forest

Decreases in fire frequency are also seriously affecting ponderosa pine (*Pinus ponderosa*) forests, a common component on about 16 million ha (40 million acres) in the western United States. Historically, the ponderosa pine ecosystem had frequent, low-intensity, surface fires that perpetuated park-like stands with grassy undergrowth (Barrett 1980). For six decades, humans attempted to exclude fire on these sites (OTA 1993). Fifty years ago, Weaver (1943) stated that complete prevention of forest fires in the ponderosa pine region had undesirable ecological effects and that already-deplorable conditions were becoming increasingly serious. Today, many ponderosa pine forests are overstocked, plagued by epidemics of insects and diseases, and subject to severe stand-destroying fires (Mutch et al. 1993).

Lodgepole Pine Forest

Like ponderosa pine forests, lodgepole pine (*Pinus contorta*) forests are experiencing a change in structure, distribution, and functioning of natural processes because of fire exclusion and increases in disease. Wildfire may be the most important factor responsible for establishment of existing stands (Wellner 1970). Historical stand-age distributions in lodgepole pine forests indicated an abundance of younger age classes resulting from periodic fires. Fire exclusion, by precluding the initiation of new stands, is responsible for a marked change in distribution of age classes in these forests (Fig. 3).

Dwarf mistletoe (*Arceuthobium americanum*), the primary disease of lodgepole pine, also has a profound effect on forest structure and function, although it occurs slowly. Data show that chronic increases of dwarf mistletoe are partly due to the exclusion of fire (Zimmerman and Laven 1984) because fire is the natural control of dwarf mistletoe and has played a major role in the distribution and abundance of current populations and infection intensities (Alexander and Hawksworth 1975). As the frequency and extent of fire have decreased in lodgepole pine stands over the last 200 years, dwarf mistletoe infection intensity and distribution are clearly increasing (Zimmerman and Laven 1984).

Southern Pinelands

In contrast to the juniper, ponderosa pine, and lodgepole pine communities, fire frequencies have not drastically decreased in the 78 million ha (193 million acres) of southern pinelands. These pinelands are composed of diverse plant communities associated with longleaf (*Pinus palustris*), slash (*P. elliotti*), loblolly (*P. taeda*), and shortleaf pines (*P. echinata*). Fire has continued on an altered basis as an ecological process in much of the southern pinelands; historically, fire burned 10%-30% of the forest annually (Wright and Bailey 1982); the southern culture never effectively excluded fire from its pinelands (Pyne 1982), although human-ignited fires have partially replaced natural fires. Consequently, the amount of fire has been reduced and the season of burns has changed from predominately growing-season to dormant-season (fall or winter) fires (Robbins and Myers 1992). Altering the burning season and frequency has significantly affected southern pineland community structure, composition, and biological diversity (Fig. 4).

Implications

The role of fire becomes more complex as it interacts with land management. Maintaining interactions between disturbance processes and ecosystem functions is emphasized in ecosystem management. It is vital for mangers to

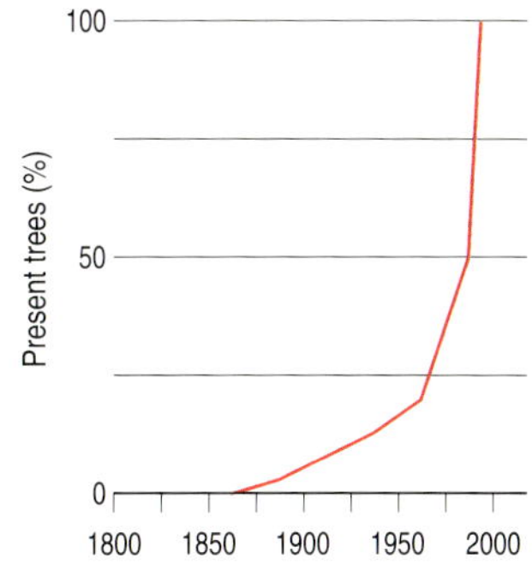

Fig. 2. Cumulative establishment of western juniper on Steens Mountain, Oregon (adapted from R.F. Miller, Eastern Oregon Agricultural Research Center, unpublished data).

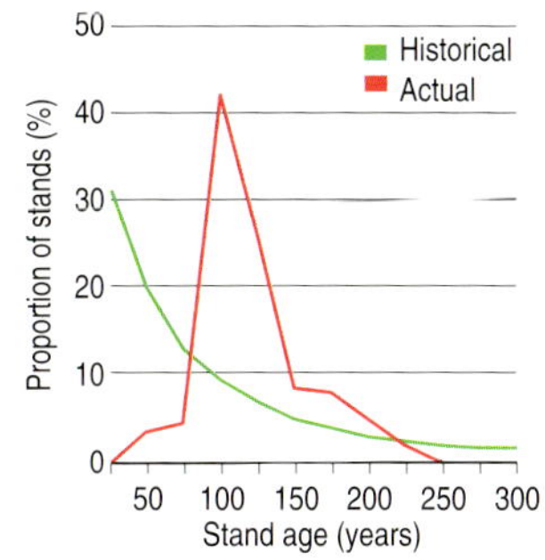

Fig. 3. Historical and actual age-class distributions of lodgepole pine forest.

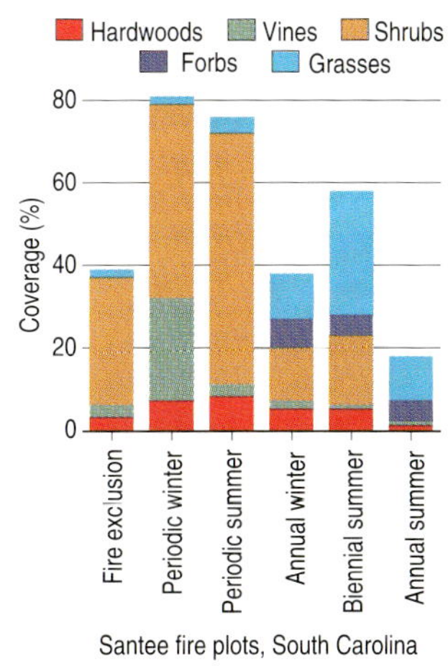

Fig. 4. Understory plant crown coverage after 30 years of burning (Waldrop et al. 1987).

recognize how society influences fire as an ecological process. In addition, managers must uniformly use information on fire history and fire effects to sustain the health of ecosystems that are both fire-adapted and fire-dependent. Managers must balance the suppression program with a program of prescribed fire applied on a landscape scale if we are to meet our stewardship responsibilities.

References

Alexander, M.E., and F.G. Hawksworth. 1975. Wildland fires and dwarf mistletoe: a literature review of ecology and prescribed burning. U.S. Forest Service Gen. Tech. Rep. RM-14. Rocky Mountain Forest Range Experiment Station, Fort Collins, CO. 12 pp.

Barrett, S.W. 1980. Indians and fire. Western wildlands 6(3):17-21.

Kilgore, B.M., and M.L. Heinselman. 1990. Pages 297-335 *in* J.C. Hendee, G.M. Stankey, and R.C. Lucas, eds. Fire in wilderness ecosystems. Wilderness management. 2nd ed. North American Press, Golden, CO.

Miller, R.F., and P.E. Waigand. 1994. Holocene changes in semi-arid pinyon-juniper woodlands: response to climate, fire and human activities in the Great Basin. Biological Science 44 (7). In press.

Mutch, R.W., S.F. Arno, J.K. Brown, C.E. Carlson, R.D. Ottmar, and J.L. Peterson. 1993. Forest health in the Blue Mountains: a management strategy for fire-adapted ecosystems. U.S. Forest Service Gen. Tech. Rep. PNW-310. 14 pp.

Nichol, A.A. 1937. The natural vegetation of Arizona. University of Arizona Tech. Bull. 68.

OTA. 1993. Preparing for an uncertain climate. 2 vols. U.S. Congress, Office of Technology Assessment. Washington, DC.

Pellant, M., and C. Hall. 1994. Distribution of two exotic grasses on public lands in the Great Basin: status in 1992. *In* Proceedings of a Symposium on Ecology, Management and Restoration of Intermountain Annual Rangelands. Intermountain Forest and Range Experiment Station, Ogden, UT.

Pyne, S.J. 1982. Fire in America: a cultural history of wildland and rural fire. Princeton University Press, Princeton, NJ. 653 pp.

Robbins, L.E., and R.L. Myers. 1992. Seasonal effects of prescribed burning in Florida: a review. Tall Timbers Research, Inc. Miscellaneous Publ. 8. 96 pp.

Waldrop, T.A., D.H. Van Lear, R.T. Lloyd, and W.R. Harms. 1987. Long-term studies of prescribed burning of loblolly pine forests of the southeastern Coastal Plain. U.S. Forest Service Southeastern Forest Experiment Station Gen. Tech. Rep. SE-45. Asheville, NC. 23 pp.

Weaver, H. 1943. Fire as an ecological and silvicultural factor in the ponderosa-pine region of the Pacific Slope. Journal of Forestry 41:7-14.

Wellner, C.A. 1970. Fire history in the northern Rocky Mountains. Pages 41-64 *in* The role of fire in the Intermountain West. Proceedings of the Intermountain Fire Research Council Symposium, Missoula, MT.

West, N.E. 1988. Intermountain deserts, shrub steppes, and woodlands. Pages 209-230 *in* M.B. Barbour and W.D. Billings, eds. North American terrestrial vegetation. Cambridge University Press, New York.

West, N.E., and M.A. Hassan. 1985. Recovery of sagebrush-grass vegetation following wildlife. Journal of Range Management 38:131-134.

Whisenant, S.G. 1990. Changing fire frequencies on Idaho's Snake River Plains: ecological and management implications. Pages 4-10 *in* E.D. McArthur, E.M. Romney, S.D. Smith, and P.T. Tueller, eds. Proceedings of a Symposium on Cheatgrass Invasion, Shrub Die-off, and Other Aspects of Shrub Biology and Management. U.S. Forest Service Gen. Tech. Rep. INT-276. Intermountain Forest and Range Experiment Station, Ogden, UT.

Wright, H.A., and S.W. Bailey. 1982. Fire ecology: United States and southern Canada. John Wiley and Sons, New York. 501 pp.

Zimmerman, G.T., and R.D. Laven. 1984. Ecological interrelationships of dwarf mistletoe and fire in lodgepole pine forests. Pages 123-131 *in* F.G. Hawksworth and R.F. Scharpf, technical coordinators. Biology of dwarf mistletoes: Proceedings of the Symposium. U.S. Forest Service Gen. Tech. Rep. RM-111. Rocky Mountain Forest Range Experiment Station, Fort Collins, CO. 131 pp.

For further information:

Gardner W. Ferry
Bureau of Land Management
Division of Fire and Aviation
Policy and Management
3905 S. Vista Ave.
Boise, ID 83705

Vegetation Change in National Parks

by
Kenneth Cole
National Biological Service

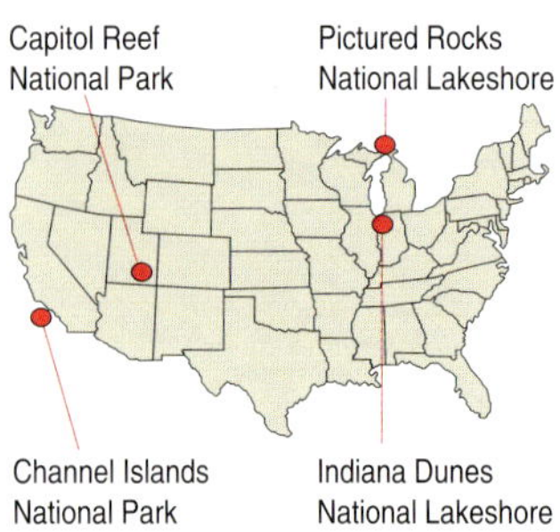

Fig. 1. Four national park units studied.

Natural ecosystems are always changing, but recent changes in the United States have been startlingly rapid, driven by 200 years of disturbances accompanying settlement by an industrialized society. Logging, grazing, land clearing, increased or decreased frequency of fire, hunting of predators, and other changes have affected even the most remote corners of the continent. Recent trends can be better understood by comparisons with more natural past trends of change, which can be reconstructed from fossil records. Conditions before widespread impacts in a region are termed "presettlement"; conditions after the impacts are "postsettlement."

Fossil plant materials from the last few thousand years are used to study past changes in many natural areas. Pollen buried in wetlands, for example, can reveal past changes in vegetation (Faegri and Iversen 1989), and larger fossil plant parts can be studied in deserts where the fossilized plant collections of packrats, called packrat middens, have been preserved (Betancourt et al. 1990).

This article summarizes the rates of vegetation change in four national park areas over the last 5,000 years as reconstructed from fossil pollen and packrat middens. These four national park areas from different ecological regions (Fig. 1) demonstrate the flexibility of these paleoecological techniques and display similar results.

Northern Indiana Prairie

A 4,500-year history of vegetation change was collected from Howes Prairie Marsh, a small marsh surrounded by prairie and oak savanna in the Indiana Dunes National Lakeshore near the southern tip of Lake Michigan. Only 40 km (25 mi) from Chicago, this area has been affected by numerous impacts from settlements but still supports comparably pristine tall-grass prairie vegetation as well as the endangered Karner blue butterfly

Courtesy W. L. Loope

Coring a small pond at Pictured Rocks National Lakeshore.

Courtesy K. Cole, NBS

Experimental prairie fire at Indiana Dunes National Lakeshore.

(*Lycaeides melissa samuelis*). Although this site has experienced more disturbances than any of the others described here, it is a most valuable site because of its many species (Wilhelm 1990) and its tall-grass prairie vegetation that has been nearly eliminated elsewhere.

The many historical impacts to this area make it a good source for studying past changes. Past amounts of pollen from the primary plant taxa are illustrated in Fig. 2. Many changes occurred before settlement, but more rapid changes occurred in the last 140 years.

Past rates of change in vegetation can be measured by summing the relative change in each plant type between successive samples and then dividing by the number of years between samples. The technique is similar to that used by Jacobson and Grimm (1986).

Although these changes had been occurring throughout the last 4,500 years, the postsettlement rates of change are at least 10 times greater than the presettlement rates of change (Fig. 3a). The rates of change have been declining over the last 50 years, but are still far greater than the presettlement rates of change.

Northern Michigan Forest

A similar analysis was carried out on pollen from a small bog (unofficial name: 12-Mile Bog) surrounded by pine forest along the southern shore of Lake Superior (Fig. 3b). This site, within Pictured Rocks National Lakeshore, was more severely affected by logging and slash burning in the 1890's than by the periodic wildfires that characterized this forest earlier, but it has been protected for the last 80 years. The magnitude of change caused by the crude logging and slash burning of the logging era was far greater than any recorded during the 2,500 years since Lake Superior receded to create the forest of white and red pine (*Pinus strobus* and *P. resinosa*).

As in the Indiana Dunes, rates of change have declined during the last 60 years, and the forest is now very similar to the forest of 2,000 years ago. Thus, although the area is still changing at a rate far above normal, it has begun to recover through protection.

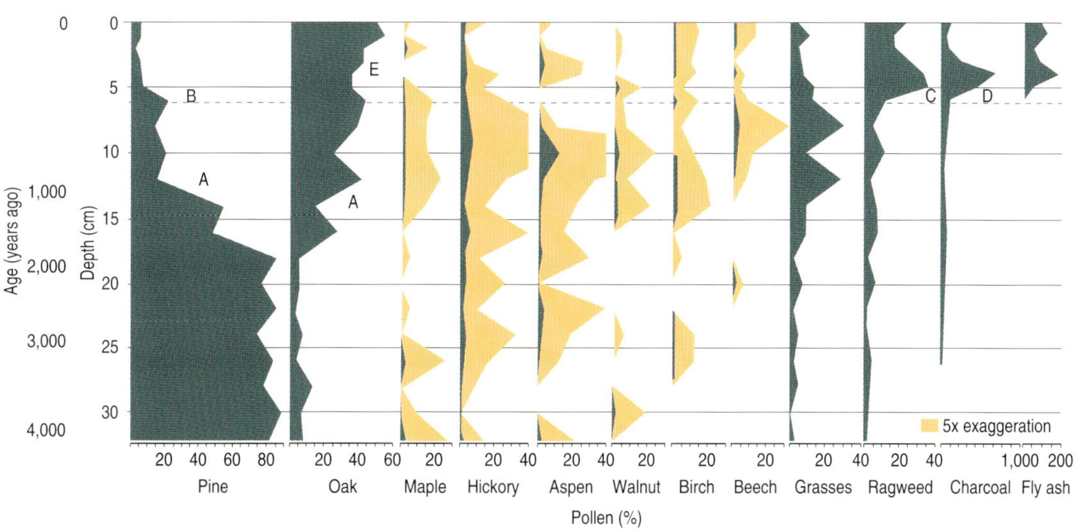

Fig. 2. Selected taxa of fossil pollen recorded from Howes Prairie in the Indiana Dunes. The percentage of total pollen representing each plant is graphed along a vertical time axis. The dotted line shows the sedimentary horizon representing settlement of the region (about A.D. 1850). Major changes indicated by letters: A — decline in pine and increases in oak and grasses due to plant succession and climate change; B — decline in pine due to logging of white pine in mid-1800's; C — increase in ragweed from cleared farm fields and increase in fly ash from the development of the steel industry in Gary, IN (22 km away) in the late 1800's (Cole et al. 1990); D — increase in charcoal particles as steam railroads ignite nearby drained wetlands and subsequent decline in charcoal as steam power ends and wildfires are controlled; and E — decline in oak as frequent fires top-kill mature trees followed by increase in oak as periodic prairie fires are extinguished.

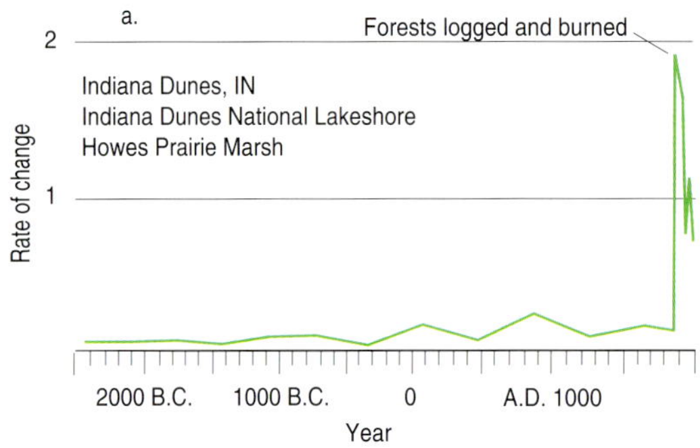

Fig. 3a. Rates of change from a tall-grass prairie and oak savanna in the Indiana Dunes, IN, based on pollen from tree species.

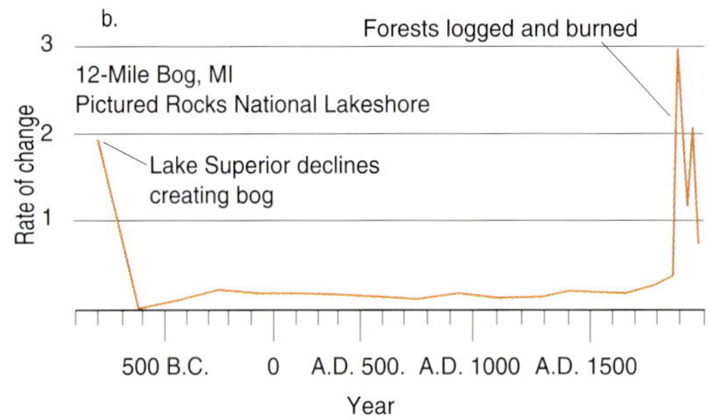

Fig. 3b. Rates of change from a boreal pine forest at Pictured Rocks, MI, based on pollen from tree species.

California Coastal Sage Scrub

Fossil pollen was analyzed from an estuary on Santa Rosa Island off the coast of southern California (Cole and Liu 1994). The semi-arid landscape around the estuary is covered with coastal sage scrub, chaparral, and grassland. This site, within Channel Islands National Park, is one of the least affected areas in this region of rapidly expanding urbanization, although the island's native plants and animals were not well adapted to withstand the grazing of the large animals introduced with the ranching era of the 1800's. This island, which had no native large herbivores, became populated with thousands of sheep, cattle, horses, goats, pigs, deer, and elk. The National Park Service is removing many of the large herbivores, although most of the island remains an active cattle ranch.

All pollen types from 33 samples spread over the last 4,600 years were analyzed. The rates of change in the pollen were similar to those observed from the other sites (Fig. 3c).

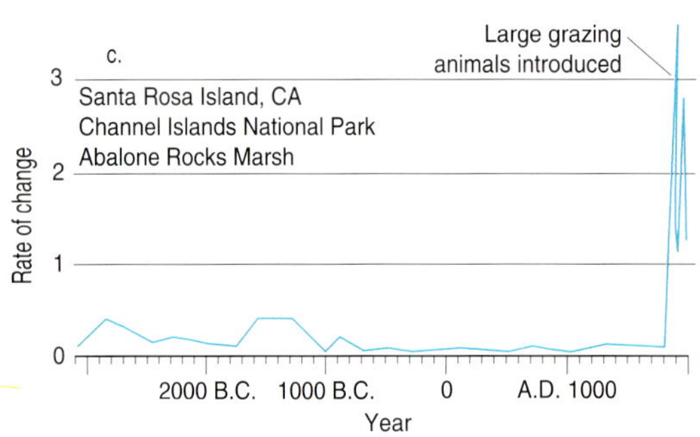

Fig. 3c. Rates of change from coastal sage scrub on Santa Rosa Island, CA, based on pollen from an estuary.

Southern Utah Desert

Because fossil pollen is usually preserved in accumulating sediments of wetlands, different paleoecological techniques are necessary in arid areas. In western North America, fossil deposits left by packrats (*Neotoma* spp.) have proven a useful source of paleoecological data (Betancourt et al. 1990). Past desert vegetation can be reconstructed by analyzing bits of leaves, twigs, and seeds collected by these small rodents and incorporated into debris piles in rock shelters or caves. These debris piles can be collected, analyzed, and radiocarbon dated.

The vegetation history of a remote portion of Capitol Reef National Park (Hartnett Draw) was reconstructed through the analysis of eight packrat middens ranging in age from 0 to 5,450 years (Cole 1995). The vegetation remained fairly stable throughout this period until the last few hundred years. The most recent deposits contain many plants associated with overgrazed areas such as whitebark rabbitbrush (*Chrysothamnus visidiflorus*), snakeweed (*Gutterezia sarothrae*), and greasewood (*Sarcobatus vermiculatus*), which were not recorded at the site before settlement.

Conversely, other plants that are extremely palatable to grazing animals were present throughout the last 5,450 years, only to disappear since settlement. Plant species preferred by sheep and cattle, such as winterfat (*Ceratoides lanata*) and rice grass (*Stipa hymenoides*), disappeared entirely, while many other palatable plant species declined in abundance after 5,000 years of comparative stability.

The past rates of vegetation change for this site were calculated in a manner similar to the fossil pollen records (Fig. 3d). Although the rate of change calculation is less precise than the fossil pollen records because there were fewer samples, the results show a similar pattern. The rate of vegetation change is highest between the two most recent records.

Although this area is still grazed by cattle today through grazing leases to private ranchers from the National Park Service, much of the

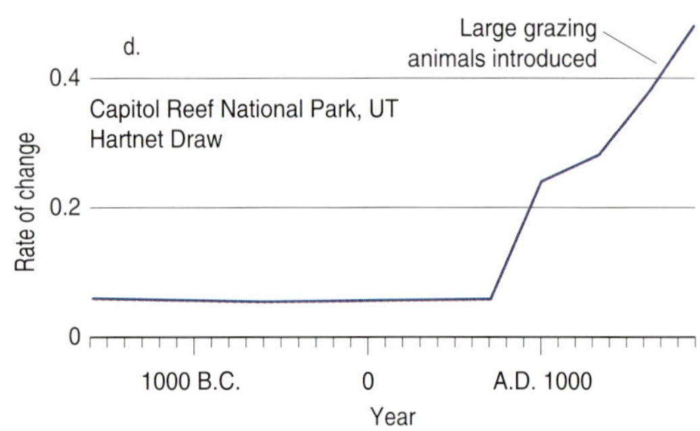

Fig. 3d. Rates of change from desert vegetation at Capitol Reef, UT, based on plant fossils in packrat middens.

severe damage was probably done by intensive sheep grazing during the late 1800's when the entire region was negatively affected by open-land sheep ranching. We cannot yet demonstrate whether the grazing effects are continuing or if the site is improving, although reinvasion of palatable species is unlikely in the face of even light grazing. Severe overgrazing is required to eliminate abundant palatable species, but once they are eliminated, even light grazing can prevent their restoration.

Courtesy K. L. Cole, NBS

Twenty-eight thousand-year-old packrat midden at Capitol Reef National Park; orange notebook is 6 inches high.

Implications

Wise land management decisions are more likely to be made if land managers understand a site and are able to place the status quo into a historical perspective. Because most of the damage to these four sites occurred before the 20th century, land managers might not even be aware of the tremendous changes that have occurred were it not for these fossil records. Since the ultimate goal for the management of many areas is to mitigate settlement impacts and return the land to its presettlement status, detailed knowledge of the effects of settlement is imperative.

In all study areas, postsettlement rates of change were at least 10 times higher than the presettlement rates of change. Thus, the changes now being observed in even remote natural ecosystems are unlike former natural changes. Some areas are continuing to change at rapid rates, while other areas, which have not been disturbed as recently, are stabilizing. The climatic warming projected for the next 50 years may exacerbate these ongoing changes, but will be only one of many variables operating in the unplanned redesign of our natural ecosystems. Land managers need to understand the nature and severity of the effects of settlement to return the land to its presettlement condition.

References

Betancourt, J.L., T.R. Van Devender, and P.S. Martin, eds. 1990. Packrat middens: the last 40,000 years of biotic change. The University of Arizona Press, Tucson. 467 pp.

Cole, K.L. 1995. A survey of the fossil packrat middens and reconstruction of the pregrazing vegetation of Capitol Reef National Park. National Park Service Res. Rep. In press.

Cole, K.L., D.R. Engstrom, R.P. Futyma, and R. Stottlemyer. 1990. Past atmospheric deposition of metals in northern Indiana measured in a peat core from Cowles Bog. Environmental Science and Technology 24:543-549.

Cole, K.L., and G. Liu. 1994. Holocene paleoecology of an estuary on Santa Rosa Island, California. Quaternary Res. 41:326-335

Faegri, K., and J. Iversen. 1989. Textbook of pollen analysis. Wiley and Sons, New York. 328 pp.

Jacobson, G.L., Jr., and E.C. Grimm. 1986. A numerical analysis of Holocene forest and prairie vegetation in central Minnesota. Ecology 67:958-966.

Wilhelm, G.S. 1990. Special vegetation of the Indiana Dunes National Lakeshore. Indiana Dunes National Lakeshore Research Program, Rep. 90-02, Porter, IN.

For further information:

Kenneth L. Cole
National Biological Service
Cooperative Park Studies Unit
University of Minnesota
Department of Forest Resources
115 Green Hall
St. Paul, MN 55108

Air Pollution Effects on Forest Ecosystems in North America

by
David L. Peterson
National Biological Service

Air pollution poses a threat to forest ecosystems in several regions of North America. Although there are isolated impacts downwind from point sources such as industrial operations, the major impacts are from regional-scale exposure to ambient ozone and acid precipitation. Acidic deposition (including sulfur and nitrogen deposition) is fairly high in the northeastern United States and southeastern Canada, although symptomatic injury and changes in forest growth have not been clearly linked to a particular pollutant. Recent evidence, however, indicates that long-term inputs of acid precipitation may be altering the chemical equilibrium of some soils, which could result in a nutritional imbalance in trees.

Elevated levels of ozone have resulted in stress in several forest ecosystems of North America: (1) those adjacent to Mexico City (extensive mortality and reduced growth); (2) those in mountains of the Los Angeles Basin in California (mortality and growth reductions); (3) those in the central and southern Sierra Nevada (some reduced growth and widespread symptomatic injury); (4) those in the Rincon Mountains of Arizona (some symptomatic injury); and (5) those in the Great Smoky Mountains (some symptomatic injury).

Recent growth reductions and changes in forest health have been reported for several locations in North America although the role of air pollution in these "declines" must be evaluated in the context of a stress complex that includes climate, stand dynamics, and site factors. Although some lichens are known to be sensitive to air pollution, there is relatively little information on the effects of air pollutants on forest species other than trees. Only if monitoring programs are implemented soon will it be possible to detect how long-term pollutant deposition affects forest health and productivity.

For further information:

David L. Peterson
National Biological Service
Cooperative Park Studies Unit
University of Washington
AR-10
Seattle, WA 98195

Air Quality in the National Park System

by
Bruce Nash
Kathy Tonnessen
National Biological Service
David Joseph
Miguel Flores
National Park Service

The National Park Service (NPS) Organic Act and the federal Clean Air Act require the NPS to protect the natural resources of the lands it manages from the adverse effects of air pollution. The NPS established a program to measure ozone—the air pollutant that is most widespread and injurious to human health and vegetation—at more than 40 monitoring sites within the National Park System.

NPS sites in southern and central California, the Great Lakes region, and the northeast and east-central United States generally record the highest ozone concentrations in the NPS network. Ozone levels exhibit strong seasonal and diurnal temporal trends, and year-to-year variation may be significant (Figure).

The 1987-91 NPS trend in maximum ozone concentrations closely resembles the corresponding trend for the entire nation. The National Biological Service (NBS) National Air Quality Research Program sponsors surveys to document ozone injury to vegetation. Current monitoring concentrates on sensitive indicator plants, including hardwoods and some herbaceous plants in the eastern United States and conifers in the West. Controlled fumigation studies have confirmed that elevated ambient ozone levels can cause decreased growth rate, decreased biomass, and premature defoliation in sensitive species such as black cherry (*Prunus serotina*), American sycamore (*Platanus occidentalis*), yellow-poplar (*Liriodendron tulipifera*), and ponderosa pine (*Pinus ponderosa*).

Acid deposition is a regional pollutant monitored at 30 NPS units as part of the National Atmospheric Deposition Program (NADP). Ten years of wet deposition (e.g., pollutants that may come down in rain or snow) data permit researchers to estimate loading of nitrate, sulfate, and hydrogen ions to sensitive ecosystems. NADP data show that the NPS units with the greatest acid loading are in the eastern United States, with Acadia, Cape Cod, Shenandoah, and Great Smoky Mountains national parks showing annual average wet deposition pH values of 4.4-4.6. These values do not reflect the contributions of cloudwater, fogwater, and dry deposition (e.g., particles and gases) to the total loading of acids, nitrogen, and sulfate to ecosystems that are sensitive to acidic inputs. NADP samplers do not measure snow efficiently and do not account for the effect of snowmelt pulses on sensitive alpine lakes and streams in the spring at high-elevation sites in the Sierra Nevada, the Cascades, and the Rocky Mountains. Research at Shenandoah National Park has shown that deposition-driven episodes in streams can result in pH levels low enough to affect native fish species.

Any assessment of ecosystem health must consider the composition of the atmosphere and its interactions with the biological and physical components of the ecosystem under investigation. Although we have some understanding of the biological effects of air pollution, more studies are necessary to ensure the protection of our natural resources.

For further information:

Bruce L. Nash
National Biological Service-AIR
National Air Quality Research Program
PO Box 25287
Denver, CO 80225

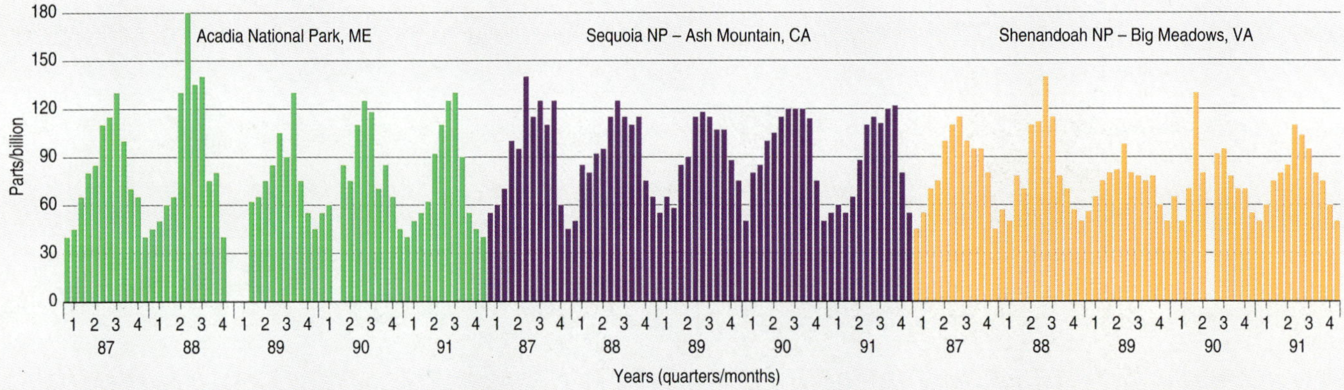

Figure. Highest daily 1-h ozone concentration per month.

Whitebark Pine: Ecosystem in Peril

by
Katherine C. Kendall
National Biological Service

Whitebark pine (*Pinus albicaulis*) is well-suited to harsh conditions and populates high-elevation forests in the northern Rocky Mountain, North Cascade, and Sierra Nevada ranges (Fig. 1a). Whitebark pine seeds are unusually large, highly nutritious, and are a preferred food for grizzly bears (*Ursus arctos*) and many other animals (Kendall and Arno 1990). These pine trees (Fig. 2) are adapted to cold, dry sites and pioneer burns and other disturbed areas. At timberline, they grow under conditions tolerated by no other tree species, thus playing an important role in snow accumulation and persistence. Because few roads occur in whitebark pine ecosystems and because the tree's wood is of little commercial interest, information on the drastic decline of this picturesque tree has only recently emerged.

Threats

Whitebark pine is threatened by an introduced disease and fire suppression. In its northern range, many whitebark pine stands have declined by more than 90% (Fig.1a). The most serious threat to the tree is from white pine blister rust (*Cronartium ribicola*), a non-native fungus that has defied control. Fewer than one whitebark pine tree in 10,000 is rust-resistant. Mortality has been rapid in areas like western Montana, where 42% of whitebark pine trees have died from the disease in the last 20 years; 89% of the remaining trees are infected with rust (Fig. 3; Keane and Arno 1993). Although drier conditions have slowed the spread of blister rust in whitebark pine's southern range, infection rates there are increasing and large die-offs are eventually expected to occur (Fig. 1b).

Before fire suppression, whitebark pine stands burned every 50-300 years. Under current management, they will burn at 3,000-year intervals. Without fire, seral whitebark pine trees are replaced by shade-tolerant conifers and become more vulnerable to insects and disease.

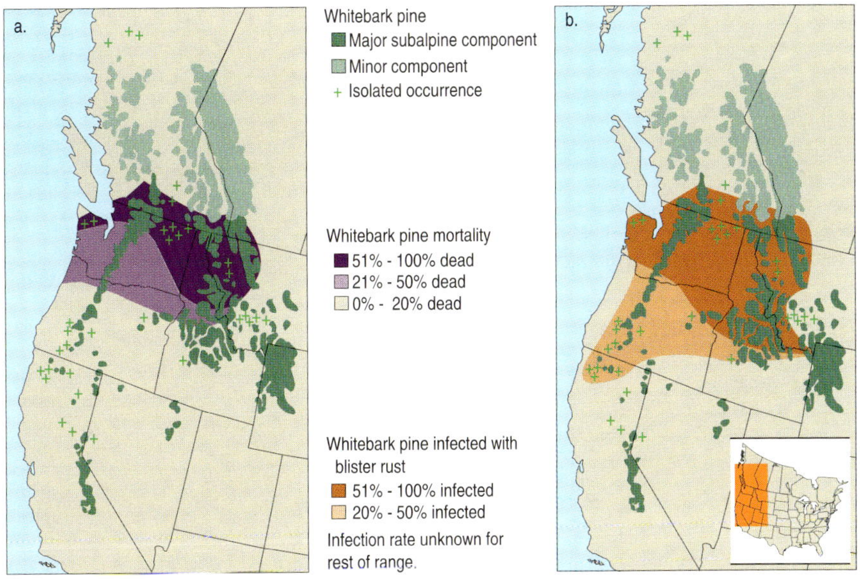

Fig. 1. **(a)** Natural distribution of whitebark pine (Arno and Hoff 1989; Olgilvie 1990) with mortality zones. Mortality level is the proportion of trees dead from all causes since presettlement. **(b)** White pine blister rust infection rates in whitebark pine. Blister rust is present but infection rates are unknown in Canada and the southern United States.

Repercussions

The alarming loss of whitebark pine has broad repercussions: mast for wildlife is diminished and the number of animals the habitat can support is reduced. Such results hinder grizzly bear recovery and may be catastrophic to Yellowstone grizzlies for whom pine seeds are a critical food. Predicted changes in whitebark pine communities include the absence of reforestation of harsh sites after disturbance and the lowering of treelines. In addition, stream flow and timing will be altered as snowpack changes with vegetation.

Implications

Whitebark pine will be absent as a functional community component until rust-resistant strains evolve. Natural selection could be speeded with a breeding program like that developed

Courtesy D. Reinhart

Fig. 2. Healthy whitebark pine stand in Yellowstone National Park not yet affected by the introduced disease, white pine blister rust.

Courtesy R. McClelland

Fig. 3. Dead whitebark pine trees in Glacier National Park.

for western white pine (*P. monticola*), which also suffers from rust. In some areas where whitebark pine is regenerating, its competitors should be eliminated. To perpetuate whitebark pine at a landscape scale, fires must be allowed to burn in whitebark pine ecosystems.

Isolated populations may become extinct where mountain pine beetle or other agents kill remaining resistant trees. To prevent loss of genetic diversity, seeds of these pines should be collected throughout the species' range and stored as insurance against catastrophic events. To guide restoration efforts, more information is needed on whitebark pine's historical distribution, trends throughout its range, and rust epidemic dynamics.

For further information:

Katherine C. Kendall
National Biological Service
Glacier Field Station
Glacier National Park
West Glacier, MT 59936

References

Arno, S.F., and R.J. Hoff. 1989. Silvics of whitebark pine (*Pinus albicaulis*). U.S. Forest Service Gen. Tech. Rep. INT-253. 11 pp.

Keane, R.E., and S.F. Arno. 1993. Rapid decline of whitebark pine in western Montana: evidence from 20-year remeasurements. Western Journal of Applied Forestry 8(2):44-47.

Kendall, K.C., and S.F. Arno. 1990. Whitebark pine—an important but endangered wildlife resource. Pages 264-273 *in* U.S. Forest Service Gen. Tech. Rep. INT-270.

Olgilvie, R.T. 1990. Distribution and ecology of whitebark pine in western Canada. Pages 54-60 *in* U.S. Forest Service Gen. Tech. Rep. INT-270.

Oak Savannas in Wisconsin

by
Richard A. Henderson
Eric J. Epstein
Wisconsin Department of Natural Resources

Oak savanna is a term given to a loosely defined, yet well-recognized, class of North American plant communities that were part of a large transitional complex of communities between the vast treeless prairies of the West and the deciduous forests of the East. This system was driven by frequent fires and possibly influenced by large herbivores such as bison and elk. A wide range of community types was found within this transitional complex; collectively, they represented a continuum from prairie to forest. The term "savanna" is generally applied to a small group of related community types in the middle portion of this continuum.

Savannas all have a partial canopy of open grown trees and a varied ground layer of prairie and forest herbs, grasses, and shrubs, as well as plants restricted to the light shading and mix of shade and sun so characteristic of savanna. Oaks were clearly the dominant trees, and, hence, the common use of the term oak savanna. Definitions of savanna tree cover range from 5% to 80% canopy; however, the lower canopy covers of 5%-50% or 5%-30% are more widely used criteria. Savanna types range from those associated with dry, gravelly, or sandy soils; those on rich, deep soils; and those on poorly drained, moist soils.

Oak savannas have probably been in North America for some 20-25 million years (Barry and Spicer 1987), expanding and contracting with climatic changes and gaining and losing species (on a geologic time scale) through evolution and extinction. For the past several thousand years, such savannas have existed as a relatively stable band of varying width and continuity, from northern Minnesota to central Texas (Figure).

At the time of European settlement (ca. 1830), oak savanna covered many millions of hectares. It varied somewhat in species composition from north to south and east to west, but

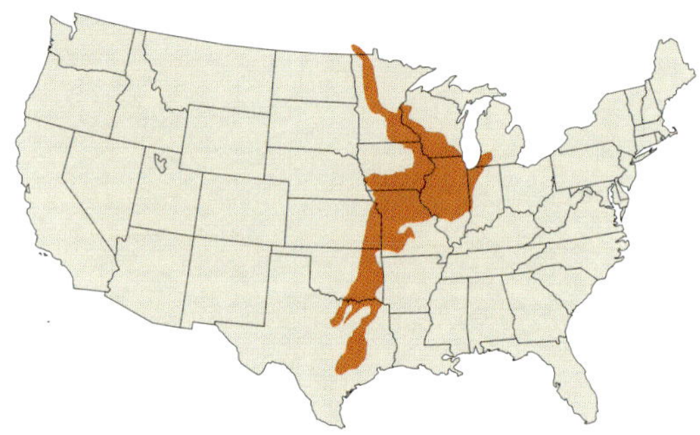

Figure. Gross range of presettlement oak savanna in central North America (adapted from Nuzzo 1986 and Smeins and Diamond 1986). The shaded area represents the general region in which oak savannas occurred, although this region was not uniformly savanna. Significant blocks of nonsavanna vegetation, such as prairie or forest, were also present within it. Nor was oak savanna totally restricted to this region. Small, disjunct outliers existed as far east as Ohio and as far west as the Dakotas.

structure and functions were probably similar throughout. In the upper Midwest (Minnesota, Wisconsin, Michigan, Iowa, Illinois, Indiana, and Missouri) there were an estimated 12 million ha (29.6 million acres) of oak savanna (Nuzzo 1986). Wisconsin's portion was 2.9 million ha (7.3 million acres; Curtis 1959). As the Midwest's rich soils were used for agriculture and fire was suppressed, this ecosystem all but disappeared from the landscape throughout its range. Today, oak savanna is a globally endangered ecosystem.

Status

In the early to mid-19th century, the oak savanna ecosystem was thoroughly fragmented and nearly totally destroyed throughout its range. Most of its acreage suffered from clearing and plowing, overgrazing, or invasion by dense shrub and tree growth caused by lack of fire, lack of grazing, or both. Consequently, oak savanna now shares equal billing with tall-grass prairie as the most threatened plant communities in the Midwest and among the most threatened in the world. Only a little more than 200 ha (500 acres) of intact examples of oak savanna vegetation are listed in the Wisconsin State Natural Heritage Inventory, or less than 0.0001 (0.01%) of the original 2.9 million ha (7.3 million acres)—a fate repeated throughout this community's entire range (Johnson 1986; Smeins and Diamond 1986). A tallying of known oak savanna sites in the upper Midwest (Missouri northward) in 1985 (Nuzzo 1986) listed only 133 sites totaling 2,600 ha (6,420 acres), or only 0.0002 (0.02%) of the estimated presettlement extent of the community. Most of what remains are dry and wet savanna types. Richer, more productive soil savanna is now nearly nonexistent.

Fortunately, most of the biota that was associated with savanna, especially the vertebrates, have either adapted to the changed landscape or have managed to survive in suboptimal habitat (e.g., the fringes of other less devastated communities, such as oak forests). This situation is precarious for many species, however, and their long-term future is doubtful. Vertebrates have been successful because major elements of the savanna structure are still well represented in various edge habitats, including wooded pastures, lawns, and woodlots. The fact that the plant species may be different in surrogate savannas has not affected savanna vertebrate species for the most part.

Oak savanna vegetation has not fared as well. Many species that were probably savanna specialists are now uncommon and are found only in the fringes and openings of oak woods, brushy areas, and lightly grazed pastures. A few examples are giant false-foxglove (*Aureolaria grandiflora*), yellow pimpernel (*Taenidia integerrima*), pale Indian-plantain (*Cacalia atriplicifolia*), New Jersey tea (*Ceanothus americanus*), sessile-leaved eupatorium (*Eupatorium sessilifolium*), and white death-camas (*Zigadenus elegans*). Two likely savanna specialists, purple milkweed (*Asclepias purpurascens*) and wild hyacinth (*Camassia scilloides*), are now listed as endangered in Wisconsin. Three others—kitten-tails (*Besseya bullii*), cream gentian (*Gentiana alba*), and Virginia lespedeza (*Lespedeza virginica*)—are listed as threatened.

Most bird species found in Wisconsin savannas are still doing well today (e.g., American robin [*Turdus migratorius*], indigo bunting [*Passerina cyanea*], and brown thrasher [*Toxostoma rufum*]). Only one oak savanna/woodland bird, the passenger pigeon (*Ecopistes migratorius*), has become extinct, and another, the wild turkey (*Meleagris gallopavo*), was extirpated but is now restored; however, both of these were lost because of unregulated hunting and not because of habitat loss.

Courtesy R. Henderson, Wisconsin Department of Natural Resources

Oak savanna.

Recently, however, a number of savanna birds have not thrived or have begun to decline throughout their range, including the northern flicker (*Colaptes auratus*), red-headed woodpecker (*Melanerpes erythrocephalus*), vesper sparrow (*Pooecetes gramineus*), northern bobwhite (*Colinus virginianus*), warbling vireo (*Vireo gilvus*), and field sparrow (*Spizella pusilla*). The orchard oriole (*Icterus spurius*) and yellow-breasted chat (*Icteria virens*) are on Wisconsin's list of special concern. The loggerhead shrike (*Lanius ludovicianus*) and barn owl (*Tyto alba*) are on Wisconsin's endangered species list, and Bell's vireo (*Vireo bellii*) is now on Wisconsin's threatened species list (D.W. Sample and M.J. Mossman, Wisconsin Department of Natural Resources, personal communication, 1994). Although loss of habitat has not been the cause of decline in all these species, it certainly is affecting many of them. The abandonment and loss of savanna/woodlot pastures in the past few decades may be playing a role in some of the recent declines of savanna bird species.

Most amphibian and reptile species that were closely associated with the historic oak savanna in Wisconsin are doing at least moderately well today, although two reptiles associated with savanna habitat are on the Wisconsin list of endangered species and are suffering from habitat loss: the western slender glass lizard (*Ophisaurus attenuatus*) and the eastern massasauga rattlesnake (*Sistrurus catenatus*). The eastern massasauga is also under consideration for federal listing.

Our knowledge of oak savanna invertebrates is limited; we know little about what species were characteristic or restricted to oak savanna, let alone their current status. Some reliable status information does exist for savanna Lepidoptera (moths and butterflies), however; of this group, the Karner blue butterfly (*Lycaeides melissa samuelis*) is listed as federally endangered while phlox flower moths (*Schinia indiana*) and tawny crescent butterflies (*Phyciodes batesii*) are under consideration for federal listing. The frosted elfin butterfly (*Incisalia irus*) is listed as threatened in Wisconsin, and four savanna skippers (*Erynnis persius*, *Hesperia leonardus*, *H. metea*, and *Atrytonopsis hianna*) and the buck moth (*Hemileuca maia*) are considered rare in the state. Other globally rare insects thought to have been part of the oak savanna ecosystem include the federally listed American burying beetle (*Nicrophorus americanus*) and the red-tailed leafhopper (*Aflexia rubranura*), which is under consideration for federal listing.

Threats

Threats to the future survival of oak savanna throughout its range can be summarized into four categories. The first, loss of recovery opportunities, can be attributed to accelerating succession to tree and shrub species that produce dense shade; a lack of recruitment and eventual death of long-lived plants surviving now only in suboptimal habitat; changes in pasturing practices through either increasing or decreasing grazing pressure; and an increasing rate of rural home building in key savanna areas. The second threat is lack of understanding about the community by the public, professional resource managers, and scientists. Resistance to the use of prescribed fire, especially in wooded areas, and lack of understanding about the importance of fire in maintaining biodiversity are the third threat; invasion by aggressive non-natives (i.e., honeysuckle, buckthorn, and reed canary grass) is the fourth.

Recovery Potential

In the absence of active management, the future of oak savanna looks bleak in Wisconsin and throughout its entire range. The increasing abandonment of lightly to moderately grazed wooded pastures and the accelerating succession of oak woodlots toward heavy shade-producing trees and shrubs are likely to lead to the further decline and possible loss of much of the remaining savanna flora and fauna, including eventual declines of the oaks themselves.

If oak savanna habitats are actively managed, however, their recovery potential in Wisconsin and throughout their range is substantial (Holtz 1985; Bronny 1989; R.A. Henderson, Wisconsin Department of Natural Resources, unpublished data). Many degraded sites in the dry and wet ends of the spectrum can be recovered with relative ease. Mesic, richer soil savannas will take more time and work, but recovery is still feasible. The native species that formerly inhabited oak savannas can be reintroduced with a reasonable amount of effort (Packard 1988), but the options available are quickly decreasing.

In Wisconsin alone, hundreds if not thousands of hectares of overgrown but still retrievable oak savanna exist on public lands and thousands more on private lands. Although Wisconsin may be above average in potential for savanna recovery, similar situations exist in other states. Much of this land, especially low productivity sites, could be restored within a few decades simply by thinning trees, brushing, and burning. Well-drained, rich soil sites, however, will require more work and time to restore. Some plant reintroductions may be necessary, but much can be accomplished with fire alone. Light grazing may also have potential as a savanna management tool.

References

Barry, A.T., and R.A. Spicer. 1987. The evolution and palaeobiology of land plants. Croom Helm, London. 309 pp.

Bronny, C. 1989. One-two punch: grazing history and the recovery potential of oak savanna. Restoration and Management Notes 7(2):73-76.

Curtis, J.T. 1959. The vegetation of Wisconsin. University of Wisconsin Press, Madison. 657 pp.

Holtz, S.L. 1985. Cutting and burning a degraded oak barrens: management techniques that stimulate natural disturbance. M.S. thesis, University of Wisconsin, Madison. 80 pp.

Johnson, F.L. 1986. Oak-hickory savannahs and transition zones: preservation status and management problems. Pages 345-347 *in* D.L. Kulhavy and R.N. Conner, eds. Wilderness and natural areas in the eastern United States: a management challenge. Stephen F. Austin State University, Nacogdoches, TX.

Nuzzo, V.A. 1986. Extent and status of Midwest oak savanna: presettlement and 1985. Natural Areas Journal 6(2):6-36.

Packard, S. 1988. Just a few oddball species: restoration and the rediscovery of the tallgrass savanna. Restoration and Management Notes 6(1):13-20.

Smeins, F.E., and D.D. Diamond. 1986. Grasslands and savannahs of east central Texas: ecology, preservation, and management. Pages 381-394 *in* D.L. Kulhavy and R.N. Conner, eds. Wilderness and natural areas in the eastern United States: a management challenge. Stephen F. Austin State University, Nacogdoches, TX.

For further information:

Richard A. Henderson
Wisconsin Department of Natural Resources
Bureau of Research
1350 Femrite Dr.
Monona, WI 53716

Courtesy J.G. Wiener, NBS

Aquatic Ecosystems

by
Science Editor
Michael J. Mac
National Biological Service
1849 C St. NW
Washington, DC 20240

Overview

Aquatic ecosystems have been especially subjected to the environmental degradation that has occurred over the last century in this country. Nearly every activity that occurs on land ultimately affects the receiving waters in that drainage. Whether it's pesticides and herbicides applied to crops, silt washed away because of vegetation removal, or even atmospheric deposition, aquatic ecosystems are a product of all local disturbances regardless of where they occur. In addition, waterways have been used for numerous activities other than providing habitat to aquatic organisms. They have been altered for transportation, diverted for agricultural and municipal needs, dammed for energy, borrowed as an industrial coolant, and straightened for convenience. These abuses have taken their toll as evidenced by worldwide declines in fisheries, monumental floods, an ever-growing list of endangered aquatic species, and communities trying to deal with finite water supplies.

The traits that make aquatic ecosystems particularly vulnerable also make them useful for monitoring environmental quality. Water serves to integrate these impacts by distributing them among the elements within aquatic ecosystems. Although dilution is occurring, subtle changes can be detected in habitats or organisms over a much larger area that may be the result of a single point source. A clean aquatic ecosystem with a healthy biological community will be indicative of the condition of the terrestrial habitat in the watershed, whereas the reverse may not necessarily be true.

This section features accounts of the alterations of aquatic habitats and their impacts on the biota. Evidence is presented documenting habitat destroyed by dams or channelization (*see* this section, Bogan et al.; Wlosinski et al.; and Wiener et al.), contaminants affecting organism health (*see* Hesselberg and Gannon; Lerczak and Sparks), wetlands affected by water-level control (*see* Wilcox and Meeker), reduced water quality (*see* Charles and Kociolek), and introductions of exotic species (*see* Hansen and Peck; Wiener et al.). These kinds of changes have caused declining biodiversity in many groups of aquatic species ranging from freshwater mussels to waterfowl.

Some encouraging trends are emerging. Persistent organic contaminants in the Great Lakes have declined (*see* Hesselberg and Gannon), and marginal water-quality improvement has been accompanied by increased diversity of the fish community (*see* Lerczak and Sparks). Despite these achievements, much needs to be done to effectively manage and conserve aquatic resources. As is evident from the reports on diatoms (*see* Charles and Kociolek),

Courtesy J.G. Wiener, NBS

Habitat loss on a stretch of the Mississippi River modified for navigation (shown here) contrasts with a diverse complex of habitats on less developed areas of the Upper Mississippi River (*see* plate previous page).

algae (*see* Moe), and protozoa (*see* Lipscomb), little is known of the national trends in their populations, diversity, or biomass. Our knowledge of these groups is poor even though they provide basic functions of photosynthesis, production, and decomposition critical to the normal functioning of aquatic ecosystems.

Without increased monitoring, some very basic attributes of aquatic systems may be unknowingly lost or severely degraded. Groups of species that seem insignificant actually are critical parts of a food web that supports valuable commercial and sport species. Subtle changes such as losses of island habitat and constant water depth or level may lead to drastic declines in productivity or diversity (*see* Wlosinski et al.; Wilcox and Meeker). The loss of some of these integral pieces of ecosystems may be impossible to restore. The unsuccessful attempt to restore self-sustaining lake trout populations in the Great Lakes, despite massive efforts, exemplifies this (*see* Hansen and Peck).

Habitat Changes in the Upper Mississippi River Floodplain

by
Joseph H. Wlosinski
Douglas A. Olsen
Carol Lowenberg
Thomas W. Owens
Jim Rogala
Mark Laustrup
National Biological Service

The U.S. Congress recognized the Upper Mississippi River (UMR) as a nationally significant ecosystem in 1986. The UMR extends northward from the confluence of the Mississippi and Ohio rivers to the Twin Cities, Minnesota, a distance of more than 1,360 km (850 mi). The floodplain (area between the bluffs) of the UMR includes 854,000 ha (2,110,000 acres) of land and water. The Mississippi River is a major migration corridor for waterfowl and provides habitat for more than 150 fish and 40 freshwater mussel species.

Since 1824 the federal government has implemented numerous changes on the UMR. The river was first modified by removing snags and then sandbars, with changes progressing to rock excavation, elimination of rapids, closing of side channels, and the construction of hundreds of wing dams, 27 navigation dams, and hundreds of kilometers of levees. Reservoirs formed by the navigation dams are known locally as pools (Fig. 1), which are numbered from north to south. Construction of the dams (mostly during the 1930's) significantly altered the northern 1,040 km (650 m) of the UMR (north of St. Louis, Missouri) by increasing the amount of open water and marsh areas. Wing dams and levees have altered aquatic habitats south of St. Louis (the open river) by reducing open-water habitats and isolating the river from much of the floodplain. Most of the changes to the river ecosystem were either designed for navigational improvements or to control the movement of river water. Here we investigate some of the habitat changes at various levels of resolution.

Spatial data were analyzed by using a geographic information system (GIS). Floodplain areas (bluff to bluff) and systemic land-cover/land-use data were obtained from Landsat Thematic Mapper data collected in 1989. Land-cover/land-use data from 1891 were created

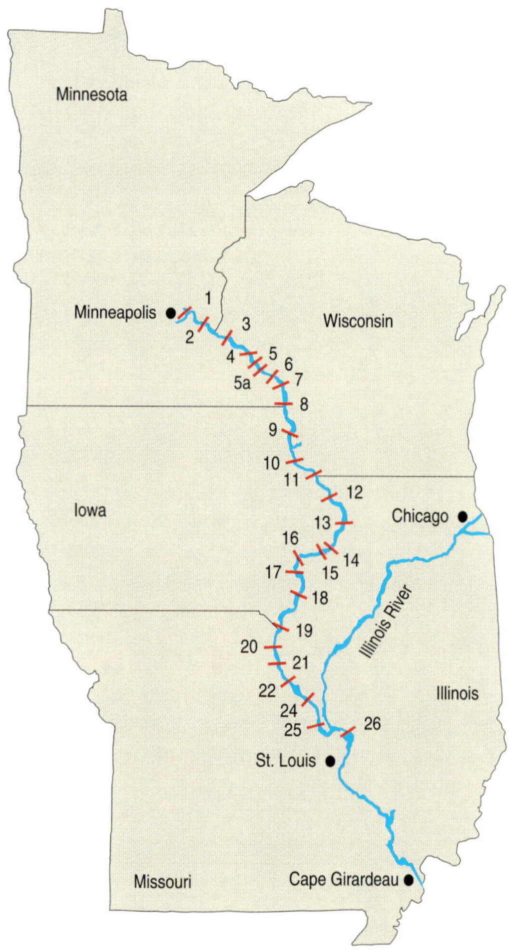

Fig. 1. The Upper Mississippi River. Numbers indicate reservoirs formed by navigation dams and known locally as pools.

from ground surveys conducted by the Mississippi River Commission. High-resolution land-cover/land-use data were created from 1:15,000 (scale) color infrared aerial photographs taken in 1989. Data for 1891 and 1989 were compared for Pools 4, 5, 8, 13, 26, and for a 64-km (40-mi) stretch of river, near Cape Girardeau, Missouri, which is not affected by navigation dams. Historical aerial photographs from 1939, 1954, 1967, and 1989 were used to measure island loss in an area just upriver of the dam in Pool 8.

Long-term daily data at three stations on the open-river portion of the UMR were analyzed to evaluate changes in the relationship between discharges and water-surface elevations.

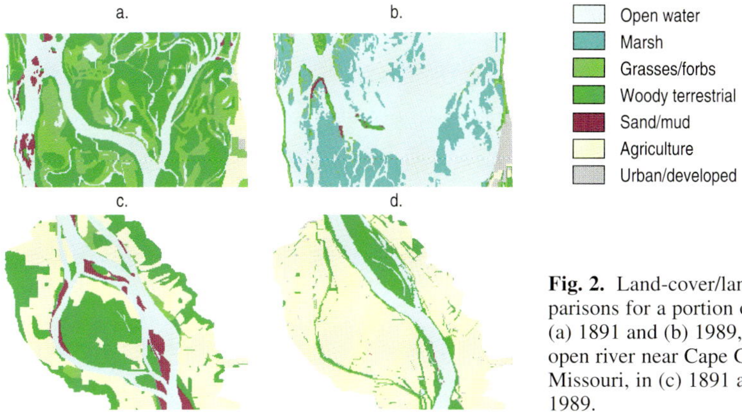

Fig. 2. Land-cover/land-use comparisons for a portion of Pool 8 in (a) 1891 and (b) 1989, and the open river near Cape Girardeau, Missouri, in (c) 1891 and (d) 1989.

Status and Trends

Comparison of the land-cover/land-use data between 1891 and 1989 in the dammed portion of the UMR showed that open water and marsh habitats generally increased, mostly at the expense of grass/forb, woody terrestrial, and agricultural classes. For example, the combined classes of open water and marsh in Pool 8 have increased from 3,600 ha (8,900 acres) in 1891 to 9,500 ha (23,430 acres) in 1989 (Figs. 2a, b). Similar increases in these two classes were found at Pools 5 and 13. In Pools 4 and 26 increases were less significant.

In many pools inundation created an impounded area with a mosaic of islands, open water, and marsh, which, in general, increased aquatic habitat for fish and wildlife. Although dam construction has benefited aquatic habitat in many pools, the reservoir aging process has reduced these benefits, especially in areas just upriver of dams. For example, island areas have been steadily eroding upriver of the dam in Pool 8 (Fig. 3). The dam that forms Pool 8 began operating in 1937, and photographs taken 2 years later showed 253 ha (624 acres) of islands. By 1989 the island area in the same location was reduced by 79% to 52 ha (129 acres).

Sedimentation is also a major concern on the UMR; rates of 1 to 3 cm/yr (0.4-1.2 in/yr) have been measured (McHenry et al. 1984). Erosion and sedimentation were both detected in comparisons between present elevation data and surveys before dam construction. Erosion was more prevalent in shallow areas and sedimentation more prevalent at greater depths. Erosion and sedimentation converge at depths of between 0.9 and 1.5 m (3 to 5 ft). This has resulted in a more homogeneous distribution of depth, which is dominated by areas 0.9 to 1.5 m (3 to 5 ft) in depth. Similar frequency distributions of water depth were observed for lower portions of Pools 8 and 13. Comparison of historical and present bottom geometry revealed the loss of elevational diversity.

In areas of the UMR unaffected by navigation dams (the 40-mi stretch of river near Cape Girardeau), there was a 28% reduction in open water and a 38% reduction in woody and terrestrial habitat between 1891 and 1989 (Figs. 2c, d). Agricultural areas increased by 6,360 ha (15,700 acres). The 1,900-ha (4,710-acre) reduction of open water can be explained by the construction of levees and wing dams (also known as pile dikes). One large side channel that existed in 1891 was cut off by construction of a levee, reducing the area of water by 550 ha (1,350 acres). In all, nearly 2,000 km (1,240 mi) of levees now isolate more than 400,000 ha (988,000 acres) from the river during all but the highest discharge rates.

Wing dams and levees, along with other changes to the watershed, have also had a major effect on habitats by changing the relationship between discharge and water-surface elevations. Wing dams have narrowed and deepened the main channel so that water elevations at low discharges are now lower than they were historically. Levees restrict flows and result in higher

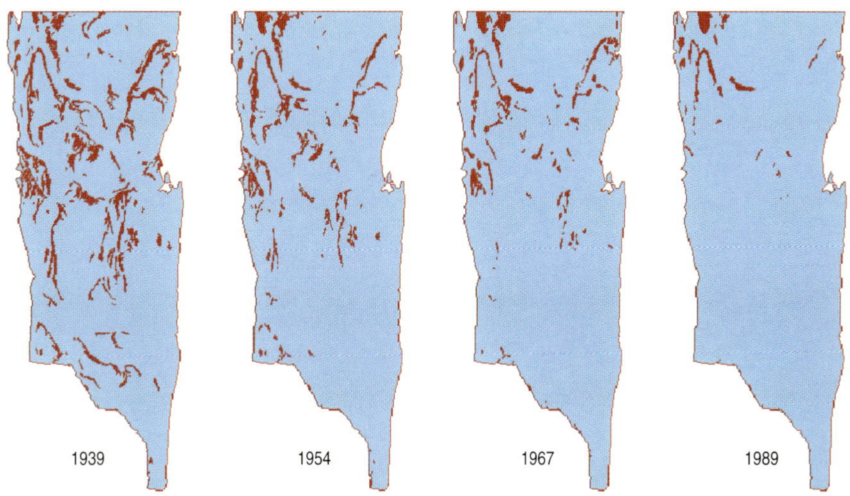

Fig. 3. Island loss that has occurred in Pool 8, in the area just upriver of the dam, since construction of the lock and dam system.

water elevations during high discharges. Water-surface elevations at relatively low discharges (60,000 cfs) have dropped about 2.4 m (8.0 ft) over the record 133-year period at St. Louis, Missouri, 0.5 m (1.5 ft) over the 52-year record at Chester, Illinois, and 1.5 m (5.0 ft) over the 60-year record at Thebes, Illinois. Water-surface elevations at relatively high discharges (780,000 cfs), however, have risen about 2.7 m (9 ft) over the record period at St. Louis, 1.5 m (5.0 ft) at Chester, and 1.1 m (3.6 ft) at Thebes.

For further information:

Joseph Wlosinski
National Biological Service
Environmental Management
Technical Center
575 Lester Ave.
Onalaska, WI 54650

Reference

McHenry, J.R., J.C. Ritchie, C.M. Cooper, and J. Verdon. 1984. Recent rates of sedimentation in the Upper Mississippi River. Pages 99-118 *in* J.G. Wiener, R.V. Anderson, and D.R. McConville, eds. Contaminants in the Upper Mississippi River. Butterworth Publishers, Boston, MA.

Biota of the Upper Mississippi River Ecosystem

by
James Wiener
Teresa Naimo
Carl Korschgen
National Biological Service
Robert Dahlgren
U.S. Fish and Wildlife Service
Jennifer Sauer
Kenneth Lubinski
Sara Rogers
National Biological Service
Sandra Brewer
University of Wisconsin-LaCrosse

The Mississippi River is one of the world's major river systems in size, habitat and biotic diversity, and biotic productivity. The navigable Upper Mississippi River, extending 1,370 km (850 mi) from St. Anthony Falls (Minnesota) to the confluence with the Ohio River, has been impounded by 27 locks and dams to enhance commercial navigation. The reach between two consecutive locks and dams is termed a "pool." The upstream portions of many pools are similar to the unimpounded river, whereas the downstream reaches are similar to reservoirs.

The Upper Mississippi contains a diverse array of wetland, open-water, and floodplain habitats, including extensive national wildlife and fish refuges. Human activities, though, have greatly altered this river ecosystem; much of the watershed is intensively cultivated, and many tributary streams deliver substantial loads of nutrients, pesticides, and sediment from farmland. Pollutants also enter the river from point sources.

We examine recent temporal trends in the abundance of several key groups of organisms in the Upper Mississippi River and show that certain flora and fauna have declined along substantial reaches of the river. Our analysis is spatially constrained by available data to the reach of river extending from Pool 2 (near Minneapolis-St. Paul, Minnesota) to Pool 19 (near Keokuk, Iowa).

Information on the abundance of selected riverine biota was obtained by compiling historical data and by censusing or sampling. Bottom-dwelling invertebrates, collectively termed benthic macroinvertebrates, were extensively studied. Some of these organisms are important in the diets of fish and wildlife and are useful as biological indicators of toxic pollution. Data on densities of fingernail clams (*Musculium transversum*) from 1973 to 1992 were obtained from regional scientists, published literature, and sampling (Wilson et al. 1994) in selected pools (pools are numbered consecutively from upstream to downstream).

In 1992 benthic macroinvertebrates were sampled in soft sediments of five reaches of the Upper Mississippi and one reach of the Illinois River to estimate densities of fingernail clams and the burrowing mayfly (*Hexagenia*). In 1975 and 1990, benthic macroinvertebrates were extensively sampled in five habitats (marsh, bay, open water, side channel, and dredged side channel) in Pool 8 of the Upper Mississippi (near La Crosse, Wisconsin) to examine changes in abundance, biomass, and community structure (Brewer 1992). Data on the unionid mussel fauna in the river were obtained from the literature and other sources (Shimek 1921; Grier and Mueller 1922; Ellis 1931a, b; Dawley 1947; Finke 1966; Coon et al. 1977; Fuller 1978, 1979; Mathiak 1979; Perry 1979; Thiel et al. 1979; Ecological Analysts, Inc. 1981; Thiel 1981; Duncan and Thiel 1983; Holland-Bartels 1990).

Wildcelery plants (*Vallisneria americana*) were sampled each August during 1980-84 and 1989-93 in quadrants along 12 0.8-km (0.5-mi) transects in Lake Onalaska, a backwater lake in Pool 7. Numbers of canvasback ducks (*Aythya valisineria*) in Pool 7 during fall migration were determined by aerial surveys. Trends in the abundance of mink (*Mustela vison*) were assessed by examining indices of mink harvest per unit of trapping effort (total harvest/total number of trappers) on the Upper Mississippi River National Wildlife and Fish Refuge and in states along the river corridor (Dahlgren 1990).

Status and Trends

Benthic Macroinvertebrates

Densities of fingernail clams declined significantly ($P \leq 0.05$) in five of eight pools examined (declines in Pools 2, 5, 7, 9, and 19; Figs. 1 and 2) along 700 km (435 mi) of river from Hastings, Minnesota, to Keokuk, Iowa. Densities in Pool 19, which had the longest historical record on fingernail clams, averaged 30,000/m^2 (2,800/ft^2) in 1985 and decreased to zero in 1990 (Fig. 2). In 1992 densities of fingernail clams were still low in sampled areas on the Upper Mississippi and Illinois rivers, averaging 5-94 individuals/m^2 (0.5-8.7/ft^2). Only

8% of 721 samples taken in 1992 had densities exceeding 100 fingernail clams/m^2 (9.3/ft^2). Corresponding mean densities of burrowing mayflies in these areas ranged from 10 to 99/m^2 (0.9-9.2/ft^2).

Wilson et al. (1994) hypothesized that the declines in fingernail clams in Pools 2 to 9 were linked to point-source pollution, and that the declines in Pool 19 were linked to low-flow conditions during drought. The causal mechanisms by which low flow influences fingernail clam abundance may involve unfavorable changes in the chemistry of sediment pore water.

In Pool 8, the structure of benthic macroinvertebrate communities changed between 1975 and 1990 in all five habitats studied. Standing crop of the benthos decreased significantly in both open-water and bay habitats, and diversity and abundance decreased in open-water habitat (Brewer 1992). These declines suggest that the standing crop of invertebrates has decreased substantially in Pool 8 because open-water habitat was 45% of the total area of the pool.

The biodiversity of the unionid mussel fauna in the Upper Mississippi River drainage has declined from about 50 to 60 species in the early 1920's to about 30 species in the mid-1980's. Many of these species are commercially important; others are threatened or endangered. Unionid mussels are further imperiled by the zebra mussel (*Dreissena polymorpha*), which recently invaded the Illinois and Upper Mississippi rivers.

Rooted Aquatic Plants

The abundance of submersed aquatic plants—including wildcelery, which produces a vegetative tuber important as food for certain migratory waterfowl—declined along extensive reaches of the Upper Mississippi River in the late 1980's. This decline has been attributed to changing environmental conditions caused by the severe midwestern drought of 1988-89. In Pool 7, the abundance of wildcelery was fairly stable during 1980-84, but declined greatly after the dry summer of 1988. In Pools 5 through 9, more than 4,000 ha (10,000 acres) of wildcelery beds were lost (C.E. Korschgen, Upper Mississippi Science Center, unpublished data). Overall, the abundance of wildcelery and many other submersed plants declined along 600 km (375 mi) of river from Pool 5 to Pool 19. Coincidentally, the abundance of the exotic plant Eurasian watermilfoil (*Myriophyllum spicatum*) has seemingly increased, particularly in locations formerly occupied by wildcelery or other native submersed plants.

Migratory Birds

Millions of migratory birds use the Mississippi River corridor during fall and spring migration. The river is critical in the life cycle of many migratory birds because of its north-to-south orientation and its nearly contiguous habitat. Diving ducks, swans, pelicans, and cormorants use the river's open waters. Dabbling ducks, geese, herons, egrets, terns, bitterns, rails, and many resident and Neotropical songbirds use the shallow riverine wetlands. Bottomland forests support migrating and nesting songbirds, and nesting raptors, herons, egrets, and waterfowl.

The primary factor affecting the use of the river ecosystem by birds is the production of food by various plants and animals. The number of birds in riverine habitats decreases rapidly if preferred food resources are unavailable. The use of Lake Onalaska (Pool 7) by canvasback ducks (*Aythya valisineria*), for example, decreased greatly when the abundances of their preferred foods, wildcelery and benthic invertebrates (Korschgen 1989), declined in the late 1980's (Fig. 3). A gradual increase in foods in 1992 resulted in increased use by canvasbacks (C.E. Korschgen, unpublished data).

Numbers of other migratory waterfowl have also decreased along the river corridor, reflecting deterioration of habitat on the breeding grounds and the river. The decrease in the abundance of fingernail clams has adversely affected waterfowl that feed heavily on the small mollusk, particularly lesser scaup (*Aythya affinis*). The peak number of lesser scaup on Pool 19 during fall migrations, for example, has decreased from 300,000-500,000 in the 1970's to fewer than 25,000 in 1993.

Mink

The abundance of mink (*Mustela vison*) on the Upper Mississippi River Refuge declined precipitously during 1959-65, remained low until about 1970, and then began to slowly increase to numbers that are now less than half those of the 1950's (Dahlgren 1990). In contrast, mink populations in the adjoining states of Iowa, Minnesota, and Wisconsin were relatively stable during this period and did not exhibit the pattern of decline and partial recovery seen in populations on the refuge. These patterns indicate that some factor unique to the river corridor, not present in the mostly agricultural watersheds of the adjoining states, caused the decline of mink populations on the refuge.

The survival and reproduction of mink are adversely affected by dietary exposure to small doses of polychlorinated biphenyls, (PCBs; Aulerich and Ringer 1977; Wren 1991). The

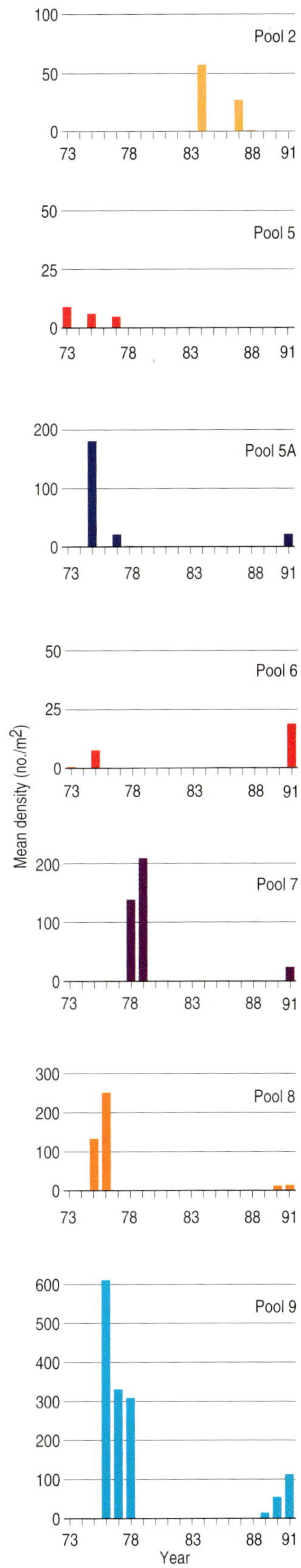

Fig. 1. Mean density of the fingernail clam *Musculium transversum* in Pools 2, 5, 5A, 6, 7, 8, and 9 of the Upper Mississippi River during 1973-91.

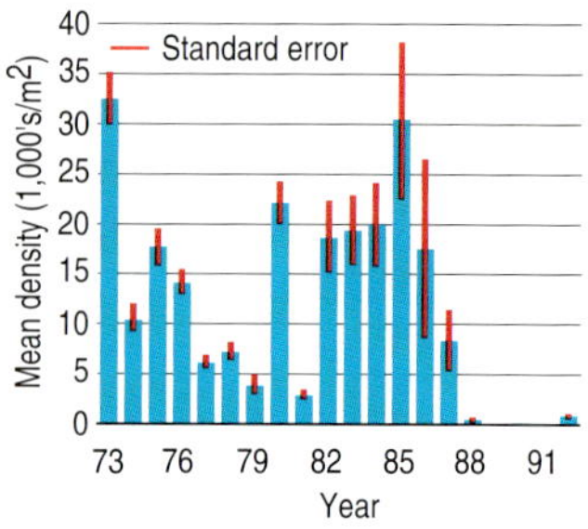

Fig. 2. Mean densities (±1 standard error) of the fingernail clam *Musculium transversum* in Pool 19 of the Upper Mississippi River during 1973-91. Mean densities that were too low to appear in the chart are 1989 (17 clams/m^2), 1990 (0), and 1991 (18 clams/m^2).

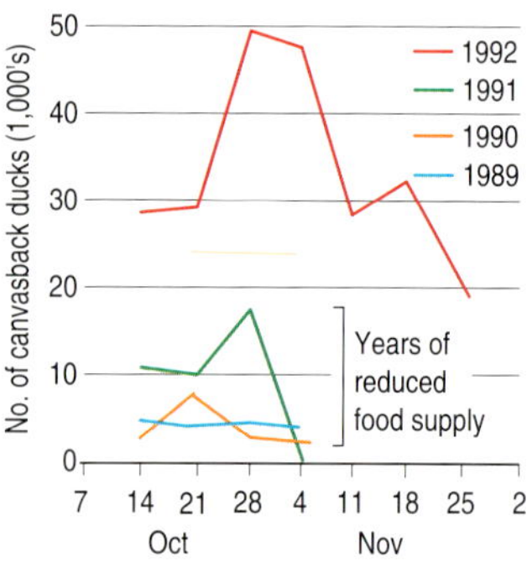

Fig. 3. Numbers of canvasback ducks (*Aythya valisineria*) in Pool 7 of the Upper Mississippi River during four consecutive fall migrations (1989-92) varied in relation to the abundance of plant and animal food in the pool.

decline of mink on the refuge coincided with the probable period of most severe PCB contamination in the river. Conversely, the partial recovery of mink populations that began in the late 1970's coincided with a period of declining PCB levels in riverine fishes (Hora 1984). R.B. Dahlgren and K.L. Ensor (U.S. Fish and Wildlife Service, personal communication) estimate that a diet containing 33% fish, having PCB concentrations similar to those in the early 1970's, would contain enough PCB to prevent reproduction in mink, based on experimental toxicity studies (Platonow and Karstad 1973). In 1989-91, PCB concentrations in mink from the Upper Mississippi River in Minnesota exceeded those in mink from all other areas of the state except Lake Superior (Ensor et al. 1993). Recent studies show that PCBs continue to enter or cycle within the riverine ecosystem and that they are transferred from the sediment to higher trophic levels via the benthic food chain (Steingraeber et al. 1994).

Ecosystem Health

The declines in these riverine flora and fauna signal a deterioration in the health of this ecosystem. In recent decades, populations of fingernail clams, unionid mussels, certain other invertebrates, submersed vegetation, migratory waterfowl, and mink have decreased along extensive reaches of the river. The Upper Mississippi is often heralded as a multiple-use resource, and human use of the river for navigation, hydropower, discharge of wastes, and other purposes may increase while inputs of sediment, nutrients, and chemicals from the watershed continue. Yet the cumulative impacts of humans may already exceed the assimilative capacity of this ecosystem.

Many complex questions concerning environmental degradation, declining flora and fauna, and human impact on this ecosystem need objective analysis and resolution. It is suspected that mink populations declined in response to PCB contamination and that fingernail clams declined in response to sediment toxicity, perhaps linked to low-flow conditions during droughts (Wilson et al. 1994). The factors causing most of the observed biotic declines are largely unknown, however, hampering the application of corrective measures. Several factors, for example, are suspected of contributing to declines in the unionid mussel fauna, including habitat modification and degradation, contaminants, overharvest, commercial and recreational navigation, and poor water quality (Williams et al. 1993). The need for scientifically based, integrated resource management of the Upper Mississippi is illustrated by the economic and ecological effects of the flood of 1993 on the river floodplain and its inhabitants. Federal and state agencies involved with resource management need integrated, proactive policies based on an understanding of the ecological structure and functioning of this complex ecosystem.

References

Aulerich, R.J., and R.K. Ringer. 1977. Current status of PCB toxicity to mink, and effect on their reproduction. Archives of Environmental Contamination and Toxicology 6:279-292.

Brewer, S.K. 1992. Community structure of benthic macroinvertebrates in Navigation Pool no. 8, Upper Mississippi River: comparisons between 1975 and 1990. M.S. thesis, University of Wisconsin-La Crosse. 84 pp.

Coon, T.G., J.W. Eckblad, and P.M. Trygstad. 1977. Relative abundance and growth of mussels (Mollusca: Eulamellibranchia) in Pools 8, 9, and 10 of the Mississippi River. Freshwater Biology 7:279-285.

Dahlgren, R.B. 1990. Fifty years of fur harvest on the Upper Mississippi River National Wildlife and Fish Refuge: consistencies, anomalies, and economics. Pages 142-160 *in* Proceedings of the 46th Annual Meeting of the Upper Mississippi River Conservation Committee, Rock Island, IL.

Dawley, C. 1947. Distribution of aquatic mollusks in Minnesota. American Midland Naturalist 38:671-697.

Duncan, R., and P.A. Thiel. 1983. A survey of the mussel densities in Pool 10 of the Upper Mississippi River. Wisconsin Department of Natural Resources Tech. Bull. 139. 14 pp.

Ecological Analysts, Inc. 1981. Survey of freshwater mussels (Pelecypoda: Unionacea) at selected sites in Pools 11 through 24 of the Mississippi River. Final Rep. 9031. U.S. Army Corps of Engineers, Rock Island, IL. 188 pp.

Ellis, M.M. 1931a. A survey of conditions affecting fisheries in the Upper Mississippi River. U.S. Bureau of Fisheries Circular 5:1-18.

Ellis, M.M. 1931b. Some factors affecting the replacement of the commercial fresh-water mussels. U.S. Bureau of Fisheries Circular 7:1-10.

Ensor, K.L., W.C. Pitt, and D.D. Helwig. 1993. Contaminants in Minnesota wildlife 1989-1991. Minnesota Pollution Control Agency, St. Paul. 75 pp.

Finke, A.H. 1966. Report of a mussel survey in Pools 4 (Lake Pepin), 5, 6, 7, and 9 of the Upper Mississippi River during 1965. Wisconsin Department of Natural Resources, La Crosse. 5 pp.

Fuller, S.L.H. 1978. Fresh-water mussels (Mollusca: Bivalvia: Unionidae) of the Upper Mississippi River: observations at selected sites within the 9-foot channel navigation project on behalf of the U.S. Army Corps of Engineers. Final Rep. 78-33. Academy of Natural Sciences, Philadelphia, PA. 400 pp.

Fuller, S.L.H. 1979. Historical and current distribution of fresh-water mussels (Mollusca: Bivalvia: Unionidae) in the Upper Mississippi River. Pages 71-119 *in* J.L. Rasmussen, ed. Proceedings of the UMRCC Symposium on Upper Mississippi River Bivalve Mollusks. Upper Mississippi River Conservation Committee, Rock Island, IL.

Grier, N.M., and J.F. Mueller. 1922. Notes on the naiad fauna of the Upper Mississippi River. II. The naiades of the Upper Mississippi drainage. Nautilus 36:46-49, 96-103.

Holland-Bartels, L.E. 1990. Physical factors and their influence on the mussel fauna of a main channel border habitat of the Upper Mississippi River. Journal of the North American Benthological Society 9:327-335.

Hora, M.E. 1984. Polychlorinated biphenyls (PCBs) in common carp (*Cyprinus carpio*) of the Upper Mississippi River. Pages 231-239 *in* J.G. Wiener, R.V. Anderson, and D.R. McConville, eds. Contaminants in the Upper Mississippi River. Butterworths, Stoneham, MA.

Korschgen, C.E. 1989. Riverine and deepwater habitats for diving ducks. Pages 157-180 *in* L.M. Smith, R.L. Peterson, and R.M. Kaminski, eds. Habitat management for migrating and wintering waterfowl in North America. Texas Tech University Press, Lubbock.

Mathiak, H.A. 1979. A river survey of the unionid mussels of Wisconsin, 1973-1977. Sand Shell Press, Horicon, WI. 75 pp.

Perry, E.W. 1979. A survey of Upper Mississippi River mussels. Pages 118-139 *in* J.L. Rasmussen, ed. A compendium of fishery information on the Upper Mississippi River. 2nd ed. Upper Mississippi River Conservation Committee, Rock Island, IL.

Platonow, N.S., and L.H. Karstad. 1973. Dietary effect of polychlorinated biphenyls on mink. Canadian Journal of Comparative Medicine 37:391-400.

Shimek, B. 1921. Mollusks of the McGregor, Iowa, Region I. Iowa Conservationist 5:1.

Steingraeber, M.T., T.R. Schwartz, J.G. Wiener, and J.A. Lebo. 1994. Polychlorinated biphenyl congeners in emergent mayflies from the Upper Mississippi River. Environmental Science and Technology 28:707-714.

Thiel, P.A. 1981. Survey of unionid mussels in the Upper Mississippi River (Pools 3-11). Tech. Bull. 124. Wisconsin Department of Natural Resources, Madison. 24 pp.

Thiel, P.A., M. Talbot, and J. Holzer. 1979. Survey of mussels in the Upper Mississippi River, Pools 3 through 8. Pages 148-156 *in* J.L. Rasmussen, ed. Proceedings of the UMRCC Symposium on Upper Mississippi River Bivalve Mollusks. Upper Mississippi River Conservation Committee, Rock Island, IL.

Williams, J.D., M.L. Warren, Jr., K.S. Cummings, J.L. Harris, and R.J. Neves. 1993. Conservation status of freshwater mussels of the United States and Canada. Fisheries 18:6-22.

Wilson, D.M., T.J. Naimo, J.G. Wiener, R.V. Anderson, M.B. Sandheinrich, and R.E. Sparks. 1994. Declining populations of the fingernail clam *Musculium transversum* in the Upper Mississippi River. Hydrobiologia. In press.

Wren, C.D. 1991. Cause-effect linkages between chemicals and populations of mink (*Mustela vison*) and otter (*Lutra canadensis*) in the Great Lakes Basin. Journal of Toxicology and Environmental Health 33:549-585.

For further information:

James Wiener
National Biological Service
Upper Mississippi Science Center
PO Box 818
La Crosse, WI 54602

Fish Populations in the Illinois River

by
Thomas V. Lerczak
Richard E. Sparks
Illinois Natural History Survey

The Illinois River is formed by the confluence of the Des Plaines and Kankakee rivers, about 80 km (50 mi) southwest of Chicago, Illinois. It then flows 439 km (273 mi) to join with the Mississippi River about 50 km (31 mi) northwest of St. Louis, Missouri (Fig. 1). The Illinois River has been extensively modified and degraded by industrial and municipal pollution for most of this century (Mills et al. 1966). The upper river reaches above the Starved Rock Dam (Fig. 1) became the most degraded because most of this pollution originated in the densely populated and heavily industrialized Chicago metropolitan area. In fact, by the late 1920's, the upper river was thought devoid of fish (Thompson 1928). Soon after this period, as pollution-control efforts began to have an effect, fish gradually returned.

Changes in the composition of a fish community in a polluted environment can be a useful index for assessing environmental health and the effectiveness of pollution control because different fish species vary in their ability to tolerate effects of pollution. In 1957 the Illinois Natural History Survey (INHS) initiated an annual electrofishing survey of the Illinois River to monitor fish populations. A central purpose of the survey was to relate changes in fish populations to environmental conditions. This article summarizes trends in fish populations of the upper Illinois River as determined from electrofishing catches from 1959 to 1993.

Status and Trends

Fish sampling was conducted at five stations in the upper Illinois River and at two stations in the Des Plaines River (Fig. 1) from late August through October. Data from these seven fixed stations were combined for analyses. At each station, fish were sampled by electrofishing for 1 hr; thus, catches are expressed as number of

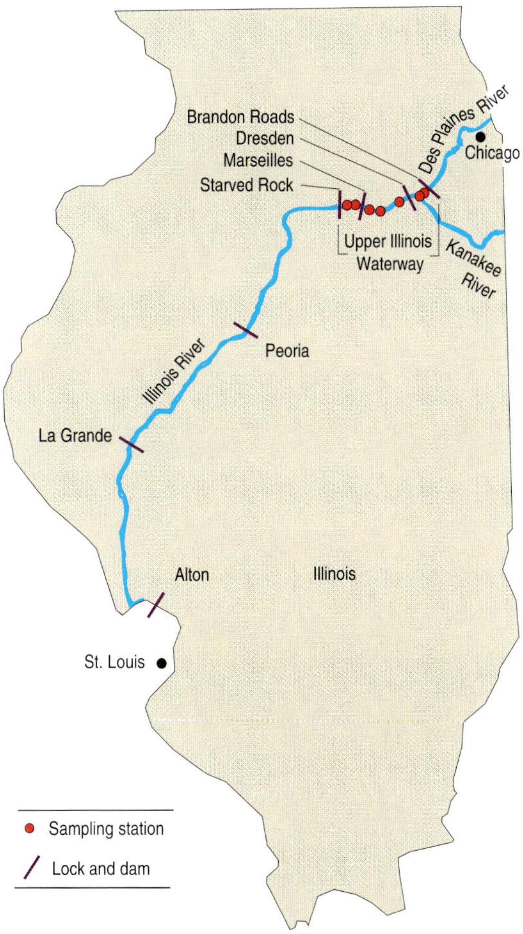

Fig. 1. The Illinois River with locations of navigation locks and dams. Locations of the Illinois Natural History Survey's upper Illinois Waterway electrofishing stations are also shown.

fish obtained per hour of sampling. Fish were stunned in an electric field, gathered with a net, measured, checked for externally visible abnormalities (sores, eroded fins, etc.), and returned to the water. The same methods and similar equipment have been used for all years of the survey to allow comparability of data among years.

Fishes of the family Centrarchidae (e.g., largemouth bass [*Micropterus salmoides*], bluegill [*Lepomis macrochirus*]) were treated as a group to simplify data analysis because they have very similar habitat requirements and are generally considered intolerant of polluted conditions. (The green sunfish [*Lepomis cyanellus*], however, is usually indicative of a stressed environment [Karr et al. 1986].) Also, because many of these fishes are piscivorous, their presence or absence will have a direct impact on overall fish community composition. Catches of common carp (*Cyprinus carpio*) and goldfish (*Carassius auratus*), both non-natives to North America, and their hydrids were analyzed separately. These two species are omnivorous habitat generalists that are tolerant of polluted waters.

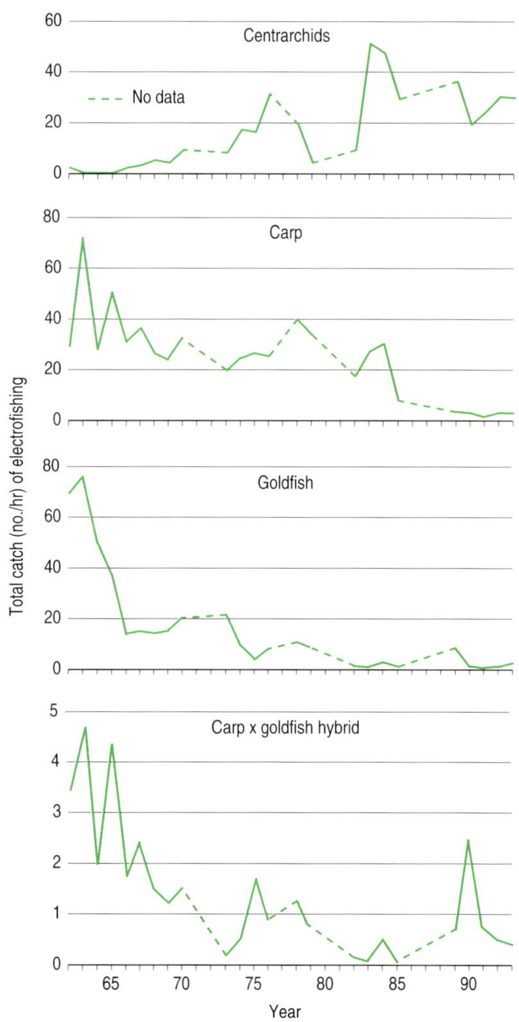

Fig. 2. Number of individual fish obtained per hour of electrofishing from the upper Illinois Waterway for fish species identified as pollution indicators. Dashed lines connect data points for years between which electrofishing was not conducted.

Sediments of the upper Illinois River contain varying amounts of toxic substances (IEPA 1992), which are thought to contribute to the incidence of abnormalities on fishes that forage in sediments while minimally affecting fishes that forage in the water column. To test this hypothesis, all fishes were assigned to one of two groups: benthic species that frequently forage in bottom sediments (e.g., common carp) and pelagic species that usually inhabit the water column (e.g., bluegill).

Substantial changes have occurred in catch rates from the seven upper Illinois Waterway stations between 1962 and 1993 (Fig. 2). Catches of centrarchids have increased (D = 327, $P < 0.001$) since the early 1960's, peaking at 52/hr in 1983. Catches of centrarchids appear to have stabilized during the last 5 years (Fig. 2), indicating their populations may have reached carrying capacity. Catches have decreased for carp ($P < 0.001$), goldfish ($P < 0.001$), and carp x goldfish hybrids ($P < 0.001$) since the early 1960's. Carp were able to maintain their numbers until the mid-1980's (Fig. 2), as larger, older individuals probably died off and smaller, younger individuals were more vulnerable to predation by piscivores. Catches of goldfish declined rather precipitously from 1963 to 1966 for an unknown reason before substantial increases in centrarchids.

Data from 1963 and 1992 were chosen for more detailed examination, those years being representative of catches from early and recent years of the electrofishing survey. In 1963, goldfish accounted for almost one-third of all fish collected per hour, followed by carp, emerald shiner (*Notropis atherinoides*), and gizzard shad (*Dorosoma cepedianum*); together these four species dominated the catch, accounting for 95.8% of all individuals collected per hour (Fig. 3). In 1992, 13 species accounted for 95.4% of all fish collected per hour: emerald shiners were most abundant followed by centrarchids; carp and goldfish were reduced to a minor component (Fig. 3). The increase in centrarchids and decrease in carp and goldfish since the early 1960's (Fig. 2) reflect a more diverse fish community in recent years (Fig. 3).

For all years when data were collected from 1959 to 1993, the percentages of fish with external abnormalities were higher on benthic fishes than on pelagic fishes, suggesting that sediments may contain significant amounts of contaminants. In fact, the IEPA (1992) identified several locations near our electrofishing stations on the upper Illinois and Des Plaines rivers as having sediments that contained elevated levels of toxicants, including mercury, lead, and PCBs. Brown et al. (1973) reported, however, that benthic fishes had a higher frequency of tumors than pelagic fishes (1.7% and 1.0%, respectively) even when collected from a relatively unpolluted Canadian watershed. Both groups of fishes, though, had higher rates of tumors in the polluted Fox River of northeastern Illinois (benthic fishes, 7.0%; pelagic fishes, 3.0%) than in the Canadian system (Brown et al. 1973). Hughes and Gammon (1987) noted that increasing pollution seems correlated with an increase in the incidence of abnormalities on fishes of the Willamette River in Oregon. Likewise, Tyler and Everett (1993) reported that bottom-dwelling barbel (*Barbus barbus*) collected from polluted rivers in England had a higher incidence of abnormalities than those collected from a clean river. Therefore, the relationship between a high incidence of abnormalities on fish and polluted waters has been well established. On the upper Illinois River, there was a marginal trend of decreasing incidence of abnormalities against years for pelagic fishes since the early 1960's ($D = 3{,}156$; $P < 0.05$), coincident with known improvements in water quality over the same period (Butts 1987), but not for benthic fishes ($D = 1{,}937$; $P = 0.23$).

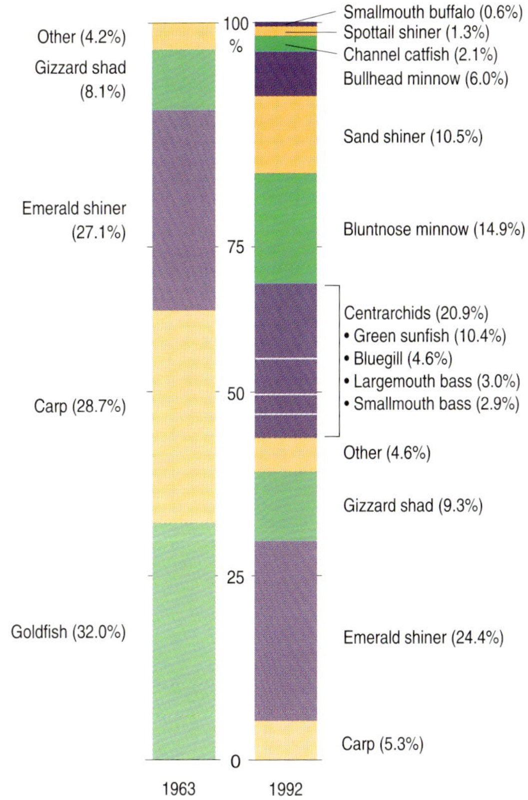

Fig. 3. Composition of catches (%) for the upper Illinois Waterway for 1963 and 1992, based on number of individuals collected per hour of electrofishing.

Conclusions

Long-term trends of fish populations in the upper Illinois River reflect improved water quality in recent years as compared with the early 1960's. This trend is consistent with data presented in other studies that showed improved water quality in the upper Illinois River (Butts 1987; Lerczak et al. 1992). The increased incidence of external abnormalities between bottom-foraging fishes compared with pelagic fishes suggests contaminated sediments (Essig 1991; IEPA 1992).

Because recovery of fish populations in the upper Illinois Waterway appears to be a response to pollution-control efforts, definite restoration goals should be identified to help guide further recovery and to determine expectations. In addition, the specific causes for the high incidence of abnormalities in benthic fishes need to be explicitly identified.

References

Brown, E.R., J.J. Hazdra, L. Keith, I. Greenspan, and J.B.G. Kwapinski. 1973. Frequency of fish tumors found in a polluted watershed as compared to nonpolluted Canadian waters. Cancer Res. 33:189-197.

Butts, T.A. 1987. Illinois River water quality. Past, present, future. Pages 195-209 *in* H. Korab, ed. Proceedings Governor's Conference on the Management of the Illinois River System: the 1990's and Beyond. Special Rep. 16. Water Resources Center, University of Illinois, Urbana.

Essig, H.W. 1991. Chemical and biological monitoring of the upper Illinois River. Pages 68-77 *in* H. Korab, ed. Proceedings 1991 Governor's Conference on the Management of the Illinois River System: the 1990's and Beyond. Special Rep. 19. Water Resources Center, University of Illinois, Urbana.

Hughes, R.M., and J.R. Gammon. 1987. Longitudinal changes in fish assemblages and water quality in the Willamette River, Oregon. Transactions of the American Fisheries Society 116:196-209.

IEPA. 1992. Illinois water quality report, 1990-1991. Illinois Environmental Protection Agency, Springfield. 405 pp.

Karr, J.R., K.D. Fausch, P.L. Angermeier, P.R. Yant, and I.J. Schlosser. 1986. Assessing biological integrity in running waters: a method and its rationale. Illinois Natural History Survey Special Publ. 5. 28 pp.

Lerczak, T.V., R E. Sparks, and K.D. Blodgett. 1992. The long-term Illinois River fish population monitoring program. Annual report to the Illinois Department of Conservation (F-101-R-3). Aquatic Ecology Tech. Rep. 92/9. Illinois Natural History Survey, Champaign. 51 pp.

Mills, H.B., W.C. Starrett, and F.C. Bellrose. 1966. Man's effect on the fish and wildlife of the Illinois River. Illinois Natural History Survey Biological Notes 57. 23 pp.

Thompson, D.H. 1928. The "knothead" carp of the Illinois River. Illinois Natural History Survey Bull. 17(8):285-320.

Tyler, C.R., and S. Everett. 1993. Incidences of gross morphological disorders in barbel (*Barbus barbus*) in three rivers in England. Journal of Fish Biology 43:739-748.

For further information:

Thomas V. Lerczak
Illinois Natural History Survey
River Research Laboratory
PO Box 590
Havana, IL 62644

Contaminant Trends in Great Lakes Fish

by
Robert J. Hesselberg
John E. Gannon
National Biological Service

The Great Lakes region is home to many large industrialized cities and extensive agricultural areas that produce and use an array of potentially toxic chemicals. Some of these chemicals entering the lakes' food chain have been related to environmental health problems including poor egg-hatching success, reproductive abnormalities, and birth defects in fish, fish-eating birds, and mammals. Tumors and other deformities in some fish and wildlife species are also attributed to exposure to toxic contaminants. In addition, fish consumption advisories are issued annually by the Great Lakes' states and the Province of Ontario for certain fish species and larger sizes of Great Lakes fish that accumulate toxic contaminants.

To measure progress in reducing chemicals in the Great Lakes ecosystem, the National Biological Service's (NBS's) Great Lakes Science Center began a contaminant trend-monitoring program in Lake Michigan in 1969. The program was expanded in 1977 to include all of the Great Lakes and additional species of fish through a cooperative agreement between the NBS Great Lakes Science Center and the U.S. Environmental Protection Agency (USEPA), Great Lakes National Program Office. Fish are sampled for this program from 12 sites. All sites were sampled annually through 1982 and thereafter were divided into odd- and even-year sampling regimes. Results from these long-term monitoring programs are extremely valuable in understanding the dynamics of contaminants, developing predictive models for contaminant trends, and determining the effectiveness of regulatory programs.

This article presents data from the top predators sampled during even years for the NBS/USEPA monitoring program, lake trout (*Salvelinus namaycush*) or walleye (*Stizostedion vitreum vitreum*, Lake Erie only). In addition, information is presented on locations in the Great Lakes where tumors and other deformities in fish have been observed, indicating potentially contaminated sediments.

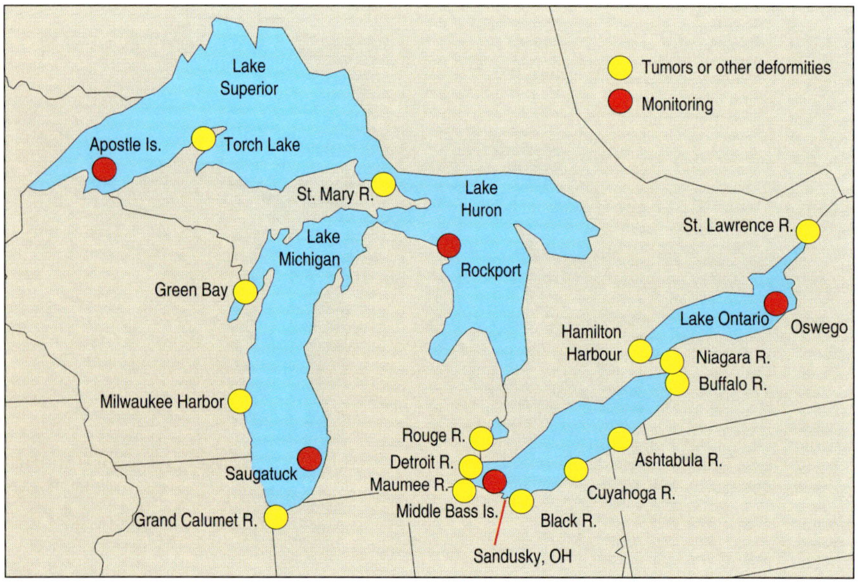

Fig. 1. Sampling sites for the NBS/USEPA Fish Contaminant Monitoring Program and "hot spots" of sediment contamination where tumors and other deformities have been detected in fish.

Methods

Lake trout from 600 to 700 mm (23.6-27.6 in) total length were collected from Lakes Superior, Michigan, Huron, and Ontario from even-year sampling sites by using gill nets (Fig. 1). Walleye from 400 to 500 mm (15.8-19.7 in) total length were collected from Lake Erie near Sandusky, Ohio, by using gill nets (Fig. 1). All fish were stored frozen until analyzed. Fish were prepared for analysis by thawing, compositing fish into five samples, and homogenizing. Contaminants were extracted and separated into nonpolar (polychlorinated biphenyls [PCBs]) and polar (DDT [sum of DDT, DDE, and DDD] and dieldrin) fractions and analyzed by a gas chromatograph equipped with an electron capture detector. Contaminants were reported as total DDT, total PCBs, and dieldrin.

Tumor surveys were conducted by the NBS Great Lakes Science Center and other agencies in highly industrialized rivers and harbors. Most of the work focused on the brown bullhead (*Ameiurus nebulosus*), a bottom-feeding fish especially exposed to tumor-causing chemicals in contaminated sediments.

Contaminant Trends

Results of DDT, PCB, and dieldrin trends during an approximately two-decade period are presented in Figs. 2-6. Data are from DeVault et al. 1985; Hesselberg et al. 1990; and DeVault and Hesselberg, in press. In general, concentrations of contaminants in fish consistently declined until the mid-1980's, but since then the downward trend has leveled off. Similar trends have been observed in fish in Canadian waters of the Great Lakes (Baumann and Whittle 1988).

Lake Michigan

Contaminants were higher in Lake Michigan lake trout than in fish of any of the other Great Lakes. Both total DDT and PCBs declined (Fig. 2), yet total PCBs did not decline after the voluntary control in 1972 but did after the mandatory ban in 1976.

In lake trout dieldrin reached a high in 1978 and a low in 1987 (Fig. 2). Dieldrin is higher in Lake Michigan fish than in fish from the other

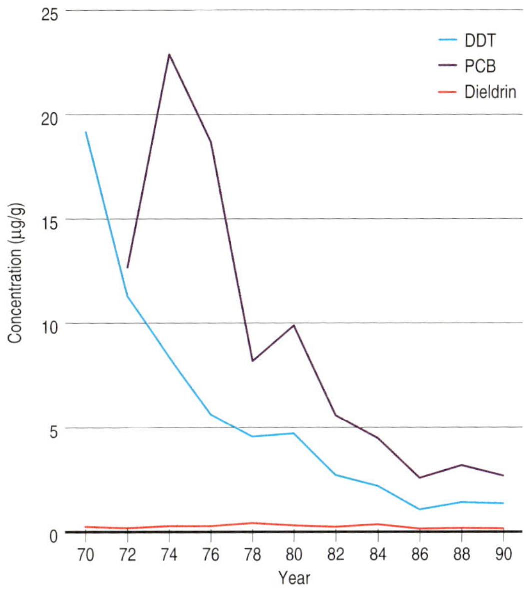

Fig. 2. Contaminant results from Lake Michigan lake trout, 1970-90.

Great Lakes, and changes in fish tissue concentrations do not follow use patterns for reasons that are not well understood.

Lake Superior

Total DDT and PCB concentrations in lake trout from Lake Superior were the lowest of all the Great Lakes and generally declined from 1977 to 1990 (Fig. 3). Dieldrin was always low and varied little from 1977 to 1990. Contaminant concentrations are lowest in Lake Superior because of the low density of agriculture and industry in the lake basin.

Lake Huron

Concentrations of total DDT and PCBs in lake trout from Lake Huron were intermediate between Lake Michigan and Lake Superior. Similar trends of declining concentrations of these chemicals were observed in Lake Huron (Fig. 4). Dieldrin concentrations were similar to Lake Superior but declined from a high in 1979 to a low by 1988. With the exception of the Saginaw Valley, both agriculture and industry are much less developed surrounding Lake Huron than Lake Michigan, thereby resulting in lower contaminants in Lake Huron fish.

Lake Ontario

The contaminants in Lake Ontario fish are relatively high (Fig. 5), second only to Lake Michigan. Trends in total DDT concentrations in lake trout from Lake Ontario were fairly constant from 1977 to 1990. Total PCBs in lake trout declined significantly from a high in 1977 to a low in 1990, a slower decline than in Lake Michigan. The relatively high contaminant concentrations in Lake Ontario fish are a result of the highly urbanized, industrial, and agricultural basin. In addition, it is the lowermost of the Great Lakes, receiving pollutants from upstream through the Niagara River. Dieldrin concentrations in lake trout from Lake Ontario reached a high in 1979 and then declined to a low by 1988.

Lake Erie

Total DDT, PCB, and dieldrin concentrations in Lake Erie walleye (Fig. 6) were lower and more similar to concentrations in lake trout in Lake Superior than those of other Great Lakes. Total DDT and PCBs peaked in 1977 and declined to a low in 1982; no consistent trend was noted for dieldrin. Low concentrations of contaminants in Lake Erie were similar to those in Lake Superior even though Lake Erie is surrounded by the largest urbanized, industrial, and agricultural basin of all the Great Lakes. Lake Erie, however, is the shallowest of all the Great Lakes and contains the highest amount of particulate matter. Contaminants flush more quickly through the shallow lake and are removed from the water column as they adhere to particulate matter and settle to the bottom. These factors work together in reducing the amount of contaminants available to fish in Lake Erie.

Contaminant Effects

Reduced reproductive success in fish-eating birds has been linked with DDT and PCBs (Giesy et al. 1994). As the concentrations of these contaminants have declined, populations of fish-eating birds such as the bald eagle (*Haliaeetus leucocephalus*) are beginning to recover in the Great Lakes basin. In lake trout, PCBs are also linked to reduced egg hatchability and may also be responsible for fry deformities and mortality (Mac et al. 1993). In spite of reductions in PCBs in lake trout in all of the Great Lakes, substantial natural reproduction occurs only in Lake Superior (Mac and Edsall 1991). The role of contaminants and other factors in lake trout reproductive problems in the other four Great Lakes is still under investigation.

Another fish health problem associated with toxic chemicals is found in Great Lakes harbors and tributaries where heavy industry was located (Baumann et al. 1991). Bottom sediments in these areas are heavily contaminated with polycyclic aromatic hydrocarbons (PAHs). Presence of liver tumors and other deformities such as lip papillomas, stubbed barbels, or skin discolorations in bottom-feeding fishes, such as the brown bullhead, have been linked to the

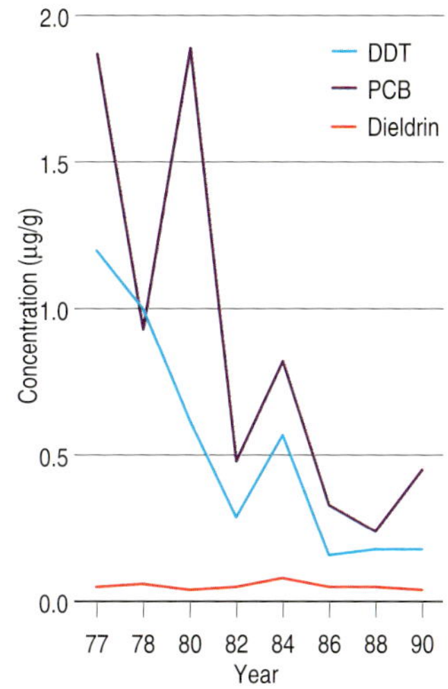

Fig. 3. Contaminant results from Lake Superior lake trout, 1977-90.

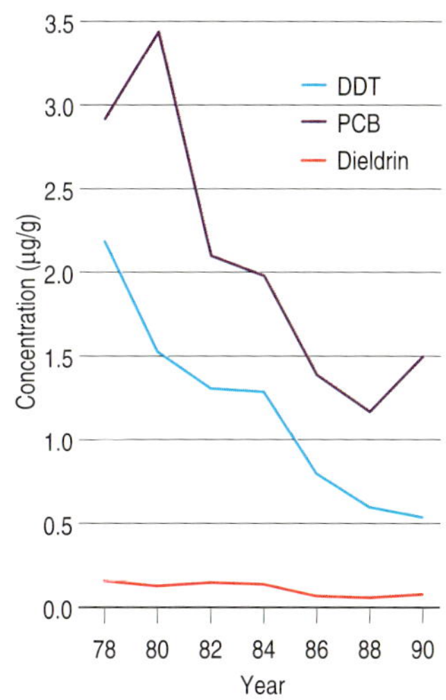

Fig. 4. Contaminant results from Lake Huron lake trout, 1977-90.

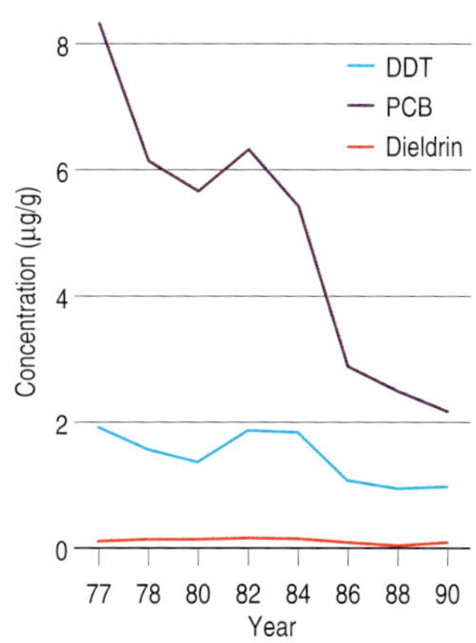

Fig. 5. Contaminant results from Lake Ontario lake trout, 1977-90.

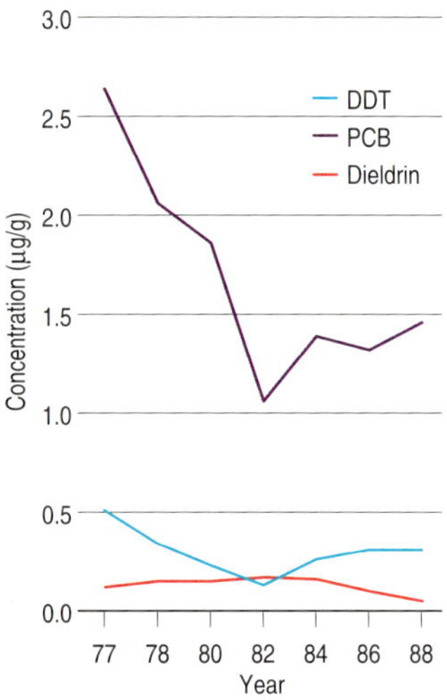

Fig. 6. Contaminant results from Lake Erie walleye, 1977-90.

presence of PAHs in the sediment (Baumann et al. 1991; Smith et al. 1994; Fig. 7). Tumors and other deformities have been detected in 15 locations (Hartig and Mikol 1992; Fig. 1).

Conclusions

The monitoring program for contaminants in Great Lakes fish has documented successful reduction of contaminants in response to usage bans for DDT and PCBs. Trends in dieldrin are less clear and concentrations of this pesticide remain especially high in Lake Michigan in comparison to the other Great Lakes. Fish communities are rebounding in some Great Lakes harbors, tributaries, embayments, and connecting channels that formerly were so contaminated that only the most pollution-tolerant organisms could survive. More reductions in contaminants are required, however. Monitoring results clearly indicate that the downward trend in contaminants leveled off in the mid-1980's, and resource-management agencies and research institutions are investigating the potential to further reduce sources of contamination in Great Lakes fish.

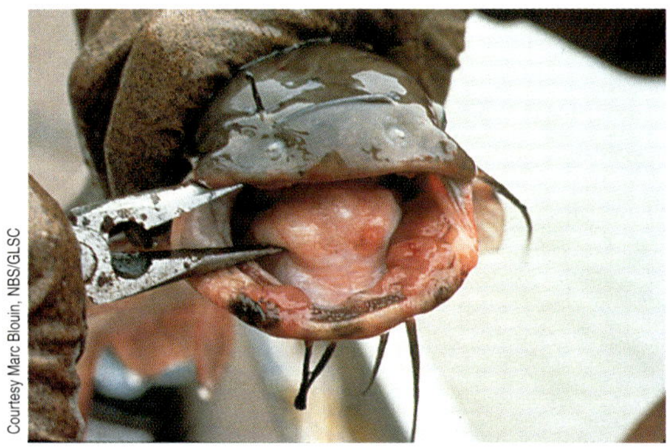

Fig. 7. Lip tumor and stubbed barbels on a brown bullhead.

Reproductive problems, tumors, and other deformities are still being detected in certain fish and wildlife populations in most of the Great Lakes. Similarly, consumption advisories recommending restrictions on eating certain species and sizes of Great Lakes fish still remain. The United States and Canada have agreed upon a virtual elimination policy for toxic contaminants under the auspices of the Great Lakes Water Quality Agreement. Remedial action plans are being developed by federal and state agencies in cooperation with local municipalities and local citizens to eliminate beneficial use impairments in the most contaminated rivers, harbors, and bays in the Great Lakes. Continued long-term monitoring of contamination in fish is required to determine the success of these programs and to guide where further corrective actions may be necessary.

For further information:

Robert J. Hesselberg
National Biological Service
Great Lakes Science Center
1451 Green Road
Ann Arbor, MI 48105

References

Baumann, P.C., M.J. Mac, S.B. Smith, and H.C. Harshbarger. 1991. Tumor frequencies in walleye (*Stizostedion vitreum*) and brown bullhead (*Ictalurus nebulosus*) and sediment contaminants in tributaries of the Laurentian Great Lakes. Canadian Journal of Fisheries and Aquatic Sciences 48:1804-1810.

Baumann, P.C., and D.M. Whittle. 1988. The status of selected organics in the Laurentian Great Lakes: an overview of DDT, PCBs, dioxins, furans, and aromatic hydrocarbons. Aquatic Toxicology 11:241-257.

DeVault, D.S., and R.J. Hesselberg. 1995. Contaminant trends in lake trout and walleye from the Laurentian Great Lakes. Journal of Great Lakes Res. 21. In press.

DeVault, D.S., W.A. Willford, R.J. Hesselberg. ,1985. Contaminant trends in lake trout (*Salvelinus namaycush*) from the upper Great Lakes. U.S. Environmental Protection Agency, Great Lakes National Program Office, Chicago, IL, EPA 905/3-85-001. 22 pp.

Giesy, J.P., J.P. Ludwig, and D.E. Tillitt. 1994. Deformities in birds of the Great Lakes region: assigning causality. Environmental Science and Technology 28:128A-135A.

Hartig, J., and G. Mikol. 1992. "How clean is clean?" An operational definition for degraded areas in the Great Lakes. Journal of Environmental Engineering and Management 2:15-23.

Hesselberg, R.J., J.P. Hickey, D.A. Nortrup, and W.A. Willford. 1990. Contaminant residues in the bloater (*Coregonus hoyi*) of Lake Michigan, 1969-1986. Journal of Great Lakes Research 16:121-129.

Mac, M.J., and C.C. Edsall. 1991. Environmental contaminants and the reproductive success of lake trout in the Great Lakes: an epidemiological approach. Journal of Toxicology and Environmental Health 33:375-394.

Mac, M.J., T.R. Schwartz, C.C. Edsall, and A.M. Frank. 1993. PCBs in Great Lakes lake trout and their eggs: relations to survival and congener composition 1979-1988. Journal of Great Lakes Research 19:752-765.

Smith, S.B., M.A. Blouin, and M.J. Mac. 1994. Ecological comparisons of Lake Erie tributaries with elevated incidence of fish tumors. Journal of Great Lakes Res. 20:701-716.

Lake Trout in the Great Lakes

by

Michael J. Hansen
National Biological Service

James W. Peck
Michigan Department of Natural Resources

Lake trout (*Salvelinus namaycush*) populations in the Great Lakes collapsed catastrophically during the 1940's and 1950's because of excessive predation by the sea lamprey (*Petromyzon marinus*) and exploitation by fisheries. The lake trout was the top-level predator in most of the Great Lakes as well as an important species harvested by commercial fisheries. Interagency efforts to restore lake trout into the Great Lakes included comprehensive control of sea lamprey populations (Smith 1971), regulation of commercial and recreational fisheries, and stocking (Eschmeyer 1968). It was hoped that without sea lamprey predation and fishery exploitation, stocked lake trout could reproduce and eventually restore wild lake trout populations in each of the Great Lakes. Lake trout restoration began during the 1950's in Lake Superior (Hansen et al. 1995), the 1960's in Lake Michigan (Holey et al. 1995), the 1970's in Lake Huron (Eshenroder et al. 1995) and Lake Ontario (Elrod et al. 1995), and the 1980's in Lake Erie (Cornelius et al. 1995).

Long-term monitoring of lake trout populations relied on catch records of commercial fisheries before the populations collapsed. Later monitoring of lake trout populations relied on assessment fisheries to measure the increase in abundance of stocked fish and, subsequently, of naturally produced fish. At present, natural reproduction by lake trout has been widespread only in Lake Superior. In contrast, lake trout reproduced in only limited areas of Lakes Huron, Michigan, and Ontario, and only in Lake Huron have progeny survived to adulthood. We describe the historical collapse and subsequent restoration of lake trout populations in U.S. waters of Lake Superior. We also describe the limited natural reproduction that has occurred in the other Great Lakes.

We compiled data describing abundance trends of lake trout in Lake Superior during 1929-93, expressed as the number of fish caught in a specified length of gill net. Data sources were for Michigan during 1929-49 (Hile et al. 1951), Michigan and Wisconsin during 1950-70 (Pycha and King 1975), and Michigan and Wisconsin during 1970-93 (Hansen et al. 1995).

Fishing was by commercial fishers during 1929-58 and by commercial-fisher contractors or state agencies during 1959-93. Lake trout populations in Lake Superior during 1929-43 sustained stable yields in commercial fisheries, providing a benchmark for judging restoration status. Therefore, lake trout abundance, expressed as a percentage of the 1929-43 average, directly compares lake trout abundance during the various phases of population collapse and recovery. Hatchery lake trout were all marked by removal of one or more fins before stocking. Thus, we show the abundance of stocked lake trout (one or more fins missing) separately from that of wild-origin lake trout (no fins missing). Comparable data are not available for Canadian waters of Lake Superior.

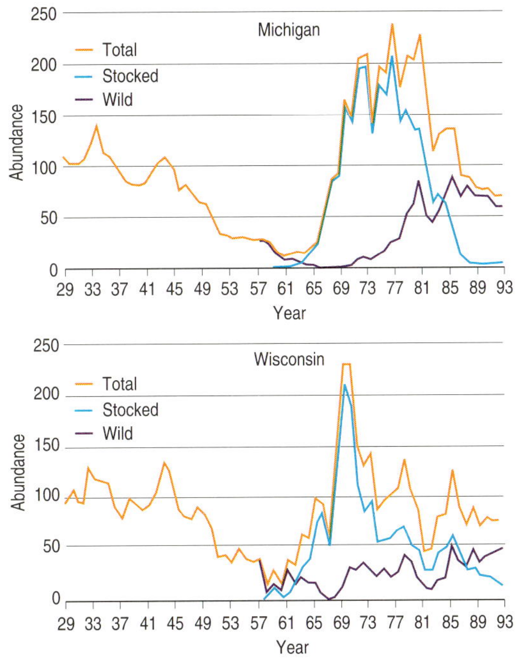

Figure. Abundance of stocked and wild lake trout (*Salvelinus namaycush*) in Michigan and Wisconsin waters of Lake Superior in 1929-93, expressed as a percentage of the 1929-43 mean (from Hansen et al. 1995).

Status and Trends

Lake Superior

Abundance of wild lake trout in Michigan declined from stable levels in the 1930's to nearly zero in the late 1960's (Figure; Hansen et al. 1995). In the 1970's and 1980's, abundance of wild lake trout increased steadily, but in the late 1980's and early 1990's decreased slowly because of increased commercial fishing and sea lamprey predation. The abundance of stocked fish increased in the late 1960's well beyond the 1929-43 average and remained there during most of the 1970's.

Abundance of stocked lake trout declined rapidly in the late 1970's and 1980's and has remained less than 10% of the 1929-43 average since 1988. Stocked lake trout reproduced in the late 1960's and produced an increased abundance of wild fish in the 1970's and 1980's.

The key to this successful natural reproduction was the presence of abundant inshore spawning grounds that inexperienced stocked lake trout easily located. Also, the decline in abundance of wild lake trout in the late 1970's and 1980's was evidently due to the earlier decline in stocked lake trout. The decline was less severe, however, because of reproduction by wild fish, the progeny of the first stocked spawners.

By 1993, 80%-90% of the lake trout in Michigan were wild, but abundance of wild lake trout was only 61% of the 1929-43 average. Fishery management agencies deferred lake trout restoration in eastern Michigan (Whitefish Bay) so that court-affirmed Native American fisheries could maximize their harvest in that area.

In Wisconsin, abundance of wild lake trout declined irregularly through 1968 and increased after that (Figure; Hansen et al. 1995). The abundance of wild lake trout in Wisconsin, even at its lowest point, remained higher than in Michigan in the late 1960's. Increased abundance in the 1970's was mostly of stocked lake trout, as in Michigan, and peak abundance also greatly exceeded the 1929-43 average. The abundance of stocked lake trout declined earlier than in Michigan, though not as much, and remained at 19% of the 1929-43 average.

Abundance of wild fish in Wisconsin increased irregularly from the 1970's through the early 1990's, but remained lower in 1993 than in Michigan and was only 53% of the

1929-43 average. Stocked lake trout were less important in the restoration of wild lake trout in Wisconsin than in Michigan. Because most spawning reefs in Wisconsin were farther offshore than in Michigan, they were not found by inexperienced stocked spawners. The increased abundance of wild lake trout in Wisconsin was largely due to reproduction by surviving wild fish in the 1960's and 1970's.

Direct measures of historical abundance do not exist for Minnesota or Ontario. The current abundance of lake trout in Minnesota is below that in Michigan and Wisconsin, but in Ontario it is probably similar to Michigan. Lake trout restoration is progressing in Minnesota but is behind that in Michigan or Wisconsin. Patterns of abundance in Minnesota since 1963 are similar to those in Michigan since 1959. Reproduction by stocked lake trout produced increased abundance of wild lake trout in Minnesota, as in Michigan (Hansen et al. 1995). Progress in lake trout restoration in Ontario is sufficient to eliminate stocking in most areas. Excessive fishery exploitation in the Michigan side of Whitefish Bay caused the deferral of lake trout stocking in the Ontario side. This deferral of lake trout restoration will continue until fishery management agencies in Michigan better regulate fishery exploitation.

Lake trout reestablished self-sustaining populations in much of Lake Superior, though few have reached former levels of abundance. Still, most of these populations are sufficiently large to support limited commercial and sport fishing. Current or proposed strategies for restoring wild lake trout in Lake Superior include controlling fishery exploitation, reducing sea lamprey populations, and reducing or eliminating stocking where self-sustaining populations exist.

Lake Michigan

Wild lake trout populations collapsed in Lake Michigan during the 1940's and the species became extirpated in the 1950's (Holey et al. 1995). Stocking began in the 1960's. The abundance of stocked lake trout increased in the late 1970's, then decreased in the northern part of the lake because of excessive fishery exploitation. Scattered evidence of lake trout reproduction, including eggs deposited on spawning grounds and newly hatched juvenile lake trout, has been found since the 1970's, although the only production of wild lake trout more than 1 year old was in Grand Traverse Bay during the late 1970's and early 1980's. Unfortunately, excessive fishery exploitation destroyed the wild lake trout produced in Grand Traverse Bay, preventing the establishment of a self-sustaining population (Holey et al. 1995). Current efforts to restore lake trout to Lake Michigan focus on stocking a variety of lake trout strains in offshore refuges that may afford protection from fishery exploitation, allowing restoration of wild populations to occur.

Lake Huron

Wild lake trout populations collapsed in Lake Huron in the 1940's and the species became extirpated in the main basin in the 1950's (Eshenroder et al. 1995). Stocking began in the 1970's. Abundance of stocked fish in southern Michigan waters increased steadily during the 1970's and 1980's, then decreased in response to reduced stocking. Abundance in northern Michigan waters increased briefly during the late 1970's and early 1980's, but decreased slowly after that because of excessive sea lamprey predation and fishery exploitation.

Natural reproduction occurred in Thunder Bay, Michigan, and South Bay, Ontario, but self-sustaining populations have not developed at either location. Restoration efforts now focus on reducing the number of sea lampreys and stocking a variety of lake trout strains on offshore reefs and in a refuge. The refuge, located in the northern part of the lake, may provide protection from fishery exploitation, and thereby may allow a self-sustaining population to become established.

Lake Erie

Wild lake trout populations collapsed in Lake Erie during the 1920's (Cornelius et al. 1995). Stocking began in the 1980's. Abundance of stocked lake trout increased steadily following initial chemical treatment of sea lampreys in 1986-87, although abundance of stocked lake trout decreased after 1990 for unexplained reasons. Current restoration efforts focus on controlling sea lampreys and stocking yearling lake trout. Research efforts focus on identifying causes of declining abundance of stocked fish and determining whether adult lake trout will aggregate at suitable spawning locations and reproduce successfully.

Lake Ontario

Wild lake trout populations collapsed in Lake Ontario between 1930 and 1960 (Elrod et al. 1995). Stocking began in the 1970's. Stocked lake trout subsequently survived to maturity, spawned, and deposited eggs that hatched into juveniles. These juveniles, however, evidently did not survive to later ages because fishery biologists have not yet discovered any older, wild-origin lake trout. Current restoration efforts focus on stocking strains of lake trout that reproduce most successfully. Research focuses on evaluating factors that limit survival of the fry, such as predation and contaminants.

References

Cornelius, F.C., K.M. Muth, and R. Kenyon. 1995. Lake trout rehabilitation in Lake Erie: a case history. Journal of Great Lakes Res. In press.

Elrod, J.H., R. O'Gorman, C.P. Schneider, T.H. Eckert, T. Schaner, J.N. Bowlby, and L.P. Schleen. 1995. Lake trout rehabilitation in Lake Ontario. Journal of Great Lakes Res. In press.

Eschmeyer, P.H. 1968. The lake trout (*Salvelinus namaycush*). U.S. Fish and Wildlife Service, Fishery Leaflet 555. 8 pp.

Eshenroder, R.L., N.R. Payne, J.E. Johnson, C.A. Bowen II, and M.P. Ebener. 1995. Lake trout rehabilitation in Lake Huron. Journal of Great Lakes Res. In press.

Hansen, M.J., J.W. Peck, R.G. Schorfhaar, J.H. Selgeby, D.R. Schreiner, S.T. Schram, B.L. Swanson, W.R. MacCallum, M.K. Burnham-Curtis, G.L. Curtis, J.W. Heinrich, and R.J. Young. 1995. Lake trout (*Salvelinus namaycush*) restoration in Lake Superior. Journal of Great Lakes Res. In press.

Hile, R., P.H. Eschmeyer, and G.F. Lunger. 1951. Status of the lake trout fishery in Lake Superior. Transactions of the American Fisheries Society 80:278-312.

Holey, M.E., R.W. Rybicki, G.W. Eck, E.H. Brown, Jr., J.E. Marsden, D.S. Lavis, M.L. Toneys, T.N. Trudeau, and R.M. Horrall. 1995. Progress toward lake trout restoration in Lake Michigan. Journal of Great Lakes Res. In press.

Pycha, R.L., and G.R. King. 1975. Changes in the lake trout population of southern Lake Superior in relation to the fishery, the sea lamprey, and stocking, 1950-70. Great Lakes Fishery Commission Tech. Rep. 28, Ann Arbor, MI. 34 pp.

Smith, B.R. 1971. Sea lampreys in the Great Lakes of North America. Pages 207-247 *in* M.W. Hardisty and I.C. Potter, eds. The biology of lampreys. Vol. 1. Academic Press, New York.

For further information:

Michael J. Hansen
National Biological Service
Great Lakes Center
1451 Green Rd.
Ann Arbor, MI 48105

Wetlands in Regulated Great Lakes

by

Douglas A. Wilcox
National Biological Service

James E. Meeker
Northland College, WI

Water levels in the Great Lakes are affected by variations in precipitation, evaporation, ice build up, internal waves (seiches), and human alterations that include modifying the connecting channels between lakes and regulating the water levels of Lake Superior and Lake Ontario. Fluctuations in water level promote the interaction of aquatic and terrestrial systems, thereby resulting in higher quality habitat and increased productivity. When the fluctuations in water levels are reduced through stabilization, shifting of vegetation types decreases, more stable plant communities develop, and species diversity and habitat value decrease (Wilcox and Meeker 1991, 1992). Although water levels in Lake Superior are regulated by structures at the outlet, water-level cycles and patterns remain fairly similar to natural conditions. Lake Ontario water levels are also regulated, but high and low water extremes have been eliminated since the mid-1970's. The effects of water-level history on wetland plant communities under the two regulation regimes were investigated by studying wetlands on each lake.

Seventeen sites on Lake Ontario and 18 on Lake Superior were sampled. Vegetation was mapped and then sampled along transects that followed elevation contours with specific water-level histories (number of years since last flooded or last dry). The histories and elevations differed between lakes. Correlations between specific elevations and accompanying plant communities were assessed across all wetlands sampled in each lake to determine the range of elevations in which the most diverse plant communities occur; these data were used to create schematic cross-sections depicting the structural habitat provided by the plant communities characteristic of each lake.

Vegetation and Water Level

At study sites on both Lakes Ontario and Superior, wetland plant communities differed at different elevations; these plant communities developed as a result of the water-level history of each elevation that was sampled. In general, plant communities at elevations that had not been flooded for many years were dominated by shrubs, grasses, and old-field plants. If flooding was more recent, small shrubs that became established after flooding were present, as were grasses, sedges, and other nonwoody plants.

The plant communities at elevations that were flooded periodically at 10- to 20-year intervals and dewatered for successive years between floods had the greatest diversity of

Courtesy J. Meeker, Northland College

Small unnamed bay near Bete Grise, Lake Superior, August 1991. Scattered lilies (*Nuphar variegata*) with submersed plants adjacent to a floating bog mat.

wetland vegetation. Dominants included grasses, sedges, rushes, short emergent plants, and submersed aquatic vegetation. At elevations that were rarely or never dewatered, submersed and floating plants were dominant, with emergent plants also occurring at some sites.

Lake Superior

Water levels on Lake Superior have been regulated for much of this century, although the range of fluctuations and the cyclic nature of high and low lake levels have not been altered

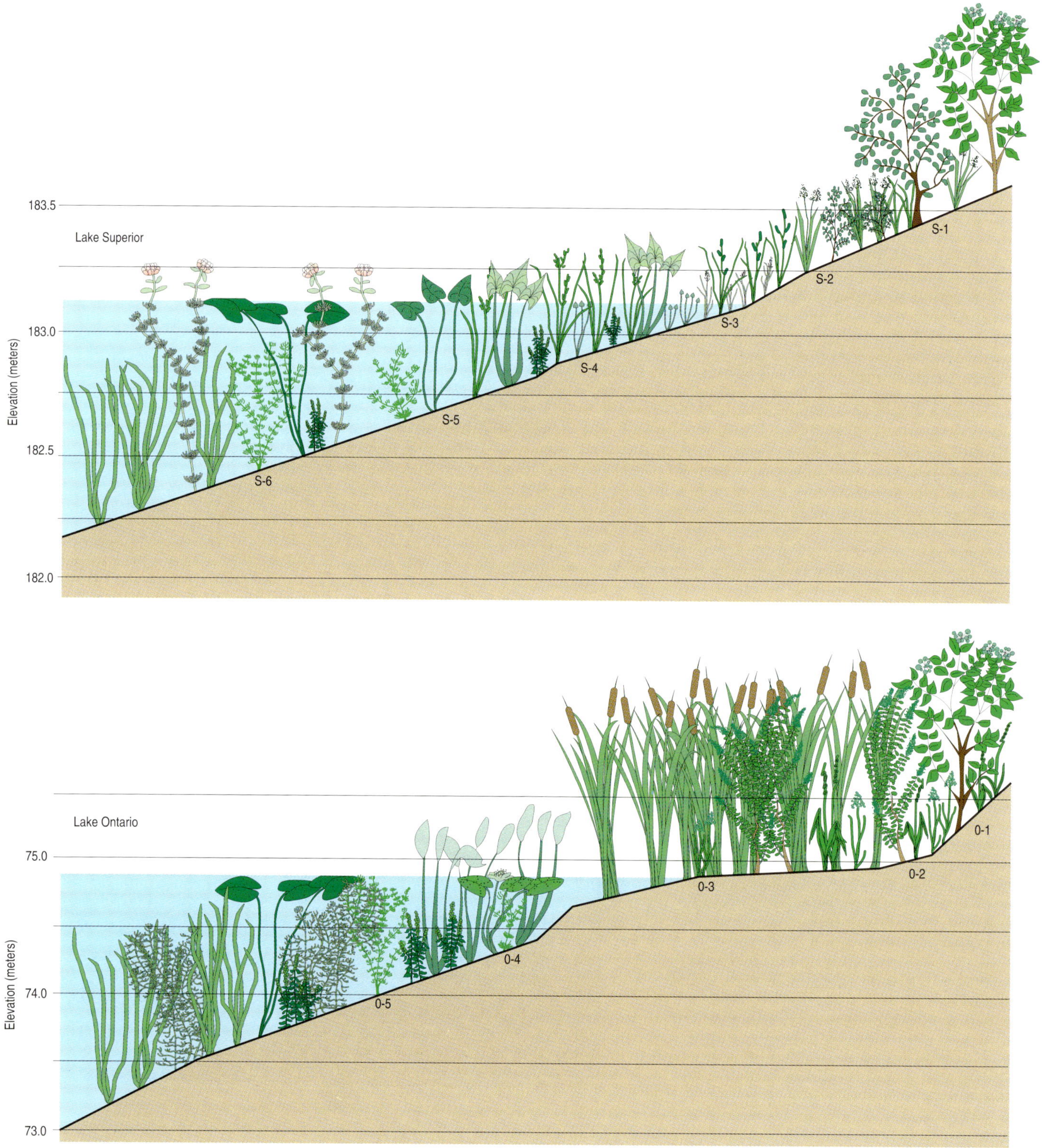

Figure. Schematic cross-sections depicting the structural habitat provided by plant communities characteristic of regulated Lakes Superior and Ontario. Elevations at which vegetation sampling was conducted are shown beneath each cross-section (benchmark: International Great Lakes Datum 1955).

substantially. More than 275 taxa were recorded in a sampling of 18 wetlands along the U.S. shoreline, 216 of which were obligate (*see* glossary) or facultative (*see* glossary) wetland species. Vegetation mapping showed the most prevalent vegetation types to be those dominated by submersed aquatic vegetation or shrubs, both of which were present in all sites and averaged about 25% of the cover. Vegetation types dominated by cattails (*Typha* sp. or other taxa plus cattails) occurred in about half the sites but averaged only about 6% of the cover. Across all sites, 27 different vegetation types were mapped.

Lake Ontario

Water levels on Lake Ontario have been regulated since 1960, when the St. Lawrence Seaway began operation. Before regulation, the range of fluctuations during the 20th century was about 2 m (6.6 ft). After regulation, the range was reduced slightly between 1960 and 1976, but low water-supply conditions in the mid-1960's and high supplies in the mid-1970's maintained much of the range. Regulation reduced the range to about 0.9 m (2.9 ft) in the years after 1976.

The lack of alternating flooded and dewatered conditions at the upper and lower edges of the wetlands resulted in establishment of extensive stands of cattail (*Typha* sp.) and domination of other areas by purple loosestrife (*Lythrum salicaria*), reed canary grass (*Phalaris arundinacea*), and various shrubs. Although more than 250 taxa were recorded in a sampling of 17 wetlands along the U.S. shoreline, only 151 were obligate or facultative wetland plants. Vegetation mapping showed the cattail-dominated vegetation type to be most prevalent, occurring at all sites and averaging about 32% of the cover. The submersed aquatic vegetation type occurred at 75% of the sites and averaged about 30% of the cover. Across all sites, 20 different vegetation types were mapped.

Habitat Structure

Differences in the species and structural types of plants at different elevations in wetlands of regulated Lakes Superior and Ontario result in different habitats for faunal organisms because the greater diversity of taxa and vegetation types in Lake Superior wetlands provides more niches for fauna than in Lake Ontario wetlands (Figure; Engel 1985; Wilcox and Meeker 1992). The prevalence of dominant cattail stands in Lake Ontario wetlands reduces habitat value there (Weller and Spatcher 1965).

Periodic high waters are necessary to reduce dominant emergent vegetation in Great Lakes wetlands; low waters are necessary to reduce dominant submersed vegetation. High waters followed by low-water years allow a diversity of plants to grow from seed on the exposed sediments, reproduce, and replenish the seed bank. Although competitive species such as cattails will again become dominant, the next high-water year will eliminate them again. When water-level fluctuations are reduced by regulation, the processes for rejuvenating wetland plant communities are lost and habitat values decrease.

References

Engel, S. 1985. Aquatic community interactions of submerged macrophytes. Wisconsin Department of Natural Resources Tech. Bull. 156. 79 pp.

Weller, M.W., and C.S. Spatcher. 1965. Role of habitat in the distribution and abundance of marsh birds. Iowa Agricultural and Home Economics Experiment Station, Ames, IA. Special Rep. 43.

Wilcox, D.A., and J.E. Meeker. 1991. Disturbance effects on aquatic vegetation in regulated lakes in northern Minnesota. Canadian Journal of Botany 69:1542-1551.

Wilcox, D.A., and J.E. Meeker. 1992. Implications for faunal habitat related to altered macrophyte structure in regulated lakes in northern Minnesota. Wetlands 12:192-203.

For further information:

Douglas A. Wilcox
National Biological Service
Great Lakes Science Center
1451 Green Rd.
Ann Arbor, MI 48105

Decline in the Freshwater Gastropod Fauna in the Mobile Bay Basin

The historical freshwater gastropod fauna of the Mobile Bay basin in Alabama, Georgia, Mississippi, and Tennessee was the most diverse in the world, comparable only to the diversity reported for the Mekong River in Southeast Asia. This fauna was represented by 9 families and about 118 species. Several families have genera endemic to the Mobile Bay basin: Viviparidae: *Tulotoma*; Hydrobiidae: *Clappia*, *Lepyrium*; Pleuroceridae: *Gyrotoma*; and Planorbidae: *Amphigyra* and *Neoplanorbis*. The greatest described species diversity was in the Pleuroceridae (76 species). The pleurocerid genera *Pleurocera*, *Leptoxis*, and *Elimia* had their greatest radiation in the Coosa River drainage.

Although this extremely diverse aquatic gastropod fauna received little attention in the past 50 years, it was actively studied during the second quarter of this century (Goodrich 1922, 1924, 1936, 1944a, 1944b). During the last 60 years, this unique gastropod fauna has declined precipitously (Table 1; Athearn 1970; Heard 1970; Stansbery 1971). More recent documentation of the decimation of this fauna was presented by Stein (1976) and Palmer (1986). The endemic genus *Tulotoma* (Figs. 1 and 2), formerly widespread in the main channel of the Alabama and Coosa rivers, was presumed extinct until recently rediscovered (Hershler et al. 1990). The pleurocerid genus *Gyrotoma*, restricted primarily to the shoals of the Coosa

by
Arthur E. Bogan
Freshwater Molluscan Research
J. Malcolm Pierson
Calera, Alabama
Paul Hartfield
U.S. Fish and Wildlife Service

River, contained six recognized species, all of which are presumed extinct (Table 2; Fig. 3).

Status and Trends

Literature records were compiled to document the gastropod species present historically. Recent surveys of the aquatic gastropod fauna of the Coosa and Cahaba river drainages in Alabama have been conducted by using standard field techniques (Bogan and Pierson 1993 a, b). Additional unpublished data (Bogan and Hartfield) are included.

Recent surveys of the aquatic gastropod fauna at about 800 sites (Table 1) have documented population declines, decreases in species' ranges, and the loss of a major portion of the gastropod diversity, especially in the Coosa River. The Coosa River drainage had at least 82 species historically (Table 1); today 26 species are presumed extinct in six genera, and four genera (*Clappia* [2 species], *Gyrotoma* [6 species], *Amphigyra* [1 species], and *Neoplanorbis* [4 species]) are presumed extinct (Tables 1 and 2). The genus *Leptoxis* has been reduced to a single species restricted to three creek tributary systems in the Coosa River.

The fauna of the Cahaba River drainage has fared much better (Table 1). Although the Cahaba River drainage does not suffer from the numerous dams and the siltation problems of the Coosa River drainage, it is heavily affected by nonpoint-source runoff, siltation, acid mine drainage, pollution from wastewater treatment

Table 1. Summary of the aquatic gastropod fauna of the river systems in the Mobile Bay basin.

Data*	Alabama River	Tombigbee R. drainage	Black Warrior R. drainage	Cahaba R. drainage	Coosa R. drainage	Talapoosa R. drainage	Mobile Bay basin total
Approximate total historical gastropod species diversity	19	8	17	36	82	8	118
Number of species found in recent surveys	3	3	7	24	30	4	80
Federally listed endangered species	1	0	0	0	1	0	1
Federal candidate species	4	1	6	16	43	2	70
Number of species presumed extinct	?	0	2	4	26	?	38
Percent decline in gastropod fauna	84%	62%	58%	33%	63%	50%	32%

* Data from Bogan and Pierson (1993 a,b), Burch (1989), and A.E. Bogan and P. Hartfield (unpublished data).

Fig. 1. Live specimens of the endangered tulotoma, *Tulotoma magnifica*, from Kelly Creek, Elmore County, Alabama, 1993.

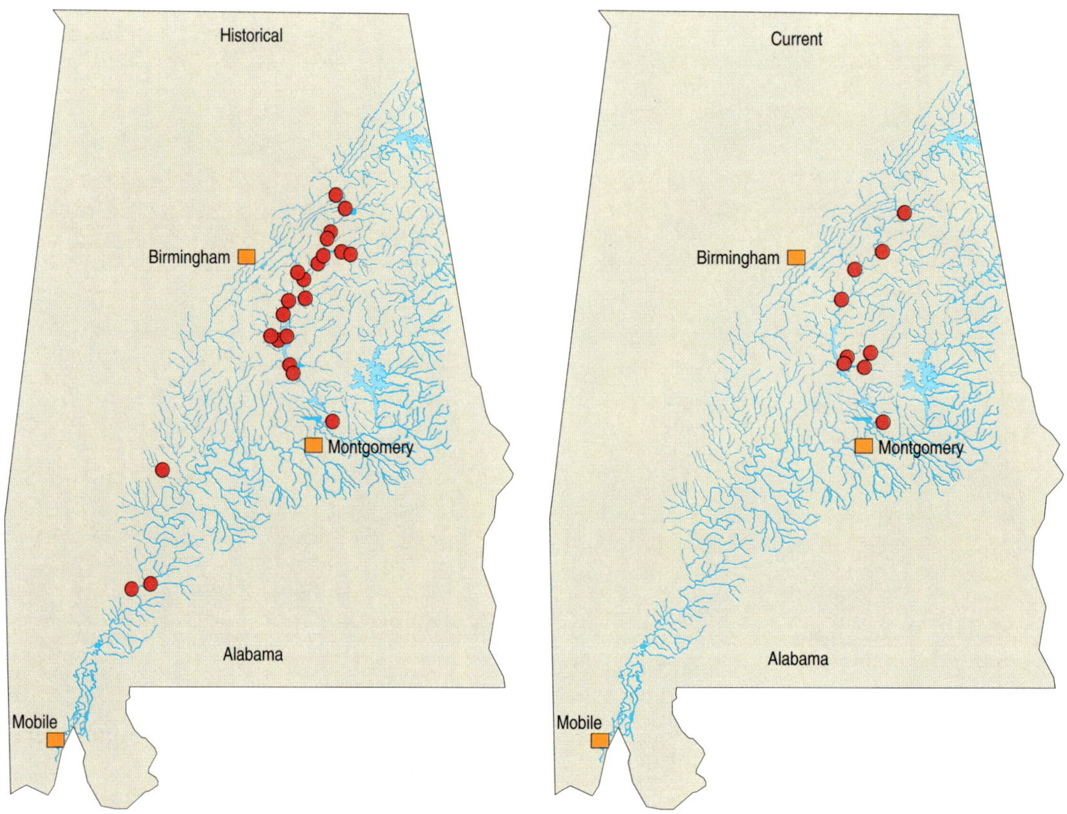

Fig. 2. Historical and current distribution of *Tulotoma magnifica*. Filled circles represent a single or two closely located collection sites (after Hershler et al. 1990). Map modified from the U.S. Geological Survey 1:500,000 scale—State of Alabama sheet (1970 ed.).

Table 2. Freshwater gastropod species presumed extinct in the Mobile Bay basin.

Family and common name	Scientific name
Hydrobiidae	
Cahaba pebblesnail	*Clappia cahabensis* Clench 1965
Umbilicate pebblesnail	*C. umbilicata* (Walker 1904)
Pleuroceridae	
Short-spire elimia	*Elimia brevis* (Reeve 1860)
Closed elimia	*E. clausa* (Lea 1861)
Fusiform elimia	*E. fusiformis* (Lea 1861)
No common name	*E. gibbera* (Goodrich 1922)
High-spired elimia	*E. hartmaniana* (Lea 1861)
Constricted elimia	*E. impressa* (Lea 1841)
Hearty elimia	*E. jonesi* (Goodrich 1936)
No common name	*E. lachryma* (Reeve 1861)
Ribbed elimia	*E. laeta* (Jay 1839)
No common name	*E. macglameriana* (Goodrich 1936)
Rough-lined elimia	*E. pilsbryi* (Goodrich 1927)
Pupa elimia	*E. pupaeformis* (Lea 1864)
Pygmy elimia	*E. pygmaea* (H.H. Smith 1936)
Cobble elimia	*E. vanuxemiana* (Lea 1843)
Excised slitshell	*Gyrotoma excisa* (Lea 1843)
Striate slitshell	*G. lewisii* (Lea 1869)
Pagoda slitshell	*G. pagoda* (Lea 1845)
Ribbed slitshell	*G. pumila* (Lea 1860)
Pyramid slitshell	*G. pyramidata* (Shuttleworth 1845)
Round slitshell	*G. walkeri* (H.H. Smith 1924)
Agate rocksnail	*Leptoxis clipeata* (H.H. Smith 1922)
Oblong rocksnail	*L. compacta* (Anthony 1854)
Interrupted rocksnail	*L. formanii* (Lea 1843)
Maiden rocksnail	*L. formosa* (Lea 1860)
Rotund rocksnail	*L. ligata* (Anthony 1860)
Lirate rocksnail	*L. lirata* (H.H. Smith 1922)
Black mudalia	*L. melanoides* (Conrad 1834)
Bigmouth rocksnail	*L. occultata* (H.H. Smith 1922)
Coosa rocksnail	*L. showalteri* (Lea 1860)
No common name	*L. torrefacta* (Goodrich 1922)
Striped rocksnail	*L. vittata* (Lea 1860)
Planorbidae	
Shoal sprite	*Amphigyra alabamensis* (Pilsbry 1906)
No common name	*Neoplanorbis carinatus* (Walker 1908)
No common name	*N. smithi* (Walker 1908)
No common name	*N. tantillus* (Pilsbry 1906)
No common name	*N. umbilicatus* (Walker 1908)

plants, and water withdrawn for domestic water use. Species such as *Lepyrium showalteri* and *Lioplax cyclostomaformis*, formerly much more widespread in the basin, are now apparently restricted to one or two shoal areas in the Cahaba River main channel. The status of the pebblesnails (Hydrobiidae) is uncertain. The former diversity of the genus *Somatogyrus* in the Coosa River has probably suffered the same fate as most of the main channel shoal-dwelling pleurocerid species—extinction. Detailed information on the distribution of the freshwater limpets (Ancylidae) is not available, but they appear to have suffered similar range restrictions.

The uncertainty expressed in the diversity of the historical gastropod fauna presented in Table 1 is indicative of our lack of information regarding all aspects of the historical gastropod fauna of the Mobile Bay basin. There are a lack of detailed data on the ecology and life history of all of the species, and a paucity of distributional information for most of the families other than the Pleuroceridae, making estimation of gastropod diversity by drainage difficult.

Courtesy A.E. Bogan

Fig. 3. Illustration of a representative species of the extinct slitshell genus *Gyrotoma* from Butting Ram Shoals, Coosa River, Alabama.

Declining species diversity can be directly linked to the inundation of the shoal areas of the rivers of the Mobile Bay basin by impoundment and siltation resulting from a variety of watershed disturbances, including 33 major dams for hydroelectric generation, locks and flood control on the major rivers of the Mobile Bay basin, and numerous smaller impoundments on tributary rivers and streams. Most gastropods inhabiting shoal areas are gill-breathing species typically grazing on the plant life growing on the rock substrate in shallow riffle and shoal areas. They formerly lived on rocks in the shallow shoal areas with highly oxygenated water. The pleurocerid gastropod fauna represented a significant portion of the invertebrate biomass living on these shoal areas.

When this habitat was impounded, the snails were not able to survive the deep, cold, and often oxygen-depleted water. Many areas not impounded have suffered because of the heavy siltation of shoal areas, smothering the plant life that formed the diet of these gastropods. Major sources of siltation include poor agricultural and silvicultural practices, lack of riparian buffer zones, and generally poor land-use practices. The drastic decline in gastropod diversity is especially evident in the Coosa River main channel where numerous species formerly found on shoals have disappeared after the damming of the river (Bogan and Pierson 1993a). Other species have had their ranges fragmented by the damming of the rivers and have become restricted to the unimpounded areas below the dams with clean current-swept gravel and bedrock outcrops.

Tulotoma magnifica (Figs. 1 and 2) is the only aquatic gastropod now federally listed as endangered; none is listed as threatened, although 104 species of aquatic gastropods from Alabama are on the federal candidate list. Most are from the Coosa and the Cahaba rivers (Table 1). Conservation and recovery of the remaining diversity will require immediate action to prevent further declines and extinctions. This will necessitate action to improve water quality across the basin and to decrease the amount of silt entering the streams and rivers. In addition, the survey of the aquatic gastropod fauna of the Mobile Bay basin is not complete, and additional fieldwork in the main channels of the larger rivers is needed, especially on the vertical limestone wall habitats.

References

Athearn, H. 1970. Discussion of Dr. Heard's paper. Malacologia 10(1):28-31.

Bogan, A.E., and J.M. Pierson. 1993a. Survey of the aquatic gastropods of the Coosa River Basin, Alabama: 1992. Final report submitted to Alabama Natural Heritage Program, Montgomery. 14 pp.

Bogan, A.E., and J.M. Pierson. 1993b. Survey of the aquatic gastropods of the Cahaba River Basin, Alabama: 1992. Final report submitted to Alabama Natural Heritage Program, Montgomery. 20 pp.

Burch, J.B. 1989. North American freshwater snails. Malacological Publications, Hamburg, MI. 365 pp.

Goodrich, C. 1922. The Anculosae of the Alabama River drainage. University of Michigan, Museum of Zoology, Miscellaneous Publ. 7:1-57.

Goodrich, C. 1924. The genus *Gyrotoma*. University of Michigan, Museum of Zoology, Miscellaneous Publ. 12:1-29.

Goodrich, C. 1936. Goniobasis of the Coosa River, Alabama. University of Michigan, Museum of Zoology, Miscellaneous Publ. 31:1-60.

Goodrich, C. 1944a. Certain operculates of the Coosa River. The Nautilus 58(1):1-10.

Goodrich, C. 1944b. Pulmonates of the Coosa River. The Nautilus 58(1):11-15.

Heard, W.H. 1970. Eastern freshwater mollusks (II). The south Atlantic and gulf drainages. Malacologia 10(1):23-27.

Hershler, R., J.M. Pierson, and R.S. Krotzer. 1990. Rediscovery of *Tulotoma magnifica* (Conrad) (Gastropoda: Viviparidae). Proceedings of the Biological Society of Washington 103(4):815-824.

Palmer, S. 1986. Some extinct molluscs of the U.S.A. Atala 13(1):1-7.

Stansbery, D.H. 1971. Rare and endangered mollusks in eastern United States. Pages 5-18 *in* S.E. Jorgenson and R.E. Sharp, eds. Proceedings of a Symposium on Rare and Endangered Mollusks (Naiads). U.S. Bureau of Sport Fisheries and Wildlife, Washington, DC.

Stein, C.B. 1976. Gastropods. Pages 21-41 *in* H. Boschung, ed. Endangered and threatened species of Alabama. Bull. Alabama Museum of Natural History 2.

For further information:

Arthur E. Bogan
Freshwater Molluscan Research
36 Venus Way
Sewell, NJ 08080

Protozoa

by
Diana Lipscomb
George Washington University

The diverse assemblage of organisms that carry out all of their life functions within the confines of a single, complex eukaryotic (*see* glossary) cell is called protozoa. *Paramecium*, *Euglena*, and *Amoeba* are well-known examples of these major groups of organisms. Some protozoa are more closely related to animals, others to plants, and still others are relatively unique. Although it is not appropriate to group them together into a single taxonomic category, the research tools used to study any unicellular organism are usually the same, and the field of protozoology has been created to carry out this research. The unicellular photosynthetic protozoa are sometimes also called algae and are addressed elsewhere. This report considers the status of our knowledge of heterotrophic protozoa (protozoa that cannot produce their own food).

Free-living Protozoa

Protozoans are found in all moist habitats within the United States, but we know little about their specific geographic distribution. Because of their small size, production of resistant cysts, and ease of distribution from one place to another, many species appear to be cosmopolitan and may be collected in similar microhabitats worldwide (Cairns and Ruthven 1972). Other species may have relatively narrow limits to their distribution.

Marine ciliates inhabit interstices of sediment and beach sands, surfaces, deep sea and cold Antarctic environments, planktonic habitats, and the algal mats and detritus of estuaries and wetlands. Our actual knowledge of salinity, temperature, and oxygen requirements of marine protozoa is poor (although some groups, such as the foraminifera, are better studied than others), and even the broadest outlines of their biogeographic ranges are usually a mystery. In general, freshwater protozoan communities are similar to marine communities except the specialized interstitial fauna of the sand is largely missing. In freshwater habitats, the foraminifera and radiolaria common in marine environments are absent or low in numbers while testate amoebae exist in greater numbers. Relative abundance of species in the marine versus freshwater habitat is unknown.

Soil-dwelling protozoa have been documented from almost every type of soil and in every kind of environment from the peat-rich soil of bogs to the dry sands of deserts. In general, protozoa are found in greatest abundance near the soil surface, especially in the upper 15 cm (6 in), but occasional isolates can be obtained at depths of a meter (yard) or more. Protozoa do not constitute a major part of soil biomass, but in some highly productive regions such as forest litter, the protozoa are a significant food source for the microinvertebrates, with a biomass that may reach 20 g/m^2 of soil surface area there.

Environmental Quality Indicators

Polluted waters often have a rich and characteristic protozoan fauna. The relative abundance and diversity of protozoa are used as indicators of organic and toxic pollution (Cairns et al. 1972; Foissner 1987; Niederlehner et al. 1990; Curds 1992). Bick (1972), for example, provided a guide to ciliates that are useful as indicators of environmental quality of European freshwater systems, along with their ecological distribution with respect to parameters such as amount of organic material and oxygen levels. Foissner (1988) clarified the taxonomy of European ciliates as part of a system for classifying the state of aquatic habitats according to their faunas.

Symbiotic Protozoa

Parasites

Protozoa are infamous for their role in causing disease, and parasitic species are among the best-known protozoa. Nevertheless, our knowledge has large gaps, especially of normally free-living protozoa that may become pathogenic in immuno-compromised individuals. For example, microsporidia comprise a unique group of obligate, intracellular parasitic protozoa. Microsporidia are amazingly diverse organisms with more than 700 species and 80 genera that are capable of infecting a variety of plant, animal, and even other protist hosts. They are found worldwide and have the ability to thrive in many ecological conditions. Until the past few years, their ubiquity did not cause a threat to human health, and few systematists worked to describe and classify the species. Since 1985, however, physicians have documented an unusual rise in worldwide infections in AIDS patients caused by four different genera of microsporidia (*Encephalitozoon*, *Nosema*, *Pleistophora*, and *Enterocytozoon*). According to the Centers for Disease Control in the United States, difficulties in identifying microsporidian species are impeding diagnosis and effective treatment of AIDS patients.

Protozoan Reservoirs of Disease

The presence of bacteria in the cytoplasm of protozoa is well known whereas that of viruses is less frequently reported. Most of these reports simply record the presence of bacteria or viruses and assume some sort of symbiotic relationship between them and the protozoa. Recently, however, certain human pathogens were shown to not only survive but also to multiply in the cytoplasm of free-living, nonpathogenic protozoa. Indeed, it is now believed that protozoa are the natural habitat for certain pathogenic bacteria. To date, the main focus of attention has been on the bacterium *Legionella pneumophila*, the causative organism of Legionnaires' disease; these bacteria live and reproduce in the cytoplasm of some free-living amoebae (Curds 1992).

Symbionts

Some protozoa are harmless or even beneficial symbionts. A bewildering array of ciliates, for example, inhabit the rumen and reticulum of ruminates and the cecum and colon of equids. Little is known about the relationship of the ciliates to their host, but a few may aid the animal in digesting cellulose.

Data on Protozoa

Bibliography

While our knowledge of recent and fossil foraminifera in the U.S. coastal waterways is systematically growing, other free-living protozoa are poorly known. There are some regional guides and, while some are excellent, many are limited in scope, vague on specifics, or difficult to use. Largely because of these problems, most ecologists who include protozoa in their studies of aquatic habitats do not identify them, even if they do count and measure them for biomass estimates (Taylor and Sanders 1991).

Parasitic protozoa of humans, domestic animals, and wildlife are better known although no attempt has been made to compile this information into a single source. Large gaps in our knowledge exist, especially for haemogregarines, microsporidians, and myxosporidians (*see* Kreier and Baker 1987).

Museum Specimens

For many plant and animal taxa, museums represent a massive information resource. This is not true for protozoa. In the United States, only the National Natural History Museum (Smithsonian Institution) has a reference collection preserved on microscope slides, but it does not have a protozoologist curator and cannot provide species' identification or verification services. The American Type Culture Collection has some protozoa in culture, but its collection includes relatively few kinds of protozoa.

Ecological Role of Protozoa

Although protozoa are frequently overlooked, they play an important role in many communities where they occupy a range of trophic levels. As predators upon unicellular or filamentous algae, bacteria, and microfungi, protozoa play a role both as herbivores and as consumers in the decomposer link of the food chain. As components of the micro- and meiofauna, protozoa are an important food source for microinvertebrates. Thus, the ecological role of protozoa in the transfer of bacterial and algal production to successive trophic levels is important.

Factors Affecting Growth and Distribution

Most free-living protozoa reproduce by cell division (exchange of genetic material is a separate process and is not involved in reproduction in protozoa). The relative importance for population growth of biotic versus chemical-physical components of the environment is difficult to ascertain from the existing survey data. Protozoa are found living actively in nutrient-poor to organically rich waters and in fresh water varying between 0°C (32°F) and 50°C (122°F). Nonetheless, it appears that rates of population growth increase when food is not constrained and temperature is increased (Lee and Fenchel 1972; Fenchel 1974; Montagnes et al. 1988).

Comparisons of oxygen consumption in various taxonomic groups show wide variation (Laybourn and Finlay 1976), with some aerobic forms able to function at extremely low oxygen tensions and to thereby avoid competition and predation. Many parasitic and a few free-living species are obligatory anaerobes (grow without atmospheric oxygen). Of the free-living forms, the best known are the plagiopylid ciliates that live in the anaerobic sulfide-rich sediments of marine wetlands (Fenchel et al. 1977). The importance of plagiopylids in recycling nutrients to aerobic zones of wetlands is potentially great.

Ecological Interactions

Because of the small size of protozoa, their short generation time, and (for some species) ease of maintaining them in the laboratory, ecologists have used protozoan populations and communities to investigate competition and predation. The result has been an extensive literature on a few species studied primarily under laboratory conditions. Few studies have been extended to natural habitats with the result that we know relatively little about most protozoa and their roles in natural communities. Intraspecific competition for common resources often results in cannibalism, sometimes with dramatic changes in morphology of the cannibals (Giese 1973). Field studies of interspecific competition are few and most evidence for such species interactions is indirect (Cairns and Yongue 1977).

References

Bick, H. 1972. Ciliated protozoa. An illustrated guide to the species used as biological indicators in freshwater biology. World Health Organization, Geneva. 198 pp.

Cairns, J., G.R. Lanza, and B.C. Parker. 1972. Pollution related structural and functional changes in aquatic communities with emphasis on freshwater algae and protozoa. Proceedings of the National Academy of Sciences 124:79-127.

Cairns, J., and J.A. Ruthven. 1972. A test of the cosmopolitan distribution of fresh-water protozoans. Hydrobiologia 39:405-427.

Cairns, J., and W.H. Yongue. 1977. Factors affecting the number of species of freshwater protozoan communities. Pages 257-303 *in* J. Cairns, ed. Aquatic microbial communities. Garland, New York.

Curds, C.R. 1992. Protozoa and the water industry. Cambridge University Press, MA. 122 pp.

Fenchel, T. 1974. Intrinsic rate increase: the relationship with body size. Oecologia 14:317-326.

Fenchel, T., T. Perry, and A. Thane. 1977. Anaerobiosis and symbiosis with bacteria in free-living ciliates. Journal of Protozoology 24:154-163.

Foissner, W. 1987. Soil protozoa: fundamental problems, ecological significance, adaptations in ciliates and testaceans, bioindicators, and guide to the literature. Progress in Protistology 2:69-212.

Foissner, W. 1988. Taxonomic and nomenclatural revision of Stádecek's list of ciliates (Protozoa: Ciliophora) as indicators of water quality. Hydrobiologia 166:1-64.

Giese, A.C. 1973. *Blepharisma*. Stanford University Press, CA. 366 pp.

Kreier, J.P., and J.R. Baker. 1987. Parasitic protozoa. Allen and Unwin, Boston, MA. 241 pp.

Laybourn, J., and B.J. Finlay. 1976. Respiratory energy losses related to cell weight and temperature in ciliated protozoa. Oecologia 44:165-174.

Lee, C.C., and T. Fenchel. 1972. Studies on ciliates associated with sea ice from Antarctica. II. Temperature responses and tolerances in ciliates from Antarctica, temperate and tropical habitats. Archive für Protistenkunde 114:237-244.

Montagnes, D.J.S., D.H. Lynn, J.C. Roff, and W.D. Taylor. 1988. The annual cycle of heterotrophic planktonic ciliates in the waters surrounding the Isles of Shoals, Gulf of Maine: an assessment of their trophic role. Marine Biology 99:21-30.

Niederlehner, B.R., K.W. Pontasch, J.R. Pratt, and J. Cairns. 1990. Field evaluation of predictions of environmental effects from multispecies microcosm toxicity test. Archives of Environmental Contamination and Toxicology 19:62-71.

Taylor, W., and R. Sanders. 1991. Protozoa. Pages 37-93 *in* J.H. Thorp and A.P. Covich, eds. Ecology and classification of North American freshwater invertebrates. Academic Press, New York.

For further information:

Diana Lipscomb
George Washington University
Department of Biological Sciences
Washington, DC 20052

Marine and Freshwater Algae

by
Richard Moe
University of California

Algae are an extremely diverse group of photosynthetic organisms that range from single-celled organisms to complex thalli (e.g., kelps). Benthic algae live attached to the bottom of a water body or to living or nonliving objects on the bottom. Planktonic algae live free-floating in the ocean and in the largest to smallest lakes and streams. Algae also occur in such varied places as the surface layers of soils and porous rocks, on the bark and leaves of trees, in snow, hot springs, and in symbiotic association with fungi to form lichens.

These organisms are important as primary producers (representing the base of the food chain or pyramid), in contributing to the fertility of soil, in providing substrate for other organisms, and in defining aquatic environments such as kelp beds and algal reefs.

The toxicity of certain marine unicellular algae can limit coastal marine fisheries (e.g., dinoflagellates in red tide). In fresh water, blooms tied to nutrient enrichment are often a major nuisance. A few species of macrophytic algae (large enough to be seen by the naked eye) are harvested from the wild for food and industrial purposes.

Knowledge of the algae of the United States is not uniform across various groups or environments. Some modern regional floras, or lists of plants (e.g., California, southeastern coast, gulf coast), are available for marine benthic macroscopic algae (Dawes 1974; Abbott and Hollenberg 1976; Schneider and Searles 1991), of which there are approximately 900 species on the Pacific coast and fewer on the Atlantic and gulf coasts (approximately 450 for the northern Atlantic coast, 350 for the southeastern Atlantic coast, and 300 for the gulf coast). Local floras are available for many places. Few species are shared between the Atlantic and Pacific coasts. Information about marine microalgae is less accessible.

Local and regional floras are available for some groups of freshwater algae (e.g., Hoshaw and McCourt 1988; Dillard 1989; Johansen 1993), but information is absent or has not been compiled for much of the country. Because no attempt has been made to produce a national flora of freshwater algae in this century, it is not possible to estimate the number of such species. Many groups of algae are cosmopolitan, however, and European monographs and floras can be useful.

In general, distribution, status, and trends of algae, even of conspicuous marine algae, are not well established. Floras usually provide ranges, but distribution of many species may be discontinuous, with various causes for the discontinuity. Filling the gaps (or confirming the discontinuities) will require considerable effort.

Although nationwide data on status and trends of North American algal populations are not readily available, scientists do know that a great deal of formerly aquatic habitat has become unavailable for algae because of landfill, reclamation, and water diversion. In addition, other habitat has been altered through farming and municipal and industrial waste discharge. In the case of reservoirs, however, one kind of aquatic habitat has been replaced by another.

Long-term information about phytoplankton is available for the Great Lakes; this information has allowed documentation of water-quality improvement in Lake Erie and analysis of the effect of the invasion of the zebra mussel (*Dreissena polymorpha*; Makarewicz 1993; Nicholls 1993). Much limnological information is available for individual water bodies or catchment basins (e.g., Brock 1985 for Lake Mendota in Wisconsin), but reconciling the different methods used when comparing separate studies is a challenge.

Interpretation of marine baseline and trend data is complicated by differences in communities over time and space (Foster et al. 1988). An example of the utility of marine baseline studies is the census of algae along the coast near Los Angeles (Dawson 1959) that showed how sewage discharge reduced algal diversity. Subsequent resurveys (Widdowson 1971) demonstrated some improvement after stricter environmental regulations were enacted. Long-term studies are available for giant kelp (*Macrocystis pyrifera*), the economically important component of southern California kelp beds. North (1971) and Foster and Schiel (1985) documented the decline of kelp beds after sewage was discharged into the ocean. They also discussed the partly successful attempts at remediation, which involved transplantation and predator control and which led to an appreciation of the complexity of organismal interactions in kelp beds.

Achieving a uniform estimate of the status of algae in North America will take considerable original observation and collection. Furthermore, different research approaches will be necessary for freshwater versus marine algae and for macrophytic algae versus microphytic algae. To determine status and trends of marine macroalgae, published literature must be compiled and analyzed. In addition, unpublished information should be obtained from herbaria and from private collections in the form of specimens, labels, and collectors' notebooks, illustrations, and checklists.

This process has been followed for west coast algae in a project by T. DeCew, the results of which are available at the Herbarium of the University of California. This project condenses the 100-year record of west coast phycology (study of algae) by using a literature review, compilation of data from specimens at west coast herbaria, and original observations. For each species a tabular representation of geographic and hydrographic range is provided. Presence or absence in different precincts along the coast and details of phenology (relations between climate and periodic biological phenomena), such as reproductive state throughout the year, are indicated. The study gives ecological information such as requirements for substrate and exposure to waves as well as the presence of epiphytes and parasites. In addition, illustrations and references to pertinent taxonomic, chemical, ecological, genetic, and physiological literature are given.

If this kind of project is done on a national scale, workers must have the necessary taxonomic training and herbarium resources must be preserved. About 100 American scientists have algal taxonomy as a principal area of interest (Anonymous 1992). Modern molecular taxonomic methods aid in the study of some groups of algae, but to progress toward a national inventory, traditional taxonomic methods must be supported.

References

Abbott, I.A., and G.J. Hollenberg. 1976. Marine algae of California. Stanford University Press, CA. 827 pp.

Anonymous. 1992. 1992 Membership Directory and Bylaws [of the Phycological Society of America]. Journal of Phycology (4, supplement).

Brock, T.D. 1985. A eutrophic lake: Lake Mendota, Wisconsin. Springer-Verlag, New York. 308 pp.

Dawes, C.J. 1974. Marine algae of the west coast of Florida. University of Miami Press, Coral Gables, FL. 201 pp.

Dawson, E.Y. 1959. A preliminary report on the marine benthic flora of southern California mainland shelf. Pages 169-264 *in* Oceanographic survey of the Continental Shelf area of southern California. California State Water Pollution Control Board Publ. 20.

Dillard, G.E. 1989. Freshwater algae of the southeastern United States. Part 1. Chlorophyceae: Volvocales, Tetrasporales and Chlorococcales. Bibliotheca Phycologica 81. 202 pp.

Foster, M.S., A.P. De Vogelaere, C. Harrold, J.S. Pearse, and A.B. Thum. 1988. Causes of spatial and temporal patterns in rocky intertidal communities of central and northern California. Memoirs of the California Academy of Sciences 9. 45 pp.

Foster, M.S., and D.R. Schiel. 1985. The ecology of giant kelp forests in California: a community profile. U.S. Fish and Wildlife Service Biological Rep. 85 (7.2). 152 pp.

Hoshaw, R.W., and R.M. McCourt. 1988. The Zygnemataceae (Chlorophyta): a twenty-year update of research. Phycologia 27:511-548.

Johansen, J.R. 1993. Cryptogamic crusts of semiarid and arid lands of North America. Journal of Phycology 29:140-147.

Makarewicz, J.C. 1993. Phytoplankton biomass and species composition in Lake Erie, 1970 to 1987. Journal of Great Lakes Res. 19:258-274.

Nicholls, K.H. 1993. Recent changes in Lake Erie (North Shore) phytoplankton: cumulative impacts of phosphorous loading reductions and the zebra mussel introduction. Journal of Great Lakes Res. 19:637-647.

North, W.J., editor. 1971. The biology of giant kelp beds (*Macrocystis*) in California. Beihefte zur Nova Hedwigia 31. 600 pp.

Schneider, C.W., and R.B. Searles. 1991. Seaweeds of the southeastern United States: Cape Hatteras to Cape Canaveral. Duke University Press, Durham, NC. 553 pp.

Widdowson, T.B. 1971. Changes in the intertidal algal flora of the Los Angeles area since the survey by E. Yale Dawson in 1957-1959. Bull. of the Southern California Academy of Sciences 70:2-16.

For further information:

Richard L. Moe
University of California
University Herbarium
Berkeley, CA 94720

Freshwater Diatoms: Indicators of Ecosystem Change

by
Donald Charles
Academy of Natural Sciences, Philadelphia
Pat Kociolek
California Academy of Sciences, San Francisco

Diatoms are photosynthetic unicellular organisms. They are found in almost all aquatic and semi-aquatic habitats and are of great ecological importance because they form an important part of the base of the food web. Although diatoms are widely distributed as a group, most species occur only in habitats with specific physical, chemical, and biological characteristics. Ecologists have long made practical use of this habitat specificity by collecting and analyzing individual species and community data to determine the quality or condition of aquatic habitats. Both long-term monitoring of specific lake and stream habitats and analysis of diatom remains (that become part of the sedimentary record of lakes) allow scientists to obtain a unique long-term historical perspective on these ecosystems. This perspective is especially valuable in assessing the long-term effects of human activities on aquatic and terrestrial ecosystems. Diatoms have been studied throughout the country, but no reasonably complete compilation or summary of these studies exists.

Diatoms are divided into two groups based on overall symmetry of the cell walls; radially symmetrical forms are informally called "centric" diatoms while bilaterally symmetrical forms are referred to as "pennate" diatoms. One remarkable aspect of these organisms is that they have cell walls made of glass (silicon dioxide). The glass cell walls are perforated and ornamented with many holes, which are usually arranged in definite patterns. The nature of these perforations as well as their orientation and densities help in the identification of diatom species. Diatom cell walls come in two pieces that fit together the way a Petri dish or pill box does. When these organisms divide, each half reproduces a "daughter" half that, because of the rigidity of the glass walls, must be smaller than the original half.

Despite the important roles diatoms play in aquatic ecosystems and their utility in evaluating and monitoring environmental change in these systems, intensive floristic or taxonomic studies on freshwater diatoms in North America have been limited. A two-volume work entitled

The Diatoms of the United States (Patrick and Reimer 1966, 1975) considered a selected number of genera, and in those genera treated only those species reported from the United States up to 1960. There are only a few regional or statewide taxonomic treatments of diatoms in the United States. The focus has been on specific habitats; areas receiving the most attention have been the Northeast, upper Midwest, the Great Lakes, and isolated areas in the West. Only a few checklists of diatom taxa exist.

Fifteen centric and 63 pennate diatom genera have been reported from fresh water. No exact species counts have been made, but about 4,000 species have been described in the literature. This number is undoubtedly a conservative estimate because in two areas where intensive research has been conducted, in Dickinson County, Iowa (around the Iowa Lakeside Laboratory), and the Laurentian Great Lakes, about 1,200 and 2,000 species, respectively, have been recognized. In the Great Lakes, nearly 10% of those species are new to science. There is still a great need to document the variety and distribution of freshwater diatoms in the United States.

Diatom assemblages provide the basis for many important assessments of trends in the status of freshwater ecosystems. These versatile indicators tell us about the acidification (*see* glossary) of lakes caused by acidic deposition, the eutrophication (*see* glossary) of lakes caused by human impacts and changing land use, improvements and declines in the quality of our rivers and streams, and changes in climate over the past thousands of years. Because diatoms are important components of the biological community and food web and are sensitive to changes in water quality, they provide information on both the biological integrity of the ecosystem and those factors likely to be causing any observed changes. Researchers are rapidly developing new techniques for using diatoms to provide even more quantitative and accurate inferences of ecosystem condition, and diatoms are being included in a growing number of local and regional-scale monitoring programs.

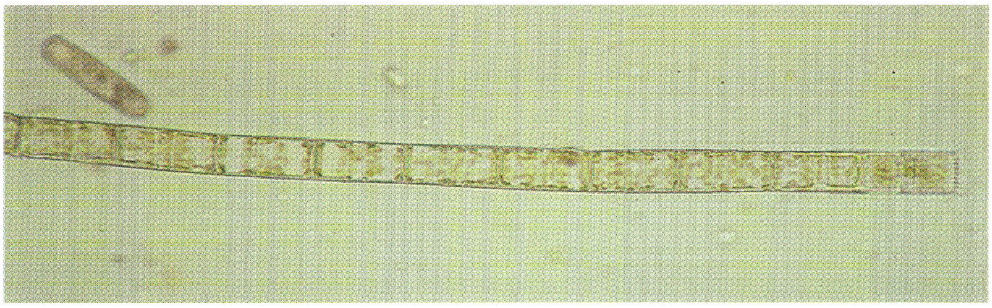

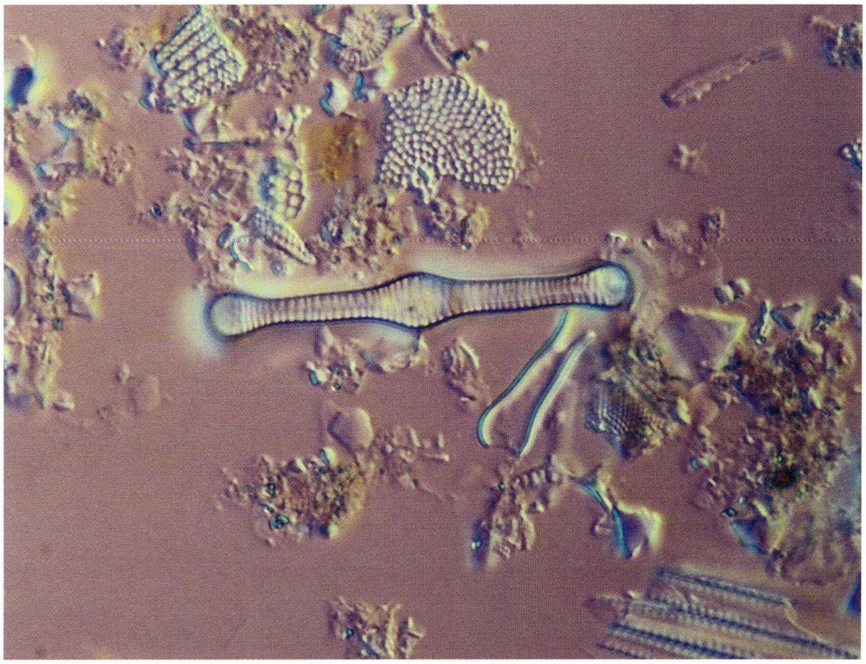

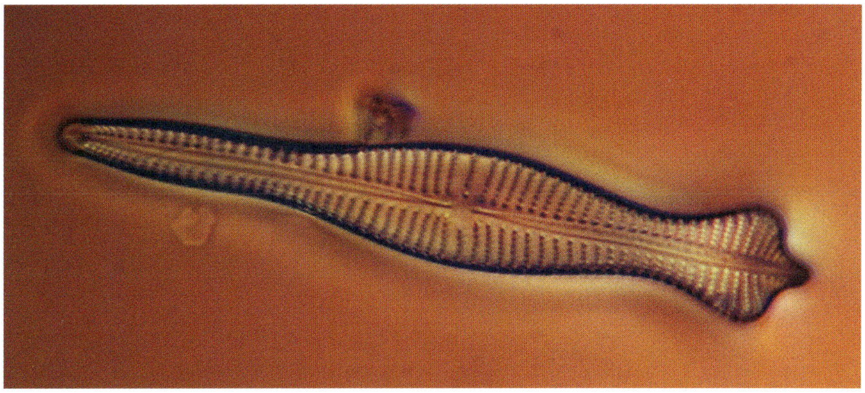

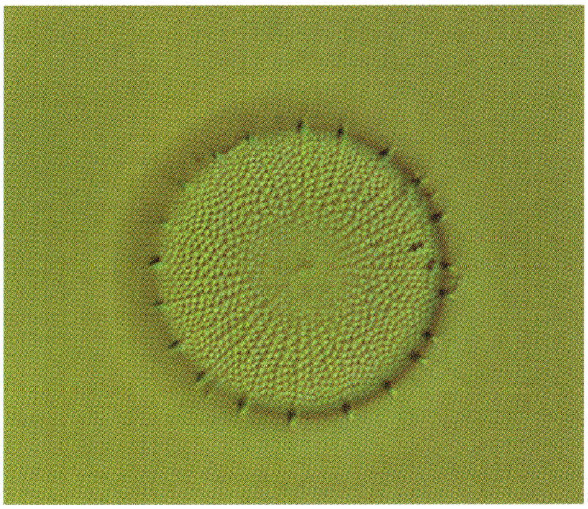

Examples of diatoms (top to bottom):

Aulacoseira sp.,
Tabellaria sp.,
Gomphonema sp., and
Stephananodiscus sp.

Photos courtesy J.P. Kociolek

Lake Acidification

The extent, magnitude, timing, and causes of lake acidification in acid-sensitive regions of the country have been inferred from analysis of diatom assemblages in the stratigraphic record of dated lake sediment cores. These paleolimnological studies show, for example, that about 25%-35% of the lakes in the Adirondack Mountains with the lowest ability to neutralize acids (acid neutralizing capacity < 400 µeq/L) have become more acidic since preindustrial times (Cumming et al. 1992). Lakes in other regions of the country have also acidified but not to the same extent (Charles et al. 1989). The amount of acidification inferred from diatoms is related to the level of atmospheric loading of strong acids and the ability of watersheds to neutralize those acids. Analysis of diatoms and sedimentary remains of other biological groups (e.g., chrysophytes, chironomids, Cladocera) reveals that acidic deposition has had significant effects on aquatic communities in many lakes. Numbers of taxa are reduced, but some acid-tolerant taxa have significantly increased in abundance.

Lake Eutrophication

Population estimates of the numbers of lakes in New England and New York that are more eutrophic now than in presettlement times are being obtained from analyses of diatom assemblages from recent and preindustrial levels of sediment cores taken as part of the U.S. Environmental Protection Agency's Surface Water component of the Environmental Monitoring and Assessment Program (EMAP; Dixit and Smol 1994). The approach of examining lake eutrophication by using diatom assemblages has been widely applied in North America and throughout the world.

Rivers and Streams

Many long-term diatom data sets exist that can inform us about trends in water quality. The monitoring program conducted by the Federal Water Pollution Control Agency in the 1960's tracked the status of major rivers throughout the country (Williams and Scott 1962). Monitoring of diatom assemblages in rivers and streams is just beginning as part of the U.S. Geological Survey's National Water Quality Assessment (NAWQA) and of the Environmental Monitoring and Assessment Program. The Academy of Natural Sciences of Philadelphia has long-term records for several rivers in the eastern United States. Many of these records show that the quality of water downstream from industrial effluent outfalls and sewage treatment plants has improved markedly, but others show worsening conditions, often due to the increased number of sources of stress along the river or in the watershed. Much more could be learned about trends by simply analyzing the immense data that already exist, especially by using new quantitative techniques developed in the past 5-10 years.

Climate Change

Diatom assemblage composition is sensitive to changes in water level, salinity, ice cover, wind-mixing patterns, and other characteristics directly and indirectly affected by climate. Paleolimnological studies of sediment cores are providing valuable data on climate change over the past hundreds to thousands of years, which are essential for understanding the nature and magnitude of ecosystem change that can be expected in future years.

Conclusions

The ability to infer ecosystem status and trends from diatoms is largely dependent on the availability of ecological data for the species occurring at study sites. The amount of such data is accumulating at an increasingly rapid rate, but it is in many separate data bases. These need to be coordinated so that users will have easier access to the data that already exist.

References

Charles, D.F., R.W. Battarbee, I. Renberg, H. Van Dam, and J.P. Smol. 1989. Paleoecological analysis of lake acidification trends in North America and Europe using diatoms and chrysophytes. Pages 207-276 *in* S.A. Norton, S.E. Lindberg, and A.L. Page, eds. Acidic precipitation. Soils, aquatic processes, and lake acidification. Vol. 4. Springer-Verlag, New York.

Cumming, B.F., J.P. Smol, J.C. Kingston, D.F. Charles, H.J.B. Birks, K.E. Camburn, S.S. Dixit, A.J. Uutala, and A.R. Selle. 1992. How much acidification has occurred in Adirondack region lakes (New York, USA) since preindustrial times? Canadian Journal of Fisheries and Aquatic Sciences 49:128-141.

Dixit, S.S., and J.P. Smol. 1994. Diatoms and indicators in the Environmental Monitoring and Assessment Program —Surface Waters (EMAP—SW). Environmental Monitoring and Assessment 31:275-306.

Patrick, R., and C.W. Reimer. 1966, 1975. The diatoms of the United States. Vols. 1 and 2. The Academy of Natural Sciences of Philadelphia. Vol. 1:688 pp. Vol. 2:213 pp.

Williams, L.G., and C. Scott. 1962. Principal diatoms of major waterways of the United States. Ecology 7:365-379.

For further information:

Donald Charles
Academy of Natural Sciences
1900 Benjamin Franklin Parkway
Philadelphia, PA 19103

Courtesy C. McIvor

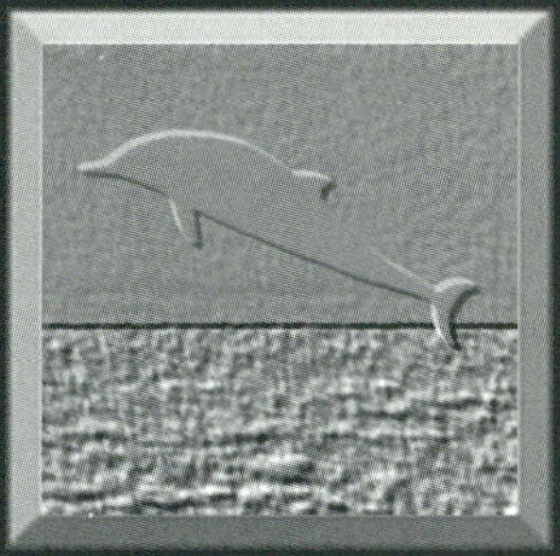

Coastal & Marine Ecosystems

by
Science Editor
Carole C. McIvor
National Biological Service
Arizona Cooperative Fish and Wildlife Research Unit
University of Arizona
Tucson, AZ 85721

Overview

The quantity and health of the nation's coastal and marine resources have declined over historical time at the species, community, and ecosystem levels. All articles in this section implicate human activities as contributing to these declines. Human impacts on the coastal and nearshore marine zone include urbanization (direct loss of habitat, lowered water quality), shoreline modification (dredging and filling, diking and impoundments), overfishing, and high-density recreational use.

Some portion of the overall downward trend is directly attributable to natural processes. Hurricanes and coastal storms can have significant negative impacts on both barrier islands (Williams and Johnston) and seagrass beds (Handley, Onuf). Rising sea level and coastal subsidence—natural processes that are likely being accelerated by anthropogenic (human-caused) activities—are responsible for coastal wetland loss in Louisiana (Johnston et al.). Rising sea level is also implicated in the erosion of barrier islands (Williams and Johnston). The inescapable conclusion is, however, that even where natural processes play a role, human impact is of equal or greater importance to the long-term health of these resources.

Despite overall declines in coastal and marine resources, there is some room for cautious optimism. Some coral reefs are far enough from human habitation that they are probably stable and not declining (Jameson). Despite changes in the relative abundances of native fish species and the introduction of exotic species in the tidal portion of the Hudson River, no native fish species have been extirpated within the period of record (1936 to 1990) (Daniels). The population trend for manatee (*Trichechus manatus*) in Florida appears stable and perhaps slightly increasing (Lefebvre and O'Shea). Recent local reversals in the decline of seagrasses have occurred in Chesapeake Bay (Pendleton) and in lower Tampa and Little Sarasota bays (Handley). These successes, however, are tempered by the realization that human populations in coastal states are projected to substantially increase soon.

It is clear from these articles that the quality and extent of our information bases for judging status and trends of our coastal and marine resources are often inadequate. Whereas the areal coverage of some ecosystems can be judged by comparison of remotely sensed data (e.g., coastal wetlands), gathering analogous information on other ecosystems or components (e.g., fishes on coral reefs) requires much smaller scale, more labor-intensive efforts. In their review of Florida Keys reef fishes, Smith-Vaniz et al. were forced to rely on a combination of human population trends and the status of Keys' reefs combined with information collected from commercial fisheries' landings to infer the health of reef fishes: no single reef site has ever been repeatedly surveyed for fish abundance over time. This example clearly demonstrates that to better judge the status of our coastal and marine resources in the future, carefully chosen and designed long-term monitoring is required.

Nearshore Fish Assemblage of the Tidal Hudson River

by
Robert A. Daniels
New York State Museum
Biological Survey Laboratory

The Hudson River drains about 45,000 km^2 (17,370 mi^2), most of it in eastern New York. Although this is a young river with a relatively small watershed at higher latitudes, the Hudson and its tributaries support a rich fish fauna of more than 200 species (Smith and Lake 1990). This fauna is a diverse mixture of native and exotic freshwater species, diadromous (migratory between fresh and salt waters) fish, and marine strays (Barnhouse et al. 1988). More than 150 of these species are reported from the tidal portion of the river that extends 243 km (151 mi) from the Battery on Manhattan Island to the Troy Lock (Fig. 1); of these, about 80 species are freshwater or diadromous forms and 50 species occur regularly in nearshore areas (Smith 1985). During the last half-century, the nearshore fauna of the tidal portion of the river has undergone two types of changes: species have been added to and deleted from the fauna and the relative abundances of the dominant species have changed.

I explore differences in the nearshore fish assemblages of the Hudson River by comparing information on the distribution and abundance of fish collected between 1936 and 1990. This comparison offers only a coarse look at change in the fish assemblage and provides little information on trends. The nearshore fish assemblage of the Hudson River is dynamic and changes on a daily, seasonal, and annual basis.

Surveys of Fish Fauna, 1936-92

The study of Hudson River fish dates to Samuel Mitchill's publication on the fish of New York (Mitchill 1815). DeKay (1842) and Bean (1903) also provided information on fish in the Hudson River, but the first synoptic survey of the fish in the river system was not undertaken until 1936 (Greeley 1937). The watershed surveys of New York conducted between 1926 and 1939 included a detailed survey of fish distribution and abundance in the lower Hudson River drainage. Fish collected during these surveys were vouchered; specimens are housed at the New York State Museum (NYSM). Beginning in the early 1970's, interest in the fish of the Hudson River increased dramatically (Limburg et al. 1986), and several long- and short-term monitoring programs began. Data collection continues in many of the long-term programs.

Courtesy R. Daniels, NY State Museum

Hudson River.

To examine change in the nearshore fish assemblage of the Hudson River, I used selected information from the 1936 watershed survey; NYSM surveys conducted between 1990 and 1992; intensive site surveys conducted between 1976 and 1979 by Lawler, Matusky and Skelly Engineers (LMS); and surveys supported by Con Edison between 1974 and 1989. Because techniques and equipment vary among the surveys, I have included in the analyses only information collected by workers using seines. Still, the size of the seines used, the mesh size, and the area sampled differ among the surveys and contribute a bias not easily quantified. Because this analysis is relatively coarse, any biases that may exist in the data should be masked. Furthermore, in most of the analyses I have made comparisons within data bases. Comparisons between data bases are used primarily with presence and absence applications.

The 1936 watershed survey collected information on fish from 112 sites in the tidal Hudson River (Fig. 1). All sampling was conducted during summer. Fish collected during this survey were identified and counted or ranked; the ranking system may have varied among the crews. To compare abundance, I assigned numbers to the ranks in the fieldbooks and compared my assigned number to the actual number of preserved fish. In 20 comparisons of each of the five ranks, the assigned number equaled or underrepresented the number preserved 73% of the time; therefore, the abundance estimates should be conservative.

Between 1990 and 1992, I collected information on fish abundance and distribution from several sites on the tidal portion of the Hudson River. Most information discussed here is from work done at four sites during the summer of 1990 (Fig. 1). These sites typified the nearshore

Painting of pumpkinseed (*Lepomis gibbosus*), a persistent species of the Hudson River.

Courtesy NY Environmental Conservation Department

habitats present along the entire main channel. LMS intensively collected fish from four sites between 1976 and 1979 (Fig. 1). Day and night sampling, using seines to collect fishes at weekly or biweekly intervals, began early in spring

Table 1. Freshwater and diadromous fishes collected from nearshore areas of the Hudson River, 1936-92. Records from 1936 are from the watershed survey of the lower Hudson River, with identifications verified, and specimens vouchered. Records from 1974-89 are from the Con Edison data base, no specimens vouchered. Records from 1990-92 are from New York State Museum (NYSM) surveys and other additional specimens, vouchered.

Species	Fish surveys			
	Watershed 1936	Con Ed 1974-80	Con Ed 1981-89	NYSM 1990-92
Longnose gar				x
American eel	x	x	x	x
Blueback herring	x	x	x	x
Alewife	x	x	x	x
American shad	x	x	x	x
Gizzard shad		x	x	x
Bay anchovy	x	x	x	x
White catfish	x	x	x	x
Brown bullhead	x	x	x	x
White sucker	x	x	x	x
Creek chubsucker	x			
Northern hog sucker	x	x	x	
Goldfish	x	x	x	
Grass carp				x
Common carp	x	x	x	x
Rudd				x
Golden shiner	x	x	x	x
Creek chub	x	x		
Fallfish	x	x	x	x
Eastern silvery minnow	x	x	x	x
Comely shiner	x	x	x	x
Emerald shiner	x	x	x	x
Bridle shiner	x	x	x	
Spottail shiner	x	x	x	x
Common shiner	x	x	x	
Spotfin shiner	x	x	x	
Fathead minnow		x	x	
Blacknose dace		x	x	
Longnose dace		x	x	
Rainbow smelt	x	x		
Redfin pickerel	x	x	x	
Northern pike		x	x	x
Chain pickerel	x	x	x	x
Banded killifish	x	x	x	x
Mummichog	x	x	x	x
Inland silverside	x	x	x	x
Fourspine stickleback	x	x	x	x
Threespine stickleback		x	x	
White perch	x	x	x	x
White bass			x	
Striped bass	x	x	x	x
Rock bass	x	x	x	x
Bluespotted sunfish	x			
Redbreast sunfish	x	x	x	x
Pumpkinseed	x	x	x	x
Bluegill	x	x	x	x
Smallmouth bass	x	x	x	x
Largemouth bass	x	x	x	x
White crappie		x	x	x
Black crappie	x	x	x	x
Tessellated darter	x	x	x	x
Yellow perch	x	x	x	x
Logperch	x	x		
Shield darter	x	x		
Freshwater drum				x
Total number of species	43	48	45	38

and continued until December. The data base from Con Edison includes information from 31,582 nearshore, shallow-water sites throughout the 243-km (151-mi) course of the lower Hudson River. These collections were made between 1974 and 1989.

Changes, 1936-90

The changes in the nearshore fish assemblages of the Hudson River that have occurred during the past six decades are illustrated in several ways. First, the component species have changed, although species richness (number of species in the assemblage) has remained relatively constant (Table 1). During the 1936 survey, the assemblage had 43 freshwater and diadromous species. Based on recent NYSM collections, the assemblage consists of 38 species. Recently introduced to the river are gamefishes such as northern pike (*Esox lucius*) and white crappie (*Pomoxis annularis*) and exotic fishes such as rudd (*Scardinius erythrophthalmus*) and grass carp (*Ctenopharyngodon idella*). Two additions from undocumented sources also included the gizzard shad (*Dorosoma cepedianum*) and freshwater drum (*Aplodinotus grunniens*). Several species that remain common in tributary streams are now extirpated or extremely rare (e.g., bridle shiner).

In addition, the relative abundance of most resident species (excluding diadromous forms) has changed (Table 2). The two dominant resident species in 1936 (spottail shiner [*Notropis hudsonius*] and white perch [*Morone americana*]) made up 34% of the individuals in the assemblage. The same two species remained dominant in the 1990 survey, but have almost doubled their relative abundance to 64% of the individuals in the assemblage. The relative abundances of an additional five persistent species have declined between the two sampling events, but only slightly. Thus, declines in relative abundance were most noticeable in the remaining species (not dominant or persistent) of the freshwater component of the river fish assemblage. In 1936, 36 species made up 26% of the catch, while the remaining species accounted for 7% in 1990.

The diadromous fishes typically dominated, by number, the nearshore assemblage during summer (Fig. 2) because of the presence of young-of-year individuals. The most common species in all samples included blueback herring (*Alosa aestivalis*), alewife (*A. pseudoharengus*), American shad (*A. sapidissima*), bay anchovy (*Anchoa mitchilli*), American eel (*Anguilla rostrata*), and striped bass (*Morone saxatilis*). The difference between 1936 and all other years was the curious near-absence of

Fig. 1. The tidal portion of the Hudson River, New York, showing areas where some fish collections have been made over the last six decades.

Table 2. The relative abundance of resident fishes (percentage) in nearshore assemblages in the summers of 1936 and 1990, lower Hudson River, New York.

Species	Survey 1936	Survey 1990
Spottail shiner	20	33
White perch	14	31
Total for two dominant species	34	64
Banded killifish	14	12
Tessellated darter	10	6
Mummichog	7	4
Pumpkinseed	5	4
Redbreast sunfish	4	3
Total for five persistent species	40	29
Fourspine stickleback	6	
Eastern silvery minnow	4	
Goldfish	3	
Fallfish		2
Bridle shiner	2	
White sucker	2	
White catfish	2	
Golden shiner	2	
Gizzard shad		1
Others	5	4
All other species	26	7

blueback herring in 1936 when this species was taken at only 4 of the 112 sites sampled. In 1990, and during the last two decades, blueback herring dominated the summer catches at nearshore sites (Fig. 2). In the nearshore fish assemblage of the Hudson River in 1974-89, the five diadromous species dominated throughout the sample period (Fig. 2).

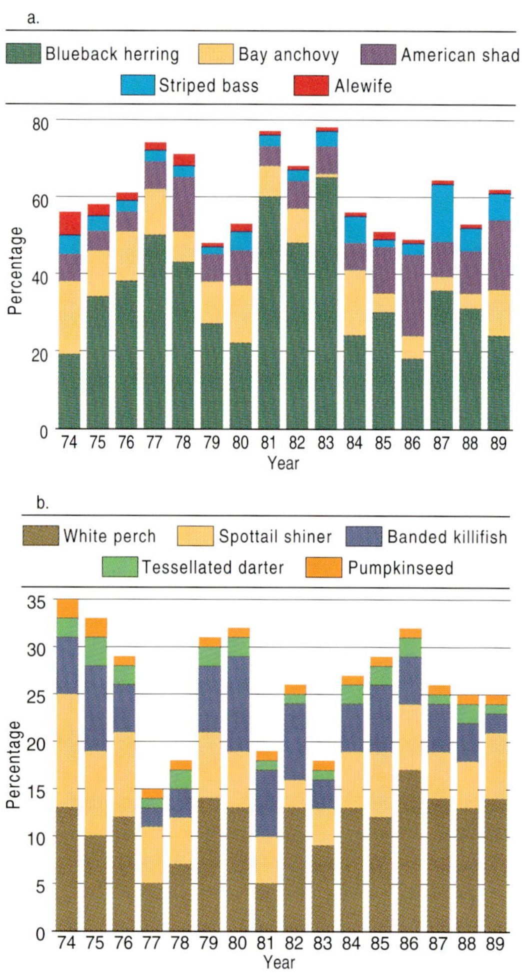

Fig. 2. Changes in the relative abundance of the 10 most numerous fishes in the nearshore fish assemblage of the Hudson River, New York, 1974-89 (data from Con Edison files). Changes in percentage abundance in (a) 5 diadromous fish species and (b) 5 resident freshwater species.

Despite fluctuations in each of the most abundant species, no obvious trends in relative abundance were apparent although the relative abundance of other species has changed. For example, the abundance of Atlantic silverside (*Menidia menidia*), a marine stray, has increased, while other species, such as two resident fish, emerald shiner (*Notropis atherinoides*) and goldfish (*Carassius auratus*), and the diadromous rainbow smelt (*Osmerus mordax*), have dramatically decreased in relative abundance (Fig. 3). Relative fish abundances exhibited site, diel, and seasonal variation.

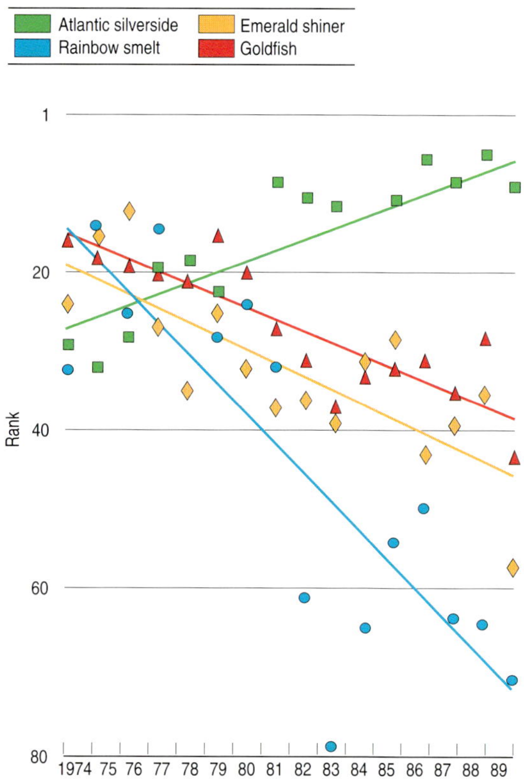

Fig. 3. Annual trends in the abundance of four fish species from the nearshore Hudson River assemblage. Rank is based on total number of fish caught during the year; the most abundant fish caught received a rank of 1, the least abundant a rank of 82.

Implications of Changes

Change in the nearshore fish assemblage of the tidal portion of the Hudson River is continuous. To identify trends in the abundance of an assemblage made up of resident freshwater and estuarine species, diadromous fishes, and marine strays, data must be collected in ways that account for the dynamic qualities of the species involved. Although the Hudson River is among the most-studied aquatic systems in North America, data necessary to confirm population trends in its fish assemblage are scant. Abundance data are best for some commercially important and protected fishes. Data on other species are often inadequate, rare, or nonexistent. Early or baseline data are often incompatible with modern surveys, and long-term data bases, although growing, are still in their early years.

Some changes appear to be trends. First, the number of fish species in the Hudson River appears to be increasing. The presence of recent entrants into the river—such as gizzard shad, rudd, grass carp, central mudminnow (*Umbra limi*), white bass (*Morone chrysops*), and freshwater drum—may create management concerns in the future.

Second, another group of fish appears to be declining, although it seems that only a few species, if any, have been extirpated. This group consists of fish that were common in the 1936 survey of the river but rare in all recent collections, including the bridle shiner (*Notropis bifrenatus*), common shiner (*Luxilus cornutus*), comely shiner (*Notropis amoenus*), spotfin shiner (*Cyprinella spiloptera*), creek chub (*Semotilus atromaculatus*), northern hog sucker (*Hypentelium nigricans*), and creek chubsucker (*Erimyzon oblongus*). These fish remain common, or at least present, in tributaries to the lower Hudson River. Their absence from the main channel may result from increasing development and loss of riparian vegetation at the mouths of many tributaries, which may isolate tributary populations from those of the main channel and lead to the creation of sub- or new populations.

The third apparent trend is that, although richness is increasing, diversity (an expression that includes the number of species and their relative abundance) in the nearshore fish assemblage has declined because of the increase in population size of the dominant species.

Studies that allow a better assessment of trends in the Hudson River fish assemblage will provide broad-based benefits. Management agencies, commercial fishing operations, and individual anglers, for example, all have an interest in the fisheries and fish of the river. Other river users, such as municipal planners and utility companies, also will gain from increased knowledge of the population trends of river-dwelling organisms because the trends reflect changes in water-quality conditions.

References

Barnhouse, L.W., R.J. Klauda, D.S. Vaughan, and R.L. Kendall, eds. 1988. Science, law, and Hudson River power plants. A case study in environmental impact assessment. American Fisheries Society Monograph 4. 347 pp.

Bean, T.H. 1903. Catalogue of the fishes of New York. New York State Museum Bull. 60, Zoology 9. 784 pp.

DeKay, J.E. 1842. Zoology of New York or the New York fauna, Part IV. Fishes. W. and A. White and J. Visscher, Albany, NY. 415 pp.

Greeley, J.R. 1937. Fishes of the area with annotated list. Pages 45-103 *in* E. Moore, ed. A biological survey of the lower Hudson watershed. Supplemental to 26th Annual Report, State of New York Conservation Department, Albany.

Limburg, K.E., M.A. Moran, and W.H. McDowell. 1986. The Hudson River ecosystem. Springer-Verlag, New York. 331 pp.

Mitchill, S.L. 1815. The fishes of New York, described and arranged. Transactions of the Literary and Philosophical Society (1814) 1:355-492.

Smith, C.L. 1985. Inland fishes of New York State. Department of Environmental Conservation, Albany. 522 pp.

Smith, C.L., and T.R. Lake. 1990. Documentation of the Hudson River fish fauna. American Museum Novitates 2981. 17 pp.

For further information:

Robert A. Daniels
New York State Museum
145 Jordan Rd.
Troy, NY 12180

Natural Resources in the Chesapeake Bay Watershed

by
Edward Pendleton
National Biological Service

The Chesapeake Bay is the nation's largest estuary; its watershed covers 165,760 km^2 (64,000 mi^2) and is occupied by 13 million people. By the 1980's, the bay's waters were enriched with nutrients from agriculture and loaded with pollutants from urban and suburban areas. The bay's submersed grasses were disappearing, fisheries 2 centuries old were in serious decline, and wetlands and other natural habitats were under continuing threats of development (Flemer et al. 1983).

In 1983 the federal government, Virginia, Maryland, Pennsylvania, the District of Columbia, and the Chesapeake Bay Commission formally declared their intent to work cooperatively to restore the natural resources of the bay. Their partnership, known as the Chesapeake Bay Program, attacked water-quality problems by adopting measures to reduce inputs of nitrogen and phosphorus from urban, industrial, and agricultural sources and to increase levels of dissolved oxygen in bay waters. Simultaneously, scientists and managers determined the status of bay species and natural habitats and began to track historical and ongoing trends.

Status and trends assumed special relevance as they were incorporated into managerial objectives and goals, or as indices of the success of programs and policies (Chesapeake Bay Implementation Committee 1988). Trends for three habitats—submersed aquatic vegetation beds, wetlands, and forests; four key aquatic species—oysters (*Crassostrea virginica*), blue crabs (*Callinectes sapidus*), striped bass (*Morone saxatilis*), and American shad (*Alosa sapidissima*); and waterfowl are summarized below. These trends represent a mixture of moderate successes and continuing challenges for managers of the bay.

Status and Trends

The status and trends of key habitats and species in the Chesapeake Bay are based on multiple annual surveys conducted by state and federal agencies. Perhaps the most comprehensive is a survey of the bay's submersed aquatic plant community; each year, the extent of submersed aquatic vegetation (SAV) is estimated by aerial photography of the entire bay and the tidal portions of its major tributaries (Orth and Moore 1983). Wetland areas are likewise estimated from aerial photographs and have been extrapolated for the watershed from a finite number of sites in various geographic strata for three time periods (Tiner and Finn 1986; Tiner et al., U.S. Fish and Wildlife Service, unpublished data). Approximately every 8 years, forested areas are estimated for each state in the bay watershed by the U.S. Department of Agriculture Forest Service's Forestland Inventory from satellite imagery (Chesapeake Bay Program 1993).

Many aquatic animal species that are surveyed annually (including those addressed here) support commercial and recreational fisheries or hunting and bird watching. Oyster, blue crab, striped bass, and American shad populations are estimated from commercial landings, and are augmented at times with surveys that are independent of fishery statistics, such as numbers of oyster spat that set each year, estimates of the biomass of spawning striped bass, or numbers of juvenile striped bass per seine haul (the young-of-year index). Waterfowl have been counted during their wintering season on the bay by the aerial Midwinter Surveys since the 1940's.

Submersed Aquatic Vegetation (SAV)

Beginning in the late 1960's and continuing into the 1970's, the distribution and abundance of a community of 20 species of submersed grasses declined throughout the bay because of nutrient enrichment, increased loads of suspended sediments, and other factors (Stevenson and Confer 1978; Orth and Moore 1983). In 1978 the first aerial survey estimated 16,500 ha (40,700 acres) of SAV in the bay (Anderson and Macomber 1980). The next year, 15,400 ha (38,000 acres) were documented (Orth et al. 1985); since that time, annual surveys have shown modest but continual increases in SAV coverage to an estimated 28,600 ha (70,600 acres; Orth et al. 1993; Fig. 1). Recent increases represent gains in brackish mid-bay regions and are tempered somewhat by slow or no SAV recovery in freshwater areas in the upper bay and by the spread of the exotic species hydrilla (*Hydrilla verticillata*) in the tidal freshwater portions of the Potomac River.

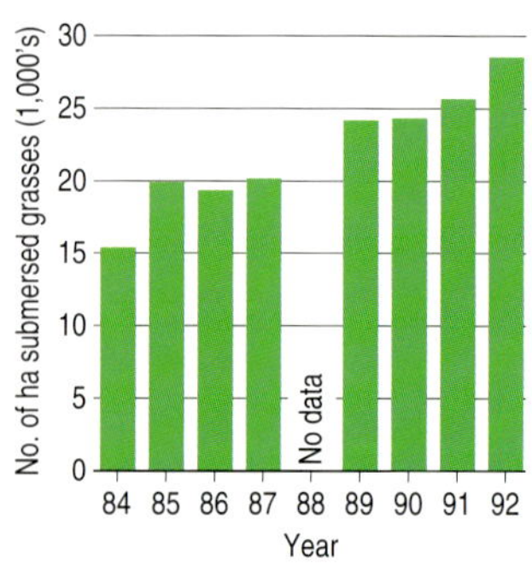

Fig. 1. Area of submersed grasses (submersed aquatic vegetation or SAV) in the Chesapeake Bay, 1984-92.

Wetlands

The status and trends for more than a million acres of wetlands in the Chesapeake Bay watershed have been estimated over two time periods, from the mid-1950's to the late 1970's and early 1980's (Tiner and Finn 1986), and from this period to 1989 (Tiner et al., USFWS, unpublished data). Dominant wetland types include nontidal forested wetlands (60% of total wetlands), nontidal shrub-scrub wetlands (10%), and salt and freshwater marshes (10% each).

Losses occurred in all of these wetland types during the period from the mid-1950's to late 1970's and early 1980's. About 9% of the watershed's salt marshes were lost to dredging, impoundment, and filling. Nontidal wetlands declined by nearly 6% as a result of being drained and converted to agriculture or impounded to form ponds, lakes, and reservoirs. During the 1980's, losses continued; the rate of marsh loss declined, while forested wetland losses increased. Overall, there was an estimated net loss of 0.5% of estuarine wetlands and a net loss of 2.0% of palustrine wetlands (roughly equal to tidal and nontidal wetlands) during the 1980's. These trends mirror historical losses over the past 200 years (Dahl 1990).

Forests

An estimated 95% of the Chesapeake Bay watershed was forested before European settlement; around 58% remains today (Chesapeake Bay Program 1993; Fig. 2). This percentage is declining for the first time in over a century because of recent forest clearing for urban and suburban development. Forest clearing has proceeded unevenly over the watershed, with some drainages intact and others as much as 85% cleared.

Oysters

Oyster landings in Chesapeake Bay have experienced a 95% decline since 1980 and are estimated to be at their lowest recorded level (Kennedy 1991; National Marine Fisheries Service, Annapolis, Maryland, unpublished data; Fig. 3a). Although reproductive success of the oyster remains high (as measured by larval oyster, or spat, set on oyster reefs and other suitable substrates; Maryland Department of Natural Resources, Oxford, Maryland, unpublished data), populations have suffered from harvest to low levels, two parasitic diseases (Dermo and MSX), habitat loss (including decreased water quality), and predation.

Blue Crabs

Blue crab populations in the Chesapeake Bay, as indicated by commercial landings data, vary from year to year, making trends less apparent than those of other bay species (Lipcius and Van Engel 1990; National Marine Fisheries Service, Annapolis, Maryland, unpublished data; Fig. 3b). Populations appear to follow a 7-12 year cycle and may be in the "trough" of this cycle at present. This perception and increasing annual harvests as fishery efforts shift to crabs from other species have prompted Maryland and Virginia to begin to regulate the blue crab fishery.

Striped Bass

Probably the most monitored fish species in the bay, striped bass populations have increased about 25% a year since 1984, after falling to low levels in the early 1980's (Gibson 1993; Fig. 3c). Increases are at least partially attributed to a moratorium on harvest from 1985 to 1989 to allow improvement of the age and sex structure of the spawning stock. The 1993 young-of-the-year index, a measure of numbers of juvenile fish entering the population, is the highest on record (National Marine Fisheries Service, Annapolis, Maryland, unpublished data) and may be related to the timing of high freshwater flows, nutrient inputs, and increases in planktonic prey (Blankenship 1994), which may interact to allow large numbers of young fish to survive after hatching.

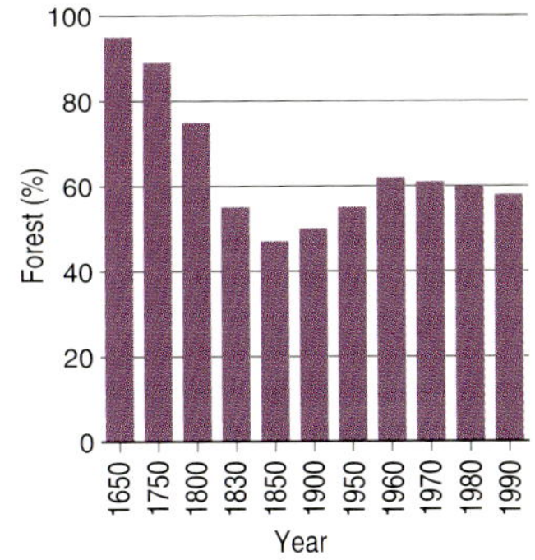

Fig. 2. Forest cover in the Chesapeake Bay watershed, 1650-1990.

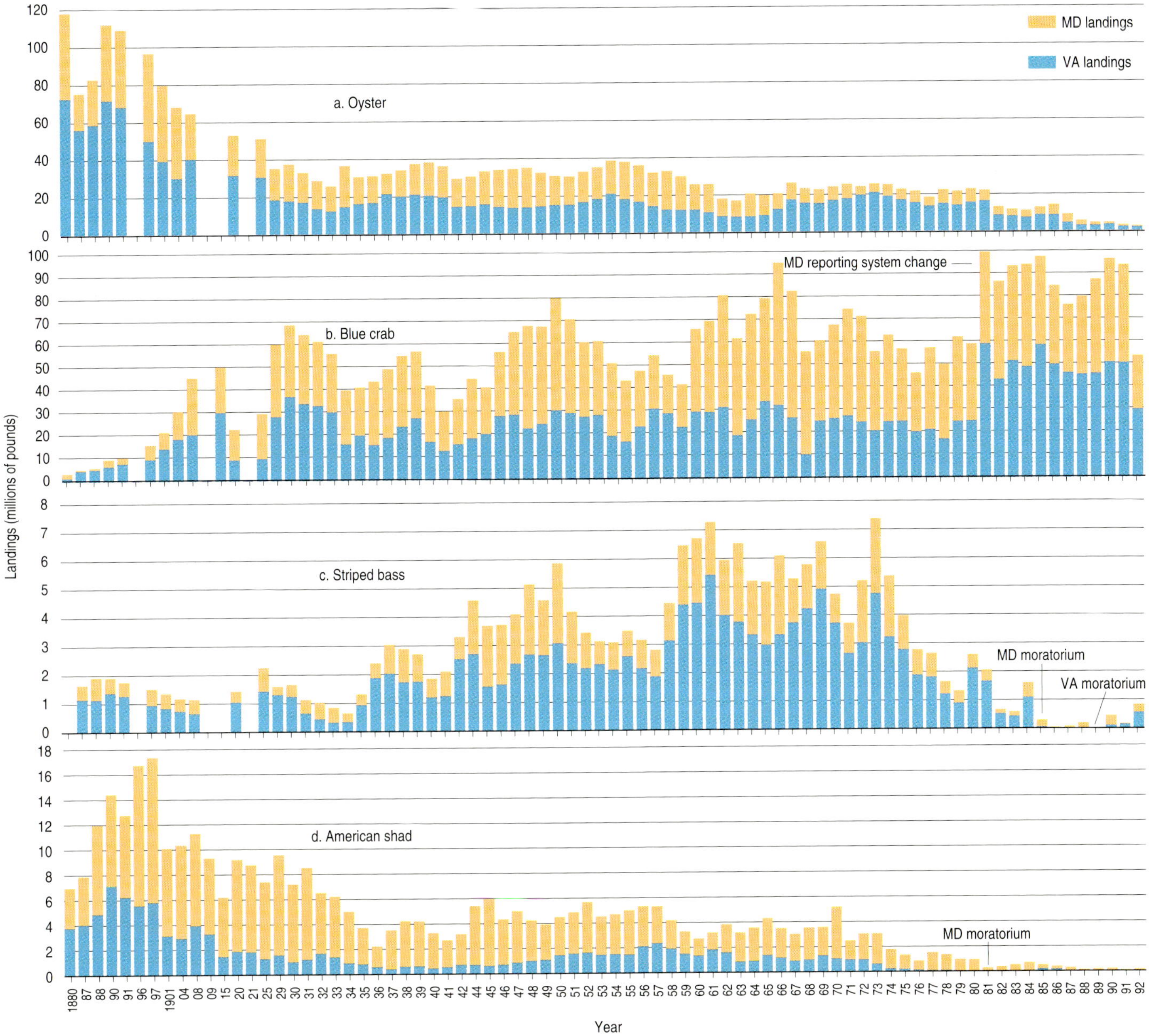

Fig. 3a-d. Commercial landings in the Chesapeake Bay of (a) oysters, 1880-1992; (b) blue crab, 1880-1992 (population increases in the 1980's are partially due to mandatory catch reporting requirements); (c) striped bass, 1887-1992; and (d) American shad, 1880-1992.

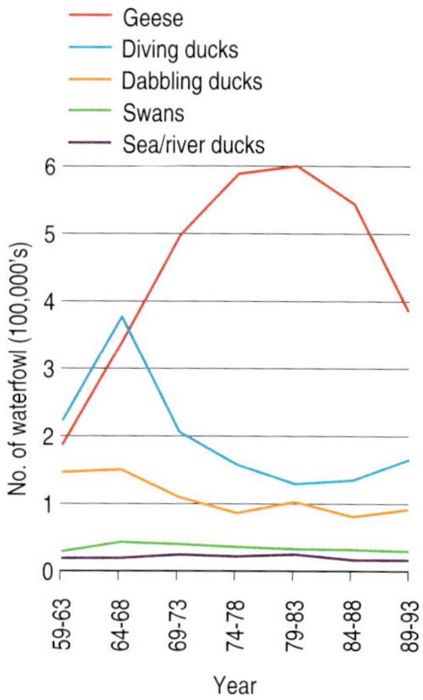

Fig. 4. Trends in waterfowl abundance in the Chesapeake Bay, based on 5-year running means, 1959-93.

American Shad

Like striped bass, American shad have declined in Chesapeake Bay in recent decades; unlike the stripers, this species has not shown a strongly positive population response despite moratoria on fishing in Maryland and Virginia. Long-term trends show a drastic decline in fishery landings to the point of almost total disappearance in the bay (National Marine Fisheries Service, Annapolis, Maryland, unpublished data; Fig. 3d). This decline has been related to blockages of spawning streams by dams, overharvest, and pollution (Blankenship 1993). Population estimates in 1992 and 1993 for the upper bay, where shad are counted during their upstream migration to the Susquehanna River, show a reversal of a recent positive trend, for reasons yet unknown.

Waterfowl

Midwinter surveys estimate an average of more than one million waterfowl along the Atlantic Flyway winter in Chesapeake Bay each year (USFWS, Chesapeake Bay Field Office, Annapolis, Maryland, unpublished data). Of the 28 species of ducks, geese, and swans represented in this total, some are declining in abundance, whereas others show increasing or variable trends in abundance (Fig. 4; Table). In general, duck numbers declined and goose populations increased since the late 1950's as submersed aquatic vegetation and other duck foods dwindled and changing farming practices left more grain in fields for geese. Recently, geese have also declined as excessive harvest and poor production on northern breeding grounds reduced their numbers. Their distribution along the Atlantic Flyway has also shifted to the north. Mallards (*Anas platyrhynchos*) and introduced mute swans (*Cygnus olor*) have shown moderate increases, but many other species, including American black duck (*Anas rubripes*), wigeon (*A. americana*), northern pintail (*A. acuta*), canvasback (*Aythya valisineria*), and redhead (*A. americana*), have declined or stabilized at population levels substantially lower than in the 1950's.

Table. Trends for waterfowl in Chesapeake Bay, based on 5-year running means from 1959 to 1993 (USFWS, Chesapeake Bay Field Office, Annapolis, Maryland, unpublished data).

Group	Trend
Swans and geese	
Tundra swan	Variable
Mute swan	Increasing
Snow goose	Increasing
Canada goose	Decreasing
Brant	Variable
Dabbling ducks	
Mallard	Increasing
Black duck	Decreasing
Gadwall	Variable
Teal (blue- and green-winged)	Variable
American wigeon	Decreasing
Northern pintail	Decreasing
Northern shoveler	Variable
Bay (diving) ducks	
Canvasback	Decreasing
Redhead	Decreasing
Ring-necked duck	Variable
Scaup	Variable
River and sea ducks	
Goldeneye	Decreasing
Bufflehead	Increasing
Ruddy duck	Decreasing
Scoter	Decreasing
Oldsquaw	Decreasing
Mergansers	Increasing

References

Anderson, R.R., and R.T. Macomber. 1980. Distribution of submerged vascular plants, Chesapeake Bay, Maryland. Environmental Protection Agency, Chesapeake Bay Program, Annapolis, MD. 126 pp.

Blankenship, K. 1994. Bay bounces back from record-setting spring "freshet." Bay Journal 3(10):1,7.

Blankenship, K. 1993. Biologists puzzled by sudden decline of East Coast shad. Bay Journal 3(9):1,6.

Chesapeake Bay Implementation Committee. 1988. The Chesapeake Bay Program: a commitment renewed. U.S. Environmental Protection Agency, Chesapeake Bay Program Office, Annapolis, MD. 88 pp.

Chesapeake Bay Program. 1993. Environmental indicators: measuring our progress. U.S. Environmental Protection Agency, Chesapeake Bay Program Office, Annapolis, MD.

Dahl, T.E. 1990. Wetlands losses in the United States 1780's to 1980's. U.S. Department of the Interior, Fish and Wildlife Service, Washington, DC. 21 pp.

Flemer, D.A., G. Mackiernan, W. Nehlsen, and V. Tippe. 1983. Chesapeake Bay: a profile of environmental change. U.S. Environmental Protection Agency, Chesapeake Bay Program Office, Annapolis, MD. 200 pp.

Gibson, M.R. 1993. Historical estimates of fishing mortality on the Chesapeake Bay striped bass stock using separable virtual population analysis applied to market class catch data. Report to the Atlantic States Marine Fisheries Commission Striped Bass Technical Committee. Atlantic States Marine Fisheries Commission, Washington, DC. 20 pp.

Kennedy, V.S. 1991. Eastern oyster. Pages 3-1 to 3-20 *in* Habitat requirements for Chesapeake Bay living resources. 2nd ed. Living Resources Subcommittee, Chesapeake Bay Program. U.S. Fish and Wildlife Service, Annapolis, MD.

Lipcius, R.M., and W.A. Van Engel. 1990. Blue crab population dynamics in Chesapeake Bay: variation in abundance (York River, 1972-1988) and stock-recruit functions. Bull. of Marine Science 46:180-194.

Orth, R.J., and K.A. Moore. 1983. Chesapeake Bay: an unprecedented decline in submerged aquatic vegetation. Science 222:51-53.

Orth, R.J., J.F. Nowak, G.F. Anderson, and J.R. Whiting. 1993. Distribution of submerged aquatic vegetation in the Chesapeake Bay and tributaries and Chincoteague Bay—1992. U.S. Environmental Protection Agency, Chesapeake Bay Program Office, Annapolis, MD. 161 pp.

Orth, R.J., J. Simons, R. Allaire, V. Carter, L. Hindman, K. Moore, and N. Rybick. 1985. Distribution of submerged aquatic vegetation in the Chesapeake Bay and tributaries—1984. EPA Final Report. Chesapeake Bay Program, Annapolis, MD. 155 pp.

Stevenson, J.C., and N.M. Confer. 1978. Summary of available information on Chesapeake Bay submerged vegetation. U.S. Fish and Wildlife Service Biological Rep. FWS/OBS-78/66. 335 pp.

Tiner, R.W., and J.T. Finn. 1986. Status and recent trends of wetlands in five mid-Atlantic states: Delaware, Maryland, Pennsylvania, Virginia, and West Virginia. U.S. Fish and Wildlife Service, Region 5, Newton Corner, MA, and U.S. Environmental Protection Agency, Region 3, Philadelphia, PA. Cooperative Tech. Publ. 40 pp.

For further information:

Edward Pendleton
National Biological Service
Leetown Science Center
1700 Leetown Rd.
Kearneysville, WV 25430

Florida Manatees

by
Lynn W. Lefebvre
Thomas J. O'Shea
National Biological Service

The endangered Florida manatee (*Trichechus manatus latirostris*) is a survivor. It is one of only three living species of manatees which, along with their closest living relative, the dugong (*Dugong dugon*), make up the Order Sirenia. This taxonomic distinctiveness reflects their evolutionary and genetic uniqueness. Sirenians are the only herbivorous marine mammals; manatees feed on seagrasses; freshwater plants, including nuisancc species such as hydrilla and water hyacinth; and even some shoreline vegetation. Because manatees depend on marine, estuarine, and freshwater ecosystems, our efforts to protect them necessitate protection of aquatic resources.

Life-history Research

Major efforts have concentrated on better quantification of Florida manatee populations, emphasizing reproduction, population size, and mortality. Most of the information on manatee reproduction (Table) comes from long-term studies based on recognizable individuals at winter aggregation sites (e.g., Rathbun et al. in press). Florida manatees are at the northern limit of the species' range and must seek warmer waters when water temperatures drop below about 20 °C. Natural springs, such as those found in Crystal River on the west coast and Blue Spring on the St. Johns River, and discharges from industrial plants provide warmwater refuges for hundreds of manatees during cold periods.

Individual manatees are recognized at these sites largely through their unique scar patterns, caused by boat strikes (Figs. 1a and 1b). National Biological Service personnel have cataloged almost 1,000 recognizable manatees and maintained their sighting histories in a computer-based system (Beck and Reid in press).

Table. Estimated population traits of the Florida manatee based on long-term life-history research (data are from the National Biological Service and the Florida Department of Environmental Protection).

Life-history trait	Data
Maximum life expectancy	60 years
Gestation	11-13 months
Litter size	1
% twins	
Blue Spring	1.79%
Crystal River	1.40%
Sex ratio at birth	1:1
Calf survival	
Blue Spring	60%
Crystal River	67%
Annual adult survival	
Atlantic coast	90%
Blue Spring	96%
Crystal River	96%
Age of first reproduction (female)	3-4 years
Mean age first reproduction (female).	5 years
Spermatogenesis (male).	2-3 years
Proportion pregnant (female)	33% salvaged carcasses
Blue Spring	41%
Proportion nursing 1st-year calves during winter season	36% (mean)
Blue Spring	30%
Crystal River	36%
Atlantic coast	38%
Calf dependency	1.2 years
Interbirth interval	2.5 years
Highest number of births	May-September
Highest frequency in mating herds	February-July
No. salvaged carcasses	2,219 (1974-93)
No. documented in ID catalog	> 950 (1975-February 1994)
Highest count (aerial surveys)	1,856 in January 1992

Courtesy J.P. Reid, NBS

Fig. 1a. Female manatee and calf. Individuals can be identified by their unique scar patterns; scars are usually the result of collisions with boats.

Courtesy R.K. Bonde, NBS

Fig. 1b. A manatee often bears scars from multiple boat collisions.

Estimates of manatee reproductive traits are similar across study sites (Table), despite large habitat differences among study areas. There is also agreement in reproductive estimates obtained from salvaged carcasses (Marmontel in press), indicating that Florida manatees have probably achieved a maximum level of reproduction (O'Shea and Hartley in press).

Aerial Surveys

The population of Florida manatees cannot be directly estimated because they are often difficult to see. They occupy waters that may be turbid or obscured by overhanging branches; they can move long distances between counting areas over a short time; and many environmental factors, particularly temperature, influence their distribution and behavior (Lefebvre et al. in press).

Three statewide aerial surveys, coinciding with maximum manatee use of winter aggregation sites, resulted in counts of 1,268 (January 1991), 1,465 (February 1991), and 1,856 (January 1992; Ackerman in press). The differences in these counts are thought to reflect the influence of different environmental conditions, not changes in population size. Manatee presence at winter aggregation sites varies within and between winters, depending upon the pattern and severity of winter cold fronts.

Garrott et al. (1994) developed a population index by using a temperature covariate to model a simple linear trend in annual aerial survey data from the winters of 1977-78 through 1991-92. Their analyses showed an increasing trend in the temperature-adjusted counts of 7%-12% annually on the Atlantic coast, but the degree to which these increases are related to true population growth is unknown. No pronounced temporal trend was detected at the largest aggregation site on the southwest coast.

While this result seems promising because it shows no evidence for major declines, it is tempered by other factors. The number of human-related manatee deaths on the Atlantic coast is more than twice as high as on the gulf coast (Ackerman et al. in press). This fact is reflected in the lower survival rate of adult manatees on the Atlantic coast than at Crystal River and Blue Spring (O'Shea and Langtimm in press). Reynolds and Wilcox (1994) found that the number of calves sighted at winter aggregation sites has decreased since 1982, and that in three recent winters, the percentage of manatees sighted that are calves has also decreased. They note that mortality of calves at or near time of birth is the fastest-growing type of manatee mortality, thus the downward trend in aerial survey calf counts is a cause for concern and further investigation.

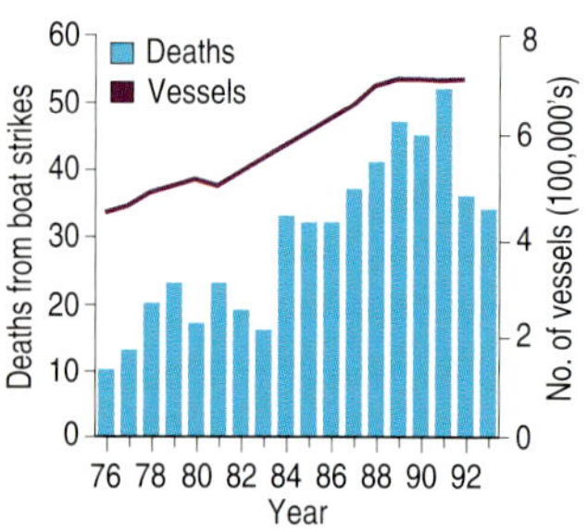

Fig. 2. Number of manatee deaths from watercraft collisions and number of Florida registered vessels from 1976-93 (data from National Biological Service and Florida Department of Environmental Protection).

Recovery Criteria

Species recovery criteria for the Florida manatee are three-fold: the population trend must be stable or increasing; mortality must be stable or declining; and threats to manatee habitat must be under control (USFWS 1989). Better population and life-history data suggest a greater potential for increase and higher numbers than previously recognized, and strong steps taken by local, state, and federal governments are increasing the number and area of sanctuaries and slow boat-speed zones. These steps may reduce mortality if they are continued and expanded, allowing the population to recover more quickly.

Management has focused on ways to reduce human-related mortality. Of greatest concern has been an increase over the years in the number of human-caused deaths, particularly those caused by collisions with boats (Fig. 2). Boat strikes account for 78% of human-related manatee mortality and 25% of all documented deaths (Wright et al. in press). A moderate reduction in the number of boat-related deaths in the last 2 years caused optimism; however, watercraft collisions accounted for 49 manatee deaths in 1994, almost matching the record number of 51 in 1991 (Fig. 2).

Habitat Threats

Habitat threats are far from under control, however. Florida has one of the fastest-growing human populations in the nation, with an estimated net gain of close to 1,000 people per day (Fernald et al. 1992). Much growth has occurred along the coast, with inevitable consequences for coastal habitats. For example, about a third of the 600,000 ha (1.5 million acres) of seagrass meadows present in coastal Florida in the 1940's no longer exist (Lewis 1987). One of the most important regions for manatees on the Atlantic coast is the Indian River Lagoon. Over the past 20 years, losses of submerged aquatic vegetation in some areas of the lagoon have exceeded 95% (Busby and Virnstein 1993). Submerged freshwater plants have also been affected adversely by increases in turbidity and nutrients.

Debris, particularly monofilament line, plastics, and unattended fishing nets and ropes, directly threatens manatees, who may ingest or become entangled in these materials (Beck and Barros 1991). Manatees are also vulnerable to natural and human-caused catastrophes, such as disease and oil spills, particularly when the animals are concentrated at winter aggregation sites.

Future

Population and life-history information suggests that the potential long-term viability of the Florida manatee population is good, provided that strong efforts are continued to curtail mortality, habitat quality is maintained or improved, and steps are taken to offset potential catastrophes.

References

Ackerman, B.B. Ongoing manatee aerial survey programs—a progress report. *In* T.J. O'Shea, B.B. Ackerman, and H.F. Percival, eds. Population biology of the Florida manatee (*Trichechus manatus latirostris*). National Biological Service, Biological Rep. Series. In press.

Ackerman, B.B., S.D. Wright, R.K. Bonde, D.K. Odell, and D.J. Banowetz. Trends and patterns in manatee mortality in Florida, 1974-1992. *In* T.J. O'Shea, B.B. Ackerman, and H.F. Percival, eds. Population biology of the Florida manatee (*Trichechus manatus latirostris*). National Biological Service, Biological Rep. Series. In press.

Beck, C.A., and N.B. Barros. 1991. The impact of debris on the Florida manatee. Marine Pollution Bull. 22(10):508-510.

Beck, C.A., and J.P. Reid. An automated photo-identification catalog for manatee life history studies. *In* T.J. O'Shea, B.B. Ackerman, and H.F. Percival, eds. Population biology of the Florida manatee (*Trichechus manatus latirostris*). National Biological Service, Biological Rep. Series. In press.

Busby, D.S., and R.W. Virnstein. 1993. Executive summary. Pages iii-viii *in* L.J. Morris and D.A. Tomasko, eds. Submerged aquatic vegetation and photosynthetically active radiation. Special Publ. SJ93-SP13, St. Johns River Water Management District, Palatka, FL.

Fernald, E.A., E.D. Purdum, J.R. Anderson, Jr., and P.A. Krafft. 1992. Atlas of Florida. University Press of Florida, Gainesville, FL. 280 pp.

Garrott, R.A., B.B. Ackerman, J.R. Cary, D.M. Heisey, J.E. Reynolds III, P.M. Rose, and J.R. Wilcox. 1994. Trends in counts of Florida manatees at winter aggregation sites. Journal of Wildlife Management 58(4):642-654.

Lefebvre, L.W., B.B. Ackerman, K.M. Portier, and K.H. Pollock. Aerial surveys for estimating manatee population size and trend—problems and prospects. *In* T.J. O'Shea, B.B. Ackerman, and H.F. Percival, eds. Population biology of the Florida manatee (*Trichechus manatus latirostris*). National Biological Service, Biological Rep. Series. In press.

Lewis, R.R., III. 1987. The restoration and creation of seagrass meadows in the Southeast United States. Pages 153-173 *in* M.J. Durako, R.C. Phillips, and R.R. Lewis III, eds. Proceedings of the Symposium on Subtropical-tropical Seagrasses of the Southeastern United States, 12 August 1985. Florida Marine Research Publications Number 42, St. Petersburg, FL.

Marmontel, M. Age and reproductive parameter estimates in female Florida manatees. *In* T.J. O'Shea, B.B. Ackerman, and H.F. Percival, eds. Population biology of the Florida manatee (*Trichechus manatus latirostris*). National Biological Service, Biological Rep. Series. In press.

O'Shea, T.J., and W.C. Hartley. Longitudinal studies of manatee reproduction and early age survival at Blue Spring, upper St. Johns River, Florida. *In* T.J. O'Shea, B.B. Ackerman, and H.F. Percival, eds. Population biology of the Florida manatee (*Trichechus manatus latirostris*). National Biological Service, Biological Rep. Series. In press.

O'Shea, T.J., and C.A. Langtimm. Adult survival estimates for Florida manatees at Crystal River, Blue Spring, and the Atlantic coast. *In* T.J. O'Shea, B.B. Ackerman, and H.F. Percival, eds. Population biology of the Florida manatee (*Trichechus manatus latirostris*). National Biological Service, Biological Rep. Series. In press.

Rathbun, G.B., J.P. Reid, R.K. Bonde, and J.A. Powell. Reproduction in free-ranging Florida manatees (*Trichechus manatus latirostris*). *In* T.J. O'Shea, B.B. Ackerman, and H.F. Percival, eds. Population biology of the Florida manatee (*Trichechus manatus latirostris*). National Biological Service, Biological Rep. Series. In press.

Reynolds, J.E., III, and J.R. Wilcox. 1994. Observations of Florida manatees (*Trichechus manatus latirostris*) around selected power plants in winter. Marine Mammal Science 10(2):163-177.

USFWS. 1989. Florida Manatee (*Trichechus manatus latirostris*) Recovery Plan. Prepared by the Florida Manatee Recovery Team for the U.S. Fish and Wildlife Service, Atlanta, GA. 98 pp.

Wright, S.D., B.B. Ackerman, R.K. Bonde, C.A. Beck, and D.J. Banowetz. Analysis of watercraft-related mortality of manatees in Florida, 1979-1991. *In* T.J. O'Shea, B.B. Ackerman, and H.F. Percival, eds. Population biology of the Florida manatee (*Trichechus manatus latirostris*). National Biological Service, Biological Rep. Series. In press.

For further information:

Lynn W. Lefebvre
National Biological Service
Sirenia Project
412 NE 16th Ave., Rm 250
Gainesville, FL 32601

Gulf of Mexico Coastal Wetlands: Case Studies of Loss Trends

by
James B. Johnston
Mary C. Watzin
John A. Barras
Lawrence R. Handley
National Biological Service

The Gulf of Mexico's coastal wetlands are of special interest because the gulf is an exceptionally productive sea that yields more than 1.1 billion kg (2.5 billion lb) of fish and shellfish annually and contains four of the top five fishery ports in the nation by weight (U.S. Environmental Protection Agency 1988). The volume of commercial shrimp landings in the gulf has been statistically related to the areal coverage of gulf coastal wetlands (and seagrass beds) that provide crucial nursery habitat to the young (Turner 1977). Coastal wetlands (particularly salt marshes and mangroves) and associated shallow waters function similarly in support of many fish species of commercial interest (Seaman 1985). The gulf wetlands are also well known for their large populations of wildlife, including shorebirds, colonial nesting birds, and 75% of the migratory waterfowl traversing the United States (Duke and Kruczynski 1992). The extensive coastal wetlands that remain along the gulf make up about half of the nation's total wetland area (NOAA 1991).

General Trends

The National Oceanic and Atmospheric Administration (NOAA 1991) examined the areal extent and distribution of gulf coast coastal wetlands in the mid-1980's by using aerial photographs and maps from 1972 to 1984 (28% from 1979 and 42% from 1980 or later). Summaries of NOAA's data are shown in the Table for three wetland categories: marshes (fresh, brackish, and salt marshes), estuarine shrub-scrub (mangroves), and freshwater forested/shrub-scrub wetlands. Even though wetland area has diminished greatly along the gulf coast during the last 30 years, about 1.3 million ha (3.3 million acres) still remain in these three categories.

Louisiana has the greatest area of coastal wetlands with 55% of the total, followed by Florida (18%), Texas (14%), and Mississippi (2%). Louisiana contains 69% of the marshes, while Florida has 97% of the estuarine scrub-shrub, most of which is mangrove. Of the three wetland types, 80% is marsh, 19% estuarine scrub-shrub, and 1% forested wetland.

Table. Total area (hectares) of selected vegetated wetlands by state for the Gulf of Mexico (from NOAA 1991).*

State	Marsh	Estuarine scrub-shrub	Fresh forested and scrub-shrub	Total	% of total**
Texas	183,900	1,100	3,000	188,000	14
Louisiana	723,500	4,100	1,900	729,500	55
Mississippi	23,800	400	-	24,200	2
Alabama	10,400	1,100	800	12,300	1
Florida	108,100	248,400	7,400	363,900	28
Total	1,049,700	255,100	13,100	1,317,900	-

* Calculated based on Fish and Wildlife Service's National Wetlands Inventory maps. Area originally reported as acres x 100; hectare = 2.471 acres.

** Fractions of percent rounded to the next highest whole percent (1.6 = 2.0%).

Because of the age of the photographs used by NOAA and because national trends suggest that the area of most wetland types is still declining (Frayer et al. 1983), the wetland statistics presented by NOAA may be overestimates. No current studies summarize coastal wetland area or loss rates for the entire Gulf of Mexico region; therefore, four case studies conducted by the National Biological Service's Southern Science Center, the U.S. Fish and Wildlife Service's National Wetland Inventory, and their partners are presented to depict status and trends from the 1950's to 1970's and the 1970's to the late 1980's. The areas chosen (Fig. 1) represent a cross-section of current trends.

Coastal Wetland Loss: Gulf of Mexico Case Studies

Galveston Bay

White et al. (1993) reported both gains and losses in Galveston Bay wetlands from the 1950's to 1989, but the net trend was one of wetland loss, going from 69,800 ha (171,000 acres) in the 1950's to 56,100 ha (138,600 acres) in 1989. The rate of loss decreased over time from about 405 ha (1,000 acres) per year between 1953 and 1979 to about 283 ha (700 acres) per year between 1979 and 1989. The rate of loss from 1979 to 1989 would probably be lower if inaccuracies in wetland interpretation of the 1979 photographs could be taken into account. In general, freshwater scrub-shrub habitats decreased in area from the 1950's to 1979 and 1989, while forested wetlands increased. Marshes (fresh and non-fresh) decreased from about 67,000 ha (165,500 acres) in the 1950's to about 52,800 ha (130,400 acres) in 1989, producing a total net marsh loss of about 21% of that resource.

The five key factors contributing most to wetlands decline in the Galveston Bay since the 1950's are (1) industrial development; (2) urbanization; (3) navigation channels; (4) flood control and multipurpose water projects to meet Houston's future water demand, especially upstream impoundments on the Trinity and San Jacinto rivers; and (5) pollution due to agricultural runoff despite the diminished acreage lost to agricultural expansion. It should be noted that human-induced subsidence due to industrial development (oil and gas activities) and urbanization (groundwater withdrawals) are considered in this analysis (D. Whitehead, U.S. Fish and Wildlife Service, personal communication).

Coastal Louisiana

Coastal wetland loss for Louisiana represents 67% of the nation's total loss. For the time period 1978-90, the loss was 177,625 ha (290,432 acres), representing an annual loss rate of 9,802 ha/yr (24,203 acres/yr) for this 12-year period; that is equal to 97.9 km^2 or 37.8 m^2/yr. For the time period 1956-78, net wetland loss was even greater, 267,800 ha (661,700 acres), representing a loss rate of 12,170 ha/yr (30,000 acres/yr); that is equal to 121.7 km^2 or 47 mi^2/yr.

Although much of this loss is only indirectly linked to human activities, most of the net current, catastrophic wetland loss is primarily the result of altered hydrology stemming from navigation, flood control, and mineral extraction and transport projects (Sasser et al. 1986; Louisiana Wetland Protection Panel 1987; Turner and Cahoon 1988). These operations do not always destroy wetlands directly, but they do amplify tidal forces in historically low-energy systems, which upsets the balance of subsidence and accretion, reduces nutrient and sediment influx, decreases freshwater retention, and increases the levels of salt, sulfate, and other substances potentially toxic to indigenous plant species (Good 1993).

Current wetland losses are concentrated in the southern Deltaic Plain (78%; Fig. 1). In this region, losses are especially severe in the fringing marshes of the Terrebonne and Barataria basins (Figs. 2 and 3). Previous losses in the Deltaic Plain occurred primarily in large areas of interior lands. In the Chenier Plain (Fig. 1), loss rates were more constant (22%); many of the larger areas of loss there seem related to impounded areas with managed water levels.

The Barataria and Terrebonne basins suffer the highest land loss rates (all land but mostly wetlands) in Louisiana (2,880 ha/yr [7,120 acres/yr] and 2,630 ha/yr [6,500 acres/yr], respectively), accounting for 64% of all land loss in the 1978-90 period. In contrast, this area accounted for only 43% of all loss in the 1956-78 period. The Mermentau and Sabine basins (Fig. 2) have the next highest loss rates (1,080 ha/yr [2,670 acres/yr] and 660 ha/yr [1,630 acres/yr]), with losses largely confined to the northern and central portions, except for shoreline erosion along the Mermentau Basin's coastline. Loss rates within the Teche-Vermilion, Mississippi, Breton Sound, and Pontchartrain basins (Fig. 2) are all less than 930 ha/yr (2,300 acres/yr), which seems to indicate more stable environments. The Atchafalaya and Pearl River basins (Fig. 2) experienced losses of less than 130 ha/yr (321 acres/yr). In summary, land loss rates in coastal Louisiana, although decreasing, remain high for the 1978-90 period.

The National Biological Service is providing future land loss updates for coastal Louisiana by using Landsat Thematic Mapper satellite imagery on a 3-year basis.

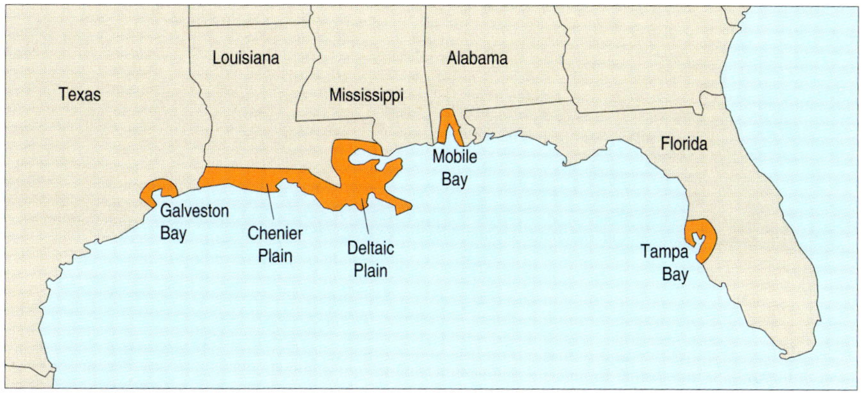

Fig. 1. Locations of wetland loss study sites along the Gulf of Mexico region.

Mobile Bay

Non-freshwater marshes surrounding Mobile Bay declined by more than 4,047 ha (10,000 acres) from 1955 to 1979, representing a loss of 35% (Roach et al. 1987). Freshwater marshes in all of coastal Alabama declined by about 69% from 1955 to 1979. More than 2,500 ha (6,200 acres) were lost during that time (Roach et al. 1987).

When comparing these data to 1988 wetland habitat maps prepared for upper Mobile Bay, it appears that in this portion of the bay no additional net loss of non-freshwater marsh has occurred since 1979. Some marsh has obviously continued to be lost in certain areas, primarily because of dredge disposal associated with navigation and industry. These losses, though, seem to have been offset by the growth of emergent marsh in existing spoil sites (Watzin et al. 1994).

The Southern Science Center's 1988 areal estimates show a substantial increase of 189 ha (467 acres) in freshwater marsh from 1979 to 1988 in upper Mobile Bay. Further investigation revealed that some of this gain was the result of the growth of emergent vegetation in existing disposal areas and in ditches along railroads and highways. Because of disparities in photointerpretation between dates, it is also quite likely that some of these differences are simply due to mapping errors and differences in mapping technique (Watzin et al. 1994).

As a result of mapping errors associated with interpreting forested and scrub-shrub wetlands in the 1956 photographs, Roach et al. (1987) had little faith in the quantitative estimate of change between 1956 and 1979 for these wetland types. The Southern Science Center's 1988 wetland area figures for forested wetlands appear relatively accurate; they indicate that about 486 ha (1,201 acres) of forested wetlands (2.7%) were lost in upper Mobile Bay between 1979 and 1988. These losses can be attributed to conversion of forested habitats to scrub-shrub areas (e.g., clearcutting associated with timber harvest), small impoundments, and commercial and residential development (Watzin et al. 1994).

Tampa Bay

Haddad (1989) reported emergent wetlands decreased from 29,000 ha (71,700 acres) in the 1950's to 23,900 ha (59,100 acres) in 1982, about an 18% loss. Mangroves decreased from 8,629 ha (21,320 acres) to 8,032 ha (19,847 acres), a decline of about 7%. Salt marshes declined from 2,063 ha (5,097 acres) to 1,423 ha (3,538 acres), or a loss of 30%. Freshwater wetlands decreased 21% from 18,335 ha (45,305 acres) to 14,440 ha (35,681 acres).

Lewis et al. (1985) estimate that 44% of the salt marsh and mangrove has been lost in Tampa Bay since the late 1800's. Although their numbers and those of Haddad (1989) are not readily

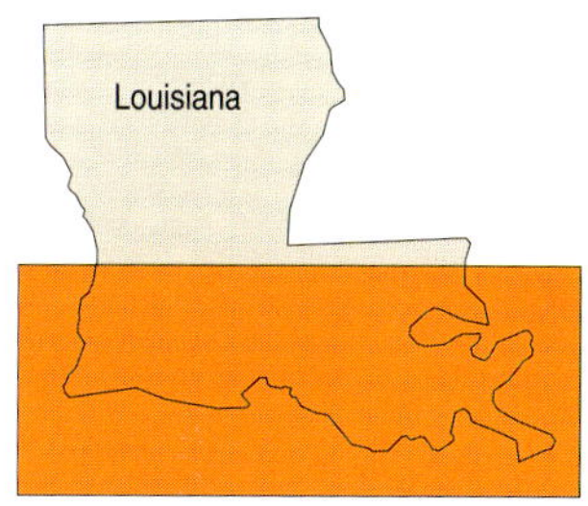

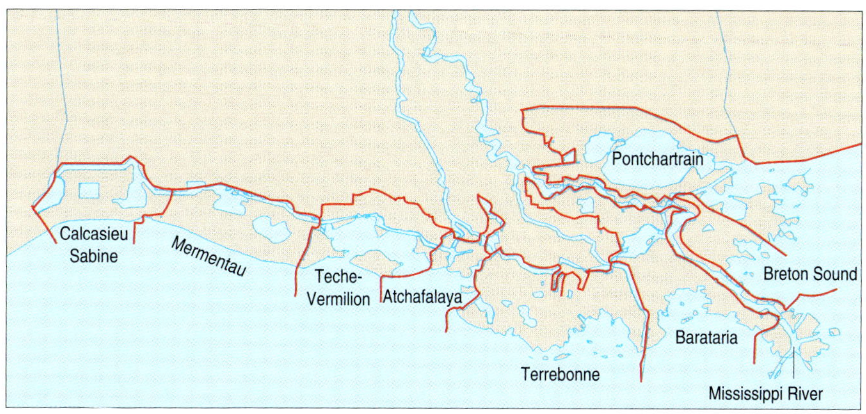

Fig. 2. Coastal Louisiana basins as defined in the Coastal Wetlands Planning, Protection, and Restoration Act Plan.

Fig. 3. Coastal landloss in Louisiana and elsewhere is analyzed by using computerized geographic information systems that produce graphics such as this map.

comparable because of differences in time frame, methodology, vegetation classification, and area mapped, the results taken together confirm that significant losses of wetland habitat have occurred. Marsh and mangrove losses are the product of dredge and fill activities that are now under strict regulatory control; although permitted dredging continues, protective measures exist to minimize loss that is not for public benefit.

Future Concerns

To protect the future of gulf coastal wetlands, status and trends over time must be continually recorded and noted in the scientific and public literature. Preliminary data from selected coastal areas studied in the 1980's show a reduced rate of wetland loss compared with earlier decades. While this is good news, the pressures of a continuously expanding human population make it unclear whether this trend will continue into the 21st century. Only additional monitoring data can answer this question.

References

Duke, T.W., and W.L. Kruczynski, eds. 1992. Status and trends of emergent and submerged vegetated habitats of the Gulf of Mexico, USA. Gulf of Mexico Program, U.S. Environmental Protection Agency, John C. Stennis Space Center, MS. 161 pp.

Frayer, W.E., T.J. Monahan, D.C. Bowden, and F.A. Graybill. 1983. Status and trends of wetlands and deepwater habitats in the coterminous United States, 1950's to 1970's. Colorado State University, Department of Forest and Wood Sciences, Fort Collins. 32 pp.

Good, B. 1993. Louisiana's wetlands: combatting erosion and revitalizing native ecosystems. Restoration and Management Notes. 11:125-133.

Haddad, K.D. 1989. Habitat trends and fisheries in Tampa and Sarasota bays. Pages 113-128 *in* Tampa and Sarasota bays: issues, resources, status, and management. National Oceanic and Atmospheric Administration, Estuary-of-the-Month Seminar Series 11.

Lewis, R.R., III, M.J. Durako, M.D. Moffler, and R.C. Phillips. 1985. Seagrass meadows of Tampa Bay. Pages 210-246 *in* S. Treat, J. Simon, R. Lewis III, and R. Whitman, Jr., eds. Proceedings Tampa Bay Area Scientific Information Symposium. Florida Sea Grant Rep. 65.

Louisiana Wetland Protection Panel. 1987. Saving Louisiana's coastal wetlands: The need for a long-term plan of action. U.S. Environmental Protection Agency, EPA-230-02-87-026. 102 pp.

NOAA. 1991. Coastal wetlands of the United States: an accounting of a national resource base. National Oceanic and Atmospheric Administration Rep. 91-3. 59 pp.

Roach, E.R., M.C. Watzin, and J.D. Scurry. 1987. Wetland changes in coastal Alabama. Pages 92-101 *in* T.A. Lowery, ed. Symposium on the Natural Resources of the Mobile Bay Estuary. Alabama Sea Grant Extension Service, Mobile, AL. MASGP-87-007.

Sasser, C.E., M.D. Dozier, J.G. Gosselink, and J.M. Hill. 1986. Spatial and temporal changes in Louisiana's Barataria Basin marshes, 1945-1980. Environmental Management 10:671-680.

Seaman, W., Jr. 1985. Florida aquatic habitat and fishery resources. Florida Chapter of the American Fisheries Society, Kissimmee. 543 pp.

Turner, R.E. 1977. Intertidal vegetation and commercial yields of penaeid shrimp. Transactions of the American Fisheries Society, Kissimmee, FL. 543 pp.

Turner, R.E., and D. Cahoon, eds. 1988. Causes of wetlands loss in the coastal Gulf of Mexico. Vol 1. Executive summary. Minerals Management Service OCS Study/MMS 87-0119.

U.S. Environmental Protection Agency. 1988. The gulf initiative: protecting the Gulf of Mexico. John C. Stennis Space Center, MS.

Watzin, M.C., S. Tucker, and C. South. 1994. Environmental problems in the Mobile Bay ecosystem: the cumulative effects of human activities. U.S. Environmental Protection Agency Tech. Rep. In press.

White, W.A., T.A. Tremblay, E.G. Wermund, and L.R. Handley. 1993. Trends and status of wetland and aquatic habitats in the Galveston Bay system, Texas. The Galveston Bay National Estuary Program. GBNEP-31. 225 pp.

For further information:

James B. Johnston
National Biological Service
Southern Science Center
700 Cajundome Blvd.
Lafayette, LA 70506

Seagrass Distribution in the Northern Gulf of Mexico

by
Lawrence R. Handley
National Biological Service

Seagrass ecosystems are widely recognized as some of the most productive benthic habitats in estuarine and nearshore waters of the gulf coast. Seagrass meadows provide food for wintering waterfowl and important spawning and foraging habitat for several species of commercially important finfish and shellfish. Physical structure provided by seagrasses affords juveniles refuge from predation and allows for attachment of epiphytes and benthic organisms. Seagrass communities also support several endangered and threatened species, including some sea turtles and manatees. Changes in seagrass distribution can reflect the health of a water body, and losses of seagrasses may signal water-quality problems in coastal waters. Losses of seagrasses in the northern Gulf of Mexico over the last five decades have been extensive—from 20% to 100% for most estuaries, with only a few areas experiencing increases in seagrasses.

Although often considered continuous around the entire periphery of the gulf, seagrasses exist only in isolated patches and narrow bands from Mobile Bay, Alabama, to Aransas Bay, Texas (Figure). This pattern of occurrence results from a combination of low salinities, high turbidity, and high wave energy in shallow waters. Seagrasses are more extensively developed from Mobile Bay to Florida Bay (Figure). Although freshwater submerged aquatic vegetation also occurs throughout gulf coast estuaries and river deltas, its distribution is not considered in this article.

Seagrass habitats in the Gulf of Mexico have declined dramatically during the past 50 years, mostly because of coastal population growth and accompanying municipal, industrial, and agricultural development. Although proximate causes of local declines can sometimes be identified, most habitat loss has resulted from widespread deterioration of water quality (Neckles 1993).

The total seagrass coverage in the shallow, clear waters in protected estuaries and nearshore waters of the Gulf of Mexico coastal states is estimated to be 1.02 million ha (2.52 million acres; Duke and Kruczynski 1992). About 693,000 ha (1.71 million acres) of seagrasses occur in waters of the Florida Big Bend and Florida Bay (Figure). The remaining 324,000 ha (800,000 acres) are within gulf estuaries, with about 95% in the estuarine areas of Florida and Texas. Florida Bay seagrass meadows occupy about 550,000 ha (1.36 million acres), while the seagrass meadows of the Florida Big Bend area cover about 300,000 ha (740,000 acres; Zieman and Zieman 1989).

Six species of seagrasses occur in the Gulf of Mexico: turtle grass (*Thalassia testudinum*), shoal grass (*Halodule wrightii*), manatee grass (*Syringodium filiforme*), star grass (*Halophila engelmanni*), *Halophila decipiens*, and widgeon grass (*Ruppia maritima*). The latter has a distribution in water with lower salinity, but is commonly reported in association with the seagrasses throughout the gulf coast.

Case Histories

Sarasota Bay

Between 1948 and 1974, South Sarasota and Roberts bays lost 193 ha (477 acres) or 25%; Dryman, Blackburn, Dona, and Roberts (a different Roberts Bay) bays lost 31 ha (77 acres) or 29%; and Lemon Bay lost 55 ha (136 acres) or 21% of seagrasses (Evans and Brungardt 1978). Losses have been attributed mainly to dredge-and-fill activities and decline in water quality (Wolfe and Drew 1990). Improved water quality in Little Sarasota Bay caused seagrasses to increase between 1948 and 1974 by 14 ha (34 acres) or 9%.

Tampa Bay

In Tampa Bay (Figure), turtle grass and shoalgrass are dominant, and widgeon grass, manatee grass, and star grass are also found. A historical estimate places 30,970 ha (76,527 acres) occurring within the shallow- water margins of Tampa Bay before human influence (ca. 1876; Lewis et al. 1985). Based on 1981 estimates of seagrass coverage, a reduction of 81% of seagrasses has occurred in Tampa Bay; 5,750 ha (14,208 acres) were present in 1981. The most striking decrease occurred between 1940 and 1963, when about 50% of the grass beds were lost (Lewis et al. 1985). During this period, Hillsborough Bay alone lost 94% of its grass beds, Old Tampa Bay lost 45%, and Tampa Bay proper lost 35%. These losses have been attributed primarily to direct dredging of grassbeds and major shoreline modifications through filling and siltation (Wolfe and Drew

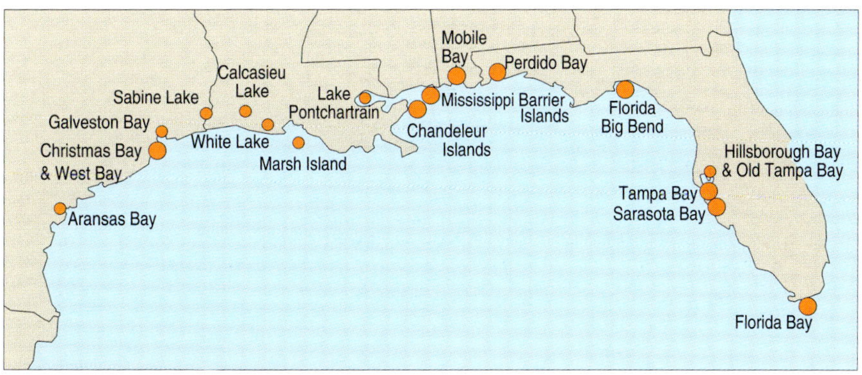

Figure. Study sites along the Gulf of Mexico region.

1990), which reduced light penetration and produced bottom sediments that are not conducive to seagrass growth and development.

Since 1963, grass beds have continued to decline in the upper bays of Tampa Bay to a point where Hillsborough Bay has lost the remaining 139 ha (343 acres) and Old Tampa Bay has lost nearly 60% (Figure). In lower Tampa Bay, grass beds have regained some area, increasing about 14% or 435 ha (1,075 acres). Tampa Bay as a whole has lost 5,984 ha (14,786 acres), or 51% of seagrasses between 1940 and 1983.

Perdido Bay

In Perdido Bay (Figure), widgeon grass and shoal grass are the predominant species. They grow in large and small beds, in numerous patches along shallow sandy reaches of the shoreline, and in large shallow flats in the lower bay and outlet. From 1940 to 1987, changes in the upper and middle parts of the bay consisted mainly of shifts in the locations of small meadows, with only minor changes in density. In the lower bay, some shifting of locations and changes in density occurred, and the coverage of seagrasses declined from 486 ha (1,201 acres) in 1940-41 to 251 ha (619 acres) in 1987. While the loss of seagrasses for the whole area was nearly 48%, some areas in U.S. Geological Survey quadrangles lost as much as 82% of the seagrasses delineated between 1940-41 and 1987. The changes in the extent of seagrasses are due to increased turbidity caused primarily by channel dredging and boat traffic; shoreline modifications; decreasing water quality and sedimentation from increasing farmlands and residential, commercial, and industrial development; and the high wave energy, overwash, sedimentation, erosion, and runoff from Hurricane Frederick in 1979.

Mississippi Gulf Coast

Along the Mississippi gulf coast, the Gulf Islands National Seashore includes most of the state's barrier islands (Figure). Manatee grass and shoal grass are the dominant seagrasses found in the shallow water on the northern side of the barrier islands, where they are protected from the high wave energy of the open gulf. Between 1956 and 1987, 416 ha (1,029 acres) of seagrasses declined to 140 ha (345 acres), a loss of 66%. The largest concentration of seagrasses was found on the north side of Horn Island, where 169 ha (417 acres) in 1956 declined to 56 ha (138 acres) by 1987, and to 6 ha (14 acres) by 1992.

Coastal Louisiana

Coastal Louisiana has a large amount of submerged aquatic vegetation but only a small portion is seagrasses (5,657 ha [13,974 acres] in 1988). Since the mid-1950's Louisiana has lost all of its seagrass in Lake Pontchartrain, in the Mississippi River Delta, behind the south coast barrier islands and Marsh Island, and in the coastal lakes (White, Calcasieu, and Sabine). The only remaining seagrass beds in coastal Louisiana exist in Chandeleur Sound behind the Chandeleur Islands. Turtle grass, shoal grass, manatee grass, widgeon grass, and star grass are present in the sandy sediments of the shallow backbarrier lagoon. These seagrass beds are virtually unaffected by human impacts because of their distance from the mainland, and they are controlled by high waves from chronic frontal passages and hurricanes causing overwash, erosion, sedimentation, changes in water depth, and turbidity. For example, Hurricane Camille in August 1969, with a storm surge of nearly 11 m (36 ft) on the Mississippi mainland, caused a loss of 530 ha (1,310 acres), or 22% of the seagrasses, on the North Islands (USGS 1:24,000 quadrangle), and a loss of 303 ha (749 acres) or 54% of the seagrasses, on Chandeleur Light (USGS 1:24,000 quadrangle).

The Chandeleur Islands (Figure) have been intensively mapped for wetland and seagrass habitats for 1978, 1982, 1987, and 1989. The areal extent of seagrasses for the Chandeleur Islands has remained relatively constant over the 11-year period, from 6,409 ha (15,831 acres) in 1978 to 5,657 ha (13,974 acres) in 1989. This constitutes a loss of only 12% of the seagrasses from 1978 to 1989, a period that had two hurricanes, two tropical storms, and countless cold fronts that influenced these islands.

Galveston Bay

In the Galveston Bay estuary (Figure), the distribution of seagrasses, predominantly shoal grass and widgeon grass, decreased in areal extent from more than 2,024 ha (5,000 acres) in the mid-1950's to about 283 ha (700 acres) in 1989, a loss of 1,471 ha (3,635 acres) or about 85% (White et al. 1993). The most significant losses were along the margins of western Galveston Bay and were related to the effects of subsidence and Hurricane Carla in 1970. In West Bay nearly 890 ha (2,200 acres) of seagrasses were completely lost, primarily through human activities including industrial, residential, and commercial development; wastewater discharges; chemical spills; and increased turbidity from boat traffic and dredging (Pulich and White 1991). In Christmas Bay, which has the largest concentration of seagrass beds in the

Galveston Bay estuarine system, seagrass areal extent declined from 121 ha (300 acres) in 1975 to 81 ha (200 acres) in 1987, but increased to 156 ha (385 acres) by 1989.

Conclusions

Losses of seagrasses in the northern Gulf of Mexico have been extensive over the last five decades, with losses varying 20%-100% for most estuaries of the northern Gulf of Mexico. Only a few locales have experienced increases in seagrasses. The high productivity of the Gulf of Mexico seagrass beds as spawning, nursery, food, and shelter areas increases the importance of the loss of this valuable habitat far beyond the areal extent of the resource. Regionwide, the loss of seagrasses is attributable to natural causes (hurricanes, cold-front storms, and increased or decreased salinities) and human-induced effects (increased turbidity and decreases in water quality resulting from dredging, boating activities, and other development pressures), which work in concert to deteriorate the environmental quality of the habitat.

References

Duke, T.W., and W.L. Kruczynski, eds. 1992. Status and trends of emergent and submerged vegetated habitats of the Gulf of Mexico. Gulf of Mexico Program, U.S. Environmental Protection Agency, Stennis Space Center, MS. 161 pp.

Evans, M., and T. Brungardt. 1978. Shoreline analysis of Sarasota County Bay systems with regard to revegetation activities. Pages 193-206 *in* D.P. Cole, ed. Proceedings of the Fifth Annual Conference on Restoration of Coastal Vegetation in Florida. Environmental Studies Center, Hillsborough Community College, Tampa.

Lewis, R.R., M.J. Durako, M.D. Moffler, and R.C. Phillips. 1985. Seagrass meadows of Tampa Bay—a review. Pages 210-246 *in* S.F. Treat, J.L. Simon, R.R. Lewis III, and R.L. Whitman, Jr., eds. Proceedings of the Tampa Bay Area Scientific Information Symposium, May 1982. University of South Florida, Tampa.

Neckles, H.A., ed. 1993. Seagrass monitoring and research in the Gulf of Mexico: draft report of a workshop held at Mote Marine Laboratory in Sarasota, Florida, January 28-29, 1992. National Biological Survey, National Wetlands Research Center, Lafayette, LA. 75 pp.

Pulich, W.M., Jr., and W.A. White. 1991. Decline of submerged vegetation in the Galveston Bay system: chronology and relationship to physical processes. Journal of Coastal Res. (4):1125-1138.

White, W.A., T.A. Tremblay, E.G. Wermund, Jr., and L.R. Handley. 1993. Trends and status of wetland and aquatic habitats in the Galveston Bay system, Texas. The Galveston Bay National Estuary Program, Publ. GBNEP-31. 225 pp.

Wolfe, S.H., and R.D. Drew, eds. 1990. An ecological characterization of the Tampa Bay watershed. U.S. Fish and Wildlife Service Biological Rep. 90(20). 334 pp.

Zieman, J.C., and R.T. Zieman. 1989. The ecology of the seagrass meadows of the west coast of Florida: a community profile. U.S. Fish and Wildlife Service Biological Rep. 85(7.25). 155 pp.

For further information:

Lawrence R. Handley
National Biological Service
Southern Science Center
700 Cajundome Blvd.
Lafayette, LA 70506

Seagrass Meadows of the Laguna Madre of Texas

by
Christopher P. Onuf
National Biological Service

A series of lagoons forms an almost continuous fringe of water behind coastal barriers for 500 km (310 mi) from Galveston Bay, Texas, to the Mexican border (Fig. 1). At the northeast end, where river discharge and precipitation greatly exceed evaporation from the embayments, fringing marshes are the dominant wetland type. Toward the southwest, freshwater inputs decrease, fringing marshes are replaced by wind-tidal flats that support highly productive algal mats during periodic inundation, and seagrasses dominate the shallow waters of the embayments (Table).

Seagrasses are so prevalent in Laguna Madre that they define the structure of the physical environment, as well as being the source of biological production for the ecosystem. Consequently, seagrass meadows serve a critical nursery function in support of the region's rich fisheries, and one waterfowl species has established an exclusive dependence on Laguna Madre and its most common seagrass. More than 75% of the world population of redhead ducks (*Aythya americana*) winters in the greater Laguna Madre ecosystem (inclusive of the

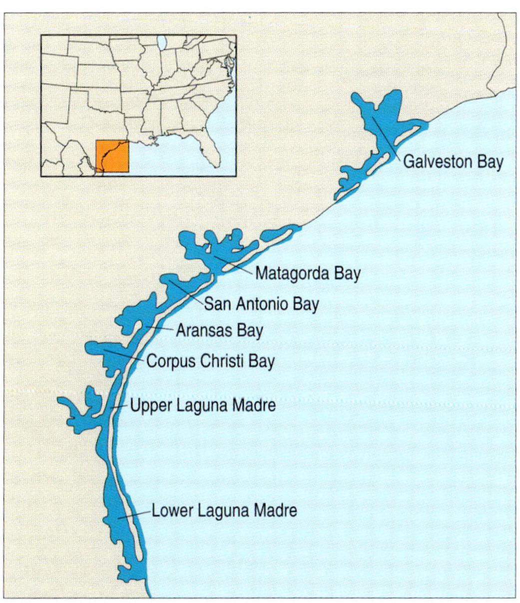

Fig. 1. Major bay systems along the Texas coast.

Table. Seagrass cover in bays of the Texas coast.

Bay system	Bottom vegetated (%)
Galveston Bay System*	0.3
Matagorda Bay System*	1.1
San Antonio Bay System**	5.0
Aransas-Copano Bay System**	5.2
Corpus Christi Bay System**	12.1
Upper Laguna Madre***	75.2
Lower Laguna Madre***	70.5

* Adair et al. 1994.
** Adair and Moore 1990.
*** Quammen and Onuf 1993.

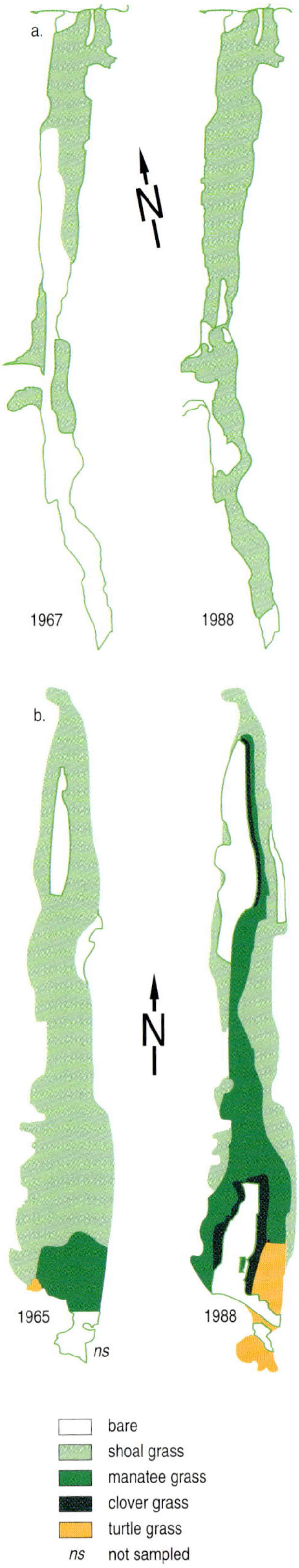

Fig. 2. Dominant cover types in the continuously submerged portions of upper (a) and lower (b) Laguna Madre.

Laguna Madre de Tamaulipas, immediately south of the delta of the Rio Grande in Mexico; Weller [1964]) and feeds almost exclusively on one species of seagrass while in residence (shoal grass, *Halodule wrightii*). Because of the degree of dependence of the redhead population on the laguna and reports of major disruptions to the laguna's seagrass community, the National Biological Service began a research program in coastal Texas.

The Texas Parks and Wildlife Department inventoried aquatic vegetation in Laguna Madre in the 1960's (Texas Parks and Wildlife Department 1965-67). In 1988 the National Wetlands Research Center (now the Souhern Science Center) resurveyed the laguna (Quammen and Onuf 1993).

Distributional Patterns

Seagrass meadows are undergoing profound change in Laguna Madre. The area of vegetated bottom in upper Laguna Madre has increased 130 km^2 (50 mi^2), from 120 km^2 (46 mi^2) to 250 km^2 (97 mi^2) between 1967 and 1988 (Quammen and Onuf 1993), an amount exceeding the total area of seagrass meadows in bays of the middle and upper Texas coast (Adair and Moore 1990; Adair et al. 1994). Concurrently, seagrass cover in lower Laguna Madre decreased by an even larger amount, 140 km^2 (54 mi^2), from 620 km^2 (239 mi^2) to 480 km^2 (185 mi^2), confined to deeper areas (Quammen and Onuf 1993).

Changes in the species composition of seagrass meadows affected even larger areas of the lower laguna (Fig. 2). Shoal grass covered 82% of the bay bottom in 1965 compared to 33% in 1988. Over the same period, cover of bay bottom by manatee grass (*Syringodium filiforme*) increased from 9% to 27% and by turtle grass (*Thalassia testudinum*) from 1% to 7%.

Factors Responsible

Processes responsible for the loss of seagrass from deep areas are different from those for the other changes. The loss of seagrass has resulted from reduced light reaching the bottom in deep areas near navigation channels because of increased turbidity caused by maintenance dredging. In 1988-89, waves generated by frequent episodes of high winds resuspended fine materials from dredge deposits and increased light attenuation for more than a year after a dredging project was completed (Onuf 1994). Since the interval between dredging projects is 2 years, the reduction in available light is essentially permanent.

Hydrological modifications of the laguna are most likely the primary cause of the expansion of seagrass cover in upper Laguna Madre and the shift in the composition of surviving seagrass meadows in lower Laguna Madre. Historically, a 20-km (12.4-mi) expanse of usually emergent flats separated the two sections of the laguna. Salinities greater than 60 ppt in the lower laguna and greater than 100 ppt in the southern part of the upper laguna were not unusual.

In 1949 the Gulf Intracoastal Waterway was completed, providing a continuous water connection between the two parts of the laguna, improving exchange with the Gulf of Mexico and moderating the salinity regime of the laguna. Since completion of the waterway, salinities have seldom reached 50 ppt in the lower laguna and 60 ppt in the upper laguna, even during extreme drought (Quammen and Onuf 1993).

Isolation from source populations of seagrass probably accounts for the slower colonization of the upper laguna than the lower laguna, after the environment became tolerable. The displacement of shoal grass by manatee grass and turtle grass after salinity moderation is consistent with the relative intolerance of those species to hypersalinity (high salinity) and their superior competitive capabilities under benign conditions. The current distributions of the three species are consistent with their relative colonizing abilities since salinity moderation: shoal grass is most widespread, manatee grass is intermediate, and turtle grass is most closely confined to its point of origin at the south end of the laguna (Quammen and Onuf 1993).

Management Implications

The dramatic decrease of shoal grass in the lower laguna is a particular concern to natural resource managers because redheads feed almost exclusively on shoal grass while in winter residence. Historically, there were several other important wintering areas for these ducks, such as Chesapeake Bay, Pamlico Sound, and Galveston Bay. The possibility existed that other areas could absorb additional birds if habitat quality in Laguna Madre deteriorated. Now, none of the alternative areas support significant winter populations of redheads, and few others do either, making the condition of Laguna Madre all the more critical for redheads.

Changes in the upper laguna since 1988 are almost certain to worsen the problem of redhead habitat deterioration. Whereas increases in the upper laguna compensated for about 40% of the losses of shoal grass in the lower laguna over the period of this analysis, a persistent phytoplankton bloom known as the brown tide has been resident in the upper laguna since 1990.

The bloom is so dense in some locations that it reduces light penetrating 1 m (3.3 ft) by more than 50% (Dunton 1994). This light reduction is leading to loss of shoal grass in the deep areas most influenced by the brown tide.

Displacement of shoal grass by manatee grass was not evident in the upper laguna in 1988 but is now. In all likelihood, the same processes responsible for the profound changes in the composition of seagrass meadows in the lower laguna will now take hold in the upper laguna. The greater isolation of the upper laguna from a source population of the invader probably accounts for the much later initiation of the replacement process than in the lower laguna.

A final factor further magnifies the importance to redheads of these changes in seagrasses of the Laguna Madre of Texas. The Laguna Madre de Tamaulipas, just south of the delta of the Rio Grande, is an integral part of the winter life-support system of redheads. In most years, more redheads overwinter in Texas than Mexico; however, in years of drought in Texas, more ducks continue south into Mexico. The large geographic extent of available habitat apparently buffers the population by increasing the probability that suitable conditions prevail somewhere in the system every year. The governor of the State of Tamaulipas, however, is now promoting the extension of the Gulf Intracoastal Waterway through the Laguna Madre in Mexico. In all likelihood, this development will reduce the support capacity of the laguna in Mexico for redheads, further increasing the reliance of the ducks on the laguna in Texas.

Modification of dredging practices in Texas and planning of waterway construction in Mexico hold the most promise for sustaining seagrasses and habitat for redheads to the maximum extent possible. At present, most dredge disposal is to submerged receiving areas along the channel, where bay resources are directly affected and wave-caused resuspension sometimes impairs water clarity for long periods after dredging. Land-based or deep-sea disposal would alleviate these problems. In Mexico, conducting an inventory of key resources, prominently including seagrasses and redheads, routing the waterway to avoid concentration areas, and implementing environmentally sound construction and disposal practices will ensure the greatest security for the wintering habitat of redheads and other resources linked to seagrass meadows.

References

Adair, S.E., and J.L. Moore. 1990. A survey of seagrass distribution of the middle Texas coast. Chapter 11 *in* S.E. Adair, J.L. Moore, W.H. Kiel, Jr., and M.W. Weller, eds. Winter ecology of redhead ducks in the gulf coast region. Final Rep., Cooperative Agreement 14-16-0009-87-909 between Texas A & M University and the U.S. Fish and Wildlife Service, National Wetlands Research Center.

Adair, S.E., J.L. Moore, and C.P. Onuf. 1994. Distributional ecology of submerged aquatic vegetation in estuaries of the upper Texas coast. Wetlands 14(2):110-121.

Dunton, K.H. 1994. Seasonal growth and biomass of the subtropical seagrass *Halodule wrightii* in relation to continuous measurements of underwater irradiance. Marine Biology 120:479-489.

Onuf, C.P. 1994. Seagrasses, dredging and light in Laguna Madre, Texas, USA. Estuarine, Coastal and Shelf Science 39:75-91.

Quammen, M.L., and C.P. Onuf. 1993. Laguna Madre: seagrass changes continue decades after salinity reduction. Estuaries 16:302-310.

Texas Parks and Wildlife Department. 1965-67. Coastal waterfowl project. Federal Aid Projects W-29-R-18 to 24. Austin, TX.

Weller, M.W. 1964. Distribution and migration of the redhead. Journal of Wildlife Management 28:64-103.

For further information:

Christopher P. Onuf
National Biological Service
Southern Science Center
Campus Box 339
6300 Ocean Dr.
Corpus Christi, TX 78412

Coastal Barrier Erosion: Loss of Valuable Coastal Ecosystems

by

S. Jeffress Williams
U.S. Geological Survey

James B. Johnston
National Biological Service

The conterminous United States has nearly 142,000 km (88,182 mi) of tidal shoreline that exists in a delicate balance with the forces of nature (Culliton et al. 1990). Much of this shoreline and the coastal barriers in particular are experiencing greatly increased pressures as a result of rapid population growth and accompanying development. Although coastal areas are highly desirable for their abundant natural resources and habitability, they are also extremely dynamic environments in which conditions hazardous to humans (e.g., erosion, flooding, pollution) may be present. In many regions, these hazards, which threaten not only humans but also valuable marine resources and even entire ecosystems, are increasing at alarming rates as coastal development, recreation, and waste disposal increase, often in direct conflict with long-term natural coastal processes. This article defines coastal barriers, summarizes their changes, and discusses the U.S. Department of the Interior's (DOI) Coastal Barrier Resources System (CBRS).

Coastal Barriers Defined

Coastal barriers are geologically recent depositional sand bodies that are highly variable in shape, size, and their response to natural processes and human alterations. They may stretch many kilometers in length and contain high sand dunes—such as the Outer Banks of North Carolina—or they may be small and isolated islands, so low in relief that they are routinely overwashed by spring tides and minor storms. Their dynamic nature means coastal barriers are constantly shifting and being modified by winds and waves, but scientific field investigations over the past several decades are revealing some disturbing trends.

Long-term survey data by the U.S. Geological Survey and others, based on analyses of archive maps, reports, and aerial photographs, demonstrate that coastal erosion is affecting each of the 30 coastal states (Figure; Williams et al. 1991a). About 80% of U.S. coastal barriers are undergoing net long-term erosion at rates of less than 1 m (3.3 ft) to as much as 20 m (65.6 ft) per year. Natural processes such as storms, rise in relative sea level, and sediment starvation (a reduction in volume of sediment transported by rivers reaching the coast), which may also be a result of human interference, are responsible for most of this erosion; but human factors such as mineral extraction, emplacement of hard coastal-engineering structures, and dredging of sand from navigation channels are now recognized as having major effects on shoreline stability (Table 1).

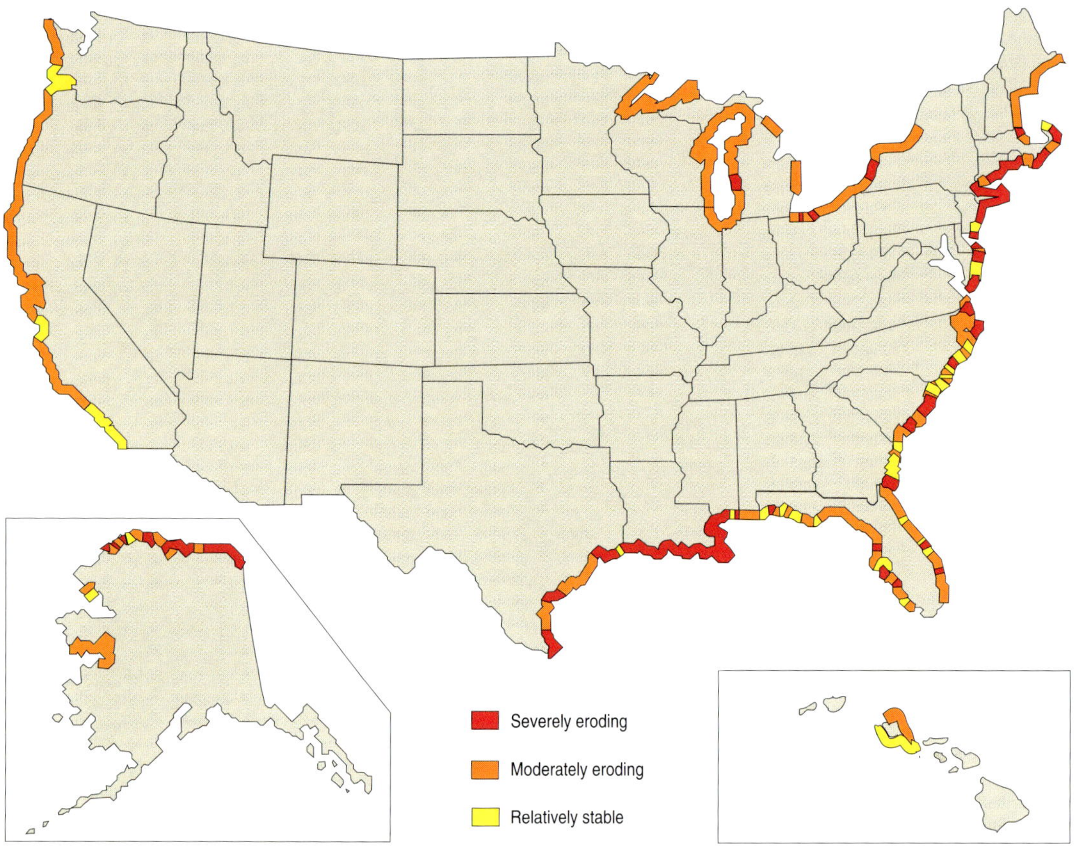

Figure. Classification of annual shoreline change around the United States (modified from U.S. Geological Survey 1985).

Table 1. Primary factors (geologic and human) affecting coastal areas ranked by decreasing relative importance (modified from Williams et al. 1991b).

Primary factors affecting coastal areas
Land subsidence (sediment compaction)
Storm impacts
Coastal processes (waves, winds, tides)
Eustatic sea-level change
Sand supply at the coast
Human activities: dredging, dams, mining, engineering structures, withdrawal of fluids (e.g., oil, gas, and water)
Regional tectonic movements

Undeveloped Coastal Barriers

Since 1982 the U.S. Fish and Wildlife Service (and now the National Biological Service) has been conducting inventories of the CBRS along the Atlantic and Gulf of Mexico coasts and the Great Lakes, as defined by the Coastal Barrier Resources Act of 1982 (Public Law 97-384) and the Coastal Barrier Improvement Act of 1990 (Public Law 101-591). (The Pacific undeveloped coastal barriers are under review by DOI as required by Section 6 of Public Law 101-591.)

The photographic inventories from aerial color infrared photographs (scales 1:12,000 to 1:65,000) provide a precise visual identification for each unit within the CBRS. Undeveloped coastal barriers are defined as areas that have less than one structure per 2 ha (4.9 acres) of fastland (areas suitable for building structures). Additionally, there are no areas in CBRS that are less than 0.4 km (0.25 mi) long. The entire barrier coastline was reviewed for inclusion into the CBRS system; inclusion into the CBRS means that the areas were ineligible for direct or indirect federal financial assistance that might support or encourage development.

The total shoreline length of the CBRS system for the United States is 2,055 km (1,276 mi), encompassing an area of about 537,000 ha (1.3 million acres; Table 2).

Table 2. Summary of undeveloped coastal barriers (F. McGilvery, U.S. Fish and Wildlife Service, Washington, DC, and National Oceanic and Atmospheric Administration, unpublished report).

State	Shoreline lengths (km)	No. CBRS units	CBRS shoreline lengths (km)	CBRS area (ha)
Maine	5,565	31	37.7	1,949
Massachusetts	2,430	79	197.0	27,301
Rhode Island	614	25	53.1	4,502
Connecticut	989	28	36.6	3,718
New York	2,960	90	167.4	26,216
New Jersey	2,867	16	16.7	3,279
Delaware	610	8	28.2	2,813
Maryland	5,104	48	45.1	2,901
Virginia	5,304	62	124.0	19,412
North Carolina	5,400	16	69.2	14,268
South Carolina	4,602	21	97.0	39,765
Georgia	3,750	11	31.6	26,085
Florida	18,811	105	304.8	115,484
Alabama	971	8	31.6	4,609
Mississippi	574	7	20.6	2,422
Louisiana	11,394	21	286.6	142,454
Texas	5,360	23	283.2	79,377
Puerto Rico	1,120	62	82.3	8,179
Virgin Islands	280	35	23.5	1,536
Ohio	320	10	13.0	1,941
Michigan	2,368	46	88.9	7,569
Wisconsin	960	7	12.2	793
Minnesota	368	1	4.8	381
Total	82,721	760	2,055.1	536,954

Comparing the CBRS maps after three surveys were conducted since 1982 shows that there have been no significant changes of CBRS unit boundaries in the United States (Frank McGilvery, USFWS, personal communication). Quite significant changes have occurred, however, in the size, shape, and character of many barriers because of natural processes.

Future

As the coastal population grows and barriers become urbanized, valuable habitats are being destroyed, and associated negative impacts such as waste disposal, pollution, and changes in freshwater and fine-grained sediment dispersal are altering entire coastal marine and maritime ecosystems. Protecting all remaining undeveloped coastal barriers should be a national priority. Some protection occurs through the Coastal Barrier Resources System, as well as other local, state, and federal programs, including acquisition, restoration, protection, and management programs.

References

Culliton, T.J., M.A. Warren, T.R. Goodspeed, D.G. Remer, C.M. Blackwell, and J.J. McDonough III. 1990. Fifty years of population change along the nation's coasts, 1960-2010. National Oceanic and Atmospheric Administration Coastal Trend Series, 2nd rep. 41 pp.

Williams, S.J., K. Dodd, and K.K. Gohn. 1991a. Coasts in crisis. U.S. Geological Survey Circular 1075. Reston, VA. 32 pp.

Williams, S.J., S. Penland, A.H. Sallenger, R.A. McBride, and J.L. Kindinger. 1991b. Geologic controls on the formation and evolution of Quaternary coastal deposits of the northern Gulf of Mexico. Pages 1082-1095 *in* N.C. Kraus, K.C. Gingrich, and D.L. Kriebel, eds. Coastal sediments '91. Vol. 1. American Society of Civil Engineers, New York.

U.S. Geological Survey. 1985. The national atlas, shoreline erosion and accretion map. U.S. Government Printing Office, Washington, DC.

For further information:

S. Jeffress Williams
U.S. Geological Survey
914 National Center
Reston, VA 22092

Reef Fishes of the Florida Keys

by

William F. Smith-Vaniz
National Biological Service

James A. Bohnsack
National Marine Fisheries Service

James D. Williams
National Biological Service

The Florida Keys are a chain of islands extending 320 km (199 mi) along the southern edge of the Florida Plateau from Biscayne Bay to the Dry Tortugas (101 km [63 mi] west of Key West). The Florida Reef Tract, a band of living coral reefs paralleling the Keys, extends from Fowey Rocks to the Marquesas and includes about 130 km (81 mi) of bank reefs and 6,000 patch reefs. For convenience, the Keys can be divided into the upper, middle, and lower Keys (Fig. 1).

The environmental and economic importance of the Florida Keys is indicated by the many protected or regulated areas, which include several national wildlife refuges, national parks, marine sanctuaries, and state-protected areas (Fig. 1). Because many recreational and commercial activities occur in nearshore habitats, these areas have high potential for environmental damage.

Relatively high rates of human population increase (28%-44%) are predicted over the next 20 years in some parts of the Keys; Monroe County, which includes all of the Keys, had a population growth of 160% during the past 40 years. Human activities associated with increased population growth may well ultimately disrupt the Florida Keys marine ecosystem and damage the area's overall economy. In recognition of this possibility, the Florida Keys National Marine Sanctuary was designated in 1990 under the Marine Protection, Research, and Sanctuaries Act, U.S. Public Law 101-605. The sanctuary includes 9,515 km^2 (3,673 mi^2) of coastal waters around the Florida Keys. The Sanctuaries and Reserves Division of the National Oceanic and Atmospheric Administration was charged with developing a comprehensive management plan and regulations to protect sanctuary resources (NOAA 1995). We focus on the current status of Florida Keys reef fishes and areas where research is needed immediately.

The Fish Fauna

The diversity and richness of fishes in the Florida Keys are unparalleled in shelf waters of the continental United States and reflect the mixing of dissimilar faunal components

Coral Reef Ecosystems

by
Stephen C. Jameson
National Oceanic and Atmospheric Administration

Coral reefs are one of the most diverse, complex, and beautiful ecosystems on earth. Coral reef ecosystems benefit humans commercially, recreationally, and environmentally (Laist et al. 1986). The abundant biological diversity of the coral reef ecosystem not only includes coral and the commercially important species associated with the reef but also thousands of other plant and animal species. Thus, the status and trends of this ecosystem are not easily evaluated.

Historically, most coral reef surveys have been limited to discrete reefs or species or have been time-limited (Rogers 1985; Dustin and Halas 1987; Bythell et al. 1992; Porter and Meier 1992; Ginsburg 1994). The status and trends of complete coral reef ecosystems around entire islands or reef tracts (e.g., the entire Florida reef tract) have never been comprehensively evaluated because of the complexity, length of time, and cost of such endeavors. Because of this lack of a comprehensive understanding of the status and trends of coral reef ecosystems under U.S. jurisdiction, this article looks at broad patterns in the status and trends of these ecosystems today with the hope of providing a useful focus for future ecosystem-based National Biological Service (NBS) coral reef endeavors.

Status and Trends

Coral reef ecosystems under U.S. jurisdiction are located in waters throughout the world (Figs. 1 and 2). These reefs can be divided into two broad categories, pristine and at risk. For references on specific areas, please contact the author.

Pristine Coral Reef Ecosystems

Pristine coral reef ecosystems are in remote locations with little or no human threats to ecosystem health. By definition, the status of these ecosystems is good and the trend in health is steady. Areas under U.S. jurisdiction with pristine coral reef ecosystems include the Flower Garden Banks in the Gulf of Mexico; the northwest Hawaiian Islands (uninhabited); Wake Island; the Northern Mariana Islands (excluding Saipan); Palmyra Island and Kingman Reef; Howland Island; Baker Island; and Jarvis Island in the Pacific Ocean (Figs. 1 and 2).

Coral Reef Ecosystems at Risk

Coral reef ecosystems at risk are near human population centers with some or all reefs experiencing local anthropogenic stress. Some important sources of stress include nutrient enrichment from sewage and agriculture, overfishing, and stress from high sedimentation caused by deforestation, agriculture, vessel traffic, and coastal runoff. The status and trends of many coral reef resources within these areas are poor (D'Elia et al. 1991; Ginsburg 1994). In addition, it is impossible to know the status and trends of these ecosystems on an island-wide or reef tract basis because of our lack of understanding of these ecosystems in any comprehensive way. Within U.S. jurisdiction, the coral reef ecosystems at risk include the Florida Reef tract, Puerto Rico, and the U.S. Virgin Islands in the western Atlantic and Caribbean; and the main Hawaiian Islands (inhabited), Johnston Atoll, Saipan (Northern Mariana Islands), and American Samoa in the Pacific Ocean (Figs. 1 and 2).

Future

The United States has abundant coral reef ecosystems. Pristine coral reef ecosystems are especially valuable as "natural" laboratories and control sites that can help us eventually understand the evolution and function of healthy coral reef ecosystems. We will not be able to clearly evaluate the status and trends of unhealthy ecosystems until we better understand pristine coral reef ecosystems. It is vital that adverse effects to these pristine areas are avoided.

Figs. 1 and 2 show that over half of all U.S. coral reef ecosystems are at risk, and some are nearly dead because of human perturbations. Swift legislative efforts and public works programs to reduce nutrients and

Fig. 1. Coral reef ecosystems under U.S. jurisdiction in the western Atlantic Ocean, Gulf of Mexico, and Caribbean Sea. Coral reef ecosystems are found on or around the Florida Reef tract, Flower Garden Banks, Puerto Rico, and U.S. Virgin Islands. Coral reef ecosystems at risk are indicated by an asterisk.

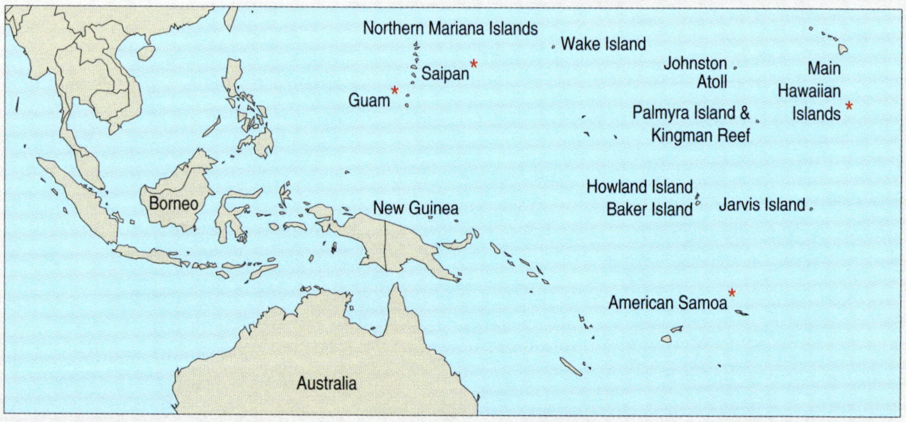

Fig. 2. Coral reef ecosystems under U.S. jurisdiction in the Pacific Ocean. Coral reef ecosystems are found around the northwest and main Hawaiian Islands, Wake Island, Johnston Atoll, Northern Mariana, Guam, Palmyra Island and Kingman Reef, Howland Island, Baker Island, Jarvis Island, and American Samoa. Coral reef ecosystems at risk are indicated by a red asterisk.

sediments may be the only way to save many of these national treasures.

References

Bythell, J.C., B. Gladfelter, and M. Bythell. 1992. Ecological studies of Buck Island Reef National Monument, St. Croix, U.S. Virgin Islands: a quantitative assessment of selected components of the coral reef ecosystem and establishment of long-term monitoring sites. Island Resource Foundation, St. Thomas. 72 pp.

D'Elia, C.F., R.W. Buddemeier, and S.V. Smith, eds. 1991. Workshop on Coral Bleaching, Coral Reef Ecosystems and Global Change: Report of Proceedings. Maryland Sea Grant College UM-SG-TS-91-03. 49 pp.

Dustin, P., and J.C. Halas. 1987. Changes in the reef-coral community of Carysfort Reef, Key Largo, Florida: 1974 to 1982. Coral Reefs 6:91-106.

Ginsburg, R.N., compiler. 1994. Proceedings of the Colloquium on Global Aspects of Coral Reefs: Health, Hazards and History, 1993. Rosenstiel School of Marine and Atmospheric Science, University of Miami. 420 pp.

Laist, D.W., T.E. Bigford, G.W. Robertson, and D.R. Gordon. 1986. Management of corals and coral ecosystems in the United States. Coastal Zone Management Journal 13(3/4):203-239.

Porter, J.W., and O.W. Meier. 1992. Quantification of loss and change in Floridian reef coral populations. American Zoologist 32:625-640.

Rogers, C.S. 1985. Degradation of Caribbean and western Atlantic coral reefs and decline of associated fisheries. Proceedings of the 7th International Coral Reef Symposium 6:491-496.

For further information:

Stephen C. Jameson
National Oceanic and Atmospheric Administration
Office of Ocean Resources Conservation and Assessment
1305 East-West Hwy.
Silver Spring, MD 20910

(Gilbert 1973) and the variety of habitats. More fish species have been reported from Alligator Reef in the upper Keys than at any single location in the Western Hemisphere (Starck 1968). These fishes consist primarily of continental, warm-temperate species characteristic of the northern Gulf of Mexico, and tropical Caribbean species, especially on the Atlantic side of the Florida Keys. Mixing of the warm-temperate and tropical Caribbean species occurs from north to south with distribution limits of individual species determined by seasonal temperature variations and the exchange of Gulf of Mexico and Atlantic Ocean waters in nearshore habitats in the middle to lower Keys. The key silverside (*Menidia conchorum*) is the only fish confined to the Florida Keys. It is not as rare as had previously been thought, and a recommendation has been made to change its official state listing from "threatened" to "special concern" (Gilbert 1992).

Two studies of single sites indicate the total diversity of Florida Keys fishes. Longley and Hildebrand (1941) listed 442 species from the Dry Tortugas, 300 of which are closely associated with coral reefs. Starck (1968) recorded 517 fish species from Alligator Reef, including 389 considered members of the reef community. The category "coral reef fish" is arbitrary, however, because a continuum exists from obligate (*see* glossary) species that spend their entire adult lives largely hidden within recesses of the reef, to opportunistic species that use many habitats. Also, most economically important reef fish are dependent on seagrasses and mangroves along the Keys and in Florida Bay for critically important nursery habitat. The availability of such habitats permits a higher density of organisms and a more complex reef community (Parrish 1989).

As more researchers, anglers, recreational scuba divers, and snorkelers have visited the Keys, an appreciation of the complex nature of reef fish communities has increased (Sale 1991). Research that uses visual census techniques has focused on the more common and readily observable reef fish. The most comprehensive census study to date (Bohnsack et al. 1987) provided a detailed quantitative description of the fish fauna of Looe Key National Marine Sanctuary for depths less than 13 m (43 ft). Quantitative studies of this kind serve as essential baseline references required for monitoring and detecting future changes in reef fish abundances and distributions. That study, with additional data from Key Largo, showed that fish faunas of the outer reefs in the Keys are diverse and complex, and their community structures are similar to well-developed reefs throughout the Caribbean.

Influences and Trends

As one of the most heavily fished areas in Florida, the Keys support extensive commercial and recreational fisheries for food, sport, and the marine aquarium trade (Bohnsack et al. 1994). A major management goal is ensuring continued sustainability of limited resources and traditional activities under rapidly increasing human population growth and exploitation of the reef fisheries. Excessive use and fishing may cause long-term harm to individual species, disrupt reef ecosystems, and damage the area's overall economy.

Demand and use of resources have increased (Fig. 2) with the growing number of residents and tourists (White 1991; Bohnsack et al. 1994). The number of registered boats has increased more than sixfold since 1965 while the number of commercial and partyboat vessels has remained stable (Bohnsack et al. 1994). Fishing success has increased, however, because of more accurate navigational aids, inexpensive electronic fish-finding equipment,

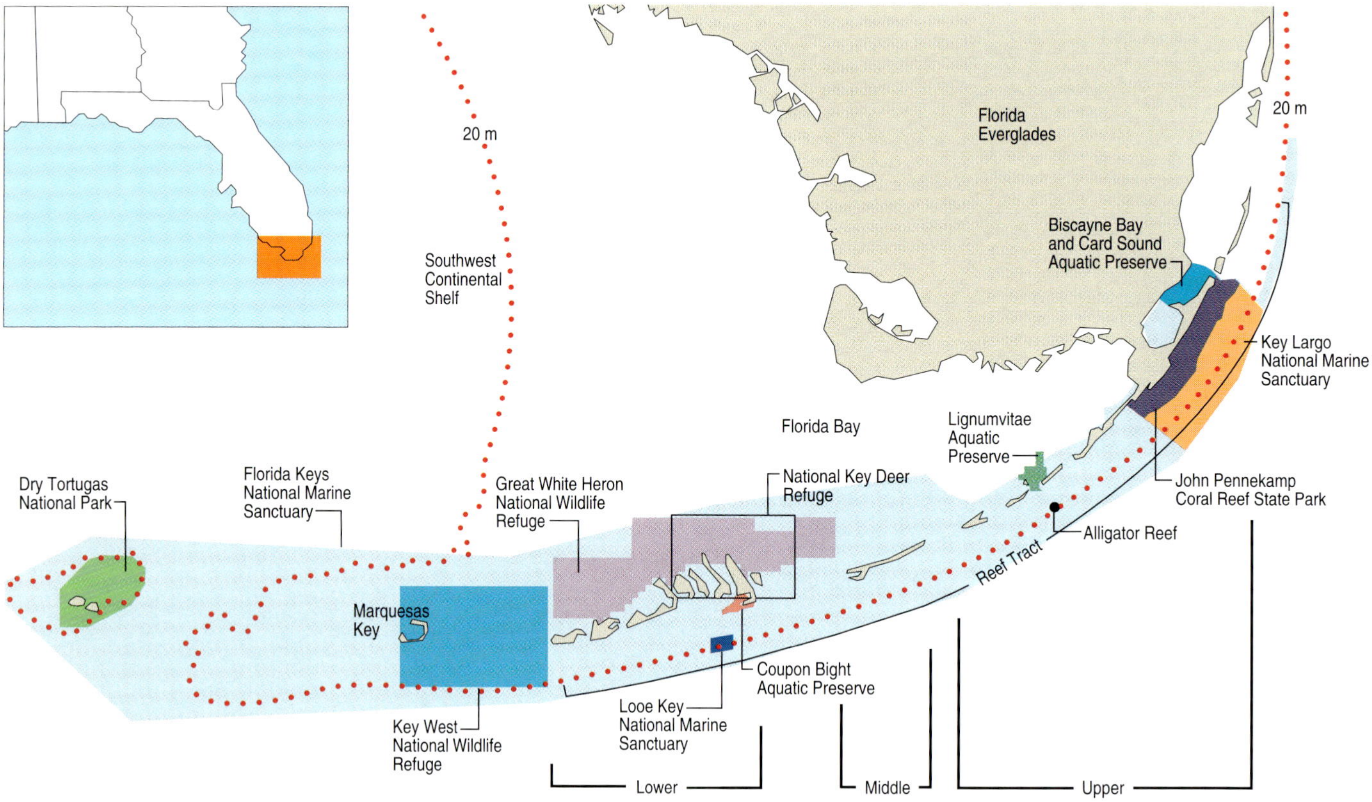

Fig. 1. The upper, middle, and lower Florida Keys. National marine sanctuaries, national parks, national wildlife refuges, and aquatic preserves are also shown (modified from maps provided by James A. Reed II, Florida Marine Research Institute). Various colors used simply to delineate designated areas.

and improved fishing gear and vessel technology. Although fishing can directly reduce stocks, other human activities also can damage resources and affect fish, including pollution, sedimentation, habitat loss from land-use practices, and vessel groundings. For example, habitat changes in Florida Bay have been attributed to water management and land-use practices in southern Florida (McIvor et al. 1994).

Fig. 2. Extensive use of resources in the Florida Keys: Looe Key (spur and groove zone).

Because of insufficient data, population trends and stock condition are impossible to determine for many species. Few fishery-independent data exist and fishery-dependent data have been limited to a relatively few years, to certain species, or to specific fishery components. Analyses are complicated because of the many species targeted, the large number of fisheries operating out of different ports, the number of different fishing methods used, and the many different fishing objectives, especially within the recreational fishery.

Some fishery trends are apparent despite data limitations. King mackerel (*Scomberomorus cavalla*) stocks collapsed in the early 1980's, but recovered somewhat after management measures were implemented. Pink shrimp (*Penaeus duorarum*) and grouper (Serranidae) landings have declined, and fisheries for queen conch (*Strombus gigas*), Nassau grouper (*Epinephelus striatus*), and jewfish (*Epinephelus itajara*) were closed because of reduced stock size. Increased landings reported for greater amberjack (*Seriola dumerili*), stone crab (*Menippe mercenaria*), blue crab (*Callinectes sapidus*), and yellowtail snapper (*Ocyurus chrysurus*) mostly reflect increased or redirected fishing efforts. For example, amberjack became commercially targeted only in the mid-1980's after king mackerel and red snapper (*Lutjanus campechanus*) landings declined. Landings of some species such as mutton snapper (*L. analis*), gray snapper (*L. griseus*), and

West Indies spiny lobster (*Panulirus argus*) have generally remained stable, despite large increases in effort (Bohnsack et al. 1994).

There is no guarantee, however, that any of these trends will continue, especially if fishing efforts increase or habitats become further impaired. For example, annual pink shrimp landings from the Dry Tortugas fluctuated around 4.5 million kg (9.9 million lb) for about 40 years before plummeting to less than half that level in the mid-1980's. Some of this decline may be a result of environmental changes caused by reduced freshwater inflow to Florida Bay (McIvor et al. 1994). Sponge and seagrass die-offs in Florida Bay may eventually reduce lobster and other fishery landings because of lost juvenile habitat. Fishery landing data will not necessarily reveal the full impact of those removals on the ecosystem or its sustainability. This is particularly true in complex tropical ecosystems such as the Florida Keys (Knowlton 1992). The annual removal of millions of kilograms of shrimp and spiny lobster is expected to affect their fish predators, while the removal of large numbers of predators may affect abundances and interactions of their prey. Fishing is a particular concern because it tends to target top predators, which are often the keystone species important for maintaining community structure (Knowlton 1992).

The widespread ecosystem changes documented in the Florida Keys and elsewhere in the Caribbean are of special concern to the long-term status of coral reef fish communities (Richards and Bohnsack 1990; Hallock et al. 1992). These changes include unexplained sea urchin mass mortalities, major coral loss and coral bleaching, shifts from coral- to algal-dominated substrates, extensive algae blooms, and numerous fish kills. Porter and Meier (1992) reported a loss of coral diversity between 1984 and 1991 at six locations and a decrease in abundance at five locations in protected areas between Miami and Key West. Although Porter and Meier (1992) could not determine the specific causes responsible for the changes, they noted that continued equal rates of loss over long periods would not allow the historical coral reef community structure of the Florida Keys to be sustained.

Algal fouling that may be related to leaching of nutrient-enriched groundwater (NOAA 1995) has recently caused severe damage to Algae Reef off Key Largo, and may be spreading to nearby Horseshoe Reef. Whether caused by increased nutrient enrichment, human alteration of historically freshwater runoff from the Everglades, reduced natural flushing effects associated with hurricanes during the last 20 years, or a combination of factors, continued deterioration of Florida Bay water quality ultimately will seriously alter the fish community structure of the bay and affect the Florida Keys ecosystem as well.

Courtesy J.A.Bohnsack

Florida Keys habitat showing representative reef fishes.

Recommendations

Realistic goals and objectives must be established to protect and restore Florida Keys ecosystems and their fish resources to allow optimal sustainable economic use while preserving biodiversity. Research efforts should focus on obtaining a better understanding of ecosystem dynamics and the effects of human interactions in order to generate and test predictive management models. Marine sanctuaries should have scientific reference sites and be used to develop strategies to reduce user conflicts. To be effective, management efforts must be international and must include cooperation between all levels of government and users. Because it is possible to love a reef to death (Fishman 1991), increased public education, understanding, awareness, and appreciation of the complex nature of reef fish communities and the effects of human activities within the Florida Keys ecosystem are especially important. Although efforts are needed to restore habitats, primary emphasis should be to prevent further habitat degradation from human activities.

Objective measures of fish populations, habitat conditions, and ecosystem function should be developed and monitored. Standard measures are needed to compare ecological impacts of different fisheries (Bohnsack et al.

1994), including better fishery and habitat data and more precise stock assessments. There is also an urgent need to develop nondestructive methods of collecting fishery-independent data. Cryptic, obligate reef fish, which have received the least attention, are likely among the best indicator species of environmental degradation because they are more sensitive to environmental changes. A comprehensive inventory of the cryptic reef fauna of the Florida Keys is also needed for baseline data in conjunction with establishment of long-term monitoring stations throughout the Keys.

References

Bohnsack, J.A., D.E. Harper, D.B. McClellan, D.L. Sutherland, and M.W. White. 1987. Resource survey of fishes within Looe Key National Marine Sanctuary. National Oceanic and Atmospheric Administration Technical Memorandum NOS MEMD 5:1-108.

Bohnsack, J.A., D.E. Harper, and D.B. McClellan. 1994. Fisheries trends from Monroe County, Florida. Bull. of Marine Science 54(3):982-1018.

Fishman, D.J. 1991. Loving the reef to death. Sea Frontiers 37(2):14-21.

Gilbert, C.R. 1973. Characteristics of the western Atlantic reef-fish fauna. Quarterly Journal of the Florida Academy of Sciences [1972] 35(2-3):130-144.

Gilbert, C.R. 1992. Key silverside, *Menidia conchorum* Family Atherinidae. Pages 213-217 *in* C.R. Gilbert, ed. Rare and endangered biota of Florida. Vol. 2. Fishes. University Press of Florida, Gainesville. 247 pp.

Hallock, P., F.E. Müller-Karger, and J.C. Halas. 1992. Coral reef decline. National Geographic Res. and Exploration 9(3):358-378.

Knowlton, N. 1992. Thresholds and multiple stable states in coral reef community dynamics. American Zoologist 32: 674-682.

Longley, W.H., and S.F. Hildebrand. 1941. Systematic catalogue of the fishes of Tortugas, Florida, with observations on color, habits, and local distribution. Papers from Tortugas Laboratory No. 34 (Carnegie Institution of Washington Publ. 535). 331 pp.

McIvor, C.C., J.A. Ley, and R.D. Bjork. 1994. Changes in freshwater inflow from the Everglades to Florida Bay including effects on biota and biotic processes: a review. Pages 117-146 *in* S.M. Davis and J.C. Ogden, eds. Everglades: the ecosystem and its restoration. St. Lucie Press, Delray Beach, FL. 826 pp.

NOAA. 1995. Florida Keys National Marine Sanctuary draft management plan/environmental impact statement. National Oceanic and Atmospheric Administration, Silver Spring, MD. Vols 1-3.

Parrish, J.D. 1989. Fish communities of interacting shallow-water habitats in tropical oceanic regions. Marine Ecology Progress Series 58:143-160.

Porter, J.W., and O.W. Meier. 1992. Quantification of loss and change in Floridian reef coral populations. American Zoologist 32:625-640.

Richards, W.J., and J.A. Bohnsack. 1990. The Caribbean Sea: a large marine ecosystem in crisis. Pages 44-53 *in* K. Sherman, L.M. Alexander, and B.D. Gold, eds. Large marine ecosystems: patterns, processes and yields. American Association for the Advancement of Science, Washington, DC. 242 pp.

Sale, P.F., ed. 1991. The ecology of fishes on coral reefs. Academic Press, Inc., New York. 754 pp.

Starck, W.A., II. 1968. A list of fishes of Alligator Reef, Florida with comments on the nature of the Florida reef fish fauna. Undersea Biology 1(1):5-36.

White, B., ed. 1991. Monroe County statistical abstract, 1991. NCS Corporation, Key West, FL. 295 pp.

For further information:

William F. Smith-Vaniz
National Biological Service
Southeastern Biological Science Center
7920 NW 71st St.
Gainesville, FL 32653

Courtesy M.L. Scott, NBS

Riparian Ecosystems

Science Editor
Austin K. Andrews
text by
Gregor T. Auble
National Biological Service
Midcontinent Ecological Science Center
Fort Collins, CO 80525

Overview

The strict definition of riparian is "streambank," but riparian ecosystems are often broadly defined to include riverine floodplains. In the broad sense, the riparian zone is both a transition and interface between riverine and upland systems. Functionally and structurally, riparian areas are different from surrounding uplands because of proximity to a water course. In the eastern United States, the upland landscape is generally moist enough to support woody vegetation while the often extensive bottomland forests comprise only those plants able to tolerate flooding and excessive moisture. In much of the West, areas near water courses are often the only places with sufficient moisture for trees. Thus, western riparian ecosystems are often relatively narrow ribbons of trees in a generally unforested landscape.

We lack good estimates of the status or historical changes in area for riparian ecosystems of the West as a whole, although we do know that they have always represented a very small fraction of the land area because of their dependence on water in a dry region. Their importance stems from the unique features that they provide, representing desirable habitat for a variety of species. Many of the same features that make these systems relatively rare and important also make them relatively sensitive.

Western riparian systems have been massively altered in the last 200 years; the history of development in the West is to a large extent one of water development. As the articles in this section illustrate, it is hard to make a hydrologic change without also altering the associated riparian ecosystem. Busch and Scott (this section) show how hydrologic changes can influence the long-term species composition by altering soil salinity and changing the nature of disturbances that create opportunities for regeneration. Some changes described in Roelle and Hagenbuck's article (this section) on the Middle Rio Grande are relatively straightforward: riparian vegetation is inundated by a reservoir or a channel narrows with lower streamflow. Other effects of hydrologic alteration are more complex and may be played out over many decades. As the authors note, the absence of a change in net area may mask dramatic shifts in the location of different vegetation types.

Although hydrology is the dominant factor shaping these ecosystems, it is not the only one. In all the riparian systems described in the following articles, invasions of non-native plants have changed the composition of the communities and the way the systems will likely respond in the future. Timber clearing, overgrazing by livestock, agricultural conversion, and urban growth are other important causes of change in these ecosystems.

Western Riparian Ecosystems

by
David E. Busch
National Park Service
Michael L. Scott
National Biological Service

In much of western North America, riparian (streamside) environments are the only part of the landscape moist enough to allow survival of trees (Fig. 1). Riparian landscapes are usually defined as ecotones or corridors between terrestrial and aquatic realms (Malanson 1993). In spite of their limited areal extent, riparian ecosystems are essential habitat for many vertebrate species and provide critical physical and biological linkages between terrestrial and aquatic environments (Gregory et al. 1991).

Because of their association with scarce surface water resources, western riparian ecosystems have long been influenced by human activities. Human-caused perturbations can alter energy and material flow in riverine ecosystems, thus modifying riparian plant communities (Brinson 1990). Among the most serious impacts to riparian ecosystems are water impoundment and diversion, groundwater pumping from alluvial aquifers, livestock grazing, land clearing for agriculture or to increase water yield, mining, road development, heavy recreational demand, fire, the elimination of native organisms (e.g., beaver [*Castor canadensis*]) or the introduction of exotics, and overall watershed degradation (Stromberg 1993).

Courtesy M.L. Scott, NBS

Fig. 1. A cottonwood-willow riparian ecosystem illustrating how trees are closely associated with a water source in an arid landscape. Arikaree River, Colorado.

Riparian ecosystems along most major western rivers have changed as the result of water development and flood control. Losses of riparian forest downstream of dams have been reported from throughout western North America (Rood and Mahoney 1990). In contrast, woodland expansion in other dam-regulated riparian ecosystems provides evidence that the interrelationships between plant communities and hydrogeomorphic processes are complex (Johnson 1994). As the result of widespread, human-induced changes in hydrology and land use, native cottonwood-willow stands are being replaced by non-native woody species such as Russian olive (*Elaeagnus angustifolia*) and tamarisk (*Tamarix ramosissima*) throughout the West (Olson and Knopf 1986; Knopf and Scott 1990; Stromberg 1993).

In this article, we contrast the roles played by natural and human-induced disturbances in structuring western riparian ecosystems. Our approach draws heavily on data from the lower Colorado and upper Missouri rivers, two large, diverse systems that showcase a range of natural and human factors influencing riparian ecosystems throughout western North America. We also focus on how water and land-use management may threaten these valuable ecological resources.

Most of the Missouri River, through the Dakotas to its confluence with the Mississippi River, is controlled by a series of large dams and reservoirs constructed between the 1930's and 1950's. These dams radically altered the magnitude, timing, and frequency of flood flows that formerly promoted regeneration and maintenance of extensive riparian cottonwood (*Populus deltoides*) forests (Johnson et al. 1976; Johnson 1992). Here, we examine the importance of flow variability and channel processes in creating and maintaining riparian cottonwood stands in one of the last relatively naturally functioning reaches of the upper Missouri River in Montana.

The lower Colorado River riparian ecosystem (Nevada, California, and Arizona) has also been affected by hydrologic change resulting from human activities. Declines in riparian forest dominated by cottonwood (*Populus fremontii*) and willow (*Salix gooddingii*) have been attributed to change in the physical environment and to the extensive invasion of tamarisk. Our evaluation of the Colorado River ecosystem centered on an investigation of surface water-groundwater linkages and how hydrologic factors affect water uptake and use in riparian trees and shrubs. We also examined how hydrologic perturbation and alteration of natural disturbance processes affect riparian community structure along the lower Colorado River.

Methods

Upper Missouri River

We intensively sampled nine sites between Fort Benton, Montana, and Fort Peck Reservoir. Sites were selected primarily to represent the range of geomorphic (*see* glossary) conditions observed within this reach. Channel movement is variously constrained through this portion of river; in some reaches the channel meanders whereas in other reaches lateral movement is limited to a narrow valley floor. Previous work demonstrates that the age structure of cottonwood populations is strongly influenced by aspects of flow that promote successful establishment. We determined the precise age and elevation of establishment of 151 plains cottonwood stems in the study site and related their years of establishment to the flow record from U.S. Geological Survey (USGS) gauges.

Lower Colorado River

We established three sites for an intensive ecophysiological analysis of riparian plant communities on the Havasu National Wildlife Refuge in the lower Colorado River floodplain. Our analyses were confined to stands of riparian vegetation that had been classified as cottonwood-willow habitats (Anderson and Ohmart 1984; Younker and Andersen 1986).

Hydrology and Riparian Ecosystem Dynamics

Reproduction and growth of riparian plant species are closely associated with peak flows and related channel processes such as meandering. Successful establishment of such plants typically occurs only in channel positions that are moist, bare, and protected from removal by subsequent disturbance (Sigafoos 1964; Everitt 1968; Noble 1979; Bradley and Smith 1986; Stromberg et al. 1991; Sacchi and Price 1992; Johnson 1994). If streamflow is diverted, young trees may die (Smith et al. 1991). Studies of plant water uptake in floodplain ecosystems indicate that maintenance of cottonwood and willow populations depends on groundwater moisture sources which, in turn, are closely linked to instream flows (Busch et al. 1992). Thus, the establishment and maintenance of riparian plant communities are a function of the interplay among surface water dynamics, groundwater, and river channel processes.

Maps and notes from the journals of Lewis and Clark (1804-06) suggest that the present distribution and abundance of cottonwoods along the Missouri River within the study reach are generally similar to presettlement conditions. Although flows through this reach are influenced by Canyon Ferry Dam on the mainstem and Tiber Dam on the Marias River, the gross seasonal timing of flows and the magnitude and frequency of daily maximum flows have not been greatly altered by dam operations. This is due in part to the dam's relatively small storage capacity and the presence of a number of unregulated tributaries below the dams. Thus, the study reach represents one, if not the last, semi-naturally functioning reach along the entire Missouri River.

In the Colorado River, the link of floodplain groundwater with instream flows is illustrated by the association of river discharge and fluctuations in water table depth in the adjacent floodplain (Fig. 2). Further evidence for this linkage comes from daily fluctuations in water table depth, which correlated closely with the Colorado River hydroperiod (Busch, unpublished data). Colorado River floodplain soils were dry. Volumetric soil moisture in the upper 1 m (3.3 ft) of the Colorado River soil profile averaged less than 4%, while that of the nearby and less heavily impacted Bill Williams River averaged 13%. Incision of stream channels, through either natural or human-induced causes, can lead to the depression of floodplain water tables (Williams and Wolman 1984). Channelization of the lower Colorado River appears to have led to floodplain groundwater declines, and this has tended to isolate riparian vegetation from its principal moisture source at or near the water table (Busch et al. 1992).

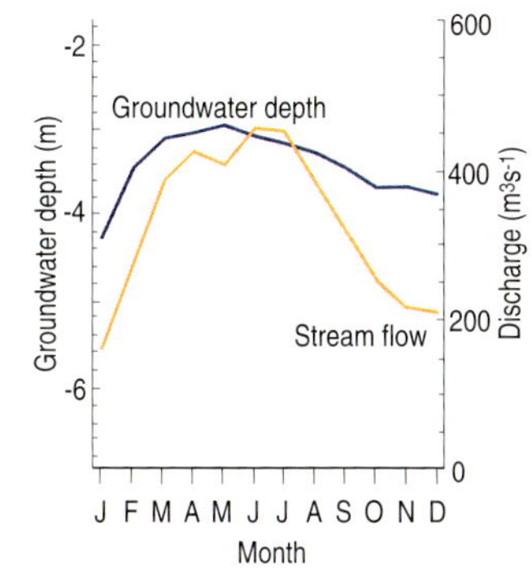

Fig. 2. Streamflow in the lower Colorado River and water table depth fluctuations in the adjacent floodplain.

Salinity and Alteration of Riparian Ecosystem Processes

In regulated rivers, a lack of flooding or infrequent groundwater incursion into surface soils can result in altered nutrient dynamics. The lack of an aqueous medium for salt dispersal may result in the elevation of soil salinity to levels that are stressful to some of the trees and shrubs native to southwestern riparian ecosystems (Busch and Smith 1995). Colorado River soils were significantly ($P < 0.05$) more saline than soils in the adjacent Bill Williams River floodplain. Salinities in Colorado River soils exceeded levels shown to inhibit germination, reduce vigor, and induce mortality in seedling cottonwood and willow (Jackson et al. 1990). Salt-tolerant species could thus benefit from elevated alluvium salinity. Evidence for salinity tolerance in both native and exotic halophytes (plants growing in salty soils or a saltwater environment) shows that arrowweed (*Tessaria sericea*) and tamarisk had significantly ($P <$ 0.05) higher leaf tissue sodium concentrations (11.2 and 18.1 mg/g [ppt], respectively) than did cottonwood (1.1 mg/g [ppt]) and willow (0.7 mg/g [ppt]).

Courtesy D.E. Busch, BOR

Colorado River study site with willow (note stress-induced canopy die back) and exotic tamarisk.

Courtesy M.L. Scott, NBS

Constrained reach of the Missouri River, Montana.

Establishment Patterns of Riparian Tree Populations

The structural diversity of riparian cottonwood and willow stands is a function of spatial and temporal patterns of occurrence. These patterns are largely determined by events during the establishment phase (Stromberg et al. 1991; Scott et al. 1993). Where stream regulation limits flooding and channel movement (e.g., the lower Colorado River), opportunities for seed germination are limited. In such systems, community structure may become less dynamic unless novel forms of disturbance such as fire increase in importance relative to the natural disturbance regime.

The magnitudes of flows associated with cottonwood establishment are influenced by local channel processes. Along the upper Missouri River, sections of meandering channel alternate with sections where lateral migration does not occur. In meandering sections, successful establishment occurs at relatively low elevations above the channel (Fig. 3a), producing several bands of even-aged trees (Bradley and Smith 1986).

If, however, lateral movement of the channel is constrained by a narrow valley, successful establishment occurs only at high elevations, often producing a single, narrow band of trees (Fig. 3b); seedlings initially established at lower positions are removed by water or ice scour. Where the channel is free to move, plant establishment occurs relatively frequently in association with both moderate and high river flows, but where the channel is constrained, plant establishment is associated with infrequent high flows in excess of 1,400 m^3/s (50,000 ft^3/s). Elimination of such high flows would largely eliminate cottonwood and willow stands from the constrained reaches of the upper Missouri River and decrease the frequency of stand establishment in the meandering reaches. From a water-management perspective, then, it is important to recognize how flow variability, including infrequent large flows, shapes the distribution and abundance of riparian tree populations.

Meandering reach of the Missouri River, Montana.

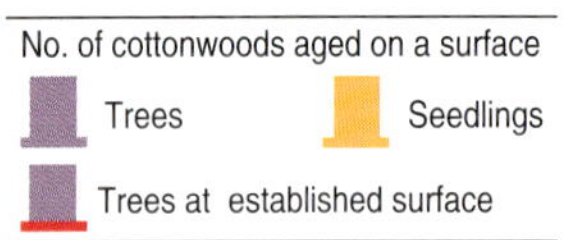

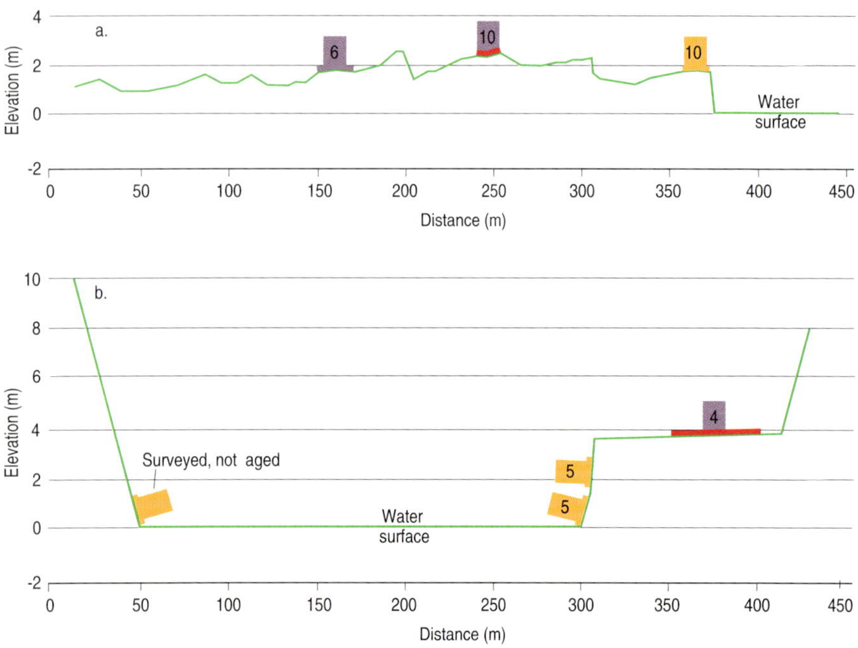

Fig. 3. Cross section in the (a) meandering channel reach, Missouri River, Montana, and (b) the constrained channel reach. All seedlings established within 10 cm (3.9 in) of the present surface (from Scott et al. 1994). For (a), at 160 m (525 ft) six trees were aged, but depth to establishment surface was measured for only one.

Disturbance Regimes and the Invasion of Non-native Species

Riparian ecosystems are dependent upon disturbance caused by occasional high flows. Along rivers where these flows have been reduced in frequency and magnitude, natural riparian ecosystems are being lost along with associated invertebrate and vertebrate species. Resource managers concerned with maintaining floodplain ecosystems need to consider ways of preserving flows that produce establishment, growth, and survival of native riparian species. If not, species such as tamarisk can exploit resources more efficiently than native riparian species, thereby altering whole ecosystem properties (Vitousek 1990). Thus, as Hobbs and Huenneke (1992) suggest, modification of the historical disturbance regimes will result in a decline in native species diversity. Although successful plant invasions are often associated with increased disturbance (Hobbs 1989; Rejmanek 1989; Hobbs and Huenneke 1992; Parker et al. 1993), in situations where the frequency or intensity of a natural disturbance is decreased, the invasion of competitively superior non-natives may be promoted (Hobbs and Huenneke 1992).

Although most riparian plants are adapted to flooding, the frequency, timing, and duration of floods may be highly altered on regulated stream reaches. Fire appears to have increased in importance relative to flooding as a form of disturbance affecting regulated southwestern rivers, including the Colorado. Colorado River cottonwood and willow canopy cover decreased only slightly following fire, but burned cottonwood-willow stands had significantly greater cover of both arrowweed and tamarisk ($P < 0.005$). Efficiency in water uptake, transport, and use are among the mechanisms responsible for superior post-fire recovery of halophytic shrubs compared with trees native to the Colorado River ecosystem (Busch and Smith 1993).

As the result of ecosystem change over the last century, cottonwoods have become rare along the lower Colorado River, and most remaining stands are dominated by senescent (i.e., in decline) individuals (Fig. 4). Although a senescent segment was also a substantial portion of the willow population, this species is still relatively abundant in stands classified as cottonwood-willow habitat. Even so, salt-tolerant or water stress-tolerant shrubs such as tamarisk and arrowweed now dominate these habitats.

Similar to tamarisk, the non-native Russian olive is a shrubby tree that has become naturalized throughout the western United States (Olson and Knopf 1986), forming extensive stands in some areas (Knopf and Olson 1984; Brown 1990), particularly where historical river flow patterns have been altered by water development, such as along the Platte River in Nebraska (Currier 1982) and the Bighorn River in Wyoming (Akashi 1988). Such conversion of riparian vegetation from native to non-native species may have profound wildlife management implications. Bird species richness and density, for example, are higher in native riparian vegetation than in habitats dominated by tamarisk or Russian-olive (Knopf and Olson 1984; Brown 1990; Rosenberg et al. 1991).

Future

The health of natural riparian ecosystems is linked to the periodic occurrence of flood flows, associated channel dynamics, and the preservation of base flows capable of sustaining high floodplain water tables. The establishment of native riparian vegetation is diminished when the frequency and magnitude of peak river flows are reduced. Water uptake and water-use patterns indicate that native trees are replaced by non-native species in riparian ecosystems where streamflows are highly modified. Although riparian ecosystems are most directly affected by altered streamflow, additional factors threaten their integrity, including groundwater pumping (Stromberg et al. 1992), grazing (Armour et al. 1991), timber harvest and land clearing (Brinson et al. 1981), and fire (Busch and Smith 1993). Studies are under way to evaluate whether exotic plants will encroach further into riparian ecosystems, given conditions predicted under global climate change scenarios.

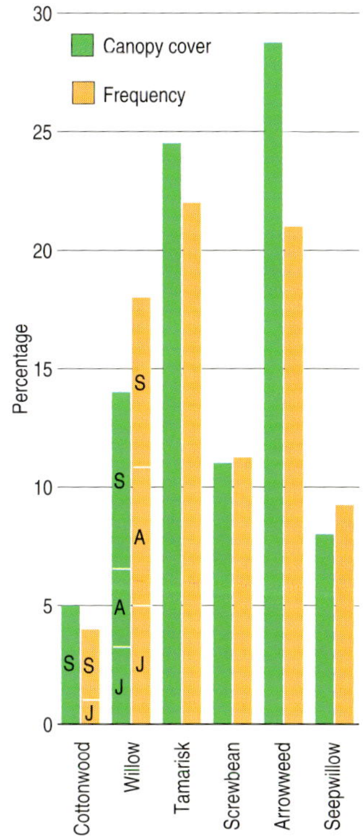

Fig. 4. Relative plant canopy cover and relative frequency percentages for woody riparian species of the lower Colorado River floodplain. Cottonwood and willow populations are grouped into juvenile (J), adult (A), and senescent (S) age segments.

References

Akashi, Y. 1988. Riparian vegetation dynamics along the Bighorn River, Wyoming. M.S. thesis, University of Wyoming, Laramie. 245 pp

Anderson, B.W., and R.D. Ohmart. 1984. Vegetation community type maps, lower Colorado River. U.S. Bureau of Reclamation, Boulder City, NV. 59 pp.

Armour, C.L., D.F. Duff, and W. Elmore. 1991. The effects of livestock grazing on riparian and stream ecosystems. Fisheries 16:7-11.

Bradley, C.E., and D.G. Smith. 1986. Plains cottonwood recruitment and survival on a prairie meandering river floodplain, Milk River, southern Alberta and northern Montana. Canadian Journal of Botany 64:1433-1442.

Brinson, M.M. 1990. Riverine forests. Pages 87-141 *in* A.E. Lugo, M. Brinson, and S. Brown, eds. Ecosystems of the world. Vol. 14. Forested ecosystems. Elsevier, Amsterdam, The Netherlands. 527 pp.

Brinson, M.M., B.L. Swift, R.C. Plantico, and J.S. Barclay. 1981. Riparian ecosystems: their ecology and status. U.S. Fish and Wildlife Service Biological Rep. 81/17. 155 pp.

Brown, C.R. 1990. Avian use of native and exotic riparian habitats on the Snake River, Idaho. M.S. thesis, Colorado State University, Fort Collins. 60 pp.

Busch, D.E., N.L. Ingraham, and S.D. Smith. 1992. Water uptake in woody riparian phreatophytes of the southwestern United States: a stable isotope study. Ecological Applications 2:450-459.

Busch, D.E., and S.D. Smith. 1993. Effects of fire on water and salinity relations of riparian woody taxa. Oecologia 94:186-194.

Busch, D.E., and S.D. Smith. 1995. Mechanisms associated with the decline and invasion of woody species in two riparian ecosystems of the southwestern U.S. Ecological Monographs. In press.

Currier, P.J. 1982. The floodplain vegetation of the Platte River: phytosociology, forest development, and seedling establishment. Ph.D. dissertation, Iowa State University, Ames. 332 pp.

Everitt, B.L. 1968. Use of the cottonwood in an investigation of the recent history of a floodplain. American Journal of Science 266:417-439.

Gregory, S.V., F.J. Swanson, W.A. McKee, and K.W. Cummins. 1991. An ecosystem perspective of riparian zones. BioScience 41:(8)540-551.

Hobbs, R.J. 1989. The nature and effects of disturbance relative to invasions. Pages 389-405 *in* J.A. Drake, H.A. Mooney, F. di Castri, R.H. Groves, F.J. Kruger, M. Rejmanek, and M. Williamson, eds. Biological invasions: a global perspective. Wiley, Chichester, England. 525 pp.

Hobbs, R.J., and L.F. Huenneke. 1992. Disturbance, diversity, and invasion: implications for conservation. Conservation Biology 6:324-337.

Jackson, J., J.T. Ball, and M.R. Rose. 1990. Assessment of the salinity tolerance of eight Sonoran Desert riparian trees and shrubs. U.S. Bureau of Reclamation, Yuma, AZ. 102 pp.

Johnson, W.C. 1992. Dams and riparian forests: case study from the upper Missouri River. Rivers 3(4):229-242.

Johnson, W.C. 1994. Woodland expansion in the Platte River, Nebraska: patterns and causes. Ecological Monographs 64:45-84.

Johnson, W.C., R.L. Burgess, and W.R. Keammerer. 1976. Forest overstory vegetation and environment on the Missouri River floodplain in North Dakota. Ecological Monographs 46:59-84.

Knopf, F.L., and T.E. Olson. 1984. Naturalization of Russian-olive: implications to Rocky Mountain wildlife. Wildlife Society Bull. 12:289-298.

Knopf, F.L., and M.L. Scott. 1990. Altered flows and created landscapes in the Platte River headwaters, 1840-1990. Pages 47-70 *in* J.M. Sweeney, ed. Management of dynamic ecosystems. North Central Section, The Wildlife Society, West Lafayette, IN. 180 pp.

Malanson, G.P. 1993. Riparian landscapes. Cambridge University Press, Cambridge, England. 296 pp.

Noble, M.G. 1979. The origin of *Populus deltoides* and *Salix* interior zones on point bars along the Minnesota River. American Midland Naturalist 102(1):59-67.

Olson, T.E., and F.L. Knopf. 1986. Naturalization of Russian-olive in the western United States. Western Journal of Applied Forestry 1:65-69.

Parker, I.M., S.K. Mertens, and D.W. Schemske. 1993. Distribution of seven native and two exotic plants in a tallgrass prairie in southeastern Wisconsin: the importance of human disturbance. American Midland Naturalist 130:43-55.

Rejmanek, M. 1989. Invasibility of plant communities. Pages 369-388 *in* J.A. Drake, H.A. Mooney, F. di Castri, R.H. Groves, F.J. Kruger, M. Rejmanek, and M. Williamson, eds. Biological invasions: a global perspective. Wiley, Chichester, England. 525 pp.

Rood, S.B., and J.M. Mahoney. 1990. Collapse of riparian poplar forests downstream from dams in western prairies: probable causes and prospects for mitigation. Environmental Management 14:451-464.

Rosenberg, K.V., R.D. Ohmart, W.C. Hunter, and B.W. Anderson. 1991. Birds of the lower Colorado River Valley. University of Arizona Press, Tucson. 416 pp.

Sacchi, C.F., and P.W. Price. 1992. The relative roles of abiotic and biotic factors in seedling demography of arroyo willow (*Salix lasiolepis*: Salicaceae). American Journal of Botany 79(4):395-405.

Scott, M.L., J.M. Friedman, and G.T. Auble. 1994. Fluvial process and the establishment of bottomland trees. Geomorphology. In press.

Scott, M.L., M.A. Wondzell, and G.T. Auble. 1993. Hydrograph characteristics relevant to the establishment and growth of western riparian vegetation. Pages 237-246 *in* H.J. Morel-Seytoux, ed. Proceedings of the Thirteenth Annual American Geophysical Union Hydrology Days. Hydrology Days Publications, Atherton, CA. 432 pp.

Sigafoos, R.S. 1964. Botanical evidence of floods and flood-plain deposition. U.S. Geological Survey Professional Paper 485A. 35 pp.

Smith, S.D., A.B. Wellington, J.L. Nachlinger, and C.A. Fox. 1991. Functional responses of riparian vegetation to streamflow diversion in the eastern Sierra Nevada. Ecological Applications 1:89-97.

Stromberg, J.C. 1993. Fremont cottonwood-Goodding willow riparian forests: a review of their ecology, threats, and recovery potential. Journal of the Arizona-Nevada Academy of Science 26:97-110.

Stromberg, J.C., D.T. Patten, and B.D. Richter. 1991. Flood flows and dynamics of Sonoran riparian forests. Rivers 2(3):221-235.

Stromberg, J.C., J.A. Tress, S.D. Wilkins, and S.D. Clark. 1992. Response of velvet mesquite to groundwater decline. Journal of Arid Environments 23:45-58.

Vitousek, P.M. 1990. Biological invasions and ecosystem processes: towards an integration of population biology and ecosystem studies. Oikos 57:7-13.

Williams, G.P., and M.G. Wolman. 1984. Downstream effects of dams on alluvial rivers. U.S. Geological Survey Professional Paper 1286. 83 pp.

Younker, G.L., and C.W. Andersen. 1986. Mapping methods and vegetation changes along the lower Colorado River between Davis Dam and the border with Mexico. U.S. Bureau of Reclamation, Boulder City, NV. 22 pp.

For further information:

David E. Busch
South Florida Natural Resources Center
National Park Service
Everglades National Park
4001 State Rd. 9336
Homestead, FL 33034

Surface Cover Changes in the Rio Grande Floodplain, 1935-89

by

James E. Roelle
National Biological Service

Warren W. Hagenbuck
U.S. Fish and Wildlife Service

Riparian (streamside) vegetation communities provide valuable habitat for wildlife, particularly in the arid and semi-arid Southwest, where such communities make up less than 1% of the landscape (Knopf et al. 1988). Agricultural conversion, urban and suburban expansion, water development, recreation, and invasion by non-native species such as Russian olive (*Elaeagnus angustifolia*) and saltcedar (*Tamarix* spp.) have severely reduced the extent and quality of these habitats. Despite such impacts, the floodplain of the Rio Grande in central New Mexico supports one of the most extensive cottonwood (*Populus fremontii*) gallery forests (bosque) remaining in the Southwest (Howe and Knopf 1991), and interest in ensuring the long-term health and viability of native communities along the Rio Grande has been steadily increasing (Crawford et al. 1993). This article documents changes between 1935 and 1989 in cover types of the floodplain of the Rio Grande in central New Mexico.

Study Area

The study area covers the historical floodplain of the Rio Grande from Velarde, New Mexico, to the narrows at Elephant Butte Reservoir, New Mexico, a distance of nearly 250 river mi (402 km; Figure). The historical floodplain in this reach encompasses more than 95,000 ha (nearly 236,000 acres); about 9,650 ha (24,000 acres) were omitted from the analysis because 1989 photography was unavailable.

Classification

Wetlands were classified according to the system used by the U.S. Fish and Wildlife Service's (USFWS) National Wetlands Inventory (Cowardin et al. 1979). Wooded riparian (nonwetland) areas were classified according to an unpublished system developed by the USFWS and the Arizona Riparian Council. The remaining uplands were classified according to a system developed by the U.S. Geological Survey (Anderson et al. 1976).

These classification systems provided more than 160 cover classes, an unmanageable number for an analysis of change. Thus, we aggregated the original classes in our geographic information system (GIS) into 11 broader types.

Expansion of saltcedar is of great concern in the Rio Grande valley in New Mexico, and

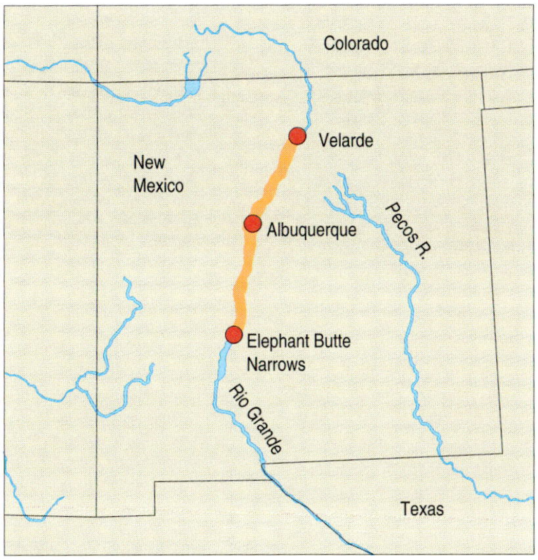

Figure. The Rio Grande study area.

separating saltcedar from other scrub-shrub types would have been desirable. Unfortunately, saltcedar could not be distinguished from other scrub-shrub types on the 1935 photography used to classify the area and was therefore included in the riparian scrub-shrub and wetland scrub-shrub classes.

Trends

Major changes in surface cover occurred on the floodplain of the Rio Grande between 1935 and 1989 (Table). Five of eight wetland or riparian types declined by about 17,000 ha (42,000 acres), including 5,453 ha (13,475 acres) of river or artificial channel; 4,015 ha (9,921 acres) of wet meadow, marsh, or pond; 2,638 ha (6,519 acres) of riparian scrub-shrub; 2,507 ha (6,195 acres) of riparian forest; and 2,482 ha (6,133 acres) of wetland scrub-shrub. Upland range also declined by 5,217 ha (12,891 acres).

The largest gains occurred in urban (11,389 ha; 28,143 acres) and agricultural (5,395 ha; 13,331 acres) cover types. Only three wetland or riparian cover types (lake, wetland forest, and dead forest or scrub-shrub) increased. Higher water levels in Elephant Butte Reservoir and construction of Cochiti Reservoir, New Mexico, produced a gain of 2,552 ha (6,306 acres) of lake. Wetland forests increased by 1,779 ha (4,396 acres). Most of this increase occurred between the levees and the stream channel, which has become narrower and straighter because of levee construction and channel stabilization. Dead forest or scrub-shrub increased by 1,197 ha (2,958 acres). Most of this mortality was at the upper end of Elephant Butte Reservoir because of high water in the mid-1980's.

The total forested area (wetland plus riparian) declined only slightly between 1935 (9,861 ha; 24,367 acres) and 1989 (9,133 ha; 22,568 acres), but this does not mean that concern for the long-term future of the woodlands is unwarranted. Only about 27% of the area forested in 1935 still supports forests, indicating that significant changes have occurred even in cases where the net change in area has been small. As noted before, much of the cottonwood forest is now confined between the levees and the river channel. The flow regime of the Rio Grande, however, has been altered significantly (e.g., lower peak flows) since most of these stands were established, and conditions favorable for germination and establishment of cottonwood now occur only rarely. Russian olive and saltcedar are likely to continue to replace cottonwood, especially under current hydrologic conditions (Howe and Knopf 1991).

Table. Surface cover changes in the Rio Grande floodplain, Velarde to Elephant Butte Reservoir, NM, 1935-89.

Cover type	Area (ha)		
	1935	1989	Change
Lake	32	2,584	+2,552
River or artificial channel	10,673	5,220	-5,453
Wet meadow, marsh, or pond	5,527	1,512	-4,015
Scrub-shrub (wetland)	9,070	6,588	-2,482
Scrub-shrub (riparian)	7,804	5,166	-2,638
Dead forest or scrub-shrub	0	1,197	+1,197
Forest (wetland)	4,683	6,462	+1,779
Forest (riparian)	5,178	2,671	-2,507
Agriculture	19,614	25,009	+5,395
Range	20,179	14,962	-5,217
Urban	3,006	14,395	+11,389

Rio Grande floodplain.

References

Anderson, J.R., E.E. Hardy, J.T. Roach, and R.E. Witmer. 1976. A land use and land cover classification system for use with remote sensor data. U.S. Geological Survey Professional Paper 964. 27 pp.

Cowardin, L.M., V.C. Carter, F.C. Golet, and E.T. LaRoe. 1979. Classification of wetlands and deepwater habitats of the United States. U.S. Fish and Wildlife Service FWS/OBS 79/81. 131 pp.

For further information:

James E. Roelle
National Biological Service
Midcontinent Ecological Science Center
4512 McMurry Ave.
Fort Collins, CO 80525

Crawford, C.S., A.C. Cully, R. Leutheuser, M.S. Sifuentes, L.H. White, and J.P. Wilber. 1993. Middle Rio Grande ecosystem: bosque biological management plan. Biological Interagency Team, U.S. Fish and Wildlife Service, Albuquerque, NM. 291 pp.

Howe, W.H., and F.L. Knopf. 1991. On the imminent decline of Rio Grande cottonwoods in central New Mexico. The Southwestern Naturalist 36(2):218-224.

Knopf, F.L., R.R. Johnson, T. Rich, F.B. Samson, and R.C. Szaro. 1988. Conservation of riparian ecosystems in the United States. Wilson Bull. 100(2):272-284.

Distribution, Abundance, and Health of

Ecoregions

The Great Plains *295*

Interior West *309*

Alaska *337*

Hawaii *361*

Doug Sokell/Tom Stack and Associates©

The Great Plains

by
Science Editor
Gary D. Willson
National Biological Service
Northern Prairie Science Center
Central Plains/Ozark Field Station
University of Missouri
316 Gentry Hall
Columbia, MO 65211

Overview

The Great Plains of North America are grasslands or former grasslands that occupy more than 200 million ha (500 million acres) of land from central Alberta, Canada, to the Texas Panhandle and eastern New Mexico and from the front range of the Rocky Mountains to the forest edge in Minnesota, Missouri, and Oklahoma. The natural plant communities dominating this landscape are known as grasslands or prairie (French for meadow) and they are composed of a rich complex of grasses and forbs. The climate, soils, and topography of the eastern Great Plains are suitable for agriculture, and consequently most of the original prairie has been converted to row crops or pasture. In the western Great Plains, large areas of intact grassland are used as rangeland. Researchers estimate that less than 1% of the original grasslands remains undisturbed by human activities (Klopatek et al. 1979).

Articles in this section focus on the effects of more than 100 years of postsettlement manipulation of the Great Plains ecosystem. For example, fire was undoubtedly an important ecological force in maintaining historical grassland landscapes and species distributions. Following fire suppression, woody plants have invaded grasslands from adjacent forest and wooded stream valleys. In addition, water management practices and the planting of farm and ranch shelterbelts have resulted in the encroachment of trees into grassland habitat. In many parts of the Great Plains today, far more woody plants exist than before agricultural development. As endemic grassland birds have declined, they have been replaced by eastern forest species moving into newly wooded habitats (Knopf; Igl and Johnson, both this section).

Native prairie fishes also have experienced significant losses in their historical distributions. Impoundments constructed on many rivers and streams of the Great Plains have fragmented populations and eliminated colonization of vacant habitat. Several prairie fishes, including the Arkansas River shiner (*Notropis girardi*) and the Arkansas River speckled chub (*Macrhybopsis aestivalis tetranemus*), have shown significant declines in their distributions and abundances (Echelle et al., this section).

The fragmentation of native grassland due to agricultural encroachment as well as the elimination of keystone species, such as bison (*Bison bison*) and the white-tailed prairie dog (*Cynomys leucurus*), have led to a general decline in prairie wildlife, although some species have adapted to human-induced changes and some have even increased in numbers. For example, native grassland birds have shown steeper, more consistent, and more geographically widespread declines than any other avian group, including Neotropical migrants (Knopf, this section). Species such as mallard (*Anas platyrhynchos*), blue-winged teal (*A. discors*), and northern pintail (*A. acuta*) are now at or near the lowest numbers ever recorded (Shaffer and Newton, this section). The primary reason for these declines in numbers is low nest success due to predation by common species such as red fox (*Vulpes vulpes;* Shaffer and Newton, this section). In other species, such as American coot (*Fulica americana*), drainage of wetlands compounded by severe drought may have played a role in depressing populations (Igl and Johnson, this section). In contrast to waterfowl, the coyote (*Canis latrans*) is increasing its range. Historical and recent trends in coyote populations and diet may reflect a response to land-use changes, especially agricultural changes and shifts in human populations on the Great Plains (Gipson and Brillhart, this section).

The Great Plains are becoming increasingly rural because of emigration of people and a shift of human populations away from farms to urban centers. Although the Great Plains encompass about 20% of the land mass of the lower 48 states, the population is only about 2% of the U.S. total. Federal agricultural land-retirement programs, such as the Soil Bank Program and the Conservation Reserve Program (CRP), devised to mediate fluctuations in the farm economy, may also help slow or reverse the declines of some grassland species. For example, recent field surveys have shown that several grassland birds that had declined in the Great Plains are much more common on CRP habitat than in cropland (Johnson and Koford, this section). In recent years numerous small to medium tracts of native grassland have been designated as preserves. These areas plus changes in agricultural practices that promote natural resource conservation (e.g., CRP) are important to protect the remaining biodiversity of the Great Plains.

References

Klopatek, E.V., R.J. Olson, C.J. Emerson, and J.L. Jones. 1979. Landuse conflicts with natural vegetation. Environmental Conservation 6:191-200.

Declining Grassland Birds

by
Fritz L. Knopf
National Biological Service

Native grasslands represent the largest vegetative province of North America. Almost 1.5 million km^2 (0.6 million mi^2) of grasslands historically occurred on the Great Plains. Although the Great Plains played a major role in the evolution of North American forest birds (Mengel 1970), the grassland avifauna itself is relatively poor with only 5% of all North American bird species believed to have evolved within the Great Plains. That group includes 12 species of birds that are considered endemic (i.e., evolved specifically within) to the grasslands, along with 20 others that have centers of evolution on the grasslands but range more widely into contiguous vegetative provinces.

The landscape of the Great Plains has undergone significant alteration from descriptions provided in early accounts. The influences have been varied with many (e.g., urbanization, mineral exploration, and defense installations) having primarily local effects on the native birds. Activities with more universal effects on the landscape have included transformation of the native grazing community, cultivation of grains and tame grasses, draining of wetlands, and woody development in the form of tree plantings in the dry central and western Great Plains (Knopf and Samson, in press). Also, ecological invasions following fire suppression in the eastern and central plains and water developments in the western plains have drastically altered historical landscapes.

Of the 435 bird species breeding in the United States, 330 have been recorded on the Great Plains. Current avian assemblages on the grasslands reflect two broad patterns of change that have occurred in the last century: native endemic species have declined in numbers (Table) while simultaneously (and rather independently) alien species have expanded their ranges (Knopf 1994).

Methods

Information on the annual status of endemic grassland birds was obtained through the Breeding Bird Surveys (1966-91), which are conducted annually during the bird breeding season at numerous sites across the nation.

Status and Trends

During the last 25 years, grassland species have shown steeper, more consistent, and more geographically widespread declines than any other behavioral or ecological guild of North American birds, including Neotropical migrants. Continental population trends of many individual species of grassland birds also declined. Excluding the wetland-associated marbled godwit (*Limosa fedoa*) and Wilson's phalarope (*Phalaropus tricolor*), 7 of the 10 endemic grassland species showed population declines during the last 26 years. Population declines of four species (mountain plover [*Charadrius montanus*], Franklin's gull [*Larus pipixcan*], Cassin's sparrow [*Aimophila cassinii*], and lark bunting [*Calamospiza melanocorys*]) are statistically significant.

Similarly, 14 of the 20 more widespread species that evolved primarily on the Great Plains declined during this period, with the declines in the eastern meadowlark (*Sturnella magna*) and 5 sparrows (grasshopper [*Ammodramus savannarum*], Henslow's [*A. henslowii*], lark [*Chondestes grammacus*], Brewer's [*Spizella breweri*], and clay-colored [*S. pallida*]) being statistically significant. Across all grassland species, populations of only the upland sandpiper (*Bartramia longicauda*) and McCown's longspur (*Calcarius mccownii*) have increased significantly since 1966.

Patterns of Bird Declines

Reasons for population declines among species within the grassland avifauna are difficult to assess. Through examining trends for those species where declines are supported statistically, the declines appear to be localized for Franklin's gull, dickcissel, Henslow's and grasshopper sparrows, lark bunting, and eastern meadowlark; these species show a significant difference in the proportion of surveys with increasing versus decreasing populations. This pattern of significant local declines for species that also are declining continentally reflects a pattern of loss of local breeding habitats.

Declines in populations of mountain plover and Cassin's and clay-colored sparrows were universal across their respective geographic ranges. The seasonal distributions and ecology of these sparrows are poorly understood. The plover is now rare on its former wintering areas in southern Texas and has a fragmented wintering distribution in California. Ongoing research on plovers indicates that declines of these species may be attributable to decline or degradation in the quality of habitats available for wintering.

Population trends for a third group of grassland species (ferruginous hawk [*Buteo regalis*]; Mississippi kite [*Ictinia mississippiensis*]; upland sandpiper; short-eared owl [*Asio flammeus*]; horned lark [*Eremophila alpestris*]; western meadowlark [*Sturnella neglecta*]; and vesper [*Pooecetes gramineus*], savannah [*Passerculus sandwichensis*], and Henslow's sparrows) show significant changes in relative abundance among surveys, even though continental numbers are stable. The geographic distributions of these species appear to be changing at present.

Table. Annual rates of change in continental populations of endemic grassland bird species, 1966-91 (Breeding Bird Survey data).

Species	No. of routes	Percentage[a]		Population[b]
		Increasing	Decreasing	
Endemics				
Ferruginous hawk	242	51.7	42.6*	+ 0.4
Mountain plover	40	45.0	50.0	- 3.6*
Long-billed curlew	222	45.5	50.0	- 0.6
Marbled godwit	181	54.7	43.1	+ 1.0
Wilson's phalarope	339	41.3	54.6	+ 0.8
Franklin's gull	225	35.1	60.4*	- 7.4*
Sprague's pipit	136	38.2	55.9	- 3.1
Cassin's sparrow	169	49.7	48.5	- 3.4*
Baird's sparrow	132	39.4	56.8	- 1.6
Lark bunting	344	39.8	57.6*	- 3.3*
McCown's longspur	66	45.5	47.0	+ 7.9*
Chestnut-colored longspur	151	42.4	54.3	+ 0.6
More widespread species				
Mississippi kite	163	58.9	40.5*	+ 0.4
Swainson's hawk	607	48.8	46.6	+ 1.2
Northern harrier	1,075	43.7	52.5	- 1.0
Prairie falcon	261	47.1	43.7*	+ 0.2
Greater prairie-chicken	47	40.0	53.2	- 10.3
Lesser prairie-chicken	8	25.0	62.5	+ 8.1
Sharp-tailed grouse	180	41.7	52.8	+ 0.9
Sage grouse	103	52.4	46.6	+ 6.2
Upland sandpiper	668	51.5	45.8*	+ 3.5*
Burrowing owl	349	43.3	51.6	- 0.2
Short-eared owl	268	38.1	57.5*	- 0.7
Horned lark	1,708	40.6	56.8*	- 0.7
Sage thrasher	230	53.0	44.8	+ 1.4
Eastern meadowlark	1,714	30.5	68.6*	- 2.2*
Western meadowlark	1,304	38.0	59.7*	- 0.6
Dickcissel	780	37.6	60.9*	- 1.7*
Green-tailed towhee	207	43.0	51.2	+ 0.3
Savannah sparrow	1,418	42.9	54.4*	- 0.6
Grasshopper sparrow	1,446	37.6	58.8*	- 4.6*
Henslow's sparrow	249	30.9	61.8*	- 4.2*
Vesper sparrow	1,473	38.5	58.2*	- 0.6
Lark sparrow	909	44.7	52.6	- 3.4*
Sage sparrow	205	38.0	58.5*	- 2.4
Brewer's sparrow	359	39.8	55.7*	- 4.1*
Clay-colored sparrow	441	43.8	52.4	- 1.5*

[a]Percentages totaling <100% include some routes with no change in numbers of birds detected. Asterisk indicates significant differences ($P < 0.05$) between number of surveys with increasing versus decreasing species populations.

[b]Annual rate (expressed as a percentage) of change in population numbers. Asterisk indicates a statistically significant ($P < 0.05$) rate of population change.

Although species associated with wetlands have certainly declined since settlement of the grasslands in the mid-1800's, Breeding Bird Survey data indicate that populations of the endemic marbled godwit and Wilson's phalarope are stable. Wetland conservation actions to benefit waterfowl have apparently stabilized populations of these two species.

Are There Fewer Birds on the Great Plains?

Many species of forest birds historically occurred west of their eastern deciduous forest habitats in streamside vegetation of the eastern Great Plains. As most endemic grassland birds have declined, they have been replaced locally by eastern species moving into windbreaks and developing riparian forests along streambeds of the short-grass prairie. The streamside forests evolved with water management practices in the west and have favored the movement of many species farther onto and across the grasslands.

At one location, Crook, Colorado, 83 species of birds in the vicinity included only 6 representatives of the Great Plains avifauna, of which only 3 species bred locally (Knopf 1986). None of those three species bred in the riparian vegetation. That riparian forest developed since 1900, and almost 90% of the native birds currently breeding in northeastern Colorado have colonized in recent times.

Causes of Declines Unknown

Ecological processes driving population trends of North American grassland birds are undescribed. As a group, grassland birds have declined more than birds of other North American vegetative associations. Unlike Neotropical migrants, which have experienced declines primarily in the northeastern deciduous forests (Robbins et al. 1989), declines in grassland species are occurring at a continental scale. For example, the decline in numbers of the mountain plover, Cassin's sparrow, and lark bunting are occurring across their ranges. The lack of understanding of the wintering ecology of grassland birds precludes optimistic projections, especially for these species experiencing widespread, geographic declines.

References

Knopf, F.L. 1986. Changing landscapes and the cosmopolitism of the eastern Colorado avifauna. Wildlife Society Bull. 14:132-142.

Knopf, F.L. 1994. Avian assemblages on altered grasslands. Studies in Avian Biology 15:247-257.

Knopf, F.L., and F.B. Samson. Conserving the biotic integrity of the Great Plains. *In* S. Johnson and A. Bouzaher, eds. Conservation of Great Plains ecosystems: current science, future options. Kluwer Academic Press, Dordrecht, Netherlands. In press.

Mengel, R.M. 1970. The North American Central Plains as an isolating agent in bird speciation. Pages 280-340 *in* W. Dort and J.K. Jones, eds. Pleistocene and recent environments of the central Great Plains. University of Kansas Press, Lawrence.

Robbins, C.S., J.R. Sauer, R.S. Greenberg, and S. Droege. 1989. Population declines in North American birds that migrate to the Neotropics. Proceedings of the National Academy of Science 86:7658-7662.

For further information:

Fritz L. Knopf
National Biological Service
Midcontinent Ecological
Science Center
4512 McMurry Ave.
Fort Collins, CO 80525

Migratory Bird Population Changes in North Dakota

by

Lawrence D. Igl
Douglas H. Johnson
National Biological Service

The status of migratory bird populations in North America has received increased attention in recent years. Much of this consideration has been on Neotropical migrants, especially those associated with eastern forests. The status of migratory bird populations in the Great Plains has received far less attention. During the past quarter-century, populations of many species of birds that breed in the northern Great Plains have increased or declined, as indicated by trends from the North American Breeding Bird Survey.

In 1967 Stewart and Kantrud (1972) conducted a survey of breeding bird populations throughout North Dakota. This study offered a rare glimpse of bird populations breeding in the northern Great Plains as well as important baseline data on breeding bird populations. These data help us evaluate relationships between birds and habitat conditions. We repeated the survey to compare bird populations in North Dakota during 1967 with those in 1992 and 1993.

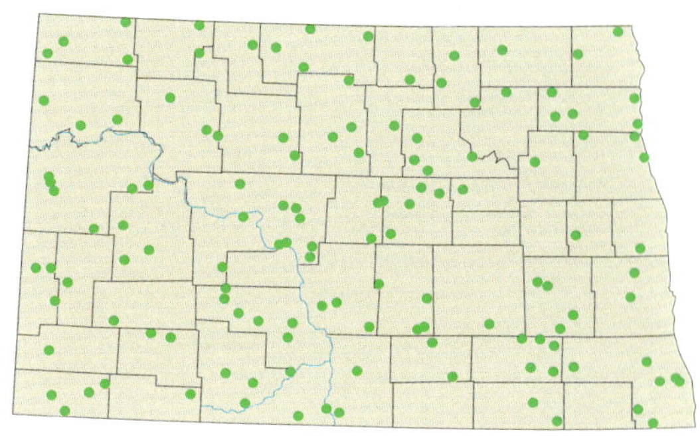

Figure. Distribution of the original 130 quarter-sections in North Dakota.

Study Areas and Methods

To aid in a direct comparison, the same 130 legal quarter-sections (64.7 ha, 160 acres) surveyed in 1967 were visited again in 1992 and 1993 (Figure). Surveys of breeding birds were conducted as similarly as possible to the methods used by Stewart and Kantrud (1972).

Each bird species was classified into one of three groups according to its migratory strategy: permanent resident (present in North Dakota year-round), short-distance migrant (winters north of the U.S.-Mexico border), and long-distance migrant (winters south of the U.S.-Mexico border). In addition, each species was categorized to a preferred breeding habitat: wetland/wet meadow, grassland/open habitat, open habitat with scattered trees, woodland/woodland-edge, shrubland, residential/habitat generalist, and other. Within each group, a mean population size was calculated and expressed as the number of indicated pairs per 100 ha (247 acres).

Status and Trends

Data were obtained on 160 breeding bird species within the 128 quarter-sections that we received permission to survey in all 3 years (Table 1), including 129 species in 1967, 144 in 1992, and 152 in 1993. Thus, about 72% of the known breeding avifauna of North Dakota (Faanes and Stewart 1982) were identified. Songbirds were the most common group,

Table 1. Distribution of species observed on 128 randomly selected quarter-sections in North Dakota in 1967, 1992, and 1993 by breeding habitat and migratory strategy.

Breeding habitat	Permanent resident	Short-distance migrant	Long-distance migrant	Total
Wetland/wet meadow	0	41	13	54
Grassland/open habitat	3	16	15	34
Open habitat with trees	1	4	3	8
Shrubland	0	4	5	9
Woodland/woodland-edge	8	15	24	47
Residential/ habitat generalist	4	1	2	7
Other	0	1	0	1
Total	16	82	62	160

accounting for about 80% of the total number of indicated pairs in each year.

Of the total number of breeding pairs of the 50 most common species in the 3 years (Table 2), the five most commonly encountered species, in order of abundance, were horned lark (*Eremophila alpestris*), chestnut-collared longspur (*Calcarius ornatus*), red-winged blackbird (*Agelaius phoeniceus*), western meadowlark (*Sturnella neglecta*), and brown-headed cowbird (*Molothrus ater*). The horned lark, the most common breeding bird species recorded each year, is a species that is most characteristic of cropland or heavily grazed prairie and which favors open areas with low sparse vegetation (Stewart 1975). The overall frequency and abundance of the brown-headed cowbird are of concern because this brood parasite has been implicated in the decline of some Neotropical migrants.

Ninety percent of the 160 species observed in the 3 years are migrants (Table 1). Moreover, migrants constitute over 95% of the indicated pairs detected in the sample units in each year. The remaining (10%) species are year-round residents in North Dakota. Of the species that migrate, 82 (51%) are short-distance migrants and 62 (39%) are long-distance (Neotropical) migrants.

The data indicate that breeding bird populations show considerable short- and long-term variability. The patterns of population change for many grassland and wetland species are remarkably similar and consistent among taxonomic groups (e.g., mallard [*Anas platyrhynchos*] versus American coot [*Fulica americana*] versus savannah sparrow [*Passerculus sandwichensis*]) and migration strategies (long-distance versus short-distance migrant; Tables 2 and 3). A common feature of these species is their dependence on grassland and wetland habitats on the breeding grounds; most breed in the northern Great Plains but winter elsewhere. Severe drought conditions in the Great Plains may have played a major role in the depressed populations (Tables 2 and 3) of some wetland and grassland species in 1992 (an extremely dry year) compared with 1967 (a near-average year) and 1993 (an extremely wet year).

Several species associated with grassland and wetland habitats (e.g., savannah sparrow and American coot) were relatively common in 1967, showed major declines in 1992, and recovered slightly in 1993 (Table 2). The fact that populations of some species (e.g., black tern [*Chlidonias niger*], Wilson's phalarope [*Phalaropus tricolor*]) remain below their 1967 levels suggests that precipitation alone may not explain all of the changes in the populations of grassland and wetland species. Drainage of wetlands, agriculture encroachment, and increased fragmentation of native prairie are also suspected in the declines of some wetland and grassland species.

Table 2. Number of indicated pairs of the 50 most common bird species observed on 128 randomly selected quarter-sections in North Dakota in 1967, 1992, and 1993.

Species	Migration strategy*	Breeding habitat**	No. indicated pairs 1967	1992	1993
Horned lark (*Eremophila alpestris*)	SDM	G/O	1,253	1,093	1,661
Chestnut-collared longspur (*Calcarius ornatus*)	SDM	G/O	1,129	602	755
Red-winged blackbird (*Agelaius phoeniceus*)	SDM	WET	945	597	710
Western meadowlark (*Sturnella neglecta*)	SDM	G/O	926	487	646
Brown-headed cowbird (*Molothrus ater*)	SDM	W/E	460	643	610
Lark bunting (*Calamospiza melanocorys*)	LDM	G/O	604	679	298
Grasshopper sparrow (*Ammodramus savannarum*)	LDM	G/O	301	402	449
Mourning dove (*Zenaida macroura*)	SDM	W/E	292	339	337
Savannah sparrow (*Passerculus sandwichensis*)	SDM	G/O	516	134	276
Clay-colored sparrow (*Spizella pallida*)	LDM	SHR	364	261	289
Vesper sparrow (*Pooecetes gramineus*)	SDM	G/O	195	224	393
Common grackle (*Quiscalus quiscula*)	SDM	O/T	140	300	299
Eastern kingbird (*Tyrannus tyrannus*)	LDM	O/T	167	321	245
Cliff swallow (*Hirundo pyrrhonota*)	LDM	G/O	152	343	226
American coot (*Fulica americana*)	SDM	WET	348	76	124
Bobolink (*Dolichonyx oryzivorus*)	LDM	G/O	216	186	172
Blue-winged teal (*Anas discors*)	LDM	WET	286	99	145
Mallard (*Anas platyrhynchos*)	SDM	WET	212	113	200
House wren (*Troglodytes aedon*)	LDM	W/E	52	219	209
Barn swallow (*Hirundo rustica*)	LDM	G/O	96	187	194
Western kingbird (*Tyrannus verticalis*)	LDM	O/T	103	194	177
House sparrow (*Passer domesticus*)	RES	RES	96	181	174
Yellow-headed blackbird (*Xanthocephalus xanthocephalus*)	LDM	WET	89	155	175
Common yellowthroat (*Geothlypis trichas*)	LDM	WET	134	91	175
American goldfinch (*Carduelis tristis*)	SDM	W/E	106	146	132
Baird's sparrow (*Ammodramus bairdii*)	LDM	G/O	170	77	125
Killdeer (*Charadrius vociferus*)	SDM	WET	105	112	142
Gadwall (*Anas strepera*)	SDM	WET	96	111	118
Marsh wren (*Cistothorus palustris*)	SDM	WET	51	113	153
American robin (*Turdus migratorius*)	SDM	GEN	67	117	123
Yellow warbler (*Dendroica petechia*)	LDM	GEN	102	90	105
Rufous-sided towhee (*Pipilo erythrophthalmus*)	SDM	W/E	94	83	118
Song sparrow (*Melospiza melodia*)	SDM	SHR	55	100	130
Upland sandpiper (*Bartramia longicauda*)	LDM	G/O	63	106	89
Northern pintail (*Anas acuta*)	SDM	WET	171	23	58
Bank swallow (*Riparia riparia*)	LDM	G/O	77	112	61
Brown thrasher (*Toxostoma rufum*)	SDM	SHR	42	85	104
Black tern (*Chlidonias niger*)	LDM	WET	118	39	39
Cedar waxwing (*Bombycilla cedrorum*)	SDM	W/E	23	149	43
Field sparrow (*Spizella pusilla*)	SDM	SHR	49	61	74
Northern shoveler (*Anas clypeata*)	SDM	WET	87	25	52
Franklin's gull (*Larus pipixcan*)	LDM	WET	22	79	56
Least flycatcher (*Empidonax minimus*)	LDM	W/E	23	52	81
Brewer's blackbird (*Euphagus cyanocephalus*)	LDM	SHR	19	47	83
Sora (*Porzana carolina*)	SDM	WET	32	41	78
Chipping sparrow (*Spizella passerina*)	LDM	W/E	22	51	70
Wilson's phalarope (*Phalaropus tricolor*)	LDM	WET	73	30	36
Lark sparrow (*Chondestes grammacus*)	LDM	G/O	40	52	40
Gray catbird (*Dumetella carolinensis*)	LDM	SHR	48	44	38
Ruddy duck (*Oxyura jamaicensis*)	SDM	WET	36	56	23

*SDM—short-distance migrant; LDM—long-distance migrant; RES—resident.
**G/O—grassland-open habitat; WET—wetland-wet meadow; W/E—woodland-woodland-edge; SHR—shrubland; O/T—open habitat with trees: GEN—residential-habitat generalist.

Table 3. Mean number of indicated breeding pairs in 128 randomly selected quarter-sections in North Dakota by year, migration strategy, and preferred breeding habitat.

Migration and habitat	Mean pairs/100 ha		
	1967	1992	1993
Migration strategy			
Permanent resident	2.6	5.7	6.1
Short-distance migrant	95.5	74.7	99.5
Long-distance migrant	43.2	52.3	45.4
Breeding habitat			
Wetland/wet meadow	37.5	24.7	32.6
Grassland/open habitat	71.7	59.3	68.3
Open habitat with trees	5.5	10.6	9.7
Shrubland	7.2	7.5	9.0
Woodland/woodland-edge	15.6	24.6	25.3
Residential/generalist	3.7	5.7	5.8
Other	0.1	0.3	0.3
Total	141.3	132.7	151.0

Federal land-retirement programs (such as the Soil Bank Program in 1967 and the Conservation Reserve Program in 1992-93) may help slow or reverse the declines of some grassland species. For example, between 1982 and 1991, the sedge wren (*Cistothorus platensis*) showed a significant decline on Breeding Bird Surveys in North Dakota. Over 50% of sedge wren breeding pairs found in all 3 years were found in these set-aside habitats.

In contrast, the populations of birds associated with woody vegetation may be less vulnerable to climatological factors such as drought. Species associated with woody vegetation have increased dramatically between 1967 and 1992-93 (Tables 2 and 3). In presettlement times, fire and grazing pressures played a major role in the formation and maintenance of the grassland landscape of the northern Great Plains. The relaxation and alteration of these pressures resulted in the encroachment of shrubs and trees into grassland habitats. Landscape fragmentation by tree plantings (e.g., farmstead windbreaks and field shelterbelts) is also suspected in the increase in species associated with woody vegetation. These conditions provided woodland and woodland-edge species with nesting opportunities that did not exist or were quite limited in presettlement times. In addition, maturation of the woody vegetation in these tree plantings may be attractive to certain species. For example, 14 of 15 species that nest in tree cavities showed increasing or stable populations in this survey.

Conservation Implications

Further analysis of habitat changes between 1967 and 1992-93 are needed to fully understand the changes in bird populations in North Dakota. Many species associated with the increasing amount of woody vegetation are common and have widespread distributions in North America (Johnson et al. 1994). On the other hand, many grassland and wetland species experienced declines and have few habitat alternatives to the Great Plains. The implication is that preservation of native grassland and wetland habitats is necessary to support breeding populations of migrants in the northern plains.

References

Faanes, C.A., and R.E. Stewart. 1982. Revised checklist of North Dakota birds. Prairie Naturalist 14:81-92.

Johnson, D.H., R.L. Kreil, G.B. Berkey, R.D. Crawford, D.O. Lambeth, and S.F. Galipeau. 1994. Influences of waterfowl management on nongame birds: the North Dakota experience. Transactions of the North American Wildlife and Natural Resource Conference 59: 293-302.

Stewart, R.E. 1975. Breeding birds of North Dakota. Tri-College Center for Environmental Studies, Fargo, ND. 295 pp.

Stewart, R.E., and H.A. Kantrud. 1972. Population estimates of breeding birds in North Dakota. Auk 89:766-788.

For further information:

Lawrence D. Igl
National Biological Service
Northern Prairie Science Center
8711 37th St. SE
Jamestown, ND 58401

Duck Nest Success in the Prairie Potholes

by
Terry L. Shaffer
Wesley E. Newton
National Biological Service

Since the early 1970's, the numbers of some waterfowl species such as mallard (*Anas platyrhynchos*), blue-winged teal (*A. discors*), and northern pintail (*A. acuta*) have reached or nearly reached the lowest ever recorded. Low nest success (the proportion of nests in which one or more eggs hatch) in key breeding areas, including the U.S. Prairie Pothole region, is partly responsible for declines in duck numbers (Klett et al. 1988; Johnson et al. 1992).

Methods

We examined status and trends of duck nest success for mallard, blue-winged teal, gadwall (*A. strepera*), northern shoveler (*A. clypeata*), and northern pintail, for one to four time periods between 1966 and 1989, and for five regions in North and South Dakota and Minnesota (Fig. 1). Nest success data originated from numerous independent studies conducted throughout the region. Some data from 1966 to 1984 were previously analyzed by Klett et al. (1988). We followed the methods of Klett et al., except we considered one additional time period (1985-89) and one additional habitat (Conservation Reserve Program lands).

Nest Success

Mallard

Data for 4,093 mallard nests showed that their nest success ranged from 6% to 20% (Fig. 2). Only 3 of 14 nest success estimates reached or exceeded 15%, the level of nest success thought necessary to maintain mallard numbers at a stable level in central North Dakota (Cowardin et al. 1985). These three areas were central South Dakota (1966-74), eastern South Dakota (1985-89), and central North Dakota (1985-89).

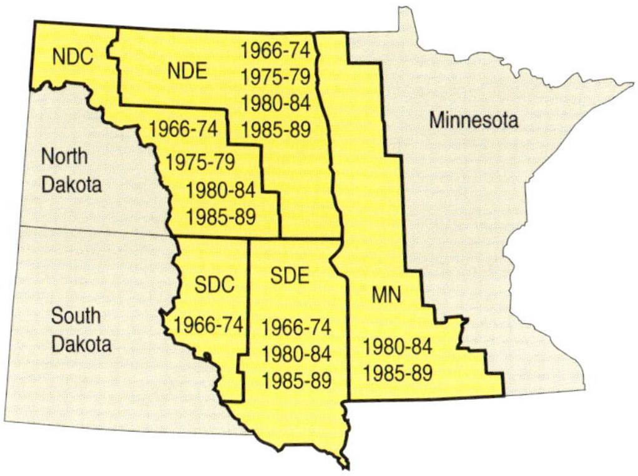

Fig. 1. Areas of the Prairie Pothole region and time periods for which estimates of duck nest success were made. See Fig. 2 for abbreviations.

Mallard nest success was relatively steady from 1966-74 to 1980-84, but increased from 1980-84 to 1985-89. In 1985-89, mallard nest success was still less than 15% in two of four regions where data were available. Predation was the major cause of nest failure, accounting for 85% of mallard nest failures in North Dakota.

Blue-winged Teal

Data for 9,819 blue-winged teal nests revealed that nest success of blue-winged teal ranged from 12% to 29% (Fig. 2). The level of nest success needed to maintain stable numbers of blue-winged teal is believed to be 20% (Klett et al. 1988). Nest success of blue-winged teal was generally below this level, except in South Dakota.

Nest success increased slightly from 1980-84 to 1985-89, but was still generally less than 20%. Predation, the principal cause of blue-winged teal nest failures, accounted for 92% of the failed nests in North Dakota.

Gadwall

Data on 3,782 gadwall nests showed that their average nest success ranged from 11% in western Minnesota (1980-84) to 26% in central South Dakota (1966-74; Fig. 2). Nest success was generally below 20%, the minimum level believed necessary to sustain populations (Klett et al. 1988).

Gadwall nest success increased from 1980-84 to 1985-89 and reached or exceeded 20% during 1985-89 in North and South Dakota. Predation was the primary cause of nest failure, accounting for 90% of failed gadwall nests in North Dakota.

Northern Shoveler

Nest success in 1,212 shoveler nests ranged from 5% in western Minnesota (1980-84) to 35% in central South Dakota (1966-74; Fig. 2). Nest success was generally below 20%, the minimum level believed necessary to sustain populations (Klett et al. 1988). In western Minnesota and eastern North Dakota, nest success was less than 10%, but it was greater than 20% in central North and South Dakota.

Shoveler nest success increased from 1980-84 to 1985-89, but was still much less than 20% in western Minnesota and eastern North Dakota. Predation was the primary cause of nest failure in all regions, and in North Dakota caused 88% of shoveler nest failures.

Northern Pintail

Data for 1,633 pintail nests revealed that their success ranged from 5% to 20% (Fig. 2). Fifteen percent is the minimum level of nest success believed necessary to sustain pintail numbers (Klett et al. 1988). Only 2 of 14 nest success estimates reached or exceeded 15%; these were for central South Dakota (1966-74) and central North Dakota (1985-89).

Within each region, pintail nest success was generally lowest in 1966-74 and highest in 1985-89. Even in 1985-89, however, nest success was much less than 15% in all regions where data were available, except central North Dakota. Predation was the major cause of nest failure; for example, in North Dakota it accounted for 81% of pintail nest failures. In addition, because pintails nest more frequently in cropland than other species (Klett et al. 1988), farming operations were also an important cause of nest failure, accounting for 16% of pintail nest failures.

Trends

Our results suggest that nest success of the five species of ducks considered here was and probably still is too low to maintain stable numbers of breeding ducks in most areas of the Prairie Pothole region. For example, even though nest success increased from 1980-84 to 1985-89, it was still below the level needed to sustain populations for most species in most regions. Except for pintails, whose nest success generally increased, we observed no consistent increases or decreases in nest success across periods. In central South Dakota in the 1966-74 period nest success was much higher than in other regions, exceeding the level needed to sustain populations. This region likely contributed a "surplus" of ducks in 1966-74 that helped make up for the "shortage" of ducks produced in other regions. Unfortunately, no data for central South Dakota have been available since then.

Predation was the primary reason for the low nest success we observed. Predator species such

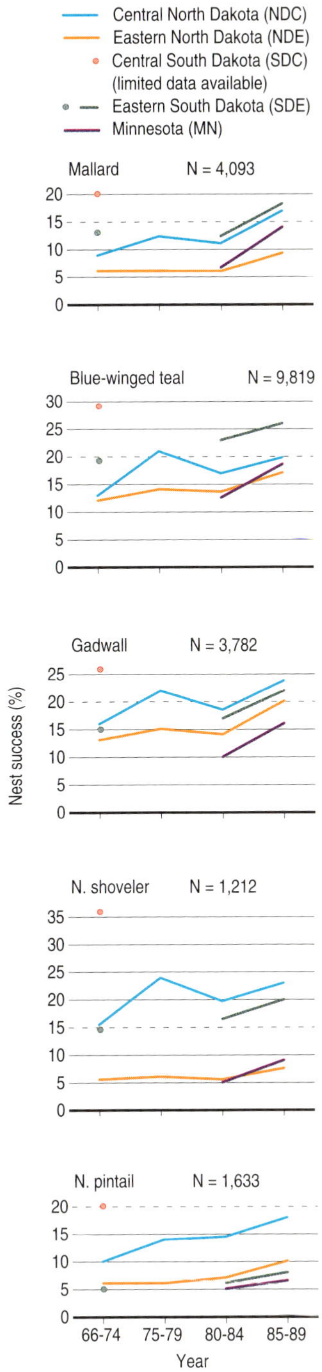

Fig. 2. Nest success (%) by period for five areas in the Prairie Pothole region for five duck species. Dashed line = level of nest success believed necessary for a stable population.

as red fox (*Vulpes vulpes*), striped skunk (*Mephitis mephitis*), and raccoon (*Procyon lotor*) are common or numerous throughout the region (Sargeant et al. 1993). Both red foxes and striped skunks are important predators of duck nests (Johnson et al. 1989), and red foxes also take many female ducks during the breeding season (Sargeant et al. 1984).

More than two-thirds of the Prairie Pothole region is in Canada. Greenwood et al. (1987) studied mallard nest success in that portion of the region from 1982 to 1985. Their findings were similar to ours: mallard nest success averaged 12% and only 7 of 31 estimates on individual areas reached or exceeded 15%. Predators caused most nest failures. The authors concluded that nest success in much of Prairie Canada in 1982-85 was too low to maintain stable numbers of breeding mallards.

The status of duck nest success in the recent past in the Prairie Pothole region seems clear. Nest success was too low for duck populations to sustain themselves. Unless steps are taken to improve duck nest success in the future, we will likely see further declines in numbers of these and possibly other waterfowl species.

For further information:

Terry L. Shaffer
National Biological Service
Northern Prairie Science Center
8711 37th St. SE
Jamestown, ND 58401

References

Cowardin, L.M., D.S. Gilmer, and C.W. Shaiffer. 1985. Mallard recruitment in the agricultural environment of North Dakota. Wildlife Monographs 92. 37 pp.

Greenwood, R.J., A.B. Sargeant, D.H. Johnson, L.M. Cowardin, and T.L. Shaffer. 1987. Mallard nest success and recruitment in Prairie Canada. Transactions of the North American Wildlife and Natural Resources Conference 52:298-309.

Johnson, D.H., J.D. Nichols, and M.D. Schwartz. 1992. Population dynamics of breeding waterfowl. Pages 446-485 *in* B.D.J. Batt, ed. The ecology and management of breeding waterfowl. University of Minnesota Press, Minneapolis.

Johnson, D.H., A.B. Sargeant, and R.J. Greenwood. 1989. Importance of individual species of predators on nesting success of ducks in the Canadian Prairie Pothole region. Canadian Journal of Zoology 67:291-297.

Klett, A.T., T.L. Shaffer, and D.H. Johnson. 1988. Duck nest success in the Prairie Pothole region. Journal of Wildlife Management 52(3):431-440.

Sargeant, A.B., S.H. Allen, and R.T. Eberhardt. 1984. Red fox predation on breeding ducks in midcontinent North America. Wildlife Monographs 89. 41 pp.

Sargeant, A.B., R.J. Greenwood, M.A. Sovada, and T.L. Shaffer. 1993. Distribution and abundance of predators that affect duck production—Prairie Pothole region. U.S. Fish and Wildlife Service Resour. Publ. 194. 96 pp.

Conservation Reserve Program and Migratory Birds in the Northern Great Plains

by
Douglas H. Johnson
Rolf R. Koford
National Biological Service

U.S. Department of Agriculture programs have mediated supply and demand of commodities and maintained the agricultural industry, but several programs have also offered various kinds of conservation benefits. The 1985 Food Security Act (Farm Bill) featured the Conservation Reserve Program (CRP), which paid farmers to plant perennial cover on highly erodible lands and to leave this land intact for a 10-year contract period. During that period we conducted two studies to determine the value of CRP fields to breeding birds in the northern Great Plains.

Methods

In one investigation, we censused breeding birds on about 400 fields in nine counties in eastern Montana, North Dakota, South Dakota, and western Minnesota (Johnson and Schwartz 1993). These four states have about 4 million ha (9.9 million acres) of CRP land, which is nearly 30% of all land included in the program. Most of these CRP fields were planted to mixtures of native and introduced grasses and legumes. We compared the average estimated density of breeding pairs in CRP fields in North Dakota with the density in croplands in a random sample of quarter-sections surveyed in the state (see Igl and Johnson, this section). We believe this is an appropriate comparison because nearly all CRP lands would have been in cropland without the program. In addition, North Dakota is the only state with comparable information about bird populations in cropland. Results are available for 1992 and 1993.

In a second investigation, we examined daily survival rates of nests (eggs and young), a key component of reproductive success, on 11 CRP fields in North Dakota and Minnesota in 1991-93. For comparison with CRP fields, we also studied an alternative habitat with a similar breeding-bird community. We studied 11 idle grassland fields on upland parts of federal Waterfowl Production Areas (WPAs); their vegetation typically is planted to mixtures of legumes and to grasses.

Bird Populations and Reproductive Success

Seventy-three different species were counted in the first study; most of these species were far more common in CRP fields than in cropland (Table 1). Differences were especially great for several grassland species that had declined markedly in the Breeding Bird Survey's Central Region of North America between 1966 (when the surveys began) and 1990. For example, lark buntings (*Calamospiza melanocorys*) and grasshopper sparrows (*Ammodramus savannarum*), whose numbers fell by about two-thirds during that period, were about 10 and 16 times more common in CRP habitat than in cropland.

The most recent Breeding Bird Surveys indicate that these grassland species, which had been declining for a long time, appear to be increasing (Reynolds et al. 1994).

Overall, daily survival rates of nests were similar in CRP fields and WPA fields (Table 2). In North Dakota there was some indication that nests of grasshopper sparrows and western meadowlarks (*Sturnella neglecta*) had higher daily survival rates in CRP fields than in WPA fields. Differences between states and among years, however, make generalizing difficult. Predation caused 80% of the nest failures.

Implications

These studies show that federal agricultural programs can have an enormous effect on wildlife resources over broad areas. In addition, with the restoration of suitable habitat, in this case mostly a mixture of introduced grasses and legumes rather than native prairie, populations of grassland birds can flourish. The similar daily survival rates of nests in CRP and WPA fields indicate that the habitat quality of CRP fields and WPA fields is roughly comparable.

More information is needed to provide a fuller picture of how the CRP is affecting trends in grassland birds. Information on temporal and spatial effects is especially useful. As CRP fields age, their attractiveness to certain species may change. Daily survival rates of nests also may change. Spatial effects are apparent in our censuses and undoubtedly exist on a wider scale. Finally, we need to integrate results from field studies with trend data from the Breeding Bird Survey.

Table 1. Densities (pairs per 100 ha) of most common birds in Conservation Reserve Program fields and in cropland fields in North Dakota, 1992-93, and trends from the Breeding Bird Survey for the Central Region of North America, 1966-90.

Species	CRP fields		Cropland		Trend
	1992	1993	1992	1993	
Lark bunting (*Calamospiza melanocorys*)	24.54	9.14	2.01	1.28	-64.5
Red-winged blackbird (*Agelaius phoeniceus*)	21.50	10.90	1.86	0.90	-1.0
Grasshopper sparrow (*Ammodramus savannarum*)	21.14	10.21	1.33	0.58	-67.5
Savannah sparrow (*Passerculus sandwichensis*)	7.28	8.60	0.12	1.48	16.3
Brown-headed cowbird (*Molothrus ater*)	7.11	5.03	2.61	2.51	-6.1
Bobolink (*Dolichonyx oryzivorus*)	7.74	3.53	2.73	2.11	-48.7
Western meadowlark (*Sturnella neglecta*)	5.43	5.26	1.28	1.16	-7.8
Clay-colored sparrow (*Spizella pallida*)	5.07	3.63	0.02	0.02	-39.6
Common yellowthroat (*Geothlypis trichas*)	2.49	1.87	0.02	0.02	-7.4
Horned lark (*Eremophila alpestris*)	2.38	0.53	19.96	29.18	-3.5
Sedge wren (*Cistothorus platensis*)	0.73	1.13	0	0	-15.5
Dickcissel (*Spiza americana*)	1.37	0.03	0.19	0	-29.4
Baird's sparrow (*Ammodramus bairdii*)	0.26	0.68	0.34	0.02	-46.6
Upland sandpiper (*Bartramia longicauda*)	0.40	0.13	0.92	0.65	145.7
Chestnut-collared longspur (*Calcarius ornatus*)	0.02	0.03	11.38	2.08	-10.7

Table 2. Daily survival rates of nests in Conservation Reserve Program fields and on Waterfowl Production Areas, North Dakota and Minnesota. Numbers of nests are in parentheses.

Species	North Dakota		Minnesota	
	CRP	WPA	CRP	WPA
American coot (*Fulica americana*)	-	-	0.92 (14)	-
Blue-winged teal (*Anas discors*)	-	-	0.94 (19)	0.95 (34)
Bobolink (*Dolichonyx oryzivorus*)	-	-	0.92 (16)	0.93 (32)
Clay-colored sparrow (*Spizella pallida*)	-	0.93 (49)	-	0.97 (24)
Grasshopper sparrow (*Ammodramus savannarum*)	0.95 (39)	0.91 (14)	0.92 (13)	-
Mallard (*Anas platyrhynchos*)	0.96 (20)	-	-	0.97 (10)
Red-winged blackbird (*Agelaius phoeniceus*)	0.92 (70)	-	0.86 (25)	-
Savannah sparrow (*Passerculus sandwichensis*)	-	-	0.86 (13)	0.95 (30)
Upland sandpiper (*Bartramia longicauda*)	-	0.99 (19)	-	-
Western meadowlark (*Sturnella neglecta*)	0.95 (20)	0.87 (14)	-	-

References

Johnson, D.H., and M.D. Schwartz. 1993. The Conservation Reserve Program and grassland birds. Conservation Biology 7:934-937.

Reynolds, R.E., T.L. Shaffer, J.R. Sauer, and B.G. Peterjohn. 1994. Conservation Reserve Program: benefit for grassland birds in the northern plains. Transactions of the North American Wildlife and Natural Resources Conference 59:328-336.

For further information:

Douglas H. Johnson
National Biological Service
Northern Prairie Science Center
8711 37th St. SE
Jamestown, ND 58401

Decline of Native Prairie Fishes

by

Anthony A. Echelle
Geffery R. Luttrell
Oklahoma State University
Robert D. Larson
New Mexico Department of Game and Fish
Alexander V. Zale
William L. Fisher
David M. Leslie, Jr.
National Biological Service

Several prairie fishes that were once widespread and abundant in riverine ecosystems of the south-central Great Plains have declined markedly in their distributions and abundances. Declines of such species likely reflect degradation of riverine ecosystems, particularly in the Arkansas River basin. At a 1994 interregional meeting, the U.S. Fish and Wildlife Service, representing various regions, considered eight riverine aquatic species in the Arkansas and Missouri river basins as Category 2 species (i.e., more data needed to determine appropriateness of listing as federally endangered or threatened species). Four of the eight species were small prairie fishes, including the Arkansas River shiner (*Notropis girardi*) and the Arkansas River speckled chub (*Macrhybopsis aestivalis tetranemus*).

We recently investigated distribution and reproductive status of the Arkansas River shiner and the Arkansas River speckled chub in relation to human alterations of river flows within the Arkansas River basin. Human impacts were identified that are detrimental to the long-term stability of native prairie fish assemblages.

Historical distributions of the Arkansas River shiner and the Arkansas River speckled chub were determined by reviewing collection records from appropriate museums. Current distributions of both species were assessed with intensive seine samples throughout historical ranges in Colorado, Kansas, New Mexico, Oklahoma, and Texas (153 collections at 116 localities for the shiner; 223 collections at 159 localities for the speckled chub). River discharges throughout the year were evaluated relative to the reproductive cycles of the fish.

Arkansas River Shiner

This shiner is endemic to the Arkansas River basin; it was widespread in the basin before

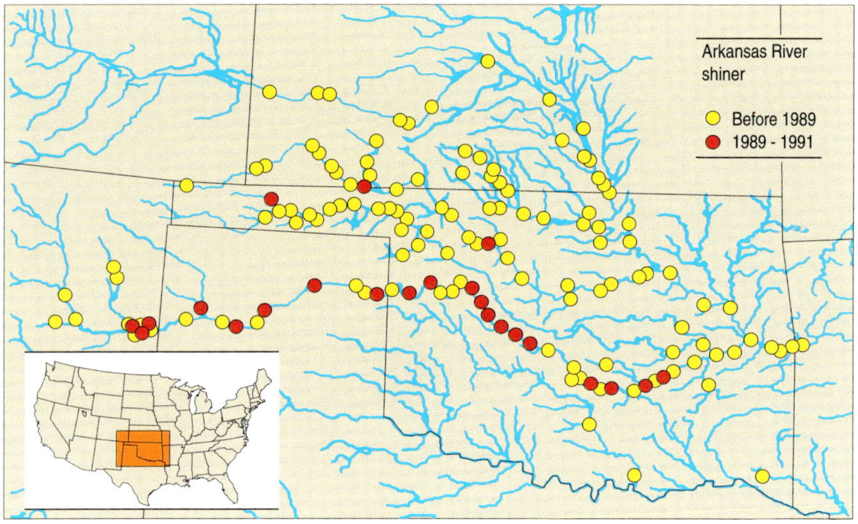

Fig. 1. Historical occurrences of the Arkansas River shiner from collections before 1989 and current occurrences from 1989 to 1991.

1985, but relative abundances varied widely. In three main tributaries of the Arkansas River (North Canadian River, Cimarron River, and Salt Fork of the Arkansas River), the shiner declined markedly between 1983 and 1985, and no specimens were collected after 1990. Our sampling between 1989 and 1991 indicated that native populations were common only in the South Canadian River in Oklahoma, Texas, and New Mexico. An introduced population (perhaps a result of bait transport) occurs in the Pecos River, New Mexico, southwest of the shiner's normal distribution (Bestgen et al. 1989). Overall, the shiner has been extirpated from about 75% of the river reaches in its historical range (Fig. 1). That, coupled with the speed with which populations became extinct in the mid-1980's, prompted action to list the shiner as threatened.

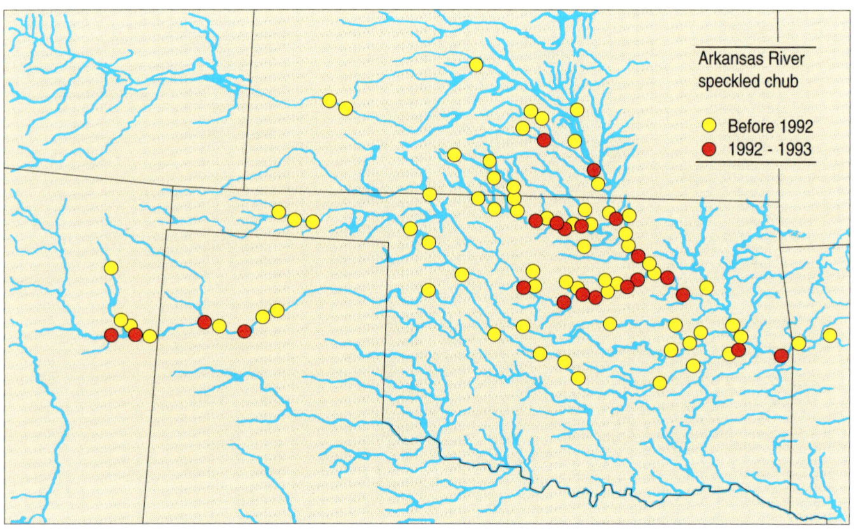

Fig. 2. Historical occurrences of the Arkansas River speckled chub from collections before 1992 and current occurrences from 1992 to 1993.

Arkansas River Speckled Chub

Historically, the speckled chub occurred throughout the Arkansas River, including the main tributaries in Arkansas, Colorado, Kansas, New Mexico, Oklahoma, and Texas. Our seining collections between 1991 and 1993, however, resulted in capture of speckled chubs at only 22 of the 159 sites sampled, indicating a marked reduction in distribution (Fig. 2). Only six stream reaches in Kansas, New Mexico, Oklahoma, and Texas support speckled chub. We believe that the species is extirpated from Arkansas and Colorado, the North Canadian and Deep Fork rivers in Oklahoma, the Salt Fork of the Arkansas River and Medicine Lodge River in Kansas, and parts of the South Canadian River. Its population in the Cimarron River in Oklahoma varied from very common in collections before 1950, absent from 1984 to 1991, and rare in 1992 and 1993.

River Flows and Reproduction

We examined duration curves of river flows from three time periods (before 1950, 1950-69, and 1970-88). Our analyses indicated that May-September river flows at most sampling sites were depressed from 1970 to 1988. Overall, 17 of 21 (81%) significant differences among river flows involved depressed flow levels from May to September.

Reproductive activity of the Arkansas River shiner extends from early May to August. The highest reproductive activity in shiners collected in 1989 occurred in June and was coincident with peak river flows. Reproductive activity in shiners in 1989 decreased as river flows declined throughout the summer. Although we do not have comparable reproductive data for the speckled chub, it is clear that it is as affected by river flows (Bottrell et al. 1964) as the shiner.

Both the shiner and the speckled chub have experienced sizeable losses (ca. 75%) in their historical distributions. Local abundances of the shiner have declined since at least the mid-1960's. The shiner and speckled chub now occur together only in the South Canadian River between two reservoirs in Texas and New Mexico and possibly in the Cimarron River in Oklahoma. Declines of these two species parallel similar declines in other native prairie fishes, such as the plains minnow (*Hybognathus placitus*; Cross and Moss 1987).

Reproduction in these two species appears dependent on periodic and intensive river flows during spring and summer when buoyant eggs are deposited directly into the current. Eggs drift in the current and hatch in 2-4 days (Moore

1944; Bottrell et al. 1964; Cross et al. 1985). In general, the south-central Great Plains is characterized by low but intense rainfall, high evaporation rates, and periodic drought (Zale et al. 1989). Such conditions likely cause great population changes year-to-year and may even cause local extinctions.

Extensive agricultural activities and resultant demands for irrigation water, coupled with the construction of numerous reservoirs in the Arkansas River basin, have degraded and restricted habitats of the shiner and speckled chub and likely other prairie fishes (Cross and Moss 1987). Successful reproduction or recruitment seems to have been impaired. Impoundments have fragmented once contiguous populations of the shiner and speckled chub to restricted river reaches with suitable habitat, effectively eliminating movements between populations and colonization of vacant habitat. Although altered flow regimes may be the ultimate explanation of the declines of these and other species, the actual pattern of decline differs between species. Overall, these declines indicate that human activities have degraded aquatic prairie ecosystems to the point of endangering parts of endemic fish assemblages.

References

Bestgen, K.R., S.P. Platania, J.E. Brooks, and D.L. Probst. 1989. Dispersal and life history traits of *Notropis girardi* (Cypriniformes: Cyprinidae) introduced into the Pecos River, New Mexico. American Midland Naturalist 122:228-235.

Bottrell, C.E., R.H. Ingersol, and R.W. Jones. 1964. Notes on the embryology, early development, and behavior of *Hybopsis aestivelis tetranemus* (Gilbert). Transactions of the American Microscopical Society 83:391-399.

Cross, F.B., and R.E. Moss. 1987. Historic changes in fish communities and aquatic habitats in plains streams of Kansas. Pages 155-165 *in* W.J. Matthews and D.C. Heins, eds. Community and evolutionary ecology of North American stream fishes. University of Oklahoma Press, Norman.

Cross, F.B., R.E. Moss, and J.T. Collins. 1985. Assessment of dewatering impacts on stream fisheries in the Arkansas and Cimarron rivers. Final Report, Kansas Fish and Game Commission, Nongame Wildlife Contract 46. 161 pp.

Moore, G.A. 1944. Notes on the early life history of *Notropis girardi*. Copeia 1944:209-214.

Zale, A.V., D.M. Leslie, Jr., W.L. Fisher, and S.G. Merrifield. 1989. The physicochemistry, flora, and fauna of intermittent prairie streams: a review of the literature. U.S. Fish and Wildlife Service Biological Report 89(5). 44 pp.

For further information:

Anthony A. Echelle
Oklahoma State University
Department of Zoology
Stillwater, OK 74078

The Coyote: An Indicator Species of Environmental Change on the Great Plains

by
Philip S. Gipson
Dennis E. Brillhart
National Biological Service

Increases or declines in wildlife populations are often the first noted indicators of widespread environmental change. Behavioral changes such as diet shifts or habitat-use also may provide sensitive indicators of environmental change. The coyote (*Canis latrans*) is an example of an opportunistic wild animal that may show both numerical and behavioral responses to environmental change.

Recent trends in populations and diets of coyotes and other canids (e.g., wolves, foxes, dogs) may reflect changes in land use, especially agricultural changes, and shifts in human populations. This article reviews both published accounts and original research to summarize how coyotes appear to have responded to changes in human populations and land use on the Great Plains.

Methods

Data presented in this paper were taken from many published sources (Sperry 1941; Young and Jackson 1951; Fichter et al. 1955; Gier 1968; Johnson and Sargeant 1977; Socolofsky and Self 1988) and from original research on coyote diets (Brillhart 1993). Although most of these studies were conducted on specific biological or social issues, we compare them to help understand human and wildlife population changes through time.

Human Population Changes

Two large-scale movements of people into the central Great Plains, from Nebraska south through Kansas and Oklahoma, occurred during the 1800's. The first large influx took place during the late 1820's, 30's, and 40's, as displaced Native American tribes were moved to the region.

Information about wildlife before 1850 is limited, but accounts suggest that bison (*Bison bison*), other big game, and wild canids were abundant when eastern Native American tribes moved into the region (Allen 1874; Cragin 1885; Mead 1899; Choate and Fleharty 1975; Bee et al. 1981). Native Americans on the Great Plains lived a subsistence lifestyle dependent upon these game animals, but even when relatively large numbers of Native Americans were moved to the region, they generally left the prairies and wildlife populations intact.

The second major influx of people occurred from 1860 through the 1880's when thousands of settlers from eastern states and Europe came to homestead or to buy land from the railroads. Settlers and market hunters killed tens of thousands of bison yearly; several million bison hides were shipped from Dodge City and other railroad communities (Socolofsky and Self 1988). Before the turn of the century, bison and elk (*Cervus elaphus*) were extirpated from the region. European settlers converted the prairies into farms, ranches, and towns. They also replaced big game with cattle, sheep, hogs, and poultry, and later waged poisoning campaigns against wolves (*Canis lupus*), coyotes, and other predators.

Since the late 1800's, a steady shift in human populations from farms to urban centers has occurred on the Great Plains. In some plains states, these changes have resulted in more people living in urban centers than in rural areas. For example, in 1880, about 90% of the Kansas population lived on farms, but by 1930, farm residents accounted for 60% of the total population (U.S. Department of Commerce 1993). Since that time, there have been further decreases in the proportion of the rural population; by 1990 about 30% lived on farms.

In addition, most farms have become larger and more highly mechanized than those 40-50 years ago. Changes also have occurred in production of domestic animals, with fewer farms today raising cattle, hogs, sheep, and chickens. Further, livestock and poultry are better cared for now and often are raised in confinement where they are unavailable to coyotes (Robel et al. 1981).

Canid Population Changes

Populations of wolves, coyotes, red foxes (*Vulpes vulpes*), swift foxes (*V. velox*), and dogs (*Canis familiaris*) on the Great Plains probably were relatively stable until settlers began arriving in the 1860's. Wolves dominated the canid social system except for the immediate area around villages, where village dogs probably dominated (Fig. 1). Because wolves are aggressive toward coyotes, coyote numbers probably were depressed (Young and Jackson 1951; Mech 1970). Mech (1994) and others have shown that the buffer zones that exist between adjacent wolf packs (about 6-7 km wide) provide refugia for deer and other animals. Coyotes may have occupied these buffer zones as well. Red and swift foxes were locally common during the 1800's, and there was probably little conflict between wolves and foxes. Because coyotes are aggressive toward foxes, fox numbers likely declined as coyote numbers increased (Johnson and Sargeant 1977).

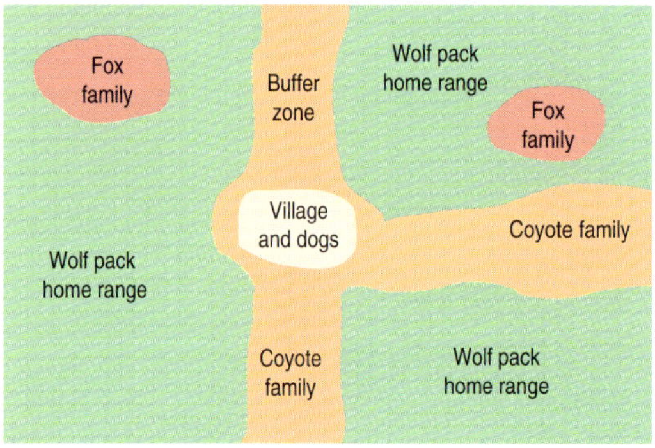

Fig. 1. Presettlement spatial relationships among home ranges of three packs of wolves, two families of coyotes, and two families of foxes near a Native American village with free-ranging dogs. Buffer zones and boundaries of wolf pack territories are dynamic, changing with availability of food and composition of wolf packs.

Coyotes increased during settlement and expanded their ranges as wolves were eliminated and bison were replaced with cattle and sheep. Coyotes may have reached their highest densities in North Dakota, and possibly other parts of the Great Plains, from about 1895 to 1915 (Johnson and Sargeant 1977).

Federal predator control started in 1915 when Congress appropriated $125,000 to organize and conduct control operations in partnership with states and local sponsors. The initial emphasis was on eliminating wolves from western and midwestern states. This wolf-control partnership was amazingly successful—almost all wolves were removed from western states by 1923 (Young and Goldman 1944). Coyotes generally increased in numbers as wolf populations declined.

Coyote populations fluctuated from 1915 to 1950, but bounty records suggest a general decline after 1915 (Gier 1968; Johnson and Sargeant 1977). In Kansas, low coyote populations were recorded from 1932 through 1940 (Cockrum 1952) and from 1954 through 1958 (Gier 1968). Compound 1080 (sodium fluoroacetate) was used to control coyotes in Kansas from 1950 through 1960; coyote numbers declined dramatically there.

Through the 1960's, coyote numbers continued to decline with increased use of Compound 1080 and other predator-control toxicants. Coyote numbers generally increased throughout the Great Plains after 1972 when the use of toxicants on federal lands was prohibited. Local fluctuations in coyote populations have occurred since 1970, largely in response to coyote fur prices and trapping and hunting.

Changing Coyote Diets

Coyote diets on the Great Plains today are markedly different than they were at the turn of the century, a likely reflection of changes in agricultural systems and human populations. Early in this century, most people on the Great Plains lived on mostly small farms and raised a variety of domestic animals. These farms usually were distributed fairly evenly across much of the region, making domestic animals widely available as prey for coyotes. Many farms suffered livestock and poultry losses from coyotes, which intensified predator-control efforts.

Studies of coyote diets on the Great Plains through the 1960's demonstrated that rabbits, rodents, and domestic animals were important food items (Sperry 1941; Fichter et al. 1955; Gier 1968). For example, in Kansas, almost 90% of the coyote diet was dominated by these three prey groups (Fig. 2), and more than half of all coyote stomachs sampled contained remains of either domestic livestock or poultry (Gier

1968). Similar patterns in consumption of rabbits, rodents, and domestic animals were evident in Nebraska, with livestock and poultry occurring in a third of all samples (Fichter et al. 1955).

Recent studies of coyote diets on the Great Plains also have shown the importance of rodents and rabbits as coyote prey (Brillhart 1993). In contrast to earlier studies, however, domestic livestock and chickens are eaten infrequently (Fig. 2); other common coyote foods today include certain insects, fruits, and wild birds.

Conclusions

Circumstantial evidence and prevailing professional opinion support our hypothesis that populations and diets of canids have changed in response to changing agricultural practices and shifts in human populations on the Great Plains. Because direct evidence is lacking to confirm these associations, research with specific testable hypotheses is needed.

Widespread changes in agricultural practices are inevitable and corresponding changes in wildlife populations should be expected. Recent changes in agricultural practices that are likely to result in changes in wildlife populations include a shift to dryland farming in formerly irrigated areas because of groundwater depletion, government regulations, and increasing energy prices. Agricultural set-aside programs authorized by the 1985 Food Security Act are positively influencing many wildlife populations, and future programs of a similar nature may benefit wildlife populations further.

References

Allen, J.A. 1874. Notes on the mammals of portions of Kansas, Colorado, Wyoming, and Utah. Part 1. On the mammals of middle and western Kansas. Bull. of the Essex Institute 6:43-66.

Bee, J.W., G.E. Glass, R.S. Hoffmann, and R.R. Patterson. 1981. Mammals in Kansas. University of Kansas, Publ. of the Museum of Natural History, Public Education Series 7:1-300.

Brillhart, D.E. 1993. Seasonal, temporal, and spatial variation in prey use by coyotes (*Canis latrans*). M.S. thesis, Kansas State University, Manhattan. 66 pp.

Cockrum, E.L. 1952. Mammals of Kansas. University of Kansas Publication of the Museum of Natural History 7:1-303.

Choate, J.R., and E.D. Fleharty. 1975. Synopsis of native, recent mammals of Ellis County, Kansas. Occasional Papers of the Museum of Texas Tech University 37:1-80.

Cragin, F.W. 1885. Notes on some mammals of Kansas, with a few additions to the list of species known to inhabit the state. Bull. of Washburn College, Laboratory of Natural History 1:42-47.

Fichter, E., G. Schildman, and J.H. Sather. 1955. Some feeding patterns of coyotes in Nebraska. Ecological Monographs 25:1-37.

Gier, H.T. 1968. Coyotes in Kansas. Kansas State College, Agricultural Experiment Station Bull. 393:1 118.

Johnson, D.H., and A.B. Sargeant. 1977. Impact of red fox predation on the sex ratio of prairie mallards. U.S. Fish and Wildlife Service Wildlife Res. Rep. 6. 56 pp.

Mead, J.R. 1899. Some natural-history notes of 1859. Transactions of the Kansas Academy of Science 16:280-281.

Mech, L.D. 1970. The wolf: the ecology and behavior of an endangered species. Natural History Press, Garden City, NY. 384 pp.

Mech, L.D. 1994. Buffer zones of territories of gray wolves as regions of intraspecific strife. Journal of Mammalogy 75:199-202.

Robel, R.J., A.D. Dayton, F.R. Henderson, R.L. Meduna, and C.W. Spaeth. 1981. Relationships between husbandry methods and sheep losses to canine predators. Journal of Wildlife Management 45:894-911.

Socolofsky, H.E., and H. Self. 1988. Historical atlas of Kansas. 2nd ed. University of Oklahoma Press, Norman, OK. 74 pp.

Sperry, C.C. 1941. Food habits of the coyote. U.S. Fish and Wildlife Service Res. Bull. 4. 70 pp.

U.S. Department of Commerce. 1993. 1990 census of population and housing. U.S. Government Printing Office, Washington, DC.

Young, S.P., and E.A. Goldman. 1944. The wolves of North America. Dover Publications, Inc., New York. 385 pp.

Young, S.P., and H.H.T. Jackson. 1951. The clever coyote. The Stackpole Company, Harrisburg, PA. 411 pp.

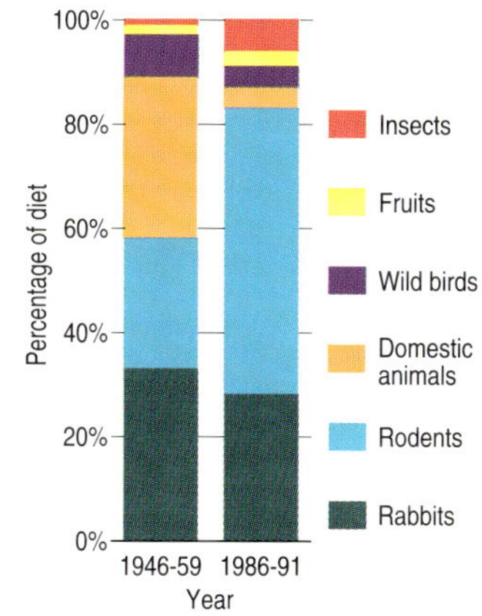

Fig. 2. Comparison of coyote diets in Kansas during the late 1940's and 1950's with diets from the late 1980's to 1991 (Gier 1968; Brillhart 1993).

For further information:

Philip S. Gipson
National Biological Service
Kansas Cooperative Fish and Wildlife Research Unit
205 Leasure Hall
Kansas State University
Manhattan, KS 66506-3501

Courtesy D. Stevens, NPS

Interior West

by
Science Editor
Thomas J. Stohlgren
National Biological Service
Natural Resource Laboratory
Colorado State University
Fort Collins, CO 80523

Overview

The articles in this section reveal the critical need for ecosystem science to direct ecosystem management in areas ranging from the Colorado Rockies, south to the Colorado Plateau, west to the Great Basin and the Pacific Northwest, and north to the Greater Yellowstone Ecosystem. Ecosystems in the Interior West are challenged by severe climatic fluctuations superimposed on rapidly changing land-use patterns and anthropogenic (human-caused) threats. Because scientists and resource managers now recognize the prohibitive cost and difficulty of a species-by-species approach to biological conservation and wise stewardship, their efforts are moving increasingly toward an ecosystem and landscape approach to conservation.

My colleagues and I begin this section by identifying and quantifying anthropogenic threats to ecosystem integrity in Rocky Mountain National Park and the Colorado Rockies (Stohlgren et al.). The article by Schullery continues this common theme by describing alarming trends in plant and animal populations in the Greater Yellowstone Ecosystem. By taking a broad view of subalpine forest dynamics in the Pacific Northwest, Peterson shows that treeline communities may be adversely influenced by rapid environmental change. Warshall examines the southwestern sky island ecosystems (the mountaintops of the Great Basin) with respect to threats from nonindigenous species, recreation and military practices, and fire-management activities.

The status and trends of many plant and animal populations are uncertain in the Interior West. Scoppettone and Rissler, however, report successful population increases of the endangered cui-ui fish (*Chasmistes cujus*) in Pyramid Lake, Nevada: the population has doubled between 1990 and 1993. Mueller and Marsh focus on how loss of critical riparian habitat through water development, pollution, and the introduction of nonindigenous species have caused population declines of the threatened and endangered razorback sucker (*Xyrauchen texanus*) and bonytail (*Gila elegans*) in the Colorado River Basin. The article by Drost and Deshler on the diversity of reptiles and amphibians on the Colorado Plateau reminds us that much inventory and monitoring work lies ahead. Van Riper III et al. also remind us that human activities in the past (e.g., pesticide use, water diversion, and the introduction of nonindigenous trout) continue to affect the status and trends of bald eagle (*Haliaeetus leucocephalus*) populations on the southern Colorado Plateau. And, Willey demonstrates that 90% of the threatened Mexican spotted owl (*Strix occidentalis lucida*) habitat on the Colorado Plateau is on timber-management sites.

The last two articles focus on restoring ecosystem integrity by reintroducing extirpated species. Singer reports that the success of restoration efforts of bighorn sheep (*Ovis canadensis*) in the Rocky Mountains is influenced negatively by their proximity to domestic sheep and by small, translocated groups of bighorn sheep that are too genetically similar. McCutchen discusses the history and status of desert bighorn (*O.c. nelsoni*) and shows that sheep translocations have been fairly successful, except in New Mexico and southern California.

It is important to note the overriding theme in this section: modern humans continue to alter ecosystem components and processes. To manage natural resources in a sustainable way to meet the needs of the American people, we must first understand the inseparable link of human and resource ecology. The perpetuation of biological diversity in the Interior West depends largely on coordinated, multiscale ecosystem science, and resource inventory and monitoring efforts.

Ecosystem Trends in the Colorado Rockies

by
Thomas J. Stohlgren
Jill Baron
National Biological Service
Timothy G.F. Kittel
University Corporation for Atmospheric Research
Dan Brinkley
Colorado State University

Fig. 1. Drastically increased urbanization in Estes Park/Rocky Mountain National Park, 1921 (above) to 1986 (below). The photographs also show, however, forest recovery from turn-of-the-century logging and human-caused fires (Veblen and Lorenz 1991).

Biological conservation is increasingly moving toward an ecosystem and landscape approach, recognizing the prohibitive cost and difficulty of a species-by-species approach (LaRoe 1993). Also, statewide (e.g., Gap Analysis Program) and national surveys (e.g., Environmental Monitoring and Assessment Program or EMAP) are conducted at a scale and level of resolution that do not meet the needs of most small land-management units that require detailed information at the ecosystem and landscape scale (Stohlgren 1994). The Colorado Rockies are an ideal outdoor laboratory for ecosystem science and management. The escalating environmental threats described in this article compelled us to design a landscape-scale assessment of the status and trends of biotic resources.

Our guiding principle is that a strong ecosystem science program provides crucial information for ecosystem management and wise stewardship. We define ecosystem science as the long-term, interdisciplinary study of ecosystem components and processes and their interactions at multiple spatial, temporal, and organizational scales, to meet management needs.

About 76% of the land adjacent to Rocky Mountain National Park is federal land. While the area has not received as much attention as the Greater Yellowstone Ecosystem, there may be as many internal and external threats to the natural resources in the area. The Colorado Rockies are an archetypal ecosystem under siege. Like many national parks, wilderness areas, wildlife refuges, and other natural areas, common threats include encroachment from urbanization and development, habitat fragmentation, fire suppression, nonindigenous species' invasion, and global change (e.g., climate change, bordering land-use changes, and air and water pollution). Since all these threats transcend ownership or stewardship borders, so have interagency concerns for conservation, inventory and monitoring, and research.

Here we identify and quantify trends that threaten ecosystem integrity in Rocky Mountain National Park and the Colorado Rockies. Our specific objectives include presenting qualitative information on vegetation change over the past 65 years, documenting quantitative trends of an ecosystem under siege, showing preliminary results of a long-term global change research program, and discussing the role of ecosystem science in assessing long-term trends in ecosystem condition.

Status and Trends

There is little doubt that the ecosystems of the Colorado Rockies have been altered significantly by humans. The density of ponderosa pine woodlands has increased (Fig. 1) as has suburban development (Veblen and Lorenz 1991). These qualitative changes are supported by qualitative measures (Fig. 2). The response of the forest from turn-of-the-century logging and fires showed a 5-fold increase in ponderosa pine bole (*see* glossary) biomass. In addition, the human population in Estes Park and the number of visitors in Rocky Mountain National Park have almost doubled since 1960. Urban development throughout the Front Range of Colorado has resulted in increased air pollution. Annual wet deposition values for nitrate, ammonium, and sulfate in the Loch Vale watershed of Rocky Mountain National Park are significantly greater than the average values of 2-4 kg/ha (about 2-4 lb/acre) in remote areas of the world (Fig. 2).

Elk and moose populations continue to increase in the park (Fig. 2) for many reasons including reduced predation (wolves have been extirpated) and hunting as well as diminished habitat and migratory corridors outside the park. Researchers are now quantifying ungulate (hooved herbivores) habitat relationships and aspen-willow community conservation. Although agricultural land use in Larimer County has declined slightly in recent years (Fig. 2), landscape and ecosystem integrity is

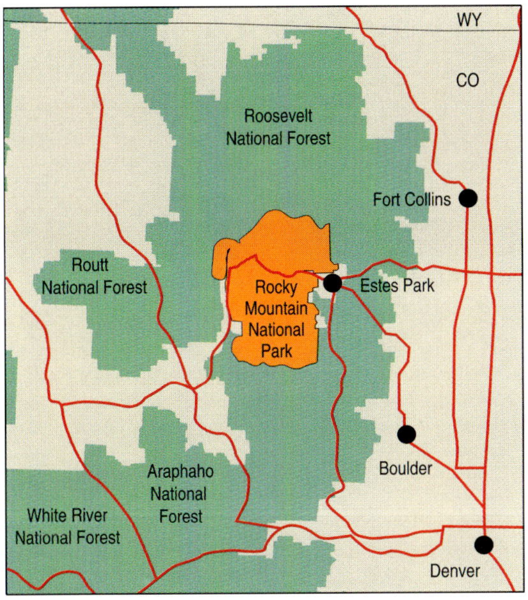

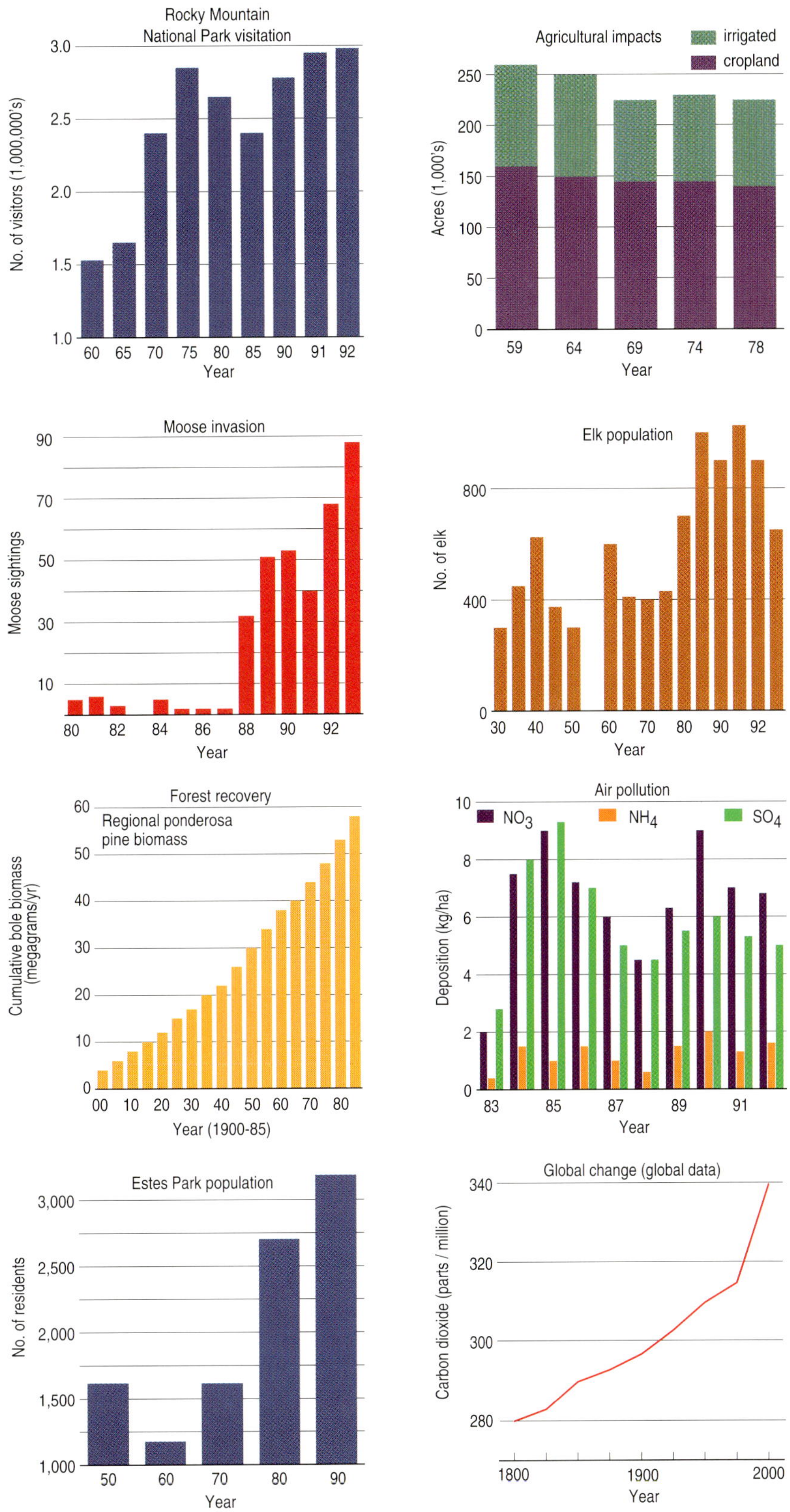

Fig. 2. Trends in Rocky Mountain National Park visitors, agricultural impacts, moose invasion, elk population, forest recovery, air pollution, Estes Park population, and global change in carbon dioxide.

challenged by fire suppression, nonindigenous species' invasions, weather modification (i.e., cloud seeding), and global climate change (Stohlgren et al. 1993).

Just as a species-by-species approach to conservation biology is prohibitively expensive, a complex of ecosystem threats cannot be addressed one by one. Our interdisciplinary approach in the Colorado Rockies is based on developing partnerships, consolidating and evaluating the status and trends in existing data, and developing a biogeographical, long-term, multiple spatial-scale monitoring program that fills information gaps and provides a scientific basis for sound ecosystem management. Preliminary results from the National Biological Service global climate change research program show significant interactions of climate, hydrological, and vegetation systems.

Mesoscale (1- to 100-km grids) climate modeling in the Front Range of the Colorado Rockies demonstrated that changes in land cover (e.g,. wild prairie to irrigated agricultural land) can lead to significant and perhaps unexpected changes in mesoscale climate. Computer modeling results indicate that the severity of summer thunderstorms in Rocky Mountain National Park is influenced by spatial patterns in albedo (*see* glossary) and surface roughness of farmlands several kilometers away (Pielke et al. 1993).

Quantifying trends in mountain hydrology and vegetation change caused by global climate change and assessing the effects of nearby cloud seeding require the development of new predictive models (Baron et al. 1994). Hydrological models are proving effective at estimating stream discharge and regional water supply.

In our long-term forest plots, we found the old-growth spruce and fir forests of the central Rocky Mountains range in biomass from

150,000 to more than 320,000 kg/ha (133,828-285,500 lb/acre) in standing biomass, and annual tree growth remains relatively high in these ancient forests. We are finding that ecotones are sensitive indicators of forest change; the forest-tundra ecotone (transitional area between distinct habitats or ecosystems) in Rocky Mountain National Park has been undergoing substantial directional change for some time (Baker et al. 1994). There is substantial evidence of seedling and sapling invasion within some previously unforested areas within the ecotone, particularly in wet areas in the patch forest zone. This filling in of the ecotone could substantially alter the ecotone environment (Baker et al. 1994). There is little evidence, however, of upward establishment of trees into tundra. To synthesize the vegetation change data, we are developing predictive vegetation change models by using geographic information systems. Our long-term study plots and transects will validate future models.

Implications

This interdisciplinary approach can be widely applied to most U.S. Department of the Interior land units and most ecosystems and will be an essential link to large-scale inventory and monitoring programs (e.g., Gap Analysis Program and EMAP). Ecosystem science is the most logical approach to determine the status and long-term trends of selected resources, populations, and ecosystems. This approach fosters discovery, standardization, linkages, and partnerships as well as coordinated inventory, monitoring, and research. New, standardized sampling protocols are being developed to accurately assess vascular plant species richness, an index of biodiversity (Stohlgren 1994).

For further information:

Thomas J. Stohlgren
National Biological Service
Rocky Mountain Field Station
Natural Resource Ecology Laboratory
Colorado State University
Fort Collins, CO 80523

References

Baker, W.L., J.J. Honaker, and P.J. Weisberg. 1994. Using aerial photography and GIS to map the forest-tundra ecotone in Rocky Mountain National Park, Colorado, for global change research. Photogrammetric Engineering and Remote Sensing. In press.

Baron, J., L. Band, S.W. Running, R.A. Pielke, and T.G.F. Kittel. 1994. Dynamic land surface/atmospheric parameterization at different spatial scales in the Colorado Rocky Mountains. *In* Proceedings of the Use of Hydrologic Models for Evaluating Climate Change Effects in Snowmelt Water Supply Basins. Agricultural Research Service. In press.

LaRoe, E.T. 1993. Implementation of an ecosystem approach to endangered species conservation. Endangered Species Update 10:3-6.

Pielke, R.A., J. Baron, T.Chase, J. Copeland, T.G.F. Kittel, T.J. Lee, R. Walko, X. Zeng. 1993. Use of mesoscale models for simulation of seasonal weather and climate change for the Rocky Mountain states. Proceedings, Second International Conference/Workshop on Integrating GIS and Environmental Modeling. Breckenridge, CO. In press.

Stohlgren, T.J. 1994. Planning long-term vegetation studies at landscape scales. *In* T. M. Powell and J. H. Steele, eds. Ecological time series. Chapman and Hall, New York. In press.

Stohlgren, T.J., J. Baron, and T. Kittel. 1993. Understanding coupled climatic, hydrological, and ecosystem responses to global climate change in the Colorado Rockies biogeographical area. Pages 184-200 *in* W.E. Brown and S.D. Veirs, Jr., eds. Partners in stewardship: Proceedings of the 7th Conference on Research and Resource Management in Parks and on Public Lands. George Wright Society, Hancock, MI.

Veblen, T.T., and D.C. Lorenz. 1991. The Colorado Front Range: a century of ecological change. University of Utah Press, Salt Lake City. 186 pp.

The Greater Yellowstone Ecosystem

by
Paul Schullery
National Park Service

Greater Yellowstone is described as the last large, nearly intact ecosystem in the northern temperate zone of the earth (Reese 1984; Keiter and Boyce 1991). Conflict over management has been controversial, and the area is a flagship site among conservation groups that aggressively promote ecosystem management (Greater Yellowstone Coalition 1992). The Greater Yellow Ecosystem (GYE) is one of the world's foremost natural laboratories in landscape ecology and geology and is a world-renowned recreational site (Knight 1994).

History

Yellowstone National Park (YNP) boundaries were arbitrarily drawn in 1872 in hopes of including all regional geothermal basins. No other landscape considerations were incorporated. By the 1970's, however, the grizzly bear's (*Ursus arctos*) range in and near YNP became the first informal minimum boundary of a theoretical Greater Yellowstone Ecosystem that included at least 1,600,000 ha (4,000,000 acres; Schullery 1992). Since then, definitions of the GYE have steadily grown larger (Fig. 1). A 1994 study listed the GYE size as 7,689,000 ha (19,000,000 acres; Clark and Minta 1994), while a 1994 speech by a Greater Yellowstone Coalition leader enlarged that to 8,000,000 ha (20,000,000 acres; Wilcox 1994).

In 1985 the House Subcommittees on Public Lands and National Parks and Recreation held a joint subcommittee hearing on Greater Yellowstone, resulting in a report by the Congressional Research Service (1986) outlining shortcomings in interagency coordination and concluding that the area's essential values were at risk.

Ecosystem Management by Species

The GYE concept has been most often advanced through concerns over individual species rather than over broader ecological principles. GYE managers must keep at least two types of "long-term" status in mind. One is the known, or at least probable, trend of a species based on historical and prehistorical information. The second type is that which has existed since the beginning of formal scientific study. Though 20 or 30 or even 50 years of information on a population may be considered long-term by some, one of the important lessons of GYE management is that even half a century is not long enough to give us a full idea of how a species may vary in its occupation of a wild ecosystem.

For example, anecdotal information on grizzly bear abundance dates to the mid-1800's (Schullery and Whittlesey 1992), and administrators have made informal population estimates for more than 70 years (Schullery 1992). From these sources, we know the species was common in the GYE when Europeans arrived, and we know that the population was not isolated before the 1930's, but is now. We do not know if bears were more or less common than now.

A 1959-70 bear study suggested a grizzly bear population size of about 175, later revised to about 229 (Craighead et al. 1974). Later estimates have ranged as low as 136 and as high as 540 (Schullery 1992); the most recent is a minimum estimate of 236 (Servheen 1993). Although the GYE population is relatively close to recovery goals, the plan's definition of recovery is controversial (Mattson and Reid 1991; Schullery 1992). Thus, even though the population may be stable or possibly increasing in the short term, in the longer term, continued habitat loss and increasing human activities may well reverse the trend.

Yellowstone cutthroat trout (*Oncorhynchus clarki bouvieri*) have suffered considerable declines since European settlement, but recently began flourishing (Varley and Schullery 1983) in some areas. Especially in Yellowstone Lake itself, long-term records indicate an almost remarkable restoration of robust populations from only three decades ago when the numbers of this fish were depleted because of excessive harvest (Gresswell and Varley 1988). Its current recovery, though a significant management achievement, does not begin to restore the species' historical abundance.

Early accounts of pronghorn (*Antilocapra americana*) in the GYE described herds of hundreds seen ranging through most major river valleys (Schullery and Whittlesey 1992). These populations were decimated by 1900, and declines continued among remaining herds. On the park's northern range, pronghorn declined from 500-700 in the 1930's to about 122 in 1968 (Houston 1982). By 1992 the herd had increased to 536 (J. Mack, National Park Service, personal communication).

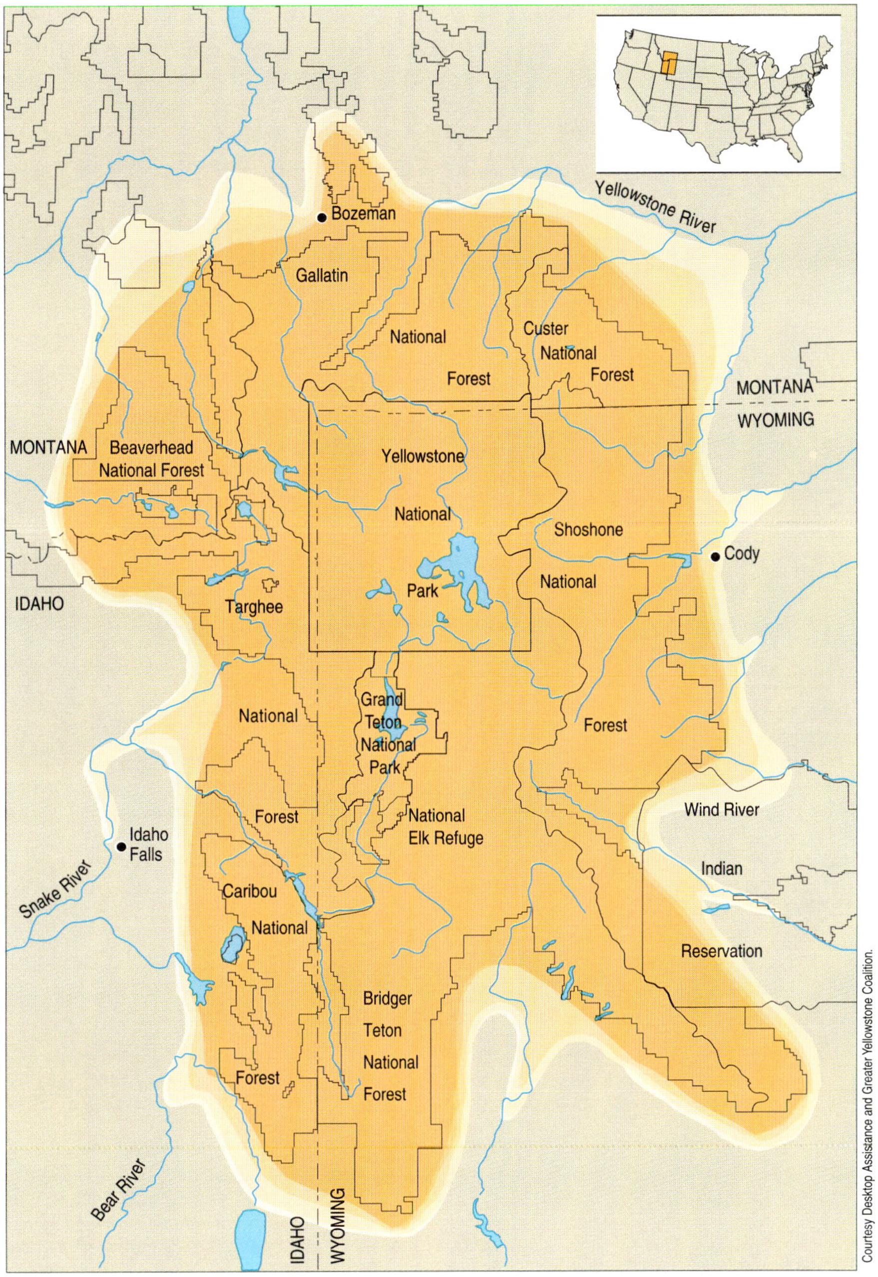

Fig. 1. Progressively lighter shading is used around the edges of a recent map of the Greater Yellowstone Ecosystem to illustrate the uncertainty that still plagues definitions of the ecosystem.

Among plants, whitebark pine (*Pinus albicaulis*) is a species of special interest, in large part because of its seasonal importance to grizzly bears, but also because its distribution could be dramatically reduced by relatively minor global warming (Blanchard and Knight 1991; Romme and Turner 1991; Fig. 2). In this case, we do not have a good long-term data set on the species, but we understand its ecology well enough to project declining future status.

Estimates of the decline of quaking aspen (*Populus tremuloides*) on YNP's northern range since 1872 range from 50% to 95% (Houston 1982; Kay 1993), and perhaps no controversy underway in the GYE more clearly reveals the need for comprehensive interdisciplinary research. Several factors are suspected in the

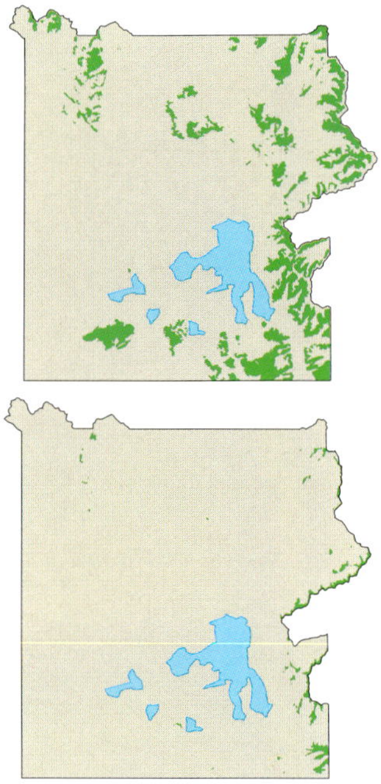

Fig. 2. Top: Current distribution of whitebark pine portrayed by a computerized geographic information system (GIS). Bottom: Distribution of whitebark pine projected by GIS analysis under a modest increase in warmth and dryness, showing a decrease of approximately 90%. (Derived from Romme and Turner [1991] by the Yellowstone GIS Laboratory, Yellowstone National Park.)

aspen's changing status, including Native American influences on numerous mammal species and on fire-return intervals before the creation of the park in 1872; European influences on fire frequency since 1886; regional climate warming; human harvests of beaver and ungulates in the first 15 years of the park's history and of wolves and other predators before 1930; human settlement of traditional ungulate migration routes north of the park since 1872; ungulate (especially elk) effects on all other parts of the ecosystem since 1900; and human influences on elk distribution in the park (Houston 1982; Schullery and Whittlesey 1992; Kay 1993).

Conclusions

Research is but one component of land-management decisions (Varley 1993). While in some respects the GYE has fulfilled the promise of early scientists who described it as one of the foremost natural laboratories on earth, both managers and researchers need more information to deal with the increasing demands on the region's resources, either in terms of raw information or in terms of an ecosystem-level understanding. In YNP, a landscape model is being developed based on a computerized geographic information system that will integrate, analyze, and display information from many disciplines (Shovic et al. 1993). Through this level of synthesis we may be able to better understand trends in the GYE.

For further information:

Paul Schullery
National Park Service
Yellowstone Center for Resources
PO Box 168
Yellowstone Park, WY 82190

References

Blanchard, B., and R. Knight. 1991. Movements of Yellowstone grizzly bears. Biological Conservation 58:41-67.

Clark, T.W., and S.C. Minta. 1994. Greater Yellowstone's future. Homestead Publishing, Moose, WY. 160 pp.

Congressional Research Service. 1986. Greater Yellowstone Ecosystem, an analysis of data submitted by federal and state agencies. U.S. Government Printing Office, Washington, DC. 210 pp.

Craighead, J., J. Varney, and F. Craighead. 1974. A population analysis of the Yellowstone grizzly bears. Montana Forest and Conservation Experiment Station Bull. 40. University of Montana, Missoula. 20 pp.

Greater Yellowstone Coalition. 1992. Inside greater Yellowstone. Greater Yellowstone Coalition, Bozeman, MT. 16 pp.

Gresswell, R., and J. Varley. 1988. Effects of a century of human influence on the cutthroat trout of Yellowstone Lake. American Fisheries Society Symposium 4:45-52.

Houston, D. 1982. The northern Yellowstone elk, ecology and management. Macmillan, New York. 474 pp.

Kay, C. 1993. Aspen seedlings in recently burned areas of Grand Teton and Yellowstone National parks. Northwest Science 67(2):94-103.

Keiter, R.B., and M.S. Boyce. 1991. The Greater Yellowstone Ecosystem, redefining America's wilderness heritage. Yale University Press, New Haven, CT. 428 pp.

Knight, D. 1994. Mountains and plains, the ecology of Wyoming landscapes. Yale University Press, New Haven, CT. 338 pp.

Mattson, D., and M. Reid. 1991. Conservation of the Yellowstone grizzly bear. Conservation Biology 5(3):364-372.

Reese, R. 1984. Greater Yellowstone, the national park and adjacent wildlands. Montana Geographic, Helena. 104 pp.

Romme, W., and M. Turner. 1991. Implications of global climate change for biogeographic patterns in the Greater Yellowstone Ecosystem. Conservation Biology 5(3):373-386.

Schullery, P. 1992. The bears of Yellowstone. High Plains Publishing, Worland, WY. 318 pp.

Schullery, P., and L. Whittlesey. 1992. The documentary record of wolves and related wildlife species in the Yellowstone National Park area prior to 1882. Pages 1-4 to 1-174 *in* J.D. Varley and W.G. Brewster, eds. Wolves for Yellowstone? Vol. 4. A report to the U.S. Congress. National Park Service, Yellowstone National Park, WY.

Servheen, C. 1993. Grizzly bear recovery plan. U.S. Fish and Wildlife Service, Missoula, MT.

Shovic, H., M. Johnson, and H. Porter. 1993. A new view of an old land. Yellowstone Science 1(2):2-6.

Varley, J. 1993. Research in Yellowstone. Bioscience 43(3):3-4.

Varley, J., and P. Schullery. 1983. Freshwater wilderness, Yellowstone fishes and their world. Yellowstone Library and Museum Association, Yellowstone National Park, WY. 132 pp.

Wilcox, L. 1994. Sustaining wilderness. Panelist presentation, Cinnabar Symposium, Museum of the Rockies, Bozeman, MT. March 25. Unpublished.

Subalpine Forests of Western North America

by
David L. Peterson
National Biological Service

Subalpine forest and meadow ecosystems are important, climatically sensitive components of mountainous regions of western North America (Peterson 1991). Changes in temperature, precipitation, snowpack, storm frequency, and fire all could affect the growth and productivity of these systems, resulting in substantial shifts in the location of ecotones (*see* glossary) between subalpine and alpine zones and montane and subalpine zones (Canaday and Fonda 1974).

Subalpine forests of western North America provide an excellent opportunity to examine response to past climate variation. Trees in the subalpine zone are frequently more than 500 years old and respond to climatic variations over annual to centuries-long time scales. The magnitude of climatic variation these forests have experienced may be compared with projections of future climate resulting from increased concentration of greenhouse gases. The population dynamics of subalpine tree species can be used to interpret climatic conditions under which trees have regenerated and can indicate how subalpine forest and meadow ecotones changed in the past. Preserved pollen and plant fossils can be used to examine subalpine vegetation distribution during different climatic periods of the Holocene (since the last ice age).

Recent literature on the potential effects of

climate change has focused on changes in the growth and distribution of low-elevation forests (e.g., Woodman 1987; Davis 1989). In western North America, most low-elevation forests are sensitive to soil moisture deficits during relatively dry summers (Peterson et al. 1991; Graybill et al. 1992). Although subalpine forests have been the subject of considerably less study, it appears that snowpack is an important limiting factor to growth, with respect to length of growing season (Graumlich 1991; Peterson 1993). Duration of snowpack also limits seedling establishment in subalpine meadows (Fonda 1976) and after disturbance by fire (Little et al. 1994). Summer temperature also positively affects the growth of mature subalpine conifers (Graumlich 1991; Peterson 1993) and negatively affects the seedlings' survival (Little et al. 1994).

Several reports document recent increases in the growth of subalpine conifer species in western North America (Innes 1991) as well as recent increases in the abundance of subalpine conifer populations at several locations. This article reviews recent reports of changes in the growth and distribution of subalpine conifers in western North America and discusses some possible causes.

Tree Growth

The first prominent report of a recent increase in growth of subalpine coniferous species was published by LaMarche et al. (1984), who reported dramatic increases in the growth rate of bristlecone pine (*Pinus longaeva, P. aristata*) and limber pine (*P. flexilis*) in California and Nevada. The extreme age of these trees, combined with the fact that radial growth has increased since 1850, makes this a particularly interesting result. The authors suggested that elevated levels of carbon dioxide associated with fossil fuel combustion may enhance the growth and productivity of these trees, perhaps through increased water-use efficiency. A more recent examination of these data corroborates the growth increase and restates that carbon dioxide fertilization is the hypothesized cause of the increase (Graybill and Idso 1993). Some disagreement exists about the factors causing the growth increase and whether the increases in growth found in these studies (which included sampling of strip-bark trees) are representative of the populations as a whole (Cooper and Gale 1986).

A subsequent study of basal area growth trends of lodgepole pine (*P. contorta*) and whitebark pine (*P. albicaulis*) at sites above 3,000-m elevation in the east-central Sierra Nevada of California also revealed that a high proportion of trees has had recent growth increases (Peterson et al. 1990), with the onset of the increase normally between 1850 and 1900, as found by LaMarche et al. (1984). Growth was particularly rapid during the past 30 years or so.

Courtesy D.L. Peterson, NBS

Many subalpine forests in western North America, such as this site in the Olympic Mountains, are currently protected in national parks and wilderness areas. Some of these areas have been experiencing increased tree growth and rapid establishment of young trees during the past century.

There are other reports of recent growth increases in subalpine conifers of western North America (Innes 1991). Jacoby (1986) found radial growth increases in lodgepole pine in the San Jacinto Mountains of southern California, but did not identify a strong causal factor despite detailed climatic analysis. Graumlich et al. (1989) found increases in the growth and productivity of Pacific silver fir (*Abies amabilis*) and mountain hemlock (*Tsuga mertensiana*) in the Cascade Mountains of Washington State, and suggested that these trends were related to increased temperature.

Recent growth increases have also been reported in European conifers (Innes 1991), such as Norway spruce (*Picea abies*; Kienast and Luxmoore 1988; Briffa 1992) and silver fir (*Abies alba*; Becker 1989), although these species are generally found below the subalpine zone. Both increased carbon dioxide (Kienast and Luxmoore 1988) and temperature (Becker 1989; Briffa 1992) have been suggested as potential causes for increased growth.

Not all studies of subalpine conifers have found recent increased growth, however. Graumlich (1991), for example, did not find increased radial growth in foxtail pine (*Pinus balfouriana*), limber pine, and western juniper (*Juniperus occidentalis*) in the Sierra Nevada. It is difficult to compare the various studies of tree growth discussed here because the studies employed a diversity of sampling and analytical techniques to evaluate growth patterns.

As noted previously, there are several potential explanations for recent increased growth in subalpine conifers. The possibility of carbon dioxide fertilization has been supported by experimental studies (Graybill and Idso 1993), but is extremely difficult to demonstrate for mature trees in the field. Increased temperature is another potential cause, but its relationship with growth is correlative and also difficult to demonstrate for mature trees. Changes in snowpack duration, which affects length of growing season, are a more likely cause of growth increases. Unfortunately, the long-term relationship of snowpack to tree growth has not been adequately investigated because snowpack data are often difficult to obtain.

Fertilization through nitrogen deposition could be another cause of growth increases. Although nitrogen deposition is relatively low in western North America, it is probably somewhat higher now than in the past because of the combustion of fossil fuels. Many subalpine forests are in sites with shallow soils and relatively low fertility, so even a small increase in nitrogen input could have some effect over several decades. Finally, the growth increases may simply be the result of normal forest stand dynamics because relatively little is known about the growth and ecological characteristics of subalpine forest ecosystems. Although the observed increases appear abnormal compared to lower elevation species, they may in fact be a normal phenomenon that reflects the natural range of variation in growth of subalpine species. Growth response to climate or other factors likely varies considerably by region (e.g., the Rocky Mountains have a continental climate, the Sierra Nevadas a Mediterranean climate) and by microsite (north aspect versus south aspect).

Patterns of Establishment

Recent increases in tree establishment in subalpine meadows have been documented in mountainous regions throughout western North America (Rochefort et al. 1994). Most locations show an expansion of the forest margin after 1890, with establishment peaks between 1920 and 1950. Additional establishment peaks have been identified on a local basis. Most investigators have concluded that increases in tree establishment are the result of a warmer climate following the Little Ice Age (Franklin et al. 1971; Kearney 1982; Heikkinen 1984; Butler 1986). It is unclear if establishment patterns signify a long-term directional change or short-term variation in relatively stable ecotones, regardless of the potential causes.

Most studies on subalpine tree establishment have been conducted in the Pacific Northwest in British Columbia in Canada and Washington and Oregon (Woodward et al. 1991; Rochefort et al. 1994) where tree invasion in subalpine meadows is widespread. Trees in this area are rapidly becoming established (Rochefort and Peterson 1991; Woodward et al. 1991), especially in meadows dominated by ericaceous species (species in the heath family such as heather and huckleberries). Much of this establishment is occurring in concavities and other places where snow would normally accumulate and inhibit germination and survival (personal observation). As trees become established, tree clumps act as black bodies to increase the absorption of radiation, snowmelt occurs progressively earlier, tree canopies intercept (and allow sublimation of) snow, and tree survival adjacent to the tree clump is further enhanced. This progression of events is termed "contagious dispersion" (Payette and Filion 1985).

Eight separate studies in the Pacific Northwest have documented large increases in populations of subalpine fir (*Abies lasiocarpa*), Pacific silver fir, mountain hemlock, subalpine larch (*Larix lyallii*), and Alaska yellow-cedar (*Chamaecyparis nootkatensis*). All these species experienced increases in establishment between 1920 and 1950. This was generally a period of lower snowpacks, which probably allowed seedlings to become established during a longer growing season. Winter precipitation limits subalpine tree growth and establishment in the Pacific Northwest, which has a maritime climate with wet winters and dry summers; high summer temperature can also limit tree establishment because shallow-rooted seedlings are subject to soil moisture stress (Little et al. 1994).

Increases in establishment of three species have been documented in the Sierra Nevada and White Mountains of California: foxtail pine, lodgepole pine, and bristlecone pine. Soil moisture stress is clearly a limiting factor in this area, which is dominated by a Mediterranean climate with very dry summers. Temporal patterns of establishment are inconsistent among the different locations in this region, and there has been little documented establishment during the past 20 years.

Studies conducted in the Rocky Mountains have documented increases in subalpine tree establishment for subalpine fir, lodgepole pine, and Engelmann spruce (*Picea engelmannii*). This region is dominated by a continental climate, with low precipitation and cold winters. Temporal patterns of establishment were more consistent in the Rocky Mountains, especially during 1940-50, a period with a warmer, wetter climate.

It is unclear whether observations of subalpine tree invasions are isolated events or part of a broad pattern in western North America. There are insufficient data from locations other than the Pacific Northwest to speculate about the geographic extent of this phenomenon.

Future Changes

Data on subalpine tree growth for western North America are too sparse to infer that growth increases are a broad regional phenomenon. Additional data from other sites are needed to quantify growth trends in subalpine species. Furthermore, consistent sampling and analytical methods should be applied so that different data sets can be compared.

Sufficient information exists, however, about long-term growth trends and shorter-term response of growth to climate to make some general predictions about potential growth under future climate scenarios. If the climate becomes warmer and drier, as predicted by general circulation models, growth rates of subalpine conifers will probably increase. This growth increase would depend on the seasonality of precipitation. A decrease in snowfall would be particularly beneficial to species such as subalpine fir and Engelmann spruce (Ettl and Peterson 1991; Peterson 1993), although warmer summer temperatures could cause summer soil moisture deficits that would be detrimental to growth. It is unknown how future growth patterns will be influenced by increased concentrations of carbon dioxide. Any potential growth changes must, of course, be considered with respect to the effects of climate change on interspecific competition and disturbance, as well as deposition of nitrogen or other nutrients.

References

Becker, M. 1989. The role of present and past vitality of silver fir forests in the Vosges Mountains of northeastern France. Canadian Journal of Forest Res. 19:1110-1117.

Briffa, K.R. 1992. Increasing productivity of "natural growth" conifers in Europe over the last century. Pages 64-71 *in* T.S. Bartholin, B.E. Berglund, D. Eckstein, and F.H. Schweingruber, eds. Tree rings and the environment. Proceedings of the International Dendrochronological Symposium, Lundqua Rep. 34. Lund University, Sweden.

Butler, D.R. 1986. Conifer invasion of subalpine meadows, Central Lemhi Mountains, Idaho. Northwest Science 60:166-173.

Canaday, B.B., and R.W. Fonda. 1974. The influence of subalpine snowbanks on vegetation pattern, reproduction, and phenology. Bull. of the Torrey Botanical Club 101:340-350.

Cooper, C., and J. Gale. 1986. Carbon dioxide enhancement of tree growth at high elevation: commentary. Science 231:859-860.

Davis, M.B. 1989. Insights from paleoecology on global change. Bull. of the Ecological Society of America 70:222-228.

Ettl, G.J., and D.L. Peterson. 1991. Growth and genetic response of subalpine fir (*Abies lasiocarpa*) in a changing environment. The Northwest Environmental Journal 7:357-359.

Fonda, R.W. 1976. Ecology of alpine timberline in Olympic National Park. Proceedings, Conference on Scientific Res. in National Parks 1:209-212.

Franklin, J.F., W.H. Moir, G.W. Douglas, and C. Wiberg. 1971. Invasion of subalpine meadows by trees in the Cascade Range, Washington and Oregon. Arctic and Alpine Res. 3:215-224.

Graumlich, L.J. 1991. Subalpine tree growth, climate, and increasing CO_2: an assessment of recent growth trends. Ecology 72:1-11.

Graumlich, L.J., L.B. Brubaker, and C.C. Grier. 1989. Long-term trends in forest net primary productivity: Cascade Mountains, Washington. Ecology 70:405-410.

Graybill, D.A., and S.B. Idso. 1993. Detecting the aerial fertilization effect of atmospheric CO_2 enrichment in tree-ring chronologies. Global Biogeochemical Cycles 7:81-95.

Graybill, D.A., D.L. Peterson, and M.J. Arbaugh. 1992. Coniferous forests of the Colorado Front Range. Pages 365-401 *in* R.K. Olson, D. Binkley, and M. Böhm, eds. Response of western forests to air pollution. Springer-Verlag, New York.

Heikkinen, O. 1984. Forest expansion in the subalpine zone during the past hundred years, Mount Baker, Washington, U.S.A. Erdkunde 38:194-202.

Innes, J.L. 1991. High-altitude and high-latitude tree growth in relation to past, present and future global climate change. The Holocene 1:174-180.

Jacoby, G.C. 1986. Long-term temperature trends and a positive departure from the climate-growth response since the 1950s in high elevation lodgepole pine from California. Pages 81-83 *in* Proceedings of the NASA Conference on Climate-vegetation Interactions. Rep. OIES-2. University Corporation Atmospheric Research. Boulder, CO.

Kearney, M.S. 1982. Recent seedling establishment at timberline in Jasper National Park, Alberta. Canadian Journal of Botany 60:2283-2287.

Kienast, F., and R.J. Luxmoore. 1988. Tree-ring analysis and conifer growth responses to increased atmospheric CO_2 levels. Oecologia 76:487-495.

LaMarche, V.C., D.A. Graybill, H.C. Fritts, and M.R. Rose. 1984. Increasing atmospheric carbon dioxide: tree ring evidence for growth enhancement in natural vegetation. Science 225:1019-1021.

Little, R.L., D.L. Peterson, and L.L. Conquest. 1994. Regeneration of subalpine fir (*Abies lasiocarpa*) following fire: effects of climate and other factors. Canadian Journal of Forest Res. 24:934-944.

Payette, S., and L. Filion. 1985. White spruce expansion at the tree line and recent climatic change. Canadian Journal of Forest Res. 15:241-251.

Peterson, D.L. 1991. Sensitivity of subalpine forests in the Pacific Northwest to global climate change. The Northwest Environmental Journal 7:349-350.

Peterson, D.L., M.J. Arbaugh, and L.J. Robinson. 1991. Regional growth changes in ozone-stressed ponderosa pine (*Pinus ponderosa*) in the Sierra Nevada, California, USA. The Holocene 1:50-61.

Peterson, D.L., M.J. Arbaugh, L.J. Robinson, and B.R. Derderian. 1990. Growth trends of whitebark pine and lodgepole pine in a subalpine Sierra Nevada forest, California, U.S.A. Arctic and Alpine Res. 22:233-243.

Peterson, D.W. 1993. Dendroecological study of subalpine conifer growth in the North Cascade Mountains. M.S. thesis, University of Washington, Seattle. 66 pp.

Rochefort, R.M., R.L. Little, A. Woodward, and D.L. Peterson. 1994. Changes in subalpine tree distribution in western North America: a review of climate and other factors. The Holocene 4:89-100.

Rochefort, R.M., and D.L. Peterson. 1991. Tree establishment in subalpine meadows of Mount Rainier National Park. The Northwest Environmental Journal 7:354-355.

For further information:

David L. Peterson
National Biological Service
Cooperative Park Studies Unit
University of Washington
Seattle, WA 98195

Woodman, J.N. 1987. Potential impact of carbon dioxide-induced climate changes on management of Douglas-fir and western hemlock. Pages 277-283 *in* W.E. Shands and J.S. Hoffman, eds. The greenhouse effect, climate change, and U.S. forests. The Conservation Foundation, Washington, DC.

Woodward, A., M.B. Gracz, and E.S. Schreiner. 1991. Climatic effects on establishment of subalpine fir (*Abies lasiocarpa*) in meadows of the Olympic Mountains. The Northwest Environmental Journal 7:353-354.

Southwestern Sky Island Ecosystems

by
Peter Warshall
University of Arizona

The "sky islands" of Arizona and New Mexico in the southwestern United States form a unique complex of about 27 mountain ranges whose boundaries, at their lowest elevation, are desert scrub, grasslands, or oak woodlands (Figs. 1 and 2; Table 1). Since the last glaciation, these forested mountain ranges have become relatively isolated from each other. Expanding desert grasslands and desert scrub in the valleys ("the sea" between the sky islands) have limited genetic interchange between populations and created environments with high evolutionary potential. The resulting sky island ecosystems support many perennial streams in an arid climate, have a high number of endemic species, and harbor most game species as well as most threatened and endangered species in the Southwest.

Fig. 1. Sky island mountain ranges of Arizona, New Mexico, and adjacent Sonora and Chihuahua (Marshall 1957). All of the labeled mountain ranges have pine-oak woodland.

The southwestern sky island "archipelago" is unique on the planet. It is the only sky-island complex extending from subtropical to temperate latitudes (compared to the Great Basin, the Venezuelan, and the African sky islands) with an exceptionally complex pattern of species of northern and southern origins. The "continents" that have been the main sources of species for the archipelago are the Sierra Madre of Mexico and the Rocky Mountains of the United States, although the flora has been influenced by the Californian, Sonoran, Intermountain, Cordilleran, and Sierra Madrean Floristic Provinces (S. McLaughlin, University of Arizona, unpublished data).

The ecosystems of each mountain range are of major interest to resource managers concerned with preserving each sky island's unique biogeography and biological diversity as well as to the public for recreation. Land uses sometimes conflict on the sky islands: camping, rock climbing, car-based tourism, military maneuvers, hunting, fishing, exotic grass and fish stocking, grazing, water-supply withdrawals, timber and fuelwood extraction, bird watching, critical habitat for threatened and endangered species, skiing, summer homes, mining, scientific research, sacred Native American ceremonies, and archaeological sites.

Most American sky islands are within the Gila River basin. About 15 additional sky islands are in Mexico and will not be discussed here. Nevertheless, the cross-border management of sky islands is important for such tasks as reintroduction of the Mexican wolf (*Canis lupus baileyi*), maintenance of disjunct populations of rare plant species, and migration of the Mexican pronghorn (*Antilocarpa americana mexicana*), if it still occurs.

Status of Information

The floras of the largest sky islands of Arizona have been inventoried (S. McLaughlin, University of Arizona, unpublished data), including most insular, endemic, and rare species. Certain inventory gaps (e.g., the Baboquivari, Galiuro, Santa Rita, Whetstone, and Patagonia mountains) exist. In addition, the

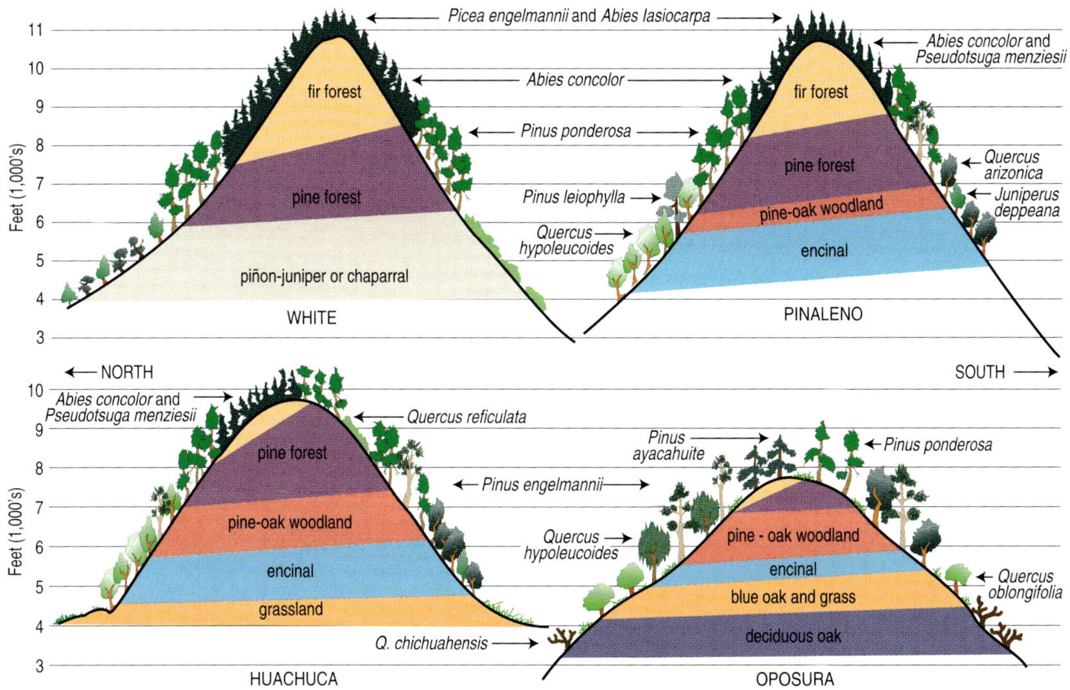

Fig. 2. Cross-sections of three sky islands showing the "stacked" biotic communities varying with latitude. The White Mountains are close to the Rocky Mountain flora and fauna. The Oposura Mountains begin to show the full development of the Sierra Madre communities (Marshall 1957).

number of sites for rare or insular plants, their abundance at each site, and other species diversity indices are lacking for many species of concern. The areal extent, age class, structural characters, and regeneration rates of the five or six biotic communities on the sky islands are poorly known, especially for oak woodlands (McPherson 1992).

Fungi have been intensely inventoried on the Chiricahuas, though only partial inventories exist for ranges of mycorrhizal hypogeous fungi, truffles, and false truffles (States 1990; Nishida et al. 1992). The lichen flora, one of the most diverse and complex in western North America, is poorly inventoried for almost all the sky islands.

The highest sky islands, except the Peloncillos and the Animas, have been intensely inventoried for all groups of insects (C. Olson, University of Arizona, personal communication). Spider and pseudoscorpion distribution is poorly understood. The larger millipedes and scorpions have been extensively collected, but the micro-millipedes, the insular flightless beetles, and the flightless grasshoppers in the upper elevations are poorly known. For instance, a 6-week survey on top of the Pinalenos yielded three new species of flightless beetles (Warshall 1986).

The land mollusks have been inventoried (Bequaert and Miller 1973), though their range extensions and taxa need review. The cienaga (wetland) mollusks are being studied by the Smithsonian. Fish, birds, amphibians, reptiles, and mammals are well-inventoried and yield continuing surprises such as the recent discovery of the Ramsey Canyon leopard frog (*Rana subaquavocalis*). Specific inventory and monitoring gaps in frequency and abundance of sensitive species remain.

Flora and Fauna

A major dividing line between the flora and fauna of southern and northern origins occurs in the sky island ecoregion. The sky island complex harbors more than 2,000 plant species. Of the more than 190 snail species in the Southwest, the sky islands support 3 endemic genera, and over 60 endemic species, including the genus *Sonorella* and the monotypic genus

Table 1. Sky island forested ecosystems of Arizona and New Mexico.

Mountains	High point	Elevation (ft)	Base (ft)	Range (ft)
Pinaleno	Mount Graham	10,720	4,000	6,720
Santa Catalina	Mount Lemmon	9,157	3,000	6,157
Rincon	Mica Mountain	8,666	3,000	5,666
Santa Rita	Mount Wrightson	9,453	4,000	5,453
Chiricahua	Chiricahua Peak	9,796	4,500	5,296
Huachucas	Miller Peak	9,466	5,000	4,466
Dos Cabezas	Dos Cabezas	8,369	4,000	4,369
Animas	Animas Peak	8,519	4,500	4,019
Baboquivari	Baboquivari Peak	7,730	3,500	4,230
Galiuro	Bassett Peak	7,650	4,000	3,650
Santa Teresa	Cottonwood Mountain	7,489	4,000	3,489
Winchester	Reiley Peak	7,631	4,400	3,231
Whetstone	Apache Peak	7,684	4,800	2,884
Dragoon	Mt. Glenn	7,519	4,700	2,819
Patagonia	Mt. Washington	7,221	4,500	2,721
Mule	Mt. Ballard	7,370	4,800	2,570
Tucson	Wasson Peak	4,687	2,200	2,487
Atascosa	Atascosa Peak	6,440	4,000	2,440
Swisshelm	Swisshelm Mountain	7,185	4,800	2,385
Peloncillo	Owl Peak	6,625	4,500	2,125
Tumacacori	(Unnamed)	5,634	3,500	2,134
Sierrita	Samaniego Peak	5,991	4,000	1,991
Mustang	(Unnamed)	6,469	4,750	1,719
Empire	(Unnamed)	5,588	4,000	1,588
Pajarito	Pajarito Peak	5,236	4,000	1,236

Table 2. Selected sensitive, rare, and endangered plants of the sky islands.

Species: Common name	Species: Scientific name	Location	Status
Agave	*Agave parviflora* ssp. *parviflora*	Patagonia/Atascosa/Santa Rita/S. Cibuta	Rare
Wild onion	*Allium gooddingii*	Catalina	Disjunct
Amsonia	*Amsonia kearneyana*	Baboquivari	Endemic/rare
Aster	*Aster potosinus*	Huachuca	Single population
Milk-vetch	*Astragalus cobrensis* var. *maguirei*	Chiricahua/Peloncillo	Insular
Milk-vetch	*A. hypoxylus*	Patagonia/Huachuca	Very rare
Zorillo	*Choisya mollis*	Atascosa	Endemic
Pincushion cactus	*Coryphantha robbinsorum*	Peloncillo	Endemic
Climbing milkweed	*Cynanchum wigginsii*	Atascosa/Patagonia/Baboquivari/Mule	Very rare
Fleabane	*Erigeron kuschei*	Chiricahua	Endemic
Fleabane	*E. heliographis*	Pinaleno	Endemic
Fleabane	*E. lemmonii*	Huachuca	Endemic
Lemon lily	*Lillium parryi*	Huachuca/Santa Rita/Chiricahua	Rare
(No common name)	*Lilaeopsis schaffneriana recurva*	Huachuca	Rare
(No common name)	*Perityle cochisensis*	Chiricahua	Endemic
Dock, sorrel	*Rumex orthoneurus*	Chiricahua/Sierra Ancha/Pinaleno/Huachuca	Unique/rare
Groundsel	*Senecio huachucanus*	Santa Rita/Huachuca	Rare
Sophora	*Sophora arizonica*	Pinaleno /Swisshelm/Whetstone/Hualapai	Vulnerable
Lady's-tresses	*Spiranthes delitescens*	Canelo	Endemic

Chaenaxis. More than 75 reptile species inhabit the sky islands, one of the most diverse herpetological regions in North America with several endemic races.

About 265 bird species occur within the sky island complex, including valley and riparian species. About 30 are of subtropical origin and have their northern limits within the sky island complex. The sky islands are the most diverse U.S. area for mammals; some 90 native mammals inhabit the area from the chaparral community to higher elevations; at least 6 are endemic subspecies.

Table 3. Various threatened, endangered, and candidate species.*

Common name	Scientific name	Location	Status
Yaqui catfish	*Ictalurus pricei*	San Bernardino Creek	E
Yaqui topminnow	*Poeciliopsis occidentalis sonoriensis*	Swisshelm	E
Apache trout	*Oncorhynchus apache*	Various reintroductions	T
Gila chub	*Gila intermedia*	Gila drainages	(T)
Rosetail chub	*G. robusta*	Galiuro	(T)
Sonora chub	*G. ditaenia*	Atascoca	E
Spikedace	*Meda fulgida*	Galiuro/Pinaleno	T
Loach minnow	*Tiaroga cobitis*	Galiuro/Pinaleno	T
Mexican stoneroller	*Campostoma ornatum*	Chiricahua	T
Barking frog	*Hylactophryne augusti*	Pajarito/Santa Rita	(E)
Chiricahua leopard frog	*Rana chiricahuensis*	Chiricahua (+)	(T)
Great Plains narrow-mouthed toad	*Gastrophryne olivacea*	Various	(C)
Mexican garter snake	*Thamnophis eques*	Various	(C)
Ridge-nosed rattlesnake	*Crotalus willardi willardi*	Huachuca, Patagonia, Santa Rita	(C)
Ridge-nosed rattlesnake	*C. willardi obscurus*	Animas; San Luis	(C)
Thick-billed parrot	*Rhynchopsitta pachyrhyncha*	Various (Chiricahua)	E
Buff-breasted flycatcher	*Empidonax fulvifrons*	Huachuca	(E)
Gray hawk	*Buteo nitidus*	Santa Cruz/San Pedro drainages	(T)
Yellow-billed cuckoo	*Coccyzus americanus*	Various gallery forests	(T)
Mexican spotted owl	*Strix occidentalis lucida*	Various	E
Northern goshawk	*Accipiter gentilis*	Various	C/T
Peregrine falcon	*Falco peregrinus*	Various	E
Sanborn's long-nosed bat	*Leptonycteris sanborni*	Various	E
Mt. Graham red squirrel	*Tamasciuris hudsonicus grahamensis*	Pinaleno	E
Mexican wolf	*Canis lupus baileyi*	Extirpated	E
Grizzly bear	*Ursus arctos*	Extirpated	T
Jaguar	*Felis onca*	Extirpated	E
Ocelot	*F. pardalis*	Extirpated	E
Mexican long-tongued bat	*Choeronycteris mexicana*	Various	T
Arizona shrew	*Sorex arizonae*	Chiricahua, Huachuca, Santa Rita	C
Red bat	*Lasiurus borealis*	Various	(C)

*This list is incomplete because of revisions since publication in 1988. Only animals inhabiting sky islands within or above the oak-pine woodlands are included. Some wetland species have not been included. The species must have been on federal lists. Various — more than two sky-island ecosystems; T — threatened; E — endangered; C — candidate in Arizona or New Mexico; ( ) — listing by Arizona Fish and Game only; federal status not yet determined.

Trends and Management

Researchers have begun measuring biological trends in six major categories concerning inventory, monitoring, preservation, and restoration that are most pertinent to sky island forested ecosystems. A discussion of these categories follows.

Endemics and Insular Species

With new investigative techniques, there has been growing respect for the genetic diversity of this area, especially late-Cenozoic and Pleistocene relict faunal populations (see examples, Tables 2, 3). For instance, recent genetic analyses on the Mt. Graham red squirrel (*Tamasciuris hudsonicus grahamensis*) and the lemon lily (*Lillium parryi*) showed that both populations are more highly divergent from closely related populations than previously thought. Increasingly, however, local and insular species have hybridized with introduced races and species; hybridization is particularly evident among fish (e.g., hybrids of the Apache trout, *Oncorhynchus apache*, with the rainbow trout, *O. mykiss*), but can also be found among white-tail deer (*Odocoileus virginianus* versus *hemionus*), pronghorn (*Antilocapra americana* versus *mexicana*), turkeys (*Meleagris merriami* versus *mexicana*), and bighorn sheep (*Ovis mexicana* versus *nelsoni*).

Selected rare, unique, threatened, and endangered species and subspecies whose critical habitat includes the sky islands are listed in Tables 2 and 3. Figure 3 compares the Coronado National Forest to other forests. The number of sensitive plant and animal species from the sky island ecoregion has increased over the last 20 years. (Sensitive means that the population's viability is of concern and requires monitoring or active protection.) The increase is, in part, the product of more detailed knowledge. For instance, a recent review of *Erigeron pringlei* split this fleabane into four species, creating a new endemic, *E. heliographis,* on the Pinalenos. Nevertheless, the Coronado Forest reports 56 sensitive plants, among the largest number reported from any national forest, including 1 on the federal endangered list and 3 candidate species. McLaughlin (University of Arizona, unpublished data) suggested that the local extirpation of six plant species in the last century was related to either global warming, habitat alteration, or both.

Seven insects are listed by the Coronado National Forest as species of concern (U.S. Forest Service, unpublished memo). About 12 fish species are considered vulnerable, including 9 federally listed and 7 living in sky island drainages (Table 3). Within this national forest, there are 11 amphibians whose population viability is of concern, though none is on the federal list; 8 dwell in the valleys or lower drainages of the sky islands. They are sensitive to upstream watershed alterations. Fourteen sky island reptiles are considered sensitive but not federally listed. There are about 55 bird species of concern, 5 federally listed, and about 20 whose population viability is of significant concern (Table 3). About 30 mammals are of concern, 4 federally listed. The grizzly bear, jaguar, ocelot, Tarahumara frog, and gray wolf have been extirpated from the sky island archipelago. Not counting the extirpated species and the 11 bats of concern, there are 13 mammal species and subspecies dwelling on the sky islands that have low populations of concern (Table 3).

Distribution

Some of the most interesting aspects of sky island ecosystems and history are why some mountains lack a particular species ("holes"), why some species skip mountain ranges or appear as an exception in an otherwise species-poor flora or fauna ("outliers"), and why some species, even mobile animals such as birds, end their distribution on a particular sky island (Warshall 1986). For example, why are there no chipmunks on the Huachucas? Why does the Mexican chickadee (*Parus sclateri*) stop on the Chiricahuas, but only 35 mi away the mountain chickadee (*P. gambeli*) inhabits the Pinalenos? Why are there no voles on the Catalinas?

Colonization of the sky islands by exotic species is increasing with over 60 non-native plants having established regenerative populations in the Arizona sky islands. Major issues include limiting introductions of buffel grass (*Pennisetum cilaris*) and exotic lovegrasses (*Eragrostis* spp.) by the U.S. Department of Agriculture, as well as controlling and restoring habitats swamped by exotic forbs such as *Euryops multifida* on the Pinalenos and Catalinas. Fifteen non-native fish species have been added to the five or six native freshwater fish families, with consequent hybridization, predation, and competition throughout springs and drainages. The Central Arizona Project has become a new corridor for exotic fish, some of which are invading the last strongholds of natives.

Three feral exotic mammals may have colonized the sky island complex. The opossum (*Didelphis* spp.) colonization is believed to be a mix of released Virginia opposum (*D. virginiana* and range-expanding Mexican opossum (*D. marsupialis*). It has been reported but not confirmed that the European ferret (*Mustela putorius*) has become feral in the Huachucas. Over the last 50 years, about 12 birds and mammals of southern origin have been recorded colonizing more northern sky islands. No animal species is known to have retreated south except the extirpated jaguar, ocelot, and thick-billed parrot. A few species such as the Abert's squirrel (*Sciuris aberti*) have been introduced for hunting and have then expanded their range. The monitoring of these changes will be an important barometer to how habitat changes, species introductions, and climate interact with ecosystem management practices.

Vertical Migration

Each sky island has a unique ecosystem with a stack of life zones ranging from arid to boreal (Fig. 2). Many species migrate vertically to feed and breed at various elevations. The Pinalenos contain the most stacked life zones in the shortest vertical distance of any mountain in North America. By traversing five biotic communities in a few hours, bears can feed on *Opuntia* (prickly pear cactus) fruit in the morning and grass roots growing in semi-alpine meadows in the afternoon. Assuring minimal viable habitat size and the appropriate forest age-class structure to support animal populations with vertical migration is an unstudied aspect of forest ecosystem management.

In addition, various biotic communities are remnants of colder climates with small relict acreage. For instance, only about 243 ha (600 acres) of spruce-fir (*Picea engelmannii-Abies lasiocarpa*) forest are left within the sky island complex. This forest type, found only on the Pinalenos, is critical habitat for the endemic and federally listed endangered Mt. Graham red squirrel, which also inhabits the transition to mixed conifer forests (Douglas fir-white fir; *Pseudotsuga menziesii-Abies concolor*) at lower elevations. These two plant associations, heavily logged and cleared, will not become mature enough to supply the minimum viable habitat to ensure the squirrel population's survival for 2 centuries. Annual growth rates, seeding rates, and regeneration cycles have become less predictable with the unknown effect of global warming and fire risk, requiring rethinking of the minimum size required for viable habitats.

Special Habitats

Special habitats and plant associations (e.g., high-elevation cienagas, limestone outcrops, perennial streams, talus slopes) create islands of habitat within the sky island ecosystem, increasing biological richness. For instance, talus slopes support a series of endemic land snails; the rock cliffs and outcrops support plant species such as the fleabanes *Erigeron lemmonii*, *E. heliographis*, and *E. pringlei*, found

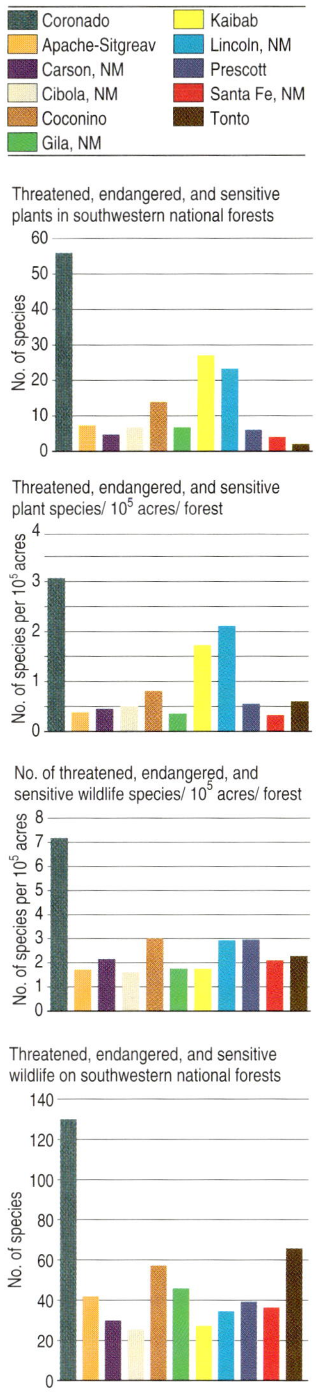

Fig. 3. Comparison of the Coronado National Forest, which administers all or part of 16 sky islands, with national forests with no sky island flora and fauna.

nowhere else in the world. The perennial streams support seven rare native fish species. Of the special habitats, the cienagas (swampy, marshy cover) and perennial streams require the most monitoring, protection, and restoration.

Special Interest Game

The densest populations of most game species are found on the sky islands. For instance, the densest populations of black bear (*Ursus americanus*) and mountain lion (*Felis concolor*) south of the Mogollon Rim are on the Pinalenos. In general, over a 20-year period, both species increased with population troughs occurring from rancher depredation and drought. White-tailed deer (*Odocoileus virginianus*) populations have increased, while mule deer are less stable. Javelina (*Tayassu tajacu*) are stable or declining, having suffered from canine distemper after the drought of 1989. Band-tailed pigeon (*Columba fasciata*), a species dependent on sky island forests, has had a long-term decline as have two subspecies of turkey.

Corridors

For many land animals, corridors of animal movement between sky islands have been through riparian zones. Increasing habitat fragmentation from increased subdivisions around the base of the sky islands is further isolating some populations, especially in the Tucson area, which separates the Santa Catalina and Rincon mountains from the Tucsons and the Santa Ritas. This structural change in migration patterns has not been studied but is believed to be the most significant threat to "safe passage" corridors between sky islands.

In summary, the single best indicator of ecosystem management has been the increasing number of threatened and endangered populations (USFS 1993). This trend requires increasing acreage of critical or otherwise protected habitat; increased monitoring and control over the introduction and spread of exotic grasses, fish, and gamebirds, and the reintroduction of locally extirpated mammals and tree species in restoration projects.

Other Issues

Fire management is planned to reduce catastrophic fires from fuel build ups, to allow natural burns required by certain species, and to increase fire suppression to maintain remnant old-growth forest biodiversity. Experimentation and debate about fire management are widespread, however. Another trend is toward the restriction of cattle to prevent overgrazing and trampling, to protect sensitive plant species, and to protect and promote recovery of wetland and riparian habitats. A third trend is toward upstream rehabilitation in specific watersheds where flooding endangers sensitive plants.

In addition, there is increasing urban pressure on the Forest Service to clear more habitat for recreation such as camping and skiing (on the Catalinas) and to expand roads into the sky islands for greater access and uses that can conflict with habitat protection (USFS 1993). Managing land use on private inholdings, on properties adjacent to public land, and on properties bordering intermountain corridors will be increasingly important.

The final trend is the unknown impact of global warming on the biseasonal (winter and summer) rainfall pattern of the southwestern sky islands. This trend is of special importance because of the large number of relict and insular species and subspecies in the region.

Because of the geological, topographic, and biological uniqueness of each sky island, the policies for each mountain range will need to be custom-designed on a watershed by watershed basis.

References

Bequaert, J.C., and W.B. Miller. 1973. The mollusks of the arid Southwest. University of Arizona Press, Tucson. 271 pp.

Marshall, J.T. 1957. Birds of the pine-oak woodland in southern Arizona and adjacent Mexico. Pacific Coast Avifauna 32. 125 pp.

McPherson, G.R. 1992. Ecology of oak woodlands in Arizona. *In* Ecology and management of oak and associated woodlands. Rocky Mountain Forest and Range Station, Gen. Tech. Rep. RM-218. Fort Collins, CO.

Nishida, F.H., W.J. Sundberg, J.A. Menge, J.S. States, R.E. Tulloss, and J. Cifuentes Blanco. 1992. Studies in the mycoflora of the Chiricahua Mountains. I. Preliminary report on species distribution, ecology, and biogeographical affinities. Pages 35-39 *in* A.M. Barton and S. Sloane, eds. 1992 Chiricahua Mountains Research Symposium, Proceedings. Southwest Parks and Monument Association, Tucson, AZ.

States, J.S. 1990. Mushrooms and truffles of the Southwest. University of Arizona Press, Tucson. 232 pp.

USFS. 1993. Executive summary, Coronado National Forest plan five year review report, October 1987 through September 1991. U.S. Forest Service, Tucson, AZ. 33 pp.

Warshall, P. 1986. Biogeography of the high peaks of the Pinalenos. Reprinted from the Environmental Data Book, U.S. Forest Service, Coronado National Forest, by Maricopa Audubon Society, Phoenix, AZ. 18 pp.

For further information:

Peter Warshall
University of Arizona
Office of Arid Lands Studies
4500 W. Speedway
Tucson, AZ 85745

Endangered Cui-ui of Pyramid Lake, Nevada

by
G. Gary Scoppettone
Peter H. Rissler
National Biological Service

Cui-ui (*Chasmistes cujus*) is a large plankton-feeding fish that only occurs in Pyramid Lake, Nevada. It was put on the federal endangered list in 1967 based on declining population and absence of reproduction. A lake dweller, cui-ui is a stream spawner. Most of this century, this sucker species was unable to access the Truckee River, Pyramid Lake's only perennial tributary, to reproduce. Water diversion from the Truckee River, as a result of the nation's first Bureau of Reclamation project (Newlands Project), reduced the lake elevation and, in most years, caused an impassable delta to form at the mouth of the Truckee River. Cui-ui live more than 40 years; it is this longevity that has allowed the species to persist for as many as 19 years with virtually no recruitment (*see* glossary) to the adult population (Scoppettone 1988).

Cui-ui is one of three remaining species of the genus *Chasmistes*. Of the three, its habitat is most intact, and it thus has the best opportunity for recovery (Scoppettone and Vinyard 1991). Each spring, cui-ui adults, most of which mature at 8-12 years of age, migrate to the mouth of the Truckee River at the south end of Pyramid Lake, where they aggregate, awaiting environmental cues and sufficient stream flow to enter the river (Scoppettone et al. 1986). This behavior provides an excellent opportunity to capture the adults for estimating population numbers and year-class (year hatched) structure. In this article we report changes in adult cui-ui population number and year-class structure from spring 1983 to spring 1993.

Status and Trends

Each spring, cui-ui are captured, anchor-tagged, and released for recapture. The proportion of tagged to untagged fish is used to estimate population number. Virtually all mature adults enter the prespawning aggregate each year (Scoppettone, unpublished data); thus an estimate of the number of adults entering the aggregate is an estimate of the entire adult population. We provide data of 4 select years (1983, 1991, 1992, and 1993) to illustrate trends between 1983 to 1994.

Courtesy G.G. Scoppettone, NBS

Cui-ui (*Chasmistes cujus*) with dorsal radio tag.

Captures of cui-ui from the prespawning aggregate have been successful enough to give us reliable estimates of the adult population. In 1982 and 1983, 3,000 adults were captured and tagged. From 1989 through 1993, captures increased markedly because of a change in capture gear and increased population. More than 100,000 cui-ui have been captured, and tags were applied to 60% of these. By spring 1993, tag returns were close to 4% of the fish captured.

The adult cui-ui population has increased 10-fold from 1983 to 1993 (Fig. 1), an increase attributed in part to unusually wet years from 1980 to 1986. During these years more than 65,000 adults entered the lower Truckee River to spawn, and produced more than 250 million cui-ui larvae for Pyramid Lake. In contrast, virtually no spawning occurred in the Truckee River from 1988 through 1992, a fact that will probably be reflected later in this decade as a downward trend in the number of adults.

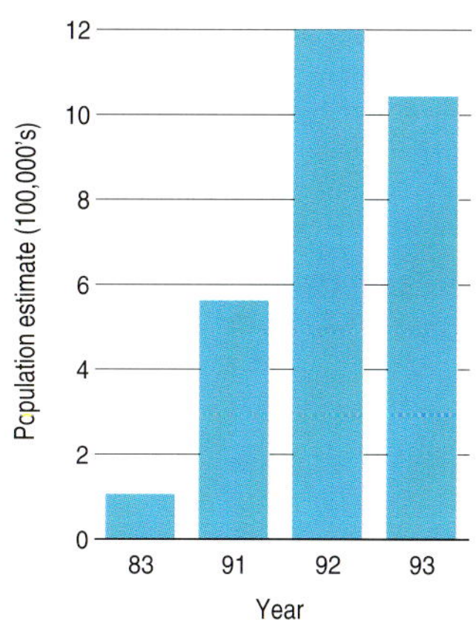

Fig. 1. Estimated population of adult cui-ui in spring 1983, 1991, 1992, and 1993.

Adult Year-class Structure

To understand cui-ui demographics and why the species is still considered endangered, it is necessary to understand its year-class structure. In 1983 when there were about 100,000 adult cui-ui in Pyramid Lake, almost 90% were from a single year class produced in 1969; the second predominate class represented about 5% of the population and was hatched in 1950 (Fig. 2). From 1950 to 1968 and from 1970 to 1979, very little recruitment occurred. The situation has improved; in 1991, the 1981 year class replaced the 1969 in predominance, and it remained so through 1993. In 1993, 400,000 of the estimated 1 million adults were fish that had been hatched in 1981. The dramatic increase in the spawning population from 1991 to 1992 is assumed to be those fish that hatched in 1981, 1982, and 1983 and finally reached adulthood.

These improvements in population numbers and year-class structure are partly attributed to several extraordinarily wet years; similar conditions may not occur with sufficient frequency to assure species recovery or preclude extinction.

In addition to the prespawning aggregate, the adult and juvenile populations have been sampled around Pyramid Lake throughout the year. Our results suggest that few juveniles hatched after 1986, and thereby provide testimony to inconsistency in cui-ui recruitment.

Spawning cui-ui.

Future Outlook

The cui-ui has an excellent prognosis for recovery. It has an approved recovery plan and supporting legislation (P.L.101-616), which provide for acquisition of water and water rights to elevate Pyramid Lake, improve fish passage over the delta, and enhance spawning flows. Plans to acquire water for Pyramid Lake and cui-ui are being developed. Cui-ui population trends over the past 10 years demonstrate the rebound potential of the species when it is provided with passage and sufficient water for reproduction. Because limited water is available for acquisition, however, Truckee River flows required for cui-ui recovery need to be precisely determined. Our monitoring of the adult cui-ui population is part of a cui-ui population dynamics study aimed at calibrating an existing Truckee River water-management model being used for cui-ui recovery. Monitoring will continue through the 1998 spawning season, at which time sufficient information should have been generated to calibrate the model.

References

Scoppettone, G.G. 1988. Growth and longevity of the cui-ui and longevity of other catostomids and cyprinids in western North America. Transactions of the American Fisheries Society 117:301-307.

Scoppettone, G.G., M. Coleman, and G.A. Wedemeyer. 1986. Life history and status of the endangered cui-ui of Pyramid Lake, Nevada. U.S. Fish and Wildlife Service Fish and Wildlife Res. 1:1-23.

Scoppettone, G.G., and G. Vinyard. 1991. Life history and management of four endangered lacustrine suckers. Pages 359-377 *in* W.L. Minckley and J.E. Deacon, eds. Battle against extinction: native fish management in the American West. The University of Arizona Press, Tucson.

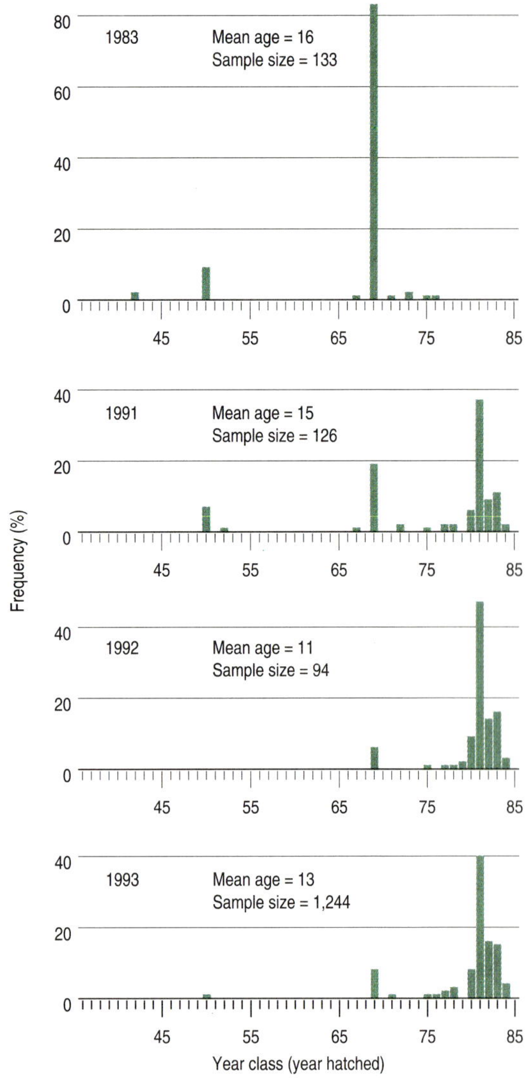

Fig. 2. Year-class structure of adult cui-ui in spring 1983, 1991, 1992, and 1993.

For further information:

G. Gary Scoppettone
National Biological Service
Reno Field Station
4600 Kietzke Ln.
Bldg. C., Rm. 120
Reno, NV 89502

Bonytail and Razorback Sucker in the Colorado River Basin

by

Gordon Mueller
National Biological Service
Paul Marsh
Arizona State University

Bonytail (*Gila elegans*) and razorback sucker (*Xyrauchen texanus*) are large river fish found only in western North America's Colorado River basin. The bonytail is nearly extinct and the razorback sucker is becoming rare.

The bonytail (Fig. 1) is a large, streamlined minnow (family Cyprinidae) that may reach 50 cm (18 in) in length and weigh up to 0.5 kg (1 lb). The razorback sucker (Catostomidae; Fig. 2) may grow to 75 cm (2.5 ft) in length and weigh up to 5 kg (10 lb). Both species have evolved a unique dorsal keel or hump, a characteristic shared by few other fish. Individual life spans approach 50 years.

Historically, both species were common and were used by Native Americans and early settlers as food and fertilizer. Physical and biological changes to their habitat and direct competition and predation from non-native fishes are responsible for their decline. Young fish no longer survive to replace adults as they die of old age.

Status

Information about these fish is found in sources ranging from scattered personal journals of early travelers to more recent biological reports and scientific literature. Bonytail and razorback sucker were first described by scientists in the late 1850's. Comprehensive studies were not conducted in the lower Colorado River until 1930, while similar investigations upstream were delayed until the 1960's because the area is rugged and remote.

Dill (1941) reported an alarming decline of endemic fish in the lower river; Miller et al. (1982) reported similar trends farther upstream. Three years after the 1973 passage of the Endangered Species Act, the Colorado River

Fishes Recovery Team was formed. The Colorado River Fishery Project was established in 1978 to recover threatened and endangered fish in 965 km (600 mi) of the upper Colorado and Green rivers. Recovery efforts intensified in 1987 with the Recovery Implementation Program. These and other projects have funded major research on the biology and habitat needs of these species.

Bonytails were historically common in the mainstem Colorado, Green, Gunnison, Yampa, Gila, and Salt rivers before the construction of large dams. Bonytail became rare in the lower river system by 1935 and suffered similar declines farther upstream by the mid-1960's. The last confirmed bonytail taken from any river was in 1985. Bonytail continue to be captured in low numbers from Lake Mohave in Arizona and Nevada, a reservoir on the Colorado River downstream of Hoover Dam.

Razorback suckers were historically common to abundant in the Colorado mainstem and portions of the Green, San Juan, Animas, Duchesne, Gila, Salt, and Verde rivers. Razorback suckers also had begun declining in the lower river by 1935, but were commercially harvested near Grand Junction, Colorado, and Phoenix, Arizona, until 1950. Numbers dramatically declined in the upper Colorado River during the 1970's and 1980's, and today the fish is very rare. The largest river population is in the Green River, Utah, and is estimated (1993) at fewer than 500 adults.

Large populations of razorback sucker developed in some newly created reservoirs in the lower river before fish predators became abundant. For example, populations that numbered into the hundreds of thousands became established in the Salton Sea, Roosevelt Lake, Saguaro Lake, Lake Havasu, Lake Mead, Lake Mohave, and Senator Wash Reservoir. Predation by non-native fishes eventually proved overwhelming, and, without recruitment (addition of individuals to a population through reproduction and immigration), populations disappeared after 40 to 50 years.

Razorback suckers are now rare except in Lake Mohave, which supports the last large population. Spawning is successful there, but as was true at older reservoirs, young razorback suckers are eaten by sunfish, bass, and other fish. The reservoir population declined by 60% between 1988 (59,500) and 1991 (23,300). Remaining suckers are expected to die by the end of the decade.

It is unlikely that the bonytail and razorback will survive in the wild. No measurable recruitment is evident in any part of the drainage and old individuals are reaching the end of their life span. Bonytail are found in less than 2% of their former range, and razorback sucker in less than 25% of their former range (Fig. 3).

Reasons for Decline

The Colorado River ecosystem has been dramatically altered by water development that transformed an erratic and turbulent river system into a series of calm reservoirs and channelized river reaches. Eight dams were built across the lower 563 km (350 mi) of the river by 1950. The historical habitat of these fish is now controlled by 44 large dams and is being drained by hundreds of miles of diversion canals. Nursery areas, critical for early life stages, have been flooded by reservoirs, and upstream migration is physically blocked by dams. Seasonally warm and turbid flows of the natural hydrology of the basin were replaced by cold, diminished reservoir releases governed by hydroelectric and downstream water demands.

Although physical habitat changes have been dramatic, subtle ecological changes may have been even more damaging. Reservoirs and cold tailwaters presented favorable conditions to develop recreational fisheries. Although the bonytail and razorback sucker were once valuable food sources, they became viewed as trash fish when more desirable sportfish (e.g., trout, catfish, and bass) became established. Resource agencies stocked and promoted recreational fisheries, often at the expense of native fishes. For example, in 1962, 723 km (450 mi) of the upper Green River was poisoned to improve trout production. Today, over 90% of all fish found in the river system are species introduced for recreational fishing. Uncounted other aquatic plants and animals, pathogens, parasites, and

Fig. 1. Bonytail (*Gila elegans*).

Fig. 2. Razorback sucker (*Xyrauchen texanus*).

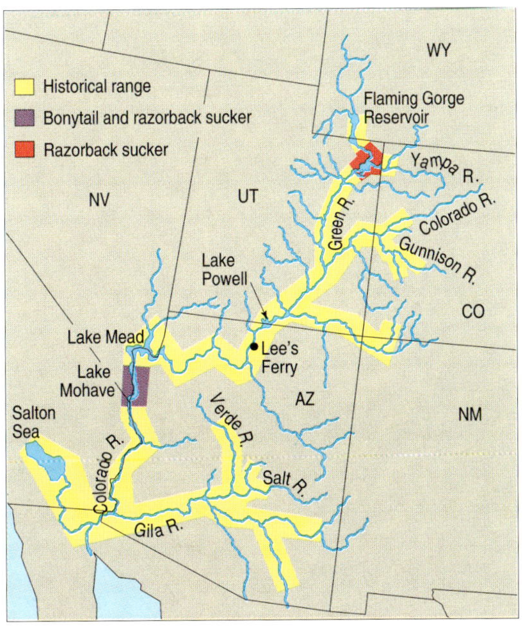

Fig. 3. Historical range and current concentrations of bonytail and razorback sucker (Minckley and Deacon 1991).

chemical contaminants were introduced and have changed the river's delicate ecosystem.

The dramatic decline prompted the listing of the bonytail as endangered in 1980, and a similar listing for the razorback sucker followed in 1991. Although both fishes are federally protected and recovery programs began over 15 years ago, these species continue to edge toward extinction. The problem lies in the complexity of the environmental and legal issues, combined with possible conflicts in land-, water-, and fishery-management philosophies. Controversy and debate have slowed, stalled, and complicated recovery effort. While sociopolitical issues of recovery are debated, old relict populations are not being aggressively protected through management and they continue to die off.

Recovery and Management

The goal of recovery is to reestablish species or enhance their ability to maintain self-perpetuating populations in native habitat, which may require both physical and biological habitat restoration. Many scientists believe recovery of bonytail and razorback sucker will take an aggressive and long-term commitment. Recovery efforts in the upper river are being intensified to restore adequate spring flows and develop nursery habitat. Stocking of bonytail and razorback sucker is being postponed until these habitat changes are made, and guidelines for stocking recreational species and possibly reducing their populations are being negotiated. Whether these actions will be sufficient to recover these fish is unknown.

While bonytail and razorback sucker are not being stocked in the upstream recovery program, they are being stocked farther downstream. A 10-year stocking program reintroduced razorback sucker into Arizona streams, but although nearly 15 million razorbacks were stocked between 1981 and 1990, the effort failed because most small suckers were believed to have been eaten by catfish and other non-native fishes. This emphasizes the need for predator removal or the stocking of larger fish.

Removal of non-native species is virtually impossible and sometimes undesirable. Larger bonytails and razorback suckers are being stocked by the Native Fish Work Group to attempt to maintain the Lake Mohave population by replacing the old population with young adults that exhibit the genetic characteristics of the remnant population. Bonytail and razorback suckers are being raised in isolated coves where other fish have been removed. Fish grow to about 30 cm (12 in) in length in a year and are then released into the reservoir. At this size, many should escape predation and could potentially survive 40 to 50 years.

Stocking is not an alternative to recovery, but if done properly, it can be used to maintain, expand, or reestablish long-lived endangered fish populations. Lake Mohave is not pristine habitat; however, maintenance of its population can help preserve genetic diversity, enhance species diversity in the reservoir, help ensure against catastrophic loss of hatchery brood stocks, and provide opportunities to study these fish in the wild.

Aggressive management of remaining populations is essential to provide the time to complete and test habitat restoration programs. If remnant populations are not saved, we stand to lose important pieces of a very complex ecological puzzle.

References

Dill, W.A. 1941. The fishery of the lower Colorado River. California Fish and Game 30:109-211.

Miller, W.H., H.M. Tyus, and C.A. Carlson, eds. 1982. Fishes of the upper Colorado River system: present and future. American Fisheries Society, Bethesda, MD. 131 pp.

Minckley, W.L., and J.E. Deacon, eds. 1991. Battle against extinction: native fish management in the American West. University of Arizona Press, Tucson. 517 pp.

For further information:

Gordon Mueller
National Biological Service
PO Box 25007 D-8220
Denver, CO 80225

Amphibian and Reptile Diversity on the Colorado Plateau

by
Charles Drost
Elena Deshler
National Biological Service

The Colorado Plateau region is an area of high uplands, cut by the dramatic canyons of the Colorado River system in northern Arizona, northeastern New Mexico, eastern Utah, and western Colorado (Figure). Habitats within the region range from upland desert in the lower stretches of the Colorado River to small areas of alpine tundra on the highest peaks. The amphibian fauna is relatively small and dominated by species adapted to dry conditions such as toads (genus *Bufo*) and spadefoot toads (genus *Scaphiopus*). Reptile species are more numerous and varied, with the spiny lizards (*Sceloporus*), whiptail lizards (*Cnemidophorus*), and garter snakes (*Thamnophis*) well-represented. The reptiles and amphibians of the area have not been well-studied, although several species are known or suspected to have suffered recent declines.

As part of an overall project to assess the completeness of biological inventory information on National Park Service lands (Stohlgren and Quinn 1992), we compiled information from species lists, literature reports, and limited field work to prepare a preliminary data base of amphibian and reptile occurrence on 25 park areas in National Park Service lands on the Colorado Plateau.

Status and Trends

The quality and completeness of amphibian

and reptile inventory information for Colorado Plateau parks vary. Grand Canyon National Park has received moderately thorough survey effort along the Colorado River corridor (Miller et al. 1982), but the canyon rim areas have received relatively little study. Parts of Glen Canyon National Recreation Area were surveyed by the University of Utah (Woodbury 1959), but this information is now 35 years old. Most other parks have had little or no thorough survey work and a few, such as Rainbow Bridge National Monument, have no inventory information at all. The information for many park species lists is based on large-scale range maps or unverified records. We found incorrect identifications or outdated taxonomy on about 10% of the species recorded. The generally poor and sometimes unreliable state of inventories in many parks echoes the results of Stohlgren and Quinn (1992) in a larger study.

Sixty-two reptile and 18 amphibian species are known from the Colorado Plateau as a whole. Most occur in one or more of the national park areas. The few species that apparently do not live in any parks, such as the mountain treefrog (*Hyla eximia*), Chiricahua leopard frog (*Rana chiricahuensis*), and narrow-headed garter snake (*Thamnophis rufipunctatus*), are primarily found in the area of the precipitous Mogollon Rim in north-central Arizona, which forms the southern boundary of the plateau. The most widespread species in the region include Woodhouse's toad (*Bufo woodhousii*; 20 areas), tiger salamander (*Ambystoma tigrinum*; 16 areas), eastern fence lizard (*Sceloporus undulatus*; 22 areas), tree lizard (*Urosaurus ornatus*; 21 areas), side-blotched lizard (*Uta stansburiana*; 22 areas), striped whipsnake (*Masticophis taeniatus*; 21 areas), pine snake (*Pituophis melanoleucus*; 22 areas), and western rattlesnake (*Crotalus viridis*; 22 areas). Some other species are much more limited; for example, the Jemez Mountains salamander (*Plethodon neomexicanus*) is known only from a small area of north-central New Mexico, including Bandelier National Monument. The painted turtle (*Chrysemys picta*) is apparently rare in the region and may have declined further; there have been no recent reports of this species from Glen Canyon National Recreation Area, where it formerly occurred. Other species, like the desert iguana (*Dipsosaurus dorsalis*) and the Gila monster (*Heloderma suspectum*), only have a small portion of their range on the Colorado Plateau (although they occur within the geographic boundaries of the area, some of these species are restricted to habitats not representative of the plateau, such as the Sonoran Desert).

Four of the 62 reptile species (7%) are listed as threatened or endangered by either individual states, the Department of the Interior, or both (Table). In contrast, 5 of the 18 amphibian species (27%) in the region are considered threatened or endangered. The high proportion of amphibians listed is due to several frog and toad species that have experienced serious population declines. One of these, the relict leopard frog (*Rana onca*) of southern Nevada, was thought extinct but has recently been rediscovered (D. Bradford, Environmental Protection Agency, Las Vegas, Nevada, personal communication). The western toad (*Bufo boreas*) has suffered drastic declines in other parts of its range (e.g., Carey 1993); its status on the Colorado Plateau is not known. The northern leopard frog (*Rana pipiens*), although not yet listed by state or federal governments, has disappeared from large areas of its range in western North America (Hayes and Jennings 1986). There are recent reports of healthy populations in a number of the perennial streams on the Colorado Plateau, but this species, in particular, needs further survey.

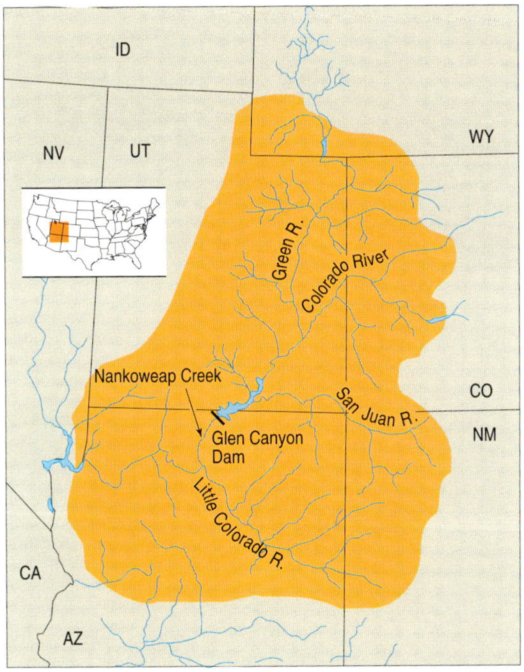

Figure. The Colorado Plateau region is cut by dramatic canyons of the Colorado River system.

Courtesy C. Drost, NBS

Northern leopard frog (*Rana pipiens*).

Table. Threatened and endangered species of the Colorado Plateau. An X indicates that a species is listed as threatened or endangered by particular states or by the U.S. Department of the Interior.

Common name	Scientific name	Threatened or endangered status				
		AZ	CO	NM	UT	Federal
Amphibians						
Jemez Mountains salamander	*Plethodon neomexicanus*			X		
Western toad	*Bufo boreas*		X	X		
Chiricahua leopard frog	*Rana chiricahuensis*	X				
Relict leopard frog	*Rana onca*	X			X	X
Spotted frog	*Rana pretiosa*				X	
Reptiles						
Desert tortoise	*Gopherus agassizii*				X	X
Chuckwalla	*Sauromalus obesus*				X	
Gila monster	*Heloderma suspectum*			X	X	
Narrow-headed garter snake	*Thamnophis rufipunctatus*			X		

Future Needs

In the arid Southwest, plant and animal communities depend on the same scarce water resources as human populations, agriculture, and industry. Amphibians and some reptiles, such as garter snakes, are directly dependent on free-flowing water and aquatic habitats. Amphibians are of further concern because of recent, unexplained losses in many areas (Barinaga 1990; Blaustein and Wake 1990). A thorough understanding of the present status, population trends, and requirements of native species is essential to avoid or lessen conflicts among competing natural resource demands.

This ongoing project provides an assessment of our current knowledge, baseline information on distribution, and a starting point for intensive studies of rare and declining species. The development of an adequate inventory, coupled with long-term population studies of particular species of concern, forms the basis for informed protection and management of local natural communities.

For further information:

Charles Drost
National Biological Service
Colorado Plateau Research Station
PO Box 5614
Flagstaff, AZ 86011

References

Barinaga, M. 1990. Where have all the froggies gone? Science 247:1033-1034.

Blaustein, A.R., and D.B. Wake. 1990. Declining amphibian populations: a global phenomenon? Trends in Ecology and Evolution 5(7):203-204.

Carey, C. 1993. Hypothesis concerning the causes of the disappearance of boreal toads from the mountains of Colorado. Conservation Biology 7(2):355-361.

Hayes, M.P., and M.R. Jennings. 1986. Decline of ranid frog species in western North America: are bullfrogs (*Rana catesbeiana*) responsible? Journal of Herpetology 20(4):490-509

Miller, D.M., R.A. Young, T.W. Gatlin, and J.A. Richardson. 1982. Amphibians and reptiles of the Grand Canyon National Park. Monograph 4, Grand Canyon Natural History Association. 144 pp.

Stohlgren, T.J., and J.F. Quinn. 1992. An assessment of biotic inventories in western U.S. National Parks. Natural Areas Journal 12(3):145-154.

Woodbury, A.M., ed. 1959. Ecological studies of the flora and fauna in Glen Canyon. Anthropological Papers, University of Utah. 229 pp.

Wintering Bald Eagles Along the Colorado River Corridor

by

Charles van Riper III
Mark K. Sogge
National Biological Service
Timothy T. Tibbits
U.S. Fish and Wildlife Service

The construction and operation of reservoirs have had a dramatic influence on wintering and migrant bald eagles (*Haliaeetus leucocephalus*; Southern 1963; Spencer 1976; Steenhof 1978; Stalmaster 1987). In contrast to reservoir-induced destruction of riverine habitat upon which many wintering bald eagles have traditionally relied, reservoirs may harbor, in some instances, new or alternative food sources (Spencer 1976; Jenkins 1992). In addition to hunting the shorelines and surface waters of reservoirs, eagles congregate below some dams in winter to feed on fish that are killed or stunned while passing through the turbines or to hunt in ice-free water (Steenhof 1978).

Commercial river guides on the Colorado River first noted winter bald eagle concentrations on the southern Colorado Plateau below Glen Canyon Dam at Nankoweap Creek in the early 1980's (Fig. 1). Before this, bald eagles were considered uncommon along the Colorado River in Grand Canyon National Park (Brown et al. 1987). A preliminary study by Brown et al. (1989) concluded that wintering bald eagles had increased in numbers, particularly below Glen Canyon Dam, because of a combination of regulated discharge of cold water from the dam and the introduction of rainbow trout (*Oncorhynchus mykiss*). Although trout were introduced by the National Park Service into tributaries of the Colorado River in the 1920's, it was not until after the dam was completed (1963) that trout numbers increased in the Colorado River. By 1988 the mouth of Nankoweap Creek had become a concentration

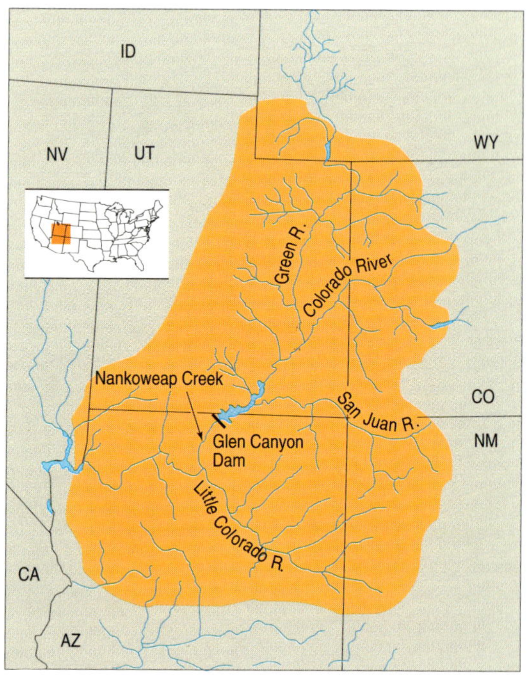

Fig. 1. The Colorado Plateau.

point for foraging because of the ease with which spawning trout could be obtained by eagles.

The concentrations of wintering and migrant bald eagles at Nankoweap are analogous to how eagles formerly concentrated at McDonald Creek in Glacier National Park, Montana (McClelland 1973). There, the introduction of non-native kokanee salmon (*Oncorhynchus nerka*) eventually attracted hundreds of migrant bald eagles (McClelland et al. 1982). The subsequent introduction of exotic zooplankton into Flathead Lake recently caused the collapse of this salmon population and ended the concentration of wintering eagles. In Grand Canyon National Park, it was felt that if the number of spawning trout remained high in the Colorado River tributaries, bald eagles might continue to concentrate there for food, as happened along McDonald Creek at Glacier National Park.

This article outlines the 1989-94 status of wintering bald eagles along the Colorado River corridor, from the Glen Canyon Dam through Grand Canyon National Park. We also discuss the trends of bald eagle numbers as determined from monitoring eagle and fish populations throughout the river corridor.

We determined the annual status of bald eagles from 1990 to 1994 by direct ground observations from the river bottom at the confluence of Nankoweap Creek and the Colorado River, and from aerial censusing flights from January through April.

Trends

Aerial Surveys

Wintering bald eagles were present each year along the Colorado River corridor from late fall (October-November) through early spring (March-April). During the 1990-91 aerial censusing surveys, peak numbers occurred in January and February, so aerial surveys in subsequent years were confined to December through March (Fig. 2). Eagles were observed on every flight, with numbers ranging from 2 (in March 1993) to 23 (in February 1991). Bald eagles were generally distributed evenly along the river corridor except in January and February, when conditions were suitable and rainbow trout were spawning in tributaries (Leibfried and Montgomery 1993). During these 2 months birds concentrated at the small tributaries.

Ground Surveys at Nankoweap

The trend of bald eagle numbers at Nankoweap Creek was for birds to closely parallel spawning trout numbers (Fig. 3). During 1990-91 we recorded the highest known bald eagle concentration at Nankoweap Creek with up to 26 eagles present on a peak day (Fig. 4). About 70-100 individuals were documented during the eagle concentration (when at least 10 eagles were present each day) from 8 February to 8 March 1990. The previous high of 18 wintering eagles was recorded at Nankoweap in February 1988 (Brown et al. 1989). The trend was for fewer numbers of trout and birds in following years (Fig 3). For example, in 1993 when spawning was extremely low in Nankoweap Creek, there were concomitantly low numbers of eagles. In 1994 spawning trout numbers were also low in the creek and few bald eagles were found in the area.

Other Areas of the Plateau

During 1992-94 when the numbers of wintering bald eagles along the Colorado River were low (Fig. 3), concentrations of bald eagles were reported at other locations on the southern Colorado Plateau. For example, in 1992, Bureau of Reclamation pilots (personal communication) noted eagle concentrations at the junction of the Green and Colorado rivers (Fig. 1). During 1993 the Arizona Game and Fish (personal communication) reported up to 20 eagles at Lake Mary, just east of Flagstaff, Arizona. These birds were feeding on some of the thousands of rainbow trout the agency had stocked into the lake during the winter. In 1994, another year of low bald eagle numbers along the Colorado River corridor, we received numerous reports from state and federal agency biologists of small eagle concentrations at elk and deer carcasses over the southern Colorado Plateau.

Status

The status of bald eagles along the Colorado River, especially in portions of Grand Canyon National Park and Glen Canyon National Recreation Area, has been improved by an increase in numbers of introduced rainbow trout. For example, at Nankoweap Creek, the trend went upward from a few birds starting in the mid-1980's to peak numbers in 1990-91. In following years (1992-94), poor rainbow trout spawning resulted in low numbers of bald eagles in this region. Creek morphology and flow conditions varied among years and influenced the availability of trout, and thus eagle numbers.

Bald eagles at Nankoweap, however, can be the largest such concentration in the southwestern United States. The 70-100 individual eagles recorded during 1990 represent what is believed to be one-fourth of the entire population of bald eagles wintering to the south of the Grand Canyon (in Arizona and northern Mexico). We expect that wintering eagles will continue to

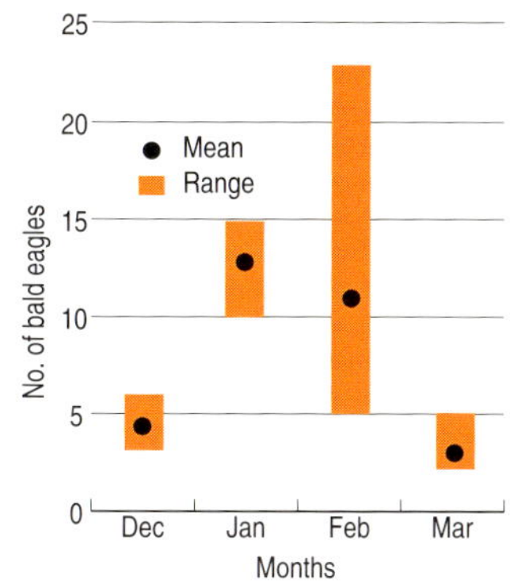

Fig. 2. Average number of bald eagles detected each month (1990-94) during aerial surveys along the Colorado River from the junction of the Little Colorado River north to Glen Canyon Dam.

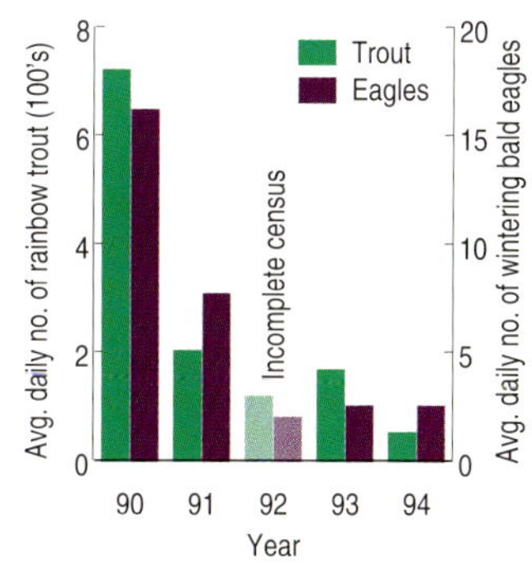

Fig. 3. Average daily numbers of rainbow trout and bald eagles at Nankoweap Creek in Grand Canyon National Park, AZ, 1990-94.

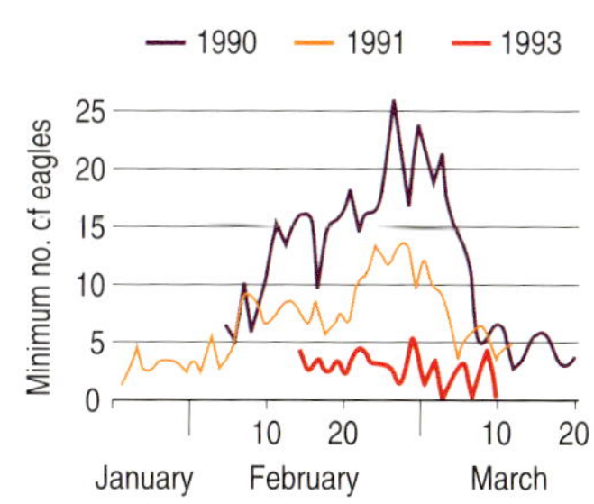

Fig. 4. Bald eagle abundance at Nankoweap in 1990, 91, and 93.

frequent this region if annual spawning trout are present.

Bald eagle counts along the Colorado River corridor during the winters of 1990-94 mirrored the bald eagle numbers at Nankoweap Creek. Their numbers peaked during late February and early March and varied greatly among years. Higher concentrations of bald eagles noted in other areas of the southern Colorado Plateau, when lower numbers were recorded along the Colorado River, suggest widespread eagle movements over the region. Bald eagles appear to concentrate in areas that have the most abundant and available food resources, and these locations change annually.

References

Brown, B.T, S.W. Carothers, and R.R. Johnson. 1987. Grand Canyon birds; historical notes, natural history, and ecology. University of Arizona Press, Tucson. 302 pp.

Brown, B.T., R. Mesta, L.E. Stevens, and J. Weisheit. 1989. Changes in winter distribution of bald eagles along the Colorado River in Grand Canyon, Arizona. Journal of Raptor Res. 23:110-113.

Jenkins, J.M. 1992. Ecology and behavior of a resident population of bald eagles. Ph.D. dissertation, University of California, Davis. 183 pp.

Leibfried, W.C., and W.L. Montgomery. 1993. Regulated flows, trout spawning, and abundance of bald eagles on the Colorado River, Grand Canyon National Park. Pages 37-48 *in* P.G. Rowlands, C. van Riper III, and M.K. Sogge, eds. Proceedings of the First Biennial Conference on Research in Colorado Plateau National Parks Transactions and Proceedings Series NPS/NRNAU/NRTP-93/10, U.S. Department of the Interior, Washington, DC.

McClelland, B.R. 1973. Autumn concentrations of bald eagles in Glacier National Park. Condor 75:121-123.

McClelland, B.R., L.S. Young, D.S. Shea, P.T. McClelland, H.L. Allen, and E.B. Spettigue. 1982. The bald eagle concentration in Glacier National Park, Montana: origin, growth, and variation in numbers. Living Bird 21:133-155.

Southern, W.E. 1963. Winter populations, behavior, and seasonal dispersal of bald eagles in northwestern Illinois. Wilson Bull. 75:42-55.

Spencer, D.A., ed. 1976. Wintering of the migrant bald eagle in the lower 48 states. National Agricultural Chemicals Association, Washington, DC. 170 pp.

Stalmaster, M.V. 1987. The bald eagle. Universe Books, New York. 227 pp.

Steenhof, K. 1978. Management of wintering bald eagles. U.S. Fish and Wildlife Service FWS/OBS-78/79. 59 pp.

For further information:

Charles van Riper III
National Biological Service
Colorado Plateau Research Station
PO Box 5614
Northern Arizona University
Flagstaff, AZ 86011

Mexican Spotted Owls in Canyonlands of the Colorado Plateau

by
David W. Willey
National Biological Service

In response to perceived threats to critical nesting habitat and lack of adequate protective regulations, the Mexican spotted owl (*Strix occidentalis lucida*) was officially listed as a threatened species under the Endangered Species Act in 1993 (Federal Register 1993). Limited information is available on the distribution of Mexican spotted owls inhabiting arid canyonlands throughout the southwestern United States (Ganey and Balda 1989). Though widely distributed, the Mexican spotted owl apparently occurs in isolated populations restricted to habitat islands (Fig. 1). Here I report findings from spotted owl surveys conducted throughout the northwest portion of the Colorado Plateau in Utah.

The Colorado Plateau Physiographic Province consists of extensive sandstone canyons interspersed by eroded valleys, upwarped plateaus, and isolated mountain ranges (Thornbury 1965). Prolonged erosional dissection produced a maze of complex watersheds within the Colorado Plateau region (Youngblood and Mauk 1985). Agency lands encompassed by the Colorado Plateau include extensive U.S. Department of Agriculture national forests and U.S. Department of the Interior Bureau of Land Management (BLM) areas, seven National Park Service national parks, two national recreation areas, several national monuments, and state-administered lands, all in the Four Corners region (Arizona, New Mexico, Colorado, and Utah) of the southwestern United States.

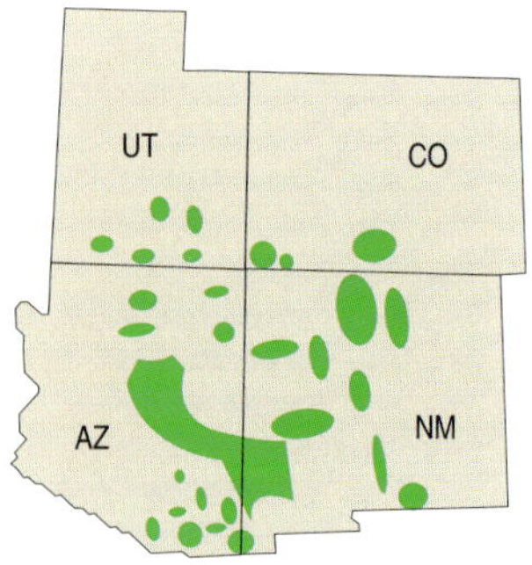

Fig. 1. Distribution of Mexican spotted owls in the southwestern United States.

These lands may function as biological refugia, providing dispersal corridors and habitat islands joining occupied and potentially suitable spotted owl habitat. In the Four Corners region, spotted owls are associated with rocky canyon terrain (i.e., canyonlands) and could be negatively affected by such activities as timber harvesting, mining, and recreation (Ganey 1988). Long-term study of spotted owl distribution and habitat use is necessary to provide information on the potential effects of human activities and to develop ecologically based conservation plans (Gutiérrez 1989).

Courtesy Steve Howe, Mills Publishing, Inc©

Mexican spotted owls (*Strix occidentalis lucida*) roosted in canyonlands in southern Utah.

Surveys

Information on Mexican spotted owl distribution within canyonlands of the Colorado Plateau was gathered by using published species accounts and conducting field surveys.

During the field surveys, individuals and

pairs of owls were located by imitating their calls with the human voice or using taped broadcasts of their calls to elicit a response from the owls (Forsman 1983). The surveys were conducted during each breeding season (1 April-31 August) from 1989 through 1993. Target areas were visited four times during the breeding season to search for owls. Spotted owl callers ("hooters") conducted searches by "hooting" at stations located on night-time survey routes placed within search areas. Hooters conducted daytime visits to sites where spotted owls were heard at night in order to find nests and count young.

Historical Records

Historical records of Mexican spotted owls on the Colorado Plateau date back to the 1920's (McDonald et al. 1990). The earliest record in the canyonlands was from Zion National Park in June 1928. A single owl was reported in August 1957, in Davis Gulch, a tributary of the Escalante River in southern Utah. Three birds were seen in July 1958, in a small side canyon of Glen Canyon National Recreation Area, and another was observed at the mouth of the Escalante River. The most northerly occurrence of a spotted owl on the Colorado Plateau was recorded in September 1958, in the Book Cliff Mountains. Spotted owls have been observed occasionally since the early 1970's throughout the canyonlands of southern Utah. Kertell (1977) detected spotted owls at six locations in Zion National Park in the early 1970's. The species accounts suggest that spotted owls were widely dispersed throughout the canyonlands of the Colorado Plateau, especially in deeply eroded sandstone gorges.

Field Survey Results

About 202,500 ha (500,000 acres) were surveyed from 1990 to 1993 on U.S. Forest Service lands, and more than 483 km (300 mi) of BLM canyons were surveyed from 1991 to 1993. Surveys were also conducted in portions of Grand Canyon, Capitol Reef, Canyonlands, and Zion national parks, as well as Natural Bridges and Navajo national monuments. Seventy-six spotted owls (26 pairs and 24 single adults) were detected at 50 locations: 6 on U.S. Forest Service lands, 12 on BLM lands, 1 on state lands, and 31 on National Park Service lands (Fig. 2).

Groups or subpopulations of owls were distributed among several landscape areas spread across the northwest portion of the Colorado Plateau including the greater Zion National Park area; the greater Capitol Reef area; the Dirty Devil River watershed; Canyonlands National Park; and near Elk Ridge and Dark Canyon on the Manti LaSal National Forest.

Mexican spotted owls were widely distributed and appeared coincident with canyon habitat. Canyon habitats on the Colorado Plateau are discontinuous and reflect the naturally fragmented topographic conditions of the plateau region. This patchy landscape could explain the patchy locations of surveyed spotted owls. A study conducted in Zion National Park found owls nesting and roosting in humid, narrow canyons with dense understories (Rinkevich 1991). Since many owls on the Colorado Plateau were found in similar habitat, the owls may be selecting these canyons because of their unique habitat features: large cliffs that provide escape cover to avoid predation, shaded roost sites to avoid high summer temperatures, patches of forest vegetation, and availability of suitable prey.

Relatively few owls were found in the canyonlands area compared with forest sites in Arizona and New Mexico; thus, canyonland owl sites may need special protection. Further surveys should be conducted across USDI lands to more accurately assess distribution and habitat of spotted owls.

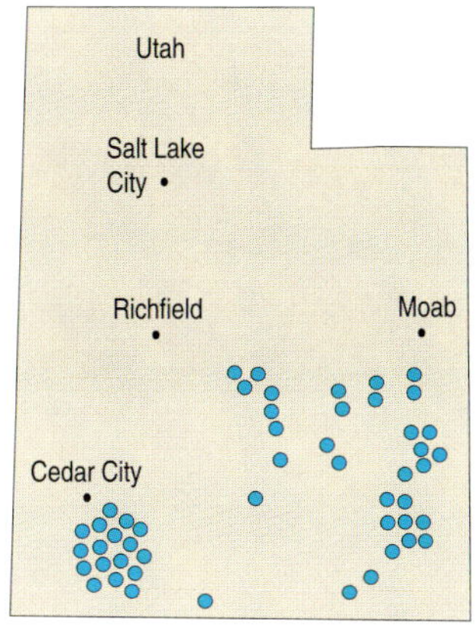

Fig. 2. Distribution of Mexican spotted owls in canyonlands of southern Utah, representing the northwest portion of the Colorado Plateau Physiographic Province.

References

Federal Register. 1993. Endangered and threatened wildlife and plants: final rule to list the Mexican spotted owl as a threatened species. Federal Register 58:14248-14271.

Forsman, E.D. 1983. Methods and materials for locating and studying spotted owls. Gen. Tech. Rep. PNW-162. U.S. Forest Service, Pacific Northwest Forest and Range Experiment Station, Portland, OR. 8 pp.

Ganey, J.L. 1988. Distribution and habitat ecology of Mexican spotted owls in Arizona. M.S. thesis, Northern Arizona University, Flagstaff. 229 pp.

Ganey, J.L., and R.P. Balda. 1989. Distribution and habitat use of Mexican spotted owls in Arizona. Condor 91:355-361.

Gutiérrez, R.J. 1989. Distribution and abundance of spotted owls in Zion National Park. National Park Service, Order PX-1200-9-C820. 69 pp.

Kertell, K. 1977. The spotted owl at Zion National Park. Western Birds 8:147-150.

McDonald, C.B., J. Andersen, J.C. Lewis, R. Mesta, A. Ratzliff, T.J. Tibbots, and S.O. Williams III. 1990. Mexican spotted owl status review. U.S. Fish and Wildlife Service, Endangered Species Rep. 20. Alberquerque, NM. 85 pp.

Rinkevich, S.E. 1991. Distribution and habitat characteristics of Mexican spotted owls in Zion National Park, Utah. M.S. thesis, Humboldt State University, Arcata, CA. 62 pp.

Thornbury, W. 1965. Regional geomorphology of the United States. John Wiley and Sons, New York. 609 pp.

Youngblood, A.P., and R.L. Mauk. 1985. Coniferous forest habitat types of central and southern Utah. Gen. Tech. Rep. INT-187. U.S. Forest Service, Intermountain Forest and Range Experiment Station, Ogden, UT. 89 pp.

For further information:

David W. Willey
National Biological Service
Colorado Plateau Research Station
Northern Arizona University
Box 5640
Flagstaff, AZ 86011

Bighorn Sheep in the Rocky Mountain National Parks

by
Francis Singer
National Biological Service

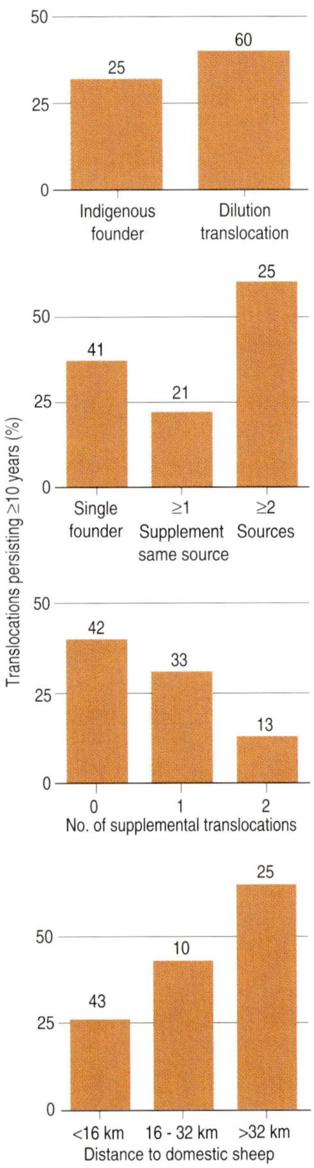

Figure. Factors contributing to persistence of bighorn sheep translocations. The numbers above the bars refer to the total number, or sample size, of populations in that category.

Current numbers of bighorn sheep (*Ovis canadensis*) are probably only 2%-8% of their numbers at the time of European settlement. The Rocky Mountain subspecies (*O.c. canadensis*) and the California subspecies (*O.c. californiana*) combined may have numbered roughly 1 million, and the desert subspecies (*O.c. nelsoni*) of the southwestern United States and Mexico also likely numbered about 1 million (Buechner 1960; Wishart 1978; Bleich et al. 1990). Unregulated harvesting, habitat destruction, overgrazing of rangelands, and diseases contracted from domestic livestock all contributed to drastic declines, the most drastic occurring from about 1870 through 1950.

Bighorn exist mostly in small, isolated populations within their former vast range. Thorne et al. (1985) found that 64% of 166 populations in the western United States contained fewer than 100 individuals. In Arizona, 88% of the populations (52 of 59) contained fewer than 100 individuals (Krausman and Leopold 1986).

Small populations of animals may be at higher risk of extirpation (Gilpin and Soulé 1986). The negative effects of small population size on bighorn were documented by Berger (1990), who reported that no indigenous populations of fewer than 50 animals survived for 5 decades, whereas all populations numbering more than 100 animals survived for the same period. Berger's (1990) published review did not consider national park populations of bighorns.

Restoration of bighorn sheep has been pursued actively by many state and federal agencies since the 1940's, although these efforts have met with only limited success, and most of the historical range of bighorns remains unoccupied. Human encroachments near bighorn populations are severe enough in some areas that the peninsular population of desert bighorns in California has been proposed for federal threatened status.

This article reviews the status of bighorns of three subspecies, the desert, Rocky Mountain, and badlands (*O.c. audobonii*), in 17 national park service units in the Rocky Mountains. Factors contributing to the success of 115 transplants of bighorn sheep that occurred over the past five decades in six Rocky Mountain states are also reviewed.

Information on the status and restoration of bighorns in the National Park Service units came from published accounts, university theses, unpublished park records, and a questionnaire mailed to state wildlife agencies and land managers in Colorado, Montana, North and South Dakota, Wyoming, and Utah. Only populations that had been translocated at least 10 years were included in the analysis.

Status in National Parks

Eighteen national park units historically contained populations of bighorn sheep. Native populations of bighorns were extirpated in all but five of the units, and populations were greatly reduced in four of these five. Only the Yellowstone's ranges remained fully occupied by bighorn during this period. Native bighorns survived but were greatly reduced in Grand Teton, Canyonlands, Glacier, and Rocky Mountain parks. The Badlands subspecies was eliminated about 1921. This subspecies originally inhabited clay badlands and low river breaks in the Dakotas, including Badlands and Theodore Roosevelt national parks.

Restoration efforts of bighorns into park units began in the late 1940's in 11 national park units where bighorns had been extirpated. Augmentations or translocations of additional subpopulations occurred in three of five national park units where bighorns had not been completely extirpated. Bighorn ranges are now considered fully or very fully occupied in two of these units, Rocky Mountain and Canyonlands parks.

Restoration of bighorns into other national park units has had only limited success. Ten national park units support persisting populations (numbering 100 or more sheep the previous 4 years); five park units have populations estimated to exceed 500 animals; and five other park units have populations of 100-200. Five other park units have fewer than 100 individuals, and two of these units support populations on the verge of extirpation (only 6-14 animals).

Translocations

Only 39% of 115 bighorn transplants in six Rocky Mountain states were rated as persisting (Figure). Sixty-four percent of transplants located more than 32 km (20 mi) from domestic sheep were persistent, but only 44% of those bighorn populations located 16 to 32 km from domestic sheep were persistent (Figure). In addition, nearly twice as many transplanted populations that were sedentary failed than populations that migrated to separate winter and summer ranges. Most translocated populations do not regain the historical migration patterns of the extirpated native population; instead, many spend the summer and winter on the same small ranges.

Limited evidence suggests that threshold population size or genetic diversity is related to persistence of bighorn transplants. Transplant persistence and genetic diversity were positively correlated to initial founder group size (the number of animals moved in the initial translocation), to multiple (versus single) source populations represented in the initial founder group, to the use of native populations as a source, and to sheep interaction with other nearby subpopulations (Fitzsimmons 1992).

Implications

Restorations into national park units are, as yet, incomplete. At present, bighorn sheep occur in small, widely scattered populations, with the smallest groups (fewer than 50 animals) seemingly at highest risk of extirpation. Thus, to achieve larger, more secure populations, restoration is necessary. To improve the chances for successful translocations, greater care must be taken; only about one-third of past translocations were persistent. Our analysis suggests that a population distant from domestic sheep improved the probability of its persistence more than any other factor. Larger founder sizes, multiple versus single sources of founders, native source groups, interactions with nearby subpopulations, and migratory tendencies also may contribute to continued persistence of translocated sheep and should be considered during translocations. Habitat suitability assessments before translocations would also probably contribute to sheep restoration success and are recommended as an integral part of any restoration.

References

Berger, J. 1990. Persistence of different-sized populations: an empirical assessment of extinction in bighorn populations. Conservation Biology 4:91-98.

Bleich, V.C., J.D. Weyhausen, and S.A. Holl. 1990. Desert-dwelling mountain sheep: conservation implications of a naturally fragmented distribution. Conservation Biology 4:383-390.

Buechner, H.K. 1960. The bighorn sheep in the United States, its past, present and future. Wildlife Monograph 4. 174 pp.

Fitzsimmons, N. 1992. Genetic factors, population histories, and horn growth in Rocky Mountain bighorn sheep. M.S. thesis, University of Wyoming, Laramie. 76 pp.

Gilpin, M.E., and M.E. Soulé. 1986. Minimum viable populations. Pages 19-34 *in* M.E. Gilpin and M.E. Soulé, eds. Conservation biology, the science of scarcity and diversity. Sinauer Association, Inc., Sunderland, MA.

Krausman, P.R., and B.D. Leopold. 1986. The importance of small populations of bighorns. Transactions North American Wildlife and Natural Resources Conference 51:52-61.

Thorne, E.T., W.O. Hickey, and S.T. Stewart. 1985. Status of California and Rocky Mountain bighorn sheep in the United States. Pages 56-81 *in* M. Hoefs, ed. Wild sheep: distribution and management and conservation of sheep in the world and closely related mountain ungulates. Special Report, Northern Wild Sheep and Goat Council, Whitehorse, Yukon.

Wishart, W. 1978. Bighorn sheep. Pages 161-171 *in* J. L. Schmidt and D.L. Gilbert, eds. Big game of North America. Stackpole Books, Harrisburg, PA.

For further information:

Francis Singer
National Biological Service
Natural Resource Ecology Laboratory
Colorado State University
Fort Collins, CO 80523

Desert Bighorn Sheep

by
Henry E. McCutchen
National Biological Service

Desert bighorn sheep (*Ovis canadensis* spp.) are subspecies of concern in the continental United States. Populations declined drastically with European colonization of the American Southwest beginning in the 1500's (Buechner 1960). At present, desert bighorn numbers are extremely low, although the overall population trend has increased since 1960.

Desert bighorn are considered good indicators of land health because the species is sensitive to many human-induced environmental problems (McCutchen 1981). In addition to their aesthetic value, desert bighorn are considered desirable animals by hunters.

The Rocky Mountain and California races of bighorn occupy the cooler western and northwestern regions of the United States. In contrast, the desert sheep races are indigenous to the hot desert ecosystems of the Southwest.

Population Trends

The number of desert bighorn in North America in pristine times is unknown but most likely was in the tens of thousands (Buechner 1960). Seton (1929) estimated the pre-Columbian numbers of all subspecies of bighorn in North America (United States, Canada, and Mexico) at 1.5-2 million. By 1960, however, the overall bighorn population in the United States, including desert bighorns, had dwindled to 15,000-18,200 (Buechner 1960). Buechner documented major declines from the 1850's to the early 1900's. These declines were attributed to excessive hunting; competition and diseases from domestic livestock, particularly domestic sheep; usurpation of watering areas and critical range by human activities; and human-induced habitat changes (Buechner 1960; Graham 1980; McCutchen 1981).

These declines were followed by a period of population stabilization that Buechner ascribed to conservation measures. The decline of desert bighorn probably mirrored the pattern of decline of the overall bighorn population. Desert bighorn population trends have been upward since the 1960's when Buechner (1960) estimated their population at 6,700-8,100. In

Courtesy H.E. McCutchen

Desert bighorn sheep (*Ovis canadensis nelsoni*).

1980 desert bighorn populations were estimated at 8,415-9,040 (Wishart 1978). Weaver (1985) conducted a state-by-state survey a few years later and estimated the U.S. desert bighorn population at 15,980. The 1993 estimate of the population is 18,965-19,040 (Table).

Table. Status and trends of desert bighorn sheep in the United States 1960-93. Estimate for 1960 by Buechner (1960). Estimates for 1993 from state wildlife agency status reports presented to the Desert Bighorn Council, Moab, Utah, 1994.

State	Population estimate by year 1960	1993
Arizona	3,000-3,500	6,000
California	2,140-2,450	4,300-4,325*
Colorado	0	475
Nevada	1,500-2,000	5,294
New Mexico	400-500	295
Texas	25	401
Utah	Remnant	2,200-2,250
Total	**7,065-8,475**	**18,965-19,040**

*In California, Nelson's bighorn (*Ovis canadensis nelsoni*) population trends are upward. Peninsular bighorn (*O.c. cremnobates*) populations are lower and are of concern.

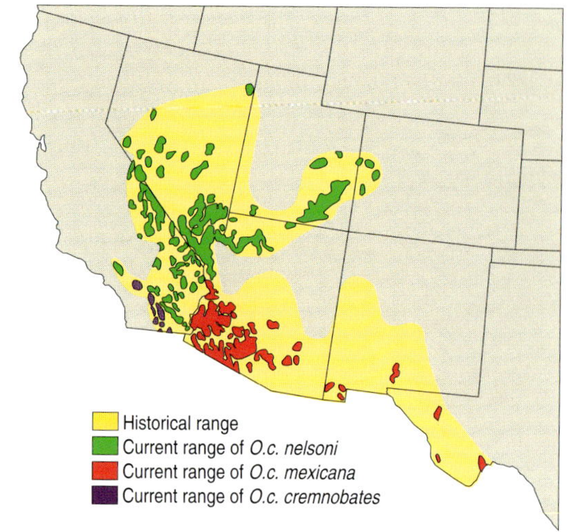

Figure. Historical range and current distribution of the three subspecies of desert bighorn in the United States (redrawn from Trefethen 1975 and Weaver 1985).

Subspecies

Cowan (1940) used morphological characters and measurements to identify three subspecies of desert bighorns (*O.c. nelsoni*, *O.c. mexicana*, and *O.c. cremnobates*) occurring in the United States. A recent reevaluation of mountain sheep races in the United States, however, suggested significant differences between the northern and southern (desert) sheep (Ramey 1993). Differences among the three desert bighorn races, however, did not support separate subspecies designations.

The distribution of desert bighorn races is uncertain, although the distribution maps of Trefethen (1975) and Weaver (1985) are accepted by mountain sheep biologists (Figure).

Status and Trends by State

Arizona

Historically, desert bighorn occurred on all mountain ranges and plateau slopes in the southern, northern, and western sections of Arizona (Russo 1956). In spite of early protection (beginning in the 1880's), researchers believed that bighorn populations declined until the 1950's (Russo 1956).

Arizona began a limited hunting program in 1953 and reintroduction programs in 1958. The Arizona Game and Fish Department conducts annual helicopter surveys. Buechner (1960) estimated the 1960 population at about 3,000-3,500. In 1993 the population had increased to an estimated 6,000 (R. Lee, Arizona Game and Fish Department, unpublished data).

California

Desert bighorn occupied desert mountains in southeast California in historical times. California protected bighorn in 1883, and by 1960 Buechner (1960) estimated the population at about 2,150-2,450 (1,800-2,100 *O.c. nelsoni* and 350 *O.c. cremnobates*). The state began transplanting in 1971 and permitted hunting beginning in 1986 (Bleich et al. 1990). In 1993 the populations were estimated at 4,300-4,325, with the breeds occupying about 50 mountain ranges (S. Torres, California Department of Fish and Game, unpublished data).

The less common peninsular bighorn (*O.c. cremnobates*) occurs in the desert mountains of southeast California from Palm Springs south to the Mexican border. From 1977 to 1993 this population declined from an estimated 1,171 to 400-425 individuals because of excessive lamb mortality (Weaver 1989; S. Torres, California Department of Fish and Game, unpublished data). In 1992 the U.S. Fish and Wildlife Service proposed listing the peninsular bighorn as endangered (Torres et al. 1993). This subspecies also occurs southward into Mexico; populations there are larger. One survey estimated a population of 780-1,170 adult bighorn in northern Baja California, Mexico (DeForge et al. 1993).

Colorado

There is no scientific evidence that desert bighorn occurred historically in Colorado, although there is habitat in the state contiguous with desert bighorn habitat in Utah. Thus, desert bighorn probably occurred in the state, and became extirpated before subspecies' determinations could be made.

The Colorado Division of Wildlife began transplanting desert bighorn in 1979. By 1993 populations containing approximately 475 bighorn had been established from the release of animals originally from Arizona and Nevada (Wolfe 1990; V. Graham, Colorado Division of Wildlife, unpublished data).

Nevada

Desert bighorn (*O.c. nelsoni*) historically occupied the central and southern portions of Nevada (McQuivey 1978). Hunting the animals was prohibited from 1901 to 1952. Transplanting programs have been successful: between 1968 and 1988 more than 800 desert bighorn were transplanted. From these animals, 21 transplanted herds have been established (Delaney 1989).

Buechner (1960) estimated the Nevada population at 1,500-2,000 in 1960. The state began annual population trend counts in 1969. In 1993 the population was estimated at 5,294 animals, occupying 45 mountain ranges (P. Cummings, Nevada Division of Wildlife, unpublished data).

New Mexico

Although desert bighorns (*O.c. mexicana*) historically occupied mountain ranges and canyons in the southern part of New Mexico, by 1930 the animals were restricted to only four mountain ranges, and by the late 1940's were found in only two (Weaver 1985).

In 1972 the state constructed the 300-ha (741 acres) Red Rock propagating enclosure and added brood stock. Transplants from the captive herd were subsequently made into the Big Hatchet, Peloncillo, and Alamo Hueco mountains (Sandoval 1979).

The San Andres Mountain population was formerly the state's largest, but declined from 200 to fewer than 25 by 1991 (Clark and Jessup 1992) because of psoroptic scabies (*Psoroptes* spp.).

Buechner estimated the New Mexican population at 400-500 in 1960. In 1993 the estimated population was 295, of which 100 were at Red Rock (A. Fisher, New Mexico Department of Game and Fish, unpublished data).

Texas

Desert bighorn (*O.c. mexicana*) appear to have occupied all the mountains in southwest Texas west of the Pecos River (Carson 1941). In 1880 the population was estimated at 1,500 animals (Kilpatric 1982); some populations still existed in the late 1930's. By the mid-1950's all bighorns had become extirpated except for a small herd of 25; excessive hunting and competition with domestic livestock are believed to have been major factors in the final decline (Buechner 1960).

In 1957 the Texas Game and Fish Department began a highly successful captive breeding and release program. By 1993 the free-ranging population was estimated at 310; 91 other sheep were in captivity (G. Calkins, Texas Parks and Wildlife Department, unpublished data).

Utah

Historically, desert bighorn (*O.c. nelsoni*) occupied canyons and ranges in southern and eastern Utah. Significant population declines occurred in the 1870's (Buechner 1960), and the state did not permit hunting of bighorn from 1899 to 1967.

In 1967 limited hunting began, and in 1973 the state started an active transplant program. Between 1973 and 1990, over 250 desert bighorn sheep were transplanted, establishing at least nine populations that augment four additional areas containing native populations (Cresto et al. 1990).

Buechner (1960) believed that only remnant populations persisted in the state. Utah, which has conducted aerial trend counts on bighorn since 1969 (Cresto et al. 1990), documented increasing populations statewide. Individual populations, however, have exhibited large increases and sudden declines. In 1993 the desert bighorn population was estimated at 2,200-2,250 (N. McKee and J. Karpowitz, Utah Division of Wildlife, unpublished data).

Future of Desert Bighorn

Since 1960 bighorn have increased in numbers, but their population levels are still low when compared with the estimates of pre-European numbers and the amount of available unoccupied habitat. The number of sheep in individual populations has fluctuated greatly. Population monitoring and efforts to restore desert bighorn must continue to ensure viable future populations.

References

Bleich, V., J.D. Weyhausen, K.R. Jones, and R.A. Weaver. 1990. Status of bighorn sheep in California, 1989 and translocations from 1971 through 1989. Desert Bighorn Council Transactions 34:24-26.

Buechner, H.K. 1960. The bighorn sheep in the United States, its past, present and future. Wildlife Monograph 4. 174 pp.

Carson, B. 1941. Man, the greatest enemy of desert bighorn mountain sheep. Bull. 21, Texas Game, Fish and Oyster Commission. Austin, TX. 23 pp.

Clark, R.K., and D.A. Jessup. 1992. The health of mountain sheep in the San Andres Mountains, New Mexico. Desert Bighorn Council Transactions 36:30-35.

Cowan, I. McT. 1940. Distribution and variation in the native sheep of North America. American Midland Naturalist 24:505-580.

Cresto, J., J. Karpowitz, and L. Seibert. 1990. Status of desert bighorn sheep in Utah, 1989. Desert Bighorn Council Transactions 34:31-32.

DeForge, J.R., S.D. Osterman, D.E. Toweill, P.E. Cyrog, and E.M. Barrett. 1993. Helicopter survey of Peninsular bighorn sheep in Northern Baja California. Desert Bighorn Council Transactions 37:24-28.

Delaney, D. 1989. Status of desert bighorn sheep in Nevada, 1988. Desert Bighorn Council Transactions 33:13-14.

Graham, H. 1980. The impact of modern man. Pages 288-309 *in* G. Monson and L. Sumner, eds. The desert bighorn, its life history, ecology and management. University of Arizona Press, Tucson.

Kilpatric, J. 1982. Status of desert bighorn sheep in Texas, 1982. Desert Bighorn Council Transactions 26:102-104.

McCutchen, H.E. 1981. Desert bighorn zoogeography and adaptation in relation to historic land use. Wildlife Society Bull. 9:171-179.

McQuivey, R.P. 1978. The desert bighorn sheep of Nevada. Nevada Department of Wildlife Biology Bull. 6:1-81.

Ramey, R.R., II. 1993. Evolutionary genetics and systematics of North American mountain sheep: implications for conservation. Ph.D. dissertation, Cornell University. 250 pp.

Russo, J.P. 1956. The desert bighorn in Arizona. Arizona Game and Fish Department Wildlife Bull. 1:1-153.

Sandoval, A.V. 1979. Bighorn sheep status report from New Mexico. Desert Bighorn Council Transactions 23:82-87.

Seton, E.T. 1929. Lives of game animals. Vol. 3. Doubleday Page and Co., New York. 780 pp.

Torres, S., V.C. Bleich, and A. Pauli. 1993. Status of bighorn sheep in Nevada, 1992. Desert Bighorn Council Transactions 37:54-55.

Trefethen, J.B., ed. 1975. The wild sheep in modern North America. Boone and Crocket Club. Winchester Press, New York. 302 pp.

Weaver, R.A. 1985. The status of the desert bighorn in the United States. Pages 82-85 *in* Manfred Hoefs, ed. Wild sheep, distribution, abundance, management and conservation of the sheep of the world and closely related mountain ungulates. Northern Wild Sheep and Goat Council Special Report. Yukon Wildlife Branch, Whitehorse, Yukon, Canada.

Weaver, R.A. 1989. Status of bighorn sheep in California, 1988. Desert Bighorn Council Transactions 33:11-12.

Wishart, W. 1978. Bighorn sheep. Pages 160-171 *in* J.L. Schmidt and D.L. Gilbert, eds. Big game of North America, ecology and management. Stackpole Books, Harrisburg, PA.

Wolfe, J. 1990. Status of bighorn sheep in Colorado, 1989. Desert Bighorn Council Transactions 34:27.

For Further Information:

Henry E. McCutchen
National Biological Service
Colorado Plateau Research Station
Northern Arizona University
PO Box 5614
Flagstaff, AZ 86011

Courtesy S.C. Amstrup, NBS

Alaska

by
William K. Seitz
National Biological Service
Alaska Science Center
1011 E. Tudor Rd.
Anchorage, AK 99503

Overview

The Alaskan ecoregion has many immense, mostly pristine ecosystems including marine waters and islands; the Arctic Coastal Plain and the Brooks Range; taiga forests and interior rivers; the extensive, treeless lowlands and deltas of the Yukon and Kuskokwim rivers; the rugged coastline with glacier-capped mountains and numerous fjords and tidewater glaciers; and coastal rain forests, bogs, and alpine tundra communities on numerous islands. This section highlights the status and trends of selected mammals and fish that inhabit these pristine ecosystems. Waterfowl, shorebirds, and seabirds are discussed in a separate chapter.

Caribou (*Rangifer tarandus*), muskox (*Ovibos moschatus*), and large mammalian predators such as the gray wolf (*Canis lupus*) and brown bear are vital components in the coastal plain tundra of the Arctic National Wildlife Refuge. All mammal populations on the refuge appear stable and healthy (McCabe et al., this section). Since 1989 the internationally shared (Canada and the United States) Porcupine caribou herd, which uses the narrow coastal plain for calving in June and July, has remained near 160,000 animals. The resident muskox population, reintroduced after being hunted to extinction in the late 1800's, now numbers nearly 720. Almost 100 brown bears (*Ursus arctos*) and 43 wolves live on the north slope of the refuge in relatively stable populations.

Arctic fisheries, of little significance in terms of commercial harvest and economic value, constitute a significantly large, locally important contribution to rural economies and provide valuable food for Alaskan Natives. Thorsteinson and Wilson document the status of Arctic cisco (*Coregonus autumnalis*), broad whitefish (*C. nasus*), least cisco (*C. sardinella*), and Dolly Varden char (*Salvelinus malma*) in the nearshore Beaufort Sea north of Prudhoe Bay.

Pacific salmon have always played a major role in the history and economy of Alaska and its commercial, sport, and subsistence fisheries. Burger and Wertheimer (this section) analyze historical and recent salmon harvest information to explore status and trends of Pacific salmon in Alaska. Total salmon harvest in Alaska was estimated at 56,000 salmon in 1878, but rose to over 21 million in 1900. After substantial population declines in the 1920's, 1960's, and 1970's, harvests in most Alaskan populations rebounded, and populations are healthy. Only populations of pink salmon (*Oncorhynchus gorbuscha*) in Prince William

Sound and chum salmon (*O. keta*) in the Kuskokwim River in western Alaska are experiencing major declines and need attention.

There is a long history of biological studies in Denali National Park and Preserve. Wolves, caribou, brown bears, moose (*Alces alces*), and Dall sheep (*Ovis dalli*) all live in this large ecosystem. The park provides scientists the opportunity to study the natural interactions of these species and serves as a baseline for comparison with areas where hunting occurs. Adams and Mech (this section) document the natural fluctuations expected in species inhabiting such a dynamic and variable environment.

Brown bears on the Kodiak Archipelago are renowned for their large size and dense aggregations along salmon-spawning streams. Barnes et al. (this section) estimate a population of more than 2,800 bears on the archipelago. Through intensive management by Alaska and the U.S. Fish and Wildlife Service, the status of the Kodiak bear population is better now than in the early 1900's.

Populations of the three marine mammals for which the Department of the Interior has management authority—polar bears (*Ursus maritimus*), Pacific walrus (*Odobenus rosmarus divergens*), and sea otters (*Enhydra lutris*)—are healthy. The estimated population of polar bears along Alaska's north coast and the Beaufort Sea is nearly 2,000 and probably larger compared to the early 1900's (Amstrup et al., this section).

About 250 years ago, more than several hundred thousand sea otters were continuously distributed from Baja California, north and west along the Pacific Rim to Kamchatka, and south along the Kuril Islands to northern Japan. When the Russian fur harvest was halted in 1911, only a few surviving colonies, likely numbering a few hundred animals or less, remained. Now, Bodkin et al. (this section) estimate more than 100,000 sea otters living throughout about 75% of their original range, illustrating the healthy recovery of a species after protection and active management.

Pacific walruses in the Bering and Chukchi seas of Alaska and Russia are an important source of meat and ivory for Native peoples of Alaska and the Chukotka Peninsula of Russia (Garner, this section). These marine mammals are also a highly visible indicator of the health of the Arctic marine ecosystem. Cooperative U.S.-Russia surveys conducted at 5-year intervals since 1975 provide estimates ranging from 246,000 walruses in 1980 to 200,000 in 1990. Even though the survey estimates have large confidence intervals, some researchers believe these surveys indicate a general decline in numbers between 1975 and 1990.

The Mentasta caribou herd, a small herd that lives in and around Wrangell-St. Elias National Park and Preserve, exhibits typical population trends and management problems found in many mountain herds in central Alaska and the Yukon Territory of Canada. This herd increased from about 2,000 caribou in the early 1970's to 3,200 in the early 1980's (Jenkins, this section). From 1989 to 1993, the herd decreased to 900 caribou, about a 24% decrease per year.

Klein (this section) documents the distribution and abundance of the tundra or Arctic hare (*Lepus timidus*) in western Alaska. The Arctic hare has long been used for food and clothing by indigenous people living in western Alaska. Arctic hares have declined in number throughout much of their range, though biologists are not sure why.

The Arctic Tundra Ecosystem in Northeast Alaska

by
Thomas R. McCabe
National Biological Service
Donald P. Garrett
U.S. Fish and Wildlife Service
Don Russell
Canadian Wildlife Service
Ken Whitten
Alaska Department of Fish and Game

The tundra of the coastal plain of the Arctic National Wildlife Refuge (ANWR; Fig. 1) represents nearly pristine, intact Arctic ecosystem. It is unique because of the close arrangement of the plants and animals occurring between the Brooks Mountains and the Beaufort Sea (Fig. 2). The Porcupine caribou (*Rangifer tarandus*) herd (PCH), which ranges between Canada and Alaska, uses the narrow Coastal Plain for calving after migrating hundreds of kilometers from its winter habitat. A now healthy muskox (*Ovibos moschatus*) population was reintroduced in 1969 after being hunted to extinction in the late 1800's. Large predators including gray wolves (*Canis lupus*), brown bear (*Ursus arctos*), and golden eagles (*Aquila chrysaetos*) are also important components as well as sensitive measures of ecosystem health.

Extensive cooperative U.S. and Canadian biological research has occurred on the Coastal Plain during the last decade because it overlies a potentially large and economically productive oilfield. The biological information resulting from these cooperative efforts will guide Congress in its decision to develop the oilfield. The information also provides an excellent measure of the status and trends of key animals in a near-pristine Arctic ecosystem (McCabe et al. 1992).

Monitoring the Ecosystem

We monitored the status and trends of caribou, muskox, and large predators to enhance our understanding of the important relationships of the Arctic ecosystem and to identify and predict the potential impacts of oil and gas development on that system.

Caribou

We periodically photographed and censused the PCH in July from 1972 to 1992. To determine when the caribou were optimally aggregated for photographing, we monitored the formation and distribution of large postcalving aggregations by using intensive aerial reconnaissance, radio tracking, and satellite telemetry.

We estimated the sex and age structure of the PCH during the postcalving aggregation period from aerial and ground counts in 1988, 1989, 1990, and 1992. We estimated annual survival of cows and calves by using aircraft and satellites to periodically track a sample of animals fitted with radio transmitters.

Muskox

We closely monitored the distribution, composition, and size of the muskox population by using radio transmitters, tags, and intensive aerial and ground surveys. We surveyed muskox at 4-8 week intervals in 1987-93 to determine their locations. An average of eight flights per year were flown. We conducted no flights from late November to late January because of severe winter weather and low light conditions.

We also determined sex and age composition from the ground counts. We conducted total counts of the population annually from 1972 to 1993 during aerial surveys and ground counts.

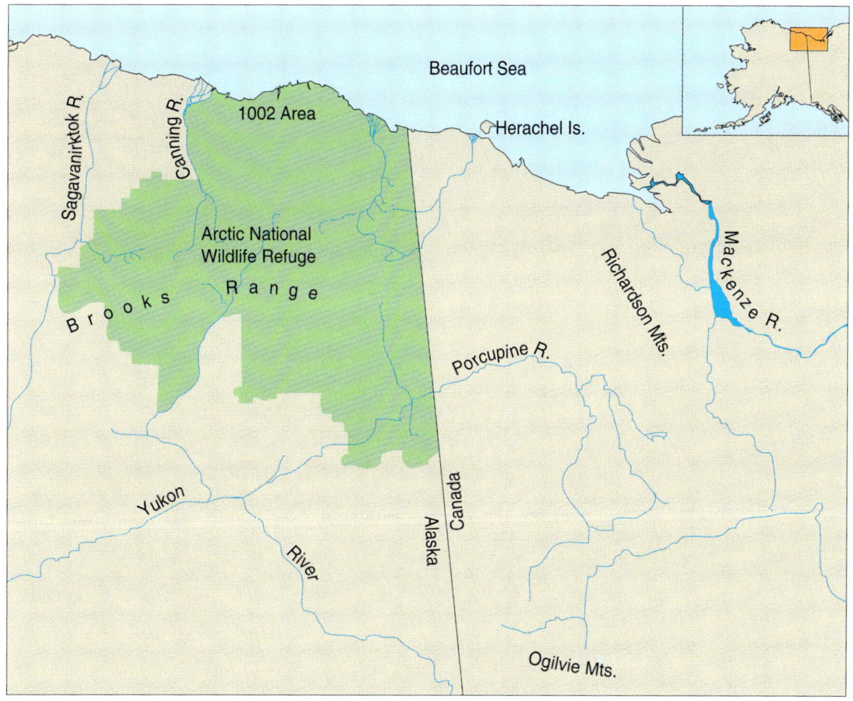

Fig. 1. Arctic National Wildlife Refuge in northeastern Alaska.

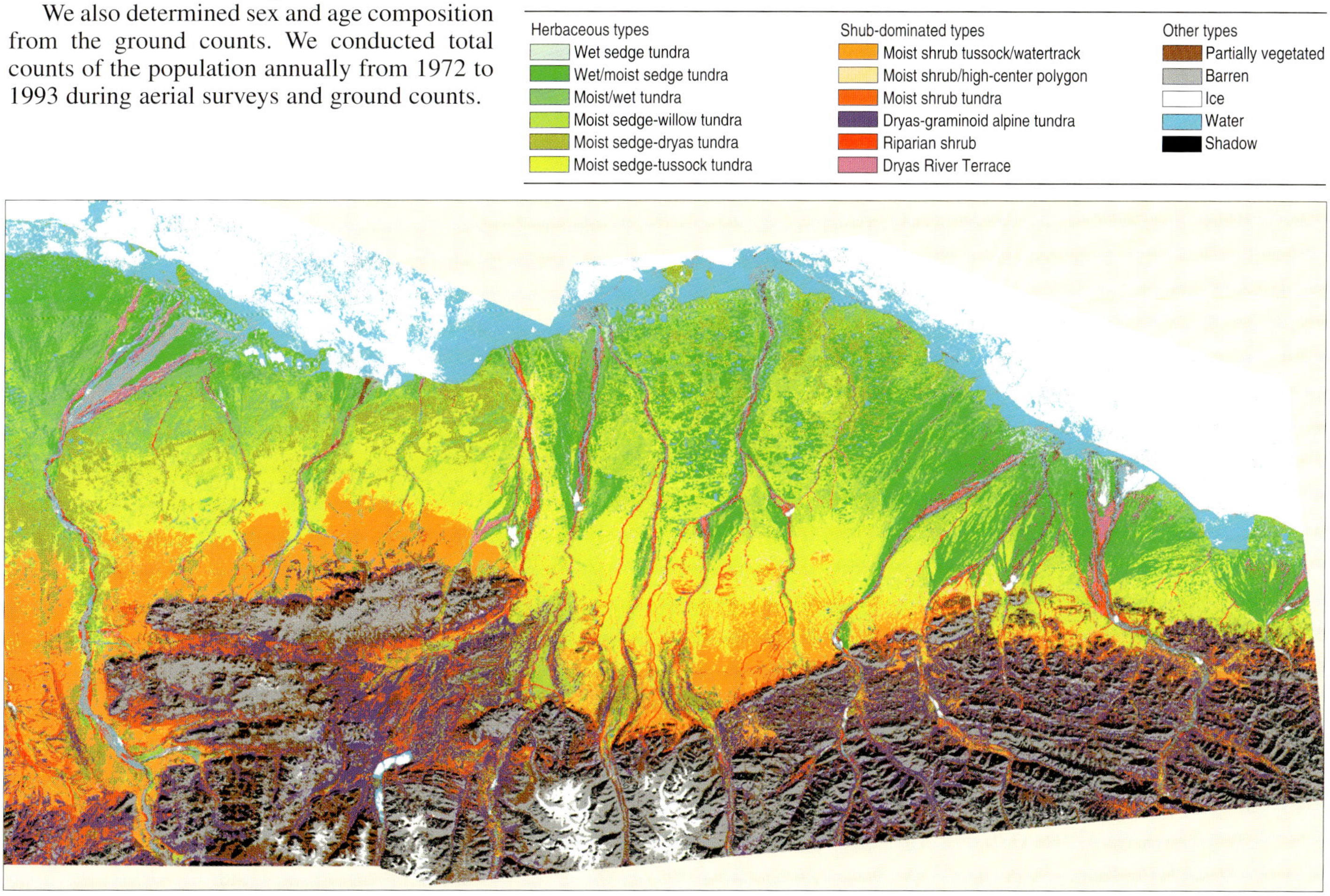

Fig. 2. Land-cover classes on the Arctic National Wildlife Refuge in northeastern Alaska.

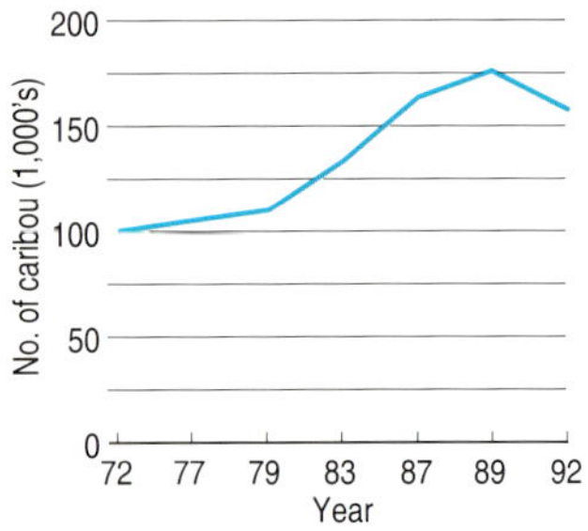

Fig. 3. Photocensus results for the Porcupine caribou (*Rangifer tarandus*) herd, 1972-92.

Predators

We determined the number of brown bears on the Coastal Plain portion of the refuge from densities recorded during extensive aerial surveys in 1983. Subsequent trends in the population were based on composition counts and survival estimates obtained from monitoring radio-tagged bears.

We located wolf dens and packs by monitoring radio-tagged animals and aerial surveys. In 1984, we made a minimum estimate of the population by recognizing individual wolves. We based trends in pack size and composition on ground observations collected at the den site.

We completed aerial surveys of golden eagle nest sites twice each year from 1988 to 1990 to monitor trends in nest occupancy and nestling production.

Status of the Arctic Ecosystem

Caribou

The PCH increased from an estimated 100,000 animals in 1972 to peak at 178,000 in 1989, then dropped to 160,000 in 1992 (Fig. 3). The growth rate averaged 4.8%/year from 1979 to 1989. Since 1989 the population has either stabilized or declined. Ratios of calves to 100 cows ranged from a low of 38 in 1971 to a high of 73 in 1983. This trend in herd productivity generally agrees with the trends in population growth.

We observed no consistent trends in the estimates of annual survival of adult females. The population dynamics of the PCH are similar to the longer term cycles observed in other barren-ground caribou herds.

Muskox

The muskox population on the Coastal Plain increased an average of 20%/year (Fig. 4). After 1986 numbers of muskox in ANWR decreased and then stabilized at about 350 animals, and numbers of muskox east and west of the refuge increased. In 1993 we observed 720 muskox, including 370 on the ANWR Coastal Plain.

Annual productivity for the muskox population on ANWR has averaged about 48 calves per 100 cows since 1985. In the highly productive years of 1984, 1985, and 1988, calf-to-cow ratios were greater than 70:100 and calves accounted for more than 21% of the total population. Age at death for five known-age cows averaged 13.8 years (range: 9-19) and annual survival averaged 88% for adult cows and 77%-78% for yearlings and calves. Changes in distribution occurred during years following winters in which biologists observed lower productivity and survival of young animals and adult cows. The dynamics and behavior of the population are typical of animals reintroduced into suitable habitat.

Predators

In 1983 we estimated that there were 108 brown bears on the north slope of ANWR. Between 1983 and 1993 estimates of survival and reproductive rates of the bear population were stable and distribution and movements of bears were consistent. This consistency suggests that the bear population is stable.

In 1984 we estimated that a minimum of 34 wolves occurred on the north slope of ANWR. A mean litter size of 4.2 during 1988-90 was consistent with the 3.0 reported for ANWR in 1984 and 4.3 in 1985. Population size appears stable.

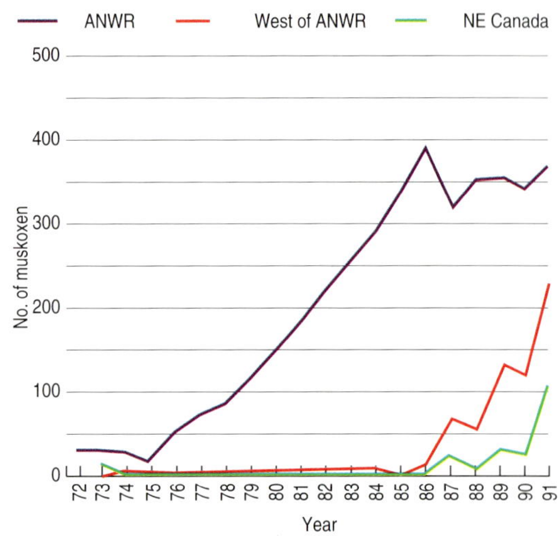

Fig. 4. Growth and stabilization of the pioneering muskox (*Ovibos moschatus*) population within the Arctic National Wildlife Refuge and adjoining areas.

Between 1988 and 1990 we observed 31 nesting attempts by golden eagles on the north slope of the ANWR between the Canning and Kongakut rivers. Twenty-seven of the 31 (87%) breeding pairs produced 33 young, resulting in 1.22 young per successful pair. The number of young remained constant from 1988 to 1990, suggesting that the ANWR eagle population is also stable.

Reference

McCabe, T.R., D.B. Griffith, N.E. Walsh, and D.D. Young. 1992. Terrestrial research: 1002 Area. Arctic National Wildlife Refuge. Interim Report, 1988-1990. U.S. Fish and Wildlife Service, Anchorage, AK. 432 pp.

For further information:

Thomas R. McCabe
National Biological Service
Fairbanks Field Station
Box 15
101 12th Ave.
Fairbanks, AK 99701

Anadromous Fish of the Central Alaska Beaufort Sea

by
Lyman K. Thorsteinson
National Biological Service
William J. Wilson
LGL Alaska Research Associates, Inc.

Today, more than 25 years after the discovery of oil in Prudhoe Bay, it is hydrocarbon resources, and not fish or wildlife, that most Americans equate with Alaska's North Slope. From national and statewide perspectives, Arctic fisheries are of little significance in terms of total landings and economic worth. But small, northern fisheries contribute to rural economies and provide necessary sources of protein for Alaska's Native people. As such, the welfare of exploited fish populations and protection of regional lifestyles are dominant environmental and sociological themes associated with the industrialization of Arctic coastlines. Fully one-third of adult Inupiat Eskimos participate in subsistence fisheries. They capture about 96,000 kg (210,000 lb) of fish annually, an amount that rivals the yearly Native harvest of bowhead whales. Commercial fishermen harvest another 40,000 fish in a fall-winter fishery.

The continued development of the Prudhoe Bay oilfields in Alaska required the construction of two solid-fill gravel causeways (West Dock in 1974-75, extended in 1976 and 1981; Endicott Causeway in 1984-85) extending several kilometers offshore (Fig. 1). These causeways can cause transient changes in local fish habitat. Biologists are concerned that fish populations may be negatively affected when causeway-induced changes in habitat quality, quantity, or availability combine with regional fishery removals. Because nearshore water circulation is wind-driven, these changes vary with wind speed, direction, and duration.

Arctic cisco (*Coregonus autumnalis*), broad whitefish (*C. nasus*), least cisco (*C. sardinella*), and Dolly Varden char (*Salvelinus malma*) are the fish of primary concern. These anadromous species have life cycles that include annual migrations from winter habitats in fresh water to summer feeding habitats in salt water.

Summer habitats are in coastal environments, which are vulnerable to industrial developments. The species have adapted to Arctic conditions through strategies that promote their welfare, including complex migrations, variable freshwater rearing periods, being long-lived with late maturity, and having low recruitment rates.

Fish Monitoring

Inventories of fish habitats, populations, and fisheries in the Alaska Beaufort Sea began in earnest during the mid-1970's. The construction of the West Dock and Endicott causeways required environmental monitoring and other research to evaluate the effects of these structures. The study area included Prudhoe Bay and 120 km (75 mi) of adjacent coastline between the Colville and Sagavanirktok river deltas (Fig. 1).

Biologists initiated fish-monitoring studies around causeways in 1981. For monitoring they incorporated common fishery techniques used to estimate population health and size (Norton 1989; Benner and Middleton 1991). Their sampling included live captures of fish along the coast, standard biological measurements, and physical assessments of fish habitat. They conducted their fieldwork between June and mid-September. Biologists have also compiled annual fishery statistics from the Colville River since 1967.

We examined three data sets: season-averaged catch rates; season-long estimates of population size from mark-recapture studies; and effort-adjusted catch rates and total harvests from the commercial fishery. Because sampling effort varied each year, we derived coastal indices of abundance from five permanent stations established in 1985. We based our counts of small Arctic ciscoes and broad whitefish on all available catch records.

We defined groups of fish of the same species that comprise the same age or size ranges (called cohorts). For Arctic cisco and broad whitefish: cohort I—age 0 (young-of-the-year), cohort II—age 1, cohort III—ages 2 and 3, and cohort IV—age 4 or older. For least cisco: cohort I less than 180 mm (7.1 in) long and cohort II at least 180 mm (7.1 in) long. For Dolly Varden char: cohort I—less than 350 mm (13.8 in) long, and cohort II—at least 350 mm (13.8 in) long.

Arctic Fish Species

Arctic Cisco

Biologists believe that Arctic cisco inhabiting the central Alaskan Beaufort Sea originate in the

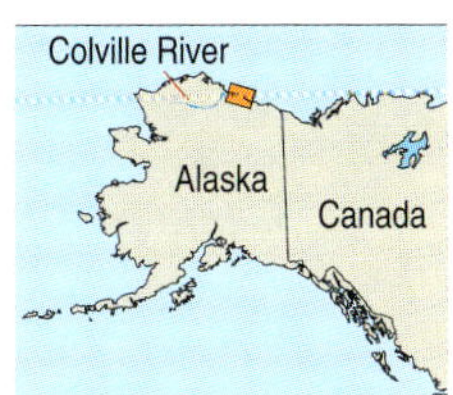

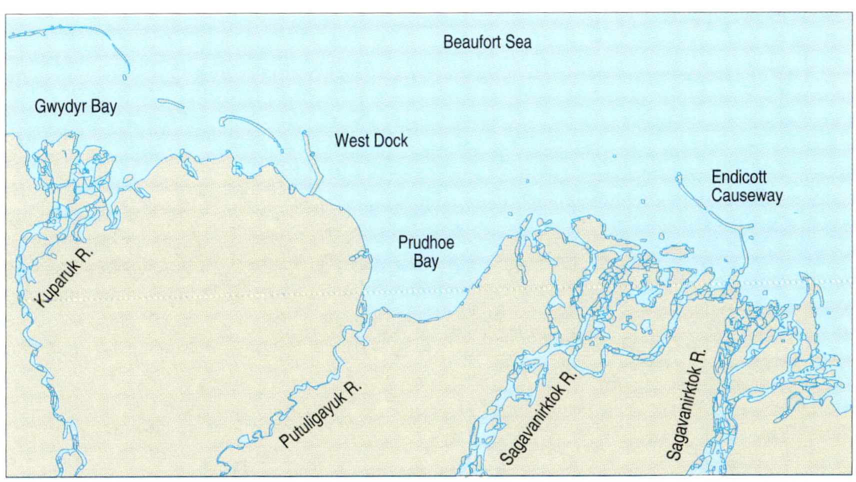

Fig. 1. Prudhoe Bay study area showing West Dock and the Endicott Causeway, Alaska.

Arctic cisco (*Coregonus autumnalis*). Courtesy R. West, USFWS

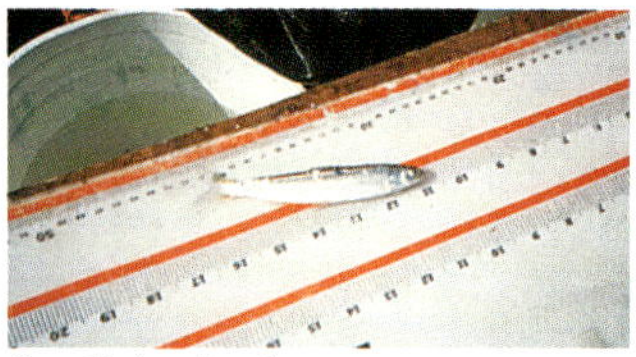

Small Arctic cisco. Courtesy W.J. Wilson, LGL Research Alaska, Inc.

Broad whitefish (*Coregonus nasus*). Courtesy R. West, USFWS

Least cisco (*Coregonus sardinella*). Courtesy R. West, USFWS

Dolly Varden char (*Salvelinus malma*). Courtesy W.J. Wilson, LGL Research Alaska, Inc.

Mackenzie River, Canada. Upon emergence, young fish are swept downstream and transported along the coast with prevailing nearshore currents. During years when easterly winds prevail, currents carry juveniles westward into Alaskan waters. Their migration is passive and recruitment varies annually. Juveniles in Alaska become mature after 7-9 years. They then return to spawn in the Mackenzie system. In Alaska, fish winter in the Colville River, and to a lesser extent, in the Sagavanirktok River deltas. Each June, young fish move to the coastal sea and are common in Prudhoe Bay. The commercial gillnet fishery is selective for 5- and 6-year-old fish.

Trends in abundance for four age groups of Arctic cisco captured in Prudhoe Bay since 1985 illustrate the cyclic nature of the species abundance by cohort in northern Alaska (Fig. 2a). The size and age structure observed in the population before 1985, and after causeway construction, generally follow predicted patterns expected from historical wind records.

The annual commercial catch of Arctic cisco from the Colville River fishery has ranged from 9,000 fish in 1979 to 72,000 in 1973 (Fig. 3). During the same period mean catch-per-unit effort (CPUE) ranged from 12 fish/net/day in 1979 to 195 fish/net/day in 1986. Biologists think that the availability of harvestable fish is due to natural mortality and interannual variations in numbers of migrants from Canada.

Broad Whitefish

Broad whitefish are indigenous to the Sagavanirktok and Colville rivers. Monitoring concentrated on the Sagavanirktok's population because of causeway construction in the river delta. Cohort analysis shows low catch rates in Prudhoe Bay from 1985 to 1987, followed by annual increases that peaked in 1990, and declining abundance thereafter (Fig. 2b). The existing data suggest a cycle of strong year-class success followed by several years of poor juvenile survival, which probably results from adult displacement of juveniles from optimal winter habitats.

A 10-year population cycle is suggested, underscoring the critical, if not limiting, nature of freshwater habitat for broad whitefish. Without long-term monitoring, the reduced abundance of juveniles in the late 1980's would have been attributed to causeways and not competition between fish at freshwater wintering sites.

Least Cisco

The center of distribution for least cisco in the Alaskan Beaufort Sea is the Colville River. Biologists have captured all cohorts of least cisco in Prudhoe Bay. No trends in CPUEs are apparent from cohort analysis (Fig. 2c), although catch rates were high in 1990, possibly because of prevailing west wind conditions that year.

Least cisco are of secondary importance in the

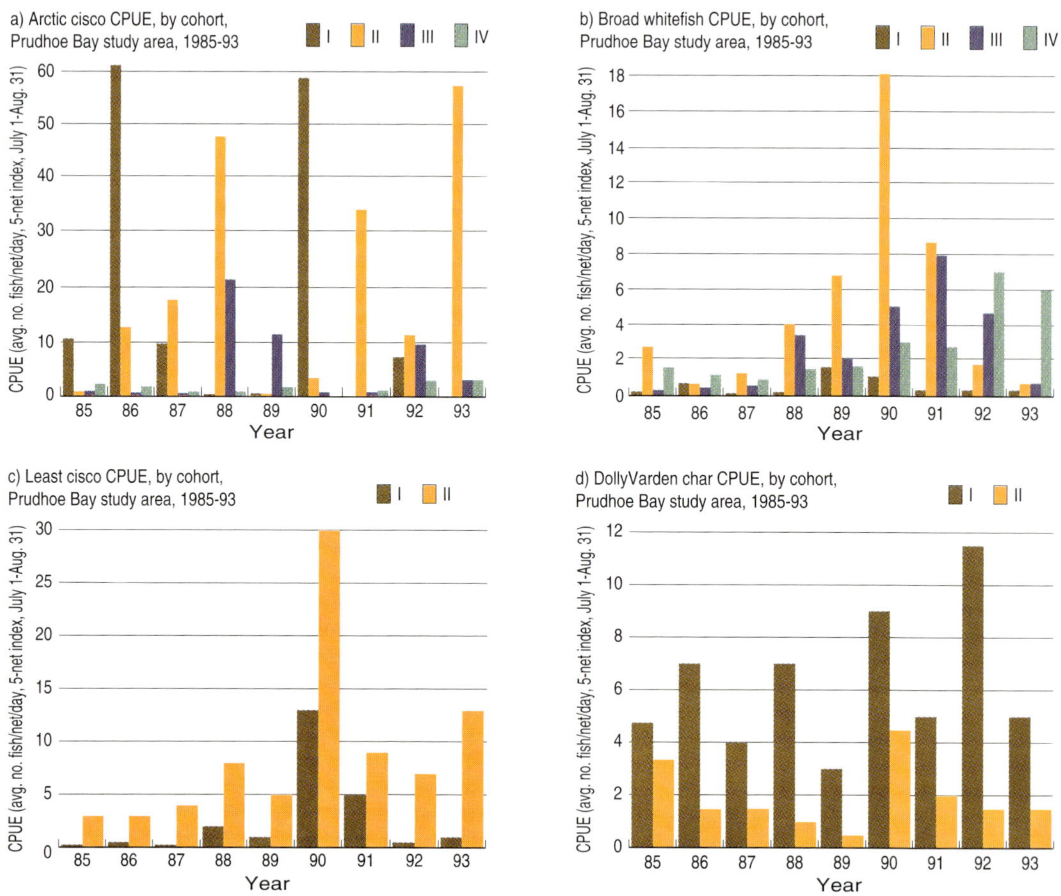

Fig. 2. Season-averaged catch rates (catch-per-unit-effort, CPUE) for (a) Arctic cisco (*Coregonus autumnalis*) cohorts I-IV, (b) broad whitefish (*C. nasus*) cohorts I-IV, (c) least cisco (*C. sardinella*) cohorts I and II, and (d) Dolly Varden char (*Salvelinus malma*) cohorts I and II, in the Prudhoe Bay study area, 1985-93 (Endicott Fish Monitoring Program, BP Exploration [Alaska], Inc.).

Colville River fishery. Annual catches have ranged from 6,000 fish in 1993 to 38,000 in 1983 (Fig. 3). Biologists believe that the annual variability observed in the catches reflects population fluctuations associated with natural mortality and fishing effects. The apparent decline in numbers of least cisco since 1991 cannot be explained by the existing data, and consequently residents of the North Slope Borough are closely monitoring this fishery.

Dolly Varden Char

Major populations of Dolly Varden char occur in the mountain streams and rivers of the eastern Brooks Range. The char is growing in importance as a recreational species; an estimated 1,000-3,500 fish are harvested annually (Alaska Department of Fish and Game 1993).

There are no apparent trends in population abundance (Fig. 2d). The Dolly Varden char is a highly mobile and tolerant species that uses freshwater, estuarine, and marine habitats. Catches tend to be highest during early and late summer when the fish are migrating near river mouths. Recent findings show that char from the eastern Alaska Beaufort Sea and Canada are present in Prudhoe Bay during summer.

Conclusions

These fish differ in their susceptibility to causeway changes in Prudhoe Bay. The broad whitefish is highly susceptible because of its more limited distribution and habitat preferences in the Sagavanirktok River delta. Young-of-the-year Arctic cisco must cross the Prudhoe Bay area to reach prime overwintering habitat; they forage in these coastal waters for several years thereafter. Continued exposure to habitat changes that affect summer habitat quality, access, or migration poses moderate risks to this species. Much of the study area is at the eastern limits of the Colville River population of least cisco and thus, at present, this species is considered at low risk from the existing causeways. Similarly, Dolly Varden char are probably at low risk because of their ability to use more offshore marine waters for feeding and migration.

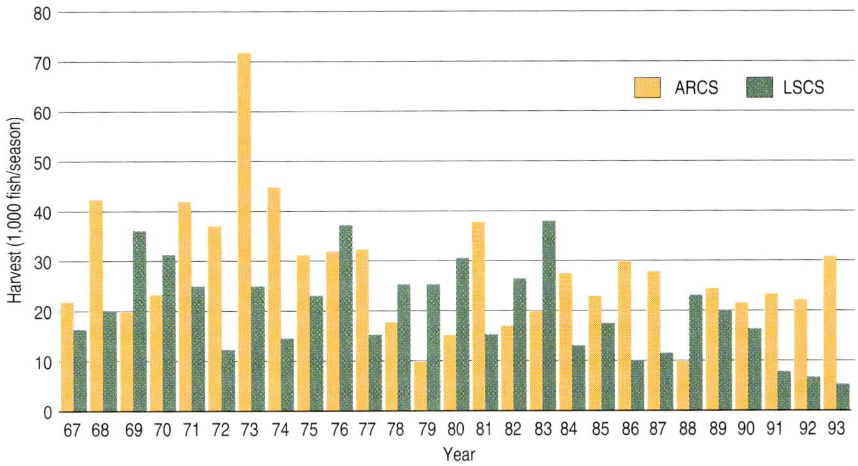

Fig. 3. Total annual commercial harvest of Arctic cisco (ARCS) and least cisco (LSCS) for the Colville River, 1967-93 (Endicott Fish Monitoring Program, BP Exploration [Alaska], Inc.).

References

Alaska Department of Fish and Game. 1993. North Slope sport fishery management plan. Sport Fish Division, Fairbanks, AK. 22 pp.

Benner, C.S., and R.W. Middleton, eds. 1991. Fisheries and oil development on the Continental Shelf. American Fisheries Society Symposium 11. 172 pp.

Norton, D.W., ed. 1989. Research advances on anadromous fish in Arctic Alaska and Canada. Biological Papers of the University of Alaska 24. 166 pp.

For further information:

Lyman K. Thorsteinson
National Biological Service
Alaska Science Center
Anchorage, AK 99503

Pacific Salmon in Alaska

by

Carl V. Burger
National Biological Service

Alex C. Wertheimer
National Marine Fisheries Service

Pacific salmon (Salmonidae) have played a major role in the history and economy of Alaska and its commercial, sport, and subsistence fisheries; Alaska currently produces about 80% of all salmon harvested in the western United States and Canada. Before commercial exploitation in the late 1800's, salmon were a main food source for Alaska's Native peoples, who subsisted by using an estimated 12 million salmon annually (Pennoyer 1988). By the end of the century, the total commercial harvest in Alaska had expanded to an estimated 56,000 salmon in 1878 but rose to more than 21 million by 1900 (Rigby et al. 1991). Since 1980 the annual commercial harvest has exceeded 100 million salmon in all but one year and is presently at a record high of more than 190 million (Fig. 1). The annual sport harvest of salmon in Alaska has averaged about 1 million fish over the past several years (Mills 1993), as has the subsistence harvest (INPFC 1992). Science-based management, "limited-entry" fishing, effective law enforcement, and establishment of fixed escapement goals for specific rivers are among the factors responsible for increased salmon abundance.

Apart from their economic, recreational, and subsistence importance, salmon are a vital link in various Alaskan ecosystems. Large populations of bears (Ursidae) and eagles (Accipitridae) in some parts of Alaska, for example, depend on late-spawning salmon as a food source before winter. Also, the carcasses of spawned-out salmon are a key element in otherwise nutrient-poor lakes and rivers. Because Alaska has a comparatively greater amount of unaltered habitat and a larger number of wild salmon stocks than do other parts of the Northwest, monitoring population status and trends is particularly important to alert managers to problems before irreversible losses occur.

We summarize trends in harvest and escapement (fish that survive sport, commercial, and

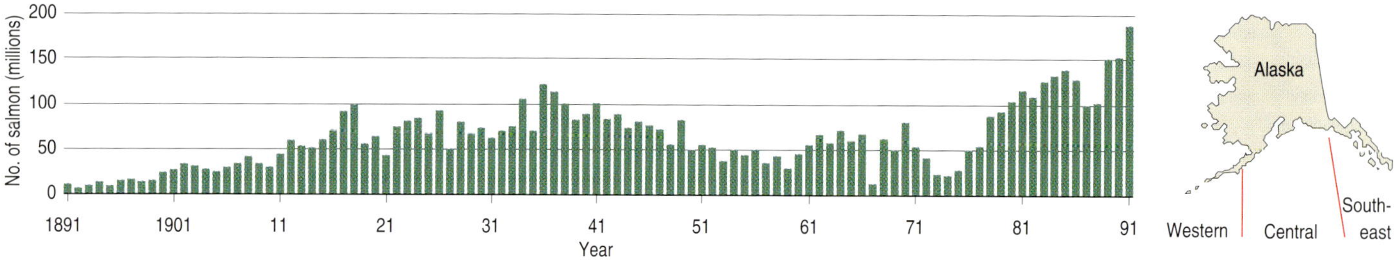

Fig. 1. Statewide commercial salmon harvest for all species of Alaskan salmon (excluding hatchery-produced fish), 1891-1991 (Rigby et al. 1991).

subsistence fishing) for five species of salmon in Alaska: pink (*Oncorhynchus gorbuscha*), chum (*O. keta*), sockeye (*O. nerka*), chinook (*O. tshawytscha*), and coho salmon (*O. kisutch*). We present historical records and data for three major regions of the state: southeastern, central, and western (Fig. 1). This summary is based on data from similar efforts completed or in progress by the Alaska Chapter of the American Fisheries Society, the Alaska Department of Fish and Game, the National Marine Fisheries Service, and the U.S. Forest Service.

The data we present originate from the Alaska Department of Fish and Game (various Area Management Reports). Information on the annual status of Alaskan salmon populations comes from numerous state and federal publications and is presented in three ways. First, we tabulate the trends in salmon escapement by species. This tabulation was done for species in central and western Alaska from 1968 to 1984 (Konkel and McIntyre 1987), for pink and chum salmon in southeast Alaska from 1960 to 1993 (Wertheimer in press), and for southeast sockeye, chinook, and coho stocks from 1960 to 1992 (C. Halupka, U.S. Forest Service, personal communication). These trend summaries do not include all populations, but are limited to those for which escapement data are readily available in a usable format.

Second, we graph the historical harvest for each species from 1891 to 1991 (Rigby et al. 1991). Because of Alaska's limited-entry fishing policy (since 1975) and the use of fixed-escapement goals, these summaries of commercial harvest may be a useful indicator of population trends.

In our third approach, we graph the escapements of pink, sockeye, chinook, and chum salmon (data for coho salmon were inadequate) in key areas of Alaska based on Department of Fish and Game Annual Management reports (1960 to 1992). This method provides an index of salmon abundance and is particularly relevant in determining sockeye salmon trends because management of this species is often based on in-season escapement enumeration. It also allows us to compare a species escapement trend in a specific area (for example, Prince William Sound) with its overall trend in other areas of Alaska. Because many Alaskan stocks are managed to meet a target escapement goal, however, a decreasing trend may not indicate a decrease in overall productivity.

Population Trends for Five Species

Pink Salmon

The trend summary for pink salmon was limited to the southeast and central regions of Alaska, where much of the harvest occurs. Most populations showed either no significant trend or were increasing in size (Table).

The plot of statewide harvest of pink salmon over time (Fig. 2a) was similar to the 100-year statewide harvest totals for all species (Fig. 1). Hatchery production of pink salmon is considerable in the central portion of Alaska and may account for up to 51% of the catch (Wertheimer in press). Statewide, a record catch occurred in 1991, when 93 million wild pink salmon and 35 million hatchery pink salmon were harvested (Fig. 2a; Wertheimer in press).

Table. Summary of trends in escapement for populations of Pacific salmon in Alaska by species and region over time. Escapement trends were classified as increasing or decreasing if the slope of the regression of escapement over time was significantly different ($P<0.05$) from zero. (NA— not available.)

Species	Years	Number of populations showing: No trend	Decrease	Increase	Source of data*
Pink salmon					
Southeast	1960-93	312	9	150	1
Central	1968-84	102	0	32	2
Western	NA	NA	NA	NA	—
Chum salmon					
Southeast	1960-93	28	5	5	1
Central	1968-84	61	11	3	2
Western	1968-84	10	2	0	2
Sockeye salmon					
Southeast	1960-92	93	10	4	3
Central	1968-84	58	0	35	2
Western	1968-84	16	0	12	2
Chinook salmon					
Southeast	1960-92	35	0	6	3
Central	1968-84	20	0	23	2
Western	1968-84	24	1	15	2
Coho salmon					
Southeast	1960-92	107	12	9	3
Central	1968-84	4	0	4	2
Western	NA	NA	NA	NA	—

*Source of data:
1—Wertheimer in press.
2—Konkel and McIntyre (1987).
3—C. Halupka, U.S. Forest Service, personal communication.

Pink salmon management in Prince William Sound is extremely complex. Record harvests of pink salmon (30-50 million fish) in Prince William Sound during 1990 and 1991 declined to 9 million in 1992. The decline in catch and recent declines in escapement (Fig. 3a) may be a result of density-dependent mortality from increased hatchery releases, environmental alterations, or changing oceanic currents. It should be noted, however, that the pink salmon escapements in Prince William Sound, Cook Inlet, and Kodiak increased in 1993 (Fig. 3a). The 1989 peak in the combined escapements for pink salmon in Cook Inlet and Kodiak reflects fishery closures related to the *Exxon Valdez* oil spill.

Chum Salmon

The trend summary for chum salmon was available for all regions of Alaska. Decreasing trends were more common than increases (Table). The statewide harvest of chum salmon attained record levels through the mid-1980's (Fig. 2b) and has generally increased in all areas of Alaska since the mid-1970's. Although the catch in western Alaska is almost all from wild populations, hatchery contributions are now

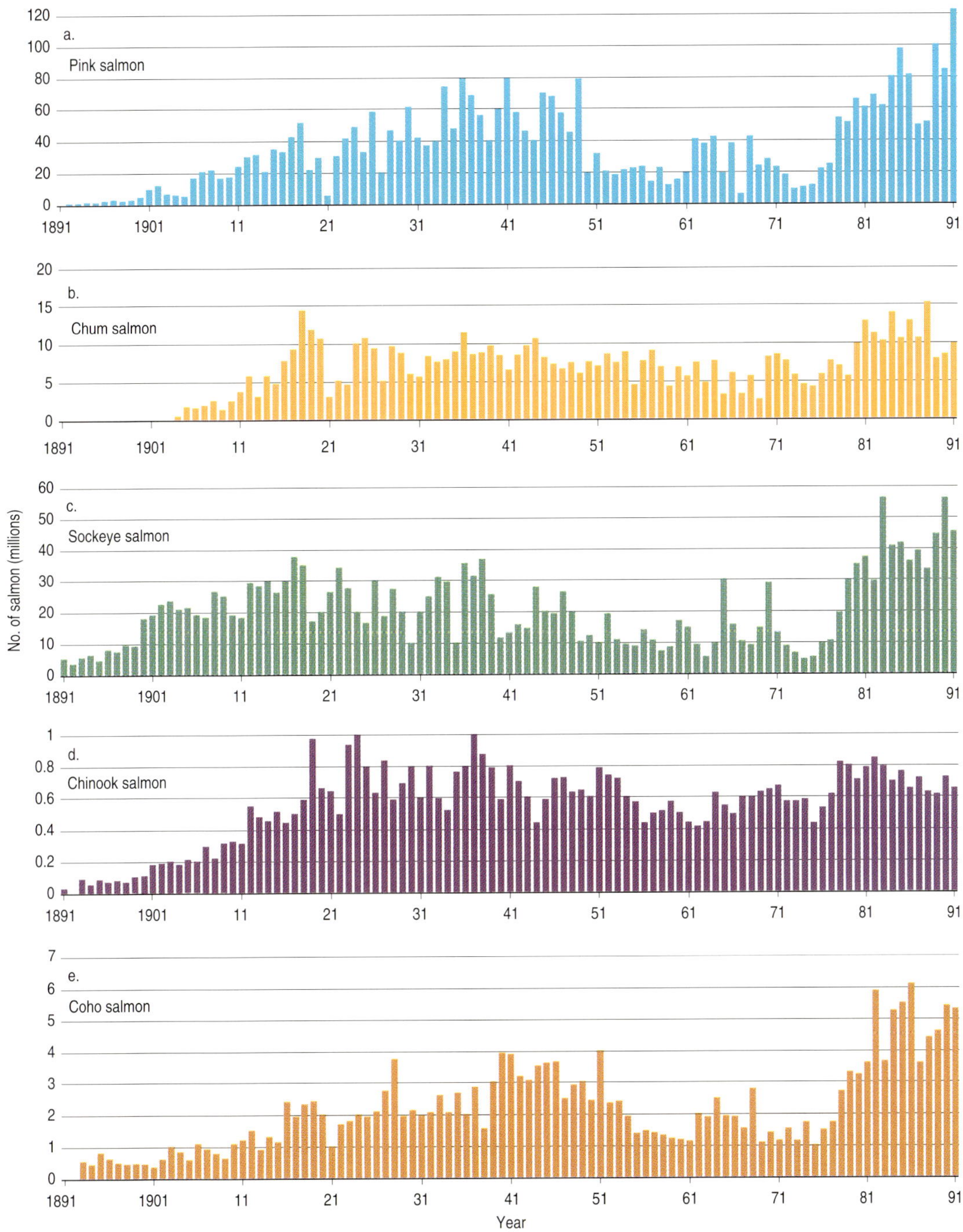

Fig. 2 a-e. Statewide commercial harvest of Alaskan salmon by species, 1891-1991 (Rigby et al. 1991).

about 12% of the catch in the central region and about 33% in southeastern Alaska (Wertheimer in press).

Chum salmon escapements (1979-93) in central and western Alaska (Fig. 3b) have generally declined, as have the escapements of fall-run chum salmon in the Yukon River. These declines have directly affected western Alaska commercial and subsistence users who depend on the chum salmon resource. Several factors could be responsible for this decline, including oceanographic change, density-dependent competition at sea with large numbers of chums released by hatcheries in Russia and Japan (Ishida et al. 1993), and interception by high seas drift-net fisheries (Olsen 1994). In addition, fishing effort has increased in recent years from expanding in-river commercial and subsistence chum salmon fisheries.

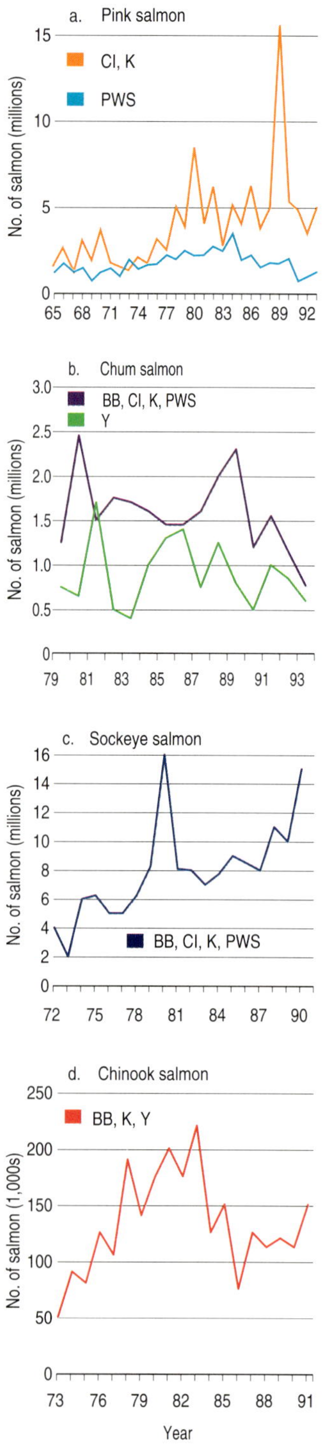

Fig. 3 a-d. Salmon escapements over time in select areas of Alaska. (BB—Bristol Bay; CI—Cook Inlet; K—Kodiak; PWS—Prince William Sound; Y—Yukon.)

Sockeye Salmon

A trend summary was possible for sockeye salmon in all regions of Alaska. Most populations were either stable or increasing (Table). Statewide sockeye salmon harvest is at a record level (Fig. 2c), and the catch throughout Alaska has risen substantially since the early 1970's (Wertheimer in press). Escapement also appears to be increasing for most populations (Fig. 3c). In addition to the few stocks in southeast Alaska that have declined, a decline in Cook Inlet sockeye salmon is predicted over the next 2 years. After many spawning adults escaped harvest when fisheries were closed in 1989 because of the *Exxon Valdez* oil spill, too many fry were produced to be supported by their habitat (D. Schmidt, Alaska Department of Fish and Game, personal communication). The resulting increase in fry mortality will probably be a factor in the abundance of Cook Inlet sockeye salmon in the immediate future.

Chinook Salmon

The trend summary for chinook salmon suggests that most populations are either stable or increasing (Table). Although present commercial harvest of chinook salmon statewide is slightly lower than the average historical level (Fig. 2d), the catch appears to be more stable than for all species combined (Fig. 1). A recent decrease in the quota for southeastern Alaska troll fisheries may be a factor in the stable catch of chinook salmon. Sport harvest of chinook salmon has increased substantially over the past several years (Mills 1993) and now exceeds 10% of the commercial catch (Wertheimer in press). Catches of chinook salmon declined in nearly all regions of Alaska in the early 1970's, rebounded through the early 1980's, and have begun to decrease since that time. High seas drift-net and trawl fisheries that target other species may be factors in the minor decline in chinook salmon harvest in western Alaska (Olsen 1994; Table). When actual escapements are plotted for several areas of Alaska, however, the trends are generally increasing (Fig. 3d).

Courtesy M. Wenger, USFWS

Sockeye salmon return to spawn in numerous Alaskan streams each summer.

Coho Salmon

A trend summary was possible for coho salmon stocks only in the southeastern and central regions of Alaska (Table). Overall, fewer data have been collected for coho than for other species of salmon because of their late run timing, smaller population sizes, and use of remote, heavily vegetated watersheds. Most populations analyzed in southeastern Alaska showed no trend: some increased and some decreased (Table). Of the eight populations examined from central Alaska, half increased and none decreased.

Statewide harvest of coho salmon is at a record level (Fig. 2e), as is the catch in all regions of Alaska (Wertheimer in press). Data were insufficient to plot coho salmon escapements in key areas of Alaska. Based on catch data alone, abundance of coho salmon is generally increasing (Wertheimer in press). For some of the populations that are declining in southeastern Alaska (Table), habitat effects associated with logging may be a factor; however, an equal number of declining populations in southeast Alaska are in pristine areas (C. Halupka, U.S. Forest Service, personal communication).

Conclusions

The population trends and escapements of pink, sockeye, chinook, and coho salmon in Alaska are generally stable or increasing based on the data analyzed. A recent decline in chum salmon escapements has occurred in central and western Alaska, the cause of which may be related to density-dependent factors and oceanic change in the marine environment. In many Alaskan streams, salmon abundance has not been determined or analysis of data is incomplete.

References

INPFC. 1992. Statistical yearbook, 1989. International North Pacific Fisheries Commission, Vancouver, BC. 123 pp.

Ishida, Y., S. Ito, M. Kaeriyama, S. McKinnell, and K. Nagasawa. 1993. Recent changes in age and size of chum salmon (*Oncorhynchus keta*) in the North Pacific Ocean and possible causes. Canadian Journal of Fisheries and Aquatic Sciences 50:290-295.

Konkel, G.W., and J.D. McIntyre. 1987. Trends in spawning populations of Pacific anadromous salmonids. U.S. Fish and Wildlife Service Tech. Rep. 9. 25 pp.

Mills, M.J. 1993. Harvest, catch, and participation in Alaska sport fisheries during 1992. Fishery Data Series Rep. 93-42. Alaska Department of Fish and Game, Anchorage. 228 pp.

Olsen, J.C. 1994. Alaska salmon resources. Pages 77-89 *in* L. Low, J. Olsen, and H. Braham, eds. Status of living marine resources off Alaska, 1993. National Oceanic and Atmospheric Administration, Tech. Memorandum NMFS-AFSC-27. U.S. Department of Commerce, Alaska Fisheries Science Center, Seattle, WA.

Pennoyer, S. 1988. Early management of Alaska fisheries. Marine Fisheries Review 50:194-197.

Rigby, P., J. McConnaughey, and H. Savikko. 1991. Alaska commercial salmon catches, 1878-1991. Alaska Department of Fish and Game Regional Information Rep. 5J91-16. 88 pp.

Wertheimer, A.C. Status of Alaska salmon. *In* D.J. Stounder, P.A. Bisson, and R.J. Naiman, eds. Pacific salmon and their ecosystems: status and future options. Chapman and Hall, Inc., New York. 18 pp. In press.

For further information:

Carl V. Burger
National Biological Service
Alaska Science Center
1011 East Tudor Rd.
Anchorage, AK 99503

Wolves and Caribou in Denali National Park, Alaska

by
Layne G. Adams
L. David Mech
National Biological Service

Management of gray wolves (*Canis lupus*) and their prey in interior Alaska has been controversial for three decades (Harbo and Dean 1983). Recently, debate was rekindled with renewed interest in wolf control to bolster two populations of caribou (*Rangifer tarandus*). Our research in Denali National Park provides insights into the declines in caribou numbers over the last few years that are the basis of recent wolf control proposals. Our observations of fluctuating populations also illustrate the complexity of managing these predator-prey systems to meet a diverse array of public interests.

Wolves and caribou are two components of the large mammal community of Denali National Park that also includes grizzly bears (*Ursus arctos*), moose (*Alces alces*), and Dall sheep (*Ovis dalli*). With the 1980 park expansion to more than 18,800 km^2 (7,300 mi^2) of central Alaska, this large mammal system became the only one of its kind that is virtually unaffected by human harvest. Therefore, Denali provides a unique opportunity to understand the natural interactions of these species and serves as a baseline for comparison with areas where hunting or other active wildlife management occurs.

We have studied Denali's wolves and caribou since 1986 to determine their numbers and status and understand their natural interactions in this protected subarctic ecosystem. Our studies began near the end of more than a decade of mostly light winter snowfalls of around 100 cm (39 in)/yr. Since winter 1988-89, we have experienced five consecutive winters with above-average snowfalls, including two record-setting years. During winters 1990-91 and 1992-93, more than 390 cm (154 in) of snow fell, four times as much as in the early years of our study. This change in snowfall had profound effects on the wildlife in central Alaska. The population trends of Denali's caribou and wolves are strong evidence of the natural fluctuations to be expected in species inhabiting such dynamic and variable environments.

Counting Caribou and Wolves

Our research has relied heavily on radiotelemetry to study the dynamics of the wild caribou and wolf in Denali. We can easily find our radio-collared study animals by using signal-receivers mounted in small airplanes (Mech 1975). Locating radio-collared wolves allows us to count their packmates, determine the number of pups born to each pack, gain information on survival and dispersal, determine the size and location of each pack's territory, and estimate the total number of wolves in our study area (Mech 1973). Regular monitoring of radio-collared caribou provides information on calf production, survival, and seasonal distribution of the herd, and makes it easier to complete aerial surveys to estimate herd size and composition (Adams et al. in press).

Population and Weather

The Denali caribou herd grew from about 1,000 in 1975 to 2,500 by 1986, during a decade of mostly below-average snowfalls, and was increasing at about 7% per year in 1986 when

Courtesy L.G. Adams, NBS

Wolf carrying week-old caribou calf.

our research began (Adams et al. in press; Figure). About 46 wolves inhabited the 10,000-km^2 (3,860-mi^2) range of the caribou herd in the early years of our study (Meier et al. in press). The number of wolves was lower than we expected based on the abundance of large prey species in Denali. Light snowfalls were favorable to caribou, and few died. Wolves preyed primarily on moose; the few caribou they took were usually very young or very old (Mech et al. in press). Times were tough for wolves, with poor production of pups and high dispersal rates for young wolves. Also, fights between packs resulted in the deaths of several wolves.

With the onset of more severe winters, beginning with winter 1988-89, wolf numbers rapidly increased to 81 wolves in just 2 years (Meier et al. in press; Figure), primarily because of higher pup production and less dispersal of young wolves. Caribou were more vulnerable to predation in the deep snow and replaced moose as the most important prey species for wolves. Losses of adult cows increased eight-fold to nearly 20% per year. Fewer than 9% of the calves survived to 4 months old, compared to nearly 60% following the light snow winters (Adams et al. in press). The caribou herd stopped growing in 1990 at about 3,300 and plummeted to 1,700 by 1993, a 50% decline in only 3 years (Figure). With declining prey, the wolves also declined to about 60 wolves within the caribou herd's range, a 23% reduction between March 1990 and March 1993.

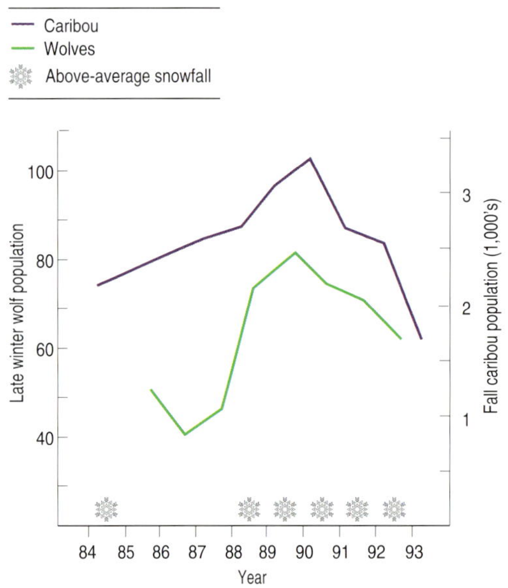

Figure. Wolf and caribou population trends in Denali National Park, Alaska, 1984-93.

The fluctuations in wolf and caribou numbers observed in Denali National Park are probably indicative of normal adjustments to the highly variable winter weather of the region. Within 8 years, the caribou herd increased by 36% and declined by 50%. At the same time, the wolves almost doubled in number and then declined halfway back to their original numbers.

The trends noted for the Denali caribou herd are representative of population trends of several mountain caribou herds throughout central Alaska, including the Chisana and Mentasta herds in the Wrangell Mountains, and the Delta and Macomb herds east of Denali Park in the Alaska Range. Unlike the Denali herd, which has been closed to hunting for nearly 20 years, these other caribou herds are important resources for subsistence and sport hunters alike. Hunting seasons have been closed for all four caribou herds because of the declines in the last few years.

These reductions in hunting opportunities have led to debates over the merits of wolf control to provide more caribou for human harvest. Arguments regarding allocation of harvestable wildlife between subsistence and sport hunters will intensify when hunting seasons are reopened. Although the future of wolves and caribou in interior Alaska is secure, natural fluctuations like those described here can be expected to generate continued controversy over the management and allocation of these important wildlife resources.

For further information:

Layne G. Adams
National Biological Service
National Park Service-Alaska Region
2525 Gambell St., Room 107
Anchorage, AK 99503

References

Adams, L.G., B.W. Dale, and L.D. Mech. Wolf predation on caribou calves in Denali National Park, Alaska. *In* L. Carbyn, S. Fritts, and D. Seip, eds. Ecology and conservation of wolves in a changing world. Canadian Circumpolar Institute, University of Alberta, Edmonton. In press.

Harbo, S.J., Jr., and F.C. Dean. 1983. Historical and current perspectives on wolf management in Alaska. Pages 51-64 *in* L.N. Carbyn, ed. Wolves in Canada and Alaska: their status, biology, and management. Canadian Wildlife Service Rep. Series 45. 135 pp.

Mech, L.D. 1973. Wolf numbers in the Superior National Forest of Minnesota. USDA Forest Service Res. Paper NC-97. 10 pp.

Mech, L.D. 1975. Current techniques in the study of elusive wilderness carnivores. Pages 315-322 *in* Proceedings 11th International Congress Game Biologists. National Swedish Environment Protection Board, Stockholm.

Mech, L.D., T.J. Meier, J.W. Burch, and L.G. Adams. Patterns of prey selection by wolves in Denali National Park, Alaska. *In* L. Carbyn, S. Fritts, and D. Seip, eds. Ecology and conservation of wolves in a changing world. Canadian Circumpolar Institute, University of Alberta, Edmonton. In press.

Meier, T.J., J.W. Burch, L.D. Mech, and L.G. Adams. Pack structure dynamics and genetic relatedness among wolf packs in a naturally regulated population. *In* L. Carbyn, S. Fritts, and D. Seip, eds. Ecology and conservation of wolves in a changing world. Canadian Circumpolar Institute, University of Alberta, Edmonton. In press.

Kodiak Brown Bears

by
Victor G. Barnes, Jr.
National Biological Service
Roger B. Smith
Alaska Department of Game and Fish
Mark S. Udevitz
National Biological Service
Jay R. Bellinger
U.S. Fish and Wildlife Service

Brown bears (*Ursus arctos middendorffi*) on the Kodiak Archipelago are famous for their large size and seasonal concentrations at salmon streams. Sport hunting of Kodiak bears has been popular since World War II. Their value as captivating subjects to observe or photograph is a more recent development that is increasing rapidly; visitors from around the world come to experience brown bears on Kodiak, adding substantially to Alaska's economy.

An equally important contribution of brown bears is their value as an indicator of ecosystem vitality. Despite high population numbers, Kodiak bears are vulnerable to the environmental effects that have seriously depleted brown bear populations in Europe and parts of North America (Cowan 1972; Servheen 1990). They are long-lived mammals that require large expanses of land to meet biological needs, and their low reproductive rate limits population recovery. Energy development, depletion of salmon resources, and recreational growth are factors that can adversely affect bears and, in doing so, signal a loss of environmental quality affecting many species.

Management of Kodiak brown bears is directed at maintaining current density, distribution, and habitat-use patterns. This goal is challenged by growing levels of commercial and private use throughout the region. An immediate concern is cabin and lodge development on 121,500 ha (300,000 acres), formerly part of the Kodiak National Wildlife Refuge, that were deeded to Alaska Natives via the Alaska Native Claims Settlement Act. Much of that Native-conveyed land is coastal or riparian habitat especially important to brown bears during summer and fall. Concurrently, recreational use of the Kodiak refuge is increasing about 10% annually (USFWS 1987). Sport fishing, bear photography, and deer and elk hunting often put bears and humans in direct conflict (Smith et al. 1989).

Timber harvest on Afognak Island, uncertain trends of salmon populations due to natural or human-caused events (e.g., *Exxon Valdez* oil spill), and hydroelectric development (Smith and Van Daele 1990) could impose additional long-term effects on localized bear populations.

Population Monitoring

Sport harvest records, available since 1950 (Troyer 1961), provide the most comprehensive information on Kodiak brown bears. In addition, biologists use aerial surveys to monitor population and habitat-use trends of brown bears on southwest Kodiak Island, an area that supports the highest bear densities and approximately 15% of Kodiak Island's bear population (Barnes et al. 1988).

We assessed status of the Kodiak bear population from estimates of density for representative study areas on northern, southwestern, and eastern Kodiak Island. We radio-collared a sample of bears on each area and estimated bear density using ratios of marked and unmarked bears observed from small aircraft (Miller et al. 1987). Brown bear abundance on other geographic units of the Kodiak Archipelago was estimated by comparing those units with the study areas.

Status and Trends

Sport Harvest Records

Excessive and localized harvest of brown bears in the mid-1960's (Fig. 1) prompted biologists to impose restrictions (season length, area closures) that dramatically reduced harvest. A sharp rise in hunting in the early 1970's produced another increase in harvest. In 1976 the Alaska Department of Fish and Game began an area permit system that distributed hunting more equitably throughout the archipelago. Since 1980 the harvest pattern has been relatively stable, with an average annual take of 163 animals (Fig. 1).

Sex composition of the sport harvest has remained relatively stable despite fluctuations in yearly harvest. From 1987 to 1993 the female portion of the harvest has ranged from 32% to 38%.

Age and skull measurements of harvested bears provide further evidence of population stability. Mean ages of males and females taken during 1981-93 (7.3 and 7.4 years, respectively) were slightly higher than during 1969-80 (6.3 and 6.8 years, respectively), but we attribute this difference to sampling variation (Fig. 2). Skull

Courtesy D. Menke, USFWS

Adult brown bear on Dog Salmon Creek, Kodiak National Wildlife Refuge, Kodiak Island, AK.

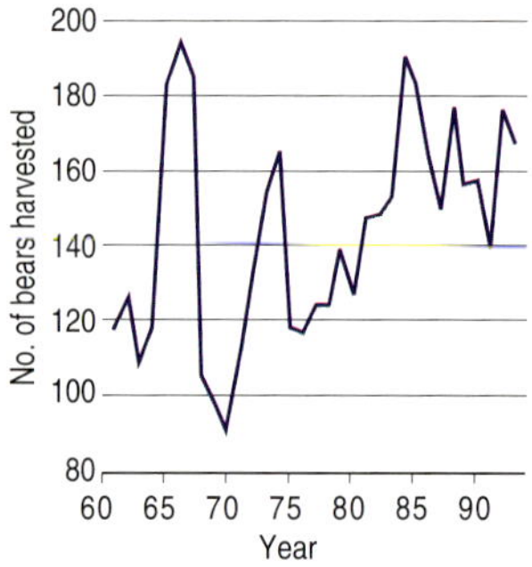

Fig. 1. Annual sport harvest of Kodiak brown bears, 1961-93.

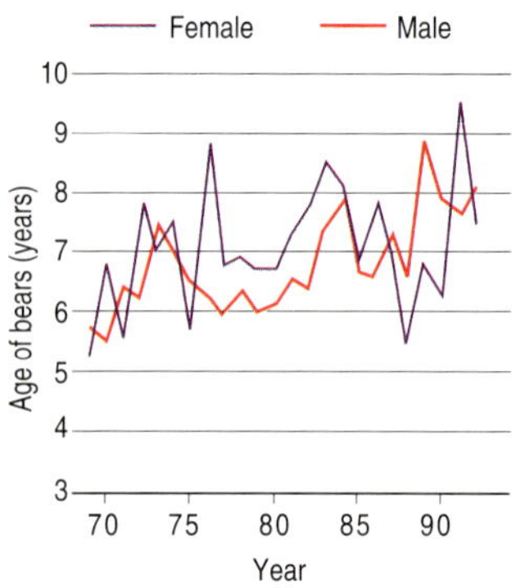

Fig. 2. Mean age of Kodiak brown bears harvested by sport hunters, 1969-92.

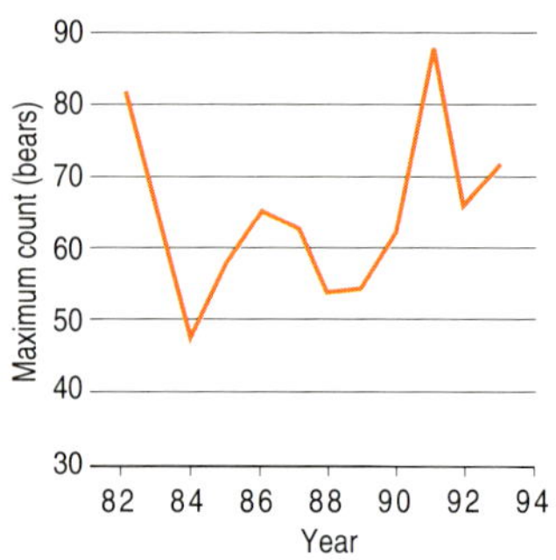

Fig. 3. Maximum counts from aerial surveys of brown bears concentrated along salmon-spawning streams on southwest Kodiak Island, 1982-93.

For further information:

Victor G. Barnes, Jr.
National Biological Service
Alaska Science Center
Kodiak Field Station
Kodiak, AK 99615

measurements (length plus width) of harvested bears, which generally indicate bear size (Glenn 1980), have remained consistent over time.

Collectively, sport hunting records point to a stable bear population on the Kodiak Archipelago. A comparison of average annual harvest and estimated population size indicates that harvest is at or near the maximum sustainable level (Miller 1990), and managers should closely monitor additional effects on the bear population arising from increased mortality or other factors.

Aerial Stream Surveys

Adjusted maximum counts from stream surveys ranged from 47 to 87 bears per survey over the past 12 years, but there has not been any consistent trend in the counts during this period (Fig. 3). The stream survey counts are used as an index to population size, but they are affected by many other factors such as timing of the surveys relative to peak bear concentrations and strength of salmon runs.

We consider estimates of composition based on the stream surveys more reliable. Annual estimates of the proportion of maternal females have varied little from the overall mean of 24% during this period. Taken together, the count and composition data suggest that the brown bear population in this area remains relatively stable.

Population Abundance

Estimates of brown bear density on three study areas on Kodiak Island ranged from 0.29 to 0.35 bears/km^2 (0.75 to 0.91 bears/mi^2). Habitats represented by the areas included precipitous mountain terrain, shrub-covered slopes, riparian zones, coastal habitat, and extensive bog and heathland flats. Extrapolating those density estimates to comparable habitats on other geographical areas provided an estimate of 2,842 bears for the Kodiak Archipelago or about 0.23 bears/km^2 (0.60 bears/mi^2). Bear density was highest at Karluk Lake (0.42 bears/km^2 [1.09 bears/mi^2]) and lowest on small, isolated islands (0.04 bears/km^2 [0.10 bears/mi^2]).

Management Considerations

Available information suggests that the status of the Kodiak brown bear population is better now than in some earlier periods. In the early 1900's bears were commercially hunted for their hides or indiscriminately killed as competitors of fisherman and ranchers (Troyer 1961; Smith et al. 1989). During the 1960's bears were killed in a controversial control program undertaken to reduce conflicts with livestock on northeast Kodiak Island (Eide 1965), and excessive sport harvest occurred on parts of southwest Kodiak Island. These events undoubtedly affected bear distribution and abundance in local areas. However, future management of brown bears and their habitat will face new problems, including accelerated timber harvest, construction of cabins on bear habitat, and additional hydroelectric development. Added to all these threats is the long-term problem of expanding recreational use. Effective management of the bear population in upcoming years will depend on inventory methods that can detect population change in a timely manner.

References

Barnes, V.G., Jr., R.B. Smith, and L.J. Van Daele. 1988. Density estimates and estimated population of brown bears on Kodiak and adjacent islands, 1987. Report to the Kodiak Brown Bear Research and Habitat Maintenance Trust. U.S. Fish and Wildlife Service and Alaska Department of Fish and Game, Kodiak. 34 pp.

Cowan, I.M. 1972. The status and conservation of bears (Ursidae) of the world—1970. Proceedings of the International Conference on Bear Research and Management 2:343-371.

Eide, S. 1965. The nature of brown bear predation on cattle, Kodiak Island, Alaska. Proceedings of the Conference of Western Association of State Game and Fish Commissioners 45:113-118.

Glenn, L.P. 1980. Morphometric characteristics of brown bears on the central Alaska Peninsula. Proceedings of the International Conference on Bear Research and Management 4:313-319.

Miller, S.D. 1990. Population management of bears in North America. Proceedings of the International Conference on Bear Research and Management 8:357-373.

Miller, S.D., E.F. Becker, and W.B. Ballard. 1987. Black and brown bear density estimates using modified capture-recapture techniques in Alaska. Proceedings of the International Conference on Bear Research and Management 7:23-35.

Servheen, C. 1990. The status and conservation of the bears of the world. International Conference on Bear Research and Management Monograph Series 2. 32 pp.

Smith R.B., V.G. Barnes, Jr., and L.J. Van Daele. 1989. Brown bear-human conflicts in the Kodiak Archipelago, Alaska. Pages 111-119 *in* M. Bromley, ed. Bear-people conflict: Proceedings of a Symposium on Management Strategies. Northwest Territories Department of Renewable Resources, Yellowknife, Canada.

Smith, R.B., and L.J. Van Daele. 1990. Impacts of hydroelectric development on brown bears, Kodiak Island, Alaska. Proceedings of the International Conference on Bear Research and Management 8:93-103.

Troyer, W.A. 1961. The brown bear harvest in relation to management on the Kodiak Islands. Transactions of the North American Wildlife and Natural Resources Conference 26:460-468.

USFWS. 1987. Kodiak National Wildlife Refuge, final comprehensive conservation plan, wilderness review and environmental impact statement. U.S. Fish and Wildlife Service, Anchorage, AK. 533 pp.

Polar Bears in Alaska

by
Steven C. Amstrup
Gerald W. Garner
George M. Durner
National Biological Service

The polar bear (*Ursus maritimus*) is the top predator of the Arctic marine ecosystem. Polar bears prey primarily on ringed seals (*Phoca hispida*) and bearded seals (*Erignathus barbatus*), which live exclusively on the sea ice (Smith and Stirling 1975; Stirling and Archibald 1977; Smith 1980), but they also can kill larger prey such as walruses (*Odobenus rosmarus*) and white whales (*Delphinapterus leucas*; Kiliaan and Stirling 1978; Fay 1982; Calvert and Stirling 1990; Stirling and Derocher 1990).

Polar bears move several thousand kilometers annually and over years occupy areas that can exceed 500,000 km^2 (nearly 200,000 mi^2; Garner et al. 1990; Amstrup and Durner, unpublished data; Fig. 1). Polar bears are circumpolar in the northern hemisphere, but they live in several largely discrete subgroups, rather than one homogeneous pan-Arctic population (Harington 1968). We used radio telemetry to show that two partially discrete subpopulations live adjacent to Alaska (Fig. 2). One subpopulation occurs largely in the Beaufort Sea of Alaska and neighboring Canada. Animals from this Beaufort Sea stock appear to spend about 25% of their time along the Chukchi Sea coast of northwestern Alaska (Amstrup and Durner, unpublished data). The Chukchi Sea subpopulation winters in the northern Bering Sea and southern Chukchi Sea adjacent to Russia and western Arctic Alaska, and its members seldom enter the Beaufort Sea (Fig. 2).

Low reproductive rates make polar bears vulnerable to excessive hunting. Yankee whalers, local resident Native people, and aerial hunters reduced numbers and local distributions of polar bears in many areas (Leffingwell 1919; Hanna 1920; Lønø 1970; Mowat 1984; Amstrup et al. 1986). Polar bears are also potentially vulnerable to industrial developments and other human activities that have increased in the Arctic recently (Lentfer 1983; Amstrup et al. 1986). Polar bears and the seals on which they prey may also be among the first species to show effects of climate warming and other global changes (Stirling and Derocher 1993).

In 1973 the five nations within whose boundaries polar bears occur negotiated the International Agreement on Conservation of Polar Bears. The agreement, ratified in 1976, prohibited the taking of polar bears by hunters in aircraft or large motor vessels, creating a de facto sanctuary in active offshore ice habitats. The agreement required each nation to conduct a research program and coordinate management and research, with other jurisdictions, for populations that overlap international boundaries.

In the United States, the agreement was implemented by enactment of the Marine Mammal Protection Act of 1972. Under the act, only Native people living along the Alaska coast were allowed to take polar bears. The act, however, removed restrictions on the take of cubs and females with cubs and the mandatory reporting requirement of the state's management program. Despite elimination of many management tools, the act required the Department of the Interior to manage polar bears within the bounds of optimum sustainable population levels.

Counting Polar Bears

We captured polar bears and marked them with ear tags and tattoos. Selected adult female polar bears also were fitted with radio collars. Captured bears were weighed and measured, and a vestigial premolar tooth was removed for age determination (Stirling et al. 1975; Hensel and Sorensen 1980). Each year, we tallied new captures and recaptures, and updated capture and reproductive histories of previously marked animals. We constructed life tables from the capture data (Seber 1973; Caughley 1977), and estimated survival rates from radio-collared bears and their young (Kaplan-Meier method; Pollock et al. 1989). We examined patterns of population size with matrix models (Leslie 1945, 1948).

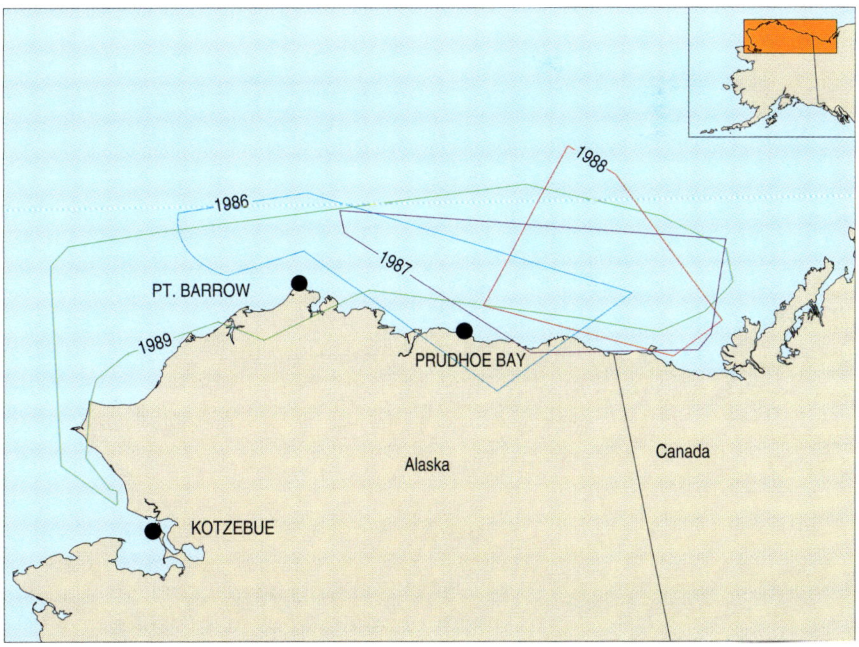

Fig. 1. Outlines of the annual activity areas for one radio-collared polar bear (*Ursus maritimus*) monitored during 4 consecutive years. The boundaries of the multiyear activity area enclosed 517,000 km^2 (about 200,000 mi^2).

Population Estimates

Recaptures were too few in the Chukchi Sea to evaluate population status for that subpopulation. Many data were available from the

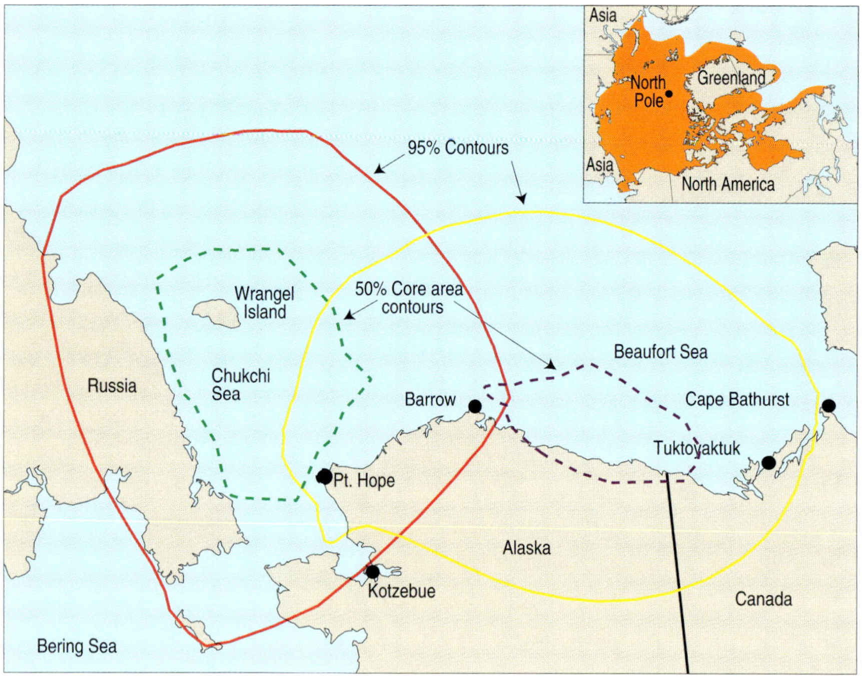

Fig. 2. Approximate bounds of the Beaufort Sea and Chukchi Sea polar bear populations. The contours for each population surround 95% and 50% of the radio relocations that were nearest the harmonic mean center of the distribution of relocations.

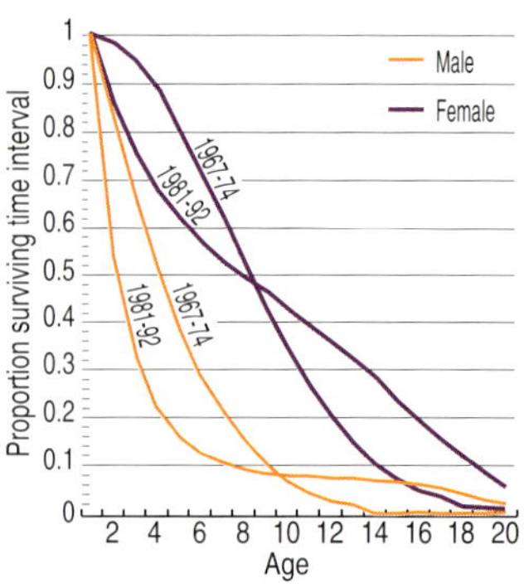

Fig. 3. Curves comparing survival of male and female polar bears in early (1967-74) and late (1981-92) periods. Note poorer survival of young and improved survival of adults, indicative of a relatively large population, in later period.

Beaufort Sea, however. We compared 986 captures and recaptures from the 1967-74 period to 1,531 captures and recaptures from the 1981-92 period to evaluate population trends. Reproduction among females commonly began at age 6 and continued until at least age 24. Numbers of cubs produced per female in both time periods were similar, but litter sizes of yearlings were larger in the first period. Differences in sampling during the two periods may have prevented effective comparisons of birth rates and of litter sizes; the age structure of the population was younger in the first period. Survival of adults, as calculated from life tables, was higher and survival of young lower in the 1981-92 period (Fig. 3). Radio-collared bears had a survival rate of 0.965 (96.5% survived), and their dependent young survived at the rate of 0.676 (67.6% survived). Of 26 radio-collared females followed until death, 22 (84%) were shot by coastal hunters.

We used a modified Petersen mark-and-recapture model (Seber 1973) to estimate there were approximately 600 females in the Beaufort Sea in 1976. Placing our calculated birth and death rates into matrix models projected growth to 900 females and 1,500 total animals in 1992. This was a realized growth rate of about 2% per year. The modified Petersen model provided an estimate of 750 females for 1986. That growth rate projected forward to 1992 indicated 850 females and just over 1,400 total animals; numbers that agreed closely with those predicted by the matrix models.

Numbers of bears captured per unit of effort, in the Beaufort Sea, also have increased, providing another indication of population growth. The few catch/effort data from the Chukchi Sea also suggest an increasing trend. There was a compensatory relationship between estimated population size in the Beaufort Sea and recruitment of subadults. Large populations of recent years recruited few juveniles, and smaller populations present in the first period recruited higher proportions of juveniles.

Implications of Growth

We are confident that the growth we detected in the Beaufort Sea population is real. A finite rate of growth of 1%-2% and a current population of approximately 1,500 are both reasonable. Increased numbers of polar bears seen along Alaska's north coast in recent years, increased encounter rates by researchers, and matrix models all suggest the population is larger now than in the recent past. This increase in numbers has occurred despite continued hunting by local resident Native people, and despite development of the nation's largest oilfield at Prudhoe Bay. The age structure and survivorship patterns of recent years suggest the population in the Beaufort Sea may be at or near the limits set by its environment.

Unfortunately, known and unknown biases in our mark and recapture data resulted in population size estimates that were associated with considerable uncertainty. The degree of fluctuation we observed in population estimates derived by the sophisticated Jolly-Seber model were biologically impossible. The estimates were more consistent in the simpler Petersen model, substantiating the observation that the trend of increase is valid, but not erasing concerns about the absolute size of the population. Less-than-perfect population estimates may not be an urgent problem if harvest is kept at a level that is known to be within long-term sustained yield (e.g., near present harvest levels). Hunting, however, already accounts for 80% of calculated annual mortality, and pressures to increase harvest are always present. Estimates of the size of the population of polar bears in the Chukchi Sea are lacking, but the catch per unit of effort during research tagging there may suggest an increase, as do observations and kills by coastal residents. Uspenski and Belikov (1991) believe there are more bears in the Chukchi Sea now than in the past despite the absence of a reliable population estimate.

Thus, the good news of apparent increases in numbers is accompanied by increased challenges for management. Those challenges can only be met by a better understanding of the dynamics of the polar bear's ecosystem. In the

Chukchi Sea, there is a pressing need for development of new methods for determining numbers and trends. This need appears more urgent in view of the likelihood that the ban on polar bear hunting in Russia, in effect since 1956, will be lifted. The bounds of optimum sustainable population levels are not known in the Beaufort or Chukchi seas, and interactions between polar bears and their prey and polar bears and sea ice, which establish these levels, are not understood. If managers are to keep polar bear numbers at optimum sustainable population levels in the face of increased harvests and other local and global perturbations, they will need more accurate and precise population estimates and an understanding of the ecosystem forces that limit polar bear population size.

References

Amstrup, S.C., I. Stirling, and J.W. Lentfer. 1986. Past and present status of polar bears in Alaska. Wildlife Society Bull. 14(3):241-254.

Calvert, W., and I. Stirling. 1990. Interactions between polar bears and overwintering walruses in the central Canadian High Arctic. International Conference of Bear Researchers and Managers 8:351-356.

Caughley, G. 1977. Analysis of vertebrate populations. John Wiley and Sons, New York. 234 pp.

Fay, F.H. 1982. Ecology and biology of the Pacific walrus, *Odobenus rosmarus divergens* Illiger. North American Fauna 74:1-279.

Garner, G.W., S.T. Knick, and D.C. Douglas. 1990. Seasonal movements of adult female polar bears in the Bering and Chukchi seas. International Conference of Bear Researchers and Managers 8:219-226.

Hanna, G.D. 1920. Mammals of the St. Matthew Islands, Bering Sea. Journal of Mammalogy 1:118-122.

Harington, C.R. 1968. Denning habits of the polar bear (*Ursus maritimus*). Canadian Wildlife Service Rep. Series 5. 33 pp.

Hensel, R.J., and F.E. Sorensen. 1980. Age determination of live polar bears. International Conference of Bear Researchers and Managers 3:93-100.

Kiliaan, H.P.L., and I. Stirling. 1978. Observations on overwintering walruses in the eastern Canadian High Arctic. Journal of Mammalogy 59:197-200.

Leffingwell, E. 1919. The Canning River region, northern Alaska. U.S. Geological Survey Professional Paper 109. 247 pp.

Lentfer, J.W. 1983. Alaskan polar bear movements from mark and recovery. Arctic 36:282-288.

Leslie, P.H. 1945. On the use of matrices in certain population mathematics. Biometrika 33:183-212.

Leslie, P.H. 1948. Some further notes on the use of matrices in population mathematics. Biometrika 35:213-245.

Lønø, O. 1970. The polar bear (*Ursus maritimus* Phipps) in the Svalbard area. Norsk Polarinstitutt Skrifter 149. 103 pp.

Mowat, F. 1984. Sea of slaughter. Atlantic Monthly Press, Boston. 322 pp.

Pollock, K.H., J.D. Nichols, C. Brownie, and J.E. Hines. 1990. Statistical inference for capture-recapture experiments. Wildlife Monographs 107:1-97.

Pollock, K.H., S.R. Winterstein, C.M. Bunck, and P.D. Curtis. 1989. Survival analysis in telemetry studies: the staggered entry design. Journal of Wildlife Management 53(1):7-15.

Seber, G.A.F. 1973. The estimation of animal abundance. Hafner Press, New York. 506 pp.

Smith, T.G. 1980. Polar bear predation of ringed and bearded seals in the land-fast ice habitat. Canadian Journal of Zoology 58(12):2201-2209.

Smith, T.G., and I. Stirling. 1975. The breeding habitat of the ringed seal (*Phoca hispida*). The birth lair and associated structures. Canadian Journal of Zoology 53:1297-1305.

Stirling, I., D. Andriashek, P. Latour, and W. Calvert. 1975. Distribution and abundance of polar bears in the eastern Beaufort Sea. Canadian Wildlife Service Tech. Rep. 2. 59 pp.

Stirling, I., and W.R. Archibald. 1977. Aspects of predation of seals by polar bears. Journal of the Fisheries Research Board of Canada 34:1126-1129.

Stirling, I., and A.E. Derocher. 1990. Factors affecting the evolution and behavioral ecology of the modern bears. International Conference of Bear Researchers and Managers 8:189-204.

Stirling, I., and A. Derocher. 1993. Possible impacts of climatic warming on polar bears. Arctic 46:240-245.

Uspenski, S.M., and S.E. Belikov. 1991. Polar bear populations in the Soviet Arctic: current state, studies, and management. Pages 93-95 *in* S.C. Amstrup and O. Wiig, eds. Polar bears. International Union for the Conservation of Nature and Natural Resources, Species Survival Committee Occasional Paper 7.

Radio-collared polar bear (*Ursus maritimus*) female and 3-month-old cub.

For further information:

Steven C. Amstrup
National Biological Service
Alaska Science Center
1011 E. Tudor Rd.
Anchorage, AK 99508

Sea Otters in the North Pacific Ocean

by
James L. Bodkin
Ronald J. Jameson
James A. Estes
National Biological Service

About 250 years ago sea otters (*Enhydra lutris)* were distributed continuously from central Baja California, north and west along the Pacific Rim to Kamchatka Peninsula in Russia, and south along the Kuril Islands to northern Japan (Kenyon 1969; Fig. 1a). Several hundred thousand sea otters may have occurred in the north Pacific region when commercial hunting began in the 18th century (Riedman and Estes 1990).

At least two attributes of the sea otter have influenced humans likely for as long as they have resided together along the coast of the north Pacific Ocean. First, sea otters rely on a dense fur, among the finest in the world, for insulation in the cold waters of the Pacific Ocean. The demand for sea otter fur led to their near extinction in the 19th century. The fur harvest, begun about 1740 and halted by international treaty in 1911, left surviving colonies, each likely numbering less than a few hundred animals, in California, south-central Alaska, and the Aleutian, Medny, and Kuril Islands (Fig. 1a). These individuals provided the nucleus for the recovery of the species. Today more than 100,000 sea otters occur throughout about 75% of their original range (Fig. 1b). Immigration has resulted in near-complete occupation of the Aleutian and Kuril archipelagos and the Alaska Peninsula. Successful translocations have resulted in viable populations in southeast Alaska, Washington, and British Columbia. Large amounts of unoccupied habitat remain along the coasts of Russia, Canada, the United States, and Mexico.

The second potential source of conflict

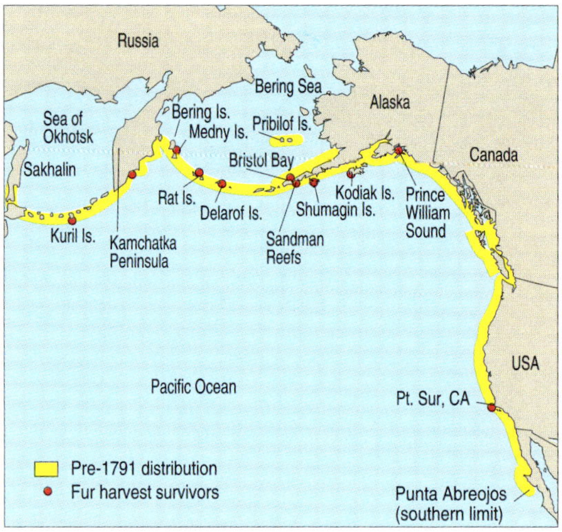

Fig. 1a. Distribution of sea otters before fur harvest began in 1741 and populations that survived the harvest, providing the nucleus for recovery of the species. Width of shaded area is not relative to sea otter habitat.

Fig. 1b. Current distribution of sea otters including locations of successful translocations.

between sea otters and humans is that sea otters prey on and often limit some benthic invertebrate populations. Because some of these invertebrates are also used by humans (Estes and VanBlaricom 1985), human perceptions about the effects of sea otter foraging on invertebrates sometimes differ. By limiting populations of herbivorous invertebrates (e.g., sea urchins [Echinoidea]) otters help maintain the integrity of kelp forest communities. At the same time, sea otter predation on other marine invertebrates can lead to direct competition with humans for resources. These interactions add complex dimensions to the conservation and management of sea otters, in large part because of wide-ranging social, ecological, and economic consequences of sea otter foraging.

Long-term data on abundance and distribution are available for relatively few sea otter populations. Here we summarize such data from three populations: Bering Island, Russia; Prince William Sound, Alaska; and Olympic Peninsula, Washington. The Bering Island population resulted from natural emigration and represents complete recovery. Prince William Sound represents near recovery of a remnant population, whereas the Washington population was established via translocations from Alaska and is just beginning to recover. We will compare growth rates and current status among these populations. Because of its unique status and growth characteristics, the California sea otter is not treated in this article.

Population Surveys

Annual skiff surveys were conducted at Bering Island from 1979 to 1993 (except 1990; Burdin et al. in press). Surveys from skiffs, airplanes, and helicopters were conducted in 1950, 1959, 1972, and 1984-85 in Prince William Sound (Johnson 1987; Irons et al. 1988). In Washington, skiff surveys augmented with ground counts were conducted from 1977 through 1987, and aerial surveys augmented with ground counts were conducted from 1989 to 1993 (Jameson et al. 1986; Jameson 1993). Instantaneous growth rates were calculated by regressing the natural logs of survey counts over time.

Population Status

Bering Island

Bering Island was recolonized by sea otters from nearby Medny Island about 1970. Growth occurred by progressive expansion around the island, with complete occupation of available habitat by 1983. The abundance of sea otters increased at an average of 22% per year, from 500 sea otters in 1979 to an estimated 3,835 in 1990 (Fig. 2). More than 20% of the population died at Bering Island during the winter of 1990-91 (Burdin et al. in press), suggesting that the number of sea otters exceeded available food resources. Little opportunity exists for emigration as the nearest unoccupied habitat is several hundred kilometers from Bering Island.

Prince William Sound

Although no surveys were conducted before 1959, at least 150 sea otters were observed in southwestern Prince William Sound in 1951 (Lensink 1962). Sea otters had spread throughout all available habitat in the sound by 1985, although growth was still apparent in the eastern part of the region (Johnson 1987). The overall growth rate in Prince William Sound between 1911 and 1985 was on average about 8% per year (Fig. 2). No density-dependent mortality event, such as observed at Bering Island, has been documented for Prince William Sound. Limited unoccupied habitat that could provide space for dispersing animals is still available both to the east and west of Prince William Sound.

Washington

In 1969 and 1970, 59 sea otters were released along the outer coast of Washington. Mortality was high, with 16 carcasses recovered after the first release (Jameson et al. 1982). Between 1977 and 1993, the population grew at an average of about 20% per year. Between 1989 and 1993, however, the average annual growth rate has been lower (12%). Unoccupied habitat currently occurs north, south, and within the present range, and continued growth is likely.

Predicted Trends

Sea otters illustrate the healthy recovery of a species following protection and active management. Rates of increase in most populations with unoccupied habitat available to them have been 17%-20% per year (Estes 1990a). As unoccupied habitats become limiting, however, density-dependent mechanisms may dramatically reduce sea otter abundance. As geographically separate populations reach equilibrium densities or as populations become so large as to create long dispersal distances to unoccupied habitats, we anticipate declining growth rates, increased mortality, and numbers of otters stabilizing near an equilibrium density. The observed trend in virtually all persisting populations since 1911 has been one of growth, with declines observed only as populations exceeded available resources (Estes 1990a, 1990b). Continued growth is expected, particularly in Washington and southeast Alaska and along the Kamchatka Peninsula.

The long-term exponential growth in many sea otter populations has allowed us to describe the process of sea otter recovery. However, as populations attain equilibrium densities and growth rates decline, evaluation of future trends will become more difficult. In addition, possible short-term changes, such as those resulting from human impacts, may remain difficult to detect. Thus, describing future population trends will require improved population- or individual-based assessment models.

At least two issues are currently relevant to sea otter conservation and management. One is competition between sea otters and humans for shellfish resources. As otters continue to reoccupy former habitat, the commercial, recreational, and subsistence harvest of species such as crabs (Crustacea), clams (Bivalvia), abalone (Gastropoda), and urchins, can be expected to decline.

Another current issue is the extent of the legal and illegal harvest for sea otter fur. Both the legal harvest by Alaska Natives and an illegal harvest in Russia have recently increased (A. Burdin, Russian Academy of Science, personal communication). Reasonable harvest guidelines and adequate inventory and monitoring programs should be established in areas with harvested populations.

Neither of these conservation issues currently appears to be precluding the continued growth of sea otter populations, but the potential to overharvest this species has been well demonstrated. Conservative management should ensure continued growth through complete recovery.

References

Burdin, A.M., V.V. Vertyankin, and V.S. Nikulin. Current status of the sea otter population at Bering Island, Commander Islands, Russia. *In* G.R. Vanblaricom, ed. Third Joint U.S./Russia Sea Otter Workshop, Petropavlovsk Kamchatski, September 1991. Biological Rep. In press.

Estes, J.A. 1990a. Growth and equilibrium in sea otter populations. Journal of Animal Ecology 59:385-401.

Estes, J.A. 1990b. Action plan for sea otters. *In* P. Foster-Turley, S. MacDonald, and C. Mason, eds. Otters: an action plan for their conservation. Chicago Zoological Society, Brookfield, IL.

Estes, J.A., and G.R. VanBlaricom. 1985. Sea otters and shellfisheries. Pages 187-235 *in* R. Beverton, J. Beddington, and D. Lavigne, eds. Conflicts between marine mammals and fisheries. Allen and Unwin, London, England.

Irons, D.B., D.R. Nysewander, and J.C. Trapp. 1988. Prince William Sound sea otter distribution in relation to population growth and habitat type. U.S. Fish and Wildlife Service. 31 pp. (Unpublished rep.)

Jameson, R.J. 1993. Results of the 1992 survey of the translocated sea otter population off the outer coast of Washington State. International Union for the Conservation of Nature and Natural Resources, Otter Specialist Group Bull. 8:2-4.

Jameson, R.J., K.W. Kenyon, S. Jeffries, and G.R. VanBlaricom. 1986. Status of a translocated sea otter population and its habitat in Washington. The Murrelet 67:84-87.

Jameson, R.J., K.W. Kenyon, A.M. Johnson, and H.W. Wight. 1982. History and status of translocated sea otter populations in North America. Wildlife Society Bull. 10:100-107.

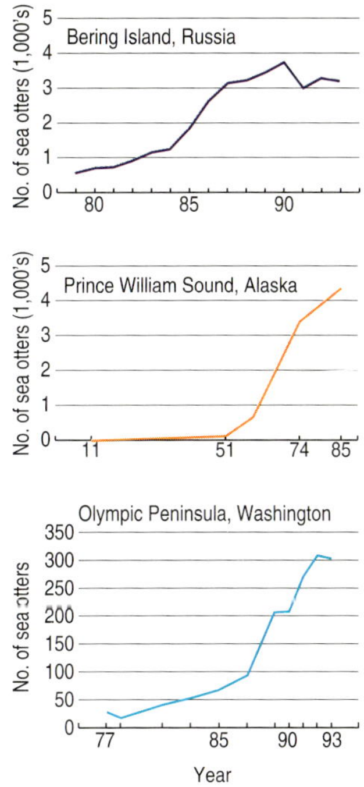

Fig. 2. Growth patterns observed in three sea otter populations in the north Pacific resulting from natural range expansion (Bering Island and Prince William Sound) or translocation (Washington).

Sea otter (*Enhydra lutris*).

For further information:

James L. Bodkin
National Biological Service
Alaska Science Center
1011 E. Tudor Rd.
Anchorage, AK 99503

Johnson, A.M. 1987. Sea otters of Prince William Sound, Alaska. U.S. Fish and Wildlife Service, Alaska Fish and Wildlife Research Center, Anchorage. 87 pp. (Unpublished rep.)

Kenyon, K.W. 1969. The sea otter in the eastern Pacific Ocean. U.S. Fish and Wildlife Service North American Fauna 68. 352 pp.

Lensink, C.J. 1962. The history and status of sea otters in Alaska. Ph.D. thesis, Purdue University, West LaFayette, IN. 188 pp.

Riedman, M.L., and J.A. Estes. 1990. The sea otter (*Enhydra lutris*): behavior, ecology and natural history. U.S. Fish and Wildlife Service Biological Rep. 90(14). 126 pp.

Pacific Walruses

by
Gerald W. Garner
National Biological Service

Pacific walruses (*Odobenus rosmarus divergens*) live in the Bering and Chukchi seas of Alaska and Russia (Figure). The population is subject to a Native subsistence harvest in Alaska and a commercial and subsistence harvest in Russia. Total annual harvest ranges between 5,500 and 10,300 walruses (Fay et al. 1989). The Marine Mammal Protection Act requires management of the population within an optimal sustainable population range, and the subsistence harvest by Alaskan Natives cannot be regulated unless the population is declared depleted.

Pacific walruses are an important source of meat and ivory for Native peoples of Alaska and the Chukotka Peninsula, Russia. The species is long-lived, has a relatively low reproductive rate, and occupies a position near the top of the marine food chain. Thus, besides being a very visible species, the walrus may be an indicator of the health of the Arctic marine ecosystem. The United States and the former Soviet Union initiated cooperative surveys throughout the entire range of the shared population in 1975 and have since conducted periodic surveys at 5-year intervals.

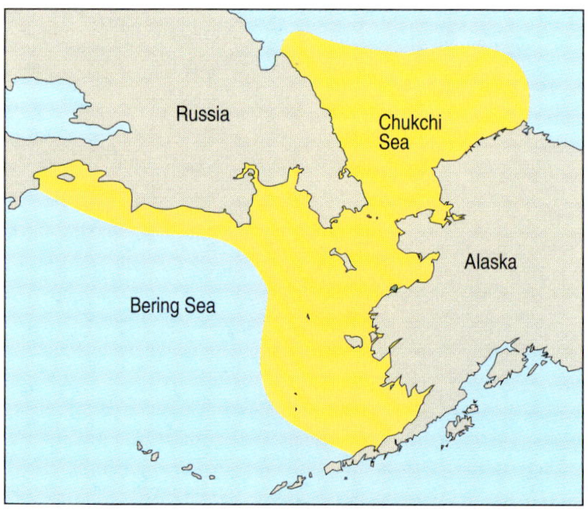

Figure. Distribution of Pacific walruses in the Bering and Chukchi seas of Alaska and Russia (Fay 1982).

U.S.-Russian Walrus Surveys

Walruses are gregarious and often form large groups when resting on sea ice or land. This behavior is called "hauling-out," and land sites where large groups traditionally congregate to rest are commonly called "haul-outs." The cooperative U.S.-Russian surveys used aerial counts of walruses on sea ice in the Russian and U.S. sectors, aerial and photographic counts at Russian land haul-outs, and ground and aerial counts at U.S. land haul-outs (Estes and Gilbert 1978; Estes and Gol'tsev 1984). Aerial surveys were conducted in the U.S. sector during 1975, 1980, and 1985, and were extended to include sea ice within the Russian sector during 1990 (Gilbert et al. 1992). Biologists altered each subsequent aerial survey to increase the precision of the estimates (Johnson et al. 1982; Gilbert 1986, 1989; Hills and Gilbert 1994).

Because of the ongoing efforts to improve the surveys, specific techniques varied among years but the basic design was to fly a series of north-south transects beginning at the edge of the polar ice pack and ending where concentration of ice was sufficient to exclude walruses. Transects were arranged systematically and stratified to achieve maximum coverage of the Chukchi Sea. Transects were located approximately between Pt. Barrow, Alaska, and the international border in 1975, 1980, and 1985, and throughout the entire Chukchi Sea during 1990. Most land haul-outs also were surveyed from aircraft, either by counts made directly by observers or from photographs. Some haul-outs were visited and counted by observers on the ground.

Biases were evident in the survey data, and lack of precision was common in all surveys (Estes and Gilbert 1978; Johnson et al. 1982; Gilbert 1989; Gilbert et al. 1992). Surveys, however, were continued because biologists believed that, despite these faults, the surveys would indicate population trends and were the best available method for assessing population size (Johnson et al. 1982; Gilbert 1986). Also, researchers recognized that an unknown and variable part of the walrus population was not available for counting because the number of walruses that were hauled out on land or ice varied significantly from day to day (Estes and Gilbert 1978; Gilbert 1989; Gilbert et al. 1992). None of these surveys used a correction factor for this unobserved fraction, and no attempts were made to classify walruses by age or sex. Even though population trends cannot yet be reliably determined by these surveys, researchers believe that long-term data from the surveys will eventually provide more definitive information about the status and trends of walrus populations.

Walrus Population Estimates

The point estimates for walrus population size were 221,000 for 1975, 246,000 for 1980, 234,000 for 1985, and 201,000 for 1990. Even though confidence intervals of these estimates were large, these estimates are considered the best information available to assess the status and trends of the Pacific walrus (Hills and Gilbert 1994). Estimates from sea ice exceeded those from land haul-outs except during 1990, when the ice pack receded much farther north and over deeper water than in most years. Because most of the large land haul-outs were in Russia, estimates there are higher than in the United States. Although these data indicate a general decline in numbers of walrus between 1975 and 1990, some biologists question the validity of this apparent decline (Hills and Gilbert 1994). Other researchers believe the population may be declining, based on various biological indices (Fay et al. 1989).

References

Estes, J.A., and J.R. Gilbert. 1978. Evaluation of an aerial survey of Pacific walruses (*Odobenus rosmarus divergens*). Journal Fisheries Research Board Canada 35:1130-1140.

Estes, J.A., and V.N. Gol'tsev. 1984. Abundance and distribution of Pacific walrus, *Odobenus rosmarus divergens*: results of the first Soviet-American joint aerial survey, autumn 1975. Pages 67-76 *in* F.J. Fay and G.A. Fedoseev, eds. Soviet-American cooperative research on marine mammals. Vol. 1: Pinnipeds. National Oceanographic and Atmospheric Administration Tech. Rep., National Marine Fisheries Service 12.

Fay, F.H. 1982. Ecology and biology of the Pacific walrus, *Odobenus rosmarus divergens* Illiger. North American Fauna 74:1-279.

Fay, F.H., B.P. Kelly, and J.L. Sease. 1989. Managing the exploitation of Pacific walruses: a tragedy of delayed response and poor communication. Marine Mammal Science 5:1-16.

Gilbert, J.R. 1986. Aerial survey of Pacific walrus in the Chukchi Sea, 1985. U.S. Fish and Wildlife Service, Anchorage, AK. 41 pp.

Gilbert, J.R. 1989. Aerial census of Pacific walruses in the Chukchi Sea, 1985. Marine Mammal Science 5:17-28.

Gilbert, J.R., G.A. Fedoseev, D. Seagars, E. Razlivalov, and A. Lachugin. 1992. Aerial census of Pacific walrus, 1990. U.S. Fish and Wildlife Service Admin. Rep. R7/MMM 92-1. 33 pp.

Hills, S., and J.R. Gilbert. 1994. Detecting Pacific walrus population trends with aerial survey—a review. Transactions of the North American Wildlife and Natural Resource Conference 59. Wildlife Management Institute, Washington, DC. In press.

Johnson, A., J. Burns, W. Dusenberry, and R. Jones. 1982. Aerial survey of Pacific walrus, 1980. U.S. Fish and Wildlife Service, Anchorage, AK. 32 pp.

For further information:

Gerald W. Garner
National Biological Service
Alaska Science Center
1011 E. Tudor Rd.
Anchorage, AK 99503

Mentasta Caribou Herd

by
Kurt Jenkins
National Biological Service

The Mentasta caribou (*Rangifer tarandus*) herd, a small herd that lives in and around Wrangell-St. Elias National Park and Preserve, Alaska, experiences population trends and management problems that are typical of many mountain herds in central Alaska and the Yukon Territory of Canada. Traditionally, the herd has been important for sport and subsistence hunting, but a recent decline in numbers led to suspension of hunting in 1992. The Alaska National Interest Lands Conservation Act authorizes the National Park Service to allow subsistence hunting throughout Wrangell-St. Elias, and sport hunting on preserve lands, provided that hunting is consistent with sound wildlife management principles and conservation of natural and healthy populations. Even though the National Park Service allows hunting, other agency mandates do not allow predator control or habitat management to enhance declining populations for hunting.

Sound information on caribou populations, gathered every year, is used to determine when hunting seasons are allowed and how many caribou can be taken by hunters. The collection of reliable data will help minimize conflicts between the dual objectives of providing hunting opportunities and maintaining natural characteristics of wildlife populations. Information on wildlife populations in national parks also provides important insights on natural population fluctuations for comparison with more actively managed wildlife on adjacent lands.

Biologists have monitored population size and composition of the Mentasta herd routinely since 1973 to provide basic information for management. They expanded monitoring and research in 1992 to improve their understanding of population-limiting factors during a period of rapid population decline.

Surveys of the Caribou Herd

Biologists have estimated the size of the Mentasta herd nearly each year since 1973 from aerial surveys conducted after the calving season. During late June, caribou congregate in high-mountain habitats or snow fields, where they are most readily visible from airplanes, and are counted by biologists.

Biologists also determine the population composition of the herd twice annually: after calving season in late June and during breeding season in early October. They classify caribou as calves, cows, or bulls. The counts in late June provide an index of early calf survival; counts in early October provide an index of summer survival of calves and proportions of bulls in the population.

In 1992 and 1993, biologists determined birth rates of cows to see whether low calf-to-cow ratios in late June resulted from low productivity. They determined birth rates by inspecting cows at close range from a helicopter during the peak of calving; observers looked for presence of calves or swollen udders, indicating cows had or would soon produce calves.

In 1993 biologists measured survival rates of calves and adult cows to help interpret causes of the rapid population decline observed in the early 1990's. They measured survival rates by fitting 39 calves and 41 adult cows with radio-collars containing mortality sensors. They located these radio-collared caribou daily during the calving season in 1993, weekly during the remainder of summer, and once every 2 months throughout winter. When biologists located carcasses of dead caribou, they inspected them as soon as possible to determine the cause of death.

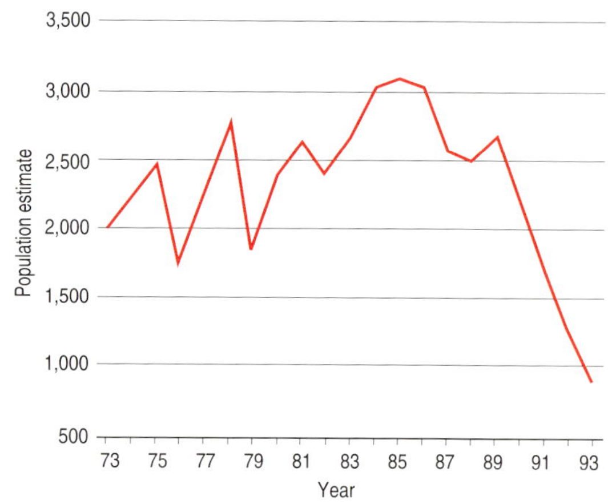

Fig. 1. Recent trends in size of the Mentasta caribou herd, 1973-93.

Population Trends

The Mentasta herd increased from about 2,000 caribou in the early 1970's to 3,200 in the early 1980's, an increase of about 5% per year (Fig. 1). Since 1989 the Mentasta herd has decreased to a low of around 900 caribou in 1993, a decrease of about 24% per year. Between 1992 and 1993 alone, the herd decreased by one-third.

This population decline appears related, in part, to changes in calf survival or production between the 1980's and 1990's. The proportion of cows with calves in late June declined from 39 calves to every 100 cows in the late 1970's and early 1980's (including a high of about 50 calves to 100 cows in 1979), to only 6 calves to every 100 cows in the early 1990's (Fig. 2). Similarly, estimates of calf-to-cow ratios in October have decreased about 90% since the 1980's.

Recent surveys of birth rates indicate that rapid declines in calf-to-cow ratios were not related to poor productivity of cows. In 1992 an estimated 81% of cows produced calves; in 1993, 70% did. Although birth rates were below average in 1993, productivity was sufficient for the herd to grow in the absence of high calf losses.

By intensively radio tracking newborn calves, biologists showed that the low calf-to-cow ratios were related to high death rates of calves. Of 39 calves radio-collared at birth, only 1 (2.5%) survived the summer. The rest were lost to predation by gray wolves (*Canis lupus*), grizzly bears (*Ursus arctos*), and wolverines (*Gulo gulo*), or they died from unknown causes.

Fig. 2. Recent trends in calf production (June) and number of calves (October) in the Mentasta caribou herd, 1973-93.

Survival rates of adult cows also were low. Of 41 cows radio-collared at the beginning of the study, only 83% (34 cows) survived 1 year. By contrast, generally 88%-96% of adult cows survive each year in stable or increasing herds.

Ongoing monitoring will increase understanding of natural fluctuations of the herd and provide information for incorporating fluctuations into a scheme for determining harvest quotas. Currently, biologists propose to allow annual harvests equal to a small percentage of the number of calves in the herd each fall, a good index of population trends. This proposal will link the harvest to patterns of herd growth and incorporate the objectives of natural populations and resource use into one workable management model.

For further information:

Kurt Jenkins
National Biological Service
Wrangell-St. Elias Field Station
PO Box 29
Glennallen, AK 99588

Tundra or Arctic Hares

by
David R. Klein
National Biological Service

The tundra or Arctic hare (*Lepus othus*; systematic studies are being conducted because some researchers classify the hare as *Lepus timidus*) now has a restricted distribution in western Alaska (Figure). It occurs in tundra habitats and also in shrub communities along streams. Its primary foods are willows, grasses, and herbaceous plants. Indigenous people, particularly in the coastal tundra of the Yukon-Kuskokwim Delta regions, the Seward Peninsula, and the Kotzebue Sound drainages, have a long history of using the tundra hare for food and clothing. The hare has declined in number throughout much of its range; biologists do not know what has caused its reduced distribution or the decrease in numbers.

Distribution Records

We obtained information on the former and present distribution and numbers of the tundra hare from historical records and reports and from interviews of state and federal wildlife biologists and local residents (Bee and Hall 1956; Murie 1959; Anderson 1974). Biologists conducted limited reconnaissance surveys on the Alaska Peninsula during 1990 and 1991, in the Yukon-Kuskokwim Delta region in 1973, and on the Seward Peninsula and in the Kotzebue region during 1985, 1986, and 1993. Field surveys continue on the Seward Peninsula and near Kotzebue, along with studies of the habitat requirements of these hares. A mail survey to determine population status throughout their distribution is being initiated through the University of Alaska-Fairbanks.

Status

Historically, the tundra hare was present in the Alaskan Arctic north of the Brooks Range (the "North Slope") from the Colville River westward (Bee and Hall 1956), but there have been no records of hares in that region since 1951 (Figure). Circumstantial evidence suggests that the tundra hare may have declined after the arrival of the snowshoe hare (*Lepus americanus*), which was not present there early in this century. The relationship may be direct through food or parasites and disease, or indirect through increased numbers of predators during snowshoe hare population highs.

The northern limit of tundra hare distribution in the coastal area of western Alaska has shrunk southward, and the hare is now absent or extremely rare north of Kotzebue. Centers of abundance are the western Seward Peninsula and the Yukon-Kuskokwim Delta region, although numbers have remained low there since population highs in the 1970's. Throughout its southern distribution on the Alaska Peninsula, tundra hare densities are currently low; high densities were last reported there in the winter of 1953-54 (Schiller and Rausch 1956). Researchers at the University of Alaska-Fairbanks are attempting to explain reasons for the tundra hare's decline.

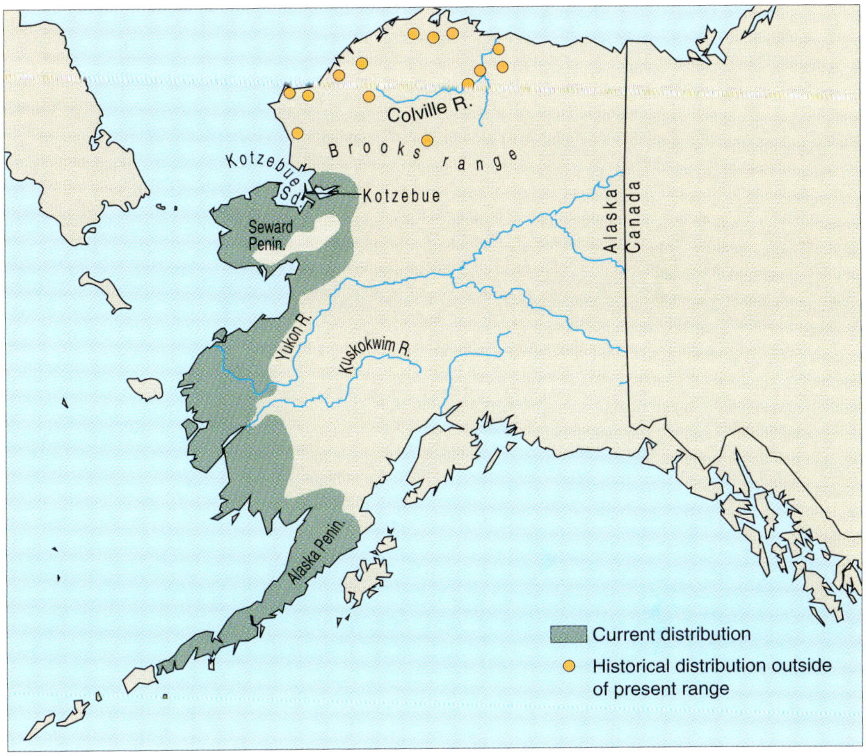

Figure. Distribution of the tundra hare, with historical records of occurrence outside of its present distribution.

Courtesy D.R. Klein, NBS

Tundra hare (*Lepus othus*).

References

Anderson, H.L. 1974. Natural history and systematics of the tundra hare (*Lepus othus* Merriam) in western Alaska. M.S. thesis, University of Alaska, Fairbanks. 106 pp.

Bee, J.W., and E.R. Hall. 1956. Mammals of northern Alaska. Miscellaneous Publ. 8, University of Kansas, Museum of Natural History, Lawrence. 309 pp.

Murie, O.J. 1959. Fauna of the Aleutian Islands and Alaska Peninsula. North American Fauna Series 61. 406 pp.

Schiller, E.L., and R. Rausch. 1956. Mammals of the Katmai National Monument, Alaska. Arctic 9:191-201.

For further information:

David R. Klein
National Biological Service
Cooperative Fish and Wildlife
Research Unit
University of Alaska-Fairbanks
Fairbanks, AK 99775

Courtesy M.J. Mac, NBS

Hawaii

by
Science Editor
J. Michael Scott
National Biological Service
Idaho Cooperative Fish and Wildlife Research Unit
Moscow, ID 83843

Overview

Of the thousands of islands in the world's oceans, none has captured the fancy and dreams of adventure more than those in the central and south Pacific. Even among those magical islands, however, the Hawaiian Archipelago stands out. Mark Twain remarked in *Roughing It*, "They are the loveliest fleet of islands that lies anchored in any ocean."

The Hawaiian Islands are geographically diverse. Stretching over some 2,200 km (364 mi) of ocean, they vary in size from hectares to thousands of square kilometers and in elevation from sun-drenched atolls less than 6 m (20 ft) above sea level up to snow-capped peaks 4,270 m (14,000 ft) high. Rainfall ranges from less than 50 cm (20 in) to more than 1,140 cm (450 in) per year. This diversity of environments and the islands' extreme isolation (more than 4,000 km [2,490 mi] to the nearest continent) have resulted in a spectacular variety of species. The Hawaiian Islands are a true showcase of evolution that has resulted in degrees of endemism (species restricted to a particular area) unmatched anywhere else in the world. Studies show that on Hawaii 46% of mosses, 70% of ferns, 91% of flowering plants, 91% of gymnosperms, 99% of terrestrial mollusks and terrestrial arthropods, 100% of land mammals, and 81% of birds are endemic at the subspecies level (Gagné 1988).

Unfortunately, loss of species in the islands has been staggering, and what remains often occupies but a fraction of its historical range. Seventy percent of the extinctions known to have occurred in the United States took place in Hawaii. The islands have lost more than 50% of their birds (Scott et al. 1986; Scott et al. 1988; Olson and James 1991; Pyle, this section; Jacobi and Atkinson, this section); perhaps 50% of their plants, 90% of their native land snails, and an unknown percentage of their terrestrial insects. Flora and fauna that evolved over millions of years have been devastated in less than 2,000 years since the arrival of humans. But despite huge losses, what remains is spectacular.

Today's unique assemblage of species is rapidly being lost. Twenty-five percent of the U.S. endangered taxa occur in the islands. The reasons for their endangerment are many, but loss of habitat and introduction of non-native species are prominent factors. Both are the result of a steadily increasing human population and the more than 4 million tourists that visit the islands annually. Few visitors realize that the lush lowland vegetation and colorful flowers they marvel at are not native to the islands, but

Hawaii Biological Survey

by
Allen Allison
Scott E. Miller
Gordon M. Nishida
Bishop Museum, Hawaii

In 1992 the Hawaii State Legislature established a biological survey at Bishop Museum, Hawaii's Museum of Natural and Cultural History. The survey conducts an ongoing natural history inventory of the archipelago and locates, identifies, evaluates, and maintains the reference collections of all native and non-native species of flora and fauna within the state. The survey works in cooperation with other agencies, including the Hawaii Heritage Program, various state agencies, and the National Biological Service.

More than 14,000 terrestrial, 300 freshwater, and 4,000 marine species inhabit Hawaii (Table 1). Bishop Museum maintains the world's largest biological collections for Hawaii (ca. 4,000,000 specimens; Table 2). Through the Hawaii Biological Survey program, and in cooperation with many partner organizations, the museum is organizing information from these collections and associated literature into comprehensive computerized data bases and conducting field surveys to document distributions of these organisms. The resulting information base has many applications in conservation, agriculture, forestry, public health, fisheries, and land management.

Table 1. Terrestrial and freshwater plant and animal species in Hawaii. In addition, another 4,000 marine organisms inhabit Hawaiian waters. Species at risk include those on the federal lists of endangered, threatened, and candidate species (not including marine).

Taxon	Species (no.)	Endemic (%)	Species at risk
Lower plants	>1,800	Few	0
Higher plants	2,143	44	359
Nematodes	>147	Few	0
Mollusks	1,100	95	60
Insects and mites	>8,800	>60	340
Fish	>25	24	1
Amphibians	4	0	0
Reptiles	13	0	0
Birds	131	43	35
Mammals	19	5	0
Total	>14,182		795

In 1992 and 1993, the Hawaii Biological Survey:

- published a summary list of the more than 8,600 species of Hawaiian insects and related arthropods;
- produced a catalog of Hawaiian land snails, including nearly 1,000 species;
- continued progress on the book series *Reef and Shore Fauna of Hawaii*, with another volume nearing completion; and
- began a collaborative project with the Smithsonian Institution and local agencies to create a data base of specimens of Hawaiian plants. Other plant projects in progress include a manual of cultivated plants in Hawaii (2,500 species treated in detail, with an additional 10,000 species evaluated); a manual of marine algae; and an updated, electronic bibliography of Hawaiian plants.

A five-stage process was developed to implement the biological survey. For each major group of plants and animals, the process involves developing a computerized literature data base; preparing summary lists of species names (checklists) based on the literature, collections, and consultation with experts; creating a data base of specimen information in our collections; creating data bases of information from other collections and other sources or establishing computer linkage to this information; and filling gaps and updating information through field surveys.

Megalagrion pacificum, a damselfly in a genus endemic to the Hawaiian Islands.

Table 2. The comprehensive collections of Bishop Museum are a core resource for the Hawaii Biological Survey. This chart indicates the relative sizes of the Hawaiian collections, plus related materials from the Pacific region and elsewhere that provide the context for understanding the Hawaiian biota.

Group	Hawaiian collections	Total collections	%
Plants (including algae, etc.)	175,000	500,000	35
Marine invertebrates	250,000	500,000	50
Mollusks	3,000,000	6,000,000	50
Insects and mites	500,000	13,000,000	4
Fish	15,000	130,000	12
Terrestrial vertebrates	20,000	85,000	24
Library		100,000	
Archives		1,100,000	

For further information:

Allen Allison
Bishop Museum
Box 19,000-A
Honolulu, HI 96817

are instead a diverse collection of alien invaders that pose the biggest threat to the integrity of Hawaii's native ecosystems.

The alien species that have had the greatest impact on Hawaii's fragile flora and fauna, however, are the ungulates (e.g., pigs, goats, sheep, and cattle), which have devastated the native plants that evolved in the absence of grazers and browsers and thus had lost any protective mechanisms they might have had. Pigs have been particularly damaging, rooting through the understory, devastating large tracts of land, and creating a seedbed for alien plants and severe erosion, especially on the steep slopes of older islands. Elimination of alien species and prevention of invasion by new non-native species are the first conservation priority in Hawaii (Stone and Scott 1985; Stone et al. 1992).

Status and trends of the amazingly diverse insects of Hawaii are described by Howarth et al. (this section). The case for using species of picture wing flies as monitors for change and evolution is made by Foote and Carson (this section), while the dramatic recovery of a plant, the Haleakala silversword (Loope and Medeiros, this section), gives hope that other species can respond to recovery efforts. There are lessons to be learned not only from our failures, but also from successes such as the silversword.

Interest in the plants and animals of Hawaii has rekindled in the last 20 years. Private, state, and federal biologists have sought to document the occurrence and abundance of species and have mounted an impressive attempt to save the remaining characters in this hotbed of evolution (*see* Vol. 38 of *BioScience* and Culliney 1988 for a review). To gain more information about Hawaii's resources, the state legislature formed the Hawaii Biological Survey in 1992, whose mission and scope are described in this section (Allison et al.).

References

Culliney, J.L. 1988. Islands in a far sea. Sierra Club Books, San Francisco, CA. 410 pp.

Gagné, W.C. 1988. Conservation priorities in Hawaiian natural systems. BioScience 38(4):264-271.

Olson, S.L., and H.F. James. 1991. Descriptions of thirty-two new species of birds from the Hawaiian Islands; Part I, Non-passeriformes. Ornithological Monographs 45:1-88.

Scott, J.M., C.B. Kepler, C. vanRiper, and S.I. Fefer 1988. Conservation of Hawaii's vanishing avifauna. BioScience 38 (4):238-253.

Scott, J.M., S. Mountainspring, F.L. Ramsey, and C.B. Kepler. 1986. Forest bird communities of the Hawaiian Islands: their dynamics, ecology and conservation. Studies in Avian Biology 9. 431 pp.

Stone, C.P., C.W. Smith, and J.T. Tunison. 1992. Alien plant invasions in native ecosystems of Hawaii management and research. University of Hawaii Press, Honolulu. 887 pp.

Stone, C.P., and J.M. Scott, eds. 1985. Hawaii's terrestrial ecosystems: preservation and management. University of Hawaii Press, HI. 584 pp.

Haleakala Silversword

by
Lloyd L. Loope
Arthur C. Medeiros
National Biological Service

The Haleakala silversword (*Argyroxiphium sandwicense* ssp. *macrocephalum*) was near extinction in the 1920's because of human vandalism and browsing by goats and cattle. The plant has increased under protection and deserves attention as the most dramatic conservation success story of the Hawaiian Islands.

The silversword is a distinctive, globe-shaped rosette plant with rigid (sword-like), succulent leaves densely covered by silver hairs. When a plant flowers at the end of its life, it produces a spectacular flowering stalk 0.5-2.0 m (1.6-6.4 ft) tall, typically with hundreds of maroon sunflowerlike flower heads. This plant receives more attention from visitors to Haleakala National Park than any other plant or animal because of its striking appearance and restricted distribution.

The Haleakala silversword is endemic to a 1,000-ha (2,471-acre) area at 2,100- to 3,000-m (6,890- to 9,843-ft) elevation in the crater and outer slopes of Haleakala Volcano, within Haleakala National Park, Maui, Hawaii. It is the most famous member of the endemic Hawaiian silversword alliance, perhaps the premier example of evolutionary adaptive radiation in plants. This morphologically diverse group comprises 28 species of herbs, vines, shrubs, trees, and rosette plants in three genera that evolved in the Hawaiian Islands from a North American tarweed (Asteraceae: Madiinae) ancestor (Robichaux et al. 1990; Baldwin et al. 1991). The monocarpic (flowers only once, at the end of its lifetime) silversword matures from seed to its final flowering stage in about 15-50 years. The plant remains a compact rosette until it sends up an erect, central flowering stalk, sets seed, and dies.

In 1992 this taxon was given threatened status by the U.S. Fish and Wildlife Service because of its extremely limited range and precarious life cycle. The other subspecies of *A. sandwicense* (ssp. *sandwicense*), endemic to Mauna Kea on the island of Hawaii, is federally listed as endangered, with fewer than 100 naturally occurring individuals.

Population Trends

The strikingly beautiful Haleakala silversword has always aroused the curiosity of human visitors to Haleakala Volcano. In pre-park days, plants were often removed by travelers to Haleakala Volcano as proof that the party had reached the summit, a practice that eventually seriously affected the silversword population. Browsing by feral goats and domestic cattle was also a significant factor in the silversword decline, but it was not a species preferred by these animals. By the 1920's, silversword numbers were so depleted that the Maui Chamber of Commerce sent a petition to Washington, DC, requesting that a serious effort be made to save the species (Loope and Crivellone 1986).

The first reliable quantitative information on silversword numbers is from the summer of 1935. In that year, Ranger S.H. Lamb tallied 1,470 plants (88 of which were flowering) on a single cinder cone (Ka Moa o Pele) within Haleakala Crater (Lamb 1935). Because about 217 plants were flowering within the crater at that time (Lamb 1935), a reasonable estimate of the total population is about 4,000 individuals.

Because silversword plants occur on otherwise barren cinder, fairly accurate counts are possible. Two studies since 1935 illustrate the trend of the silversword population over about 60 years of protection. Methods are described in the original reports (Kobayashi 1973, 1993; Loope and Crivellone 1986).

On Ka Moa o Pele, a single cinder cone where the largest number of plants were in 1935, the silversword population had increased from 1,470 to 6,528 plants as of 1991 (Fig. 1).

Elsewhere in Haleakala Crater, the silversword has increased in numbers and extent, large local populations having developed in areas where few plants occurred in 1935. A census of the entire silversword population has been attempted four times since 1971, with the following results: 1971: 43,262 (Kobayashi 1973); 1979-80: 35,000 (Kobayashi 1993); 1982: 47,640 (Loope and Crivellone 1986); and 1991:

Courtesy A.C. Medeiros, NBS

Haleakala silversword (*Argyroxiphium sandwicense* ssp. *macrocephalum*).

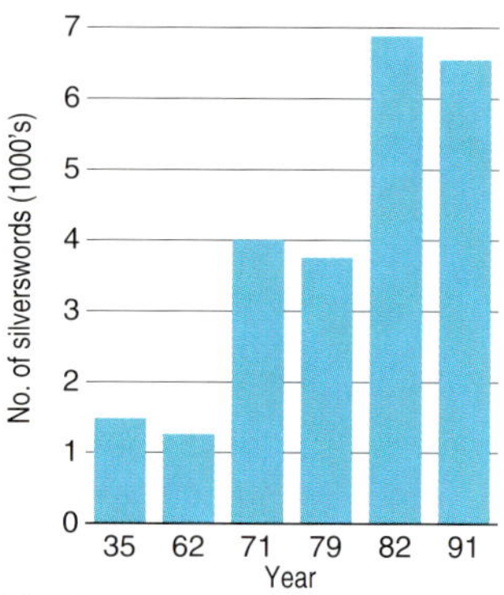

Fig. 1. Number of silversword plants counted by investigators on a single cinder cone, Ka Moa o Pele, within Haleakala Crater (Loope and Crivellone 1986; Kobayashi 1993).

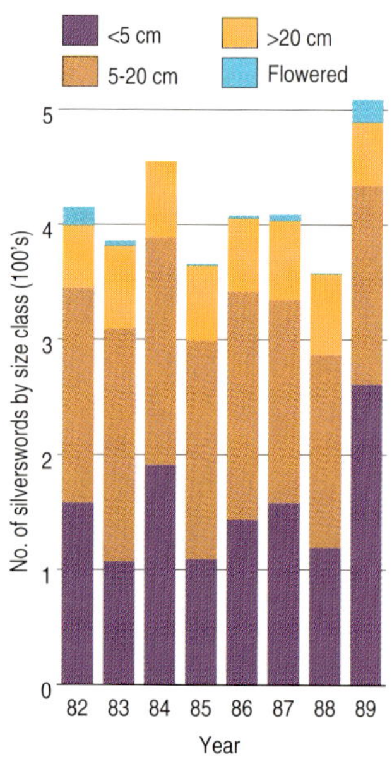

Fig. 2. Number by diameter classes of Haleakala silversword in fixed monitoring plots, 1982-89. Summary of data from eleven 5 m x 20 m (16.4 x 65.6 ft) plots in representative sites in Haleakala Crater (Loope and Medeiros 1994).

64,800 (Kobayashi 1993). The current population of Haleakala silversword is about 16 times larger than the estimated population in 1935.

Annual trends in 11 fixed plots, 5 m x 20 m (16.4 x 65.5 ft), from 1982 through 1989, suggest occurrence of substantial annual fluctuations in the recruitment and survival of seedlings (Loope and Crivellone 1986; Loope and Medeiros 1994; Fig. 2).

Data on Silversword Flowering

The Haleakala silversword flowers from June to September, with annual numbers of flowering plants varying dramatically from year to year. Reliable counts of flowering plants were made in 1935 (217 flowered) and in 1941 (815 flowered; Loope and Crivellone 1986). Numbers recorded in recent years have ranged from zero in 1970 to 6,632 in 1991. The environmental stimulus for flowering or nonflowering of silversword within a given annual flowering season is still unknown. An apparent relationship of the 1991 mass flowering event to stratospheric alteration by the eruption of Pinatubo Volcano in the Philippines is intriguing.

Threats

As a result of management within Haleakala National Park, the most serious former threats to the Haleakala silversword have been virtually eliminated: human vandalism and browsing by goats and cattle. To date, no introduced plant species competes significantly with silversword. Cooperative interagency efforts are being made to exclude the non-native mullein (*Verbascum thapsus*) and fountain grass (*Pennisetum setaceum*) from becoming established on Maui; since these plants occupy similar habitat on other Hawaiian Islands, they might compete with silverswords.

The greatest threat to the silversword appears to be potential loss of endemic pollinators because of the invasion of silversword habitat by the Argentine ant (*Iridomyrmex humilis*). This ant species occupies two disjunct areas between 2,070 m (6,792 ft) and 2,850 m (9,351 ft) elevation in Haleakala National Park, with a total area of 175 ha (432 acres; Cole et al. 1992). Because queens are unable to fly, the spread of this species is relatively slow. This alien ant species negatively affects the locally endemic arthropod fauna (Cole et al. 1992), including pollinators that evolved in the absence of ant predation. A marked expansion in the ant's range was noted in 1993, especially at higher elevations (Medeiros et al. 1994). Unless this ant species is controlled, it could cause potentially catastrophic effects on locally endemic biota, including the silversword, which is associated with several endemic insect species (Loope and Crivellone 1986) and which requires cross-pollination for successful seed set (Carr et al. 1986). Experimental control efforts are under way.

Trends

Recovery of the Haleakala silversword is one of the most dramatic single-species conservation success stories known. The primary factor contributing to its decline, human vandalism, was effectively addressed by the National Park Service beginning in the 1930's. Over the past 60 years the species has steadily recovered through protection within Haleakala National Park. It is increasing in numbers and expanding its range. Continued protection from human vandalism and feral ungulates, such as goats and cattle, is essential, and potential threats from the Argentine ant and alien plants must be addressed. Given the plant's limited range and precarious life cycle, the long-term prognosis for survival of this species appears remarkably favorable.

For further information:

Lloyd L. Loope
National Biological Service
Haleakala National Park
PO Box 369
Makawao, Maui, HI 96768

References

Baldwin, B.G., D.W. Kyhos, J. Dvorak, and G.D. Carr. 1991. Chloroplast DNA evidence for a North American origin of the Hawaiian silversword alliance. Proceedings of the National Academy of Science 88:1840-1843.

Carr, G.D., E.A. Powell, and D.W. Kyhos. 1986. Self-incompatibility in the Hawaiian Madiinae (Compositae): an exception to Baker's rule. Evolution 40:430-434.

Cole, F.R., A.C. Medeiros, L.L. Loope, and W.W. Zuehlke. 1992. Effects of the Argentine ant (*Iridomyrmex humilis*) on the arthropod fauna of high-elevation shrubland, Haleakala National Park, Maui, Hawaii. Ecology 73:1313-1322.

Kobayashi, H.K. 1973. Ecology of the silversword, Haleakala Crater, Hawaii. Final Rep. HALE-N-3. 124 pp.

Kobayashi, H.K. 1993. Census report on the Haleakala silversword *Argyroxiphium sandwicense* dc. (Compositae) ssp. *macrocephalum* (Gray) Meyrat for 1980 and 1991. Hawaii Natural History Association and Haleakala National Park. 12 pp.

Lamb, S.H. 1935. First progress report, silversword project, March 1935. Second progress report, silversword project, July 1935. National Park Service.

Loope, L.L., and C.F. Crivellone. 1986. Status of the silversword in Haleakala National Park: past and present. Cooperative National Park Studies Unit, University of Hawaii at Manoa, Department of Botany, Tech. Rep. 58. 33 pp.

Loope, L.L., and A.C. Medeiros. 1994. Impacts of biological invasion on the management and recovery of rare plants in Haleakala National Park, Maui, Hawaii. Pages 143-158 *in* M. Bowles and C.J. Whelan, eds. Restoration of endangered species. Cambridge University Press, Cambridge, UK.

Medeiros, A.C., F.R. Cole, and L.L. Loope. 1994. Patterns of expansion of an invading Argentine ant (*Iridomyrmex humilis*) population in Haleakala National Park, Maui, Hawaii. (Abstract) Bull. of the Ecological Society of America 75:150-151

Robichaux, R.H., G.D. Carr, M. Liebman, and R.W. Pearcy. 1990. Adaptive radiation of the silversword alliance (Compositae: Madiinae): ecological, morphological, and physiological diversity. Annals of the Missouri Botanical Garden 77:64-72.

Insects of Hawaii

by
Francis G. Howarth
Gordon Nishida
Bishop Museum, Hawaii
Adam Asquith
U.S. Fish and Wildlife Service

Insects are the dominant animals in most terrestrial ecosystems, especially on isolated oceanic islands where many larger animals are absent. In Hawaii, many of the original colonizing species evolved into perhaps 10,000 or more new species and adapted to live in the diverse island habitats. In addition to their importance as pollinators of native plants, recyclers of nutrients in ecosystems, and food for native birds and other animals, insects are also excellent subjects for evolutionary research. The isolation and habitat diversity of the Hawaiian Islands make them wonderful natural laboratories for studying ecology and evolution. Many important research projects have featured Hawaiian insects, such as the native *Drosophila* (*see* Foote and Carson, this section) and crickets (Otte 1989).

Because insects are important components of ecosystems, insect surveys can be used to assess the health of native ecosystems, and reserve managers often need to be able to determine the status of insects to properly manage other natural resources. Such assessments, however, are daunting tasks: although about 5,100 native insect species have been described in Hawaii, probably at least as many more remain undescribed or unknown. In addition, about 2,600 insect species have been established through human activities. Many native species are declining from the combined effects of invasive non-native organisms and human alteration of habitats.

Information on the status of Hawaiian insects came from a data base compiled at the Bishop Museum of all published records on the taxonomy, biology, and distribution of Hawaiian arthropods (Nishida 1992). Further information on the status and trends of selected rare species was obtained from label data of preserved specimens, especially those in the research collections at Bishop Museum and University of Hawaii, Honolulu, as well as from personal communications and observations of researchers in the field. Population surveys are in progress to determine the status and trends of a few insect groups such as the damselflies (*Megalagrion*; Polhemus 1993) and cave species.

Insects of Hawaii

Only 16 out of 30 insect orders recognized worldwide are represented in the native fauna. Another 11 orders have become established through human activities (Figs. 1 and 2). The beetles (Coleoptera), flies (Diptera), bees and wasps (Hymenoptera), and moths (Lepidoptera) are the largest groups in the Hawaiian Islands. Most native species are found on the high, main islands, but each of the northwest Hawaiian Islands harbors a few interesting species (Fig. 3). Oahu currently has the most known species, but this stems from collecting bias because most entomologists have lived and worked on Oahu. Maui and Kauai, in particular, should have comparable numbers. Western Maui, for example, was missed in the early insect surveys, and its insect fauna remains poorly known. About 63% of the species occur on only one island, and many have extremely restricted ranges within their island. This limited distribution and lack of information on how many species there are and where they survive have important consequences in planning for their conservation.

Trends

Profound changes are occurring in the Hawaiian insect fauna. Increasing contact with the outside world has broken the isolation that allowed the evolution of native species. The changing composition of the Hawaiian insect

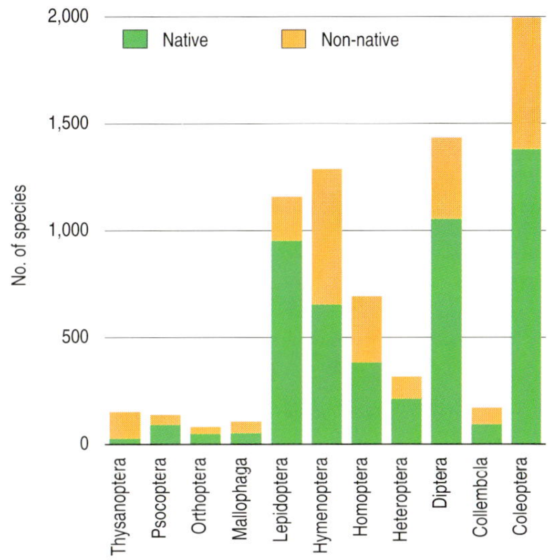

Fig. 1. Comparison of native and non-native insects in the larger orders (i.e., represented by more than 75 species) in Hawaii. Source: Hawaiian Terrestrial Arthropod Database, February 1994.

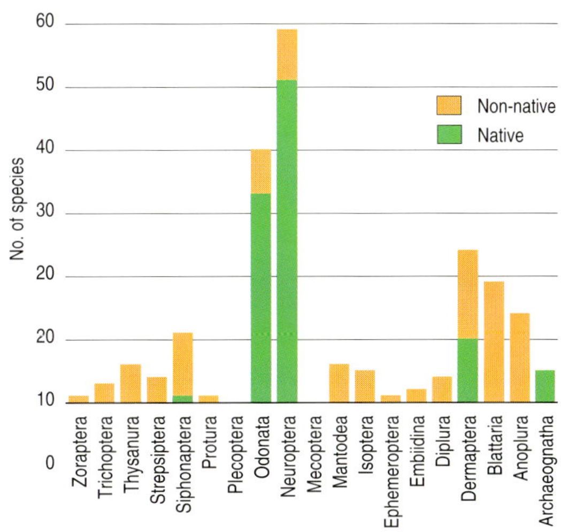

Fig. 2. Comparison of native and non-native insects in the smaller orders (i.e., represented by fewer than 75 species) in Hawaii. Source: Hawaiian Terrestrial Arthropod Database, February 1994.

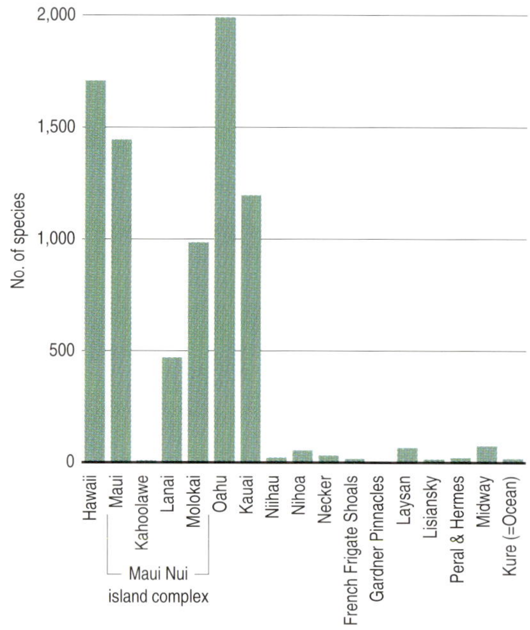

Fig. 3. The island distribution of native insect species.

fauna is readily apparent from the contrast between historical collections and reports (e.g., Perkins 1913; Zimmerman 1948) and more recent records and surveys. This change is particularly obvious in lowland areas where land conversion (Cuddihy and Stone 1990) and the establishment of alien species have eliminated or drastically reduced the abundance and diversity of native arthropods. For example, Asquith and Messing (1993) found that less than 10% of the insect fauna of a lowland agricultural area on Kauai is composed of native species, and, at a low-elevation site on the island of Hawaii, even the arthropod community on the native tree ‘Ohi ‘a lehua (*Metrosideros polymorpha*) is composed primarily of alien species (Gagné 1979).

At higher elevations in more intact vegetative communities, invasive alien arthropods have become dominant in some guilds, such as honey bees as pollinators and millipedes and isopods as detritivores. The effects of predatory species, such as the Argentine ant (*Linepithema humilis*; Cole et al. 1992) and the western yellowjacket (*Vespula pensylvanica*; Gambino et al. 1990), in the decline of some native groups are well documented.

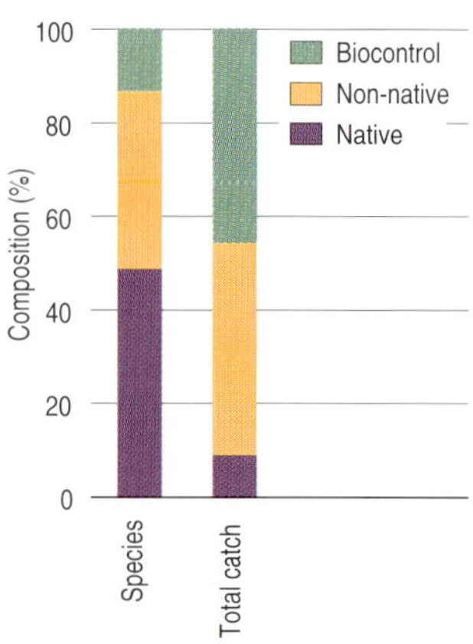

Fig. 4. Species and total catch composition of wasps in the families Braconidae and Ichneumonidae in a high-elevation mesic forest on the Hawaiian island of Kauai. Wasps were sampled from January 1992 to January 1993 by using a Malaise trap.

Introduced Parasites

Although the native Hawaiian faunas naturally bear some pressure from parasitoids, with endemic taxa of wasps in the families Ichneumonidae, Bethylidae, Diapriidae, Eucoiliidae, and Eulophidae, and the fly family Pipunculidae, the taxonomic composition and therefore the ecology of parasitism itself have been altered by the addition of alien species. For example, the Hawaiian Islands originally had no native species of braconid wasps, but now harbor 76 species in 42 genera (Nishida 1992). Many of these parasitoids are not confined to disturbed habitats or alien hosts. By using a Malaise trap (a tentlike net left in place to capture flying insects), A. Asquith and M. Kido (USFWS and University of Hawaii, Kauai, unpublished data) recently sampled the parasitic wasp community in a high-elevation native mesic forest on the island of Kauai for a full year. Of the 17 species of Braconidae and Ichneumonidae captured, all but one are parasitoids of moth larvae and pupae, and all but two are known to attack native Hawaiian moths. No known species of native ichneumonid wasp in this forest is extinct, and the endemic taxa still contribute the most to species diversity (Fig. 4). Human activities, however, have essentially doubled the number of species parasitizing native Lepidoptera. Furthermore, parasitoid abundance in this community is dominated by non-native species (Fig. 4), with less than 1 in 10 parasitoids being native to Hawaii. On a numbers of individuals per species basis, the two species introduced for biological control (*Eriborus sinicus* and *Meteorus laphygmae*) are more invasive in this forest than the supposed inadvertently introduced species of parasitoids. Not only have these introductions increased the number of species and individuals parasitizing Hawaii's native Lepidoptera, but also the new species have searching and immobilizing behaviors to which the native fauna is unaccustomed.

Populations of the native stink bug (Pentatomidae) genera *Coleotichus* and *Oechalia* dramatically decreased after 1962 following the purposeful introductions of a tachinid fly and several scelionid wasps for biological control of the non-native pest, the southern green stink bug (*Nezara viridula*). The koa bug (*Coleotichus blackburniae*) is the largest and most spectacular native Hawaiian true bug and was, until the early 1970's, common on all the major Hawaiian islands, including within the city of Honolulu, where it could frequently be found on introduced acacia trees. Numerous specimens were deposited in the insect collections of the Bishop Museum and University of Hawaii every decade from 1890 to 1970, but very few specimens have been seen since 1978. Because the koa bug is conspicuous, and its rarity has been publicized (Howarth 1991), its population decline seems real and not an artifact of survey effort.

Since the koa bug is gregarious and hundreds of individuals could be collected from a single tree, it was used as an alternate host for rearing introduced parasites before their release. Thus, circumstantial evidence implicates these biological-control introductions in the demise of these native bugs. Recent observations suggest that small populations of the koa bug still survive on most of the major islands, but quantitative status surveys and protection for this insect may need to be initiated to ensure its continued existence.

These examples support the arguments of Gagné and Howarth (1985) and Howarth (1991) that alien parasitoids are the major factor contributing to the decline and extinction of many native insect species. Lepidopteran caterpillars were an important food source for native forest birds and other native organisms; thus, their decline may affect other parts of the forest community. The ability of non-native arthropods to invade intact native communities demonstrates that conservation efforts aimed at habitat preservation, or the selection and management of nature reserves based on plant diversity or endemism, may not provide sufficient protection for some insects and their associated biota because of the continued emphasis on biological control and insufficient quarantine control in Hawaii. The effect of invasive alien arthropods means that we could save the forest and still lose the bugs, but we would eventually lose the forest as well because of the loss of pollinators and other functional groups of insects.

Extinctions

With at least 50% of Hawaii's native birds (Stone 1989) and mollusks (Solem 1990) extinct, it is likely that Hawaii has also lost a significant proportion of its terrestrial arthropod fauna. While 36 arthropod species are recognized as extinct by the U.S. Fish and Wildlife Service, populations of two species, a damselfly (*Megalagrion nesiotes*) and a sphinx moth (*Manduca blackburni*), have recently been rediscovered. The lack of intensive surveys for most of Hawaii's rare arthropods makes their status equivocal and weakens arguments for the allocation of conservation resources for these animals.

One of the few areas in Hawaii where arthropod extinctions are reasonably well documented is on Laysan Island in the northwestern part of the chain. While only 3.8 km^2 (1.5 mi^2) in size, it harbored at one time a native arthropod fauna of at least 77 taxa with at least 14 endemic species (Conant et al. 1984). With intensive surveys during the 1960's and 1980's, we now know that 35% of Laysan's endemic species are extinct. Other evidence of arthropod extinctions comes from those species associated with endangered or extinct plants. In 1917 a new species of *Proterhinus* weevil was collected from the last remaining tree of *Hibiscadelphus giffardianus* on the island of Hawaii. While the tree has been given a reprieve from extinction by propagation of individuals from seed, the weevil, which breeds in dying branches, was doomed with the death of the last wild tree. Many Hawaiian insect groups are similarly extremely host-specific; for example, some species of long-horned beetles (*Plagithmysus*), with 139 known species, and leaf bugs (*Nesiomiris*), with 50 species, occur on rare hosts and face a similar fate.

Survey Needs

Waiting for confirmations of extinctions or the discoveries of relict populations is ineffective, reactive conservation and will not preserve Hawaii's remaining arthropods. We need to identify species early in their decline or at least before they slide beyond recovery (Howarth and Ramsay 1991). This report is limited to the insects, but other native invertebrates deserve mention, including the spiders and relatives (arachnids), sandhoppers and relatives (crustaceans) (Howarth and Mull 1992; Nishida 1992), and mollusks (Solem 1990; Cowie et al., in press). The worms and smaller invertebrate groups are even less well-known than the arthropods.

The urgency and effectiveness of status surveys are exemplified by one being conducted for Hawaii's damselflies. On the island of Oahu alone, two damselfly species are believed extinct, and three additional taxa are severely reduced from their historical ranges and in danger of extinction. For example, sometime between 1983 and 1985, *Megalagrion nigrohamatum nigrolineatum* disappeared from its usual haunts along streams near Honolulu. Surveys begun in 1990 have found it in only three isolated localities near the headwaters of Oahu streams. This represents a greater than 99% reduction in range in a decade. Most of its former habitat still appears suitable and the reasons for its decline are uncertain, but researchers suspect the decline results from the effects of non-native species, as well as habitat destruction (Polhemus 1993).

Status surveys of additional selected groups of arthropods should be a top priority so that appropriate conservation measures can be planned. Studies on the systematics of Hawaiian biota, including descriptions of new species, are also urgently needed. Whether a population represents a native or non-native species or 1, 10, or 20 closely related species has bearing on effective conservation strategies in reserves (Howarth and Ramsay 1991).

References

Asquith, A., and R.H. Messing. 1993. Contemporary Hawaiian insect fauna of a lowland agricultural area on

Courtesy W.P. Mull

The koa bug (*Coleotichus blackburniae*), the largest Hawaiian true bug (three-fourths inch long), was common, but few have been seen since 1978.

Kaua'i: implications for local and island-wide fruit fly eradication programs. Pacific Science 47:1-16.
Cole, F.R., A.C. Medeiros, L.L. Loope, and W.W. Zuehlke. 1992. Effects of the Argentine ant on arthropod fauna of Hawaiian high-elevation shrubland. Ecology 73:1313-1322.
Conant, S., C.C. Christensen, P. Conant, W.C. Gagné, and M. Lee Goff. 1984. The unique terrestrial biota of the northwestern Hawaiian Islands. Pages 77-94 *in* Proceedings of a Symposium on Resource Investigations in the Northwestern Hawaiian Islands. University of Hawaii, Sea Grant College Program, Honolulu.
Cowie, R.H., N.L. Evenhuis, and C.C. Christensen. Catalog of the native land and freshwater molluscs of the Hawaiian Islands. Bishop Museum Bull. in Zoology. In press.
Cuddihy, L.W., and C.P. Stone. 1990. Alteration of the native Hawaiian vegetation: effects of humans, their activities and introductions. University of Hawaii Press, Honolulu. 138 pp.
Gagné, W.C. 1979. Canopy-associated arthropods in Acacia koa and *Metrosideros* tree communities along an altitudinal transect on Hawaii Island. Pacific Insects 21:56-82.
Gagné, W.C., and F.G. Howarth. 1985. Conservation status of endemic Hawaiian Lepidoptera. Pages 74-84 *in* Proceedings of the 3rd Congress of European Lepidopterology Cambridge, 1982. Karlsruhe: Societas Europaea Lepidopterologica.
Gambino, P., A.C. Medeiros, and L.L. Loope. 1990. Invasion and colonization of upper elevations on East Maui (Hawaii) by *Paravespula pensylvanica* (Hymenoptera: Vespidae). Annals of the Entomological Society of America 83:1088-1095.
Howarth, F.G. 1991. The environmental impacts of classical biological control. Annual Review of Entomology 36:485-509.
Howarth, F.G., and W.P. Mull. 1992. Hawaiian insects and their kin. University of Hawaii Press, Honolulu. 160 pp.
Howarth, F.G., and G.W. Ramsay. 1991. The conservation of island insects and their habitats. Pages 71-107 *in* N.M. Collins and J.A. Thomas, eds. The conservation of insects and their habitats. Academic Press, London.
Nishida, G.M., ed. 1992. Hawaiian terrestrial arthropod checklist. Bishop Museum Tech. Rep. 1. 262 pp.
Otte, D. 1989. Speciation in Hawaiian crickets. Pages 482-526 *in* D. Otte and J.A. Endler, eds. Speciation and its consequences. Sinauer Associates, Sunderland, MA.
Perkins, R.C.L. 1913. Introduction to fauna Hawaiiensis. Vol. 1. D. Sharp, ed. Cambridge University Press, Cambridge. xv-ccxxvii.
Polhemus, D.A. 1993. Damsels in distress: a review of the conservation status of Hawaiian *Megalagrion* damselflies (Odonata: Coenagrionidae). Aquatic Conservation: Marine and Freshwater Ecosystems 3:343-349.
Solem, A. 1990. How many Hawaiian land snail species are left? and what we can do for them. Bishop Museum Occasional Papers 30:27-40.
Stone, C.P 1989. Native birds. Pages 96-102 *in* C.P. Stone and D.B. Stone, eds. Conservation biology in Hawaii. University of Hawaii Press, Honolulu.
Zimmerman, E.C. 1948. Insects of Hawaii. Vol. 1. University of Hawaii Press, Honolulu. 206 pp.

For further information:

Francis G. Howarth
Bishop Museum
Department of Natural Sciences
PO Box 19000-A
Honolulu, HI 96817

Drosophila as Monitors of Change in Hawaiian Ecosystems

by
David Foote
National Biological Service
Hampton L. Carson
University of Hawaii

Most of the world's named species of plants and animals are insects, which are the dominant component of the biological diversity of most terrestrial and aquatic ecosystems (Hammond 1992). In Hawaii, there are probably more than 10,000 native insect species, whose functions in Hawaiian ecosystems include prey for forest birds, pollinators of native plants, and decomposers associated with the cycling of plant nutrients.

The population trends and distributions for most Hawaiian insects are unknown and cannot possibly be determined for more than a small minority of species. To successfully develop management strategies to monitor and preserve our biological heritage, focal taxa or "indicator species" need to be identified and used to develop the biological information necessary for making management decisions (Quinn and Karr 1993). The data discussed here demonstrate how one well-studied group of Hawaiian insects, the Hawaiian *Drosophila* (Pomace flies), may serve as a focal insect taxon. The species diversity, underlying genetic diversity, and evolutionary history of this group have been described in detail. Their sensitivity to direct and indirect environmental change has also been demonstrated. These attributes make them an ideal model species to monitor and understand changes in patterns of biological diversity associated with human impacts on native ecosystems in Hawaii.

Background

Hawaiian *Drosophila* and, in particular, the large "picture-winged" species within the genus, are unique among living organisms because different levels of biological diversity within a single large, closely related group of species can be characterized by researchers. Polymorphisms (*see* glossary) due to inverted chromosomal segments have been used to assess genetic variation within and between species. The banding patterns of all five major chromosome arms among 106 species of Hawaiian picture-winged *Drosophila* have yielded a 5 million-year-old phylogeny (*see* glossary) that is rooted to species on the island of Kauai (Carson 1992). This work on the evolutionary history of Hawaiian *Drosophila* augments an extensive systematic treatment of the genus (Hardy 1965; Kaneshiro 1976).

More recent genetic surveys have complemented research on chromosome variation. These include the description of nuclear and mitochondrial DNA sequences and extensive fieldwork that describes genetic variation within and among populations and species of Hawaiian *Drosophila* for allozymes (*see* glossary) and quantitative traits (Carson et al. 1981; DeSalle and Hunt 1987). Attention has focused on characters thought to play an important role in speciation, the process underlying the diversification of *Drosophila* in Hawaii.

Molecular and cytological (*see* glossary) research has been paralleled by ecological research on the natural breeding sites of these species on specific Hawaiian plants, such as olapa (Araliaceae) and ohawai (lobelioids). Extensive surveys have determined that Hawaiian *Drosophila* are specialized microbivores (*see* glossary) that complete their life cycle in the decaying tissue of over 40 families of Hawaiian plants (Carson and Kaneshiro 1976). This ecological information is the most detailed for any group of native Hawaiian insects, and the combined phylogenetic and ecological data provide a firm foundation for further study of the position and function of these insects in ecosystems.

Current ecological studies focus on quantifying species diversity over ecosystem gradients and evaluating long-term trends in population sizes (Carson 1986; Foote and Carson, unpublished data). Among the patterns observed are changes in *Drosophila* community structure associated with invasions of nonindigenous plants and animals in Hawaii. One dominant trend is the increasing representation of the recently introduced cosmopolitan species of *Drosophila* in wet forest communities disturbed by feral pigs and alien weeds. A second pattern is the apparent decline of certain guilds of endemic picture-wing *Drosophila* and their host plants over a 20-year period of observation.

Close to one-fifth of the world's known *Drosophila* fauna are endemic to Hawaii. Invasions by nonindigenous *Drosophila* are adding to the diversity of the group. This abundance of species is increasingly useful for tracking biological diversity at several levels, from changes in chromosome inversion frequencies over altitudinal clines to the measurement of long-term changes in community structure.

Status and Trends

Methods

Beginning in 1971, as part of the International Biological Program, the relative frequencies of populations of 14 species of picture-wing *Drosophila* were measured in the Olaa Forest at Hawaii Volcanoes National Park on the island of Hawaii (Fig. 1). The fly populations were surveyed earlier by using baits placed on tree trunks, vines, and tree ferns, but the most recent survey (1992-93) used tree fern stipes exclusively. Since 1980 the surveys have employed nondestructive sampling where individuals are identified by unique wing and thorax markings in the field (Carson 1986; Foote and Carson, unpublished data).

Since 1982 four fenced feral pig exclosures have been constructed in rain forests where *Drosophila* surveys have been undertaken.

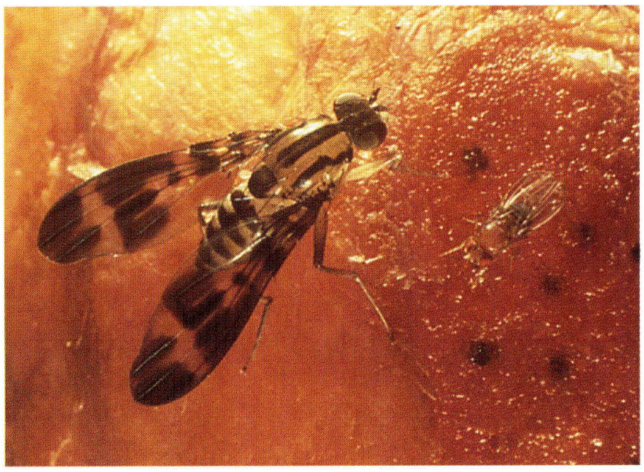

Courtesy W.P. Mull

An endemic Hawaiian picture-wing *Drosophila* (*D. conspicua*-left) perched on a *Clermontia* fruit next to a cosmopolitan member of the species complex that includes *D. simulans* (right), one of the common non-indigenous species in Hawaiian rain forests. The picture-wings are close to the size of common house flies, giants in comparison to their mainland relatives.

These exclosures average about 300 ha (740 acres) in size. The impact on *Drosophila* communities of removing non-native pigs has been evaluated through the comparison of recently introduced cosmopolitan and endemic flies attracted to baits in different-aged exclosures and adjacent forest where feral pigs are still active (Foote et al., unpublished data).

Using Chromosomes To Trace Evolutionary History

There are 491 described Hawaiian species in the family Drosophilidae. Most of the species belong to one of two genera, *Drosophila* and *Scaptomyza*. Among the Hawaiian species, 124 have been genetically surveyed, including 106 of 111 picture-wing species in the genus *Drosophila* (Carson 1992). Most species are single-island endemics, reflecting the forces of geographical isolation imposed by this volcanic archipelago.

Inversion polymorphisms (*see* glossary) have been detected within or between populations of about one-third of the species and their frequencies have been measured over environmental gradients in several well-studied species on the island of Hawaii (Carson 1992). Variations in the frequency of different polymorphisms along a gradient are used as an indicator of the role of natural selection in maintaining genetic variation. One such genetic gradient occurs among populations of *Drosophila*

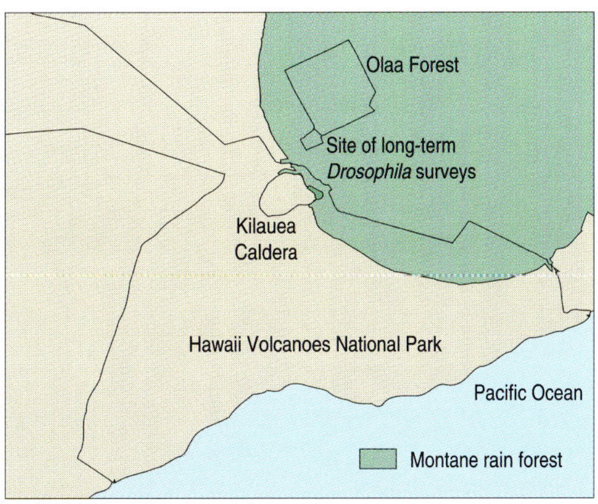

Fig. 1. Hawaii Volcanoes National Park, including the Olaa Forest, where population surveys of picture-wing *Drosophila* have been carried out.

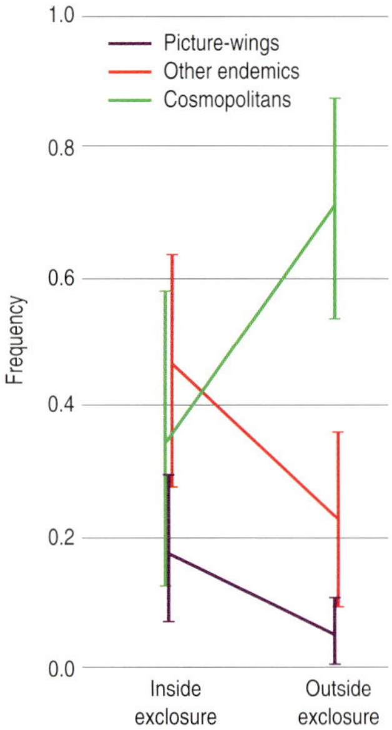

Fig. 2. A comparison of the average relative frequencies of Hawaiian picture-wing flies, other endemic *Drosophila*, and recently introduced cosmopolitan species inside and outside of a 7-year-old 240-ha (593 acre) fenced pig exclosure in Hawaii Volcanoes National Park. Frequencies are based on observations at 837 bait stations set up over five survey periods in 1992 and 1993 along four 2,400-m (7,875-ft) transects. The data are expressed as a percentage of total observations within a survey period.

silvestris above Kilauea Volcano and reflects the recolonization of habitat destroyed by two explosive eruptions within the last 2,100 years. The surfaces of Kilauea Volcano are covered by new lava flows at a rate of about 90% per 1,000 years. The population biology of Hawaiian *Drosophila* and other endemic insects has been one of continual local extinction of and recolonization by populations on single volcanoes over thousands of years (Carson et al. 1990).

Dominance of Cosmopolitan *Drosophila* in Disturbed Habitat

The dominance of non-native plant and animal species in Hawaii associated with human activity and the subsequent loss of endemic species have long been recognized (Perkins 1913). Rain forests above 1,000-m (3,280-ft) elevation provide habitat for much of the remaining native biota. Most of these wet forests occur on the "Big Island" of Hawaii where about 30% of the 500,000 ha (1.2 million acres) of upland native woodland is rain forest (Jacobi and Scott 1985). While this wet forest vegetation is among the most intact in the state, invasions by alien species have seriously degraded components of the understory.

Non-native ungulates (cattle, goats, pigs, etc.) cause major problems. Feral pigs, in particular, feed upon and uproot native tree ferns, shrubs, and herbs. They also actively consume fleshy fruits of non-native plants and thereby spread their seeds. As a consequence, feral pigs help establish non-native plants that can permanently alter native communities (Cuddihy and Stone 1990; Stone et al. 1992).

Twenty-five years ago fenced feral ungulate exclosures were first tested by Hawaii Volcanoes National Park as a method of promoting natural wet forest restoration. In the park, including approximately one-third of the Olaa Forest, exclosures encompassing as much as 800 ha (1,980 acres) are now used to manage tracts of montane wet forest (Stone et al. 1992).

When intensive research on Hawaii *Drosophila* was initiated over 30 years ago, it was apparent that habitats altered by human activity had greatly reduced populations of endemic *Drosophila* (Carson 1967). The construction of large fenced feral pig exclosures has provided an opportunity to measure changes in *Drosophila* community composition associated with this one particular agent of disturbance in wet forests. A significant increase in the frequency of cosmopolitan species of *Drosophila* has been measured in wet forest habitat disturbed by feral pigs and associated non-native plants (Fig. 2). In areas with high pig densities, many host plants for endemic *Drosophila* are reduced to those few individuals growing as epiphytes above the reach of pigs.

In contrast, many alien plants that thrive in pig-disturbed areas are species that produce fleshy fruits eaten by the pigs. These fruit-bearing non-native plants, such as banana poka (*Passiflora mollissima*) and yellow Himalayan raspberry (*Rubus ellipticus*), also support large populations of introduced cosmopolitan *Drosophila*, such as *D. immigrans*, *D. simulans*, and *D. suzukii* (known collectively to geneticists as "yellow flies"), that breed primarily on rotting fruit (Foote et al., unpublished data).

Long-term Changes Among Populations of Picture-wing *Drosophila*

Changes in the relative proportions of different species of endemic picture-wing flies in Olaa Forest between 1971 and 1993 are shown in Fig. 3. These proportions are based upon observations of individual picture-wings, totaling 1,222 in 1971, 1,467 in 1981, and 1,294 in 1992. A general decline in overall picture-wing diversity is suggested by the observation that 4 species out of 14 were missing from one or more of the more recent surveys. There has also been a change in the relative abundances of species within the group. For example, two of the most common species of picture-wing flies from the original survey, *D. murphyi* and *D. setosimentum*, are now much less common.

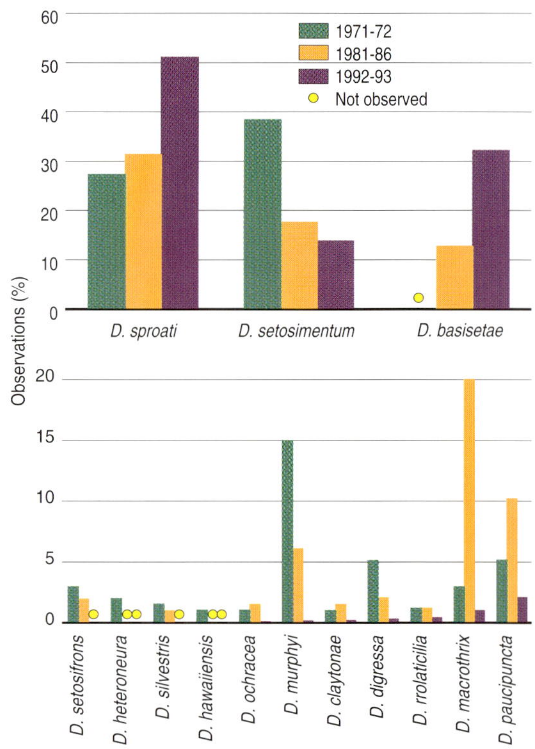

Fig. 3. A comparison of the relative frequencies of 14 endemic species of picture-wing *Drosophila* over three periods of observation from 1971 to 1993 in Olaa Rain Forest in Hawaii Volcanoes National Park. The data are expressed as a percentage of total observations within a survey period.

Population trends are suggested by analysis of guilds of picture-wing flies that breed on specific host plants. One example is the increase in the relative frequency of observations of *D. sproati*, a species that appears to breed exclusively in rotting bark of one of the most common trees in this rain forest, the endemic member of the Aralia family, Olapa (*Cheirodendron trigynum*). In contrast, a long-term decline is evident in the guild of four species that breed primarily in rotting bark of native lobelioids in the genus *Clermontia*. There has been a concordant reduction in frequency of all four species over the last two decades and two of the four species are now missing from this site (Fig. 4).

There is historical evidence that the decline in the latter group is a consequence of the reduction of host plant populations. For example, an important host of the two picture-wings that appear locally extinct, *Clermontia hawaiiensis*, has been extirpated from at least one nearby forest with a long history of disturbance by feral pigs and cattle.

Another factor may have been the invasion of Olaa Forest by alien western yellowjackets (*Vespula pensylvanica*) in the early 1980's. The wasps have become dominant predators of other insects and may have contributed to the decline of picture-wing *Drosophila* by feeding on larvae that are particularly exposed on *Clermontia* (Carson 1986; Foote and Carson, unpublished data). These and other potential causative agents of the changes in the community structure of *Drosophila* need further investigation.

Drosophila as a Focal Taxon To Monitor Ecosystem Change

Because of the extensive data that exist on the population genetics and evolutionary relationships within the picture-wing *Drosophila*, the potential consequences of disruption of the *Drosophila* community can be examined at several levels. For instance, the loss of a local population of *Drosophila silvestris* (one of the four species that breeds in *Clermontia*) occurs at the lower end of an altitudinal cline in inversion frequencies among populations that extend above Kilauea Volcano (Carson et al. 1990). A continued decline of *D. silvestris* populations at lower elevations has the potential to change inversion frequencies. Such long-term data may prove useful in evaluating why different populations or species may not respond similarly to greenhouse stresses associated with global climate change (Hoffman and Blows 1993).

The potential influence of species extinctions of picture-wing *Drosophila* on specific lineages of subgroups (Fig. 2) can also be evaluated. For example, the guild of lobelioid-associated picture-wing flies undergoing a decline in Olaa Forest is made up of species from three separate lineages of Hawaiian *Drosophila*. Among fossil records of Hawaiian birds, an analogous situation may have occurred with the extinction of several groups of flightless geese and geeselike terrestrial herbivores (Olson and James 1991). The monitoring of focal groups of Hawaiian insects may well complement our understanding gained from vertebrates on how changes in Hawaiian ecosystems selectively favor certain taxa that make up the contemporary species diversity found in the islands.

Lastly, data from active resource management programs, such as the construction of feral pig exclosures at Hawaii Volcanoes National Park, suggest that the population declines of certain Hawaiian *Drosophila* are reversible, even in the case of local extinction. Many host populations are recovering in Olaa Forest following removal of pigs, and nearby populations of many of the missing picture-wing species persist. This is a rain forest that has twice been devastated by volcanic eruptions and has recovered. We are testing an old evolutionary tradition on these islands as we encourage the recolonization of this protected habitat by its former occupants from nearby populations.

The 30 years of intensive research on Hawaiian *Drosophila* will not be readily repeated for even a small fraction of the remaining species that make up the biological diversity present in Hawaii. The fact that Hawaiian *Drosophila* flies have received so much attention is due in part to the fact that *Drosophila* are readily observed and sampled (at baits) in native forests. The sensitivity of such groups to a wide range of human impacts needs to be evaluated. Taxa, such as Hawaiian *Drosophila*, that can be monitored in a cost-effective manner and yield statistically reliable data need to be exploited as potential indicators of the impact of environmen tal change on the vast majority of species about which we know too little to manage intelligently.

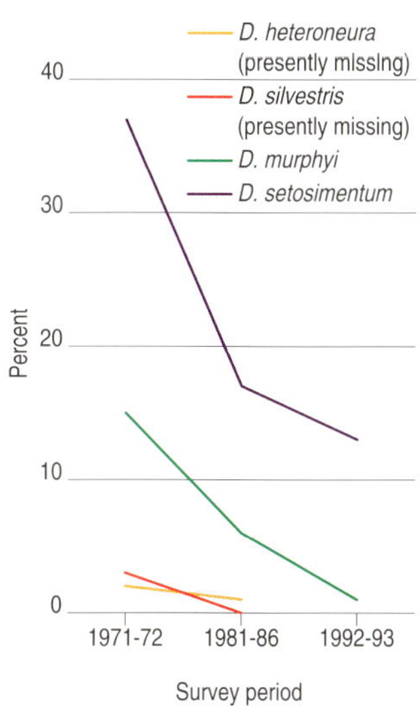

Fig. 4. Long-term trends within one host-specific guild of picture-wing *Drosophila* that breed in rotting bark of native lobelioids in the genus *Clermontia*.

References

Carson, H.L. 1967. Genetics and evolution of Hawaiian Drosophilidae: preliminary report of collections made in Kipahulu Valley, Maui, cooperatively with the Kipahulu Valley Expedition, August, 1967. *In* R.E. Warner, ed. Scientific report of the Kipahulu Valley Expedition: Maui, Hawaii, 2 August-31 August, 1967. The Nature Conservancy, Honolulu, HI.

Carson, H.L. 1986. *Drosophila* populations in the Ola'a tract, Hawaii Volcanoes National Park, 1971-1986. Proceedings Hawaii Volcanoes National Park Natural Science Conference 6:3-9.

Carson, H.L. 1992. Inversions in Hawaiian *Drosophila*. Pages 407-439 *in* C.B. Krimbas and J.R. Powell, eds. *Drosophila* inversion polymorphism. CRC Press, Ann Arbor, MI.

Carson, H.L., E.M. Craddock, W.E. Johnson, L.J. Newman, Y.K. Paik, W.W.M. Steiner, and K.C. Sung. 1981. Genetic studies of natural populations. Pages 438-470 *in* D. Mueller-Dombois, K.W. Bridges, and H.L. Carson, eds. Island ecosystems. Hutchinson-Ross, Stroudsburg, PA.

Carson, H.L., and K.Y. Kaneshiro. 1976. *Drosophila* of Hawaii: systematics and ecological genetics. Annual Review of Ecology and Systematics 7:311-345.

Carson, H.L., J.P. Lockwood, and E.M. Craddock. 1990. Extinction and recolonization of local populations on a growing shield volcano. Proceedings of the National Academy of Sciences 87:7055-7057.

Cuddihy, L.W., and C.P. Stone. 1990. Alteration of native Hawaiian vegetation. University of Hawaii Press, Honolulu. 138 pp.

DeSalle, R., and J.A. Hunt. 1987. Molecular evolution in Hawaiian Drosophilids. Trends in Ecology and Evolution 2:212-216.

Hammond, P. 1992. Species inventory. Pages 17-39 *in* B. Groombridge, ed. Global biodiversity: status of the earth's living resources. Chapman Hall, London.

Hardy, D.E. 1965. Insects of Hawaii. Vol. 12. Diptera: Cyclorrhapha II, Series Schizophora, Section Acalypterae I, Family Drosophilidae. University of Hawaii Press, Honolulu. 814 pp.

Hoffmann, A.A., and M.W. Blows. 1993. Evolutionary genetics and climate change: will animals adapt to global warming? Pages 165-178 *in* P.M. Kareiva, J.G. Kingsolver, and R.B. Huey, eds. Biotic interactions and global change. Sinauer and Associates, Sunderland, MA.

Jacobi, J.D., and J.M. Scott. 1985. An assessment of the current status of native upland habitats and associated endangered species on the island of Hawaii. Pages 1-22 *in* C.P. Stone and J.M. Scott, eds. Hawaii's terrestrial ecosystems: preservation and management. Cooperative National Park Resources Studies Unit, University of Hawaii, Manoa.

Kaneshiro, K.Y. 1976. A revision of the generic concepts in the biosystematics of Hawaiian Drosophilidae. Proceeding of the Hawaiian Entomological Society 22:255-278.

Olson, S.L., and H.F. James. 1991. Descriptions of thirty-two new species of birds from the Hawaiian Islands: Part 1, Non-passeriformes. Ornithological Monographs 45:1-88.

Perkins, R.C.L. 1913. Introduction. Vol. 1. Fauna Hawaiiensis. D. Sharp, ed. Cambridge University Press, Cambridge.

Quinn, J.F., and J.R. Karr. 1993. Habitat fragmentation and global change. Pages 451-463 *in* P.M. Kareiva, J.G. Kingsolver, and R.B. Huey, eds. Biotic interactions and global change. Sinauer and Associates, Sunderland, MA.

Stone, C.P., L.W. Cuddihy, and J.T. Tunison. 1992. Responses of Hawaiian ecosystems to removal of feral pigs and goats. Pages 666-704 *in* C.P. Stone, C.W. Smith, and J.T. Tunison, eds. Alien plant invasions in native ecosystems of Hawaii: management and research. University of Hawaii Press, Honolulu.

For further information:

David Foote
National Biological Service
Cooperative Parks Study Unit
PO Box 52
Hawaii National Park, HI 96718

Birds of Hawaii

by
Robert L. Pyle
Bishop Museum, Hawaii

The wild birds inhabiting Hawaii are unique and known worldwide. Native breeding birds rank among the world's highest in endemism, endangerment, and extinction, and Hawaii's total bird life contains a higher proportion of non-native species than perhaps any other area of comparable size. Interest in Hawaii's birds centers on the status and trends of its populations, understanding their ecological requirements, and developing measures to protect and conserve their remaining populations, which are dwindling at an alarming rate.

The unique nature of Hawaii's bird life results primarily from isolation. The Hawaiian Islands, a linear archipelago extending some 2,650 km (1,646 mi) from Kure to Hawaii, is 4,000 km (2,484 mi) from the nearest point in North America and 3,400 km (2,111 mi) from Asia. The wild colonizers, individual birds or small groups out over the ocean, were the first to stumble on Hawaii, where they remained to live and breed. This process has been going on for millions of years, with two species repeating the same process within the past 15 years. Then came evolution of new species in situ, as many of these original colonizers changed through adaptation and filled unused ecological niches in these young islands. During the past 2,000 years, humans began inhabiting the islands, bringing with them some birds that otherwise would never have reached Hawaii on their own. In addition, some strong-flying species that regularly migrate long distances have found Hawaii and developed annual migration patterns that bring them to the islands for part of each year during the nonbreeding season.

Current Status

It is convenient to categorize the wild birds of Hawaii into residents and visitors. Resident species remain permanently in Hawaii; visitors regularly come to Hawaii for only part of each year. Each group can be further divided. Residents are either native species that arrived or evolved here naturally or alien species brought in by humans. Visiting birds either come to Hawaii to breed or breed elsewhere and come during nonbreeding season. True pelagic species, which spend all their time at sea except when breeding, are considered to have visited Hawaii if they have occurred in Hawaii's offshore waters within the 200-nautical mile zone.

Native Residents

Native resident species may be either indigenous, meaning that others of the same species or subspecies reside elsewhere in the world, or endemic, meaning that they are found nowhere else. The latter may be endemic at subspecies level, at species level, or at genus or higher level. For example, endemic at subspecies level means that others of its species are found elsewhere, but the subspecies occurs only in Hawaii.

Alien Residents

Polynesians first settled in Hawaii roughly 2,000 years ago (Kirch 1982). Only one bird species brought by the early Polynesians still survives in Hawaii as an established alien species, the red junglefowl (*Gallus gallus*), ancestor of the domestic chickens. The

Hawaiians called it *moa*, not to be confused with the huge, extinct, flightless birds in New Zealand of the same name. How many other bird species may have been brought by the Polynesians and failed to become established is unknown.

Since Captain Cook first visited Hawaii in 1778, alien bird species have been brought to the islands in a steady stream. Only a few have been successful in establishing a continuing breeding population. Of the 54 alien species now considered to have established populations, fully half have origins in Asia (Fig. 1). Fewer are from North America and Africa; a few have come from Australia and South America. Among the continents, only Antarctica is not represented by an established alien species. Penguins have indeed been brought to Hawaii, and one thriving population is in captivity today. But were they to escape, they would not find sufficient krill and ice to maintain a wild population in the islands.

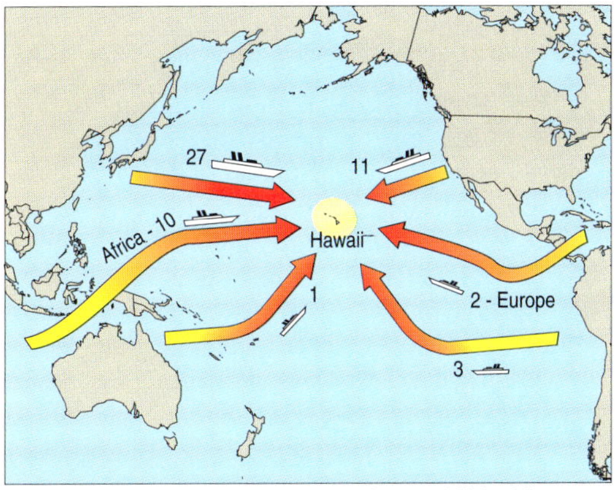

Fig. 1. Origins of the 54 alien species established in Hawaii.

Breeding Visitor Species

Visiting species that come to Hawaii to breed are basically pelagic, that is, living on the open ocean. They come to land to breed, but depart again as soon as parental duties end. Many go to the food-rich boundaries of ocean currents just north of the equator, but some species range throughout the North Pacific (Fig. 2). None appears at any other land during nonbreeding season. First-year birds of most species remain at sea for 3 years or more before returning to breed. Breeding visitors are the albatrosses, shearwaters, petrels, terns, and some tropicbirds. Other seabird species, including boobies, frigatebirds, and noddies, are classed as residents since they remain based at their breeding areas throughout the year, going to sea usually for only a few days at a time.

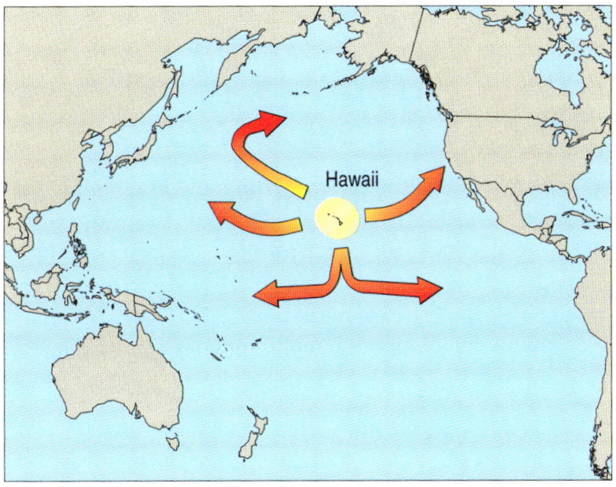

Fig. 2. Dispersal of the 13 breeding visitor species when not breeding.

Nonbreeding Visitor Species

A great many birds that breed elsewhere depart their nesting grounds after chick rearing is finished, some wandering freely and others following traditional migration routes. Some species, notably the familiar Pacific golden-plover (*Pluvialis fulva*) and some other shorebirds and ducks, have developed migration patterns that bring large numbers to Hawaii regularly each year, with some individuals even coming to the same plot of ground each winter (Johnson et al. 1981). For other species, just a few individuals turn up each year. For still others, an individual or two may be reported in only a few years out of ten. A number of species have been recorded in Hawaii fewer than a dozen times, perhaps only once or twice. All regular visitors and most others are strong flyers, accustomed to making long migration flights annually, or are larger birds able to store enough energy to reach Hawaii on their own. Almost all are waterbirds. Only nine species of passerine landbirds are known to have straggled to Hawaii, and most of these have been recorded only one or two times each. *Note that absolutely no species of small landbird migrates regularly to Hawaii, either for breeding or in nonbreeding season.* Virtually all nonbreeding visitor species nest in the northern hemisphere, most of them in the far north (Fig. 3). A few shearwaters and petrels, a skua, and the great crested tern (*Sterna bergii*) are the only exclusively southern hemisphere nesters that have straggled to Hawaii.

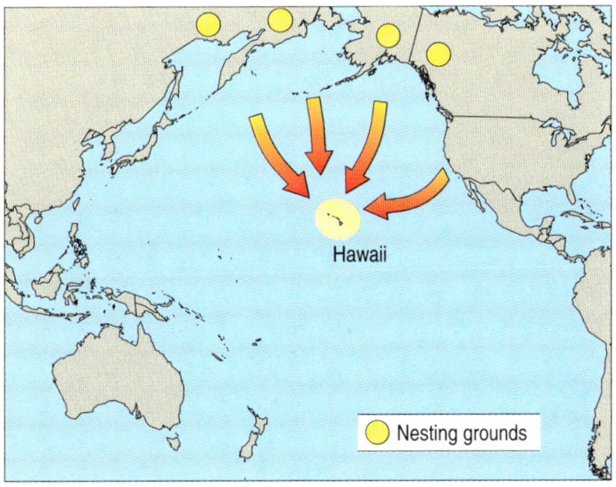

Fig. 3. Origins of the 155 visitor species that do not breed in Hawaii.

All Species

Of the 272 species resident now or recorded as visitors (Pyle 1992), roughly 40% are permanent residents, about equally divided between native and alien species (Fig. 4). The breeding visitors, all seabirds, are relatively few. The remaining 55% of the species are nonbreeding visitors. This large percentage for nonbreeding visitor species is not surprising, since these include many species that have wandered to Hawaii as individual stragglers. But in terms of total individuals, the picture is reversed. The nonbreeding visitor species account for probably the fewest individuals, while the breeding visitor seabirds have much larger populations in their huge nesting colonies in the unpopulated Northwest Hawaiian Islands. But the largest of all in total population are the alien residents, which include the ubiquitous Japanese white-eye (*Zosterops japonicus*), zebra dove (*Geopelia striata*), and other residents found almost everywhere in the main populated islands.

Birds known to have been in Hawaii in the past, but which are no longer there, can be summarized as follows: 16 species (resident-native) have become extinct since Captain Cook's visit; 35 or more species (subfossils, probably native residents) were extinct before Captain Cook's visit; and about 150 species are alien introductions not established. Adding these to the 272 species here now constitutes about 475 species of birds known to have occurred in Hawaii.

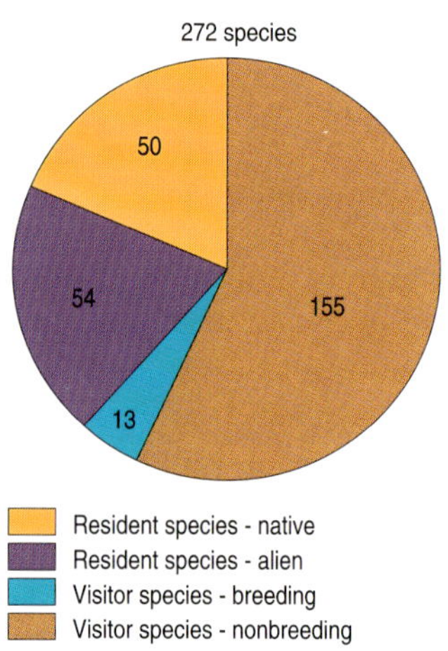

Fig. 4. Bird species currently resident in or visiting Hawaii.

Trends

Native Landbirds

Meaningful estimates of total populations of landbirds in Hawaii are difficult to derive. Native species have been confined, at least since Captain Cook's visit, to thickly vegetated and wet higher elevation forests on steep slopes or occupied by deep muddy bogs. Not surprisingly, naturalists over the years could make no real estimates of landbird populations for the island group or even for an individual island, despite the relatively small total areas that were occupied by many of these endemic species.

It was not until the Hawaii Forest Bird Survey in the late 1970's to early 1980's that thoroughly planned fieldwork was conducted, leading to the first comprehensive population estimates for native Hawaiian landbirds. Pioneering techniques for field surveys in such terrain and for statistical analysis were used to obtain population estimates for the native landbird species on all forested islands except O'ahu and Ni'ihau (Scott et al. 1986, 1988). For O'ahu Island, Shallenberger's surveys during the latter 1970's in the Koolau Mountains (Shallenberger and Vaughn 1978) and in the Waianae Mountains (unpublished) have been the most comprehensive.

More recently, Ellis et al. (1993) estimated populations for each native forestbird on each Hawaiian island, based on information available at the end of 1992. These are not directly comparable with the earlier estimates derived from field surveys. However, these estimates and numerous other less comprehensive surveys over the years involving some species on some islands do reinforce a general consensus that Hawaiian forestbirds have declined steadily both in the long term during the past century and in the short term in the past decade.

Resident Waterbirds and Visitors

The Hawaii Division of Forestry and Wildlife has conducted statewide counts of wetland birds semiannually during recent decades. These have included resident wetland species (not the seabirds) and nonbreeding visitor species. Variations in these population counts over the years reflect changes in available wetland habitat, thoroughness of coverage, and possibly some irregular interisland movements. Engilis and Pratt (1993) analyzed these statewide counts for the resident species during 1978-87. Data from earlier surveys covering only certain islands and using less rigorous counting techniques are not readily comparable. Longer-term historical trends in populations of four endangered wetland species are being examined for the Hawaii Wetland Bird Recovery Plan now in preparation by the Recovery Team for the U.S. Fish and Wildlife Service (A. Engilis, personal communication).

The breeding visitor seabirds gather to nest in large colonies in the Northwest Hawaiian Islands. Gross estimates of population numbers for these species made in the 1960's and again in the late 1970's are not comparable for trends analysis because of varying techniques used in attempting to arrive at meaningful numbers in these huge colonies totaling in the millions. Harrison (1990) discussed the difficulties involved in making representative counts and finds no evidence of long-term trends in species numbers, although some wide fluctuations occurred earlier this century. One notable feature has been the return of the Laysan albatross (*Diomedea immutabilis*) as a breeding visitor to Kaua'i and O'ahu in the main Hawaiian Islands. Since 1977, a steady increase in numbers now measured in the hundreds has local interest but has had a rather small effect on the total statewide population of millions.

Extinction

The rate of extinction within Hawaii's endemic birds is by far the highest in the United States and is approached worldwide only within a few other isolated island groups. At the time of Captain Cook's visit in 1778 some 93 species and subspecies of native birds were breeding in Hawaii, as determined by subsequent discovery and scientific description. In the ensuing two centuries, at least 23 of these have gone extinct (A.O.U. 1983) and another 13 are imperiled. Recent discoveries of the bones of prehistorically vanished species now reveal a vast array of former birds that became extinct long before Captain Cook arrived. Thirty-five of these have already been scientifically described (Olson and James 1991) and must represent only a small fraction of the forms of birds that existed prehistorically in Hawaii.

Extinctions over the past 200 years (Fig. 5) show a disproportionate number of bird species vanishing during the 1890's, a decade concluding a period of intense discovery and collecting of Hawaii's birds. A similar large decline in the 1980's represents nine forms not reported since then. Ralph and van Riper (1985) discussed the factors that have contributed to the decline in Hawaiian bird populations since the arrival of the Polynesians.

Aliens

An early listing of the alien species in Hawaii was that of Caum (1933), who identified 92 species as alien introductions. These may be categorized as established, not established, or uncertain. Most (75%) of the alien species established in 1933 are still present (Pyle 1992). Few of those deemed uncertain or not established in 1933 have persisted until today. Introductions continued during the 1940's and 1950's but thereafter were severely curtailed by stronger governmental restrictions on importation of wild birds. Of the 54 alien species considered established in Hawaii today, 31 (57%) had been introduced more than 60 years ago, and 23 (43%) have been introduced and have become established since 1933.

Conclusion

Hawaii's birds comprise four groups: native and alien resident species, and breeding and nonbreeding visitor species. Factors affecting population levels differ markedly among the groups. Although current status of species within all groups is fairly well understood, assessing meaningful trends for species is difficult for lack of comparable quantitative data on statewide populations over time.

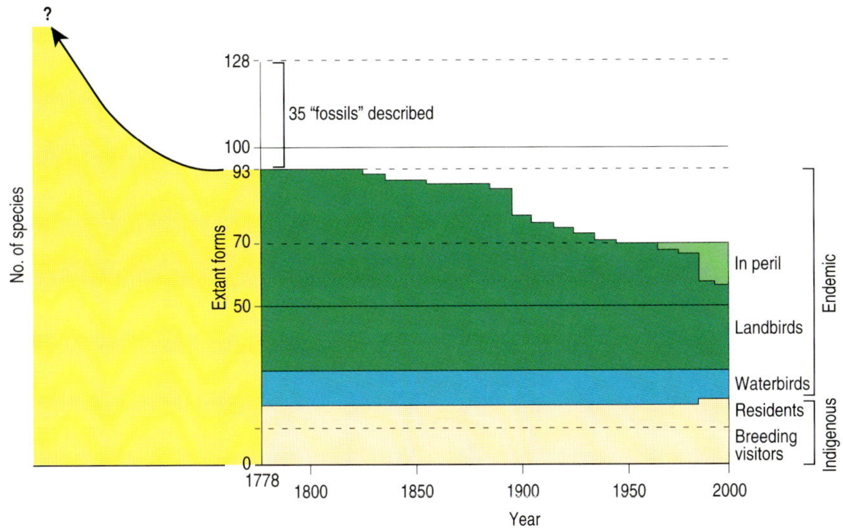

Fig. 5. Extinction of native breeding birds since 1778. Steps mark the decade of the last record for each form considered extinct (A.O.U. 1983). The 70 forms shown as currently existing include 13 in peril, with steps marking the decades of their last known records. Yellow represents prehistoric forms.

References

A.O.U. 1983. Check-list of North American birds, 6th ed. [with supplements through 1993]. American Ornithologists Union, Washington, DC. 877 pp.

Caum, E.L. 1933. The exotic birds of Hawaii. B.P. Bishop Museum Occasional Papers 10(9). Honolulu, HI.

Ellis, S., C. Kuehler, R. Lacy, K. Hughes, and U.S. Seal. 1993. Hawaiian forest birds conservation and assessment management plan. Final report. Captive Breeding Specialist Group, IUCN. Apple Valley, MN. 142 pp.

Engilis, A., and T.K. Pratt. 1993. Status and population trends of Hawaii's native waterbirds. Wilson Bull. 105:142-158.

Harrison, C.S. 1990. Seabirds of Hawaii—natural history and conservation. Cornell University Press, Ithaca, NY. 249 pp.

Johnson, O.W., P.M. Johnson, and P.L. Bruner. 1981. Wintering behavior and site-faithfulness of golden plovers on Oahu. 'Elepaio 41:123-130.

Kirch, P. 1982. The impact of the prehistoric Polynesians on the Hawaiian ecosystem. Pacific Science 36:1-14.

Olson, S.L., and H.F. James. 1991. Descriptions of thirty-two new species of birds from the Hawaiian Islands. Ornithological Monographs 45 and 46. American Ornithologists Union, Washington, DC.

Pyle, R.L. 1992. Checklist of the birds of Hawaii—1992. 'Elepaio 52(8):53-62.

Ralph, C.J., and C. van Riper III. 1985. Historical and cultural factors affecting Hawaiian native birds. Pages 43-70 *in* S.A. Temple, ed. Bird conservation 2. University of Wisconsin Press, Madison.

Scott, J.M., C.B. Kepler, C. van Riper III, and S.I. Fefer. 1988. Conservation of Hawaii's vanishing avifauna. Bioscience 38(4):238-253.

Scott, J.M., S. Mountainspring, F.L. Ramsey, and C.B. Kepler. 1986. Forest bird communities of the Hawaiian Islands. Studies in Avian Biology 9, Cooper Ornithological Society, Lawrence, KS. 431 pp.

Shallenberger, R.J., and G.K. Vaughn. 1978. Avifaunal survey in the central Ko'olau Range, O'ahu. Ahuimanu Productions, Honolulu. 106 pp.

For further information:

Robert L. Pyle
Bishop Museum
Department of Natural Sciences
741 N. Kalaheo Ave.
Kailua, HI 96734

Hawaii's Endemic Birds

by
James D. Jacobi
Carter T. Atkinson
National Biological Service

The endemic landbirds of Hawaii, particularly the Hawaiian honeycreepers, an endemic subfamily of the cardueline finches, are one of the world's most dramatic examples of adaptive radiation and speciation (*see* glossary) in island ecosystems (Freed et al. 1987; Scott et al. 1988). From what is believed to have been a single successful colonization of the Hawaiian Archipelago by an ancestral species from North America, the honeycreepers evolved into a diverse array of species and subspecies of birds with bills ranging from thick, seed-eating beaks of the palila (*Loxioides bailleui*), to small insectivorous bills as seen on the `amakihi (*Hemignathus virens*), woodpecker-like adaptations of the `akiapola`au (*H. munroi*), and large, decurved nectar-feeding bills of the `i`iwi (*Vestiaria coccinea*).

In addition to the honeycreepers, the historically documented endemic Hawaiian avifauna included three seabirds, several waterfowl, two raptors, and perching birds that include a species of crow, and representatives of Old World flycatchers, honeyeaters, and thrushes. In all, at least 71 endemic species and subspecies of Hawaiian birds existed at the time of Captain Cook's arrival in the Hawaiian Islands in 1778. Now, however, 76% of the Hawaiian birds are either extinct or endangered, and several of the remaining unlisted species are showing significant population declines.

The arrival of humans to the Hawaiian Islands—starting with the Polynesians more than 1,500 years ago and continuing following European contact—drastically changed many natural ecosystems, leading not only to the extinction of many plant and animal species, but also to a significant reduction in both range and abundance for many other taxa. Originally, the Hawaiian birds were found in all habitat zones on each island, but today few native forest birds are found below 610-m (2,000-ft) elevation, and many of the wetland areas that once provided abundant habitat for waterbirds have been destroyed.

Of the historically documented 71 taxa of endemic Hawaiian birds, 23 are now extinct, and 30 of the remaining 48 species and subspecies are listed as endangered or threatened by the U.S. Fish and Wildlife Service (USFWS 1992), many with few or only single populations remaining (Fig. 1; Table 1; Table 2). Studies of recently discovered fossil bird bones have further identified nearly 40 additional species of Hawaiian birds never seen alive by the post-Cook naturalists; many of these became extinct after the Polynesians arrived (Olson and James 1982; H. James, Smithsonian Institution, personal communication).

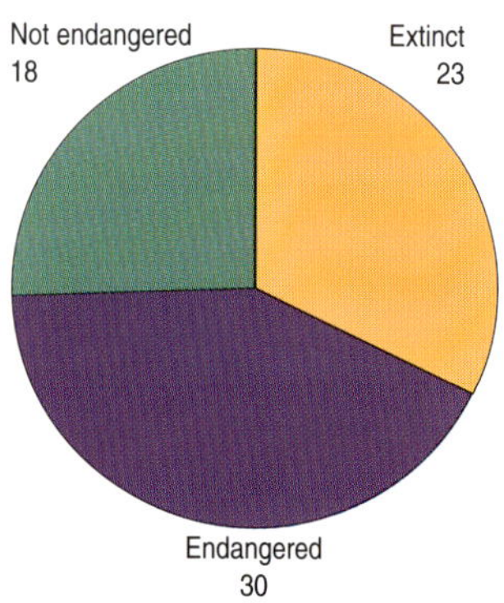

Fig. 1. Current status of endemic Hawaiian bird species known to exist at the time of Western contact (1778).

Table 1. Historically known endemic Hawaiian birds that are now extinct.

Common name	Scientific name	Island distributions
Hawaiian rail	*Porzana sandwichensis*	Hawaii
Laysan rail	*P. palmeri*	Laysan
Laysan millerbird	*Arcrocephalus f. familiaris*	Laysan
`Amaui	*Myadestes oahuensis*	Oahu
Oloma`o	*M. l. lanaiensis*	Lanai
Oahu `o`o	*Moho apicalis*	Oahu
Hawaii `o`o	*M. nobilis*	Hawaii
Kioea	*Chaetoptila angustipluma*	Hawaii
Lanai hookbill	*Dysmorodrepanis munroi*	Lanai
Lesser koa-finch	*Rhodacanthis flaviceps*	Hawaii
Greater koa-finch	*R. palmeri*	Hawaii
Kona grosbeak	*Chloridops kona*	Hawaii
Greater `amakihi	*Hemignathus sagittirostris*	Hawaii
Oahu `akialoa	*H. obscurus ellisianus*	Oahu
Lanai `akialoa	*H.o. lanaiensis*	Lanai
Hawaii `akialoa	*H.o. obscurus*	Hawaii
Oahu nukupu`u	*H.l. lucidus*	Oahu
Lanai creeper	*Paroreomyza m. montana*	Lanai
Oahu `akepa	*Loxops c. rufus*	Oahu
`Ula-`ai-hawane	*Ciridops anna*	Hawaii
Hawaii mamo	*Drepanis pacifica*	Hawaii
Black mamo	*D. funerea*	Molokai
Laysan honeycreeper	*Himatione sanguinea freethii*	Laysan
Total number extinct		23

Reasons for the Decline

Many factors have been suggested to explain the decline of Hawaiian bird species since human colonization (Ralph and van Riper 1985; Scott et al. 1988). The most important and plausible of these include habitat loss (Berger 1981; Kirch 1982; Olson and James 1982; Jacobi and Scott 1985), susceptibility to introduced avian diseases (Warner 1968; Ralph and van Riper 1985; van Riper et al. 1986), predation by introduced mammals (Atkinson 1977), and competition from introduced birds (Mountainspring and Scott 1985) and arthropods (Perkins 1903; Banko and Banko 1976). Although no one factor is believed to be the single cause for the loss or decline of the Hawaiian birds, many biologists believe that habitat loss and avian diseases have had the greatest effect on native birds.

Habitat Loss

Habitat loss from forest removal and development in the Hawaiian Islands started when large tracts of mostly lower elevation land were cleared for agriculture by the first Hawaiian colonists. After European and American settlers arrived, starting in the late 18th century, habitat loss increased dramatically as agriculture and ranching expanded. Today, less than 40% of the land surface of Hawaii is covered with native-dominated vegetation (Jacobi 1990; S. Gon, The Nature Conservancy of Hawaii, unpublished data). Some of the most significant loss of habitat has occurred below 610 m (2,000 ft)

Table 2. Status of extant species and subspecies of endemic Hawaiian birds. Data summarized from Scott et al. (1986), Harrison (1990), Ellis et al. (1992), Engilis and Pratt (1993), and J. Jacobi (unpublished data).

Species	Island distributions[a]	Listing status[b]	Estimated population	Population data source[c]	No. of populations	Trend[d]	Comments
Dark-rumped petrel *Pterodroma phaeopygia sandwichensis*	All?	E	400-600	2	5 ?	Unk	Upland nester; stable on Ma with predator control
Newell's shearwater *Puffinus newelii*	All?	T	4,000-6,000	2	4 ?	Unk	Upland nester; susceptible to predators
Nene (Hawaiian goose) *Branta sandvicensis*	Ka,Ma,Ha	E	< 500	3, 5	4	S ?	Recently introduced population on Ka increasing
Koloa (Hawaiian duck) *Anas wyvilliana*	Ka,Oa,Ha (Ma, Mo)	E	2,500	4	3	S ?	Problem of hybridization with mallards on Oa
Laysan duck *A. laysanensis*	NW	E	< 50	3	1	D	Population declined from > 600 to < 50 since 1990
Hawaiian hawk *Buteo solitarius*	Ha	E	1,500-2,500	3, 5	1	S	Occupies native and non-native forest
Hawaiian moorhen *Gallinula chlorpus sandwicensis*	Ka,Oa (Mo,Ma,Ha)	E	Unk	4	2	Unk	Difficult species to survey; no reliable population count data
Hawaiian coot *Fulica americana alai*	All	E	2,000-4,000	4	5	S ?	Population fluctuations may result from addition of migrating birds
Hawaiian stilt *Himantopus mexicanus knudseni*	All	E	1,200-1,600	4	6	I ?	Increase on Oa and Ma due to habitat managment
Hawaiian noddy *Anous minutus melanogenys*	All		Many thousands	2	?	S ?	Frequent along ocean cliffs of all islands
Pueo (Hawaiian owl) *Asio flammeus sandwichensis*	All	C	3,500 ?	3	7	D ?	No detailed data on population trends
`Alala (Hawaiian crow) *Corvus hawaiiensis*	Ha	E	< 20	5	1	D	An additional 18 birds in captivity
Nihoa millerbird *Acrocephalus familiaris kingi*	NW	E	< 300	3	1	S	Extremely vulnerable; entire population on one island
Kauai `elepaio *Chasiempis sandwichensis sclateri*	Ka		> 20,000	1, 5	1	D ?	Relatively common in suitable habitat
Oahu `elepaio *C.s. gayi*	Oa	C	200 ?	5	2	D	Few birds seen in recent years
Hawaii `elepaio *C.s. sandwichensis*	Ha		> 200,000	1, 5	3	S ?	Widespread, even in mid-elevations
Kama`o *Myadestes myadestinus*	Ka	E	< 50 ?	1, 5	1	EX ?	Not seen in over 5 years; possibly extinct
Oloma`o *M. lanaiensis rutha*	Mo	E	< 10 ?	1, 5	1	EX ?	Not seen in over 5 years; possibly extinct
`Oma`o *M. obscurus*	Ha		>170,000	1, 5	1	S	Extinct from Kona and Kohala sections of island
Puaiohi *M. palmeri*	Ka	E	< 50 ?	1, 5	1	D	Extremely rare; seen in 1994 survey
Bishop's `o`o *Moho bishopi*	Ma (Mo)	EX?	?	1, 5	?	EX ?	Extinct? No recent black bird populations
Kauai `o`o *M. braccatus*	Ka	E	< 10 ?	1, 5	1	EX ?	Not seen since 1986
Laysan finch *Telespyza cantans*	NW	E	10,000	3	1	S	Translocated population on Perl and Hermes Reef
Nihoa finch *T. ultima*	NW	E	1,000-3,000	3	1	S	Extremely vulnerable; entire population on one island
`O`u *Psittirostra psittacea*	All (Oa,Mo,La Ma)	E	< 50 ?	1, 5	2	D, EX ?	Rapid decline on Ka and Ha; last seen in 1986 on Ha
Palila *Loxioides bailleui*	Ha	E	> 3,000	1, 5	1	S ?	Habitat extremely vulnerable to fire
Maui parrotbill *Pseudonestor xanthophrys*	Ma	E	< 500	1, 5	1	D	Recent decline in distribution
Kauai `amakihi *Hemignathus virens stejnegeri*	Ka		> 15,000	1, 5	1	S ?	Population seems stable
Oahu `amakihi *H.v. chloris*	Oa		20,000-60,000	5	2	S	Also in areas with non-native vegetation
Maui `amakihi *H.v. wilsoni*	Ma,Mo (La)		> 45,000	1, 5	4	S	Last seen on La in 1970's
Hawaii `amakihi *H.v. virens*	Ha		> 800,000	1, 5	Many	S	Population seems stable
`Anianiau *H. parvus*	Ka		15,000-25,000	1, 5	1	S	Common in suitable habitat
Kauai `akialoa *H. procerus*	Ka	E	< 10 ?	1, 5	1	D, EX ?	Not seen in 30 years
Kauai nukupu`u *H. lucidus hanapepe*	Ka	E	< 10 ?	1, 5	1	D, EX ?	Last seen in 1986
Maui nukupu`u *H.l. affinus*	Ma	E	< 10 ?	1, 5	1	D, EX ?	Not seen in nearly 10 years
`Akiapola`au *H. munroi*	Ha	E	< 1,500	1, 5	4	D	Nearly extinct on upper slopes of Mauna Kea

—continued—

Table 2. Continued.

Species	Island distributions[a]	Listing status[b]	Estimated population	Population data source[c]	No. of populations	Trend[d]	Comments
Kauai creeper *Oreomystis bairdi*	Ka	C	800-1,000	1, 5	1	D	Population now concentrated in central Alakai region
Hawaii creeper *O. mana*	Ha	E	12,500	1, 5	3	S ?	Uncommon but appears stable
Oahu creeper *Paroreomyza maculata*	Oa	E	< 10 ?	5	1	D, EX ?	Last seen in 1985
Molokai creeper *P. flammea*	Mo	E	< 10 ?	1, 5	1	D, EX ?	Not seen since the 1960's
Maui creeper *P. montana newtoni*	Ma		35,000	1, 5	2	S	Common in wet and mesic forests >1,220-m elevation
Kauai \`akepa *Loxops coccineus caeruleirostris*	Ka	C	> 3,000	1, 5	1	D	Population has been declining over past 10 years
Maui \`akepa *L.c. ochraceus*	Ma	E	< 10 ?	1, 5	1	D, EX ?	Not seen during past 10 years
Hawaii \`akepa *L.c. coccineus*	Ha	E	14,000	1, 5	3	D ?	Locally common in upper-elevation forests
\`I\`iwi *Vestiaria coccinea*	All		> 350,000	1, 5	7	S/D	Appears stable on Ka, Ma, and Ha; rare or EX elsewhere
\`Akohokohe *Palmaria dolei*	Ma (Mo)	E	>3,500	1, 5	1	S ?	Locally common above 1,220 m; EX on Molokai
\`Apapane *Himatione s. sanguinea*	All		> 1,300,000	1, 5	7	S	Very common above 1,220 m, less common below
Po\`o-uli *Melamprosops phaeosoma*	Ma	E	< 50	1, 5	1	D	Last seen in 1993; extremely rare
Total number of extant species	48						

[a]Islands: All — all major Hawaiian Islands; Ha — Hawaii; Ka — Kauai; La — Lanai; Ma — Maui; Mo — Molokai; NW — Northwest Hawaiian Islands; Oa — Oahu; species extinct where island listed in parentheses.
[b]Listing status: C — candidate for listing; E — endangered; EX — extinct; T — threatened species.
[c]Population and trend data source: 1 — Scott et al. (1986); 2 — Harrison (1990); 3 — Ellis et al. (1992); 4 — Engilis and Pratt (1993); 5 —recent survey data (J. Jacobi, unpublished data).
[d]Trend: D — declining; EX? — possibly extinct; I — increasing; S — stable; Unk — unknown.

elevation, where less than 10% of the native vegetation remains. In addition to direct clearing, all remaining native plant communities are further degraded by disturbance and competition from introduced plants and animals.

The current ranges of most Hawaiian forest birds appear closely tied to the distribution of forests dominated by native tree species. It is unclear whether this association is due to feeding specialization on native plants, or if other factors, such as disease or predators, restrict native birds from disturbed habitats. The only real exception to this is the Oahu \`amakihi (*Hemignathus virens chloris*), which recently appears to be colonizing habitats dominated by introduced plant species around Honolulu.

Avian Disease

The accidental introduction of *Culex* mosquitoes in the early 19th century, and the importation and widespread release of domestic fowl, gamebirds, and cage birds with their accompanying diseases, are believed responsible for the establishment of avian pox virus and malaria (*Plasmodium relictum*) in Hawaiian forest bird populations (Warner 1968; van Riper et al. 1986). The concurrent fragmentation of native forests probably hastened the spread of mosquitoes and exotic birds into forest habitats, exposing native birds to avian pox (Perkins 1893; Henshaw 1902) and malaria.

Warner (1968) first identified pox and malaria as major pathogens of native forest birds. Van Riper et al. (1986) demonstrated that the highest incidence of malaria occurs in wet midelevation forests (between 900 m [3,000 ft] and 1,500 m [5,000 ft]) where populations of *Culex* mosquitoes overlap with highly susceptible native birds. Current investigations support these observations. Surveys for other disease agents identified a number of potentially pathogenic parasites and bacteria, but none has been implicated as a significant cause of mortality (van Riper and van Riper 1985).

Introduced Predators

While introduced rats (*Rattus* spp.), cats (*Felis catus*), dogs (*Canis familiaris*), and mongooses (*Herpestes auropunctatus*) have seriously affected nesting waterbirds, less information exists on the significance of these predators in restricting the distribution and abundance of upland forest birds in Hawaii (Atkinson 1977; Griffin et al. 1989). Several projects have begun in Hawaii to develop adequate control strategies for introduced predators and to monitor the response of forest bird populations to the reduction or elimination of these predators.

Competition and Food

Competition for nesting and food resources by introduced birds and food resource limitation by introduced arthropods (e.g., ants or wasps) are the two most difficult of the limiting factors hypotheses to evaluate. Although a study by Mountainspring and Scott (1985) found a negative association between several native and introduced bird species pairs, much more work is needed to understand the significance of these relationships. Similarly, preliminary evidence suggests that arthropods such as the introduced yellowjacket wasps (*Vespula* spp.) and several species of ants may seriously deplete the resident arthropods that many native birds eat, particularly during nesting (P. Banko, NBS, personal communication).

Current Status

Table 2 summarizes the most recent information on the status of endemic Hawaiian bird species. The population size for many forest birds comes from the Hawaii Forest Bird Survey, 1976-81 (Scott et al. 1986). While most of these numbers are more than 15 years old, they represent a distribution and abundance baseline upon which subsequent surveys can be based. The trend information in Table 2 is based on population surveys conducted during the past 15 years.

Seabirds

Three seabird species are endemic to Hawaii: the endangered dark-rumped petrel (*Pterodroma phaeopygia sandwichensis*), the threatened Newell's shearwater (*Puffinus newelii*), and the Hawaiian noddy (*Anous minutus melanogenys*). The first two relatively rare species nest in upland forest or subalpine and alpine sites. As with all of the ground-dwelling or nesting birds, the dark-rumped petrel and Newell's shearwater are extremely susceptible to predation by cats, dogs, rats, and mongooses during their long nesting period. A successful predator-control program in nesting areas for the dark-rumped petrel in Haleakala National Park on Maui has resulted in a significant increase in petrel productivity. Recently discovered nesting areas for the dark-rumped petrel and Newell's shearwater on the island of Hawaii offer similar opportunities to use predator control to reestablish significant breeding colonies for these species in upland habitats.

Waterbirds

Historically, the Hawaiian avifauna includes six waterbird species, five of which are typically found in and around fresh-, brackish-, and saltwater impoundments and estuaries (Engilis and Pratt 1993). The sixth species, the nene or Hawaiian goose (*Branta sandwichensis*), though occasionally found around water, most typically occurs in upland sites.

Continued loss of habitat and predation are the two biggest threats to the remaining Hawaiian waterbirds. Although the Hawaiian coastal zone formerly contained many large wetland areas, few remain. For example, the resort area known as Waikiki Beach was an extensive wetland that was drained in the early 1900's. Because introduced predators are a major threat to waterbirds in Hawaii, predator control has become essential in all waterbird-management programs.

An intensive captive propagation and release program has kept the nene from extinction. This ground-nesting goose, however, is extremely vulnerable to predation by introduced mongooses, cats, dogs, and possibly rats and is not able to sustain wild populations in most areas (Stone et al. 1983). A recently established population on the island of Kauai appears to be thriving, probably mostly because of the absence of mongooses on this island.

Both duck species endemic to Hawaii are endangered. The Laysan duck (*Anas laysanensis*) is known only from Laysan Island, a small atoll about halfway up the northwest Hawaiian Island chain. Although population levels have been as high as 600 birds over the past 25 years, they dropped to fewer than 50 during 1993 (T. Work, NBS, personal communication). Species confined to such a small geographical area are extremely vulnerable to natural disasters (e.g., hurricane damage) or human-related impacts (e.g., introduction of disease or predators to the island).

The koloa or Hawaiian duck (*A. wyvilliana*), formerly found on all major Hawaiian Islands, is now relatively rare, with small populations on Kauai, Oahu, and Hawaii. It, too, is extremely vulnerable to predators. Additionally, because koloa on Oahu are hybridizing with feral populations of the closely related mallard (*A. platyrhynchos*; Engilis and Pratt 1993), a mallard-control program has been recommended to protect the native koloa populations from genetic alteration.

Courtesy J. Jeffrey Photography©

\`I\`iwi (*Vestiaria coccinea*). The long sickle-bill of the \`i\`iwi enables it to feed on nectar from flowers and to probe for insects.

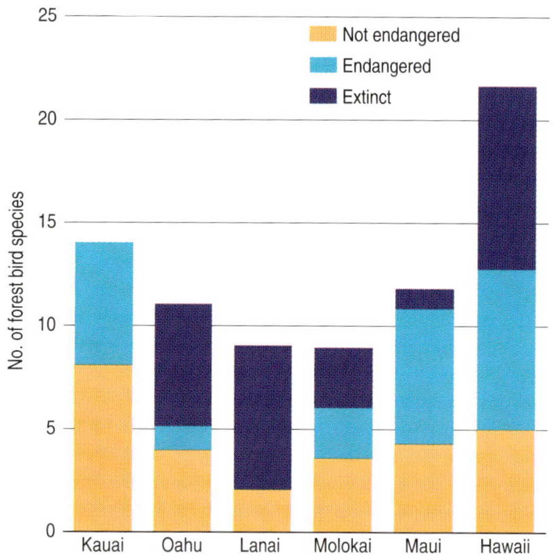

Fig. 2. Summary of status of endemic forest birds on the major Hawaiian Islands.

In addition to the waterbirds, the two rail species endemic to Hawaii are now extinct. One species, the Laysan rail (*Porzana palmeri*), known only from Laysan Island, became extinct after introduced rabbits nearly totally defoliated this small atoll in the early 1900's (Berger 1981). The other rail species endemic to Hawaii (*P. sandwichensis*) was extremely rare in the late 1700's when Western naturalists first began to document the Hawaiian birds. This species was probably extinct by the early 1900's (Berger 1981).

Forest Birds

Forest birds constitute the largest group of Hawaiian birds, with 60 species and subspecies described since Western contact. Several species of passerines known from the Northwest Hawaiian Islands are also included with the forest bird group, although none of these atolls has any forest habitat.

Both the greatest number of species and the number of losses of species of Hawaiian birds are found in the forest bird group. Of 60 endemic species and subspecies of Hawaiian forest birds, 22 are believed extinct, an additional 23 are endangered or threatened (USFWS 1992), and 4 are candidate species for listing (Table 2). Thirteen of the endangered forest birds have estimated populations of less than 50 individuals; 10 of these species have not been sighted during the past 10 years and may be extinct. The island of Kauai, which seemed to be the only island with all historically known bird species still extant, now has five species that may be extinct (Fig. 2). Surveys in 1993 and 1994 resulted in finding only one of the endangered forest bird species, the puaiohi or small Kauai thrush (*Myadestes palmeri*).

Only 11 Hawaiian forest bird taxa are considered relatively stable, but several populations of these species, particularly the \`i\`iwi, have experienced recent declines. The \`oma\`o or Hawaii thrush (*M. obscurus*) has relatively robust populations on the windward side of the island of Hawaii, but is extirpated in the wet forests of both the leeward (Kona) side of the island and in the Kohala region.

Conservation Outlook

While the prospects for survival of all remaining Hawaiian bird species appear limited, conservation efforts to further the chances of survival of even some of the rarest species can be enhanced by using techniques such as translocation, predator and disease vector control, and captive propagation in conjunction with habitat-management programs.

Avian Disease Research and Management

Since 1992 the National Biological Service's disease studies have focused on determining the effect of pox and malaria transmission on the island of Hawaii and whether significant changes in the prevalence and distribution of these diseases have occurred since van Riper and colleagues completed their work in the late 1970's. Major new efforts to develop strategies for monitoring transmission of these diseases in remote forest habitats and for controlling vector populations are in progress.

In 1992 NBS scientists witnessed a major pox and malaria epidemic in midelevation forest birds on the island of Hawaii. These birds are highly susceptible to malaria. Results of experimental infections with isolates of malaria from wild birds demonstrated that a minimal infective dose, equivalent to the bite of a single malaria-infected mosquito, was sufficient to kill 90% of juvenile \`i\`iwi under experimental conditions. The high susceptibility of this species could explain its disappearance during the past 20 years from many midelevation forests where it was previously common.

Strategies for breaking the cycle of vector-transmitted diseases include intensive environmental management to reduce mosquito breeding sites, chemical and biological control agents, genetic manipulation of the vector population, and release of sterile male mosquitoes. In addition, removal of feral ungulates from critical forest habitats may reduce available breeding sites and mosquito densities to levels too low to support disease transmission, but this needs to be evaluated under controlled conditions. Efforts by land managers in Hawaii to fence and control feral ungulates will provide an opportunity to coordinate disease research with management.

Additional Research and Management

Conservation programs in Hawaii need to have both species and ecosystem components.

Species actions include intensive site-management programs (e.g., predator control, disease and vector control, food supplementation, detailed ecological research, nest manipulation), coupled with translocation and state-of-the-art captive propagation and reintroduction. These strategies are being applied to the critically endangered `alala or Hawaiian crow (*Corvus hawaiiensis*). During 1993 the remaining wild population of 12 `alala was augmented with the release of 5 juvenile birds hatched in captivity from eggs removed from wild nests. Limited nesting success in the remaining three wild pairs prompted a "double-clutching" (*see* glossary) strategy to increase egg productivity and allow for artificial incubation and hatching. Two other birds hatched from artificial incubation of wild-laid eggs were added to a captive breeding flock; the 1994 season yielded five chicks from wild nests and four new birds from captive breeding.

Habitat and ecosystem management are also essential to conserve the remaining Hawaiian birds, as well as for recovery of rare and listed species. Unless we can better protect the natural ecosystems in Hawaii today, the already enormous list of endangered and extinct species known from the Hawaiian Islands will grow and species that are still common will also decline.

Courtesy J. Jeffrey Photography©

Palila (*Loxioides bailleui*) perched in a mamane (*Sophora chrysophylla*) tree. This endemic forest bird feeds primarily on the immature seeds of the mamane.

References

Atkinson, I.A.E. 1977. A reassessment of factors, particularly *Rattus rattus* L., that influenced the decline of endemic forest birds in the Hawaiian Islands. Pacific Science 31:109-133.

Banko, W.E., and P.C. Banko. 1976. Role of food depletion by foreign organisms in historical decline of Hawaiian forest birds. Pages 29-34 *in* C.W. Smith, ed. Proceedings. First Conference in Natural Sciences, Hawaii Volcanoes National Park. Cooperative National Park Resources Studies Unit, Department of Botany, University of Hawaii, Honolulu. 243 pp.

Berger, A.J. 1981. Hawaiian birdlife. 2nd ed. University Press of Hawaii, Honolulu. 274 pp.

Ellis, S. C. Kuehler, R. Lacy, K. Hughes, and U.S. Seal. 1992. Hawaiian Forest Birds Conservation Assessment and Management Plan, Final Report. Proceedings of the Hawaiian Forest Birds Conservation Assessment and Management Plan Workshop, 7-12 December 1992, Hilo, Hawaii. Published by the Captive Breeding Specialist Group, IUCN-The World Conservation Union/Species Survival Commission. 142 pp.

Engilis, A., Jr., and T. Pratt. 1993. Status and population trends of Hawaii's native waterbirds. Wilson Bull. 105:142-158.

Freed, L.A., S. Conant, and R.C. Fleischer. 1987. Evolutionary ecology and radiation of Hawaiian passerine birds. Trends in Ecology and Evolution 2:196-203.

Griffin, C.R., C.M. King, J.A. Savidge, F. Cruz, and J.B. Cruz. 1989. Effects of introduced predators on island birds: contemporary case histories from the Pacific. Pages 687-698 *in* H. Ouellet, ed. Proceedings of the XIX Ornithological Congress. Vol. 1.

Harrison, C.S. 1990. Seabirds of Hawaii. Comstock Publishing Associates, Cornell University Press, Ithaca, NY. 249 pp.

Henshaw, H.W. 1902. Birds of the Hawaiian possessions. T.G. Thrum, Publisher, Honolulu, HI. 146 pp.

Jacobi, J.D. 1990. Distribution maps, ecological relationships, and status of native plant communities on the island of Hawaii. Ph.D. dissertation, University of Hawaii, Honolulu. 291 pp.

Jacobi, J.D., and J.M. Scott. 1985. An assessment of the current status of upland habitats and associated endangered species on the island of Hawaii. Pages 3-22 *in* C.P. Stone and J.M. Scott, eds. Hawaii's terrestrial ecosystems, preservation and management. Cooperative National Park Resources Studies Unit, University of Hawaii, Honolulu. 584 pp.

Kirch, P.V. 1982. The impact of the prehistoric Polynesians on the Hawaiian ecosystem. Pacific Science 36:1-14.

Mountainspring, S., and J.M. Scott. 1985. Interspecific competition among Hawaiian forest birds. Ecological Monographs 55:219-239.

Olson, S.L., and H.F. James. 1982. Fossil birds from the Hawaiian Islands: evidence for wholesale extinction by man before western contact. Science 217:633-635.

Perkins, R.C.L. 1893. Notes on collecting in Kona, Hawaii. Ibis 1893:101-114.

Perkins, R.C.L. 1903. Vertebrata (Aves). Pages 365-466 *in* D. Sharp, ed. Fauna Hawaiiensis. Vol. 1, Part 4. The University Press, Cambridge, England.

Ralph, C.J., and C. van Riper III. 1985. Historical and current factors affecting Hawaiian native birds. Pages 7-42 *in* S.A. Temple, ed. Bird conservation 2. The International Council for Bird Preservation, United States Section. The University of Wisconsin Press, Madison, WI.

Scott, J.M., C.B. Kepler, C. van Riper III, and S.I. Fefer. 1988. Conservation of Hawaii's vanishing avifauna. BioScience 38:238-253.

Scott, J.M., S. Mountainspring, F.L. Ramsey, and C.B. Kepler. 1986. Forest bird communities of the Hawaiian Islands: their dynamics, ecology, and conservation. Studies in Avian Biology 9. 431 pp.

Stone, C.P., R.L. Walker, J.M. Scott, and P.C. Banko. 1983. Hawaiian goose research and management—where do we go from here? 'Elepaio 44:11-15.

USFWS. 1992. Endangered and threatened wildlife and plants. 50 Code of Federal Regulations 17.11 and 17.12. 38 pp.

van Riper, C., III, and S.G. van Riper. 1985. A summary of known parasites and diseases recorded from the avifauna of the Hawaiian Islands. 1985. Pages 298-371 *in* C.P. Stone and J.M. Scott, eds. Hawaii's terrestrial ecosystems: preservation and management. Cooperative National Park Resources Studies Unit, University of Hawaii, Honolulu. 584 pp

van Riper, C., III, S.G. van Riper, M.L. Goff, and M. Laird. 1986. The epizootiology and ecological significance of malaria in Hawaiian land birds. Ecological Monographs 56:327-344.

Warner, R.E. 1968. The role of introduced diseases in the extinction of the endemic Hawaiian avifauna. Condor 70:101-120.

For further information:

James D. Jacobi
National Biological Service
Hawaii Field Station
PO Box 44
Hawaii National Park, HI 96718

PART 3

Special Issues

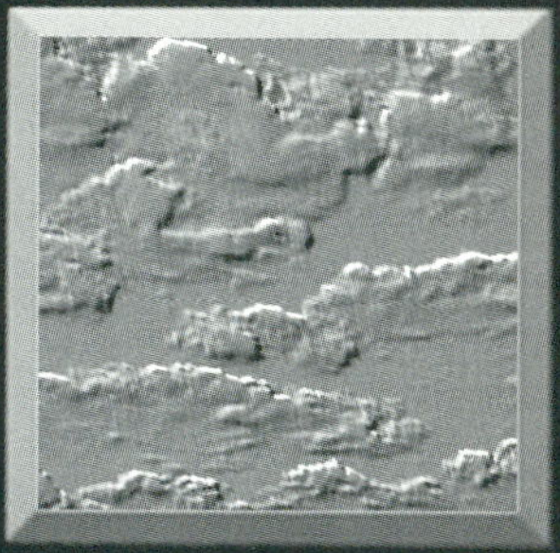

Courtesy A.L. Batik

Global Climate Change

Edward T. LaRoe
Senior Science Editor
National Biological Service

Overview

Scientists have long recognized climate, especially temperature and precipitation, as one of the major ecological forces affecting the abundance, location, and ecological health of living organisms. This relationship is so strong that in many cases, if biologists know what plants and animals are present in an area, they can approximate the climate of the area. Quantifying these relationships will allow scientists to predict the ecological consequences of global climate change.

Recently the scientific community reached a remarkable consensus on the likelihood and magnitude of global climate change, describing a likely scenario of a 3°C (5.4°F) average global warming, significant changes in the patterns and abundance of precipitation, and 0.6-m (1.9-ft) sea-level rise in the next 60 years (Houghton et al. 1990; LaRoe 1991). These changes will occur faster than previous change in geologic history and are therefore expected to have greater ecological impact.

Because of the strong relation between climate and ecosystem health and distribution, the U.S. Global Change Research Program has as a major component the monitoring of plants and animals to detect, understand, and ultimately predict the effects of global climate change on living resources (CEES 1990). The National Biological Service's research includes several projects to monitor the effects of climate change on animal and plant populations and ecosystems. Not only will the results of these projects allow a better understanding of the ecological effects of climate change, but they will also give an early, clear indication of the onset and magnitude of climate change because living resources may be sensitive indicators of global change.

Determining if long-term change in a species' population abundance or distribution was caused by specific climate changes is an extremely difficult scientific problem for two reasons: first, both climate and biological factors vary greatly from year to year, and these annual variations often mask long-term trends, making them difficult to detect. Second, several factors such as habitat loss, hunting pressure, competition with other native species and non-native species, and contaminants are simultaneously affecting species' population size and distribution along with climate change so that it is difficult to determine definitively the effect of any one cause.

Some species of plants and animals already may be affected by one type of global climate change: global warming. Much of the evidence for this, however, is anecdotal or poorly documented. For example, some cold-intolerant species such as opossums (*Didelphis* spp.) and armadillos (*Dasypus novemcinctus*) have expanded their range significantly northward during the last 50 years, and some heat-sensitive species, such as white birch (*Betula papyrifera*), have receded northward during the same period. Data from some recent studies also suggest that global warming may be influencing the distribution or physiology of other plants and animals. Although these data are not sufficient to determine cause and effect relationships, they are intriguing enough to identify future research needs.

The articles that follow all investigate interesting trends between one aspect of climate change—global warming—and plant and animal behavior. Root and Weckstein document long-term change in the winter distribution of some birds; global warming is one possible explanation for these changes. LaRoe and Rusch's article shows change in onset of hatching behavior in selected populations of geese; and Oglesby and Smith's contribution shows a long-term trend in migratory behavior of some birds and in blooming of some plants. Finally, Morse et. al. use existing models to provide a preliminary assessment of patterns of plant vulnerability to climate change.

All four articles are subject to the complexities common to most work on global change; all the trends show dramatic year-to-year variation in response to short-term temperature changes and all have multiple possible explanations; and while all show intriguing statistical correlations, none demonstrates a cause-and-effect relationship. Moreover, these trends do not affect all species, because different species have different sensitivities to temperature and because global climate change is not the only factor affecting species. As discussed in Root and Weckstein's article, a number of competing hypotheses can be used to explain these changes. Nonetheless, together these articles suggest that global warming should be considered as a contributing factor.

References

CEES. 1990. Our changing planet: the FY1991 research plan of the U.S. Global Change Research Program. A report by the Committee on Earth and Environmental Sciences. U.S. Geological Survey, Reston, VA. 169 pp.

Houghton, J.T., G.J. Jenkins, and J.J. Ephraums, eds. 1990. Climate change, the IPCC scientific assessment. Report prepared for the Intergovernmental Panel on Climate Change by Working Group 1. Cambridge University Press, England. 364 pp.

LaRoe, E.T. 1991. The effects of global climate change on fish and wildlife resources. Transactions of the North American Wildlife and Natural Resources Conference 56:171-176.

Changes in Winter Ranges of Selected Birds, 1901-89

by
Terry L. Root
Jason D. Weckstein
University of Michigan

Over time the ranges of species expand and contract, and abundance patterns shift. Ranges can expand when suitable new habitat becomes available or when population pressure forces migration to new areas. Contractions can occur when populations decline and individuals abandon less-than-ideal habitats, which are often along the edges of species' ranges.

We wish to compare historical and recent range and abundance patterns of selected wintering birds, categorize the type of changes that occurred, and speculate on possible causes of the changes. We found range expansions in most birds examined; only a few species exhibited contractions, and patterns of abundance shifted in almost all species.

Sources of Data

We used data collected by volunteers for the National Audubon Society's Christmas Bird Counts. Wing (1947) summarized data from 1901-40 (from winter 1900-01 to winter 1939-40), which included 6,853 censuses. We obtained data for 32,167 censuses from the U.S. Fish and Wildlife Service for 1960-89, excluding those for 1969, which were not available at time of analysis. For more information on how we used these data, *see* Root and Weckstein (1994).

Changes in Ranges

We found extensive changes in the ranges and abundance patterns of the birds we examined. Environmental changes that facilitate rearrangements in species' ranges and abundances can be due to natural factors, such as hurricanes transporting cattle egrets (*Bubulcus ibis*) to North America (Bock and Lepthien 1976). In the fairly recent past, however, such changes have been primarily precipitated by humans, including breaking sod in the prairies for farming, which allowed the western spread of American robins (*Turdus migratorius*; Bent 1949), and building cooling ponds for waste heat from power plants, which provided open water for various wintering ducks in the northern states (Root 1988a).

Over the last several decades most ecological studies examining range and abundance changes have focused primarily on investigating direct natural and human-induced effects of habitat change. Consequently, by reading the literature one gets the impression that such changes are the most common and most important.

Indirect effects of habitat change, however, are probably just as common and important, and perhaps even more so, although obtaining clear obvious evidence for indirect effects is difficult given the fact that other factors are changing at the same time. One such effect is the biotic response to the abiotic changes induced by human disturbance. A good example is changes in birds' ranges in response to increasing temperature.

Range Expansion

One way to examine the possible importance of global warming on changing ranges is examining possible physiological mechanisms constraining birds' ranges to warm areas. Previous work has shown that 50 species of songbirds (e.g., sparrows and warblers) have range boundaries apparently dictated by average minimum January temperatures (Root 1988b). Ongoing studies of a few of these key species have shown significant changes in the location of northern range boundaries from year to year, and these correspond to annual climate changes.

Preliminary studies on northern cardinals (*Cardinalis cardinalis*) suggest that the lack of stored fat, which is needed to fuel increased metabolic rates in colder areas (Root 1991), is the primary factor restraining this bird's range. Consequently, as the earth warms, we expect birds with ranges restricted by low temperatures to readily expand their ranges. Such expansions may indeed be already occurring.

Successfully managed birds show extensive range expansions. Up to 1940, the mute swan (*Cygnus olor*) was recorded only in Pennsylvania and Michigan (Fig. 1a), but since then, programs to introduce and establish it—primarily in parks—have allowed it to spread to 19 states (Fig. 1b).

The wild turkey (*Meleagris gallopavo*) shows even a more dramatic change (Fig. 2). It originally occurred in the Southwest and in all the states east of the 100th meridian, except for North Dakota (Schorger 1966). Hunting pressures, habitat loss, and disease spread by domestic poultry all contributed to its dramatic range contraction (Schorger 1966; Hewitt 1967; Lewis 1973). From 1901 to 1940 it was recorded in only 10 states (Fig. 2a). Turkeys were reintroduced into all but three states within its original range and introduced into all the states outside its original range (Fig. 2b). Obviously, management has had a major effect on this gamebird.

Similarly, people may have contributed to a change in both ranges and abundances of various seed-eating birds (Fig. 3). On average, a third of the households in North America provide about 60 lb of bird food a year, with the average being even higher in New England (Ehrlich et al. 1988). Consequently, feeders may have contributed to the expansion of winter ranges of some birds into the northeastern part of the country (e.g., mourning dove [*Zenaida macroura*] Fig. 3; tufted titmouse [*Parus bicolor*]; northern cardinal; and evening grosbeak [*Coccothraustes vespertinus*]).

Habitat change due to logging may have contributed to the extensive and recent range changes of the barred owl (*Strix varia;* Fig. 4), which tends to prefer mixed-aged forests. Before 1972 no northern populations of this owl were reported west of the 100th meridian (Root 1988a). The recent expansion is of concern because this owl's range is now partly sympatric with that of the endangered northern spotted owl (*S. occidentalis caurina*), which prefers ancient forests. The consequences of competition between these two species are not understood well yet, but nesting sites, foraging, and diet are similar, particularly in the Northwest (Taylor and Forsman 1976). Anecdotal evidence, however, suggests the larger, more aggressive barred owl may be able to displace the smaller spotted owl (Sharp 1989).

Other raptors (e.g., northern harrier [*Circus cyaneus*] and ferruginous hawk [*Buteo regalis*]) have also significantly expanded their ranges. In particular, the golden eagle (*Aquila chrysaetos*) has moved east, while the bald eagle (*Haliaeetus leucocephalus*; Fig. 5) has spread into the center of the continent.

Over the years humans have strongly influenced the expansion of the bald eagle's range through water-management programs (Root 1988a). Large lakes and impoundments built in the 1930's, locks placed on major waterways,

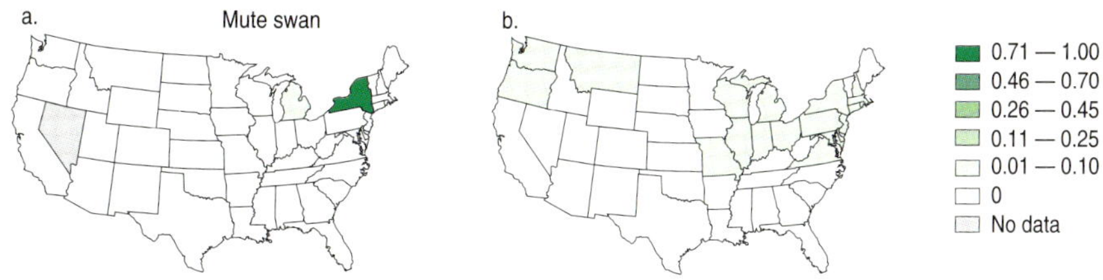

Fig. 1. Range and abundance patterns of the mute swan. (a) Data from 1901 to 1940, (b) Data from 1960 to 1989 (except 1969).

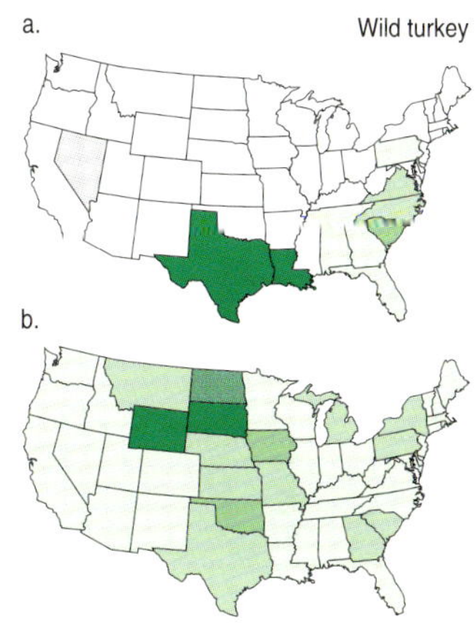

Fig. 2. Range and abundance patterns of the wild turkey. (a) Data from 1901 to 1940, (b) Data from 1960 to 1989 (except 1969).

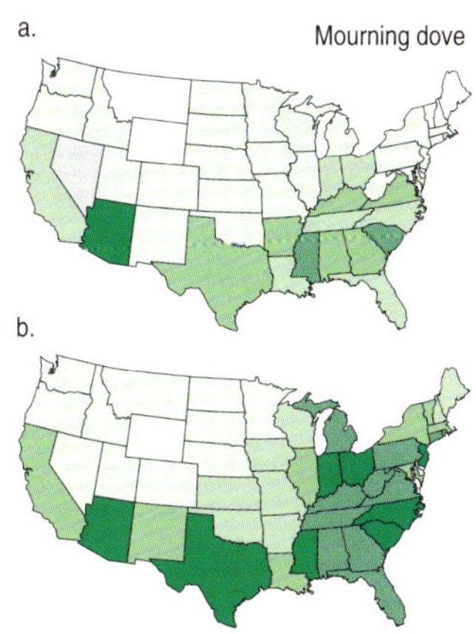

Fig. 3. Range and abundance patterns of the mourning dove. (a) Data from 1901 to 1940, (b) Data from 1960 to 1989 (except 1969).

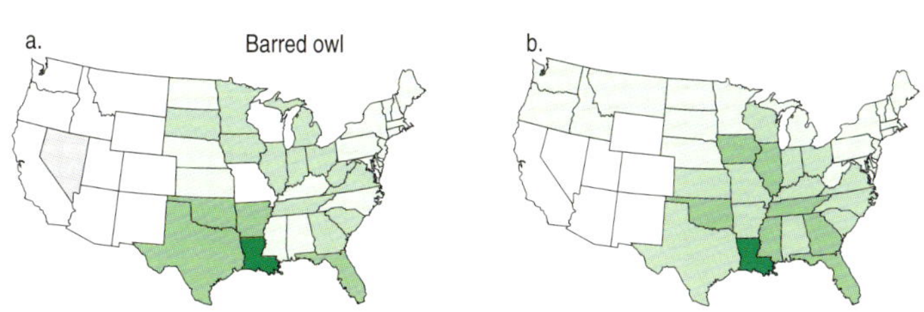

Fig. 4. Range and abundance patterns of the barred owl. (a) Data from 1901 to 1940. (b) Data from 1960 to 1989 (except 1969).

Changes in Nesting Behavior of Arctic Geese

by
Edward T. LaRoe
Donald H. Rusch
National Biological Service

As part of the joint United States-Canada efforts to monitor populations of Arctic geese and to provide data necessary to set hunting regulations, scientists have recorded not only goose population levels, but also nesting behavior. MacInnes et al. (1990) analyzed data from four long-term studies of five different Arctic goose populations. These studies documented the date the eggs hatched and the clutch size (number of eggs per nest) over 35 years (Fig. 1).

The dates of nest initiation and hatch are clearly affected by climate and are delayed by cold weather. The records not only show wide fluctuations from year to year in response to annual variations in climate, but also demonstrate a consistent trend toward earlier hatching over the period (Fig. 2). Young Arctic geese today, on the average, hatch about 30 days earlier than they did 35 years ago; during the same time, average clutch size has shrunk (Fig. 3). MacInnes et al. (1990) suggest the change in nest date is a result of climatic amelioration, that is, warming (although whether from a long-term trend or short-term cycle is unclear), and the change in clutch size is a result of habitat deterioration.

Reference

MacInnes, C.D., E.H. Dunn, D.H. Rusch, F. Cooke, and F.G. Cooch. 1990. Advancement of goose nesting dates in the Hudson Bay Region, 1951-86. Canadian Field Naturalist 104:295-297.

For further information:

Donald H. Rusch
National Biological Service
226 Russell Labs
1630 Linden Drive
Madison, WI 53706

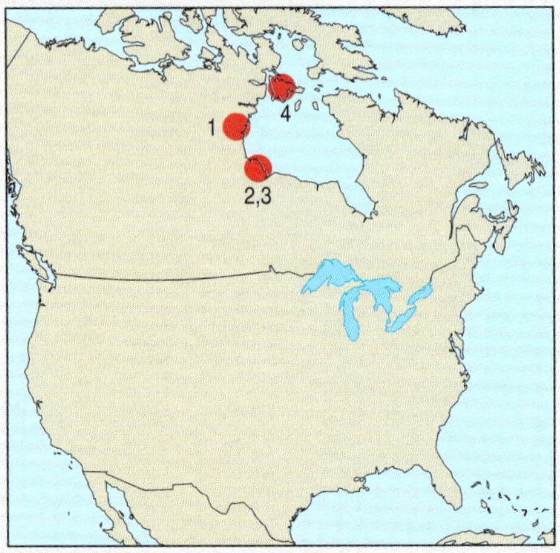

Fig. 1. Location of four monitoring sites. Sites chosen represent localities with information for 5 or more years. Site 1–McConnell River (60° 50'N, 94° 25'W; snow goose [*Chen caerulescens*] and small Canada goose [*Branta canadensis*]); 2– La Pérouse Bay (58° 24'N, 94° 24'W; snow goose); 3–Cape Churchill (58° 25'N, 93°W; medium Canada goose); and 4–Southhampton Island (63° 60'N, 86°W; snow goose).

Courtesy E.T. LaRoe

Canada goose (*Branta canadensis*).

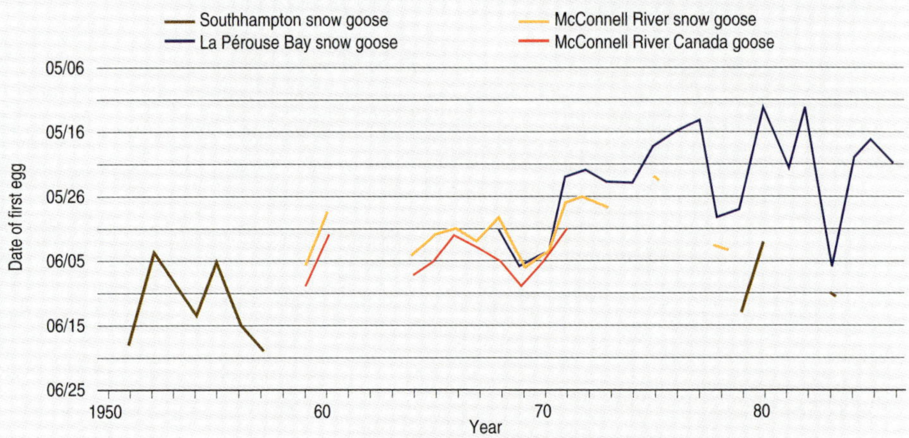

Fig. 2. Date of first egg. Although all sites displayed large fluctuations, both the date of first egg and the mean (average) hatch date became significantly earlier during the period 1951-86. There were no significant differences in the slopes of trends among sites or species.

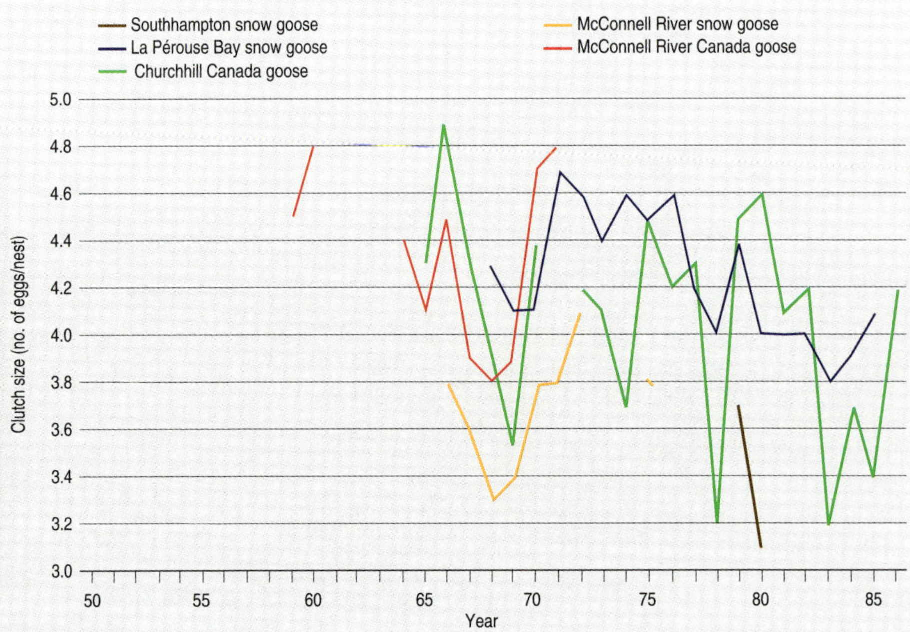

Fig. 3. Clutch size of nests of snow and Canada geese at Hudson Bay.

and numerous hydroelectric plants built with cooling ponds provide open water in winter, which this eagle needs for hunting (Southern 1963).

The winter abundance of the bald eagle throughout most of the contiguous United States dropped by about a third from 1957 to 1970 because of the use of persistent insecticides (e.g., DDT) and habitat destruction (Brown 1975). Since World War II, population declines in the East have been blamed on habitat destruction due to human disturbances (waterfront housing and outdoor recreation; Sprunt 1969). Shooting by ranchers from small planes from the late 1930's to the early 1960's could have depressed their abundance during this period and later (USFWS 1992).

Range Contractions

Of the 58 species examined, only 4 showed range contractions. This result could have been partly an artifact of our sample: we did not examine species that have very restricted ranges. It may also be due to our methods of examination because species had to abandon entire states, not just part of them, before we recorded a contraction. Of the four species showing range contractions, one is the brown-headed cowbird (*Molothrus ater*) and the other three depend on open water: pied-billed grebe (*Podilymbus podiceps*), northern pintail (*Anas acuta*), and common merganser (*Mergus merganser*).

The contraction of the northern pintail is of particular concern (Fig. 6). This game species has been extensively managed, yet estimates of its breeding population have shown a fivefold decrease since the mid-1900's (USFWS 1992). The reasons for this large decline are not yet understood.

Conclusion

The data collected by volunteers for the National Audubon Society's Christmas Bird Counts provide excellent information to examine the ranges and abundance patterns of wintering North American birds over both a very broad spatial scale and a long temporal scale. The changes that we found were primarily due to human activity, both purposeful (e.g., management of game species) and accidental. Some of these changes could be viewed as being beneficial (e.g., water management programs increasing bald eagle numbers), while others could be viewed as negative (e.g., logging allowed barred owls to invade spotted owl territories).

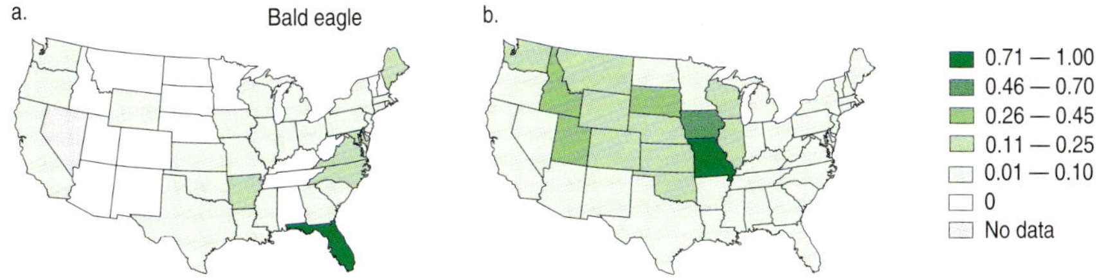

Fig. 5. Range and abundance patterns of the bald eagle. (a) Data from 1901 to 1940. (b) Data from 1960 to 1989 (except 1969).

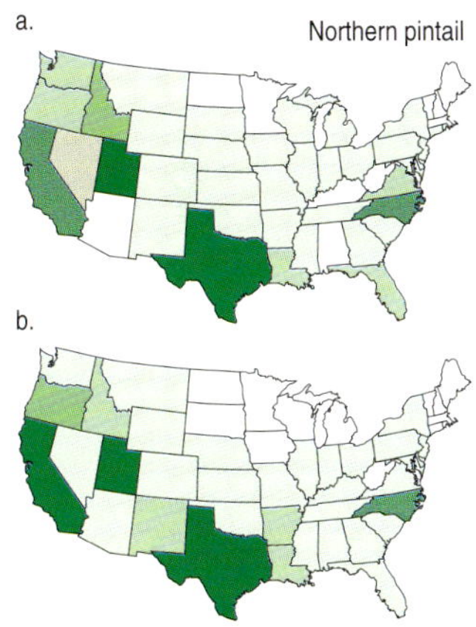

Fig. 6. Maps showing range and abundance patterns of the northern pintail. (a) Data from 1901 to 1940. (b) Data from 1960 to 1989 (except 1969).

References

Bent, A.C. 1949. Life histories of North American thrushes, kinglets, and their allies. United States National Museum Bull. 196. 454 pp.

Bock, C.E., and L.W. Lepthien. 1976. Population growth in the cattle egret. Auk 93:164-166.

Brown, W.H. 1975. Winter population trends in the bald eagle. American Birds 29:12-14.

Ehrlich, P.R., D.S. Dobkin, and D. Wheye. 1988. The birder's handbook. Simon and Schuster, New York. 785 pp.

Hewitt, O.H. 1967. The wild turkey and its management. Wildlife Society, Washington, DC. 589 pp.

Lewis, J.C. 1973. The world of the wild turkey. J.B. Lippincott Co., Philadelphia, PA. 158 pp.

Root, T.L. 1988a. Atlas of wintering North American birds. University of Chicago Press, IL. 312 pp.

Root, T.L. 1988b. Environmental factors associated with avian distributional boundaries. Journal of Biogeography 15:489-505.

Root, T.L. 1991. Positive correlation between range sizes and body size: a possible mechanism. Proceedings of the Twentieth International Ornithological Congress 2:817-825.

Root, T.L., and J.D. Weckstein. 1994. Changes in distribution patterns of select wintering North American birds from 1901 to 1989. Pages 191-201 *in* J.R. Jehl, Jr. and N.K. Johnson, eds. A century of avifaunal change in western North America. Studies in Avian Biology 15.

Schorger, A.W. 1966. The wild turkey: its history and domestication. University of Oklahoma Press, Norman. 625 pp.

Sharp, D.V. 1989. Range extension of the barred owl in western Washington and first breeding record on the Olympic Peninsula. Journal of Raptor Research 23:179-180.

Southern, W.E. 1963. Winter populations, behavior, and seasonal dispersal of bald eagles in northwestern Illinois. Wilson Bull. 75:42-55.

Sprunt, A., IV. 1969. Population trends of the bald eagle in North America. Pages 347-351 *in* J. Hickey, ed. Peregrine falcon populations: their biology and decline. University of Wisconsin Press, Madison.

Taylor, A.L., and E.D. Forsman. 1976. Recent range extensions of the barred owl in Western North America, including the first records for Oregon. Condor 78:560-561.

USFWS. 1992. Status of waterfowl and fall flight forecast. U.S. Fish and Wildlife Service, Office of Migratory Bird Management. 30 pp.

Wing, L. 1947. Christmas census summary 1900-1939. State College of Washington, Pullman. Mimeograph.

For further information:

Terry L. Root
University of Michigan
School of Natural Resources and Environment
430 E. University
Ann Arbor, MI 48109

Climate Change in the Northeast

by
Ray T. Oglesby
Charles R. Smith
Cornell University

Climate is a principal determinant of biological distributions and of patterns that characterize the seasonal physiology and behavior of many organisms (Gates 1993). Consequently, a changing climate should elicit responses in these biological properties. Detecting and characterizing such changes are logical early steps in assessing the significance of climate change to species and ecosystems (Schwartz 1990). Most published work on this subject involves species and ecosystem modeling based on known physiological and behavioral traits of selected species. This article presents evidence from an array of phenological data suggesting that climate change is occurring and that its biological effects may already be of considerable magnitude. (Phenological data are those associated with the relationship between climate and periodic phenomena like bird migration and flowering.)

Most research intended to explore possible effects of climate change on vegetation has understandably focused on agricultural and forest plants. Our approach, however, focuses on examining historical trends at the regional level and identifying species of potential value as climate change indicators. With the assumption, based on climate models, that unidirectional warming is already occurring and will probably accelerate over the next few decades, we began to search for evidence of biological responses among very different groups of organisms. Specifically, dates of the first return of spring-migrating birds and of the first bloom of spring wildflowers in the Northeast were sought in long-term (50 or more years), continuous, reliable records.

We computerized and analyzed two major and several minor long-term data sets from handwritten records from three New York State locations (Fig. 1). An especially rich source was records from the Cayuga Bird Club at Ithaca. Highly reliable observers recorded first spring sightings of migratory birds from 1903 to the present in the Cayuga Lake basin of central New York as delineated by Wiegand and Eames (1925). A second source of extensive, high-quality information was records for dates of first spring arrival for migrating birds and dates of first bloom for spring wildflowers at the Mohonk Preserve, an upland site in the mid-Hudson Highlands region of southeastern New York; these records extend from the late 1920's onward. Both sites are expected to continue generating comparable data sets. A third data set includes dates of first spring arrival for Louisiana waterthrush (*Seiurus motacilla*) and solitary vireo (*Vireo solitarius*) in western New York (1952 to present) on the Allegheny Plateau.

Our general approach to data analysis has been in the form of species plots with date of first arrival or first bloom as the vertical axis and sequence of years as the horizontal axis (Figs. 2-4).

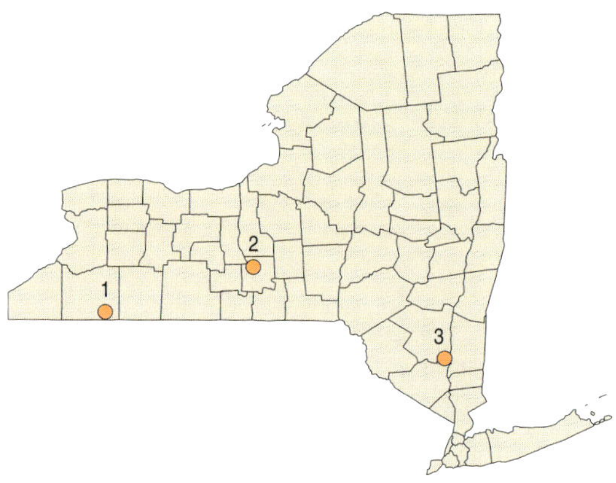

Fig. 1. New York locations from which phenological data were obtained. 1–Allegheny Plateau (birds); 2–Cayuga Lake basin, Ithaca (birds); and 3–Hudson Highlands (flowering plants and birds).

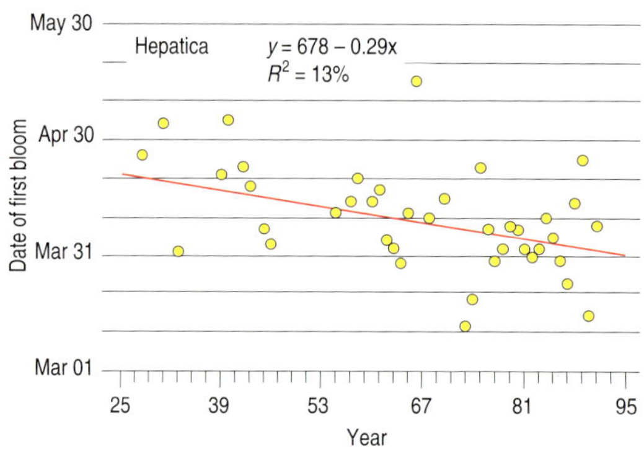

Fig. 2. Trend for hepatica (*Hepatica acutiloba*) from the Hudson Highlands (Mohonk Preserve) of southeastern New York, showing tendency for earlier spring blooming. The negative slope of the trend is significant at $P < 0.05$.

Status and Trends

Flowering Plants

Phenological data were examined for 15 species of spring wildflowers on time of first blooming at the Hudson Highlands site (Fig. 2). Six species of wildflowers all exhibited significantly earlier ($P \leq 0.05$) rather than later blooming (averaging -19.8 days/50 yr; $R^2 = 0.26$). The remaining nine species showed no significant patterns of change. We only can speculate why six species exhibited such a pronounced change and nine others did not. Clues may be obtained when existing data for other plant species at this site are examined. For example, the set of species showing earlier blooming appears to include plants typically found in more open locations where soil temperature would show the earliest and most rapid response to warming. One woody shrub, common witch-hazel (*Hamamelis virginiana*), which blooms in early fall, also showed a significant trend toward earlier bloom.

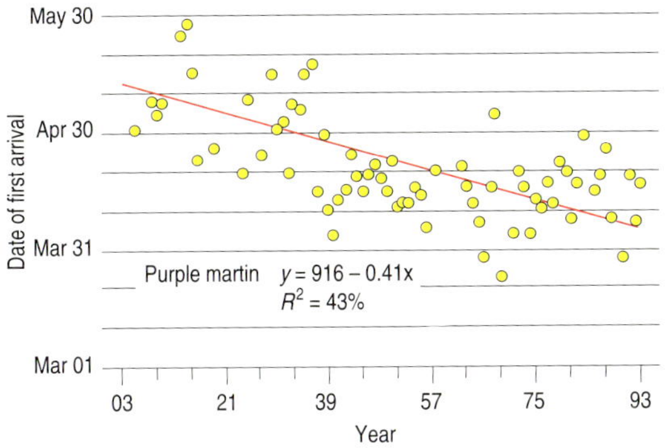

Fig. 3. Trend for purple martin (*Progne subis*) from the Cayuga Lake basin of central New York showing a clear tendency for earlier spring arrival. The negative slope of the trend is significant at $P < 0.005$.

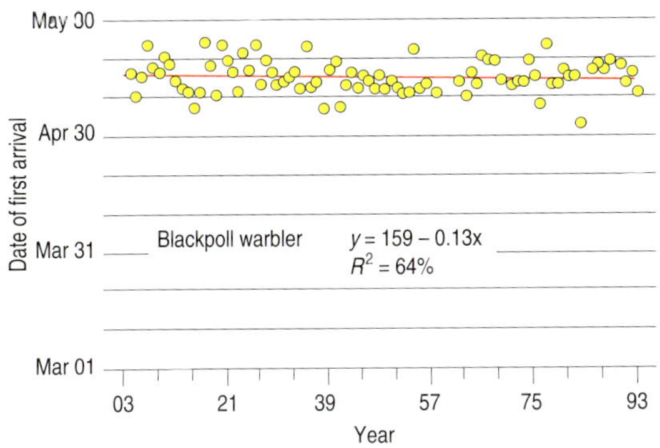

Fig. 4. Trend for blackpoll warbler (*Dendroica striata*) from the Cayuga Lake basin of central New York, showing no tendency for earlier spring arrival. The slope of the trend is not significantly different from zero.

Migratory Landbirds

Records of first sightings of spring migrants from the Cayuga Lake basin of central New York are from 1903 to the present and represent 130 species. More than 5,000 first arrival dates were examined for 76 species of migratory landbirds, excluding waterfowl, hawks, and shorebirds. Thirty-nine species (51%) showed 90-year trends of significantly earlier ($P \leq 0.05$) arrival, with an average shift of about 5.5 days/50 yr (for example, see Fig. 3). Thirty-five species (46%) showed no significant change ($P \leq 0.05$) in dates of first spring arrival (Fig. 4). Louisiana waterthrush and mourning warbler (*Oporornis philadelphia*) were the only species showing significant trends ($P \leq 0.05$) toward later spring arrival dates.

Of those species showing significantly earlier arrival dates, 85% are long-distance Neotropical migrants, which normally overwinter in Central and South America. For the Mohonk Preserve data, 6 of 10 migratory bird species examined showed significant trends toward earlier arrival. For the western Allegheny Plateau, first-arrival data, covering a 40-year period beginning in 1952, were available for two species of birds, Louisiana waterthrush and solitary vireo. Both species tended to earlier arrival.

Conclusions

The trends reported here toward earlier arrival dates for migratory birds and earlier blooming dates for spring wildflowers are concurrent with patterns of climatic warming and consistent with what might be expected in the context of global warming. At the same time, local changes in land cover, with the forested area of the region increasing by more than 30% since 1900, may provide greater amounts of suitable habitat for attracting and holding migrating landbirds, thereby contributing to observed patterns of change in migratory behavior.

It is noteworthy that only two bird species examined, and no plant species, showed trends to either later spring arrival or later blooming. If explanations of trends for only one of these two major groups were sought, alternative explanations could be advanced, such as expansion of bird ranges due to changes in land use (*see* Root and Weckstein, this section). In addition, a recent examination of dates of fall departure for migrating birds in Germany (Gatter 1992) shows later fall departures. Such fall trends would be expected in the context of climatic warming and agree with the spring trends we report.

Given the patterns reported here, climate change is the one variable affecting diverse groups of organisms that offers a rational and parsimonious explanation for the observed changes in timing of migration in birds and blooming in plants we and others have observed. Research either planned or in progress includes analyzing additional data sets as well as more sophisticated statistical analysis; determining the species most appropriate for monitoring climate change; finding and analyzing data sets that describe the phenology of other taxa; and possibly extending the study to other locations.

References

Gates, D.M. 1993. Climate change and its biological consequences. Sinauer Associates, New York. 280 pp.

Gatter, W. 1992. Zugzeiten und zugmuster im herbst: einfluss des treibhauseffekts auf den vogelzug. Journal für Ornithologie 133:427-436.

Schwartz, M.D. 1990. Detecting the onset of spring: a possible application of phenological models. Climate Res. 1:23-29.

Wiegand, K.M., and A.J. Eames. 1925. The flora of the Cayuga Lake basin, New York. Cornell University Agricultural Experiment Station Memoir 92. 491 pp.

For further information:

Ray T. Oglesby
Cornell University
Department of Natural Resources
New York Cooperative Fish and
Wildlife Research Unit
Ithaca, NY 14853

Potential Impacts of Climate Change on North American Flora

by
Larry E. Morse
Lynn S. Kutner
The Nature Conservancy
John T. Kartesz
North Carolina Botanical Garden

Climate change is a natural phenomenon that has occurred throughout the history of the earth. The frequency and magnitude of climate change have varied substantially during and between glacial periods, and temperatures on both global and local scales have been both substantially warmer and colder than present-day averages (Ruddiman and Wright 1987; Pielou 1991; Peters and Lovejoy 1992). While potential magnitudes of local and global climate change are of concern, it is the predicted rate of temperature change that poses the greatest threat to biodiversity. The ability of species to survive rapid climate changes may partially depend on the rate at which they can migrate to newly suitable areas.

In the next few centuries climate may change rapidly because of human influences. The concentrations of "greenhouse" gases in the atmosphere are being altered by activities such as carbon dioxide emission from burning fossil fuels. Models of climate change (IPCC 1990, 1992) predict an increase in mean global temperature of about 1.5-4.5°C (2.7-8.1°F) in the next century. Temperature changes suggested by general circulation models would present natural systems with a warmer climate than has been experienced during the last 100,000 years. While this would be a substantial change from the current climate, the *rate* of climate change is the greatest determinant of the impact on biological diversity. Future climate change due to human influences could occur many times faster than any past episode of global climate change (IPCC 1990, 1992; Schneider et al. 1992).

The strong association between distributions of plant species and climate suggests that rapid global climatic changes could alter plant distributions, resulting in extensive reorganization of natural communities (Graham and Grimm 1990). Climate changes could also lead to local extirpations of plant populations and species extinctions. The effects of global climate change are likely to vary regionally, depending on factors such as proximity to oceans and mountain ranges. Alteration of the amount and timing of precipitation and evaporation would affect soils and habitats; freshwater ecosystems are likely to be vulnerable to these changes in hydrology (Carpenter et al. 1992). Even minor fluctuations in the availability of water can radically affect habitat suitability for many wetland plant species. Rapid, large-scale shifts in temperature, precipitation, and other climate patterns could have broad ecological effects, presenting major challenges to the conservation of biodiversity.

Analysis of Potential Effects

An analysis conducted by The Nature Conservancy on the potential effects of climate change on the native vascular flora of North America (Morse et al. 1993) provides a preliminary assessment of patterns of plant species' vulnerability. For this preliminary analysis, we made several simplifying assumptions about the relationships between plants and climate to estimate the viable climate "envelopes" for each of over 15,000 native vascular plant species in North America recognized in the checklist by Kartesz (1994).

The principal assumptions are that climate determines the range of plant species; mean annual temperature adequately approximates climate; species distribution appears to be in equilibrium with present climate; and a species' current climate envelope is equivalent to its tolerance of climate variation. Together, these assumptions state that the current distribution of each species is greatly influenced by climate and that temperature adequately represents climate.

Clearly, each of the above assumptions are not actually met for all native vascular plant species. For example, precipitation and soil moisture are extremely important determinants of range limits in some regions. These simplified temperature envelopes, however, allow the initial identification of broad patterns of species' vulnerability to climate change.

In the analysis, the mean temperature was uniformly increased in 1°C (1.8°F) increments up to an increase of 20°C (36°F) above current mean annual temperatures (Fig. 1). Many species would be vulnerable to climate change in all scenarios of uniform temperature increase. With a mean global warming of 3°C

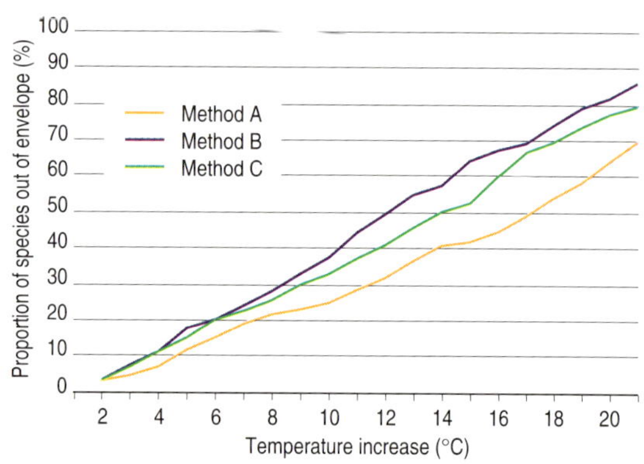

Fig. 1. The proportion of native vascular plant species that were entirely out of their climate envelopes as a function of the increase in temperature above mean annual temperature. Three methods were used to determine climate envelopes (A, B, C).

(5.4°F), 7% to 11% of 15,148 native vascular plant species in North America (about 1,060 to 1,670 species) could be entirely out of their climate envelopes. These species would thus be vulnerable to extinction unless they can migrate rapidly enough or can persist despite climate change. In comparison, about 90 plant species in North America are believed to have gone extinct in the last two centuries (Russell and Morse 1992).

Rarity and Vulnerability

Of the native vascular plant species studied, about 4,100 (27%) are considered rare by The Nature Conservancy (*see* article by Stein et al., p. 399, for definitions of ranking system for rarity). These species occur at fewer than 100 sites or are comparably vulnerable. Our analysis shows that these rare plants are likely to be further affected by climate change. In this analysis, about 10%-18% of the rare species would be vulnerable to a mean 3°C (5.4°F) temperature increase. In contrast, only 1% to 2% of the common species appear vulnerable under these conditions. These results imply that numerous rare vascular plant species could be additionally threatened by climate change. Early warnings of species' vulnerability to a rapidly changing climate might allow the development and implementation of new conservation strategies before a crisis occurs, thus improving the success rate for the protection of rare plants while minimizing the cost.

Regional Patterns of Vulnerability

Based on the uniform 3°C (5.4°F) mean increase in temperature used for this preliminary climate change impacts analysis, there appear to be regional patterns to the proportion of potentially vulnerable species in each state or province (Fig. 2). In this initially simplified analysis, the southeastern states have the highest percentage of species out of their climate envelopes, while the Great Plains states and provinces may experience proportionally fewer species losses. The relatively high proportion of species vulnerability in the Southeast may be due in part to the presence in state floras of Appalachian Mountain species at their southern range limits. Many of these species are already rare in states along their southern range limits and are likely to be lost from the local floras if the climate warms.

Global warming models, however, suggest that the temperature and precipitation changes in the interior of the continent will be far greater than in coastal regions. In the Great Plains, some models suggest increases in summer temperatures by 4-7°C (7.2-12.6°F), accompanied by dramatic decreases in precipitation. Future analyses that incorporate regional changes in climate projected by models will further refine our understanding of regional patterns of plant species' vulnerability to climate change.

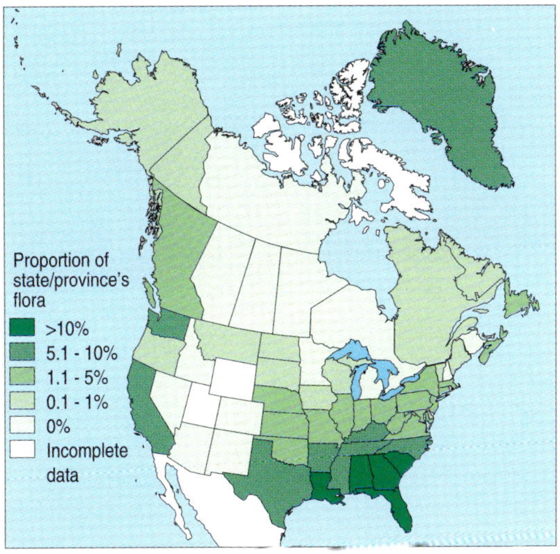

Fig. 2. The proportion of species that would be out of their climate envelope in each state or province with a +3°C (+5.4°F) temperature change.

Dispersal and Persistence of Vascular Plants

The survival of species during periods of changing climate will be determined in part by their abilities to disperse to new sites or to persist in place. For this analysis, a dispersal-ability scale was used to assess the potential for different species to migrate. The scale is based on characteristics important to species mobility such as pollination mechanisms, dispersal mechanisms, reproductive characteristics, degree of self-compatibility, growth form, trophic type, and number of populations. Biological factors likely to increase species mobility include wind pollination, at least partial self-compatibility, dispersal of propagules by wind or birds, and a short generation time. Characteristics such as dependence on specific pollinators (e.g., yucca and yucca moth), dispersal by ants, or a long generation time reduce the chances for successful rapid dispersal and establishment. By using these criteria, most of the species studied appear to have an intermediate dispersal potential.

The species in this analysis that would be vulnerable in a +3°C (5.4°F) climate appear to have characteristics that limit long-distance dispersal (Fig. 3). This suggests that the plants potentially most vulnerable to climate change may be those forced to adapt in place to new conditions. In general, rare plants and narrow endemics will be particularly endangered by climate change. These plants often have restricted ranges, a reduced seed source, and may depend on specific microclimatic conditions for

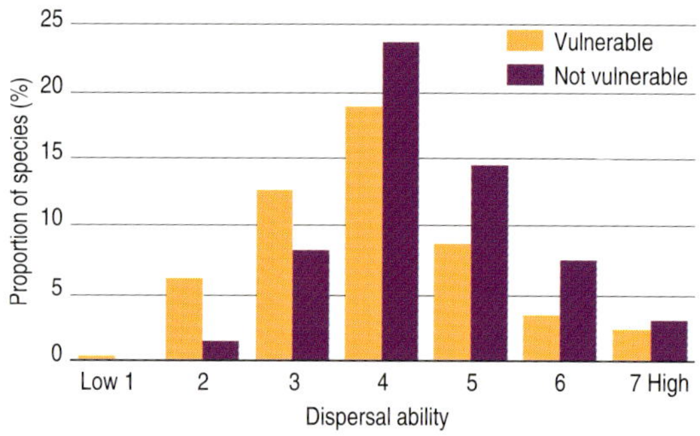

Fig. 3. The proportion of species on the dispersal-ability scale that are out of their climate envelopes (vulnerable) or in their climate envelopes (not vulnerable) with a +3°C (+5.4°F) temperature change. Full data were available for 8,668 species.

survival. Rare plants would thus potentially have trouble migrating to comparable new sites, regardless of their ability to disperse. For example, Boott's rattlesnake-root (*Prenanthes boottii*) and mountain avens (*Geum peckii*), endemic to alpine habitats in the northeastern United States, would be particularly sensitive to global warming.

Migration Rate

During the warming at the end of the last glacial period, plant migration rates, as calculated from the fossil pollen record, ranged from about 5 to 150 km (3-90 mi) per century (Shugart et al. 1986). Human-caused climate change may occur at rates more than five times faster than any changes since the last glacial maximum, including the period of most rapid deglaciation (Overpeck et al. 1991). Various studies have suggested that such rapid climate changes would require shifts of plant ranges of up to 500 km (300 mi) within the next century, exceeding the known rates of migration for many plant species (Davis 1984; Davis and Zabinski 1992).

Since species respond individually to climate change, migration rates will vary within and among natural communities. It is unlikely that entire biological communities would move together in response to climate changes (Graham and Grimm 1990). Some plants may respond rapidly to changes; others may survive for several generations in place or persist as long-lived clones despite significant climate change. The fossil record provides evidence of decade- or even century-long time lags in species migration (Davis 1989). The process of changing community composition in response to climate change has been documented in the fossil record through the disassociation and reassembly of plant and animal taxa (Graham and Grimm 1990). This variation in species assemblages displays the transitory nature of former as well as existing and future community types.

Temperature extremes and changes in the frequency and severity of local disturbances may have a greater influence on the survival of plant species at particular locations than small shifts in the average climate. More frequent droughts, fires, and pest and pathogen outbreaks are predicted to act in conjunction with climate change to significantly transform the landscape (Peters 1992). This prediction is supported by paleoecological evidence that altered disturbance regimes can intensify the effects of climate change on plants and increase the amount of overall vegetational change (Davis 1989).

Threats by Weedy Exotics

With global climate change, some exotic weeds may be favored over native species. Many weeds are able to expand relatively quickly, posing serious threats to existing species and overall biodiversity (Schwartz 1992). Many weedy species are widespread, prolific, fast-growing annuals capable of establishing in disturbed habitats and are often favored by disturbances. Climate-induced changes could expose native plants to nonnative competitors for the first time (Peters 1992), stressing the balance established between native plants and their habitat. Exotic weeds may become a greater problem in the management of many preserves and natural areas.

Landscape Fragmentation

The potentially rapid rates of warming, combined with habitat loss and fragmentation from human development, suggest that many species will not adjust as successfully to climate change as in the past. Most native plant species exist in a highly fragmented landscape that further separates appropriate habitat patches, increasing the dependence of many species on relatively rare events of long-distance dispersal. Furthermore, species often must disperse across hostile habitats, including roads, cities and suburbs, and farmland (Peters 1992). Finally, plants would need to establish themselves in landscapes where many of the open or disturbed areas have been colonized by aggressive weedy exotics.

Climate Change and Conservation Planning

Rapid climate change could place novel demands and constraints on plant species conservation. Vulnerability to climate change could affect selection and design of new preserves and management procedures in existing preserves, especially in southern or low-elevation portions of species' ranges. Management of species threatened by climate change could involve restoration and transplantation of species among preserves or into new locations. Actions such as removal of exotic species or hydrological controls may not be qualitatively different than those that are currently required of land managers, but climate change may increase the intensity and frequency of threats from exotic species, drought, and fire. In view of the unpredictable and potentially devastating effects of global climate change on species' viability and distribution, conservation strategies such as propagation of critical species outside of their natural range to provide materials for reintroductions are likely to become increasingly important to preserve biological diversity.

References

Carpenter, S.R., S.G. Fisher, N.B. Grimm, and J.F. Kitchell. 1992. Global change and freshwater ecosystems. Annual Review of Ecology and Systematics 23:119-140.

Davis, M.B. 1984. Climatic instability, time lags, and community disequilibrium. Pages 269-284 *in* J. Diamond and T.J. Case, eds. Community ecology. Harper and Row, New York.

Davis, M.B. 1989. Insights from paleoecology on global change. Ecological Society of America Bull. 70(4):222-228.

Davis, M.B., and C. Zabinski. 1992. Changes in geographical range resulting from greenhouse warming effects on biodiversity in forests. Pages 298-308 *in* R.L. Peters and T.L. Lovejoy, eds. Global warming and biological diversity. Yale University Press, New Haven, CT.

Graham, R.W., and E.C. Grimm. 1990. Effects of global climate change on the patterns of terrestrial biological communities. Trends in Ecology and Evolution 5(9):289-292.

IPCC. 1990. Climate change: the Intergovernmental Panel on Climate Change scientific assessment. World Meteorological Organisation and United Nations Environment Programme, Geneva. 270 pp.

IPCC. 1992. 1992 Intergovernmental Panel on Climate Change supplement. World Meteorological Organisation and United Nations Environment Programme, Geneva. 70 pp.

Kartesz, J.T. 1994. A synonymized checklist of the vascular flora of the United States, Canada, and Greenland. 2nd ed. Timber Press, Portland, OR. 622 pp.

Morse, L.E., L.S. Kutner, G.D. Maddox, J.T. Kartesz, L.L. Honey, C.M. Thurman, and S.J. Chaplin. 1993. The potential effects of climate change on the native vascular flora of North America: a preliminary climate-envelopes analysis. Rep. TR-103330, Electric Power Research Institute, Palo Alto, CA.

Overpeck, J.T., P.J. Bartlein, and T. Webb III. 1991. Potential magnitude of future vegetation change in eastern North America: comparison with the past. Science 254:692-695.

Peters, R.L. 1992. Conservation of biological diversity in the face of climate change. Pages 15-30 *in* R.L. Peters and T.L. Lovejoy, eds. Global warming and biological diversity. Yale University Press, New Haven, CT.

Peters, R.L., and T.L. Lovejoy, eds. 1992. Global warming and biological diversity. Yale University Press, New Haven and London. 386 pp.

Pielou, E.C. 1991. After the ice age: the return of life to glaciated North America. University of Chicago Press, Chicago and London. 366 pp.

Ruddiman, W.F., and H.E. Wright, Jr., eds. 1987. North America and adjacent oceans during the last deglaciation. The Geological Society of America, Boulder, CO. 501 pp.

Russell, C.A., and L.E. Morse. 1992. Plants. Biodiversity Network News [The Nature Conservancy] 5(2):4.

Schneider, S.H., L. Mearns, and P.H. Gleick. 1992. Climate-change scenarios for impact assessment. Pages 38-55 *in* R.L. Peters and T.L. Lovejoy, eds. Global warming and biological diversity. Yale University Press, New Haven, CT.

Schwartz, M.W. 1992. Modelling effects of habitat fragmentation on the ability of trees to respond to climatic warming. Biodiversity and Conservation 2:51-61.

Shugart, H.H., M.Y. Antonovsky, P.G. Jarvis, and A.P. Sandford. 1986. CO_2, climatic change and forest ecosystems. Pages 475-521 *in* B. Bolin, B.R. Döös, J. Jäger, and R.A. Warrick, eds. The greenhouse effect, climatic change, and ecosystems. John Wiley and Sons, New York.

For further information:

Larry E. Morse
The Nature Conservancy
1815 N. Lynn St.
Arlington, VA 22209

Courtesy R. Hayes/ H. Weiss, BLM

Human Influences

by
Science Editor
Wayne Willford
National Biological Service
1849 C St. NW
Washington, DC 20240

Overview

The following articles are directed at neither a specific species nor an ecosystem, but at human activities that affect living resources nationally and internationally. These broad-scale effects on, and changes in, ecosystem health are frequently the result of local or regional actions and land-use practices that collectively have effects across the nation.

The first article (Stein et al.) examines the significance of federal lands as refugia for the protection and conservation of endangered species. Stein et al. (box) then describe a system used to rank species by their need for conservation measures to prevent their endangerment and future extinction. In the article by Friend, we learn about the history of diseases in waterfowl, the trends in disease outbreak, and how the loss of wetlands and the discharge of water can be associated with these disease outbreaks. Dein et al. describe the use of propagation and translocation (transplanting species to a certain area) to recover or augment threatened or endangered species as well as recreational species. Dein et al. also examine the secondary consequences of such efforts on the transfer and spread of disease to wildlife, domestic animals, and humans. Cumulatively, these articles broaden the focus of status and trends assessments beyond individual species and ecosystems, and begin to reveal the interrelatedness between species, ecosystems, and human activities.

The remainder of the articles focus on the effects of pollution that results from human activities such as agricultural, industrial, and municipal development. The articles pay special attention to monitoring of pollution because the effects of pollution are excellent examples of the links between ecosystem health and the health of organisms, including humans, that depend on those systems.

The first article on pollution (Turgeon and Robertson) describes toxic contaminants in fish and mollusks from U.S. coastal waters. Next, Schmitt and Bunck describe the trends of chemical residues in fish and wildlife from across the nation during the past 25 years. The note by Glaser emphasizes how birds are being affected by the "new family" of pesticides in use across the United States.

Schreiber discusses the adverse impacts of acid deposition (acid rain) on sensitive species and ecosystems and the influence of recent regulatory efforts to control this form of pollution. Everson and Graber describe the results of a long-term study on the effects of acid rain on forest watersheds, the secondary impacts on water chemistry because of leaching of nutrients from soils, and the influence of a forest fire on the process. Allen discusses the agricultural effects on wildlife habitat.

Overall, these varied articles introduce the ways in which large-scale assessments of status and trends in the health and condition of biota also provide an excellent indication of overall ecosystem health, particularly in relation to the less visible effects such as long-term, subtle declines due to diseases and pollution.

Significance of Federal Lands for Endangered Species

by
Bruce A. Stein
The Nature Conservancy
Tom Breden
Association for Biodiversity Information
Richard Warner
The Nature Conservancy

The federal government has overall trust responsibilities for species listed as threatened or endangered under the Endangered Species Act (ESA). The options available for managing and protecting these species, however, are directly related to the ownership of the lands on which the species are found. This article provides information about the presence and numbers of federally listed species on federal lands and the responsibility of federal land managers to care for these species. Our analysis helps evaluate the potential and actual role of federal land-management agencies in the overall protection of threatened and endangered species (Natural Heritage Data Center Network 1993).

Natural Heritage Programs—a partnership between state and federal agencies and The Nature Conservancy—gather and manage a variety of information linking both biological and nonbiological factors of relevance to biodiversity conservation. Central to this effort is the inventory of all known occurrences for species of conservation concern, including all federally listed endangered or threatened species. An occurrence is defined as an example of a species at a specific location representing a habitat capable of sustaining the survival of that species. What constitutes an occurrence depends on the biology of the particular species, but most often reflects a mappable and geographically distinct population or subpopulation. Pertinent information is documented for each occurrence, such as the biological health and population trends of the occurrence, habitat quality, protection or management status, and land ownership.

Heritage Programs in all 50 states queried their data bases for all documented occurrences of federally listed species in their jurisdiction and reported the class of landowner or type of managing agency. (Note: "Species" under the ESA includes subspecies as well as full species; in the strictest taxonomic sense these collectively would be referred to as "taxa.") Only occurrences observed since 1973 were included in the analysis.

Karner blue butterfly (*Lycaeides melissa samuelis*), an endangered species found partially on federal lands.

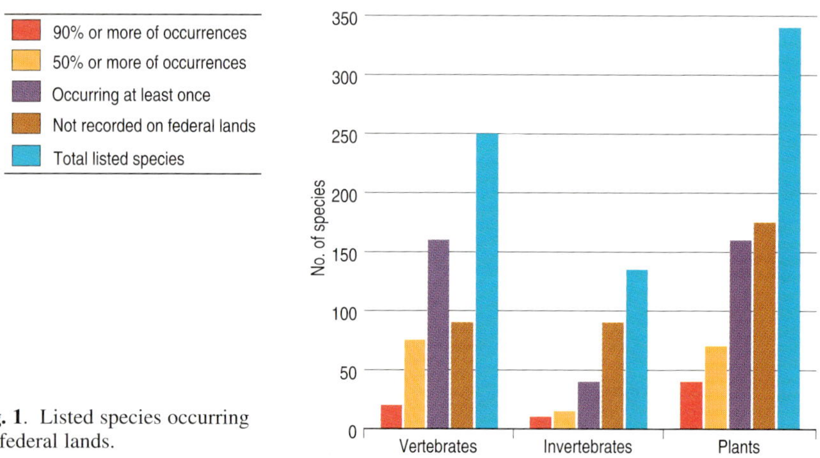

Fig. 1. Listed species occurring on federal lands.

While the Heritage Programs are the most comprehensive source for such locality information on rare species and reviewed about 350,000 occurrence records for this analysis, this information is incomplete for four reasons: (1) Heritage Programs may not be aware of all occurrences, and indeed, many populations for species of concern may yet be discovered; (2) most programs have a data-entry backlog; (3) not all data centers have completely recorded the land ownerships for all their occurrence records; and (4) species occurrences in lakes and rivers are generally not recorded as under the jurisdiction of a federal agency except where they are entirely included in such areas as national parks or wildlife refuges. On the other hand, in many states more is known about the status of listed species on federal lands than on state or private lands. This imbalance in the available data, largely the result of federally funded inventories on federal lands, will tend to overstate the proportion of a species' range or population on federal lands.

Species on Federal Lands

This analysis includes 344 plant, 254 vertebrate, and 130 invertebrate species found in the 50 United States that as of March 1993 were

Status of U.S. Species: Setting Conservation Priorities

by
Bruce A. Stein
Lawrence L. Master
Larry E. Morse
Lynn S. Kutner
Melissa Morrison
The Nature Conservancy

Resources available for conservation of species and ecosystems invariably are in short supply relative to the needs for those resources. Targeting conservation and management actions toward those species and ecosystems in greatest need, and where opportunities for success are greatest, requires clearly established priorities. Accordingly, setting priorities is a necessary prerequisite for effective biodiversity conservation and ecosystem management.

Many systems and methods for setting priorities and determining endangerment status have been developed, including those used by the U.S. Fish and Wildlife Service, the IUCN (World Conservation Union; formerly International Union for the Conservation of Nature and Natural Resources), and many individual states. Among the most widely applied systems is the biodiversity status-ranking system developed and used by the Natural Heritage Network and The Nature Conservancy (Master 1991; Morse 1993; Stein 1993). This ranking system has been designed to evaluate the biological and conservation status of plant and animal species and within-species taxa as well as ecological communities. For status-ranking purposes, collectively these are all referred to as "elements" of natural diversity (Jenkins 1988).

Status ranks are based primarily on objective factors relating to a species' rarity, population trends, and threats. Four aspects of rarity are considered: number of individuals, number of populations or occurrences, rarity of habitat, and size of geographic range. Ranks are assigned according to a rigorous and standardized process, with all supporting information documented in Heritage Program data bases. Ranking is based on an approximately logarithmic scale, ranging from (1) critically imperiled to (5) demonstrably secure (Table). Typically species with ranks from one to three would be considered of conservation concern and broadly overlap with species that might be considered for review under the Endangered Species Act or similar state or international statutes.

For conservation priorities to be set at local as well as rangewide scales, ranking is carried out at three hierarchical levels: subnational (e.g., state), national, and global. Thus, a species may be relatively common and secure globally (G4), but within a given state may be critically imperiled (S1). The combined rank within that state (G4/S1) allows local priorities to be set within a global context.

Table. Definition of biodiversity status ranks.

Rank	Definition
GX	Presumed extinct; not located despite searches
GH	Of historical occurrence; possibly extinct but some expectation of rediscovery
G1	Critically imperiled; typically 5 or fewer occurences or 1,000 or fewer individuals
G2	Imperiled; typically 6 to 20 occurrences or 1,000 to 3,000 individuals
G3	Rare or uncommon but not imperiled; typically 21 to 100 occurrences or 3,000 to 10,000 individuals
G4	Uncommon but not rare; apparently secure, but with cause for some long-term concern; usually more than 100 occurrences or 10,000 individuals
G5	Common; demonstrably widespread, abundant, and secure

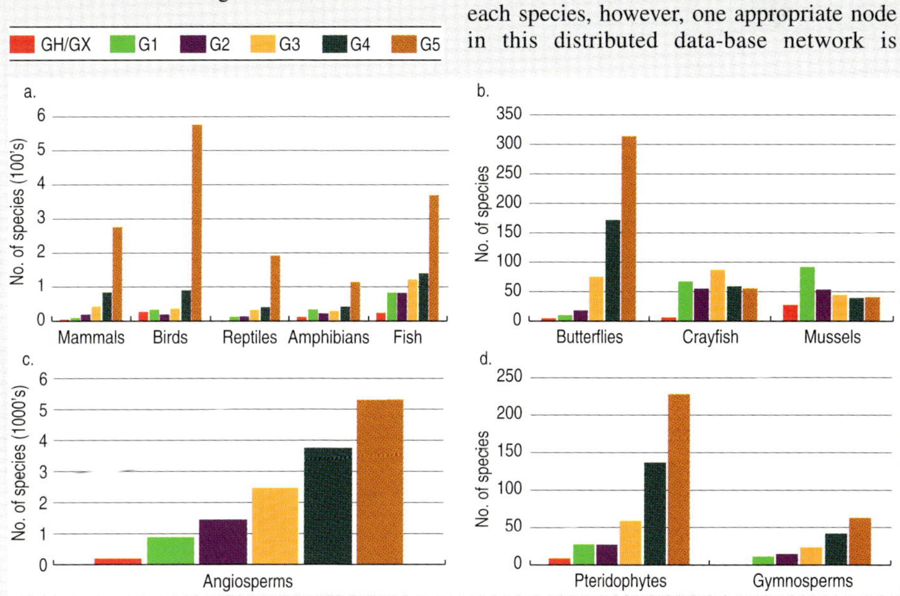

Figure. GH/GX - potentially extinct; G1 to G5 rank species from rarest (G1) to most common (G5). a - status ranks of U.S. vertebrate species (fish include freshwater only); b - status ranks of selected invertebrate groups: native U.S. species of butterflies, crayfish, and freshwater mussels; c - status ranks of native U.S. flowering plant species; and d - status ranks of native U.S. fern and conifer species.

The natural world is extremely dynamic, due to both intrinsic ecological factors and increasing human influences. At the same time, our knowledge of the distribution, abundance, and basic biology of species and ecological communities is imperfect, but continually improving. For these reasons, biodiversity status ranks must be viewed as working hypotheses based on the best available information. Ranks are continually reevaluated and refined as new populations are discovered, known populations are extirpated, or new or better information is available on overall status, trends, or threats. Indeed, ranks initially assigned to some poorly known species may reflect the inadequate state of knowledge about the organism more than its actual biological status. The very process of assigning these ranks and documenting the gaps in our understanding, however, works as a powerful tool in setting priorities for additional inventory and research. The increased inventory attention accorded highly ranked (i.e., rare) species tends to improve understanding of the species' distribution and status, often showing the species to be more common or secure than previously known and, therefore, of lesser concern from a conservation perspective.

The Natural Heritage Network is a distributed data-base network operating on the principle of shared information-management concepts and shared responsibilities. Each state is responsible for assessing the status of each species and natural community within its jurisdiction and for assigning and documenting a state rank for those elements. For each species, however, one appropriate node in this distributed data-base network is responsible for maintaining the current global rank based on a combination of the state and national-level status ranks together with other available information.

Global-level ranks have been assigned to all U.S. vertebrate species, selected groups of invertebrates (including all federally listed, proposed, and candidate species), all vascular plant species, and selected nonvascular plant species (e.g., many lichens and bryophytes). Preliminary global ranks also have been assigned to all rare terrestrial natural communities for the United States.

A summary of status ranks is presented in the Figure for the approximately 2,500 species of native U.S. vertebrates, for more than 1,200 invertebrate species from several groups for which complete data sets are available (butterflies, crayfish, and freshwater mussels), for the approximately 16,300 species of native U.S. flowering plants, and for the approximately 675 species of ferns and conifers native to the United States. These ranks are the result of collaborative work with Heritage Programs, conservation data centers, The Nature Conservancy's scientific staff, and many other state, federal, and private cooperators.

References

Jenkins, R.E., Jr. 1988. Information management for the conservation of biodiversity. Pages 231-239 *in* E.O. Wilson, ed. Biodiversity. National Academy Press, Washington, DC.

Master, L. 1991. Assessing threats and setting priorities for conservation. Conservation Biology 5(4):559-563.

Morse, L.E. 1993. Standard and alternative taxonomic data in the multi-institutional Natural Heritage Data Center Network. Pages 69-79 *in* F.A. Bisby, G.F. Russell, and R.J. Pankhurst, eds. Designs for a global plant species information system. Oxford University Press, Oxford.

Stein, B.A. 1993. Towards common goals: collections information in conservation databases. Association for Systematics Collections Newsletter 21(1):1-6.

For further information:

Bruce A. Stein
The Nature Conservancy
1815 N Lynn St.
Arlington, VA 22209

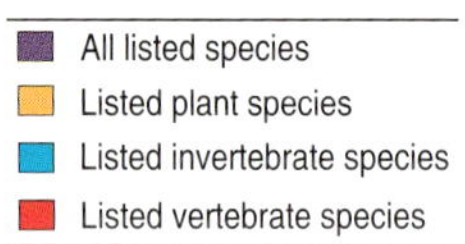

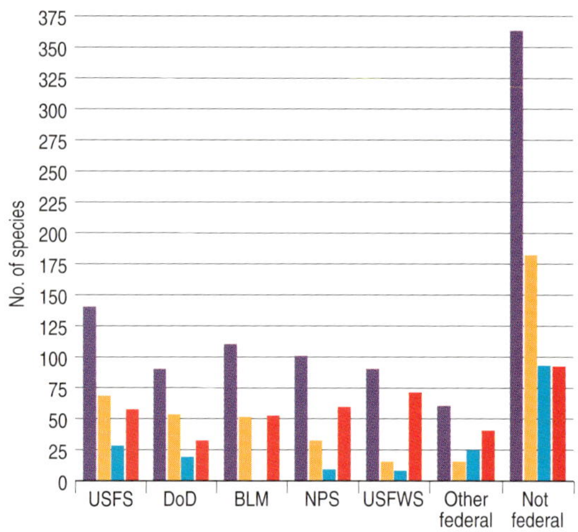

Fig. 2. Listed species occurring by jurisdiction on federal land. USFS—U.S. Forest Service; DoD—Department of Defense; BLM—Bureau of Land Management; NPS—National Park Service; USFWS—U.S. Fish and Wildlife Service.

listed as endangered or threatened under the ESA. About 50% of all federally listed threatened and endangered species occur at least once on federal lands (Fig. 1). The other half of the listed species are found on either state and local public lands, tribal lands, or private lands. About 25% of listed species have more than half of their known occurrences on federal lands and can benefit substantially from federal land-management protection and recovery actions. About 12% of listed species are found almost exclusively on federal lands, with 90%-100% of their known occurrences restricted to lands under federal management.

Number of Listed Species

About 100 species occur on land managed by most major federal agencies (Fig. 2). With nearly 150 species, the U.S. Forest Service (USFS) harbors the largest number of federally listed species of any federal agency. This larger number reflects not only the size and geographic extent of the landholdings, but also the intensive biological surveys that have been conducted on many national forests.

Occurrences of Listed Species

Of the 24,573 occurrences of federally listed species recorded by the Natural Heritage Network nationwide, 36% are found on federal lands (Fig. 3). The USFS, with 16% of the total, has the largest number of occurrences followed by the Bureau of Land Management (8%) and the Department of Defense (4%). Both the U.S. Fish and Wildlife Service and the National Park Service have 3% of known occurrences on their lands.

The average number of occurrences per listed species varies markedly among plants, vertebrates, and invertebrates. There are an average of 34 occurrences per listed species for all species combined, 17 occurrences per plant species, 13 occurrences per invertebrate species, and 67 occurrences per vertebrate species. The broad distribution of some federally listed vertebrates heavily influences these figures, however. Just 12 vertebrate species (e.g., bald eagle) account for 12,121 occurrences, representing 49% of the total for all species. Excluding these 12 vertebrates, the average number of occurrences for vertebrates drops to 20 and the overall average to 17.

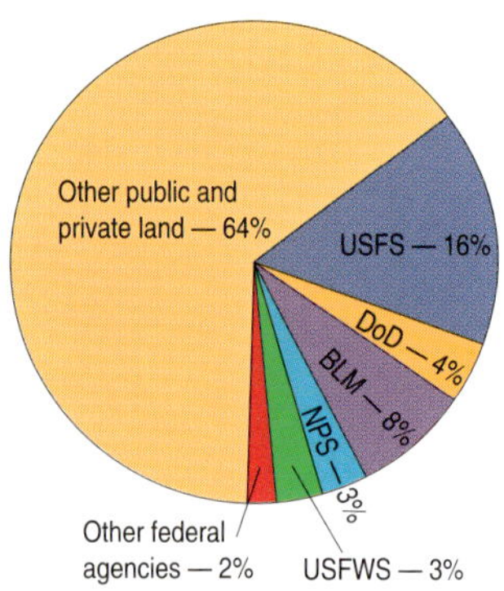

Fig. 3. Occurrences of all listed species by land ownership category. See Fig. 2 for definitions of abbreviations.

Conclusions

This analysis puts in perspective the relevance of federal land management for the protection of federally listed threatened and endangered species. Agencies that manage federal lands have substantial responsibilities and opportunities for protecting listed species, particularly those that are found exclusively, or mostly, on federal lands. An example of a plant entirely restricted to federal lands is Ruth's golden aster (*Pityopsis ruthii*), and an animal found exclusively on federal lands is the Laysan duck (*Anas laysanensis*), which lives only on national wildlife refuges.

Federal agencies can also provide substantial, permanent conservation of the listed species that have more than half of all occurrences on their lands. For example, most populations of both the white-haired goldenrod (*Solidago albopilosa*) and white birds-in-a-nest (*Macbridea alba*) are on national forest lands.

For many listed species with less than 50% of their occurrences on federal land, federal agencies may still be able to provide important protection and recovery opportunities. For example, the Karner blue butterfly (*Lycaeides melissa samuelis*), with less than one-quarter of its occurrences on federal lands, still can substantially benefit from federal management actions, such as restoration of the pine and oak savanna habitat on which this butterfly depends.

This study found, however, that fully 50% of federally listed species are not known to occur on federal lands, and that for all listed species, 64% of known occurrences are on nonfederal lands. This strongly points to the need for developing and strengthening federal efforts for protecting these species through cooperative efforts and incentive programs with state and local agencies, private conservation organizations, and private landholders.

Reference

Natural Heritage Data Center Network. 1993. Perspectives on species imperilment (revised printing). The Nature Conservancy, Arlington, VA. 40 pp.

For further information:

Bruce A. Stein
The Nature Conservancy
1815 N. Lynn St.
Arlington, VA 22209

Increased Avian Diseases With Habitat Change

by
Milton Friend
National Biological Service

Changes in disease patterns and trends reflect changing relationships between the affected species (host) and the causes of disease (agent). Host-agent interactions are closely linked to environmental factors that either enhance or reduce the potential for disease to occur. As a result, wildlife disease patterns and trends are, to a substantial extent, indicators of environmental quality and changing host-agent interactions within the environment being evaluated. The types, distribution, and frequency of diseases causing major avian die-offs have changed greatly during the 20th century. Too little is known to assess the changes of most avian diseases that result in chronic attrition rather than major die-offs, or about those that affect reproductive success, reduce body condition, or affect survival in other indirect ways. Nevertheless, the changing patterns and trends in highly visible avian diseases provide notice of problems needing attention.

Information on the status of disease in wild birds was obtained from National Wildlife Health Center (NWHC) evaluations of the cause of death for more than 30,000 carcasses from across the United States during the past two decades, reports of avian mortality received from collaborators, the scientific literature, and NWHC field investigations of bird mortality. Comprehensive assessments of causes of wild bird mortality, magnitude of losses, and geographic distribution of specific diseases are not possible from these data, although we can identify general relationships for waterfowl and some other species.

Changes in Disease Patterns

The occurrence of disease involves three factors: a susceptible host, presence of an agent capable of causing disease, and suitable environmental conditions for contact between the host and agent in a manner that results in disease. Environment is often the dominant factor in this relationship (Fig. 1).

Avian Botulism

The most dramatic example of geographic expansion of a noninfectious indigenous disease is avian botulism, caused by the bacterium *Clostridium botulinum*. In 1914 a Bureau of Biological Survey researcher began investigating catastrophic die-offs that had begun in 1910 and in which millions of waterbirds along the Great Salt Lake, Utah, had died. Later studies revealed that avian botulism was responsible for those die-offs. Historically, avian botulism was referred to as "western duck disease" because of its rather limited geographical distribution of occurrence (Kalmbach and Gunderson 1934; Fig. 2)

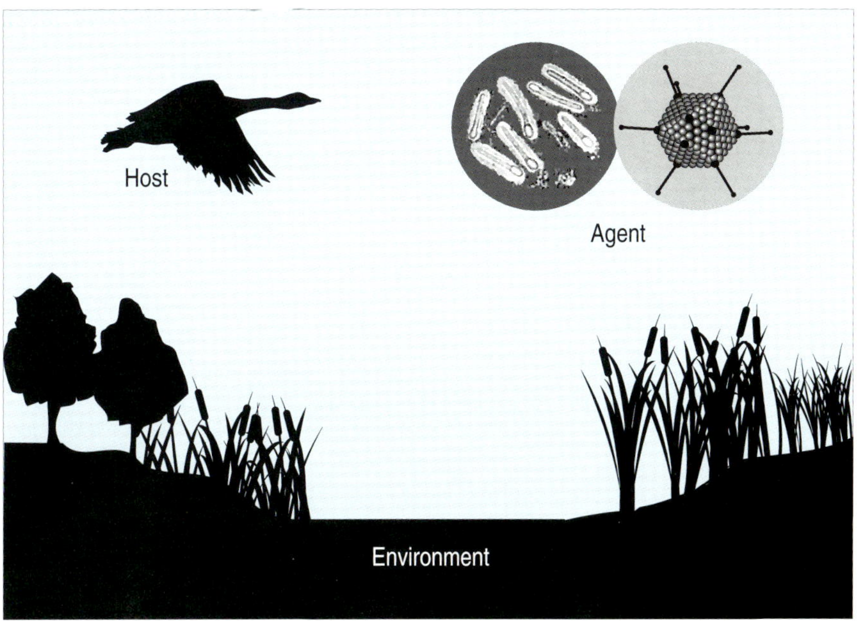

Fig. 1. Common factors required for disease to occur. Environmental factors greatly influence occurrence of disease by changing the amount and type of host-agent interactions.

Avian botulism now occurs all over the United States (Fig. 2) and in many other countries as well. Because of the visibility of massive die-offs, avian botulism is probably the best-documented nonhunting waterfowl mortality (Stout and Cornwell 1976). The continued reporting of avian botulism die-offs since the early 1900's makes researchers suspect that much of the disease's geographic expansion is of recent origin. Also, most (15 of 21) initial outbreaks of avian botulism in countries other than in North America have occurred since 1970.

Avian Cholera

Avian cholera, caused by the bacterium *Pasteurella multocida*, has been recognized as an important infectious disease of domestic poultry in the United States since at least 1867 (Rhoades and Rimler 1991). Therefore, it is noteworthy that a 1930 evaluation of the status of waterfowl commented on the lack of documentation of avian cholera in wild waterfowl (Phillips and Lincoln 1930). In 1944, however, the disease was documented in wild waterfowl in the United States (Quortrup et al. 1946). Limited geographical expansion of avian cholera in wild waterfowl occurred during the 1940's and 1950's, and sporadic occurrences were documented at a few new locations during the 1960's (Fig. 3). By the end of the 1960's, though, avian cholera was reported as established in the Central and Pacific flyways. Outbreaks in the Mississippi Flyway were unusual, and only two outbreaks had occurred in the Atlantic Flyway. With the exception of a single instance during the breeding season, outbreaks occurred in winter (Stout and Cornwell 1976). During the 1970's, avian cholera became established as a major cause of waterfowl mortality in all four flyways within the United States and as a recurring cause of waterfowl mortality in Canada (Fig. 3). Geographic expansion of die-off locations continues, and outbreaks now occur during all seasons of the year (Friend 1987).

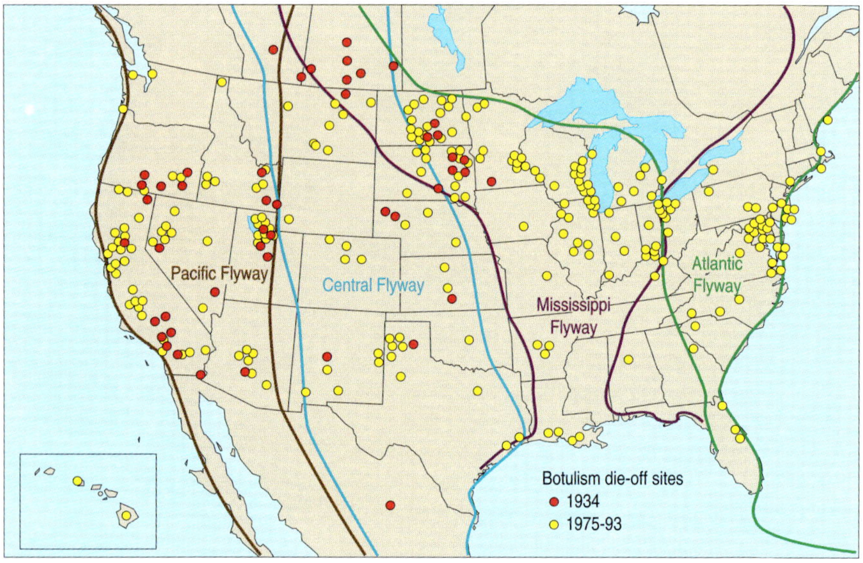

Fig. 2. Known distribution of "western duck sickness" (avian botulism) in North America, 1934 (Kalmback and Gunderson 1934), and from 1975 to 1993.

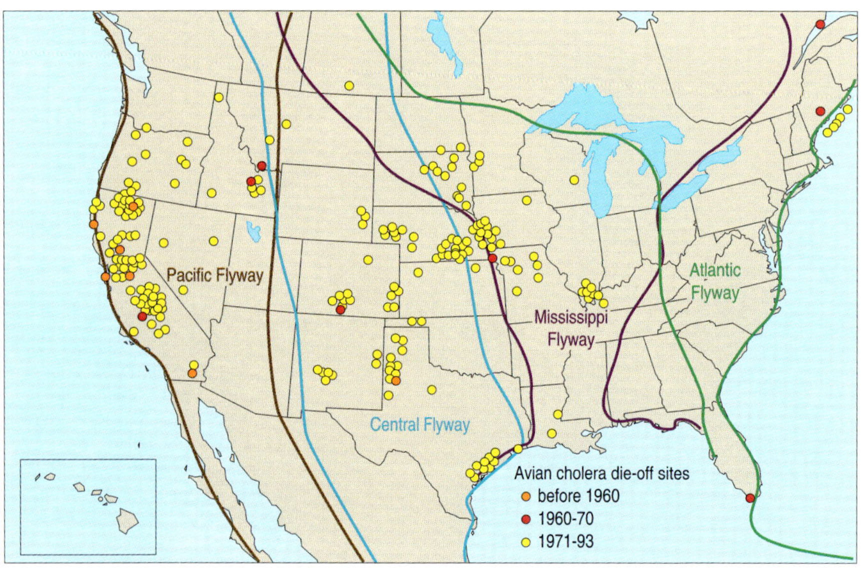

Fig. 3. Geographic distribution of avian cholera in wild waterfowl within the United States, before 1960 (first outbreak in 1944), during the 1960's, and after 1970, when disease spread (National Wildlife Health Center files).

Duck Plague

Duck plague is another emerging disease of North American waterfowl. This herpesvirus infection first appeared on the North American continent in 1967 when it caused large-scale losses in the domestic duck industry and losses of a small number of wild waterfowl (Leibovitz and Hwang 1968). The first major die-off involving wild waterfowl occurred during January 1973 at the Lake Andes National Wildlife Refuge in South Dakota (Friend and Pearson 1973). Duck plague has expanded throughout North America since the initial outbreak, along with an increasing number of outbreaks in each decade (Fig. 4). Nearly all occurrences of duck plague have involved nonmigratory waterfowl (captive, tame, and resident waterfowl that do not undergo traditional migratory movements). A February 1994 outbreak in the Finger Lakes region of New York State involving mallards (*Anas platyrhynchos*) and American black ducks (*A. rubripes*) is the first major outbreak involving migratory waterfowl since the January 1973 Lake Andes outbreak.

Other Diseases

Other diseases affecting wild birds are newly recognized, are occurring with increasing frequency, or have expanded their geographic

Table 1. Changes in patterns of diseases affecting wildlife.

Disease	Cause	Species	First occurrence in species		Major expansion	Current status
			Year	Location		
Avian botulism	Bacteria	Waterbirds	1910	UT	1970's	Widespread; major problem
Avian cholera	Bacteria	Waterbirds	1944	TX,CA	1970's	Widespread; major problem
Duck plague	Virus	Waterfowl	1967	NY	1990's	Expanding
Avian pox	Virus	Waterfowl	1978	AK	1990's	Expanding
		Bald eagle	1978	AK	1980's	Stable
Salmonellosis	Bacteria	Songbirds	Historical	MA	1980's	Major problem of urban environments (bird feeders)
Canine parvovirus	Virus	Wild carnivores	1978	TX	1980's	Widely distributed
Canine heartworm	Parasite	Wild carnivores	Historical	Southeastern U.S.	1960's	Continuing spread northward from southeastern U.S.
Fibropapilloma	Unknown	Marine turtles	1938	FL	1980's	Major geographic spread; increased frequency of occurrence

Table 2. Emerging diseases of wildlife.

Disease	Cause	Species	Occurrence in species		Ecosystem type
			Year	Location	
Inclusion body disease of cranes	Virus	Exotic cranes	1978	WI	Terrestrial (captive-propagation flock)
Eastern equine encephalitis	Virus	Whooping crane	1984	MD	Terrestrial (captive-propagation flock)
Mycotoxicosis (Trichothecenes)	Fungus	Sandhill crane	1982	TX	Agricultural (peanuts)
Coccidioidomycosis	Fungus	California sea otter	1976	CA	Marine
Upper respiratory disease syndrome	Bacteria?	Desert tortoise	1987	NV	Desert
Avian tuberculosis	Bacteria	Whooping crane	1982	CO	Mixed; wetlands and agricultural fields
Neoplasia	Unknown	Mississippi sandhill crane	1975	MS	Mixed; wetlands and agricultural fields
Velogenic Newcastle disease	Virus	Double-crested cormorants	1990	Canada	Aquatic
			1992	Great Lakes	
Woodcock reovirus	Virus	American woodcock	1989	NJ	Forest
Seal plague	Virus	Marine mammals	1987	Russia	Inland sea
			1988	Europe	Marine
			1990	U.S.	Marine

occurrence during the 20th century. Changes in disease patterns in wild birds are consistent with such changes in other species, including humans, and reflect environmental changes that foster the eruption of disease and the spread of infectious agents (Tables 1 and 2).

Magnitude of Losses

Changes over time in the frequency of wild bird die-offs and losses from disease cannot be precisely determined because no appropriate data base exists. Also, changes in surveillance and reporting confounds interpretation of existing data. Nevertheless, with the exception of rare catastrophic events, available information suggests that substantially greater numbers of wild birds are dying from diseases now than in earlier periods of the 20th century. The yearly average of 55,066 reported waterfowl deaths from disease during 1930-64 (Friend 1992) has been exceeded or nearly exceeded by single events since 1964. Several disease outbreaks, for example, have killed between 25,000 and 100,000 waterfowl; die-offs of 5,000 to 10,000 waterfowl are common.

The number of avian die-offs in the NWHC data base with reported mortality of 1,000 or more is sufficient to support the contention of increased numbers of birds dying from disease compared with the period before 1965. The annual number of avian die-offs is an additional indicator of the relatively high frequency of avian disease during the late part of the 20th century (Fig. 5).

Habitat and Human Interactions

Causes of major bird die-offs during the past decade and their geographic distribution are shown in Table 3. Composite data indicate a relation between bird concentrations and the occurrences of avian disease. Those states with large concentrations of migratory birds on migrational staging and wintering areas tend to have the most disease outbreaks. Preliminary assessments suggest that habitat quantity and quality are important factors in this relation.

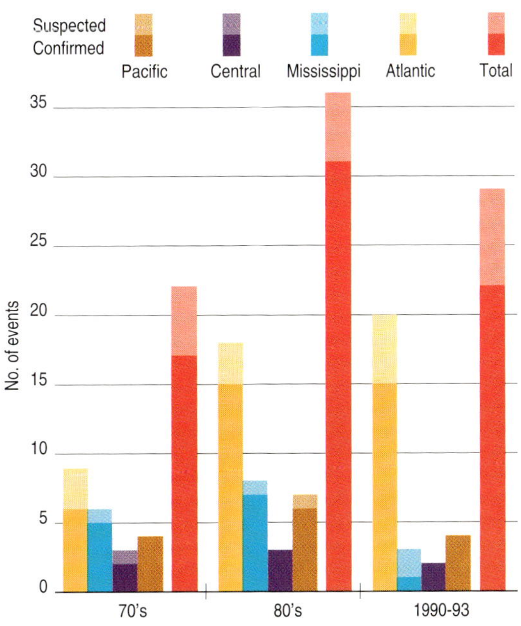

Fig. 4. Number of duck plague outbreaks in waterfowl in the United States, by flyway (suspect cases have pathology consistent with duck plague but lack isolation of the virus to confirm the diagnosis), and total number of outbreaks by decade.

Table 3. Geographic distribution of major (>500 birds) die-offs of wild birds by cause, 1983-93.

Disease	Cause	States (number of events)		
		<5	5-9	>9
Aspergillosis	Fungus	CO(1), VT(1)		
Avian botulism	Bacteria	AR(1), CO(2), KS(2), LA(2), MD(2) MI(1), MN(1), NM(1), NY(1), OH(2) OK(1), OR(4), WI(3)	ID(7) NB(5) NV(5)	CA(80), MT(14) ND(25), SD(14) UT(11)
Avian cholera	Bacteria	CO(3), IA(4), ID(1), ME(1), MN(3) OR(1), NM(4), NV(1), SD(1)	MO(5)	CA(21), NB(13) TX(18)
Chlamydiosis	Bacteria	ND(1)		
Erysipelas	Bacteria	MD(1)		
Mycotoxicosis	Fungus	TX(3)		
Myocardiopathy	Unknown	CA(1)		
Necrotic enteritis	Bacteria?	ND(3), SD(1)		
Newcastle disease	Virus	MN(1), ND(2), SD(1)		
Salmonellosis	Bacteria	CA(2), VT(1), WA(1)		
Toxicosis	Environmental contaminants	AZ(1), CA(2), IL(1), NV(1), VA(1)		
Trichomoniasis	Parasite	CA(1), NM(1)		

Wetland losses, for example, are well documented and clearly a contributing factor in the spread of avian cholera. California, a focal point for the occurrence and spread of avian cholera in waterfowl, lost 91% of its historical wetland acreage by 1980 (Dahl 1990). A similar situation exists in the Rainwater Basin of Nebraska where avian cholera first appeared in waterfowl in 1975; this area has subsequently become a focal point for spread to other areas. About 90% of historical wetland acreage within the Rainwater Basin has been lost (Farrar 1980). The association between wetland losses and spread of infectious disease is due to the interactive factors that aid disease transmission.

Habitat (environment) loss often results in crowding birds on the remaining habitat, thereby enhancing the potential for transmission of infectious agents. Movement patterns of infected birds can spread the disease to other locations and populations and help establish the disease as a recurring problem.

High concentrations of birds for prolonged periods of time on limited habitat often degrade the quality of habitat through fecal contamination and damage to vegetation. Deposition and survival of pathogenic parasites and microbes are aided by such environmental conditions and can result in enhanced disease maintenance and spread.

Habitat degradation due to human-caused factors is also important. For example, although the bacterium responsible for avian botulism is a common inhabitant of wetland substrates, the production of the botulinum toxin that causes botulism is dependent on specific environmental factors such as ambient temperature, pH, oxygen depletion, and other factors (Locke and Friend 1987). Discharges into wetlands of sewage, agricultural chemicals, and poultry wastes from factories have frequently been associated with eruptions of avian botulism, although cause and effect relationships have not been clearly established.

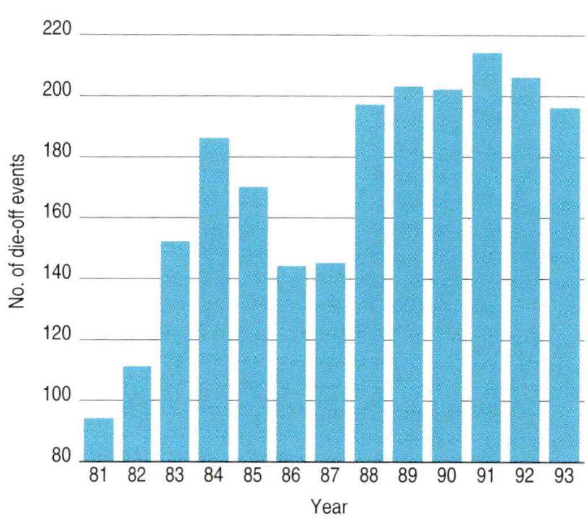

Fig. 5. Number of wild bird die-off events listed in the National Wildlife Health Center data base, 1981-93.

Prevention of Avian Diseases

Diseases affecting wild birds can be prevented and controlled despite the challenges of dealing with species and populations that are often highly mobile and spend much or all of their lives in remote areas.

Methodical monitoring and surveillance programs are needed to provide early detection of emerging problems so that intervention can begin when problems are most manageable. Accurate diagnostic assessments of the causes of morbidity and mortality are essential for focusing control efforts. Also, greater emphasis is needed on studies of disease ecology to provide enhanced understanding of host-agent-environment relationships for specific diseases. This information serves to identify weak links where disease-control and prevention efforts will be most effective. Molecular biology and the associated field of genetic engineering will greatly assist these efforts.

References

Dahl, T.E. 1990. Wetland losses in the United States 1780's to 1980's. U.S. Fish and Wildlife Service, Washington, DC. 13 pp.

Farrar, J. 1980. The Rainwater Basin . . . Nebraska's vanishing wetlands. Nebraska Game and Parks Commission, Lincoln. 15 pp.

Friend, M. 1987. Avian cholera. Pages 69-82 *in* M. Friend and C.J. Laitman, eds. Field guide to wildlife diseases. Vol. 1. General field procedures and diseases of migratory birds. U.S. Fish and Wildlife Service Resour. Publ. 167.

Friend, M. 1992. Environmental influences on major diseases. Transactions of the North American Wildlife and Natural Resour. Conference 57:517-525.

Friend, M., and G.L. Pearson. 1973. Duck plague: the present situation. Proceedings of the Western Association of State Game and Fish Commissioners 53:315-325.

Kalmbach, E.R., and M.F. Gunderson. 1934. Western duck sickness: a form of botulism. U.S. Department of Agriculture Tech. Bull. 411. 81 pp.

Leibovitz, L., and J. Hwang. 1968. Duck plague on the American continent. Avian Diseases 12:361-378.

Locke, L.N., and M. Friend. 1987. Avian botulism. Pages 83-93 *in* M. Friend and C.J. Laitman, eds. Field guide to wildlife diseases. Vol. 1. General field procedures and diseases of migratory birds. U.S. Fish and Wildlife Service Resour. Publ. 167.

Phillips, J.G., and F.C. Lincoln. 1930. American waterfowl. Houghton Mifflin Company, Cambridge, MA. 312 pp.

Quortrup, E.R., F.B. Queen, and L.T. Merovka. 1946. An outbreak of pasteurellosis in wild ducks. Journal of the American Veterinary Medical Association 108:94-100.

Rhoades, K.R., and R.B. Rimler. 1991. Pasteurellosis. Pages 145-162 *in* B.W. Calnek, H.J. Barnes, C.W. Beard, W.H. Reid, and H.W. Yoder, Jr., eds. Diseases of poultry. 9th ed. Iowa State University Press, Ames.

Stout, I. J., and G.W. Cornwell. 1976. Nonhunting mortality of fledged North American waterfowl. Journal of Wildlife Management 40:681-693.

For further information:

Milton Friend
National Biological Service
National Wildlife Health Center
6006 Schroeder Rd.
Madison, WI 53711

Captive Propagation, Introduction, and Translocation Programs for Wildlife Vertebrates

by
Joshua Dein
Kathryn Converse
National Biological Service
Christy Wolf
University of Wisconsin

Captive propagation, introduction, and translocation (relocation) programs for many animals have been undertaken by federal, state, and private agencies for more than 20 years. These programs help aid the recovery of endangered and threatened species, reestablish lost species, augment declining populations, increase recreational opportunities, reduce nuisance species, and introduce non-native species. Davidson and Nettles (1992) discuss translocation as a component of successful early restorations of game species including wild turkey (*Meleagris gallopavo*) and white-tailed deer (*Odocoileus virginianus*), and recovery of endangered species such as the peregrine falcon (*Falco peregrinus*). Despite some successes, the total number of translocations that occur yearly is unknown, as is the success and effects of these programs, because there is rarely appreciable monitoring after release (Griffith et al. 1989; Gogan 1990). This report focuses on trends in the use of translocation programs and disease transmission following translocation of wildlife vertebrates other than fish.

In the absence of a national data base on wildlife translocations, a search for publications with information on translocations was performed by using *Wildlife Review* and the U.S. Fish and Wildlife Reference Service CD-ROM data bases for the 20-year period, 1971-91. In addition, personnel from multiple federal, state, and private agencies that conduct propagation and translocation programs were contacted for supplemental information and literature. Increasing numbers of books (Neilsen and Brown 1988), journals (Ullrey 1993), and meetings (Junge 1992; Wolff and Seal 1992) discuss wildlife translocations and many contain information on the effects of translocations on animals and their environment.

Trends

Of 292,628 citations reviewed, 1,431 addressed translocations. There were relatively high percentages of citations that included translocation programs in the early 1970's and again in the late 1980's with a general increasing trend overall (Fig. 1). Although the number of publications probably underestimates the true extent of translocation programs, it does demonstrate the trend of continued interest, research, and publication over the past 20 years.

Griffith et al. (1989) published a comprehensive survey that estimated an average of 515 translocations per year (414 programs) of terrestrial vertebrates occurring in the United States, Canada, New Zealand, and Australia between 1973 and 1989; 98% were conducted in the United States and Canada. Birds were most frequently (59%) translocated (Fig. 2); 92% of the translocations involved game species, 7% endangered and threatened species, and 1% nongame species (Griffith et al. 1993). Of the 261 translocations in the United States reported by Griffith et al. (1993), wild species were most frequently translocated, and the Southeast had the greatest number of translocations (Table 1).

In 1985 Boyer and Brown (1988) surveyed the 50 state conservation agencies; 29 confirmed they were translocating mammals (56% native game species, 5% nongame species, and 5% endangered species). In addition, 19 states reported that mammals were translocated by private agencies in their states.

A 1993 follow-up to the Griffith et al. (1993) survey suggests that many of the 414 programs originally surveyed were still releasing animals (C. Wolf, University of Wisconsin, unpublished data). The average duration of these translocation programs was 4.8 years, an increase from

Courtesy International Crane Foundation

Captive-reared whooping crane chicks released in Kissimmee Prairie, Florida, 1993.

the 3 years estimated by Griffith et al. (1989). Boyer and Brown (1988) reported that 40 states projected either no change or an increase in translocation activity.

It is impossible to estimate the total number of animals released throughout the United States, but Maryland provides an example of an ongoing and intensive propagation and release program for mallard ducks to augment the natural population. The state released 409,838 mallards from 1967 to 1991. An estimated 100,000-150,000 ducks per year are also released in Maryland by private parties onto regulated shooting areas (L. Hindman, Maryland Department of Natural Resources, personal communication).

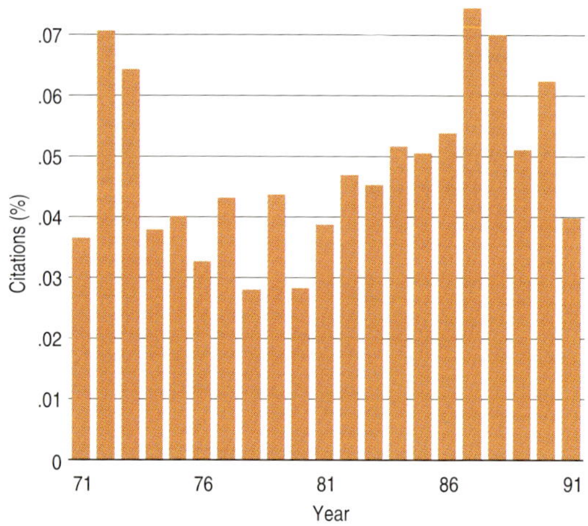

Fig. 1. Percentage of citations relating to translocations, 1971-91. Sample size = 292,628.

Disease

Every animal represents a living micro-ecosystem containing bacteria, viruses, fungi, and parasites. Wildlife scientists now recognize the translocation of a wild animal never represents movement of a single species (Davidson and Nettles 1992). Unless health-monitoring programs for source populations are in place, the risk is greater that hazardous disease agents may be moved and released into new environments along with the species of interest.

Table 1. Percentages of translocations by geographic area and source of translocated animals, 1973-86 (Griffith et al. 1993).

Geographic area U.S. regions	Wild (%)	Captive (%)*	Total no. translocations
Northwest	85	15	53
Southwest	88	12	24
Central	82	18	40
Southeast	62	38	61
Northeast	89	11	46
Rocky Mountain	81	19	37

*May include some wild-caught animals.

This threat also exists in reverse, of course: animals for which a significant amount of money has been spent on their production may be decimated by a disease agent existent at the release site. Success of a whooping crane (*Grus americana*) reintroduction program in Idaho was limited by disease, and the current whooping crane reintroductions in Florida face similar challenges.

Williams et al. (1992) documented the importance of overcoming disease problems to have a successful reintroduction program. Table 2 summarizes other documented incidents of disease introduction into new environments via animal translocations. These diseases have substantial effects on wildlife, domestic animals, and humans.

Conclusions

Data presented here show a consistent if not increasing trend in the number of translocation programs between 1971 and 1991. Multiple disease problems have been documented in animals moved in similar programs. We are alarmed because many of these programs will continue and probably increase in the future, and because most programs do not monitor or follow up to detect ecosystem change caused by translocations. Griffith et al. (1989) found only 27% of the agencies that responded to their survey followed specific protocols for collecting and recording information during translocations. These data indicate a need for a national

Raccoon Rabies: Example of Translocation, Disease

The spread of raccoon (*Procyon lotor*) rabies in the eastern United States is an excellent example of disease transmission through the movement of animals (Winkler and Jenkins 1991).

The first reported raccoon rabies occurred in Florida in the early 1950's. Between the 1950's and 1977, rabies spread by raccoon-to-raccoon transmission primarily within Florida and Florida to Georgia.

More than 3,500 raccoons were trapped in Florida and transported to Virginia between 1977 and 1981 to restock raccoon populations and provide increased hunting opportunities. Rabid raccoons were confirmed in shipments of animals to Virginia and North Carolina from Florida.

After these shipments, between 1977 and 1994, mid-Atlantic rabies outbreaks spread north to Virginia, Maryland, Pennsylvania, District of Columbia, Delaware, New York, New Jersey, Connecticut, Massachusetts, Vermont, and New Hampshire, and south to North Carolina (Jenkins et al. 1988; Rupprect and Smith 1995).

Studies of rabies viruses have shown great similarity between rabies virus isolates from raccoons collected between 1950 and 1977 in Florida and Georgia and the mid-Atlantic outbreaks (Winkler and Jenkins 1991). There is increased public health concern about human contact with the rabid animals that are common in densely populated urban and suburban areas.

Table 2. Some diseases transmitted by or to translocated animals in the United States (adapted from Woodford 1993).

Species translocated	Source*	Disease or agent	Release area	Species affected	Reference
Desert tortoise	Pet shops, w/c	*Mycoplasma*	Mojave Desert	Desert tortoise	Jacobson et al. 1991
Whooping crane	MD, c/b	Avian tuberculosis	ID	Whooping crane	Snyder et al. 1991
Waterfowl	Various, c/b	Duck plague	Various	Waterfowl	Brand 1987
Wild turkey	Various, w/c, c/b	*Mycoplasma*	Various	Wild turkey	Davidson et al. 1982
Parrot	Central, S. Am., w/c	Newcastle disease	CA	Domestic poultry, pet birds	Utterback 1973
Raccoon	FL, w/c	Rabies	VA	Raccoon; 6 other spp.	Winkler and Jenkins 1991
	TX, w/c	Parvovirus	WV	Skunk, raccoon	Nettles et al. 1980
Red fox	OH, other states	*Echinococcus multilocularis*	SC	Unknown	Davidson et al. 1992
Bighorn sheep	Los Angeles Co., CA, w/c	Contagious ecthyma (Orf)	Ventura Co., CA	Human	Jessup et al. 1991
	ID, w/c	Scabies (mange mite)	OR	Human	Thorne et al. 1992
Tule elk	CA, w/c	Paratuberculosis	Pt. Reyes, CA	Tule elk (from contact with domestic cattle)	Jessup et al. 1981
Elk, caribou	Various, w/c	Brainworm	Various, w/c	Elk, caribou (from contact with wild white-tailed deer)	Samuel et al. 1992

*w/c—wild caught, c/b—captive bred.

data base and monitoring program for propagation, introduction, and translocation programs. This data base should be readily available to managers planning similar programs and should provide a mechanism for assessing the positive and negative effects of these programs.

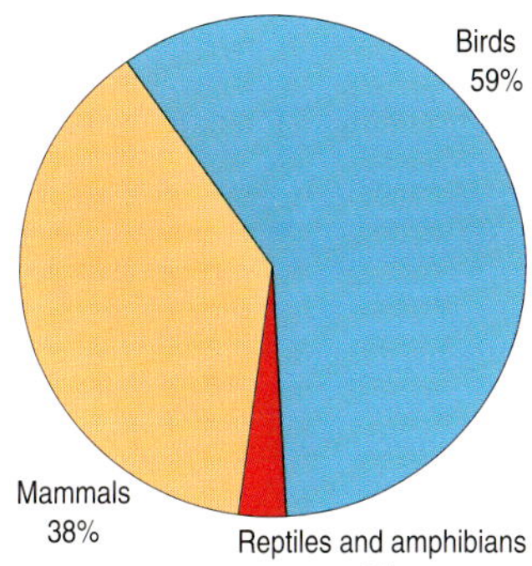

Fig. 2. Percentage of translocated animals by classes, 1971-86; average number of translocations per year was 515 (Griffith et al. 1993).

References

Boyer, D.A., and R.D. Brown. 1988. A survey of translocations of mammals in the United States, 1985. Pages 1-11 *in* L. Neilsen and R.D. Brown, eds. Translocation of wild animals. Wisconsin Humane Society, Milwaukee, WI, and Caesar Kleberg Wildlife Research Institute, Kingsville, TX.

Brand, C.J. 1987. Duck plague. Pages 117-127 *in* M. Friend, ed. Field guide to wildlife diseases. U.S. Fish and Wildlife Service Resour. Publ. 167.

Davidson, W.R., M.J. Appel, G.L. Doster, O.E. Baker, and J.F. Brown. 1992. Diseases and parasites of red foxes, gray foxes, and coyotes from commercial sources selling to fox-chasing enclosures. Journal of Wildlife Diseases 28:581-589.

Davidson, W.R., and V.F. Nettles. 1992. Relocation of wildlife: identifying and evaluating disease risks. Transactions of the North American Wildlife and Natural Resour. Conference 57:466-473.

Davidson, W.R., V.F. Nettles, C.E. Couvillion, and H.W. Yoder, Jr. 1982. Infectious sinusitis in wild turkeys. Avian Diseases 26:402-405.

Gogan, P.J.P. 1990. Considerations in the reintroduction of native mammalian species to restore natural ecosystems. Natural Areas Journal 10:210-217.

Griffith, B., J.M. Scott, J.W. Carpenter, and C. Reed. 1989. Translocation as a species conservation tool: status and strategy. Science 245:477-480.

Griffith, B., J.M. Scott, J.W. Carpenter, and C. Reed. 1993. Animal translocations and potential disease transmission. Journal of Zoo and Wildlife Medicine 24:231-236.

Jacobson, E.R., J.M. Gaskin, M.B. Brown, R.K. Harris, C.H. Gardiner, L. La Pointe, H.P. Adams, and C. Reggiardo. 1991. Chronic upper respiratory tract disease of free-living desert tortoises. Journal of Wildlife Diseases 27:296-316.

Jenkins, S.R., B.D. Perry, and W.G. Winkler. 1988. Ecology and epidemiology of raccoon rabies. Review of Infectious Diseases 10:S620-S625.

Jessup, D.A., B. Abbas, D. Behymer, and P. Grogan. 1981. Paratuberculosis in tule elk in California. Journal of the American Veterinary Medical Association 179:1253-1254.

Jessup, D.A., R.K. Clark, and N.D. Kock. 1991. Contagious ecthyma in free-ranging desert bighorn sheep: Iatrogenic and zoonotic cases. Pages 215-217 *in* Proceedings of the American Association of Zoo Veterinarians.

Junge, R.E., ed. 1992. Proceedings of the Joint Conference of the American Association of Zoo Veterinarians and the American Association of Wildlife Veterinarians. Oakland, CA. 404 pp.

Neilsen, L., and R.D. Brown, eds. 1988. Translocation of wild animals. Wisconsin Humane Society, Milwaukee, and Caesar Kleberg Wildlife Research Institute, Kingsville, TX. 333 pp.

Nettles, V.N., J.E. Pearson, G.A. Gustafson, and J.L. Blue. 1980. Parvovirus infection in translocated raccoons. Journal of the American Veterinary Medical Association 177:787-789.

Rupprect, C.E., and J.F. Smith. 1995. Raccoon rabies: the reemergence of an epizootic in a densely populated area. Seminars in Virology 5. In press.

Samuel, W.M., M.J. Pybus, D.A. Welch, and C.J. Wilke. 1992. Elk as a potential host for meningeal worm: implications for translocation. Journal of Wildlife Management 56:629-639.

Snyder, S.B, M.J. Richard, R.C. Drewein, N. Thomas, and J.P. Thilsted. 1991. Pages 74-80 *in* Diseases of whooping cranes seen during annual migration of the Rocky Mountain flock. Proceedings of the American Association of Zoo Veterinarians.

Thorne, E.T., M.M. Miller, D.A. Jessup, and D.L. Hunter. 1992. Pages 18-25 *in* Disease as a consideration in translocating wild animals. Proceedings of the American Association of Zoo Veterinarians.

Ullrey, D.E., ed. 1993. Journal of Zoo and Wildlife Medicine 24:229-424.

Utterback, W.W. 1973. Epidemiology of VVND in southern California. Proceedings of the United States Animal Health Association 76: 280-287.

Williams, E.S., E.T. Thorne, D. Kwaitkowski, and R. Oakleaf. 1992. Overcoming disease problems in the black-footed ferret recovery program. Pages 474-485 *in* R.E. McCabe, ed. Transactions of the Fifty-seventh North American Wildlife and Natural Resources Conference. Charlotte, NC.

Winkler, W.G., and S.R. Jenkins. 1991. Raccoon rabies. Pages 325-340 *in* G.M. Baer, ed. The natural history of rabies, 2nd ed. CRC Press, Boca Raton, FL.

Wolff, P.L., and U.S. Seal, eds. 1992. International Conference on Implications of Infectious Diseases for Captive Propagation and Reintroduction Programs of Threatened Species. Oakland, CA.

Woodford, M.H. 1993. International disease implications for wildlife translocation. Journal of Zoo and Wildlife Medicine 24:265-270.

For further information:

Joshua Dein
National Biological Service
National Wildlife Health Center
6006 Schroeder Rd.
Madison, WI 53711

Contaminants in Coastal Fish and Mollusks

by
Donna Turgeon
Andrew Robertson
National Oceanic and Atmospheric Administration

Historically, U.S. coastal fish and shellfish have been plentiful, supporting native carnivores such as ospreys, bald eagles, striped bass, sharks, sea lions, porpoises, and whales in ecosystem food webs. Since the 1960's, however, the capacity of coastal ecosystems to produce abundant fish and shellfish has declined. Increasingly frequent reports of closures of shellfish beds and bathing beaches, contamination of living resources and habitats by toxic chemicals, decreases in commercial fish stocks, shallow-water strandings of porpoises and pilot whales, losses of wetland habitat, and spread of toxic and nuisance algal blooms indicate there has been widespread environmental degradation.

As part of the national response to concerns over the deteriorating health of our coastal ecosystems, several federal and state programs monitor changes in the levels of toxic chemicals in select organisms at coastal sites. In general, contaminant levels have been found to be holding steady or, in the case of several contaminants, decreasing in coastal areas over the past few years, reversing the trends of contaminant increases that occurred in the first two-thirds of this century.

The National Oceanic and Atmospheric Administration (NOAA) initiated its National Status and Trends (NS&T) Program in 1984 with its National Benthic Surveillance Project. The project monitors concentrations of about 70 chemical contaminants (Table) in fish livers and sediments and investigates some of the effects of these chemicals on fish (e.g., liver tumors, reproductive impairment, fin loss) from nearshore waters of the Atlantic, Gulf of Mexico, and Pacific coasts, including Alaska. In 1986, the NS&T Program began its Mussel Watch Project to monitor concentrations of the same contaminants in the tissues of bivalve mollusks (primarily mussels and oysters) and sediments. More than 350 coastal sites (Fig. 1) of the continental United States, off Hawaii, several Caribbean islands, and in the Great Lakes are regularly monitored. Since 1986 the NS&T Program also has conducted intensive studies of the magnitude and extent of contaminant effects on selected indicator species from the most contaminated U.S. estuaries (Wolfe et al. 1993).

Table. Chemicals measured in the National Oceanic and Atmospheric Administration's National Status and Trends Program.

Chemicals measured		
DDT and its metabolites	**Polycyclic aromatic hydrocarbons**	**Major elements**
2,4'-DDD	*2-ring*	Aluminum
4,4'-DDD	Biphenyl	Iron
2,4'-DDE	Naphthalene	Manganese
4,4'-DDE	1-Methylnaphthalene	Silicon
2,4'-DDT	2-Methylnaphthalene	
4,4'-DDT	2, 6-Dimethylnaphthalene	
	1,6,7-Trimethylnaphthalene	
Tetra-, tri-, di-, and monobutyltins	*3-ring*	**Trace elements**
	Flourene	Antimony
	Phenanthrene	Arsenic
	1-Methylphenanthrene	Cadmium
Chlorinated pesticides other than DDT	Anthracene	Chromium
Aldrin	Acenapthene	Copper
Chlordanes	Acenaphthylene	Lead
cis-Chlordane		Mercury
trans-Nonachlor	*4-ring*	Nickel
Heptachlor	Fluoranthene	Selenium
Heptachlor epoxide	Pyrene	Silver
Dieldrin	Benz[*a*]anthracene	Tin
Hexachlorobenzene	Chrysene	Zinc
Lindane (gamma-HCH)		
Mirex	*5-ring*	
	Benzo[*a*]pyrene	
	Benzo[*e*]pyrene	**Toxaphene at some sites**
	Perylene	
Polychlorinated biphenyls	Dibenz[*a,h*]anthracene	**Related parameters**
PCB congeners 8, 18, 28,	Benzo[*b*]fluoranthene	Grain size
44, 52, 66, 77, 101, 105,	Benzo[*k*]fluoranthene	Total organic carbon (TOC)
118, 126, 128, 138, 153,		*Clostridium perfringens* spores
179, 180, 187, 195, 206, 209	*6-ring*	
	Benzo[*g,h,i*]perylene	
	Indeno[1,2,3-*cd*]pyrene	

Methods

All sites are located away from known point sources and dumpsites (Lauenstein et al. 1993). Sites are sampled every 1-2 years for biota and less frequently for sediments. Sediments for chemical analyses are collected from the top 2 cm (0.75 in) of grab samples. Mollusks (e.g., oysters, clams, and mussels) are dredged from deep subtidal zones or hand-collected in intertidal to shallow subtidal zones. Fish are sampled by otter trawl tows in depths of 1-70 m (3-230 ft). Details of sampling protocols and methods of analysis are described elsewhere (Lauenstein and Cantillo 1993).

Sediment contaminant concentrations have been adjusted for particle size to account for differences in concentrations due to variations in physical properties of absorption surfaces among large-grained sands, fine muds, and sediment mixtures (NOAA 1988). Measurements for individual chemicals have been combined for groups of related compounds. Thus total PCB (polychlorinated biphenyl) is based on the sum of the concentrations of 20 PCBs; total PAH (polycyclic aromatic hydrocarbon) is the sum of 24 PAH compounds; total DDT (1,1'-[2,2,2-trichloroethylidene]bis[4-chlorobenzene]) is the sum of the concentrations of DDT and its metabolites; and total chlordane is the sum of the concentrations of two major constituents of chlordane mixtures (*cis*-chlordane and *trans*-nonachlor) and two minor components (heptachlor and heptachlorepoxide).

The primary species of mollusks monitored are the eastern oyster (*Crassostrea virginica*), the northeastern and west coast species of mussels (*Mytilus edulis*, *M. trossulus*, and *M.*

californianus), and the Great Lakes zebra mussel species (*Dreissena polymopha*). The six primary fish species monitored nationwide are winter flounder (*Pleuronectes americanus*), spot (*Leiostomus xanthurus*), Atlantic croaker (*Micropogonias undulatus*), flathead sole (*Hippoglossoides elassodon*), white croaker (*Genyonemus lineatus*), and starry flounder (*Platichthys stellatus*).

Status and Trends

Contaminants in Sediments (1984-90), Bivalves (1986-90), and Fish (1984-88)

Nationwide, the highest concentrations of the chemicals measured in coastal sediments were found near urbanized areas of the Northeast (New York City, Boston, and Baltimore) and the west coast (San Diego, Los Angeles, and Seattle). An NS&T inventory revealed that more than 90% of the coastal and estuarine areas have concentrations below "High" (the geometric mean plus one standard deviation of all NS&T site means). The greatest number of sites with concentrations greater than five times the "High" ("5 x High") was near densely populated areas in poorly flushed water bodies (e.g., harbors and intracoastal waterways) of the Northeast and Gulf of Mexico. The most common chemicals in the inventory at these "5 x High" levels were metals in decreasing frequency: mercury, cadmium, tin, and silver. Total PAH was the organic compound group most commonly found in the "5 x High" range, a finding important to the consumption of fish and mollusks taken near highly contaminated sites. According to the U.S. Environmental Protection Agency, 22 states had advisories warning against consumption of fish and shellfish from coastal waters in 1992 (Fish Consumption Database 1993).

Mollusks accumulate many organic and inorganic contaminants. Although tissue concentrations of organic contaminants did not vary by species, tissue concentrations of several inorganic contaminants were species-dependent (O'Connor 1993). Mercury was used to illustrate the national spatial distribution of contamination in mollusks because differences in tissue concentrations are not species-dependent (Fig. 2). Mercury concentrations in mollusks from sites with corresponding high sediment concentrations off Texas, Florida, and California were among the highest measured. The highest concentrations of organic contaminants in molluscan tissues were found at sites with corresponding high sediment concentrations, near Boston, New York City, Mobile, San Diego, San Francisco, and Los Angeles (O'Connor 1992).

Levels of silver, lead, and the organic compounds (total DDT, total chlordane, and total PCBs) in fish livers have been found to be positively correlated with sediment concentrations (i.e., high levels of contaminants in sediments and high levels in fish livers from the same site; Turgeon et al. 1993). Figure 3 illustrates the distribution of mean concentrations of lead and total DDT in the livers of fish from sites along the east, Gulf of Mexico, and west coasts. Concentrations of lead were highest in winter flounder from Casco Bay in Maine, in Atlantic croaker from the Elizabeth River in Virginia, and in white croaker from San Diego and San Pedro bays in California. Concentrations of total DDT were highest in white croaker from San Pedro Bay and winter flounder from the Hudson-Raritan Estuary in New York and New Jersey.

Contamination assessments have been made for selected regions and compared with sites nationwide (e.g., Turgeon et al. 1989; Gottholm and Turgeon 1992; Gottholm et al. 1993). For example, mean concentrations of total DDT in sediment from 213 sites, mussel tissue from 111 sites, and fish liver from 118 sites nationwide were compared to mean concentrations at 14 Hudson-Raritan Estuary and coastal New Jersey sites (Gottholm et al. 1993). Among these sites, concentrations in sediments were all above the national median, whereas concentrations in mussel tissues were at or above the median at most sites. Advisories (Fish Consumption Database 1993) were in effect warning against consumption of fish and shellfish collected from much of the Hudson-Raritan Estuary area. NS&T-sampled sites with high levels of contaminants were within the fish-consumption advisory area (Fish Consumption Database 1993) and a health advisory area (New York State Department of Health 1993).

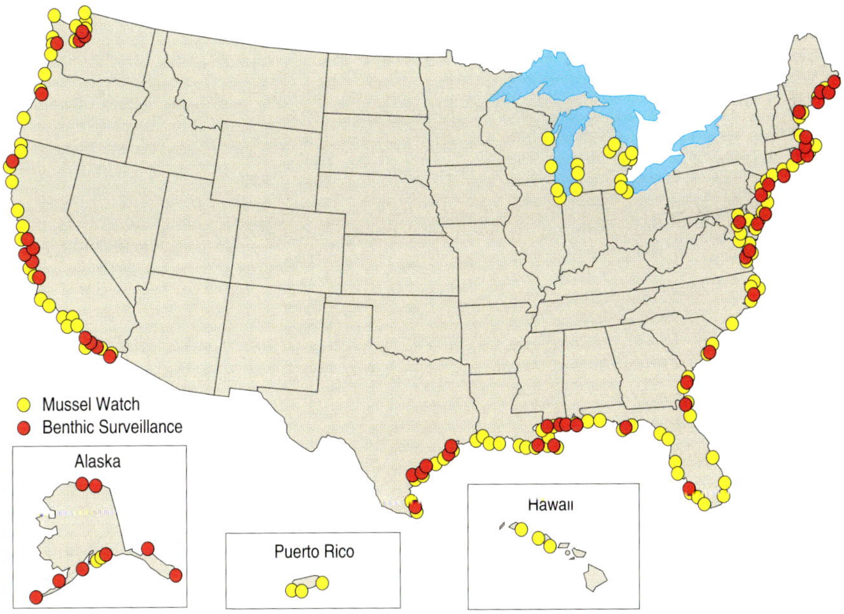

Fig. 1. The National Oceanic and Atmospheric Administration's National Status and Trends Program monitors more than 350 sites nationwide for chemical contaminants in fish livers, bivalve molluscan tissues, and associated sediments. Yellow symbols depict Mussel Watch Project (molluscan and sediment monitoring) and red symbols depict National Benthic Surveillance Project (fish and sediment monitoring) sites.

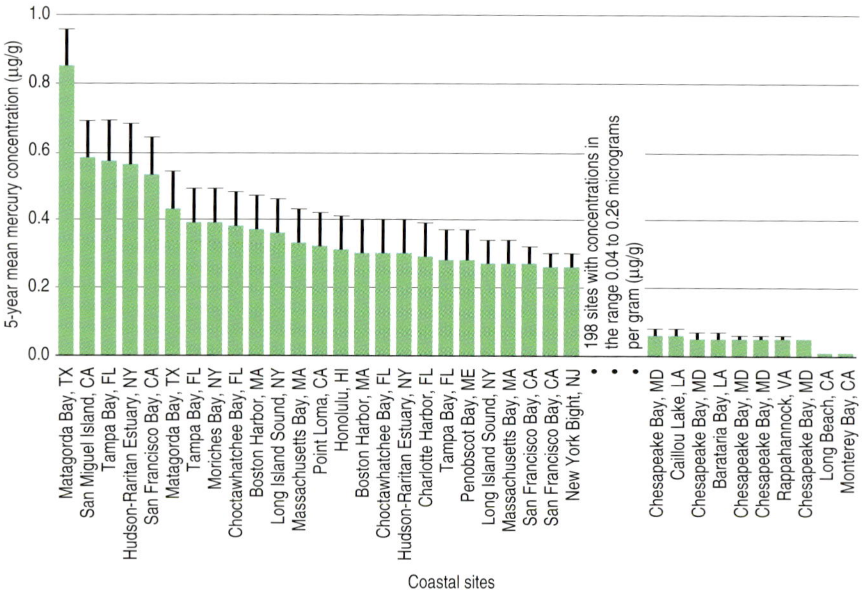

Fig. 2. Mercury in molluscan tissues from the National Oceanic and Atmospheric Administration's National Status and Trends Program coastal sites. Bar graphs are 5-year (1986-90) mean concentrations (mercury measurements in micrograms per gram dry weight) and standard deviations ("T" symbols on bars).

Contaminants in Surface Sediments, Bivalves, and Fish

Recent national and regional trends for certain contaminants have been identified in data from fish livers (McCain et al. 1992), surface sediments (O'Connor 1990; Wade et al. 1992), and molluscan tissues (O'Connor 1992). A statistical test (Spearman rank-correlation) was used to detect trends in annual mean concentrations of 14 chemicals at 141 sites with 4 or 5 years of molluscan monitoring data (O'Connor 1992). Among 1,974 chemical-site combinations, there were 239 occurrences (152 decreasing and 87 increasing) with strong correlations between those concentrations and time. The hypothesis was offered that similar trends among sites in an area corroborate that the trend is real and areawide. For example, at nine Long Island Sound (New York) sites, decreases occurred in copper at six sites, cadmium at five, silver at four, and zinc at three. An apparent 20-year decreasing trend in annual concentration of total PCB in mussels has been recorded at Palos Verdes, California.

Contaminants Determined from Dated Sediment Cores

Trends in contamination can also be detected from contaminant profiles in dated estuarine and coastal sediment cores. Since 1989 the NS&T Program has sponsored projects that use sediment cores to reconstruct the history of contamination in U.S. coastal waters (Hudson-Raritan Estuary, Long Island Sound, Chesapeake Bay, Savannah River, southern California Basin, San Francisco Bay, and Puget Sound). In 1994 sediment cores were analyzed for sites in the Mississippi River Delta and Galveston Bay. Generally, results show a slow increase in contamination in the late 1800's, followed by an acceleration of pollution in the mid-1900's. Maximum contamination was reached around the mid-1970's, and in most areas a decrease has been observed for anthropogenic contaminants (e.g., antimony, lead, DDT, and PCBs; Valette-Silver and O'Connor 1989; Valette-Silver et al. 1994).

Selected Studies in Highly Contaminated Coastal Areas

Liver neoplasms (cancerous tumors) were found in 10 fish species collected from 1984 to 1988 from sites near urban centers along the west and northeast coasts (Turgeon et al. 1992). Scientists concluded that the contaminants most likely to be factors in the development of these tumors were the PAHs, PCBs, and DDTs (Myers et al. 1993).

Although incidences of cancerous tumors are generally low in fish from U.S. coastal waters, other liver disease conditions, some of which may progress to neoplasms, occur more frequently in areas where contaminants are high. Neoplasms and pre-neoplasms (pre-cancerous tumors) were found in up to 15% of the winter flounder from sites in Boston Harbor (Murchelano and Wolke 1991). Along the west coast, neoplasm incidences are well below 10% in most fish species (Myers et al. 1993). Relatively high incidences of nontumorous disease conditions occur in fish from contaminated sites. For example, in English sole (*Parophrys vetulus*) from Elliott Bay, Washington, incidences of 42% for specific degeneration and necrosis (SDN) of liver cells and 13% proliferative disorders (cells duplicating out of control) have been recorded; and in white croaker from San Pedro Outer Harbor, California, 22% SDN and 7% for proliferative disorders have been found (Varanasi et al. 1989; Myers et al. 1993). At Morris Cove, a highly contaminated site in New Haven, Connecticut, up to 90% of the cells in winter flounder livers have been found to be vacuolated cells (large areas of apparently empty, nonfunctioning cells; Gronlund et al. 1991).

Although fin erosion (fish with reduced fins or in extreme stages of disease with no fins) has been found in all species at all sites, this condition is still unusual, except in a few highly contaminated areas. Eroded fins occurred in 27% of the black croaker (*Cheilotrema saturnum*) and 22% of barred sand bass (*Paralabrax nebulifer*) from the West Harbor site in San Diego Bay, California (McCain et al. 1989). Up to 90% of Atlantic croaker, 100% of sand seatrout (*Cynoscion arenarius*), and 17% of spot sampled from the Houston Ship Channel at Green Bayou, Texas, experienced fin loss due to disease (P. Hanson, National Marine Fisheries Service, personal communication).

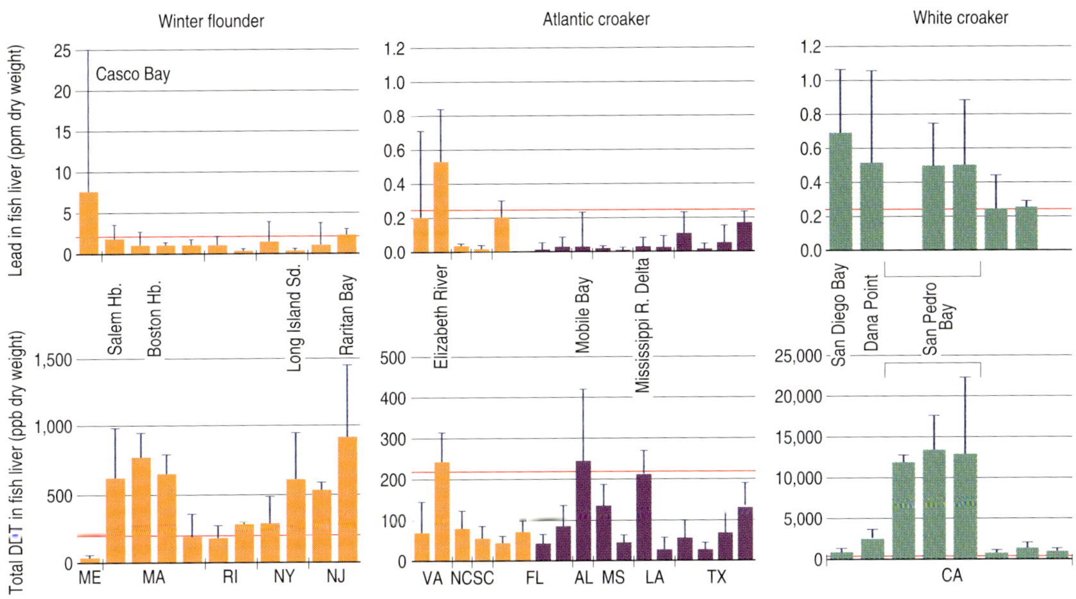

Fig. 3. Lead (Pb) and total DDT in fish livers from the National Oceanic and Atmospheric Administration's National Status and Trends Program coastal sites. Bar graphs are 5-year (1984-88) mean concentrations (Pb measurements in parts per million dry weight; total DDT in parts per billion dry weight) and standard deviations ("T" symbols on bars).

Reproductive impairment occurred in fish from Eagle Harbor and Duwamish Waterway in Puget Sound, San Francisco and San Pedro bays, and in Morris Cove. Significantly lower levels of estradiol (a reproductive hormone) and vitellogenin (yolk protein critical to the development of fertile eggs for reproduction) have been found in English sole from contaminated sites in Puget Sound than those at relatively clean sites (Johnson et al. 1989). Also, a significant proportion of fish from contaminated sites failed to produce yolked eggs and undergo normal ovarian development. Moreover, fewer English sole spawned from the Duwamish Waterway (54%) in comparison with those from Port Susan during the 1987 and 1988 reproductive seasons (Casillas et al. 1991).

White croaker from a site near Los Angeles and kelp bass (*Paralabrax clathratus*) from San Pedro Bay had lower reproductive succcess than those from less contaminated sites at Dana Point and Santa Catalina Island (Cross and Hose 1989). In this study, the percentage of spawning fish was 24%-68% lower, batch fecundity (number of eggs produced) was 36%-44% lower, and the proportion of eggs fertilized was 14%-45% lower at the contaminated site. Gonadally mature female starry flounders from an urbanized central San Francisco Bay site off Berkeley had a reduced proportion of floating eggs and poorer fertilization success than those captured at a site in northern San Pablo Bay (Spies and Rice 1988). In Long Island Sound, embryo abnormalities were most frequent and hatching success was lowest in female winter flounders from more contaminated sites near Milford and New Haven; larvae were smallest off Deer Island, a highly contaminated site in Boston Harbor (Nelson et al. 1991). Thus, we conclude that at contaminated sites the observed lower reproductive success of sampled benthic fish could have long-term effects on spawning populations from contaminated sediments.

Impacts of Contaminants on Coastal Ecosystems

Results from NS&T monitoring have helped define the body burdens of toxic contaminants, evaluate trends in these contaminant data, and assess certain related biological effects in fish and shellfish from more than 350 estuarine and coastal sites over the past 10 years. Contaminant levels have been found to be quite low in most U.S. coastal areas, although substantially elevated levels of a number of contaminants have been measured near most major urban centers. Some biological effects associated with contaminants have been found near these same urban centers, but there is little indication of widespread acute biological effects on benthic fishes and bivalves associated with toxic contaminants throughout most of the U.S. coastal waters. Overall, contaminant levels have been stable or, for a few contaminants, decreasing in the past few years.

References

Casillas, E., D. Misitano, L.L. Johnson, L.D. Rhodes, T.K. Collier, J.E. Stein, B.B. McCain, and U. Varanasi. 1991. Inducibility of spawning and reproductive success of female English sole (*Parophrys vetulus*) from urban and nonurban areas of Puget Sound, Washington. Marine Environmental Res. 31:99-122.

Cross, J.N., and J.E. Hose. 1989. Reproductive impairment in two species of fish from contaminated areas off southern California. Oceans '89 Conference Proceedings 2:382-384.

Fish Consumption Database. 1993. State fish consumption advisories and bans listed in the Fish Consumption Database. U.S. Environmental Protection Agency. 215 pp.

Gottholm, B.W., and D.D. Turgeon. 1992. Toxic contaminants in the Gulf of Maine. NOAA/ORCA Special Rep. 12 pp. Rockville, MD.

Gottholm, W.B., M.H. Harmon, and D.D. Turgeon. 1993. Assessment of chemical contamination in the Hudson-Raritan Estuary and coastal New Jersey area. NOAA/ORCA/CMBAD Special Rep. Silver Spring, MD. 20 pp.

Gronlund, W.D., S.-L. Chan, B.B. McCain, R.C. Clark, Jr., M.S. Myers, J.E. Stein, D.W. Brown, J.T. Landahl, M.M. Krahn, and U. Varanasi. 1991. Multidisciplinary assessment of pollution at three sites in Long Island Sound. Estuaries 14(3):299-305.

Johnson, L., E. Casillas, D. Misitano, T.K. Collier, B.B. McCain, and U. Varanasi. 1989. Bioindicators of reproductive impairment in female English sole (*Parophrys vetulus*) exposed to environmental contaminants. Oceans '89 Conference Proceedings 2:391-396.

Lauenstein, G.G., and A.Y. Cantillo, eds. 1993. Sampling and analytical methods of the NS&T Program National Benthic Surveillance and Mussel Watch projects. Comprehensive descriptions of elemental analytical methods. National Oceanic and Atmospheric Administration Tech. Memorandum 71.

Lauenstein, G.G., M.R. Harmon, and B.W. Gottholm. 1993. National Status and Trends Program: monitoring site descriptions (1984-1990) for the National Mussel Watch and Benthic Surveillance Projects. National Oceanic and Atmospheric Administration Tech. Memorandum 70. 306 pp.

McCain, B.B., S.-L. Chan, M.M. Krahn, D.W. Brown, M.S. Myers, J.T. Landahl, S. Pierce, R.C. Clark, Jr., and U. Varanasi. 1989. Results of the National Benthic Surveillance Project (Pacific coast): 1987. Oceans '89 Conference Proceedings 2:590-596.

McCain, B.B., S.-L. Chan, M.M. Krahn, D.W. Brown, M.S. Myers, J.T. Landahl, S. Pierce, R.C. Clark, Jr., and U. Varanasi. 1992. Chemical contamination and associated fish diseases in San Diego Bay. Environmental Science and Technology 26(4):725-733.

Murchelano, R.A., and R.E. Wolke. 1991. Neoplasms and nonneoplastic liver lesions in winter flounder, *Pseudopleuronectes americanus*, from Boston Harbor, Massachusetts. Environmental Health Perspectives 90:17-26.

Myers, M.S., C.M. Stehr, O.P. Olson, L.L. Johnson, B.B. McCain, S.-L. Chan, and U. Varanasi. 1993. National Benthic Surveillance Project: Pacific coast. Fish histopathology and relationships between toxicopathic lesions and exposure to chemical contaminants for cycles I to V (1984-88). National Oceanic and Atmospheric Administration Tech. Memorandum NMFS-NWFSC-6. 160 pp.

Nelson, D.A., J.E. Miller, D. Rusanowsky, R.A. Greig, G.R. Sennefelder, R. Mercaldo-Allen, C. Kuropat, E. Gould, F.P. Thurberg, and A. Calabrese. 1991. Comparative reproductive success of winter flounder in Long Island Sound: a three-year study (biology, biochemistry, and chemistry). Estuaries 14(3):318-331.

New York State Department of Health. 1993. Health advisory: Chemicals in sportfish and game, 1992-1993. Albany, NY. 9 pp.

NOAA. 1988. National Status and Trends Program: a summary of selected data on chemical contaminants in sediments collected during 1984, 1985, 1986, and 1987. U.S. Department of Commerce, NOAA Tech. Memorandum 49. 22 pp.

O'Connor, T.P. 1990. Coastal environmental quality in the United States, 1990: chemical contamination in sediment and tissues. A special National Oceanic and Atmospheric Administration twentieth anniversary report. NOAA/NOS/OAD, Silver Spring, MD. 34 pp.

O'Connor, T.P. 1992. Mussel Watch: recent trends in coastal environmental quality: results from the first five years of the NOAA Mussel Watch Project. Special publ. NOAA/ORCA/CMBAD, Silver Spring, MD. 29 pp.

O'Connor, T.P. 1993. The National Oceanic and Atmospheric Administration (NOAA) National Status and Trends Mussel Watch Program: national monitoring of chemical contamination in the coastal United States. Pages 331-349 *in* C.R. Cothern and N.P. Ross, eds. Environmental statistics, assessment, and forecasting. Lewis Publishers, Boca Raton, FL.

Spies, R.B., and D.W. Rice, Jr. 1988. Effects of organic contaminants on reproduction of the starry flounder *Platichthys stellatus* in San Francisco Bay. II. Reproductive success of fish captured in San Francisco Bay and spawned in the laboratory. Marine Biology 98:191-200.

Turgeon, D.D., S.B. Bricker, and T.P. O'Connor. 1992. National Status and Trends Program: chemical and biological monitoring of U.S. coastal waters. Pages 425-457 *in* D.H. McKenzie, D.E. Hyatt, and V.J. McConald, eds. Ecological indicators. Vol. 1. Elsevier Applied Science, New York.

Turgeon, D.D., B.W. Gottholm, K.D. McMahon, and A. Robertson. 1989. National Status and Trends Program for marine environmental quality: toxic contaminants in Long Island Sound. NOAA/OOMA Special Rep., Rockville, MD. 7 pp.

Turgeon, D.D., B.W. Gottholm, D.A. Wolfe, and A. Robertson. 1993. The National Status and Trends Program National Benthic Surveillance Project—contaminants in fish tissues. Pages 3474-3483 *in* O.T. Magoon, S. Wilson, H. Converse, and L.T. Tobin, eds. Proceedings of Coastal Zone '93, the Eighth Symposium on Coastal and Ocean Management. American Society of Civil Engineers, New York.

Valette-Silver, N. and T.P. O'Connor. 1989. Contamination history of marine, lacustrine and riverine sediments: a review. EOS 70(43):1150.

Valette-Silver, N., T.P. O'Connor, and A. Robertson. 1994. Chronology of the contamination in recent sediments of three U.S. estuaries: results from the National Oceanic and Atmospheric Administration National Status and Trends Program. EOS 75(16):199.

Varanasi, U., S.L. Chan, B.B. McCain, J.T. Landahl, M.H. Schiewe, R.C. Clark, D.W. Brown, M.S. Myers, M.M. Krahn, W.D. Gronlund, and W.D. MacLeod, Jr. 1989. National Benthic Surveillance Project: Pacific coast. Part II: Technical presentation of the results for cycles I to III (1984-86). National Oceanic and Atmospheric Administration Tech. Memorandum NMFS F/NWC-170. 159 pp.

Wade, T.L., E.N. Powell, T.J. Jackson, and J.M. Brooks. 1992. Processes controlling temporal trends in Gulf of Mexico oyster health and contaminant concentrations. Pages 223-229 *in* Proceedings of the Marine Technological Society '92, Washington, DC.

Wolfe, D.A., E.R. Long, and A. Robertson. 1993. The NS&T Intensive Bioeffects Surveys: design strategies and preliminary results. Pages 298-312 *in* O.T. Magoon, S. Wilson, H. Converse, and L.T. Tobin, eds. Proceedings of Coastal Zone '93, the Eighth Symposium on Coastal and Ocean Management. American Society of Civil Engineers, New York.

For further information:

Donna Turgeon
National Oceanic and Atmospheric Administration
National Status and Trends Program
N/ORCA 22, Bldg. SSMC4
1305 East-West Highway
Silver Spring, MD 20910

Persistent Environmental Contaminants in Fish and Wildlife

by
C.J. Schmitt
C.M. Bunck
National Biological Service

The publication of *Silent Spring* (Carson 1962) highlighted the potential for dichlorodiphenyl trichloroethane (DDT) and other pesticides that persist in the environment to accumulate in and to harm fish, wildlife, and the ecosystems upon which they depend. The federal government responded in the mid-1960's by establishing a multi-agency program to monitor the concentrations of pesticides and, later, other long-lived toxic contaminants in all segments of the environment.

The U.S. Fish and Wildlife Service (USFWS) participated in this program by periodically measuring contaminant concentrations in freshwater fish and birds (Johnson et al. 1967). Fish were selected for monitoring aquatic ecosystems because of their tendency to accumulate pesticides and other contaminants. The European starling (*Sturnus vulgaris*) was selected for monitoring contaminant levels in terrestrial habitats because of its varied diet and wide geographic distribution. Following a successful pilot study (Heath and Prouty 1967), the wings of hunter-killed ducks were used to monitor contaminants in duck populations of the major flyways, and thereby to also provide an assessment of contaminant levels in wetlands. The USFWS maintained this National Contaminant Biomonitoring Program into the 1980's, with the objective of continuing the documentation of temporal and geographic trends in contaminant concentrations (Prouty and Bunck 1986; Bunck et al. 1987; Schmitt and Brumbaugh 1990; Schmitt et al. 1990).

Status and Trends

During the two decades spanned by USFWS contaminant monitoring, the use of persistent insecticides such as DDT was greatly curtailed, and concentrations in fish and wildlife declined. In the environment, DDT breaks down gradually into several different toxic metabolites, of which dichlorodiphenylethylene (DDE) is the most stable and most toxic. A downward trend was clearly evident for DDE in all three networks (Fig. 1), indicating that the total DDT burden in North America declined. In fish, DDE increased from about 70% of total DDT in 1976 to about 74% in 1986 (Fig. 2).

As existing DDT is metabolized, DDE increases proportionally if DDT inputs are reduced; the proportional change evident in fish therefore provides additional evidence of reduced inputs to North American ecosystems. A similar trend toward increasing percentages of DDE relative to DDT has been noted elsewhere (Aguillar 1984), indicating that the global DDT burden is also declining.

In the United States, the bioaccumulation (*see* glossary) of DDT led to eggshell thinning in fish-eating birds such as the bald eagle (*Haliaeetus leucocephalus*). The resulting decline in recruitment of young to bald eagle populations caused the near extirpation and subsequent listing of this species as endangered in the conterminous states (Federal Register 1978). The downward trend of DDT concentrations documented in fish, starlings, and duck wings (Figs. 1 and 2) was paralleled by declining DDE concentrations in bald eagle eggs, and eagle eggshell thickness increased (Wiemeyer et al. 1993). Corresponding increases in recruitment have led to bald eagles repopulating many areas (Fig. 2), and reclassification of the bald eagle from endangered to threatened has been proposed for most of the conterminous states (Federal Register 1994).

In addition to the effects of DDT and its metabolites on eggshell thickness, these compounds, as well as PCBs (polychlorinated biphenyls) and other contaminants, are reported to interfere with other reproductive and maturation processes in fish and wildlife (e.g., Fry and Toone 1981). Although overall concentrations have declined, residues of DDT, other insecticides, and PCBs remain widespread, and problem areas are still evident. In the United States, concentrations of DDT (mostly as DDE) remain highest in fish and wildlife from areas in the South, Southwest, and Northwest where DDT was used to protect cotton and orchards from insects; in the Northeast, where it was used to control mosquitos; and near former centers of DDT production and formulation. Areas affected by former production centers include northern Alabama, near the former Red Stone Arsenal—now Wheeler National Wildlife Refuge (O'Shea et al. 1980); and the Arkansas, Tombigbee, Alabama, and Tennessee rivers (Fig. 3).

Concentrations of other persistent insecticides that are no longer in widespread use, such as heptachlor, dieldrin, endrin, and chlordane, have also declined in all three networks (Prouty and Bunck 1986; Bunck et al. 1987; Schmitt et al. 1990). Nevertheless, residues of chlordane remain sufficiently high in fish from some areas of the Midwest to warrant the issuance of human consumption advisories by state health agencies. Concentrations are also high in Hawaii, where chlordane and other chemically similar compounds were used against termites and agricultural pests, as they were in the Midwest.

Chlordane is a mixture of structurally similar compounds that decompose at different rates over time. The composition of the chlordane mixture present in fish has changed during the 1980's in a manner indicative of an overall

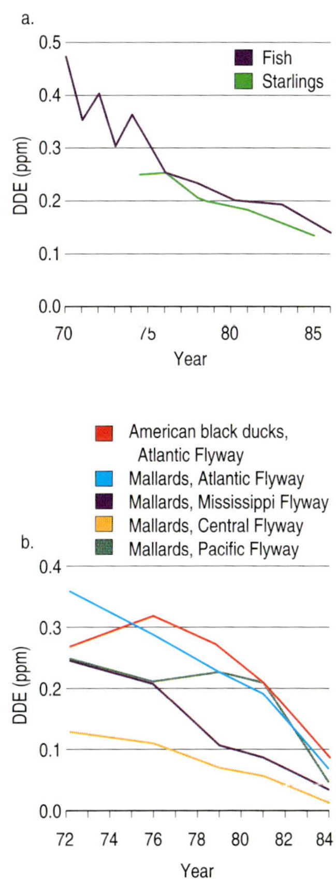

Fig. 1. Mean concentrations of DDE in U.S. Fish and Wildlife Service monitoring networks: (a) fish and starlings and (b) flyway populations of mallards and American black ducks.

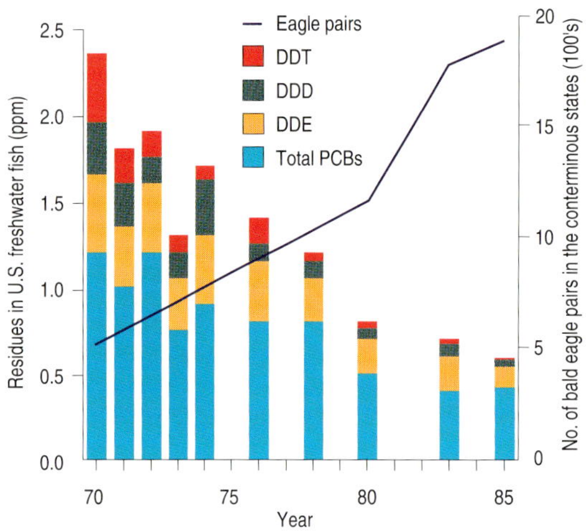

Fig. 2. Mean concentrations of DDT and its primary metabolites, DDE and DDD (TDE—dichlorodiphenyldichloroethane), and of total polychlorinated biphenyls (PCBs), in fish,1970-86. Also shown are the estimated number of bald eagle pairs in the conterminous United States during the same period (Federal Register 1994).

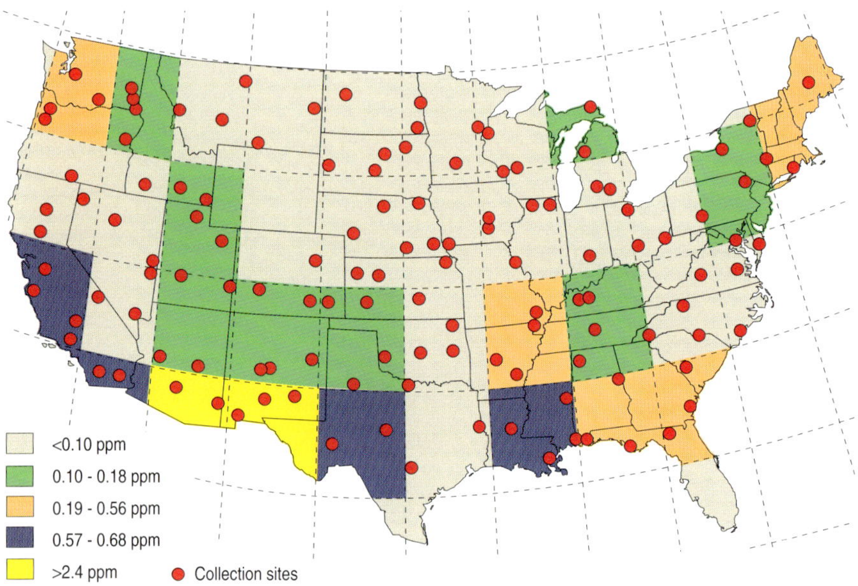

Fig. 3. Geographic distribution of DDE residues in starlings collected in 1985. Also shown are boundaries of the 5-degree (latitude and longitude) sampling blocks and collection sites.

decline (Schmitt et al. 1990). Concentrations of toxaphene, an insecticide that replaced DDT in cotton farming and many other applications, have also declined in fish since 1980, when its registration was canceled (Schmitt et al. 1990). Toxaphene does not accumulate in birds and was not measured in either starling or duck-wing samples.

Polychlorinated biphenyls (PCBs) are also complex mixtures of chemicals. Comprising as many as 209 different compounds, various PCB formulations were used historically as lubricants, hydraulic fluids, and fire retardants; as heat transfer agents in electrical equipment, including fluorescent light ballasts; and as a component of carbonless copy papers. Much like DDT, many PCBs are persistent and toxic. Large quantities were discharged directly to waterways, including Lakes Michigan and Ontario and the Hudson, Mississippi, Kanawha, and Ohio rivers. PCBs are also often present in landfills and urban runoff. These discharge and disposal patterns are reflected in the geographic trends evident for PCBs in fish and wildlife; greatest concentrations generally occur in the urban-industrial regions of the Midwest and Northeast (Fig. 4). By 1980, the direct discharge of PCBs to waterways had been greatly restricted, and total PCB concentrations generally declined in U.S. fish and wildlife (Bunck et al. 1987; Schmitt et al. 1990). Residual PCBs nevertheless remain a problem in some areas, as evidenced by human consumption advisories in effect for fish from the Great Lakes, Lake Champlain, the Hudson River, and elsewhere.

Some highly toxic PCBs are long-lived and are selectively accumulated by aquatic organisms. Fish samples collected in 1988 from some regions, especially the Great Lakes, still contained toxic PCBs at concentrations great enough to be harmful to fish-eating birds (C.J. Schmitt, National Biological Service, unpublished data, 1993). Indeed, PCBs and other contaminants in Great Lakes fish are believed to limit the reproduction of bald eagles and other fish-eating birds, mink (*Mustela vison*), and river otters (*Lutra canadensis*) in coastal areas of the Great Lakes (Wren 1991; Giesy et al. 1994). PCBs, along with DDE and other contaminants, including chlorinated dioxins, may also be involved in the failure of lake trout (*Salvelinus namaycush*) to reproduce naturally in Lake Michigan (USFWS 1981; Spitsbergen et al. 1991). In spite of discharge restrictions, the concentrations of PCBs and chemically similar compounds in the Great Lakes will likely remain elevated because of atmospheric transport and the internal cycling of contaminants already present in the lakes.

The primary sources of mercury to U.S. waters were discharges from chemical facilities that manufactured caustic soda (sodium hydroxide). These discharges have been regulated since the 1970's. Other historical sources included paper mills, gold and silver mines, and the production and use of mercury-containing pesticides. Concentrations of mercury in fish declined significantly from 1969 through 1974 as a result of restrictions on these historical uses, but concentrations have not changed appreciably since 1974. Concentrations in fish from heavily contaminated waters, such as Lake St. Clair, declined the most (Schmitt and Brumbaugh 1990). Despite these declines, fish consumption advisories remain in effect for some waters. Recent findings have highlighted the importance of atmospheric transport and the accumulation of mercury in natural sinks, such as Lake Champlain (e.g., Driscoll et al. 1994) and the Everglades, in the maintenance of elevated concentrations (Zillioux et al. 1993).

Lead concentrations in fish declined from 1976 to 1986 (Schmitt and Brumbaugh 1990), paralleling a trend reported for U.S. rivers (Smith et al. 1987). This decline has been attributed to reductions in the lead content of gasoline and to discharge restrictions at smelters and other industrial sources (Smith et al. 1987).

Selenium is a trace element required by plants and animals; it is toxic at high concentrations. Concentrations of selenium in fish declined in some areas of the United States. In some parts of the West, however, where concentrations were historically elevated, levels either increased or remained unchanged (Schmitt and Brumbaugh 1990). Selenium is a natural component of soils and is present at high concentrations in some arid areas of the U.S. West. The dissolution of selenium and other potentially toxic elements from soils and their accumulation in ecosystems are accelerated by irrigation. Elevated selenium concentrations, induced by irrigation, are responsible for the widely publicized wildlife deaths and deformities at Kesterson National Wildlife Refuge in California (Lemly 1993).

In general, U.S. concentrations of persistent contaminants that accumulate in fish and wildlife are lower now than at any time for which accurate data exist, although problem areas remain. These results imply that direct inputs of many toxic substances to the environment have been reduced through the regulation of industrial discharges and pesticide use. Declining concentrations of DDT and other contaminants in North America have permitted the return of predatory birds, such as bald eagles, to some areas from which they had been eliminated (Fig. 2).

The persistence of contaminant problems, despite curtailment of direct discharges to waterways and restrictions on the uses of persistent pesticides, has highlighted the importance of global and ecosystem processes such as atmospheric transport and internal cycling. The accumulation of selenium in California, and mercury in the Everglades, has resulted from natural processes—the leaching of elements from soils and vegetation. The rates of these processes have been accelerated by irrigation and other activities associated with agriculture. Atmospheric transport also represents an important source of PCBs to the Great Lakes; it has also been linked to the accumulation of mercury in Lake Champlain (*see* Glaser, this section; Baker et al. 1993) and other northeastern lakes (Driscoll et al. 1994).

The exposure of migratory birds such as peregrine falcons (*Falco peregrinus*) to contaminants on their wintering grounds outside of the United States (Henny et al. 1982), where DDT and other persistent compounds are still used, also remains a problem. Moreover, the curtailment of organochlorine pesticide use in North America has led to increasing reliance on so-called soft pesticides—highly toxic organophosphate, carbamate, and synthetic pyrethroid compounds—that are difficult to monitor because they are short-lived and do not accumulate. Evidence of the increasing use and potential adverse effects of these chemicals is highlighted by increasing occurrences of wildlife mortality attributable to them (*see* Glaser, this section). Additionally, chemical analysis has demonstrated the presence of highly toxic contaminants such as the chlorinated dioxins. No long-term monitoring data exist for these compounds, which may affect fish and wildlife at extremely low concentrations (Giesy et al. 1994). New approaches and technologies, capable of detecting chemical exposure and its effects at all levels of biological organization, will be required to monitor and assess highly toxic chemicals and those that do not accumulate in fish and wildlife before concentrations reach harmful levels.

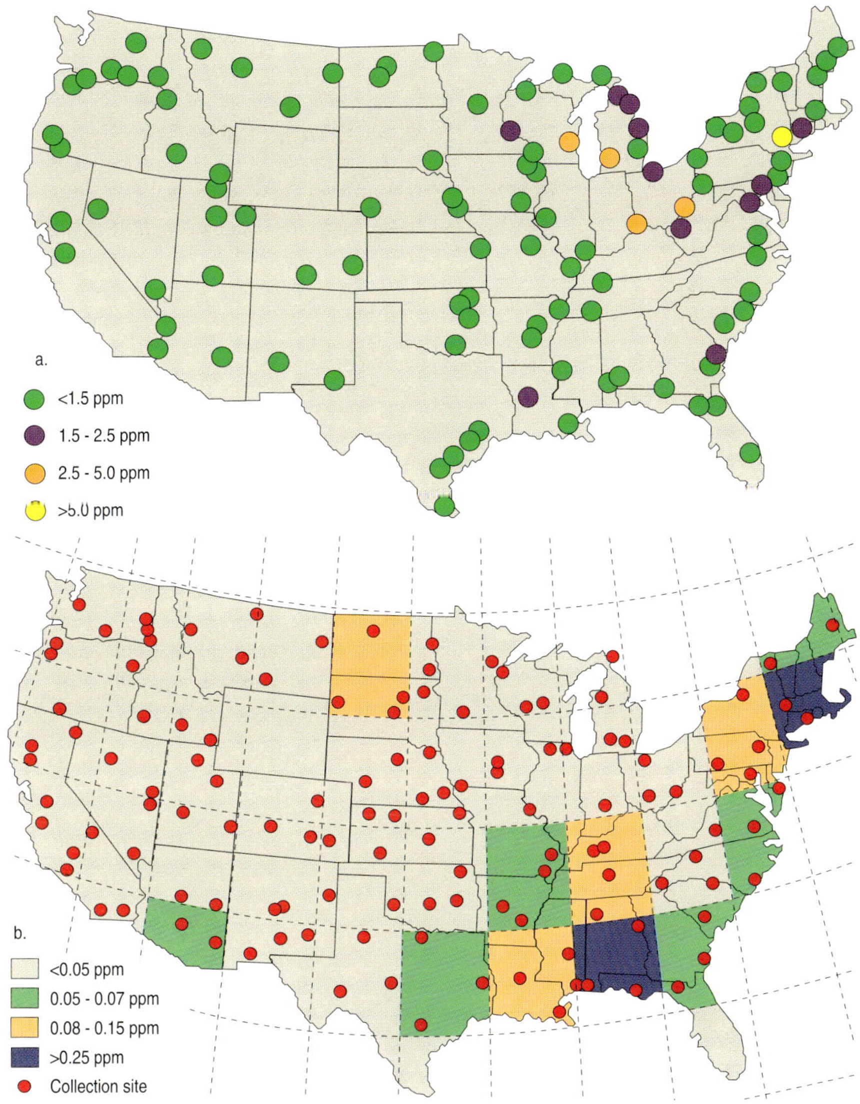

Fig. 4. Geographic distribution of PCB residues in U.S. Fish and Wildlife Service monitoring networks: (a) PCB concentrations in fish collected in 1986 from the indicated sites. Not shown are stations in Alaska and Hawaii, at which PCB concentrations were < 1.5 parts per million (ppm) at all sites; (b) PCBs in starlings collected in 1985. Also shown are boundaries of the 5-degree (latitude and longitude) sampling blocks and collection sites.

References

Aguillar, A. 1984. Relationships of DDE/DDT in marine mammals to the chronology of DDT input into the ecosystem. Canadian Journal of Fisheries and Aquatic Science 21:840-844.

Baker, J.E., T.M. Church, S.J. Eisenreich, W.J. Fitzgerald, and J.R. Scudlark. 1993. Relative atmospheric loadings of toxic contaminants and nitrogen to the Great Waters. A report to the Great Waters Program. U.S. Environmental Protection Agency, Research Triangle Park, NC. 142 pp.

Bunck, C.M., R.M. Prouty, and A.J. Krynitsky. 1987. Residues of organochlorine pesticides and polychlorobiphenyls in starlings (*Sturnus vulgaris*) from the continental United States, 1982. Environmental Monitoring and Assessment 8:59-75.

Carson, R. 1962. Silent spring. Houghton-Mifflin, Boston, MA. 368 pp.

Driscoll, C.T., C. Yan, C.L. Schofield, R. Munson, and J. Holsapple. 1994. The mercury cycle and fish in the Adirondack lakes. Environmental Science and Technology 28:136A-143A.

Federal Register. 1978. Determination of certain bald eagle populations as endangered or threatened. Federal Register 43:6230-6233. [Also see Federal Register 1990; 55:4209-4212.]

Federal Register. 1994. Endangered and threatened wildlife and plants; reclassify the bald eagle from endangered to threatened in most of the lower 48 states. Federal Register 59:35584-35594.

Fry, D.M., and C.K. Toone. 1981. DDT-induced feminization of gull embryos. Science 213:922-924.

Giesy, J.P., J.P. Ludwig, and D.E. Tillitt. 1994. Deformities in birds of the Great Lakes Region. Environmental Science and Technology 28:128A-135A.

Heath, R.G., and R.M. Prouty. 1967. Trial monitoring of pesticides in wings of mallards and black ducks. Bull. of Environmental Contamination and Toxicology 2:101-110.

Henny, C.J., F.R. Ward, K.E. Riddle, and R.M. Prouty. 1982. Migratory peregrine falcons, *Falco peregrinus*, accumulate pesticides in Latin America during winter. Canadian Field-Naturalist 96:333-338.

Johnson, R.E., T.C. Carver, and E.H. Dustman. 1967. Indicator species near top of food chain chosen for assessment of pesticide base levels in fish and wildlife—clams, oysters, and sediment in estuarine environment. Pesticide Monitoring Journal 1:7-13.

Lemly, A.D. 1993. Guidelines for evaluating selenium data from aquatic monitoring and assessment studies. Ecotoxicology and Environmental Safety 26:181-204.

O'Shea, T., W.J. Fleming III, and E. Cromartie. 1980. DDT contamination at Wheeler National Wildlife Refuge. Science 209:509-510.

Prouty, R.M., and C.M. Bunck. 1986. Organochlorine residues in adult mallard and black duck wings, 1981-82. Environmental Monitoring and Assessment 6:49-57.

Schmitt, C.J., and W.G. Brumbaugh. 1990. National Contaminant Biomonitoring Program: concentrations of arsenic, cadmium, copper, lead, mercury, selenium, and zinc in freshwater fishes of the United States, 1976-1984. Archives of Environmental Contamination and Toxicology 19:731-747.

Schmitt, C.J., J.L. Zajicek, and P.L. Peterman. 1990. National Contaminant Biomonitoring Program: residues of organochlorine chemicals in freshwater fishes of the United States, 1976-1984. Archives of Environmental Contamination and Toxicology 19:748-782.

Smith, R.A., R.B. Alexander, and M.G. Holman. 1987. Water quality trends in the nation's rivers. Science 235:1607-1615.

Spitsbergen, J.M., M.K. Walker, J.R. Olson, and R.E. Peterson. 1991. Pathologic alterations in early life stages of lake trout, *Salvelinus namaycush*, exposed to 2,3,7,8-tetrachlorodibenzo-*p*-dioxin as fertilized eggs. Aquatic Toxicology 19:41-72.

USFWS. 1981. Chlorinated hydrocarbons as a factor in the reproduction and survival of lake trout (*Salvelinus namaycush*) in Lake Michigan. U.S. Fish and Wildlife Service Tech. Paper 105. 42 pp.

Wiemeyer, S.N., C.M. Bunck, and C.J. Stafford. 1993. Environmental contaminants in bald eagle eggs—1980-84—and further interpretations of relationships to productivity and shell thickness. Archives of Environmental Contamination and Toxicology 24:213-227.

Wren, C.D. 1991. Cause-effect linkages between chemicals and populations of mink (*Mustela vison*) and otter (*Lutra canadensis*) in the Great Lakes basin. Journal of Toxicology and Environmental Health 33:549-585.

Zillioux, E.J., D.B. Porcella, J.M. Benoit. 1993. Mercury cycling and effects in freshwater wetland ecosystems. Environmental Toxicology and Chemistry 12:2245-2264.

For further information:

C.J. Schmitt
National Biological Service
Midwest Science Center
4200 New Haven Rd.
Columbia, MO 65201

Wildlife Mortality Attributed to Organophosphorus and Carbamate Pesticides

by
Linda C. Glaser
National Biological Service

Organophosphorus (OP) and carbamate pesticides are used widely in agricultural and residential applications as insecticides, herbicides, fungicides, and rodenticides. This family of chemicals replaced the organochlorine pesticides banned for use in the United States in the 1970's. Unlike organochlorine pesticides, which are long-lived in the environment and cause biological damage when they accumulate in an organism's system over time, OP and carbamate pesticides are short-lived in the environment and fast-acting on their "target pest." Direct mortality of wildlife from organochlorine pesticides was uncommon (Hayes and Wayland 1975); however, mortality is the primary documented effect on wildlife from OP and carbamate pesticides (Grue et al. 1983). Organophosphorus and carbamate pesticide toxicity is not specific to a target "pest," and lethal effects are seen in nontarget organisms; birds appear to be the most sensitive class of animals affected by these pesticides.

Organophosphorus and carbamate pesticides primarily affect the nervous system by inhibiting acetylcholinesterase (AChE) enzyme activity. This enzyme's main function in the nervous system is to break down the neurotransmitter acetylcholine. When AChE is altered by OP and carbamate pesticides, it cannot perform this breakdown function and acetylcholine accumulates. Acetylcholine accumulation increases nerve impulse transmission and leads to nerve exhaustion and, ultimately, failure of the nervous system. When the nervous system fails, muscles do not receive the electrical input they require to move. The respiratory muscles are the most critical muscle group affected, and respiratory paralysis is often the immediate cause of death.

Documentation of Poisoning

Virtually no reported findings of dead or affected birds are based on planned surveys or follow-ups to specific pesticide applications. In fact, there is often no suspicion of OP or carbamate pesticide poisoning because it is only after necropsy and laboratory testing that the poisoning is revealed. A cholinesterase (ChE) screening test compares brain ChE activity (primarily acetylcholinesterase activity) in a bird suspected of being poisoned with the ChE activity of normal birds of the same species. Enzyme activity reduced 20% or more is considered evidence of exposure to a cholinesterase-inhibiting compound; a reduction greater than 50% is evidence of lethal exposure (Ludke et al. 1975). In these incidents the cholinesterase-inhibiting compounds are OP and carbamate pesticides, and specific OP and carbamate compounds may be identified by chemical analysis of esophagus or stomach contents.

Effects on Wildlife

Wildlife mortality attributed to OP and carbamate pesticides has been documented for at least two decades, and the number of incidents recorded since 1980 is increasing (Fig. 1). In this article, 207 separate mortality incidents related to an OP or carbamate pesticide are described. These incidents occurred in nonendangered wildlife from 1980 to 1993. Of the 207 mortalities, a specific chemical compound was identified as the cause of death in 124 incidents and 19 different compounds were detected. Of the specific compounds identified, 4 were carbamates and 15 were OP compounds (Table). Carbamates were responsible in 31 mortalities while OP compounds were responsible in 93. On the basis of inhibited ChE activity in the brain, carbamate and OP pesticides were suspected as the cause of 64 additional incidents. In 19 unconfirmed reports, 5 had 20%-40% brain ChE inhibition, exposure levels not considered high enough to be lethal. The remaining 14 had a history suggesting pesticide exposure, but a diagnostic evaluation was not made.

Table. Specific compounds identified in organophosphorus and carbamate pesticide-related wildlife mortality incidents, 1980-93.

Carbamates	Organophosphorus compounds
Carbofuran	Chlorpyrifos
Methiocarb	Diazinon
Oxamyl	Dicrotophos
Aldicarb	Dimethoate
	Disulfoton
	Famphur
	Fenamiphos
	Fensulfothion
	Fenthion
	Fonofos
	Methamidiphos
	Monocrotophos
	Parathion
	Phorate
	Phosphamidon

Thousands of birds representing more than 50 species including waterfowl, passerines, colonial waterbirds, shorebirds, gulls, raptors, and others have been killed in these incidents. A die-off incident can involve a few birds of one species or hundreds of birds of a variety of species. Gross necropsy findings in birds dying from OP and carbamate toxicity were minimal. Lung edema and hyperemia (*see* glossary) were the predominant findings when lesions were observed. Mammals such as Virginia opossum (*Didelphis virginiana*), raccoon (*Procyon lotor*), and coyote (*Canis latrans*) were occasionally involved.

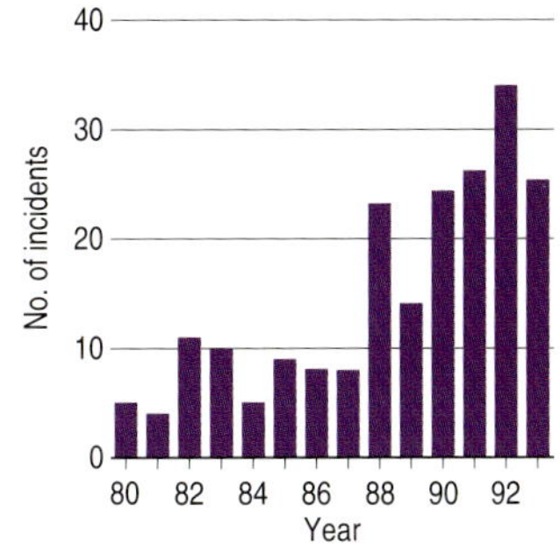

Fig. 1. Number of organophosphorus and carbamate pesticide-related wildlife mortality incidents, 1980-93.

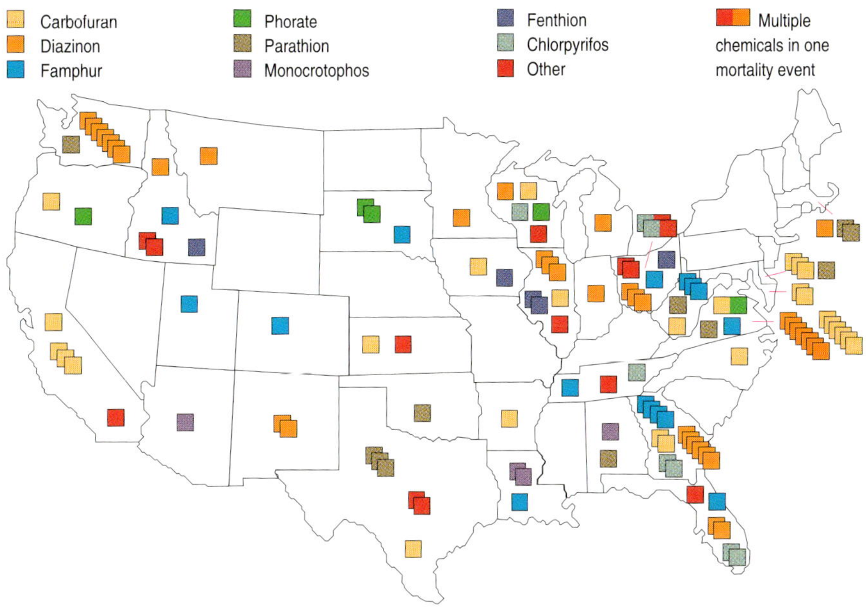

Fig. 2. Location by state of organophosphorus and carbamate compounds in pesticide-related wildlife mortality incidents, 1980-93.

The geographic distribution of mortality associated with specific compounds varied, although multiple incidents where the same compound was identified occurred within states (Fig. 2). Of the 124 deaths where a specific pesticide was identified, 64 had a known pesticide application (Fig. 3). The application varied from use on agricultural crops or livestock (agricultural) to lawn care or other uses in residential areas (residential) and on golf courses. Other known applications did not fall into these three categories and are primarily incidents of intentional baiting with grain.

Documentation of wildlife mortality in this manner has supported restrictions on the use of some OP and carbamate pesticides, such as the removal of diazinon from use for turf applications and limiting the use of granular carbofuran. Studies are under way to determine the sublethal effects of these chemicals (Grue et al.

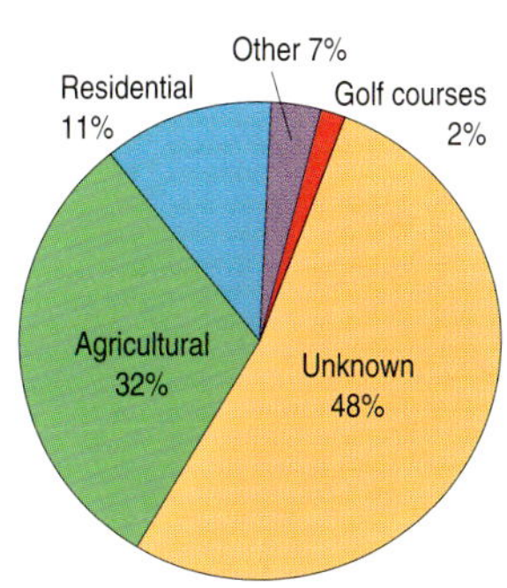

Fig. 3. General type of application associated with organophosphorus and carbamate pesticide-related wildlife mortality incidents, 1980-93.

1991; Hart 1993); preliminary findings indicate that OP and carbamate pesticides cause alterations in behavior and physiology and could affect survival in the wild. The total effect of carbamate and organophosphorus pesticides to wildlife is still unknown.

For further information:

Linda C. Glaser
National Biological Service
National Wildlife Health Center
6006 Schroeder Rd.
Madison, WI 53711

References

Grue, C.E., W.J. Fleming, D.G. Busby, and E.F. Hill. 1983. Assessing hazards of organophosphate pesticides to wildlife. Pages 200-220 *in* Transactions of the 48th North American Wildlife and Natural Resources Conference. The Wildlife Management Institute, Washington, DC.

Grue, C.E., A.D.M. Hart, and P. Mineau. 1991. Biological consequences of depressed brain cholinesterase activity in wildlife. Pages 151-209 *in* P. Mineau, ed. Cholinesterase-inhibiting insecticides. Elsevier Science Publishers, Amsterdam, The Netherlands.

Hart, A.D.M. 1993. Relationships between behavior and the inhibition of acetylcholinesterase in birds exposed to organophosphorus pesticides. Environmental Toxicology and Chemistry 12:321-336.

Hayes, W.J., Jr., and J. Wayland. 1975. Effects on wildlife. Chapter 11 *in* Toxicology of pesticides. Williams and Wilkins Company, Baltimore, MD.

Ludke J.L., E.F. Hill, and M.P. Dieter. 1975. Cholinesterase (ChE) response and related mortality among birds fed ChE inhibitors. Archives of Environmental Contamination and Toxicology 3:1-21.

Acidic Deposition ("Acid Rain")

by
Kent Schreiber
National Biological Service

Acidic deposition, or "acid rain," describes any form of precipitation, including rain, snow, and fog, with a pH of 5.5 or below (Note: pH values below 7 are acidic; vinegar has a pH of 3). It often results when the acidity of normal precipitation is increased by sulfates and nitrates that are emitted into the atmosphere from burning fossil fuels. This form of airborne contamination is considered harmful, both directly and indirectly, to a host of plant and animal species.

Although acid rain can fall virtually anywhere, ecological damages in environmentally sensitive areas downwind of industrial and urban emissions are a major concern. This includes areas that have a reduced capacity to neutralize acid inputs because of low alkalinity soils and areas that contain species with a low tolerance to acid conditions. To determine the distribution of acidic deposition and evaluate its biological effects, research and monitoring are being conducted by the federal government with support from states, universities, and private industry.

The national extent of the acid rain problem has been estimated by sampling water from 3,000 lakes and 500 streams (Irving 1991), representing more than 28,000 lakes and 56,000 stream reaches with a total of 200,000 km (125,000 mi). Some particularly sensitive areas, such as the Adirondack Mountain region, have been more intensively sampled and the biota examined in detail for effects from acidity.

To identify trends in aquatic ecosystems, present and historical survey data on water chemistry and associated biota are compared. In lakes, the chemical and biological history and pH trends may be inferred or reconstructed in some cases by examining assemblages of fossil diatoms and aquatic invertebrates in the sediment layers. In terrestrial ecosystems, vegetation damage is surveyed and effects of acidic deposition to plants and animals are determined from laboratory and field exposure experiments. Natural variation in populations and the complex interactions between acidity and other ecosystem components make it difficult to extend many of the research findings to populations or communities. Acidity can also modify ecosystem processes such as decomposition and the flow of nutrients. Therefore, models are often used to predict such effects by combining information on individual species' effects, population distributions, and the patterns and amounts of acidic deposition.

Status and Trends

Aquatic Species

Research in the United States, Scandinavia, and Canada has demonstrated that acidity affects the physiology, reproduction, food resources, and habitat of aquatic species. Laboratory experiments and field surveys have shown that sensitive aquatic species, ranging from plankton and aquatic invertebrates at the bottom of the food chain to fish at the top (Figure), decrease in numbers with increased acidity (i.e., decreased pH). Some reductions in sensitive species may be partially offset by increases in more acid-tolerant species, resulting in little change in the total number of organisms in the community even though the diversity of species may change.

Melting snow, which accumulates the winter's deposition of acidic materials, and episodes of spring rainfall can be especially damaging to sensitive streams and lakes. The acidity that flushes from the surrounding landscape often enters the aquatic ecosystem at a time of important reproductive activity in fish and invertebrates. Acid conditions leach aluminum from the watershed soils, creating toxic levels for aquatic organisms in the lakes and streams that receive the runoff. Acidity also increases the availability and toxicity of other metals, such as mercury, that may be present in the aquatic environment (Longcore et al. 1993).

The Adirondack region of New York, one of the most extensively studied areas in the United States, has exhibited some of the most evident

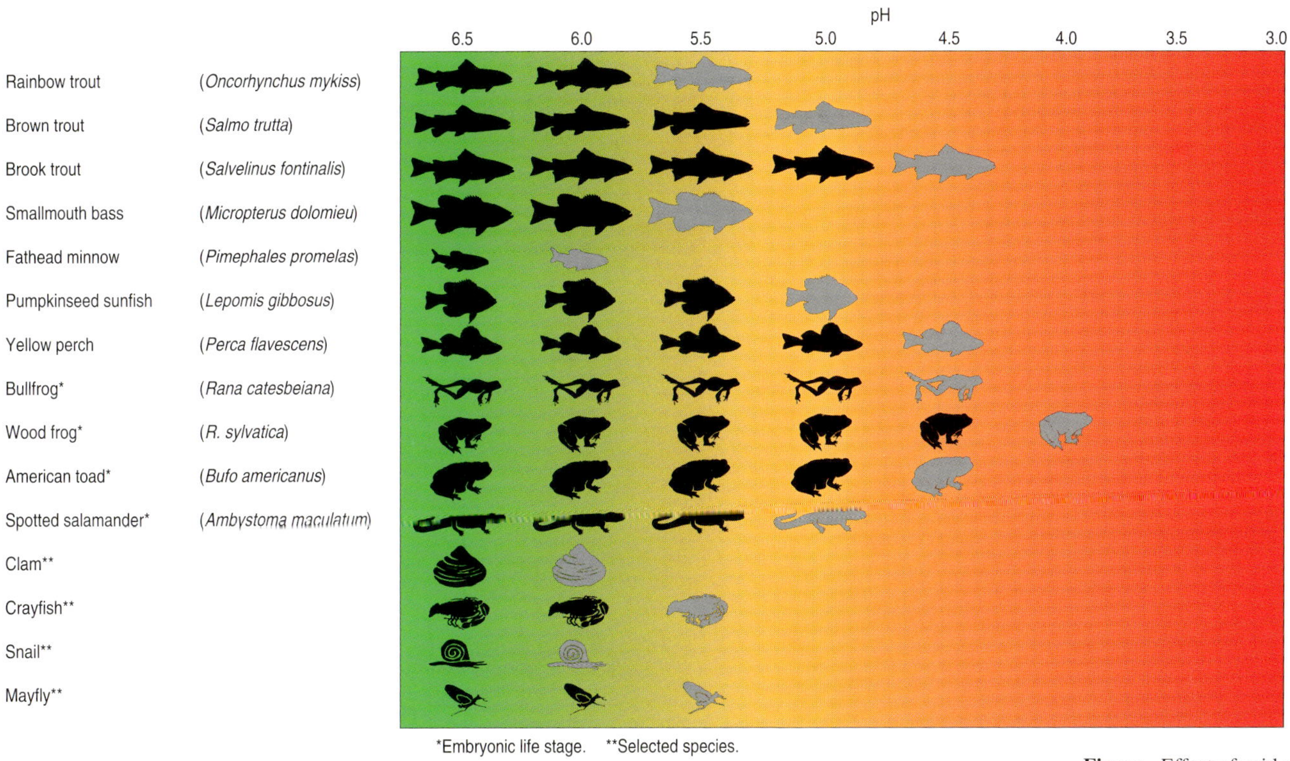

Figure. Effect of acid rain on some aquatic species. As acidity increases (and pH decreases) in lakes and streams, some species are lost (gray).

effects from acidification. Comparison of information for 274 lakes surveyed between 1929 and 1934 and again between 1975 and 1985 showed 80% of the lakes had declined in their capacity to neutralize acidity (Driscoll et al. 1991). Surveys for fish in 1,469 Adirondack lakes during 1984-87 showed 24% without fish, with a high percentage (61%) of these fishless lakes located in the southwestern portion of the region where buffering of acid input is limited by the local geology. These lakes are generally small, shallow, and at relatively high elevations. Although fish species demonstrated a wide range of tolerance to acidity, studies found that the number of species in a lake declined as pH declined. Comparisons with historical information demonstrated that populations of brook trout (*Salvelinus fontinalis*), a relatively acid-tolerant species, had disappeared from 44 of 409 lakes (11%), and acid-sensitive minnows were lost from 33 of 170 lakes (19%) surveyed (Baker et al. 1993).

Surveys in the mid-Appalachian and mid-Atlantic Coastal Plain regions indicate many streams that are vulnerable to acidic deposition. Acidic deposition influences an estimated 3,000 km (1,865 mi) of trout streams in Pennsylvania (Carline et al. 1992). Of 344 streams surveyed in western Virginia, nearly all (93%) are considered sensitive and nearly half are considered extremely sensitive because of their low buffering capacity. Ten percent of the surveyed streams in this area are acidic (Cosby et al. 1991). Waters farther south in the Blue Ridge province presently show little or no effects from atmospheric acidity, but the potential for damage exists because of their low buffering capacity. Evidence suggests that the ability of the watersheds to neutralize the acidic input is declining, and future acidification of surface waters is a continuing concern.

In the extreme south, Florida has the distinction of having the greatest number of acidic lakes of any U.S. region; however, many of the fish species commonly found in these lakes are tolerant of low pH levels. Other regions, such as the upper Midwest and western states, have waters with low buffering capacity and, although existing information does not indicate biological problems, they remain vulnerable to acidity.

Acidification of the aquatic environment can also affect vertebrate species other than fish. For example, studies show acidic deposition can affect the diet, foraging, distribution, and reproduction of bird species that depend on the aquatic environment (Table). Such indirect effects are often difficult to interpret, but they could potentially lead to fundamental changes in the ecosystem.

Terrestrial Species

Acidic deposition affects terrestrial wildlife species by damaging habitat and by reducing or contaminating food sources through uptake of toxic levels of metals (Schreiber and Newman 1988). Species such as amphibians, which

Table. Bird species studies that either did (yes) or did not (no) yield evidence that acidic deposition affected the birds (modified from Longcore et al. 1993).

Species	Diet/foraging		Breeding distribution		Reproductive measures	
	Yes	No	Yes	No	Yes	No
Common loon (*Gavia immer*)	x		x**	x**	x	x
Arctic loon (*Gavia arctica*)						x
Common merganser (*Mergus merganser*)			x		x	
Belted kingfisher (*Ceryle alcyon*)			x			
Osprey (*Pandion haliaetus*)	x		x		x	
American black duck (*Anas rubripes*)	x		x		x*	
Common goldeneye (*Bucephala clangula*)			x*			
Ring-necked duck (*Aythya collaris*)	x				x	
Eastern kingbird (*Tyrannus tyrannus*)				x	x	
Tree swallow (*Tachycineta bicolor*)	x			x	x	

*Beneficial effect.
**Evidence of both an effect and no effect.

require both aquatic and terrestrial environments, are perhaps most at risk. For example, in the acid-sensitive areas of eastern Canada, 16 of the 17 amphibian species have more than 50% of their ranges affected by acidic deposition (Clark 1992). Monitoring amphibian populations could provide a biological indication of changes in acid deposition (Freda et al. 1991).

Forest damage attributed to acid deposition, such as the maple dieback, can change the biomass of invertebrates available to birds. Species such as the red-eyed vireo (*Vireo olivaceus*) and least flycatcher (*Empidonax minimus*), which forage in the overstory, may have fewer prey because of habitat loss. Other species, though, including the wood thrush (*Hylocichla mustelina*) and ovenbird (*Seiurus aurocapillus*), which are associated with shrubs and ground-feeding, may benefit from an increased biomass of invertebrates in their foraging areas (DesGranges 1987). Such effects could result in changes in the ecosystem; however, little direct evidence of population-level changes is available.

Future Conditions

Legislation to reduce emissions that form acid rain has been enacted in both the United States and Canada. There is evidence that acidic deposition in some areas has started to decline and that water quality has improved (Gunn and Keller 1990). Monitoring over a decade at 81 selected sites in the Northeast and upper Midwest has shown that most of the lakes and streams there have decreased in sulfate levels, coinciding with the general decrease (about 11%) in national emissions of sulfur dioxides (NAPAP 1993). Results from modeling the effects of 30 years of emission controls (i.e., 1980 to 2010) on 2,500 affected lakes in the Adirondacks suggest improvements will occur in water chemistry and fish habitat in up to 150 lakes (Rubin et al. 1992).

Control measures take time to implement, and it is too early to determine their overall ecological effects. Episodes of acidification continue to adversely affect fish populations and invertebrates. To prevent loss of fisheries and aquatic biota in some severely affected localities, limestone, a neutralizing agent, is being applied to reduce acidity levels (Olem 1991). It is important to continue monitoring the status of species and populations in sensitive areas to evaluate the effect of emission controls and to ensure healthy ecosystems.

For further information:

R. Kent Schreiber
National Biological Service
Leetown Science Center
1700 Leetown Rd.
Kearneysville, WV 25430

References

Baker, J.P., W.J. Warren-Hicks, J. Gallagher, and S.W. Christensen. 1993. Fish population losses from Adirondack lakes: the role of surface water acidity and acidification. Water Resour. Res. 29:861-874.

Carline, R.F., W.E. Sharpe, and C.J. Gagen. 1992. Changes in fish communities and trout management in response to acidification of streams in Pennsylvania. Fisheries 17:33-38.

Clark, K.L. 1992. Monitoring the effects of acidic deposition on amphibian populations in Canada. Occasional Papers of the Canadian Wildlife Service 76:63-66.

Cosby, B.J., P.F. Ryan, J.R. Webb, G.M. Hornberger, and J.N. Galloway. 1991. Mountains of western Virginia. Pages 297-318 *in* D.E. Charles, ed. Acidic deposition and aquatic ecosystems. Springer-Verlag, New York.

DesGranges, J.-L. 1987. Forest birds as biological indicators of the progression of maple dieback in Quebec. Pages 249-257 *in* A.W. Diamond and F.C. Fillion, eds. The value of birds. International Council for Bird Preservation Tech. Publ. 6. Cambridge, England.

Driscoll, C.T., R.M. Newton, C.P. Gubala, J.P. Baker, and S.W. Christensen. 1991. Adirondack Mountains. Pages 133-202 *in* D.E. Charles, ed. Acidic deposition and aquatic ecosystems. Springer-Verlag, New York.

Freda, J., W.J. Sadinski, and W.A. Dunson. 1991. Long-term monitoring of amphibian populations with respect to the effects of acidic deposition. Water, Air, and Soil Pollution 55:445-462

Gunn, J.M., and W. Keller. 1990. Biological recovery of an acid lake after reductions in industrial emissions of sulphur. Nature 345:431-433.

Irving, P.M., ed. 1991. Acidic deposition: state of science and technology. National Acid Precipitation Assessment Program, Washington, DC. 265 pp.

Longcore, J.R., H. Boyd, R.T. Brooks, G.M. Haramis, D.K. McNicol, J.R. Newman, K.A. Smith, and F. Stearns. 1993. Acidic depositions: effects on wildlife and habitats. Wildlife Society Tech. Rev. 93-1. 42 pp.

NAPAP. 1993. 1992 report to Congress. National Acid Precipitation Assessment Program, Office of the Director, Washington, DC. 130 pp.

Olem, H. 1991. Liming acidic surface waters. Lewis Publishers, Chelsea, MI. 331 pp.

Rubin, E.S., M.J. Small, C.N. Bloyd, and M. Henrion. 1992. Integrated assessment of acid-deposition effects on lake acidification. Journal of Environmental Engineering 118:120-134.

Schreiber, R.K., and J.R. Newman. 1988. Acid precipitation effects on forest habitats: implications for wildlife. Conservation Biology 2:249-259.

Atmospheric Deposition and Solute Transport in a Montane Mixed-Conifer Forest System

by
Daniel A. Everson
David M. Graber
National Biological Service

In 1981 a long-term, cooperative study of ecosystems in the southern Sierra Nevada was begun to address concerns over high levels of air pollutants. Atmospheric pollutants are generated throughout California and because of topography, wind patterns, and a Mediterranean climate, they eventually concentrate in the San Joaquin Valley, west of the study area. Baseline ecosystem data—chemical and biological—were collected to determine basic system structure and function. This collection was followed by long-term measurements of pollutants to assess their present and potential effects on terrestrial and aquatic ecosystems. Studies included measurement of precipitation chemistry, dry deposition, stream hydrology, aquatic chemistry and biology, soil chemistry, meteorology, nutrient fluxes, watershed response to fire disturbance, and vegetation structure and dynamics.

Methods

Research was designed to take advantage of the striking elevation gradient by including measurements at three core areas:

Elk Creek is a low-elevation, 750-m (2,460-ft) foothill site dominated by chamise chaparral (*Adenostoma fasciculatum*). Precipitation averages 66 cm (36 in) annually, nearly all falling as rain in winter. Precipitation chemistry and volume are collected weekly from a site at Ash Mountain, 3 km (1.9 mi) south of the site. An intermittent first-order tributary, Chamise Creek, is sampled when possible.

Log Creek is a mid-elevation 2,100-m (6,890-ft) montane mixed conifer forest site dominated numerically by white fir (*Abies concolor*); however, giant sequoia (*Sequoiadendron giganteum*) contribute the greatest basal area. Mean annual precipitation is 100 cm (39.4 in); more than 85% falls as snow during the winter. Precipitation chemistry and volume are collected weekly. Paired watersheds, Tharp's and Log creeks, are sampled biweekly.

Emerald Lake is a subalpine 3,000-m (9,840-ft) cirque (*see* glossary), largely treeless but including lodgepole pine (*Pinus contorta*), western white pine (*Pinus monticola*), foxtail pine (*Pinus balfouriana*), and red fir (*Abies magnifica*). Annual precipitation varied between 70 cm (28 in) and 300 cm (118 in) in the past decade, nearly all as snow during the winter. Precipitation depth is estimated by using snow-water equivalents; precipitation chemistry is collected as the opportunity arises.

Inputs

Precipitation and discharge at all three sites vary greatly from year to year. As a result, annual ion input and export also vary considerably in each watershed. There was a general increase in precipitation and a decrease in ion concentration with elevation. Most precipitation falls during the winter as snow above 1,800 m (5,900 ft).

Unlike the eastern United States, where the major source of acidification of lakes and streams is sulfur deposition, the southern Sierra Nevada is considered to be most exposed to nitrogen. At the Log Creek and Elk Creek sites, over the sampling period the mean loading of nitrogen, expressed as NO_3^-, was 16.77 kg ha^{-1} yr^{-1}, with 74% contributed from wet deposition. The mean loading of sulfur (S), expressed as SO_4^{2-}, was 5.24 kg ha^{-1} yr^{-1}, with 67% contributed by wet deposition (Table). The dry deposition input estimates are conservative because the dry deposition sampling site appears more shielded from pollutant inputs than the wet sampling site.

The mean precipitation pH at the Log Creek site was 5.25, and at the Elk Creek site 5.37; there is not a chronic acid rain problem in the area. The frequency and volume of summer storms were fairly constant. Wet-deposition ion loading at Log Creek and Elk Creek was similar. Elk Creek, a somewhat more polluted site, received 57% as much precipitation as Log Creek, but ions were proportionately more concentrated (Blanchard and Tonneson 1993), yielding equivalent loading. Dry deposition loading in the Elk Creek site was not measured.

Emerald Lake received about 99% of its precipitation in the form of snow, with a mean pH of 5.3, meaning the site is only slightly acidic. Concentrations of individual ions were extremely dilute, usually less than 5 μEq L^{-1}. Mean wet deposition loading of nitrogen and sulfur, expressed as NO_3^- and SO_4^{2-}, was 2.15 kg ha^{-1} yr^{-1} and 0.78 kg ha^{-1} yr^{-1}, respectively (estimated from Dozier et al. 1987). No reliable estimate of annual dry deposition flux is available at Emerald Lake. Like Log Creek, Emerald Lake has no chronic acid precipitation problem.

Table. Log Creek site inputs of nitrogen and sulfur (kg ha^{-1} yr^{-1}).

Year*	NO^3		NH^4		Total N		SO^4		S	
	Wet	Dry	Wet	Dry	Wet	Dry**	Wet	Dry	Wet	Dry***
85	4.98	-	2.39	-	2.98	-	2.89	-	0.97	-
86	10.52	-	2.08	-	3.99	-	8.62	-	2.88	-
87	4.87	3.23	1.66	-	2.39	0.73	6.38	1.79	2.13	0.73
88	5.39	3.80	1.89	-	2.68	0.86	3.12	1.79	1.04	0.60
89	4.48	4.82	1.42	-	2.11	1.09	2.57	2.07	0.86	0.69
90	4.88	5.71	2.24	-	2.84	1.29	1.84	1.68	0.61	0.56
91	8.41	4.11	2.09	-	3.52	0.93	2.23	0.75	0.74	0.25
92	5.46	-	2.23	-	2.96	-	2.26	-	0.75	-
93	4.12	-	1.11	-	1.79	-	2.02	-	0.67	-
				Mean	**2.81**	**0.98**			**1.18**	**0.57**
				Standard deviation	**0.64**	**0.19**			**0.74**	**0.17**

* Hydrologic year is October 1 to September 30.
Wet deposition from National Atmospheric Deposition Program site.
Dry deposition from Wolverton National Oceanic and Atmospheric Administration site.
** Dry deposition N is HNO^3 and NO^3 reported as NO^3.
*** Dry deposition S is SO^2 and SO^4 reported as SO^4.

Trends in Input

Between 1981 and 1989, 29 precipitation events were highly acidic (pH values less than 4.5), with 22 of these events occurring in low-volume summer storm events from May through September. Since 1989 only one storm, in November 1991, had a pH below 4.5. A rise in mean precipitation pH was also recorded during this period at both the low-elevation and mid-elevation sites. The Elk Creek site mean precipitation pH rose from 5.23 before 1989 to 5.68, while the Log Creek site rose from 5.12 to 5.39. A beneficial downward trend in the total annual loading of sulfur occurred (Table), which was poorly explained by variation in precipitation ($r = 0.397$). The frequency of storm events with very high (> 2 mg L^{-1}) concentrations of sulfur was also reduced from 25 events from 1983 to 1988 to 6 events in the following 5 years. No apparent trend for nitrogen was seen. Emerald Lake precipitation chemistry data are insufficient to infer trends.

Effects on Stream and Lake Chemistry

Stream discharge peaks in February at the Elk Creek site, April at Log Creek, and June at Emerald Lake. The first is a direct response to maximum rainfall; the other two discharges reflect the peaks of snowmelt.

In the mixed-conifer Log Creek site, more than 99% of the nitrogen deposited was conserved. Mean discharge of nitrogen, expressed as NO_3^-, is 0.09 kg ha^{-1} yr^{-1}, virtually all of which was derived from the melting snowpack. Seventy-four percent of the annual loading of sulfur was conserved, with a mean discharge, expressed as SO_4^{2-}, of 0.93 kg ha^{-1} yr^{-1}, again mostly derived from melting snowpack. Acid-neutralizing capacity (ANC), expressed as HCO_3^- (bicarbonate), was many times greater than annual acidic loading at a mean of 320 μEq L^{-1}. At these levels, most sulfur and nitrogen are being conserved by the biota in the ecosystem, and the ecosystem's ability to neutralize acid is generally good except when extreme events occur.

Tharp's Creek, one of the paired watersheds in the mid-elevation Log Creek site, was burned by prescription in October of 1990, producing striking changes in stream output chemistry that continued through 1993. Although net retention of NO_3^- continued, discharge of NO_3^- increased from a pre-burn mean of 0.04 kg ha^{-1} yr^{-1} to 1.59 kg ha^{-1} yr^{-1} in the 3 years following the burn. Thus, the ability to retain nitrogen was decreased, and the system leaked nitrogen and sulfur. The SO_4^{2-} outputs exceeded inputs following the fire, increasing from a pre-burn mean of 0.35 kg ha^{-1} yr^{-1} to over 6.63 kg ha^{-1} yr^{-1}. Acid-neutralizing capacity also rose from a pre-burn mean of 21.16 kg ha^{-1} yr^{-1} to over 37.16 kg ha^{-1} yr^{-1}.

At the Elk Creek site, mean alkalinity (or ANC) expressed as HCO_3^- was 310.0 μEq L^{-1} with a mean stream pH of 6.61 (neutral). The creek flows intermittently, mostly between January and March. The water is cloudy with suspended clay particles, and debris flows are common after heavy rains.

The outflow of Emerald Lake had a mean pH of 6.17 (neutral) and mean HCO_3^- of only 30 μEq L^{-1}. Episodic acidification on the order of days to weeks was recorded at Emerald Lake under two scenarios: (1) during dirty summer storms, when buffering capacity was overwhelmed by low-pH stormwater flashing into the lake, and (2) during snowmelt, when NH_4^+, NO_3^-, and SO_4^{2-} were preferentially eluted from the snowpack, causing an acidic pulse (Williams and Melack 1991).

Discussion

We have found no long-term chronic acidification of lakes and streams in our study area, even though the hundreds of lakes in the region are considered to be the most poorly buffered in the western United States (Landers et al. 1987).

Emerald Lake, typical of subalpine lakes in the region, currently generates enough ANC, mostly through cation exchange (Williams et al. 1993a), to buffer two to five times the current annual acidic inputs (Sickman and Melack 1989). Very clean winter air and the large-volume, extremely dilute snowpack it produced were significant factors in maintaining the buffering capacity. Dirty summer storms and spring snowmelt that concentrate NO_3^- and SO_4^{2-} and deliver them quickly to the lake have caused episodic acidification, a phenomenon that we are studying. An increase in the frequency of storm events during the summer and fall would likely be harmful to subalpine lake basins if the chemistry of these events were to remain the same.

The low concentrations of nitrogen and sulfur in stream water indicate that neither reached saturation in soils and plants of the mixed-conifer forest (Williams et al. 1993b). The buffering capacities of low- and mid-elevation sites were many times greater than acidic inputs. But human-caused nitrogen and sulfur have been shown to stimulate the growth of ponderosa pines (*Pinus ponderosa*), which in turn increases vulnerability to high local ozone levels and potential for damages (Temple et al. 1992).

Precipitation and snowmelt pH levels were not low enough to mobilize aluminum (Dozier et al. 1987). Bradford et al. (1992) found that present pH levels in Sierran lakes and streams are not sufficient, directly or through aluminum mobilization, to affect indigenous amphibians.

The measured reductions in sulfur loading and decrease in frequency of low-pH storm events suggest that reduced pollutant emissions may be having a positive effect on the air quality of the southern Sierra Nevada. Continued monitoring and further research will be needed to determine if this trend will continue, given the tremendous rate of population growth in California.

References

Blanchard, C.L., and K.A. Tonnesen. 1993. Precipitation-chemistry measurements from the California Acid Deposition Monitoring Program, 1985-1990. Atmospheric Environment 27A(11):1755-1763.

Bradford, D.F., C. Swanson, and M.S. Gordon. 1992. Effects of low pH and aluminum on two declining species of amphibians in the Sierra Nevada, California. Journal of Herpetology 26:369-377.

Dozier, J., J.M. Melack, D. Marks, K. Elder, R. Kattelman, and M. Williams. 1987. Snow deposition, melt, runoff, and chemistry in a small alpine watershed, Emerald Lake basin, Sequoia National Park. Final report to California Air Resources Board, Contract A3-106-32. 155 pp.

Landers, D.H., J.M. Ellers, D.F. Brakke, W.S. Overton, P.E. Kellar, M.E. Silverstein, R.D. Schonbrod, R.E. Crowe, R.A. Linthurst, J.M. Omernik, S.A. Teague, and E.P. Meier. 1987. Characteristics of lakes in the western United States. Vol. 1. Population descriptions and physico-chemical relationships. U.S. Environmental Protection Agency EPA-600/3-86/054a. 176 pp.

Sickman, J.O., and J.M. Melack. 1989. Characterization of year-round sensitivity of California's montane lakes to acidic deposition. Final report to California Air Resources Board, Contract A5-203-32. 104 pp.

Temple, P.J., G.H. Riechers, and P.R. Miller. 1992. Foliar injury responses of ponderosa pine seedlings to ozone, wet and dry acidic deposition, and drought. Environmental and Experimental Botany 32(2):101-113.

Williams, M.W., A.D. Brown, and J.M. Melack. 1993a. Geochemical and hydrologic controls on the composition of surface water in a high-elevation basin, Sierra Nevada, California. Limnology and Oceanography 38(4):775-797.

Williams, M.W., and J.M. Melack. 1991. Solute chemistry of snowmelt and runoff in an alpine basin, Sierra Nevada. Water Resour. Res. 27:1575-1588.

Williams, M.W., J.M. Melack, and D.A. Everson. 1993b. Export of major ionic solutes from two subalpine watersheds before and after fire. Poster presented at Annual Meeting of American Geophysical Union, San Francisco, CA.

For further information:

David M. Graber
National Biological Service
Sequoia and Kings Canyon
Field Station
Three Rivers, CA 93271

Agricultural Ecosystems

by

Arthur W. Allen
National Biological Service

Approximately 45% of the U.S. land area is used for agricultural purposes, with 191 million ha (472 million acres) in cropland and 238 million ha (587 million acres) in range or pasture (Knutson et al. 1990). American agriculture has become the most productive in the world based on technology and increased specialization. Energy, machinery, agrochemicals, and irrigation are principal components of modern American agriculture, all of which potentially affect farm and off-farm environmental quality. In addition, government policies have pervasively affected U.S. agriculture, often precluding producers from responding to changing market conditions or affecting adoption of farm practices that potentially improve environmental quality (National Research Council 1989; Reichelderfer 1990).

Energy and technology have propelled American agriculture from pioneering conversion of the landscape to intensive, high-yield, monocultural production. The composition of agriculture in terms of farm numbers, size, and methods of production have changed dramatically throughout this century. The effects of the agricultural industry on the diversity, distribution, and abundance of wildlife continue to be profound.

Larger, more economically efficient producers that could tolerate smaller profit margins have absorbed the assets of smaller, less successful operations. In 1991 the U.S. human population on farms was less than one-tenth of what it was in 1920 (Haynes 1991). As the number of farms decreased by two-thirds during this same period, farm size increased. In response to fewer farms and the need to increase production efficiency, fields have become larger, crop diversity has decreased, crop rotation patterns have become simpler and less frequent, and agrochemicals play a major role in crop production. Over the last 30 years, these elements have had significant effects on environmental quality within agricultural ecosystems.

The Conservation Title of the Food Security Act of 1985, commonly referred to as the Farm Bill, was formulated in a time of commodity surpluses, economic stress within the agricultural community, and increasing public concern about environmental quality. The Conservation Reserve Program (CRP), a cornerstone of the 1985 Farm Bill, was enacted to remove highly erosive cropland from production. This legislation reflects an effective integration of economic support to the agricultural community with environmental policies advocated by a strong coalition of organizations representing a wide spectrum of the American public. The CRP has provided substantial benefits to wildlife populations across the nation. To appreciate the CRP's significance to wildlife, we must remember that tremendous changes in agriculture have influenced the abundance and quality of habitat in this century (Soil and Water Conservation Society 1994).

World War II, for example, brought an increased demand for American agricultural products. New technologies adopted in the post-war period reduced production costs and further escalated farm output. Tractors and farm machinery became more powerful and efficient. Time and energy savings decreased the amount of human labor needed to work larger fields. Advances in biological and chemical technologies further increased agricultural efficiency and crop yields. The use of nitrogen fertilizer increased from 197 million metric tons (217 million tons) in 1940 to 6,765 million metric tons (7,459 million tons) in 1970 (Haynes 1991). By the early 1970's, crop yields had skyrocketed to new records.

American agriculture entered the world market in the 1970's in response to increased global demands for agricultural products. American farmers expanded production by cultivating existing croplands more intensively and bringing new, less fertile and more fragile lands into production. The 1980's arrived with the farm industry in crisis due to overproduction, increased costs for fuels and fertilizers, elevated interest rates, declining land values, and decreased demand for export sales. The agricultural economic predicament, as well as heightened public concern about environmental quality, set the stage for the 1985 Farm Bill and establishment of the CRP.

Agricultural Effects on Wildlife Habitat

The effects of modern agriculture on wildlife are indisputable, ranging from habitat elimination to long-term effects of agrochemicals on water quality and reproductive success of ground-nesting birds (Capel et al. 1993). Habitat diversity in agricultural ecosystems has declined drastically as a consequence of the elimination of hay and pasture needed by draft animals and a shift to crop monocultures. In many regions, wetland drainage, consolidation of fields and farms, and elimination of fencerows and idle areas have reduced habitat diversity even further, thereby diminishing the ability of agricultural ecosystems to sustain viable populations of wildlife. The amount of undisturbed grass-dominated cover and noncropped areas has decreased, resulting in lower availability of habitat and higher losses to predators for many nongame and game species of wildlife. In many agricultural regions, crucial wildlife habitat components such as undisturbed grassland have become dissected into small, isolated patches, or spatially segregated tracts. Increased agrochemical use has been implicated in the long-term decline of species such as the northern bobwhite (*Colinus virginianus*).

Monocultures, with minimal rotations between crops, have accelerated soil erosion and led to a greater dependence on chemical fertilizers and pesticides (Bender 1984) resulting in surface and groundwater contamination (Ribaudo 1989). Larger, heavier equipment used for tillage, planting, application of agrochemicals, and harvesting contributes to increased soil compaction and decreased soil tilth (suitability), further contributing to erosion. Agriculture has become the largest single nonpoint source of water pollution, delivering not only soil particles but also absorbed and dissolved nutrients and pesticides (National Research Council 1989).

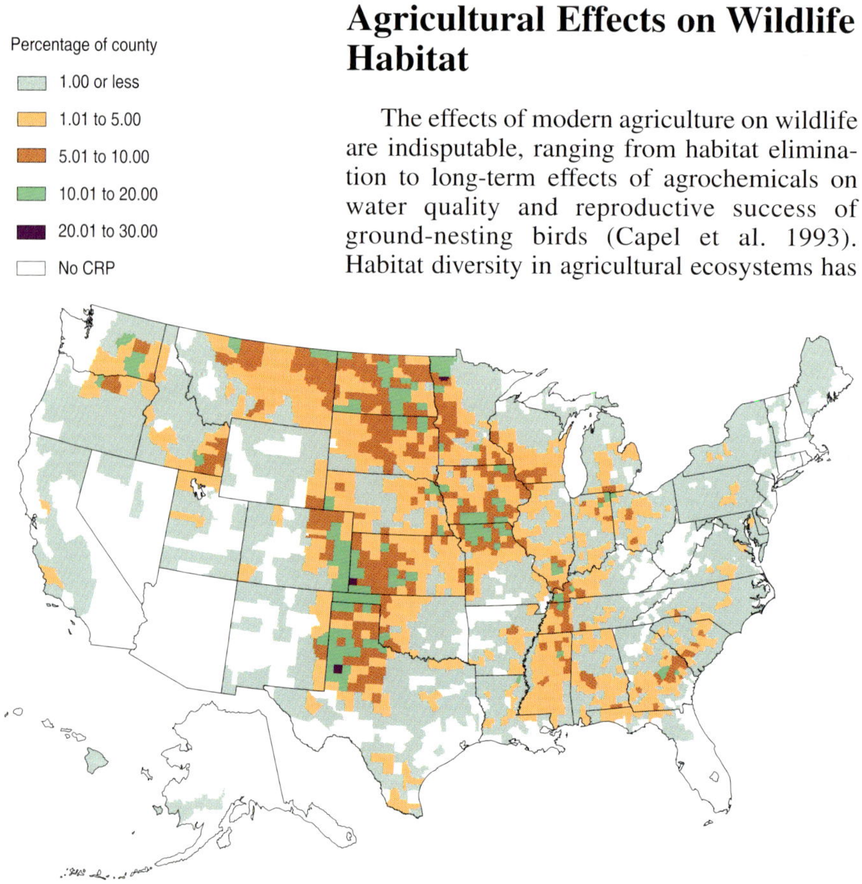

Fig. 1. Percentage of county area enrolled in the Conservation Reservation Program (CRP) through July 1992.

The Conservation Reserve Program (CRP)

About 14.7 million ha (36.4 million acres) were removed from production for a minimum of 10 years during the 12 sign-up periods of the CRP (U.S. Department of Agriculture 1993). The percentage of each county area enrolled in the CRP is illustrated in Fig. 1. Grasses represent the vast majority of cover established on retired acres (Table). The most significant benefit to wildlife from the CRP is the more than 13 million ha (32 million acres) of grass interspersed with lands remaining in production. This grass cover has enhanced the quality and distribution of habitat for nongame and game species in both terrestrial and aquatic ecosystems. To document CRP-derived benefits to habitat quality, a cooperative study between the International Association of Fish and Wildlife Agencies and the U.S. Fish and Wildlife Service was initiated in 1987 (Farmer et al. 1988).

From 1987 to 1993 fish and wildlife agency personnel from 31 states collected vegetation

Table. Conservation Reserve Program (CRP) acres contracted in dominant conservation practices through the 12th sign-up period (U.S. Department of Agriculture 1993).

Conservation practice	Acres	Percent of all CRP acres
CP1 Tame grass	21,385,848	58.68
CP2 Native grass	8,459,343	23.21
CP3 Tree planting	2,321,290	6.37
CP4 Wildlife habitat	1,972,563	5.41
CP10 Existing grass	2,010,940	5.52
CP12 Wildlife food plots	18,449	0.05
CP13 Filter strips	52,931	0.15
All other conservation practices	221,408	0.61
Total	**36,442,772**	**100.00**

data in 501 counties under guidance of the U.S. Fish and Wildlife Service's National Ecology Research Center in Fort Collins, Colorado (now part of the National Biological Service). Study sites were based on environmental conditions and dominant agricultural practices before CRP enrollment. CRP fields in each region were sampled based on the conservation practice established (e.g., tame grasses, native grasses) and the year the CRP contract began.

Vegetation data were collected before spring growth (pre-greenup) and during midsummer (July-August) after the peak of the growing season. Visual obstruction readings (VOR) provide a simultaneous measure of vegetation height and density. Pre-greenup VOR are used to assess the amount of residual vegetation (i.e., dead material remaining from the previous growing season), and is an important indicator of habitat quality for ground-nesting birds, which often establish nests before significant growth of the current year's vegetation. Midsummer VOR provide evidence of the amount and quality of vegetative cover present during the peak growing season.

Data pertaining to vegetation height and density in CRP fields planted to tame and native grasses in the Southern Plains, Northern Plains, and Midwest regions portray some of the wildlife habitat realized under this program (Figs. 2 and 3). Pre-greenup VOR were essentially 0.0 cm in all fields before establishment of CRP cover, indicating an absence of reproductive and protective cover for grassland-dependent species.

Pre-greenup VOR in tame grasses (Fig. 2a) showed increased height and density of residual vegetation in the Southern Plains and Midwest regions for the first 4 years after establishment. In the Northern Plains, the vegetation response was more immediate, reaching a maximum of 31 cm (12.2 in) only 2 years after planting. Within all three regions, however, tame grass VOR eventually showed a declining or stable trend.

In comparison, pre-greenup VOR in native grasses indicated a slower response in the first 2 years after planting (Fig. 2b); however, these fields eventually showed greater height and density of vegetation across all three regions. In the Midwest, VOR in native grasses reached a maximum value of 65 cm (25.5 in) 5 years after planting.

Midsummer VOR in tame grasses showed rapid increases in value for the first 2 years after planting (Fig. 3a). In the Northern Plains, however, VOR decreased from 60 cm (23.5 in) to 36 cm (14.2 in) by 6 years after planting. Relatively constant increases in VOR values for tame grasses were evident in the Southern Plains and Midwest regions. Midsummer VOR measurements in native grasses indicated a slower response in terms of height and vegetation density (Fig. 3b). With the exception of the Southern Plains, long-term VOR values are generally higher for native grasses than tame grasses across all regions (Fig. 3b).

CRP Benefits

The CRP has provided substantial wildlife habitat with millions of acres planted to tame and native grasses. These millions of acres of high-quality habitat have, in turn, provided benefits to populations of both nongame and game wildlife (Allen 1993a, 1993b). Tame grasses appear to provide greater cover in initial years after planting, but height and density of residual and midsummer vegetation begin to decline in a few years. Conversely, native grasses appear to take longer to provide substantial benefits but provide habitat of higher quality for longer periods. Eventually, regardless of species composition, some type of management (e.g., burning, mowing, limited grazing, or haying) will be required to maintain stand vigor and long-term habitat quality in grass-dominated CRP fields.

Although the CRP's highest priorities were to reduce deficiency payments, decrease soil erosion, and provide economic support to the agricultural community, it also provides benefits to wildlife and their habitat. The restoration of more than 14 million ha (36 million acres) of cropland to long-term cover has provided an essential element of habitat stability and has helped repair the widespread deterioration in habitat and environmental quality experienced across the agricultural landscape. In contrast to the U.S. Department of Agriculture's annual set-aside programs, which typically have negative effects on wildlife (Berner 1988), the key beneficial element of the CRP to wildlife is the long-term provision of relatively undisturbed vegetation cover dispersed across agricultural ecosystems. Based on the propensity of American agriculture to overproduce selected commodity crops, the return of most CRP acres to production can be expected to result in

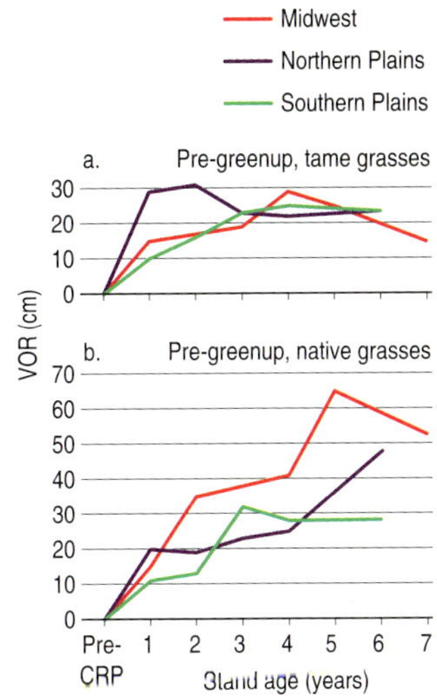

Fig. 2. Relationships between pre-greenup Visual Obstruction Readings (VOR) and stand age in Conservation Reserve Program fields planted with (a) tame grasses and (b) native grasses. VOR data were collected in spring before current season's vegetation growth (1 cm—0.4 in).

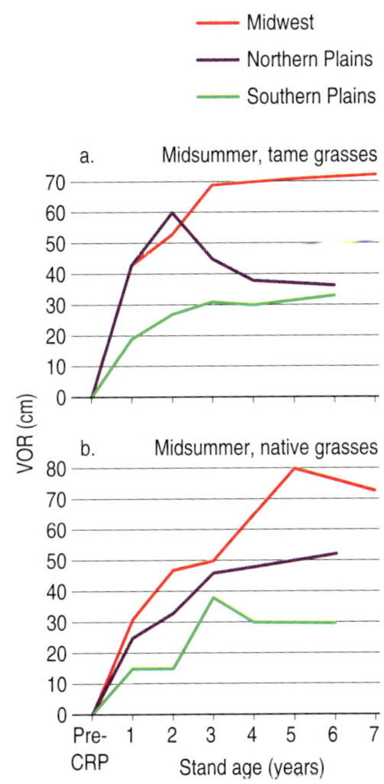

Fig. 3. Relationships between midsummer Visual Obstruction Readings (VOR) and stand age in Conservation Reserve Program fields planted with (a) tame grasses and (b) native grasses. VOR data were collected in midsummer after peak growth (1 cm—0.4 in).

greater dependence on annual set-aside programs that potentially have drastic negative effects on wildlife and environmental quality.

The CRP has provided environmental benefits, particularly in terms of water quality stemming from reduced amounts of soil erosion and reduced applications of agrochemicals. Agricultural production, environmental quality, and viable populations of wildlife in agricultural ecosystems are not mutually exclusive objectives. The Food Security Act of 1985 and the CRP have successfully integrated environmental and agricultural policies, providing public benefits on a national scale. Recent surveys indicate that most lands enrolled in the CRP will return to crop production upon the program's termination in 1995 (Dicks 1994). If this does occur and if remaining lands are subjected to uncontrolled haying and grazing, the many benefits to wildlife and environmental quality realized over 10 years will be lost.

References

Allen, A.W. 1993a. Wildlife habitat criteria in relation to future use of Conservation Reserve Program lands. Pages 41-88 *in* Proceedings of the Great Plains Agricultural Council Annual Meeting. June 2-4. Rapid City, SD.

Allen, A.W. 1993b. Regional and state perspectives on Conservation Reserve Program contributions to wildlife habitat. U.S. Fish and Wildlife Service Federal Aid Rep., National Ecology Research Center, Fort Collins, CO. 28 pp.

Bender, M. 1984. Industrial versus biological traction. Pages 87-105 *in* W. Jackson, W. Berry, and B. Coleman, eds. Meeting the expectations of the land: essays in sustainable agriculture and stewardship. North Point Press, San Francisco.

Berner, A.H. 1988. Federal pheasants—impact of federal agricultural programs on pheasant habitat, 1934-1985. Pages 45-93 *in* D.L. Hallett, W.R. Edwards, and G.V. Burger, eds. Pheasants: symptoms of wildlife problems on agricultural lands. Northcentral Section of the Wildlife Society, Bloomington, IN.

Capel, S., J.A. Crawford, R.J. Robel, L.W. Burger, Jr., and N.W. Southerton. 1993. Agricultural practices and pesticides. Pages 172-173 *in* K.E. Church and T.V. Donley, eds. Quail Symposium. Kansas Dept. Wildlife and Parks.

Dicks, M.R. 1994. Proceedings of the NCT-163 Post Conservation Reserve Program land use conference. USDA/CSRS/Project 92110381. January 10-11, Denver, CO.

Farmer, A.H., F.L. Hays, and R.P. Webb. 1988. Effects of the Conservation Reserve Program on wildlife habitat: a cooperative monitoring study. Transactions of the Fifty-third North American Wildlife and Natural Resources Conference 53:232-238.

Haynes, R.P. 1991. Science, technology, and the farm crisis. Pages 121-129 *in* C.V. Blatz, ed. Ethics and agriculture: an anthology on current issues in world context. University of Idaho Press, Moscow.

Knutson, R.D., J.B. Penn, and W.T. Boehm. 1990. Agricultural and food policy. Prentice Hall, Englewood Cliffs, NJ. 437 pp.

National Research Council. 1989. Alternative agriculture. National Academy Press, Washington, DC. 448 pp.

Reichelderfer, K. 1990. Environmental protection and agricultural support: are trade-offs necessary? Pages 201-230 *in* K. Allen, ed. Agricultural policies in a new decade. Resources for the Future and National Planning Association, Washington, DC.

Ribaudo, M.O. 1989. Water quality benefits from the Conservation Reserve Program. U.S. Department of Agriculture, Resources and Technology Division, Economic Research Service. Agricultural Economic Rep. 606. 30 pp.

Soil and Water Conservation Society. 1994. When conservation and reserve program contracts expire: the policy options Conference Proceedings, Feb. 10-11, 1994, Arlington, VA. Soil and Water Conservation Society, Ankeny, IA. 143 pp.

U.S. Department of Agriculture. 1993. Conservation Reserve Program, 12th sign-up statistics. Agricultural Stabilization and Conservation Service, Washington, DC. 16 pp.

For further information:

Arthur W. Allen
National Biological Service
Midcontinent Ecological Science Center
4512 McMurry Ave.
Fort Collins, CO 80525

Courtesy G.H. Rodda

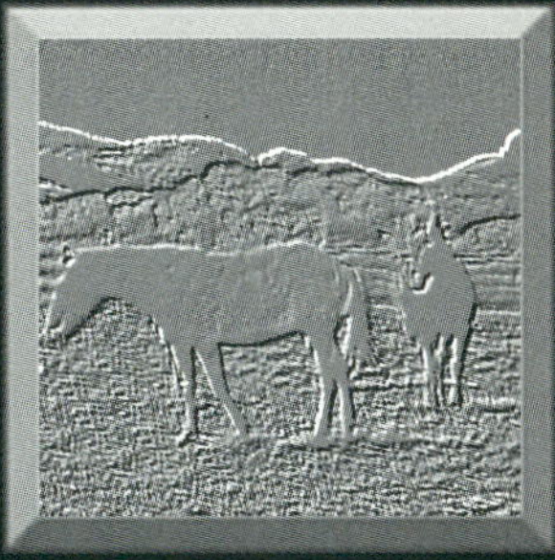

Non-native Species

by
Science Editor
Hiram W. Li
National Biological Service
Oregon Cooperative Fishery Research Unit
Oregon State University
Corvallis, OR 97331

Overview

Introduced species evolved elsewhere and have been transported and purposefully or accidentally disseminated by humans. Many synonyms are used to describe these species: alien, exotic, non-native, and nonindigenous. The spread of non-native species during the last century has been unprecedented in Earth's history, with the speed and scale of these infestations more rapid than natural invasions. The spread of non-native species in human-disturbed habitats reflects a deterioration of the North American landscape.

Introduced species disrupt the functioning of native ecosystems upon which humans depend. Many non-native species become pests by rapidly dispersing into communities in which they have not evolved, and by displacing native species because of evolutionary mismatches. For example, non-native species contributed to 68% of the fish extinctions in the past 100 years, and the decline of 70% of the fish species listed in the Endangered Species Act (Lassuy 1994).

As several articles indicate, the economic cost incurred because of non-native species reaches millions, or even billions, of dollars. Non-native species damage agricultural crops and rangelands, contribute to the decline of commercially important fishes, spread diseases that affect domestic animals and humans, and disrupt vital ecosystem functions.

Some species that have become pests were first introduced to "create" a desired landscape; these non-natives include exotic game animals, fish, and decorative plants. Mack and Thompson (1982), for example, traced the widespread dissemination of 139 weedy, non-native plants in the United States to seed catalogues and the commercial seed trade of the 19th century. Similarly, feral (wild) domestic animals such as mustangs are a major problem on public lands, and sound management of such animals has been impeded by romantic images of America's past.

Accidental introductions through human travel is a theme repeated in several articles, indicating that cargo traffic (ship, air, land) is a major vector of non-native species and should be monitored as world trade increases. The zebra mussel (*Dreissena polymorpha*) is the most notorious hitchhiker, but introductions through ballast water are not isolated to the Laurentian Great Lakes. My colleagues and I recently found that 11 exotic benthic invertebrates have become established in Oregon estuaries. Similarly, dinoflagellates causing red tide toxins have spread into Australian waters through cargo traffic. The importation of raw logs from New Zealand and Siberia endanger Pacific Northwest forests through forest pests hitchhiking in the bark and wood (J. Lattin, Oregon State University, personal communication). It is clear that international cargo traffic must be monitored to reduce the spread of non-native species.

Although this section only briefly mentions disease, it may be one of the most important problems caused by non-native species. After Columbus landed in the New World, for example, 95% of the Native tribes became extinct because their people were susceptible to European microbes (Diamond 1992). Likewise, exotic diseases have devastated populations of aquatic organisms worldwide, killed many native trees, and exterminated much of Hawaii's avifauna. Non-native species are the primary vector for these diseases; for instance, the spread of fish diseases worldwide resulted from the unprecedented transfer of non-native fishes for hatchery production.

It is clear from the small sampling of articles here that changes caused by non-native species are widespread and profound. We present different case histories representative of a myriad of management problems today. New problems continually arise, however, because humans deliberately and accidentally release non-native species and encourage their invasion through massive disturbances of the landscape, thereby mitigating against native species' resistance to invaders by stressing native populations. These articles should make it clear that although non-native species are costly to manage, manage them we must.

References

Diamond, J. 1992. The arrow of disease. Discover 13 (10):64-73.

Lassuy, D.R. 1994. Aquatic nuisance organisms: setting national policy. Fisheries (Bethesda) 19(4):14-17.

Mack, R.N., and J.N. Thompson. 1982. Evolution in step with few large, hooved mammals. American Naturalist 119:757-773.

Non-native Aquatic Species in the United States and Coastal Waters

by

Christina Bjergo
U.S. Coast Guard

Charles Boydstun
National Biological Service

Michael Crosby

Steve Kokkanakis
National Oceanic and Atmospheric Administration

Richard Sayers, Jr.
U.S. Fish and Wildlife Service

Since the European colonization of North America, many non-native aquatic species have been introduced into the United States and adjacent waters. The harm caused by recent introductions, particularly by the zebra mussel (*Dreissena polymorpha*), and concern about a possible increase in the number of unintentional introductions resulted in passage of the Nonindigenous Aquatic Nuisance Prevention and Control Act of 1990. This statute mandates development and implementation of a comprehensive national program to prevent and respond to problems caused by the unintentional introduction of nonindigenous aquatic species into waters of the United States. This article presents an overview of nonindigenous aquatic species, a summary of potential pathways of introduction, and response strategies.

Presence and Distribution

Non-native aquatic species in the United States and coastal waters include species from many plant and animal taxa and span the entire country (Figure). That this problem is extensive is clear by the numbers: 139 nonindigenous species are now established in the Great Lakes (Mills et al. 1993); 32 species of nonindigenous marine organisms were collected from one small Oregon estuary (Carlton 1991); 96 nonindigenous sponges, worms, crustaceans, and other invertebrates are now found in San Francisco Bay (Carlton 1979); and more than half of Hawaii's free-living species are nonindigenous (U.S. Congress 1993). The rate of nonindigenous species' introductions into the Great Lakes has increased in spurts since 1810, largely in response to an expanding human population, development in the basin, and increased transoceanic shipping.

Benefits and Costs

Nonindigenous aquatic species have been both beneficial and problematic. Beneficial aspects include enhancing recreational opportunities such as sport-fishing; providing reliable, high-quality food via aquaculture and mariculture; and aesthetically improving the human environment via the aquarium industry. Recreational fishing contributed an estimated $24 billion in expenditures and $69.4 billion in economic output in 1991 (SFI 1994).

Problems associated with nonindigenous aquatic species are primarily related to ecological issues, such as their effects on indigenous species, and financial issues, such as economic losses caused by biofouling of water-intake pipes. For example, nonindigenous species were cited as a contributing cause in the extinction of 27 species and 13 subspecies of North American fishes over the past 100 years (Miller et al. 1989). Federal, state, and local governments, as well as industry, have often borne significant costs related to nonindigenous aquatic species. From 1906 to 1991, estimated losses associated with 79 aquatic and terrestrial nonindigenous species were roughly $97 billion (Table 1), and worst-case estimates for 15 potential high-impact nonindigenous species project future economic losses of another $134 billion (U.S. Congress 1993).

Table 1. Estimated cumulative losses to the United States from selected categories of harmful nonindigenous species, 1906-91 (U.S. Congress 1993).

Category	Species analyzed (no.)	Cumulative loss est. ($ millions, 1991)	Species not analyzed (no.)
Plants	15	603	-
Terrestrial vertebrates	6	225	> 39
Insects	43	92,658	> 330
Fish	3	467	> 30
Aquatic invertebrates	3	1,207	> 35
Plant pathogens	5	867	> 44
Other	4	917	-
Total	79	96,944	> 478

Introduction and Dispersal

Many non-native aquatic species have entered the country in infested stock for aquaculture or fishery enhancement. For example, the introduction of the Pacific oyster (*Crassostrea gigas*) to the west coast in the 1920's brought with it a Japanese snail (*Ocenebra japonica*) that preys on native oysters, a flatworm (*Pseudostylochus ostreophagus*), and possibly also a copepod parasite (*Mytilicola orientalis*). An Asian tapeworm (*Bothriocephalus opsarichthydis*) was found in several species of native fish in the 1970's following its introduction via infected grass carp (*Ctenopharyngodon idella*). A non-native freshwater snail (*Potamopyrgus antipodarum*) that probably escaped from a fish aquaculture facility now threatens indigenous mollusks of the Snake River region.

The aquarium industry is a significant entry and dispersal pathway for non-native aquatic species. Hydrilla (*Hydrilla verticillata*), an aquatic weed that causes a major navigation hazard, is believed to have been released by aquarium dealers in an attempt to create a domestic source of the plant (Williams 1980). At least three snail species entered U.S. waters when individual snails were discarded by aquarium dealers or their customers over the past few decades. Since 1980, releases from aquaria were the source of at least seven nonindigenous fish species that are now established, and the aquarium fish industry is believed the source of at least 27 nonindigenous fish species now established in the continental United States (Courtenay and Williams 1992; U.S. Congress 1993).

Another major introduction and dispersal pathway for non-native aquatic species is via ballast water discharge. Since many ports are infested with non-native aquatic species, ballasting operations often bring these species, as well as indigenous species, into the ballast tanks of a vessel. These organisms are then transported around the world within the ballast tanks. When a vessel unloads or picks up cargo, the operator often empties the ballast tanks, thus introducing these organisms into new environments. This mode of introduction is probably responsible for the introduction of zebra mussels, ruffe (*Gymnocephalus cernuus*), and the spiny water flea (*Bythotrephes cederstroemi*) into the Great Lakes (U.S. Congress 1993).

Many non-native aquatic species are intentionally imported as pets, for aquaculture, or to supplement recreational fishing. State and federal natural resource agencies have intentionally introduced a variety of non-native aquatic species to enhance recreational and commercial interests (e.g., brown trout [*Salmo trutta*], carp, and Pacific oyster). Some animals (e.g., water fleas, freshwater shrimp, crayfish, and others; Wildlife Nurseries, Inc. 1989) can be purchased through the mail and introduced outside their natural range. Many tropical aquarium species now found in Florida's waters escaped from aquaculture facilities (Courtenay and Williams 1992). The Aquatic Nuisance Species (ANS) Task Force suggests that it is inevitable that cultured species will eventually escape confinement and enter U.S. waterways.

Assessment and Monitoring

Efforts to assess or monitor non-native aquatic species are, at best, fragmented. Generally, these species are not monitored until they reach nuisance status, such as purple loosestrife (*Lythrum salicaria*) or zebra mussels have, and no broad, nationally coordinated program exists for detecting new species. A nationally coordinated effort for providing timely notification to appropriate entities of the detection and dispersal of all non-native aquatic species is needed. There is currently no definitive evidence to suggest that rates of introduction for non-native aquatic species are increasing or decreasing (Table 2).

Table 2. Number of new species of foreign origin established per decade (U.S. Congress 1993).

Category	1940-50	1950-60	1960-70	1970-80	1980-90
Terrestrial vertebrates	3	11	13	3	No data
Fish	2	15	18	5	12
Mollusks	5	5	6	10	4
Plant pathogens	3	5	4	16	7

Research Strategies

Three main research strategies are used to limit the damages caused by nonindigenous aquatic species: prevention, control, and detection and monitoring. Prevention relies on the identification and elimination of pathways through which nonindigenous ANS enter the nation's waters. Although prevention should be the first line of defense, it is unlikely to be

Purple loosestrife (*Lythrum salicaria*)*

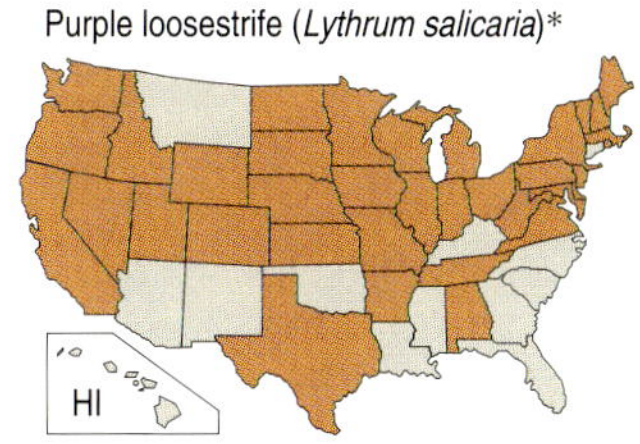

Zebra mussel (*Dreissena polymorpha*)*

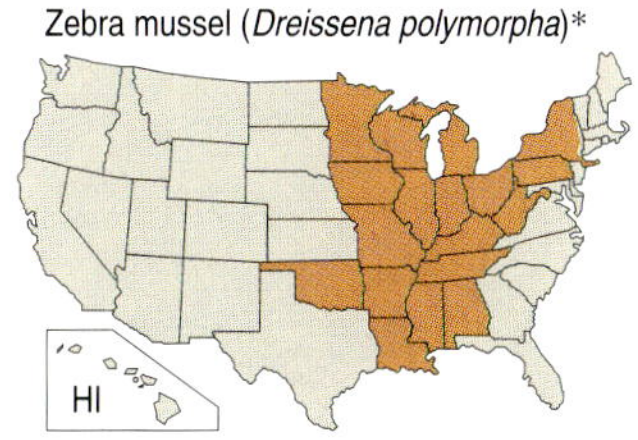

Asian clam (*Corbicula fluminea*)**

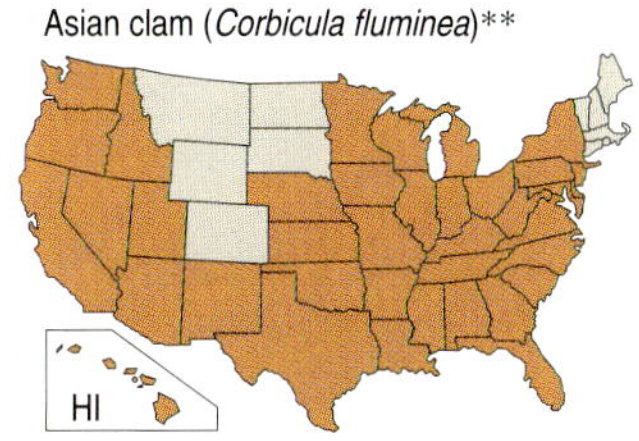

Common carp (*Cyprinus carpio*)**

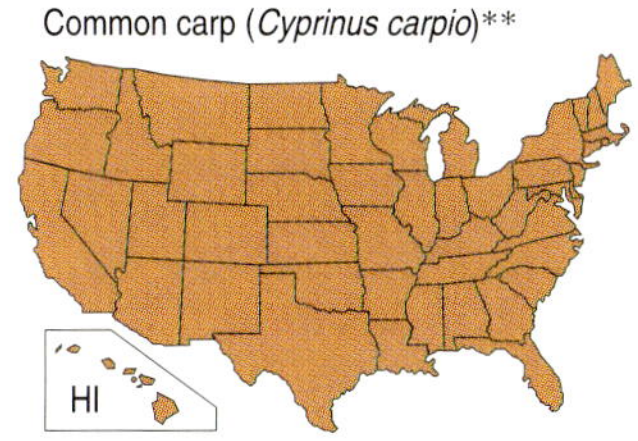

Figure. Distribution of purple loosestrife, Asian clam, zebra mussel, and carp in the United States (shading indicates species presence).

*U.S. Congress 1993.

**U.S. Department of the Interior, National Biological Service 1994. Non-native aquatic species data base.

100% effective and can never eliminate all threats from nonindigenous aquatic species. Therefore, rapid response and control techniques must be identified and in place to control and limit damages caused by nonindigenous ANS. This approach is being used to control ruffe.

Control is intended to reduce the effects of nonindigenous aquatic species through eradication, reduction in numbers to tolerable levels, and exclusion from sensitive areas. Three general control methods exist to prevent the spread of these species: chemical, biological, and physical. Proper evaluation and use of selective chemicals may provide effective control of non-native aquatic species with an apparent minimum of ecological hazard or other side effects. Increasing concern exists, however, about the long-term environmental safety and impacts of chemicals used to control nonindigenous aquatic species. Efforts to control sea lamprey (*Petromyzon marinus*) in the Great Lakes are a prime example of chemical control. This control has been highly successful in reducing the population size of an invading species, but carries an enormous price tag: more than $10 million annually (U.S. Congress 1993).

Carefully planned biological-control programs may provide rapid, cost-effective control and pose negligible ecological problems. The success rate for biological-control programs typically ranges from 16% to 36% (Meyers et al. 1989) and improperly screened biological-control agents have themselves become nuisance species in the past (e.g., blue tilapia [*Tilapia aurea*]; McClelland 1992).

Although often very expensive, physical control of aquatic nuisance species can be an appropriate technique in certain circumstances. Physical control has been used to control nuisance aquatic weeds like Eurasian watermilfoil (*Myriophyllum spicatum*).

Since no single method is likely to provide the necessary level of control, a comprehensive, integrated control strategy combining techniques is usually necessary for an effective control program. Few, if any, control methods are without some environmental risk. When properly used, and with continual monitoring for effectiveness and ecological side effects, environmentally sound control of at least some aquatic nuisance species can be achieved, as in the Great Lakes sea lamprey control program.

Detection and monitoring strategies serve as early warning systems that first identify new invasions and then track ranges and populations. This strategy complements or integrates prevention and control to allow for early intervention and assessment of management actions. The capability for early detection of new invasions will allow managers to implement strategies for limiting their spread and reducing negative effects. Timely detection of non-native aquatic species that are or could become nuisances can also help identify gaps in prevention procedures. Monitoring of those organisms will not only allow rapid response if harmful situations arise but will also allow verification or repudiation of assumptions that may have been made during assessments before intentional releases.

Because of extremely limited resources, cooperative ventures and collaborations between agencies are essential for collecting monitoring information. The Detection and Monitoring Committee of the ANS program is developing a national network to coordinate and provide information regarding occurrences of known nonindigenous aquatic species. This network is intended to provide managers and researchers with an important tool for determining the status of a particular nonindigenous aquatic species, its potential and known effects, and proven or potential control techniques.

By and large, three interrelated problems associated with nonindigenous ANS remain unsolved: (1) determining levels of acceptable risk; (2) setting thresholds or other variables above which more formal decision making and costly approaches for control are invoked; and (3) identifying trade-offs in terms of costs and economic ramifications in the face of uncertainty as to probable success in controlling ANS. Current federal methods and programs to identify risks of potentially harmful nonindigenous aquatic species have many shortfalls—including long response times.

Summary

Nonindigenous aquatic species are widespread in the United States. While many of these organisms have been intentionally introduced, many others dispersed via unintended introductions. The potential for ecological and economic harm resulting from introductions of nonindigenous aquatic species can be large. For example, zebra mussels seem to be jeopardizing a number of native North American mussel species (Williams et al. 1993) and could result in economic losses in excess of $3 billion (U.S. Congress 1993). The actual extent of problems associated with non-native aquatic species remains largely unknown. The ability to detect new species and limit their dispersal before they become problematic is critical if we are to limit future nonindigenous species problems.

References

Carlton, J.T. 1979. Introduced invertebrates of San Francisco Bay. Pages 427-444 *in* T.J. Conomos, ed. San Francisco Bay: the urban estuary. American Association of the Advancement of Science, Pacific Division, San Francisco, CA.

Carlton, J.T. 1991. Man's role in changing the face of the ocean: biological invasions and implications for conservation of near-shore environments. Conservation Biology 3(3):265-273.

Courtenay, W.R., and J.D. Williams. 1992. Dispersal of exotic species from aquaculture sources, with emphasis on freshwater fishes. Pages 49-81 *in* A. Rosenfield and R. Mann, eds. Dispersal of living organisms into aquatic ecosystems. Maryland Sea Grant, College Park.

McClelland, M. 1992. Invasion of the bio-snatchers. Florida Environments 6(2):5,13.

Meyers, J.H., C. Higgins, and E. Kovacs. 1989. How many insect species are necessary for the biological control of insects? Environmental Entomology 18:541-547.

Miller, R.R., J.D. Williams, and J.E. Williams. 1989. Extinctions of North American fishes during the last century. Fisheries 14(6):22-38.

Mills, E.L., J.H. Leach, J.T. Carlton, and C.L. Secor. 1993. Exotic species in the Great Lakes: a history of biotic crises and anthropogenic introductions. Journal of Great Lakes Research 19(1):1-54.

SFI. 1994. Sport Fishing Institute releases report on economics of sport fishing. Sport Fishing Institute Bull. 452:6, March/April 1994.

U.S. Congress, Office of Technology Assessment. 1993. Harmful non-indigenous species in the United States. OTA-F-565. U.S. Government Printing Office, Washington, DC. 391 pp.

Wildlife Nurseries, Inc. 1989. What brings them in. Catalogue of animals and plants for wildlife habitat. Oshkosh, WI.

Williams, J.D., M.L. Warren, Jr., K.S. Cummings, J.L. Harris, and R.J. Neves. 1993. Conservation status of freshwater mussels of the United States and Canada. Fisheries 18(9):6-22.

Williams, M.C. 1980. Purposefully introduced plants that have become noxious or poisonous weeds. Weed Science 28(3):300-305.

For further information:

Richard Sayers, Jr.
U.S. Fish and Wildlife Service
1849 C Street, NW, Mailstop
ARLSQ-820
Washington, DC 20240

Nonindigenous Fish

by
Charles Boydstun
Pam Fuller
James D. Williams
National Biological Service

Within the United States alone, humans have intentionally or unintentionally introduced more than 4,500 species of terrestrial and aquatic species to areas outside their historical range (U.S. Congress 1993). Although many terrestrial introductions are viewed as beneficial to humans because of economic and social considerations, all but a few intentional aquatic introductions have proven to be mixed blessings (Courtenay and Williams 1992; Steirer 1992; U.S. Congress 1993). No unintentional aquatic introductions have been considered beneficial (Steirer 1992); instead, their environmental consequences are generally harmful and sometimes catastrophic (Taylor et al. 1984; U.S. Congress 1993).

Both intentional and unintentional introductions have enabled nonindigenous fish to become temporary, and often permanent, residents in nearly every U.S. aquatic system. Complete eradication or exclusion is neither economically plausible nor socially justified (U.S. Congress 1993); therefore, nonindigenous fish are and will continue to be components of these aquatic systems. Because nonindigenous fish have the potential to alter significantly the U.S. aquatic ecosystems during the next century and beyond, their interactions within the aquatic community must be monitored and analyzed to ensure that effective management actions are taken before a crisis arises.

To help document the consequences of nonindigenous fish introductions, the National Biological Service monitors the status and distribution of these organisms in U.S. waters (Williams and Jennings 1991). Since 1978, reports and specimens of various nonindigenous fish have been collected, verified, and entered in a geographic information system, which is a computerized mapping and data base system.

Obtaining qualitative and quantitative information on nonindigenous fish for a national assessment requires cooperation by many agencies, organizations, and individuals (Boydstun and Benson 1992). We collect much of our ecological and geographical data using a voluntary reporting form. Historical accounts are gathered through review of both scientific and other literature, including natural resource agency publications that often provide accounts of nonindigenous fish, stockings, and discoveries. For our purposes, we established a historic cut-off date for usable nonindigenous fish reports at 1800.

We limited this analysis to only reports of nonindigenous fish from open waters identifiable to species level and recognizable nonindigenous hybrids.

Status of Nonindigenous Fish

We have collected more than 11,000 reports that document 404 unique fish species or hybrids introduced outside their native ranges within U.S. waters. This diverse group of 67 families of fish includes species from every continent except Antarctica. Of the 404 species, 252 (62%) are native to the United States but found outside their native ranges, and 152 (38%) are from other countries. Nonindigenous hybrid fish represent roughly 5% (19) of the total 404 nonindigenous fish species.

Our total is considerably higher than the 127 nonindigenous fish (70 U.S. and 57 non-U.S.) reported in the United States in 1992 by the Office of Technology Assessment (U.S. Congress 1993). Courtenay and Williams (1992) reported 99 exotic (non-U.S.) nonindigenous fish species in the contiguous U.S. waters in 1992, of which 46 were established as sustaining populations. The disparity between our results and these estimates is most influenced by our intent to include all reported nonindigenous fish that have been found within the United States since 1800, regardless of their current status.

Game and associated forage fish are the most widely distributed nonindigenous fish. These include the salmonids (salmon and trout), ictalurids (catfish), centrarchids (bass and sunfish), percids (walleye and sauger), and cyprinids (minnows). The two most widely distributed nonindigenous fish species are goldfish (*Carassius auratus*) and common carp (*Cyprinus carpio*). Both have been reported or collected from all states except Alaska (Table). Goldfish introductions are the result of the release of bait and aquarium fish and forage fish stocking for game fish. Widespread distribution of common carp is primarily due to the stocking program of the U.S. Fish Commission in the late 1800's and early 1900's and later use of juvenile carp as bait.

Reported Occurrences

All 50 states have reported nonindigenous fish from their open waters (Fig. 1). When considering total diversity of nonindigenous fish species, the top five states are California (114), Texas (96), Florida (96), North Carolina (83), and Nevada (82). In fact, of the total 404 species, 312 (77%) are reported as occurring or having been found within the 11 states crossing or below the 35th parallel (e.g., Hawaii, California, Arizona, New Mexico, Texas, Oklahoma, Arkansas, Louisiana, Alabama, Georgia, and Florida). Although Hawaii was historically without any native freshwater fish, it now has 52 nonindigenous freshwater fish species.

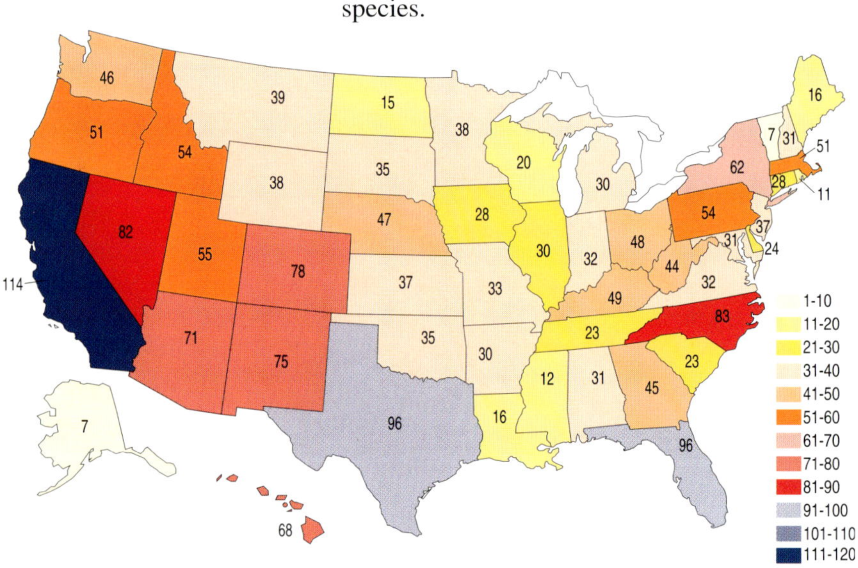

Fig 1. The number of nonindigenous fish species reported by state, 1800-1994. Some species may not be established or have been eradicated.

Table. Nonindigenous fish introduced into 10 or more states, 1800-1994.

Common name (scientific name)	No. of states reported outside native range
Goldfish (*Carassius auratus*)	49
Common carp (*Cyprinus carpio*)	49
Brown trout (*Salmo trutta*)	47
Rainbow trout (*Oncorhynchus mykiss*)	47
Grass carp (*Ctenopharyngodon idella*)	44
Largemouth bass (*Micropterus salmoides*)	41
Walleye (*Stizostedion vitreum*)	40
Smallmouth bass (*Micropterus dolomieu*)	38
Brook trout (*Salvelinus fontinalis*)	36
White crappie (*Pomoxis annularis*)	36
Bluegill (*Lepomis macrochirus*)	33
Northern pike (*Esox lucius*)	33
Striped bass (*Morone saxatilis*)	32
Green sunfish (*Lepomis cyanellus*)	31
Black crappie (*Pomoxis nigromaculatus*)	31
Yellow perch (*Perca flavescens*)	29
Channel catfish (*Ictalurus punctatus*)	29
Coho salmon (*Oncorhynchus kisutch*)	28
Rock bass (*Ambloplites rupestris*)	26
Lake trout (*Salvelinus namaycush*)	26
Threadfin shad (*Dorosoma petenense*)	26
Western mosquitofish (*Gambusia affinis*)	25
Fathead minnow (*Pimephales promelas*)	24
Rainbow smelt (*Osmerus mordax*)	23
Chinook salmon (*Oncorhynchus tshawytscha*)	22
White bass (*Morone chrysops*)	22
Atlantic salmon (*Salmo salar*)	22
Golden shiner (*Notemigonus crysoleucas*)	21
Redear sunfish (*Lepomis microlophus*)	20
Muskellunge (*Esox masquinongy*)	20
Sockeye salmon (*Oncorhynchus nerka*)	19
Pumpkinseed (*Lepomis gibbosus*)	19
Blue catfish (*Ictalurus furcatus*)	19
Alewife (*Alosa pseudoharengus*)	18
Tench (*Tinca tinca*)	18
Rudd (*Scardinius erythrophthalmus*)	18
American shad (*Alosa sapidissima*)	17
Brown bullhead (*Ameiurus nebulosus*)	15
Chain pickerel (*Esox niger*)	15
Flathead catfish (*Pylodictis olivaris*)	15
Black bullhead (*Ameiurus melas*)	15
Spotted bass (*Micropterus punctulatus*)	15
Warmouth (*Lepomis gulosus*)	15
Lake whitefish (*Coregonus clupeaformis*)	14
Cutthroat trout (*Oncorhynchus clarki*)	14
White catfish (*Ameiurus catus*)	14
Bighead carp (*Aristichthys nobilis*)	13
Arctic grayling (*Thymallus arcticus*)	13
Mozambique tilapia (*Tilapia mossambica*)	13
Redbreast sunfish (*Lepomis auritus*)	13
Guppy (*Poecilia reticulata*)	12
Piranha (*Serrasalmus* spp.)	12
Blue tilapia (*Tilapia aurea*)	12
Tiger muskellunge (*Esox lucius* x *masquinongy*)	12
Golden trout (*Oncorhynchus aguabonita*)	11
White perch (*Morone americana*)	10
Green swordtail (*Xiphophorus helleri*)	10
Sauger (*Stizostedion canadense*)	10
Redbelly tilapia (*Tilapia zilli*)	10

Trends

The first fish translocation effort began in the early 1870's with an attempt to introduce several eastern species to the west coast and to stock chinook salmon in the East. Fish that were introduced to the West included eels, brook and lake trout, lake whitefish, northern pike, striped bass, American shad, yellow perch, catfish, bullheads, sunfish, black bass, and crappies. Most of these introductions resulted in established populations that still persist today. At this same time brown trout, tench, and carp were being stocked throughout the country. A resurgence of stocking occurred around 1950 when many state agencies began stocking game fish. The popularity of home aquaria and the availability of foreign fish have also contributed to an increase in the number of species introduced in the past 40 years (Courtenay and Williams 1992; Fig. 2).

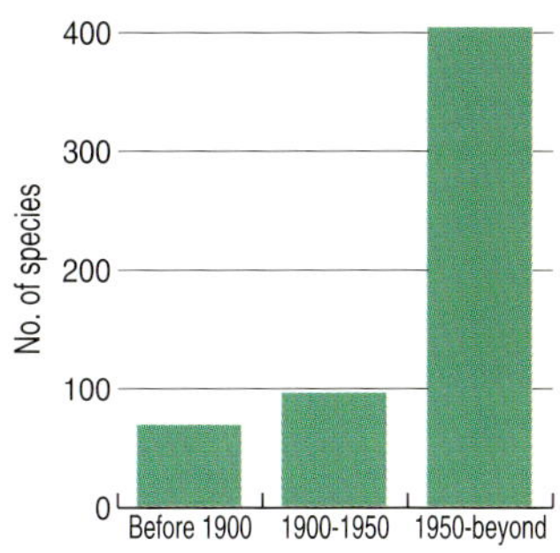

Fig. 2. Diversity of fish introductions over time.

The Future

The presence of nonindigenous fish will continue to alter U.S. aquatic resources. These species compete with or prey on native game and nongame fish, often with severe negative effects on aquatic ecosystems. Nonindigenous fish that survive the initial introduction and subsequently become established are often tolerant of adverse or altered environmental conditions, including habitat disturbance. This tolerance has been used to justify nonindigenous fish introductions rather than to restore disrupted environments. The environmental tolerance of nonindigenous fish combined with increasing habitat disruption in streams and lakes assures their continued dispersal into formerly unoccupied areas. If the introduction and establishment of nonindigenous fish continue at their present rates, distribution and survival of native aquatic organisms could be drastically affected. These introductions can also profoundly change biological diversity and composition of habitats and ecosystems, which could result in substantially increased rates of extinction of native aquatic species.

References

Boydstun, C.P., and A. Benson. 1992. Nonindigenous report (1992:1): zebra mussel (*Dreissena polymorpha*) sites in the United States and Canada. National Fisheries Research Center, Gainesville, FL. 10 pp.

Courtenay, W.R., and J.D. Williams. 1992. Dispersal of exotic species from aquaculture sources, with emphasis on freshwater fish. Pages 49-81 *in* A. Rosenfield and R. Mann, eds. Dispersal of living organisms into aquatic ecosystems. Maryland Sea Grant, College Park.

Steirer, F.S., Jr. 1992. Historical perspective on exotic species. Pages 1-4 *in* M.R. DeVoe, ed. Introductions and transfers of marine species. South Carolina Sea Grant Consortium, Hilton Head Island.

Taylor, J.N., W.R. Courtenay, Jr., and J.A. McCann. 1984. Known impacts of exotic fish in the continental United States. Pages 322-373 *in* W.R. Courtenay, Jr., and J.R. Stauffer, Jr., eds. Distribution, biology and management of exotic fish. The Johns Hopkins University Press, Baltimore, MD.

U.S. Congress, Office of Technology Assessment. 1993. Harmful non-indigenous species in the United States. U.S. Government Printing Office OTA-F-565, Washington, DC. 391 pp.

Williams, J.D., and D.P. Jennings. 1991. Computerized data base for exotic fish: the western United States. California Fish and Game 77(2):86-93.

For further information:

Charles Boydstun
National Biological Service
Southeastern Biological Science Center
7920 NW 71st St.
Gainesville, FL 32653

Non-native Reptiles and Amphibians

by

Michael James McCoid
Caesar Kleberg Wildlife Research Institute

Interest in established, non-native species of reptiles and amphibians in the United States (including territories and possessions) has been increasing the past quarter-century. Concerns regarding the interactions of introduced and native species have driven this interest (Wilson and Porras 1983). Most successful introductions have taken place in the southern tier of states (California to Florida) and on islands. This success rate is probably due, in part, to favorable environmental conditions. Movements by indigenous peoples to islands also may have substantially augmented existing faunas. For example, in American Samoa, virtually the entire terrestrial reptile fauna may have been introduced by the original human colonizers (T.D. Schwaner, Alabama School of Science and Math, personal communication). Since many species of reptiles and amphibians on islands could be considered as introduced, the scope of this report, for islands, is restricted to those introductions that occurred after contact with western societies and for the mainland United States, within the past century. A review of both successful and unsuccessful reptile and amphibian introductions in North America is presented by Smith and Kohler (1977).

Of the documented 53 established non-native amphibian and reptile species (Table), at least 5—spectacled caiman (*Caiman crocodilus*), marine toad (*Bufo marinus*), African clawed frog (*Xenopus laevis*), bullfrog (*Rana catesbeiana*), and brown tree snake (*Boiga irregularis*)—have been established at least 30 years and have been sufficiently monitored to enable preliminary assessment of impacts on the native biota. The marine toad is established in Florida, Hawaii, the Territories of Guam, U.S. Virgin Islands, and American Samoa, and the Commonwealths of Puerto Rico and of the Northern Mariana Islands, where it is regarded as a nuisance species. The spectacled caiman is established in Puerto Rico and Florida, where it may be negatively affecting vertebrates. The African clawed frog is established in Arizona and California, but is not demonstrating any apparent negative effects on native vertebrates. The bullfrog is widely established in western North America, Hawaii, and Puerto Rico, and is implicated in restricting the range of native North American ranid frogs and the Mexican garter snake (*Thamnophis eques*). The brown tree snake is established on Guam and is identified as the agent in the extirpation of native forest-dwelling birds and small reptiles.

Courtesy G. Rodda, NBS

Marine toad (*Bufo marinus*).

Table. Established exotic species of amphibians and reptiles in the United States (including territories and possessions).

Scientific name (common name)	Area (reference) [a]
Frogs and toads	
Bufo marinus (marine toad)[b]	FL (Ashton and Ashton 1988; Wilson and Porras 1983), MP (Rodda et al.1991), GU McCoid 1993), HI (McKeown 1978), AS (Amerson et al. 1982), PR[c], VI[c]
Dendrobates auratus (poison-dart frog)	HI (McKeown 1978)
Eleutherodactylus coqui (common coqui)	FL, LA, VI (Conant and Collins 1991)
E. planirostris (greenhouse frog)	FL, LA (Conant and Collins 1991)
Litoria fallax (eastern dwarf treefrog)	GU, MP? (McCoid 1993)
Osteopilus septentrionalis (Cuban treefrog)	FL (Ashton and Ashton 1988; Conant and Collins 1991)
Rana catesbeiana (bullfrog)[d]	HI, PR, western U.S. except ND and MN (Conant and Collins 1991; McKeown 1978)
R. pipiens (northern leopard frog)[d]	CA (Stebbins 1985)
R. rugosa (wrinkled frog)	HI (McKeown 1978)
Xenopus laevis (African clawed frog)	CA, AZ, NC?[e],VA?[f] (McCoid and Fritts 1980b)
Salamanders	
Ambystoma tigrinum (tiger salamander)[d]	CA, AZ (Stebbins 1985)
Lizards	
Ameiva ameiva (South American ground lizard or giant ameiva)	FL (Ashton and Ashton 1988; Conant and Collins 1991)
Anolis carolinensis (green anole)[d]	CA (Bury and Luckenbach 1976), GU (McCoid 1993), HI (McKeown 1978), MP (Rodda et al. 1991)
A. chlorocyanus (Hispaniolan green anole)	FL[g]
A. cristellatus (Puerto Rican crested anole)	FL (Ashton and Ashton 1988; Conant and Collins 1991)
A. cybotes (large-headed anole)	FL (Ashton and Ashton 1988; Conant and Collins 1991)
A. distichus (bark anole)	FL (Ashton and Ashton 1988; Conant and Collins 1991)
A. equestris (knight anole)	FL (Ashton and Ashton 1988; Conant and Collins 1991)
A. garmani (Jamaican giant anole)	FL (Ashton and Ashton 1988; Conant and Collins 1991)
A. sagrei (brown anole)	FL, TX (Ashton and Ashton 1988; Conant and Collins 1991), LA (Thomas et al. 1990)
Basiliscus vittatus (brown basilisk)	FL (Ashton and Ashton 1988; Conant and Collins 1991)
Carlia fusca (brown four-fingered skink)	GU (McCoid 1993), MP (Rodda et al. 1991)
Chamaeleo jacksonii (Jackson's chameleon)	HI (McKeown 1978)
Cnemidophorus lemniscatus (South American whiptail)	FL (Ashton and Ashton 1988; Conant and Collins 1991)
Cyrtopodion scabrum (roughtail gecko)	TX (Conant and Collins 1991)
Ctenosaura pectinata (Mexican spiny-tailed iguana)	FL, TX (Conant and Collins 1991)
Gekko gekko (tokay gecko)	FL (Ashton and Ashton 1988; Conant and Collins 1991)
Gonatodes albogularis (yellow-headed gecko)	FL (Ashton and Ashton 1988; Conant and Collins 1991)
Hemidactylus frenatus (house gecko)	GU (McCoid 1993), MP (Rodda et al. 1991), HI (McKeown 1978), AS (Amerson et al. 1982), FL (Meshaka et al. 1994).
H. garnotii (fox gecko)	FL (Ashton and Ashton 1988; Conant and Collins 1991)
H. mabouia (cosmopolitan house gecko)	FL (Butterfield et al. 1993; Lawson et al. 1991)
H. turcicus (Mediterranean house gecko)	AZ (Stebbins 1985), NM (Painter et al. 1992), AR (Paulissen and Buchanan 1990), NV (Saethre and Medica 1993), FL (Ashton and Ashton 1988), TX, OK, LA, AL, MS, GA, PR (Conant and Collins 1991)
Iguana iguana (green iguana)	FL (Conant and Collins 1991), HI (McKeown 1978), PR[c]
Lamprolepis smaragdina (green tree skink)	MP (Rodda et al. 1991)
Leiocephalus carinatus (carinate curly-tailed lizard)	FL (Ashton and Ashton 1988; Conant and Collins 1991)
L. schreibersii (Schreiber's curly-tailed lizard)	FL (Conant and Collins 1991)
Lampropholis delicata (rainbow skink)	HI (Baker 1979)
Phelsuma sp. (day gecko)	HI[h]
Phrynosoma cornutum (Texas horned lizard)[i]	LA, FL, GA (Ashton and Ashton 1988; Conant and Collins 1991)
Podarcis muralis (common wall lizard)	OH (Conant and Collins 1991)
P. sicula (Italian wall lizard)	NY, KS (Conant and Collins 1991)
Sphaerodactylus argus (ocellated dwarf gecko)	FL (Ashton and Ashton 1988; Conant and Collins 1991)
S. elegans (ashy gecko)	FL (Ashton and Ashton 1988; Conant and Collins 1991)
S. notatus (reef gecko)[j]	FL (Conant and Collins 1991)
Snakes	
Boiga irregularis (brown tree snake)	GU (Fritts 1988), MP?[k]
Rhamphotyphlops braminus (Brahminy blind snake)	FL (Ashton and Ashton 1988; Conant and Collins 1991)
Turtles	
Apalone spinifera (spiny softshell turtle)[d]	CA (Stebbins 1985), AZ[l]
Chelydra serpentina (snapping turtle)[d]	CA, NV?, UT? (Bury and Luckenbach 1976; Stebbins 1985), AZ[l]
Palea steindachneri (wattle-necked softshell turtle)	HI (Ernst and Barbour 1989)
Pelodiscus sinensis (Chinese softshell turtle)	GU (McCoid 1993), HI (McKeown 1978)
Terrapene carolina (common box turtle)[d]	GU? (McCoid 1993)
Trachemys scripta (common slider)[d]	GU, MP? (McCoid 1993), HI (McKeown 1978), CA, AZ (Stebbins 1985)
Alligators and crocodiles	
Caiman crocodilus (spectacled caiman)	FL, PR (Ashton and Ashton 1988; Conant and Collins 1991)

[a] Area abbreviations conform to the United States Postal Service; uncommon usages are Commonwealth of the Northern Mariana Islands — MP, Guam —GU, American Samoa — AS, U.S. Virgin Islands — VI, Puerto Rico — PR.
[b] Native range in United States is extreme southern Texas.
[c] R. Henderson, Milwaukee Public Museum, personal communication.
[d] Native range is eastern North America.
[e] M. McCoid, Texas A & M University-Kingsville, unpublished data.
[f] R. Tinsley, University of Bristol, personal communication.
[g] J. Collins, University of Kansas, personal communication.
[h] L. Nakahara, Hawaii Department of Agriculture, personal communication.
[i] Native range is south-central United States.
[j] Native range is Florida Keys.
[k] T. Fritts, National Biological Service, personal communication.
[l] C. Schwalbe, University of Arizona, personal communication.

Case Studies

Spectacled Caiman

The spectacled caiman has been established in southern Florida for about 30 years (Ellis 1980). There are few published accounts of this species in Florida, but one (Ellis 1980) indicated that these animals eat fish, amphibians, and mammals. This information, coupled with the species' ability to tolerate crowding in bodies of water and relatively rapid maturation, suggests that impacts on native alligators (*Alligator mississippiensis*) might be expected (C.M. Sekerak, University of South Florida, personal communication). Studies in the species' native range (J. Dixon, Texas A&M University, personal communication), however, suggest that the spectacled caiman does not co-occur with larger species of crocodilians, perhaps because of their predation on the smaller caimans. Since the American alligator reaches a larger size than the spectacled caiman, it is possible that the American alligator will deter the caiman from substantially expanding its range.

Marine Toad

The marine toad, native to the tropical New World, is widely introduced and now has a virtually circumtropical range (Zug and Zug 1979). Populations were originally established for insect control, but the species itself became a pest. Information from Australia (Tyler 1989) indicates that ingestion of marine toads, because they have highly toxic skin glands, results in deaths of native reptiles, birds, and mammals. Observations on Guam, where the marine toad has been established since 1937 (McCoid 1993), indicate that poisonings of pet dogs and cats by biting or mouthing marine toads are relatively common (R. Dorner, Marianas Veterinary Clinic, personal communication). On Guam, the island-wide decline of a large varanid lizard is attributed to its predation on the introduced toad (McCoid et al. 1994). In Florida, where the marine toad has been established since 1955, poisonings of pets (Ashton and Ashton 1988) and declines of native amphibians in areas of co-occurrence with the marine toad are reported (J. Rossi, Jacksonville University, personal communication). In a laboratory situation, a native toad (*Bufo americanus*) was behaviorally dominated and excluded from feeding by marine toads (Boice and Boice 1970). There is a literature survey on the marine toad that includes information on extralimital populations (Lawson 1987).

African Clawed Frog

Despite initial fears of the effect of the African clawed frog on aquatic California vertebrates (St. Amant 1975), a subsequent study (McCoid and Fritts 1980a) indicated that these fears may be unwarranted because the only vertebrates found in stomach analyses were immature African clawed frogs and an introduced fish species. Other studies (McCoid and Fritts 1980b, 1993) characterize populations as living primarily in temporary and artificial bodies of water, where most native aquatic vertebrates are expected to be absent. Recently, populations in southern California may have declined because of drought (McCoid et al. 1993). Although African clawed frogs have been established in California since the mid-1960's (McCoid and Fritts 1980b), impacts on native invertebrates, their primary food source, are unassessed.

Bullfrog

Although precise dates of introductions of the bullfrog into many areas of western North America are not well known (Bury and Whelan 1984), the earliest introduction occurred in 1896 (Hayes and Jennings 1986). Impacts on native ranid frogs, however, are well documented and may account for range restrictions of native ranids (Moyle 1973; Hayes and Jennings 1986; Stuart and Painter 1993). Recent information indicates that the Mexican garter snake is also declining because of predation by bullfrogs (*see* Rosen and Schwalbe, this section).

Brown Tree Snake

Since the introduction of the brown tree snake on Guam about 40 years ago, the snake has reached enormous densities (Rodda et al. 1992) and is implicated in the demise of the entire native forest-dwelling bird community (Savidge 1987) and some of the larger lizard species (Rodda and Fritts 1992). Additional impacts include disruption of electrical power (Fritts et al. 1987), predation on domesticated animals (Fritts and McCoid 1991), and human health risks (Fritts et al. 1990, 1994). There are several overviews of the brown tree snake problem on Guam (Fritts 1988; McCoid 1991; also *see* Fritts and Rodda, this section).

Discussion and Summary

Exotic species of reptiles and amphibians are established in the following areas of the United States (Table): Florida (30 species), Hawaii (12), Guam (9), Commonwealth of the Northern Mariana Islands (8), California (6), Louisiana (5), Puerto Rico (5), Texas (4), and Arizona (3). All other areas combined have 9 species. Many of these introductions are due to released or escaped pets.

The ability to assess impacts of exotics on native species may be related, in part, to the length of time that the exotic has been established. For example, deleterious impacts by the brown tree snake on Guam were not noticed by biologists until about 25 years after initial colonization (Savidge 1987). Thus, short-term studies of many non-native reptiles and amphibians may not reveal impacts on native biota. Of the five long-term infestations discussed earlier, only the African clawed frog seems to have not affected the native vertebrate biota. The four detrimental case studies suggest, however, the trend that introduced reptiles and amphibians, like many other introductions, negatively affects established biota. Importantly though, populations of most introduced species of reptiles and amphibians remain unstudied and long-term effects are largely unassessed.

References

Amerson, A.B., Jr., W.A. Whistler, and T.D. Schwaner. 1982. Wildlife and wildlife habitat of American Samoa. 1. Environment and ecology. U.S. Fish and Wildlife Service, Washington, DC. 119 pp.

Ashton, R.E., Jr., and P.S. Ashton. 1988. Handbook of reptiles and amphibians of Florida. Parts 1-3. Windward Publishing, Miami, FL. 558 pp.

Baker, J.K. 1979. The rainbow skink, *Lampropholis delicata*, in Hawaii. Pacific Science 33:207-212.

Boice, R., and C. Boice. 1970. Interspecific competition in captive *Bufo marinus* and *Bufo americanus* toads. Journal of Biological Psychology 12:32-36.

Bury, R.B., and R.A. Luckenbach. 1976. Introduced amphibians and reptiles in California. Biological Conservation 10:1-14.

Bury, R.B., and J.A. Whelan. 1984. Ecology and management of the bullfrog. U.S. Fish and Wildlife Service, Res. Publ. 155. 23 pp.

Butterfield, B.P., B. Hauge, and W.E. Meshaka, Jr. 1993. The occurrence of *Hemidactylus mabouia* on the United States mainland. Herpetological Review 24:111-112.

Conant, R., and J.T. Collins. 1991. A field guide to reptiles and amphibians. Eastern and Central North America. Houghton Mifflin Company, Boston, MA. 450 pp.

Ellis, T.M. 1980. *Caiman crocodilus*: an established exotic in south Florida. Copeia 1980:152-154.

Ernst, C.H., and R.W. Barbour. 1989. Turtles of the world. Smithsonian Institution Press, Washington, DC. 313 pp.

Fritts, T.H. 1988. The brown tree snake, *Boiga irregularis*, a threat to Pacific islands. U.S. Fish and Wildlife Service Biological Rep. 88(31). 36 pp.

Fritts, T.H., and M.J. McCoid. 1991. Predation by the brown tree snake *Boiga irregularis* on poultry and other domesticated animals on Guam. The Snake 23:75-80.

Fritts, T.H., M.J. McCoid, and R.L. Haddock. 1990. Risks to infants on Guam from bites of the brown tree snake (*Boiga irregularis*). American Journal of Tropical Medicine and Hygiene 42:607-611.

Fritts, T.H., M.J. McCoid, and R.L. Haddock. 1994. Symptoms and circumstances associated with bites by the brown tree snake (*Boiga irregularis*) on Guam. Journal of Herpetology 28:27-33.

Fritts, T.H., N.J. Scott, and J.A. Savidge. 1987. Activity of the arboreal brown tree snake (*Boiga irregularis*) on Guam as determined by electrical power outages. The Snake 19:51-58.

Hayes, M.P., and M.R. Jennings. 1986. Decline of ranid frog species in western North America: are bullfrogs (*Rana catesbeiana*) responsible? Journal of Herpetology 20:490-509.

Lawson, R., P.G. Frank, and D.L. Martin. 1991. A gecko new to the United States herpetofauna, with notes on geckoes of the Florida Keys. Herpetological Review 22:11-12.

Lawson, W.J. 1987. The cane toad *Bufo marinus*. A bibliography. Australian Environmental Studies Working Paper 1/87. Griffith University, Queensland. 39 pp.

McCoid, M.J. 1991. Brown tree snake (*Boiga irregularis*) on Guam: a worst case scenario of an introduced predator. Micronesica Supplement 3:63-69.

McCoid, M.J. 1993. The new herpetofauna of Guam, Mariana Islands. Herpetological Review 24:16-17.

McCoid, M.J., and T.H. Fritts. 1980a. Notes on the diet of a feral population of *Xenopus laevis* (Pipidae) in California. Southwestern Naturalist 25:272-275.

McCoid, M.J., and T.H. Fritts. 1980b. Observations of feral populations of *Xenopus laevis* (Pipidae) in southern California. Bull. of the Southern California Academy of Sciences 79:82-86.

McCoid, M.J., and T.H. Fritts. 1993. Speculations on colonizing success of the African clawed frog, *Xenopus laevis* (Pipidae), in California. South African Journal of Zoology 28:59-61.

McCoid, M.J., R.A. Hensley, and G.J. Witteman. 1994. Factors in the decline of *Varanus indicus* on Guam, Mariana Islands. Herpetological Review 25:60-61.

McCoid, M.J., K. Pregill, and R.M. Sullivan. 1993. Possible decline of *Xenopus* populations in southern California. Herpetological Review 24:29-30.

McKeown, S. 1978. Hawaiian reptiles and amphibians. Oriental Publishing Company, Honolulu, HI. 80 pp.

Meshaka, W.E., Jr., B.P. Butterfield, and J.B. Hauge. 1994. Reproductive notes on the introduced gecko *Hemidactylus mabouia* in southern Florida. Herpetological Natural History 2:109-110.

Moyle, P.B. 1973. Effects of introduced bullfrogs, *Rana catesbeiana*, on the native frogs of San Joaquin Valley, California. Copeia 1973:18-22.

Painter, C.W., P.W. Hyder, and G. Swinford. 1992. Three species new to the herpetofauna of New Mexico. Herpetological Review 23:62.

Paulissen, M.A., and T.M. Buchanan. 1990. *Hemidactylus turcicus* (Mediterranean gecko). Herpetological Review 21:22.

Rodda, G.H., and T.H. Fritts. 1992. The impact of the introduction of the colubrid snake *Boiga irregularis* on Guam's lizards. Journal of Herpetology 26:166-174.

Rodda, G.H., T.H. Fritts, and P.J. Conry. 1992. Origin and population growth of the brown tree snake, *Boiga irregularis*, on Guam. Pacific Science 46:46-57.

Rodda, G.H., T.H. Fritts, and J.D. Reichel. 1991. The distributional patterns of reptiles and amphibians in the Mariana Islands. Micronesica 24:195-210.

Saethre, M.B., and P.A. Medica. 1993. *Hemidactylus turcicus* (Mediterranean gecko). Herpetological Review 24:154-155.

St. Amant, J.A. 1975. Exotic visitor becomes permanent resident. Terra 13:22-23.

Savidge, J.A. 1987. Extinction of an island avifauna by an introduced snake. Ecology 68:660-668.

Smith, H.M., and A.J. Kohler. 1977. A survey of herpetological introductions in the United States and Canada. Transactions of the Kansas Academy of Science 80:1-24.

Stebbins, R.C. 1985. A field guide to western reptiles and amphibians. The Easton Press, Norwalk, CT. 336 pp.

Stuart. J.N., and C.W. Painter. 1993. *Rana catesbeiana* (bullfrog). Cannibalism. Herpetological Review 24:103.

Thomas, R.A., P.J. Thomas, J.O. Coulson, and T. Coulson. 1990. *Anolis sagrei sagrei* (brown anole). Herpetological Review 21:22.

Tyler, M. 1989. Australian frogs. Viking O Neill Publishers, Ringwood, Victoria. 220 pp.

Wilson, L.D., and L. Porras. 1983. The ecological impact of man on the south Florida herpetofauna. University of Kansas Museum of Natural History Special Publ. 9. 89 pp.

Zug, G.R., and P.B. Zug. 1979. The marine toad, *Bufo marinus*: a natural history resumé of native populations. Smithsonian Contributions to Zoology 284. 58 pp.

For further information:

Michael James McCoid
Caesar Kleberg Wildlife Research Institute
Campus Box 218
Texas A & M University Kingsville
Kingsville, TX 78363

Non-native Birds

by
Chandler S. Robbins
National Biological Service

Two of the three most common nesting species in North America today are birds whose ancestors were brought here from Europe. Some non-native birds are more conspicuous than others, so comparisons are only relative, but according to the two largest continental surveys, non-native species (excluding house finches) constitute, on average, about 6% of the bird population during the summer months (Breeding Bird Survey [BBS]) and about 8% in winter (Christmas Bird Count). Percentages vary considerably by habitat and geographic location.

Many exotic bird species were introduced to the United States by European colonists who missed the familiar birds of their homeland and tried to establish populations of familiar Old World species. Farmers also saw opportunities for pest control by birds such as starlings and house sparrows, but they did not anticipate the degree to which these exotic species would outcompete native birds for nesting sites. Most introductions, however, were by sporting or hunting organizations and state game departments that wished to provide more hunting opportunities.

Competition between exotic and native species has been particularly severe on islands. In the Hawaiian Islands, introduced songbird species far exceed native ones. Visitors to Honolulu, for example, see only exotic songbirds unless they hike mountain trails in search of the few remaining endemic species. MacArthur and Wilson (1967) predicted that for every new species colonized or introduced on an island, an average of one species will become extinct. Even Puerto Rico has breeding populations of about 20 kinds of exotic songbirds, far outnumbering the endemics.

The best-known introductions in North America are those that were highly successful: the house sparrow (*Passer domesticus*), European starling (*Sturnus vulgaris*), rock dove or common pigeon (*Columba livia*), ring-necked pheasant (*Phasianus colchicus*), mute swan (*Cygnus olor*), gray or Hungarian partridge (*Perdix perdix*), and the chukar (*Alectoris chukar*). They readily adapted to their new environments, and most have prospered here for more than 100 years.

Data Sources

Before the mid-20th century, information on the distribution and population trends of exotic birds came primarily from scattered accounts in the literature, from state bird books, and from the Audubon Christmas Bird Count (CBC). Since 1966 in the eastern states and Canada, and 1968 in the West, the BBS (Robbins et al. 1986) has provided information on geographic distribution, relative abundance, and population trends for all but the rarest species. A condensed summary of BBS trends of exotic species (Table) based on as many as 2,500 fifty-stop roadside transects per year is presented for the three major regions of the continent.

History and Status

Cattle Egret (*Bubulcus ibis*)

The only records of intentional release of this African species in the United States are from Hawaii, where the bird was deliberately introduced on five major islands in July and August 1959 to control flies around homes and cattle (Breeze 1959). These birds were obtained in Florida, where they arrived in the early 1940's from South America by way of the West Indies. The species had been known from British Guiana since the 1870's (American Ornithologists' Union 1983), but no firm documentation of its arrival there from Africa is known. The species' spread across the continental United States is well documented by the BBS (Table) and the CBC. The cattle egret is highly migratory, and many of the American birds winter in Latin America. Cattle egrets feed primarily in pastures with cattle. Concerns that cattle diseases might be carried across international boundaries have so far lacked documentation, but populations are being monitored and movements of banded birds are being tracked.

Waterfowl

Many species of exotic waterfowl have found their way into the wild through intentional introductions and by escaping from captivity. The large, heavy-bodied muscovy duck (*Cairina moschata*) from Mexico, in both natural and white plumage, is a common sight in many city parks, but is less commonly found in the wild where it must forage for itself. Many European and some Asiatic ducks, and even a few exotic geese, escape from private collections, especially during storms. Because the number of these individuals is small, their populations have not been monitored.

Table. Population trends from the Breeding Bird Survey, 1966-92. Dashes under "Population trend" indicate insufficient data; dashes under "Significance" indicate no significant changes.

Species	Population trend		Significance*	
	1966-79	1980-92	1966-92	1966-92
Cattle egret				
East	Increase	Stable**	Stable**	-
Center	Increase	Decrease	Increase	-
Continent	Increase	Decrease	Increase	-
Mute swan				
East	-	Increase	Increase	*P* < 0.05
Continent	-	Increase	Increase	*P* < 0.05
Ring-necked pheasant				
East	Decrease	Increase	Decrease	-
Center	Stable**	Decrease	Stable**	-
West	Stable**	Decrease	Decrease	-
Continent	Decrease	Stable**	Decrease	-
Gray partridge				
East	Decrease	-	Decrease	-
Center	Increase	Decrease	Increase	-
West	-	Decrease	Decrease	-
Continent	Increase	Stable**	Increase	-
Rock dove				
East	Increase	Stable**	Increase	-
Center	Increase	Decrease	Stable**	-
West	Increase	Increase	Increase	-
Continent	Increase	Decrease	Increase	-
European starling				
East	Decrease	Stable**	Decrease	*P* < 0.01
Center	Increase	Decrease	Stable**	-
West	Increase	Decrease	Decrease	-
Continent	Stable**	Stable**	Decrease	*P* < 0.05
House sparrow				
East	Decrease	Decrease	Decrease	*P* < 0.001
Center	Stable**	Decrease	Decrease	*P* < 0.001
West	Increase	Decrease	Stable**	-
Continent	Stable**	Decrease	Decrease	*P* < 0.001

*Probability values are given for those species with a significant continuing decline for the entire survey period.
**Less than 1% change per year on average.

The mute swan of Eurasia is the notable exception. Introduced on Long Island and the lower Hudson Valley of New York in the late 1800's (Bump 1941), this swan is now locally common around Long Island Sound, in the Maryland portion of Chesapeake Bay, and in the Great Lakes region. Small populations thrive in other localities. Recent population increases (Table) and the resulting destruction of submerged aquatic vegetation needed by native waterfowl are causing concern.

Upland Game Birds

Most exotic birds imported for release are nonmigratory gallinaceous species: pheasants and francolins from Asia and partridges from Europe. Most were released to provide more hunting opportunities. Some states, such as Oregon, still have active introduction programs.

The only two Old World species to have become established widely enough to be monitored by the BBS are the ring-necked pheasant and the gray (Hungarian) partridge (Table). The first successful release of ring-necked pheasants was the release of 199 pairs in the Willamette Valley of Oregon in 1881 (Bump and Robbins 1966). Ring-necked pheasants have become an important game species in the northern states, but have had detrimental effects on remnant populations of the greater prairie chicken (*Tympanuchus cupido*; Vance and Westemeier 1979). Gray partridges have been in America since 1908-09, when nearly 40,000 birds, mostly wild-trapped in Hungary, were released in the United States and Canada (Bump and Robbins 1966).

Doves

The domestic pigeon or rock dove was first introduced from Europe by French settlers in the early 1600's (Schorger 1952). Now they are one of the most noticeable birds in American cities and farming communities. Countless thousands are still reared annually by pigeon fanciers who use them for homing and racing competitions, and each year the feral population is supplemented by captive-reared individuals that fail to return home. Rock dove populations were ignored by scientists and bird watchers before the Breeding Bird Survey began in 1966 and were not reported on Christmas Bird Counts until 1974. The population appears to have stabilized following a sharp increase in the 1960's and 1970's (Table).

Introductions of several other dove species have been successful locally, especially in the mild climates of Florida, California, and Hawaii, but these species are not sufficiently widespread to be monitored by existing surveys. The spotted or lace-necked dove (*Streptopelia chinensis*) of eastern Asia was well established in the Hawaiian Islands before 1900, and local populations have been established in southern California since 1917 (Willet 1933). The species now also occurs on St. Croix in the Virgin Islands (Raffaele 1989).

Likewise, the small barred or zebra dove (*Geopelia striata*) was brought to the Hawaiian Islands in 1922, and by 1936-37 it was common on all the major islands except Hawaii. Ten years later the Hawaiian population was estimated at 237,000 birds (Schwartz and Schwartz 1949).

The ringed turtle-dove or Barbary dove (*Streptopelia risoria*) has been domesticated so long that its origin is uncertain. Small populations are established in southern California, eastern Texas, Florida, and Puerto Rico. Occasional individuals occur each year in more northern states. A close relative, the Eurasian collared-dove (*S. decaocto*), has bred in southern Florida since the late 1970's (Smith 1987) and has been found as far north as Louisiana and Georgia. Its rapid spread across Europe in the past few decades suggests its potential for rapid expansion in America.

Parrots

Many species of parrots imported for the cagebird trade have escaped, especially at ports of entry. The budgerigar (*Melopsittacus undulatus*) from Australia and the canary-winged parakeet (*Brotogeris versicolurus*) from South America have established populations in southern Florida and Puerto Rico, while the parakeet has become established in Los Angeles County, California. Of greater concern to orchardists has been the survival and reproduction in more northern states of monk parakeets (*Myiopsitta monachus*) from temperate South America (Bull 1975). Control measures have eliminated most populations of this exotic species in the United States.

Songbirds

Berger (1981) includes accounts of 37 exotic songbird species that maintain breeding populations in Hawaii, and Raffaele (1989) lists 19 that are breeding or probably breeding in Puerto Rico. Fewer nest on the U.S. mainland. The two most notorious species that dominate the environment and have negative effects on native species are the house sparrow and European starling, both of which compete with native birds for nesting cavities.

One hundred house sparrows from England established the first breeding population in New York City in 1851-52. Additional introductions helped the population spread westward to the Mississippi River by 1870, and by 1910, this species was established across the continent (Robbins 1973). Their numbers continued to expand until the automobile replaced the horse and the supply of waste grain was markedly reduced. Their decrease since the mid-1960's is well documented by the BBS (Table).

Sixty European starlings released in New York City in April 1890 (Cruickshank 1942) were the ancestors of the millions that now occupy the American countryside. Although these birds consume enormous quantities of noxious insects and weed seeds, they are serious competitors with native species for nesting cavities and food. Fortunately, their populations seem to have peaked and are now declining (Table).

The house finch (*Carpodacus mexicanus*), native to the western states, is an adaptable species that has rapidly colonized the East since the illegal release of the species on Long Island, New York, in the early 1940's. The birds now breed in every eastern state.

Migratory Immigrants

In addition to birds intentionally released in North America, two migratory species, the cattle egret (already discussed) from Africa and the parasitic shiny cowbird (*Molothrus bonariensis*) from South America, have invaded via the West Indies in recent decades. Shiny cowbirds, which lay their eggs in the nests of other songbirds, may be as real a threat to the reproductive success of native North American species as they have been to the yellow-shouldered blackbird (*Agelaius xanthomus*) in Puerto Rico (Wiley 1985). Shiny cowbirds have been found as far north as Maine and as far west as Texas and Oklahoma.

Future Concerns

The North American avifauna has developed over millions of years, changing as climatic conditions altered habitats. New species evolved; others became extinct. Today, human influences are speeding extinction rates without any comparable increase in evolution of new species. Introducing aggressive exotic species often results in unforeseen problems, including extinction of native species.

References

American Ornithologists' Union. 1983. Check-list of North American birds. 6th ed. Allen Press, Lawrence, KS. 877 pp.

Berger, A.J. 1981. Hawaiian birdlife. 2nd ed. University of Hawaii Press, Honolulu. 260 pp.

Breeze, P.L. 1959. Information on the cattle egret, a bird new to Hawaii. 'Elepaio 20:33-34.

Bull, J. 1975. Introduction to the United States of the monk parakeet—a species with pest potential. International Council for Bird Preservation Bull. 12:98.

Bump, G. 1941. The introduction and transplantation of game birds and mammals into the state of New York. Transactions of the North American Wildlife Conference 5:409-420.

Bump, G., and C.S. Robbins. 1966. The newcomers. Pages 343-353 *in* A. Stefferud, ed. Birds in our lives. U.S. Fish and Wildlife Service, Washington, DC.

Cruickshank, A.D. 1942. Birds around New York City. American Museum of Natural History Handbook Series 13. New York. 489 pp.

MacArthur, R.H., and E.O. Wilson. 1967. The theory of island biogeography. Princeton University Press, NJ. 203 pp.

Raffaele, H.A. 1989. A guide to the birds of Puerto Rico and the Virgin Islands. Princeton University Press, NJ. 254 pp.

Robbins, C.S. 1973. Introduction, spread, and present abundance of the house sparrow in North America. Pages 3-9 *in* S.C. Kendeigh, ed. A symposium on the house sparrow (*Passer domesticus*) and European tree sparrow (*Passer montanus*) in North America. Ornithological Monograph

For further information:

Chandler S. Robbins
National Biological Service
Patuxent Wildlife Research Center
11410 American Holly Dr.
Laurel, MD 20708-4015

14, American Ornithologists' Union. Allen Press, Lawrence, KS.

Robbins, C.S., D. Bystrak, and P.H. Geissler. 1986. The Breeding Bird Survey: its first fifteen years, 1965-1979. U.S. Fish and Wildlife Service Res. Publ. 157. 196 pp.

Schorger, A.W. 1952. Introduction of the domestic pigeon. Auk 69:462-463.

Schwartz, C.W., and E.R. Schwartz. 1949. A reconnaissance of the game birds in Hawaii. Board of Commissioners of Agriculture and Forestry, Hilo, HI. 168 pp.

Smith, P.W. 1987. The Eurasian collared-dove arrives in the Americas. American Birds 41(5):1371-1379.

Vance, D.R., and R.L. Westemeier. 1979. Interactions of pheasants and prairie chickens in Illinois. Wildlife Society Bull. 7(4):221-225.

Wiley, J.W. 1985. Shiny cowbird parasitism in two avian communities in Puerto Rico. Condor 87:165-176.

Willet, G. 1933. A revised list of the birds of southwestern California. Cooper Ornithological Club, Pacific Coast Avifauna 21. 204 pp.

Non-native Animals on Public Lands

by
Charles A. Drost
Gary M. Fellers
National Biological Service

Non-native plants and animals have become part of our surroundings, in cities, agricultural areas, and wildlands. While there are many beneficial purposes for non-native animals, such as for food and sport hunting and as agricultural animals, the introduction of some has had major negative economic consequences (Palmer 1899), and adverse effects on native wildlife, plants, and habitats. The British ecologist Charles Elton, in a major review of introduced species, described the increasing number of invasions as constituting "one of the great historical convulsions in the world's flora and fauna" (Elton 1958, p. 31).

Non-native species are significant problems on large areas of state and federal public lands, and areas set aside to protect native plant and animal communities are not immune to such harm. Science and conservation journals have devoted entire issues to the threats posed by non-native plants and animals in nature reserves (e.g., Usher et al. 1988). In a compilation of threats to U.S. national parks, non-native plants and animals were the most often reported threat, and were reported by the most areas; feral cats (*Felis catus*), feral dogs (*Canis familiaris*), and wild pigs (*Sus scrofa)* were the non-native animals cited most often (NPCA 1977). Non-native species present serious threats, but at the same time, coordinated efforts on public lands offer the best possibility for controlling some harmful non-native species, and protecting both native plant and animal communities and human interests and needs.

We compiled information on non-native animals on public and private land-management areas by conducting a mail survey to assess their occurrence and management status in land-management areas. Survey results represent contributions from 937 national parks, national forests, national wildlife refuges, Bureau of Land Management field areas, and state and private land-management areas. The results reflect those species that land managers considered of greatest concern, and their general distribution on public lands. Non-native invertebrate animals, particularly forest insects and agricultural pests, cause severe economic and environmental damage as well (OTA 1993), but were not the focus of this survey.

Distribution and Effects

The forests, parks, refuges, and other areas that responded to the surveys identified 205 non-native animal species as species of management concern. As a group, non-native mammals were most often reported by land managers as problem species, accounting for 60% (823 of 1,370) of the reports received (Table 1). Twenty-eight non-native mammal species were listed for the areas surveyed, with feral cats and dogs and wild pigs reported most often (Table 2). Feral cats and dogs are nearly ubiquitous (Figure) and are of concern because they prey on native birds and mammals (Van't Woudt

Table 1. Non-native species reported from U.S. national forests, parks, refuges, and other land-management areas. "Species introduced" is the total number of non-native species of each group that are known to have been brought into the United States (fish from Courtenay and Stauffer 1984; amphibians and reptiles from Smith and Kohler 1977; birds from Long 1981; mammals from Lever 1985). "Established" is the number of species that have established successful long-term populations. "Species reported" is the number of species noted in mail surveys sent to U.S. land-management areas, and "Number of reports" is the number of areas reporting each species.

	Fish	Amphibians	Reptiles	Birds	Mammals
Species introduced	104	27	67	119	40
Established	41	14	35	56	35
Species reported in this survey	40	3	4	19	28
Number of reports	272	24	6	245	823

Table 2. Non-native animal species most commonly reported in national forests, parks, and other U.S. land-management areas.

Name: Common	Name: Scientific	No. of areas
Cat (feral)	*Felis catus*	180
Dog (feral)	*Canis familiaris*	123
Pig	*Sus scrofa*	100
European starling	*Sturnus vulgaris*	93
Carp	*Cyprinus carpio*	56
Cow	*Bos taurus*	35
Horse	*Equus caballus*	31
Nutria	*Myocastor coypus*	29
Rainbow trout	*Oncorhynchus mykiss*	28
Burro	*Equus asinus*	25
Goat	*Capra hircus*	25
Brown trout	*Salmo trutta*	23
Brook trout	*Salvelinus fontinalis*	21
Red fox	*Vulpes vulpes*	11
Rock dove	*Columba livia*	28

1990). Wild pigs were reported primarily in the southeastern United States, California, and Hawaii; despite their status as game in most areas, they pose serious threats to native plant communities and rare plant species by their foraging and digging (Singer 1981; Stone and Loope 1987). Wild horses (*Equus caballus*) are primarily present in the western United States and on the barrier islands of the east coast. Although they may damage native vegetation, wild horses are generally protected as part of the historic scene.

After mammals, non-native fish were listed most often as problem non-native species. For all areas combined, we received 272 reports representing a total of 40 non-native fish species. Non-native trout (introduced to augment local fisheries) and common carp (*Cyprinus carpio*) were reported most. Introduced trout include species from other parts of the United States (e.g., eastern brook trout, *Salvelinus fontinalis*, introduced in many areas of the West) and species from other areas of the world (primarily European brown trout, *Salmo trutta*). Introduced trout may decimate susceptible native fish populations, lead to the loss of native varieties through interbreeding, and deplete amphibians and aquatic invertebrates in waters originally without fish (Taylor et al. 1984; Larson and Moore 1985). Most areas reporting problems or threats from non-native trout are in the western United States (Figure). Carp have been introduced in waters throughout much of the United States, but most areas reporting them as serious pests were wetland-management districts and wildlife refuges along the Mississippi, Missouri, and Columbia river systems.

We received 245 reports of non-native birds from survey respondents. Although many bird species have been introduced into the United States (Table 1), many failed to become established or remained restricted to areas where introduced. Only 19 species were reported as causing significant damage. European starling (*Sturnus vulgaris*) and rock dove (common pigeon, *Columba livia*) were reported most often, primarily in developed areas.

Only three non-native amphibian species and four non-native reptiles were reported. These species (e.g., marine toad, *Bufo marinus*) are primarily a problem in tropical and subtropical areas of southern Florida and Hawaii and some U.S. territories.

Seventy-three of the species identified in the surveys had been targeted for control or eradication. Feral cats were the subject of the greatest number of management projects (138 areas). Seventy-eight areas were conducting or had completed projects to control wild pigs, while 60 areas listed management for feral dogs, 41 for wild horses, 35 for cows (*Bos taurus*), and 35 for feral burros (*Equus asinus*).

Non-mammalian species were less often targets for control. Thirty-four areas, primarily U.S. Fish and Wildlife Service areas, listed control or eradication programs for carp. Other fish subject to control were introduced rainbow trout (*Oncorhynchus mykiss*; 22 areas) and brook trout (20 areas) in streams in western North America. Fewer projects were listed for birds. European starlings were the target of most controls (15 areas). A few areas listed control projects for non-native invertebrates. Most common were fire ants (*Solenopsis* spp., 14 areas) and gypsy moths (*Lymantria dispar;* 9 areas).

This survey highlights widespread and serious concerns about the effects of introduced species on native plant and animal communities. Geographically, this was true for areas across most of the United States except Alaska, where survey respondents generally reported few problems with non-native species, possibly because of the extreme climate of that area. Even there, however, non-native species can be a serious threat in local areas; some nesting waterfowl and seabirds on island wildlife refuges are severely affected by predation from introduced Arctic foxes (*Alopex lagopus*).

Some of the greatest adverse impacts of non-native species have been in freshwater communities and on islands. Introduced fish have caused calamitous changes in the Great Lakes, decimating both the natural community of the lakes and the commercial fishery that depends on these inland seas (Lawrie 1970; Eck and Wells 1987). Adverse effects of introduced fish, especially predaceous species, on native fish, amphibians, and invertebrates are a recurrent pattern (Taylor et al. 1984; Moyle 1986). Introduced brown trout, in particular, are serious predators on native salmonids in the United States. In spite of their small size, introduced western mosquitofish (*Gambusia affinis*) may eliminate other small, native fishes through competition or predation; they may also prey heavily on the young of food and game fish and also on aquatic amphibian larvae (Meffe et al. 1983).

Non-native species introduced to islands have caused the greatest harm to terrestrial plant and animal communities. Areas specifically responding to our surveys included the national seashores on the barrier islands of the east coast and Gulf of Mexico, the National Park Service on the California Channel Islands, and national parks and wildlife refuges on the Hawaiian Islands. It is generally considered that long-isolated island plants and animals are poorly adapted to cope with introduced predators, competitors, and disease organisms, and all of these island areas have suffered serious damage from

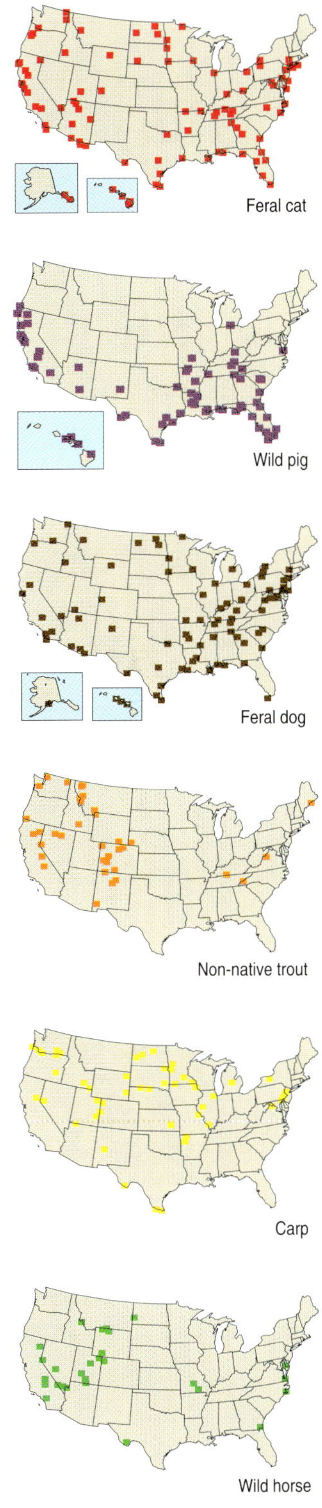

Figure. Distribution of several non-native animal species on public lands as reported by land managers responding to mail surveys: feral cat, wild pig, feral dog, non-native trout, carp, and wild horse.

introduced herbivores such as goats (*Capra hircus*), pigs, and Old World rabbits (*Oryctolagus cuniculus*), and introduced predators such as feral cats, rats, and mongooses (*Herpestes auropunctatus*; Stone 1985; Brockie et al. 1988). At the same time, these island areas have had some of the greatest success at controlling and managing non-native species. Feral goats, pigs, rabbits, and cats have been eliminated from some of the Channel Islands, allowing native plant and animal communities to begin to recover, and Hawaiian parks and refuges have successfully protected parts of their unique flora and fauna through aggressive and innovative control and exclusion measures against non-natives (Stone and Loope 1987).

References

Brockie, R.E., L.L. Loope, M.B. Usher, and O. Hamann. 1988. Biological invasions of island nature reserves. Biological Conservation 44:9-36.

Courtenay, W.R., Jr., and J.R. Stauffer, Jr. 1984. Distribution, biology, and management of exotic fishes. Johns Hopkins University Press, Baltimore, MD. 430 pp.

Eck, G., and L. Wells. 1987. Recent changes in Lake Michigan's fish community and their probable causes, with an emphasis on the role of the alewife (*Alosa pseudoharengus*). Canadian Journal of Fisheries and Aquatic Science 44:53-60.

Elton, C.S. 1958. The ecology of invasions by animals and plants. Chapman and Hall, London, UK. 181 pp.

Larson, G.L., and S.E. Moore. 1985. Encroachment of exotic rainbow trout into stream populations of native brook trout in the southern Appalachian Mountains. Transactions of the American Fisheries Society 114:195-203.

Lawrie, A.H. 1970. The sea lamprey in the Great Lakes. Transactions of the American Fisheries Society 99:766-775.

Lever, C. 1985. Naturalized mammals of the world. Longman Inc., New York. 487 pp.

Long, J.L. 1981. Introduced birds of the world. Universe Books, New York. 528 pp.

Meffe, G.K., D.A. Hendrickson, W.L. Minckley, and J.N. Rinne. 1983. Factors resulting in decline of the endangered Sonoran topminnow *Poeciliopsis occidentalis* (Atheriniformes: Poeciliidae) in the United States. Biological Conservation 25:135-159.

Moyle, P.B. 1986. Fish introductions into North America: patterns and ecological impact. Pages 27-43 *in* H.A. Mooney and J.A. Drake, eds. Ecology of biological invasions of North America and Hawaii. Springer-Verlag, New York.

NPCA. 1977. Summary of National Parks and Conservation Association survey of feral animals in the national park system. National Parks and Conservation Magazine 51:16-20.

NPS. 1991. Natural resources management guideline, NPS-77. U.S. Department of the Interior, National Park Service. 675 pp.

OTA. 1993. Harmful non-indigenous species in the United States. U.S. Congress, Office of Technology Assessment. 391 pp.

Palmer, T.S. 1899. The danger of introducing noxious animals and birds. Pages 87-110 *in* U.S. Department of Agriculture yearbook, 1898. USDA, Washington, DC.

Singer, F.J. 1981. Wild pig populations in the national parks. Environmental Management 5:263-270.

Smith, H.M., and A.J. Kohler. 1977. A survey of herpetological introductions in the United States and Canada. Transactions of the Kansas Academy of Science 80:1-24.

Stone, C.P. 1985. Alien animals in Hawaii's native ecosystems: toward controlling the adverse effects of introduced vertebrates. Pages 251-297 *in* C.P. Stone and J.M. Scott, eds. Hawaii's terrestrial ecosystems: preservation and management. Cooperative National Parks Resources Studies Unit, University of Hawaii, Honolulu.

Stone, C.P., and L.L. Loope. 1987. Reducing negative effects of introduced animals on native biotas in Hawaii: what is being done, what needs doing, and the role of national parks. Environmental Conservation 14:245-258.

Taylor, J.N., W.R. Courtenay, and J.A. McCann. 1984. Known impacts of exotic fish introductions in the continental United States. Pages 322-373 *in* W.R. Courtenay, Jr. and J.R. Stauffer, eds. Distribution, biology and management of exotic fishes. Johns Hopkins University Press, Baltimore, MD.

Usher, M.B., F.J. Kruger, I.A.W. Macdonald, L.L. Loope, and R.E. Brockie. 1988. The ecology of biological invasions into nature reserves: an introduction. Biological Conservation 44:1-8.

Van't Woudt, B.D. 1990. Roaming, stray, and feral domestic cats and dogs as wildlife problems. Pages 291-295 *in* L.R. Davis and R.E. Marsh, eds. Proceedings of the Fourteenth Vertebrate Pest Conference, University of California, Davis.

For further information:

Charles A. Drost
National Biological Service
Colorado Plateau Research Station
Northern Arizona University
PO Box 5614
Flagstaff, AZ 86011

Exotic Species in the Great Lakes

by

Thomas A. Edsall
National Biological Service
Edward L. Mills
Cornell University Biological Field Station
Joseph H. Leach
Ontario Ministry of Natural Resources

Global transfer of exotic organisms is one of the most pervasive and perhaps least recognized effects of humans on aquatic ecosystems of the world. Such transfers to new environments may lead to loss of species diversity and the extensive alteration of the native community. These changes, in turn, may have broad economic and social effects on the human communities that rely on the system for food, water, or recreation. Here we describe the exotic aquatic species that have become established in the Great Lakes and discuss their entry mechanisms or routes, the timeline of introduction, their geographic origins or sources, and their effects on the ecosystem of the Great Lakes. A recent review (Mills et al. 1993) provides the basis for much of this report.

Introductions of Species

Since the early 1800's, at least 139 new aquatic organisms have become established in the Great Lakes (Fig. 1); most are aquatic or wetland plants (42%), fishes (18%), and algae (17%). Introduced species of mollusks, oligochaetes, crustaceans, flatworms, bryozoans, cnidarians, and disease pathogens combined represent 22% of the total. All entered the Great Lakes basin by major mechanisms or routes (Fig. 2) including shipping (41 exotic species); unintentional releases (40 new species); ship or barge canals, along railroads or highways, or deliberate releases (17 species); unknown entry vectors (14 species); and multiple entry mechanisms (27 species).

The rate of introduction of exotic species increased markedly since the 1800's, as human activity in the Great Lakes basin increased. Almost one-third of the introductions to the Great Lakes were reported in the past 30 years. The first introductions of aquatic plants occurred when ships discharged solid ballast in the late 1800's. The opening of the St. Lawrence Seaway in 1959 greatly increased the number of ocean-going vessels entering the Great Lakes and dramatically increased the entry of exotic species by ships. Deliberate releases declined after the 1800's, and entry by canal increased slightly through 1959; entry by railroad and highway occurred mostly in the 1800's, and unintentional releases were consistently high since the late 1800's.

Origins of Introduced Species

Although most exotic species established in the Great Lakes are native to Eurasia (55%) and the Atlantic coast (13%), Great Lakes populations of many of these exotic species may have been established from sources outside their original native range. Purple loosestrife (*Lythrum salicaria*), Eurasian watermilfoil (*Myriophyllum spicatum*), and the Asiatic clam (*Corbicula fluminea*) are examples of Eurasian organisms that invaded the Great Lakes from source populations established outside their native ranges. Invading Atlantic coast species, such as sea lamprey (*Petromyzon marinus*) and white perch (*Morone americana*) probably entered through the Erie and Welland canals. Pacific salmon (*Oncorhynchus* spp.), rainbow trout (*O. mykiss*), brown trout (*Salmo trutta*), alewife (*Alosa pseudoharengus*), and rainbow smelt (*Osmerus mordax*) are examples of species that were introduced directly into the Great Lakes basin from populations in their original native ranges.

Effects of Introductions

The ecological and economic effects of the introduced fish species have been large. Of the 25 introduced fish species established in the Great Lakes, nearly half have had substantial effects. The extension of the range of the sea lamprey since the 1830's contributed to the decline of several fish species and severely damaged the sport and commercial fisheries of the Great Lakes. Millions of dollars are spent annually on sea lamprey control. The lake trout was the major predator species in the four lower Great Lakes, and its extermination by the sea lamprey allowed the alewife to move quickly through the lakes and experience almost unrestrained population growth. This growth was followed by massive die-offs of alewives, which polluted shorelines and blocked the intake pipes of water treatment plants and other industries. The alewife probably also suppressed native coregonines (*Coregonus* spp.), yellow perch (*Perca flavescens*), emerald shiner (*Notropis atherinoides*), and rainbow smelt. Eventually the alewife became an important prey for trout and salmon.

The ruffe (*Gymnocephalus cernuus*), a small, perchlike fish, reached the St. Louis River estuary in Lake Superior in ballast water in the early to mid-1980's. Ruffe abundance increased rapidly and in 1993, 61% (by number) of the fish caught in 440 bottom-trawl tows in the estuary were ruffe (J.H. Selgeby, National Biological Service, personal communication). The ruffe is spreading to other parts of the lake and has the potential to occupy at least 6.6 million ha (16.3 million acres) of Great Lakes' habitat that is suitable for use by native percid fishes, including the economically important walleye (*Stizostedion vitreum*) and yellow perch (Edsall et al. 1993). The effect of ruffe on native Great Lakes percids has not been demonstrated, but yellow perch numbers in the St. Louis River estuary declined markedly as ruffe abundance increased. There is concern that the ruffe has the potential to adversely affect percid abundance in other areas of the Great Lakes.

The common carp (*Cyprinus carpio*) was stocked in the 1870's, but it never became popular and by the 1890's was considered a problem because of its negative effects on more favored fish species and on waterfowl habitat. The stockings of Pacific salmon and rainbow and brown trout had profound and permanent ecological effects on the fish fauna through competition and predation. These salmonids now support a major element of the fishery in the Great Lakes, valued at more than $6 billion annually (GLFC 1992).

Of the fish disease pathogens introduced into the Great Lakes, *Glugea hertwigi*, a protozoan, caused extensive mortality in rainbow smelt in Lakes Erie and Ontario in the 1960's

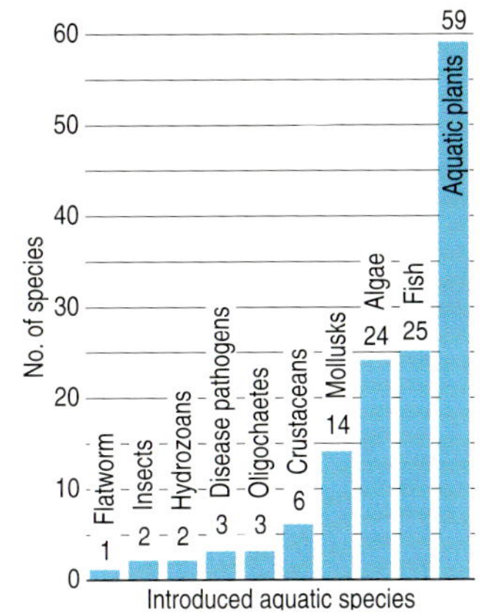

Fig. 1. Introduced aquatic species established in the Great Lakes. The number of species in each category is given above the bars.

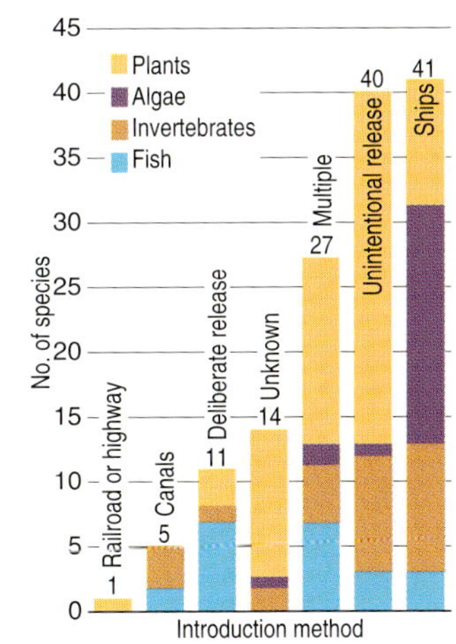

Fig. 2. Entry mechanisms or routes of exotic species established in the Great Lakes. The number of species in each category is given above the bars.

Courtesy T.A. Edsall

The introduced ruffe (*Gymnocephalus cernuus*).

and 1970's. A second pathogen, bacterial kidney disease, has been implicated in the massive mortalities of Pacific salmon in Lake Michigan in recent years (MDNR 1992). Two other introduced pathogens cause salmon whirling disease and furunculosis, but they occur mainly in fish hatcheries where crowding makes fish vulnerable to outbreaks of disease.

The arrival of the zebra mussel (*Dreissena polymorpha*) in Lake Erie in 1986 (Leach 1992) set the stage for long-term changes in pelagic and benthic communities in the Great Lakes and in the economic and social future of lake users. The zebra mussel may cause substantial changes in the food chain, resulting in a probable reduction in the overall production of fish in the Great Lakes. Zebra mussels also foul private vessels and structures, and nautical and littoral structures, including water intakes, in the Great Lakes. The zebra mussel has spread to southern Ontario in Canada; its westward range extension includes the Mississippi River and some of its tributaries from the river's headwaters near St. Paul, Minnesota, to its mouth at New Orleans, Louisiana. Negative ecological, economic, and societal effects are expected from these and future range expansions.

Introduced plant species outnumber all other groups of introduced organisms, but the effects of only a few of these are known. Purple loosestrife has spread throughout the Great Lakes basin and is replacing the cattail (*Typha latifolia*) and other wetland native plants. Purple loosestrife has no food value for wildlife and is making wetlands less suitable as wildlife habitat. Eurasian watermilfoil has also had a substantial effect in lakes in the Great Lakes basin. Massive beds of the plant often make boating and swimming impossible and reduce fish and invertebrate populations. Some introduced species of algae have become dominant members of the algal community of the Great Lakes. Their ecological impacts are generally unknown, but one, *Stephanodiscus binderanus*, has caused water-quality problems on several occasions.

The ecological effects of the introduced crustaceans, oligochaetes, bryozoans, cnidarians, and flatworms are largely unknown. Historically, the ecological and economic risks associated with these groups have not been as high as those posed by other plants and animals. The recently introduced spiny water flea (*Bythotrephes cederstroemi*), a predatory zooplankter, has rapidly expanded in the Great Lakes. Its ecological effect is unknown, but its establishment in Lake Michigan coincided with observed changes in the zooplankton community characteristic of those caused by an invertebrate predator.

Conclusions

The ecological, social, and economic effects of exotic species in the Great Lakes continue to be enormous. Serious effects have been documented for only a fraction of the species introduced into the Great Lakes. However, most introduced species have not been thoroughly studied to determine their effects on the ecosystem. Introduced species exist at almost every level in the food chain, and their effects must certainly pervade the entire aquatic community of the Great Lakes. We believe that as long as human-mediated transfer mechanisms persist and habitat alterations that stress native aquatic communities are allowed to occur, the Great Lakes ecosystem will also be at substantial risk from new, undesirable, exotic species.

References

Edsall, T.A., J.H. Selgeby, T.J. DeSorcie, and J.R.P. French III. 1993. Growth-temperature relation for young-of-the-year ruffe. Journal of Great Lakes Res. 19:630-633.

GLFC. 1992. Ruffe in the Great Lakes: a threat to North American fisheries. Great Lakes Fishery Commission, Ann Arbor, MI. 144 pp.

Leach, J.H. 1992. Impacts of zebra mussels (*Dreissena polymorpha*) on water quality and fish spawning reefs in western Lake Erie. Pages 381-397 *in* T.F. Nalepa and D.W. Schloesser, eds. Zebra mussels: biology, impacts, and control. CRC Press, Boca Raton, FL.

MDNR. 1992. Dingell-Johnson Annual Report. Project F-53-R-8, Study 471-2. Michigan Department of Natural Resources, Lansing. 10 pp.

Mills, E.L., J.H. Leach, J.T. Carlton, and C.L. Secor. 1993. Exotic species in the Great Lakes: a history of biotic crises and anthropogenic introductions. Journal of Great Lakes Res. 19:1-54.

For further information:

Thomas A. Edsall
National Biological Service
Great Lakes Science Center
1451 Green Rd.
Ann Arbor, MI 48105

Zebra Mussels in Southwestern Lake Michigan

by
Tammy Keniry
J. Ellen Marsden
Illinois Natural History Survey

The zebra mussel (*Dreissena polymorpha*) is a European species that was accidentally introduced into North America. It has had a tremendous impact on freshwater ecosystems of the United States and Canada. Since the zebra mussel was first discovered in Lake St. Clair in 1988, it has spread to each of the Great Lakes and to the major river systems of central and eastern United States. Communities along the affected lakes and rivers rely on these waters for drinking, industrial water supplies, transportation, commercial fishing and shelling, and recreation. Rapidly expanding populations of zebra mussels could ultimately affect many of these activities, in addition to changing the structure of the ecosystem.

By firmly attaching to hard surfaces, zebra mussels have clogged water-intake pipes and fouled hard-shelled animals such as clams and snails. In addition, zebra mussels have reduced plankton populations as colonies of mussels filter large volumes of water for food (e.g., Holland 1993), potentially depleting food resources of larval and planktivorous fishes such as smelt, chub, and alewife (*Alosa pseudoharengus*). Transfer of suspended material to the lake bottom in mussel waste products also leads to increased water clarity (Reeders et al. 1992) and increased growth of aquatic plants, a phenomenon already observed in some of the shallower harbors of Chicago. Although clear water is often considered aesthetically pleasing, this clarity indicates that drastic changes have occurred at the base of the food web and that energy flow through the ecosystem has been altered.

The first live zebra mussel was discovered in Lake Michigan near Chicago in 1989. We documented the subsequent establishment of the zebra mussel in southern Lake Michigan by monitoring larval and adult zebra mussels in 1991-93. Monitoring was conducted primarily along the Illinois and Indiana shorelines; limited sampling occurred along the southern Wisconsin shoreline. We also quantified the initial effects of the invasion on water clarity and native fauna.

Zebra Mussel Densities

Larval zebra mussels were present at all sampling locations during 1991-93; however, the number of sampling locations decreased from 8 to 3 over the 3 years. Peak numbers were collected each year at Burns Harbor, Indiana, where the highest average density was 37,044 veligers/m^3 (1,049/ft^3) in 1991; 74,493/m^3 (2,109/ft^3) in 1992; and 42,099/m^3 (1,192/ft^3) in 1993.

Attached zebra mussels were found in quite low numbers (less than 150/m^2 or 14/ft^2) in 1991 at one Wisconsin and four Illinois locations sampled by divers. The maximum density in 1991 (up to 2,389/m^2 or 222/ft^2) was recorded on concrete blocks in the intake channel of an Indiana power plant inaccessible to divers. By 1992, sampling at 2 Wisconsin and 4 Illinois sites revealed that the population had exploded, with a *minimum* average density of 57,115/m^2

Courtesy D.W. Schlosser, NBS

Zebra mussel (*Dreissena polymorpha*) on fragile papershell mussel (*Leptodea fragilis*).

Invasion of the Zebra Mussel in the United States

by
Amy J. Benson
Charles P. Boydstun
National Biological Service

Passage of the Nonindigenous Aquatic Nuisance Prevention and Control Act of 1990 called for a national program to control and reduce the risk of further introductions of nonindigenous aquatic nuisance species. This legislation specifically addressed the non-native zebra mussel (*Dreissena polymorpha*), which is expected to affect two-thirds of the nation's waterways.

The zebra mussel, a European species, was first discovered in Lake St. Clair in June 1988 and is now well established in North America. Zebra mussel introductions through ballast water may be responsible for many other introductions to the Great Lakes as well.

Aside from economic impacts, there could also be severe biological impacts. Plankton populations are directly affected by zebra mussels because of the tremendous filtering capacity of large mussel colonies; this could potentially shift system energetics and reduce available food resources for higher organisms. Biologists in the Great Lakes region believe that zebra mussels have already had an effect on the ecology of Lake St. Clair (Griffiths 1993); increased water clarity there potentially could cause a shift in the fish species composition. There has also been a detrimental effect on native mussel populations in Lake Erie since the arrival of zebra mussels (Masteller and Schloesser 1991). Native freshwater mussels are affected when zebra mussel larvae settle and attach on native mussels, covering them so completely that they can no longer carry out life processes. In addition, zebra mussels reduce the amount of food and possibly oxygen available to native mussels.

One important part of the nonindigenous program is to monitor the zebra mussel's distribution and provide technical assistance to other federal agencies, states, and the private sector. The National Biological Service's Southeastern Biological Science Center (SBSC) in Gainesville, Florida, monitors the zebra mussel as part of this program. By using the zebra mussel as a prototype species, personnel at SBSC also began developing a national geographic information system (GIS) to organize a coherent set of nonindigenous aquatic species data.

Federal, state, and private cooperators supplied us with information, resulting in the most complete digital data set of zebra mussel sightings in North America (Boydstun and Benson 1992). The locations of sightings were then entered into a data base. Since July 1991, between the United States and Canada we have collected more than 1,000 records of zebra mussel occurrences going back to their discovery in 1988 in Lake St. Clair.

Types of Observations

Zebra mussels are observed and collected by artificial substrate samplers, plankton nets, and inspection of pipes and water intakes. In the Great Lakes pipes and water intakes at power plants, water-treatment facilities, and various industries pump lake water into their plants. Zebra mussels clog these water pipelines, causing serious mechanical problems. The U.S. Coast Guard found zebra mussels on navigational buoys in the Great Lakes during routine inspections; these buoys now serve as an artificial substrate sampler, giving us hundreds of records each winter. Zebra mussels have also been collected inadvertently while sampling for fish when using gill nets or when collecting native mussels. The incidental finds account for many important sightings in newly expanded areas.

Range Expansion

Since the first zebra mussel was sighted in 1988 (Fig. 1), the species quickly colonized regions in all five Great Lakes by 1990. Currently, they have been reported in the waterways of 19 states and 2 Canadian provinces (Fig. 2). They are established in the Great Lakes and the following rivers: Mississippi, Arkansas, Illinois, Ohio, Tennessee, Cumberland, Hudson, Susquehanna, Ottawa, Niagara, Mohawk, Genesee, Kanawha, and St. Lawrence. Established colonies exist throughout the lower Great Lakes (Erie, Ontario, and St. Clair) wherever there is suitable habitat. Lake Huron has populations in Saginaw Bay and at the southern end of the lake where it flows into the St. Clair River. There are also a few isolated populations around the lake and in the Georgian Bay area. Zebra mussels are abundant in most of the southern portion of Lake Michigan's shoreline from Sheboygan, Wisconsin, to Frankfort, Michigan. The northern portion of the lake has populations in Green Bay, Traverse Bay,

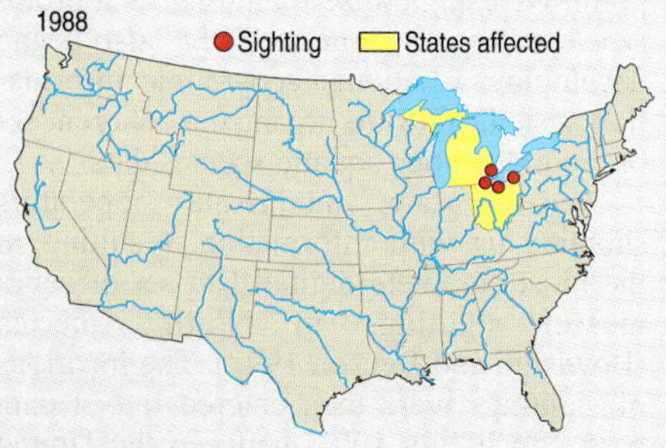

Fig. 1. States with zebra mussel sightings in inland or adjacent waters, 1988. In 1989, they spread to Lake Superior, Lake Michigan, and Lake Ontario (National Biological Service, unpublished data).

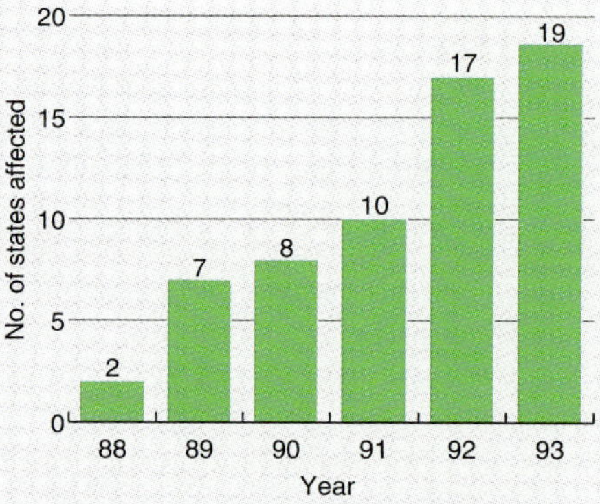

Fig. 2. Numbers of states affected by zebra mussels since their arrival in the United States in the mid-1980's.

and in the lake at Escanaba and St. Ignace, Michigan. Zebra mussels have also been found in 11 inland lakes in Michigan. Lake Superior is the only Great Lake where zebra mussels are not spreading quickly. Since the first sightings in Duluth Harbor in October 1989, they have been found only in Thunder Bay (Canada), Sault Ste. Marie, and Marquette, Michigan.

The first sighting in the Mississippi River was in Alton, Illinois, on 10 September 1991. Two days later a single zebra mussel was found about 764 km (475 mi) upstream at La Crosse, Wisconsin. In January 1992, mussels were found at Clarksville, Missouri; Oquawka, Illinois; and Genoa, Wisconsin. In July 1992, mussels were reported near Winona, Minnesota. By early 1993 (Fig. 3), almost every lock and dam in the Upper Mississippi River north of Dubuque, Iowa, had zebra mussels. The Lower Mississippi River was colonized more recently in the later part of 1992 and early 1993. Mussels were collected in the river at Greenville and Vicksburg, Mississippi, in 1992. By the end of June 1993, zebra mussels were collected in Louisiana at Shaw, Lettsworth, St. Francisville, New Orleans, and Berwick.

Vectors

It is important to be aware of the spread of nonindigenous species, especially ones with the potential to be an ecological menace such as the zebra mussel. The natural means of dispersal is larval drift downstream. Aside from natural mechanisms, canals and barge traffic in navigable rivers are suspected as major vectors for dispersal. In April 1992, a barge dry-docked for repairs at Hartford, Illinois, had more than 1,000 zebra mussels attached to a section of exposed hull (Keevin et al. 1992). The total number of zebra mussels on the entire hull could not be determined. The barge's log book showed that it had traveled 20,558 km (12,777 mi) up and down the Mississippi

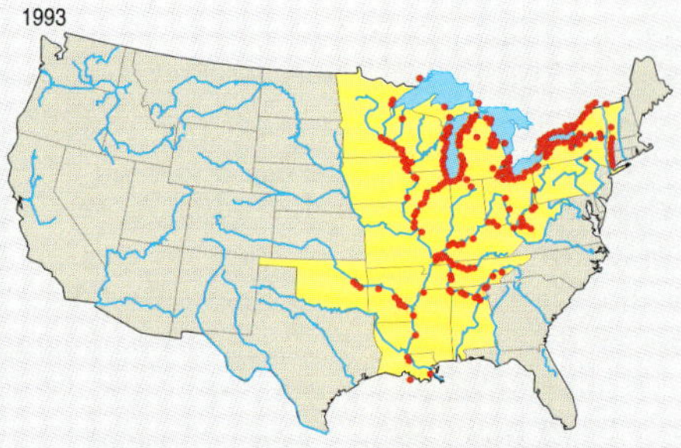

Fig. 3. States with zebra mussel sightings in inland or adjacent waters in 1993. The range has extended west of the Mississippi River into Oklahoma by way of the Arkansas River (National Biological Service, unpublished data).

River from Minnesota to Louisiana in just over 1 year before dry-docking. This documented long-distance transport of live mussels gives credibility to the assumption that barge traffic has been a primary dispersal mechanism in navigable waters. Zebra mussels can also be dispersed overland, especially by human activities such as recreational boating. Dead zebra mussels from Lake Erie were found on a boat trailer entering California (D. Peterson, California Department of Water Resources, personal communication).

References

Boydstun, C.P., and A.J. Benson. 1992. Nonindigenous report (1992:1): zebra mussel (*Dreissena polymorpha*) sightings in the United States and Canada. National Fisheries Research Center, Gainesville, FL. 10 pp.

Griffiths, R.W. 1993. The changing environment of Lake St. Clair. Proceedings of the Third International Zebra Mussel Conference, Toronto, Canada.

Keevin, T., R. Yarbrough, and A. Miller. 1992. Inadvertent transport of live zebra mussels on barges: experiences in the St. Louis District, Spring 1992. Zebra Mussel Research Tech. Notes ZMR-1-07. U.S. Army Corps of Engineers Waterways Experiment Station, Vicksburg, MS. 4 pp.

Masteller, E.C., and D.W. Schloesser. 1991. Infestation and impact of zebra mussels on the native unionid population at Presque Isle State Park, Erie, Pennsylvania. Page 20 *in* Proceedings of the Second Annual Zebra Mussel Research Conference, Rochester, NY.

For further information:

Amy J. Benson
National Biological Service
Southeastern Biological Science Center
7920 NW 71st St.
Gainesville, FL 32606

(5,306/ft^2) near Glencoe, Illinois. The *maximum* average density in 1992 was 267,885/m^2 (24,885/ft^2) at a Waukegan site that 1 year previously had only 25 mussels/m^2 (2/ft^2) (Marsden et al. 1993). Densities at two Illinois locations remained high in 1993, with average densities of 224,428/m^2 (20,858/ft^2) at Waukegan and 52,428/m^2 (4,870/ft^2) at Lake Forest.

High reproductive success during 1991 was clearly responsible for the huge increase in the number of attached mussels during 1992. It is interesting that although 1992 levels of reproduction were generally twice as high as in 1991, the population increase did not continue in 1993 at the two locations sampled.

Water Clarity

Water visibility (using a secchi disk) increased from a maximum depth of 4 m (13 ft) in 1990, to 6 m (20 ft) in 1991, to 10 m (33 ft) in 1992. Water remained clear in 1993, with a maximum depth at disappearance of 9.5 m (31 ft). At the site for which data are most consistently available (Waukegan), minimum water visibility measurements during 1991-93 were higher than any measured values during 1990. This trend should be interpreted with caution given the natural variability in water clarity values. The data suggest, however, that the water clarity of southern Lake Michigan may be increasing due to colonization of the lake by massive numbers of zebra mussels. This trend has been documented in other recently colonized lakes, such as Lake Erie (Leach 1992).

Impacts on Snails

Most native snails we collected were colonized by one or more zebra mussels. *Stagnicola* was the most common genus collected in nonquantitative samples. In 1991, 72% of these snails had attached zebra mussels, with an average of 1.6 mussels per snail. By 1992, 99% of *Stagnicola* were fouled, with the average number increasing to 3.7 zebra mussels per individual snail. *Elimia* snails dominated the quantitative samples from rocky areas. In 1992, 99% of 94 *Elimia* were fouled with mussels; in 1993 divers failed to find any live *Elimia* at the Waukegan reef.

Conclusions

In the Great Lakes and associated river systems, populations of native clams are threatened because of the colonization of their shells with massive numbers of zebra mussels (Mackie 1991). Our data indicate that snails are also being used as substrate for mussel attachment in Lake Michigan. As grazers, snails are an important part of the bottom community. They are also a source of food for fishes such as yellow perch (*Perca flavescens*), sunfish, and whitefish (Scott and Crossman 1973). Given the limited knowledge of the role of snails in Lake Michigan and other large lakes, it is not possible to fully anticipate the effects of reduced or decimated snail populations.

The rapid increase in zebra mussel densities we observed in the open waters of the lake was reflected in their colonization of municipal and industrial water-intake pipes. In 1991 and 1992 facilities drawing raw water from Lake Michigan began treatment programs to reduce infestation of intake pipes. The cost of retrofitting plants in Chicago and northern Illinois shoreline communities had totaled $1,778,000 by 1992 (Nelson 1992). This value does not include chemical costs, or increased personnel costs as workers dealt with mussel-related problems. In addition to economic costs of retrofitting and chemical treatments, Lake Michigan has an increased ecological risk of accidental chemical spills or leakages.

Zebra mussels also affect the aesthetic and recreational value of the lake. Boat owners are concerned about zebra mussels fouling boat hulls and engine cooling systems, and windrows of broken shells have begun to appear along Lake Michigan beaches.

The economic impact of zebra mussels is not limited to industrial and recreational interests, however. Native clams from the Illinois River are shipped to Japan for use in the cultured pearl industry; in 1991 the value of this resource was $1.4 million annually. The infestation of clams by zebra mussels has increased dramatically, resulting in significant clam mortality. Commercial shelling on the Illinois River was recently banned, following a drop in harvest from over 454,000 kg (1 million lb) in 1991 to 67,646 kg (149,000 lb) in 1993 (Don Duffert, Illinois Department of Conservation, personal communication).

Zebra mussels are a permanent addition to the Lake Michigan ecosystem and connected waters. Chemical and mechanical controls for zebra mussels are only useful in localized areas such as intake pipes and other artificial structures, but not in the open waters of the lake. Ultimately, zebra mussel populations will exceed the capacity of the environment to support them, after which their numbers will likely decline. Native predators such as freshwater drum (*Aplodinotus grunniens*), diving ducks, and crayfish may also keep mussel populations in check in some areas. The adverse effects of zebra mussels on human activities and native aquatic species will never be totally eliminated, but eventually they may become a more tolerable nuisance.

References

Holland, R.E. 1993. Changes in the planktonic diatoms and water transparency in Hatchery Bay, Bass Island area, western Lake Erie since the establishment of the zebra mussel. Journal of Great Lakes Res. 19:617-624.

Leach, J.H. 1992. Impacts of the zebra mussel (*Dreissena polymorpha*) on water quality and fish spawning reefs in western Lake Erie. Pages 381-395 *in* T.F. Nalepa and D.W. Schloesser, eds. Zebra mussels: biology, impacts, and control. Lewis Publishers, Boca Raton, FL.

Mackie, G.L. 1991. Biology of the exotic zebra mussel, *Dreissena polymorpha*, in relation to native bivalves and its potential impact on Lake St. Clair. Hydrobiologia 219:251-268.

Marsden, J.E. 1992. Standard protocol for monitoring and sampling zebra mussels. Illinois Natural History Survey Biological Notes 138. 40 pp.

Marsden, J.E., N. Trudeau, and T. Keniry. 1993. Zebra mussel study on Lake Michigan. Final report to the Illinois Department of Conservation. Illinois Natural History Survey Tech. Rep. 93/14. 51 pp.

Nelson, S. 1992. A pound of cure for a ton of mussels. Aquaticus: Journal of the Shedd Aquarium 23:28-29.

Reeders, H.H., A. bij de Vaate, and R. Noordhuis. 1992. Potential of the zebra mussel (*Dreissena polymorpha*) for water quality management. Pages 439-451 *in* T.F. Nalepa and D.W. Schloesser, eds. Zebra mussels: biology, impacts, and control. Lewis Publishers, Boca Raton, FL.

Scott, W.B., and E.J. Crossman. 1973. Freshwater fishes of Canada. Bull. 184. Fisheries Research Board of Canada, Ottawa. 966 pp.

For further information:

Tammy Keniry
Illinois Natural History Survey
Lake Michigan Biological Station
400 17th St.
Zion, IL 60099

Africanized Bees in North America

by
Michael R. Kunzmann
National Biological Service
Stephen L. Buchmann
John F. Edwards
Steven C. Thoenes
Eric H. Erickson
U.S. Department of Agriculture

The honeybee genus *Apis* likely has the greatest breadth of pollen diet of any insect and, because of its human-caused cosmopolitan distribution, the species directly affects the reproductive biology of about 25% of the world's flowering plants (Schmalzel 1980; Buchmann et al. 1992). This situation has profound consequences for agribusiness, native plants and animals, and ecosystems. In 1956, bee geneticist Warwick E. Kerr imported queen bees of an African race (*Apis mellifera scutellata*) into Brazil to breed a more productive honeybee that was better adapted to the Neotropical climate and vegetation (Kerr 1967). The following year, 26 of Kerr's Africanized honeybee queens were inadvertently released into the surrounding forest (Winston 1987). Since then, the Africanized hybrids have been expanding their range northward, with an average rate of between 330 and 500 km (200 and 300 mi) each year (Fig. 1).

The first U.S. Africanized honeybee colony was reported in October 1990, at Hidalgo, Texas, along the international boundary. By fall 1993, Africanized honeybees (AHBs) had extended their territory north and west into numerous counties of Arizona, New Mexico, and Texas (Fig. 2). Since the first U.S. AHB swarm was detected, the rate of spread has accelerated to over 600 km (375 mi) per year in the southwestern United States (Guzman-Novoa and Page 1994).

European honeybees (EHBs) were introduced into North America as early as the 16th century by Spanish conquistadors and missionaries (Brand 1988). Today, one of the three most common subspecies or races of the EHB, the Italian honeybee (*A.m. ligustica*), is nearly pandemic throughout North America because of its popularity with professional and hobbyist beekeepers. As a consequence, these non-native bees have become naturalized and have been a part of the North American arthropod biota for about 3,500 bee generations, or at least the past

Courtesy J.O. Schmidt, USDA

Africanized honeybees swarm outside a trap in Costa Rica.

200 years (Buchmann et al. 1992). European honeybees are commonly seen visiting agricultural food crops, cultivated flowers, and roadside wildflowers to gather nectar and pollen. They are even common in areas far from human population centers. These bees are also the preferred, "managed" pollinator for over 100 U.S. agricultural crops (e.g., fruits, vegetables, and some nuts), most of which depend on or benefit from insect pollination. The value of these pollination services by EHBs is estimated at $5-$10 billion annually in the United States (Southwick and Southwick 1992).

Africanized and European honeybees represent divergent subspecies within the *mellifera* species of the genus *Apis*. Both have nearly the same biochemistry, morphology, genetics, diet, and reproductive and other behaviors. Their diet includes pollen and spores from most seed plants. Both EHBs and AHBs are social bees living in perennial colonies. They are active on most days collecting nectar, water, pollen, and plant resins for their subsistence. These honeybees "hoard" excess honey as energy-rich carbohydrate reserves in hexagonal wax combs. Energy from honey consumption partially supports brood-rearing and, most importantly, supplies the energy necessary for foraging flights by thousands of adult worker bees.

Africanized and European honeybees exhibit different foraging strategies (largely tropical versus temperate attributes). Africanized honeybee colonies in Africa, and now in much of the Neotropics, are attuned to finding and exploiting isolated mass-flowering tropical trees, and also use pollen and nectar from the nocturnal flowers of bat-pollinated flowering plants. Some tropical *Apis* species even migrate to follow nectar and pollen flows across the floral landscape. Consequently, these bees depend on increased colony mobility (reproductive swarming and abandoning the hive) as behavioral responses to seasonal floral richness or dearths. EHBs are better at hoarding vast amounts of honey and surviving long, cold winters.

Although preliminary evidence for behavioral differences between the two races have been documented in the Neotropics (French Guiana, Venezuela, Panama; *see* reviews by Taylor 1977; Seeley 1985; Roubik 1989), the behavioral ecology of AHBs and their interactions with EHBs and thousands of species of native U.S. bees remain largely unknown. Africanized honeybees have slightly shorter developmental times than do European bees, enabling them to produce more bees per unit time compared with EHBs. Africanized bees will also accept smaller cavities to nest in than European bees. This behavior increases potential competition for nesting sites with birds and other animals and also increases the potential for greater numbers of honeybee colonies in an area. Africanized honeybees commonly abandon their hives, often 15%-30% annually or even much greater in some localities. Absconding colonies may travel as far as 170 km (about 100 mi) before selecting a new nesting site (USDA 1994). Thus they have been able to rapidly colonize new areas in the Neotropics.

The most often-discussed characteristic separating the two races is the AHBs' propensity to vigorously defend their colony and nest site. Although all honeybees respond to threats to their colonies, AHBs respond more quickly and in much greater numbers than do EHBs. In comparison to EHBs, greater numbers of AHBs will pursue intruders for much greater distances to defend their colonies. Recent research reported that 3 to 4 times as many AHBs responded and left 8 to 10 times more stings in a black leather measuring target in stinging experiments (USDA 1994).

Biochemical comparisons of AHB and EHB venoms indicate they are nearly identical. Nineteen stings per 1 kg (2.2 lb) of human victim body weight is the predicted median lethal dose (Schumacher et al. 1992). Massive stinging incidents by AHBs are more likely to result in toxic envenomation. Reported 1993 stinging incidents in Mexico have involved more than 60 human fatalities (one death per 1.4 million). From 1988 to 1992, the Mexican national African Bee Program eliminated 117,000 AHB swarms in densely populated urban areas (Guzman-Novoa and Page 1994). To date, the worst U.S. stinging incident occurred in July 1992, when a 44-year-old man mowing his lawn experienced a massive bee attack resulting in 800-1,000 stings (McKenna 1992).

Fig. 1. Migration of Africanized honeybees.

Ecological Implications

Competition among nectar- and pollen-feeding invertebrate and vertebrate pollinators, resource partitioning, insect and plant community interactions, and ecosystem processes are affected by introduced EHBs and AHBs, with important short- and long-term ecological and perhaps evolutionary consequences. The influence of exotic honeybees on individual species or communities of native tropical (or temperate) plants or animals can only have one of three outcomes: the native species will suffer, benefit, or remain more or less unaffected. The key to understanding these seemingly obvious outcomes is, however, based on obtaining sufficient information to delineate the very complex short- and long-term competitive dynamics between introduced bees, native bees and pollinators, and native plants in diverse, interacting, natural communities.

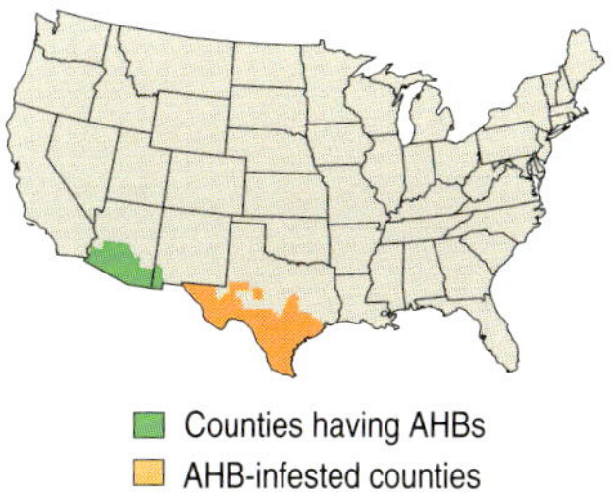

Fig. 2. Confirmed presence of Africanized honeybees in (colored) counties of Arizona, New Mexico, and Texas, January 15, 1994.

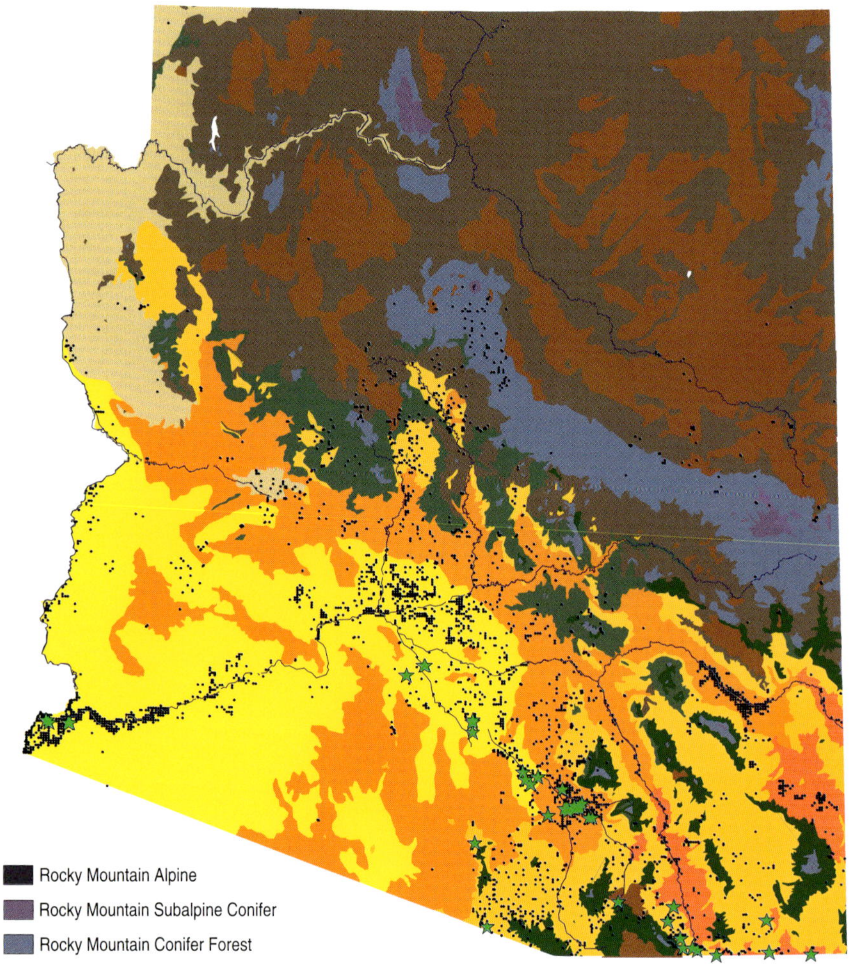

Fig. 3. Known honeybee locations in Arizona displayed with vegetation classes; derived from Brown et al. (1979).

One observational and manipulative competition study between honeybees, bumblebees, solitary bees, and ants was at midelevations in the Santa Catalina Mountains in the Sonoran Desert near Tucson, Arizona (Schaffer et al. 1983). Dramatic shifts in abundance of ants and bumblebees were detected when honeybees were present (introduced) or sealed inside their hives. The researchers suggested that direct competition between introduced honeybees and native hymenopteran floral visitors was caused by honeybees numerically dominating the site. Initial evidence seems to indicate that honeybees seek out and preempt the most profitable habitats and partially exclude native bees indirectly by rapidly reducing the standing crop of plant nectar and pollen (*Agave* in this study).

Both species of non-native bees forage vast expanses of territory containing native and non-native floral resources. Estimates of the amount of terrain foraged annually by an average-sized honeybee colony in New York hardwood forests (Visscher and Seeley 1982) are 80-100 km^2 (30-40 mi^2). Forage area estimates for AHB colonies living in lowland Panamanian rain forests (Roubik 1989) are 200-300 km^2 (75-115 mi^2), although 90% of these foraging flights are completed within 5 km (3 mi) of the nest (Visscher and Seeley 1982). Even given this restrictive caveat, the amount of "bee pasture" grazed by these aerial herbivores is immense.

In studying honeybee colonies foraging in temperate forests in New York State, Visscher and Seeley (1982) found that these cold-hardy EHB colonies amassed 15-30 kg (33-66 lb) of pollen and 60-80 kg (132-176 lb) of honey each year. To collect this amount of food, a colony must dispatch tens of thousands of foragers on many millions of foraging bouts with the bees flying 20-30 million km (12-19 million mi) overall. Similar studies of AHBs in Panama (Roubik 1989) determined that AHBs placed more emphasis on pollen collection. The Sonoran Desert of northern Mexico and southern Arizona is perhaps one of the richest areas in the world in floral resources because of the relative high plant diversity and the many fair-weather days for worker-bee foraging.

Many important nectar- and pollen-producing plants visited by AHBs bloom at night and are pollinated by bats. Africanized honeybees find and exploit these rich flowers at first light, and we predict that saguaros and other columnar cacti will be heavily used as food plants for AHBs in Arizona. Early Arizona data for AHB colonies illustrate that most AHB colonies have been found in the subtropical climate zones in Sonoran desertscrub.

Determining which plants are used primarily for nectar versus pollen, or both, depends on direct observations of bees on flowers or indirectly by identifying pollen grains in stored nest samples of honey. In Panama, Roubik (1989) found that AHB colonies harvested pollen from at least 142-204 flowering plant species in a forest containing about 800-1,000 species. European honeybees collected pollen or nectar from about 185 plant species from a secondary forest and agricultural area in Mexico (Villanueva 1984). These studies suggest that honeybees are using about 25% of the local flora, but intensively use far fewer species at any given time (Roubik 1989). In Arizona EHBs will often harvest pollen from more than 60 species annually, but of these, only 10-15 are harvested heavily and consistently from year to year (Buchmann et al. 1992). Because of their pollen herbivory and reproductive contact with so many plants, there can be serious long-term ecological and evolutionary consequences of these interactions that we simply do not yet understand.

Ecological Monitoring

Although we have made a case for potential serious, competitive displacement of food resources by honeybees to the exclusion of some native bees and pollinators, there is a little-appreciated yet unique ecological application for using EHB colonies (*A. mellifera*) as short- and long-term local and regional monitoring devices of vegetation diversity, plant productivity, flowering phenology, precipitation, climate, and general ecosystem health. No expensive equipment is required since the bees do all the "fieldwork." In addition, floral changes in landscapes can be determined from the rich "fossilized" source of pollen dietary information in old, dark brood combs or in 75- to 100-year-old "debris middens" in the Sonoran Desert (Buchmann et al. 1992). Long-term records (some spanning decades) for certain beekeeping locations are invaluable aids to beekeepers, ecologists, and resource managers for ecological evaluation and monitoring.

To validate any AHB range-expansion prediction or to measure potential effects on native pollinators or ecosystem components, we must monitor the bees and evaluate habitats on national and local scales. Information must be collected, integrated, and shared by researchers, individuals, and agencies. Public-and-private-sector partnerships have been developed to exchange AHB information and develop monitoring protocols.

Researchers use geographic information systems (GIS) and global positioning systems (GPS) technologies to track the locations of known AHB and EHB colonies; delineate honeybee habitat parameters such as preferred vegetation community, climatic zone, elevation, and distance to water; investigate potential ecological consequences to native bees and other nectar-dependent species; monitor and detect habitat productivity changes; and develop computer models to illustrate and predict preferred AHB habitats and potential ecological consequences (Fig. 3).

The Future

Knowing how far north AHBs will spread is critical in predicting their ecological effects. There is general agreement that they have a climatic limit, but precise limits of their U.S. range expansion is disputed. Some researchers suggest that AHBs will disperse almost as far north as Canada; others propose that they will go no farther than the U.S. southwestern and southeastern corners. In all likelihood, AHBs will become established as a dominant ecosystem forager in the southern third of the United States, where EHB overwintering behavior is less critical for survival. If conditions are favorable, however, the AHBs may expand into marginally productive or colder habitats in higher latitudes or elevations.

While the ecological range limits and economic consequences of non-native AHB migration into the United States are not precisely known, researchers agree that honeybees are economically important, and that sufficient biological information exists to develop adequate inventory and monitoring programs. Added benefits to honeybee monitoring programs are also important because bee colonies can also serve as excellent indicators of flowering plant productivity, ecosystem stability, and relative ecological health.

References

Brand, D.D. 1988. The honeybee in New Spain and Mexico. Journal of Cultural Geography 9:71-81.

Brown, D.E., C.H. Lowe, and C.P. Pase. 1979. A digitized classification system for the biotic communities of North America, with community (services) and association examples for the Southwest. Journal of the Arizona Nevada Academy of Science 14 (Supplement 1):1-16.

Buchmann, S.L., M.K. O'Rourke, C.W. Shipman, S.C. Thoenes, and J.O. Schmidt. 1992. Pollen harvest by honeybees in Saguaro National Monument: potential effects on plant reproduction. Pages 149-156 *in* C.P. Stone and E.S. Bellantoni, eds. Proceedings of the Symposium on Research in Saguaro National Monument, Tucson, AZ.

Guzman-Novoa, E., and R. Page. 1994. The impact of Africanized bees on Mexican beekeeping. American Bee Journal 134:101-106.

Kerr, W.E. 1967. The history of the introduction of African bees to Brazil. South African Bee Journal 39:3-5.

McKenna, W.R. 1992. Killer bees: what the allergist should know. Pediatric Asthma, Allergy and Immunology 6(4):19-26.

Roubik, D.W. 1989. Ecology and natural history of tropical bees. Cambridge University Press, England. 514 pp.

Schaffer, W.M., D.W. Zeh, S.L. Buchmann, S. Kleinhaus, M.V. Schaffer, and J. Antrim. 1983. Competition for nectar between introduced honeybees and native North American bees and ants. Ecology 64:564-577.

Schmalzel, R.J. 1980. The diet breadth of *Apis* (Hymenoptera: Apidae). M.S. thesis, University of Arizona, Tucson. 79 pp.

Schumacher, M.J., J.O. Schmidt, N.B. Egen, and K.A. Dillion. 1992. Biochemical variability of venoms from individual European and Africanized honeybees (*Apis mellifera*). The Journal of Allergy and Clinical Immunology 90:59-65.

Seeley, T.D. 1985. Honeybee ecology: a study of adaptation in social life. Monographs in Behavior and Ecology, Princeton University Press, NJ. 201 pp.

Southwick, E.E., and L. Southwick, Jr. 1992. Economic value of honeybees (*Hymenoptera: Apidae*) in the United States. Journal of Economic Entomology 85(3):621.

Taylor, O.R. 1977. The past and possible future spread of Africanized honeybees in the Americas. Bee World 58:19-30.

USDA. 1994. African honeybee fact sheet. U.S. Department of Agriculture.

Villanueva, R. 1984. Plantas de importancia apicola en el ejido de Plan del Rio. Veracruz, Mexico. Biotica 9:279-340.

Visscher, P.K., and T.D. Seeley. 1982. Foraging strategy of honeybee colonies in a temperate deciduous forest. Ecology 63:1790-1801.

Winston, M.L. 1987. The biology of the honeybee. Harvard University Press, Cambridge, MA. 281 pp.

For further information:

Michael R. Kunzmann
National Biological Service
Cooperative Park Studies Unit
University of Arizona
125 Biological Sciences E., Bldg. 43
Tucson, AZ 85721

Bullfrogs: Introduced Predators in Southwestern Wetlands

by
Philip C. Rosen
University of Arizona
Cecil R. Schwalbe
National Biological Service

In the American Southwest, much of the native fish fauna is facing extinction (Minckley and Deacon 1991); frogs in California (Fellers and Drost 1993) and frogs and garter snakes in Arizona (Schwalbe and Rosen 1988) are also in critical decline. Habitat destruction and introduced predators appear to be primary causes of native frog declines (Jennings and Hayes 1994), and habitat modification often yields ponds and lakes especially suitable for introduced species. Introduced bullfrogs (*Rana catesbeiana*) have been blamed for amphibian declines in much of western North America (e.g., Hayes and Jennings 1986; Leonard et al. 1993; Vial and Saylor 1993). Extensive cannibalism by bullfrogs renders them especially potent predators at the population level. The tadpoles require only perennial water and grazeable plant material; hence, transforming young can sustain a dense adult bullfrog population even if alternate prey are depleted. This may increase the probability that native species may be extirpated by bullfrog predation.

Introduced predatory fishes are apparently an important cause of frog declines (Hayes and Jennings 1986). They have been strongly implicated in one important case of decline of native ranid frog (family Ranidae, the "true" frogs; Bradford 1989). Some introduced crayfish may also be devastating in some areas (Jennings and Hayes 1994). In our study region, however, neither introduced fishes nor crayfish are dominant. We present results that sustain a "bullfrog hypothesis" for some native ranid declines, and we present our study as an example of how evidence accumulates to support such a hypothesis.

In 1985 we began documenting historical localities for wetland herpetofaunas (reptiles and amphibians), based on museum records and personal interviews, then revisited these and additional areas to determine current species' status. Results of this process, plus circumstantial evidence, suggested that the bullfrog was a primary cause for declines of leopard frogs and garter snakes in southern Arizona (Schwalbe and Rosen 1988).

Courtesy J.N. Carr, Arizona Game and Fish Dept.

Fig. 1. The worm has turned! In this unstaged photograph taken at Parker Canyon Lake, Cochise County, Arizona, 1964, an introduced bullfrog is swallowing a Mexican garter snake, normally a frog-eating species. Such predation appears to be destroying remaining populations of this garter snake in the United States.

In 1986-89 and 1992-93 we conducted removal censuses of bullfrogs at San Bernardino National Wildlife Refuge (SBNWR), Cochise County, Arizona. We simultaneously monitored native Chiricahua leopard frogs (*R. chiricahuensis*) and Mexican garter snakes (*Thamnophis eques*) at the sites of bullfrog removal. A control site, with no bullfrog removal, was established in comparable habitat at Buenos Aires National Wildlife Refuge (BANWR), Pima County, Arizona.

Evidence for Bullfrog Effects

Bullfrogs ate garter snakes, including Mexican garter snakes (Fig. 1), as well as numerous frogs, including young bullfrogs and the last observed leopard frogs on our intensive study areas. In addition, these frogs ate other frogs and snakes, lizards, fish, birds, and mammals in addition to many invertebrates (*see* also Bury and Whelan 1984).

We currently know of no examples of overlap between populations of the native leopard frogs *R. chiricahuensis* and *R. yavapaiensis* and bullfrogs in southern Arizona. Leopard frogs were abundant at both SBNWR and BANWR before bullfrog proliferation, and as recently as 1981, bullfrogs and leopard frogs were both still widespread at SBNWR (D. Lanning, The Arizona Nature Conservancy, unpublished data). Leopard frogs apparently were extirpated from our SBNWR study area by 1989.

In 1993-94 relict populations of Chiricahua leopard frogs (2-20 adults each) were found 5, 10, and 19 km (3.1, 6.2, and 11.8 mi) east of SBNWR. These populations are in areas not occupied by bullfrogs in habitats that may dry too frequently for non-native predators (personal observations), as seen in native frogs of the central valley of California (Hayes and Jennings 1988). These recent findings near SBNWR further support the bullfrog hypothesis in southeastern Arizona.

Checkered garter snakes (*Thamnophis marcianus*) are semi-terrestrial and coexist in abundance with bullfrogs. The highly aquatic Mexican garter snake, however, has only small, apparently declining populations where its habitat overlaps with that of bullfrogs. Because the bullfrog is also highly aquatic, its effects on the Mexican garter snake have been greater than on the checkered.

Although Mexican garter snakes do reproduce where they occur with bullfrogs, few young survive (Fig. 2). Once the young snakes outgrow vulnerability to bullfrog predation, they survive well; young adults marked in 1986-88 have been recovered at ages 7-10 in 1993, equaling and exceeding known ages for garter snakes in the wild (Fitch 1965). All of the larger, older Mexican garter snakes have damaged tails from repeated bullfrog bites, and the largest and oldest one was found dying in 1993 with gross inflammation of the tail. It appears that without successful reproduction by some of these old snakes, the study population will shortly disappear.

Bullfrog Removal Experiments

Before 1993 intensive bullfrog removals were conducted two to three times per year at SBNWR. At one study pond, 854 large (80+ mm body length) bullfrogs had been removed from about 0.2 ha (0.5 acre) of habitat. After the 3 to 4 active-season months between removals, we saw a 50%-80% rebound toward preremoval numbers, and we observed weak evidence of positive effects on native leopard frogs and garter snakes (Schwalbe and Rosen 1988). Because a bullfrog can have as many as 20,000 eggs per clutch and has multiple clutches each year, the bullfrog was clearly uncontrollable at our initial level of effort.

Starting in 1993, we increased our efforts to remove bullfrogs from SBNWR by eliminating adult bullfrogs and catching juveniles as they matured.

Discussion

If adult-free bullfrog populations are attained at SBNWR during 1994, we predict that this will result in successful recruitment of juvenile Mexican garter snakes. We propose to translocate leopard frogs from nearby areas into fenced, newly created, bullfrog-free ponds. A primary objective is to have at least one natural area to save genetic stock of the local leopard frogs.

The SBNWR, with its numerous highly productive water sources, was probably a historical regional metapopulation (a set of populations connected by immigration and emigration) center (Gilpin and Hanski 1991) for leopard frogs. During times of drought, it was likely the mainstay of the species in the San Bernardino Valley system. Some of the unexplained frog declines in western North America (Cary 1993) may ultimately be traceable to catastrophic, localized extinctions in such refugia (Sjögren 1991; Bradford et al. 1993). An observation of probable rapid migratory spread by an introduced leopard frog species in Arizona (12 km/yr; Platz et al. 1990) suggests that individuals do disperse enough to consider metapopulation models. Information related to metapopulation phenomena could markedly enhance management for leopard frogs.

It is notable that the checkered garter snake, with an evolutionary background of geographical overlap with bullfrogs, succeeds with introduced bullfrogs in the West. Similarly, the accidentally introduced and rapidly spreading Rio Grande leopard frog (*Rana berlandieri*) in Arizona (Platz et al. 1990) also evolved with bullfrogs. In fact, this leopard frog is spreading into areas where the endemic Yavapai leopard frog (*R. yavapaiensis*) has been extirpated, probably by introduced predators as well as habitat alteration (Vitt and Ohmart 1978; Jennings and Hayes 1994).

Conclusion

Introduced predators such as the bullfrog can have devastating effects on faunas that evolved without equivalent predatory types. The bullfrog, as an exotic in the absence of key original enemies (the basses, pikes, snapping turtles, and water snakes of the eastern United States), attains tremendous population densities. Such non-native predators, in core population areas of native species, can lead to regional extinctions, and may account for some unexplained amphibian declines.

We now have abundant documentation that introduced predators, especially fish, crayfish, and bullfrogs, have caused major declines of frogs and other species in western North America. In Arizona, current trends suggest that inaction could lead to disappearance of three of five native leopard frog species within a decade. We urge, in addition to simply monitoring declines, active management where appropriate, within a controlled and documented framework. There is a pressing need for a practical, successful, and vigorously supported management strategy to preserve genetic stocks and restore habitats of native ranid frogs.

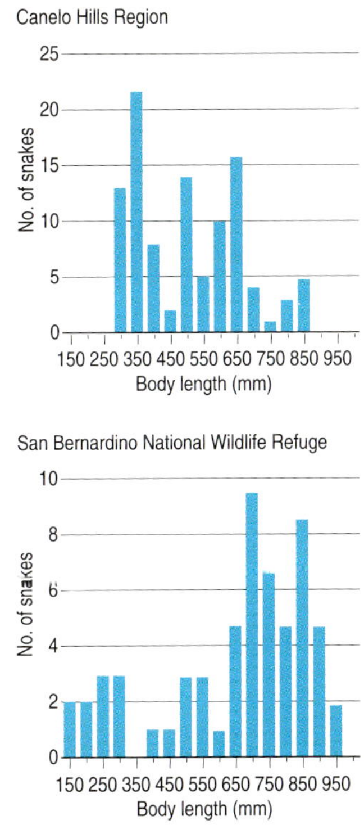

Fig. 2. Population structure of the Mexican garter snake. Numerous young snakes (200-700 mm, 1-3 years old) show successful reproduction in apparently intact populations (top), whereas bullfrog-affected populations (bottom) are composed mainly of older (700-1,000 mm, 3+ years old) snakes.

References

Bradford, D. 1989. Allotopic distribution of native frogs and introduced fishes in high Sierra Nevada lakes of California: implication of the negative effect of fish introduction. Copeia 989:775-778.

Bradford, D.F., F. Tabatabai, and D.M. Graber. 1993. Isolation of remaining populations of the native frog, *Rana muscosa*, by introduced fishes in Sequoia and Kings Canyon national parks, California. Conservation Biology 7:882-888.

Bury, R.B., and J.A. Whelan. 1984. Ecology and management of the bullfrog. U.S. Fish and Wildlife Service Resour. Publ. 155. 23 pp.

Cary, C. 1993. Hypothesis concerning the causes of the disappearance of boreal toads from the mountains of Colorado. Conservation Biology 7:355-362.

Fellers, G.M., and C.A. Drost. 1993. Disappearance of the Cascades frog *Rana cascadae* at the southern end of its range, California, USA. Biological Conservation 65:177-181.

Fitch, H.S. 1965. An ecological study of the garter snake, *Thamnophis sirtalis*. University of Kansas Publications of the Museum of Natural History 15:493-564.

Gilpin, M.E., and I. Hanski. 1991. Metapopulation dynamics. Academic Press, London. 336 pp.

Hayes, M.P., and M.R. Jennings. 1986. Decline of ranid frog species in western North America: are bullfrogs (*Rana catesbeiana*) responsible? Journal of Herpetology 20:490-509.

Hayes, M.P., and M.R. Jennings. 1988. Habitat correlates of distribution of the California red-legged frog (*Rana aurora draytonii*) and the foothill yellow-legged frog (*Rana boylei*): implications for management. Pages 144-158 *in* R.C. Szaro, K.E. Severson, and D.R. Patton, eds. Management of amphibians, reptiles, and small mammals in North America. U.S. Forest Service Gen. Tech. Rep. RM-166, Fort Collins, CO.

Jennings, M.R., and M.P. Hayes. 1994. Decline of native ranids in the desert southwest. *In* P.R. Brown and J.W. Wright, eds. Proceedings of the Conference on the Herpetology of the North American Deserts. Southwestern Association of Herpetologists Special Publ. 5. Los Angeles, CA. In press.

Leonard, W.P., H.A. Brown, L.L.C. Jones, K.R. McAllister, and R.M. Storm. 1993. Amphibians of Washington and Oregon. Seattle Audubon Society, Seattle, WA. 168 pp.

Minckley, W.L., and J.E. Deacon, eds. 1991. Battle against extinction: native fish management in the American West. University of Arizona Press, Tucson. 517 pp.

Platz, J.E., R.W. Clarkson, J.C. Rorabaugh, and D.M. Hillis. 1990. *Rana berlandieri*: recently introduced populations in Arizona and southeastern California. Copeia 1990:324-333.

Schwalbe, C.R., and P.C. Rosen. 1988. Preliminary report on effects of bullfrogs on wetland herpetofauna in southeastern Arizona. Pages 166-173 *in* R.C. Szaro, K.E. Severson, and D.R. Patton eds. Management of amphibians, reptiles, and small mammals in North America. U.S. Forest Service Gen. Tech. Rep. RM-166, Fort Collins, CO.

Sjögren, P. 1991. Extinction and isolation gradients in metapopulations: the case of the pool frog (*Rana lessonae*). Pages 135-147 *in* M.E. Gilpin and I. Hanski, eds. Metapopulation dynamics. Academic Press, London, UK.

Vial, J.L., and L. Saylor. 1993. The status of amphibian populations: a compilation and analysis. International Union for the Conservation of Nature and Natural Resources, Species Survival Commission, Declining Amphibian Populations Task Force Working Document 1. 98 pp.

Vitt, L.J., and R.D. Ohmart. 1978. Herpetofauna of the lower Colorado River: Davis Dam to the Mexican border. Proceedings of the Western Foundation of Vertebrate Zoology 2:35-72.

For further information:

Philip C. Rosen
University of Arizona
Department of Ecology and
Evolutionary Biology
Tucson, AZ 85721

Invasions of the Brown Tree Snake

by
T.H. Fritts
G.H. Rodda
National Biological Service

Around 1950, populations of the brown tree snake (*Boiga irregularis*) were introduced on Guam, a previously snake-free island. This introduction was the result of post-World War II traffic carrying military materials from the South Pacific region (Savidge 1987; Rodda et al. 1992). It resulted in major ecological changes and the loss of several bird and lizard species from the island starting in the 1970's and extending to the late 1980's. The severity of ecological damages resulting from this introduced snake may have been increased by the presence of other nonindigenous species, which served as alternative prey as native species declined.

The brown tree snake dispersed throughout Guam in the 1950's, 1960's, and 1970's, reaching high populations that resulted in devastating levels of predation on most native and introduced vertebrates (Savidge 1987; Engbring and Fritts 1988; Rodda et al. 1992). At the peak of the snake's irruption on Guam, densities probably exceeded 100 snakes/ha (40 snakes/acre), but following depletion of many of Guam's birds and mammals, snake densities appear to have fallen to 20-50 snakes/ha (8-20 snakes/acre; Rodda et al. 1992).

In the face of the loss of native forest birds and drastic reductions in other bird, mammal, and reptile species, the snake subsisted on smaller lizard prey and on introduced species, including lizards (*Hemidactylus frenatus* and *Carlia* cf. *fusca*), domestic poultry and cage birds, rodents (*Rattus* spp. and *Mus musculus*), house shrews (*Suncus murinus*), Eurasian tree sparrows (*Passer montanus*), and Javanese turtle doves (*Streptopelia bitorquata*). Thus, the reduction of snake densities that might have been expected after the loss of native prey species was limited because the snake could subsist on alternative introduced prey.

Brown tree snake (*Boiga irregularis*).

Courtesy G.H. Rodda

Species Lost from Guam

Since the arrival of the snake on Guam, the island has lost most of its indigenous forest vertebrates (Fig. 1). Too few baseline data are available to unequivocally determine the degree to which the snake is responsible for these losses, but several kinds of evidence create a strong case for the snake's role in the extirpation of many bird species (Savidge 1987, 1988; Conry 1988; Engbring and Fritts 1988) and several lizard species (Rodda et al. 1991). Additionally, some evidence exists that the snake played a role in the disappearance and decline of Guam's native mammals, three bat species (Wiles 1987), but no direct information is available for the two bat species that disappeared before 1980. The evidence clearly shows, however, that Guam has experienced a remarkably complete loss of its vertebrate fauna.

Even with all of the vertebrates at risk from the snake, the pattern of species' losses has followed a size gradient that is consistent with the snake's dietary habits (Engbring and Fritts 1988; Fritts 1988). Small birds, small mammals, and medium-sized lizards disappeared first and seem to have been most heavily affected. Contrary to what might have been expected, the most abundant bird species were affected first. We cannot determine if the abundance of the prey led to more effective search images for the snakes or if the ecological characteristics of the species and the habitats occupied contributed to this prey difference. The surviving native species and those that lasted the longest in the wild all exhibited extreme sizes (i.e., larger or smaller than those most affected) or some other trait that has minimized their vulnerability to snake predation.

Examples of these traits include large size: Mariana flying fox (*Pteropus mariannus*), Marianas crow (*Corvus kubaryi*), and Indian monitor lizard (*Varanus indicus*); urban dwelling: Micronesian starling (*Aplonis opaca*), mourning gecko (*Lepidodactylus lugubris*), and stump-toed gecko (*Gehyra mutilata*); cavity nesting: Micronesian starling and Micronesian kingfisher (*Halcyon cinnamomina*); cave ceiling roosting: gray swiftlet (*Aerodramus vanikorensis*); and extremely small size: mourning gecko and Marianas blue-tailed skink (*Emoia caeruleocauda*). All surviving endotherm populations (birds and mammals) consist of fewer than 1,000 individuals, and long-term population viabilities are in doubt for most of these groups on Guam.

Small lizards are much more numerous and have better long-term prospects even though evidence exists of localized extinctions caused by temporary surges in snake populations. The big tree gecko (*Gehyra oceanica*) has virtually disappeared since 1985, but its smaller congener (species in the same genus), the stump-toed gecko, persists in forested habitats in low numbers (Rodda et al. 1991). Some small introduced lizard species (mourning gecko, common house gecko, *Hemidactylus frenatus*, and brown four-fingered skink, *Carlia* cf. *fusca*) have expanded into new habitats in the absence of other species; they therefore maintain larger population levels on Guam even though they experience heavy predation by snakes.

The relative abundance of the Marianas blue-tailed skink has dropped markedly as the brown four-fingered skink increased after its arrival in Guam in the early 1950's (Fig. 2). Effects of predation by the snake and interactions between introduced lizards are evident in the relative abundances of lizard families, with the primarily arboreal gekkonids declining while the primarily terrestrial and more predation-resistant skinks have increased. Even introduced rodents and shrews show declines due to predation by snakes; trapping success for rodents and shrews was significantly reduced in 1984-85 compared to that of 1962-64 (Savidge 1987).

Risks of Dispersal from Guam

The many brown tree snakes on Guam make it probable that they may disperse as passive stowaways in ship and air traffic to other islands and the U.S. mainland (Fritts 1987, 1988; McCoid and Stinson 1991). To date, stowaway brown tree snakes have arrived in the northern Marianas Islands (Saipan, Rota, and Tinian); Marshall Islands (Kwajalein Atoll); Cocos Island near Guam; Okinawa; Diego Garcia in the Indian Ocean; Oahu Island, Hawaii; and Corpus Christi, Texas (Fritts 1988; unpublished manuscript). Verified and probable sightings of brown tree snakes span 1949-94 and show that dispersal of the brown tree snake is not uncommon. The apparent surge in the 1990's probably reflects better reporting of stowaway incidents rather than increased dispersal.

Risks of Damages from Further Colonizations

The islands adjacent to Guam are the northern Marianas, which have vertebrate faunas that are similar to Guam's, including some of the same introduced species. Like Guam, the northern Marianas have no native snakes. Thus, prey bases similar to those on Guam and capable of supporting high population levels of brown tree snakes exist in the northern Marianas, and species losses can be anticipated if the snake becomes established. For example, of 27 native resident bird species on the main islands of the northern Marianas (Saipan, Tinian, and Rota), 20 are shared with the original fauna of Guam and an additional 7 species are closely related to birds known from Guam. Guam and the northern Marianas also share five introduced bird species (Engbring et al. 1986). Six species of birds are federally listed as endangered or threatened in the northern Marianas, and all of these are conspecific (of or relating to the same species) or closely related to birds that have disappeared from Guam or declined significantly there (Engbring and Ramsey 1984; U.S. Department of the Interior 1990). Of 20 species of terrestrial amphibians and reptiles presently or formerly known from Guam and Cocos Island, 15 are shared with the northern Marianas, 8 native and 7 introduced (Rodda et al. 1991). Thus, the northern Marianas not only share the ecological vulnerabilities that led to mass extirpations on Guam, but also the bulk of the remaining habitat for Marianas' native species is on islands that have received stowaway snakes from Guam.

Hawaii suffered major losses in its vertebrate fauna after the arrival of the Polynesians and again after contact with Europeans. The state originally had 59 passerine bird species, but only 38 survived into historical times. Fifty species of passerines have been introduced in Hawaii, and those birds make up most of the land birds present today. At least 30 species of birds native to Hawaii are federally listed as threatened or endangered. One bird species native to Guam, the gray swiftlet, is established on Oahu (Moulton and Pimm 1986). Of the 14

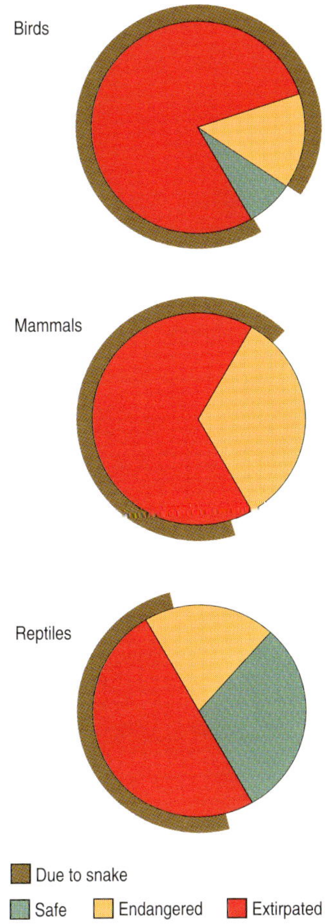

Fig. 1. Status in 1993 of Guam's native forest vertebrates (those present in 1950) with estimates of the degree to which decline was due to the introduction of the brown tree snake.

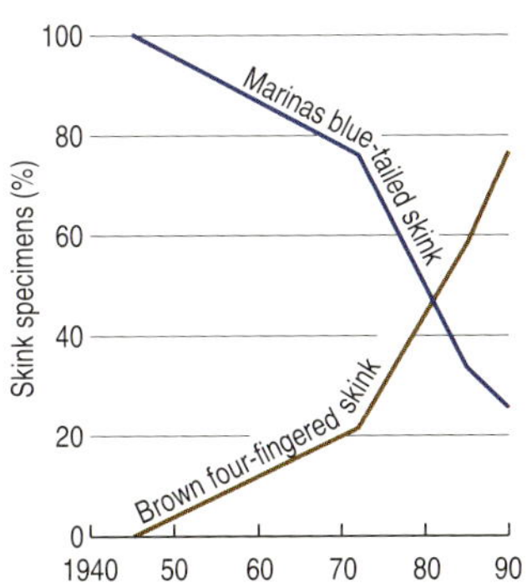

Fig. 2. Changes in the proportions of specimens of two common skinks in museum collections from Guam in four samples spanning 1945-90: Marianas blue-tailed skink (*Emoia caeruleocauda*) and brown four-fingered skink (*Carlia* cf. *fusca*).

reptile species present in Hawaii (all introduced), 8 are known as native or introduced species on Guam. Many of these introduced species are locally abundant and attain high population levels in Hawaii. All these factors show how capable the brown tree snake is in exploiting elements of the native and introduced fauna of Hawaii and in attaining high population levels in Hawaii and on other Pacific islands on which it may become established.

The effects of the brown tree snake extend beyond ecological damages; the snakes frequently climb on electrical transmission lines causing faults and disrupting electrical supplies, enter urban and residential areas where they consume poultry and pets, and bite humans causing trauma and serious health risks for small children (Fritts 1988).

For further information:

T.H. Fritts
National Biological Service
National Museum of
Natural History
MRC 111
10th and Constitution, NW
Washington, DC 20560

References

Conry, P.J. 1988. High nest predation by brown tree snakes on Guam. Condor 90:478-482.

Engbring, J., and T.H. Fritts. 1988. Demise of an insular avifauna: the brown tree snake on Guam. Transactions of the Western Section of the Wildlife Society 24:31-37.

Engbring, J., and F.L. Ramsey. 1984. Distribution and abundance of the forest birds of Guam: results of a 1981 survey. U.S. Fish and Wildlife Service FWS/OBS 84(20). 54 pp.

Engbring, J., F.L. Ramsey, and V.J. Wildman. 1986. Micronesian forest bird survey, 1982: Saipan, Tinian, Agiguan, and Rota. U.S. Fish and Wildlife Service, Honolulu, HI. 143 pp.

Fritts, T.H. 1987. Movements of snakes via cargo in the Pacific region. 'Elepaio 47:17-18.

Fritts, T.H. 1988. The brown tree snake, *Boiga irregularis*, a threat to Pacific Islands. U.S. Fish and Wildlife Service Biological Rep. 88(31). 36 pp.

McCoid, M.J., and D.W. Stinson. 1991. Recent snake sightings in the Mariana Islands. 'Elepaio 51:36-37.

Moulton, M.P., and S.L. Pimm. 1986. Species introductions to Hawaii. Pages 231-249 *in* H.A. Mooney and J.A. Drake, eds. Ecology of biological invasions of North America and Hawaii. Springer-Verlag, New York.

Rodda, G.H., T.H. Fritts, and P.J. Conry. 1992. Origin and population growth of the brown tree snake, *Boiga irregularis*, on Guam. Pacific Science 46:46-57.

Rodda, G.H., T.H. Fritts, and J.D. Reichel. 1991. The distributional patterns of reptiles and amphibians in the Mariana Islands. Micronesica 24:195-210.

Savidge, J.A. 1987. Extinction of an island forest avifauna by an introduced snake. Ecology 68:660-668.

Savidge, J.A. 1988. Food habits of *Boiga irregularis*, an introduced predator on Guam. Journal of Herpetology 22:275-282.

U.S. Department of the Interior. 1990. Endangered and threatened species recovery program. Report to Congress. Washington, DC. 406 pp.

Wiles, G.J. 1987. Current research and future management of Marianas fruit bats (Chiroptera:Pteropidae) on Guam. Australian Mammalogy 10:93-95.

Wild Horses and Burros on Public Lands

by
Tom Pogacnik
Bureau of Land Management

On December 15, 1971, Congress passed legislation to protect, manage, and control wild horses (*Equus caballus*) and burros (*E. asinus*) on public lands. The Wild Free-Roaming Horses and Burros Act (Public Law 92-195) described these animals as fast-disappearing symbols of the historic and pioneer spirit of the West. The Bureau of Land Management (BLM) and the U.S. Forest Service are charged with administering the law, which specifies how wild horses and burros are to be managed on the range and how excess animals are to be disposed. Section 3.(a) requires the Secretary of the Interior to manage wild free-roaming horses and burros in a manner designed to achieve and maintain a thriving natural ecological balance on public lands. This section also specifies requirements for inventorying, monitoring, establishing appropriate management levels, making removals, placing excess animals, and establishing criteria for destruction of animals.

Although these animals were once considered endangered by the nearly unrestrained onslaught of the mustangers and others, they have thrived under federal protection (Fig. 1). With few predators and with protection from humans, wild horse and burro populations on BLM-administered lands (where most of the animals are located) quickly grew until control of the populations and the effect on their habitat became a major concern.

Courtesy T. Pogacnik, BLM

Wild horses (*Equus caballus*).

The act requires that BLM maintain a current inventory of wild horses and burros on certain public lands. At present, BLM censuses each of the 196 herd-management areas on a rotating basis, usually every 3 years, using census techniques based on research published by the National Academy of Sciences (1982). Censuses in 1993 identified a nationwide population of 46,500 wild horses and burros (Fig. 2). Accuracy for the 1993 census ranged from 85% to 99% on wild horses and 75% to 88% on wild burros.

Annual population growth in wild horse herds varies from 5% to 25%, depending on range and environmental conditions, with 15% being a long-term average. At this rate of increase, wild horse populations may double in 5 years. The annual growth in wild burro populations has not been determined, but their reproductive capacity may be similar to that of wild horse herds.

The act specifies that wild horses and burros may be managed only on lands where they existed on December 15, 1971, the time of the act's passage. The population of wild horses and burros within those 1971 areas of use was estimated at 17,000 animals; however, at that time no formal inventory policies or procedures existed to census populations. The BLM now has 269 herd areas, 196 within which wild horses and burros are managed to some extent and 73 from which all wild horses and burros will be removed.

Wild horse and burro herd areas occupy almost 43 million acres (17.4 million ha) of public and private land in Arizona (about 4 million acres or 1.6 million ha), California (6 million+ acres or 2.4 million ha), Colorado (800,000+ acres or 324,000 ha), Idaho (450,000+ acres or 182,250 ha), Montana (55,000+ acres or about 22,275 ha), Nevada (nearly 19 million acres or nearly 8 million ha), New Mexico (nearly 150,000 acres or 60,750 ha), Oregon (nearly 4 million acres or 1.6 million ha), Utah (2.5 million acres or 1 million ha), and Wyoming (nearly 6 million acres or 2.4 million ha) (BLM 1993).

Within most herd areas, wild horses and burros graze with domestic livestock and a variety of indigenous wildlife species. Because they are generalist species, wild horses and burros inhabit a variety of habitats and vegetative communities.

The BLM's land-use planning process and evaluation of current inventory and monitoring data are used to determine a population level that maintains a thriving natural ecological balance with other uses. The act directs BLM to achieve appropriate population levels by removals, humane destruction, or other options, including antifertility methods.

BLM no longer destroys healthy excess wild horses and burros. Since 1973, when the first removals occurred, BLM has removed 141,762 wild horses and burros from public land and placed 122,627 animals into private care through the Adopt-A-Horse program.

Removing excess animals from populations that exceed appropriate numbers is expensive, has restricted BLM's attempts to pursue other management alternatives, and therefore has often allowed populations to increase dramatically. When populations reached crisis proportions, funding was increased and large numbers of excess animals were removed from the range and placed with private citizens through the adoption program. The number of animals removed often was greater than the number that could be adopted, resulting in high costs for feeding and veterinary services while animals were held pending adoption.

In June 1992 the Director of BLM approved the Strategic Plan for the Management of Wild Horses and Burros on Public Lands (BLM 1992). This plan represents BLM's first comprehensive policy for addressing wild horse and burro management. To reduce the frequency of removals, the plan recommends the use of antifertility management to slow population growth to a level where removals are only required on a cycle of 5 or more years instead of the current 3-year cycle. Pending the availability of practical and cost-effective fertility-control techniques, selective removal of animals based on age or sex is being used to reduce the growth rate in wild horse populations. The negative aspects of selective removal include the difficulty of predicting results through computer modeling and the extensive monitoring needed to ensure that age and sex ratios have not been altered to a level that could threaten the herd. Selective removals for controlling population growth are considered a temporary management option until research on immunocontraception is completed and can be implemented.

The BLM supports research on the use of immunocontraception for controlling wild horse population growth. Successful immunocontraceptive antigens have been developed; researchers are now trying to develop a system that would inhibit reproduction for 2 to 3 years (J.F. Kirkpatrick, Deacones Medical Research Institute, Billings, personal communication).

Before the passage of the act, wild horses and burros were often captured and destroyed as nuisances or were sold for profit, chiefly for use in commercial products. The methods employed in their capture and destruction were often less than humane. As public awareness of these animals grew, so too did support for federal legislation to protect them from inhumane treatment.

Public interest in the wild horse and burro program continues to direct implementation of the act. Since the act's passage in 1971, there have been 44 district court suits and in excess of 200 appeals of BLM decisions to the Interior Board of Land Appeals.

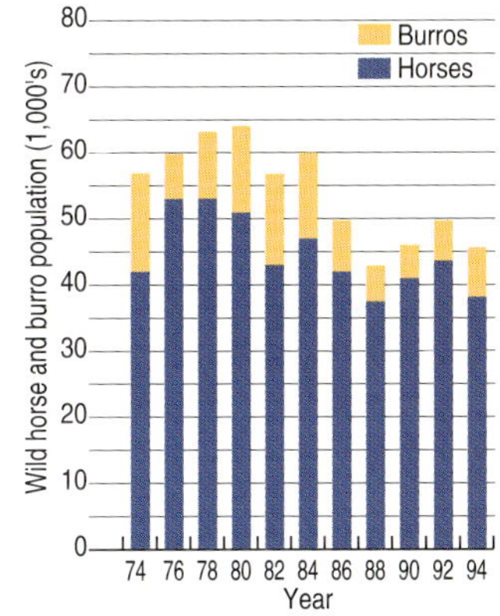

Fig. 1. Wild horse and burro population trends in BLM-administered lands since passage of the Wild Free-Roaming Horses and Burros Act of 1971.

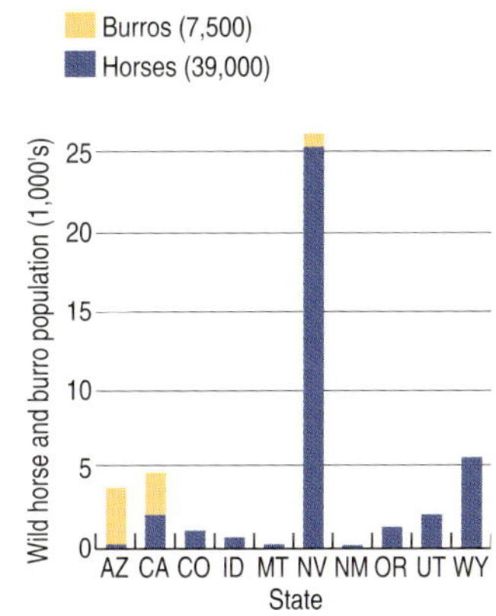

Fig. 2. Wild horses and burros in 1993: population by state.

References

BLM. 1992. Strategic plan for the management of wild horses and burros on public lands. Bureau of Land Management, Washington, DC. 12 pp.

For further information:

Tom Pogacnik
Bureau of Land Management
Wild Horse and Burro National
Program Office
850 Harvard Way
PO Box 12000
Reno, NV 89520

BLM. 1993. Ninth report to Congress on the administration of the Wild Free-Roaming Horse and Burro Act. Bureau of Land Management, Washington, DC. 70 pp.

National Academy of Sciences. 1982. Wild and free-roaming horses and burros: final report of the Committee on Wild Free-Roaming Horses and Burros Board on Agriculture and Renewable Resources National Research Council. National Academy Press, Washington, DC. 80 pp.

Purple Loosestrife

by
Richard Malecki
National Biological Service

Courtesy R. Malecki, NBS

Purple loosestrife (*Lythrum salicaria*).

Purple loosestrife (*Lythrum salicaria*) is an exotic wetland perennial introduced to North America from Europe in the early 19th century (Stuckey 1980). By the 1930's, the plant was well established along the New England seaboard. The construction of inland canals and waterways in the 1880's favored the expansion of purple loosestrife into interior New York and the St. Lawrence River Valley (Thompson et al. 1987). The continued expansion of loosestrife has coincided with increased development and use of road systems (Thompson et al. 1987), commercial distribution of the plant for horticultural purposes, and regional propagation of seed for bee forage (Pellet 1977). The plant now occurs in dense stands throughout the northeastern United States, southeastern Canada, the Midwest, and in scattered locations in the western United States and southwestern Canada. Newly created irrigation systems in many of the western states have supported its further spread.

Purple loosestrife is a classic example of an introduced species whose distribution and spread have been enhanced by the absence of natural enemies and the disturbance of natural systems, primarily by human activity. Although noted for the beauty of its late summer flowers, which also provide a nectar source for bees, loosestrife has few other redeeming qualities. Its invasion into a wetland system results in suppression of the native plant community and the eventual alteration of the wetland's structure and function (Thompson et al. 1987). Large, monotypic stands not only jeopardize various threatened and endangered plants and wildlife, such as Long's bulrush (*Scirpus longii*) in Massachusetts (Coddington and Field 1978), small spikerush (*Eleocharis parvula*) in New York (Rawinski 1982), and the bog turtle (*Clemmys muhlenbergii*) in the northeastern United States (Bury 1979), but they also eliminate natural foods and cover essential to many wildlife, including waterfowl (Rawinski and Malecki 1984).

Purple loosestrife has many traits that enabled it to become a nuisance in North America. A single, mature plant can produce more than 2.5 million seeds annually; these seeds are long-lived (Welling and Becker 1990) and easily dispersed by water and in mud, adhering to aquatic wildlife, livestock, and people (Thompson et al. 1987). Established plants are tall (about 2 m or 6.5 ft) with 30-50 stems forming wide-topped crowns that dominate the herbaceous canopy. A strong rootstock serves as a storage organ, providing resources for growth in spring and regrowth if the aboveground shoots are cut, burned, or killed by application of foliar herbicides. No native herbivores or pathogens in North America are known to suppress purple loosestrife (Hight 1990).

No effective method is available to control loosestrife, except in small localized stands that can be intensively managed. In such isolated areas, the plant can be eliminated by uprooting by hand and ensuring that all vegetative parts are removed. Other control techniques include water-level manipulation, mowing or cutting, burning, and herbicide application (Malecki and Rawinski 1985). Although these controls can eliminate small and young stands, they are costly, require continued long-term maintenance, and in the case of herbicides, are nonselective and environmentally degrading.

The most promising control measure for purple loosestrife is the application of classical biological weed-control procedures that use natural enemies like insects, mites, nematodes, and pathogens to reduce weed densities to tolerable levels. Results of insect surveys and screening tests conducted with the U.S. Department of Agriculture's Agriculture Research Service and the International Institute of Biological Control in Europe have identified five beetle species as potential control agents for purple loosestrife. Each species showed enough host specificity for purple loosestrife to be introduced with no ill effects to native North American plants.

Efforts are under way to rear large numbers of these insect species for further distribution and establishment in other states and provinces. A petition to introduce two of these beetles is under review by the USDA's Animal and Plant Health Inspection Service. Initial collection of these insects in Europe for release into the United States is planned for 1994.

A cooperative state and federal program for the biological control of purple loosestrife focuses on an international environmental weed problem that cannot be controlled by conventional means. With support from federal and state agencies we have brought together an international scientific advisory staff to participate in and oversee the selection, screening, and introduction of an insect predator community that will provide a long-lasting biological control mechanism for loosestrife, and which will also develop a corresponding program of research and evaluation.

Purple loosestrife is now a naturalized weed that always will be a part of most North American wetlands. Researchers hope that introducing select insects will result in replacing monotypic stands of loosestrife by native vegetation and an overall decrease in the occurrence of the plant. We predict a reduction of purple loosestrife abundance over the next 15-20 years to about 10% of its current level over about 90% of its North American range (Malecki et al. 1993).

References

Bury, R.B. 1979. Review of the ecology and conservation of the bog turtle, *Clemmys muhlenbergii*. U.S. Fish and Wildlife Service Spec. Sci. Rep. Wildlife 219. 9 pp.

Coddington, J., and K.G. Field. 1978. Rare and endangered vascular plant species in Massachusetts. U.S Fish and Wildlife Service, Washington, DC. 62 pp.

Hight, S.D. 1990. Available feeding niches in populations of *Lythrum salicaria* L. (purple loosestrife) in the northeastern United States. Proceedings of the International Symposium of Biological Control of Weeds 7:269-278.

Malecki, R., B. Blossey, S. Hight, D. Schroeder, L. Kok, and J. Drea. 1993. Biological control of purple loosestrife. BioScience 43:680-686.

Malecki, R., and T.J. Rawinski. 1985. New methods for controlling purple loosestrife. New York Fish and Game Journal 32:9-19.

Pellet, M. 1977. Purple loosestrife spreads down river. American Bee Journal 117:214-215.

Rawinski, T.J. 1982. The ecology and management of purple loosestrife (*Lythrum salicaria*) in central New York. M.S. thesis, Cornell University, Ithaca, NY. 88 pp.

Rawinski, T.J., and R.A. Malecki. 1984. Ecological relationships among purple loosestrife, cattail, and wildlife at the Montezuma National Wildlife Refuge. New York Fish and Game Journal 31:81-87.

Stuckey, R.L. 1980. Distributional history of *Lythrum salicaria* (purple loosestrife) in North America. Bartonia 47:3-20.

Thompson, D.Q., R.L. Stuckey, and E.B. Thompson. 1987. Spread, impact, and control of purple loosestrife (*Lythrum salicaria*) in North American wetlands. U.S. Fish and Wildlife Service Res. Rep. 2. 55 pp.

Welling, C.H., and R.L. Becker. 1990. Seed bank dynamics of *Lythrum salicaria* L.: implications for control of this species in North America. Aquatic Botany 38:303-309.

For further information:

Richard Malecki
National Biological Service
New York Cooperative Fish and Wildlife Research Unit
Fernow Hall
Cornell University
Ithaca, NY 14853

Courtesy R. Hayes and H. Weiss, BLM

Habitat Assessments

Michael D. Jennings
National Biological Service
Idaho Cooperative Fish and Wildlife Research Unit
University of Idaho
Moscow, ID 83843

Overview

Articles in this section address the development, interpretation, and analysis of ecological information over very large geographic regions and are characterized by the huge undertaking to assemble and manipulate the data required. These articles illuminate the imperative need for information at multiple levels of both geographic scale (site, small watershed, state, regional, national) and biotic organization (population, species, natural community, landscape, and biome).

A systematic approach toward the development of science-based ecological information at multiple scales and across large areas has been lacking from our management of natural resources. Significant gains in achieving an environmentally sustainable society with an acceptable standard of living can be had by addressing this issue, and the articles in this section point the way.

Edwards and Stoms and Davis present some early results of the National Gap Analysis Project (see box by Scott et al., this section). Edwards shows that less than 10% of the vegetation cover types in Utah are represented within conservation lands. There is no assurance that the 90% of vegetation types (or habitat types) not represented in conservation lands will not be eliminated by changes in land use. In a world where the demand for raw materials is increasing and the rate of land-use change is rapid, adequate representation of habitat types in conservation lands is important if we are to prevent extinctions. The lack of adequate representation of habitat types in conservation lands is also the situation described by Stoms and Davis in the next article. They show that while almost 10% of the total surface area of southwestern California is managed to protect native biological diversity, most of this land is at high elevations. Natural communities at low elevations, such as coastal sage scrub and California walnut woodlands, are in considerable danger of extinction.

Shaw and Jennings describe the Multi-Resolution Land Characteristics Database, which is the first effort to provide consistent, direct, and integrated observations of large-area ecosystems, producing basic as well as interpreted information for a range of purposes. This effort includes the land-cover types of agricultural and urban areas as well as natural areas. With access to these data sets, policy decisions as well as daily management choices may, for the first time, be regularly examined in a biogeographic context covering the entire distribution of a natural feature of concern (such as a particular habitat type). Some of these data are already available in digital format over the Internet.

Loveland and Hutcheson compare the most current picture of general vegetation patterns (taken from the weather satellite, which is at an altitude of 833 km, or 517 mi, above Earth) with a map of what the vegetation may have been like before European settlement. In addition to providing some idea of the difference between the land cover of today and, hypothetically, presettlement land cover, they show conceptually the value of being able to make these kinds of comparisons. The authors carefully point out the limitations of each of the maps they use, then they walk the reader through how such a comparison is made. Because of the coarse geographic scale used, only general patterns can be shown and the results of this comparison are more meaningful when used to estimate large-area carbon flux, for example, than for calculating changes in biological diversity. The importance of this article is not in the results of the comparison but in the concepts of using large-area land-cover data to assess the past and present trends of landscape-level ecological conditions and processes.

Wilen's article cites studies showing that half of the nation's wetlands have been converted to uplands since colonial times. He demonstrates that the apparent slowing trend in overall wetland loss is deceptive because qualitative changes that do not show up as a net loss of wetlands are occurring in different types of wetlands. For example, in recent times, vast tracts of forested wetlands have been converted to other wetland forms, such as wet meadows. This is especially important because of their complex functions, such as flood control and pollution abatement, as well as their providing critical wildlife habitat. By using data from the National Wetlands Inventory, Wilen shows that overall, wetlands are losing their diversity. Without systematic science-based efforts like the NWI to map our natural resources, there can be no meaningfully coherent information for making decisions about how to manage them.

Because the dynamics of larger systems (e.g., landscapes) constrain the behavior and occurrence of the smaller systems that they encompass (e.g., populations or species), by means that are independent of the smaller systems, conservation efforts implemented at the levels of populations or species cannot be effective when systemwide changes are occurring at the landscape level. Environmental changes that were formerly limited to affecting populations and species are now manifest at scales by which natural community and landscape systems function. Therefore, if we are to make significant progress in slowing the loss of our biological heritage, the basis for solving problems and implementing decisions must be predicated on information derived from multiple scales of geographic resolution as well as of biotic organization.

Gap Analysis: A Geographic Approach to Planning for Biological Diversity

by
J. Michael Scott
Edward T. LaRoe
Michael D. Jennings
National Biological Service

The Gap Analysis Program (GAP) is an approach to protecting the nation's biological diversity based on a collaborative effort among citizens, businesses, nonprofit groups, universities, and local, state, and federal agencies. More than anything, GAP is a method of developing information about biological diversity that will enable individuals, planners, managers, and policy makers to make informed decisions. Species and habitats not adequately represented within conservation areas constitute gaps in programs meant to prevent species from becoming extinct. By providing information before extinction crises, GAP seeks proactive rather than reactive solutions.

The questions that GAP asks are: How can we prevent the components of biological diversity from becoming endangered with extinction before they reach social and economic crises? What is the present conservation status of all species and their habitats, not just those currently endangered?

To answer these questions on a state-by-state basis, people with expertise in geography, sociology, economics, zoology, botany, statistics, and ecology cooperate in mapping the distributions of dominant natural vegetation (as habitat types), and the distributions of each vertebrate species. Nationwide standards are used so that the maps of one state will fit with the maps of adjacent states. Because these maps are standardized across the United States, yet based on state and local information, they provide a critical framework for ecosystem management that is integrated across the private and public sectors. For example, these maps help define areas with the highest species diversity as well as how these areas match up with present conservation areas.

In the process of mapping land cover, GAP provides most states not only with computerized maps of existing conditions throughout the state (most for the first time ever), but also with maps of these same conditions across contiguous states, thereby providing context for what occurs within the state. The GAP is not a substitute for detailed studies of any particular site; instead it provides information, focus, and direction for management decisions at the ecosystem level. GAP is now under way in 33 states and consists of more than 200 cooperating organizations nationwide. It is coordinated by the National Biological Service.

For further information:

Gap Analysis Program
National Biological Service
Idaho Cooperative Fish and Wildlife Research Unit
University of Idaho
Moscow, ID 83843

Furthermore, the mechanisms, or the "emergent properties," by which an ecological system operates cannot be identified by a simple aggregation of its smaller components nor by a reduction of its larger components (Allen and Starr 1982; O'Neill et al. 1986). To adequately characterize an ecosystem, it must be observed as a functioning whole rather than only inferred by reducing it to its component parts and then re-aggregating the information discovered about the components. For ecosystems that cover large areas, observation is difficult, perhaps impossible, without using aerial photography and satellite imagery along with computerized systems that can handle the large amounts of information for analysis.

There are four requisites to the effective management of biological diversity, soil, water, and natural processes across large landscapes: standardized definitions of the resources; replicative scientific methods for inventories that must go beyond lists of species to include natural communities and their processes; a high-quality environmental information system with easy access for all; and the expertise to usefully synthesize the information (Jennings and Reganold 1991). The National Wetlands Inventory, Gap Analysis, and the Multi-Resolution Land Characteristics Database are achieving these requisites.

References

Allen, T.F., and T.B. Starr. 1982. Hierarchy: perspectives for ecological complexity. University of Chicago Press, IL. 310 pp.

Jennings, M.D., and J.P. Reganold. 1991. A theoretical basis for managing environmentally sensitive areas. Environmental Conservation 18(3):211-218.

O'Neill, R.V., D.L. DeAngelis, J.B. Waide, and T.F.H. Allen. 1986. A hierarchical concept of ecosystems. Princeton University Press, NJ. 253 pp.

Protection Status of Vegetation Cover Types in Utah

by
Thomas C. Edwards, Jr.
National Biological Service

Maintaining biological diversity must be done at all levels of an ecosystem, not just for endangered species (Noss 1991; Scott et al. 1991). The Gap Analysis Program is one proactive approach for assessing the current status of biodiversity at all levels. By using computerized mapping techniques called geographic information systems (GIS) to identify "gaps" in biodiversity protection, gap analysis provides a systematic approach for evaluating how biological diversity can be protected in given areas. If problems are identified through gap analysis, appropriate management action can be taken, including establishing new preserves or changing land-use practices (Edwards et al. 1993; Scott et. al 1993; Edwards and Scott 1994).

Our gap analysis includes three primary GIS layers: distribution of actual vegetation cover types; land ownership; and distributions of terrestrial vertebrates as predicted from the distribution of vegetation and from observations. By using the GIS, map overlays of animal distribution and land ownership are compared to estimate the relative extent of protection afforded each vertebrate species. Gap analysis functions organize biological information by using the data base to provide the context for other, more detailed studies.

In this article, we apply gap analysis to assess the protection status of mapped vegetation cover types in Utah. We briefly describe the process used to model and map vegetation cover types and how this process was linked with land ownership to provide an estimate of the level of protection afforded each vegetation cover type in Utah. A central tenet of gap analysis is that the degree of conservation protection afforded a given area can be determined by ownership and management. To assess protection, we used land ownership maps; each ownership was assigned one of four management status codes (Table 1). For Utah, 38 vegetation cover types and land-cover classes were modeled by using Landsat Thematic Mapper satellite data (Table 2). How much land is necessary to protect biodiversity or certain species is problematic. We arbitrarily define adequate protection as requiring at least 10% of a vegetation cover type in status category 1 or 2.

Table 1. Management status codes applied to Utah land ownership (Scott et al. 1993).

Code	Description
1	An area having an active management plan in operation to maintain a natural state and within which natural disturbances (e.g., fire, floods) are allowed to proceed without interference or are mimicked through management.
2	An area generally managed for natural values, but which may receive use that degrades the quality of existing natural communities.
3	Most nondesignated public lands. Legal mandates prevent the permanent conversion of natural habitat types to anthropogenic habitat types and confer protection to federally listed endangered and threatened species.
4	Private or public lands without an existing easement or irrevocable management agreement to maintain native species and natural communities and which are managed for intensive human use.

Status of Lands

State and federal public lands make up roughly 71% of the 21,979,000 ha (54,288,130 acres) of Utah (Table 3). Land protection status reflects this public control over lands (Table 3). Only 1,554 ha (3,833 acres) of the state's land are considered status 1 lands; these are owned exclusively by The Nature Conservancy. The area in status code 2 is 874,736 ha (3.98%; 2,160,605 acres); the area considered status code 3 is 15,464,474 ha (70.36%; 38,197,251

Table 2. Protection status of mapped vegetation cover types in Utah.

Cover type	Status code								
	Status 1		Status 2		Status 3		Status 4		Total
	ha	(%)	ha	(%)	ha	(%)	ha	(%)	ha
Open water	61	(<0.1)	23,447	(2.7)	737,835	(86.3)	93,652	(11.0)	854,995
Spruce-fir	0	(0)	95,733	(19.3)	374,611	(75.4)	26,738	(5.4)	497,082
Ponderosa pine	0	(0)	2,281	(4.8)	39,677	(83.5)	5,577	(11.7)	47,535
Lodgepole pine	0	(0)	28,901	(12.6)	191,103	(83.3)	9,330	(4.1)	229,334
Mountain fir	0	(0)	17,376	(6.7)	198,893	(76.2)	44,862	(17.2)	261,131
Juniper	11	(<0.1)	83,467	(5.3)	1,248,615	(79.1)	246,778	(15.6)	1,578,871
Pinyon pine	66	(<0.1)	22,320	(3.4)	528,572	(81.4)	97,996	(15.1)	648,954
Pinyon-juniper	641	(<0.1)	78,067	(3.9)	1,619,168	(80.4)	316,571	(15.7)	2,014,447
Mountain mahogany	0	(0)	0	(0)	167	(62.1)	102	(37.9)	269
Aspen	0	(0)	20,660	(2.8)	475,551	(64.7)	238,769	(32.5)	734,980
Oak	88	(<0.1)	43,158	(5.4)	460,617	(58.1)	289,542	(36.5)	793,405
Maple	0	(0)	8,588	(11.4)	30,155	(40.1)	36,488	(48.5)	75,231
Mountain shrub	0	(0)	17,812	(8.6)	139,128	(67.3)	49,786	(24.1)	206,726
Sagebrush	43	(<0.1)	32,334	(1.5)	1 ,579,658	(73.5)	536,498	(25.0)	2,148,533
Sagebrush/perennial grass	0	(0)	50,818	(3.0)	863,295	(50.9)	781,212	(46.1)	1,695,325
Grassland	0	(0)	20,580	(2.4)	539,027	(62.8)	298,461	(34.8)	858,068
Alpine	0	(0)	33,542	(41.4)	46,270	(57.2)	1,123	(1.4)	80,935
Dry meadow	0	(0)	3,019	(1.4)	122,521	(56.6)	91,118	(42.1)	216,658
Wet meadow	0	(0)	40	(0.7)	3,956	(68.3)	1,793	(31.0)	5,789
Barren	0	(0)	67,922	(11.8)	451,191	(78.4)	56,378	(9.8)	575,491
Lodgepole pine/aspen	0	(0)	24	(0.4)	5,408	(92.2)	435	(7.4)	5,867
Ponderosa pine/mountain shrub	0	(0)	7,694	(3.4)	196,052	(86.0)	24,145	(10.6)	227,891
Spruce-fir/mountain shrub	0	(0)	104	(2.5)	3,320	(79.4)	756	(18.1)	4,180
Mountain fir/mountain shrub	0	(0)	117	(1.3)	6,271	(67.2)	2,943	(31.5)	9,331
Aspen/conifer	0	(0)	57	(0.4)	9,766	(71.9)	3,767	(27.7)	13,590
Mountain riparian	0	(0)	1,612	(4.2)	17,895	(46.2)	19,205	(49.6)	38,712
Lowland riparian	149	(0.3)	908	(1.8)	12,605	(24.7)	37,445	(73.3)	51,107
Lava	0	(0)	0	(0)	259	(100)	0	(0)	259
Agriculture	494	(0.1)	6,208	(0.7)	19,647	(2.1)	908,905	(97.2)	935,254
Urban	0	(0)	88	(0.1)	5,233	(3.6)	139,338	(96.3)	144,659
Salt desert scrub	0	(0)	98,802	(2.2)	3,660,972	(80.6)	779,929	(17.2)	4,539,703
Desert grassland	0	(0)	9,307	(1.0)	662,639	(73.8)	225,936	(25.2)	897,882
Blackbrush	2	(<0.1)	84,091	(8.8)	708,021	(74.1)	162,854	(17.1)	954,968
Creosote-bursage	0	(0)	308	(0.8)	36,084	(76.1)	10,925	(23.1)	47,317
Greasewood	0	(0)	1,316	(1.3)	73,890	(75.4)	22,840	(23.3)	98,046
Pickleweed barrens	0	(0)	3,990	(0.9)	385,163	(89.3)	42,401	(9.8)	431,554
Wetland	0	(0)	9,967	(18.5)	10,470	(19.5)	33,390	(62.0)	53,827

Table 3. Land ownership and protection status in Utah by major category.

Ownership	Total area		Status 1		Status 2		Status 3		Status 4	
	ha	(%)	ha	(%)	ha	(%)	ha	(%)	ha	(%)
Federal	14,006,997	(63.7)	0	(0)	699,692	(5.0)	13,307,095	(95.0)	210	(<0.1)
Native American	942,363	(4.3)	0	(0)	0	(0)	0	(0)	942,363	(100.0)
Private	4,699,145	(21.4)	1,554	(<0.1)	1,935	(<0.1)	0	(0)	4,695,656	(99.9)
State	1,666,700	(7.6)	0	(0)	173,109	(10.4)	1,493,591	(89.6)	0	(0)
Water	663,792	(3.0)	0	(0)	0	(0)	663,792	(100.0)	0	(0)

acres). The remaining 5,638,229 ha (25.65%; 13,926,440 acres) are status 4 lands. By far, most lands in Utah are nondesignated public lands subject to multiple-use guidelines (i.e., status 3). Based on the 10% rule, only 6 of the 37 mapped vegetation cover types are protected as status 1 or status 2 lands (Table 1). Four of these six cover types are timber or other high-elevation cover types. The remaining two cover types are wetlands and barrens areas with less than 5% vegetation.

A common perception is that there currently exist sufficient protected lands that preserve and maintain biological diversity. Our analyses indicate that while some cover types are protected, most of the mapped cover types in Utah have less than 10% of their area protected. Our analyses also indicate that the Utah lands that are protected are more of a random product than a systematic approach to protecting the diversity of vegetation cover types. A more reasoned approach to the management of lands for the conservation of biological resources should include a systematic evaluation of the geographic distribution of resources.

References

Edwards, T.C., Jr., J.M. Scott, C.G. Homer, and R.D. Ramsey. 1993. Gap analysis: a geographic approach for assessing national biological diversity. Natural Resources and Environmental Issues 2:65-72.

Edwards, T.C., Jr., and J.M. Scott. 1994. Use of gap analysis as a tool for the management of biodiversity. Proceedings of the 21st Congress of the International Union of Game Biologists. In press.

Homer, C.G., T.C. Edwards, Jr., D.H. Ramsey, and K.H. Price. 1993. Use of remote sensing methods in modeling sage grouse winter habitat. Journal of Wildlife Management 57:78-84.

Noss, R.F. 1991. From endangered species to biodiversity. Pages 227-246 *in* K. Kohm, ed. Balancing on the brink of extinction: the Endangered Species Act and lessons for the future. Island Press, Washington, DC.

Scott, J.M., B. Csuti, K. Smith, J.E. Estes, and S. Caicco. 1991. Gap analysis of species richness and vegetation cover: an integrated biodiversity conservation strategy. Pages 282-297 *in* K. Kohm, ed. Balancing on the brink of extinction: the Endangered Species Act and lessons for the future. Island Press, Washington, DC.

Scott, J.M., F. Davis, B. Csuti, R. Noss, B. Butterfield, S. Caicco, C. Groves, T.C. Edwards, Jr., J. Ulliman, H. Anderson, F. D'Erchia, and R.G. Wright. 1993. Gap analysis: a geographic approach to protection of biological diversity. Wildlife Monographs 123.

For further information:

Thomas C. Edwards, Jr.
National Biological Service
Utah Cooperative Fish and Wildlife Research Unit
Department of Fisheries and Wildlife
Utah State University
Logan, UT 84322

Biodiversity in the Southwestern California Region

by
David M. Stoms
Frank Davis
University of California-Santa Barbara

The Gap Analysis Program (GAP), coordinated by the National Biological Service, provides a regional screening of elements of biodiversity (plant communities and wildlife species) to identify elements most at risk and to identify general areas of highest concentrations of the at-risk elements. Data collection and analysis have been completed for southwestern California, the first of 10 regions to be analyzed in the state. This region covers roughly 8% of the land area of California, spanning the southern coast from Point Conception to the U.S.-Mexico border and from the western edge of the Sonoran and Mojave deserts to the Pacific Ocean. Urban growth has been exceptionally rapid in this region at the expense of species and habitats, particularly in the Coastal Plain. This article summarizes the gap analysis of this region and identifies plant communities and wildlife species considered at risk. Further details can be found in Davis et al. (1994).

Land Management Status

In this analysis we defined three levels of management to determine the protection status of elements of biodiversity. Level 1 represents areas managed for the long-term protection of biodiversity, such as wilderness areas, research natural areas, state parks, and some private preserves. Level 2 includes publicly owned lands not specifically designated for Level 1 management, and Level 3 contains lands with no formal management for biodiversity.

The amount of Level 1 areas managed to preserve biodiversity is 9.6% of the region, mostly in national forest wilderness areas. Other public lands managed at Level 2 account for another 30%, while the remaining 60% is private land. Lower elevations, where most urban and agricultural development occurs, are predominately private land. Government agencies manage most higher elevation lands, that is, lands at 1,500-2,500 m (4,920-8,200 ft), 25% of which is managed at Level 1.

Vegetation Status

A team from the University of California, Santa Barbara (UCSB) produced a map of actual vegetation. The California Natural Diversity Data Base staff has identified some plant communities of special concern; they generally have less than 10% of their distribution in Level 1 areas or over 70% of the mapped distribution in privately owned Level 3 areas. We used these criteria to identify other plant communities that are at risk.

Communities restricted largely to the lower elevations, like coastal sage scrub (Figure) and non-native annual grasslands, are at considerable risk (Table 1). Although grasslands are dominated by non-native species, they can be rich in native plant species and are habitat to many animal species. Roughly 88% of areas below 500 m (1,640 ft) have no formal protection status; most low-elevation land has already been converted to agricultural or urban uses, and most remaining low-elevation land is zoned for future urbanization.

Especially alarming is the condition of the California black walnut woodlands. The southern variety of this species is endemic to this region and its current distribution is highly fragmented and reduced compared with its original distribution. Sagebrush steppe shrubland, although widespread elsewhere in California, appears vulnerable in this region. A significant proportion of the sagebrush steppe habitat is on Level 2 lands, and conservation concern for these communities can probably be adequately

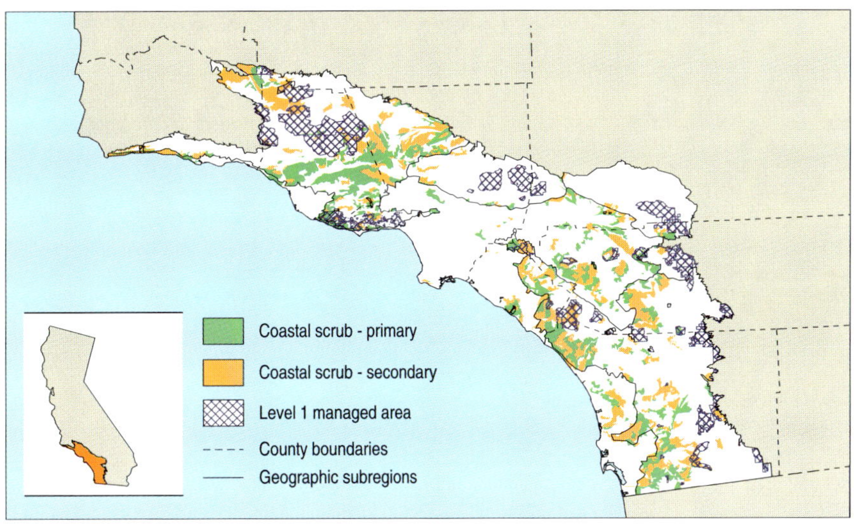

Figure. Gap analysis of coastal sage scrub in the southwestern region of California. Highlighted are landscapes where coastal sage scrub is the primary and secondary upland vegetation type.

Table 1. Natural communities identified as at risk by using Gap Analysis Program criteria. The list is ordered from highest to lowest percentage of the community that occurs on Level 3 private lands.

Natural community	Private lands %
Valley oak woodland *	94
Valley needlegrass grassland *	93
California walnut woodland	89
Coastal sage-chaparral scrub	82
Digger pine-oak woodland *	76
Non-native grassland	73
Coastal sage scrub	71
Coast live oak woodland	71
Coast live oak forest	70
Engelmann oak woodland	66
Southern mixed chaparral	62
Redshanks chaparral	42
Big sagebrush scrub	38
Upper Sonoran manzanita chaparral	24
Southern interior cypress forest *	22
Mojavean pinon woodland *	6
Northern juniper woodland	4

*Mapped distribution totals less than 50 km^2 (19.3 mi^2).

addressed by the public land managing agencies. Many oak woodlands appear to be at risk now or will be within the next one or two decades. Most of the chaparral communities seem reasonably secure; they are generally found on steeper slopes, largely on public lands, and in areas with 10%-20% Level 1 status.

Wildlife Status

Detailed field-based maps of the distribution of wildlife do not exist for all species and would be too difficult to compile in the time available. Biologists do know, however, what habitats most wildlife species prefer. We combined this knowledge, contained in the California Wildlife-Habitat Relationships (WHR) data base (Airola 1988), with the vegetation map to identify suitable habitat for all native wildlife species in the region. These predictions do not guarantee that a species occurs at a given location, only that suitable habitat exists. Threatened and endangered species usually had less than 15% of their distribution in Level 1 areas. We used this proportion as our criterion for identifying what other species breeding in the region are at highest risk.

Forty-two wildlife species are at highest risk from inadequate habitat protection (Table 2). Basically, the number of at-risk species is relatively uniform throughout San Bernardino, western Riverside, San Diego, and eastern Orange counties. The western half of the region in Los Angeles, Ventura, and Santa Barbara counties has fewer species that are at risk although some of these species may only occur in the western half; thus, this area should not be dismissed as less critical to preserving biodiversity until a comprehensive nature reserve network is designed.

Table 2. Wildlife species considered at risk based on Gap Analysis Program criteria.

Scientific name	Common name
Amphibians	
Batrachoseps pacificus	Pacific slender salamander
B. stebbinsi	Tehachapi slender salamander
Bufo microscaphus	Southwestern toad
Rana muscosa	Mountain yellow-legged frog
Reptiles	
Clemmys marmorata	Western pond turtle
Sceloporus orcutti	Granite spiny lizard
Phrynosoma coronatum	Coast horned lizard
Xantusia henshawi	Granite night lizard
Cnemidophorus hyperythrus	Orange-throated whiptail
Anniella pulchra	California legless lizard
Lichanura trivirgata	Rosy boa
Crotalus ruber	Red diamond rattlesnake
Birds	
Elanus caeruleus	Black-shouldered kite
Haliaeetus leucocephalus	Bald eagle
Aquila chrysaetos	Golden eagle
Coccyzus americanus	Yellow-billed cuckoo
Asio otus	Long-eared owl
Archilochus alexandri	Black-chinned hummingbird
Calypte costae	Costa's hummingbird
Empidonax difficilis	Western flycatcher
Tachycineta thalassina	Violet-green swallow
Polioptila caerulea	Blue-gray gnatcatcher
P. californica	California gnatcatcher
Sialia mexicana	Western bluebird
Lanius ludovicianus	Loggerhead shrike
Vireo bellii	Bell's vireo
V. vicinior	Gray's vireo
Dendroica petechia	Yellow warbler
Icteria virens	Yellow-breasted chat
Guiraca caerulea	Blue grosbeak
Aimophila ruficeps	Rufous-crowned sparrow
Amphispiza belli	Sage sparrow
Passerculus sandwichensis	Savannah sparrow
Ammodramus savannarum	Grasshopper sparrow
Agelaius tricolor	Tricolored blackbird
Mammals	
Tamias obscurus	California chipmunk
Perognathus longimembris	Little pocket mouse
P. alticola	White-eared pocket mouse
P. fallax	San Diego pocket mouse
Dipodomys agilis	Pacific or agile kangaroo rat
D. stephensi	Stephens' kangaroo rat
D. merriami	Merriam's kangaroo rat

Future Plans and Concerns

Implementation of protective measures should occur soon. Land-management agencies are the appropriate parties to set land acquisition priorities and to change existing management practices. The Southern California Association of Governments, for example, has used the GAP data base to identify natural communities of greatest concern throughout its six-county planning area as part of its Regional Comprehensive Plan Open Space Element. Multi-species conservation plans are also using biodiversity and land-management data to select and design a network of nature reserves to protect adequate habitat over large regions.

References

Airola, D.A. 1988. Guide to the California Wildlife Habitat Relationships System. California Department of Fish and Game, Sacramento. 74 pp.

Davis, F.W., D.M. Stoms, P. Stine, A. Hollander, M. Borchert, M. Bueno, K. Beardsley, and V. Gray. 1994. Gap analysis of the Southwestern California Region. Report submitted to the National Fish and Wildlife Foundation and the Southern California Edison Company. Center for Remote Sensing and Environmental Optics and Department of Geography, University of California, Santa Barbara. 107 pp.

For further information:

David M. Stoms
University of California
Department of Geography
Santa Barbara, CA 93106

Federal Data Bases of Land Characteristics

by
Denice M. Shaw
Environmental Protection Agency
Michael Jennings
National Biological Service

At the federal level of government, there is a clear need for developing comprehensive and consistent land-cover and land-characteristics information for the United States. Increased attention to environmental research and planning that addresses spatial context and relationships requires baseline land characteristics across a range of spatial and temporal scales. The demand for this information parallels advances in computer and other technologies, such as geographic information systems (GIS) and remote sensing, which permit the processing, analysis, and management of this type and volume of data.

To initiate this effort for the federal government, four ecological and environmental research and monitoring programs have formed a partnership with the U.S. Geological Survey's (USGS) Earth Resources Observation System (EROS) Data Center to design, develop, and test a Multi-Resolution Land Characteristics (MRLC) monitoring program. The overall objective of MRLC is to develop a land-characteristics monitoring system that provides a baseline of multi-scale environmental characteristics and mechanisms for monitoring, identifying, and assessing environmental change. In addition, the MRLC program is developing a national land-cover data set based on Landsat Thematic Mapper satellite imagery. Partners of the MRLC program are described below.

EPA: EMAP

The Environmental Monitoring and Assessment Program (EMAP), managed by the U.S. Environmental Protection Agency (EPA) Office of Research and Development, is a research, monitoring, and assessment effort to report on the condition of our nation's ecosystems. The EMAP is developing and using ecological indicators for wetlands, surface waters, the Great Lakes, agroecosystems, arid ecosystems, forests, and estuaries. For the EMAP, land-cover information is critical to determine sample locations, resource extent, and potential human-caused stress. When fully implemented, the EMAP will provide comparable, high-quality data on the condition of our nation's ecological resources at regional and national scales.

NBS: GAP

The National Biological Service's (NBS's) Gap Analysis Program (GAP) provides a regional and national overview of the distribution and protection status of biological diversity by producing comprehensive and synoptic biogeographic data. Analysis involves using GIS technology to compare the distributions of vegetation and native vertebrate species with land ownership and management. One of the central questions GAP addresses is how well native species are represented in areas managed for their long-term sustainability.

USGS: NAWQA

The National Water Quality Assessment (NAWQA) program of the USGS is designed to describe the status and trends in the quality of the nation's groundwater and surface-water resources and to link these status and trends with an understanding of the natural and human factors that affect water quality. The program integrates information about water quality at a wide range of spatial scales, from local to national, and focuses on water-quality conditions that affect large areas of the nation or that occur frequently within small areas.

USGS: EROS Data Center

The EROS Data Center is a data-management, systems-development, and research center of the USGS. Established in the early 1970's to receive, process, and distribute data from the National Aeronautics and Space Administration's experimental Landsat satellites (Fig. 1), it houses the world's largest collection of space and aircraft imagery of the Earth. It manages

Courtesy D.M. Shaw, EPA

Fig. 1. Landsat Thematic Mapper image of Philadelphia and New York City, taken May 20, 1991.

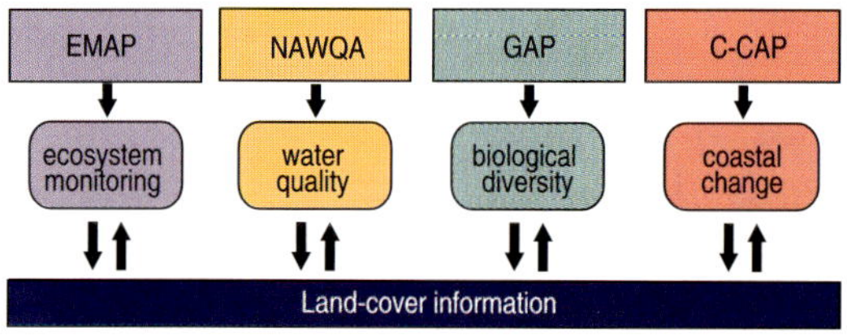

Fig. 2. Land-cover requirements of the Multi-Resolution Land Characteristics consortium land.

global Earth observational data, including the development and operation of advanced systems for receiving, processing, distributing, and applying land-related earth science, mapping, and other geographic data and information.

NOAA: C-CAP

The National Oceanographic and Atmospheric Administration's (NOAA's) Coastal Change Analysis Program (C-CAP) develops a comprehensive, nationally standardized information system to assess changes in wetlands and adjacent uplands in U.S. coastal regions. It uses satellite sensors to detect change in coastal emergent wetlands (mainly tidal marshes) and adjacent uplands and uses aerial photography to detect change in submerged aquatic vegetation. The ultimate goal of the program is to monitor coastal areas every 1 to 5 years, depending on the rate and magnitude of change in each region.

Approach

Collaboration among these programs is the most efficient approach (Fig. 2). Thus, the MRLC generates these data according to common standards for content, format, accuracy, and management; traditionally, environmental data collected for federal ecological studies have not been gathered according to standard or common methods, resulting in data that are not easily shared and in work that is duplicative. The MRLC provides partner programs and others with a data base that is collected according to consistent methods where possible.

For further information:

Denice M. Shaw
EMAP Center/EPA
MD75
RTP, NC 27711

Monitoring Changes in Landscapes from Satellite Imagery

by
Thomas R. Loveland
U.S. Geological Survey
H.L. Hutcheson
South Dakota State University

It has been said that "a model without data has no predictive power" (Rasool 1992). The need to model the extent, condition, and trends in biological resources is a central element for most environmental assessments. Whether the issues involve biological diversity or the effects of changing biogeochemical cycles, accurate baseline data are essential to the environmental monitoring and modeling of future environmental conditions.

Methods and tools for monitoring natural vegetation at the level of plots to small sites-from a single square meter to millions of square meters are well developed and widely used (Küchler and Zonneveld 1988), but at the national level there is a lack of comprehensive environmental data from which we can assess national patterns of environmental diversity. The early western explorers conducted extensive surveys of regional geological, topographic, and ethnographical resources but did not collect enough detailed biological data that could provide us with a starting point for understanding the environmental transformations that have taken place since the nation was founded. More recently, Klopatek et al. (1979) tried to assess the modification of natural vegetation in the United States but concluded that the exercise was difficult because recent land-use changes were typically undocumented. As a result, assessments of current environmental conditions are too frequently based on decades-old data.

Current Estimates of Vegetation Patterns

Perhaps the best estimate of vegetation patterns of the conterminous United States before European settlement is from Küchler's potential natural vegetation (Küchler 1964). His map of the potential natural vegetation divides the country into 116 potential vegetation types. He defines potential natural vegetation as the vegetation that would exist today if humans were removed from the scene and if the resulting plant succession were telescoped into a single moment.

There are, however, limitations in using potential natural vegetation as an indicator of pre-European settlement vegetation patterns, including problems related to the coarse scale of the Küchler map (1:3,168,000), the processes of succession, and the determination of climax vegetation types (Klopatek et al. 1979). Küchler, for example, attempted to show the potential climax stage of vegetation, although some ecosystems never reached climax because of natural controls such as fire. Küchler also pointed out the difficulties and the assumptions in using the terms "natural" and "original" vegetation. His map, however, probably represents the best approximation available today of the continent's vegetation before European settlement.

The most current picture of national land-cover vegetation patterns is from a 1990 data set produced by the U.S. Geological Survey (USGS; Loveland et al. 1991). The USGS land-cover data were interpreted from 1990 satellite imagery from the Advanced Very High Resolution Radiometer (AVHRR) sensor aboard the National Oceanic and Atmospheric Administration's polar-orbiting meteorological satellites. The USGS map of land cover is limited in its use for local applications because of the coarse ground resolution of AVHRR data and its subsequent inability to distinguish vegetation structure, seral stages, and exotic versus natural vegetation. It does, though, provide a picture of vegetation and land-cover patterns at the national level. For example, in the lower 48 states, about 38% of the land is forested, 29% is rangeland or grassland, and 23% is agricultural land. While the USGS land-cover study did not identify urban lands, information from the Defense Mapping Agency's Digital Chart of the World shows that at least 14,500 km^2 (5,655 mi^2) or 1.0% of the conterminous United States is urbanized (Danko 1992).

It must be noted that a comprehensive assessment of accuracy of the 1990 land-cover map has not been completed, although an independent study shows that the classification of forest lands is within 4% of the estimate of the U.S. Forest Service (Turner et al. 1993). Comparisons with selected state land-cover maps and U.S. Department of Agriculture crop area statistics have also shown general correspondence between land-cover estimates at the national level (Merchant et al. 1995).

Change in Natural Vegetation

The estimated extent of change in the natural vegetation since European settlement is derived by comparing Küchler's potential natural vegetation (Küchler 1964) with the 1990 land-cover data set produced by the USGS (Loveland et al. 1991). Both potential natural vegetation (Fig. 1) and 1990 land cover (Fig. 2) have been generalized to show six vegetation groups: needleleaf forest, broadleaf forest, mixed forest, grassland, shrubland, and grassland-shrubland. Note that the 1990 land-cover classification does not distinguish between natural and altered vegetation (e.g., an even-age tree plantation is mapped as forest even though it does not have the ecological value or function of a natural forest). The 1990 land-cover map (Fig. 2) also includes four additional categories: urban areas, cropland, cropland-woodland mosaics, and cropland-grassland mosaics.

A representation of the percentage of land modified from its natural state by either cultivation or urban development was produced by calculating the percentage of 1990 agriculture and urban lands found within each Küchler

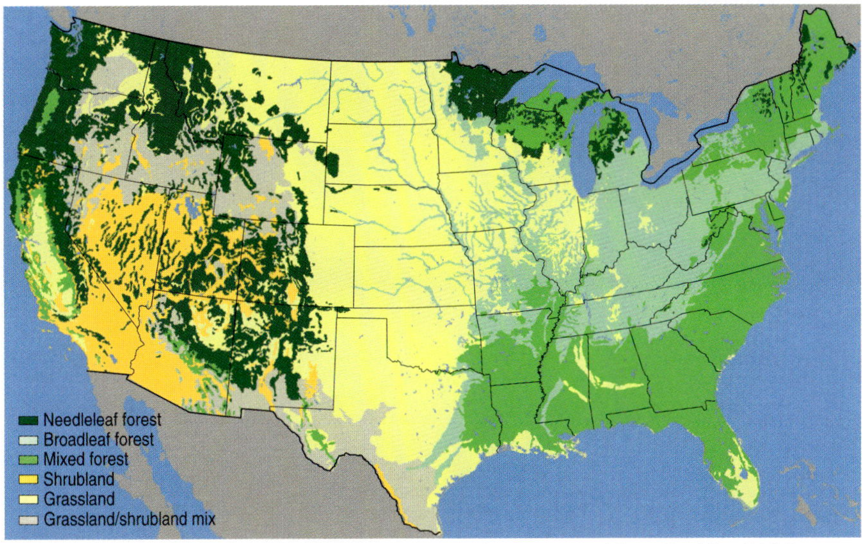

Fig. 1. Grouped categories of potential natural vegetation aggregated from Küchler (1964).

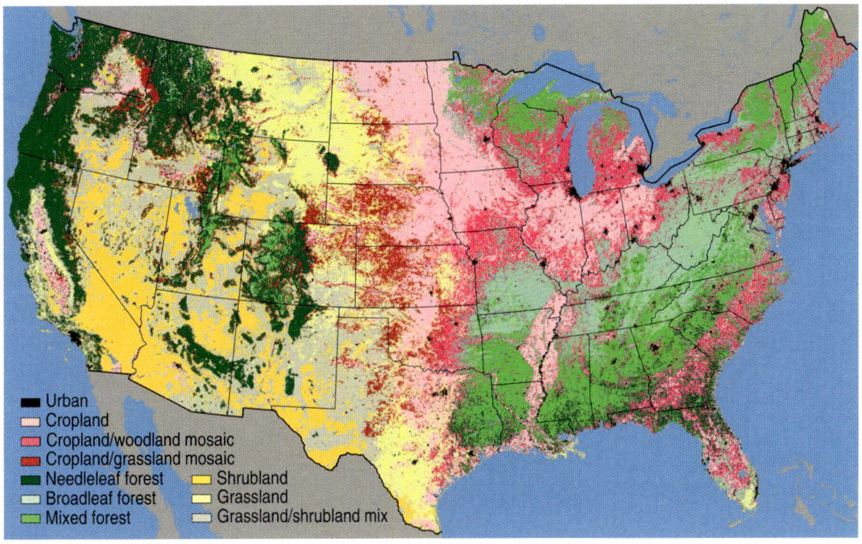

Fig. 2. Grouped categories of 1990 land cover depicting 1990 conterminous U.S. land cover that was developed from 1990 AVHRR imagery.

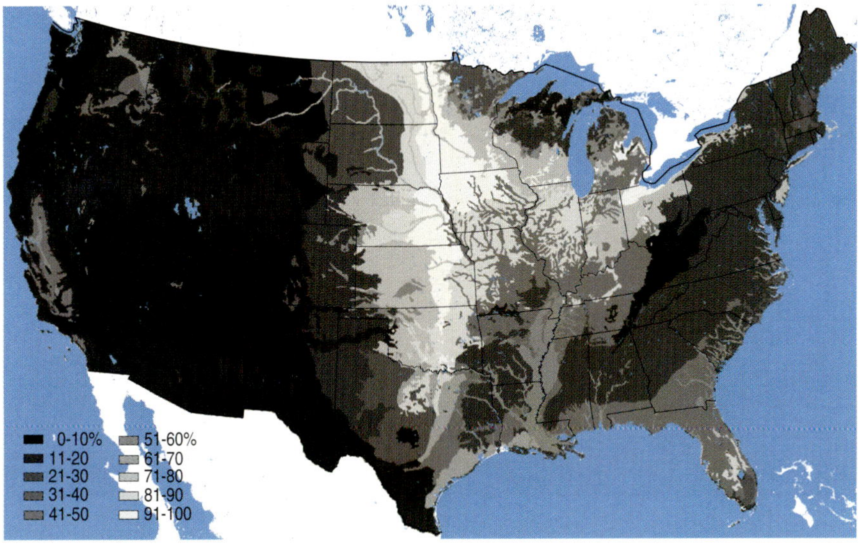

Fig. 3. Percentage of Küchler's potential natural vegetation types (Küchler 1964) that have been converted to agricultural and urban land cover. The lighter tones represent the higher levels of human modification. Percentages of modification are displayed as deca-percentiles.

Landsat MSS Images

Landsat Multispectral Scanner (MSS) images of the Dallas-Fort Worth area in 1974 and 1989 indicate that in this region urbanization has been the cause of land-cover change (Fig. 1). The population of the metropolitan area grew by an estimated 1.25 million during this 15-year period. The substantial conversion of cropland, woodlands, and grasslands to urban land uses resulted from the trend toward migration to sun belt cities and increased job opportunities.

Landsat MSS images can also display landscape transformations resulting from natural events such as the eruption of Mount Saint Helens in southern Washington and the subsequent recovery of vegetation (Fig. 2). The 1973 image represents the region in its "original state." The 1983 image displays the large denuded landscape north of the

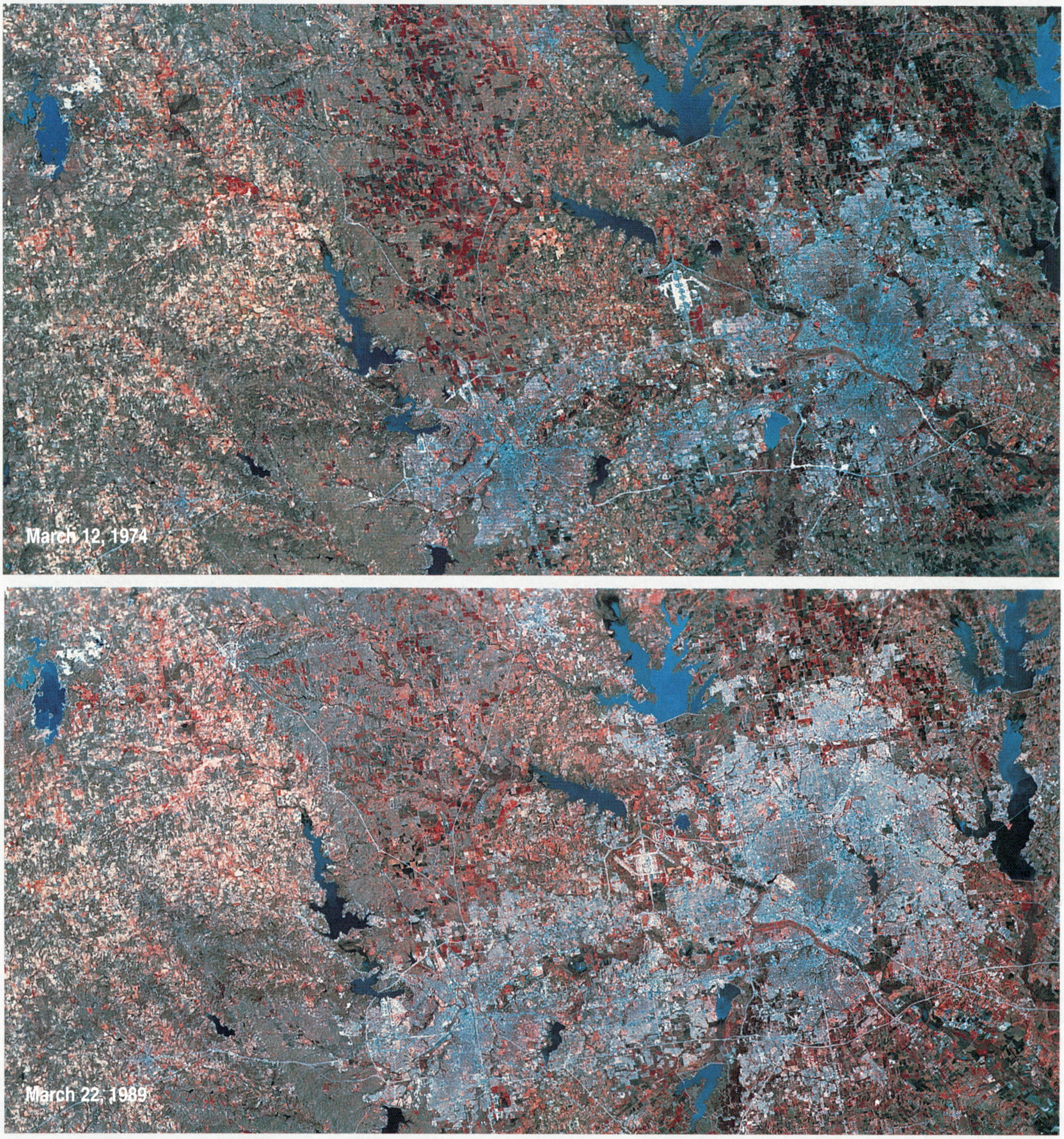

Fig. 1. The 1974 and 1989 Landsat MSS images of Dallas-Fort Worth, Texas. Expanded urban areas are clearly identifiable in the 1989 image and are particularly evident around Dallas-Fort Worth International Airport in the center of the image.

volcano shortly after its eruption May 18, 1980. The 1988 image shows revegetation of the northern slopes, and a landscape gradually recovering to a new "natural" state. A feature of this set of images is the small bluish rectangular patches surrounding Mount Saint Helens, representing areas that have been logged by clear-cutting.

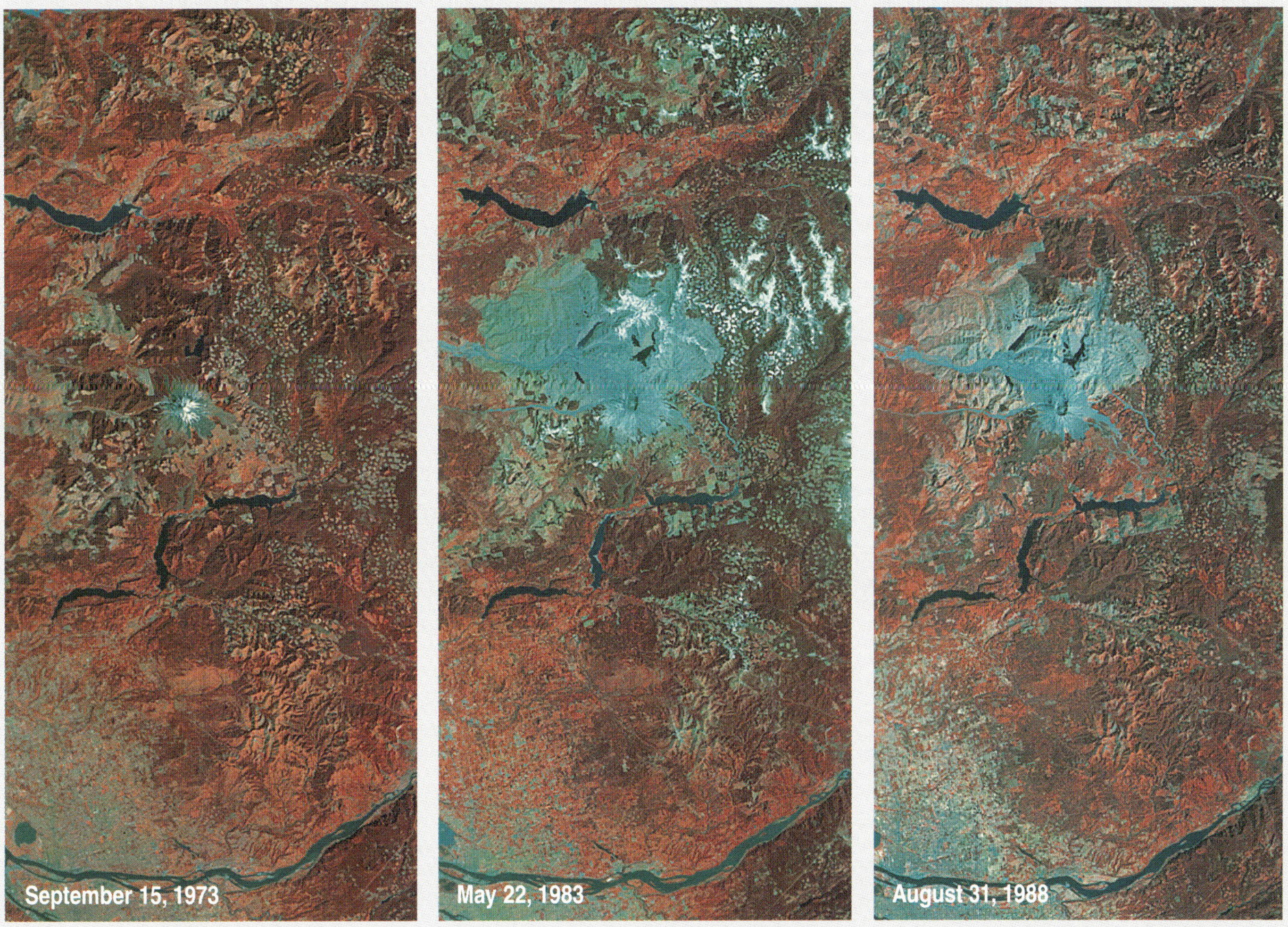

Fig. 2. Landsat MSS images of the Mount Saint Helens area in southern Washington in 1973, 1983, and 1988. The 1973 image shows the area before eruption. The area north of the crater in the image with the bluish color was most devastated by the 1980 eruption. In the 1988 image the light pink color in the blow-out area shows vegetation regrowth.

vegetation type (Fig. 3). Although more than 61% of the conterminous United States is covered with the same dominant vegetation as Küchler suggests, the percentage varies considerably by region. Almost 92% of the western forests region remains covered with tree species, while only 29% of the central and eastern grasslands region remains as grasslands.

It must be understood that a low percentage of agricultural or urban lands in a region does not imply that the landscape exists in a pristine, natural state. In some cases, the "natural" vegetation may be altered substantially by local land-use practices such as grazing and logging or changed by the introduction or invasion of non-native vegetation. Küchler (1964) recognized overgrazing as having long altered the central grassland. He also mentioned Kentucky bluegrass (*Poa pratensis*) as an exotic that has become the dominant grassland in regions including the Black Hills of South Dakota. As a result, many areas that are not affected by agriculture or urbanization are far from their natural state and do not perform the same ecological role as did the original ecosystem. The coarse nature of the AVHRR data and the lack of detailed baseline data on original vegetation conditions do not allow for the detection of these important landscape qualities. While these assessments have limitations, the comparisons represent the type of analysis and monitoring that can be done with a properly designed operational vegetation monitoring system.

The areas with the highest percentage of land modified from its natural condition are in the central United States. With one exception, the most intensively cultivated areas coincide with Küchler's grassland or mixed grassland-

Table 1. Küchler vegetation types least modified by urbanization and agricultural developments.

Type and location	Unaltered %
Grama-tobosa prairie (Arizona, New Mexico)	99.80
Trans-pecos shrub savanna (Texas, New Mexico)	98.82
Oak-juniper woodland (Arizona, New Mexico, Texas)	98.67
Southeastern spruce-fir forest (southern Appalachia)	98.18
Silver fir-Douglas fir (Oregon, Washington)	97.08
Cedar-hemlock pine forest (northern Rocky Mountains)	97.30
Grama-tobosa shrub-steppe (Arizona, New Mexico)	97.14
Creosote bush-tarbush (Arizona, New Mexico)	96.04
Chaparral (California)	96.03
Blackbrush (Utah, Arizona)	95.67
Montane chaparral (California)	95.36
Redwood forest (California, Oregon)	94.70
Mixed mesophytic forest (Pennsylvania, West Virginia, Ohio, Kentucky, Tennessee)	94.57

Table 2. Küchler vegetation types most affected by urbanization, their locales, and associated urban areas.

Type and location	Urbanized %
Fescue oatgrass (western slopes of northern coast ranges, California, San Francisco)	24.00
Subtropical pine forest (southern Florida, Miami)	21.07
Coastal sagebrush (coastal regions of southern California, Los Angeles)	15.87
Pine-cypress forest (coastal California)	6.10
Northeastern oak-pine forest (coastal New England to New Jersey, New York, Newark, Philadelphia)	5.86

Table 3. Selected grassland types arranged by percentage cultivation.

Grassland type and location	Cultivation %
Bluestem prairie (North Dakota and Minnesota southward to Oklahoma)	90.28
Wheatgrass-bluestem-needlegrass (North Dakota, South Dakota, Nebraska)	82.43
Bluestem-grama prairie (Kansas, Nebraska, Colorado, Oklahoma)	76.24
Nebraska sandhills prairie (Nebraska, South Dakota)	73.32
Wheatgrass-needlegrass (North Dakota, South Dakota, Montana, Wyoming, Colorado)	32.71
Grama-buffalograss (New Mexico, Colorado, Wyoming, Nebraska, Kansas, Oklahoma, Texas)	25.75
Wheatgrass-grama-buffalograss (South Dakota)	10.21
Grama-needlegrass-wheatgrass (Wyoming, Montana)	7.39

forest types. The exception is the elm-ash forest south and west of Lake Erie (91.03% cropland). This vegetation type covers a relatively small area (23,103 km^2; 9,010 mi^2). The principal vegetation type that is now more than 90% cropland or mixed cropland is Küchler's bluestem prairie, which covers 271,990 km^2 (106, 076 mi^2), 3.5% of the conterminous United States. The 1990 land-cover data indicate that 90.28% is predominately cropland.

The least cultivated of Küchler's types are grama-tobosa prairie (0.18%), trans-pecos shrub savanna (0.28%), creosote bush (0.60%), and blackbrush (0.66%). These four types are all part of the western shrub and woodland group. In the eastern United States, the most "uncultivated" of Küchler's types is the mixed-mesophytic forest (5.07%), which covers an area of 496,790 km^2 (193,748 mi^2) and has been noted as having the highest species diversity of all the eastern broadleaf forests (Braun 1950).

There are several other ways to evaluate the differences in the 1990 landscape versus the potential natural state. For example, certain Küchler types retain the highest percentages of areas not covered by agriculture or urban land cover (Table 1), although these areas may be presently highly disturbed by logging, road building, strip mining, grazing, or other activities. With the exception of the mixed-mesophytic forest and the relatively small southeastern spruce-fir forest, these are all in the western part of the country. Vegetation types from the Küchler map that have the highest percentage of urbanization on the USGS map (Table 2) are relatively small and are all coastal. Some, like coastal sagebrush, are types that are considered threatened (*see* Stoms and Davis, this section). For selected grassland types of the Great Plains and central lowlands, there is a decrease in percentage of cultivation from east to west (Table 3), reflecting the role of annual precipitation in conversion of grassland areas to cultivation.

Comparing forested areas from the USGS map with the Küchler map would indicate that about 57% of the potential forested area is currently covered by tree species (Turner et al. 1993). The potential impacts of these changes are significant. For example, the loss of forest cover since before European settlement (43%) has increased both albedo and carbon dioxide levels. A rise in albedo has been shown to cause a decrease in mesoscale rainfall (Charney et al. 1975). Increases in irrigated agriculture can result in a decrease in albedo, which can cause an increase in mesoscale rainfall (Barnston and Shickedanz 1984). Also, a shift from forest to grasses results in a decrease in primary productivity by a factor of two, thus reducing the rate of atmospheric carbon fixation.

Continuing Transformations

The comparison of 1990 land cover with potential natural vegetation illustrates the magnitude of change that has possibly occurred in the past 250 years. Changes in the landscape are not exclusive to that period, however; in fact, the 1990 view of United States land cover is already becoming outdated in some regions as natural and human forces continue to transform the landscape. For example, a comparison of 1970's and 1980's satellite images from the Landsat Multispectral Scanner (MSS; *see* box) shows that significant changes in some areas selected for examination are taking place. Landsat MSS images have been acquired over most of the United States since July 1972. With approximately 80 m x 80 m (260 ft x 260 ft) resolution, they provide a means to map in more detail the changes that have occurred in the past 22 years.

Future Possibilities

The vignettes presented here illustrate both the potential and the limitations associated with modeling and monitoring of environmental conditions and processes with satellite images. Clearly, baseline data are an essential starting point for these applications. Also needed is a sound framework from which baseline data can be collected, calibrated, and used in a monitoring system to target and assess environmental changes.

Remote-sensing images from orbiting satellites can play an important role in the collection of baseline vegetation data and in monitoring their status. Coarse-resolution data such as 1-km (0.62-mi) AVHRR imagery offer a means to view landscapes with daily frequency, thereby allowing the monitoring of vegetation condition both within a growing period and between years. Over a long period, AVHRR may provide a means for monitoring the subtle changes in the vegetation that may relate to such events as long-term drought. AVHRR data are not adequate for assessing the effects of more local changes. Landscape changes at the local level will be better understood with higher resolution imagery such as that provided by Landsat systems. Improved data from the sensors planned as part of the National Aeronautics and Space Administration's (NASA) Mission to Planet Earth's Earth Observing System will likely provide even better remote sensing systems for environmental monitoring.

Many components needed for a national environmental monitoring system already exist. A robust system that provides mechanisms for targeting and quantifying changes in the landscape will need to include both the synoptic overview capabilities from Earth-orbiting satellites and detailed site-specific observations of biological processes. The National Biological Service's Gap Analysis Program (GAP) provides an essential high-resolution inventory of habitat and natural vegetation for the United States by using Landsat Thematic Mapper imagery with 30 m x 30 m (98 ft x 98 ft) resolution along with substantial amounts of ancillary information such as field reconnaissance and air photos (Scott et al. 1993). Regional monitoring of the stressors to the natural systems is needed to improve the predictive capabilities of an operational monitoring system. Those systems, tied together with an integrated sampling and assessment framework, could provide a synergistic means for long-term environmental monitoring.

References

Barnston, A.G., and P.T. Schickedanz. 1984. The effect of irrigation on warm season precipitation in the southern Great Plains. Journal of Climatology and Applied Meteorology 23:865-888.

Braun, E.L. 1950. Deciduous forests of eastern North America. Blaskiston Co., Philadelphia, PA. 596 pp.

Charney, J., P.H. Stone, and W.J. Quick. 1975. Drought in the Sahara: a biogeophysical feedback mechanism. Science 187:434-435.

Danko, D.M. 1992. The digital chart of the world. GeoInfo Systems (2)1:29-36.

Klopatek, J.M., R.J. Olson, C.J. Emerson, and J.L. Joness. 1979. Land-use conflicts with natural vegetation in the United States. Environmental Conservation (6)3:191-199.

Küchler, A.W. 1964. Potential natural vegetation of the conterminous United States: a map and manual. American Geographical Society Special Publ. 36. Princeton Polychrome Press, Princeton, NJ. 116 pp.

Küchler, A.W., and I.S. Zonneveld, eds. 1988. Vegetation mapping. Kluwer Academic, Norwell, MA. 635 pp.

Loveland, T.R., J.W. Merchant, D.O. Ohlen, and J.F. Brown. 1991. Development of a land-cover characteristics database for the conterminous U.S. Photogrammetric Engineering and Remote Sensing (57)11:1453-1463.

Merchant, J.W., L. Yang, and W. Yang. 1995. Validation of continent-scale land cover data bases developed from AVHRR data. Pages 63-72 *in* Proceedings: Pecora 12 Symposium on Land Information from Space-based Systems. American Society of Photogrammetry and Remote Sensing, Bethesda, MD.

Rasool, S.I., ed. 1992. Requirements for terrestrial biospheric data for IGBP core projects. International Geosphere-Biosphere Programme-Data and Information System Working Paper 2. Université de Paris VI, Paris, France. 48 pp.

Scott, M.J., F. Davis, B. Csuti, R. Noss, B. Butterfield, C. Groves, H. Anderson, S. Caicco, F. D'Erchia, T. Edwards, J. Ulliman, and R.G. Wright. 1993. Gap analysis: a geographic approach to protection of biological diversity. Wildlife Monographs 123. 41 pp.

Turner, D.P., G. Koerper, H. Gucinski, C. Peterson, and R. Dixon. 1993. Monitoring global change: comparison of forest cover estimates using remote sensing and inventory approaches. Environmental Monitoring and Assessment 26:295-305.

For further information:

Thomas R. Loveland
EROS Data Center
U.S. Geological Survey
Sioux Falls, SD 57198

The Nation's Wetlands

by
Bill O. Wilen
U.S. Fish and Wildlife Service

The national interest in wetlands is set forth in the findings of the Emergency Wetlands Resources Act of 1986:

> The Congress finds that wetlands play an integral role in maintaining the quality of life through material contributions to our national economy, food supply, water supply and quality, flood control, and fish, wildlife, and plant resources, and thus to the health, safety, recreation, and economic well-being of all citizens of the Nation.

The act requires the Secretary of the Interior to map the nation's wetlands, develop a national digital wetlands data base, and report to Congress on the status and trends of wetlands within the conterminous United States. The U.S. Fish and Wildlife Service (USFWS) has delivered three reports to Congress (Frayer et al. 1983; Dahl 1990; Dahl and Johnson 1991). The reports show that half of the nation's wetlands have been converted to uplands since colonial times (Dahl 1990), and that although the rate of

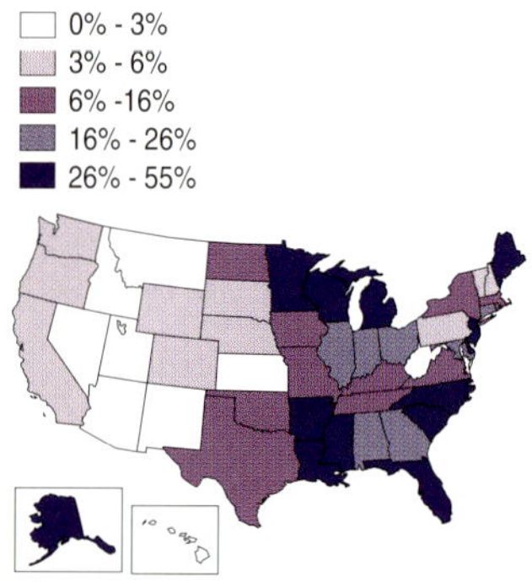

Fig. 1. Surface-area percentage of wetlands in each state: 1780's (Dahl 1990).

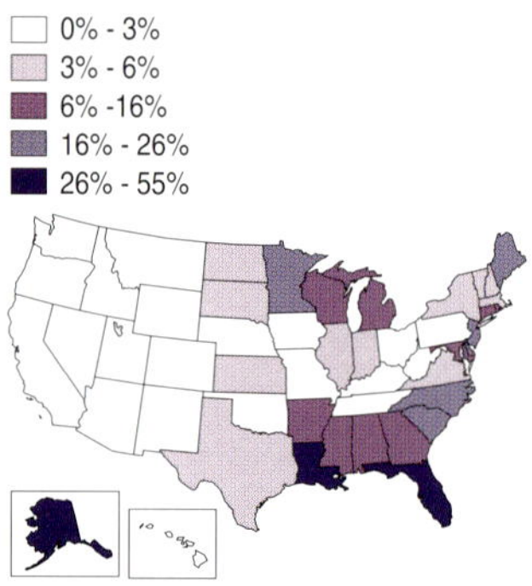

Fig. 2. Surface-area percentage of wetlands in each state: 1980's (Dahl 1990).

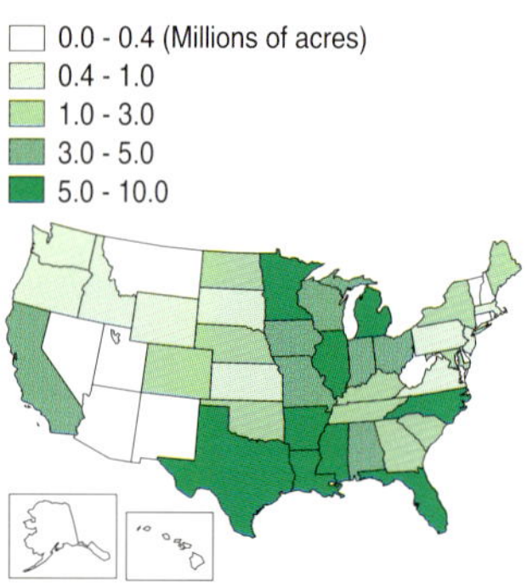

Fig. 3. Wetland acreage loss by state (Dahl 1990).

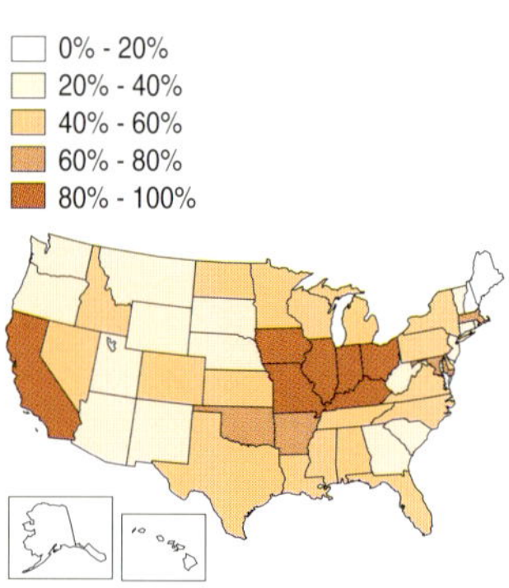

Fig. 4. Surface-area percentage of wetland base loss by state (Dahl 1990).

conversion has slowed, wetland losses continue to outdistance gains (Frayer et al. 1983; Dahl and Johnson 1991).

The quality of the remaining wetlands continues to be an unanswered question. Presidential candidate George Bush's 1988 No-Net-Loss campaign promise was adopted by the federal government as a policy goal. It was expanded by President Clinton in his August 25, 1993, policy statement, "Protecting America's Wetlands: A Fair, Flexible, and Effective Approach," to include a long-term goal of increasing the quality and quantity of the nation's wetlands resource base. Here we present a brief overview of wetlands, their definition, distribution and abundance, dynamics, functions, values, and future.

Wetland Descriptions and Definitions

The United States encompasses an area of about 931 million ha (2.3 billion acres) extending from above the Arctic Circle to the Virgin Islands and spanning the North American continent, and includes the Hawaiian Islands as well as Puerto Rico. Within this broad area, regional variations in climate, topography, hydrology, geology, soils, and vegetation create diverse wetland habitats ranging from the tundra in Alaska to the tropical rain forests of Hawaii to isolated wetlands in the arid Southwest.

Cowardin et al. (1979) defined wetlands as

> lands where saturation with water is the dominant factor determining the nature of soil development and the types of plant and animal communities living in the soil and on its surface. The single feature that most wetlands share is soil or substrate that is at least periodically saturated with or covered by water. The water creates severe physiological problems for all plants and animals except those that are adapted for life in water or in saturated soil. (p. 3)

There are three widely used definitions of wetlands. All use three parameters: hydrology, hydric soil (wetland soils), and hydrophytic vegetation (wetland plants). The USFWS's definition is ecological whereas the definitions used by the U.S. Environmental Protection Agency, the U.S. Army Corps of Engineers, and the Soil Conservation Service are regulatory. All three, however, endorse and use the same interagency wetland plant list, *National List of Plant Species That Occur in Wetlands* (Reed 1988), and wetland soils list, *Hydric Soils of the United States* (SCS 1991).

Regulators are concerned with establishing a definitive line to delineate wetlands from uplands and with placing the wetlands into administrative or regulatory categories. In contrast, the USFWS and the National Biological Service (NBS) are concerned with ecological characterization and mapping the biological extent of both vegetated and nonvegetated wetlands found on soils and substrates. The biological extent of wetlands should be established by scientists using biological criteria. Likewise, policy makers should establish regulations for the subset of wetlands that needs regulating. The subset of wetlands to be regulated and the degree of regulation have changed and will change over time based on our understanding of the functions and values of wetlands, wetlands scarcity, our ever-changing social values, and the political climate.

The USFWS classification system was developed to provide uniformity in concepts and terminology for wetlands. It is hierarchical, moving from systems at the broadest level through subsystems, classes and subclasses, to modifiers describing hydrology (water regime), soils, and water chemistry, and special modifiers relating to human activities.

These categories are used to form wetland types for mapping. More than 2,500 wetland types are commonly used on National Wetlands Inventory maps nationwide. Counties will have from 10 to 400 types, with an average of 100. These wetland types describe ecological units that have certain homogeneous natural attributes. The USFWS's National Wetlands Inventory maps are available for 84% of the conterminous United States, 28% of Alaska, and all of Hawaii.

Distribution and Abundance

The distribution of wetlands has changed dramatically since the 1780's (Figs. 1 and 2). In addition, the percentage of the landscape occupied by wetlands varies markedly from state to state (i.e., Alaska, where 43.3% of the landscape is covered by wetlands as compared with nine states where 1% or less of the landscape is covered by wetlands). The wetland areal loss by states tells one story (Fig. 3) and the percentage of the wetland base lost by states tells another (Fig. 4). Wetlands occupy 11.9% of the landscape of the United States, which is about 5% of the conterminous United States, 43% of Alaska, and 1% of Hawaii.

Wetland Dynamics

The three status and trends reports to Congress provide estimates of net wetland gains or losses; they do not examine wetland quality as a result of disturbance. Wetlands are constantly being disturbed. Even when a wetland is not converted to upland, its successional stage is often pushed back to an earlier stage. For example, between the mid-1970's and mid-1980's, forested wetlands suffered tremendous loss from agriculture and "other" uses. (The category of "other" includes all wetland areas converted to upland where the ultimate land use could not be determined.) Thousands of hectares of forested wetlands were converted to emergent, scrub-shrub, and nonvegetated wetlands. Likewise, thousands of hectares of scrub-shrub wetlands were converted to the "other" category and the agricultural land-use category. These losses were nearly offset by the conversion of forested wetlands to scrub-shrub wetlands. Despite these gains to the scrub-shrub category, however, there was an overall net loss of scrub-shrub wetlands during the study period.

The net gain of thousands of hectares of freshwater emergent wetlands is similarly deceptive. The thousands of hectares that were lost to agricultural, "other," and urban land uses were more than offset by the conversion of forested wetlands and scrub-shrub wetlands to freshwater emergent wetlands. The area of nonvegetated wetlands (primarily ponds) increased by several million hectares. Most of these gains, however, resulted from construction of ponds on uplands not used for agricultural production, but additional thousands of hectares were built on former agricultural lands. This category also experienced gains from converted forested wetlands and scrub-shrub wetlands.

Functions and Values

The functions and values of the nation's wetlands are nearly as diverse as the wetlands themselves (Table), but include flood protection and plant, fish, and wildlife habitat.

All wetlands do not perform all functions. Some functions tend to be compatible, such as flood control and water purification. Other functions tend to be incompatible, such as flood control and food chain support. In addition, wetlands of a given type do not have the same effectiveness in performing a given function. For example, the effectiveness of a given forested wetland for flood control depends on its size, shape, location in the watershed, and so forth. Because wetlands are constantly being affected by disturbance, their effectiveness in performing functions constantly changes. Thus, the effectiveness of a wetland area as wildlife habitat can be improved or degraded by the creation, maintenance, or destruction of vegetated corridors; the ratio of vegetated wetland to upland areas; buffer zones; and plants that provide for wildlife food and habitat. Uplands can and do perform some of the functions performed by wetlands, such as sediment trapping. But because wetlands are situated in the low points of the landscape or are adjacent to streams, rivers, lakes, and oceans, they are more able to perform these functions. In many cases, wetlands are the last line of defense for the protection of surface water quality.

Table. Major wetland functions and values documented in the National Wetlands Inventory "Wetlands Values Database."

Functions and values	Examples
Biogeochemical processes	Carbon cycling, sulfur cycling
Food chain	Detritus production, food source, nutrient cycling, nutrient export, primary production
Habitat	Amphibians, fish, furbearers, insects, mammals, nongame birds, reptiles, shellfish, shorebirds, waterfowl, endangered species
Hydrology	Erosion control, flood control, flow stabilization, groundwater discharge, groundwater recharge, saltwater intrusion prevention, storm dampening
Socioeconomic	Aesthetics, agricultural crops, aquaculture, archaeological, commercial harvest, cultural, educational, energy source (peat), food, forage, heritage, hunting and trapping, indicator species, medicinal, open space, natural products, recreation, research, timber, wastewater treatment, water supply
Water quality	Chemical and nutrient absorption, pollution filtering, oxygen production, sediment trapping

Some wetland functions and values can be replaced by artificial substitutes; for example, flood-control values of wetlands can be replaced by dams, ditches, levees, floodwalls, and reservoirs. Other wetland functions, however, cannot be performed by uplands or replaced by artificial substitutes. An especially important function of wetlands is supporting rich plant diversity. Although wetlands occupy only about 5% of the surface area of the conterminous United States, 6,728 plant species (31% of the U.S. flora) occur in wetlands (Reed 1988). Of these plants, half are restricted to, or usually occur in, wetlands. Thus, wetlands provide critical habitat for a high percentage of the U.S. flora.

Some argue that we cannot afford to maintain the remaining 40 million ha (99 million acres) of wetlands in the conterminous United States because of our increasing population, living standards, and competition for resources. Others argue that wetlands must occupy a greater percentage of the nation's landscape. In the conterminous United States, non-federal rural land occupies nearly 75% of the landscape and contains more than 75% of the nation's wetlands (USDA 1989). Wetlands comprise nearly 6% of the rural non-federal landscape. Specifically, wetlands occupy roughly 1% of cropland, 2% of rangeland, 5% of pastureland, 12% of forestland, and 31% of other rural land (USDA 1989).

Future

Although our understanding of wetlands is imperfect, it is clear we have more information upon which to make public policy decisions on wetlands than we have for many other ecosystems. The challenge for policy makers is to avoid ecologically irreversible choices that would diminish the wealth of future generations while promoting economic development and improving income distribution.

For further information:

Bill O. Wilen
U.S. Fish and Wildlife Service
National Wetlands Inventory
Project
4401 N. Fairfax Dr.
Rm. 400
Arlington, VA 22203

References

Cowardin, L.M., V. Carter, F.C. Golet, and E.T. LaRoe. 1979. Classification of wetlands and deepwater habitats of the United States. U.S. Fish and Wildlife Service FWS/OBS-79/31. 131 pp.

Dahl, T.E. 1990. Wetland losses in the United States: 1780's to 1980's. U.S. Fish and Wildlife Service, Washington, DC. 21 pp.

Dahl, T.E., and C.F. Johnson. 1991. Status and trends of wetlands in the conterminous United States: mid-1970's to mid-1980's. U.S. Fish and Wildlife Service, Washington, DC. 28 pp.

Frayer, W.E., T.J. Monahan, D.C. Bowden, and F.A. Graybill. 1983. Status and trends of wetlands and deepwater habitats in the conterminous United States: 1950's to 1970's. Department of Forest and Weed Sciences, Colorado State University, Fort Collins. 32 pp.

Reed, P.B., Jr. 1988. National list of plant species that occur in wetlands: national summary. U.S. Fish and Wildlife Service Biological Rep. 88(24). 244 pp.

SCS. 1991. Hydric soils of the United States. USDA Soil Conservation Service Miscellaneous Publ. 1491. n.p.

USDA. 1989. Summary report, 1987 National Resources Inventory U.S. Department of Agriculture Statistical Bull. 790. 37 pp.

Glossary & Index

Glossary

All definitions (except those followed by an asterisk) are from *The Dictionary of Ecology and Environmental Science*, edited by Henry W. Art, published by Henry Holt and Company, Inc., copyright 1993 by Storey Communications, Inc., Pownal, Vermont, and are used with permission.

acidification. The process of making a substance acidic, lowering its pH or making it "sour."

adaptive radiation. The evolutionary divergence of a species into a variety of different forms, usually as an ancestral form encounters new resources or habitats.

adventive. A species that is not native and has been introduced into the area but has not become permanently established.

agent. Something that produces or is capable of producing an effect; an active or efficient cause or a chemically, physically, or biologically active principle.

albedo. Fraction of light reflected by a surface, such as ice, or by an entire planet. Studying a planet's albedo can help determine the composition of its surface.

allele. One of a pair or series of genes that occupies a specific physical position in a specific chromosome; any of the alternative forms of a given gene.*

allozymes. One of several possible forms of an enzyme that is the product of a particular allele at a given gene locus.

anadromous. Describing a fish life cycle in which adult individuals travel upriver from the sea to spawn, usually returning to the area where they were born. Salmon and shad are anadromous species.

anthropogenic. Caused by human action, such as changes in vegetation, an ecosystem, or an entire landscape.

architomy. Reproduction by fission followed by bodily reorganization.

bioaccumulation. The absorption and concentration of toxic chemicals in living organisms. Heavy metals and pesticides, such as DDT, are stored in the fatty tissues of animals and passed along to predators of those animals. The result is higher and higher concentrations of the pesticide in fatty tissue, eventually reaching harmful levels in predators at the top of the food chain, such as eagles. Also called biomagnification.

bioassay. Testing the strength of a drug or other substance by examining its effects on a living organism and comparing it with those of a standard substance.

biodiversity. Number and variety of living organisms; includes genetic diversity, species diversity, and ecological diversity.

biome. Regional land-based ecosystem type such as a tropical rainforest, tiaga, temperate deciduous forest, tundra, grassland, or desert. Biomes are characterized by consistent plant forms and are found over a large climatic area.*

bole. Trunk of a tree above the root collar and extending along the main axis.*

broth. A fluid culture medium.*

cirque. A deep steep-walled basin high on a mountain usually shaped like half a bowl and often containing a small lake.*

clutch. The group of eggs laid at one time by either a bird or a reptile.

community. All the groups of organisms living together in the same area, usually interacting or depending on each other for existence. Also called biological community.

confidence intervals. An interval formulated to have specific probability of containing the real value of an unknown parameter. A 95 percent confidence interval has a 95 percent probability of containing the parameter being estimated.

cytological. Describing cytology, a branch of biology dealing with the structure, function, multiplication, pathology, and life history of cells.*

diadromous. Adjective describing organisms that migrate between fresh and salt water, such as eels and carp.

double clutching. When an egg-laying individual produces two clutches of eggs in the same season.*

ecosystem. A functioning unit of nature that combines biotic communities and the abiotic environments with which they interact. Ecosystems vary greatly in size and characteristics.*

ecotones. A transitional area between two (or more) distinct habitats or ecosystems, which may have characteristics of both or its own distinct characteristics. The edge of a woodland, next to a field or lawn, is an ecotone, as are some savanna areas between forests and grasslands.

ectomycorrhiza(e) or **ectotrophic mycorrhizae.** A symbiotic condition between a fungus and the root of a plant in which the fungus forms a sheath around the root. Some hyphae connecting to this sheath penetrate the host root and spread through the soil surrounding the roots. Ectomycorrhizae form between tree species and basidiomyctete fungi.

endemic. Indigenous to, and restricted to, a particular area; also, an endemic plant or animal.

endomycorrhiza(e) or **endotrophic mycorrhizae.** A symbiotic condition between a fungus and the root of the plant in which the fungal hyphae (root like structures) grow between and within the cells of the plant roots, benefiting both the fungi and the plants. Many orchids and members of the heath family (Ericacea) cannot survive without endotrophic mycorrhizae.

eukaryotic. Describing eukaryotes, organisms composed of one more cells containing visibly evident nuclei and organelles.*

eutrophication. The process by which a body of water acquires a high concentration of nutrients, especially phosphates and nitrates, which typically promote excessive growths of algae. As the algae die and decompose, high levels of organic matter and the decomposing organisms deplete the water of available oxygen, causing the death of other organisms, such as fish. Eutrophication is a natural, slow-aging process for a water body, but human activity greatly speeds up the process.

extinction. The dying out of a species, or the condition of having no remaining living members; also the process of bringing about such a condition.

extirpation. Eradication; the loss or removal of a species from one or more specific areas, but not from all areas.

facultative. Capable of existing under different conditions or using different modes for nutrition. Facultative parasites are organisms that can function either as parasites or as saprophytes (decomposers). Facultative wetland plants can occur in either wetlands or uplands, although they are more abundant in the former.

fauna. All the animals of a particular region or a particular era. For example, the fauna of New Zealand.

flora monogram. A systematic treatise on or a list of the plants of an area, habitat, or period.*

genetic drift. Random fluctuations in gene frequency occurring in isolated populations from generation to generation. Genetic drift is the result of chance combinations of different characteristics.*

geomorphic. Of or relating to the form or surface features of the earth or other celestial bodies such as the moon.*

heterotrophic. Describing consumers, organisms that cannot synthesize food from inorganic materials and therefore must use the bodies of other organisms as a source of energy and body-building materials.*

heterozygosity. A measure of genetic diversity in a population, as measured by the number of heterozygous loci across individuals.*

heterozygous. The situation in which an individual has two different alleles at a given gene locus.*

host. An organism that supports a parasite, often to its own detriment.

hydroperiod. The time during which a soil is waterlogged.*

hyperemia. Excess of blood in a body part.*

inbreeding depression. A decline in desirable characteristics such as fertility, general vigor, or yield produced by repeatedly crossing related organisms (inbreeding). Inbreeding depression can be seen in some specimens of purebred pets.

invertebrates. Animals without backbones or internal bony skeletons. All animals except for the phylum Chordata (vertebrates) fall into this category, including insects, crustaceans, worms, corals, and mollusks.

microbivore. An organism that feeds on microorganisms.

morphologic. Describing the form and structure of an organism or any of its parts.*

mycorrhizae. The symbiotic relationship between the mycelia of some species of fungi and the roots or other structures of some flowering plants. The fungal mycelia help the plant absorb minerals and in return absorb energy compounds produced by the plant. Many tree species such as beech cannot grow without their associated mycorrhizae.

Neotropical. Adjective used to describe migrating birds that winter in the Neotropics.*

obligate. Restricted to one particularly characteristic mode of life or biologically essential for survival.

Oceania. The islands and archipelagos of the central and south Pacific.*

outbreeding depression. Reduced fitness in offspring resulting from breeding between individuals from genetically distinct populations.*

pelagic. Living in or relating to the open sea, especially surface waters to the middle depths. Krill and the whales that feed upon them are examples of pelagic animals.

phenotypical. Describing phenotype, the physical expression (outward appearance) of a trait of an organism, which may be due to genetics, environment, or an interaction of the two.*

phylogeny. The evolutionary history or development of a species or higher grouping of organisms.

polymorphism. 1) The existence of more than one form or type in a species, beyond simple gender differences. Social insects such as honeybees with queens, drones, and workers demonstrate polymorphism. 2) Another term for pleomorphism, the occurrence of distinct forms during the life cycle of a plant or animal, such as the caterpillar and pupa that precede the adult.

primiparous. Describing an individual bearing a first offspring or that has borne only one offspring.*

propagules. Structures (cuttings, seeds, or spores) that propagate a plant.*

recruitment. The addition of individuals to a population through reproduction and immigration.

red data book. A catalog published by the International Union for the Conservation of Nature and Natural Resources (IUCN) that lists rare species and those in danger of extinction.

rumen. The first stomach division in animals known as ruminant that chew cud, such as cows and goats. Rumen is also the term for the first stomach division in whales and in dolphins.

seral. Of, relating to, or characteristic of an ecological sere (a series of ecological communities that follow one another in the course of the biotic development of an area or formation from pioneer stage to climax).*

speciation. The evolutionary development of new species, usually as one population separates into two different populations no longer capable of interbreeding.

species. A naturally occurring population or group of potentially interbreeding populations that is reproductively isolated (i.e., cannot exchange genetic material) from such other groups. This definition does not apply to asexually reproducing forms such as many types of Monera or Protista, etc.

subalpine. Describing the region, the climate, the vegetation, or all three found just below alpine regions, usually on mountainsides at 1300 to 1800 meters in elevation. Subalpine vegetation is that just below the treeline, often dominated by pine or spruce trees.

synonomy. The scientific names that have been used in different publications to designate a taxonomic group, such as species. Also a list of names.*

voucher. A specimen used for comparison in order to identify or verify species.*

wetland species. Organisms found in wetlands, lands transistional between aquatic and terrestrial ecosystems that are covered with water for at least part of the year.*

Index

Note: Page numbers in italics refer to a table or figure.

Abert's squirrel (*Sciuris aberti*), 321
Abies
  alba (silver fir), 315
  amabilis (Pacific silver fir), 315
  concolor (white fir), 421
  lasiocarpa (subalpine fir), 316, 317, 321
  magnifica (red fir), 421
Academy of Natural Sciences in Philadelphia (ANSP), 163, 164, 258
Acadian flycatcher (*Empidonax virescens*), *20*
Accipiter
  cooperii (Cooper's hawk), *67*
  gentilis (northern goshawk), *66*, 67, *320*
  striatus (sharp-shinned hawk), 67
Acer rubrum var *drummondii* (red maple), 216
acetylcholine, 416
acetylcholinesterase (AChE), 416
acid deposition (acid rain), 418-20
  air pollution and, 421
  aquatic ecosystems, 420
  aquatic species and, 418-19
  future conditions, 420
  melting snow and, 418
  in national park system, 228
  terrestrial species and, 418, 419-20
acidification, diatoms and, 257-58
Acipenser transmontanus (white sturgeon), 154-56
acorn woodpecker (*Melanerpes formicivorus*), *20*
Acrocephalus familiaris kingi (Nihoa millerbird), *377*
Acrocercops arbutella (leaf miner moth), *160*
Acroneuria
  abnormis (stonefly), *186*
  evoluta (stonefly), *186*
  filicis (stonefly), 185, *186*
  frisoni (stonefly), *186*
  internata (stonefly), *186*
  perplexa (stonefly), 185, *186*
Acronicta
  lanceolaria (owlet moth), 166
  radcliffei (owlet moth), *166*
Acrostichum danaeifolium, *203*
Actinonaias ligamentina [carinata] (mucket), *180*
Actitis macularia (spotted sandpiper), 59, *62*
Adenostoma fasciculatum (chamise chaparral), 421-22
Adirondack Mountains, acid deposition (acid rain) and, 418-19, 420
adult survival, birds, 23-26
Advanced Very High Resolution Radiometer (AVHRR), 469, 473
Aegolius
  acadicus (northern saw-whet owl), *67*
  funereus (boreal owl), *67*
Aerodramus vanikorensis (gray swiftlet), 455
Aeronautes saxatalis (white-throated swift), *20*
Aethia
  cristatella (crested auklet), 51
  pusilla (least auklet), 51
Aflexia rubranura (red-tailed leafhopper), 232
African bee (*Apis mellifera scutellata*), 448
African Bee Program (Mexico), 449
African clawed frog (*Xenopus laevis*), 433, *434*, 435, 436
Africanized honeybee (AHB), 448-51
  distribution of, 451
  ecological implications, 449-50
  monitoring, 450-51
  sting fatalities, 449
agate rocksnail (*Leptoxis clipeata*), *251*
agave (*Agave parviflora parviflora*), *320*
Agave parviflora parviflora (agave), *320*
Agelaius
  phoeniceus (red-winged blackbird), *21*, 299, *303*
  tricolor (tricolored blackbird), *21*, *466*
  xanthomus (yellow-shouldered blackbird), 439
agile (Pacific) kangaroo rat (*Dipodomys agilis*), *466*
Agnetina flavescens (stonefly), *186*
Agosia chrysogaster (longfin dace), *150*
agricultural ecosystems, 423-26
  Conservation Reserve Program (CRP) and, 423-26
  cropland, 423-26
  grassland, 424
  habitat change and, 108, 109, 115-16
  land used for, 423
  monoculture, 424
  productivity, 423, 424
  runoff, wetland loss and, 270
  wetland dynamics and, 475
  wildlife habitats and, 423
Agroperina lutosa, *166*
Agrotis stigmosa (owlet moth), 167
AIDS, 253
Aimophila
  cassinii (Cassin's sparrow), *21*, 297, 298
  ruficeps (rufous-crowned sparrow), *466*
air pollution, *See also* environmental quality
  in forest ecosystems, 227
  lichen and, 195, 196
  macrofungi decline and, 193
  in mixed-conifer forest system, 421-23
  in national park system, 228
  plant and fungi species decline and, 190
Aix sponsa (wood duck), 35
'akepa
  Hawaii (*Loxops coccineus coccineus*), *378*
  Kauai (*Loxops coccineus caeruleirostris*), *378*
  Maui (*Loxops coccineus ochraceus*), *378*
  Oahu (*Loxops c. rufus*), *376*
'akialoa
  Hawaii (*Hemignathus obscurus obscurus*), *376*
  Kauai (*Hemignathus procerus*), *377*
  Lanai (*Hemignathus obscurus lanaiensis*), *376*
  Oahu (*Hemignathus obscurus ellisianus*), *376*
'akiapola'au (*Hemignathus munroi*), 376, *377*
'akohokohe (*Palmaria dolei*), *378*
Alabama cavefish (*Speoplatyrhinus poulsoni*), *143*
Alabama red-bellied turtle (*Pseudemys alabamensis*), *119*
Alabama River, freshwater gastropod fauna, 249
'alala (Hawaiian crow) (*Corvus hawaiiensis*), *377*, 381
Alaska Department of Fish and Game, 344, 350
Alaska Lepidoptera Survey, 161
Alaska Maritime National Wildlife Refuge, 49
Alaska National Interest Lands Conservation Act, 357
Alaska Native Claims Settlement Act, 349
Alaskan ecoregion, 337-59
  Arctic nesting geese, 30-34
  Arctic tundra, 338-41
  Beaufort Sea, anadromous fish, 341-43
  coastal wetlands, 33
  Denali National Park, wolves and caribou, 347-48
  habitats, 32-34
  interior wetlands, 33
  Kodiak brown bears, 349-50
  Mentasta caribou, 357-58
  north Pacific Ocean, sea otters, 353-55
  overview, 337-38
  Pacific salmon, 343-47
  Pacific walruses, 356-57
  polar bears, 350-53
  seabirds, 49-52
  tundra or Arctic hare, 359
  vascular plants, 202
Alaska yellow-cedar (*Chamaecyparis nootkatensis*), 316
Alasmidonta
  marginata (elktoe), *180*
  viridis (slippershell mussel), *180*
Albany pine barrens, insects, 166-67
albatross
  Laysan (*Diomedea immutabilis*), 374
  short-tailed (*Diomedea albatrus*), 50
albedo, forest cover loss and, 472
Alca torda (razorbill), *54*, 55
Alces alces (moose), 310, *311*, 338, 347, 348
alcids (Alcidae), 16, 35, 46-48
alder flycatcher (*Empidonax alnorum*), *20*
aldicarb, *417*
aldrin, *408*
Alectoria sarmentosa, 196
Aleutian Canada goose (*Branta canadensis leucopareia*), 28, 30, 31, 32
Aleutian Islands, seabirds, 51
alewife (*Alosa pseudoharengus*), 261, *262*, *432*, 443, 445
Alextoris chukar (chukar), 437
algae
  benthic, 255

  fouling, 283
  freshwater, 255-56
  macrophytic, 255
  marine, 255-56
  non-native species, 444
  planktonic, 255
  sewage discharge and, 255
  Stephanodiscus binderanus, 444
  taxonomy, 256
  toxicity of, 255
  unicellular, 255
Algae Reef, 283
alien species. *See* non-native species
alligator, American (*Alligator mississippiensis*), 118, 127-28, 435
Alligator mississippiensis (American alligator), 118, 127-28, 435
Alligator Reef, 281
alligator snapping turtle (*Macroclemys temminckii*), 119
Allium gooddingii (wild onion), *320*
Allocapnia
  forbesi (stonefly), *186*
  granulata (stonefly), 186, *186*
  illinoensis (stonefly), 185, *186*
  mystica (stonefly), *186*
  nivicola (stonefly), 185, *186*
  recta (stonefly), *186*
  rickeri (stonefly), *186*
  smithi (stonefly), 185, *186*
  vivipara (stonefly), *186*, 187
Allonais paraguayensis, 176
Alloperla
  caudata (stonefly), 185, *186*
  roberti (stonefly), 185, *186*
Alopex lagopus (Arctic fox), 28, 51, 441
Alosa
  aestivalis (blueback herring), 261, 262
  pseudoharengus (alewife), 261, *262*, *432*, 443, 445
  sapidissima (American shad), 261, *262*, 263, *265*, 266, *432*
Alvord cutthroat trout (*Oncorhynchus clarki* spp.), *143*
'amakihi (*Hemignathus virens*), 376
  greater (*Hemignathus sagittirostris*), *376*
  Hawaii (*Hemignathus virens virens*), *377*
  Kauai (*Hemignathus virens stejnegeri*), *377*
  Maui (*Hemignathus virens wilsoni*), *377*
  Oahu (*Hemignathus virens chloris*), *377*, 378
'amaui (*Myadestes oahuensis*), *376*
Amazona, 85, *407*
  vittata (Puerto Rican parrot), 83-85
amber darter (*Percina antesella*), *143*
amberjack, greater (*Seriola dumerili*), 282
Amblema plicata plicata (threeridge), *180*
Ambloplites rupestris (rock bass), *261*, *432*
Ambystoma
  cingulatum (flatwoods salamander), 129
  macrodactylum croceum (Santa Cruz long-toed salamander), 124
  maculatum (spotted salamander), *419*
  tigrinum (tiger salamander), 327, *434*
Ameiurus
  catus (white catfish), *261*, *262*, *432*
  melas (black bullhead), *432*
  nebulosus (brown bullhead), 242, *244*, *432*
ameiva, giant (*Ameiva ameiva*), *434*
Ameiva ameiva (South American ground lizard or giant ameiva), *434*
American alligator (*Alligator mississippiensis*), 118, 435
  in Florida, 127-28
  hunting, 128
  night-light counts, 127
  nuisance control program, 128
American avocet (*Recurvirostra americana*), *59*, *62*
American badger (*Taxidea taxus*), in Illinois, 108-9
American black duck (*Anas rubripes*), 35, *36*, 266, 402, *413*, *420*
American Bryological and Lichenological Society, 196
American burying beetle (*Nicrophorus americanus*), 232
American chaffseed (*Schwalbea americana*), 207
American chestnut (*Castanea dentata*), 213
American coot (*Fulica americana*), *19*, 296, 299, *303*
American crocodile (*Crocodylus acutus*), 127
American crow (*Corvus brachyrhynchos*), *20*, *22*
American eel (*Anguilla rostrata*), 261
American Fisheries Society (AFS), Endangered Species Committee, imperiled fish listings, 142-44
American golden-plover (*Pluvialis dominica*), 58, *59*, *62*, 64
American goldfinch (*Carduelis tristis*), *21*, *22*, *299*
American kestrel (*Falco sparverius*), *67*
  Florida subspecies (*Falco sparverius paulus*), *67*
American Ornithological Union (AOU), 7
American oystercatcher (*Haematopus palliatus*), *59*, *62*
American peregrine falcon (*Falco peregrinus anatum*), *67*, 69
American pipit (*Anthus rubescens*), 22
American redstart (*Setophaga ruticilla*), *20*
American robin (*Turdus migratorius*), *20*, 231, *299*, 386
American shad (*Alosa sapidissima*), 261, *262*, 263, *265*, 266, *432*
American Somoa, non-native species, 433
American swallow-tailed kite (*Elanoides forficatus*), 67
American sycamore (*Platanus occidentalis*), 228
American tree sparrow (*Spizella arborea*), *22*, 23
American Type Culture Collection, protozoa, 253
American white pelican (*Pelecanus erythrorhynchos*), *19*, 53, *54*
American wigeon (*Anas americana*), 35, *36*, *266*
American woodcock (*Scolopax minor*), 58, *59*, *62*
Amistad gambusia (*Gambusia amistadensis*), *143*
Ammodramus
  bairdii (Baird's sparrow), *21*, *299*, *303*
  henslowii (Henslow's sparrow), 297
  savannarum (grasshopper sparrow), *21*, 297, *299*, 302, 303, *466*
Ammodytes hexapterus (sandlance), 51
Amoeba, 252
amphibians, 124-26
  acid deposition (acid rain) and, 419-20
  causes of decline, 125-26
  Colorado Plateau, 326-28
  ecological importance of, 124
  in endangered longleaf pine ecosystem, 129-31
  endemic species, 124
  faunal comparisons, 124-25
  monitoring needs, 126
  non-native species, 433-36
  population fluctuations, 126
  protection of, 118
  in southwestern sky mountains, 321
  status and trends, 117-18
  widespread species, 124-25
Amphigyra, 249, 250
  alabamensis (shoal sprite), *251*
Amphinemura
  delosa (stonefly), *186*
  nigritta (stonefly), 185, *186*
  varshava (stonefly), *186*
amphipod, troglobitic (*Gammarus troglophilus*), 177
Amphispiza
  belli (sage sparrow), *21*, *466*
  bilineata (black-throated sparrow), *21*
amsonia (*Amsonia kearneyana*), *320*
Amsonia kearneyana (amsonia), *320*
anadromous fish, of Beaufort Sea, central Alaska, 341-43
Anas
  acuta (northern pintail), *4*, 16, *19*, 35, 38-39, 266, *266*, 296, *299*, 300, 301, 389
  americana (American wigeon), 35, *36*, *266*
  clypeata (northern shoveler), *19*, 35, *36*, *266*, *299*, 300, 301
  crecca (green-winged teal), 35, *36*, *266*
  discors (blue-winged teal), *19*, 35, *36*, *266*, 296, *299*, 300, 301, *303*
  fulvigula (mottled duck), *19*, 35, *36*
  laysanensis (Laysan duck), *377*, 379, 400
  platyrhynchos (mallard), *19*, 35, *36*, 266, 296, 299, 300-301, *303*, 379, 402, 406, *413*
  rubripes (American black duck), 35, *36*, 266, 402, *413*, *420*
  strepera (gadwall), *19*, 35, *36*, *266*, *299*, 300, 301
  wyvilliana (Hawaiian duck or koloa), *377*, 379
Anchoa mitchilli (bay anchovy), 261, *262*
anchovy, bay (*Anchoa mitchilli*), 261, *262*
ancient murrelet (*Synthliboramphus antiquus*), 46, 51
Ancylidae (freshwater limpets), 251
Andropogon
  gerardii, 221
  virginicus (broomsedge), 206
angiosperms, 201
Anguilla rostrata (American eel), 261
anhingas, 53-54
'anianiau (*Hemignathus parvus*), *377*
Animal Damage Control Program, U.S. Department of Agriculture, 100
Animal and Plant Health Inspection Service, U.S. Department of Agriculture (USDA), 458
Anniella pulchra (California legless lizard), *466*
Anodontoides ferussacianus (cylindrical papershell), *180*

anole
  bark (*Anolis distichus*), *434*
  brown (*Anolis sagrei*), *434*
  green (*Anolis carolinensis*), *434*
  Hispaniolan green (*Anolis chlorocyanus*), *434*
  Jamaican giant (*Anolis garmani*), *434*
  knight (*Anolis equestris*), *434*
  large-headed (*Anolis cybotes*), *434*
  Puerto Rican crested (*Anolis cristellatus*), *434*
Anolis
  carolinensis (green anole), *434*
  chlorocyanus (Hispaniolan green anole), *434*
  cristellatus (Puerto Rican crested anole), *434*
  cybotes (large-headed anole), *434*
  distichus (bark anole), *434*
  equestris (knight anole), *434*
  garmani (Jamaican giant anole), *434*
  sagrei (brown anole), *434*
Anomogyna badicollis, *166*
Anous minutus melanogenys (Hawaiian noddy), *377*, 379
Anser
  albifrons frontalis (greater white-fronted goose), 30, 31, 33
  albifrons gambeli (tule white-fronted goose), 30, 31
ANSP (Academy of Natural Sciences in Philadelphia), 163, 164
ant, 161
  Africanized honeybees and, 450
  Argentine (*Iridomyrmex humilis*), 364, 366
  fire (*Solenopsis* spp.), 441
  red fire (*Solenopsis invicta*), 131
antelope, pronghorn (*Antilocapra americana*), 313, 320
Anthus
  rubescens (American pipit), 22
  spragueii (Sprague's pipit), *20*
Antilocapra americana (pronghorn antelope), 313, 320
  mexicana (Mexican pronghorn), 318
anurans, 124
AOU. *See* American Ornithological Union (AOU)
Apache trout (*Oncorhynchus apache*), 147, *150*, 320, *320*
Apalone spinifera (spiny softshell turtle), *119*, *434*
'apapane (*Himatione s. sanguinea*), *378*
Aphelocoma coerulescens (scrub jay), *20*
aphids, 161
Aphriza virgata (surfbird), *62*, *64*
Apis mellifera (honey bee), 85
  scutellata (African bee), 448
Aplodinotus grunniens (freshwater drum), 261, 263, 447
aplomado falcon (*Falco femoralis septentrionalis*), *67*
Aplonis opaca (Micronesian starling), 455
aquarium industry, non-native species and, 429
aquatic ecosystems, 233-58, *See also* coastal and marine ecosystems; water quality
  acid deposition (acid rain) and, 418-19, 420
  alterations, 149
  Colorado River Basin, 149-41
  freshwater diatoms, 256-58
  Great Lakes, 247-49
    fish contaminant trends, 242-44
    lake trout, 244-46
  habitat loss/change, 178, 234-36
  Illinois River fish, 239-41
  Mobile Bay basin, freshwater gastropod fauna, 249-52
  non-native species, 428-30
  overview, 233-34
  protozoa, 252-54
  trends, 233-34
  Upper Mississippi River, 234-38
aquatic insects
  as indicators of environmental quality, 182-84
  larval deformities, 183
  Mississippi River, 183-84
  Ohio River, 182-83, 184
Aquatic Nuisance Species (ANS) Task Force, 429
  Detection and Monitoring Committee, 430
aquatic vegetation
  marine and freshwater algae, 255-56
  non-native species, 443
  in regulated Great Lakes wetlands, 247-49
  submersed, navigation dams and, 237
  vascular plants, 207
  weeds, 429
  zebra mussel (*Dreissena polymorpha*) and, 445
Aquila chrysaetos (golden eagle), *67*, 68, 338, 340, 387, *466*
Aratinga canicularis (orange-fronted parakeet), 85
Arceuthobium americanum (dwarf mistletoe), 223
Archilochus alexandri (black-chinned hummingbird), *466*
Arcrocephalus f. familiaris (Laysan millerbird), *376*
Arctic-breeding shorebirds, 64
Arctic cisco (*Coregonus autumnalis*), 337, 341-43
Arctic fox (*Alopex lagopus*)
  Alaskan seabirds and, 51, 441
  Canada geese and, 28
Arctic geese
  Alaskan populations, 30-34
  inventory, 31
  nesting behavior changes, global climate change and, 388
Arctic grayling (*Thymallus arcticus*), *432*
Arctic (tundra) hare (*Lepus othus* or *Lepus timidus*), 338, 359
  distribution records, 359
  status, 359
Arctic loon (*Favia arctica*), *420*
Arctic National Wildlife Refuge, 33-34, 337, 338-40
Arctic peregrine falcon (*Falco peregrinus tundrius*), *67*, 69
Arctic tern (*Sterna paradisaea*), 46, 51
Arctiidae (tiger moths), 172, *173*
Ardea herodias (great blue heron), 53, *54*, 55
Arenaria
  interpres (ruddy turnstone), 59, *62*
  melanocephala (black turnstone), *62*, 64
Argentine ant, (*Iridomyrmex humilis*), 364, 366
Argyrostrotis quadrifiliaris (owlet moth), 166
Argyroxiphium sandwicense
  ssp. *macrocephalum* (Haleakala silversword), 362, 363-64
  ssp. *sandwicense* (Haleakala silversword), 363
Aristichthys nobilis (bighead carp), *432*
Arizona, Africanized honeybees, 450, 451
Arizona Riparian Council, 290
Arizona shrew (*Sorex arizonae*), *320*
Arizona-Sonora Desert Museum (ASDM), 140
Arkansas River, prairie fish, 303-5
Arkansas River shiner (*Notropis girardi*), 295, 303-5
Arkansas River speckled chub (*Macrhybopsis aestivalis tetranemus*), 295, 303-5
armadillo (*Dasypus novemcinctus*), 386
arogos skipper (*Atrytone arogos iowa*), 175
arribadas, Kemp's ridley, 123
arrowweed (*Tessaria sericea*), 287, 288
arsenic, Tarahumara frog (*Rana tarahumarae*) and, 140
Artemisia spp. (intermountain sagebrush), 222
Ascaphus truei (tailed frog), 125
Asclepias purpurascens (purple milkweed), 231
ascomycetes (truffles), 193
ash-throated flycatcher (*Myiarchus cinerascens*), *20*
ashy gecko (*Sphaerodactylus elegans*), *434*
ashy storm-petrel (*Oceanodroma homochroa*), 44
Asian clam (*Corbicula fluminea*), 178, *429*, 443
Asian tapeworm (*Bothriocephalus opsarichthydis*), 429
Asio
  flammeus (short-eared owl), *67*, 297
  flammeus sandwichensis (Pueo or Hawaiian owl), *377*
  otus (long-eared owl), *67*, *466*
aspen, quaking (*Populus tremuloides*), 313
Aspergillosis, *404*
Association of Natural Heritage Programs and Conservation Data Centers (Natural Heritage Network), 218
aster
  Aster potosinus, *320*
  Ruth's golden (*Pityopsis ruthii*), 400
Asterella echinella, *199*
Aster potosinus (aster), *320*
Astragalus, 201
  cobrensis var. *maguirei* (milk-vetch), *320*
  hypoxylus (milk-vetch), *320*
Athene cunicularia (burrowing owl), *67*
Atlantic croaker (*Micropogonias undulatus*), 409, *411*
Atlantic Flyway
  avian cholera, 402
  Canada goose, 27, 28-30
  canvasback, 40, 42
  northern pintail, 38-39
  regions of, 28
Atlantic puffin (*Fratercula arctica*), 55
Atlantic salmon (*Salmo salar*), *432*
Atlantic silverside (*Menidia menidia*), 261, *262*
Atlantic Slope hydrologic region, 145
Atlantic white cedar (*Chamaecyparis thyoides*), bogs, 216
Atlas of the Flora of the Great Plains, 203
Atlas of United States Trees, 201
Atlas of Vascular Plants of Utah, 202

atmospheric carbon fixation, forest cover loss and, 472
Atriplex patula, 220
Atrytone arogos iowa (arogos skipper), 175
Atrytonopsis hianna (savanna skipper), 232
Attaneuria ruralis (stonefly), *186*
Audubon Christmas Bird Count (CBC). *See* Christmas Bird Count (CBC)
Audubon's crested caracara (*Caracara plancus audubonii*), in Florida, 82-83
Audubon's warbler (*Dendroica coronata auduboni*), *20*
auk (Family Alcidae), 50
auklet
  Cassin's (*Ptychoramphus aleuticus*), 46, 51
  crested (*Aethia cristatella*), 51
  least (*Aethia pusilla*), 51
  rhinoceros (*Cerorhinca monocerata*), 46
Aulacoseira sp., *257*
Aureolaria grandiflora (giant false-foxglove), 231
Auriparus flavicep (verdin), *20*
Austurina plagiata (gray hawk), *67*
avens, mountain (*Geum peckii*), 394
AVHRR (Advanced Very High Resolution Radiometer), 469, 473
avian diseases, 402, *403*, *407*
  avian botulism, 401-2, *403*, 404
  avian cholera, 402, *403*, 404
  avian pox virus, 378, 380, *403*
  avian tuberculosis, *403*, *407*
  changes in disease patterns, 401-3
  duck plague, 402, *403*, *407*
  habitat change and, 401-4
  Hawaiian birds and, 378, 380
  host-agent interactions, 401
  human interactions and, 403-4
  magnitude of losses, 403
  major die-offs, *404*
  prevention of, 404
  salmonellosis, *403*
  western duck disease, 401
avocet, American (*Recurvirostra americana*), *59*, *62*
Aythya
  affinis (lesser scaup), *19*, 35, *36*, 237
  americana (redhead), 35, *36*, 266, 275-77
  collaris (ring-necked duck), 35, *36*, *266*, *420*
  marila (greater scaup), 35, *36*
  valisineria (canvasback), 16, 35, 40-42, 236, 237, *238*, 266

Babbitt, Bruce, 9
badger, American (*Taxidea taxus*), 108-9
badlands bighorn sheep (*Ovis canadensis audobonii*), 332
Baird's sandpiper (*Calidris bairdii*), 59, *62*
Baird's sparrow (*Ammodramus bairdii*), *21*, *299*, *303*
Balaena mysticetus (bowhead whale), 95, *96*
baldcypress (*Taxodium distichum*), 216
bald eagle (*Haliaeetus leucocephalus*), 16, 44, 66-67, 243, 309, 387, 389, 415, *466*
  causes of death, 68
  eggshell thickness, 413
  persistent contaminants and, 413
  wintering, Colorado River corridor, 328-30
Bald Eagle Protection Act (1940), 66
ballast water discharge, non-native species and, 429
Baltimore oriole (*Icterus galbula*), *21*
banana poka (*Passiflora mollissima*), 370
band-tailed pigeon (*Columba fasciata*), *19*, 322
bandwing grasshopper (*Pardalophora haldemani*), 164
Banff National Park, 99
bank swallow (*Riparia riparia*), *20*, *299*
Baptisa australis (fabaceae), *203*
Barbary dove (ringed turtle-dove) *Streptopelia risoria*, 438
barbel (*Barbus barbus*), 241
Barbour's map turtle (*Graptemys barbouri*), *119*, *120*
Barbus barbus (barbel), 241
bark anole (*Anolis distichus*), *434*
barking frog (*Hylactophryne augusti*), *320*
barn owl (*Tyto alba*), *67*, 231
barn swallow (*Hirundo rustica*), *20*, *299*
barred owl (*Strix varia*), *67*, 387
barred sand bass (*Paralabrax nebulifer*), 410
bar-tailed godwit (*Limosa lapponica*), *62*
Barton Springs salamander (*Eurycea sosorum*), 125
Bartramia longicauda (upland sandpiper), *19*, 58-59, *62*, 297, *299*, *303*
basidiomycetes (false truffles), 193
Basiliscus vittatus (brown basilisk), *434*
basilisk, brown (*Basiliscus vittatus*), *434*
bass, 145
  barred sand (*Paralabrax nebulifer*), 410
  Florida largemouth (*Micropterus salmoides floridanus*), 148
  Guadalupe (*Micropterus treculi*), 147
  kelp (*Paralabrax clathratus*), 411
  largemouth (*Micropterus salmoides*), 148, 240, *261*, *432*
  northern largemouth (*Micropterus salmoides salmoides*), 148
  rock (*Ambloplites rupestris*), *261*, *432*
  smallmouth (*Micropterus dolomieu*), 147, *261*, *419*, *432*
  spotted (*Micropterus punctulatus*), 147, *432*
  striped (*Morone saxatilis*), 261, *262*, 263, 265, *432*
  white (*Morone chrysops*), *261*, 263, *432*
bat
  gray (*Myotis grisescens*), 176
  Indiana (*Myotis sodalis*), 97-98, 176
  Mexican long-tongued (*Choeronycteris mexicana*), *320*
  red (*Lasiurus borealis*), *320*
  Sanborn's long-nosed (*Leptonycteris sanborni*), *320*
  southeastern (*Myotis austroriparius*), 176
Batrachoseps
  pacificus (Pacific slender salamander), *466*
  stebbinsi (Tehachapi slender salamander), *466*
bay anchovy (*Anchoa mitchilli*), 261, *262*
bay-breasted warbler (*Dendroica castanea*), *20*
BBIRD. *See* Breeding Biology Research and Monitoring Database (BBIRD)
BBS. *See* Breeding Bird Survey (BBS); Bureau of Biological Survey (BBS)
beaked whale, 94
beakgrain (*Diarrhena americana obovata*), 209
bear
  black (*Ursus americanus*), 100-102, 109, 322
  brown (*Ursus arctos*), 104, 337, 338, 340
  Florida (*Ursus americanus floridanus*), 102
  grizzly (*Ursus arctos*), 103-5, 228, 229, 312, 313, *320*, 321, 338, 347, 358
  Kodiak brown (*Ursus arctos middendorffi*), 349-50
  Louisiana (*Ursus americanus luteolus*), 102
  polar (*Ursus maritimus*), 338, 351-53
bearded seal (*Erignathus barbatus*), 351
Beaufort Sea
  anadromous fish, 341-43
  polar bear (*Ursus maritimus*), 351-53
bee, 161
  African (*Apis mellifera scutellata*), 448
  Africanized honeybee (AHB), 448-51
  bumble, 450
  European honeybee (EHB), 448-51
  honeybee (*Apis mellifera*), 85
  Italian honeybee (*Apis mellifera ligustica*), 448
beetle
  American burying (*Nicrophorus americanus*), 232
  Coleoptera, 161
  Plagithmysus, 367
Bell's vireo (*Vireo bellii*), 231, *466*
belted kingfisher (*Ceryle alcyon*), *420*
benthic algae, 255
Bering Island, Russia, sea otter (*Enhydra lutris*), 354, *355*
Besseya bullii (kitten-tails), 231
Bewick's wren (*Thryomanes bewickii*), *20*
Big Creek Reserve, University of California, 169-70
bighead carp (*Aristichthys nobilis*), *432*
bighorn sheep
  Ovis canadensis, 309, 332-33, *407*
    badlands (*Ovis canadensis audobonii*), 332
    California (*Ovis canadensis californiana*), 332
    desert
      Ovis canadensis cremnobates, 334
      Ovis canadensis mexicana, 320, 334, 335
      Ovis canadensis nelsoni, 309, 320, 332, 333-35
    restoration of, 332-33
    in Rocky Mountain national parks, 332-33
    Rocky Mountain (*Ovis canadensis canadensis*), 332
    translocations, 332-33
bigmouth rocksnail (*Leptoxis occultata*), *251*
big sagebrush scrub, *465*
Big Spring spinedace (*Lepidomeda mollispinis pratensis*), *150*
Bill Williams River, 287
biodiversity, *See also* genetic diversity; species diversity
  altered, Colorado River Basin, 151-52
  conserving, 6-7
  coral reef ecosystems, 283
  defined, 6

edge effect and, 160
exotic weeds and, 394
identifying gaps in, 463-64
Illinois stonefly, 184-87
importance of, 6
insects, 168
managing, 94, 463
microfungi, 191
native vascular plants, 208
non-native species and, 152
planning for, using Gap Analysis Program (GAP), 462
plants and fungi, 189
protecting through Gap Analysis Program (GAP), 465-66
protection of vegetation type status, 463-64
in southwestern California, 465-66
status ranks, 399-400
in terrestrial ecosystems, 213
biogeographic mapping, 8
bioindicators
aquatic insects, 182-84
Drosophila (Pomance flies), 368, 371
insects, 365
protozoa, 253
seabirds, 49
biological control programs, for non-native species, 430
biological pollution, 160
freshwater mussels and, 179, 181-82
in Lake Erie, 179
biological surveys, *See also* inventory and monitoring programs
early, 8
BIOTA (Biosystematic Information on Terrestrial Arthropods), 161
birds, *See also* colonial waterbirds; landbirds; migratory birds; seabirds; shorebirds; songbirds; waterfowl
adult survival, 23-26
avian diseases, 370, 380, 401-4, *407*
Breeding Bird Survey (BBS), 15, 17-21, 53, 58, 63, 64, 73-74, 303, 437
breeding productivity, 23-26
changes in winter ranges of, global climate change and, 386-89
determining status and trends, 15-17
extinct species, 375
grassland, 18, 396-98
Hawaiian, 372-75
endangered, 376
endemic, 376-81
extinct, 375, *376*
information sources, 5
nest success, 25, 300-302
non-native species, 379, 437-39
overview, 15-17
population health, 23
population trends 1966-92, 17-21
of prey, 16, 65-69
reproductive success, 302-3
shrubland, 18
in southwestern sky islands, 320
translocation programs, 405
woodland, 18
birds-in-a-nest, white (*Macbridea alba*), 401
Bishop Museum of Natural and Cultural History, 202, 362, 365
Bishop's 'o'o (*Moho bishopi*), *377*
bison (*Bison bison*), 295, 305
Bison bison (bison), 295, 305
Bitterroot ecosystem, grizzly bear, 103, 105
bivalves, contaminants in, 409-11
black-and-white warbler (*Mniotilta varia*), *20*
black-backed gull, great (*Larus marinus*), *19*, *54*, 55
black bear (*Ursus americanus*), 100-102, 109, 322
hunting, 100, 101, 102
sanctuaries, 102
black-bellied plover (*Pluvialis squatarola*), 58, *59*, *62*
blackberry, wild (*Rubus argutus*), 206
black-billed magpie (*Pica pica*), *20*
blackbird
Brewer's (*Euphagus cyanocephalus*), *21*, *299*
red-winged (*Agelaius phoeniceus*), *21*, 299, *303*
tricolored (*Agelaius tricolor*), *21*, *466*
yellow-headed (*Xanthocephalus xanthocephalus*), *21*, *299*
yellow-shouldered (*Agelaius xanthomus*), 439
black brant, Pacific (*Branta bernicla nigricans*), 30, 31, 32
black bullhead (*Ameiurus melas*), *432*
Blackburnian warbler (*Dendroica fusca*), *20*
black-capped chickadee (*Parus atricapillus*), *20*
black cherry (*Prunus serotina*), 228
black-chinned hummingbird (*Archilochus alexandri*), *466*
black crappie (*Pomoxis nigromaculatus*), *261*, *432*
black-crested titmouse (*Parus bicolor atricristatus*), *20*
black croaker (*Cheilotrema saturnum*), 410
black-crowned night-heron (*Nycticorax nycticorax*), *54*
black duck, American (*Anas rubripes*), 35, *36*, 266, 402, *413*, *420*
blackfin cisco (*Coregonus nigripinnis*), *143*
black-footed ferret (*Mustela nigripes*), 106-8
black hawk, common (*Buteogallus anthracinus*), *67*
black-headed grosbeak (*Pheucticus melanocephalus*), *20*
black-legged kittiwake (*Rissa tridactyla*), 49-50
black mamo (*Drepanis funerea*), *376*
blackmouth shiner (*Notropis melanostomus*), *143*
black mudalia (*Leptoxis melanoides*), *251*
black-necked stilt (*Himantopus mexicanus*), *19*, *59*, *62*
black oystercatcher (*Haematopus bachmani*), *62*
blackpoll warbler (*Dendroica striata*), *20*, *391*
black rat (*Rattus rattus*), Puerto Rican parrot and, 85
black sandshell (*Ligumia recta*), *180*
black scoter (*Melanitta nigra*), *36*
black-shouldered kite (*Elanus caeruleus*), *466*
blackside dace (*Phoxinus cumberlandensis*), *143*
black skimmer (*Rynchops niger*), 45, 55
black storm-petrel (*Oceanodroma melania*), 44
black swift (*Cypseloides niger*), *20*
black-tailed gnatcatcher (*Polioptila melanura*), *20*
black-tailed prairie dog (*Cynomys ludovicianus*), 107
black tern (*Chlidonias niger*), *19*, 46, *54*, *55*, 56, 299
black-throated gray warbler (*Dendroica nigrescens*), *20*
black-throated green warbler (*Dendroica virens*), *20*
black-throated sparrow (*Amphispiza bilineata*), *21*
black turnstone (*Arenaria melanocephala*), *62*, 64
black vulture (*Coragyps atratus*), *19*, *67*
black walnut woodlands, 465
bladderwort, large floating (*Utricularia inflata* Walter), 209
Blanding's turtle (*Emydoidea blandingii*), *119*
blazingstar, dwarf (*Liatris cylindracea*), *172*
blind salamander, Texas (*Typhlomolge rathbuni*), 124
blind snake, Brahminy (*Rhamphotyphlops braminus*), *434*
blister rust, white pine (*Cronartium ribicola*), 229
blueback herring (*Alosa aestivalis*), 261, 262
bluebird
eastern (*Sialia sialis*), *20*, *22*
mountain (*Sialia currucoides*), *20*
western (*Sialia mexicana*), *466*
blue butterfly, Karner (*Lycaeides melissa samuelis*), 166, 167, *172*, 224- 25, 232, *398*, 401
blue catfish (*Ictalurus furcatus*), *432*
blue crab (*Callinectes sapidus*), 263, 264-65, 282
bluegill (*Lepomis macrochirus*), 240, *261*, *432*
bluegrass, Kentucky (*Poa pratensis*), 471
blue-gray gnatcatcher (*Polioptila caerulea*), *20*, *466*
blue grosbeak (*Guiraca caerulea*), *20*, *466*
bluehead sucker
Catostomus discobolus discobolus, *150*
Zuni (*Catostomus discobolus yarrowi*), *143*
blue heron
great (*Ardea herodias*), 53, *54*, 55
little (*Egretta caerulea*), *19*
blue jay (*Cyanocitta cristata*), *20*
blue pike (*Stizostedion vitreum glaucum*), *143*
bluestem prairie, 472
blue-tailed skink, Marianas (*Emoia caerulocauda*), 455
blue tilapia (*Tilapia aurea*), 430, *432*
blue-winged teal (*Anas discors*), *19*, 35, *36*, *266*, *296*, *299*, 300, 301, *303*
boa, rosy (*Lichanura trivirgata*), *466*
boat-tailed grackle (*Quiscalus major*), *21*
bobcat (*Lynx rufus*), 193
bobolink (*Dolichonyx oryzivorus*), *21*, *299*, *303*
bobwhite, northern (*Colinus virginianus*), *19*, 231, 424
bog turtle (*Clemmys muhlenbergii*), 119, 458
Boiga irregularis (brown tree snake), 433, *434*, 435, 436, 454-56
Bombycilla cedrorum (cedar waxwing), *20*, 21-22, 23, *299*
bonytail (*Gila elegans*), *150*, 151, 309, 324-26
Colorado River basin, 324-26
Boott's rattlesnake-root (*Prenanthes boottii*), 394
boreal owl (*Aegolius funereus*), *67*
Bos taurus (cow), *440*, 441
bot fly (*Philornis* spp.), Puerto Rican parrot and, 85
Bothriocephalus opsarichthydis (Asian tapeworm), 429
bottlenose dolphin (*Tursiops truncatus*), *94*, 96
Bouteloua gracilis, *202*
bowhead whale (*Balaena mysticetus*), 95, *96*

box turtle, common (*Terrapene carolina*), 119, *434*
Brachyramphus
  brevirostris (Kittlitz's murrelet), 50
  marmoratus (marbled murrelet), 43, 47, 50
bracket fungi, 192
braconid wasps, Hawaii, 366
Brahminy blind snake (*Rhamphotyphlops braminus*), *434*
brainworm, *407*
Brandt's cormorant (*Phalacrocorax penicillatus*), 44
brant, Pacific black (*Branta bernicla nigricans*), 30, 31, 32
Branta
  bernicla nigricans (Pacific black brant), 30, 31, 32
  canadensis (Canada goose), 16, *19*, 26-33, *266*
    canadensis, 28-29
    fulva (Vancouver Canada goose), 28, 30-31, 32
    hutchinsii, 28
    interior, 28, 29
    leucopareia (Aleutian Canada goose), 28, 30, 31, 32
    maxima (giant Canada goose), 26, 27, 29
    minima (cackling Canada goose), 26, 28, 30, 31, 33
    moffitti, 27, 28, 29
    occidentalis, 28
    taverneri (Taverner's Canada goose), 28, 33
  sandvicensis (nene or Hawaiian goose), *377*, 379
Breeding Biology Research and Monitoring Database (BBIRD), 16, 24
  breeding productivity, 25-26
  data provided by, 24
  habitat-specific differences, 25
Breeding Bird Counts, 437
Breeding Bird Survey (BBS), 15, 17
  colonial waterbirds, 53
  common ravens, 73-74
  Great Plains migratory birds, 303
  methods, 17
  non-native species, 437
  population trends 1966-92, 17-21
  shorebirds, 58, 63, 64
Breeding Population and Habitat Survey
  ducks, 34-35
  northern pintail estimates, 38
breeding productivity, birds, 23-26
breeding seabirds, 43-48
  Alaska, 49-52
Breeding Waterfowl and Habitat Surveys, 40
Brephos infans (geometer), 166
Brewer's blackbird (*Euphagus cyanocephalus*), *21*, *299*
Brewer's sparrow (*Spizella breweri*), *21*, 297
bridle shiner (*Notropis bifrenatus*), *261*, *262*, 263
bristlecone pine
  (*Pinus aristata*), 315, 316
  (*Pinus longaeva*), 315, 316
bristle-thighed curlew (*Numenius tahitiensis*), *62*, 64
Bristol Bay Lowlands, 33
British Lichenological Society, 196
broadleaf forest, 469
broad-tailed hummingbird (*Selasphorus platycercus*), *20*
broad whitefish (*Coregonus nasus*), 337, 341, 342
broad-winged hawk (*Buteo platypterus*), *67*
  Puerto Rican (*Buteo platypterus brunnescens*), *67*
Bromus tectorum (cheatgrass), 222
bronzed cowbird (*Molothrus aeneus*), *21*
brood parasitism, southwestern willow flycatcher, 90, 91
brook trout (*Salvelinus fontinalis*), 419, *432*, *440*, 441
broomsedge (*Andropogon virginicus*), 206
Brotogeris versicolurus (canary-winged parakeet), 439
brown anole (*Anolis sagrei*), *434*
brown basilisk (*Basiliscus vittatus*), *434*
brown bear (*Ursus arctos*), 104, 337, 338, 340
  Kodiak (*Ursus arctos middendorffi*), 349-50
brown bullhead (*Ameiurus nebulosus*), 242, *244*, *432*
brown-crested flycatcher (*Myiarchus tyrannulus*), *20*
brown four-fingered skink (*Carlia fusca*), *434*
brown-headed cowbird (*Molothrus ater*), *21*, 25, 299, *303*, 389
  southwestern willow flycatcher and, 90, 91
brown-headed nuthatch (*Sitta pusilla*), *20*, *22*
brown pelican (*Pelecanus occidentalis*), 16, 43, 44, 53
  eastern, *54*
brown thrasher (*Toxostoma rufum*), *20*, 231, *299*
brown tide, 276-77
brown towhee (*Pipilo fuscus*), *21*
brown tree snake (*Boiga irregularis*), 433, *434*, 435, 436, 454-56
brown trout (*Salmo trutta*), *419*, 429, *432*, *440*, 441, 443
bryophytes, 197-98
  distribution, 197-98
  ecological roles, 197
  floristic inventories, 198-200
  New York, 209-10
  status, 198
Bubo virginianus (great horned owl), *67*
Bubulcus ibis (cattle egret), *19*, 386, 437, *438*, 439
Bucephala
  albeola (bufflehead), 36, *266*
  clangula (common goldeneye), *36*, *266*, *420*
buck moth (*Hemileuca maia*), 232
budgerigar (*Melopsittachus undulatus*), 439
Buenos Aires National Wildlife Refuge (BANWR), 452-53
buff-breasted flycatcher (*Empidomax fulvifrons*), *320*
buff-breasted sandpiper (*Tryngites subruficollis*), 59, *62*, 64
buffel grass (*Pennisetum cilaris*), 321
bufflehead (*Bucephala albeola*), 36, *266*
Bufo, 326
  americanus (toad), 435
  boreas (western toad), 125, 327, *328*
  marinus (marine toad), 433, *434*, 435, 441
  microscaphus (southwestern toad), *466*
  woodhousii (Woodhouse's toad), 327
bullfrog (*Rana catesbeiana*), 133, *419*, 433, *434*, 435
  control of, 453
  in southwestern wetlands, 452-53
bullhead, 145
  black (*Ameiurus melas*), *432*
  brown (*Ameiurus nebulosus*), 242, *244*, *432*
Bullock's oriole (*Icterus galbula bullockii*), *21*
bulrush, Long's (*Scirpus longii*), 458
bumblebee, Africanized honeybees and, 450
bunting
  indigo (*Passerina cyanea*), *20*, 231
  lark (*Calamospiza melanocorys*), *21*, 297, 298, *299*, 302, *303*
  lazuli (*Passerina amoena*), *20*
  painted (*Passerina ciris*), *20*
Bureau of Biological Survey (BBS), 401
  activities, *8*
  history, 7-9
Bureau of Land Management (BLM), *5*, 135, 214, 400
  distribution of, 215
  raven predation control, 74
  Strategic Plan for the Management of Wild Horses and Burros on Public Lands, 457
  wild horses and burros management, 456-57
burro, feral (or wild) (*Equus asinus*), *440*, 441, 456-57
burrowing birds, 51
burrowing mayfly
  Hexagenia spp., 181, 236
  Hexagenia bilineata, 183-84
  Hexagenia limbata, 183-84
  Pentagenia vittigera, 183
burrowing owl (*Athene cunicularia*), *67*
burying beetle, American (*Nicrophorus americanus*), 232
bush, creosote, 472
bushtit (*Psaltriparus minimus*), *20*
Buteo
  albicaudatus (white-tailed hawk), *67*
  albonotatus (zone-tailed hawk), *67*
  brachyurus (short-tailed hawk), *67*
  jamaicensis (red-tailed hawk), *67*
  lagopus (rough-legged hawk), *67*
  lineatus (red-shouldered hawk), *67*
  nitidus (gray hawk), *67*, *320*
  platypterus (broad-winged hawk), *67*
    brunnescens (Puerto Rican broad-winged hawk), *67*
  regalis (ferruginous hawk), 67, 69, 297, 387
  solitarius (Hawaiian hawk), *67*, *377*
  swainsoni (Swainson's hawk), *67*
Buteogallus anthracinus (common black hawk), *67*
butterfly, 161, *See also* Lepidoptera
  arogos skipper (*Atrytone arogos iowa*), 175
  Dakota skipper (*Hesperia dacotae*), 174, 175
  elfin (*Incisalia henrici*), 167
  endangered species, 171
  Fourth of July Butterfly Count (FJC), 171-72
  frosted elfin (*Incisalia irus*), 167, 232
  habitat change and, 160, 174
  haying management and, 175
  inventories, 168-70
  Karner blue (*Lycaeides melissa samuelis*), 166, *172*, 224-25, 232, *398*, 401
  monarch (*Danaus plexippus*), *171*, 172

ottoe skipper (*Strytone arogos iowa*), 175
painted lady (*Vanessa cardui*), 171-72
poweshiek skipper (*Oarisma poweshiek*), 175
regal fritillary (*Speyeria idalia*), 174, *175*
sky islands, 172-73
tall-grass prairie, 174-76
tawny crescent (*Phyciodes batesii*), *166*, 232
western, 172-73
Bythotrephes cederstroemi (spiny water flea), 429, 444

Cabinet-Yaak ecosystem, grizzly bear, 104-5
Cacalia atriplicifolia (pale Indian-plantain), 231
cackling Canada goose (*Branta canadensis minima*), 26, 28, 30, 31, 33
cactus
pincushion (*Coryphantha robbinsorum*), *320*
prickly pear (*Opuntia*), 321
cactus ferruginous pygmy owl (*Gaucidium brasilianum cactorum*), 69
cactus wren (*Campylorhynchus brunneicapillus*), *20*
caddisfly, 182, 183
cadmium
in coastal sediments, 409
Tarahumara frog (*Rana tarahumarae*) and, 140
Cagle's map turtle (*Graptemys caglei*), *119*
Cahaba pebblesnail (*Clappia cahabensis*), *251*
Cahaba River, freshwater gastropod fauna, 250-52
caiman, spectacled (*Caiman crocodilus*), 433, *434*, 435
Caiman crocodilus (spectacled caiman), 433, *434*, 435
Cairina moschata (muscovy duck), 437-38
Calamospiza melanocorys (lark bunting), *21*, 297, 298, *299*, 302, *303*
Calcarius
mccownii (McCown's longspur), *21*, 297
ornatus (chestnut-collared longspur), *21*, 299, *303*
Calidris
alba (sanderling), 59, *62*
alpina (dunlin), 59, *62*
articola (dunlin), 64
pacifica (dunlin), 63
bairdii (Baird's sandpiper), *59*, *62*
canutus (red knot), 59, *62*
fuscicollis (white-rumped sandpiper), 59
himantopus (stilt sandpiper), *59*, *62*
maritima (purple sandpiper), *59*, *62*
mauri (western sandpiper), *59*, *62*
melanotos (pectoral sandpiper), *59*, *62*
minutilla (least sandpiper), 59, *62*
ptilocnemis (rock sandpiper), *62*
pusilla (semipalmated sandpiper), 59, *62*
California
African clawed frog (*Xenopus laevis*), 435
breeding seabirds, 43-48
coastal sage scrub, vegetation change, 226
Department of Fish and Game, 131
Fish and Game Code, 80
native ranid frogs, 131-34
southwestern, biodiversity in, 465-66
California Academy of Sciences, 131
California bighorn sheep (*Ovis canadensis californiana*), 332
California chipmunk (*Tamias obscurus*), *466*
California condor (*Gymnogyps californianus*), 16, *67*, 80-81
California gnatcatcher (*Polioptila californica*), *466*
California gull (*Larus californicus*), *19*, 45
California Insect Survey, 161
California least tern (*Sterna antillarum browni*), 56, 74
California legless lizard (*Anniella pulchra*), *466*
California Natural Diversity Data Base, 465
California quail (*Callipepla californica*), *19*
California red-legged frog (*Rana aurora draytonii*), 132
California sea lion (*Zalophus californianus*), 96
California spotted owl (*Strix occidentalis occidentalis*), *67*
California thrasher (*Toxostoma redivivum*), *20*
California towhee (*Pipilo californicus*), *21*
California, University of
Big Creek Reserve, 169-70
Santa Barbara (UCSB), 465
California walnut woodlands, *465*
California Wildlife-Habitat Relationships (WHR) data base, 466
Callinectes sapidus (blue crab), 263, 264-65, 282
Callipepla
californica, *19*
gambelii (Gambel's quail), *19*
squamata (scaled quail), *19*
Callorhinus ursinus (northern fur seal), *95*, 96
Calyptae costae (Costa's hummingbird), *466*
Camassia scilloides (wild hyacinth), 231
Camnula pellucida (clear-winged), 164
Campostoma ornatum (Mexican stoneroller), *320*
Campylorhynchus brunneicapillus (cactus wren), *20*
Canada goose (*Branta canadensis*), 16, *19*, 26-33, *266*
Alaskan populations, 30, 33
Aleutian (*Branta canadensis leucopareia*), 28, 30, 31, 32
Atlantic Flyway, 27, 28-30
cackling (*Branta canadensis minima*), 26, 28, 30, 31, 33
dusky, 32
harvest rate, Atlantic Flyway, 29
human activity and, 27
hunting of, 26-27, 28
hutchinsii, 28
migratory population, 28-30
nesting behavior changes, global climate change and, *388*
North American population, 26-28
population changes, Atlantic Flyway, 29-30
resident populations, 30
spring surveys, 26
status and trends, 26-28
Taverner's (*Branta canadensis taverneri*), 28, 33
Vancouver (*Branta canadensis fulva*), 28, 30-31, 32
wintering areas, 29-30
Canada warbler (*Wilsonia canadensis*), *20*
Canadian Wildlife Service
duck survey, 35
seabird monitoring program, 53
canary grass, reed (*Phalaris arundinacea*), 249
canary-winged parakeet (*Brotogeris versicolurus*), 439
canine distemper, in black-footed ferrets, 106
canine heartworm, *403*
canine parvovirus, *403*
Canis
familiaris (dog), 306, 378, 440, 441
latrans (coyote), 45, 193, 296, 305-7, 417
lupus (gray wolf), 98-100, 109, 306, 310, 321, 337, 338, 340, 347-48, 358
baileyi (Mexican wolf), 98, 99, 318, *320*
irremotus (northern Rocky Mountain wolf), 98
lycaon (timber wolf), 98
nubilus, 98
occidentalis, 98
rufus (red wolf), 98
cannon net, for capturing wild turkey, 71
Cantharellus cibarius (chanterelle), *193*
canvasback (*Aythya valisineria*), 16, 35, *36*, 40-42, 236, 237, *238*, 266
habitat trends, 41-42
hunting, 41
mortality sources, 41
sex ratio, 40
survival rates, 40-41
wetland availability and, 37
Cape Fear shiner (*Notropis mekistocholas*), *143*
capelin (*Mallotus villosus*), 51
Cape May warbler (*Dendroica tigrina*), *20*
Cape Thompson, murres, 50
Capitol Reef National Park
packrat middens, *227*
vegetation change, 226, *227*
Capra hircus (goat), *440*, 441
Caprimulgus carolinensis (chuck-will's-widow), *20*
captive propagation, introduction, and translocation programs, 405-7
caracara
Audubon's crested (*Caracara plancus audubonii*), 82-83
crested (*Caracara plancus*), 16, *67*
Caracara
plancus (crested caracara), 16, *67*
plancus audubonii (Audubon's crested caracara), 82-83
Carassius auratus (goldfish), 240, 261, *262*, 432, *432*
carbamate pesticides
distribution of mortality, 417
documentation of poisoning by, 416-17
monitoring, 415
wildlife mortality and, 416-18
carbofuran, *417*
carbon cycle, 6
carbon dioxide fertilization, subalpine forest growth and, 315-16
carbon fixation, atmospheric, forest cover loss and, 472
Carcharodon carcharias (great white shark), 112

cardinal, northern (*Cardinalis cardinalis*), *20*, *22*, 387
Cardinalis
  cardinalis (northern cardinal), *20*, *22*, 387
  sinuatus (Pyrrhuloxia), *20*
Carduelis
  pinus (pine siskin), *21*
  psaltria (lesser goldfinch), *21*
  tristis (American goldfinch), *21*, *22*, *299*
Caretta caretta (loggerhead), *119*, 121, 122-23
Carex spp., 201
caribou (*Rangifer tarandus*), 337, 338-40, *407*
  counting, 347
  gray wolf (*Canis lupus*) and, 347-48
  hunting, 357
  Mentasta herd, 357-58
  population trends, 358
  Porcupine herd, 337, 338-40
  surveys, 357-58
  weather and, 347-48
carinate curly-tailed lizard (*Leiocephalus carinatus*), *434*
Carlia
  cf. *fusca* (lizard), 454
  fusca (brown four-fingered skink), *434*
Carolina chickadee (*Parus carolinensis*), *20*, *22*
Carolina gopher frog (*Rana capito capito*), 129
Carolina locust (*Dissosteira carolina*), 164
Carolina wren (*Thryothorus ludovicianus*), *20*
carp
  bighead (*Aristichthys nobilis*), *432*
  common (*Cyprinus carpio*), 240, *261*, 429, 432, *440*, 441, 443
  grass (*Ctenopharyngodon idella*), 261, 263, 429, *432*
Carpodacus
  cassinii (Cassin's finch), *21*
  mexicanus (house finch), *21*, 439
  purpureus (purple finch), *21*
Cascades frog (*Rana cascadae*), 132
Casmerodius albus (great egret), *19*
Caspian tern (*Sterna caspia*), 45
Cassin's auklet (*Ptychoramphus aleuticus*), 46, 51
Cassin's finch (*Carpodacus cassinii*), *21*
Cassin's kingbird (*Tyrannus vociferans*), *20*
Cassin's sparrow (*Aimophila cassinii*), *21*, 297, 298
Castanea dentata (American chestnut), 213
cat (*Felis catus*), 378, 440, 441
Catahoula Lake, canvasbacks, 42
catalpa, northern (*Catalpa speciosa*), 206
Catalpa speciosa (northern catalpa), 206
catbird, gray (*Dumetella carolinensis*), *20*, *299*
catfish
  blue (*Ictalurus furcatus*), *432*
  bullhead (family Ictaluridae), 145, 242, *244*, *432*
  channel (*Ictalurus punctatus*), 151, *432*
  flathead (*Pylodictis olivaris*), 151, *432*
  madtom (*Noturus* spp.), 145
  pygmy madtom (*Noturus stanauli*), *143*
  Scioto (*Noturus trautmani*), 146
  white (*Ameiurus catus*), *261*, *262*, *432*
  Yaqui (*Ictalurus pricei*), *320*
Cathartes aura (turkey vulture), *19*, *67*
Catharus
  fuscescens (veery), *20*, 25
  guttatus (hermit thrush), *20*
  minimus (gray-cheeked thrush), *20*
  ustulatus (Swainson's thrush), *20*
Catocala pretiosa (owlet moth), 166
Catoctin Mountain National Park, 113
Catoptrophorus semipalmatus (willet), *19*, 58, *59*, *62*
Catostomus
  clarki
    clarki (desert sucker), *150*
    intermedius (White River sucker), *143*, *150*
    Meadow Valley sucker, *150*
  commersoni (white sucker), 151, *261*, *262*
  discobolus
    discobolus (bluehead sucker), *150*
    yarrowi (Zuni or Zuni bluehead sucker), *143*, *150*
  insignis (Sonora sucker), *150*
  latipinnis (flannelmouth sucker), *150*
  Little Colorado sucker, *150*
  platyrhynchus (mountain sucker), *150*
cattail
  Typha spp., 249
  Typha latifolia, 444
cattle egret (*Bubulcus ibis*), *19*, 386, 437, *438*, 439
cavefish
  Alabama (*Speoplatyrhinus poulsoni*), *143*
  spring (*Forbesichthys agassiz*), 177
caves
  bats and, 97-98
  Illinois, biota, 176-77
Cayuga Bird Club, 390
Cayuga Lake, 390, 391
Ceanothus americanus (New Jersey tea), 231
cedar
  Alaska yellow (*Chamaecyparis nootkatensis*), 316
  Atlantic white (*Chamaecyparis thyoides*), 216
cedar waxwing (*Bombycilla cedrorum*), *20*, 21-22, 23, *299*
celery, wild (*Vallisneria americana*), 42, 236, 237
Centers for Disease Control, 253
Central Arizona Project, 321
Central Flyway
  avian cholera, 402
  canvasbacks, 40
  northern pintails, 38-39
Central Management Unit, mourning doves, 72
central mudminnow (*Umbra limi*), 263
Central United States Plant Inventory Database (CUSPID), 203
Centrarchidae family, 240
centric diatoms, 256, 257
Centro Ecologico de Sonora, 140
Cepphus columba (pigeon guillemot), 46, 51
Ceratoides lanata (winterfat), 226
Cerorhinca monocerata (rhinoceros auklet), 46
Cervus elaphus (elk), 115-16, 305, 310, *311*, *407*
Ceryle alcyon (belted kingfisher), *420*
Chaenaxis, 320
Chaetaglaea cerata (owlet moth), 167
Chaetoptila angustipluma (kioea), *376*
Chaetura pelagica (chimney swift), *20*
chaffseed, American (*Schwalbea americana*), 207
chain pickerel (*Esox niger*), *261*, *432*
Chamaea fasciata (wrentit), *20*
Chamaecyparis
  nootkatensis (Alaska yellow-cedar), 316
  thyoides (Atlantic white cedar), 216
Chamaeleo jacksonii (Jackson's chameleon), *434*
chameleon, Jackson's (*Chamaeleo jacksonii*), *434*
chamise chaparral (*Adenostoma fasciculatum*), 421-22
channel catfish (*Ictalurus punctatus*), 151, *432*
Channel Islands National Park, 226
  non-native species, 441-42
  vegetation change, 226
channelization, riparian ecosystem dynamics and, 287
chanterelle (*Cantharellus cibarius*), *193*
chaparral, 466
  chamise (*Adenostoma fasciculatum*), 421-22
  redshanks, *465*
  southern mixed, *465*
  Upper Sonoran manzanita, *465*
char, Dolly Varden (*Salvelinus malma*), 337, 341, *342*, 343
Charadriiformes, 54, 55-56
Charadrius
  alexandrinus (snowy plover), 58, *59*, *62*
  hiaticula (common ringed plover), *62*
  melodus (piping plover), 16, 58, *59*, *62*, 77-79
  montanus (mountain plover), 58, *59*, *62*, 63, 297, 298
  semipalmatus (semipalmated plover), 58, *59*, *62*
  vociferus (killdeer), *19*, 58, *59*, *62*, *299*
  wilsonia (Wilson's plover), 58, *59*, 61, *62*
Chasiempis sandwichensis
  gayi (Oahu 'elepaio), *377*
  sandwichensis (Hawaii 'elepaio), *377*
  sclateri (Kauai 'elepaio), *377*
Chasmistes
  brevirostris (shortnose sucker), *143*
  liorus mictus (June sucker), *143*
Chasmistes cujus (cui-ui), 309, 323-24
chat, yellow-breasted (*Icteria virens*), *20*, 231, *466*
cheatgrass (*Bromus tectorum*), 222
checkered garter snake (*Thamnophis marcianus*), 452, 453
Cheilotrema saturnum (black croaker), 410
Cheirodendron trigynum (Olapa), 371
Chelonia mydas (green sea turtle), *119*
Chelydra serpentina (snapping turtle), *434*
Chen
  caerulescens caerulescens (lesser snow goose), 30, 33
  caerulescens (snow goose), *266*, *388*
  canagica (emperor goose), 30, 31, 33

Chenier Plain, wetland loss, 270, *271*
cherry, black (*Prunus serotina*), 228
Chesapeake Bay
  turtles, 121, 122
  water quality degradation, canvasback and, 42
  watershed, natural resources, 263-66
Chesapeake Bay Program, 263
chestnut (*Castanea dentata*), American, 213
chestnut-backed chickadee (*Parus rufescens*), *20*
chestnut blight, 191
chestnut-collared longspur (*Calcarius ornatus*), *21*, 299, *303*
chestnut-sided warbler (*Dendroica pensylvanica*), *20*
Cheyenne Bottoms Wildlife Management Area, 78
chickadee
  black-capped (*Parus atricapillus*), *20*
  Carolina (*Parus carolinensis*), *20*, *22*
  chestnut-backed (*Parus rufescens*), *20*
  Mexican (*Parus sclateri*), 321
  mountain (*Parus gambeli*), *20*, 321
chicken, prairie (*Tympanuchus cupido*), 438
Chihuahua chub (*Gila nigrescens*), *143*
Chihuahuan raven (*Corvus cryptoleucus*), *20*
chimney swift (*Chaetura pelagica*), *20*
Chinese softshell turtle (*Pelodiscus sinensis*), *434*
chinook salmon (*Oncorhynchus tshawytscha*), 344, *345*, 346, *432*
chipmunk, California (*Tamias obscurus*), *466*
chipping sparrow (*Spizella passerina*), *21*, *299*
Chiricahua leopard frog (*Rana chiricahuensis*), 139, *320*, 327, *328*, 452, 453
chlamydiosis, *404*
Chlidonias niger (black tern), *19*, 46, *54*, *55*, 56, 299
chlordane, 408
  in fish, 409
  persistance of, 413-14
Chloridops kona (Kona grosbeak), *376*
chlorinated dioxins, 415
chlorinated pesticides, *408*
chlorpyrifos, *417*
Choeronycteris mexicana (Mexican long-tongued bat), *320*
Choisya mollis (zorillo), *320*
cholinesterase-inhibiting compounds, 417
cholinesterase (ChE) screening, 417
Chondestes grammacus (lark sparrow), *21*, 297, *299*
Chondrohierax uncinatus (hook-billed kite), *67*
Chordeiles
  acutipennis (lesser nighthawk), *19*
  minor (common nighthawk), *19*
Christmas Bird Count (CBC), 15, 17, 386, 389, 437
  Alaskan seabirds, 50
  colonial waterbirds, 53
  non-native species, 437
  shorebirds, 58, 64
  songbirds, winter population trends, 21-23
Chrysemys picta (painted turtle), 327
Chrysothamnus visidiflorus (whitebark rabbitbrush), 226
chub
  Chihuahua (*Gila nigrescens*), *143*
  creek (*Semotilus atromaculatus*), *261*, 263
  Fish Creek Springs tui (*Gila bicolor euchila*), *143*
  Gila (*Gila intermedia*), *150*, *320*
  humpback (*Gila cypha*), *150*, 151
  Independence Valley tui (*Gila bicolor isolata*), *143*
  least (*Iotichthys phlegethontis*), *143*
  Oregon (*Oregonichthys crameri*), *143*
  pahranagat (*Gila robusta jordani*), *150*
  rosetail (*Gila robusta*), *320*
  roundtail (*Gila robusta robusta*), *150*
  slender (*Erimystax cahni*), 146
  Sonora (*Gila ditaenia*), *320*
  thicktail (*Gila crassicauda*), *143*
  Virgin (*Gila seminuda*), *150*
chubsucker, creek (*Erimyzon oblongus*), *261*, 263
chuckwalla (*Sauromalus obesus*), 74, *328*
chuck-will's-widow (*Caprimulgus carolinensis*), *20*
chukar (*Alextoris chukar*), 437
Chukchi Sea
  Pacific walrus (*Odobenus rosmarus divergens*), 356
  polar bear (*Ursus maritimus*), 351-53
chum salmon (*Oncorhynchus keta*), 338, 344, 345-46
Chytonix sensilis, 167
Cicindela patruela, *166*
Ciconiiformes, *54*, 55
cienaga (wetland) mollusks, in southwestern sky islands, 319
cinnamon teal, *36*
Circus cyaneus (northern harrier), *67*, 387
Ciridops anna ('ula-'ai-hawane), *376*
cisco
  Arctic (*Coregonus autumnalis*), 337, 341-43
  blackfin (*Coregonus nigripinnis*), *143*
  deepwater (*Coregonus johannae*), *143*
  least (*Coregonus sardinella*), 337, 341, 342, 343
  longjaw (*Coregonus alpenae*), *143*
Cistothorus
  palustris (marsh wren), *299*
  platensis (sedge wren), 300, 303
Citheronia
  imperialis (giant silkworm moth), 166
  sepulchralis (giant silkworm moth), 166
citophilus mealybug (*Pseudococcus calceolariae*), *160*
citrus conversion, Audubon's crested caracara (*Caracara plancus audubonii*) and, 83
Cladinia
  mitis, *195*
  rangiferina (reindeer moss), 195
clam, *419*
  Asian (*Corbicula fluminea*), 178, *429*, 443
  contaminants in, 408
  fingernail (*Musculium transversum*), 236-37, 238
  zebra mussel (*Dreissena polymorpha*) and, 181, 447
Clangula hyemalis (oldsquaw), 36, *266*
Clappia, 249, 250
  cahabensis (Cahaba pebblesnail), *251*
  umbilicata (umbilicate pebblesnail), *251*
clawed frog, African (*Xenopus laevis*), *434*
clay-colored sparrow (*Spizella pallida*), *21*, 297, *299*, *303*
Clean Air Act, 228
Clean Water Act, 217
Clear Creek gambusia (*Gambusia heterochir*), 143
clear-winged grasshopper (*Camnula pellucida*), 164
Clemmys
  guttata (spotted turtle), 119
  insculpta (wood turtle), *119*
  marmorata (western pond turtle), *119*, *466*
  muhlenbergii (bog turtle), 119, 458
Clermontia, 371
  fruit, *369*
  hawaiiensis, 371
cliff swallow (*Hirundo pyrrhonota*), *20*, *299*
climate change, *See also* global climate change; temperature change
  conservation planning and, 394-95
  diatoms and, 258
  flowering plants and, 391
  human activities and, 394
  landscape fragmentation and, 394
  migratory landbirds and, 391
  in the Northeast, 390-91
  plants and, 392-95
  species migration and, 394
  subalpine forest growth and, 315-17
  warming, 227
  wetland loss and, 217
climate modeling, Colorado Rockies, 311
climbing milkweed (*Cynanchum wigginsii*), *320*
Clioperla clio (stonefly), *186*
closed elimia (*Elimia clausa*), *251*
Clostridium botulinum, 401
clover
  running buffalo (*Trifolium stoloniferum*), 208
  Trifolium microcephalum, 208
Clupea harengus (herring), 51
Cnemidophorus
  hyperythrus (orange-throated whiptail), *466*
  lemniscatus (South American whiptail), *434*
  whiptail lizard, 326
Coachella Valley fringe-toed lizard (*Uma inornata*), 118, 137-38
Coachella Valley Preserve system, 137-38
Coastal Barrier Improvement Act of 1990 (Public Law 101 591), 278
Coastal Barrier Resources Act of 1982 (Public Law 97-384), 278
Coastal Barrier Resources System (CBRS), U.S. Department of Interior (DOI), 277-79
Coastal Change Analysis Program (C-CAP), 468
coastal and marine ecosystems, 259-84, *See also* aquatic ecosystems
  barrier erosion, 277-79
  coastal storms and, 259
  contaminants in, 408-11

fish harvests and, 52
Florida manatees, 267-69
human impacts on, 259 60
natural resources, Chesapeake Bay watershed, 263-66
nearshore fish assemblage, tidal Hudson River, 260-63
non-native species, 428-30
overview, 259-60
reef fishes, Florida Keys, 279-84
seagrass
Laguna Madre, Texas, 275-77
northern Gulf of Mexico, 273-75
wetlands, Gulf of Mexico, 269-72
Coastal Plain, longleaf pine ecosystem, 129-31
coastal sage-chaparral scrub, *465*
coastal sage scrub, 226, *465*
coast horned lizard (*Phrynosoma coronatum*), *466*
coast live oak woodlands, *465*
cobble elimia (*Elimia vanuxemiana*), *251*
coccidioidomycosis, *403*
Coccothraustes vespertinus (evening grosbeak), *21*, 387
Coccyzus americanus (yellow-billed cuckoo), *19*, *320*, *466*
Cochiti Reservoir, 291
cod (*Gadus macrocephalus*), 51
coho salmon (*Oncorhynchus kisutch*), 344, *345*, 346, *432*
Colaptes auratus
cafer (red-shafted flicker), *20*
northern flicker, 231
yellow-shafted flicker, *20*
Coleoptera (beetles), 161
Coleotichus
blackburniae (koa bug), 366-67
stink bug, 366
Colinus virginianus (northern bobwhite), *19*, 231, 424
collared-dove, Eurasian (*Streptopelia decaocto*), 438-39
colonial waterbirds, 53-56
Charadriiformes, 54, 55-56
Ciconiiformes, *54*, 55
Pelecaniformes, 53
status and trends, 16
Colorado cutthroat trout (*Oncorhynchus clarki pleuriticus*), *150*
Colorado Desert, desert tortoise (*Gopherus agassizii*) in, 135-36
Colorado Desert fringe-toed lizard (*Uma notata*), 137
Colorado Division of Wildlife, 334-35
Colorado Plateau
amphibian and reptile diversity, 326-28
canyonlands, Mexican spotted owls (*Strix occidentalis lucida*), 330-31
Colorado Plateau Research Station, 90
Colorado River Basin
bonytail and razorback sucker, 324-26
ecosystem alterations, 149
fish, 149-52, 324-26
non-native species, 151
riparian ecosystems, 287-89
southwestern willow flycatcher, 90-91
species diversity alterations, 151-52
wintering bald eagles (*Haliaeetus leucocephalus*), 328-30
Colorado River Fishery Project, 325
Colorado River Fishes Recovery Team, 324-25
Colorado Rockies
ecosystem trends, 310-12
human activities and, 310-11
Colorado squawfish (*Ptychocheilus lucius*), *150*, 151
Columba
fasciata (band-tailed pigeon), *19*, 322
livia (rock dove or common pigeon), *19*, 437, 438, *440*, 441
Columbia River Basin
habitat change, 156
white sturgeon (*Acipenser transmontanus*), 154-56
Columbina passerina (common ground dove), *19*
combshell, southern (*Epioblasma penita*), *178*
comely shiner (*Notropis amoenus*), *261*, 263
commercial fishing
Beaufort Sea, central Alaska, 341
Great Lakes, 244-46
marine ecosystem and, 52
seabirds and, 51
turtles and, 122, 123
common black hawk (*Buteogallus anthracinus*), *67*
common box turtle (*Terrapene carolina*), 119, *434*
common carp (*Cyprinus carpio*), 240, *261*, 429, 432, *432*, *440*, 441, 443
common coqui (*Eleutherodactylus coqui*), *434*
common eider, *36*
common goldeneye (*Bucephala clangula*), *36*, *266*, *420*
common grackle (*Quiscalus quiscula*), *21*, *299*
common ground dove (*Columbina passerina*), *19*
common loon (*Gavia immer*), *420*
common merganser (*Mergus merganser*), *266*, 389, *420*
common murre (*Uria aalge*), 46-47, 50, *51*
common nighthawk (*Chordeiles minor*), *19*
common pigeon (rock dove) (*Columba livia*), *19*, 437, 438, *440*, 441
common raven (*Corvus corax*), *20*, 73-74, *75*
common ringed plover (*Charadrius hiaticula*), *62*
common shiner (*Luxilus cornutus*), *261*, 263
common slider (*Trachemys scripta*), 119, *434*
common snipe (*Gallinago gallinago*), *19*, 59, *62*
common tern (*Sterna hirundo*), *53*, *54*, 56
common wall lizard (*Podarcis muralis*), *434*
common witch-hazel (*Hamamelis virginiana*), 390
common yellowthroat (*Geothlypis trichas*), *20*, *299*, *303*
Compound 1080 (sodium fluoroacetate), 306
conch, queen (*Strombus gigas*), 282
condor, California (*Gymnogyps californianus*), 16, *67*, 80-81
conservation
objectives of, 6
planning, climate change and, 394-95
Conservation Reserve Program (CRP), 217, 296, 300, 302-3, 423-26
agriculture and, 423-26
dabbling ducks and, 37
northern plains migratory birds and, 302-4
percentage of county area enrolled in, 424
visual obstruction readings (VOR), 425
wildlife and, 423-26
Conservation Title, Food Security Act of 1985, 423
contagious dispersion, 316
contagious ecthyma, *407*
contaminants
bald eagle and, 413
in bivalves, 409-11
in clams, 408
in coastal and marine ecosystems, 408-11
European starling (*Sturnus vulgaris*) and, 413, *415*
in fish, 242-44, 408-11, 413-15
Great Lakes, 242-44
Hawaii, 413
mercury, 414, 415
in mollusks, 408-11
monitoring, *5*, 413-15
in oysters, 408
persistent, 413-15
in sediments, 408, 409-11
in shellfish, 408-9
Contopus
borealis (olive-sided flycatcher), *20*
sordidulus (western wood-pewee), *20*
virens (eastern wood-pewee), *20*
Cooper's hawk (*Accipiter cooperii*), *67*
Coosa River, freshwater gastropod fauna, 249-52
Coosa rocksnail (*Leptoxis showalteri*), *251*
coot
American (*Fulica americana*), *19*, 299, *303*
Hawaiian (*Fulica americana alai*), *377*
copper smelters, Tarahumara frog (*Rana tarahumarae*) and, 138, 140
coqui, common (*Eleutherodactylus coqui*), *434*
Coragyps atratus (black vulture), *19*, *67*
coral reef ecosystems, 259, 280-81
at-risk, 280
fish, Florida Keys, 279-84
loss of diversity, 283
pristine, 280
Corbicula fluminea (Asian clam), 178, *429*, 443
coregonine (*Coregonus* spp.), 443
Coregonus, 443
alpenae (longjaw cisco), *143*
artedi (lake herring), 179
autumnalis (Arctic cisco), 337, 341-43
clupeaformis (lake whitefish), *432*
johannae (deepwater cisco), *143*
nasus (broad whitefish), 337, 341, 342
nigripinnis (blackfin cisco), *143*
sardinella (least cisco), 337, 341, 342, 343
cormorant, 44-45, 53-54
Brandt's (*Phalacrocorax penicillatus*), 44
double-crested (*Phalacrocorax auritus*), 16, 44, 54

  pelagicus (*Phalacrocorax pelagicus*), 44
  red-faced (*Phalacrocorax urile*), 51
Corvus
  brachyrhynchos (American crow), *20*, *22*
  corax (common raven), 16, *20*, 73-74, *75*
  cryptoleucus (Chihuahuan raven), *20*
  hawaiiensis ('alala or Hawaiian crow), *377*, 380
  kubaryi (Marianas crow), 455
  ossifragus (fish crow), *20*
Coryphantha robbinsorum (pincushion cactus), *320*
cosmopolitan house gecko (*Hemidactylus mabouia*), *434*
Costa's hummingbird (*Calyptae costae*), *466*
cottonwood (*Populus fremontii*), 286-89, 290
Cottus
  bairdii (mottled sculpin), *150*
  beldingii (Paiute sculpin), *150*
  echinatus (Utah Lake sculpin), *143*
  greenei (Shoshone sculpin), *143*
cow (*Bos taurus*), *440*, 441
cowbird
  bronzed (*Molothrus aeneus*), *21*
  brown-headed (*Molothrus ater*), *21*, 25, 90, 91, 299, *303*, 389
  shiny (*Molothrus bonariensis*), 439
coyote (*Canis latrans*), 109, 193, 296, 305-7, 417
  California gull and, 45
  diet, 306-7
  population change, 306
crab
  blue (*Callinectes sapidus*), 263, 264-65, 282
  stone (*Menippe mercanaria*), 282
crane
  Mississippi sandhill (*Grus canadensis pulla*), 16-17, 75-77
  sandhill (*Grus canadensis*), *19*
  whooping (*Grus americana*), *405*, 406, *407*
crappie
  black (*Pomoxis nigromaculatus*), *261*, *432*
  white (*Pomoxis annularis*), 261, *432*
Crassostrea
  gigas (Pacific oyster), 429
  virginica (eastern oyster), 263, 264, *265*, 408
Craters of the Moon National Monument, Idaho, insect survey, 161
crayfish, *419*
cream gentian (*Gentiana alba*), 231
creek chub (*Semotilus atromaculatus*), *261*, 263
creek chubsucker (*Erimyzon oblongus*), *261*, 263
creek heelsplitter (*Lasmigona compressa*), *180*
creeper
  Hawaii (*Oreomystis mana*), *378*
  Kauai (*Oreomystis bairdi*), *378*
  Lanai (*Paroreomyza m. montana*), *376*
  Maui (*Paroreomyza montana newtoni*), *378*
  Molokai (*Paroreomyza flammea*), *378*
  Oahu (*Paroreomyza maculata*), *378*
Crenichthys
  baileyi
    albivallis (Preston springfish), *143*, *150*
    baileyi (White River springfish), *143*, *150*
    grandis (Hiko springfish), *150*
    moapae (Moapa springfish), *150*
    thermophilus (Moorman springfish), *143*, *150*
  nevadae (Railroad Valley springfish), *143*, *150*
creosote bush, 472
crested anole, Puerto Rican (*Anolis cristellatus*), *434*
crested auklet (*Aethia cristatella*), 51
crested caracara (*Caracara plancus*), 16, *67*
  Audubon's (*Caracara plancus audubonii*), 82-83
crested flycatcher, great (*Myiarchus crinitus*), *20*
crested tern, great (*Sterna bergii*), 373
croaker
  Atlantic (*Micropogonias undulatus*), 409, *411*
  black (*Cheilotrema saturnum*), 410
  white (*Genyonemus lineatus*), 409, 411
crocodile, American (*Crocodylus acutus*), 127
crocodilians, 117
Crocodylus acutus (American crocodile), 127
cropland, 423-26, *See also* agricultural ecosystems
  productivity of, 423, 424
crossbill
  red (*Loxia curvirostra*), *21*
  white-winged (*Loxia leucoptera*), *21*
Crotalus
  adamanteus (eastern diamondback rattlesnake), 129
  ruber (red diamond rattlesnake), *466*
  viridis (western rattlesnake), 327
  willardi
    obscurus (ridge-nosed rattlesnake), *320*
    willardi (ridge-nosed rattlesnake), *320*
crow
  American (*Corvus brachyrhynchos*), *20*, *22*
  fish (*Corvus ossifragus*), *20*
  Hawaiian ('alala) (*Corvus hawaiiensis*), *143*, *377*
  Marianas (*Corvus kubaryi*), 455
Cryphonectria parasitica, 191
cryptic reef fish, 284
Cryptobranchus alleganiensis (hellbender), 125
Ctenitis
  sloanei, *204*
  submarginalis, *204*
Ctenopharyngodon idella (grass carp), 261, 263, 429, *432*
Ctenosaura pectinata (Mexican spiny-tailed iguana), *434*
Cuban treefrog (*Osteopilus septentrionalis*), *434*
cuckoo, yellow-billed (*Coccyzus americanus*), *19*, *320*, *466*
cui-ui (*Chasmistes cujus*), 309, 323-24
Culex (mosquito), 378, 380
cup fungi, 192
curlew
  bristle-thighed (*Numenius tahitiensis*), *62*, 64
  Eskimo (*Numenius borealis*), *62*, 63
  long-billed (*Numenius americanus*), *19*, *59*, *62*
curly-tailed lizard
  carinate (*Leiocephalus carinatus*), *434*
  Schreiber's (*Leiocephalus schreibersii*), *434*
curve-billed thrasher (*Toxostoma curvirostre*), *20*
CUSPID (Central United States Plant Inventory Database), 203
cutthroat trout (*Oncorhynchus clarki*), *432*
  Alvord (*Oncorhynchus clarki* spp.), *143*
  Colorado (*Oncorhynchus clarki pleuriticus*), *150*
  westslope (*Oncorhynchus clarki lewisi*), 153-54
  Yellowstone (*Oncorhynchus clarki bouvieri*), 153-54, 313
Cyanocitta
  cristata (blue jay), *20*
  stelleri (Steller's jay), *20*
cyclic variations, 4
Cyclonaias tuberculata (purple wartyback), *180*
Cygnus olor (mute swan), 266, 387, 437, 438
cylindrical papershell (*Anodontoides ferussacianus*), *180*
Cynanchum wigginsii (climbing milkweed), *320*
Cynomys, 106-8
  leucurus (white-tailed prairie dog), 295
  ludovicianus (black-tailed prairie dog), 107
Cynoscion arenarius (sand seatrout), 410
cypress
  baldcypress (*Taxodium distichum*), 216
  southern interior forest, *465*
Cyprinella
  lutrensis (red shiner), 151
  spiloptera (spotfin shiner), *261*, 263
Cyprinodon
  diabolis (Devils Hole pupfish), 143, *143*
  macularius (desert pupfish), *143*
  macularius macularius (desert pupfish), *150*
  Monkey Springs pupfish, *150*
Cyprinus carpio (common carp), 240, *261*, 429, 432, *440*, 441, 443
Cypseloides niger (black swift), *20*
Cyrtomium falcatum, *201*
Cyrtopodion scabrum (roughtail gecko), *434*
Cyrtopogon laphriformis, 166

dabbling ducks
  factors protecting, 37
  population, 35, 36
  wetland loss and, 36
dace
  blackside (*Phoxinus cumberlandensis*), *143*
  Kendall Warm Springs (*Rhinichthys osculus thermalis*), *150*
  Las Vegas (*Rhinichthys deaconi*), *150*
  longfin (*Agosia chrysogaster*), *150*
  Meadow Valleys speckled (*Rhinichthys osculus* ssp.), *150*
  Moapa (*Moapa coriacea*), *150*
  mountain redbelly (*Phoxinus oreas*), *146*
  Preston speckled (*Rhinichthys osculus* ssp.), *150*
  speckled (*Rhinichthys osculus osculus*), *150*
  White River speckled (*Rhinichthys osculus* ssp.), *150*
Dakota skipper (*Hesperia dacotae*), 174, 175
Dallas-Fort Worth, Landsat Multispectral Scanner (MSS) images, 470
Dall sheep (*Ovis dalli*), 338, 347

dams
  Colorado River basin, bonytail and razorback sucker and, 325
  habitat change and, 234-36
  riparian ecosystem dynamics and, 287
  water quality and, 51-52
damselfly
  Megalagrion
    nesiotes, 367
    nigrohamatum nigrolineatum, 367
    pacificum, 362
Danaus plexippus (monarch butterfly), *171*, 172
Darapsa versicolor (sphinx moth), 166
dark-rumped petrel (*Pterodroma phaeopygia sandwichensis*), *377*, 379
darter, 145
  amber (*Percina antesella*), *143*
  johnny (*Etheostoma nigrum*), 143
  least (*Etheostoma microperca*), 146
  Maryland (*Etheostoma sellare*), 146
  sharphead (*Etheostoma acuticeps*), *143*
  tangerine (*Percina aurantiaca*), *146*
Dasypus novemcinctus (armadillo), 386
data sets
  long-term, *4*
  short-term, *4*
day gecko (*Phelsuma* sp.), *434*
DDE pesticides
  monitoring, 413
  raptors and, 67
DDT pesticides
  California sea otter and, 112
  chemicals measured in National Status and Trends Program, *408*
  colonial waterbirds and, 53-54
  in fish, 409, *411*
  Great Lakes, 242-44
  monitoring, 413-15
  persistence of, 413-15
  raptors and, 66, 67, 69
  sediment concentrations, 408
death-camas, white (*Zigadenus elegans*), 231
Declining Amphibian Populations Task Force, 118
Deep Creek, California, insect survey, 161
deepwater cisco (*Coregonus johannae*), *143*
deer
  management plans, 112
  white-tailed (*Odocoileus hemionus*), 320
  white-tailed (*Odocoileus virginianus*), 93-94, 112-15, 320, 322, 405
deer mouse (*Peromyscus maniculatus*), 109
deertoe (*Truncilla truncata*), *180*
Defense Mapping Agency, Digital Chart of the World, 469
Delaware Bay (DELBAY) surveys, 58
Delphinapterus leucas (white whale), 351
Deltaic Plain, wetland loss, 270, *271*
Deltistes luxatus (Lost River sucker), *143*
Denali National Park and Preserve, 338
  caribou (*Rangifer tarandus*), 347-48
  gray wolf (*Canis lupus*), 347-48
Dendrobates auratus (poison-dart frog), *434*
Dendrocygna spp. (whistling ducks), 35
Dendroica
  castanea (bay-breasted warbler), *20*
  coronata (myrtle warbler), *20*
  coronata auduboni (Audubon's warbler), *20*
  fusca (Blackburnian warbler), *20*
  magnolia (magnolia warbler), *20*
  nigrescens (black-throated gray warbler), *20*
  occidentalis (hermit warbler), *20*
  pensylvanica (chestnut-sided warbler), *20*
  petechia (yellow warbler), *20*, *299*, *466*
  pinus (pine warbler), *20*, *22*
  striata (blackpoll warbler), *20*, *391*
  tigrina (Cape May warbler), *20*
  townsendi (Townsend's warbler), *20*
  virens (black-throated green warbler), *20*
Dermochelys coriacea (leatherback), *119*, 121, 123
desert bighorn sheep
  Ovis canadensis cremnobates, 334
  Ovis canadensis mexicana, 320, 334, 335
  Ovis canadensis nelsoni, 309, 320, 332, 333-35
desert iguana (*Dipsosaurus dorsalis*), 327
desert pupfish (*Cyprinodon macularius*), *143*
  desert pupfish (*Cyprinodon macularius macularius*), *150*
desert sucker (*Catostomus clarki clarki*), *150*
desert tortoise (*Gopherus agassizii*), 16, 74, 119, *328*, *407*
  causes of death, 136
  in Mojave and Colorado deserts, 135-36
  population density, 135-36
Desert Tortoise Recovery Team, 135
Desmognathus spp. (dusky salamander), 130
Detection and Monitoring Committee, Aquatic Nuisance Species (ANS) Task
  Force, 430
Devils Hole pupfish (*Cyprinodon diabolis*), 143, *143*
diadromous fish
  Hudson River, 261
  rainbow smelt (*Osmerus mordax*), 261
diamondback rattlesnake, eastern (*Crotalus adamanteus*), 129
diamondback terrapin (*Malaclemys terrapin*), *119*, 120
diamond rattlesnake, red (*Crotalus ruber*), *466*
Diarrhena americana obovata (beakgrain), 209
diatoms
  centric, 256, 257
  climate change and, 258
  freshwater, ecosystem change and, 256-58
  lake acidification and, 257-58
  lake eutrophication and, 258
  pennate, 256, 257
  river and stream water quality and, 258
Diatoms of the United States, The, 257
diazinon, *417*
dibutyltin, *408*
dichlorodiphenylethylene (DDE). *See* DDE pesticides
dicholorodiphenyl trichlororethane (DDT). *See* DDT pesticides
dickcissel (*Spiza americana*), *20*, 303
dicrotophos, *417*
Didelphis
  marsupialis (opossum), 321
  virginiana (opossum), 321, 386, 417
dieldrin, *408*, 413
  Great Lakes, 242-44
differential grasshopper (*Melanoplus differentialis*), 164
digger pine-oak woodlands, *465*
Digital Chart of the World, Defense Mapping Agency, 469
dimethoate, *417*
dinoflagellates, 255, 427
Diogmites symmachus (robber fly), 162
Diomedea
  albatrus (short-tailed albatross), 50
  immutabilis (Laysan albatross), 374
Dipodomys
  agilis (Pacific or agile kangaroo rat), *466*
  merriami (Merriam's kangaroo rat), *466*
  stephensi (Stephens' kangaroo rat), *466*
Dipsosaurus dorsalis (desert iguana), 327
Diptera (flies), 161
disc snail, Pleistocene (*Discus macclintocki*), 176-77
Discula destructiva, 191
Discus macclintocki (Pleistocene disc snail), 176-77
diseases, *See also* avian diseases
  changes in patterns of, *403*
  emerging, *403*
  fish pathogens, 443-44
  non-native species and, 428
  translocation-related, 406-7
dispersal-ability scale, for vascular plants, 393
Dissosteira carolina (Carolina locust), 164
disulfoton, *417*
diversity, *See also* biodiversity
  genetic, 147-48, 319-21
  species, 151-52, 251, 368
diving ducks, *36*
dock sorrel (*Rumex orthoneurus*), *320*
dog (*Canis familiaris*), 306, 378, 440, 441
dogwood anthracnose, 191
Dolichonyx oryzivorus (bobolink), *21*, *299*, *303*
Dolly Varden char (*Salvelinus malma*), 337, 341, *342*, 343
dolphin, 96
  bottlenose (*Tursiops truncatus*), *94*, 96
  spotted, *94*, *95*
  whitesided, *94*
domestic animals, feral. *See* feral domestic animals
domestic poultry, avian cholera, 402
Dorosoma
  cepedianum (gizzard shad), 240, 261, *262*, 263
  petenense (threadfin shad), *432*
double-clutching, 381
double-crested cormorant (*Phalacrocorax auritus*), 16, 44, 54

Douglas fir (*Pseudotsuga menziesii*), 116, 193, 321
dove
  common ground (*Columbina passerina*), *19*
  Eurasian collared- (*Streptopelia decaocto*), 438-39
  mourning (*Zenaida macroura*), 16, *19*, 71-73, *299*, 387
  Philippine turtle (*Streptopelia bitorquata*), 454
  ringed turtle- or Barbary (*Streptopelia risoria*), 438
  rock (common pigeon) (*Columba livia*), *19*, 437, 438, *440*, 441
  small barred or zebra (*Geopelia striata*), 374, 438
  spotted or lace-necked (*Streptopelia chinensis*), 438
  white-winged (*Zenaida asiatica*), *19*
dowitcher
  long-billed (*Limnodromus scolopaceus*), *59*, *62*
  short-billed (*Limnodromus griseus*), 59, *62*
downy woodpecker (*Picoides pubescens*), *20*
dredging, seagrass and, 277
Dreissena polymorpha (zebra mussel), 42, 160, 178, 181, 237, 255, 409, 427, 428, 429, 444, 445-47
Drepanis
  funerea (black mamo), *376*
  pacifica (Hawaii mamo), *376*
Drosophila (Pomance flies), 368-71
  as an ecological indicator, 368, 371
  in disturbed habitat, 370
  diversity of, 368
  Drosophila
    conspicua (Hawaiian picture-wing), *369*
    immigrans, 370
    murphyi, 371
    setosimentum, 371
    silvestris, 369-70, 371
    simulans, 370
    sproati, 371
    suzukii, 370
  inversion polymorphism in, 369
  long-term changes, 370-71
  non-native species and, 369
  polymorphisms in, 368
drum, freshwater (*Aplodinotus grunniens*), 261, 263, 447
Dry Domain, shorebirds, 60-61
Drymarchon corais couperi (eastern indigo snake), 129
duck
  American black (*Anas rubripes*), 35, *36*, 266, 402, *413*, *420*
  Breeding Population and Habitat Survey, 34-35
  canvasback (*Aythya valisineria*), 16, 35, *36*, 37, 40-42, 236, 237, *238*, 266
  declines, 16
  Hawaiian (*Anas wyvilliana*), *377*
  hunting regulations, 37
  Laysan (*Anas laysanensis*), *377*, 379, 400
  Midwinter Survey, 34-35
  mottled (*Anas fulvigula*), *19*, 35
  muscovy (*Cairina moschata*), 437-38
  nest success, 300-302
  North American, 34-37
  pintail (*Anas acuta*), 16
  redhead (*Aythya americana*), 35, *36*, 266, 275-77
  ring-necked (*Aythya collaris*), 35, *266*, *420*
  ruddy (*Oxyura jamaicensis*), *266*, *299*
  status and trends, 16
  wetland loss and, 36-37
  whistling (*Dendrocygna* spp.), 35
  wood (*Aix sponsa*), 35
dugong (*Dogong dugon*), 267
Dumetella carolinensis (gray catbird), *20*, *299*
dunlin
  Calidris alpina, 59, *62*
    articola, 64
    pacifica, 63
dusky Canada goose, 32
dusky flycatcher (*Empidonax oberholseri*), *20*
dusky gopher frog (*Rana capito sevosa*), 129
dusky salamander (*Desmognathus* spp.), 130
dwarf blazingstar (*Liatris cylindracea*), *172*
dwarf gecko, ocellated (*Sphaerodactylus argus*), *434*
dwarf mistletoe (*Arceuthobium americanum*), 223
dwarf treefrog, eastern (*Litoria fallax*), *434*
Dysmorodrepanis munroi (Lanai hookbill), *376*

eagle
  bald (*Haliaeetus leucocephalus*), 16, 44, 66-67, 68, 309, 328-30, 387, 389, 413, 415, *466*
  causes of death, 68
  golden (*Aquila chrysaetos*), *67*, 68, 338, 340, 387, *466*
  mortality factors, 16
Earth Resources Observation System (EROS), 467
  Data Center, 467-68
eastern bluebird (*Sialia sialis*), *20*, *22*
eastern box turtle (*Terrapene carolina*), 119
eastern brown pelican, *54*
eastern diamondback rattlesnake (*Crotalus adamanteus*), 129
eastern dwarf treefrog (*Litoria fallax*), *434*
eastern equine encephalitis, *403*
eastern fence lizard (*Sceloporus undulatus*), 327
eastern indigo snake (*Drymarchon corais couperi*), 129
eastern kingbird (*Tyrannus tyrannus*), *20*, *299*, *420*
eastern lampmussel (*Lampsilis siliquoidea*), *180*
Eastern Management Unit, mourning doves, 72
eastern massasauga rattlesnake (*Sistrurus catenatus*), 231
eastern meadowlark (*Sturnella magna*), *21*, *22*, 297
eastern oyster (*Crassostrea virginica*), 263, 264, *265*, 408
eastern phoebe (*Sayornis phoebe*), *20*
eastern pondmussel (*Ligumia nasuta*), *180*
eastern screech-owl (*Otus asio*), *67*
Eastern Timber Wolf Recovery Plan, 98
Eastern Timber Wolf Recovery Team, 100
eastern United States
  rare terrestrial ecological communities, 219-20
  vascular plants, 203
eastern wood-pewee (*Contopus virens*), *20*
Echinococcus multilocularis, *407*
Echinoidea (sea urchins), 283, 354
ecological communities
  as a coarse conservation unit, 218
  rare, 218-21
Economic Entomology and Mammalogy, Division of, U.S. Department of Agriculture, 8
economic ornithology, 7-8
Ecopistes migratorius (passenger pigeon), 231
ecoregions, 5
ecosystems
  change, freshwater diatoms and, 256-58
  defined, 5
  detecting long-term trends, 4
  emergent properties of, 463
  inventory and monitoring programs, 2
  management, red-cockaded woodpecker and, 88
  status and trend studies, 1-2
ecosystem science, 310
ecotones, 314
ecthyma, contagious, *407*
ectomycorrhizae, 193
edge effect, biodiversity and, 160
eel, American (*Anguilla rostrata*), 261
egret
  cattle (*Bubulcus ibis*), *19*, 386, 437, *438*, 439
  great (*Casmerodius albus*), *19*
  reddish (*Egretta rufescens*), *54*, 55
  snowy (*Egretta thula*), *54*, 55
Egretta
  caerulea (little blue heron), *19*
  rufescens (reddish egret), *54*, 55
  thula (snowy egret), *54*, 55
eider, 36
  common, *36*
  king, *36*
  spectacled, *36*
  Steller's, *36*
Eisenhower National Historic Site, 113
Elaeagnus angustifolia (Russian olive), 286, 289, 290, 291
Elanoides forficatus (American swallow-tailed kite), *67*
Elanus
  caeruleus (black-shouldered kite), *466*
  caeruleus (white-tailed kite), *67*
Elassoma sp. (spring pygmy sunfish), *143*
electrofishing survey, Illinois River, 239-41
elegant tern (*Sterna elegans*), 45
Eleocharis parvula (small spikerush), 220, 458
'elepaio
  Hawaii (*Chasiempis sandwichensis sandwichensis*), *377*
  Kauai (*Chasiempis sandwichensis sclateri*), *377*
  Oahu (*Chasiempis sandwichensis gayi*), *377*
Elephant Butte Reservoir, 291
Eleutherodactylus
  coqui (common coqui), *434*
  planirostris (greenhouse frog), *434*

elfin butterfly
  frosted (*Incisalia irus*), 167, 232
  Incisalia henrici, 167
elf owl (*Micrathene whitneyi*), *67*
elimia
  closed (*Elimia clausa*), *251*
  cobble (*Elimia vanuxemiana*), *251*
  constricted (*Elimia impressa*), *251*
  fusiform (*Elimia fusiformis*), *251*
  hearty (*Elimia jonesi*), *251*
  high-spired (*Elimia hartmaniana*), *251*
  pupa (*Elimia pupaeformis*), *251*
  pygmy (*Elimia pygmaea*), *251*
  ribbed (*Elimia laeta*), *251*
  rough-lined (*Elimia pilsbryi*), *251*
  short-spire (*Elimia brevis*), *251*
Elimia, 249
  brevis (short-spire elimia), *251*
  clausa (closed elimia), *251*
  fusiformis (fusiform elimia), *251*
  gibbera, *251*
  hartmaniana (high-spired elimia), *251*
  impressa (constricted elimia), *251*
  jonesi (hearty elimia), *251*
  lachryma, *251*
  laeta (ribbed elimia), *251*
  macglameriana, *251*
  pilsbryi (rough-lined elimia), *251*
  pupaeformis (pupa elimia), *251*
  pygmaea (pygmy elimia), *251*
  vanuxemiana (cobble elimia), *251*
elk (*Cervus elaphus*), 115-16, 305, 310, *311*, *407*
  hunting, 115, 116
Elk Creek, atmospheric pollutants, 421-22
elktoe (*Alasmidonta marginata*), *180*
Elliptio dilatata (spike), *180*
elm-ash forest, 472
El Niño conditions
  cormorants and, 44-45
  murres and, 46, 47
Elops affinis (machete), *150*
EMAP. *See* Environmental Monitoring and Assessment Program (EMAP)
Emerald Lake, atmospheric pollutants, 421-22
emerald shiner (*Notropis atherinoides*), 240, 261, *262*, 443
Emergency Wetlands Resources Act of 1986, 473
emergent properties, of ecosystems, 463
Emilia (snail), 447
Emoia caerulocauda (Marianas blue-tailed skink), 455
emperor goose (*Chen canagica*), 30, 31, 33
Empidonax
  alnorum (alder flycatcher), *20*
  difficilis (Pacific-slope flycatcher), *20*
  difficilis (western flycatcher), *466*
  flaviventris (yellow-bellied flycatcher), *20*
  fulvifrons (buff-breasted flycatcher), *320*
  hammondii (Hammond's flycatcher), *20*
  minimus (least flycatcher), *20*, *299*, 420
  oberholseri (dusky flycatcher), *20*
  traillii (willow flycatcher), 17, *20*
    extimus (southwestern willow flycatcher), 89-91
  virescens (Acadian flycatcher), *20*
Emydoidea blandingii (Blanding's turtle), *119*
Encephalitozoon, 253
endangered species, *See also* rare species; threatened species
  California condor, 80
  freshwater mussels, 178
  habitat protection and, 466
  inventory and monitoring programs, *5*
  lichen, 195
  Mississippi sandhill crane (*Grus canadensis pulla*), 75
  number on federal lands, 400
  occurrences of, 398, 400
  piping plover, 77
  seabirds, 50, 56
  significance of federal lands for, 398-401
  translocation programs, 405
Endangered Species Act, 398, 400
  black bear, 102
  black-footed ferret, 106
  butterflies, 171
  California sea otter, 110
  gray wolf, 99
  grizzly bear, 103, 105
  Habitat Conservation Plans, 137
  marine mammals, 94
  Mexican spotted owl (*Strix occidentalis lucida*), 330
  plants and fungi, 190
  turtles, 119
  vascular plants, 206
Endangered Species Committee, American Fisheries Society (AFS), imperiled fish listings, 142-44
endemic species, Hawaii, 361
Endicott Causeway, anadromous fish and, 341
endrin, *408*, 413
Englemann oak woodlands, *465*
Englemann spruce (*Picea engelmannii*), 316, 317
English sole (*Parophrysvetulus*), 410, 411
Enhydra lutris (sea otter), 110-12, 338, 353-56
Enterocytozoon, 253
Entoloma salmoneum, *192*
Entomology, Division of, U.S. Department of Agriculture, 7-8
Environmental Monitoring and Assessment Program (EMAP), 310, 312, 467
  U.S. Environmental Protection Agency, 258
environmental monitoring. *See* inventory and monitoring programs
environmental quality, *See also* air pollution; water quality
  American alligator (*Alligator mississippiensis*) and, 128
  aquatic insects as indicators of, 182-84
  Drosophila (Pomance flies) as indicators of, 368, 371
  insects as indicators of, 365
  protozoa as indicators of, 253
  seabirds as indicators of, 49
  tree decline and, 205
Epinephelus
  itajara (jewfish), 282
  striatus (Nassau grouper), 282
Epioblasma
  penita (southern combshell), *178*
  torulosa rangiana (northern riffleshell), *180*
  triquetra (snuffbox), *180*
Equus
  asinus (burro), *440*, 456-57
  caballus (horse), *440*, 441, 456-57
Eragrostis spp. (lovegrasses), 321
Eremophila alpestris (horned lark), *20*, 297, 299, *303*
Eretmochelys imbricata (hawksbill), *119*, 121, 123
Eriborus sinicus, 366
ericaceous species, 316
Erigeron
  heliographis (fleabane), 320, 321
  kuschei (fleabane), *320*
  lemmonii (fleabane), *320*, 321
  pringlei (fleabane), 320, 321
Erignathus barbatus (bearded seal), 351
Erimystax cahni (slender chub), 146
Erimyzon oblongus (creek chubsucker), *261*, 263
erosion
  of coastal barriers, 277-79
  navigation dam resevoirs and, 235
EROS. *See* Earth Resources Observation System (EROS)
Erynnis
  brizo, *166*
  persius (savanna skipper), *166*, 232
erysipelas, *404*
Eschrichtius robustus (grey whale), 95
Eskimo curlew (*Numenius borealis*), *62*, 63
Esox
  lucius (northern pike), *432*
  lucius x masquinongy (tiger muskellunge), *432*
  masquinongy (muskellunge), *432*
  niger (chain pickerel), *261*, *432*
Estes Park, 310
Etheostoma
  acuticeps (sharphead darter), *143*
  microperca (least darter), 146
  nigrum (johnny darter), 143
  sellare (Maryland darter), 146
Eubalaena glacialis (right whale), *94*, 95-96
Eudocimus albus (white ibis), *19*, *54*, 55
Euglena, 252
Eugraphe subrosea (owlet moth), 166
Eumetopias jubatus (northern sea lion or Steller sea lion), 52, *95*, 96
eupatorium, sessile-leaved (*Eupatorium sessilifolium*), 231
Eupatorium sessilifolium (sessile-leaved eupatorium), 231
Euphagus cyanocephalus (Brewer's blackbird), *21*, *299*
Eurasian collared-dove (*Streptopelia decaocto*), 438-39

Eurasian tree sparrow (*Passer montanus*), 454
Eurasian watermilfoil (*Myriophyllum spicatum*), 237, 430, 443, 444
European ferret (*Mustela putorius*), 321
European honeybee (EHB), 448-51
European starling (*Sturnus vulgaris*), *20*, 413, 437, *438*, 439, *440*, 441
  contaminant monitoring and, 413, *415*
Eurycea sosorum (Barton Springs salamander), 125
Euryops multifida (forb), 321
eutrophication, diatoms and, 257, 258
evening grosbeak (*Coccothraustes vespertinus*), *21*, 387
Everglades, 414, 415
excised slitshell (*Gyrotoma excisa*), *251*
Eximacris superbum Hebard, 164-65
exotic species. *See* non-native species
extinction
  grasshoppers, 164-65
  habitat loss and, 452
  Hawaiian species, 361, 367, 371
    birds, 371, 375
    insects, 367, 371
  non-native species and, 427
  predators and, 452
  rate of, 6 7
  species conservation priorities, 399-400
  vascular plants, 207-8
extirpation, 207
Exxon Valdez oil spill, 346

fabaceae (*Baptisa australis*), *203*
Falco
  columbarius (merlins), *67*
  femoralis septentrionalis (aplomado falcon), *67*
  mexicanus (prairie falcon), *67*
  peregrinus
    anatum (American peregrine falcon), *67*, 69
    peregrine falcon, 16, *320*, 405, 415
    tundrius (arctic peregrine falcon), *67*, 69
  rusticolus (gyrfalcon), *67*
  sparverius (American kestrel), *67*
    paulus (Florida subspecies), *67*
falcon, 69
  American peregrine (*Falco peregrinus anatum*), *67*, 69
  aplomado (*Falco femoralis septentrionalis*), *67*
  arctic peregrine (*Falco peregrinus tundrius*), *67*
  peregrine (*Falco peregrinus*), 16, *320*, 405, 415
  prairie (*Falco mexicanus*), *67*
false-foxglove, giant (*Aureolaria grandiflora*), 231
false truffles (basidiomycetes), 193
famphur, *417*
Farm Bill, Food Security Act of 1985, 302, 307, 423, 426
fathead minnow (*Pimephales promelas*), *261*, *432*
Favia arctica (Arctic loon), *420*
fawnsfoot (*Truncilla donaciformis*), *180*
federal data bases, land, 467-68
Federal Land Policy and Management Act, *5*
federal lands
  land protection status, 463-64
  significance for endangered species, 398-401
Federal Water Pollution Control Agency, 258
Felis
  catus (cat), 378, 440, 441
  concolor (mountain lion), 109, 322
  onca (jaguar), *320*, 321
  pardalis (ocelot), *320*, 321
fenamiphos, *417*
fence lizard, eastern (*Sceloporus undulatus*), 327
fensulfothion, *417*
fenthion, *417*
feral domestic animals, 427
  burro (*Equus asinus*), *440*, 441, 456-57
  cat (*Felis catus*), 378, 440, 441
  dog (*Canis familiaris*), 306, 378, 440, 441
  pig (*Sus scrof*), 362, 370, 371, 440, 441
fern, 200
  Acrostichum danaeifolium, *203*
  Ctenitis
    sloanei, *204*
    submarginalis, *204*
  Cyrtomium falcatum, *201*
  Hawaii, 201-2
  Phanerophlebia auriculata, *204*
  Pityrogramma
    rittata, *203*
    trifoliata, *202*
  Polystichum lonchitis, *202*
ferret
  black-footed (*Mustela nigripes*), 106-8
  European (*Mustela putorius*), 321
ferruginous hawk (*Buteo regalis*), 67, 69, 297, 387
ferruginous pygmy owl (*Glaucidium brasilianum cactorum*), *67*, 69
fertilizer, nitrogen, 424
fibropapilloma, *403*
field sparrow (*Spizella pusilla*), *21*, *22*, 231, *299*
finch
  Cassin's (*Carpodacus cassinii*), *21*
  house (*Carpodacus mexicanus*), *21*, 439
  Laysan (*Telespyza cantans*), *377*
  Nihoa (*Telespyza ultima*), *377*
  purple (*Carpodacus purpureus*), *21*
fin erosion, 410
Finger Lakes, 402
fingernail clam (*Musculium transversum*), 236-37, 238
fin whale, *94*, *95*
fir
  Douglas (*Pseudotsuga menziesii*), 116, 193, 321
  Pacific silver (*Abies amabilis*), 315
  red (*Abies magnifica*), 421
  silver (*Abies alba*), 315
  subalpine (*Abies lasiocarpa*), 316, 317, 321
  white (*Abies concolor*), 321, 421
fire
  Great Plains and, 300
  management
    pine barrens, 167
    southwestern sky islands, 322
  riparian ecosystems, 289
  suppression
    rare forest types, 220
    whitebark pine (*Pinus albicaulis*) ecosystem, 229
  tall-grass prairie, 174, 175
  terrestrial ecosystems, 213-14
fire-adapted ecosystems, 214, 222-24
  altered fire regimes, 222-24
  lodgepole pine forest, 223
  ponderosa pine forests, 223
  sagebrush-grass steppe, 222-23
  southern pinelands, 223
  western juniper woodlands, 223
fire ant (*Solenopsis* spp.), 441
  red (*Solenopsis invicta*), 131
fish, 141-56, *See also* freshwater fish
  acid deposition (acid rain) and, 419
  Alaskan seabirds and, 51-52
  anadromous, of Beaufort Sea, central Alaska, 341-43
  coastal, contaminants in, 408-11
  Colorado River basin, 149-52
  contaminants in, 242-44, 409-11, 413-15
  disease pathogens, 443-44
  fin erosion in, 410
  freshwater, imperiled, 142-47
  habitat change, 141, 143
  Illinois River populations, 239-41
  lead in, *411*, 414-15
  liver neoplasm in, 410
  managed populations, 141, 147-48
  mercury contamination, 414, 415
  nearshore assemblage, tidal Hudson River, 260-63
  non-native species, 153-54, *429*, 431-33, 441
  overview, 141-42
  persistent contaminants in, 413-15
  prairie, 295, 303-5
  reef, Florida Keys, 279-84
  reproductive impairment, 411
  selenium concentrations, 415
  in southwestern sky mountains, 320-21
  stocking, loss of genetic integrity through, 147-48
Fish Creek Springs tui chub (*Gila bicolor euchila*), *143*
fish crow (*Corvus ossifragus*), *20*
fishing
  commercial
    Beaufort Sea, central Alaska, 341
    Great Lakes, 244-46
    marine ecosystem and, 52
    seabirds and, 51
    turtles and, 122, 123

recreational, 325, 428, 443
subsistence, Beaufort Sea, central Alaska, 341
FJC (Fourth of July Butterfly Count), 171-72
flammulated owl (*Otus flammeolus*), *67*
flannel moth (*Megalopyge crispata*), 166
flannelmouth sucker (*Catostomus latipinnis*), *150*
flathead catfish (*Pylodictis olivaris*), 151, *432*
flathead sole (*Hippoglossoides elassodon*), 409
flattened musk turtle (*Stemotherus depressus*), *119*
flatwoods salamander (*Ambystoma cingulatum*), 129
flatworm (*Pseudostylochus ostreophgus*), 429
flea, spiny water (*Bythotrephes cederstroemi*), 429, 444
fleabane
Erigeron
heliographis, 320, 321
kuschei, *320*
lemmonii, *320*, 321
pringlei, 320, 321
flicker
northern (*Colaptes auratus*), 231
red-shafted (*Colaptes auratus cafer*), *20*
yellow-shafted (*Colaptes auratus*), *20*
floater, giant (*Pyganodon grandis*), *180*
floating bladderwort, large (*Utricularia inflata* Walter), 209
flood control
rare forest types and, 220
western riparian ecosystems and, 286
wetlands and, 217, 270, 475
floodplains, 285, *See also* riparian ecosystems
change, Rio Grande, 290-91
habitat change, 234-36
riparian ecosystem dynamics and, 287
Flora of Alaska and Neighboring Territories, 202
Flora of the Great Plains, 203
Flora of New England, The, 203
Flora of North America, 198, 201
Florida
acid lakes, 419
American alligator (*Alligator mississippiensis*) in, 127-28
Audubon's crested caracara (*Caracara plancus audubonii*), 82-83
Game and Fresh Water Fish Commission, 127
spectacled caiman (*Caiman crocodilus*), 435
Florida Bay, 273
turtles, 121, 122
Florida bear (*Ursus americanus floridanus*), 102
Florida Big Bend, 273
Florida Keys, reef fishes, 279-84
Florida Keys National Marine Sanctuary, 279
Florida largemouth bass (*Micropterus salmoides floridanus*), 148
Florida manatee (*Trichechus manatus latirostris*), 267-69
aerial surveys, 268
future, 269
habitat threats, 268
recovery criteria, 268
Florida pine snake (*Pituophis melanoleucus mugitus*), 129
Florida Reef Tract, 279
Florida State Collectiomn of Arthropods, 161
Florida subspecies, American kestrel (*Falco sparverius paulus*), *67*
flounder
starry (*Platichthys stellatus*), 409, 411
winter (*Pleuronectes americanus*), 409, *411*
flowering plants, 200, 201
climate change and, 390
fluted-shell (*Lasmigona costata*), *180*
fly
Diptera, 161
Drosophila (Pomance flies), 367-71
Hawaii, 366
robber (*Diogmites symmachus*), 162
flycatcher, 18
Acadian (*Empidonax virescens*), *20*
alder (*Empidonax alnorum*), *20*
ash-throated (*Myiarchus cinerascens*), *20*
brown-crested (*Myiarchus tyrannulus*), *20*
buff-breasted (*Empidonax fulvifrons*), *320*
dusky (*Empidonax oberholseri*), *20*
great crested (*Myiarchus crinitus*), *20*
Hammond's (*Empidonax hammondii*), *20*
least (*Empidonax minimus*), *20*, *299*, 420
olive-sided (*Contopus borealis*), *20*
Pacific-slope (*Empidonax difficilis*), *20*
scissor-tailed (*Tyrannus forficatus*), *20*
southwestern willow (*Empidonax traillii extimus*), 89-91
western (*Empidonax difficilis*), *466*
willow (*Empidonax traillii*), 17, *20*
yellow-bellied (*Empidonax flaviventris*), *20*
flying fox, Mariana (*Pteropus mariannus*), 455
flying squirrel, northern (*Glaucomys sabrinus*), 192, 193
fonofos, *417*
food habitat studies, 8
Food Security Act of 1985 (Farm Bill), 302, 307, 426
Conservation Title, 423
foothill yellow-legged frog (*Rana boylii*), 132, *133*
forb
Euryops multifida, 321
Mentha pipertia, 176
Forbesichthys agassiz (spring cavefish), 177
forested wetlands, 475
restoration, 217
size estimates, 216
Southern, 216-17
Forest Inventory and Analysis program, U.S. Department of Agriculture Forest Service (USFS), 214
forest land
acid deposition (acid rain) and, 420
air pollution effects, 227
characteristics of, 214-15
distribution of, 215
environmental degradation and, 205
fragmentation, bird breeding productivity and, 25
growth, 215
growth-removal ratios, 215
harvest, 215
loss of, 472
mixed-conifer, atmospheric deposition and solute transport in, 421-23
mortality, 215
ownership of, 215
sky island ecosystems, 318-22
subalpine, 314-17
timberland, 215
U.S. resources, 214-15
forestry practices, red-cockaded woodpecker and, 87, 88
fork-tailed storm-petrel (*Oceanodroma furcata*), 44
Forster's tern (*Sterna forsteri*), 45, *54*, 56
fossil plant materials, 224-27
pollen, 224-27
fountain grass (*Pennisetum setaceum*), 364
four-fingered skink, brown (*Carlia fusca*), *434*
Fourth of July Butterfly Count (FJC), 171-72
fox
Arctic (*Alopex lagopus*), 28, 51, 441
Mariana flying (*Pteropus mariannus*), 455
red (*Vulpes vulpes*), 45, 51, 296, 302, 306, *407*, *440*
fox gecko (*Hemidactylus garnotii*), *434*
foxglove, giant false- (*Aureolaria grandiflora*), 231
fox sparrow (*Passerella iliaca*), *21*
foxtail pine (*Pinus balfouriana*), 315, 316, 421
Fractercula corniculata (horned puffin), 51
fragile papershell mussel (*Leptodea fragilis*), *180*, *445*
Franklinia alatamaha (Franklinia), 208
Franklinia (*Franklinia alatamaha*), 208
Franklin's gull (*Larus pipixcan*), *19*, 45, *54*, 56, 297, *299*
Fratercula
arctica (Atlantic puffin), 55
cirrhata (tufted puffin), 46, 51
freshwater algae, 255-56
freshwater diatoms, 256-58
freshwater drum (*Aplodinotus grunniens*), 261, 263, 447
freshwater fish, *See also* fish
habitat change, 146
imperiled, 142-47
managed populations, 141, 147-48
Southeastern, 144-46
freshwater gastropod fauna, Mobile Bay basin, 249-52
freshwater limpets (Ancylidae), 251
freshwater mussels, 177-79, *See also* mussel
biological pollution and, 179, 181-82
conservation of, 179
current status, 181-82
decline, 178-79
determining status, 177-78
endangered species, 178
historical status, 180-81
Lake Huron-Lake Erie corridor, 179-82
shells, uses of, 177
freshwater protozoa, 252

freshwater snail (*Potamopyrgus antipodarum*), 429
fringe-toed lizard, 137-38
  Coachella Valley (*Uma inornata*), 118, 137-38
  Colorado Valley (*Uma notata*), 137
  Mojave Valley (*Uma scoparia*), 137
fritillary, regal (*Speyeria idalia*), *166*, 174, *175*
frog, 117, 118, 124
  African clawed (*Xenopus laevis*), 433, *434*, 435, 436
  barking (*Hylactophryne augusti*), *320*
  bullfrog (*Rana catesbeiana*), 433, *434*, 435, 452-53
  California red-legged (*Rana aurora draytonii*), 132
  Carolina gopher (*Rana capito capito*), 129
  Cascades (*Rana cascadae*), 132
  Chiricahua leopard (*Rana chiricahuensis*), 139, *320*, 327, *328*, 452, 453
  Cuban treefrog (*Osteopilus septentrionalis*), *434*
  dusky gopher (*Rana capito sevosa*), 129
  eastern dwarf treefrog (*Litoria fallax*), *434*
  foothill yellow-legged (*Rana boylii*), 132, *133*
  gopher (*Rana capito*), 125
  greenhouse (*Eleutherodactylus planirostris*), *434*
  leopard (*Rana* spp.), 125, 140, 452-53
  mountain yellow-legged (*Rana muscosa*), 133, *134*, *466*
  native ranid, in California, 131-34
  northern leopard (*Rana pipiens*), 133-34, 327, *434*
  northern red-legged (*Rana aurora aurora*), 132
  poison-dart (*Dendrobates auratus*), *434*
  ranid (Ranidae), 452
  relict leopard (*Rana onca*), 327
  Rio Grande leopard (*Rana berlandieri*), 133, 453
  spotted (*Rana pretiosa*), 132-33, *328*
  tailed (*Ascaphus truei*), 125
  Tarahumara (*Rana tarahumarae*), 118, 138-40, 321
  wood (*Rana sylvatica*), *419*
  wrinkled (*Rana rugosa*), *434*
  Yavapai leopard (*Rana yavapaiensis*), 133, 139, 452-53
frosted elfin butterfly (*Incisalia irus*), 167, 232
Fulica
  americana (American coot), *19*, 299, *303*
  americana alai (Hawaiian coot), *377*
fulmar (*Fulmarus glacialis*), 51
Fundulus albolineatus (whiteline topminnow), 146
FUNGI.ARS-GRIN.GOV, 191
fungus, 189
  biological role of, 189, 190-91, 192
  bracket, 192
  cup, 192
  declines, 193
  Internet information, 191
  lichen, 194-96
  macrofungi, 192-94
  microfungi, 190-91
  polypore, 192
  in southwestern sky islands, 319
  species loss, 189
fur seal, northern (*Callorhinus ursinus*), *95*, 96
Fusconaia flava (Wabash pigtoe), *180*
fusiform elimia (*Elimia fusiformis*), *251*

Gadus macrocephalus (cod), 51
gadwall (*Anas strepera*), *19*, 35, *36*, *266*, *299*, 300, 301
Gallinago gallinago (common snipe), *19*, 59, *62*
Gallinula chlorpus sandwicensis (Hawaiian moorhen), *377*
Gallus gallus (red junglefowl), 372-73
Galveston Bay
  seagrass distribution, 274-75
  wetland loss, 270
Gambel's quail (*Callipepla gambelii*), *19*
gambusia
  Amistad (*Gambusia amistadensis*), *143*
  Clear Creek (*Gambusia heterochir*), 143
  San Marcos (*Gambusia georgei*), *143*
Gambusia
  affinis (western mosquitofish), *432*, 441
  amistadensis (Amistad gambusia), *143*
  georgei (San Marcos gambusia), *143*
  heterochir (Clear Creek gambusia), 143
game species
  non-native birds, 438
  surveys, *5*
Gammarus
  acherondytes, 176
  minus, 176
  pseudolimneaus, 176
  troglophilus (troglobitic amphipod), 177
Gap Analysis Program (GAP), 6, 310, 312, 461, 462, 463, 467, 473
  land protection status, 463-64
  protection of biodiversity through, 465-66
garter snake, 132, 326, 328
  checkered (*Thamnophis marcianus*), 452, 453
  Mexican (*Thamnophis eques*), *320*, 433, 435, 452-53
  narrow-headed (*Thamnophis rufipunctatus*), 327, *328*
  western terrestrial (*Thamnophis elegans*), 133
Gastrophyrne olivacea (Great Plains narrow-mouthed toad), *320*
gastropod fauna, freshwater, Mobile Bay basin, 249-52
gecko
  ashy (*Sphaerodactylus elegans*), *434*
  cosmopolitan house (*Hemidactylus mabouia*), *434*
  day (*Phelsuma* sp.), *434*
  fox (*Hemidactylus garnotii*), *434*
  house (*Hemidactylus frenatus*), *434*, 455
  Mediterranean house (*Hemidactylus turcicus*), *434*
  mourning (*Lepidodactylus lugubris*), 455
  ocellated dwarf (*Sphaerodactylus argus*), *434*
  reef (*Sphaerodactylus notatus*), *434*
  roughtail (*Cyrtopodion scabrum*), *434*
  stump-toed (*Gehyra mutilata*), 455
  tokay (*Gekko gekko*), *434*
  tree (*Gehyra oceanica*), 455
  yellow-headed (*Gonatodes albogularis*), *434*
Gehyra
  mutilata (stump-toed gecko), 455
  oceanica (tree gecko), 455
Gekko gekko (tokay gecko), *434*
gene flow, 147
general circulation models, 392
Generic Flora of the Southeastern United States, 203-4
genetic diversity, *See also* biodiversity; species diversity
  managed populations and, 147-48
  in southwestern sky islands, 319-21
genetic variability, 147
gentian, cream (*Gentiana alba*), 231
Gentiana alba (cream gentian), 231
Genyonemus lineatus (white croaker), 409, 411
geographic information systems (GIS), 467
  floodplain area analysis via, 234-35
  identifying gaps in biodiversity protection with, 463-64
  zebra mussel (*Dreissena polymorpha*) and, 445
geometer (*Brephos infans*), 166
Geopelia striata (small barred or zebra dove), 374, 438
George Miksch Avian Research Center, 81
Geothlypis trichas (common yellowthroat), *20*, *299*, *303*
Gettysbury National Military Park, 113
Geum
  peckii (mountain avens), 394
  triflorum (prairie smoke), 210
giant ameiva (*Ameiva ameiva*), *434*
giant anole, Jamaican (*Anolis garmani*), *434*
giant Canada goose (*Branta canadensis maxima*), 27
Giant Canada Goose, The (Hanson), 27
giant false-foxglove (*Aureolaria grandiflora*), 231
giant floater (*Pyganodon grandis*), *180*
giant kelp (*Macrocystis pyrifera*), 255
giant sequoia (*Sequoiadendron giganteum*), 421
giant silkworm moth
  Citheronia imperialis, 166
  Citheronia sepulchralis, 166
Gila
  bicolor
    euchila (Fish Creek Springs tui chub), *143*
    isolata (Independence Valley tui chub), *143*
  crassicauda (thicktail chub), *143*
  cypha (humpback chub), *150*, 151
  ditaenia (Sonora chub), *320*
  elegans (bonytail), *150*, 151, 309, 324-26
  intermedia (Gila chub), *150*, *320*
  nigrescens (Chihuahua chub), *143*
  robusta
    jordani (pahranagat chub), *150*
    robusta (roundtail chub), *150*
    rosetail chub, *320*
  seminuda (Virgin chub), *150*
Gila chub (*Gila intermedia*), *150*, *320*
Gila elegans (bonytail), *150*, 151, 309, 324-26
Gila monster (*Heloderma suspectum*), 327, *328*
Gila River, 149, 150
Gila River basin, sky islands, 318
Gila topminnow (*Poeciliopsis occidentalis*), *143*, *150*

Gila trout (*Oncorhynchus gilae*), *150*
gill nets, seabirds and, 51
GIS. *See* geographic information systems (GIS)
gizzard shad (*Dorosoma cepedianum*), 240, 261, *262*, *263*
Glacier National Park, 99
  cutthroat trout, 153-54
  non-native fish, 153-54
  wintering bald eagles (*Haliaeetus leucocephalus*), 329
Glaucidium
  brasilianum cactorum (ferruginous pygmy owl), *67*, 69
  gnoma (northern pygmy owl), *67*
Glaucomys sabrinus (northern flying squirrel), 193
glaucous-winged gull (*Larus glaucescens*), *19*, 45, 51
Glen Canyon National Recreation Area, 327
  wintering bald eagles (*Haliaeetus leucocephalus*), 329
global climate change, 385-95, *See also* climate change; temperature change
  changes in the Northeast, 390-91
  Colorado Rockies, 311
  determining impacts of, 385
  flora and, 392-95
  nesting behavior of Arctic geese and, 389
  overview, 385-86
  southwestern sky islands and, 322
  subalpine forest growth and, 315-17
  warming, 227, 322, 385-86, 392
  wetland loss and, 217
  winter ranges of birds and, 386-89
globally rare community types, 219
globeflower, spreading (*Trollius laxus*), *209*
Glugea hertwigi (protozoan), 443
gnatcatcher
  black-tailed (*Polioptila melanura*), *20*
  blue-gray (*Polioptila caerulea*), *20*, *466*
  California (*Polioptila californica*), *466*
goat (*Capra hircus*), *440*, 441
godwit
  bar-tailed (*Limosa lapponica*), *62*
  Hudsonian (*Limosa haemastica*), *59*, *62*
  marbled (*Limosa fedoa*), *19*, *59*, *62*, 396
golden aster, Ruth's (*Pityopsis ruthii*), 400
golden-crowned kinglet (*Regulus satrapa*), *20*, *22*
golden eagle (*Aquila chrysaetos*), *67*, 338, 340, 387, *466*
  causes of death, 68
goldeneye, 36, *266*
  Barrow's, *36*
  common (*Bucephala clangula*), *36*, *420*
golden-fronted woodpecker (*Melanerpes aurifrons*), *20*
golden-plover
  American (*Pluvialis dominica*), 58, *59*, *62*, 64
  Pacific (*Pluvialis fulva*), *62*, 373
goldenrod, white-haired (*Solidago albopilosa*), 401
golden shiner (*Notemigonus crysoleucas*), *261*, *262*, *432*
golden trout (*Oncorhynchus aguabonita*), *432*
goldfinch
  American (*Carduelis tristis*), *21*, *22*, *299*
  lesser (*Carduelis psaltria*), *21*
goldfish (*Carassius auratus*), 240, 261, *262*, 432, *432*
Gomphonema sp., *257*
Gonatodes albogularis (yellow-headed gecko), *434*
goose
  Arctic, 30-34, 388
  Canada (*Branta canadensis*), *19*, 26-33, *266*
    Aleutian (*Branta canadensis leucopareia*), 28, 30, 31, 32
    cackling (*Branta canadensis minima*), 26, 28, 30, 31, 33
    giant (*Branta canadensis maxima*), 26, 27, 29
    interior, 28, 29
    moffitti, 27, 28, 29
    occidentalis, 28
    Taverner's (*Branta canadensis taverneri*), 33
    Vancouver (*Branta canadensis fulva*), 28, 30-31, 32
  dusky Canada, 32
  emperor (*Chen canagica*), 30, 31, 33
  greater white-fronted (*Anser albifrons frontalis*), 30, 31, 33
  Hawaiian or nene (*Branta sandvicensis*), *377*, 379
  lesser snow (*Chen caerulescens caerulescens*), 30, 33
  monitoring, 15-16
  snow (*Chen caerulescens*), *266*, *388*
  tule white-fronted (*Anser albifrons gambeli*), 30, 31
  Yukon-Kuskokwim Delta, 31
gopher frog (*Rana capito*), 125, 130
  Carolina (*Rana capito capito*), 129
  dusky (*Rana capito sevosa*), 129
gopher tortoise (*Gopherus polyphemus*), 119, 129, 130
Gopherus
  agassizii (desert tortoise), 16, 74, 119, 135-36, *407*
  berlandieri (Texas tortoise), *119*
  polyphemus (gopher tortoise), 119, 129
goshawk, northern (*Accipiter gentilis*), *66*, 67, *320*
grackle
  boat-tailed (*Quiscalus major*), *21*
  common (*Quiscalus quiscula*), *21*, *299*
  great-tailed (*Quiscalus mexicanus*), *21*
grama-tobosa prairie, 472
Grand Bay National Wildlife Refuge, 76
Grand Canyon National Park, 327
  southwestern willow flycatcher, 89-91
  wintering bald eagles (*Haliaeetus leucocephalus*), 328, 329
granite night lizard (*Xantusia henshawi*), *466*
granite spiny lizard (*Sceloporus orcutti*), *466*
Graptemys, 120
  barbouri (Barbour's map turtle), *119*, *120*
  caglei (Cagle's map turtle), *119*
  flavimaculata (yellow-blotched map turtle), *119*
  oculifera (ringed map turtle), *119*
grass, 201
  buffel (*Pennisetum cilaris*), 321
  clover (*Halophila engelmanni*), 273
  fountain (*Pennisetum setaceum*), 364
  Kentucky bluegrass (*Poa pratensis*), 471
  manatee (*Syringodium filiforme*), 273, 274, 276, 277
  reed canary (*Phalaris arundinacea*), 249
  rice (*Stipa hymenoides*), 226
  shoal (*Halodule wrightii*), 273, 274, 276
  star, 273, 274
  turtle (*Thalassia testudinum*), 273, 274, 276
  widgeon (*Ruppia maritima*), 273, 274
grass carp (*Ctenopharyngodon idella*), 261, 263, 429, *432*
grasshopper, 163-65
  bandwing (*Pardalophora haldemani*), 164
  clear-winged (*Camnula pellucida*), 164
  Eximacris superbum, 164-65
  flying species, 165
  management implications, 165
  Marginatus group, species *Melanoplus*, 163, 164
  migratory (*Melanoplus sanguinipes*), 164
  natural range increases, 163-64
  nonflying species, 165
  Pardalophora apiculata, 164
  Pleistocene island species, 165
  range reductions and extinctions, 164-65
  sky island species, 163, 165
  taxonomic status, 163
  Trimerotropic
    huroniana, 164
    maritima, 164
  two-striped (*Melanoplus bivittatus*), 164
  unnatural range increases, 164
grasshopper sparrow (*Ammodramus savannarum*), *21*, 297, *299*, 302, 303, *466*
grasslands, 469, 471-72, *See also* Great Plains
  birds, 18, 396-98
  conversion, shorebirds and, 63, 64
  Great Plains, 295
  non-native, *465*
  retired agricultural land, 424-25
  tall-grass prairie butterfly community, 174-76
grassland-shrubland, 469
gray bat (*Myotis grisescens*), 176
gray catbird (*Dumetella carolinensis*), *20*, *299*
gray-cheeked thrush (*Catharus minimus*), *20*
gray hawk
  Austurina plagiata, *67*
  Buteo nitidus, *67*, *320*
gray-headed junco (*Junco hyemalis caniceps*), *21*
gray jay (*Perisoreus canadensis*), *20*
grayling, Arctic (*Thymallus arcticus*), *432*
gray owl, great (*Strix nebulosa*), *67*
gray (or Hungarian) partridge (*Perdix perdix*), 437, 438
Gray's Manual of Botany, 203
gray snapper (*Lutjanus griseus*), 282-83
Gray's vireo (*Vireo vicinior*), *466*
gray swiftlet (*Aerodramus vanikorensis*), 455
gray warbler, black-throated (*Dendroica nigrescens*), *20*
gray wolf (*Canis lupus*), 98-100, 109, 321, 337, 338, 340, 358
  caribou (*Rangifer tarandus*) and, 347-48
  counting, 347
  factors impeding recovery, 99

Lake Superior region, 99
Mexican (*Canis lupus baileyi*), 318, *320*
outlook, 100
reclassification of, 98-99
weather and, 347-48
western U.S., 99
greasewood (*Sarcobatus vermiculatus*), 226
great black-backed gull (*Larus marinus*), *19*, *54*, 55
great blue heron (*Ardea herodias*), 53, *54*, 55
great crested flycatcher (*Myiarchus crinitus*), *20*
great crested tern (*Sterna bergii*), 373
great egret (*Casmerodius albus*), *19*
greater 'amakihi (*Hemignathus sagittirostris*), *376*
greater amberjack (*Seriola dumerili*), 282
greater koa-finch (*Rhodacanthis palmeri*), *376*
greater scaup (*Aythya marila*), 35, *36*
greater white-fronted goose (*Anser albifrons frontalis*), 30, 31, 33
greater yellowlegs (*Tringa melanoleuca*), 59, *62*
Greater Yellowstone Ecosystem, 312-14
ecosystem management by species, 312-14
landscape model, 314
great gray owl (*Strix nebulosa*), *67*
great horned owl (*Bubo virginianus*), *67*
Great Lakes
commercial fishing, 244-46
discharge of PCBs into, 414
fish contaminant trends, 242-44
high waters, 249
lake trout (*Salvelinus namaycush*), 244-46
low waters, 249
non-native species, 428, 430, 442-47
regulated, wetlands in, 247-49
vascular plants, 203
water levels, 247-49
water quality improvements, canvasback and, 42
zebra mussel (*Dreissena polymorpha*), 409, 445-47
Great Lakes National Program Office, U.S. Environmental Protection Agency (USEPA), 242
Great Lakes Science Center, National Biological Service (NBS), 242
Great Lakes Water Quality Agreement, 244
Great Plains, 295-307
Conservation Reserve Program (CRP), and, 296, 302-3
coyote, 305-7
duck nest success, 300-302
grassland birds, 296-98
migratory birds, 298-300, 302-3
native prairie fishes, 295, 303-5
overview, 295-96
piping plover threats, 78-79
vascular plants, 203
Great Plains narrow-mouthed toad (*Gastrophyrne olivacea*), *320*
Great Salt Plains National Wildlife Refuge, 78
Great Smoky Mountains National Park, 196
black bear, 102
great-tailed grackle (*Quiscalus mexicanus*), *21*
great white shark (*Carcharodon carcharias*), California sea otter and, 112
grebe, pied-billed (*Podilymbus podiceps*), 389
green anole
Anolis carolinensis, *434*
Hispaniolan (*Anolis chlorocyanus*), *434*
greenhouse frog (*Eleutherodactylus planirostris*), *434*
greenhouse gases, 392
green iguana (*Iguana iguana*), *434*
green sea turtle (*Chelonia mydas*), *119*
green stink bug, southern (*Nezara viridula*), 366
green sunfish (*Lepomis cyanellus*), 240, *432*
green swordtail (*Xiphophorus helleri*), *432*
green-tailed towhee (*Pipilo chlorurus*), *20*
green tree skink (*Lamprolepis smaragdina*), *434*
green turtle (*Chelonia mydas*), 121, 122-23
green warbler, black-throated (*Dendroica virens*), *20*
green-winged teal (*Anas crecca*), 35, *36*, *266*
grey whale (*Eschrichtius robustus*), 95
grizzly bear (*Ursus arctos*), 103-5, 228, 229, 312, 313, *320*, 321, 338, 347, 358
hunting, 103, 105
grosbeak
black-headed (*Pheucticus melanocephalus*), *20*
blue (*Guiraca caerulea*), *20*, *466*
evening (*Coccothraustes vespertinus*), *21*, 387
Kona (*Chloridops kona*), *376*
pine (*Pinicola enucleator*), *21*
rose-breasted (*Pheucticus ludovicianus*), *20*
ground dove, common (*Columbina passerina*), *19*
groundfish, Alaskan seabirds and, 51
ground lizard, South American (*Ameiva ameiva*), *434*
groundsel (*Senecio huachucanus*), *320*
ground squirrel (*Spermophilus* spp.), 109
groundwater, riparian ecosystem dynamics and, 287
grouper, 282
Nassau (*Epinephelus striatus*), 282
Grus
americana (whooping crane), *405*, 406, *407*
canadensis (sandhill crane), *19*
pulla (Mississippi sandhill crane), 75-77
Gryotoma, 249-50
Guadalupe bass (*Micropterus treculi*), 147
Guam
marine toad (*Bufo marinus*), 435
non-native species, 454-56
guillemot, pigeon (*Cepphus columba*), 46, 51
Guiraca caerulea (blue grosbeak), *20*, *466*
Gulf Coast Joint Venture, North American Waterfowl Management Plan, 217
Gulf Intracoastal Waterway, 276, 277
Gulf of Mexico
canvasbacks, 42
coastal wetlands, 269-72
northern, seagrass distribution, 273-75
gull, 16, 51
California (*Larus californicus*), *19*, 45
Franklin's (*Larus pipixcan*), *19*, 45, *54*, 56, 297, *299*
glaucous-winged (*Larus glaucescens*), *19*, 45, 51
great black-backed (*Larus marinus*), *19*, *54*
Heermann's (*Larus heermanni*), 45
herring (*Larus argentatus*), *19*, 53, *54*, 55-56
laughing (*Larus atricilla*), *19*
ring-billed (*Larus delawarensis*), *19*, 45, *54*, 56
western (*Larus occidentalis*), 45
gull-billed tern (*Sterna nilotica*), 46, *54*, 56
Gulo gulo (wolverine), 358
guppy (*Poecilia reticulata*), *432*
Gutterezia sarothrae (snakeweed), 226
Gymnocephalus cernuus (ruffe), 429, 430, 443
Gymnoderma lineare, 196
Gymnogyps californianus (California condor), 16, *67*, 80-81
Gymnorhinus cyanocephalus (pinyon jay), *20*
gymnosperms, 200, 201
gypsy moth (*Lymantria dispar*), 441
gyrfalcon (*Falco rusticolus*), *67*
Gyrotoma, 251
excisa (excised slitshell), *251*
lewisii (striate slitshell), *251*
pagoda (pagoda slitshell), *251*
pumila (ribbed slitshell), *251*
pyramidata (pyramid slitshell), *251*
walkeri (round slitshell), *251*

habitat assessment, 461-75
biodiversity, southwestern California, 465-66
land characteristics, federal data bases, 467-68
landscape change, monitoring with satellite imagery, 468-73
overview, 461-63
vegetation cover types, Utah, 463-64
wetlands, 473-75
Habitat Conservation Plan, 137
habitat loss/change, *See also* human influences
agriculture and, 423, 424
aquatic ecosystems, 178, 234-36
avian diseases and, 401-4
Columbia River basin, 156
edge effect, 160
extinction and, 452
fragmentation, *146*
Hawaiian bird deline and, 376, 378
human influences and, 404
insects and, 161
inventory and monitoring programs, *5*
invertebrates and, 159-60
mammals and, 93-94
prairie, 174, 175
protection against, endangered and threatened species and, 466
reptiles and amphibians and, 118, 131
Upper Mississippi River (UMR) floodplain, 234-36
wetlands, 216, 236-38, 473-74, 475
winter ranges of birds and, 386-87

Haematopus
bachmani (black oystercatcher), *62*
palliatus (American oystercatcher), *62*
Halcyon cinnamomina (Micronesian kingfisher), 455
Haleakala National Park, 363-64, 379
Haleakala silversword
Argyroxiphium sandwicense ssp. *macrocephalum*, 362, 363-64
flowering, 363, 364
population trends, 363-64
recovery of, 362
threats to, 364
trends, 364
Argyroxiphium sandwicense ssp. *sandwicense*, 363
Haleakala Volcano, 363
Haliaeetus leucocephalus (bald eagle), 16, 44, 66-67, 68, 309, 328-30, 387, 389, 413, 415, *466*
Halodule wrightii (shoal grass), 273, 274, 276
Halophila
decipiens, 273
engelmanni (star grass), 273
Hamamelis virginiana (common witch-hazel), 390
Hammond's flycatcher (*Empidonax hammondii*), *20*
Hanson, Harold C., 27
Haploperla brevis (stonefly), 185, *186*
harbor porpoise (*Phocoena phocoena*), *94*, *95*, 96
harbor seal (*Phoca vitulina*), 52, *94*, 96
hardwoods, 215
hare
snowshoe (*Lepus americanus*), 359
tundra or Arctic (*Lepus othus* or *L. timidus*), 338, 359
harelip sucker (*Moxostoma lacerum*), 146
harlequin duck (*Histrionicus histrionicus*), 35, 36
harrier, northern (*Circus cyaneus*), *67*, 387
Harris' hawk (*Parabuteo unicinctus*), 67
haul outs, Pacific walrus (*Odobenus rosmarus divergens*), 356
Havasu National Wildlife Refuge, 287
Hawaii, 361-81
birds, 372-75
avian disease and, 378, 380
breeding visitor species, 373
competition among, 378-79
conservation outlook, 380-81
current status, 372-74
endemic, 376-81
extinction, 375
forest birds, 379-80
habitat loss, 376, 378
native species, 372, 374
non-native, 372-73, 375, 379, 437, 455-56
seabirds, 379
visitor species, 373, 374
waterbirds, 374, 379-80
Dropsophila (Pomance flies), 367-71
endemic species, 361
extinctions, 367, 371
fern, 201-2
Haleakala silversword (*Argyroxiphium sandwicense* ssp. *macrocephalum*), 363-64
insects, 365-67, 368-71
extinctions, 367, 371
introduced parasites, 366-67
non-native species, 365, 366-67, 379
survey needs, 367
non-native species, 361-62, 364, 379, 428, 441-42
birds, 372-73, 375, 379, 437, 455-56
insects, 365, 366-67, 379
predators, 379
ungulates, 370
overview, 361-63
persistent contamination in, 413
rain forests, 370
species at risk, *362*
species extinction, 361
vascular plants, 201-2, 206
Hawaii 'akepa (*Loxops coccineus coccineus*), *378*
Hawaii 'akialoa (*Hemignathus obscurus obscurus*), *376*
Hawaii 'amakihi (*Hemignathus virens virens*), *377*
Hawaiian coot (*Fulica americana alai*), *377*
Hawaiian crow ('alala) (*Corvus hawaiiensis*), *377*, 381
Hawaiian duck (koloa) (*Anas wyvilliana*), *377*, 379
Hawaiian goose (nene) (*Branta sandvicensis*), *377*, 379
Hawaiian hawk (*Buteo solitarius*), *67*, *377*
Hawaiian monk seal, *95*
Hawaiian moorhen (*Gallinula chlorpus sandwicensis*), *377*
Hawaiian noddy (*Anous minutus melanogenys*), *377*, 379
Hawaiian owl (pueo) (*Asio flammeus sandwichensis*), *377*
Hawaiian picture-wing *Drosophila* (*Drosophila conspicua*), *369*
Hawaiian rail (*Porzana sandwichensis*), *376*, 380
Hawaiian stilt (*Himantopus mexicanus knudseni*), *377*
Hawaiian Wetland Bird Recovery Plan, 374
Hawaii Biological Survey, 362
Hawaii creeper (*Oreomystis mana*), *378*
Hawaii Division of Forestry and Wildlife, 374
Hawaii 'elepaio (*Chasiempis sandwichensis sandwichensis*), *377*
Hawaii Forest Bird Survey, 374, 379
Hawaii Heritage Program, 362
Hawaii mamo (*Drepanis pacifica*), *376*
Hawaii 'o'o (*Moho nobilis*), *376*
Hawaii thrush ('oma'o) (*Myadestes obscurus*), 380
Hawaii, University of, 365
Hawaii Volcanoes National Park, 369, 370, 371
hawk, 16
broad-winged (*Buteo platypterus*), *67*
common black (*Buteogallus anthracinus*), *67*
Cooper's (*Accipiter cooperii*), *67*
ferruginous (*Buteo regalis*), 67, 69, 297, 387
gray (*Buteo nitidus* or *Austurina plagiata*), *67*, *320*
Harris' (*Parabuteo unicinctus*), 67
Hawaiian (*Buteo solitarius*), *67*, *377*
Puerto Rican broad-winged (*Buteo platypterus brunnescens*), *67*
red-shouldered (*Buteo lineatus*), *67*
red-tailed (*Buteo jamaicensis*), *67*
rough-legged (*Buteo lagopus*), *67*
sharp-shinned (*Accipiter striatus*), 67
short-tailed (*Buteo brachyurus*), *67*
Swainson's (*Buteo swainsoni*), *67*
white-tailed (*Buteo albicaudatus*), *67*
zone-tailed (*Buteo albonotatus*), *67*
hawkmoth
Sphingidae, 172, *173*
spurge (*Hyles euphorbiae*), 162
hawk owl, northern (*Surnia ulula*), *67*
hawksbill (*Eretmochelys imbricata*), *119*, 121, 123
haying management, butterfly population and, 175
health, population, 23
hearty elimia (*Elimia jonesi*), *251*
heavy metal posioning, Tarahumara frog (*Rana tarahumarae*), 139, 140
heelsplitter
creek (*Lasmigona compressa*), *180*
pink (*Potamilus alatus*), *180*, *181*, *182*
white (*Lasmigona complanata complanata*), *180*
Heermann's gull (*Larus heermanni*), 45
hellbender (*Cryptobranchus alleganiensis*), 125
Heloderma suspectum (Gila monster), 327, *328*
Hemaris gracilis (sphinx moth), 166
Hemidactylus
frenatus (house gecko), *434*, 455
frenatus (lizard), 454
garnotii (fox gecko), *434*
mabouia (cosmopolitan house gecko), *434*
turcicus (Mediterranean house gecko), *434*
Hemignathus
lucidus
affinus (Maui nukupu'u), *377*
hanapepe (Kauai nukupu'u), *377*
zlucidus (Oahu nukupu'u), *376*
munroi ('akiapola'au), 376, *377*
obscurus
ellisianus (Oahu 'akialoa), *376*
lanaiensis (Lanai 'akialoa), *376*
obscurus (Hawaii 'akialoa), *376*
parvus ('anianiau), *377*
procerus (Kauai 'akialoa), *377*
sagittirostris (greater 'amakihi), *376*
virens
'amakihi, 376
chloris (Oahu 'amakihi), *377*, 378
stejnegeri (Kauai 'amakihi), *377*
virens (Hawaii 'amakihi), *377*
wilsoni (Maui 'amakihi), *377*
Hemileuca maia (buck moth), 232
hemlock, mountain (*Tsuga mertensiana*), 315
Henslow's sparrow (*Ammodramus henslowii*), 297
Hepatica acutiloba (hepatica), 390
hepatica (*Hepatica acutiloba*), 390
hepatics, 197

heptachlor, 413
herbaceous ecosystems, rare, 219, 220
hermit thrush (*Catharus guttatus*), *20*
hermit warbler (*Dendroica occidentalis*), *20*
heron, 55
 black-crowned night- (*Nycticorax nycticorax*), *54*, 55
 great blue (*Ardea herodias*), 53, *54*
 little blue (*Egretta caerulea*), *19*
Herpestes auropunctatus (mongoose), 378, 441
herpetofauna, 117-18
 in endangered longleaf pine ecosystem, 129-31
herring
 blueback (*Alosa aestivalis*), 261, 262
 Clupea harengus, 51
 lake (*Coregonus artedi*), 179
herring gull (*Larus argentatus*), *19*, 53, *54*, 55-56
Hesperia
 dacotae (Dakota skipper), 174, 175
 leonardus (savanna skipper), 232
 metea (savanna skipper), 232
Heteroptera, 161
Heteroscelus incanus (wandering tattler), *62*
heterotropic protozoa, 252-54
heterozygosity, 75
hexachlorobenzene, *408*
Hexagenia, 181, 236
 bilineata (burrowing mayfly), 183-84
 limbata (burrowing mayfly), 183-84
Hibiscadelphus giffardianus, 367
hickorynut (*Obovaria olivaria*), *180*
hickorynut, round (*Obovaria subrotunda*), *180*
high-spired elimia (*Elimia hartmaniana*), *251*
Hiko springfish (*Crenichthys baileyi grandis*), *150*
Himalayan raspberry, yellow (*Rubus ellipticus*), 370
Himantopus
 mexicanus (black-necked stilt), *19*, *59*, *62*
 mexicanus knudseni (Hawaiian stilt), *377*
Himatione sanguinea
 freethii (Laysan honeycreeper), *376*
 sanguinea ('apapane), *378*
Hippoglossoides elassodon (flathead sole), 409
Hirundo
 pyrrhonota (cliff swallow), *20*, *299*
 rustica (barn swallow), *20*, *299*
Hispaniolan green anole (*Anolis chlorocyanus*), *434*
Histrionicus histrionicus (harlequin duck), 35, 36
H.J. Andrews Experimental Forest, Orgeon, insect survey, 161
Homohadena badistriga (owlet moth), *166*, 167
Homoptera (aphids, leaf hoppers, scale insects), 161
honeybee
 Africanized, 448-51
 Apis mellifera, 85
 European, 448-51
 Italian (*Apis mellifera ligustica*), 448
honeycreeper, 376
 Laysan (*Himatione sanguinea freethii*), *376*
hooded warbler (*Wilsonia citrina*), *20*
hookbill, Lanai (*Dysmorodrepanis munroi*), *376*
hook-billed kite (*Chondrohierax uncinatus*), *67*
hooters (Mexican spotted owl callers), 331
horned lark (*Eremophila alpestris*), *20*, 297, 299, *303*
horned lizard
 coast (*Phrynosoma coronatum*), *466*
 Texas (*Phrynosoma cornutum*), *434*
horned owl, great (*Bubo virginianus*), *67*
horned puffin (*Fractercula corniculata*), 51
hornwort, 197
 floristic inventories, 198, 199-200
horse (*Equus caballus*), *440*, 441, 456-57
Horsehoe Reef, 283
house finch (*Carpodacus mexicanus*), *21*, 439
house gecko
 cosmopolitan (*Hemidactylus mabouia*), *434*
 Hemidactylus frenatus, *434*, 455
 Mediterranean (*Hemidactylus turcicus*), *434*
house mouse (*Mus musculus*), 454
house shrew (*Subcus murinus*), 454
house sparrow (*Passer domesticus*), *21*, *299*, 437, *438*, 439
house wren (*Troglodytes aedon*), *20*, *299*
Howes Prairie Marsh, 224-25
Hudsonian godwit (*Limosa haemastica*), *59*, *62*
Hudson River, nearshore fish assemblage, 260-63
 changes, 261-62
 implications of changes, 262-63
 surveys, 260-61
human influences, 397-426, *See also* habitat loss/change; pesticides
 acid deposition (acid rain), 418-20
 agricultural ecosystems, 423-26
 amphibians and reptiles, 118
 atmospheric deposition and solute transport, in montane mixed-conifer forest, 421-23
 avian diseases, habitat change and, 401-4
 Canada goose (*Branta canadensis*), 27
 captive propagation, introduction, and translocation programs, 404-7
 climate change, 394
 coastal erosion, 278, 279
 coastal and marine ecosystems, 259-60
 Colorado Rockies, 310-11
 contaminants in coastal fish and mollusks, 408-11
 coral reef ecosystems, 283
 endangered species, federal lands and, 398-401
 Florida Keys ecosystem, 279
 freshwater fish, 142
 habitat loss and, 404
 Interior West ecosystems, 309
 mammals, 93-94
 non-native species and, 427
 organophosphorus and carbamate pesticides, 416-18
 overview, 397-98
 persistent environmental contaminants, 411-15
 riparian ecosystems, 286, 287
 seagrass loss, 273, 275
 terrestrial ecosystems, 214
 west coast breeding seabirds, 43
 wetland loss and, 270
 winter ranges of birds, 386-87, 389
Humid Temperate Domain, shorebirds, 60-61
hummingbird
 black-chinned (*Archilochus alexandri*), *466*
 broad-tailed (*Selasphorus platycercus*), *20*
 Costa's (*Calyptae costae*), *466*
 rufous (*Selasphorus rufus*), *20*
humpback chub (*Gila cypha*), *150*, 151
humpback whale, *94*, *95*
Hungarian (gray) partridge (*Perdix perdix*), 437, 438
hunting
 American alligator (*Alligator mississippiensis*), 128
 black bear (*Ursus americanus*), 100, 101, 102
 Canada goose (*Branta canadensis*), 26-27, 28
 canvasback (*Aythya valisineria*), 41
 caribou (*Rangifer tarandus*), 357
 duck, 37
 grizzly bear (*Ursus arctos*), 103, 105
 Kodiak brown bear (*Ursus arctos middendorffi*), 349-50
 North American elk (*Cervus elaphus*), 115, 116
 polar bear (*Ursus maritimus*), 351
 sea otter (*Enhydra lutris*), 355
 shorebirds and, 57
 wild turkey (*Meleagris gallopavo*), 71
Huperzia haleakalae (lycopod), 206
hurricanes
 coastal and marine ecosystems and, 259
 Puerto Rican parrot and, 84-85
hyacinth, wild (*Camassia scilloides*), 231
Hybognathus placitus (plains minnow), 304
Hydnum repandum, 193
hydraulic change, western riparian ecosystems and, 285-91
Hydric Soils of the United States, 474
hydrilla (*Hydrilla verticillata*), 264, 429
Hydrilla verticillata (hydrilla), 264, 429
Hydrobatidae (storm-petrels), 43-44
Hydrobiidae (pebblesnails), 249, 251
hydrology
 regions, southeastern U.S., 144-45
 riparian ecosystem dynamics and, 287
Hydroperla
 crosbyi (stonefly), *186*
 fugitans (stonefly), *186*
Hylactophryne augusti (barking frog), *320*
Hyla eximia (mountain treefrog), 327
Hyles euphorbiae (spurge hawkmoth), 162
Hylocichla mustelina (wood thrush), *20*, 25, 420
Hymenoptera (ants, bees, wasps, sawflies), 161
Hypentelium nigricans (northern hog sucker), *261*, 263

ibis, 55
white (*Eudocimus albus*), *19*, *54*, 55
white-faced (*Plegadis chihi*), *54*, 55
ichneumonid wasps, Hawaii, 366
Ictalurus
furcatus (blue catfish), *432*
pricei (Yaqui catfish), *320*
punctatus (channel catfish), 151, *432*
Icteria virens (yellow-breasted chat), *20*, 231, *466*
Icterus
galbula (Baltimore oriole), *21*
galbula bullockii (Bullock's oriole), *21*
parisorum (Scott's oriole), *21*
spurius (orchard oriole), *21*, 231
Ictinia mississippiensis (Mississippi kite), *67*, 297
iguana
desert (*Dipsosaurus dorsalis*), 327
green (*Iguana iguana*), *434*
Mexican spiny-tailed (*Ctenosaura pectinata*), *434*
Iguana iguana (green iguana), *434*
'i'iwi (*Vestiaria coccinea*), 376, *378*, 379, 380
Illinois
American badger, 108-9
caves and springs, 176-77
stonefly, 184-87
Illinois Environmental Protection Agency (IEPA), 240
Illinois Natural History Survey (INHS), 161, 183, 185, 239
Illinois River
electrofishing survey, 239-41
fish populations, 239-41
pollution of, 239
immunocontraception, for controlling wild horses, 457
inbreeding depression, 147
Incisalia
henrici (elfin butterfly), 167
irus (frosted elfin butterfly), 167, 232
inclusion body disease of cranes, *403*
Independence Valley tui chub (*Gila bicolor isolata*), *143*
Indiana bat (*Myotis sodalis*), 97-98, 176
Indiana Bat Recovery Team, 97
Indiana Dunes National Lakeshore, 224-25, *226*
Indian monitor lizard (*Varanus indicus*), 455
Indian-plantain, pale (*Cacalia atriplicifolia*), 231
Indian River Lagoon, 269
indigo bunting (*Passerina cyanea*), *20*, 231
indigo snake, eastern (*Drymarchon corais couperi*), 129
industrial development, wetland loss and, 270
inland salt marsh, 220
insecticides, *See also* DDE pesticides; DDT pesticides; organochlorine pesticides; pesticides
aquatic insects and, 182
insects
Albany pine barrens, 166-67
aquatic, 182-84
diversity and abundance of, 161-62
extinctions, 367
habitat change and, 161
Hawaiian, 365-67, 368-71, 379
immature stages, 161
as indicators of ecosystem health, 365
non-native species, 365, 366-67, 379, *429*
in southwestern sky islands, 319, 320
surveys, 161-62
Insects of Virginia, Blacksburg, 161
Institute for Bird Populations, 24
interior cypress forest, southern, *465*
interior least tern (*Sterna antillarum athalassos*), *54*, *55*, 56
Interior Plateau hydrologic region, 145
Interior West ecosystems, 309-35
Colorado Plateau
amphibian and reptile diversity, 326-28
Mexican spotted owl, 330-31
Colorado River Basin
bonytail and razorback sucker, 324-26
wintering bald eagles, 328-30
Colorado Rockies, 310-12
desert bighorn sheep, 333-35
Greater Yellowstone Ecosystem, 312-14
human activities and, 309
overview, 309-10
Pyramid Lake, Nevada, cui-ui, 323-24
Rocky Mountain National Parks, bighorn sheep, 332-33
southwestern sky island ecosystems, 318-22
subalpine forests, 314-17
Intermountain Flora, 201, 202
Intermountain region, vascular plants, 202
intermountain sagebrush (*Artemisia*spp.), 222
International Association of Fish and Wildlife Service Agencies, 424
International Biological Program, 369
International Legume Data Information System (ILDIS), 201
International Paper Company, 217
International Shorebird Survey (ISS), 58, 64
International Union for the Conservation of Nature and Natural Resources (IUCN, now World Conservation Union), 399
Species Survival Commission (IUCN/SSC), 101
International Whaling Commission (IWC), 95
Internet, microfungi information on, 191
interrupted rocksnail (*Leptoxis formanii*), *251*
introduced species. *See* non-native species
Inupiat Eskimos, 341
inventory and monitoring programs, *5*
examples, *5*
forest, 214-15
international, 4
national, 4, 473
non-native species, 431-32
purpose of, 2
satellite imagery for, 468-73
inversion polymorphism, in *Dropsophila* (Pomance flies), 369
invertebrates, 159-87
habitat change and, 159
metamorphosis, 159
overview, 159-60
Iotichthys phlegethontis (least chub), *143*
Iridomyrmex humilis (Argentine ant), 364, 366
Isogenoides varians (stonefly), 185, *186*
Isoperla
bilineata (stonefly), *186*, 187
burksi (stonefly), 185, *186*
conspicua (stonefly), 185, *186*
decepta (stonefly), *186*
longiseta (stonefly), *186*
marlynia (stonefly), 185, *186*
mohri (stonefly), *186*
nana (stonefly), *186*
richardsoni (stonefly), *186*
Italian honeybee (*Apis mellifera ligustica*), 448
Italian wall lizard (*Podarcis sicula*), *434*
IUCN (World Conservation Union, formerly International Union for the Conservation of Nature), 399
Ixoreus naevius (varied thrush), *20*
Izembek Lagoon, habitats, 32-33

Jackson County, Mississippi, Mississippi sandhill crane population, 76
Jackson's chameleon (*Chamaeleo jacksonii*), *434*
jaguar (*Felis onca*), *320*, 321
Jamaican giant anole (*Anolis garmani*), *434*
Japanese snail (*Ocenebra japonica*), 429
Japanese white-eye (*Zosterops japonicus*), 374
javelina (*Tayassu tajacu*), 322
jay
blue (*Cyanocitta cristata*), *20*
gray (*Perisoreus canadensis*), *20*
pinyon (*Gymnorhinus cyanocephalus*), *20*
scrub (*Aphelocoma coerulescens*), *20*
Steller's (*Cyanocitta stelleri*), *20*
Jemez Mountains salamander (*Plethodon neomexicanus*), 327, *328*
Jepson Manual, The, Higher Plants of California, 202
jewfish (*Epinephelus itajara*), 282
johnny darter (*Etheostoma nigrum*), 143
junco
gray-headed (*Junco hyemalis caniceps*), *21*
Oregon (*Junco hyemalis oregonus*), *21*, *22*
slate-colored (*Junco hyemalis*), *21*, *22*
Junco
hyemalis (slate-colored junco), *21*, *22*
hyemalis caniceps (gray-headed junco), *21*
hyemalis oregonus (Oregon junco), *21*, *22*
June sucker (*Chasmistes liorus mictus*), *143*
junglefowl, red (*Gallus gallus*), 372-73
juniper
single-seeded (*Juniperus monosperma*), 223
Utah (*Juniperus osteosperma*), 223
western (*Juniperus occidentalis*), 223, 315
woodlands
altered fire regimes, *222*, 223

northern, *465*
Juniperus
monosperma (single-seeded juniper), 223
occidentalis (western juniper), 223
osteosperma (Utah juniper), 223

kama'o (*Myadestes myadestinus*), *377*
kangaroo rat
Merriam's (*Dipodomys merriami*), *466*
Pacific or agile (*Dipodomys agilis*), *466*
Stephens' (*Dipodomys stephensi*), *466*
Karner blue butterfly (*Lycaeides melissa samuelis*), 166, 167, *172*, 224-25, 232, *398*, 401
karst limestone regions, Illinois, biota, 176-77
Kauai 'akepa (*Loxops coccineus caeruleirostris*), *378*
Kauai 'akialoa (*Hemignathus procerus*), *377*
Kauai 'amakihi (*Hemignathus virens stejnegeri*), *377*
Kauai creeper (*Oreomystis bairdi*), *378*
Kauai 'elepaio (*Chasiempis sandwichensis sclateri*), *377*
Kauai nukupu'u (*Hemignathus lucidus hanapepe*), *377*
Kauai thrush, small (puaiohi) (*Myadestes palmeri*), 380
Kauai 'o'o (*Moho braccatus*), *377*
kelp, giant (*Macrocystis pyrifera*), 255
kelp bass (*Paralabrax clathratus*), 411
Kemp's ridley (*Lepidochelys kempii*), *119*, 121, 123
Kendall Warm Springs dace (*Rhinichthys osculus thermalis*), *150*
Kentucky bluegrass (*Poa pratensis*), 471
Kentucky warbler (*Oporornis formosus*), *20*
kestrel
American (*Falco sparverius*), *67*
American (Florida subspecies), (*Falco sparverius paulus*), *67*
Keterson National Wildlife Refuge, 415
Key Largo, 281
key silverside (*Menidia conchorum*), 281
kidneyshell (*Ptychobranchus fasciolaris*), *180*
Kilauea Volcano, 370
killdeer (*Charadrius vociferus*), *19*, 58, *59*, *62*, *299*
kingbird
Cassin's (*Tyrannus vociferans*), *20*
eastern (*Tyrannus tyrannus*), *20*, *299*, *420*
western (*Tyrannus verticalis*), *20*, *299*
king eider, *36*
kingfisher
belted (*Ceryle alcyon*), *420*
Micronesian (*Halcyon cinnamomina*), 455
kinglet
golden-crowned (*Regulus satrapa*), *20*, *22*
ruby-crowned (*Regulus calendula*), *20*, *22*
king mackerel (*Scomberomorus cavalla*), 282
Kinosternon
flavescens (yellow mud turtle), *119*
hirtipes (Mexican rough-footed mud turtle), *119*
kioea (*Chaetoptila angustipluma*), *376*
Kirwin National Wildlife Refuge, 78
kite
American swallow-tailed (*Elanoides forficatus*), *67*
black-shouldered (*Elanus caeruleus*), *466*
hook-billed (*Chondrohierax uncinatus*), *67*
Mississippi (*Ictinia mississippiensis*), *67*, 297
snail (*Rostrhamus sociabilis*), 66, *67*
white-tailed (*Elanus caeruleus*), *67*
kitten-tails (*Besseya bullii*), 231
kittiwake, 16
in Alaska, 49-50
black-legged (*Rissa tridactyla*), 49-50
red-legged (*Rissa brevirostris*), 50
Kittlitz's murrelet (*Brachyramphus brevirostris*), 50
knight anole (*Anolis equestris*), *434*
knot, red (*Calidris canulus*), *59*, *62*
koa bug (*Coleotichus blackburniae*), 366-67
koa-finch
greater (*Rhodacanthis palmeri*), *376*
lesser (*Rhodacanthis flaviceps*), *376*
Kodiak brown bear (*Ursus arctos middendorffi*), 349-50
aerial stream surveys, 350
management considerations, 350
population abundance, 350
population monitoring, 349
sport harvest records, 349-50
Kodiak National Wildlife Refuge, 349
kokanee salmon (*Oncorhynchus nerka*), 154, 329
koloa (Hawaiian duck) (*Anas wyvilliana*), *377*, 379
Kona grosbeak (*Chloridops kona*), *376*
Kootenai River, white sturgeon (*Acipenser transmontanus*), 155
Küchler, A. W., potential natural vegetation map, 468-72

Laccaria laccata, *192*
laced-necked (spotted) dove (*Streptopelia chinensis*), 438
lady's-tresses (*Spiranthes delitescens*), *320*
Laguna Madre, Texas
seagrass meadows, 275-77
turtles, 121, 122
Laguna Madre de Tamaulipas, 277
Lake Andes National Wildlife Refuge, 402
Lake Champlain, 414
Lake Erie
fish contaminant trends, 243
freshwater mussels, 179-82
lake trout (*Salvelinus namaycush*), 244-46
pollution of, 179
lake herring (*Coregonus artedi*), 179
Lake Huron
fish contaminant trends, 243
freshwater mussels, 179-82
lake trout (*Salvelinus namaycush*), 244-46
Lake Michigan
fish contaminant trends, 242-43
lake trout (*Salvelinus namaycush*), 244-46
zebra mussel (*Dreissena polymorpha*), 445-47
Lake Mohave, bonytail and razorback sucker, 325, 326
Lake Ontario
fish contaminant trends, 243
lake trout (*Salvelinus namaycush*), 244-46
water levels, 249
lakes
acidification of, 257-58, 418-19
disturbances to, canvasback ducks and, 41-42
eutrophication of, diatoms and, 257, 258
Lake St. Clair, 414
freshwater mussels, 180-81
Lake Superior
fish contaminant trends, 243
lake trout (*Salvelinus namaycush*), 244-46
region, gray wolf population, 99
water levels, 248-49
lake trout (*Salvelinus namaycush*), 154, 242, *243*, 414, *432*
Great Lakes, 242, 244-46
stocked, 245-46
lake whitefish (*Coregonus clupeaformis*), *432*
lampmussel
eastern (*Lampsilis siliquoidea*), *180*
wavy-rayed (*Lampsilis fasciola*), *180*
lamprey, sea (*Petromyzon marinus*), 244, 430, 443
Lamprolepis smaragdina (green tree skink), *434*
Lampropholis delicata (rainbow skink), *434*
Lampsilis
fasciola (wavy-rayed lampmussel), *180*
ovata (pocketbook), *180*
siliquoidea (eastern lampmussel), *180*
Lanai 'akialoa (*Hemignathus obscurus lanaiensis*), *376*
Lanai creeper (*Paroreomyza m. montana*), *376*
Lanai hookbill (*Dysmorodrepanis munroi*), *376*
landbirds
adult survival, 23-26
breeding productivity, 23-26
land cover, *See also* vegetation
change, 461, 469-73
classes, protection of, 463-64
data sets, 461
federal data bases, 467-68
fragmentation, climate change and, 394
mapping, 469-73
monitoring with satellite imagery, 468-73
satellite photos of, 462
types, 461-62, 469
landfill practices, gulls and, 56
land mollusks, in southwestern sky islands, 319
land ownership, biodiversity protection and, 463
land protection status, 463-64, 465-66
management and, 465-66
Landsat Multispectral Scanner (MSS), 470-71, 472
Landsat Thematic Mapper, 463, 467-68, 473
landscape model, Greater Yellowstone Ecosystem, 314
Lanius
ludovicianus (loggerhead shrike), *20*, 231, *466*
ludovicianus mearnsi (San Clemente Island loggerhead shrike), 74
larch, subalpine (*Larix lyallii*), 316

large floating bladderwort (*Utricularia inflata* Walter), 209
large-headed anole (*Anolis cybotes*), *434*
largemouth bass (*Micropterus salmoides*), 240, *261*, *432*
Florida (*Micropterus salmoides floridanus*), 148
managed populations, 148
northern (*Micropterus salmoides salmoides*), 148
Larix lyallii (subalpine larch), 316
lark, horned (*Eremophila alpestris*), *20*, 297, 299, *303*
lark bunting (*Calamospiza melanocorys*), *21*, 297, 298, *299*, 302, *303*
lark sparrow (*Chondestes grammacus*), *21*, 297, *299*
Larus
argentatus (herring gull), *19*, 53, *54*, 55-56
atricilla (laughing gull), *19*
californicus (California gull), *19*, 45
delawarensis (ring-billed gull), *19*, 45, *54*, 56
glaucescens (glaucous-winged gull), *19*, 45, 51
heermanni (Heermann's gull)), 45
marinus (great black-backed gull), *19*, *54*, 55
occidentalis (western gull), 45
pipixcan (Franklin's gull), *19*, 45, *54*, 56, 297, *299*
Lasiurus borealis (red bat), *320*
Lasmigona
complanata complanata (white heelsplitter), *180*
compressa (creek heelsplitter), *180*
costata (fluted-shell), *180*
Las Vegas dace (*Rhinichthys deaconi*), *150*
laughing gull (*Larus atricilla*), *19*
Lawler, Matusky and Skelly Engineers (LMS), 260
Laysan albatross (*Diomedea immutabilis*), 374
Laysan duck (*Anas laysanensis*), *377*, 379, 400
Laysan finch (*Telespyza cantans*), *377*
Laysan honeycreeper (*Himatione sanguinea freethii*), *376*
Laysan millerbird (*Arcrocephalus f. familiaris*), *376*
Laysan rail (*Porzana palmeri*), *376*, 380
lazuli bunting (*Passerina amoena*), *20*
Leach's storm-petrel (*Oceanodroma leucorhoa*), 43-44
lead poisoning
canvasbacks, 41, 42
eagles, 68
fish concentrations, *411*, 414-15
leaf bug (*Nesiomiris*), 367
leaf-footed bug (*Thasus* sp.), *161*
leafhopper, 161
red-tailed (*Aflexia rubranura*), 232
leaf miner moth (*Acrocercops arbutella*), *160*
least auklet (*Aethia pusilla*), 51
least chub (*Iotichthys phlegethontis*), *143*
least cisco (*Coregonus sardinella*), 337, 341, 342, 343
least darter (*Etheostoma microperca*), 146
least flycatcher (*Empidonax minimus*), *20*, *299*, 420
least sandpiper (*Calidris minutilla*), 59, *62*
least tern
California (*Sterna antillarum browni*), 56, 74
interior (*Sterna antillarum athalassos*), *54*, *55*, 56
Sterna antillarum, 43, 45, *54*, 56
leatherback (*Dermochelys coriacea*), *119;* 121, 123
Legionella pneumophila, 253
Legionnaires' disease, 253
legless lizard, California (*Anniella pulchra*), *466*
legumes, 201
Leiocephalus
carinatus (carinate curly-tailed lizard), *434*
schreibersii (Schreiber's curly-tailed lizard), *434*
Leiostomus xanthurus (spot), 409
lemon lily (*Lillium parryi*), 320, *320*
leopard frog, 125, 140, 452-53
Chiricahua (*Rana chiricahuensis*), 139, *320*, 327, *328*, 452, 453
northern (*Rana pipiens*), 133-34, 327, *434*
Ramsey Canyon (*Rana subaquavocalis*), 319
relict (*Rana onca*), 327, *328*
Rio Grande (*Rana berlandieri*), 133, 453
Yavapai (*Rana yavapaiensis*), 133, 139, 452-53
Lepidochelys
kempii (Kemp's ridley), *119*, 121, 123
olivacea (olive ridley), *119*
Lepidodactylus lugubris (mourning gecko), 455
Lepidomeda
albivallis (White River spinedace), *143*, *150*
altivelis (pahranagat spinedace), *150*
mollispinis
mollispinis (Virgin spinedace), *150*
pratensis (Big Spring spinedace), *150*
vittata (Little Colorado spinedace), *150*
Lepidoptera, 161, *See also* butterfly; moth
Albany pine barrens, 166
Fourth of July Butterfly Count (FJC), 171-72
inventories, 168-70
light trapping, 169, 170
microlepidoptera, 168
state lists, 169
tall-grass prairie community, 174-76
trends, 168-70
western, 172-74
Lepomis
auritus (redbreast sunfish), *261*, *262*, *432*
cyanellus (green sunfish), 240, *432*
gibbosus (pumpkinseed sunfish), *261*, *262*, *419*, *432*
gulosus (warmouth), *432*
macrochirus (bluegill), 240, *261*, *432*
microlophus (redear sunfish), *432*
Leptodea fragilis (fragile papershell mussel), *180*, *445*
Leptodictyum riparium, 176
Leptonycteris sanborni (Sanborn's long-nosed bat), *320*
Leptoxis, 249, 250
clipeata (agate rocksnail), *251*
compacta (oblong rocksnail), *251*
formanii (interrupted rocksnail), *251*
formosa (maiden rocksnail), *251*
ligata (rotund rocksnail), *251*
lirata (lirate rocksnail), *251*
melanoides (black mudalia), *251*
occultata (bigmouth rocksnail), *251*
showalteri (Coosa rocksnail), *251*
torrefacta, *251*
vittata (striped rocksnail), *251*
Lepus
americanus (snowshoe hare), 359
othus (tundra or Arctic hare), 359
timidus (tundra or Arctic hare), 338, 359
Lepyrium, 249
lespedeza, Virginia (*Lespedeza virginica*), 231
Lespedeza virginica (Virginia lespedeza), 231
lesser goldfinch (*Carduelis psaltria*), *21*
lesser koa-finch (*Rhodacanthis flaviceps*), *376*
lesser nighthawk (*Chordeiles acutipennis*), *19*
lesser scaup (*Aythya affinis*), *19*, *36*, 237
lesser snow goose (*Chen caerulescens caerulescens*), 30, 33
lesser yellowlegs (*Tringa flavipes*), 59, *62*
Leucolepis acanthoneuron, *199*
Leuctra
rickeri (stonefly), *186*
sibleyi (stonefly), 185, *186*
tenuis (stonefly), 185, *186*
levees, habitat change and, 234, 235-36
Liatris cylindracea (dwarf blazingstar), *172*
Liatus, 221
Lichanura trivirgata (rosy boa), *466*
lichen, 194-96
air pollution and, 195, 196, 227
biological roles, 195
declines, 190
endangered species, 195
growth of, 195
overcollecting, 196
in southwestern sky islands, 319
status, 195
trends, 195-96
lichen deserts, 195-96
light trapping, Lepidoptera, 169, 170
Ligumia
nasuta (eastern pondmussel), *180*
recta (black sandshell), *180*
Lilaeopsis schaffneriana recurva, *320*
lilliput (*Toxolasma parvus*), *180*
Lillium parryi (lemon lily), 320, *320*
lily
lemon (*Lillium parryi*), 320, *320*
Nuphar variegata, *247*
limber pine (*Pinus flexilis*), 315
Limnodromus
griseus (short-billed dowitcher), 59, *62*
scolopaceus (long-billed dowitcher), *59*, *62*
Limosa
fedoa (marbled godwit), *19*, *59*, *62*, 396
haemastica (Hudsonian godwit), *59*, *62*
lapponica (bar-tailed godwit), *62*
limpet, freshwater (Ancylidae), 251

Lincoln's sparrow (*Melospiza lincolnii*), *21*
lindane, *408*
lion, mountain (*Felis concolor*), 109, 322
Lioplax
  cyclostomaformis, 251
  showalteri, 251
lirate rocksnail (*Leptoxis lirata*), *251*
Liriodendron tulipifera (yellow-poplar), 228
Lithophane
  georgii (owlet moth), 166
  lepida (owlet moth), 166
  semiusta (owlet moth), 166
  thaxteri (owlet moth), 166
Litoria fallax (eastern dwarf treefrog), *434*
little blue heron (*Egretta caerulea*), *19*
Little Colorado spinedace (*Lepidomeda vittata*), *150*
Little Colorado sucker (*Catostomus* sp.), *150*
little pocket mouse (*Perognathus longimembris*), *466*
live oak
  coast woodlands, *465*
  mourning doves and, 72
liver neoplasm, in coastal fish, 410
liverwort, 197
  floristic inventories, 198, 199-200
lizard, 117
  California legless (*Anniella pulchra*), *466*
  carinate curly-tailed (*Leiocephalus carinatus*), *434*
  Carlia cf. *fusca*, 454
  Coachella Valley fringe-toed (*Uma inornata*), 118, 137-38
  coast horned (*Phrynosoma coronatum*), *466*
  Colorado Valley fringe-toed (*Uma notata*), 137
  common wall (*Podarcis muralis*), *434*
  eastern fence (*Sceloporus undulatus*), 327
  fringe-toed (*Uma* spp.), 137-38
  granite night (*Xantusia henshawi*), *466*
  granite spiny (*Sceloporus orcutti*), *466*
  Hemidactylus frenatus, 454
  Indian monitor (*Varanus indicus*), 455
  Italian wall (*Podarcis sicula*), *434*
  Mojave Valley fringe-toed (*Uma scoparia*), 137
  orange-throated whiptail (*Cnemidophorus hyperythrus*, *466*
  Schreiber's curly-tailed (*Leiocephalus schreibersii*), *434*
  side-blotched (*Uta stansburiana*), 327
  South American ground (*Ameiva ameiva*), *434*
  South American whiptail (*Cnemidophorus lemniscatus*), *434*
  spiny (*Sceloporus* sp.), 326
  Texas horned (*Phrynosoma cornutum*), *434*
  tree (*Urosaurus ornatus*), 327
  western slender glass (*Ophisaurus attenuatus*), 231
  whiptail (*Cnemidophorus* sp.), 326
loach minnow
  Rhinichthys cobitis, *143*, *150*
  Tiaroga cobitis, *320*
Lobaria scrobiculata, 196
loblolly pine (*Pinus taeda*), 223
lobster, West Indies spiny (*Panulirus argus*), 283
locust (insect)
  Carolina (*Dissosteira carolina*), 164
  Rocky Mountain (*Melanoplus spretus*), 164
locust (tree) (*Robinia pseudoacacia*), 167
lodgepole pine (*Pinus contorta*), 315, 316, 421
  altered fire regimes, *222*, 223
Log Creek, atmospheric pollutants, 421-22
loggerhead (*Caretta caretta*), *119*, 121, 122-23
loggerhead shrike (*Lanius ludovicianus*), *20*, 231, *466*
  San Clemente Island (*Lanius ludovicianus mearnsi*), 74
Lomatium peckianum, 208
long-billed curlew (*Numenius americanus*), *19*, *59*, *62*
long-billed dowitcher (*Limnodromus scolopaceus*), *59*, *62*
long-eared owl (*Asio otus*), *67*, *466*
longfin dace (*Agosia chrysogaster*), *150*
longjaw cisco (*Coregonus alpenae*), *143*
longleaf pine (*Pinus palustris*), 17, 223
  ecosystem
    loss of, 129-30
    red-cockaded woodpecker, 87
    reptiles and amphibians, 129-31
long-nosed bat, Sanborn's (*Leptonycteris sanborni*), *320*
Long's bulrush (*Scirpus longii*), 458
longspur
  chestnut-collared (*Calcarius ornatus*), *21*, 299, *303*
  McCown's (*Calcarius mccownii*), *21*, 297
long-term trends, detecting, 4
long-toed salamander, Santa Cruz (*Ambystoma macrodactylum croceum*), 124
long-tongued bat, Mexican (*Choeronycteris mexicana*), *320*
Lonicera dioica, 167
Looe Key National Marine Sanctuary, 281
loon
  Arctic (*Favia arctica*), *420*
  common (*Gavia immer*), *420*
loosestrife, purple (*Lythrum salicaria*), 249, 429, 443, 444, 458-59
Lophodytes cucullatus (sea duck), 35
Los Angeles Zoo, 81
Los Padres National Forest, 80
Lost River sucker (*Deltistes luxatus*), *143*
Louisiana
  coastal wetland loss, 270-71
  seagrass distribution, 274
Louisiana bear (*Ursus americanus luteolus*), 102
Louisiana waterthrush (*Seiurus motacilla*), 390, 391
lovegrasses (*Eragrostis* spp.), 321
Lower Apalachicola hydrologic region, 145
Lower Mississippi Basin hydrologic region, 145
Lower Mississippi River Alluvial Floodplain (LMRAF), wetland losses, 216-17
Lower Mobile Basin hydrologic region, 145
Loxia
  curvirostra (red crossbill), *21*
  leucoptera (white-winged crossbill), *21*
Loxioides bailleui (palila), 376, *377*
Loxops coccineus
  caeruleirostris (Kauai 'akepa), *378*
  coccineus (Hawaii 'akepa), *378*
  ochraceus (Maui 'akepa), *378*
  rufus (Oahu 'akepa), *376*
lumber, importing, non-native species and, 428
lupine (*Lupinus perennis*), 167
Lupinus perennis (lupine), 167
Luquillo Mountains, Puerto Rican parrot, 84, 85
Lutjanus
  analis (mutton snapper), 282-83
  campechanus (red snapper), 282
  griseus (gray snapper), 282-83
Lutra canadensis (river otter), 414
Luxilus cornutus (common shiner), *261*, 263
Lycaeides melissa samuelis (Karner blue butterfly), 166, 167, *172*, 224-25, 232, *398*, 401
lycopod (*Huperzia haleakalae*), 206
Lymantria dispar (gypsy moth), 441
Lynx rufus (bobcat), 193
Lythrum salicaria (purple loosestrife), 249, 429, 443, 444, 458-59

Macbridea alba (white birds-in-a-nest), 401
McCown's longspur (*Calcarius mccownii*), *21*, 297
MacGillivray's warbler (*Oporornis tolmiei*), *20*
machete (*Elops affinis*), *150*
mackerel, king (*Scomberomorus cavalla*), 282
Macrhybopsis aestivalis tetranemus (Arkansas River speckled chub), 295, 303-5
Macroclemys temminckii (alligator snapping turtle), 119
Macrocystis pyrifera (giant kelp), 255
macrofungi, 192-94
  biological role of, 192
  declines, 190, 193
  fruiting, 193-94
  species diversity, 192
macrophytic algae, 255
madtom catfish
  Noturus spp., 145
  pygmy (*Noturus stanauli*), *143*
  Scioto (*Noturus trautmani*), 146
magnolia warbler (*Dendroica magnolia*), *20*
Magnuson Fisheries Conservation and Management Act (MMPA), 94-96
magpie, black-billed (*Pica pica*), *20*
maiden rocksnail (*Leptoxis formosa*), *251*
Malaclemys terrapin (diamondback terrapin), *119*, 120
malaria (*Plasmodium relictum*), 378
mallard (*Anas platyrhynchos*), *19*, 35, *36*, 266, 296, 299, 300-301, *303*, 379, 402, 406, *413*
Mallotus villosus (capelin), 51
mammals, 93-116
  habitat changes, 93-94
  inventory and monitoring programs, 93
  marine, 94-96
  in southwestern sky mountains, 321

mamo
black (*Drepanis funerea*), *376*
Hawaii (*Drepanis pacifica*), *376*
managed populations
anadromous fish, 147-48
freshwater fish, 141, 147-48
manatee (*Trichechus manatus*), 259
manatee grass (*Syringodium filiforme*), 273, 274, 276, 277
Manduca blackburni (sphinx moth), 367
mange mite (scabies), *407*
Mannia fragrans, 199
mantidfly (*Mantispa interrupta*), *161*
Mantispa interrupta (mantidfly), *161*
Manual of the Flowering Plants of Hawaii, 201-2
Manual of the Grasses of the United States, 201
Manual of the Vascular Flora of the Carolinas, The, 204
Manual of Vascular Plants of Northeastern United States and Adjacent
Canada, 203
Manual of the Vascular Plants of Texas, 203
manzanita chaparral, Upper Sonoran, *465*
maple
dieback, acid deposition (acid rain) and, 420
red (*Acer rubrum* var *drummondii*), 216
mapleleaf (*Quadrula quadrula*), *180*
mapping, wetlands, 474
MAPS. *See* Monitoring Avian Productivity and Survivorship (MAPS)
map turtle
Barbour's (*Graptemys barbouri*), *119*, *120*
Cagle's (*Graptemys caglei*), *119*
ringed (*Graptemys oculifera*), *119*
yellow-blotched (*Graptemys flavimaculata*), *119*
marbled godwit (*Limosa fedoa*), *19*, *59*, *62*, 396
marbled murrelet (*Brachyramphus marmoratus*), 43, 47, 50
Margarops fuscatus (pearly-eyed thrasher), 85
Marginatus group, species *Melanoplus*, 163, 164
Mariana flying fox (*Pteropus mariannus*), 455
Marianas blue-tailed skink (*Emoia caerulocauda*), 455
Marianas crow (*Corvus kubaryi*), 455
marine algae, 255-56
unicellular, 255
marine ecosystems. *See* coastal and marine ecosystems
Marine Mammal Protection Act of 1972, 351, 356
California sea otter, 110
marine mammals, 94-96
Alaska, 338
dolphins and porpoises, 96
seals and sea lions, 96
whales, 95-96
Marine Protection, Research, and Sanctuaries Act (Public Law 101-605), 279
marine protozoa, 252
marine toad (*Bufo marinus*), 433, *434*, 435, 441
marine turtles, 119, 120
protection of, 123
in the Southeast, 121-23
status and trends, 121-22
Maritimes Shorebird Surveys (MSS), 58
marsh wren (*Cistothorus palustris*), *299*
martin, purple (*Progne subis*), *20*, *391*
Maryland darter (*Etheostoma sellare*), 146
massasauga rattlesnake, eastern (*Sistrurus catenatus*), 231
Masticophis taeniatus (striped whipsnake), 327
Maui 'akepa (*Loxops coccineus ochraceus*), *378*
Maui 'amakihi (*Hemignathus virens wilsoni*), *377*
Maui creeper (*Paroreomyza montana newtoni*), *378*
Maui nukupu'u (*Hemignathus lucidus affinus*), *377*
Maui parrotbill (*Pseudonestor xanthrophrys*), *377*
mayfly, 183, *419*
burrowing
Hexagenia, 181, 236
bilineata, 183-84
limbata, 183-84
Pentagenia, *vittigera*, 183
meadow, subalpine, 314-17
meadowlark
eastern (*Sturnella magna*), *21*, *22*, 297
western (*Sturnella neglecta*), *21*, *22*, 297, 299, 303, *303*
winter population trends, 21
Meadow Valley speckled dace (*Rhinichthys osculus* ssp.), *150*
Meadow Valley sucker (*Catostomus clarki* ssp.), *150*
mealybug, citophilus (*Pseudococcus calceolariae*), *160*
Meda fulgida (spikedace), *150*, *320*
Mediterranean house gecko (*Hemidactylus turcicus*), *434*
medusahead (*Taeniatherum caput-medusae*), 222
Meeteetse prairie dog complexes, 106, 107
Megalagrion
nesiotes (damselfly), 367
nigrohamatum nigrolineatum (damselfly), 367
pacificum (damselfly), 362
Megalopyge crispata (flannel moth), 166
Melamprosops phaeosoma (po'o-uli), *378*
Melanerpes
aurifrons (golden-fronted woodpecker), *20*
carolinus (red-bellied woodpecker), *20*
erythrocephalus (red-headed woodpecker), *20*, 231
formicivorus (acorn woodpecker), *20*
Melanitta
fusca (white-winged scoter), *36*
nigra (black scoter), *36*
perspicillata (surf scoter), *36*
Melanoplus, 163-65
bivittatus (two-striped grasshopper), 164
Marginatus group, 163, 164
Montanus group, 165
sanguinipes (migratory grasshopper), 164
spretus (Rocky Mountain locust), 164
Meleagris
gallopavo (wild turkey), 16, 70-71, 231, 387, 405, *407*
merriami (turkey), 320
mexicana (turkey), 320
ocellata (ocellated turkey), 70
Melopsittachus undulatus (budgerigar), 439
Melospiza
georgiana (swamp sparrow), *21*
lincolnii (Lincoln's sparrow), *21*
melodia (song sparrow), *21*, *22*, *299*
Menidia
conchorum (key silverside), 281
menidia (Atlantic silverside), 261, *262*
Menippe mercanaria (stone crab), 282
Mentasta caribou herd (*Rangifer tarandus*), 338, 357-58
Mentha pipertia, 176
Mephitis mephitis (striped skunk), 302
mercury
in coastal sediments, 409
fish contamination, 414, 415
in Illinois cave streams, 177
in mollusks, 409, *410*
sources, 414
merganser
common (*Mergus merganser*), 35, *36*, *266*, 389, *420*
hooded, *36*
population, 35-36
red-breasted (*Mergus serrator*), *19*, 35, *36*
Mergus
merganser (common merganser), 35, *36*, *266*, 389, *420*
serrator (red-breasted merganser), *19*, 35, *36*
merlins (*Falco columbarius*), *67*
Merriam, Dr. C., 8, 9
Merriam's kangaroo rat (*Dipodomys merriami*), *466*
mesoscale rainfall, forest cover loss and, 472
Metarrhanthis apiciaria, *166*
Meteorus laphygmae, 366
methamidiphos, *417*
methiocarb, *417*
Metrosideros polymorpha ('Ohi 'a lehua), 366
Mexican chickadee (*Parus sclateri*), 321
Mexican garter snake (*Thamnophis eques*), *320*, 433, 435, 452-53
Mexican long-tongued bat (*Choeronycteris mexicana*), *320*
Mexican pronghorn (*Antilocarpa americana mexicana*), 318
Mexican rough-footed mud turtle (*Kinosternon hirtipes*), *119*
Mexican spiny-tailed iguana (*Ctenosaura pectinata*), *434*
Mexican spotted owl
Strix occidentalis lucida, *67*, 309
in Colorado Plateau canyonlands, 330-31
historical records, 331
surveys, 330-31
Strix occidentalis mexicanus, *320*
Mexican stoneroller (*Campostoma ornatum*), *320*
Mexican wolf (*Canis lupus baileyi*), 98, 99, 318, *320*
Mexican Wolf Recovery Plan, 98
Michigan Flora, 203
Micranthemum micranthemoides (Nuttall's mudwort), 208
Micrathene whitneyi (elf owl), *67*
microbes, 6

microfungi, 190-91
  diversity of, 191
  Internet information, 191
  survey and inventory of needs, 191
microlepidoptera, 168
Micronesian kingfisher (*Halcyon cinnamomina*), 455
Micronesian starling (*Aplonis opaca*), 455
Micropogonias undulatus (Atlantic croaker), 409, *411*
Micropterus
  dolomieu (smallmouth bass), 147, *261*, *419*, *432*
  punctulatus (spotted bass), 147, *432*
  salmoides (largemouth bass), 148, 240, *261*, *432*
    floridanus (Florida largemouth bass), 148
    salmoides (northern largemouth bass), 148
  treculi (Guadalupe bass), 147
microsporidia, 253
Middleton Island
  kittiwakes, 50
  murres, 50
  seabirds, 51
midge, 182, 183
Midwest, acid lakes, 419
midwestern United States, rare terrestrial ecological communities, 220
Midwinter Survey, ducks, 34-35
Midwinter Waterfowl Inventories (MWI), 40
migratory birds, 15-17
  avian diseases and, 403-4
  climate change and, 391
  determining status and trends, 15-17
  inventory and monitoring programs, *5*
  Mississippi River navigation dams and, 237
  neotropical, 18
  non-native species, 439
  northern plains, Conservation Reserve Program (CRP) and, 302-4
  population changes, North Dakota, 298-300
  shorebirds, 61
  short-distance, 18
  surveys, *5*
Migratory Bird Treaty, 27, 57
migratory grasshopper (*Melanoplus sanguinipes*), 164
Miksch, George, Avian Research Center, 81
mildew, 190-91
milk-vetch
  Astragalus cobrensis var. *maguirei*, *320*
  Astragalus hypoxylus, *320*
milkweed
  climbing (*Cynanchum wigginsii*), *320*
  purple (*Asclepias purpurascens*), 231
millerbird
  Laysan (*Arcrocephalus f. familiaris*), *376*
  Nihoa (*Acrocephalus familiaris kingi*), *377*
Mimus polyglottos (northern mockingbird), *20*, *22*
mink (*Mustela vison*), 236, 237-38, 414
  canvasback predation, 41
minnow, 145
  acid-sensitive, 419
  fathead (*Pimephales promelas*), *261*, *432*
  loach
    Rhinichthys cobitis, *143*, *150*
    Tiaroga cobitis, *320*
  plains (*Hybognathus placitus*), 304
mirex, *408*
Mission to Plant Earth, Earth Observing System, 473
Mississippi Flyway
  avian cholera, 402
  canvasbacks, 40, 42
  northern pintails, 38-39
Mississippi Gulf Coast, seagrass distribution, 274
Mississippi kite (*Ictinia mississippiensis*), *67*, 297
Mississippi River, aquatic insects, 183-84
Mississippi sandhill crane (*Grus canadensis pulla*), 16-17, 75-77
  Jackson County, Mississippi population, 76
  reintroductions of, 76
Mississippi Sandhill Crane National Wildlife Refuge, 75
Missouri River, riparian ecosystems, 286-88
mistletoe, dwarf (*Arceuthobium americanum*), 223
mixed chaparral, southern, *465*
mixed forest, 469
  conifer, 421-23
  mesophytic, 472
mixed grasslands, 471-72
Mniotilta varia (black-and-white warbler), *20*
moa (red junglefowl, *Gallus gallus*), 372-73
Moapa coriacea (Moapa dace), *150*
Moapa dace (*Moapa coriacea*), *150*
Moapa springfish (*Crenichthys baileyi moapae*), *150*
Mobile Bay basin
  coastal wetland loss, 271
  freshwater gastropod fauna, 249-52
mockingbird, northern (*Mimus polyglottos*), *20*, *22*
Moho
  apicalis (Oahu 'o'o), *376*
  bishopi (Bishop's 'o'o), *377*
  braccatus (Kauai 'o'o), *377*
  nobilis (Hawaii 'o'o), *376*
Mohonk Preserve, 390, 391
Mojavean pinon woodlands, *465*
Mojave Desert, desert tortoise (*Gopherus agassizii*) in, 135-36
Mojave Valley fringe-toed lizard (*Uma scoparia*), 137
mold, 190-91
mollusks
  coastal, contaminants in, 408-11
  mercury concentrations in, 409, *410*
  non-native species, 178, *429*
  in southwestern sky islands, 319
Molokai creeper (*Paroreomyza flammea*), *378*
Molothrus
  aeneus (bronzed cowbird), *21*
  ater (brown-headed cowbird), *21*, 25, 90, 91, 299, *303*, 389
  bonariensis (shiny cowbird), 439
monarch butterfly (*Danaus plexippus*), *171*, 172
mongoose (*Herpestes auropunctatus*), 378, 441
Monitoring Avian Productivity and Survivorship (MAPS), 16, 24
  breeding productivity, 25-26
  data provided by, 24
  habitat-specific differences, 25
monitoring programs. *See* inventory and monitoring programs
monitor lizard, Indian (*Varanus indicus*), 455
Monkey Springs pupfish (*Cyprinodon* sp.), *150*
monk parakeet (*Myiopsitta monachus*), 439
monk seal, Hawaiian, *95*
monobutyltin, *408*
monocrotophos, *417*
Mono Lake, California gull, 45
moorhen, Hawaiian (*Gallinula chlorpus sandwicensis*), *377*
Moorman springfish (*Crenichthys baileyi thermophilus*), *143*, *150*
moose (*Alces alces*), 310, *311*, 338, 347, 348
morel, 192
Morone
  americana (white perch), 261, *262*, *432*, 443
  chrysops (white bass), *261*, 263, *432*
  saxatilis (striped bass), 261, *262*, 263, 265, *432*
mortality, organophosphorus and carbamate pesticides and, 416-18
mosquito (*Culex*), 378, 380
mosquitofish, western (*Gambusia affinis*), *432*, 441
Mosquito Lagoon, Florida, turtles, 121, 122
moss, 197
  floristic inventories, 198-99
  Leptodictyum riparium, 176
  Leucolepis acanthoneuron, *199*
  New York, 209-10
  Ozobryum ogalalense, 198
  peat (*Sphagnum*), 198
  peat (*Sphagnum angermanicum*), 210
  reindeer (*Cladinia rangiferina*), 195
moth, 161, *See also* Lepidoptera
  buck (*Hemileuca maia*), 232
  flannel (*Megalopyge crispata*), 166
  giant silkworm (*Citheronia imperialis*), 166
  gypsy (*Lymantria dispar*), 441
  habitat change and, 160
  Hawaii, 366
  hawkmoths (Sphingidae), 172, *173*
  inventories, 168-70
  leaf miner (*Acrocercops arbutella*), *160*
  owlet
    Acronicta lanceolaria, 166
    Agroperina lutosa, 166
    Agrotis stigmosa, 167
    Argyrostrotis quadrifiliaris, 166
    Catocala pretiosa, 166
    Chaetaglaea cerata, 167
    Eugraphe subrosea, 166
    Homohadena badistriga, *166*, 167
    Lithophane
      georgii, 166

lepida, 166
semiusta, 166
thaxteri, 166
Platypolia anceps, 166
Psectraglaea carnosa, *166*, 167
Pyreferra ceromatica, 166
Xestia (Anomogyna) badicollis, 166
Xylena
cineritia, 166
thoracica, 166
Xylotype capax, 166
silkmoths (Saturniidae), 172, *173*
sky islands, 172-73
sphinx
Darapsa versicolor, 166
Hemaris gracilis, 166
Manduca blackburni, 367
Proserpinus juanita, *174*
tiger (Arctiidae), 172, *173*
western, 172-74
white phlox flower (*Schinia indiana*), 232
mottled duck (*Anas fulvigula*), *19*, 35, *36*
mottled sculpin (*Cottus bairdii*), *150*
mountain avens (*Geum peckii*), 394
mountain bluebird (*Sialia currucoides*), *20*
mountain chickadee (*Parus gambeli*), *20*, 321
mountain hemlock (*Tsuga mertensiana*), 315
mountain lion (*Felis concolor*), 109, 322
mountain plover (*Charadrius montanus*), 58, *59*, *62*, 63, 297, 298
mountain quail (*Oreortyx pictus*), *19*
mountain redbelly dace (*Phoxinus oreas*), *146*
mountain sucker (*Catostomus platyrhynchus*), *150*
mountain treefrog (*Hyla eximia*), 327
mountain whitefish (*Prosopium williamsoni*), *150*
mountain yellow-legged frog (*Rana muscosa*), 133, *134*, *466*
Mount Desert Island, Maine, insect survey, 161
Mount Saint Helens, Landsat Multispectral Scanner (MSS) images, 470-71
mourning dove (*Zenaida macroura*), 16, *19*, 71-73, *299*, 387
Mourning Dove Callcount Survey, 71
mourning gecko (*Lepidodactylus lugubris*), 455
mourning warbler (*Oporornis philadelphia*), *20*, 391
mouse
deer (*Peromyscus maniculatus*), 109
house (*Mus musculus*), 454
little pocket (*Perognathus longimembris*), *466*
San Diego pocket (*Perognathus fallax*), *466*
white-eared pocket (*Perognathus alticola*), *466*
Moxostoma lacerum (hairlip sucker), 146
Mozambique tilapia (*Tilapia mossambica*), *432*
MRLC. *See* Multi-Resolution Land Characteristics (MRLC)
MSS (Landsat Multispectral Scanner), 470-71, 472
Mt. Graham red squirrel (*Tamasciuris hudsonicus grahamensis*), 320
mucket (*Actinonaias ligamentina [carinata]*), *180*
mudalia, black (*Leptoxis melanoides*), *251*
mudminnow, central (*Umbra limi*), 263
mud turtle
Mexican rough-footed (*Kinosternon hirtipes*), *119*
yellow (*Kinosternon flavescens*), *119*
mudwort, Nuttall's (*Micranthemum micranthemoides*), 208
Mugil cephalus (striped mullet), *150*
mullein (*Verbascum thapsus*), 364
mullet, striped (*Mugil cephalus*), *150*
Multi-Resolution Land Characteristics (MRLC), 461, 463, 467, 468
murre, 16
in Alaska, 50
common (*Uria aalge*), 46-47, 50, *51*
thick-billed (*Uria lomvia*), 50, *51*
Uria spp., 49
murrelet, 16
ancient (*Synthliboramphus antiquus*), 46, 51
Kittlitz's (*Brachyramphus brevirostris*), 50
marbled (*Brachyramphus marmoratus*), 43, 47, 50
Xantus' (*Synthliboramphus hypoleucus*), *46*, 47-48
muscovy duck (*Cairina moschata*), 437-38
Musculium transversum (fingernail clam), 236-37, 238
mushroom, 192
muskellunge
Esox masquinongy, *432*
tiger (*Esox lucius x masquinongy*), *432*
muskox (*Ovibos moschatus*), 337, 338-40
musk turtle, flattened (*Sternotherus depressus*), *119*
Mus musculus (house mouse), 454
mussel, 160, *See also* freshwater mussels
black sandshell (*Ligumia recta*), *180*
contaminants in, 408
creek heelsplitter (*Lasmigona compressa*), *180*
cylindrical papershell (*Anodontoides ferussacianus*), *180*
deertoe (*Truncilla truncata*), *180*
eastern lampmussel (*Lampsilis siliquoidea*), *180*
eastern pondmussel (*Ligumia nasuta*), *180*
elktoe (*Alasmidonta marginata*), *180*
fawnsfoot (*Truncilla donaciformis*), *180*
fluted-shell (*Lasmigona costata*), *180*
fragile papershell (*Leptodea fragilis*), *180*, *445*
freshwater, 177-82
giant floater (*Pyganodon grandis*), *180*
Great Lakes zebra species (*Dreissena* spp.), 409
hickorynut (*Obovaria olivaria*), *180*
kidneyshell (*Ptychobranchus fasciolaris*), *180*
lilliput (*Toxolasma parvus*), *180*
mapleleaf (*Quadrula quadrula*), *180*
mucket (*Actinonaias ligamentina [carinata]*), *180*
Mytilus
californianus, 408-9
edulis, 408-9
trossulus, 408-9
northern riffleshell (*Epioblasma torulosa rangiana*), *180*
Ohio pigtoe (*Pleurobema cordatum*), *180*
paper pondshell (*Utterbackia imbecillis*), *180*
pimpleback (*Quadrula pustulosa pustulosa*), *180*
pink heelsplitter (*Potamilus alatus*), *180*
pink papershell (*Potamilus ohiensis*), *180*
pocketbook (*Lampsilis ovata*), *180*
polychlorinated biphenyls (PCBs) in, 410
pondhorn (*Uniomerus tetralasmus*), *180*
Potamilus alatus, *181*, *182*
purple wartyback (*Cyclonaias tuberculata*), *180*
rainbow (*Villosa iris*), *180*
rayed bean (*Villosa fabalis*), *180*
round hickorynut (*Obovaria subrotunda*), *180*
round pigtoe (*Pleurobema coccineum*), *180*
salamander (*Simpsonaias ambigua*), *180*
slippershell (*Alasmidonta viridis*), *180*
snuffbox (*Epioblasma triquetra*), *180*
southern combshell (*Epioblasma penita*), *178*
spike (*Elliptio dilatata*), *180*
squawfoot (*Strophitus undulatus*), *180*
threehorn wartyback (*Obliquaria reflexa*), *180*
threeridge (*Amblema plicata plicata*), *180*
Wabash pigtoe (*Fusconaia flava*), *180*
wavy-rayed lampmussel (*Lampsilis fasciola*), *180*
white heelsplitter (*Lasmigona complanata complanata*), *180*
zebra (*Dreissena polymorpha*), 42, 160, 178, 181, 237, 255, 427, 428, 429, 430, 444, 445-47
Mussel Watch Project, 408
Mustela
nigripes (black-footed ferret), 106-8
putorius (European ferret), 321
vison (mink), 41, 236, 237-38, 414
mute swan (*Cygnus olor*), 266, 387, 437, 438
mutton snapper (*Lutjanus analis*), 282-83
Myadestes
lanaiensis
lanaiensis (oloma'o), *376*
rutha (oloma'o), *377*
myadestinus (kama'o), *377*
oahuensis ('amaui), *376*
obscurus ('oma'o or Hawaii thrush), *377*, 380
palmeri (puaiohi or small Kauai thrush), *377*, 380
mycelium, 192
Mycoplasma agassizii, 136, *407*
mycorrhizae, 192, 193, 194
mycotoxicosis, *404*
Mycteria americana (wood stork), *54*, 55
Myiarchus
cinerascens (ash-throated flycatcher), *20*
crinitus (great crested flycatcher), *20*
tyrannulus (brown-crested flycatcher), *20*
Myiopsitta monachus (monk parakeet), 439
myocardiopathy, *404*
Myocastor coypus (nutria), *440*
Myotis
austroriparius (southeastern bat), 176
grisescens (gray bat), 176
sodalis (Indiana bat), 97-98, 176

Myriophyllum, spicatum (Eurasian watermilfoil), 237, 430, 443, 444
myrtle warbler (*Dendroica coronata*), *20*
Mytilicola orientalis), 429
Mytilus
  californianus, 408-9
  edulis, 408-9
  trossulus, 408-9

NABA (North American Butterfly Association), 171
NADP (National Atmospheric Deposition Program), 228
Nankoweap Creek, wintering bald eagles (*Haliaeetus leucocephalus*), 328, 329
narrow-headed garter snake (*Thamnophis rufipunctatus*), 327, *328*
narrow-mouthed toad, Great Plains (*Gastrophyrne olivacea*), *320*
NASA. *See* National Aeronautics and Space Administration (NASA)
Nashville warbler (*Vermivora ruficapilla*), *20*
Nassau grouper (*Epinephelus striatus*), 282
National Academy of Sciences, 456
National Aeronautics and Space Administration (NASA)
  Landsat Multispectral Scanner (MSS), 470-71, 472
  Landsat satellite mapping, 463, 467-68, 473
  Mission to Planet Earth, Earth Observing System, 473
National Air Quality Research Program, National Biological Service (NBS), 228
National Atmospheric Deposition Program (NADP), 228
National Audubon Society, *5*
  Christmas Bird Count (CBC), 15, 17, 21-23, 50, 53, 58, 64, 386, 389, 437
  colonial waterbird protection, 53
National Benthic Surveillance Project, 408
National Biological Service (NBS), 4, *5*, 271, 385
  amphibian population monitoring, 118
  Gap Analysis Program (GAP), 462, 465, 467, 473
  Great Lakes Science Center, 242
  history of, 7-9
  name change, 9
  National Air Quality Research Program, 228
  non-native species monitoring, 431
  Southeastern Biological Science Center (SBSC), 445
  Southern Science Center, 270, 271, 276
  wetlands definition, 474
National Biological Survey (NBS)
  created, 9
  name change, 9
National Contaminant Biomonitoring Program, 413
National Council of the Pulp and Paper Industry for Air and Stream Improvement, 217
National Ecology Research Center, 425
National Gap Analysis Program (GAP). *See* Gap Analysis Program (GAP)
National Insect Collection, National Museum of Natural History, 162
National List of Plant Species That Occur in Wetlands, 474
National Marine Fisheries Service (NMFS), *5*, 94-96
National Migratory Bird Harvest Information Program, 71
National Museum of Natural History, National Insect Collection, 162
National Natural History Museum, protozoa, 253
National Oceanic and Atmospheric Administration (NOAA), *5*, 94, 96, 269-70
  Advanced Very High Resolution Radiometer (AVHRR), 469, 473
  Coastal Change Analysis Program (C-CAP), 468
  National Benthic Surveillance Project, 408
  National Status and Trends (NS&T) Program, 408-11
  Sanctuaries and Reserves Division, 279
national parks
  air quality in, 228
  non-native species, 440
  vegetation change in, 224-28
National Park Service (NPS), 113, 400
National Park Service (NPS) Organic Act, 228
National Science Foundation (NSF), 193
National Status and Trends (NS&T) Program, 408-11
  Mussel Watch Project, 408
National Tropical Botanical Garden, 202
National Water Quality Assessment (NAWQA), 258, 467
National Wetlands Inventory, 4, *5*, 207, 270, 290, 462, 463, 474, *475*
  wetland functions and values, *475*
National Wetlands Research Center, 276
National Wildlife Health Center (NWHC), 401-4
national wildlife refuge system, dabbling ducks and, 37
National Wild Turkey Federation, 71
Native Americans
  Inupiat Eskimos, 341
  whale harvesting rules, 95
  Yupik Natives, 32
Native Fish Work Group, 326
Natural Heritage Network, *5*, 218, 219, 399, 400
  Central Scientific Databases, 206
Natural Heritage Programs, 200, 206-7, 398
Natural Resource Conservation Service, *5*
Nature Conservancy, 4, 118, 119, 190, 206-8, 392, 398, 399, 463
  science regions, 219
Nature Conservancy/Natural Heritage Network classification system, 219
navigation channels, wetland loss and, 270
navigation dams, habitat change and, 234-36
NAWQA. *See* National Water Quality Assessment (NAWQA)
NBS. *See* National Biological Service (NBS); National Biological Survey (NBS)
necrotic enteritis, *404*
needlegrass grasslands, valley, *465*
needleleaf forest, 469
Nemocapnia carolina (stonefly), 185, *186*
Nemoura trispinosa (stonefly), *186*
nene (Hawaiian goose) (*Branta sandvicensis*), *377*, 379
Neoperla clymene (stonefly), *186*
Neoplanorbis, 249, 250
  carinatus, *251*
  smithi, *251*
  tantillus, *251*
  umbilicatus, *251*
Neotropical migrants, 298
  brown-headed cowbird (*Molothrus ater*) and, 299
  climate change and, 391
  population trends, 18, 89
Nesiomiris (leaf bug), 367
nesting behavior, Arctic geese, global climate change and, 388
nest success
  birds, 25
  canvasbacks, 41
  ducks, Prairie Pothole region, 36-37, 300-302
Neviusia spp. (snow-wreath), 202
Newcastle disease, *404*, *407*
Newell's shearwater (*Puffinus newelii*), *377*, 379
New Jersey tea (*Ceanothus americanus*), 231
News of The Lepidopterists' Society, 168
newt, striped (*Notophthalmus perstriatus*), 129, 130
New York
  moss, 209-10
  plant waifs, 209
  vascular plants, 209-10
New York State Geological and Natural History Survey, 209
New York State Heritage Program, 210
New York State Museum (NYSM), 209, 260, 261
Nezara viridula (southern green stink bug), 366
Nicrophorus americanus (American burying beetle), 232
nighthawk
  common (*Chordeiles minor*), *19*
  lesser (*Chordeiles acutipennis*), *19*
night-heron, black-crowned (*Nycticorax nycticorax*), *54*, 55
night-light counts, of American alligator (*Alligator mississippiensis*) in, 127
night lizard, granite (*Xantusia henshawi*), *466*
Nihoa finch (*Telespyza ultima*), *377*
Nihoa millerbird (*Acrocephalus familiaris kingi*), *377*
nitrogen
  atmospheric deposition of, 421-22
  deposition, subalpine tree growth and, 316
  fertilizer, 424
NOAA. *See* National Oceanic and Atmospheric Administration (NOAA)
noddy, Hawaiian (*Anous minutus melanogenys*), *377*, 379
Nonindigenous Aquatic Nuisance Prevention and Control Act of 1990, 428, 445
nonindigenous species. *See* non-native species
non-native species, 427-59
  Africanized bees, 448-51
  algae, 444
  American Somoa, 433
  amphibians, 433-36
  aquarium industry and, 429
  aquatic, 428-30, 443

ballast water discharge and, 429
benefits of, 428, 431, 440
biodiversity and, 152
birds, 379, 437-39
brown tree snake (*Boiga irregularis*), 433, *434*, 435, 436, 454-56
bullfrogs (*Rana catesbeiana*), 452-53
Channel Islands, 441-42
Colorado River Basin, 151
controlling, 326, 430
costs of, 427, 428, *429*, 431, 440, 445-47
diseases and, 428
distribution of, 435, 440-43
Dropsophila (Pomance flies) and, 369
early detection of, 430
effects of, 440-42
extinctions and, 427
fish, 153-54, *429*, 431-33, 441
future of, 433
game birds, 438
Glacier National Park, 153-54
grasslands, *465*
Great Lakes, 428, 442-47
Guam, 454-56
Hawaii, 361-62, 364, 365, 366-67, 370, 379, 428, 437, 441-42, 455-56
insects, 365, 366-67, 379, *429*
introduction and dispersal of, 429
inventory and monitoring programs, 431-32
lumber and, 428
mollusks, 178, *429*
monitoring, 430, 431-32
national parks, 440
number of, 431
Oregon, 428
overview, 427-28
pig (*Sus scrof*), 362, 370, 371, 440, 441
plant pathogens, *429*
plants, 190, 394, *429*, 444, 458-59
on public lands, 440-42, 456-57
Puerto Rico, 439
purple loosestrife (*Lythrum salicaria*), 458-59
for recreational fishing, 325-26
recreational fishing, 428, 443
reptiles, 433-36
research strategies, 429-30
riparian ecosystems, 286, 288-89
San Francisco Bay, 428
snails, 429
Snake River, 429
southwestern sky islands, 321
terrestrial vertebrates, *429*
translocation, 405, 432-33
trout, northern leopard frog and, 133
vascular plants, 190, 203, 206
weeds, 394
wild horses and burros on public lands, 456-57
zebra mussel (*Dreissena polymorpha*), 160, 445-47
North American Breeding Bird Survey. *See* Breeding Bird Survey (BBS)
North American Butterfly Association (NABA), 171
North American Fauna series, 8
North American Waterfowl Management Plan, 37, 217
canvasback population goal, 40
Gulf Coast Joint Venture, 217
North Dakota, migratory birds, 298-300
northeastern United States, climate change, 390-91
northern bobwhite (*Colinus virginianus*), *19*, 231, 424
northern cardinal (*Cardinalis cardinalis*), *20*, *22*, 387
northern catalpa (*Catalpa speciosa*), 206
Northern Continental Divide, grizzly bear, 103-5
northern flicker (*Colaptes auratus*), 231
northern flying squirrel (*Glaucomys sabrinus*), 192, 193
northern fur seal (*Callorhinus ursinus*), *95*, 96
northern goshawk (*Accipiter gentilis*), *66*, 67, *320*
northern harrier (*Circus cyaneus*), *67*, 387
northern hawk owl (*Surnia ulula*), *67*
northern hog sucker (*Hypentelium nigricans*), *261*, 263
northern juniper woodlands, *465*
northern largemouth bass (*Micropterus salmoides salmoides*), 148
northern leopard frog (*Rana pipiens*), 133-34, 327, *434*
northern mockingbird (*Mimus polyglottos*), *20*, *22*
northern parula (*Parula americana*), *20*
northern pike (*Esox lucius*), 261, *432*
northern pintail (*Anas acuta*), 16, *19*, 35, 266, 296, *299*, 300, 301, 389
breeding population, *4*
decline of, 38-39
wetland availability and, 36
northern pygmy owl (*Glaucidium gnoma*), *67*
northern red-legged frog (*Rana aurora aurora*), 132
northern riffleshell (*Epioblasma torulosa rangiana*), *180*
northern right whale, *94*, *95*
northern Rocky Mountain wolf (*Canis lupus irremotus*), 98
Northern Rocky Mountain Wolf Recovery Plan, 98
northern rough-winged swallow (*Stelgidopteryx serripennis*), *20*
northern saw-whet owl (*Aegolius acadicus*), *67*
northern sea lion (Steller sea lion) (*Eumetopias jubatus*), 52, *95*, 96
northern shoveler (*Anas clypeata*), *19*, 35, *36*, *266*, *299*, 300, 301
northern spotted owl (*Strix occidentalis caurina*), *67*, 193, 387
northern waterthrush (*Seiurus noveboracensis*), *20*
Norway spruce (*Picea abies*), 315
Nosema, 253
Notemigonus crysoleucas (golden shiner), *261*, *262*, *432*
Notophthalmus perstriatus (striped newt), 129, 130
Notropis
amoenus (comely shiner), *261*, 263
atherinoides (emerald shiner), 240, 261, *262*, 443
bifrenatus (bridle shiner), *261*, *262*, 263
girardi (Arkansas River shiner), 295, 303-5
hudsonius (spottail shiner), 261, *262*
mekistocholas (Cape Fear shiner), *143*
melanostomus (blackmouth shiner), *143*
Noturus, 145
stanauli (pygmy madtom), *143*
trautmani (Scioto madtom), 146
NS&T (National Status and Trends (NS&T) Program), 408-11
nukupu'u
Kauai (*Hemignathus lucidus hanapepe*), *377*
Maui (*Hemignathus lucidus affinus*), *377*
Oahu (*Hemignathus l. lucidus*), *376*
Numenius
americanus (long-billed curlew), *19*, *59*, *62*
borealis (Eskimo curlew), *62*, 63
phaeopus (whimbrel), 59, *62*
tahitiensis (bristle-thighed curlew), *62*, 64
Nuphar variegata (lily), *247*
nuthatch
brown-headed (*Sitta pusilla*), *20*, *22*
red-breasted (*Sitta canadensis*), *20*
nutria (*Myocastor coypus*), *440*
Nuttall's mudwort (*Micranthemum micranthemoides*), 208
Nuttall's woodpecker (*Picoides nuttallii*), *20*
NWHC. *See* National Wildlife Health Center (NWHC)
Nyctea scandiaca (snowy owl), *67*
Nycticorax nycticorax (black-crowned night-heron), *54*, 55
Nyssa
aquatica (tupelo), 216
Sylvatica var *biflora* (tupelo), 216

Oahu 'akepa (*Loxops c. rufus*), *376*
Oahu 'akialoa (*Hemignathus obscurus ellisianus*), *376*
Oahu 'amakihi (*Hemignathus virens chloris*), *377*, 378
Oahu creeper (*Paroreomyza maculata*), *378*
Oahu 'elepaio (*Chasiempis sandwichensis gayi*), *377*
Oahu nukupu'u (*Hemignathus l. lucidus*), *376*
Oahu 'o'o (*Moho apicalis*), *376*
oak
live, mourning doves and, 72
scrub (*Quercus ilicifolia*), 166, 167
oak savanna
recovery potential, 232
status, 230-32
threats to, 232
in Wisconsin, 230-32
oak woodlands, 466
coast live, *465*
Englemann, *465*
valley, *465*
Oarisma poweshiek (poweshiek skipper), 175
Obliquaria reflexa (threehorn wartyback), *180*
oblong rocksnail (*Leptoxis compacta*), *251*
Obovaria
olivaria (hickorynut), *180*
subrotunda (round hickorynut), *180*
occurrences, of endangered or threatened species, 398, 400
Oceanodroma
furcata (fork-tailed storm-petrel), 44

  homochroa (ashy storm-petrel), 44
  leucorhoa (Leach's storm-petrel), 43-44
  melania (black storm-petrel), 44
ocellated dwarf gecko (*Sphaerodactylus argus*), *434*
ocellated turkey (*Meleagris ocellata*), 70
ocelot (*Felis pardalis*), *320*, 321
Ocenebra japonica (Japanese snail), 429
Ocyurus chrysurus (yellowtail snapper), 282
Odobenus, rosmarus divergens (Pacific walrus), 338, 351, 356-57
Odocoileus
  hemionus (white-tailed deer), 320
  virginianus (white-tailed deer), 93-94, 112-15, 320, 322, 405
Oechalia (stink bug), 366
'Ohi 'a lehua (*Metrosideros polymorpha*), 366
Ohio pigtoe (*Pleurobema cordatum*), *180*
Ohio River, aquatic insects, 182-83, 184
oil spills
  murres and, 47
  seabirds and, 50, 51
Olaa Forest, 369-71
Olapa (*Cheirodendron trigynum*), 371
oldsquaw (*Clangula hyemalis*), 36, *266*
Old World rabbit (*Oryctolagus cuniculus*), 441
oligochaete worms, 176
olive, Russian (*Elaeagnus angustifolia*), 286, 289, 290, 291
olive ridley (*Lepidochelys olivacea*), *119*
olive-sided flycatcher (*Contopus borealis*), *20*
oloma'o
  Myadestes lanaiensis lanaiensis, *376*
  Myadestes lanaiensis rutha, *377*
Olympic National Park, 200
Olympic Peninsula, Washington, sea otter (*Enhydra lutris*), 354, 355
'oma'o (Hawaii thrush) (*Myadestes obscurus*), *377*, 380
Oncorhynchus
  aguabonita (golden trout), *432*
  apache (Apache trout), 147, *150*, 320
  clarki (cutthroat trout), *432*
    Alvord cutthroat trout, *143*
    bouvieri (Yellowstone cutthroat trout), 153-54, 313
    lewisi (westslope cutthroat trout), 153-54
    pleuriticus (Colorado cutthroat trout), *150*
  gilae (Gila trout), *150*
  gorbuscha (pink salmon), 337-38, 344-45, *346*
  keta (chum salmon), 338, 344, 345-46
  kisutch (coho salmon), 344, *345*, 346, *432*
  mykiss (rainbow trout), 147, 153, 320, 328-29, *419*, *432*, *440*, 441, 443
  nerka (kokanee salmon), 154, 329
  nerka (sockeye salmon), 344, *345*, 346, *432*
  Pacific salmon, 337-38, 343-47, 443, 444
  tshawytscha (chinook salmon), 344, *345*, 346, *432*
onion, wild (*Allium gooddingii*), *320*
'o'o
  Bishop's (*Moho bishopi*), *377*
  Hawaii (*Moho nobilis*), *376*
  Kauai (*Moho braccatus*), *377*
  Oahu (*Moho apicalis*), *376*
OP. *See* organophosphorus (OP) pesticides
Ophisaurus attenuatus (western slender glass lizard), 231
Oporornis
  formosus (Kentucky warbler), *20*
  philadelphia (mourning warbler), *20*, 391
  tolmiei (MacGillivray's warbler), *20*
opossum
  Didelphis marsupialis, 321
  Didelphis virginiana, 321, 386, 417
Opuntia (prickly pear cactus), 321
orange-crowned warbler (*Vermivora celata*), *20*
orange-fronted parakeet (*Aratinga canicularis*), 85
orange-throated whiptail lizard (*Cnemidophorus hyperythrus*), *466*
orchard oriole (*Icterus spurius*), *21*, 231
Oregon
  breeding seabirds, 43-48
  non-native species, 428
Oregon chub (*Oregonichthys crameri*), *143*
Oregonichthys crameri (Oregon chub), *143*
Oregon junco (*Junco hyemalis oregonus*), *21*, *22*
Oreomystis
  bairdi (Kauai creeper), *378*
  mana (Hawaii creeper), *378*
Oreortyx pictus (mountain quail), *19*
Oreoscoptes montanus (sage thrasher), *20*
organochlorine pesticides
  banning of, 415, 416
  raptors and, 66
organophosphorus (OP) pesticides
  distribution of mortality, 417
  documentation of poisoning by, 416-17
  wildlife mortality and, 416-18
oriole
  Baltimore (*Icterus galbula*), *21*
  Bullock's (*Icterus galbula bullockii*), *21*
  orchard (*Icterus spurius*), *21*, 231
  Scott's (*Icterus parisorum*), *21*
ornithology, economic, 7-8
Oryctolagus cuniculus (Old World rabbit), 441
Osmerus mordax (rainbow smelt), 261, *262*, *432*, 443
osprey (*Pandion haliaetus*), 16, 66, *67*, *420*
Osteopilus septentrionalis (Cuban treefrog), *434*
otter
  river (*Lutra canadensis*), 414
  sea (*Enhydra lutris*), 110-12, 338, 353-56
ottoe skipper (*Strytone arogos iowa*), 175
Otus
  asio (eastern screech-owl), *67*
  flammeolus (flammulated owl), *67*
  kennicottii (western screech-owl), *67*
  nudipes newtoni (Virgin Islands screech-owl), *67*
  trichopsis (whiskered screech-owl), *67*
'o'u (*Psittirostra psittacea*), *377*
outbreeding depression, 147
ovenbird (*Seiurus aurocapillus*), *20*, 25, 420
overfishing, coastal and marine ecosystems and, 259
overgrazing
  land cover change and, 470
  in southwestern sky islands, 322
Ovibos moschatus (muskox), 337, 338-40
Ovis
  canadensis (bighorn sheep), 309, 332-33, *407*
    audobonii (badlands bighorn sheep), 332
    californiana (California bighorn sheep), 332
    canadensis (Rocky Mountain bighorn sheep), 332
    cremnobates (desert bighorn sheep), 334
    mexicana (desert bighorn sheep), 320, 334, 335
    nelsoni (desert bighorn sheep), 309, 320, 332, 333-35
  dalli (Dall sheep), 338, 347
owl, 16
  barn (*Tyto alba*), *67*, 231
  barred (*Strix varia*), *67*, 387
  boreal (*Aegolius funereus*), *67*
  burrowing (*Athene cunicularia*), *67*
  California spotted (*Strix occidentalis occidentalis*), *67*
  eastern screech- (*Otus asio*), *67*
  elf (*Micrathene whitneyi*), *67*
  ferruginous pygmy (*Glaucidium brasilianum cactorum*), *67*, 69
  flammulated (*Otus flammeolus*), *67*
  great gray (*Strix nebulosa*), *67*
  great horned (*Bubo virginianus*), *67*
  Hawaiian (*Asio flammeus sandwichensis*), *377*
  long-eared (*Asio otus*), *67*, *466*
  Mexican spotted (*Strix occidentalis lucida*), *67*, 309, *320*, 330-31
  northern hawk (*Surnia ulula*), *67*
  northern pygmy (*Glaucidium gnoma*), *67*
  northern saw-whet (*Aegolius acadicus*), *67*
  northern spotted, *Strix occidentalis caurina*, *67*, 193, 387
  short-eared (*Asio flammeus*), *67*, 297
  snowy (*Nyctea scandiaca*), *67*
  spotted (*Strix occidentalis*), 16, *67*, 69
  Virgin Islands screech- (*Otus nudipes newtoni*), *67*
  western screech- (*Otus kennicottii*), *67*
  whiskered screech- (*Otus trichopsis*), *67*
owlet moth
  Acronicta lanceolaria, 166
  Agroperina lutosa, 166
  Agrotis stigmosa, 167
  Argyrostrotis quadrifiliaris, 166
  Catocala pretiosa, 166
  Chaetaglaea cerata, 167
  Eugraphe subrosea, 166
  Homohadena badistriga, *166*, 167
  Lithophane
    georgii, 166
    lepida, 166
    semiusta, 166
    thaxteri, 166
  Platypolia anceps, 166
  Psectraglaea carnosa, *166*, 167

Pyreferra ceromatica, 166
Xestia (Anomogyna) badicollis, 166
Xylena
cineritia, 166
thoracica, 166
Xylotype capax, 166
Oxyura jamaicensis (ruddy duck), 36, *266*, *299*
oyster
contaminants in, 408
eastern (*Crassostrea virginica*), 263, 264, *265*, 408
Pacific (*Crassostrea gigas*), 429
oystercatcher, 61
American (*Haematopus palliatus*), *59*, *62*
black (*Haematopus bachmani*), *62*
Ozobryum ogalalense (moss), 198
ozone
forest ecosystems and, 227
national park system and, 228

Pacific black brant (*Branta bernicla nigricans*), 30, 31, 32
Pacific Flyway, 30
avian cholera, 402
canvasback, 40, 42
northern pintails, 38-39
Pacific golden-plover (*Pluvialis fulva*), *62*, 373
Pacific kangaroo rat (*Dipodomys agilis*), *466*
Pacific Northwest, subalpine forests, 316
Pacific oyster (*Crassostrea gigas*), 429
Pacific salmon (*Oncorhynchus* spp.), 337-38, 443, 444
in Alaska, 343-47
Pacific silver fir (*Abies amabilis*), 315
Pacific slender salamander (*Batrachoseps pacificus*), *466*
Pacific-slope flycatcher (*Empidonax difficilis*), *20*
Pacific walrus (*Odobenus rosmarus divergens*), 338, 351, 356-57
haul outs, 356
population estimates, 357
U.S.-Russia surveys, 356
packrat middens, 224, 226
paddlefish (*Polyodon spathula*), 143, 145
pagoda slitshell (*Gyrotoma pagoda*), *251*
PAH (polycyclic aromatic hydrocarbon), 408, 409
pahranagat chub (*Gila robusta jordani*), *150*
pahranagat speckled dace (*Rhinichthys osculus velifer*), *150*
pahranagat spinedace (*Lepidomeda altivelis*), *150*
PAHs. *See* polycyclic aromatic hydrocarbons (PAHs)
painted bunting (*Passerina ciris*), *20*
painted lady butterfly (*Vanessa cardui*), 171-72
painted turtle (*Chrysemys picta*), 327
Paiute sculpin (*Cottus beldingii*), *150*
Palaropus tricolor (Wilson's phalarope), *59*
Palea steindachneri (wattle-necked shoftshell turtle), 434
pale Indian-plantain (*Cacalia atriplicifolia*), 231
palila (*Loxioides bailleui*), 376, *377*, *381*
pallid sturgeon (*Scaphirhynchus albus*), *143*
Palmaria dolei ('akohokohe), *378*
palm warbler, *22*
palustrine wetlands, losses, 216
Pamlico Sound, North Carolina, turtles, 121, 122
Panama, Africanized honeybees, 450
Pandion haliaetus (osprey), 16, 66, *67*, *420*
Panulirus argus (West Indies spiny lobster), 283
paper pondshell (*Utterbackia imbecillis*), *180*
papershell
cylindrical (*Anodontoides ferussacianus*), *180*
fragile (*Leptodea fragilis*), *180*, *445*
pink (*Potamilus ohiensis*), *180*
Papilio multicaudata (two tailed swallow-tail), *173*
Parabuteo unicinctus (Harris' hawk), 67
Paracapnia angulata (stonefly), 185, *186*
Paragnetina
kansensis (stonefly), *186*
media (stonefly), 185, *186*
parakeet
canary-winged (*Brotogeris versicolurus*), 439
monk (*Myiopsitta monachus*), 439
orange-fronted (*Aratinga canicularis*), 85
Paralabrax
clathratus (kelp bass), 411
nebulifer (barred sand bass), 410
Paramecium, 252
parasites, 253
parasitic wasps, Hawaii, 366
parathion, *417*
paratuberculosis, *407*
Pardalophora
apiculata, 164
haldemani (bandwing grasshopper), 164
Parophrysvetulus (English sole), 410, 411
Paroreomyza
flammea (Molokai creeper), *378*
maculata (Oahu creeper), *378*
montana
montana (Lanai creeper), *376*
newtoni (Maui creeper), *378*
parrot
non-native species, 439
Puerto Rican (*Amazona vittata*), 17, 83-85
Puerto Rican parrot and, 85
thick-billed (*Rhynchopsitta pachyrhyncha*), *320*
translocation diseases, *407*
parrotbill, Maui (*Pseudonestor xanthrophrys*), *377*
Partners for Wildlife Program, U.S. Fish and Wildlife, 217
partridge, gray or Hungarian (*Perdix perdix*), 437, 438
parula, northern (*Parula americana*), *20*
Parula americana (northern parula), *20*
Parus
atricapillus (black-capped chickadee), *20*
bicolor (tufted titmouse), *20*, 21, *22*, 387
bicolor atricristatus (black-crested titmouse), *20*
carolinensis (Carolina chickadee), *20*, *22*
gambeli (mountain chickadee), *20*, 321
inornatus (plain titmouse), *20*
rufescens (chestnut-backed chickadee), *20*
sclateri (Mexican chickadee), 321
parvovirus, *407*
passenger pigeon (*Ecopistes migratorius*), 231
Passer
domesticus (house sparrow), *21*, *299*, 437, *438*, 439
montanus (Eurasian tree sparrow), 454
Passerculus sandwichensis (savannah sparrow), *21*, *22*, 297, 299, *303*, *466*
Passerella iliaca (fox sparrow), *21*
Passerina
amoena (lazuli bunting), *20*
ciris (painted bunting), *20*
cyanea (indigo bunting), *20*, 231
Passiflora
lutea (yellow passionflower), 208
mollissima (banana poka), 370
passionflower, yellow (*Passiflora lutea*), 208
Pasteurella multocida, 402
pasture, 423
Patuxent Wildlife Research Center (PWRC), 75, 76, 106
Pawnee Grasslands, Colorado, insect survey, 161
PCBs. *See* polychlorinated biphenyls (PCBs)
pearly-eyed thrasher (*Margarops fuscatus*), Puerto Rican parrot and, 85
peat moss
Sphagnum, 198
Sphagnum angermanicum, 210
pebblesnail
Cahaba (*Clappia cahabensis*), *251*
Hydrobiidae, 251
umbilicate (*Clappia umbilicata*), *251*
pectoral sandpiper (*Calidris melanotos*), *59*, *62*
pelagic cormorant (*Phalacrocorax pelagicus*), 44, 51
Pelecaniformes, 53-54
Pelecanus
erythrorhynchos (American white pelican), *19*, 53, *54*
occidentalis (brown pelican), 16, 43, 44, 53
pelican, 44, 53-54
American white (*Pelecanus erythrorhynchos*), 19, 53, *54*
brown (*Pelecanus occidentalis*), 16, 43, 44, 53
eastern brown, *54*
white, 44
Pelican Island, Florida, 53
Pellia megalospora, 200
Pelodiscus, sinensis (Chinese softshell turtle), *434*
Penaeus duorarum (pink shrimp), 282, 283
Penicillium notatum, 190
Peninsular hydrologic region, 145
pennate diatoms, 256, 257
Pennisetum
cilaris (buffel grass), 321
setaceum (fountain grass), 364
Pennsylvania, acid lakes, 419
Pentagenia vittigera (burrowing mayfly), 183
Pentatomidae (stink bug), 366

Perca flavescens (yellow perch), *261*, *419*, *432*, 443, 447
perch, 145
  white (*Morone americana*), 261, *262*, *432*, 443
  yellow (*Perca flavescens*), *261*, *419*, *432*, 443, 447
perching ducks, *36*
Percina
  antesella (amber darter), *143*
  aurantiaca (tangerine darter), *146*
Perdido Bay, seagrass distribution, 274
Perdix perdix (gray or Hungarian partridge), 437, 438
peregrine falcon (*Falco peregrinus*), 16, *320*, 405, 415
  American (*Falco peregrinus anatum*), *67*, 69
  arctic (*Falco peregrinus tundrius*), *67*, 69
Perisoreus canadensis (gray jay), *20*
Perityle cochisensis, *320*
Perlesta decipiens (stonefly), *186*
Perlinelia
  drymo (stonefly), *186*
  ephrye (stonefly), *186*
Perognathus
  alticola (white-eared pocket mouse), *466*
  fallax (San Diego pocket mouse), *466*
  longimembris (little pocket mouse), *466*
Peromyscus maniculatus (deer mouse), 109
persistent contaminants, in fish and wildlife, 413-15
Pestalotia natans, 205
pesticides, *See also* DDE pesticides; DDT pesticides; organochlorine pesticides; poisoning
  chlorinated, *408*
  colonial waterbirds and, 53-54, 56
  Indiana bats and, 98
  persistent, 413-15
  raptors and, 66-69
  soft, 415
  wildlife mortality and, 416-18
petrel
  dark-rumped (*Pterodroma phaeopygia sandwichensis*), *377*, 379
  Leach's storm- (*Oceanodroma leucorhoa*), 43-44
petrochemicals, toxic, American alligator (*Alligator mississippiensis*) and, 128
Petromyzon marinus (sea lamprey), 244, 430, 443
Phaeognathus hubrichti (Red Hills salamander), 124, 129, 131
Phagocata gracilis, 176
Phainopepla nitens (phainopepla), *20*
phainopepla (*Phainopepla nitens*), *20*
Phalacrocorax
  auritus (double-crested cormorant), 16, 44, 54
  pelagicus (pelagic cormorant), 44, 51
  penicillatus (Brandt's cormorant), 44
  urile (red-faced cormorant), 51
Phalaris arundinacea (reed canary grass), 249
phalarope
  red (*Phalaropus fulicaria*), *59*, *62*
  red-necked (*Phalaropus lobatus*), *59*, *62*
  Wilson's (*Phalaropus tricolor*), 59, *62*, 299, 396
Phalaropus
  fulicaria (red phalarope), *59*, *62*
  lobatus (red-necked phalarope), *59*, *62*
  tricolor (Wilson's phalarope), 59, *62*, 299, 396
Phanerophlebia auriculata, *204*
Phasianus colchicus (ring-necked pheasant), *19*, 437, 438
pheasant, ring-necked (*Phasianus colchicus*), *19*, 437, 438
Phelsuma sp. (day gecko), *434*
phenology, 256, 390
Pheucticus
  ludovicianus (rose-breasted grosbeak), *20*
  melanocephalus (black-headed grosbeak), *20*
Philadelphia vireo (*Vireo philadelphicus*), *20*
Philippine turtle dove (*Streptopelia bitorquata*), 454
Philornis spp. (bot fly), 85
Phoca
  hispida (ringed seal), 351
  vitulina (harbor seal), 52, *94*, 96
Phocoena phocoena (harbor porpoise), *94*, *95*, 96
phoebe, eastern (*Sayornis phoebe*), *20*
phorate, *417*
phosphamidon, *417*
Phoxinus
  cumberlandensis (blackside dace), *143*
  oreas (mountain redbelly dace), *146*
Phrynosoma
  cornutum (Texas horned lizard), *434*
  coronatum (coast horned lizard), *466*
Phyciodes batesii (tawny crescent butterfly), *166*, 232
phycology, 256
Phytophthora infestans, 191
phytoplankton, 255
Pica pica (black-billed magpie), *20*
Picea
  abies (Norway spruce), 315
  engelmannii (Englemann spruce), 316, 317
  engelmannii (spruce), 321
  rubens (red spruce), 205
  sitchensis, *204*
pickerel, chain (*Esox niger*), *261*, *432*
Picoides
  borealis (red-cockaded woodpecker), 17, 86-88
  nuttallii (Nuttall's woodpecker), *20*
  pubescens (downy woodpecker), *20*
Pictured Rocks National Lakeshore, 225, *226*
picture-wing *Drosophila*, 368-71
pied-billed grebe (*Podilymbus podiceps*), 389
pig (*Sus scrof*), 440, 441
  native Hawaii species and, 362, 370, 371, 440, 441
pigeon
  band-tailed (*Columba fasciata*), *19*, 322
  common (rock dove) (*Columba livia*), *19*, 437, 438, *440*, 441
  passenger (*Ecopistes migratorius*), 231
pigeon guillemot (*Cepphus columba*), 46, 51
pigtoe
  Ohio (*Pleurobema cordatum*), *180*
  round (*Pleurobema coccineum*), *180*
  Wabash (*Fusconaia flava*), *180*
pike
  blue (*Stizostedion vitreum glaucum*), *143*
  northern (*Esox lucius*), 261, *432*
pile dikes, habitat change and, 235
pilot whale, *94*
Pimephales promelas (fathead minnow), *261*, *432*
pimpernel, yellow (*Taenidia integerrima*), 231
pimpleback (*Quadrula pustulosa pustulosa*), *180*
Pinatubo Volcano, 364
pincushion cactus (*Coryphantha robbinsorum*), *320*
pine
  bristlecone
    Pinus aristata, 315, 316
    Pinus longaeva, 315, 316
  foxtail (*Pinus balfouriana*), 315, 316, 421
  limber (*Pinus flexilis*), 315
  loblolly (*Pinus taeda*), 223
  lodgepole (*Pinus contorta*), 223, 315, 316, 421
  longleaf (*Pinus palustris*), 17, 129-31, 223
  pitch- (*Pinus rigida*), 166, 167
  ponderosa (*Pinus ponderosa*), 223, 228, 422
    Laccaria laccata and, *192*
  red, (*Pinus resinosa*), 225
  shortleaf (*Pinus echinata*), 223
  slash (*Pinus elliotti*), 223
  western white (*Pinus monticola*), 230, 421
  white, (*Pinus strobus*), 225
  whitebark (*Pinus albicaulis*), 213, 228-30, 313, *314*, 315
pine barrens
  Albany, 166-67
  management, 167
pine grosbeak (*Pinicola enucleator*), *21*
pine-oak woodlands, digger, *465*
Pine Ridge Indian Reservation, 107
pine siskin (*Carduelis pinus*), *21*
pine snake (*Pituophis melanoleucus*), 327
  Florida (*Pituophis melanoleucus mugitus*), 129
pine warbler (*Dendroica pinus*), *20*, *22*
Pinicola enucleator (pine grosbeak), *21*
pink heelsplitter (*Potamilus alatus*), *180*, *181*, *182*
pink papershell (*Potamilus ohiensis*), *180*
pink salmon (*Oncorhynchus gorbuscha*), 337-38, 344-45, *346*
pink shrimp (*Penaeus duorarum*), 282, 283
pinon woodlands, Mojavean, *465*
pintail, northern (*Anas acuta*), *4*, 16, *19*, 35, 36, 38-39, 266, 296, *299*, 300, 301, 389
Pinus
  albicaulis (whitebark pine), 213, 228-30, 313, *314*, 315
  aristata (bristlecone pine), 315, 316
  balfouriana (foxtail pine), 315, 316, 421
  contorta (lodgepole pine), 223, 315, 316, 421
  echinata (shortleaf pine), 223
  elliotti (slash pine), 223
  flexilis (limber pine), 315
  longaeva (bristlecone pine), 315, 316

monticola (western white pine), 230, 421
palustris (longleaf pine), 17, 223
ponderosa (ponderosa pine), 223, 228, 422
rigida (pitch-pine), 166, 167
taeda (loblolly pine), 223
pinyon jay (*Gymnorhinus cyanocephalus*), *20*
Pipilo
californicus (California towhee), *21*
chlorurus (green-tailed towhee), *20*
erythrophthalmus (rufous-sided towhee), *20*, *299*
fuscus (brown towhee), *21*
piping plover (*Charadrius melodus*), 16, 58, *59*, *62*, 77-79
status and trends, 77-78
threats to, 78-79
pipit
American (*Anthus rubescens*), 22
Sprague's (*Anthus spragueii*), *20*
Piranga
ludoviciana (western tanager), *20*
olivacea (scarlet tanager), *20*
rubra (summer tanager), *20*
piranha (*Serrasalmus* spp.), *432*
pitch-pine (*Pinus rigida*), 166, 167
Pituophis melanoleucus (pine snake), 327
Pityopsis ruthii (Ruth's golden aster), 400
Pityrogramma
rittata, *203*
trifoliata, *202*
Plagiochasma, 199
plagiopylid ciliates, 254
Plagithmysus (beetle), 367
Plagopterus argentissimus (woundfin), *150*
plains minnow (*Hybognathus placitus*), 304
plain titmouse (*Parus inornatus*), *20*
plankton, 445
zebra mussel (*Dreissena polymorpha*) and, 445
planktonic algae, 255
Planorbidae, 249
Plant List of Attributes, Nomenclature, Taxonomy, and Symbols, 201
plant pathogens, non-native species, *429*
plants, 189-210, *See also* aquatic vegetation; vascular plants; vegetation
angiosperms, 201
biological role of, 189
climate change and, 392-95
flowering, 200, 201
gymnosperms, 200, 201
major groups, 201
migration rates, 394
non-native species, 190, *429*, 444, 458-59
overview, 189-90
pteridophytes, 201
regional, 201-4
species loss, 189
vascular, 200-210, 392-95
weeds, 394, 429
wildflowers, 390
plant waifs, New York, 209
Plasmodium relictum (malaria), 378
Platanus occidentalis (American sycamore), 228
Platichthys stellatus (starry flounder), 409, 411
Platypolia anceps (owlet moth), 166
Plegadis chihi (white-faced ibis), *54*, 55
Pleistocene disc snail (*Discus macclintocki*), 176-77
Pleistophora, 253
Plethodon neomexicanus (Jemez Mountains salamander), 327, *328*
Pleurobema
coccineum (round pigtoe), *180*
cordatum (Ohio pigtoe), *180*
Pleurocera, 249
Pleuroceridae, 249
Pleuronectes americanus (winter flounder), 409, *411*
plover, 58
American golden- (*Pluvialis dominica*), 58, *59*, *62*, 64
black-bellied (*Pluvialis squatarola*), 58, *59*, *62*
common ringed (*Charadrius hiaticula*), *62*
mountain (*Charadrius montanus*), 58, *59*, *62*, 63, 297, 298
Pacific golden- (*Pluvialis fulva*), *62*, 373
piping (*Charadrius melodus*), 16, 58, *59*, *62*, 77-79
semipalmated (*Charadrius semipalmatus*), 58, *59*, *62*
snowy (*Charadrius alexandrinus*), 58, *59*, *62*
Wilson's (*Charadrius wilsonia*), 58, *59*, 61, *62*
Pluvialis
dominica (American golden-plover), 58, *59*, *62*, 64
fulva (Pacific golden-plover), *62*, 373
squatarola (black-bellied plover), 58, *59*, *62*
Poanes viator (skipper), 166
Poa pratensis (Kentucky bluegrass), 471
pocketbook (*Lampsilis ovata*), *180*
pocket mouse
little (*Perognathus longimembris*), *466*
San Diego (*Perognathus fallax*), *466*
white-eared (*Perognathus alticola*), *466*
Podarcis
muralis (common wall lizard), *434*
sicula (Italian wall lizard), *434*
Podilymbus podiceps (pied-billed grebe), 389
Poecilia reticulata (guppy), *432*
Poeciliopsis
occidentalis (Gila topminnow), *143*, *150*
occidentalis sonoriensis (Yaqui topminnow), *320*
poison-dart frog (*Dendrobates auratus*), *434*
poisoning, *See also* DDE pesticides; DDT pesticides; organochlorine
pesticides; pesticides
by carbamate pesticides, 416-17
heavy metal, 139, 140
lead, 41, 42, 68, *411*, 414-15
by organophosphorus (OP) pesticides, 416-17
polar bear (*Ursus maritimus*), 338
in Alaska, 351-53
counting, 351
hunting, 351
implications of growth, 352-53
population estimates, 351-52
Polar Domain, shorebirds, 60-61, 64
Polioptila
caerulea (blue-gray gnatcatcher), *20*, *466*
californica (California gnatcatcher), *466*
melanura (black-tailed gnatcatcher), *20*
pollen, fossil, 224-27
pollock, walleye (*Theragra chalcogramma*), 51, 52
polychlorinated biphenyls (PCBs)
atmospheric transport of, 415
California sea otter and, 112
distribution of, *415*
in fish, 409
Great Lakes, 242-44
mink and, 237-38
in mussels, 410
National Status and Trends Program, *408*
persistence of, 414
raptors and, 67
reproductive and maturity processes and, 413
sediment concentrations, 408
uses of, 414
polycyclic aromatic hydrocarbons (PAHs), 408, 409
Great Lakes, 243-44
polymorphism, in *Dropsophila* (Pomance flies), 368
Polyodon spathula (paddlefish), 143, 145
polypore fungi, 192
Polystichum lonchitis, *202*
Pomance flies (*Drosophila*), 368-71
Pomoxis
annularis (white crappie), 261, *432*
nigromaculatus (black crappie), *261*, *432*
ponderosa pine (*Pinus ponderosa*), 223, 228, 422
altered fire regimes, *222*, 223
Laccaria laccata and, *192*
woodlands, Colorado Rockies, 310
pondhorn (*Uniomerus tetralasmus*), *180*
pondmussel, eastern (*Ligumia nasuta*), *180*
pondshell, paper (*Utterbackia imbecillis*), *180*
pond turtle, western (*Clemmys marmorata*), *119*, *466*
Pooecetes gramineus (vesper sparrow), *21*, *22*, 231, 297, *299*
pools (navigation dam reservoirs)
biota, 236-38
habitat change and, 234-36
po'o-uli (*Melamprosops phaeosoma*), *378*
poplar, yellow- (*Liriodendron tulipifera*), 228
population health, birds, 23
Populus
fremontii (cottonwood), 286-89, 290
tremuloides (quaking aspen), 313
Porcupine caribou herd, 337, 338-40

porpoise, 96
  harbor (*Phocoena phocoena*), *94*, *95*, 96
Porzana
  carolina (sora), *299*
  palmeri (Laysan rail), *376*, 380
  sandwichensis (Hawaiian rail), *376*, 380
postsettlement conditions
  defined, 224
  rates of change, 227
Potamilus
  alatus (pink heelsplitter), *180*, *181*, *182*
  ohiensis (pink papershell), *180*
Potamopyrgus antipodarum (freshwater snail), 429
pothole margins, northern pintails and, 39
poultry, avian cholera, 402
poweshiek skipper (*Oarisma poweshiek*), 175
prairie, *See also* Great Plains
  bluestem, 472
  grama-tobosa, 472
  Great Plains, 295
  habitat loss, 174, 175
  tall-grass, 174-76, 221
prairie chicken (*Tympanuchus cupido*), 438
prairie dog (*Cynomys* spp.), 106-8
  black-tailed (*Cynomys ludovicianus*), 107
  ecosystem, 107-8
  white-tailed (*Cynomys leucurus*), 295
prairie falcon (*Falco mexicanus*), *67*
prairie fishes, 295, 303-5
Prairie Pothole region
  duck nest success, 300-302
  duck population, 34, 36-37
prairie smoke (*Geum triflorum*), 210
Prenanthes boottii (Boott's rattlesnake-root), 394
presettlement conditions
  defined, 224
  rates of change, 227
Preston speckled dace (*Rhinichthys osculus* ssp.), *150*
Preston springfish (*Crenichthys baileyi albivallis*), *143*, *150*
prethroid compounds, synthetic, 415
Pribilof Islands, red-legged kittiwake, 50
prickly pear cactus (*Opuntia*), 321
Prince's Island, Cassin's auklet, 46
Prince William Sound, Alaska
  seabirds, 50, 51
  sea otter (*Enhydra lutris*), 354, *355*
Procyon lotor (raccoon), 37, 41, 302, 406, *407*, 417
productivity, birds, 23-26
Progne subis (purple martin), *20*, *391*
Promachus bastardii, *166*
pronghorn antelope
  Antilocapra americana, 313, 320
  Mexican (*Antilocapra americana mexicana*), 318
Proserpinus juanita (spinx moth), *174*
Prosopium williamsoni (mountain whitefish), *150*
Prostoia completa (stonefly), 185, *186*
Proterhinus sp. (weevil), 367
protozoa, 252-54
  data on, 253
  disease reservoirs, 253
  ecological role of, 252-54
  environmental quality indicators, 253
  free-living, 252, 254
  freshwater, 252
  Glugea hertwigi, 443
  growth and distribution of, 254
  heterotropic, 252-54
  marine, 252
  parasitic, 253
  soil-dwelling, 252
  symbiotic, 253
  water quality and, 253
Prudhoe Bay, anadromous fish, 341
Prunus serotina (black cherry), 228
Psaltriparus minimus (bushtit), *20*
Psectraglaea carnosa (owlet moth), *166*, 167
Pseudemys
  alabamensis (Alabama red-bellied turtle), *119*
  rubriventris (red-bellied turtle), *119*
Pseudococcus calceolariae (citophilus mealybug), *160*
Pseudonestor xanthrophrys (Maui parrotbill), *377*
Pseudostylochus ostreophgus (flatworm), 429
Pseudotsuga menziesii (Douglas fir), 116, 193, 321
Psittirostra psittacea ('o'u), *377*
psoroptic scabies (*Psoroptes* spp.), 335
pteridophytes, 201
Pterodroma phaeopygia sandwichensis (dark-rumped petrel), *377*, 379
Pteronarcys pictetii (stonefly), *185*, *186*
Pteropus mariannus (Mariana flying fox), 455
Ptychobranchus fasciolaris (kidneyshell), *180*
Ptychocheilus lucius (Colorado squawfish), *150*, 151
Ptychoramphus aleuticus (Cassin's auklet), 46, 51
puaiohi (small Kauai thrush) (*Myadestes palmeri*), *377*, 380
public lands
  inventory and monitoring programs, *5*
  land protection status, 463-64
  non-native species on, 440-42, 456-57
Public Laws
  92-195, Wild Free-Roaming Horses and Burros Act, 456
  97-384, Coastal Barrier Resources Act (1982), 278
  100-233 (1985), "Swampbuster" provision, 217
  101-591, Coastal Barrier Improvement Act (1990), 278
  101-605, Marine Protection, Research, and Sanctuaries Act, 279
pueo (Hawaiian owl) (*Asio flammeus sandwichensis*), *377*
Puerto Rican broad-winged hawk (*Buteo platypterus brunnescens*), *67*
Puerto Rican crested anole (*Anolis cristellatus*), *434*
Puerto Rican parrot (*Amazona vittata*), 17, 83-85
Puerto Rican Parrot Recovery Plan, 84
Puerto Rico, 83-85
  non-native birds, 439
puffball, 192
puffin
  Atlantic (*Fratercula arctica*), 55
  horned (*Fractercula corniculata*), 51
  tufted (*Fratercula cirrhata*), 46, 51
Puffinus newelii (Newell's shearwater), *377*, 379
pumpkinseed sunfish (*Lepomis gibbosus*), *261*, *262*, *419*, *432*
pupa elimia (*Elimia pupaeformis*), *251*
pupfish
  desert
    Cyprinodon macularius, *143*
    Cyprinodon macularius macularius, *150*
  Devils Hole (*Cyprinodon diabolis*), 143, *143*
  Monkey Springs (*Cyprinodon* sp.), *150*
purple finch (*Carpodacus purpureus*), *21*
purple loosestrife (*Lythrum salicaria*), 249, 429, 443, 444, 458-59
  controlling, 458-59
purple martin (*Progne subis*), *20*, *391*
purple milkweed (*Asclepias purpurascens*), 231
purple sandpiper (*Calidris maritima*), *59*, *62*
purple wartyback (*Cyclonaias tuberculata*), *180*
Pyganodon grandis (giant floater), *180*
pygmy elimia (*Elimia pygmaea*), *251*
pygmy madtom (*Noturus stanauli*), *143*
pygmy owl
  ferruginous (*Glaucidium brasilianum cactorum*), *67*, 69
  northern (*Glaucidium gnoma*), *67*
pygmy sunfish, spring (*Elassoma* sp.), *143*
Pylodictis olivaris (flathead catfish), 151, *432*
Pyramid Lake, Nevada, cui-ui (*Chasmistes cujus*), 323-24
pyramid slitshell (*Gyrotoma pyramidata*), *251*
Pyreferra ceromatica (owlet moth), 166
Pyrrhuloxia (*Cardinalis sinuatus*), *20*

Quadrula
  pustulosa pustulosa (pimpleback), *180*
  quadrula (mapleleaf), *180*
quail
  California (*Callipepla californica*), *19*
  Gambel's (*Callipepla gambelii*), *19*
  mountain (*Oreortyx pictus*), *19*
  scaled (*Callipepla squamata*), *19*
quaking aspen (*Populus tremuloides*), 313
queen conch (*Strombus gigas*), 282
Quercus ilicifolia (scrub oak), 166, 167
Quiscalus
  major (boat-tailed grackle), *21*
  mexicanus (great-tailed grackle), *21*
  quiscula (common grackle), *21*, *299*

rabbit, Old World (*Oryctolagus cuniculus*), 441
rabbitbrush, whitebark (*Chrysothamnus visidiflorus*), 226
rabies, transmission through raccoons (*Procyon lotor*), 406, *407*
raccoon (*Procyon lotor*), 302, 417
  canvasback predation, 41
  duck nest success and, 37

transmission of rabies through, 406, *407*
rail
Hawaiian (*Porzana sandwichensis*), *376*, 380
Laysan (*Porzana palmeri*), *376*, 380
Railroad Valley springfish (*Crenichthys nevadae*), *143*, *150*
rainbow (*Villosa iris*), *180*
Rainbow Bridge National Monument, 327
rainbow skink (*Lampropholis delicata*), *434*
rainbow smelt (*Osmerus mordax*), 261, *262*, *432*, 443
rainbow trout (*Oncorhynchus mykiss*), 147, 153, 320, 328-29, *419*, *432*, *440*, 441, 443
rainfall, forest cover loss and, 472
rain forests, Hawaii, 370, 371
Rainwater Basin, 404
Ramalina thrausta, 196
Ramsey Canyon leopard frog (*Rana subaquavocalis*), 319
Rana
aurora
aurora (northern red-legged frog), 132
draytonii (California red-legged frog), 132
berlandieri (Rio Grande leopard frog), 133, 453
boylii (foothill yellow-legged frog), 132, *133*
capito (gopher frog), 125, 130
capito (Carolina gopher frog), 129
sevosa (dusky gopher frog), 129
cascadae (Cascades frog), 132
catesbeiana (bullfrog), 133, *419*, 433, *434*, 435, 452-53
chiricahuensis (Chiricahua leopard frog), 139, *320*, 327, 452, 453
leopard frogs, 125, 452-53
muscosa (mountain yellow-legged frog), 133, *134*, *466*
onca (relict leopard frog), 327, *328*
pipiens (northern leopard frog), 133-34, 327, *434*
pretiosa (spotted frog), 132-33, *328*
rugosa (wrinkled frog), *434*
subaquavocalis (Ramsey Canyon leopard frog), 319
sylvatica (wood frog), *419*
tarahumarae (Tarahumara frog), 118, 138-40, 321
yavapaiensis (Yavapai leopard frog), 133, 139, 452-53
rangeland, 423
Rangifer tarandus (caribou), 337, 338-40, 347-48, 357-58, 359, *407*
ranid frog (Ranidae), 452
raptors, 16, 65-69
rare species, *See also* endangered species; threatened species
inventory and monitoring programs, *5*
rare terrestrial ecological communities, 218-21
classification of, 219
distribution of, 219-20
fire suppression and, 220
flood control and, 220
protection of, 220-21
ranking of, 219
raspberry, yellow Himalayan (*Rubus ellipticus*), 370
rat
agile kangaroo (*Dipodomys agilis*), *466*
black (*Rattus rattus*), 85
Merriam's kangaroo (*Dipodomys merriami*), *466*
Pacific kangaroo (*Dipodomys agilis*), *466*
Rattus spp., 378, 454
Stephens' kangaroo (*Dipodomys stephensi*), *466*
rattlesnake
eastern diamondback (*Crotalus adamanteus*), 129
eastern massasauga(*Sistrurus catenatus*), 231
red diamond (*Crotalus ruber*), *466*
ridge-nosed
Crotalus willardi obscurus, *320*
Crotalus willardi willardi, *320*
western (*Crotalus viridis*), 327
rattlesnake-root, Boott's (*Prenanthes boottii*), 394
Rattus, 378, 454
rattus (black rat), 85
raven (*Corvus corax*), 16
Chihuahuan (*Corvus cryptoleucus*), *20*
common (*Corvus corax*), *20*, 73-74, *75*
rayed bean (*Villosa fabalis*), *180*
razorback sucker (*Xyrauchen texanus*), *143*, *150*, 151, 309, 324-26
razorbill (*Alca torda*), *54*, 55
recovery, goals of, 326
Recovery Implementation Program, 325
Recovery Team, U.S. Fish and Wildlife Service, 374
recreational fishing
introduced species, 325
non-native species, 428, 443
recreational use, coastal and marine ecosystems and, 259
Recurvirostra americana (American avocet), *59*, *62*
red bat (*Lasiurus borealis*), *320*
red-bellied turtle
Alabama (*Pseudemys alabamensis*), *119*
Pseudemys rubriventris, *119*
red-bellied woodpecker (*Melanerpes carolinus*), *20*
redbelly dace, mountain (*Phoxinus oreas*), *146*
redbelly tilapia (*Tilapia zilli*), *432*
red-breasted merganser (*Mergus serrator*), *19*, *36*
red-breasted nuthatch (*Sitta canadensis*), *20*
redbreast sunfish (*Lepomis auritus*), *261*, *262*, *432*
red-cockaded woodpecker (*Picoides borealis*), 17, 86-88
causes of decline, 87
distribution, 87
recovery efforts, 88
Red-cockaded Woodpecker Recovery Plan, 88
red crossbill (*Loxia curvirostra*), *21*
reddish egret (*Egretta rufescens*), *54*, 55
redear sunfish (*Lepomis microlophus*), *432*
red-eyed vireo (*Vireo olivaceus*), *20*, 25, 420
red-faced cormorant (*Phalacrocorax urile*), 51
red fir (*Abies magnifica*), 421
red fire ant (*Solenopsis invicta*), 131
red fox (*Vulpes vulpes*), 296, 302, 306, *407*, *440*
Alaskan seabirds and, 51
terns and, 45
redhead duck (*Aythya americana*), 35, *36*, 266, 275-77
red-headed woodpecker (*Melanerpes erythrocephalus*), *20*, 231
Red Hills salamander (*Phaeognathus hubrichti*), 124, 129, 131
red junglefowl (*Gallus gallus*), 372-73
red knot (*Calidris canutus*), *59*, *62*
red-legged frog
California (*Rana aurora draytonii*), 132
northern (*Rana aurora aurora*), 132
red-legged kittiwake (*Rissa brevirostris*), 50
red maple (*Acer rubrum* var *drummondii*), 216
red-necked phalarope (*Phalaropus lobatus*), *59*, *62*
red phalarope (*Phalaropus fulicaria*), *59*, *62*
red pine (*Pinus resinosa*), 225
red-shafted flicker (*Colaptes auratus cafer*), *20*
redshanks chaparral, *465*
red shiner (*Cyprinella lutrensis*), 151
red-shouldered hawk (*Buteo lineatus*), *67*
red snapper (*Lutjanus campechanus*), 282
red spruce (*Picea rubens*), 205
red squirrel, Mt. Graham (*Tamasciuris hudsonicus grahamensis*), 320
redstart, American (*Setophaga ruticilla*), *20*
red-tailed hawk (*Buteo jamaicensis*), *67*
red-tailed leafhopper (*Aflexia rubranura*), 232
red tide, 255, 427
red-winged blackbird (*Agelaius phoeniceus*), *21*, 299, *303*
red wolf (*Canis rufus*), 98
reed canary grass (*Phalaris arundinacea*), 249
reef fishes, Florida Keys, 279-84
reef gecko (*Sphaerodactylus notatus*), *434*
Reef and Shore Fauna of Hawaii, 362
reforestation, wetlands, 217
regal fritillary (*Speyeria idalia*), *166*, 174, *175*
Regional Comprehensive Plan Open Space Element, 466
Regulus
calendula (ruby-crowned kinglet), *20*, *22*
satrapa (golden-crowned kinglet), *20*, *22*
reindeer moss (*Cladinia rangiferina*), 195
relict leopard frog (*Rana onca*), 327, *328*
reproductive impairment
in fish, 411
PCBs and, 413
reptiles
Colorado Plateau, 326-28
in endangered longleaf pine ecosystem, 129-31
non-native species, 433-36
protection of, 118
in southwestern sky islands, 320, 321
status and trends, 117-18
reservoirs, navigation dam, 234-36
resident game species, inventory and monitoring programs, *5*
Resources Planning Act, *5*
Rhamphotyphlops braminus (Brahminy blind snake), *434*
Rhinichthys
cobitis (loach minnow), *143*, *150*
deaconi (Las Vegas dace), *150*

osculus
    Meadow Valleys speckled dace), *150*
    osculus (speckled dace), *150*
    Preston speckled dace, *150*
    thermalis (Kendall Warm Springs dace), *150*
    velifer (pahranagat speckled dace), *150*
    White River speckled dace, *150*
rhinoceros auklet (*Cerorhinca monocerata*), 46
rhizoids, 197
Rhodacanthis
  flaviceps (lesser koa-finch), *376*
  palmeri (greater koa-finch), *376*
Rhyacotriton spp. (torrent salamander), 125
Rhynchopsitta pachyrhyncha (thick-billed parrot), *320*
ribbed elimia (*Elimia laeta*), *251*
ribbed slitshell (*Gyrotoma pumila*), *251*
rice grass (*Stipa hymenoides*), 226
ridge-nosed rattlesnake
  Crotalus willardi obscurus, *320*
  Crotalus willardi willardi, *320*
ridley
  Kemp's (*Lepidochelys kempii*), *119*, 121, 123
  olive (*Lepidochelys olivacea*), *119*
Ridley Creek State Park (Pennsylvania), 113
riffleshell, northern (*Epioblasma torulosa rangiana*), *180*
right whale (*Eubalaena glacialis*), *94*, 95-96
ring-billed gull (*Larus delawarensis*), *19*, 45, *54*, 56
ringed map turtle (*Graptemys oculifera*), *119*
ringed plover, common (*Charadrius hiaticula*), *62*
ringed seal (*Phoca hispida*), 351
ringed turtle-dove (Barbary dove) (*Streptopelia risoria*), 438
ring-necked duck (*Aythya collaris*), 35, *36*, *266*, *420*
ring-necked pheasant (*Phasianus colchicus*), *19*, 437, 438
Rio Grande, floodplain, 290-91
Rio Grande leopard frog (*Rana berlandieri*), 133, 453
riparian ecosystems, 285-91
  classification, 290-91
  dams and, 234-36
  disturbance regimes, 288-89
  flows, 289, 291
  future of, 289
  human activities and, 286, 287
  hydrology and, 287
  lower Colorado River, 287-89
  non-native species, 286, 288-89
  Rio Grande floodplain, 290-91
  salinity, 287, 288
  southwestern willow flycatcher, 89-91
  tree population patterns, 288
  trends, 291
  upper Missouri River, 286-88
  western, 285, 286-91
Riparia riparia (bank swallow), *20*, *299*
Rissa family (kittiwakes), 16
  brevirostris (red-legged kittiwake), 50
  tridactyla (black-legged kittiwake), 49-50
river otter (*Lutra canadensis*), 414
rivers. *See* riparian ecosystems; streams; waterways
robber fly (*Diogmites symmachus*), 162
robin, American (*Turdus migratorius*), *20*, 231, *299*, 386
Robinia pseudoacacia (locust), 167
rock bass (*Ambloplites rupestris*), *261*, *432*
rock dove (common pigeon) (*Columba livia*), *19*, 437, 438, *440*, 441
rock sandpiper (*Calidris ptilocnemis*), *62*
rocksnail
  agate (*Leptoxis clipeata*), *251*
  bigmouth (*Leptoxis occultata*), *251*
  Coosa (*Leptoxis showalteri*), *251*
  interrupted (*Leptoxis formanii*), *251*
  lirate (*Leptoxis lirata*), *251*
  maiden (*Leptoxis formosa*), *251*
  oblong (*Leptoxis compacta*), *251*
  rotund (*Leptoxis ligata*), *251*
  striped (*Leptoxis vittata*), *251*
rock wren (*Salpinctes obsoletus*), *20*
Rocky Mountain bighorn sheep (*Ovis canadensis canadensis*), 332
Rocky Mountain Elk Foundation, 115
Rocky Mountain locust (*Melanoplus spretus*), 164
Rocky Mountain National Park, ecosystem trends, 310-12
Rocky Mountains
  ecosystem trends, 310-12
  human activities and, 310-11
  sky islands, 318
  vascular plants, 202-3
Rocky Mountain wolf, northern (*Canis lupus irremotus*), 98
roseate tern (*Sterna dougallii*), *54*, *55*, 56
rose-breasted grosbeak (*Pheucticus ludovicianus*), *20*
rosetail chub (*Gila robusta*), *320*
Rostrhamus sociabilis (snail kite), 66, *67*
rosy boa (*Lichanura trivirgata*), *466*
rotund rocksnail (*Leptoxis ligata*), *251*
rough-footed mud turtle, Mexican (*Kinosternon hirtipes*), *119*
rough-legged hawk (*Buteo lagopus*), *67*
rough-lined elimia (*Elimia pilsbryi*), *251*
roughtail gecko (*Cyrtopodion scabrum*), *434*
rough-winged swallow, northern (*Stelgidopteryx serripennis*), *20*
round hickorynut (*Obovaria subrotunda*), *180*
round pigtoe (*Pleurobema coccineum*), *180*
round slitshell (*Gyrotoma walkeri*), *251*
roundtail chub (*Gila robusta robusta*), *150*
route regression, 17
Royal Entomological Society, 161
Rubus
  argutus (wild blackberry), 206
  ellipticus (yellow Himalayan raspberry), 370
ruby-crowned kinglet (*Regulus calendula*), *20*, *22*
rudd (*Scardinius erythrophthalmus*), 261, 263, *432*
ruddy duck (*Oxyura jamaicensis*), 36, *266*, *299*
ruddy turnstone (*Arenaria interpres*), 59, *62*
ruffe (*Gymnocephalus cernuus*), 429, 430, 443
rufous-crowned sparrow (*Aimophila ruficeps*), *466*
rufous hummingbird (*Selasphorus rufus*), *20*
rufous-sided towhee (*Pipilo erythrophthalmus*), *20*, *299*
Rumex orthoneurus (dock sorrel), *320*
running buffalo clover (*Trifolium stoloniferum*), 208
Ruppia maritima (widgeon grass), 273, 274
Russian olive (*Elaeagnus angustifolia*), 286, 289, 290, 291
rust, 190-91
  white pine blister (*Cronartium ribicola*), 229
Ruth's golden aster (*Pityopsis ruthii*), 400
Rynchops niger (black skimmer), 45, 55

sagebrush
  -grass steppe, altered fire regimes in, 222-24
  intermountain (*Artemisia* spp.), 222
  scrub, *465*
  steppe, 465
sage-chaparral scrub, coastal, *465*
sage scrub
  coastal, *465*
    vegetation change, 226
sage sparrow (*Amphispiza belli*), *21*, *466*
sage thrasher (*Oreoscoptes montanus*), *20*
St. Lawrence Seaway, 443
salamander, 117, 124
  Barton Springs (*Eurycea sosorum*), 125
  dusky (*Desmognathus* spp.), 130
  flatwoods (*Ambystoma cingulatum*), 129
  Jemez Mountains (*Plethodon neomexicanus*), 327, *328*
  Pacific slender (*Batrachoseps pacificus*), *466*
  Red Hills (*Phaeognathus hubrichti*), 124, 129, 131
  Santa Cruz long-toed (*Ambystoma macrodactylum croceum*), 124
  spotted (*Ambystoma maculatum*), *419*
  Tehachapi slender (*Batrachoseps stebbinsi*), *466*
  Texas blind (*Typhlomolge rathbuni*), 124
  tiger (*Ambystoma tigrinum*), 327, *434*
  torrent (*Rhyacotriton* spp.), 125
salamander mussel (*Simpsonaias ambigua*), *180*
salinity, of riparian ecosystems, 287, 288
Salix gooddingii (willow), 286, 287
Salmo
  salar (Atlantic salmon), *432*
  trutta (brown trout), *419*, 429, *432*, *440*, 441, 443
salmon
  Atlantic (*Salmo salar*), *432*
  chinook (*Oncorhynchus tshawytscha*), 344, *345*, 346, *432*
  chum (*Oncorhynchus keta*), 338, 344, 345-46
  coho (*Oncorhynchus kisutch*), 344, *345*, 346, *432*
  kokanee (*Oncorhynchus nerka*), 154, 329
  Pacific (*Oncorhynchus* spp.), 337-38, 343-47, 443, 444
  pink (*Oncorhynchus gorbuscha*), 337-38, 344-45, *346*
  sockeye (*Oncorhynchus nerka*), 344, *345*, 346, *432*
salmonellosis, *404*
Salmonidae (whitefish), 179
Salpinctes obsoletus (rock wren), *20*
saltcedar (*Tamarix* spp.), 290-91
salt marsh, inland, 220

Salvelinus
fontinalis (brook trout), 419, *432*, *440*, 441
malma (Dolly Varden char), 337, 341, *342*, 343
namaycush (lake trout), 154, 242, *243*, 244-46, 414, *432*
San Bernadino National Wildlife Refuge (SBNWR), 452-53
Sanborn's long-nosed bat (*Leptonycteris sanborni*), *320*
San Clemente Island loggerhead shrike (*Lanius ludovicianus mearnsi*), 74
Sanctuaries and Reserves Division, National Oceanic and Atmospheric
Administration (NOAA), 279
sand bass, barred (*Paralabrax nebulifer*), 410
sand dune ecosystem, Coachella Valley Preserve system, 137-38
sanderling (*Calidris alba*), 59, *62*
sandhill crane (*Grus canadensis*), *19*
Mississippi (*Grus canadensis pulla*), 16-17, 75-77
San Diego pocket mouse (*Perognathus fallax*), *466*
San Diego Wild Animal Park, 81
sandlance (*Ammodytes hexapterus*), 51
sandpiper, 58-59
Baird's (*Calidris bairdii*), 59, *62*
buff-breasted (*Tryngites subruficollis*), 59, *62*, 64
least (*Calidris minutilla*), 59, *62*
pectoral (*Calidris melanotos*), *59*, *62*
purple (*Calidris maritima*), *59*, *62*
rock (*Calidris ptilocnemis*), *62*
semipalmated (*Calidris pusilla*), 59, *62*
solitary (*Tringa solitaria*), *59*, *62*
spotted (*Actitis macularia*), 59, *62*
stilt (*Calidris himantopus*), *59*, *62*
upland (*Bartramia longicauda*), *19*, 58-59, *62*, 64, 297, *299*, *303*
western (*Calidris mauri*), *59*, *62*
white-rumped (*Calidris fuscicollis*), 59, *62*
sand seatrout (*Cynoscion arenarius*), 410
sandshell, black (*Ligumia recta*), *180*
San Francisco Bay
canvasbacks, 42
non-native species, 428
San Joaquin Valley, 421
San Juan Mountains, grizzly bear, 103, 105
San Marcos gambusia (*Gambusia georgei*), *143*
Santa Barbara Island, Xantus' murrelet, 47
Santa Catalina Mountains, Africanized honeybees, 450
Santa Cruz long-toed salamander (*Ambystoma macrodactylum croceum*), 124
Santa Rosa Island, vegetation change, *226*
sapsucker, yellow-bellied (*Sphyrapicus varius*), *20*
Sarasota Bay, seagrass distribution, 273
Sarcobatus vermiculatus (greasewood), 226
satellite imagery, monitoring vegetation with, 467-73
sauger (*Stizostedion canadense*), *432*
Sauromalus obesus (chuckwalla), 74, *328*
savanna, oak, in Wisconsin, 230-32
savannah sparrow (*Passerculus sandwichensis*), *21*, *22*, 297, 299, *303*, *466*
savanna skipper
Atrytonopsis hianna, 232
Erynnis persius, 232
Hesperia
leonardus, 232
metea, 232
sawfly, 161
saw-whet owl, northern (*Aegolius acadicus*), *67*
Sayornis phoebe (eastern phoebe), *20*
scabies (mange mite), *407*
psoroptic (*Psoroptes* spp.), 335
scaled quail (*Callipepla squamata*), *19*
scale insects, 161
Scaphiopus sp. (spadefoot toad), 326
Scaphirhynchus albus (pallid sturgeon), *143*
Scaptomyza sp., 369
Scardinius erythrophthalmus (rudd), 261, 263, *432*
scarlet tanager (*Piranga olivacea*), *20*
scaup
greater (*Aythya marila*), 35, *36*
lesser (*Aythya affinis*), *19*, 35, *36*, 237
Sceloporus
orcutti (granite spiny lizard), *466*
spiny lizard, 326
undulatus (eastern fence lizard), 327
Schinia indiana (white phlox flower moth), 232
Schofieldia spp., 200
Schreiber's curly-tailed lizard (*Leiocephalus schreibersii*), *434*
Schwalbea americana (American chaffseed), 207
Scioto madtom (*Noturus trautmani*), 146
Scirpus
longii (Long's bulrush), 458
maritimus, 220
scissor-tailed flycatcher (*Tyrannus forficatus*), *20*
Scleria cilata, *201*
Scolopax minor (American woodcock), 58, *59*, *62*
Scomberomorus cavalla (king mackerel), 282
scoter, 36, *266*
black (*Melanitta nigra*), *36*
surf (*Melanitta perspicillata*), *36*
white-winged (*Melanitta fusca*), *36*
Scott Paper Company, 217
Scott's oriole (*Icterus parisorum*), *21*
screech-owl
eastern (*Otus asio*), *67*
Virgin Islands (*Otus nudipes newtoni*), *67*
western (*Otus kennicottii*), *67*
whiskered (*Otus trichopsis*), *67*
scrub
big sagebrush, *465*
coastal sage, *465*
coastal sage-chaparral, *465*
scrub jay (*Aphelocoma coerulescens*), *20*
scrub oak (*Quercus ilicifolia*), 166, 167
scrub-shrub
riparian, 291
wetlands, 475
sculpin
mottled (*Cottus bairdii*), *150*
Paiute (*Cottus beldingii*), *150*
Shoshone (*Cottus greenei*), *143*
Utah Lake (*Cottus echinatus*), *143*
Scytalidium sp., 205
seabirds
in Alaska, 49-52
breeding, in California, Oregon, and Washington, 43-48
colonial waterbirds, 53-56
diet, 51-52
factors affecting, 51-52
as indicators of ecological change, 49
monitoring needed, 48
nest sites, 49
status and trends, 16
survival data, 49
threatened and endangered species, 50
sea duck (*Lophodytes cucullatus*), 35
population trends, 35, 36
seagrass
distribution, northern Gulf of Mexico, 273-75
ecosystems, 273
Laguna Madre, Texas, 275-77
seal, 96
bearded (*Erignathus barbatus*), 351
harbor (*Phoca vitulina*), 52, *94*, 96
Hawaiian monk, *95*
northern fur (*Callorhinus ursinus*), 96
ringed (*Phoca hispida*), 351
sea lamprey (*Petromyzon marinus*), 244, 430, 443
sea level rise, coastal and marine ecosystems and, 259
sea lion, 96
California (*Zalophus californianus*), 96
northern or Steller (*Eumetopias jubatus*), 52, *95*, 96
seal plague, *403*
sea otter (*Enhydra lutris*), 110-12, 338, 353-56
future of, 355
legal and illegal harvest, 355
management of, 355
population status, 354-55
population surveys, 354
seaside sparrow, *22*
seatrout, sand (*Cynoscion arenarius*), 410
sea turtle, green (*Chelonia mydas*), *119*
sea urchin (Echinoidea), 283, 354
sedge, 201
sedge wren (*Cistothorus platensis*), 300, 303
sediments
contaminants in, 408, 409-11
freshwater fish and, 146
navigation dam resevoirs and, 235
sediment starvation, coastal erosion and, 278
sediment trapping, wetlands and, 475
Seiurus
aurocapillus (ovenbird), *20*, 25, 420

motacilla (Louisiana waterthrush), 390, 391
noveboracensis (northern waterthrush), *20*
Selasphorus
platycercus (broad-tailed hummingbird), *20*
rufus (rufous hummingbird), *20*
selenium, fish concentrations, 415
Selkirk ecosystem, grizzly bear, 103-5
semipalmated plover (*Charadrius semipalmatus*), 58, *59*, *62*
semipalmated sandpiper (*Calidris pusilla*), 59, *62*
Semotilus atromaculatus (creek chub), *261*, 263
Senecio huachucanus (groundsel), *320*
sequoia, giant (*Sequoiadendron giganteum*), 421
Sequoiadendron giganteum (giant sequoia), 421
Seriola dumerili (greater amberjack), 282
Serranidae (grouper), 282
Serrasalmus spp. (piranha), *432*
Sespe condor sanctuary, 80
sessile-leaved eupatorium (*Eupatorium sessilifolium*), 231
Setophaga ruticilla (American redstart), *20*
sewage discharge, algae and, 255
shad
American (*Alosa sapidissima*), 261, *262*, 263, *265*, 266, *432*
gizzard (*Dorosoma cepedianum*), 240, 261, *262*, 263
threadfin (*Dorosoma petenense*), *432*
shark, great white (*Carcharodon carcharias*), 112
sharphead darter (*Etheostoma acuticeps*), *143*
sharp-shinned hawk (*Accipiter striatus*), 67
shearwater, Newell's (*Puffinus newelii*), *377*, 379
sheep
bighorn (*Ovis canadensis*), 309, 332-33, *407*
badlands (*Ovis canadensis audobonii*), 332
California (*Ovis canadensis californiana*), 332
desert
Ovis canadensis cremnobates, 334
Ovis canadensis mexicana, 320, 334, 335
Ovis canadensis nelsoni, 309, 320, 332, 333-35
Rocky Mountain (*Ovis canadensis canadensis*), 332
Dall (*Ovis dalli*), 338, 347
shellfish, coastal, contaminants in, 408-11
shiner
Arkansas River (*Notropis girardi*), 295, 303-5
blackmouth (*Notropis melanostomus*), *143*
bridle (*Notropis bifrenatus*), *261*, *262*, 263
Cape Fear (*Notropis mekistocholas*), *143*
comely (*Notropis amoenus*), *261*, 263
common (*Luxilus cornutus*), *261*, 263
emerald (*Notropis atherinoides*), 240, 261, *262*, 443
golden (*Notemigonus crysoleucas*), *261*, *262*, *432*
red (*Cyprinella lutrensis*), 151
spotfin (*Cyprinella spiloptera*), *261*, 263
spottail (*Notropis hudsonius*), 261, *262*
shiny cowbird (*Molothrus bonariensis*), 439
Shipsa rotunda (stonefly), 185, *186*
shoal grass (*Halodule wrightii*), 273, 274, 276
shoal sprite (*Amphigyra alabamensis*), *251*
shorebirds
Arctic-breeding, 64
east of the Rocky Mountains, 57-59
ecological domains, 60-61
future trends, 64
habitat alternation and, 63-64
migratory, 61
phalaropes, 59
plovers, 58
sandpipers, 58-59
staging sites, 57-58
status and trends, 16
western, 60-64
shoreline
erosion, 277-79
modification, 259
short-billed dowitcher (*Limnodromus griseus*), 59, *62*
short-distance migratory birds, population trends, 18
short-eared owl (*Asio flammeus*), *67*, 297
shortleaf pine (*Pinus echinata*), 223
shortnose sucker (*Chasmistes brevirostris*), *143*
short-spire elimia (*Elimia brevis*), *251*
short-tailed albatross (*Diomedea albatrus*), 50
short-tailed hawk (*Buteo brachyurus*), *67*
short-term data sets, *4*
short-term variations, 4
Shoshone sculpin (*Cottus greenei*), *143*
shoveler, northern (*Anas clypeata*), *19*, 35, *36*, *266*, *299*, 300, 301
shrew
Arizona (*Sorex arizonae*), *320*
house (*Subcus murinus*), 454
shrike
loggerhead (*Lanius ludovicianus*), *20*, 231, *466*
San Clemente Island shrike (*Lanius ludovicianus mearnsi*), 74
shrimp, pink (*Penaeus duorarum*), 282, 283
shrubland, 469
birds, population trends, 18
shrub savanna, trans pecos, 472
Sialia
currucoides (mountain bluebird), *20*
mexicana (western bluebird), *466*
sialis (eastern bluebird), *20*, *22*
side-blotched lizard (*Uta stansburiana*), 327
Sierra Madre, sky islands, 318
Sierra Madre Occidental, Tarahumara frog (*Rana tarahumarae*), 138
Sierra Nevada, atmospheric deposition and solute transport in, 421-23
Silent Spring (Carson), 413
silkmoths (Saturniidae), 172, *173*
silkworm moth
giant
Citheronia imperialis, 166
Citheronia sepulchralis, 166
Silphium laciniatum, 221
siltation, freshwater fish and, 146
silver, in coastal sediments, 409
silver fir
Abies alba, 315
Pacific (*Abies amabilis*), 315
silverside
Atlantic (*Menidia menidia*), 261, *262*
key (*Menidia conchorum*), 281
silversword
Haleakala
Argyroxiphium sandwicense ssp. *macrocephalum*, 362, 363-64
Argyroxiphium sandwicense ssp. *sandwicense*, 363
Simpsonaias ambigua (salamander mussel), *180*
single-seeded juniper (*Juniperus monosperma*), 223
Sirenia, 267
siskin, pine (*Carduelis pinus*), *21*
Sisquoc condor sanctuary, 80, 81
Sistrurus catenatus (eastern massasauga rattlesnake), 231
Sitta
canadensis (red-breasted nuthatch), *20*
pusilla (brown-headed nuthatch), *20*, *22*
skimmer
black (*Rynchops niger*), 45, 55
west coast, 45-46
skink
brown four-fingered (*Carlia fusca*), *434*
green tree (*Lamprolepis smaragdina*), *434*
Marianas blue-tailed (*Emoia caerulocauda*), 455
rainbow (*Lampropholis delicata*), *434*
skipper
arogos (*Atrytone arogos iowa*), 175
Dakota (*Hesperia dacotae*), 174, 175
ottoe (*Strytone arogos iowa*), 175
Poanes viator, 166
poweshiek (*Oarisma poweshiek*), 175
savanna
Atrytonopsis hianna, 232
Erynnis persius, 232
Hesperia
leonardus, 232
metea, 232
skunk, striped (*Mephitis mephitis*), 302
sky islands
butterflies and moths, 172-73
endemic and insular species, 320-21
grasshopper, 163, 165
non-native species, 321
southwestern ecosystems, 318-22
slash pine (*Pinus elliotti*), 223
slate-colored junco (*Junco hyemalis*), *21*, *22*
slender chub (*Erimystax cahni*), 146
slender glass lizard, western (*Ophisaurus attenuatus*), 231
slender salamander
Pacific (*Batrachoseps pacificus*), *466*
Tehachapi (*Batrachoseps stebbinsi*), *466*
slider, common (*Trachemys scripta*), 119, *434*

slippershell mussel (*Alasmidonta viridis*), *180*
slitshell
  excised (*Gyrotoma excisa*), *251*
  pagoda (*Gyrotoma pagoda*), *251*
  pyramid (*Gyrotoma pyramidata*), *251*
  ribbed (*Gyrotoma pumila*), *251*
  round (*Gyrotoma walkeri*), *251*
  striate (*Gyrotoma lewisii*), *251*
small barred (zebra) dove (*Geopelia striata*), 374, 438
small Kauai thrush (puaiohi) (*Myadestes palmeri*), 380
smallmouth bass (*Micropterus dolomieu*), 147, *261*, *419*, *432*
small spikerush (*Eleocharis parvula*), 220, 458
smelt, rainbow (*Osmerus mordax*), 261, *262*, *432*, 443
Smithsonian Institution, 202, 362
smut, 190-91
snail, *419*
  Emilia, zebra mussel (*Dreissena polymorpha*) and, 447
  Japanese (*Ocenebra japonica*), 429
  non-native species, 429
  Pleistocene disc (*Discus macclintocki*), 176-77
  Potamopyrgus antipodarum, 429
  in southwestern sky islands, 319-20
  Stagnicola, zebra mussel (*Dreissena polymorpha*) and, 447
snail kite (*Rostrhamus sociabilis*), 66, *67*
snake, 117
  Brahminy blind (*Rhamphotyphlops braminus*), *434*
  brown tree (*Boiga irregularis*), 433, *434*, 435, 436, 454-56
  checkered garter (*Thamnophis marcianus*), 452, 453
  eastern diamondback rattlesnake (*Crotalus adamanteus*), 129
  eastern indigo (*Drymarchon corais couperi*), 129
  Florida pine (*Pituophis melanoleucus mugitus*), 129
  garter (*Thamnophis* spp.), 132, 326, 328
  Mexican garter (*Thamnophis eques*), *320*, 433, 435, 452-53
  narrow-headed garter(*Thamnophis rufipunctatus*), 327, *328*
  pine (*Pituophis melanoleucus*), 327
  red diamond rattlesnake (*Crotalus ruber*), *466*
  stiped whipsnake (*Masticophis taeniatus*), 327
  western terrestrial garter (*Thamnophis elegans*), 133
Snake River
  non-native species, 429
  white sturgeon (*Acipenser transmontanus*), 156
snakeweed (*Gutterezia sarothrae*), 226
snapper
  gray (*Lutjanus griseus*), 282-83
  mutton (*Lutjanus analis*), 282-83
  red (*Lutjanus campechanus*), 282
  yellowtail (*Ocyurus chrysurus*), 282
snapping turtle
  alligator (*Macroclemys temminckii*), 119
  Chelydra serpentina, *434*
snipe, common (*Gallinago gallinago*), *19*, 59, *62*
snow goose
  Chen caerulescens, *266*, *388*
  lesser (*Chen caerulescens caerulescens*), 30, 33
snowpack duration, subalpine tree growth and, 316
snowshoe hare (*Lepus americanus*), 359
snow-wreath (*Neviusia* spp.), 202
snowy egret (*Egretta thula*), *54*, 55
snowy milkweed (*Asclepias speciosa*), 171
snowy owl (*Nyctea scandiaca*), *67*
snowy plover (*Charadrius alexandrinus*), 58, *59*, *62*
snuffbox (*Epioblasma triquetra*), *180*
sockeye salmon (*Oncorhynchus nerka*), 344, *345*, 346, *432*
soft pesticides, 415
softshell turtle
  Chinese (*Pelodiscus sinensis*), *434*
  spiny (*Apalone spinifera*), *119*, *434*
  wattle-necked (*Palea steindachneri*), *434*
softwoods, 215
Soil Bank Program, 296, 300
Soil Conservation Service, 201
soil-dwelling protozoa, 252
sole
  English (*Parophrysvetulus*), 410, 411
  flathead (*Hippoglossoides elassodon*), 409
Solenopsis
  fire ant, 441
  invicta (red fire ant), 131
Solidago albopilosa (white-haired goldenrod), 401
solitary sandpiper (*Tringa solitaria*), *59*, *62*
solitary vireo (*Vireo solitarius*), *20*, 390, 391
Somatogyrus, 251
songbirds
  non-native species, 439
  winter population trends, 21-23
song sparrow (*Melospiza melodia*), *21*, *22*, *299*
Sonora chub (*Gila ditaenia*), *320*
Sonoran Desert
  Africanized honeybees, 450, 451
  desert tortoise (*Gopherus agassizii*) in, 135-36
Sonora sucker (*Catostomus insignis*), *150*
Sonorella, 319
Sophora arizonica (sophora), *320*
sophora (*Sophora arizonica*), *320*
sora (*Porzana carolina*), *299*
Sorex arizonae (Arizona shrew), *320*
Sorghastrum nutans, 221
sorrel, dock (*Rumex orthoneurus*), *320*
South American ground lizard (*Ameiva ameiva*), *434*
South American whiptail lizard (*Cnemidophorus lemniscatus*), *434*
southeastern bat (*Myotis austroriparius*), 176
Southeastern Biological Science Center (SBSC), 445
Southeastern Regional Flora Information System, 204
southeastern United States
  freshwater fish, 144-46
  hydrologic regions, 144-45
Southern Appalachian Region, black bear, 102
Southern Appalachians hydrologic region, 145
Southern California Association of Governments, 466
southern combshell (*Epioblasma penita*), *178*
southern green stink bug (*Nezara viridula*), 366
Southern Science Center, 276
southern United States
  forested wetlands, 216-18
  interior cypress forest, *465*
  mixed chaparral, *465*
  pinelands
    altered fire regimes, 222, 223
    red-cockaded woodpecker and, 86-87
  rare terrestrial ecological communities, 220
  vascular plants, 203-4
South Farallon Islands
  Cassin's auklet, 46
  cormorants, 44-45
  murres, 47
  pigeon guillemot, 46
  rhinoceros auklet, 46
  storm-petrels, 44
  western gull, 45
southwestern toad (*Bufo microscaphus*), *466*
southwestern United States
  riparian ecosystems, 290-91
  sky island ecosystems, 318-22
  vascular plants, 202-3
southwestern willow flycatcher (*Empidonax traillii extimus*), 89-91
  brood parasitism, 90, 91
Soyedina vallicularia (stonefly), 185, *186*
spadefoot toads (*Scaphiopus* sp.), 326
sparrow
  American tree (*Spizella arborea*), *22*, *23*
  Baird's (*Ammodramus bairdii*), *21*, *299*, *303*
  black-throated (*Amphispiza bilineata*), *21*
  Brewer's (*Spizella breweri*), *21*, 297
  Cassin's (*Aimophila cassinii*), *21*, 297, 298
  chipping (*Spizella passerina*), *21*, *299*
  clay-colored (*Spizella pallida*), *21*, 297, *299*, *303*
  Eurasian tree (*Passer montanus*), 454
  field (*Spizella pusilla*), *21*, *22*, 231, *299*
  fox (*Passerella iliaca*), *21*
  grasshopper (*Ammodramus savannarum*), *21*, 297, 302, 303, *303*, *466*
  Henslow's (*Ammodramus henslowii*), 297
  house (*Passer domesticus*), *21*, *299*, 437, *438*, 439
  lark (*Chondestes grammacus*), *21*, 297, *299*
  Lincoln's (*Melospiza lincolnii*), *21*
  rufous-crowned (*Aimophila ruficeps*), *466*
  sage (*Amphispiza belli*), *21*, *466*
  savannah (*Passerculus sandwichensis*), *21*, *22*, 297, 299, *303*, *466*
  seaside, *22*
  song (*Melospiza melodia*), *21*, *22*, *299*
  swamp (*Melospiza georgiana*), *21*
  vesper (*Pooecetes gramineus*), *21*, *22*, 231, 297, *299*
  white-crowned (*Zonotrichia leucophrys*), *21*, *22*
  white-throated (*Zonotrichia albicollis*), *21*, *22*
  winter population trends, 21

species, *See also* endangered species; rare species; threatened species
  biodiversity, 6-7
  conservation, priorities for, 399-400
  detecting long-term trends, 4
  genetic variability in, 147
  inventory and monitoring programs, 2
  loss, Guam, 454-55
  migration, climate change and, 394
  status and trend studies, 1-2
species diversity, *See also* biodiversity; genetic diversity
  altered, Colorado River Basin, 151-52
  Hawaii, 368
  Mobile Bay basin, watershed disturbances and, 251
Species Survival Commission, 118
speckled chub, Arkansas River (*Macrhybopsis aestivalis tetranemus*), 295, 303-5
speckled dace
  Meadow Valleys (*Rhinichthys osculus* ssp.), *150*
  pahranagat (*Rhinichthys osculus velifer*), *150*
  Preston (*Rhinichthys osculus* ssp.), *150*
  Rhinichthys osculus osculus, *150*
  White River (*Rhinichthys osculus* ssp.), *150*
spectacled caiman (*Caiman crocodilus*), 433, *434*, 435
spectacled eider, *36*
Speoplatyrhinus poulsoni (Alabama cavefish), *143*
Spermophilus spp. (ground squirrel), 109
Speyeria idalia (regal fritillary), *166*, 174, *175*
Sphaerodactylus
  argus (ocellated dwarf gecko), *434*
  elegans (ashy gecko), *434*
  notatus (reef gecko), *434*
Sphagnum
  angermanicum Melin (peat moss), 210
  peat moss, 198
sphinx moth
  Darapsa versicolor, 166
  Hemaris gracilis, 166
  Manduca blackburni, 367
  Proserpinus juanita, *174*
Sphyrapicus varius (yellow-bellied sapsucker), *20*
spike (*Elliptio dilatata*), *180*
spikedace (*Meda fulgida*), *150*, *320*
spikerush, small (*Eleocharis parvula*), 220, 458
spinedace
  Big Spring (*Lepidomeda mollispinis pratensis*), *150*
  Little Colorado (*Lepidomeda vittata*), *150*
  pahranagat (*Lepidomeda altivelis*), *150*
  Virgin (*Lepidomeda mollispinis mollispinis*), *150*
  White River (*Lepidomeda albivallis*), *143*, *150*
spinner, whitebelly, *95*
spiny lizard
  granite (*Sceloporus orcutti*), *466*
  Sceloporus sp., 326
spiny lobster, West Indies (*Panulirus argus*), 283
spiny softshell turtle (*Apalone spinifera*), *119*
spiny-tailed iguana, Mexican (*Ctenosaura pectinata*), *434*
spiny water flea (*Bythotrephes cederstroemi*), 429, 444
Spiranthes delitescens (lady's-tresses), *320*
Spiza americana (dickcissel), *20*, 303
Spizella
  arborea (American tree sparrow), *22*, 23
  breweri (Brewer's sparrow), *21*, 297
  pallida (clay-colored sparrow), *21*, 297, *299*, *303*
  passerina (chipping sparrow), *21*, *299*
  pusilla (field sparrow), *21*, *22*, 231, *299*
Sporobolus heterolepis, 221
spot (*Leiostomus xanthurus*), 409
spotfin shiner (*Cyprinella spiloptera*), *261*, 263
spottail shiner (*Notropis hudsonius*), 261, *262*
spotted bass (*Micropterus punctulatus*), 147, *432*
spotted dolphin, *94*, *95*
spotted (laced-necked) dove (*Streptopelia chinensis*), 438
spotted frog (*Rana pretiosa*), 132-33, *328*
spotted owl (*Strix occidentalis*), 16, *67*, 69
  California (*Strix occidentalis occidentalis*), *67*
  Mexican (*Strix occidentalis lucida*), *67*, 309, *320*, 330-31
  northern (*Strix occidentalis caurina*), *67*, 193, 387
spotted salamander (*Ambystoma maculatum*), *419*
spotted sandpiper (*Actitis macularia*), 59, *62*
spotted turtle (*Clemmys guttata*), 119
Sprague's pipit (*Anthus spragueii*), *20*
spreading globeflower (*Trollius laxus*), *209*
spring cavefish, *Forbesichthys agassiz*, 177
springfish
  Hiko (*Crenichthys baileyi grandis*), *150*
  Moapa (*Crenichthys baileyi moapae*), *150*
  Moorman (*Crenichthys baileyi thermophilus*), *143*, *150*
  Preston (*Crenichthys baileyi albivallis*), *143*, *150*
  Railroad Valley (*Crenichthys nevadae*), *143*, *150*
  White River (*Crenichthys baileyi baileyi*), *143*
spring pygmy sunfish (*Elassoma* sp.), *143*
springs, Illinois, biota, 176-77
sprite, shoal (*Amphigyra alabamensis*), *251*
spruce
  Englemann (*Picea engelmannii*), 316, 317, 321
  Norway (*Picea abies*), 315
  red (*Picea rubens*), 205
spruce-fir forest, 472
spurge hawkmoth (*Hyles euphorbiae*), 162
squawfish, Colorado (*Ptychocheilus lucius*), *150*, 151
squawfoot (*Strophitus undulatus*), *180*
squirrel
  Abert's (*Sciuris aberti*), 321
  ground (*Spermophilus* spp.), 109
  Mt. Graham red (*Tamasciuris hudsonicus grahamensis*), 320
  northern flying (*Glaucomys sabrinus*), 192, 193
staging sites, shorebirds, 57-58
Stagnicola (snail), 447
star grass, 273, 274
starling
  European (*Sturnus vulgaris*), *20*, 413, 437, *438*, 439, *440*, 441
  Micronesian (*Aplonis opaca*), 455
starry flounder (*Platichthys stellatus*), 409, 411
state lands, land protection status, 463-64
status studies, purpose of, 1-2
Stelgidopteryx serripennis (northern rough-winged swallow), *20*
Steller sea lion (northern sea lion) (*Eumetopias jubatus*), 52, *95*, 96
Steller's eider, *36*
Steller's jay (*Cyanocitta stelleri*), *20*
Stephananodiscus sp., *257*
Stephanodiscus binderanus (algae), 444
Stephens' kangaroo rat (*Dipodomys stephensi*), *466*
steppe, sagebrush-grass, altered fire regimes in, 222-24
Sterna
  antillarum (least tern), 43, 45, *54*, 56
    athalassos (interior least tern), *54*, *55*, 56
    browni (California least tern), 56, 74
  bergii (great crested tern), 373
  caspia (Caspian tern), 45
  dougallii (roseate tern), *54*, *55*, 56
  elegans (elegant tern), 45
  forsteri (Forster's tern), 45, *54*, 56
  hirundo (common tern), *53*, *54*, 56
  nilotica (gull-billed tern), 46, *54*, 56
  paradisaea (arctic tern), 46, 51
Sternotherus depressus (flattened musk turtle), *119*
stifftails, *36*
stilt
  black-necked (*Himantopus mexicanus*), *19*, *59*, *62*
  Hawaiian (*Himantopus mexicanus knudseni*), *377*
stilt sandpiper (*Calidris himantopus*), *59*, *62*
stink bug
  Coleotichus, 366
  Oechalia, 366
  Pentatomidae, 366
  southern green (*Nezara viridula*), 366
Stizostedion
  canadense (sauger), *432*
  vitreum
    glaucum (blue pike), *143*
    vitreum (walleye), 242, *243*
    walleye, *432*, 443
stocking
  loss of genetic integrity through, 147-48
  recovery and, 326
stone crab (*Menippe mercanaria*), 282
stonefly, 182, 183
  Acroneuria
    abnormis, *186*
    evoluta, *186*
    filicis, 185, *186*
    frisoni, *186*
    internata, *186*
    perplexa, 185, *186*
  Agnetina flavescens, *186*
  Allocapnia
    forbesi, *186*

granulata, 186, *186*
illinoensis, 185, *186*
mystica, *186*
nivicola, 185, *186*
recta, *186*
rickeri, *186*
smithi, 185, *186*
vivipara, *186*, 187
Alloperla
caudata, 185, *186*
roberti, 185, *186*
Amphinemura
delosa, *186*
nigritta, 185, *186*
varshava, *186*
Attaneuria ruralis, *186*
Clioperla clio, *186*
Haploperla brevis, 185, *186*
Hydroperla
crosbyi, *186*
fugitans, *186*
Illinois, 184-87
Isogenoides varians, 185, *186*
Isoperla
bilineata, *186*, 187
burksi, 185, *186*
conspicua, 185, *186*
decepta, *186*
longiseta, *186*
marlynia, 185, *186*
mohri, *186*
nana, *186*
richardsoni, *186*
Leuctra
rickeri, *186*
sibleyi, 185, *186*
tenuis, 185, *186*
Nemocapnia carolina, 185, *186*
Nemoura trispinosa, *186*
Neoperla clymene, *186*
Paracapnia angulata, 185, *186*
Paragnetina
kansensis, *186*
media, 185, *186*
Perlesta decipiens, *186*
Perlinelia
drymo, *186*
ephrye, *186*
Prostoia completa, 185, *186*
Pteronarcys pictetii, *185*, *186*
Shipsa rotunda, 185, *186*
Soyedina vallicularia, 185, *186*
Strophopteryx fasciata, *186*
Taeniopteryx
burksi, *186*, 187
lita, *186*
metequi, *186*
nivalis, *186*
parvula, *186*
Zealeuctra
claaseni, *186*
fraxina, 185, *186*
narfi, 185, *186*
stoneroller, Mexican (*Campostoma ornatum*), *320*
stork, wood (*Mycteria americana*), *54*, 55
storm-petrel, 43-44
ashy (*Oceanodroma homochroa*), 44
black (*Oceanodroma melania*), 44
fork-tailed (*Oceanodroma furcata*), 44
Leach's (*Oceanodroma leucorhoa*), 43-44
Strategic Plan for the Management of Wild Horses and Burros on Public
Lands, 457
streams, *See also* riparian ecosystems; waterways
atmospheric pollution and, 421-23
banks, 285
water quality, diatoms and, 258
Streptopelia
bitorquata (Philippine turtle dove), 454
chinensis (spotted or lace-necked dove), 438
decaocto (Eurasian collared-dove), 438-39
risoria (ringed turtle-dove or Barbary dove), 438
striate slitshell (*Gyrotoma lewisii*), *251*
striped bass (*Morone saxatilis*), 261, *262*, 263, 265, *432*
striped mullet (*Mugil cephalus*), *150*
striped newt (*Notophthalmus perstriatus*), 129, 130
striped rocksnail (*Leptoxis vittata*), *251*
striped skunk (*Mephitis mephitis*), 302
striped whipsnake (*Masticophis taeniatus*), 327
Strix
nebulosa (great gray owl), *67*
occidentalis (spotted owl), 16, *67*, 69
caurina (northern spotted owl), *67*, 193, 387
lucida (Mexican spotted owl), *67*, 309, *320*, 330-31
occidentalis (California spotted owl), *67*
varia (barred owl), *67*, 387
Strombus gigas (queen conch), 282
Strophitus undulatus (squawfoot), *180*
Strophopteryx fasciata (stonefly), *186*
Strytone arogos iowa (ottoe skipper), 175
stump-toed gecko (*Gehyra mutilata*), 455
sturgeon, 145
pallid (*Scaphirhynchus albus*), *143*
white (*Acipenser transmontanus*), 154-56
Sturnella
magna (eastern meadowlark), *21*, *22*, 297
neglecta (western meadowlark), *21*, *22*, 297, 299, 303
Sturnus vulgaris (European starling), *20*, 413, 437, *438*, 439, *440*, 441
subalpine forests, 314-17
fir (*Abies lasiocarpa*), 316, 317
future of, 317
larch (*Larix lyallii*), 316
patterns of establishment, 316-17
tree growth, 315-16
subalpine meadows, 314-17
tree establishment in, 316
Subcus murinus (house shrew), 454
submerged aquatic vegetation
Chesapeake Bay, 264
navigation dams and, 237
subsistence fishing, Beaufort Sea, central Alaska, 341
sucker, 145
bluehead (*Catostomus discobolus discobolus*), *150*
cui-ui (*Chasmistes cujus*), 323-24
desert (*Catostomus clarki clarki*), *150*
flannelmouth (*Catostomus latipinnis*), *150*
hairlip (*Moxostoma lacerum*), 146
June (*Chasmistes liorus mictus*), *143*
Little Colorado (*Catostomus* sp.), *150*
Lost River (*Deltistes luxatus*), *143*
Meadow Valley (*Catostomus clarki* ssp.), *150*
mountain (*Catostomus platyrhynchus*), *150*
northern hog (*Hypentelium nigricans*), *261*, 263
razorback (*Xyrauchen texanus*), *143*, *150*, 151, 309, 324-26
shortnose (*Chasmistes brevirostris*), *143*
Sonora (*Catostomus insignis*), *150*
Suni (*Catostomus discobolus yarrowi*), *150*
white (*Catostomus commersoni*), 151, *261*, *262*
White River (*Catostomus clarki intermedius*), *143*, *150*
Zuni (or Zuni bluehead) (*Catostomus discobolus yarrowi*), *143*
sulfur, atmospheric deposition of, 421-23
summer tanager (*Piranga rubra*), *20*
sunfish
green (*Lepomis cyanellus*), 240, *432*
pumpkinseed (*Lepomis gibbosus*), *261*, *262*, *419*, *432*
redbreast (*Lepomis auritus*), *261*, *262*, *432*
redear (*Lepomis microlophus*), *432*
spring pygmy (*Elassoma* sp.), *143*
surfbird (*Aphriza virgata*), *62*, *64*
surf scoter (*Melanitta perspicillata*), *36*
Surnia ulula (northern hawk owl), *67*
Sus scrof (wild pig), 440, 441
Swainson's hawk (*Buteo swainsoni*), *67*
Swainson's thrush (*Catharus ustulatus*), *20*
swallow
bank (*Riparia riparia*), *20*, *299*
barn (*Hirundo rustica*), *20*, *299*
cliff (*Hirundo pyrrhonota*), *20*, *299*
northern rough-winged (*Stelgidopteryx serripennis*), *20*
tree (*Tachycineta bicolor*), *20*, *420*
violet-green (*Tachycineta thalassina*), *20*, *466*
swallow-tail, two tailed (*Papilio multicaudata*), *173*
swallow-tailed kite, American (*Elanoides forficatus*), *67*
"Swampbuster" provision, Public Law 100-233 (1985), 217
Swamp Lands Acts, 217
swamp sparrow (*Melospiza georgiana*), *21*
swan, mute (*Cygnus olor*), 266, 387, 437, 438

swift
black (*Cypseloides niger*), *20*
chimney (*Chaetura pelagica*), *20*
white-throated (*Aeronautes saxatalis*), *20*
swiftlet, gray (*Aerodramus vanikorensis*), 455
swordtail, green (*Xiphophorus helleri*), *432*
sycamore, American (*Platanus occidentalis*), 228
sylvanti plague, in prairie dogs, 106, 107
symbiotic protozoa, 253
Synthliboramphus
antiquus (ancient murrelet), 46, 51
hypoleucus (Xantus' murrelet), 47-48
Syringodium filiforme (manatee grass), 273, 274, 276, 277
Systematic Entomology Laboratory, 161

Tabellaria sp., *257*
Tachycineta
bicolor (tree swallow), *20*, *420*
thalassina (violet-green swallow), *20*, *466*
Taeniatherum caput-medusae (medusahead), 222
Taenidia integerrima (yellow pimpernel), 231
Taeniopteryx
burksi (stonefly), *186*, 187
lita (stonefly), *186*
metequi (stonefly), *186*
nivalis (stonefly), *186*
parvula (stonefly), *186*
tailed frog (*Ascaphus truei*), 125
tall-grass prairie, 221
butterfly community, 174-76
distribution, 175-76
fire, 175
management, 174, 175-76
tamarisk
Tamarix, southwestern willow flycatcher and, 89
Tamarix ramoisissima, 286, 288, 289
Tamarix
ramoisissima (tamarisk), 286
saltcedar, 290-91
Tamasciuris hudsonicus grahamensis (Mt. Graham red squirrel), 320
Tamias obscurus (California chipmunk), *466*
Tampa Bay
coastal wetland loss, 271-72
seagrass distribution, 273-74
tanager
scarlet (*Piranga olivacea*), *20*
summer (*Piranga rubra*), *20*
western (*Piranga ludoviciana*), *20*
tangerine darter (*Percina aurantiaca*), *146*
tapeworm, Asian (*Bothriocephalus opsarichthydis*), 429
Tarahumara frog (*Rana tarahumarae*), 118, 138-40, 321
heavy metal poisoning, 139, 140
population decline, 138-40
Tarahumara Frog Reestablishment Oversight Group, 140
tattler, wandering (*Heteroscelus incanus*), *62*
Taverner's Canada goose (*Branta canadensis taverneri*), 28, 33
tawny crescent butterfly (*Phyciodes batesii*), *166*, 232
Taxidea taxus (American badger), 108-9
Taxodium distichum (baldcypress), 216
Tayassu tajacu (javelina), 322
tea, New Jersey (*Ceanothus americanus*), 231
teal
blue-winged (*Anas discors*), *19*, 35, *36*, *266*, 296, *299*, 300, 301, *303*
cinnamon, *36*
green-winged (*Anas crecca*), 35, *36*, *266*
Tehachapi slender salamander (*Batrachoseps stebbinsi*), *466*
Telespyza
cantans (Laysan finch), *377*
ultima (Nihoa finch), *377*
temperature change, *See also* climate change; global climate change
vascular plants and, 392-95
tench (*Tinca tinca*), *432*
Tennessee warbler (*Vermivora peregrina*), *20*
tern, 51
arctic (*Sterna paradisaea*), 46, 51
black (*Chlidonias niger*), *19*, 46, *54*, *55*, 56, 299
California least (*Sterna antillarum browni*), 56, 74
Caspian (*Sterna caspia*), 45
common (*Sterna hirundo*), *53*, *54*, 56
elegant (*Sterna elegans*), 45
Forster's (*Sterna forsteri*), 45, *54*, 56
great crested (*Sterna bergii*), 373
gull-billed (*Sterna nilotica*), 46, *54*, 56
interior least (*Sterna antillarum athalassos*), *54*, *55*, 56
least (*Sterna antillarum*), 43, 45, 56
roseate (*Sterna dougalli*), *54*, *55*, 56
west coast, 45-46
Terrapene, carolina (common box turtle), 119, *434*
terrapin, diamondback (*Malaclemys terrapin*), *119*, 120
terrestrial ecosystems
acid deposition (acid rain) and, 418, 419-20
biological diversity in, 213
classification systems, 219
fire-adapted, 214, 222-24
forest resources, 241-45
human activities and, 214
imperiled, 213
introduced diseases and, 213
National Park vegetation change, 224-28
natural fire regimes, 213-14
non-native vertebrates, *429*
oak savannas, 230-32
overview, 213-14
rare communities, 218-21
southern forested wetlands, 216-17
whitebark pine (*Pinus albicaulis*), 228-30
terrestrial garter snake, western (*Thamnophis elegans*), 133
Teshekpuk Lake Special Area (TLSA), habitats, 33
Tessaria sericea (arrowweed), 287, 288
tetrabutyltin, *408*
Texas blind salamander (*Typhlomolge rathbuni*), 124
Texas Game and Fish Department, 335
Texas horned lizard (*Phrynosoma cornutum*), *434*
Texas Parks and Wildlife Department, 276
Texas tortoise (*Gopherus berlandieri*), *119*
Thalassia testudinum (turtle grass), 273, 274, 276
thalli, 255
Thamnophis, 132, 326
elegans (western terrestrial garter snake), 133
eques (Mexican garter snake), *320*, 433, 435, 452, 453
marcianus (checkered garter snake), 452
rufipunctatus (narrow-headed garter snake), 327, *328*
Tharp's Creek, 422
Thasus sp. (leaf-footed bug), *161*
Theragra chalcogramma (walleye pollock), 51-52
thick-billed murre (*Uria lomvia*), 50, *51*
thick-billed parrot (*Rhynchopsitta pachyrhyncha*), *320*
thicktail chub (*Gila crassicauda*), *143*
thrasher
brown (*Toxostoma rufum*), *20*, 231, *299*
California (*Toxostoma redivivum*), *20*
curve-billed (*Toxostoma curvirostre*), *20*
pearly-eyed (*Margarops fuscatus*), 85
sage (*Oreoscoptes montanus*), *20*
threadfin shad (*Dorosoma petenense*), *432*
threatened species, *See also* endangered species; rare species
Audubon's crested caracara (*Caracara plancus audubonii*), 82
habitat protection and, 466
number on federal lands, 400
occurrences of, 398, 400
seabirds, 50
surveys, *5*
threehorn wartyback (*Obliquaria reflexa*), *180*
threeridge (*Amblema plicata plicata*), *180*
thrush
gray-cheeked (*Catharus minimus*), *20*
hermit (*Catharus guttatus*), *20*
small Kauai (puaiohi) (*Myadestes palmeri*), 380
Swainson's (*Catharus ustulatus*), *20*
varied (*Ixoreus naevius*), *20*
wood (*Hylocichla mustelina*), *20*, 25, 420
Thryomanes bewickii (Bewick's wren), *20*
Thryothorus ludovicianus (Carolina wren), *20*
Thymallus arcticus (Arctic grayling), *432*
Tiaroga cobitis (loach minnow), *320*
tiger moth
Arctiidae, 172, *173*
Gnophaela vermiculata, *173*
tiger muskellunge (*Esox lucius x masquinongy*), *432*
tiger salamander (*Ambystoma tigrinum*), 327, *434*
tilapia
blue (*Tilapia aurea*), 430, *432*
Mozambique (*Tilapia mossambica*), *432*
redbelly (*Tilapia zilli*), *432*

Tilapia
aurea (blue tilapia), 430, *432*
mossambica (Mozambique tilapia), *432*
zilli (redbelly tilapia), *432*
timberland, *See also* forest land
characteristics of, 215
forestry practices, 87, 88
removals from, 215
timber wolf (*Canis lupus lycaon*), 98-99
tin, in coastal sediments, 409
Tinca tinca (tench), *432*
titmouse
black-crested (*Parus bicolor atricristatus*), *20*
plain (*Parus inornatus*), *20*
tufted (*Parus bicolor*), *20*, 21, *22*, 387
toad, 124
Bufo sp., 326
Bufo americanus, 435
Great Plains narrow-mouthed (*Gastrophyrne olivacea*), *320*
marine (*Bufo marinus*), 433, *434*, 435, 441
southwestern (*Bufo microscaphus*), *466*
spadefoot (*Scaphiopus* sp.), 326
western (*Bufo boreas*), 125, 327, *328*
Woodhouse's (*Bufo woodhousii*), 327
tokay gecko (*Gekko gekko*), *434*
Tolyposporium niveum, 190
topminnow
Gila (*Poeciliopsis occidentalis*), *143*, *150*
whiteline (*Fundulus albolineatus*), 146
Yaqui (*Poeciliopsis occidentalis sonoriensis*), *320*
torrent salamander (*Rhyacotriton* spp.), 125
Torreya taxifolia, 205
tortoise, 119
desert (*Gopherus agassizii*), 16, 74, 119, 135-36, *328*, *407*
gopher (*Gopherus polyphemus*), 119, 129, 130
Texas (*Gopherus berlandieri*), *119*
towhee
brown (*Pipilo fuscus*), *21*
California (*Pipilo californicus*), *21*
green-tailed (*Pipilo chlorurus*), *20*
rufous-sided (*Pipilo erythrophthalmus*), *20*, *299*
Townsend's warbler (*Dendroica townsendi*), *20*
toxaphene, 414
toxicosis, *404*
toxic substances, *See also* contaminants
aquatic insects and, 183
Toxolasma parvus (lilliput), *180*
Toxostoma
curvirostre (curve-billed thrasher), *20*
redivivum (California thrasher), *20*
rufum (brown thrasher), *20*, 231, *299*
Trachemys scripta (common slider), 119, *434*
Traffic USA, 101
translocation programs, 405-7
disease transmission and, 406, *407*
non-native fish, 405, 432-33
trends, 405-6
trans-pecos shrub savanna, 472
treefrog
Cuban (*Osteopilus septentrionalis*), *434*
eastern dwarf (*Litoria fallax*), *434*
mountain (*Hyla eximia*), 327
tree gecko (*Gehyra oceanica*), 455
tree lizard (*Urosaurus ornatus*), 327
tree skink, green (*Lamprolepis smaragdina*), *434*
tree snake, brown (*Boiga irregularis*), 433, *434*, 435, 436, 454-56
tree sparrow
American (*Spizella arborea*), *22*, 23
Eurasian (*Passer montanus*), 454
tree swallow (*Tachycineta bicolor*), *20*, *420*
trend studies, purpose of, 1-2
tributyltin, *408*
Trichechus manatus (manatee), 259
trichomoniasis, *404*
tricolored blackbird (*Agelaius tricolor*), *21*, *466*
Trifolium
microcephalum, 208
stoloniferum (running buffalo clover), 208
Trimerotropic
huroniana, 164
maritima, 164
Tringa
flavipes (lesser yellowlegs), 59, *62*
melanoleuca (greater yellowlegs), 59, *62*
solitaria (solitary sandpiper), *59*, *62*
Trionyx, *steindachneri* (wattle-necked softshell turtle), *434*
troglobitic amphipod (*Gammarus troglophilus*), 177
Troglodytes
aedon (house wren), *20*, *299*
troglodytes (winter wren), *20*
Trollius laxus (spreading globeflower), *209*
trout
Alvord cutthroat (*Oncorhynchus clarki* spp.), *143*
Apache (*Oncorhynchus apache*), 147, *150*, 320, *320*
brook (*Salvelinus fontinalis*), 419, *432*, *440*, 441
brown (*Salmo trutta*), *419*, 429, *432*, *440*, 441, 443
Colorado cutthroat (*Oncorhynchus clarki pleuriticus*), *150*
cutthroat (*Oncorhynchus clarki*), *432*
Gila (*Oncorhynchus gilae*), *150*
golden (*Oncorhynchus aguabonita*), *432*
lake (*Salvelinus namaycush*), 154, 242, *243*, 244-46, 414, *432*
non-native, northern leopard frog and, 133
rainbow (*Oncorhynchus mykiss*), 147, 153, 320, 328-29, *419*, *432*, *440*, 441, 443
westslope cutthroat (*Oncorhynchus clarki lewisi*), 153-54
Yellowstone cutthroat (*Oncorhynchus clarki bouvieri*), 153-54
Truckee River, cui-ui (*Chasmistes cujus*), 323-24
truffles
ascomycetes, 193
false (basidiomycetes), 193
Truncilla
donaciformis (fawnsfoot), *180*
truncata (deertoe), *180*
Tryngites subruficollis (buff-breasted sandpiper), 59, *62*, 64
Tsuga mertensiana (mountain hemlock), 315
tufted puffin (*Fratercula cirrhata*), 46, 51
tufted titmouse (*Parus bicolor*), *20*, 21, *22*, 387
tui chub
Fish Creek Springs (*Gila bicolor euchila*), *143*
Independence Valley (*Gila bicolor isolata*), *143*
tule white-fronted goose (*Anser albifrons gambeli*), 30, 31
Tulotoma, 249
magnificia, *250*, 252
tundra ecosystem
monitoring, 338-40
Northeast Alaska, 338-40
status, 340
tundra (Arctic) hare (*Lepus othus* or *Lepus timidus*), 338, 359
tupelo
Nyssa aquatica, 216
Nyssa Sylvatica var *biflora*, 216
Turdus migratorius (American robin), *20*, 231, *299*, 386
turkey
Meleagris merriami, 320
Meleagris mexicana, 320
ocellated (*Meleagris ocellata*), 70
wild (*Meleagris gallopavo*), 16, 70-71, 231, 387, 405, *407*
turkey vulture (*Cathartes aura*), *19*, *67*
turnstone
black (*Arenaria melanocephala*), *62*, 64
ruddy (*Arenaria interpres*), 59, *62*
Tursiops truncatus (bottlenose dolphin), *94*, 96
turtle, 117, 118-20
Alabama red-bellied (*Pseudemys alabamensis*), *119*
alligator snapping (*Macroclemys temminckii*), 119
Barbour's map (*Graptemys barbouri*), *119*, *120*
Blanding's (*Emydoidea blandingii*), *119*
bog (*Clemmys muhlenbergii*), 119, 458
Cagle's map (*Graptemys caglei*), *119*
Chinese softshell (*Pelodiscus sinensis*), *434*
commerical fisheries, 122
common box (*Terrapene carolina*), 119, *434*
common slider (*Trachemys scripta*), 119, *434*
conservation of, 120
diamondback terrapin (*Malaclemys terrapin*), *119*, 120
flattened musk (*Sternotherus depressus*), *119*
Graptemys spp., 120
green (*Chelonia mydas*), 121, 122-23
green sea (*Chelonia mydas*), *119*
hawksbill (*Eretmochelys imbricata*), 121, 123
Kemp's ridley (*Lepidochelys kempii*), *119*, 121, 123
leatherback (*Dermochelys coriacea*), 121, 123
loggerhead (*Caretta caretta*), 121, 122-23
marine, 119, 120, 121-23
Mexican rough-footed mud (*Kinosternon hirtipes*), *119*
painted (*Chrysemys picta*), 327
red-bellied (*Pseudemys rubriventris*), *119*
ringed map (*Graptemys oculifera*), *119*

slider (*Trachemys scripta*), 119
snapping (*Chelydra serpentina*), *434*
spiny softshell (*Apalone spinifera*), *119*, *434*
spotted (*Clemmys guttata*), 119
status and trends, 119-20
value of, 118
wattle-necked softshell (*Palea steindachneri*), *434*
western pond (*Clemmys marmorata*), *119*, *466*
wood (*Clemmys insculpta*), *119*
yellow-blotched map (*Graptemys flavimaculata*), *119*
yellow mud (*Kinosternon flavescens*), *119*
turtle-dove
Javanese (*Streptopelia bitorquata*), 454
ringed (Barbary dove) (*Streptopelia risoria*), 438
turtle excluder devices (TEDs), 123
turtle grass (*Thalassia testudinum*), 273, 274, 276
two-striped grasshopper (*Melanoplus bivittatus*), 164
two tailed swallow-tail (*Papilio multicaudata*), *173*
Tyler State Park (Pennsylvania), 113
Tympanuchus cupido (prairie chicken), 438
Typha
cattail, 249
latifolia (cattail), 444
Typhlomolge rathbuni (Texas blind salamander), 124
Tyrannus
forficatus (scissor-tailed flycatcher), *20*
tyrannus (eastern kingbird), *20*, *299*, *420*
verticalis (western kingbird), *20*, *299*
vociferans (Cassin's kingbird), *20*
Tyto alba (barn owl), *67*, 231

'ula-'ai-hawane (*Ciridops anna*), *376*
Uma
inornata (Coachella Valley fringe-toed lizard), 118, 137-38
notata (Colorado Valley fringe-toed), 137
scoparia (Mojave Valley fringe-toed lizard), 137
umbilicate pebblesnail (*Clappia umbilicata*), *251*
Umbra limi (central mudminnow), 263
ungulates, native Hawaii species and, 362, 370, 380
Uniomerus tetralasmus (pondhorn), *180*
United Nations Educational, Scientific and Cultural Organization (UNESCO), 219
upland game birds, non-native species, 438
upland sandpiper (*Bartramia longicauda*), *19*, 58-59, *62*, 64, 297, *299*, *303*
Upper Cook Inlet, habitats, 33
Upper Mississippi River (UMR) ecosystem
biota, 236-38
floodplain habitat change, 234-36
Upper Mississippi River National Wildlife and Fish Refuge, 236
upper respiratory disease syndrome, *403*
Upper Sonoran manzanita chaparral, *465*
upstream rehabilitation, in southwestern sky islands, 322
urbanization
coastal and marine ecosystems and, 259
southwestern sky islands and, 322
wetland loss and, 270
Uria
aalge (common murre), 46-47, 50, *51*
lomvia (thick-billed murre), 50, *51*
Urosaurus ornatus (tree lizard), 327
Ursus
americanus
black bear, 100-102, 109, 322
floridanus (Florida bear), 102
luteolus (Louisiana bear), 102
arctos
brown bear, 337, 338, 340
grizzly bear, 103-5, 228, 229, 312, 313, *320*, 321, 338, 347, 358
middendorffi (Kodiak brown bear), 349-50
maritimus (polar bear), 338, 351-53
U.S. Army Corps of Engineers, wetlands definition, 474
U.S. Bureau of Land Management. *See* Bureau of Land Management
U.S. Bureau of Reclamation, 90
U.S. Coast Guard, 446
U.S. Department of Agriculture (USDA), 201, 302
Animal Damage Control Program, 100
Animal and Plant Health Inspection Service, 458
crop area statistics, 469
Division of Economic Entomology and Mammalogy, 8
Division of Entomology, 7-8
Forest Service (USFS), Forest Inventory and Analysis program, 214
set-aside programs, 425
U.S. Department of Defense, 400
U.S. Department of the Interior (DOI)
Bureau of Biological Survey (BBS) tranferred to, 9
Coastal Barrier Resources System (CBRS), 277-79
U.S. Environmental Protection Agency (USEPA), *5*
Environmental Monitoring and Assessment Program (EMAP), 258, 467
Great Lakes National Program Office, 242
wetlands definition, 474
U.S. Fish Commission, 432
U.S. Fish and Wildlife Program, 217
U.S. Fish and Wildlife Reference Service CD-ROM data bases, 405
U.S. Fish and Wildlife Service (USFWS), 9, 206-8, 218, 303, 338, 386, 399, 400, 424
duck survey, 35
endangered freshwater fish, 145
inventory of Arctic geese, 31
National Contaminant Biomonitoring Program, 413
National Ecology Research Center, 425
National Wetlands Inventory, 4, 207, 270, 290, 462, 463, 474, *475*
Partners for Wildlife Program, 217
Recovery Team, 374
responses to wetland loss, 473-74
wetlands definition, 474
U.S. Forest Service (USFS), 214, 400
wild horses and burros and, 456
U.S. Geological Survey (USGS), 290
Earth Resources Observation System (EROS), 467-68
land cover data, 469
National Water Quality Assessment (NAWQA), 258, 467
U.S. Global Change Research Program, 385
U.S. National Fungus Collections, 191
Usnea longissima, 196
U.S. Soil Conservation Service, *5*
Utah, land protection status, 463-64
Utah Flora, 202
Utah juniper (*Juniperus osteosperma*), 223
Utah Lake sculpin (*Cottus echinatus*), *143*
Uta stansburiana (side-blotched lizard), 327
Utricularia inflata Walter (large floating bladderwort), 209
Utterbackia imbecillis (paper pondshell), *180*

valley needlegrass grasslands, *465*
valley oak woodlands, *465*
Vallisneria americana (wildcelery), 42, 236, 237
Vancouver Canada goose (*Branta canadensis fulva*), 28, 30-31, 32
Vanessa cardui (painted lady), 171-72
Varanus indicus (Indian monitor lizard), 455
Varichaetadrilus angustipenis, 176
varied thrush (*Ixoreus naevius*), *20*
Vascular Flora of the Southeastern United States, 201, 203
vascular plants, 200-210
Alaska, 202
dispersal-ability scale, 393
eastern U.S., 203
endangered species, 206
environmental degradation and, 205
geographic distribution, 206
Great Lakes, 203
Great Plains, 203
Hawaii, 201-2
Intermountain region, 202
migration rates, 394
native, 205-8
New York, 209-10
possibly extinct species, 207-8
rare species, 206-7
Rocky Mountains, 202-3
southern U.S., 203-4
southwestern U.S., 202-3
species loss, 207
temperature change and, 392-95
threats to diversity, 208
western U.S., 202
wetland species, 207
Vascular Plants of the Pacific Northwest, 202
veery (*Catharus fuscescens*), *20*, *25*

vegetation, *See also* aquatic vegetation; land cover; plants; vascular plants
federal data bases, 467-68
general types, 469-73
monoitoring with satellite imagery, 467-73
status, management of, 465-66
in Utah, protection of, 463-64
wetlands, 474, 475
vegetation patterns
changes in national parks, 224-28
changes since European settlement, 469-72
current estimates, 468-69
future monitoring, 473
satellite photos of, 462
Velogenic Newcastle disease, *403*
Verbascum thapsus (mullein), 364
verdin (*Auriparus flavicep*), *20*
Vermivora
celata (orange-crowned warbler), *20*
peregrina (Tennessee warbler), *20*
ruficapilla (Nashville warbler), *20*
vesper sparrow (*Pooecetes gramineus*), *21*, *22*, 231, 297, *299*
Vespula (yellowjacket wasp), 378-79
pensylvanica (western yellowjacket wasp), 366, 371
Vestiaria coccinea ('i'iwi), 376, *378*, 379, 380
Villosa
fabalis (rayed bean), *180*
iris (rainbow), *180*
violet-green swallow (*Tachycineta thalassina*), *20*, *466*
vireo
Bell's (*Vireo bellii*), 231, *466*
Gray's (*Vireo vicinior*), *466*
Philadelphia (*Vireo philadelphicus*), *20*
red-eyed (*Vireo olivaceus*), *20*, 25, 420
solitary (*Vireo solitarius*), *20*, 390, 391
warbling (*Vireo gilvus*), *20*, 231
white-eyed (*Vireo griseus*), *20*
Vireo
bellii (Bell's vireo), 231, *466*
gilvus (warbling vireo), *20*, 231
griseus (white-eyed vireo), *20*
olivaceus (red-eyed vireo), *20*, 25, 420
philadelphicus (Philadelphia vireo), *20*
solitarius (solitary vireo), *20*, 390, 391
vicinior (Gray's vireo), *466*
Virgin chub (*Gila seminuda*), *150*
Virginia, acid lakes, 419
Virginia lespedeza (*Lespedeza virginica*), 231
Virgin Islands screech-owl (*Otus nudipes newtoni*), *67*
Virgin spinedace (*Lepidomeda mollispinis mollispinis*), *150*
visual obstruction readings (VOR), 425
Vivparidae, 249
Vulpes vulpes (red fox), 45, 51, 296, 302, 306, *407*, *440*
vulture, 16
black (*Coragyps atratus*), *19*, *67*
turkey (*Cathartes aura*), *19*, *67*

Wabash pigtoe (*Fusconaia flava*), *180*
wading birds, colonial waterbirds, 53-56
Waikiki Beach, 379
walleye
Stizostedion vitreum, *432*, 443
Stizostedion vitreum vitreum, 242, *244*
walleye pollock (*Theragra chalcogramma*), 51, 52
wall lizard
common (*Podarcis muralis*), *434*
Italian (*Podarcis sicula*), *434*
walnut woodlands
black, 465
California, *465*
walrus, Pacific (*Odobenus rosmarus divergens*), 338, 351, 356-57
wandering tattler (*Heteroscelus incanus*), *62*
wapiti (*Cervus elaphus*), 115-16
warbler, 18
Audubon's (*Dendroica coronata auduboni*), *20*
bay-breasted (*Dendroica castanea*), *20*
black-and-white (*Mniotilta varia*), *20*
Blackburnian (*Dendroica fusca*), *20*
blackpoll (*Dendroica striata*), *20*, *391*
black-throated gray (*Dendroica nigrescens*), *20*
black-throated green (*Dendroica virens*), *20*
Canada (*Wilsonia canadensis*), *20*
Cape May (*Dendroica tigrina*), *20*
chestnut-sided (*Dendroica pensylvanica*), *20*
hermit (*Dendroica occidentalis*), *20*
hooded (*Wilsonia citrina*), *20*
Kentucky (*Oporornis formosus*), *20*
MacGillivray's (*Oporornis tolmiei*), *20*
magnolia (*Dendroica magnolia*), *20*
mourning (*Oporornis philadelphia*), *20*, 391
myrtle (*Dendroica coronata*), *20*
Nashville (*Vermivora ruficapilla*), *20*
orange-crowned (*Vermivora celata*), *20*
palm, *22*
pine (*Dendroica pinus*), *20*, *22*
Tennessee (*Vermivora peregrina*), *20*
Townsend's (*Dendroica townsendi*), *20*
Wilson's (*Wilsonia pusilla*), *20*
yellow (*Dendroica petechia*), *20*, *299*, *466*
yellow-rumped, *22*
warbling vireo (*Vireo gilvus*), *20*, 231
warmouth (*Lepomis gulosus*), *432*
wartyback
purple (*Cyclonaias tuberculata*), *180*
threehorn (*Obliquaria reflexa*), *180*
Washington state
breeding seabirds, 43-48
sea otter (*Enhydra lutris*), 354, 355
wasp, 161
Hawaii, 366
western yellowjacket (*Vespula pensylvanica*), 366, 371
waterbirds
avian botulism and, 401-2
changes in patterns of diseases affecting, *403*
colonial. *See* colonial waterbirds
water development
Colorado River basin, bonytail and razorback sucker and, 325
western riparian ecosystems and, 285, 286
water flea, spiny (*Bythotrephes cederstroemi*), 429, 444
waterfowl
avian cholera, 402
duck plague, 402
monitoring, 15-16
non-native species, 437-38
translocation diseases, *407*
Waterfowl Parts Collection Survey, northern pintail estimates, 38
Waterfowl Protection Areas (WPAs), 302-3
water levels, Great Lakes, vegetation and, 247-49
watermilfoil, Eurasian (*Myriophyllum spicatum*), 237, 430, 443, 444
water quality, *See also* aquatic ecosystems; environmental quality
agricultural runoff and, 270
clarity, zebra mussel (*Dreissena polymorpha*) and, 445, 447
diatoms and, 257-58
fish habitat change, 141, 143
habitat change, 146
Illinois River, 239
monitoring, 233
National Water Quality Assessment (NAWQA), 467
protecting, 146
protozoa and, 253
species diversity and, 251
zebra mussel (*Dreissena polymorpha*) and, 445, 447
water supply systems, zebra mussel (*Dreissena polymorpha*) and, 446, 447
waterthrush
Louisiana (*Seiurus motacilla*), 390, 391
northern (*Seiurus noveboracensis*), *20*
waterways
acid deposition (acid rain) and, 418-19
atmospheric pollution and, 421-23
damming
habitat change and, 234-36, 287, 325
water quality and, 51-52
discharge of PCBs into, 414
habitat loss, 178
water-level regulation policies, piping plover and, 78-79
water quality, diatoms and, 258
wattle-necked softshell turtle (*Palea steindachneri*), *434*
wavy-rayed lampmussel (*Lampsilis fasciola*), *180*
waxwing, cedar (*Bombycilla cedrorum*), *20*, 21-22, 23, *299*
weeds
aquatic, 429
temperature change and, 394
weevil (*Proterhinus* sp.), 367
west coast, breeding seabirds, 43-48
West Dock causeway, andromous fish and, 341
western bluebird (*Sialia mexicana*), *466*

western duck disease, 401
western flycatcher (*Empidonax difficilis*), *466*
western gull (*Larus occidentalis*), 45
western juniper (*Juniperus occidentalis*), 223, 315
western kingbird (*Tyrannus verticalis*), *20*, *299*
Western Management Unit, mourning doves, 72
western meadowlark (*Sturnella neglecta*), *21*, *22*, 297, 299, 303, *303*
western mosquitofish (*Gambusia affinis*), *432*, 441
western pond turtle (*Clemmys marmorata*), *119*, *466*
western rattlesnake (*Crotalus viridis*), 327
western sandpiper (*Calidris mauri*), *59*, *62*
western screech-owl (*Otus kennicottii*), *67*
western slender glass lizard (*Ophisaurus attenuatus*), 231
western tanager (*Piranga ludoviciana*), *20*
western terrestrial garter snake (*Thamnophis elegans*), 133
western toad (*Bufo boreas*), 125, 327, *328*
western United States, *See also* Interior West
  rare terrestrial ecological communities, 220
  riparian ecosystems, 285-91
  subalpine forests, 314-17
  vascular plants, 202
western white pine (*Pinus monticola*), 230, 421
western wood-pewee (*Contopus sordidulus*), *20*
western yellowjacket wasp (*Vespula pensylvanica*), 366
West Indies spiny lobster (*Panulirus argus*), 283
westslope cutthroat trout (*Oncorhynchus clarki lewisi*), in Glacier National Park, 153-54
wet deposition, 228
wetland loss
  avian diseases and, 403-4
  canvasback ducks and, 41-42
  causes of, 217
  Gulf of Mexico, 269-72
  habitat change, 236-38
  mapping and, 462
  Prairie Pothole region, duck population and, 36-37
  responses to, 473-74
  shorebirds and, 60, 63-64
  southern forested wetlands, 216-17
  wetlands dynamics and, 474-75
Wetland Reserve Program, 217
wetlands, 473-75
  Chesapeake Bay, 264
  classification, 290-91
  descriptions and definitions, 474
  distribution and abundance, 474
  dynamics, 474-75
  Emergency Wetlands Resources Act of 1986, 473
  functions and values, 475
  future of, 475
  mapping, 271-72, 462, 474
  National Wetlands Inventory, 4, *5*, 207, 270, 290, 462, 463, 474, *475*
  palustrine, 216
  purple loosestrife (*Lythrum salicaria*), 458-59
  in regulated Great Lakes, 247-49
  restoration, 217
  shorebirds and, 61
  size estimates, 216
  southern forested, 216-17
  southwestern, bullfrog (*Rana catesbeiana*) in, 452-53
  trends, 291
  vascular plants, 207
  vegetation, 474, 475
  water levels and, 247-49
Wetlands Protection Act (1984) (Florida), 128
whale, 95-96
  beaked, 94
  bowhead (*Balaena mysticetus*), 95, *96*
  fin, *94*, *95*
  grey (*Eschrichtius robustus*), 95
  humpback, *94*, *95*
  Native American harvesting rules, 95
  northern right, *94*, *95*
  pilot, *94*
  right (*Eubalaena glacialis*), 95-96
  white (*Delphinapterus leucas*), 351
Wheeler National Wildlife Refuge, 413
whimbrel (*Numenius phaeopus*), 59, *59*, *62*
whipsnake, striped (*Masticophis taeniatus*), 327
whiptail lizard
  Cnemidophorus sp., 326
  orange-throated (*Cnemidophorus hyperythrus*), *466*
  South American (*Cnemidophorus lemniscatus*), *434*
whiskered screech-owl (*Otus trichopsis*), *67*
whistling ducks (*Dendrocygna* spp.), 35
whitebark pine (*Pinus albicaulis*), 313, *314*, 315
  ecosystem, 213, 228-30
whitebark rabbitbrush (*Chrysothamnus visidiflorus*), 226
white bass (*Morone chrysops*), *261*, 263, *432*
whitebelly spinner, *95*
white birds-in-a-nest (*Macbridea alba*), 401
white catfish (*Ameiurus catus*), *261*, *262*, *432*
white cedar, Atlantic (*Chamaecyparis thyoides*), bogs, 216
white crappie (*Pomoxis annularis*), 261, *432*
white croaker (*Genyonemus lineatus*), 409, 411
white-crowned sparrow (*Zonotrichia leucophrys*), *21*, *22*
white death-camas (*Zigadenus elegans*), 231
white-eared pocket mouse (*Perognathus alticola*), *466*
white-eye, Japanese (*Zosterops japonicus*), 374
white-eyed vireo (*Vireo griseus*), *20*
white-faced ibis (*Plegadis chihi*), *54*, 55
white fir (*Abies concolor*), 321, 421
whitefish, 179
  broad (*Coregonus nasus*), 337, 341, 342
  lake (*Coregonus clupeaformis*), *432*
  mountain (*Prosopium williamsoni*), *150*
white-fronted goose
  greater (*Anser albifrons frontalis*), 30, 31, 33
  tule (*Anser albifrons gambeli*), 30, 31
white-haired goldenrod (*Solidago albopilosa*), 401
white heelsplitter (*Lasmigona complanata complanata*), *180*
white ibis (*Eudocimus albus*), *19*, *54*, 55
whiteline topminnow (*Fundulus albolineatus*), 146
white pelican, 44
  American (*Pelecanus erythrorhynchos*), *19*, 53, *54*
white perch (*Morone americana*), 261, *262*, *432*, 443
white phlox flower moth (*Schinia indiana*), 232
white pine
  Pinus strobus, 225
  western (*Pinus monticola*), 230, 421
white pine blister rust (*Cronartium ribicola*), 229
White River speckled dace (*Rhinichthys osculus* ssp.), *150*
White River spinedace (*Lepidomeda albivallis*), *143*, *150*
White River springfish (*Crenichthys baileyi baileyi*), *143*, *150*
White River sucker (*Catostomus clarki intermedius*), *143*, *150*
white-rumped sandpiper (*Calidris fuscicollis*), 59, *62*
whitesided dolphin, *94*
white sturgeon (*Acipenser transmontanus*)
  Columbia River Basin, 154-56
  Kootenai River, 155
  Snake River, 156
white sucker (*Catostomus commersoni*), 151, *261*, *262*
white-tailed deer
  Odocoileus hemionus, 320
  Odocoileus virginianus, 93-94, 112-15, 320, 322, 405
white-tailed hawk (*Buteo albicaudatus*), *67*
white-tailed kite (*Elanus caeruleus*), *67*
white-tailed prairie dog (*Cynomys leucurus*), 295
white-throated sparrow (*Zonotrichia albicollis*), *21*, *22*
white-throated swift (*Aeronautes saxatalis*), *20*
Whitewater River, Coachella Valley fringe-toed lizard population, 137-38
white whale (*Delphinapterus leucas*), 351
white-winged crossbill (*Loxia leucoptera*), *21*
white-winged dove (*Zenaida asiatica*), *19*
white-winged scoter (*Melanitta fusca*), *36*
whooping crane (*Grus americana*)
  captive-reared, *405*
  diseases related to translocation of, *407*
widgeon grass (*Ruppia maritima*), 273, 274
wigeon, American (*Anas americana*), 35, *36*, 266
wild blackberry (*Rubus argutus*), 206
wild burro (*Equus asinus*), *440*, 441
  control of, 457
  on public lands, 456-57
wildcelery (*Vallisneria americana*), 42, 236, 237
wildflowers, climate change and, 390
Wild Free-Roaming Horses and Burros Act (Public Law 92-195), 456
wild horse (*Equus caballus*), *440*, 441
  control of, 457
  on public lands, 456-57
wild hyacinth (*Camassia scilloides*), 231
wildlife mortality, organophosphorus and carbamate pesticides and, 416-18
Wildlife Review, 405

wild onion (*Allium gooddingii*), *320*
wild pig (*Sus scrof*), 440, 441
wild turkey (*Meleagris gallopavo*), 16, 70-71, 387, 405, *407*
  hunting, 71
  restoration of, 71
willet (*Catoptrophorus semipalmatus*), *19*, 58, *59*, *62*
Williamsonia lintneri, *166*
willow (*Salix gooddingii*), 286, 287
willow flycatcher (*Empidonax traillii*), 17, *20*
  southwestern (*Empidonax traillii extimus*), 89-91
Wilsonia
  canadensis (Canada warbler), *20*
  citrina (hooded warbler), *20*
  pusilla (Wilson's warbler), *20*
Wilson's phalarope (*Phalaropus tricolor*), 59, *62*, 299, 396
Wilson's plover (*Charadrius wilsonia*), 58, *59*, 61, *62*
Wilson's warbler (*Wilsonia pusilla*), *20*
wing dams, habitat change and, 234, 235-36
winterfat (*Ceratoides lanata*), 226
winter flounder (*Pleuronectes americanus*), 409, *411*
winter population trends, songbirds, 21-23
winter wren (*Troglodytes troglodytes*), *20*
Wisconsin, oak savanna, 230-32
Wisconsin State Natural Heritage Inventory, 231
witch-hazel, common (*Hamamelis virginiana*), 390
Witless Bay, Atlantic puffin, 55
wolf
  control of, 306
  gray (*Canis lupus*), 98-100, 109, 306, 310, 321, 337, 338, 340, 347-48, 358
  Mexican (*Canis lupus baileyi*), 98, 99, 318, *320*
  northern Rocky Mountain (*Canis lupus irremotus*), 98
  red (*Canis rufus*), 98
  timber (*Canis lupus lycaon*), 98-99
wolverine (*Gulo gulo*), 358
woodcock, American (*Scolopax minor*), 58, *59*, *62*
woodcock reovirus, *403*
wood duck (*Aix sponsa*), 35, 36
wood frog (*Rana sylvatica*), *419*
Woodhouse's toad (*Bufo woodhousii*), 327
woodland birds, population trends, 18
woodpecker
  acorn (*Melanerpes formicivorus*), *20*
  downy (*Picoides pubescens*), *20*
  golden-fronted (*Melanerpes aurifrons*), *20*
  Nuttall's (*Picoides nuttallii*), *20*
  red-bellied (*Melanerpes carolinus*), *20*
  red-cockaded (*Picoides borealis*), 17, 86-88
  red-headed (*Melanerpes erythrocephalus*), *20*, 231
wood-pewee
  eastern (*Contopus virens*), *20*
  western (*Contopus sordidulus*), *20*
wood stork (*Mycteria americana*), *54*, 55
wood thrush (*Hylocichla mustelina*), *20*, 25, 420
wood turtle (*Clemmys insculpta*), *119*
woody vegetation, bird populations and, 300
World Center for Birds of Prey, 81
World Conservation Union, 118
woundfin (*Plagopterus argentissimus*), *150*
Wrangell-St. Elias National Park and Preserve, Mentasta caribou herd, 357-58
wren
  Bewick's (*Thryomanes bewickii*), *20*
  cactus (*Campylorhynchus brunneicapillus*), *20*
  Carolina (*Thryothorus ludovicianus*), *20*
  house (*Troglodytes aedon*), *20*, *299*
  marsh (*Cistothorus palustris*), *299*
  rock (*Salpinctes obsoletus*), *20*
  sedge (*Cistothorus platensis*), 300, 303
  winter (*Troglodytes troglodytes*), *20*
wrentit (*Chamaea fasciata*), *20*
wrinkled frog (*Rana rugosa*), *434*

Xanthocephalus xanthocephalus (yellow-headed blackbird), *21*, *299*
Xantusia henshawi (granite night lizard), *466*
Xantus' murrelet (*Synthliboramphus hypoleucus*), 47-48
Xenopus laevis (African clawed frog), 433, *434*, 435, 436
Xestia (Anomogyna) badicollis (owlet moth), 166
Xiphophorus helleri (green swordtail), *432*
Xylena
  cineritia (owlet moth), 166
  thoracica (owlet moth), 166
Xylotype capax (owlet moth), 166
Xyrauchen texanus (razorback sucker), *143*, *150*, 151, 309, 324-26

Yaqui catfish (*Ictalurus pricei*), *320*
Yaqui topminnow (*Poeciliopsis occidentalis sonoriensis*), *320*
Yavapai leopard frog (*Rana yavapaiensis*), 133, 139, 452-53
yellow-bellied flycatcher (*Empidonax flaviventris*), *20*
yellow-bellied sapsucker (*Sphyrapicus varius*), *20*
yellow-billed cuckoo (*Coccyzus americanus*), *19*, *320*, *466*
yellow-blotched map turtle (*Graptemys flavimaculata*), *119*
yellow-breasted chat (*Icteria virens*), *20*, 231, *466*
yellow-cedar, Alaska (*Chamaecyparis nootkatensis*), 316
"yellow flies," 370
yellow-headed blackbird (*Xanthocephalus xanthocephalus*), *21*, *299*
yellow-headed gecko (*Gonatodes albogularis*), *434*
yellow Himalayan raspberry (*Rubus ellipticus*), 370
yellowjacket wasp (*Vespula* spp.), 378-79
  western (*Vespula pensylvanica*), 366, 371
yellow-legged frog
  foothill (*Rana boylii*), 132, *133*
  mountain (*Rana muscosa*), 133, *134*, *466*
yellowlegs
  greater (*Tringa melanoleuca*), 59, *62*
  lesser (*Tringa flavipes*), 59, *62*
yellow mud turtle (*Kinosternon flavescens*), *119*
yellow passionflower (*Passiflora lutea*), 208
yellow perch (*Perca flavescens*), *261*, *419*, *432*, 443, 447
yellow pimpernel (*Taenidia integerrima*), 231
yellow-poplar (*Liriodendron tulipifera*), 228
yellow-rumped warbler, *22*
yellow-shafted flicker (*Colaptes auratus*), *20*
yellow-shouldered blackbird (*Agelaius xanthomus*), 439
Yellowstone cutthroat trout
  Oncorhynchus clarki, 313
  Oncorhynchus clarki bouvieri, 153-54
Yellowstone National Park, 312-14
  elk (*Cervus elaphus*), 116
  grizzly bear (*Ursus arctos*), 103-5
yellowtail snapper (*Ocyurus chrysurus*), 282
yellowthroat, common (*Geothlypis trichas*), *20*, *299*, *303*
yellow warbler (*Dendroica petechia*), *20*, *299*, *466*
Yukon-Kuskokwim Delta (YKD)
  geese, 31
  habitats, 32
Yupik Natives, 32

Zalophus californianus (California sea lion), 96
Zealeuctra
  claaseni (stonefly), *186*
  fraxina (stonefly), 185, *186*
  narfi (stonefly), 185, *186*
zebra (small barred) dove (*Geopelia striata*), 374, 438
zebra mussel (*Dreissena polymorpha*), 42, 160, 178, 181, 237, 255, 409, 427, 428, 429, 444
  control of, 445, 447
  densities, 445, 447
  distribution of, 445-46
  native mussel population and, 430
  range expansion, 446
  snails and, 447
  in southwestern Lake Michigan, 445-47
  water clarity and, 445, 447
Zenaida
  asiatica (white-winged dove), *19*
  macroura (mourning dove), 16, *19*, 71-73, *299*, 387
Zigadenus elegans (white death-camas), 231
zinc, Tarahumara frog (*Rana tarahumarae*) and, 140
zone-tailed hawk (*Buteo albonotatus*), *67*
Zonotrichia
  albicollis (white-throated sparrow), *21*, *22*
  leucophrys (white-crowned sparrow), *21*, *22*
zooplankton, 444
zorillo (*Choisya mollis*), *320*
Zosterops japonicus (Japanese white-eye), 374
Zuni (or Zuni bluehead) sucker (*Catostomus discobolus yarrowi*), *150*